Advanced CHEMISTRY

Michael Clugston
Rosalind Flemming

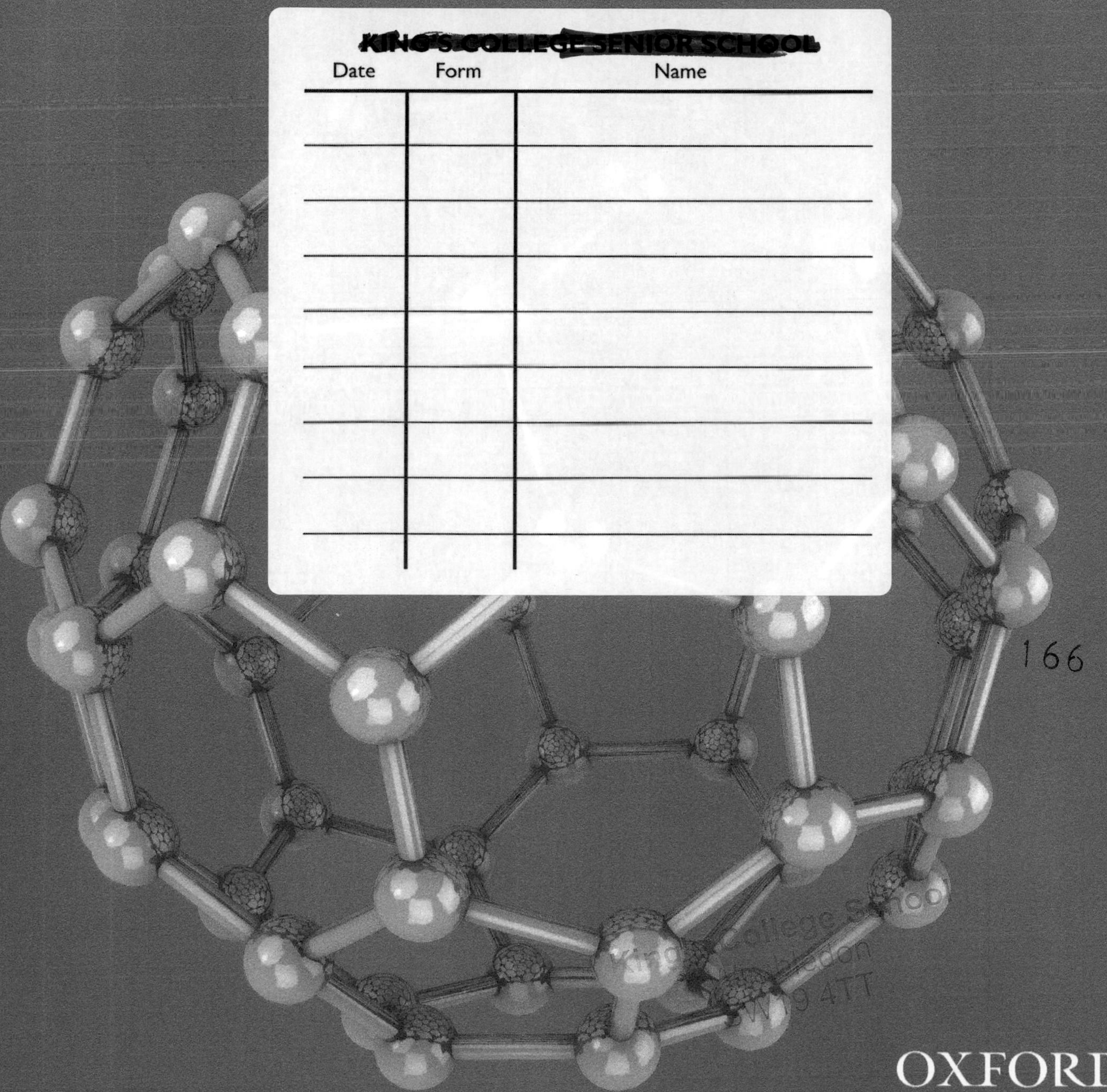

OXFORD
UNIVERSITY PRESS

OXFORD
UNIVERSITY PRESS

Great Clarendon Street, Oxford OX2 6DP

Oxford University Press is a department of the University of Oxford.
It furthers the University's objective of excellence in research, scholarship, and education by publishing worldwide in

Oxford New York

Auckland Cape Town Dar es Salaam Hong Kong Karachi
Kuala Lumpur Madrid Melbourne Mexico City Nairobi
New Delhi Shanghai Taipei Toronto

With offices in

Argentina Austria Brazil Chile Czech Republic France Greece
Guatemala Hungary Italy Japan Poland Portugal Singapore
South Korea Switzerland Thailand Turkey Ukraine Vietnam

Oxford is a registered trade mark of Oxford University Press
in the UK and in certain other countries

First published 2000
Reprinted with revisions 2013

British Library Cataloguing in Publication Data

Data available

ISBN: 978-0-19-839291-0

10 9

Typeset in New Aster

by Mark Walker Design, Plymouth, Devon

Artwork by Mark Walker

Printed in Great Britain by Ashford Colour Press Ltd.

Paper used in the production of this book is a natural, recyclable product made from wood grown in sustainable forests. The manufacturing process conforms to the environmental regulations of the country of origin.

Introduction

Advanced Chemistry has been written specially for the A-level specifications introduced in September 2000. Our central aim in writing this book has been to make chemistry accessible and stimulating without sacrificing rigour. With this in mind we have:

- organized content into double-page spreads so that you can find your way around the book easily and identify the information that you are looking for quickly (some spreads are clearly for use in the early terms of the course, while a few go beyond A level with the new Advanced Extension candidates in mind);
- given great emphasis to the visual presentation of concepts, by using hundreds of photographs and original diagrams, including images created with innovative molecular modelling software;
- researched original sources to ensure the accuracy of the data used in diagrams and tables;
- explained quantitative methods carefully and given extra maths help in boxes and in a dedicated 'Mathematics toolbox' appendix.

In addition, there are content summaries to help you to review topics effectively, hundreds of practice questions, and over fifty pages of examination questions to enable you to check your understanding and develop good exam technique.

We would like to acknowledge the many valuable contributions made to the early drafts of this book by Peter Atkins, one of today's great chemistry educators. We thank Peter for his helpful advice.

MJC would like to thank the Governors, Headmaster, colleagues, and students at Tonbridge School for their enthusiastic support during the writing of this book.

RF would like to thank her colleague David Dunevein at Abingdon College for his support and encouragement when she embarked on this project.

Both of us are very grateful to our respective families for their patience and support throughout.

We hope that you enjoy using our book and that it helps you to a greater understanding of chemistry, a subject that is central to our lives and our future.

Second edition 2013

We have taken the opportunity to update the content to ensure that it provides coverage of all the main A-level Chemistry specifications (AQA, Edexcel, and OCR) as revised for first examination at A2 in 2010, as well as the Pre-U. These changes have mainly been accomplished through the addition of margin boxes, although a couple of spreads required more extensive reworking.

These updated specifications have also decided to include current developments that continue to show the central role chemistry has in society. This 'cutting edge' material also appears in margin boxes, though they are distinguished by being outlined in red. A small health warning attaches to these 'cutting edge' boxes: the topics covered are necessarily rapidly changing and it may turn out that something new comes along during the lifetime of this edition which makes the information we supply outdated.

Mike Clugston
Ros Flemming

Contents

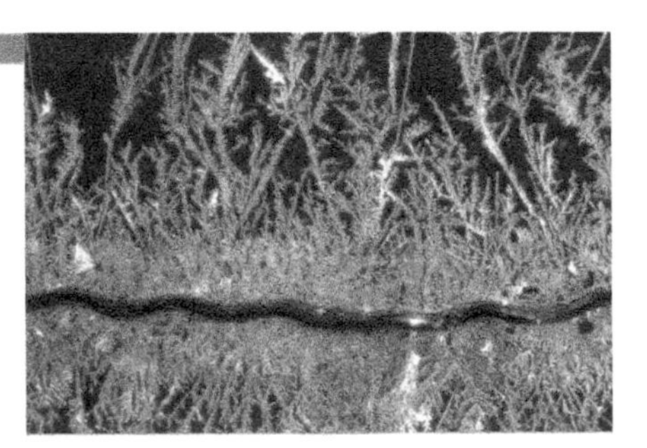

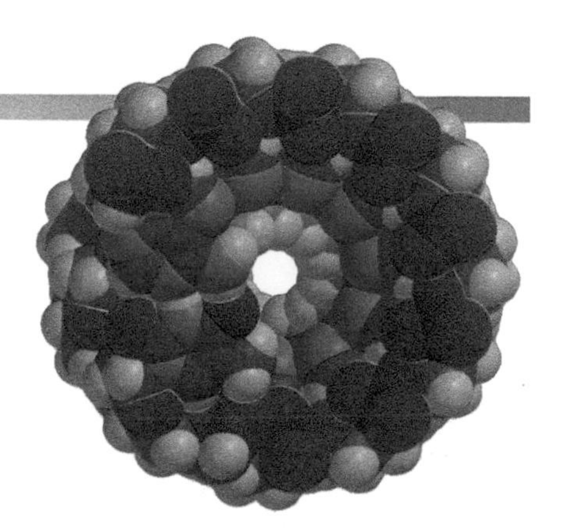

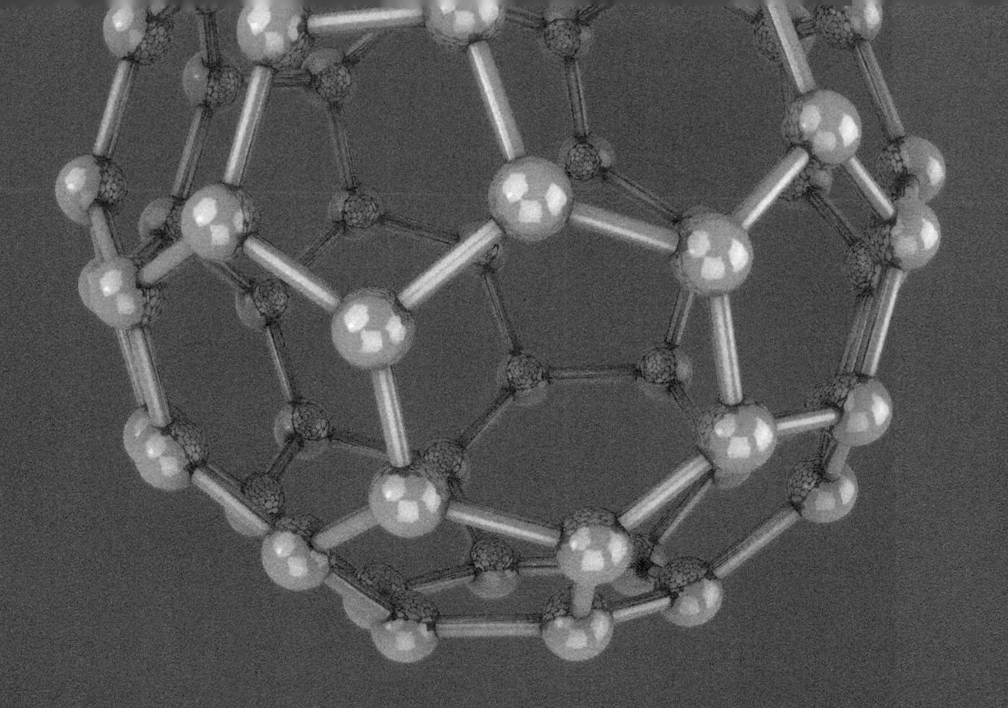

Physical **CHEMISTRY**

Physical chemistry seeks to uncover the underlying principles of chemistry. Our ideas were transformed by the discovery just over a hundred years ago of the electron (see chapter 2). Recently, such spectacular images as those on the right have provided experimental confirmation of the wave nature of electrons when they are within atoms. This idea led to a thorough understanding of the structure of atoms and of the bonding between atoms (see chapters 3–5).

Bonding theory explains the structure of elements and compounds; structure and bonding together can explain the physical properties of substances, such as their melting and boiling points and their conductivities (see chapters 6 and 7). Gases (chapter 8) complete our survey of the behaviour of individual chemicals on their own.

We begin our study of chemical reactions in chapter 9 with an account of the masses and volumes of chemicals that react. This is followed by a discussion of the energy changes that accompany chemical reactions (chapter 10). The focus then shifts in a set of four chapters (11–14) to an investigation of competition between substances. This involves the general study of equilibrium followed by an examination of the two major classes of chemical reaction: acid–base and redox reactions. These four chapters conclude with an overview of the nature of spontaneous change towards equilibrium.

The final chapter in the physical chemistry section of the book (chapter 15) considers chemical kinetics. A reaction that is predicted by thermodynamics to be spontaneous may be too slow to be of any economic value. To investigate whether any particular substance is stable or not, we need to consider both thermodynamic stability and kinetic stability.

This image shows a ring of 48 iron atoms on a copper surface. The electrons in the surface scatter from the iron atoms. The ring of iron atoms forms a boundary, or 'corral', which traps electrons in its interior. The trapped electrons occupy the quantum states of the corral. Quantum corrals provide us with an opportunity to visualize the quantum behaviour of electrons in small confining structures.

1 Patterns in chemistry

Human beings are naturally inquisitive. We have progressed from the Bronze Age to our current state of technological ability in just 5000 years. Throughout that time we have constantly observed the world around us and asked the questions: 'Why does that happen?' and 'How can I control what happens?' We try to answer these questions by carrying out experiments and searching for patterns in what we see and measure. Chemists are concerned with the study of matter – the structure of materials and how they interact. Chemistry aims to explain patterns in the behaviour of materials by formulating rules, theories, and laws to reveal the underlying nature of matter.

AN INTRODUCTION TO CHEMISTRY

OBJECTIVES

- The origins and scope of chemistry

Chemistry is the study of the elements and the compounds formed when they bond with each other. The subject is subdivided into three main branches: physical chemistry, inorganic chemistry, and organic chemistry. Physical chemistry is concerned with how the chemical structure of a substance affects its properties. It includes the study of chemical bonding and the structures of solids, liquids, and gases. Physical chemistry also investigates energy changes that accompany chemical reactions and how fast reactions happen. Inorganic chemistry is concerned with describing the properties and reactions of all elements and compounds other than those of carbon. Organic chemistry is concerned with the chemistry of carbon compounds. Its study centres particularly on the ability of carbon atoms to join with each other to form rings and long chains. There are also related branches of chemistry such as biochemistry, chemical engineering, and geochemistry.

The origins of chemistry

Iron extraction uses chemical techniques discovered more than 3000 years ago. Converting iron into steel became much cheaper in the nineteenth century.

The earliest uses of chemical processes were in the extraction of metals such as copper and iron, the firing of clays and sands to make ceramics and glasses, and the extraction and use of dyes. People were able to do all these things, but how could they explain what they saw happening? When thinking about the material world, people have always tried to find patterns and describe rules of behaviour. For example, when faced with sorting out a box of jumbled objects, the first thing anyone might try to do is sort them into groups of similar things. In the case of a collection of coloured marbles, a suitable procedure would be to sort them into different colours or sizes. The same was true of people at the dawn of civilization as they searched for a way of sorting out and explaining the non-living world around them.

At first, in the fifth century BC, their sorting was very broad. They concluded that there were four main categories of substance: fire, air, water, and earth. Each of these four categories was composed of pairs of the four fundamental qualities: hot, cold, wet, and dry. Everything in the whole of the non-living world had to fit one of these categories or a combination of them. At a simple level this classification works: mountains, deserts, rocks, and houses are all 'earthy'; all liquids are 'watery'; all gases are 'airy'; and all flames are 'fiery'. This scheme can also explain changes in materials. For example, when (cold–wet) water is boiled, the result is a form of (hot–wet) air, i.e. steam.

The early alchemists

The systematic study of chemistry as a subject started in Egypt about 1700 years ago. Writings by Zosimos (c. AD 300) describe chemical experiments and chemical apparatus. For the next thousand years, most chemical exploration sought ways of changing (or transmuting) cheap base metals such as lead into the precious metals, gold or silver. This work on transmutation was called alchemy (from the Arabic *al-kimia*: *al* = the, *kimia* = art of transmuting metals).

They did not succeed in their search but, almost accidentally, they developed many sound techniques of chemical manipulation. Modern chemistry was born during the seventeenth century when some early scientists started to investigate the mechanisms by which substances were changed. Alchemists continued their work during this period of transition, but their use of ghostly spirits and 'uncorporeal bodies' to explain their findings was gradually seen to make no sense.

Paracelsus

Theophrastus Bombast von Hohenheim (c. 1493–1541), a Swiss physician who called himself Paracelsus, attempted to move alchemy beyond simply striving to transmute metals. He declared that an equally valid aim for alchemists was to try to cure illness using chemicals as remedies. At that time four fluids of the human body called *humours* were thought to determine a person's physical and mental state. However, Paracelsus believed illness arose from specific external causes rather than from an imbalance of the *humours*.

The discovery of phosphorus in 1669 resulted from the alchemical investigations of Hennig Brandt of Hamburg. Brandt's starting material was urine. Phosphorus emits light (it is luminescent) by reaction with air; its name comes from the Greek for 'bringer of light'.

The foundation of modern chemistry

During the seventeenth century, the study of chemistry began to concentrate on the preparation, isolation, and use of new substances. In 1661, Robert Boyle wrote a landmark book called *The Sceptical Chymist*, in which he attacked the idea of the four categories (fire, air, water, and earth) and introduced the modern concept of chemical elements. During the following two centuries, Boyle's ideas took root and slowly developed. In 1789, in his *Elementary Treatise on Chemistry*, Antoine Lavoisier published a list of 33 chemical elements, many of which we recognize today. At that time, an element was thought of as a substance that cannot be broken down into two or more simpler substances. We know now that an **element** is a substance that contains only one sort of atom, and that an **atom** is the smallest particle of an element that can exist independently. Most of our modern chemical understanding rests on these two simple and fundamental points.

Antoine Lavoisier was the founder of modern chemistry. He showed that air contains oxygen and that water is a compound of oxygen and hydrogen. In 1789, using Boyle's definition of an element, he drew up the first table of chemical elements. This portrait is by Jacques-Louis David.

SUMMARY

- An element is a substance that cannot be broken down into two or more simpler substances.
- An element is a substance that contains only one sort of atom.
- An atom is the smallest particle of an element that can exist independently.

1.2

O B J E C T I V E S

- Nineteenth-century attempts at classification

Elements: the search for patterns

If you were a chemist working during the 1800s, you would have been living in exciting times. New elements were being discovered at an amazing rate. In 1807 Humphry Davy used electrolysis to isolate the new metals sodium and potassium. In 1808 he isolated the metals calcium, strontium, and barium. In 1810, Davy went on to show that chlorine is an element similar to iodine. During this period, chemists also began to investigate the *quantities* in which elements reacted with each other. From these investigations, each element was assigned an atomic mass.

Atomic mass

During the nineteenth century, the atomic masses of elements were calculated relative to hydrogen. One atom of this element was used as the arbitrary unit of mass. On this scale, oxygen had an atomic mass of 16, indicating that one atom of oxygen had a mass equal to the total mass of 16 atoms of hydrogen. The atomic mass for each element was calculated from carefully measuring the quantities of reactants and products involved in chemical reactions.

Atomic masses are now measured relative to one-twelfth of the mass of one atom of carbon-12, as we shall see later.

Döbereiner's triads

With about fifty elements clearly identified, chemists tried to group together elements that resembled each other. Following the process of classification started by the Ancient Greeks, they were looking for an underlying theory that would arrange elements into groups and explain their properties. The first real success came in 1817 when Johann Döbereiner noted that the metals calcium, strontium, and barium were very alike.

A decade later, once bromine had been discovered, he saw that the non-metals chlorine, bromine, and iodine were also very similar. Döbereiner also noted that the atomic mass of the middle element of each three was approximately the average of the atomic masses of the outer two elements. However, he was not able to suggest why this was so. Döbereiner believed that elements could be arranged in threes, or **triads** as he called them; but he could not find enough triads to construct a convincing theory to explain his classification.

Element	Atomic mass
Calcium	40.1
Strontium	87.6
Barium	137.3
Sulphur	32.1
Selenium	79.0
Tellurium	127.6
Chlorine	35.5
Bromine	79.9
Iodine	126.9

Three of Döbereiner's triads. In each triad, the atomic mass of the middle element falls approximately mid-way between the atomic masses of the outer two elements.

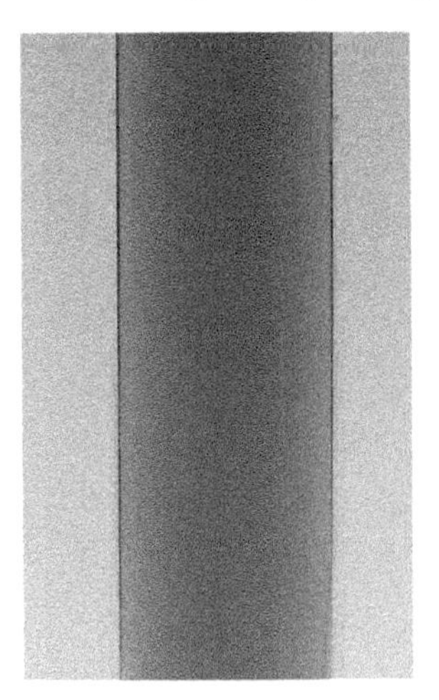

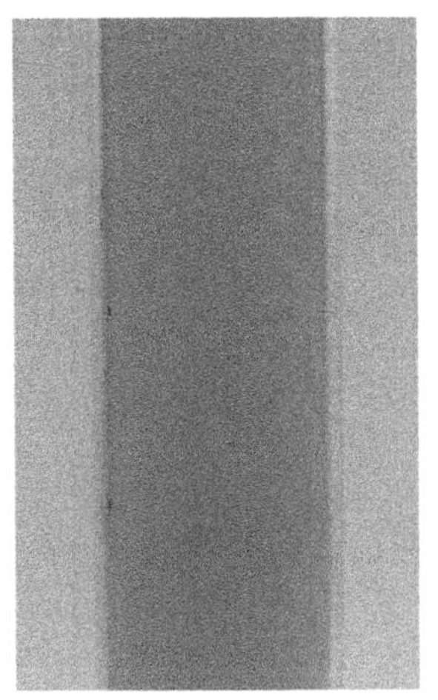

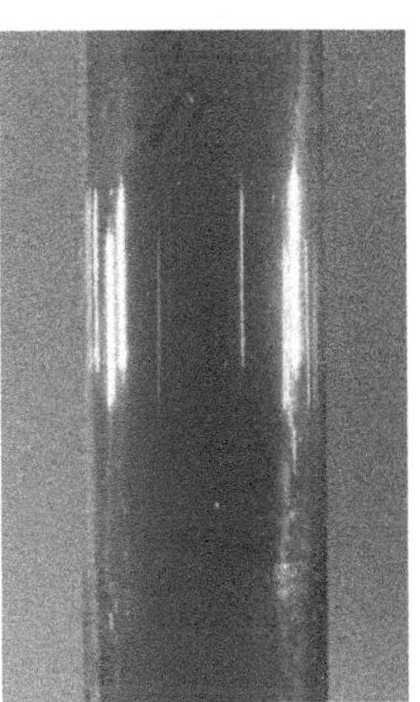

A Döbereiner triad. Chlorine, bromine, and iodine are all non-metals. Bromine (left) is a liquid with a brown vapour; chlorine (centre) is a gas; iodine (right) is a solid with a purple vapour. The atomic mass of bromine (79.9) is approximately equal to the average of the atomic masses of chlorine (35.5) and iodine (126.9) (the calculated average is 81.2).

Alexandre Béguyer de Chancourtois

In 1862, the French geologist Alexandre Béguyer de Chancourtois arranged the names of the elements in order of increasing atomic mass in a helical pattern around a cylinder. This procedure divided the elements into vertical columns. Each column contained some elements with similar properties. For example, lithium, sodium, and potassium appeared in one column; and oxygen, sulphur, selenium, and tellurium in another. (One problem was that he included some compounds and alloys thought to be elements at the time.) Béguyer de Chancourtois concluded that: 'The properties of substances are the properties of numbers.' His work was ignored by the majority of chemists, mainly because the diagram explaining his idea was omitted from the published paper.

William Odling

In 1864, William Odling published an article entitled 'On the Proportional Numbers of the Elements'. He grouped certain elements together and noticed that, for 'well-defined groups', their sequences of chemical properties and their sequences of atomic masses went in parallel with each other. Odling constructed a table of the elements to illustrate these relationships.

Newlands' octaves

John Newlands was the chief chemist in a London sugar refinery. In 1865 he noticed that, if the elements were written in order of their atomic masses, similar chemical properties appeared at every eighth element. Newlands likened this behaviour to a musical scale and suggested that the elements obeyed a *law of octaves*. Newlands had glimpsed the correct underlying pattern to the recurring properties of the elements, but he did not take the idea far enough. The real father of the modern periodic table was a Russian, Dmitri Mendeleyev.

John Newlands

Like Odling, he was born in a district of London called Southwark. Newlands' law of octaves represented an important step on the path to a systematic classification of the elements. However, his work was ridiculed.

When he presented his ideas to the Chemical Society in London, he was asked 'whether he had ever examined the elements according to the order of their initial letters?', and he was told that his work was 'not adapted for publication'. He did not receive the credit he deserved.

In later years it became clear that Newlands had anticipated Mendeleyev's 1869 periodic law. Belatedly, Newlands was awarded the Davy Medal of the Royal Society in 1887. A plaque was unveiled in 1998 at Elephant and Castle, London, to commemorate the centenary of his death.

SUMMARY

- Döbereiner identified groups containing three elements with similar properties. He called these groups *triads*.
- The triads included calcium–strontium–barium and chlorine–bromine–iodine.
- Newlands drew up a list of elements in order of increasing atomic mass and noted that every eighth element fell into a group with similar chemical properties.

PRACTICE

1 Look through this spread and list:

a the metals mentioned, and

b the non-metals mentioned.

2 Look up the following names and words in a large encyclopedia or on a computer. Arrange them into a sequence that shows the development of chemistry.

a Bronze Age **b** Iron Age

c Alchemy **d** Elements

e Robert Boyle **f** Antoine Lavoisier

g Joseph Priestley **h** Paracelsus

i John Dalton.

1.3 Mendeleyev's periodic table

OBJECTIVES

- How Mendeleyev constructed the first periodic table

Towards the end of the 1860s, chemists were on the verge of proposing a grand unifying model that would account for the properties of the elements. The 62 separate elements known in 1869 would soon be grouped and classified according to a set of clear rules. Newlands and others had glimpsed the underlying pattern in the properties of the elements. However, their understanding was incomplete and fragmented. It was the Russian Dmitri Mendeleyev who collected the elements together in a table that revealed the periodic (repeating) pattern in their properties.

The periodic law

We can arrange glass marbles systematically by referring to their colour or size. Mendeleyev arranged the 62 elements then known by referring to their atomic masses. Starting with hydrogen, he wrote out the elements in horizontal rows, in order of increasing atomic mass. This gathered elements with similar properties below one another in vertical groups. Mendeleyev's stroke of genius was to group elements according to their properties, even if he had to move some elements out of the strict sequence of increasing atomic mass. He looked at the arrangement of the elements in his table and was then able to state his periodic law, as follows:

- 'The elements, if arranged according to their atomic masses, show an evident periodicity of properties.'

A biographical note

Dmitri Ivanovich Mendeleyev was born in Tobolsk, Siberia, in 1834, the youngest of fourteen children. Mendeleyev became the most celebrated chemist of his generation, once the predictions he had made were verified.

Mendeleyev had divorced his first wife and then married a young art student. According to Russian Orthodox Law, he was a bigamist. No action was taken as Tsar Alexander II rejected criticism of his behaviour by saying: 'Yes, Mendeleyev has two wives, but I have only one Mendeleyev.'

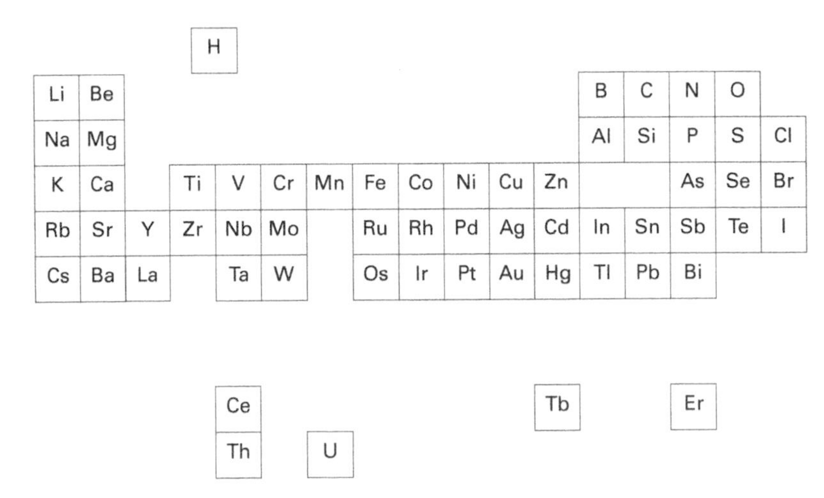

A modern outline of the periodic table incorporating the 62 elements known to Mendeleyev in 1869.

The shape of the periodic table

The illustration above shows the 62 elements known to Mendeleyev, positioned on an outline of a modern form of the periodic table. His original table successfully classified all the elements in Groups I to VII. The elements now referred to as 'transition metals' were scattered throughout the table in regions called 'subgroups'. These elements were later assigned to their own specific area created at the centre of the table.

Changing the order to fit the properties

The problem with arranging the elements in strict order of atomic mass is that the pattern does not match the properties of all the elements. For example, iodine ends up in the wrong place, away from bromine and chlorine. Mendeleyev had the courage to use his knowledge of the properties of the elements to bend his own rule; he simply exchanged some positions on the basis that the atomic masses known at the time might be inaccurate. Mendeleyev also had the foresight to realize that some elements had yet to be discovered.

The undiscovered elements

Having suggested that there must be undiscovered elements, Mendeleyev *left gaps* for them in his table in order to preserve the principle of periodicity. Most significantly, he went on to predict in detail the chemical properties that these unknown elements would have. There was a gap in his original table between silicon and tin, now occupied by germanium. Mendeleyev inserted an element he called 'eka-silicon' into this gap and he predicted its properties by inspection of the properties of the other elements in the group. The properties of germanium, isolated by Clemens Winkler in 1886, were in close agreement with Mendeleyev's predictions for eka-silicon. It is a further testimony to the brilliance of Mendeleyev's ideas that the later discovery of a completely new group of elements, the noble gases, did not disrupt his overall scheme.

Property	Eka-silicon, E (predicted)	Germanium, Ge (observed)
Atomic mass	72	72.6
Density	5.5 g cm^{-3}	5.35 g cm^{-3}
Oxide		
nature	white solid	white solid
formula	EO_2	GeO_2
density	4.7 g cm^{-3}	4.23 g cm^{-3}
Chloride		
boiling point	below 100 °C	84 °C
formula	ECl_4	$GeCl_4$
density	1.9 g cm^{-3}	1.84 g cm^{-3}

The properties of eka-silicon suggested by Mendeleyev compared with the actual properties of germanium.

This stamp commemorates the centenary of the introduction of the periodic table by Dmitri Mendeleyev. Gallium (Ga), unknown in 1869, is identified.

Lothar Meyer

As Mendeleyev was completing his periodic table, Lothar Meyer in Germany was also working with the concept of periodicity. He constructed a plot of the atomic volumes of the elements against their atomic masses. He calculated the atomic volume of each element by dividing its atomic mass by its density. The overall shape of the plot consisted of a series of periodically repeating peaks and troughs, illustrating the periodic nature of the atomic volumes of the elements.

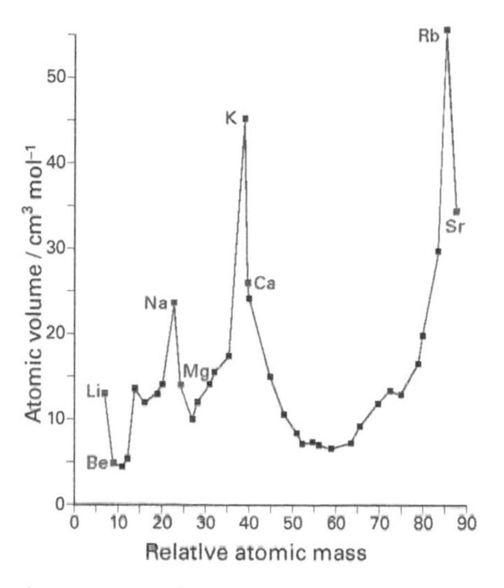

Lothar Meyer's curve of atomic volume against atomic mass would have looked similar to this one. Notice the elements at the peaks of the curve – lithium, sodium, potassium, and rubidium.

SUMMARY

- Mendeleyev was able to obtain groups containing similar elements by arranging the elements in order of increasing atomic mass.
- He left gaps or altered the order of the elements, to make better sense of their properties.
- Elements discovered later fitted in to the gaps in the table; their properties matched those predicted by Mendeleyev.

PRACTICE

1 Chemists greeted Mendeleyev's periodic law, which is only one sentence long, with a great deal of enthusiasm and approval. Why did it create so much excitement?

2 Why did the later discovery of the noble gases (helium, neon, argon, krypton, xenon, and radon) not affect the overall arrangement of Mendeleyev's periodic table?

3 Copy the outline of the periodic table given opposite. Add to it the symbols of the elements about whose chemistry you already know something. Is your chemical knowledge scattered randomly across the table or is there an underlying pattern?

4 You may have completed most of a whole period in question 3. Run your eye across it from left to right. Do you notice any trend in a property of the elements? Describe any trend that you see.

1.4

OBJECTIVES

- The overall shape of the modern periodic table
- Groups of elements
- Periods of elements
- The metal/non-metal divide

Group numbering

The groups in the periodic table have traditionally been numbered with Roman numerals I to VIII, and that is the numbering system used in this book. (Note that some books use the Arabic numerals 1 to 8 for the groups, which can lead to confusion with the period numbers 1 to 7.)

By international agreement, modern forms of the periodic table number the groups 1 to 18.

Atomic mass and atomic number

Mendeleyev's original periodic table attempted to list elements in order of increasing atomic mass.

The modern periodic table lists elements by their atomic number, rather than by atomic mass. The atomic number of an element is the number of protons in its nucleus. Using atomic numbers avoids the problems encountered by Mendeleyev in trying to establish an order. Atomic number is explained fully in the next chapter.

Gold is extremely malleable. This medallion was made by the Vikings.

THE MODERN PERIODIC TABLE

A modern periodic table can appear very complicated. It is covered in unfamiliar chemical symbols; it also has a particular shape. If you can learn to read the table as a musician reads a musical score, then an enormous amount of information becomes available to you. In order to start to understand and make use of the periodic table, it is helpful first to define some regions within it. The vertical columns of the table are called **groups**, which are numbered I to VIII. Groups contain elements that have similar chemical properties to each other. The horizontal rows are called **periods**, which are numbered from 1 to 7.

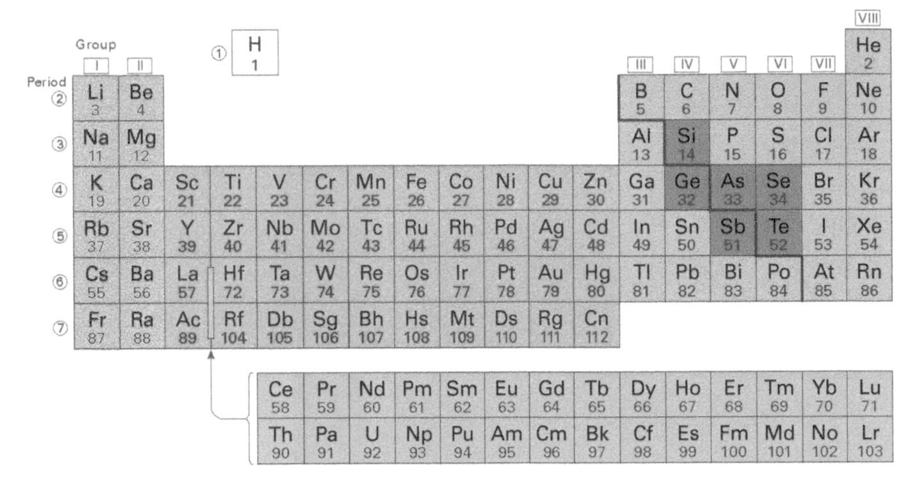

The standard modern form of the periodic table. The deeper green colour identifies the metalloids, see opposite. Md (element 101) is the symbol for mendelevium.

Metals and non-metals

The position of an element in the periodic table gives an indication of its overall properties. For example, run your eye across the elements in Period 3 (sodium to argon) of the periodic table above. Notice that there is a change from metals to non-metals. Overall, the majority of the elements in the table are metals. The relatively few non-metals are found on the right-hand side of the table.

Metallic elements

All the metallic elements, except mercury, are shiny solids. Metals are good electrical conductors, whereas most non-metals do not conduct electricity. Metals are *malleable*: they may be bent or pressed into different shapes. They are also *ductile*: they may be stretched to form, for example, thin wires. The most typically metallic elements are found in Groups I and II. Group I elements are called the alkali metals and include sodium and potassium. Group II elements include magnesium and calcium.

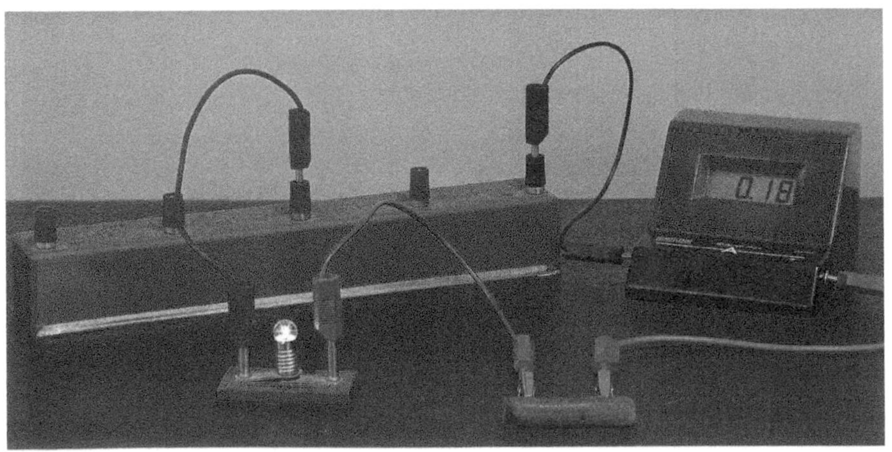

The alkali metal lithium, like all metals, is a good conductor of electricity.

Non-metallic elements

Most non-metals are electrical insulators, having an electrical conductivity at least one billion times smaller than that of a typical metal. They are also brittle when solid, and often have low melting and boiling points compared to typical metals. The most typically non-metallic elements are found in Group VII. These elements are called the **halogens** and include chlorine and bromine. Group VIII contains the **noble gases**, which include helium and neon.

Sulphur is a typical non-metal. It melts at 113 °C, is brittle, and is an electrical insulator. Compare these properties with those of a typical metal.

Metalloids

Look again at the right-hand side of the periodic table opposite. Notice the stepped line drawn rather like a staircase. This line roughly marks the division between metallic and non-metallic elements, with metals to its left and non-metals to its right. While Groups I and II contain metals and Groups VII and VIII contain non-metals, Groups III to VI contain both sorts of element. Group IV, for example, contains the non-metal carbon and the metals tin (Sn) and lead (Pb). Metallic character increases as atomic number increases down the group. The elements silicon (Si) and germanium (Ge) are intermediate in nature between metals and non-metals. Together with arsenic (As), antimony (Sb), selenium (Se), and tellurium (Te), they are called **metalloids**. Because of their intermediate electrical conductivity, silicon and germanium are classed as *semiconductors*.

> **Glenn Seaborg**
>
> Glenn Seaborg, the first person ever honoured in his lifetime with an element named after him (seaborgium, element 106), died in 1999. He could write his address in chemical elements: lawrencium, berkelium, californium, americium (most of which he had discovered).

The 'transition' elements

The periodic table contains elements other than those in Groups I to VIII. The central regions of Periods 4, 5, and 6 contain transition elements such as titanium, iron, copper, silver, and gold. These elements are metallic. Note that two rows of metals are separated out at the bottom of the periodic table, the *lanthanides* Ce to Lu and the *actinides* Th to Lr.

Glenn T. Seaborg pointing to seaborgium. The elements on either side had temporary names which have now been permanently renamed by international agreement. Hahnium (Ha) is now called dubnium (Db); nielsbohrium (Ns) is now shortened to bohrium (Bh).

SUMMARY

- The modern periodic table lists elements in order of their atomic number.
- Elements are arranged in vertical groups (I–VIII) and horizontal periods (1–7).
- Most elements are metals.
- Metals conduct electricity, and are malleable and ductile.
- Non-metals are found towards the top right-hand side of the periodic table.
- Most non-metals are insulators, are brittle when solid, and have lower melting and boiling points than metals.
- The diagonal band of metalloids in the periodic table acts like a fuzzy frontier between the metals and the non-metals.
- The six metalloids are silicon, germanium, arsenic, antimony, selenium, and tellurium.

PRACTICE

1 Give the names of the elements corresponding to the symbols: Li, Mg, Si, Br, K, Ag, Sn, F.

2 Give the symbol for each of the following elements: germanium, sulphur, calcium, gold, krypton, antimony, lead.

3 For the elements in questions 1 and 2, say whether each is a metal, a non-metal, or a metalloid.

4 Name a non-metal in Group III and a metal in Group VI.

5 Give the symbols and names of the elements corresponding to the atomic numbers: 23, 13, 56, 36, 38, 7.

2 The nuclear atom

In the previous chapter we discussed some of the early ideas about sorting and classifying elements. In this chapter we look in more detail at what we think an element actually is. We know that all matter is made up of very small particles called atoms. The idea of atoms originated with the Greek philosophers Leucippus and Democritus during the fifth century BC, but the concept remained ignored and undeveloped until it was reintroduced in the early nineteenth century by John Dalton. By measuring the masses of the elements taking part in chemical reactions, he was able to provide *indirect* evidence that matter is made up of atoms. It is only over the past few years that *direct* evidence for this has been obtained, using the scanning tunnelling microscope (STM).

Ideas about atoms

OBJECTIVES

- Matter is atomic rather than continuous
- Early ideas about atoms

For the past 2500 years people have asked the following question, in the form of a thought experiment: 'Can you go on cutting a piece of matter into ever smaller pieces, or is there a limit?' Zeno of Elea (born about 490 BC) thought that the answer was: 'Yes. Matter is continuous and fills space completely, rather like jelly.' In about 420 BC, Democritus took the opposite view. He held that matter is divided into small particles with empty space – a vacuum – between them. He also said that these particles, *atoms*, are hard and are in constant motion. He had no evidence to back up his assertions; they seemed to him to be simply the better idea.

A set of sixth-century multiplication tables gives the size of the 'smallest particle' as equivalent to 10^{-10} m, which is close to the modern value for the diameter of a typical atom. However, all these ideas about atoms were speculative, which means they were based solely on guessing and thinking.

Dalton was born in 1766 and left school at the age of twelve. He was fascinated by the weather, which he meticulously recorded for 57 years. As a life-long Quaker he shunned fame and would have been embarrassed to know that 40 000 people filed past his coffin as it lay in Manchester Town Hall. This is one of the earliest photographs of a scientist to exist (a Daguerrotype by Dancer).

John Dalton

John Dalton, a teacher from Manchester, is credited with developing the modern theory of the atom. He presented the idea in detail in his book *New System of Chemical Philosophy* in 1808. Following a long series of experiments Dalton suggested that:

- different elements have atoms which differ in mass;
- each element is characterized by the mass of its atoms.

The basis of Dalton's atomic theory is that all matter – elements, compounds, and mixtures – is composed of extremely small particles called atoms. Four *postulates* describe how these atoms behave:

1. The atoms in a given element are all of the same kind.
2. A compound contains atoms of two or more elements combined together in fixed proportions.
3. An atom retains its identity during a chemical reaction.
4. During a chemical reaction, the atoms in the reacting substances rearrange to form the products of the reaction.

Dalton also introduced the idea of chemical symbols. He was the first to find the correct formula for carbon dioxide, which he wrote down in symbols as ○●○. The open circles represent oxygen atoms and the filled circle represents a carbon atom. His system of circular symbols is no longer used. The current familiar system based on letters was devised by the Swede Jacob Berzelius in 1811.

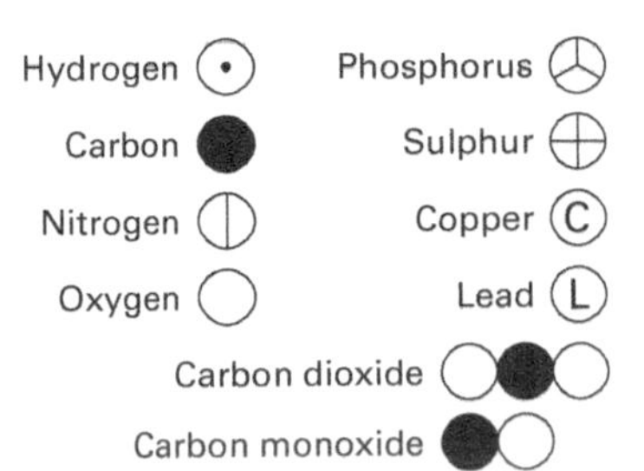

Dalton's symbols for some atoms and two compounds. His formulae for carbon dioxide and carbon monoxide are correct, but he was not always so successful!

The STM – direct evidence for atoms?

The scanning tunnelling microscope (STM) was invented in 1981. It creates images of surfaces, with a resolution that is sufficiently high to detect individual atoms. We perceive everyday objects around us by using our brains to analyse reflected light entering our eyes. The STM constructs an image by using a computer to analyse the electric current flowing between a surface and a very fine probe. More details of the STM are given in the box.

Principles of STM operation

Electricity does not usually escape from the surface of conducting wires, but if a second conductor is brought close enough to the wire, electrons cross the gap by a process called *tunnelling*. A tiny but measurable current flows.

The STM uses a minute needle-like probe, tipped with tungsten, to scan the surface being examined. A very small potential difference is applied across the gap between the probe and the surface. When the probe encounters a bump on the surface, the gap becomes smaller and the current increases; when the probe encounters a dip, the gap becomes larger and the current decreases. When all the data have been collected, the computer generates a contour map of the surface. A very fine probe is able to detect individual atoms in the surface being scanned.

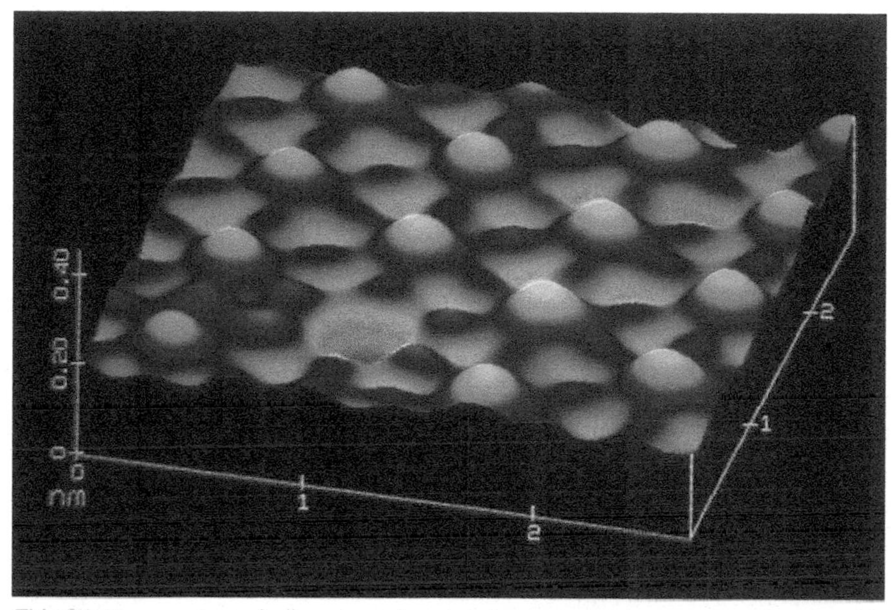

This STM image shows iodine atoms (on a platinum surface) as a series of peaks with pink tops. The colour was added to the image during computer analysis of the data from the probe. Note the 'vacancy' where an iodine atom is missing.

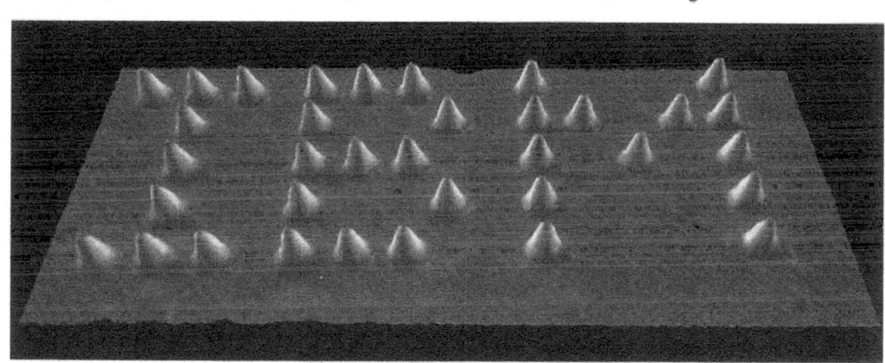

The STM image above of xenon atoms manipulated on a nickel surface must have pleased the sponsors of the research. The same company produced the image to the right using iron atoms on copper which contains the Kanji characters for 'atom' (the literal translation resembles 'original child').

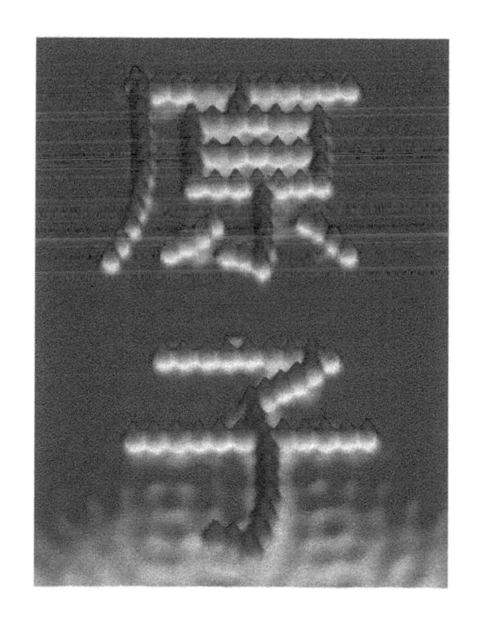

SUMMARY

- Matter is made up of atoms.
- The atoms in a given element are all of the same kind.
- A compound contains atoms of two or more elements combined in fixed proportions.
- Atoms are very small – about 10^{-10} m in diameter.

Discussion point

Would you say that the STM enables us to *see* individual atoms?

PRACTICE

1 Dalton's atomic theory is based on four postulates. Give an example for each one of these, to explain what they mean.

2 The diameter of a typical atom is 10^{-10} m. How many atoms would cover the area of a printed full stop? You can assume that a full stop is 0.1 mm in diameter and that atoms are spherical.

2.2

The discovery of the electron

OBJECTIVES

- Early evidence for atomic structure
- The negatively charged electron

Dalton's atomic theory of matter steadily gained acceptance during the second half of the nineteenth century, but there was still a long way to go before the modern theory became fully developed. Following Dalton's work it was at first thought that atoms could not be broken down into anything simpler. (In fact, the word 'atom' is derived from the Greek *atomos*, meaning 'indivisible'.) The first indication that atoms have an internal structure came with the discovery of the electron. The mass of the electron is far smaller than the mass of any atom so, the reasoning went, ordinary neutral atoms must also contain a massive positively charged part. We mainly owe our modern ideas on atoms to the experiments and thinking of three people: the Englishman J. J. Thomson, the New Zealander Ernest Rutherford, and the young Englishman Henry Moseley. The last two of these will take centre stage in the next spread.

J. J. Thomson and the electron

During the 1870s, physicists carried out experiments into the electrical conductivity of gases. They used a piece of apparatus called a 'discharge tube', which was a long glass tube fitted with a metal electrode at each end. When the gas inside was at very low pressure, a green glow appeared in the glass at the end of the tube furthest from the cathode (the negative electrode). Scientists proved that the glow was caused by invisible rays originating from the cathode and travelling away from it in straight lines. In 1895 the French physicist Jean Perrin placed a metal cylinder inside the tube to collect the 'cathode rays'. The charge on the cylinder showed that cathode rays are negatively charged.

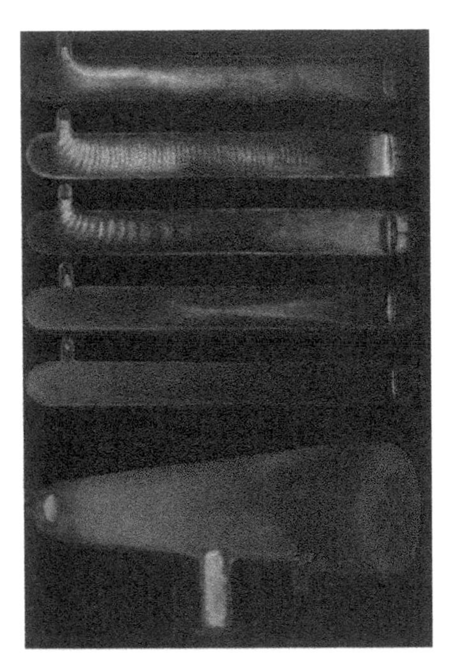

Cathode rays (electrons) stream from the cathode. As the pressure in the tube is reduced, a changing pattern of colourful lights occurs. At low pressure, the inner surface of the glass glows. A metal cross placed in the path of the electrons casts a sharp shadow, showing that they travel in straight lines.

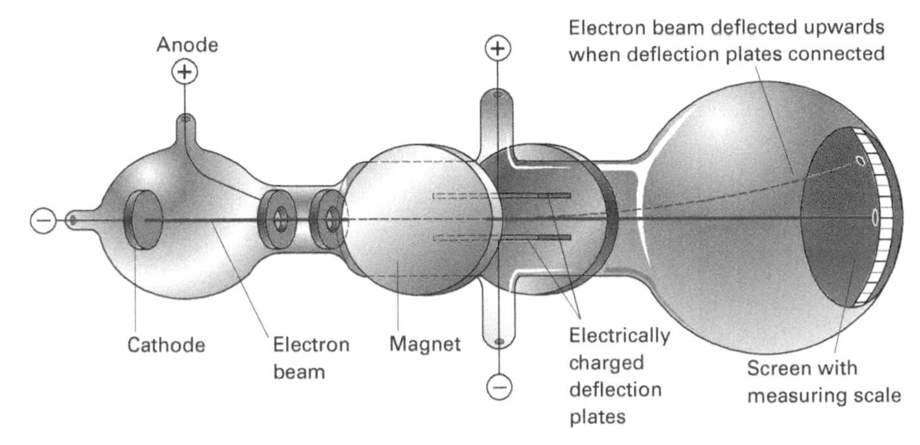

Thomson's apparatus for determining the charge/mass ratio (e/m) of electrons. The rays (electrons) from the cathode pass through a slit in the anode and exit from the slit as a narrow beam. The beam is deflected by a vertical electric field and a horizontal magnetic field, whose strengths are known. The beam strikes a fluorescent screen and appears as a dot on the measuring scale.

In 1897, J. J. Thomson deflected cathode rays with both electric and magnetic fields, and used his results to measure the ratio of their charge to their mass. He reasoned that, if cathode rays have mass, then they must be composed of a stream of particles. We now call these particles **electrons**. Thomson assumed the charge to be the same as the smallest charge observed during electrolysis, and so calculated the mass of an electron to be nearly 2000 times smaller than the mass of a hydrogen atom.

Joseph John (J. J.) Thomson (1856–1940) with the apparatus he used to determine the charge/mass ratio of the electron.

The 'plum pudding' model of the atom

Atoms have no overall charge; they are **neutral**. So, there must be a part of the atom that is positively charged, to balance the negative charge of the electrons. At the end of the nineteenth century, scientists understood that the atom had positively and negatively charged components, and that the negatively charged components had a very low mass.

J. J. Thomson proposed a model that described atoms as negatively charged electrons embedded in a sphere of positive charge. It became known as the 'plum pudding' model because the electrons were spread randomly throughout the positive charge like the dried fruit in a pudding. This model was soon shown to be inadequate.

The name 'electron'

The name 'electron' had been coined in 1891 by Johnstone Stoney, who suggested that this name would be appropriate for a fundamental electrically charged particle within atoms. Thomson called the very light particles found in this cathode-ray experiments 'corpuscles'. It was Hendrik Lorentz (see spread 14.4) who pointed out (in 1899) that Thomson had in fact found Stoney's electrons.

SUMMARY

- The electron has a negative charge; its mass is very small compared to that of an atom.
- Cathode rays are electrons.

PRACTICE

1 Look at the diagram of Thomson's e/m apparatus. Describe what would happen if you could reverse the + and – connections to the deflection plates.

2.3

OBJECTIVES

- The nucleus contains protons
- Atomic number as a basis for the periodic table

THE NUCLEUS AND PROTONS

Answers to questions often suggest further questions. In response to the question, 'What is inside atoms?', Thomson showed that they contain electrons. In this spread we explain how Rutherford subsequently demonstrated that a massive nucleus lies at their centre. A further conclusion was that the electrons surrounding the nucleus define the overall size of an atom. Of course, chemists were not content to let the matter rest at that point. They started thinking 'What is inside the nucleus?' Henry Moseley made the crucial breakthrough.

The Geiger–Marsden experiment

In 1909, Ernest Rutherford suggested to his colleague Hans Geiger that their young student Ernest Marsden might start a little research project. The project involved measuring the deflection of alpha particles as they struck a thin sheet of gold foil less than one micrometre (1 μm) thick. Alpha particles were known to be positively charged. Predictions based on Thomson's model of the atom suggested that the alpha particles would be deflected by a few degrees from the straight-ahead direction when they passed through the gold foil.

Geiger and Marsden made two unexpected observations. First, the vast majority of the alpha particles were deflected by less than one degree, showing that they must have passed through essentially empty space. Secondly, a few (about 1 in 8000) actually bounced *back* towards the source. Rutherford remarked that this was 'almost as incredible as if you had fired a 15-inch shell at a piece of tissue paper and it had bounced back and hit you'.

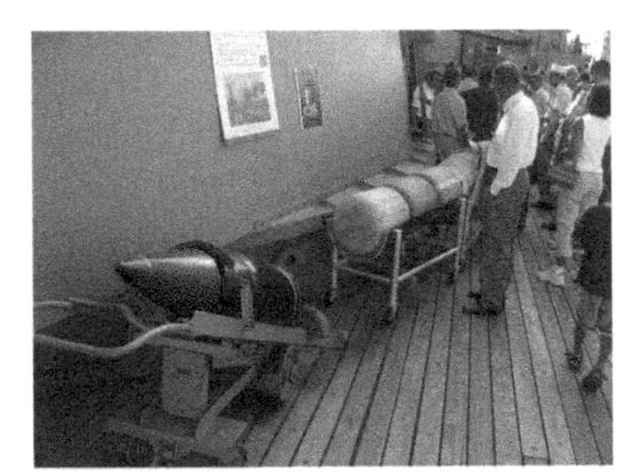

A (dummy) 16-inch shell on board the battleship USS Missouri.

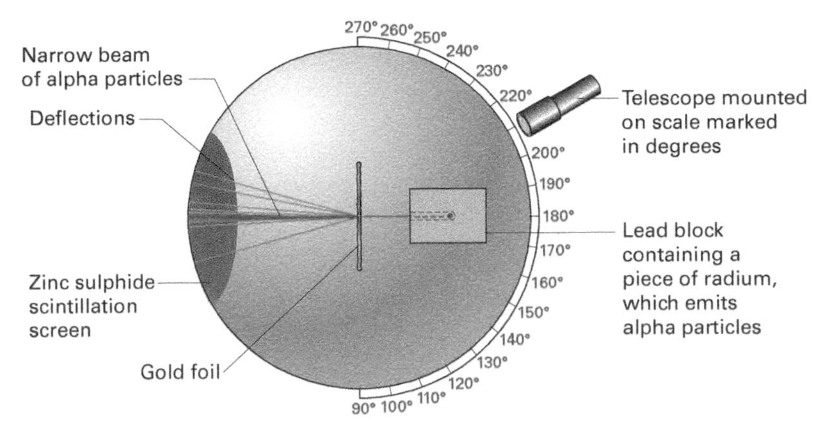

The Geiger–Marsden apparatus. A radioactive substance emits high-speed alpha particles. After passing through the foil, they produce visible scintillations (flashes of light) on a fluorescent screen. The angle of deviation of an alpha particle from the straight-ahead direction can be measured from the position of the flash on the screen.

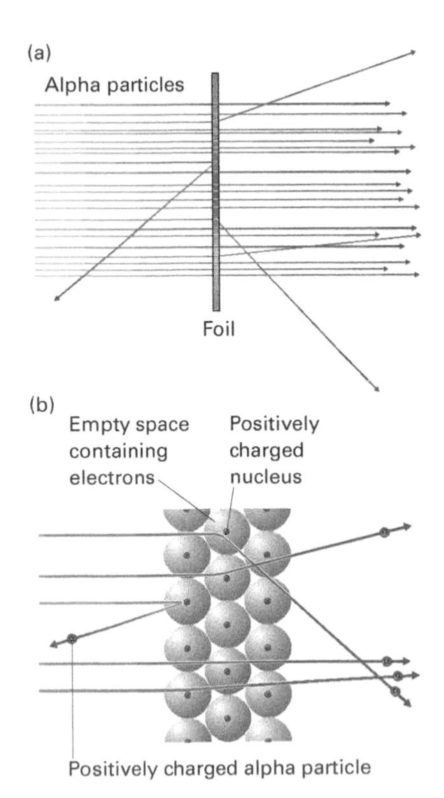

Diagram (a) summarizes the results of Geiger and Marsden and (b) shows Rutherford's interpretation of them. The size of the nucleus is greatly exaggerated for clarity.

The atom: mostly empty space

A new model of the atom was needed to account for the results of the Geiger–Marsden experiment. In 1911 Rutherford proposed that the atom has at its centre a very small positively charged **nucleus**, which contains almost all the mass of the atom. This nucleus is tiny and the rest of the atom is mostly empty space. (If you magnified an atom to the size of a football stadium, its nucleus would be about the size of a pea.) So, most of the alpha particles passed through empty space. Only very rarely would one travel close enough to the nucleus to be repelled strongly by the dense concentration of positive charge. Sufficient electrons surround the nucleus of an atom to balance the charge of the nucleus and to make the atom neutral overall.

Positive particles called protons

Between 1917 and 1921, Rutherford bombarded six different elements with alpha particles. He discovered that the nuclei of boron, nitrogen, fluorine, sodium, aluminium, and phosphorus all gave out the same positive particle, which was identical to the nucleus of the hydrogen atom. Because this was the first particle found in the nucleus, he called it the 'proton' (from the Greek *protos*, meaning 'first'). Rutherford concluded that protons made up the positive part of the nuclei of all elements. The proton carries a positive charge of exactly the same magnitude as the negative charge on the electron.

Radioactivity

α particles are helium nuclei;

β particles are electrons.

Exposure to radioctivity should be avoided where possible as radiation can damage human tissue. See spread 15.7 for an explanation of half-life.

The number of protons

The number of electrons in an atom equals the number of protons in its nucleus. The number of protons is called the **atomic number** (also called **proton number**) of the element concerned. In 1913 Henry Moseley found that when he bombarded elements with high-speed electrons, they emitted X-rays. He observed that the frequency of the emitted X-rays depended on the element used. Moseley plotted a graph of atomic number against the square root of the X-ray frequency. The result was a straight-line plot.

Ordering the elements

The idea that the elements should be ordered by atomic number was first suggested by the Dutch lawyer and amateur theoretical physicist Antonius van den Broek (1870–1926).

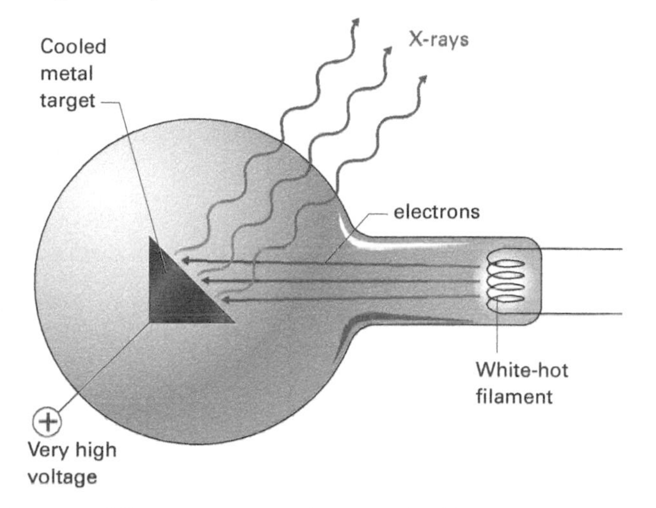

In a typical X-ray apparatus, a heated wire filament emits electrons. They accelerate towards a metal target, which has a very high positive voltage (typically 25 000 volts).

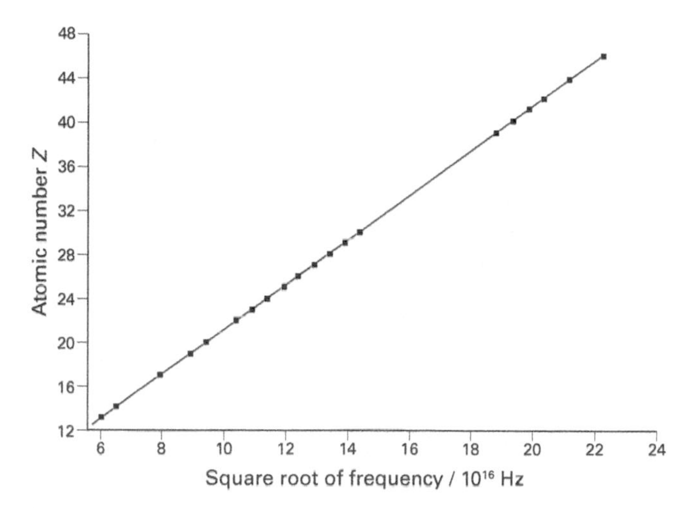

The plot of Moseley's results shows that there is a linear dependence between the atomic number of the element concerned and the square root of the frequency of the emitted X-rays.

It is easy to miss the importance of this result. Remember that Mendeleyev had arranged the elements in his periodic table in order of increasing relative atomic mass. He had then been forced to change the positions of some elements (notably iodine) to ensure that his vertical groups contained elements with similar properties. Moseley's graph listed the elements in their final order in the periodic table. Atomic number is therefore a more fundamental property of an element than its relative atomic mass. The atomic number of an element distinguishes it from all others because it specifies the number of protons in the nucleus.

A recreation of Moseley's study as it would have looked in 1913. His original apparatus is preserved in the Museum of the History of Science in Broad Street, Oxford.

SUMMARY

- The atom has at its centre a very small, positively charged nucleus.
- Almost all the mass of an atom is concentrated in the nucleus.
- The nucleus of an atom contains protons.
- The proton has a positive charge.
- The charge on a proton is equal in magnitude, and opposite in sign, to the charge on an electron.
- The total positive charge on the nucleus equals the total negative charge of the electrons.
- The atomic number of an element equals the number of protons in its nucleus.
- The atomic number of an element distinguishes that element from all others.

Henry Moseley 1887–1915

Moseley was killed in the Gallipoli Campaign during the First World War. The War Office only later understood the magnitude of the loss to science (and hence possibly to the war effort). From that point onwards, valuable scientists were never again allowed to serve in the Front Line.

2.4

OBJECTIVES

- The discovery of the neutron
- Mass number
- Comparison of electrons, protons, and neutrons
- Isotopes

THE NUCLEUS AND NEUTRONS

Having discovered the proton, Rutherford was immediately faced with another problem. He could explain the *charge* on the nucleus of an atom in terms of the number of protons in it. However, there were always (except in the case of hydrogen) too few protons to explain the *mass* of the nucleus. He suggested that there must be another, uncharged, particle present in the nucleus. Rutherford introduced the term **neutron** for these particles in 1921, but they remained a theoretical speculation.

The neutron

Experimental evidence for the neutron was found in 1932, when James Chadwick bombarded the element beryllium with alpha particles. This bombardment produced a highly penetrating stream of particles, which could pass through many centimetres of solid lead and which was not deflected by electric or magnetic fields. Chadwick decided that the stream must consist of particles with almost the same mass as protons but with no charge. Chadwick had detected the neutrons postulated earlier by Rutherford. Protons and neutrons are collectively known as **nucleons** because they are both found in the nucleus.

The winning team – Ernest Rutherford and research colleagues at the Cavendish Laboratory, Cambridge, in 1920. James Chadwick (discoverer of the neutron) is at the extreme left of the middle row. J. J. Thomson (electron) and Ernest Rutherford (proton) are three and four places, respectively, to Chadwick's left. G. P. Thomson (see spread 4.4) is sitting next to Chadwick.

Protons, neutrons, and mass number

An atom consists of a nucleus surrounded by electrons. The nucleus is positively charged because it contains positively charged protons. All atoms of the same element contain the same number of protons. The number of protons in an atom defines its atomic number and hence its identity as an element. Atoms of the same element all have the same atomic number.

Neutrons have almost the same mass as protons, so they contribute to the mass of an atom. The **mass number** (also called **nucleon number**) of an atom is defined as the sum of the numbers of protons and neutrons in the atom's nucleus. For example, the nucleus of a fluorine atom contains 9 protons and 10 neutrons. So, the atomic number of fluorine is 9 and its mass number is 19 (9 + 10).

The strong interaction

Why do nuclei not fly apart? They contain protons packed together in a very small space, but the protons do not seem to follow the rule that 'like charges repel, unlike charges attract'. The explanation is that neutrons and protons participate in an even stronger *attractive* force. This **strong interaction** is about 100 times stronger than charge repulsion or attraction, but only operates over very short distances in the nucleus (about 10^{-15} m).

Full symbols for atoms

An atom is defined by its atomic number. Chemists often incorporate the atomic number and the mass number into the symbol for an element. The full symbol for fluorine is ${}^{19}_{9}F$. The subscript 9 is the atomic number, and the superscript 19 is the mass number. From this full symbol you can work out that an atom of fluorine consists of 9 protons, 10 neutrons, and 9 electrons. As a further short-hand, the subscript (for the atomic number) can be omitted; thus you can write the symbol for fluorine as ${}^{19}F$.

- The number of electrons in a neutral atom always equals the number of protons.
- The number of neutrons equals the mass number minus the atomic number.

	Electron	Proton	Neutron
Relative charge	−1	+1	0
Relative mass	1/1840	1	1*

*The charges and approximate relative masses of electrons, protons, and neutrons. *The mass of the neutron is 1.00138 times the mass of the proton.*

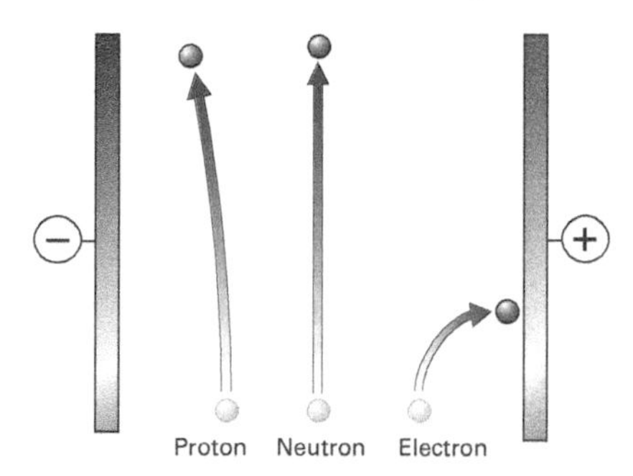

In an electric field, a neutron is not deflected, being neutral. A proton is deflected in the opposite direction to an electron; it is deflected less because it is heavier.

Neutrons and isotopes

Atoms of the same element always have the same number of protons, but may have different numbers of neutrons. Atoms with the same number of protons but different numbers of neutrons are called **isotopes**. A **nuclide** is an isotope with a specified mass number. For example, naturally occurring chlorine consists of two isotopes, ${}^{35}Cl$ (17 protons and 18 neutrons) and ${}^{37}Cl$ (17 protons and 20 neutrons). The proportion of each isotope in a sample is called its **relative abundance**. The relative abundance of ${}^{35}Cl$ is 75.8% and of ${}^{37}Cl$ is 24.2%.

Isotopes and relative atomic mass

Most elements exist naturally as two or more different isotopes. The mass of an element therefore depends on the relative abundances of all the isotopes present in the sample. The **relative atomic mass** A_r of an element is defined as the mass of one atom of that element relative to 1/12th the mass of one atom of carbon-12 (i.e. ${}^{12}C$ = exactly 12). The relative atomic mass is the average of the masses of the stable isotopes of the element, weighted to take into account the relative abundance of each isotope. (Note that the term **relative isotopic mass** refers to the mass of a specific isotope.) For the example of chlorine given above, the relative atomic mass is calculated as follows:

$$A_r(\text{Cl}) = \left(\frac{75.8}{100} \times 35\right) + \left(\frac{24.2}{100} \times 37\right) = 35.5$$

So the relative atomic mass of chlorine is 35.5.

Symbols

The full symbols for protons, neutrons, and electrons are ${}^{1}_{1}p$, ${}^{1}_{0}n$, and ${}^{0}_{-1}e$.

Magnesium – three isotopes

The calculation of relative atomic mass is similar when the element has more than two isotopes. For example, naturally occurring magnesium has the following isotopic composition:

${}^{24}Mg$ = 79.0%, ${}^{25}Mg$ = 10.0%, ${}^{26}Mg$ = 11.0%

The relative atomic mass of magnesium is calculated as:

$$A_r(\text{Mg}) = \left(\frac{79.0}{100} \times 24\right) + \left(\frac{10.0}{100} \times 25\right) + \left(\frac{11.0}{100} \times 26\right) = 24.3$$

Relative isotopic mass

The mass number of an isotope, the sum of the numbers of protons and neutrons, is an integer. Relative isotopic masses are, however, *not* integers, because protons and neutrons have very slightly different masses (see table above).

SUMMARY

- The neutron has zero charge and almost the same mass as the proton.
- Protons and neutrons are called nucleons because they are found in the nucleus of an atom.
- The mass number is the sum of the numbers of protons and neutrons in the nucleus of an atom.
- Isotopes of an element have the same number of protons but different numbers of neutrons; they therefore have different masses.
- A nuclide is an isotope with a specified mass number.
- The relative atomic mass of an element is the average mass of one atom of the element relative to 1/12th the mass of one atom of carbon-12.

PRACTICE

1 Give the number of protons, neutrons, and electrons in:

a ${}^{35}Cl$ b ${}^{24}Mg$ c ${}^{235}U$.

2 Give the atomic number and mass number for each of the isotopes in question 1.

3 Calculate the relative atomic mass of silicon with the following isotopic composition:

${}^{28}Si$ = 92.2%, ${}^{29}Si$ = 4.7%, ${}^{30}Si$ = 3.1%.

3 Masses of atoms and the mole

Chemists are interested in atoms and how they react together. They want to know about two main aspects of chemical reactions: the *qualitative* and the *quantitative*. Examples of qualitative questions are: 'What happens if I heat substance *x* with substance *y*?' and 'Why does that mixture explode when I drop it on the floor?' Quantitative questions are concerned with the 'how much' aspects of chemistry. A typical question would be:
'What mass of substance *x* will react with 10 g of substance *y*?'
Quantitative questions are particularly important in the chemical industry: adding too much of a reagent will result in unnecessary cost and may also contaminate the product. The aim of this chapter is to show how you can calculate the quantities of reactants and products involved in chemical reactions. To do this you need to understand relative formula masses and the concept of the *mole*.

FINDING THE MASSES OF ATOMS

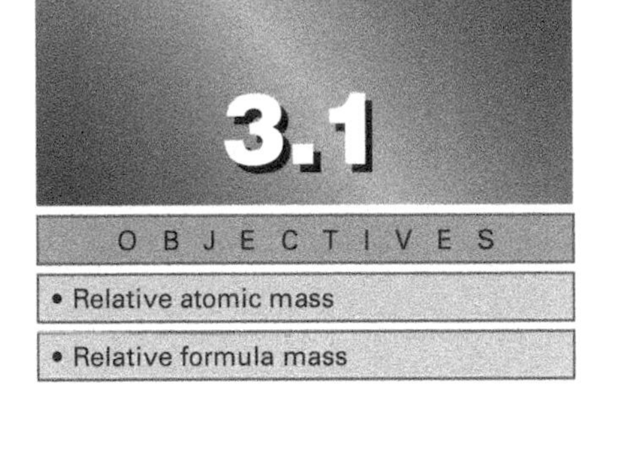

OBJECTIVES

- Relative atomic mass
- Relative formula mass

During an earlier science course, you may have seen the following spectacular but *very* dangerous demonstration. Mix iron(III) oxide powder with aluminium powder in a clay crucible. Insert a strip of magnesium ribbon to act as a fuse, light the end of the ribbon, and move back to a safe distance. When the reaction has died down, a red-hot lump of iron will be found amongst the shattered remains of the crucible. The balanced chemical equation below shows the number of atoms involved and how they react together:

$$Fe_2O_3(s) + 2Al(s) \rightarrow 2Fe(l) + Al_2O_3(s)$$

The equation shows *how many atoms* react, but it does not indicate directly the *masses* that react. To find the answer to this question, we need to consider the masses of atoms themselves.

Iron(III) oxide and aluminium react to produce molten iron, which can weld railway lines together (the Thermit process). The correct masses must be used for a successful reaction.

Relative atomic mass

The masses of individual atoms are too small to be used in calculations on chemical reactions. So instead, the mass of an atom is expressed relative to a chosen standard atomic mass:

- The relative atomic mass of an element is the average mass of one atom of the element relative to 1/12th the mass of one atom of carbon-12.

The symbol for relative atomic mass is A_r.

On this scale, A_r for carbon-12 is *exactly* 12. The value of A_r for carbon given in a data book is 12.01 (to two decimal places). This figure illustrates the meaning of the phrase 'average mass of one atom of the element' in the definition above. Carbon occurs naturally as a mixture of carbon-12 (98.9%) and carbon-13 (1.1%). If the A_r value for a specific isotope is quoted, the name given is *relative isotopic mass* (see final spread in previous chapter).

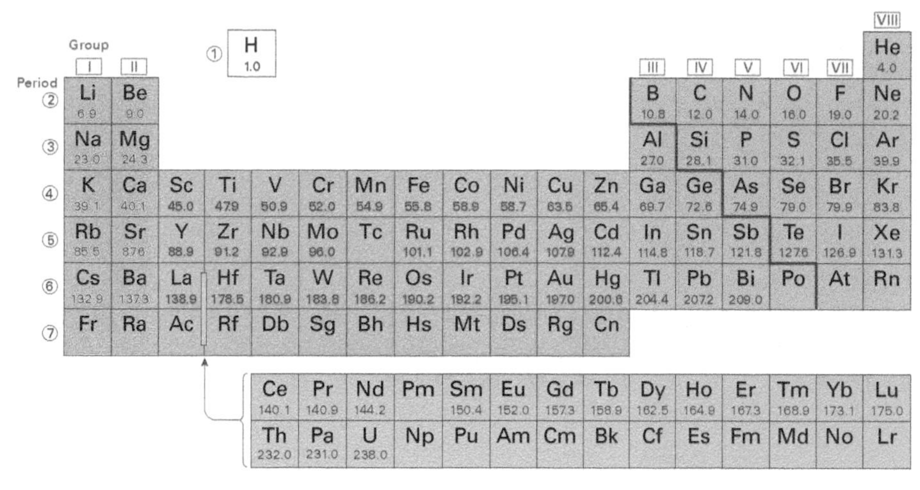

Periodic table of the elements showing relative atomic masses. Values are left out if an element has no stable isotope.

Relative atomic masses are known with great precision, usually to at least three decimal places. The values in the table above are consistently given to one decimal place. It is important to understand that, because A_r *compares* the masses of atoms, it does not itself have units. The scale is *relative*; one mass is divided by another mass and so the ratio has no units.

Relative formula mass

Atoms combine to form compounds. The formula of a compound shows the ratio in which the atoms combine. For example, the formula of water is H_2O; two atoms of hydrogen combine with one atom of oxygen. The formula of aluminium oxide is Al_2O_3; two atoms of aluminium combine with three atoms of oxygen. The idea of relative atomic masses can be extended to compounds:

- The **relative formula mass** of a compound is the sum of the relative atomic masses of all the atoms present in its formula.

The symbol for relative formula mass is M_r.

From the table above, A_r for hydrogen is 1.0, for oxygen is 16.0, and for aluminium is 27.0. The relative formula mass for water is:

$(2 \times 1.0) + (1 \times 16.0) = 18.0$

The relative formula mass for aluminium oxide is:

$(2 \times 27.0) + (3 \times 16.0) = 102.0$

These calculations allow us to compare the masses of atoms and compounds with each other. Calculating the actual masses that react depends on the idea of the *mole*, which is described later in this chapter. The concept of the 'mole' cannot be used successfully without a grasp of relative formula mass, so try the questions below to check your understanding.

Relative molecular mass

'Relative molecular mass' is the traditional name used for 'relative formula mass'. The modern name is better because the traditional term 'relative molecular mass' really refers only to compounds composed of molecules.

SUMMARY

- Relative atomic mass has no units.
- The relative formula mass (M_r) is the sum of the relative atomic masses of all the atoms present in the formula of a compound.

PRACTICE

1 Write down the relative atomic masses of: beryllium; boron; calcium; carbon; iron; manganese; phosphorus; sodium; xenon.

2 Calculate the relative formula masses of:

a N_2; CO_2; $CaCO_3$; H_2SO_4.

b ammonia; sulphur dioxide; methane; nitric acid.

3.2 THE MASS SPECTROMETER

OBJECTIVES

- The mass spectrometer
- Interpreting mass spectra

Ions

A **positive ion** has more protons than electrons, whereas a **negative ion** has more electrons than protons.

The relative atomic mass of an element is the average mass of one atom of the element relative to 1/12th the mass of one atom of carbon-12. In order to calculate the average mass, the masses of the isotopes of the element must be known, together with their relative abundances. These values are found using an instrument called the *mass spectrometer*.

The mass spectrometer

The diagram below illustrates the operating principles of a **time-of-flight (TOF)** mass spectrometer. TOF mass spectrometers only became possible with the advent of fast computers with at least 100 gigabytes of memory.

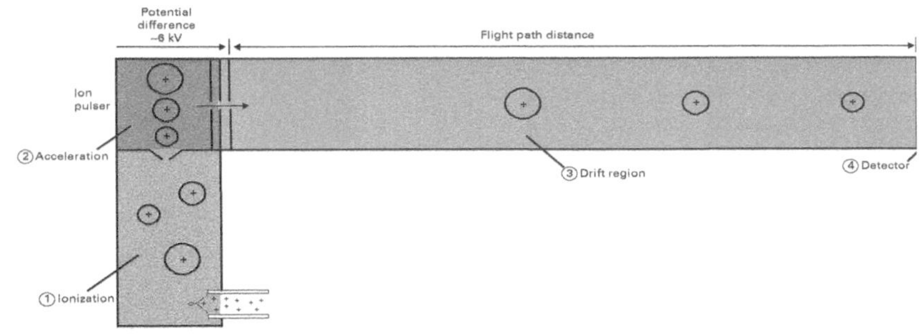

Other MS versions

Large biological molecules can be dislodged from a solid sample by a stream of rapidly moving atoms, called fast atom bombardment (**FAB**). They can be ionized by matrix-assisted laser desorption ionization (**MALDI**). MALDI-TOF can measure molecular masses up to more than 100 000.

There are four main steps:

1. *Ionization.* Some instruments use **electron ionization** (EI) in which a vaporized sample is ionized by impact of a beam of high-energy electrons from an 'electron gun' (a heated wire filament), see figure on opposite page. The electron beam knocks out an electron from the atom or molecule, forming a positive ion. The high energy involved in the impact can cause a molecule to undergo fragmentation: this is discussed in detail in spread 31.5. Modern instruments often use **electrospray ionization** (ESI) in which the sample in a solvent such as aqueous methanol is sprayed through a fine capillary held at a high potential (1 to 4 kV) aided by a flow of warm nitrogen gas, producing charged aerosol droplets. The ionization chamber, at atmospheric pressure, is separated by a small orifice from the high vacuum of the rest of the instrument. Droplets, swept through the orifice, become smaller as the solvent evaporates — they eventually 'explode' to create individual ions. A trace of ethanoic acid is often added to aid protonation, so a species X produces an XH^+ ion. Fragmentation rarely occurs with electrospray ionization, which is a 'soft' ionization technique.
2. *Acceleration.* The ions pass into the acceleration chamber — a stack of plates, each with a central hole, apart from the back plate. To start the ion's flight to the detector, a high-voltage pulse is applied to the back plate. The *electric field* produced accelerates the ions to such high kinetic energy that random thermal motion is negligible — all ions of the same charge have the same kinetic energy. Therefore ions of the same **mass-to-charge ratio** (***m/z***) have the same speed.
3. *Drift region.* The ions are allowed to drift in a *field-free* region typically 30 to 100 cm long. Ions having the same kinetic energy have the same value for $\frac{1}{2}mv^2$, where m is the mass and v the speed of the ion; so an ion of lower mass will have a higher speed. Since lighter ions travel faster, they reach the end of the drift region before heavier ions do. The different times at which the ions reach the detector, after calibration with samples of accurately known mass, give the m/z ratios of the ions.
4. *Detector/data analysis.* The stream of ions entering the detector makes an electric current: a *microchannel plate* acts as an electron multiplier, amplifying this current. The output of the detector is sampled, often by an analogue-to-digital converter (ADC), about once every nanosecond. The magnitude of the current depends on the number of ions entering the detector at a specific time. The relative abundances of different isotopes are calculated by comparing the magnitudes of their currents.

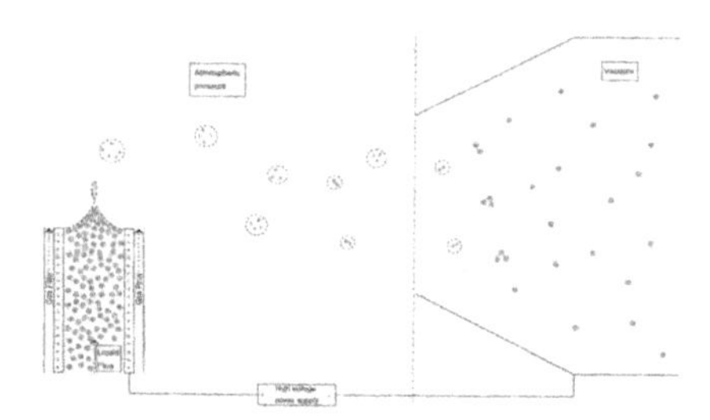

Electrospray ionization.

The need for a vacuum

The space inside a mass spectrometer is connected to a vacuum pump. The ions under investigation must be able to move freely. The machine would not work properly if the ions collided with the molecules of oxygen and nitrogen that are present in the atmosphere. So a vacuum is needed inside the apparatus.

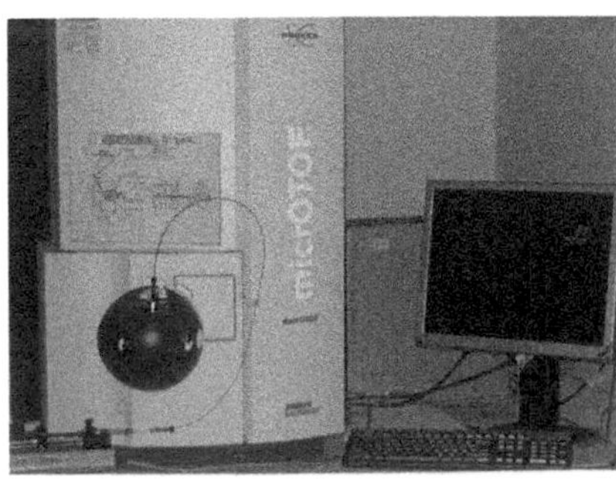

A time-of-flight mass spectrometer. See spread 31.5.

Interpreting a mass spectrum

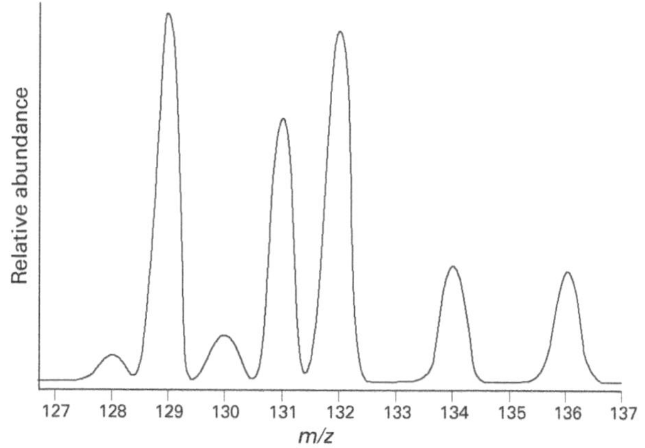

The mass spectrum for naturally occurring xenon. The horizontal axis is the mass-to-charge ratio m/z. As all ions are assumed to be unipositive, this measures the isotopic mass. The vertical axis shows the relative abundance.

A commercial mass spectrometer like this can be used to produce the spectrum shown to the left.

The mass spectrum (plural = spectra) for lead gives the following information:

Isotope	Detector current/arbitrary units
204	0.16
206	2.72
207	2.50
208	5.92

The total detector current is (0.16 + 2.72 + 2.50 + 5.92) = 11.30.
The relative abundance of lead-206 is therefore $\frac{2.72}{11.30} \times 100\% = 24.1\%$.
Similar calculations give the following results:

Isotope	Relative abundance/%
204	1.4
206	24.1
207	22.1
208	52.4

The relative atomic mass of lead can now be calculated:

$$A_r = \left(\frac{1.4}{100} \times 204\right) + \left(\frac{24.1}{100} \times 206\right) + \left(\frac{22.1}{100} \times 207\right) + \left(\frac{52.4}{100} \times 208\right)$$

$$= 207.2$$

Data presentation

Most modern mass spectrometers are equipped with microprocessors, which provide a numerical display of the masses of the isotopes and their relative abundances. Simpler machines display results in the form of a trace called a *mass spectrum*. Both types of display give the information needed to calculate the relative atomic mass of an element.

Space probes

Mass spectrometers are frequent passengers on space probes. The Viking spacecraft found that the atmosphere of Mars was mostly carbon dioxide.

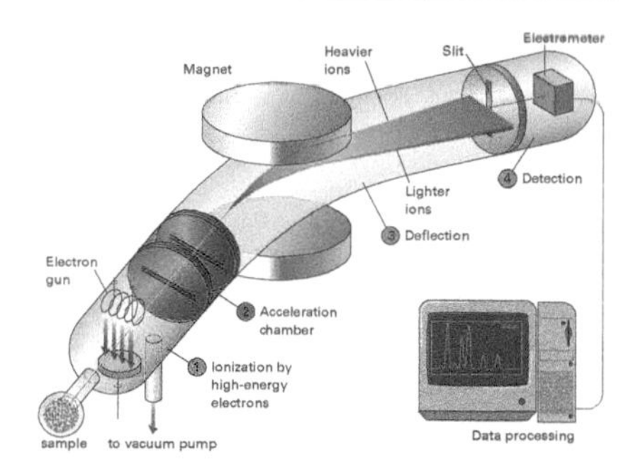

This older mass spectrometer used electron ionization (and separated the ions by deflection in a magnetic field).

SUMMARY

- The mass spectrometer produces ions by electron ionization or electrospray ionization – the ions are accelerated by an electric field. They drift at different speeds through a field-free region before arriving at the detector at different times.
- The magnitudes of the detector currents of different isotopes are proportional to their relative abundances.
- The mass spectrometer can be used to determine relative atomic masses.

PRACTICE

1 A mass spectrometer provided the information on the right for a sample of naturally occurring germanium.

Calculate the relative atomic mass for germanium.

Isotope	Detector current/arbitrary units
70	6.83
72	9.13
73	2.60
74	12.17
76	2.60

3.3

OBJECTIVES

- The mole
- Avogadro constant
- Molar mass
- Amount of substance

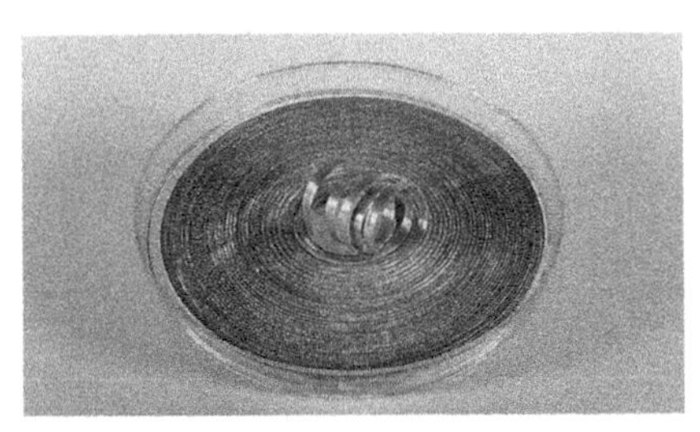

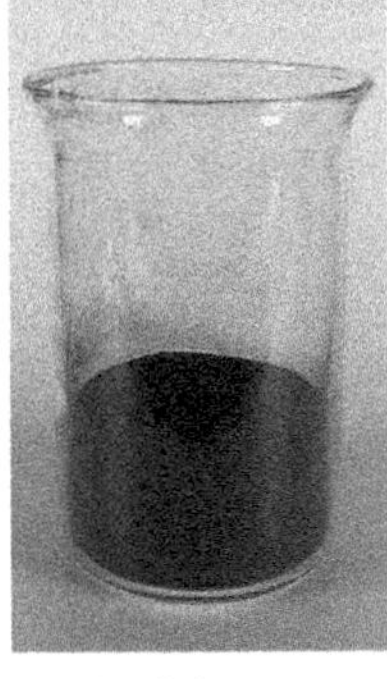

Each of these samples consists of one mole of an element. Each sample contains 6.02×10^{23} atoms. From top to bottom, the elements are magnesium, copper, mercury (which is very dense), carbon, and iodine.

A very big number

One mole (1 mol) of particles contains 6.02×10^{23} of those particles. This number of grains of sand would cover the surface of the Earth to a depth of about 2 metres. The mole lives up to the meaning of its name: 'massive heap'.

THE MOLE

Chemical reactions change reactants into products. The atoms that make up the reactants rearrange to form the products. When carrying out a chemical reaction, the correct quantity of each reactant must be used for all the reactants to change into products. However, the balanced chemical equation for a reaction only gives precise information about the numbers of atoms involved. It is not possible to count out individual atoms: they are too small to be of any use in the measurement of reacting masses. Instead, chemists use the concept of the *mole*, which allows them to count atoms by weighing them.

The mole

A bank clerk (and the bank's customers!) would find it very tedious to count out individual coins to confirm that a customer had paid in the correct number of coins. Instead of counting them out individually, the clerks know the mass of a chosen number of each coin. For example, they count ten-pence pieces in collections of 50, and fifty-pence pieces in collections of 20. They can then weigh a bag of coins to confirm that the correct number of coins is present.

In a similar fashion, atoms are too small to be counted individually on a routine basis. So, chemists also count atoms by weighing a collection of them. The connection between the microscopic world of atoms and the everyday world of balances is that the mass of a particular fixed number of atoms is known.

The number of atoms in exactly 12 g of carbon-12 is chosen as the standard, because carbon-12 is the standard chosen for relative atomic mass:

- The number of atoms in exactly 12 g of carbon-12 is called one mole.
- The number of atoms per mole is called the **Avogadro constant**.

As a unit, we use the abbreviation 'mol' for mole. The symbol L (or N_A) is used for the Avogadro constant.

The mass of a single carbon-12 atom has been found (using a mass spectrometer) to be 1.993×10^{-23} g, and so the Avogadro constant L has the value

$$L = \frac{\text{mass per mole of } ^{12}\text{C}}{\text{mass of one atom of } ^{12}\text{C}} = \frac{12\,\text{g mol}^{-1}}{1.993 \times 10^{-23}\,\text{g}} = 6.02 \times 10^{23}\,\text{mol}^{-1}$$

The relative atomic mass of carbon-12 is exactly 12, by definition; that of magnesium-24 is 24. So one atom of magnesium-24 has twice the mass of one atom of carbon-12. However many atoms are chosen, a given number of magnesium-24 atoms always has twice the mass of the same number of atoms of carbon-12. So one mole of magnesium-24 atoms has twice the mass of one mole of carbon-12 atoms.

Molar mass

The mass per mole of an atom is called its **molar mass**, M. The molar mass has the same numerical value as the relative atomic mass, *but it also has units* of grams per mole (g mol^{-1}).

The concept of the mole can be extended to compounds. Once again, the mass per mole of the compound is called its **molar mass**. The molar mass of a compound has the same numerical value as the relative formula mass, but it also has units of grams per mole (g mol^{-1}). One mole of any object always contains 6.02×10^{23} copies of that object.

There is nothing unusual in counting using a named collection of items: farmers count eggs in dozens (12); and shopkeepers sell sheets of paper in reams (500). Now we see that chemists count atoms using moles. The mole is simply the chemist's equivalent of the dozen or the ream. The numbers involved are much, much larger, but the principle is the same.

Amount of substance

The physical quantity **amount of substance** (symbol n) is measured in moles. **One mole** (1 mol) is the amount of any substance that contains the same number of particles as there are atoms in exactly 12 g of carbon-12. The particles (atoms, molecules, ions, etc.) need to be carefully specified.

For example, 1 mol C contains one mole of carbon atoms; 1 mol H_2O contains one mole of water molecules; 1 mol NaCl contains one mole of sodium chloride ion pairs, i.e. one mole of sodium ions and one mole of chloride ions. In each case, the *formula* indicates the species concerned.

Calculations: mass, molar mass, and amount in moles

The amount in moles (n) in a sample, the mass of the sample (m), and its molar mass (M) are related by the expression:

$$\text{amount in moles} = \frac{\text{mass}}{\text{molar mass}} \qquad \text{or} \qquad n = \frac{m}{M}$$

Suppose a sample of carbon has a mass of 3.0 g. The molar mass of carbon C is 12.0 g mol^{-1}. The amount in moles of carbon in the sample may be calculated by substitution into the above expression:

$$n = \frac{m}{M} = \frac{3.0\,\text{g}}{12.0\,\text{g mol}^{-1}} = 0.25\,\text{mol}$$

That is, 3.0 g of carbon contains 0.25 mol of carbon atoms.

The expression for amount in moles may be rearranged in two ways to make mass, or molar mass, the subject:

$$\text{mass} = \text{amount in moles} \times \text{molar mass} \qquad \text{or} \qquad m = nM$$

and

$$\text{molar mass} = \frac{\text{mass}}{\text{amount in moles}} \qquad \text{or} \qquad M = \frac{m}{n}$$

SUMMARY

- One mole is the amount of any substance that contains the same number of particles as there are atoms in exactly 12 g of carbon-12.
- The number of particles per mole (6.02×10^{23} mol^{-1}) is the Avogadro constant.
- The molar mass of a substance is its mass per mole. It has the same numerical value as the relative formula mass.

Amount

Because the word 'amount' is used so imprecisely in everyday speech, we will emphasize the scientific use by using the term **amount in moles**.

New definition of one mole

IUPAC issued a new definition of one mole in 2018 as the amount of substance containing exactly $6.02214076 \times 10^{23}$ particles. The new definition helpfully emphasizes that the amount of substance is concerned with counting particles rather than measuring masses. However, this change is expected to make a negligible difference in everyday usage.

Mass calculation

What is the mass of 2.00 mol of neon [A_r(Ne) = 20.2]?

We have

amount in moles = 2.00 mol

molar mass = 20.2 g mol^{-1}

So

mass = amount in moles × molar mass

= (2.00 mol) × (20.2 g mol^{-1})

= 40.4 g

Molar mass calculation

A pure sample of lithium has a mass of 1.39 g and contains 0.200 mol of lithium. What is the molar mass of lithium?

We have

mass = 1.39 g

amount in moles = 0.200 mol

So

$$\text{molar mass} = \frac{\text{mass}}{\text{amount in moles}} = \frac{1.39\,\text{g}}{0.200\,\text{mol}} = 6.95\,\text{g mol}^{-1}$$

PRACTICE

1 Calculate the mass in grams of:
   - a 2 mol of carbon atoms [A_r(C) = 12.0]
   - b 0.5 mol of magnesium atoms [A_r(Mg) = 24.3]
   - c 0.01 mol of aluminium atoms [A_r(Al) = 27.0]
   - d 5 mol of sodium atoms [A_r(Na) = 23.0]
   - e 5 mol of sodium ions Na^+.

2 Calculate the mass in kilograms of:
   - a 1000 mol of iron atoms [A_r(Fe) = 55.8]
   - b 1×10^4 mol of tungsten atoms [A_r(W) = 183.8]
   - c 1×10^{-3} mol of hydrogen atoms [A_r(H) = 1.0].

3 Calculate the amount in moles (to 3 sig. figs) of atoms in:
   - a 23.0 g of sodium
   - b 62.0 g of phosphorus [A_r(P) = 31.0]
   - c 10.0 g of calcium [A_r(Ca) = 40.1].

3.4

O B J E C T I V E S

- Diatomic molecules
- Covalent compounds
- Ionic compounds

Finding molar masses

Some elements exist as separate atoms, e.g. He and Ne. Some elements exist as molecules consisting of two or more atoms joined together, e.g. O_2 and S_8. **Compounds** consist of two or more atoms of *different* elements bonded together, e.g. HCl, NH_3, NaBr, and H_2SO_4. The molar mass of any of these substances may be derived from its chemical formula. This spread shows you how.

Diatomic molecules

Elements such as neon consist of separate atoms. Therefore 1 mol Ne contains one mole of neon atoms. Elements such as oxygen (O_2), hydrogen (H_2), and chlorine (Cl_2) exist as diatomic molecules. A **diatomic molecule** contains two atoms. 1 mol O_2 contains 6.02×10^{23} oxygen *molecules*. It is also true to say that one mole of oxygen molecules contains two moles of oxygen atoms, because each oxygen molecule contains two atoms of oxygen.

Iodine consists of I_2 molecules. The relative atomic mass of iodine is 126.9. The molar mass of I_2 is therefore $2 \times 126.9 = 253.8\ g\ mol^{-1}$
Note that this sample is twice the size of that shown in the previous spread: this is one mole of I_2 **molecules**.

Worked examples on calculating the molar mass of a diatomic molecule

We shall look at two examples, oxygen O_2 and chlorine Cl_2.

The relative atomic mass of oxygen is 16.0.
1 mol of oxygen atoms O therefore has a mass of 16.0 g.
1 mol of oxygen *molecules* O_2 has a mass of
$2 \times (16.0\ g) = 32.0\ g$.

By similar reasoning,
1 mol of chlorine molecules Cl_2 [$A_r(Cl) = 35.5$] has a mass of
$2 \times (35.5\ g) = 71.0\ g$.
It follows that the molar mass of chlorine Cl_2 is $71.0\ g\ mol^{-1}$.

A ball-and-stick model of an iodine molecule.

Covalent compounds

Covalent compounds consist of atoms bonded together to form molecules. Examples include carbon dioxide CO_2, water H_2O, and ammonia NH_3. The chemical formula of a covalent compound identifies the numbers of atoms of each element making up each molecule. The formula of carbon dioxide is CO_2, indicating that each molecule of carbon dioxide consists of one carbon atom bonded to two oxygen atoms.

Worked examples on calculating the molar mass of a covalent compound

We shall look at two examples, carbon dioxide CO_2 and ammonia NH_3. One mole (1 mol) of carbon dioxide molecules contains 6.02×10^{23} carbon dioxide molecules. Each of these molecules contains one atom of carbon bonded to two atoms of oxygen.

The molar mass of carbon C is $12.0\ g\ mol^{-1}$.
The molar mass of oxygen O is $16.0\ g\ mol^{-1}$.
The molar mass of carbon dioxide CO_2 is therefore
$12.0\ g\ mol^{-1} + 2 \times (16.0\ g\ mol^{-1}) = 44.0\ g\ mol^{-1}$.
So, the mass of 1 mol CO_2 is 44.0 g.

By similar reasoning,
NH_3 has a molar mass of
$14.0\ g\ mol^{-1} + (3 \times 1.0\ g\ mol^{-1}) = 17.0\ g\ mol^{-1}$.

That very big number again!

Note that 44.0 g of carbon dioxide and 17.0 g of ammonia contain the same number of molecules (6.02×10^{23} molecules).

Ionic compounds

Ionic compounds consist of oppositely charged ions. Metal ions are positively charged and non-metal ions are negatively charged. For example, sodium chloride consists of sodium ions Na^+ and chloride ions Cl^-. The formula NaCl tells you that sodium chloride consists of equal numbers of sodium and chloride ions. The formula for calcium chloride is $CaCl_2$, which tells you that a sample of calcium chloride contains twice as many chloride ions as calcium ions. Ionic compounds do not consist of molecules, so it is incorrect to speak about 'sodium chloride molecules'.

Worked examples on calculating the molar mass of an ionic compound

Again we consider two examples, sodium chloride NaCl and calcium bromide $CaBr_2$.

1 mol NaCl consists of one mole of sodium ions Na^+ and one mole of chloride ions Cl^-. Ions form when atoms gain or lose electrons. The mass of an electron is negligible compared to the masses of the protons and neutrons in an atom, so the mass of an ion is regarded as being the same as the mass of its parent atom.

The molar mass of the sodium ion Na^+ is $23.0\,g\,mol^{-1}$.
The molar mass of the chloride ion Cl^- is $35.5\,g\,mol^{-1}$.
The molar mass of sodium chloride NaCl is therefore
$23.0\,g\,mol^{-1} + 35.5\,g\,mol^{-1} = 58.5\,g\,mol^{-1}$.
So, the mass of 1 mol NaCl is 58.5 g.

By similar reasoning,
the molar mass of calcium bromide $CaBr_2$ is
$40.1\,g\,mol^{-1} + 2 \times (79.9\,g\,mol^{-1}) = 199.9\,g\,mol^{-1}$.

One mole of some ionic compounds. The green solid is hydrated nickel(II) chloride $NiCl_2{\cdot}6H_2O$, the pink solid is hydrated cobalt(II) chloride $CoCl_2{\cdot}6H_2O$, and the blue solid is hydrated copper(II) sulphate $CuSO_4{\cdot}5H_2O$. The orange solid is potassium dichromate(VI) $K_2Cr_2O_7$ and the white solid is sodium chloride NaCl.

Water of crystallization

Water of crystallization is the definite amount of water chemically combined in a crystal when it is formed. There are three examples in the photograph on the left; other examples are found in spreads 10.2 and 18.6. If hydrated copper(II) sulphate is heated strongly, the water is lost and the resulting white solid is **anhydrous** copper(II) sulphate, $CuSO_4$.

SUMMARY

- The molar mass of a compound has the same numerical value as its relative formula mass.

PRACTICE

1 Calculate the molar masses of the following substances:
 a ammonia NH_3
 b ethanol CH_3CH_2OH
 c sodium sulphate Na_2SO_4
 d trioxygen (ozone) O_3.

2 Calculate the masses of:
 a 2.0 mol of ammonia
 b 0.20 mol of ethanol
 c 2.5 mol of sodium bromide NaBr.

3 Calculate the amount in moles of each of the following:
 a 2.57 g of sulphur S_8
 b 19.5 g of sodium chloride
 c 4.9 g of sulphuric acid.

PRACTICE EXAM QUESTIONS

1

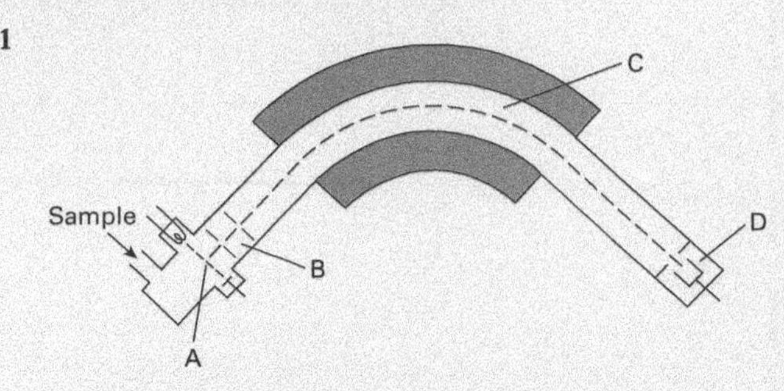

The simplified diagram shows the path of a $^{20}Ne^+$ ion through a mass spectrometer.

a Give the number of neutrons and the number of electrons in this ion. [2]

b Name the processes which occur in regions A, B, C, and D in the mass spectrometer. [4]

c **i** On a copy of the diagram of the mass spectrometer sketch the path which would be followed by a $^{21}Ne^+$ ion introduced into the spectrometer at the same time as the $^{20}Ne^+$ ion shown.

ii Explain why the paths travelled by the two ions differ. [2]

d The relative abundances of the two neon isotopes in a sample are 91.0% ^{20}Ne and 9.0% ^{21}Ne. Calculate a value for the relative atomic mass of neon. [2]

2 **a** Give the meaning of the terms *mass number* of an isotope and *relative atomic mass* of an element. [2]

b Before atoms are deflected in a mass spectrometer they must be ionized and then accelerated.

i Explain briefly why the atoms need to be ionized.

ii What method is used to accelerate the ions? [2]

c The diagram below shows the mass spectrum of an element which has been ionized by removal of only one electron from each atom (forming mononuclear ions).

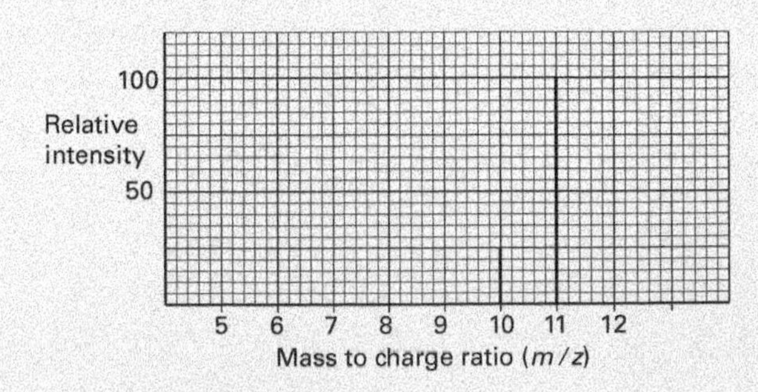

i Use the mass spectrum and your Periodic Table to identify this element. Give the symbol for one isotope of the element with its mass number and atomic number.

ii Use the information given on the mass spectrum to calculate a value for the relative atomic mass of the element. [5]

d Describe briefly how the spectrum in part **c** would differ if the ions were produced by removing two electrons from each atom and give a reason for your answer. [3]

3 The table below shows some accurate relative atomic masses.

Atom	^{1}H	^{12}C	^{6}Li
Relative atomic mass	1.0078	12.0000	6.0149

a Why is ^{12}C the only atom with a relative atomic mass which is an exact whole number? [1]

b Calculate the mass of 1 mol of $^{1}H^+$ ions. The mass of a single electron is 9.1091×10^{-28} g. [Avogadro's number, L, is 6.0225×10^{23} mol^{-1}] [2]

c **i** Explain briefly the process by which a sample is ionized in a mass spectrometer.

ii Give **one** reason why it is important to use the minimum possible energy to ionize a sample in a a mass spectrometer.

iii After ionization and before deflection, what happens to the ions in a mass spectrometer; how is this achieved? [5]

d Why is it a good approximation to consider that the relative atomic mass of the $^{6}Li^+$ ion, determined in a mass spectrometer, is the same as that of ^{6}Li? [1]

4 **a** A proton, a neutron, and an electron all travelling at the same velocity enter a magnetic field. State which particle is deflected the most and explain your answer. [2]

b Give two reasons why particles must be ionized before being analysed in a mass spectrometer. [2]

c A sample of boron with a relative atomic mass of 10.8 gives a mass spectrum with two peaks, one at $m/z = 10$ and one at $m/z = 11$. Calculate the ratio of the heights of the two peaks. [2]

d Compound **X** contains only boron and hydrogen. The percentage by mass of boron in **X** is 81.2%. In the mass spectrum of **X** the peak at the largest value of m/z occurs at 54.

i Use the percentage by mass data to calculate the empirical formula of **X**.

ii Deduce the molecular formula of **X**. [4]

(See also chapter 9.)

5 **a** State the meaning of the term *atomic number*. [1]

b What is the function of the electron gun and magnet in a mass spectrometer? [2]

c The mass spectrum of a pure sample of a noble gas has peaks at the following m/z values.

m/z	10	11	20	22
Relative intensity	2.0	0.2	17.8	1.7

i Give the complete symbol, including mass number, and atomic number for one isotope of this noble gas.

ii Give the species which is responsible for the peak at *m*/*z* = 11.

iii Use appropriate values from the data above to calculate the relative atomic mass of this sample of noble gas. [6]

6 a Define the term *mole*. [2]

b A carbonate of metal **M** has the formula M_2CO_3.

The equation for the reaction of M_2CO_3 with hydrochloric acid is given below

$M_2CO_3 + 2HCl \rightarrow 2MCl + CO_2 + H_2O$

0.245 g of M_2CO_3 was found to exactly neutralize 23.6 cm^3 of hydrochloric acid of concentration of 0.150 mol dm^{-3}. Carry out the following calculations and hence deduce the identity of **M**.

i Calculate the number of moles of hydrochloric acid. [1]

ii Calculate the number of moles of M_2CO_3. [2]

iii Calculate the relative molecular mass of M_2CO_3. [1]

iv Calculate the relative atomic mass of metal **M** and hence deduce its identity. [4]

c Here is a simplified diagram of a mass spectrometer which could also be used to determine the relative atomic mass of metal **M**.

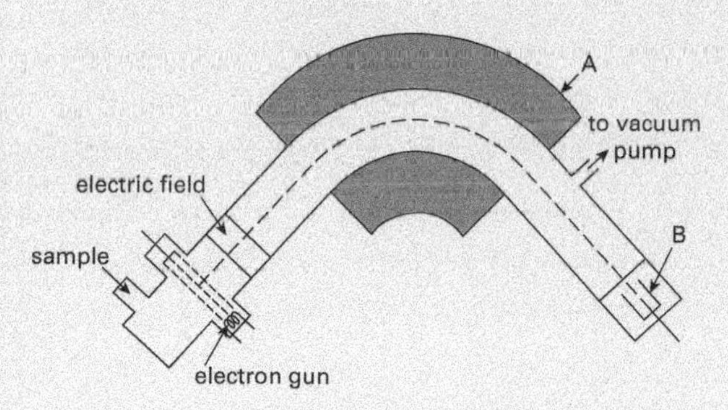

Give the names of the parts labelled **A** and **B**, and state the purpose of the parts labelled *electron gun* and *electric field*. [4]

(See also chapter 9.)

7 a Complete the following table, giving the name, relative mass, and relative charge of each of the fundamental sub-atomic particles:

Particle	Relative mass	Relative charge
	1.0	
neutron		
		-1

[3]

b The element magnesium (atomic number 12) has three isotopes of mass numbers 24, 25, and 26, having abundances of 78.6%, 10.1%, and 11.3%, respectively.

Explain the meaning of the terms:

i atomic number, [1]

ii isotopes. [1]

c Before magnesium atoms can be deflected in a mass spectrometer, they must be ionized and then accelerated.

i Briefly explain why the atoms need to be ionized. [1]

ii By what method are the ions accelerated? [1]

iii By what means are the ions deflected? [1]

d Define the term *relative atomic mass*. [2]

e Use the information given in **b** to:

i calculate the relative atomic mass of magnesium, giving your answer to two decimal places; [2]

ii draw the mass spectrum for magnesium on graph paper. [3]

4 Electrons in atoms

In the early years of the twentieth century, Rutherford's model of the atom consisted of a positively charged nucleus surrounded by a cluster of negatively charged electrons. It is well known that opposite charges attract each other, so how do the electrons remain apart from the nucleus? Rutherford's student, Niels Bohr, suggested that electrons orbit around the nucleus, rather like the planets around the Sun. The Bohr model allowed the structure of the hydrogen atom to be worked out. In the course of this chapter, you will come to understand that a whole new way of thinking about matter and energy is required, as the Bohr model failed for other atoms.

USING LIGHT TO FIND OUT ABOUT ATOMS

OBJECTIVES

- Light and electromagnetic radiation
- Emission of electromagnetic radiation by atoms
- Measurement of the wavelength of light
- Planck's equation

The modern theory of the atom rests on one main observed fact: atoms can emit light if they have absorbed energy (become excited). The wavelength of the light is characteristic of the element concerned. For example, sodium salts colour a Bunsen flame yellow; a discharge tube containing neon gas glows red. However, before attempting to link the colour of the radiated light to the internal structure of the atom, you must first understand something about the nature of light.

Light and electromagnetic radiation

Light is a form of electromagnetic radiation. **Electromagnetic radiation** is energy, associated with electric and magnetic fields, travelling as waves. A wave is described by its frequency (f) and its wavelength (λ).
The frequency and wavelength of light are related by the equation:

$$c = f\lambda$$

where c is the speed at which the waves are travelling (the speed of light, $2.998 \times 10^8\ \mathrm{m\,s^{-1}}$). The unit of wavelength is the metre (m); the unit of frequency is the hertz (Hz). The whole range of frequencies of electromagnetic radiation is called the **electromagnetic spectrum**.

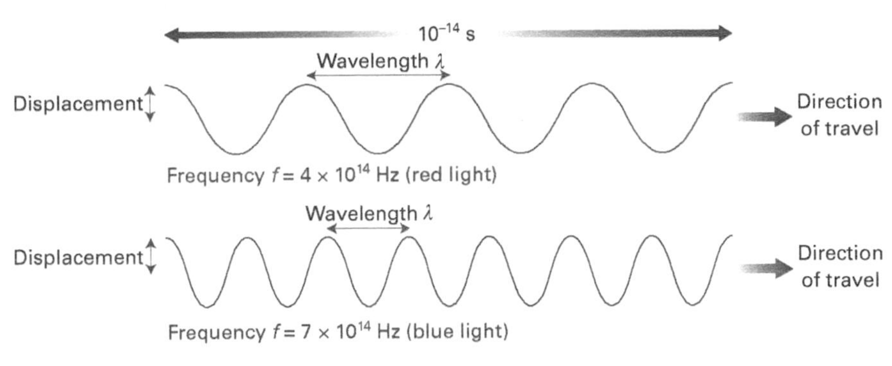

A transverse wave (such as a water wave or an electromagnetic wave) moves from one place to another at a characteristic speed. Its displacement is at right angles to its direction of travel. The wavelength λ is the distance between successive peaks (points of maximum displacement). The frequency f is the number of complete cycles of the wave that pass a stationary point in one second.

Using a spectrometer to measure wavelength

A light bulb emits light with a continuous range of wavelengths. The mixing together of this range of wavelengths produces the 'white' light that we see. Light from a coloured Bunsen flame or a discharge tube consists of a mixture of distinct, separate wavelengths. A **spectrometer** is an instrument that separates this light into its constituent wavelengths.

A diffraction grating in the spectrometer bends light of a particular wavelength λ through a specific angle. The lower the wavelength, the greater will be the angle of deviation. When the light is viewed through the telescope, each specific wavelength λ appears as a thin vertical line of coloured light. The telescope can be moved around the turntable to measure the angle of deviation for each line. We can then calculate the wavelength, and hence the frequency, of each line from its angle of deviation.

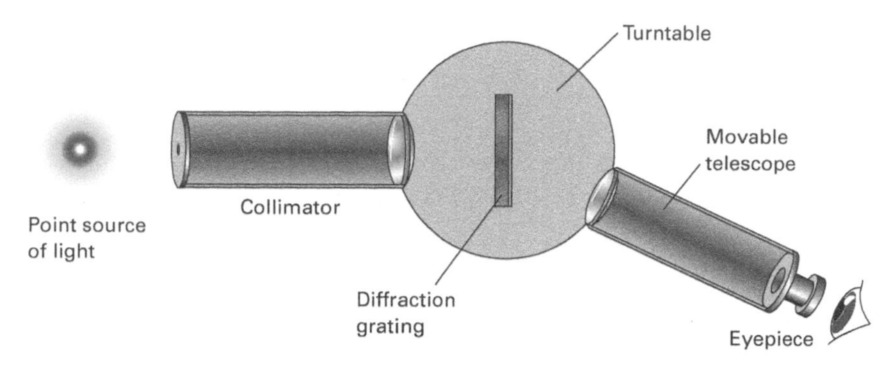

The main components of a spectrometer. The collimator produces a parallel beam of light from the point source. The diffraction grating consists of parallel slits scratched equal distances apart on a glass surface. The angle of deviation depends on the wavelength of the light and the distance between neighbouring slits.

Emission spectra

If you rotate the telescope of a spectrometer on its turntable, a number of separate coloured lines pass across your field of vision. The complete set of these lines laid out against a wavelength (or frequency) scale is a spectrum. When the spectrum is the result of light being *emitted* from *atoms*, it is called an **atomic emission spectrum**. An external energy source increases the energy of the atoms; and they lose energy as they emit electromagnetic radiation.

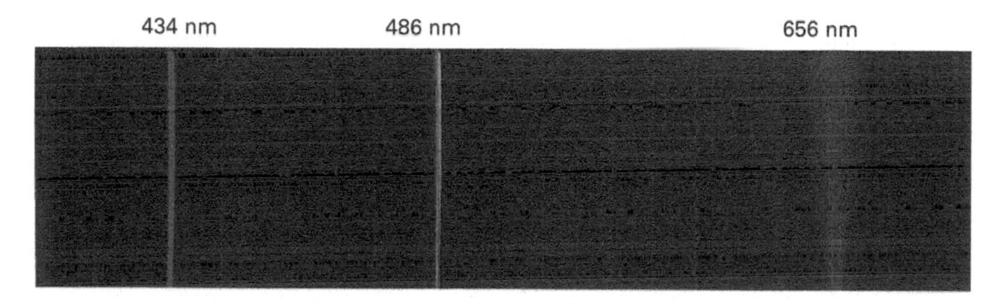

The emission spectrum of hydrogen. The wavelengths of the lines have values around 5×10^{-7} m (500×10^{-9} m, 500 nanometres, 500 nm). The corresponding frequencies are of the order of 6×10^{14} Hz.

Frequency and energy of light

Light is a form of energy. The lower the wavelength (and hence the higher the frequency), the greater is the energy of the light. Frequency f (hertz, Hz) and energy E (joules, J) are connected by an equation introduced by Max Planck in 1900:

$$E = hf$$

where h is the Planck constant (6.626×10^{-34} J s).

Measurements from a spectrometer show that the atoms in a Bunsen flame or discharge tube emit light at fixed wavelengths and hence fixed frequencies. Planck's equation indicates that energy is *directly proportional* to frequency; so these atoms emit *energy* in fixed quantities. Each fixed quantity of energy emitted is like a bundle or packet of energy. It is called a **quantum** (plural = quanta) of energy and corresponds to electromagnetic radiation of a specific frequency. Physicists needed to make a whole new range of models for the structure of the atom to fit in with this idea of quantized energy. We shall look at these in the following spreads.

SUMMARY

- Atoms can take in energy and then emit it as electromagnetic radiation at specific frequencies.
- Atoms emit fixed quantities of energy in packets called quanta.

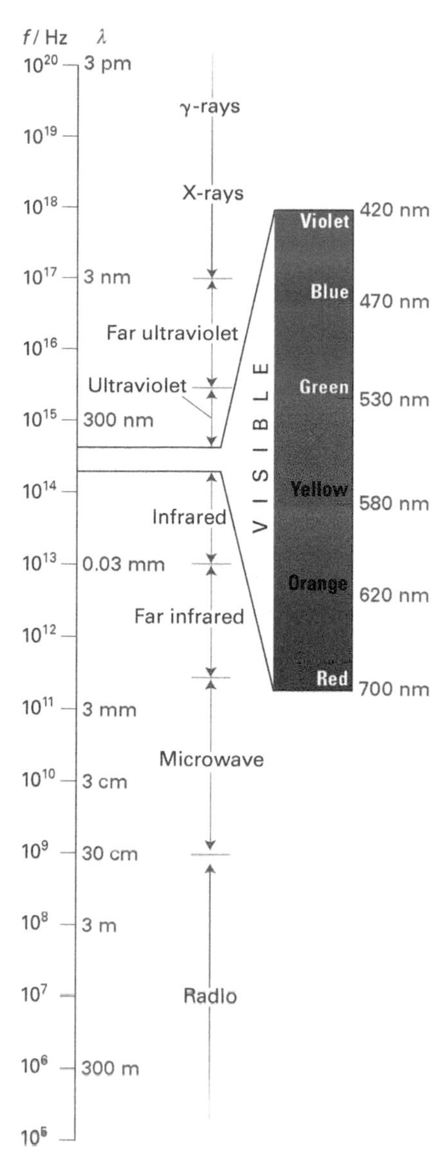

The electromagnetic spectrum.

Scientific methodology

Scientists observe something happening (a phenomenon), take measurements to gather data, and analyse the data in an attempt to explain the cause of the phenomenon. They often create a model, which behaves in a particular way to produce the observed effects. The modern model of the atom evolved over the past century.

The reflective thinker, writer, and TV personality Jacob Bronowski stated in his book *The Ascent of Man* that: 'The inside of the atom is invisible, but there is a window in it – a stained-glass window: the spectrum of the atom.' Deducing the structure of the atom from emission spectra is similar to deducing the existence of water and clouds from observing a rainbow.

4.2

OBJECTIVES

- Interpreting emission spectra
- The Bohr model of the atom
- Quantized energy levels (shells)
- Principal quantum numbers

The hydrogen spectrum and shells

Atoms are far too small for us to see their structure directly. We have to devise pictures and models that explain the results of experiments we carry out on them. This process is rather like poking at an elephant with sticks through the bars of a darkened cage. We then have to use the evidence gathered on the ends of our sticks to draw the elephant! Our unknown specimen is not an elephant, but the hydrogen atom. It is contained inside a discharge tube, and its outer skin is defined by its surrounding electrons. We poke at it with electrical energy and look at the electromagnetic radiation that results. We then try to translate this information into a picture of the atom in our minds.

The Balmer series

The Balmer series is the most straightforward part of the hydrogen emission spectrum to study because it occurs in the visible region of the electromagnetic spectrum. Each line in the series represents electromagnetic radiation of a specific single wavelength. (The purple line, for example, is at 434.05 nm.) The energy of the radiation may be calculated by using Planck's equation:

$E = hf$

The hydrogen spectrum

An atom of hydrogen consists of just one proton with one surrounding electron. The emission spectrum of hydrogen is relatively simple compared to those of other elements. The complete spectrum of hydrogen consists of separate series of distinct wavelengths concentrated in the ultraviolet, visible, and infrared regions of the electromagnetic spectrum. The six series found are named after their discoverers. In order of increasing wavelength they are the Lyman series (ultraviolet), Balmer series (visible), Paschen, Brackett, Pfund, and Humphreys series (infrared). Each of these series is called a **line spectrum** because the film record from the spectrometer appears as a pattern of separate thin vertical lines.

Quantized energy levels

In 1913, Niels Bohr introduced his model of the hydrogen atom. He *assumed* that the electron within the hydrogen atom will *not* absorb or radiate energy so long as it stays in one of a number of circular **orbits.** His model of the atom was designed to explain the observation that the electromagnetic radiation emitted by an excited hydrogen atom has specific energies. These energies are fixed, or quantized. Bohr suggested that the energy of an electron in an atom must also be quantized, so the electron can only have certain discrete **energy levels** rather than a continuous range of possible energies. Each of these energy levels may be occupied by an electron of the appropriate energy.

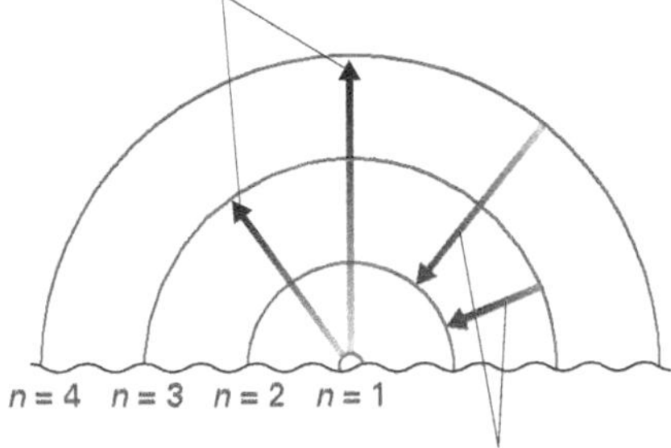

If an atom collides with another atom, or absorbs radiation, this can increase the energy of an electron within the atom. The electron moves into a higher energy level. The atom emits electromagnetic radiation when the electron moves to a lower energy level.

When an atom is excited by absorbing energy, an electron jumps up to a higher energy level. In the Bohr model, the electron is then circling at a greater distance from the nucleus. The excited atom can emit energy in the form of electromagnetic radiation as the electron falls back down to a lower energy level. When an electron moves from one energy level to another, this is called an **electronic transition**. The emitted energy can be seen as a line in the spectrum (as viewed through a spectrometer, for example). If the electronic energy levels were not quantized but could have any value, a *continuous* spectrum rather than a line spectrum would result.

The difference in energy ΔE between the two energy levels in this electronic transition, E(higher) and E(lower), is equal to the energy of the emitted radiation, E(radiation):

$$E(\text{radiation}) = E(\text{higher}) - E(\text{lower}) = \Delta E$$

If you combine this equation with Planck's equation, you can see that the frequency (f) of the radiation emitted depends on the energy level difference (ΔE) of the particular electronic transition:

$$\Delta E = hf$$

So, electronic transitions between energy levels result in emission of radiation of different frequencies and therefore produce different lines in the spectrum.

Differences

The Greek capital letter delta Δ is often used (as here) to describe a difference between two quantities. Here ΔE is a difference in energy.

Shells

Bohr labelled each of the energy levels in the hydrogen atom with a number called the **principal quantum number**, n. The energy level closest to the nucleus is labelled $n = 1$. The next energy levels are $n = 2$, $n = 3$, and so on. Each of these energy levels is called a **shell**. The principal quantum number defines the energy of the electron in a given shell. In an unexcited hydrogen atom, the electron is in the energy level $n = 1$. This state of lowest energy for the atom is called the **ground state**.

Bohr showed that the series in the high-energy ultraviolet region (the Lyman series) arises from electronic transitions from higher energy levels to the energy level $n = 1$. Each line in the Lyman series is due to electrons returning from a *particular* higher energy level to the energy level $n = 1$. The Balmer series arises from electronic transitions from higher energy levels to the energy level $n = 2$. Each line in the Balmer series is due to electrons returning from a *particular* higher energy level to the energy level $n = 2$.

Series in the infrared

The various series in the low-energy infrared region are caused by electrons returning to the energy levels $n = 3$ (Paschen series), $n = 4$ (Brackett series), $n = 5$ (Pfund series), and $n = 6$ (Humphreys series).

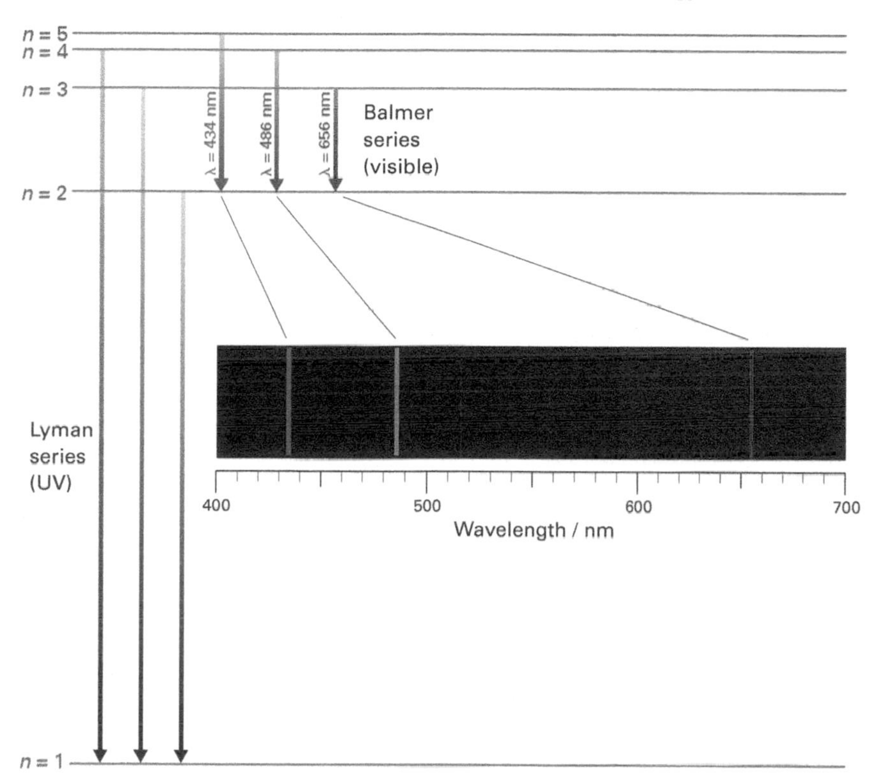

The Lyman series of lines results from electronic transitions from higher energy levels down to energy level n = 1. The Balmer series results from electronic transitions down to energy level n = 2.

The convergence limit

The separate lines in a series become closer together as their wavelength decreases, i.e. as their frequency (and energy) increases. At the high-frequency end of the series, the lines are so close together that they form a continuous band of radiation, known as a **continuum**.

The start of this continuum, beyond which separate lines cannot be distinguished, is called the **convergence limit**. The convergence limit corresponds to the point at which the energy of an electron within the atom is no longer quantized. At that point, the nucleus has lost all influence over the electron; the atom has become **ionized**.

For the Lyman series, the convergence limit represents the ionization of the hydrogen atom:

$H(g) \rightarrow H^+(g) + e^-(g)$

SUMMARY

- Electrons within atoms occupy fixed energy levels.
- The energy of an electron in an atom is quantized; the electron may have only certain energies.
- Energy levels with the same principal quantum number are in the same shell.
- Atoms emit electromagnetic radiation when electrons move from a higher to a lower energy level.

PRACTICE

1 Draw an energy level diagram to show the electronic transitions responsible for the lowest-energy spectral lines in the Paschen series.

2 In no more than 60 words each, explain the meanings of the following terms:

a Quantized

b Principal quantum number

c Line spectrum

d Convergence limit.

4.3 The sodium spectrum and subshells

OBJECTIVES

- Shells
- s, p, and d subshells

Bohr's model of the atom used the idea of quantized electronic energy levels (shells) to explain the sequence of lines in the emission spectrum of hydrogen. He labelled the shells $n = 1$, $n = 2$, etc. in order of increasing energy. The Bohr model explains the emission spectrum of the hydrogen atom in terms of electronic transitions between shells. However, the emission spectrum of sodium is a good deal more complex. To explain this spectrum, there must be subdivisions of the Bohr shells, called *subshells*. The structure of the atom must be more complicated than Bohr thought.

The sodium spectrum

The hydrogen spectrum is simple throughout most of the visible region (corresponding to the Balmer series for electronic transitions to energy level $n = 2$). There is a single red line at 656 nm and then no further line until 486 nm. This red line originates from electronic transitions from energy level $n = 3$ to $n = 2$.

The sodium emission spectrum is more complex than the very simple hydrogen emission spectrum. If we look in particular at the region where wavelengths are greater than 550 nm, we find that, whereas there is a *single* line for hydrogen, there are *three* lines fairly close together for sodium, i.e.

- a green line at 569 nm;
- a yellow line at 589 nm;
- an orange line at 616 nm.

This observation suggests that there are more energy levels available in sodium than in hydrogen. In the sodium atom, the shells must be composed of **subshells**, each of which is at a different energy. It turns out that:

- For the shell $n = 1$, there is only one subshell, labelled 1s.
- For the shell $n = 2$, there are two subshells, labelled 2s and 2p. The 2p subshell is at a higher energy than the 2s subshell.
- For the shell $n = 3$, there are three subshells, labelled 3s, 3p, and 3d. These subshells increase in energy in the order 3s < 3p < 3d.

Wavelength

A wave of wavelength λ_1 has a *longer* wavelength than a wave of wavelength λ_2 if the numerical value of λ_1 is greater than the value of λ_2, i.e. the yellow sodium emission at 589 nm has a longer wavelength than the green emission at 569 nm.

f subshells

For the shells $n = 4$ and higher, there are *four* subshells: s, p, d, and f. The f subshell will be ignored for the moment.

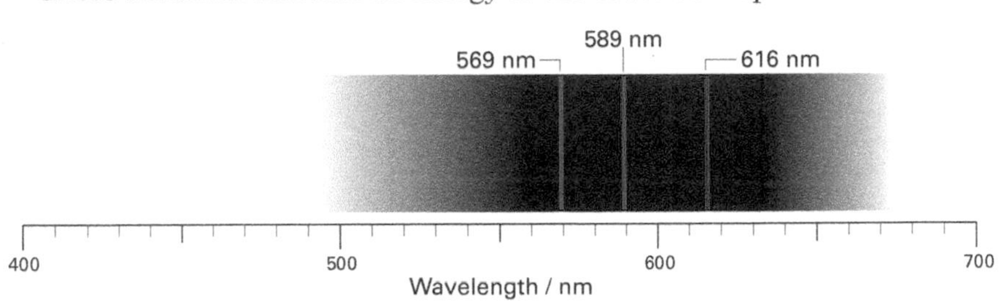

The visible region of the sodium emission spectrum. Note the green, yellow, and orange lines referred to in the text. There are numerous other lines at wavelengths below 550 nm. Lines at wavelengths greater than 616 nm are due to sodium ions Na^+ and not to sodium atoms themselves.

s, p, and d subshells – history

The letters used to label the subshells came from the descriptions of the series observed in the sodium spectrum. The series of lines had been given names that reflected their character in some way, e.g. 'sharp' because they were sharp. Each series arose from transitions in which the electron fell from a particular subshell. So it was natural to name the subshells after the associated series. The '**s**harp series' arose from transitions in which the electron fell from an s subshell. The '**p**rincipal series' (in which the electron fell from a p subshell) was so named because these lines also occurred in the absorption spectrum of sodium. The '**d**iffuse series' (in which the electron fell from a d subshell) was so named because of the characteristic visible difference from the sharp series.

Absorption spectra

If continuous radiation (electromagnetic radiation of all wavelengths) passes through the vapour of an element, lines of certain wavelengths will be absorbed by the atoms and removed from the radiation. Looking through a spectrometer, you would see a series of black lines where wavelengths have been absorbed, against the background of continuous radiation. This is an **absorption spectrum.** The wavelengths of these lines correspond to the quantized energy taken in by the atoms to promote electrons from lower to higher energy levels. For example, excited hydrogen atoms in the photosphere of the Sun cause a dark line at 656 nm in the solar spectrum.

Interpreting the sodium spectrum

We can explain the sodium atomic emission spectrum by assigning each line in the spectrum to an electronic transition on an energy level diagram. The diagram shows the energy levels corresponding to the subshells in the sodium atom. It is called a **Grotrian diagram**, after its originator, Walter Grotrian.

The most intense line, the yellow emission at 589 nm, is caused by an electronic transition from the 3p energy level down to the 3s energy level. Expressed more simply, the yellow line is the result of a 3p to 3s transition.

The orange line at 616 nm is the result of a 5s to 3p transition. This line is at a slightly longer wavelength than the yellow line. The frequency is therefore lower, corresponding to the smaller energy gap seen on the Grotrian diagram.

The green line at 569 nm is the result of a 4d to 3p transition. This line is at a slightly shorter wavelength than the yellow line. The frequency is therefore higher, corresponding to the larger energy gap seen on the Grotrian diagram.

At even shorter wavelength are emissions at 515 nm (6s to 3p), 498 nm (5d to 3p), and so on. At even longer wavelength, the 3d to 3p transition causes an intense line at 819 nm, which is in the infrared.

The Bohr model of the atom assumes that electrons are particles moving in orbits around the nucleus. This model became inadequate when it was understood in the 1920s that the electron can also behave as a *wave*. This startling discovery is explained in the next spread.

The ideas introduced in the next spread **will not be tested in exams** but are essential to explain why we now focus on the idea of *electron density*.

SUMMARY

- Shells are labelled 1, 2, 3, etc. in order of increasing energy.
- Subshells are labelled s, p, d, and f.

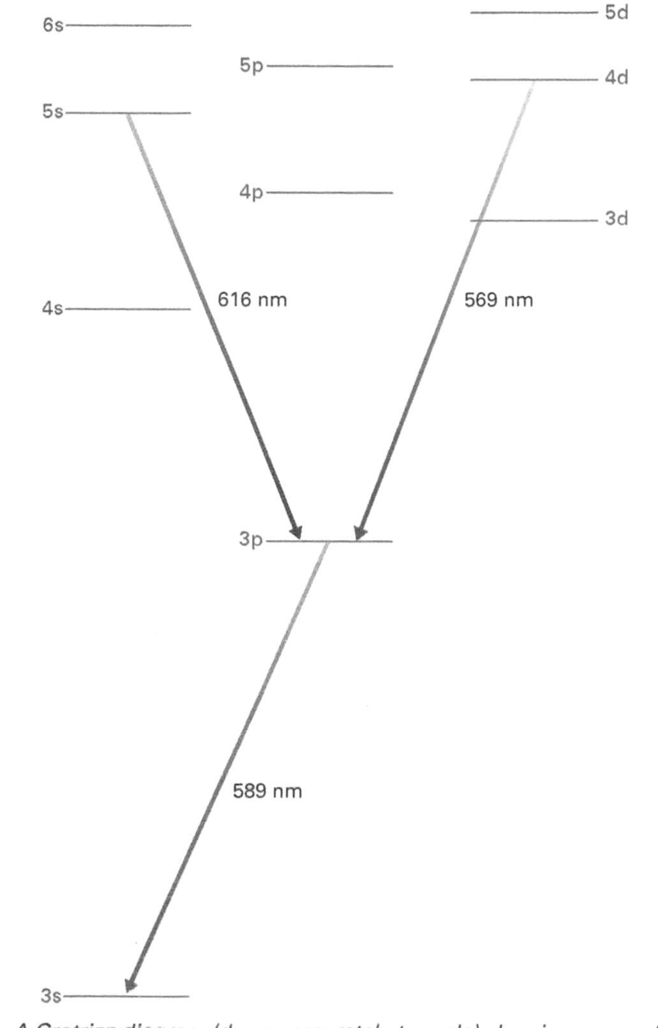

A Grotrian diagram (drawn accurately to scale) showing some of the electronic transitions responsible for the emission spectrum of sodium. Notice how the energies of the subshells get closer to each other as the principal quantum number increases.

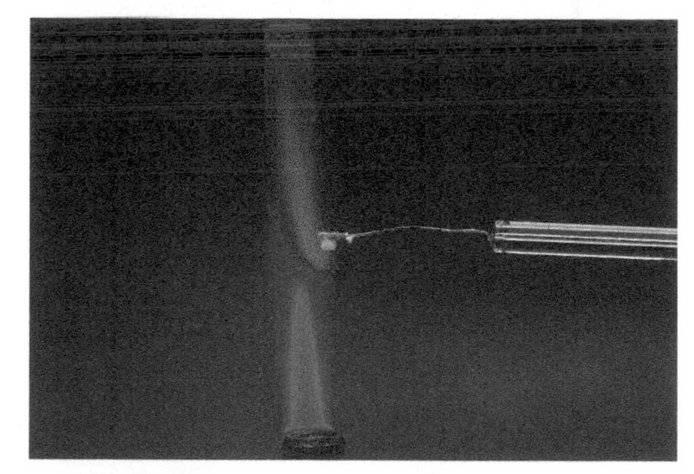

The test for sodium in the laboratory makes use of the bright yellow line in its spectrum: when heated in a flame, sodium colours the flame yellow.

PRACTICE

1 What subshells can occur in each of the following shells?
  a $n = 1$
  b $n = 2$
  c $n = 3$
  d $n = 4$.

2 Look at the Grotrian diagram for sodium, and arrange the following subshells of the sodium atom in order of *increasing* energy:
  3s, 2p, 1s, 2s, 3d, 4s, 4p, 3p.

3 Look at the Grotrian diagram for sodium, and estimate the wavelength of the emission resulting from a 5d to 4p transition.

4.4

Wave mechanics

OBJECTIVES

- Wave–particle duality of the electron
- The uncertainty principle
- The Schrödinger equation
- Probability
- Atomic orbitals

Bohr attempted to construct a model of the atom which accounted for the data obtained from emission spectra. He assumed the electron to be a solid particle of matter and tried to describe its motion in terms of the mathematical equations of classical mechanics. Bohr's physical model of the atom was only partially convincing. In constructing it he ignored the inconvenient physical law which states that an orbiting charged particle such as an electron should lose energy and spiral in towards the nucleus. A completely new way of thinking was required to describe objects as small as atoms. The 1920s finally saw the emergence of our modern view of the atom and the branch of mathematics needed to describe it – wave mechanics (sometimes called quantum mechanics).

The problem with the Bohr model

Any object moving in a circle must experience a force holding it in the circle; this force is called a **centripetal force**. This force produces an acceleration towards the centre of the circle. This applies equally to the Earth orbiting the Sun and to Bohr's electron orbiting the nucleus.

The difference between the astronomical model and Bohr's atomic model is that the electron is *charged*. The laws of electromagnetism explain that an *accelerating* charged particle should radiate electromagnetic radiation. The orbiting electron should gradually lose energy and spiral into the nucleus.

The electron: particle and wave

In 1924, Louis de Broglie presented a revolutionary idea in his doctoral thesis. He predicted that electrons, which were thought at the time to be particles, would also possess wave-like properties. One of the properties of waves is that they can be diffracted. This property means that when

A wave spreads out when it passes through a small aperture. (The size of the aperture must be of the same order as the wavelength of the wave.) The waves from two or more apertures interfere with each other, forming a diffraction pattern of alternating maxima and minima.

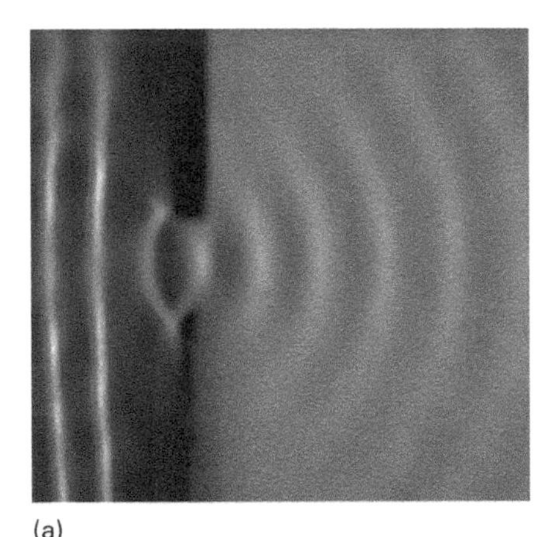

(a)

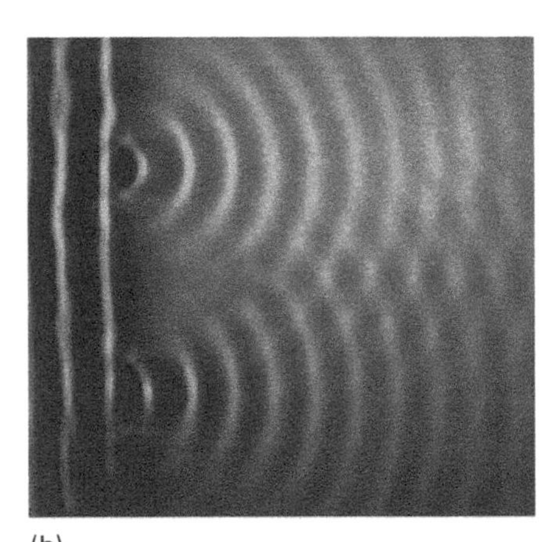

(b)

The Thomsons

George Paget Thomson was the son of Joseph John (J. J.) Thomson, who discovered the electron. J. J. received the 1906 Nobel Prize in Physics for his work. G. P. shared the 1937 Nobel Prize with Clinton Davisson for discovering electron diffraction.

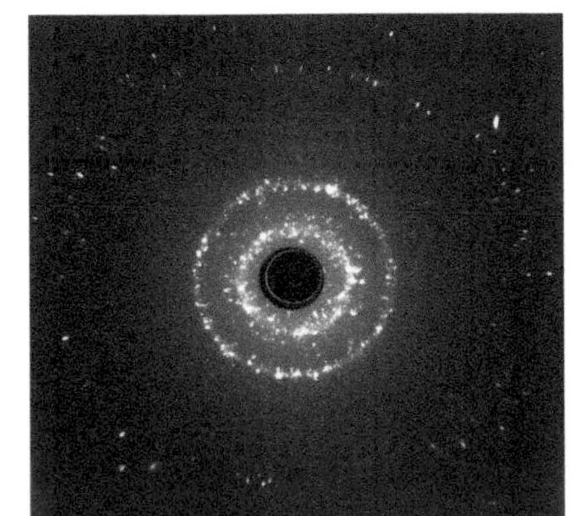

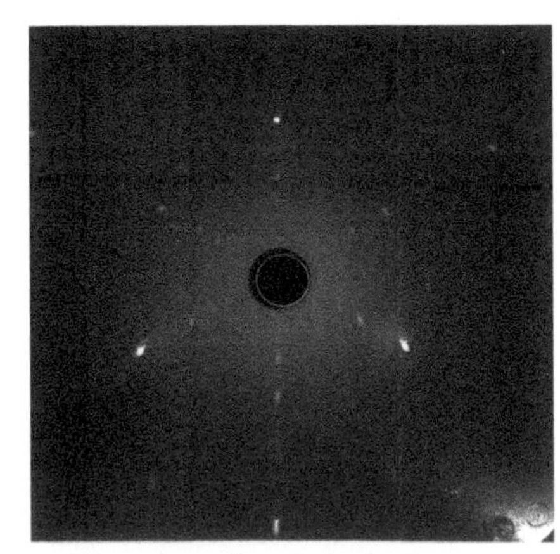

(a) The diffraction pattern produced by a beam of X-rays passing through aluminium foil. The rows of atoms in the metal act like the slits in a diffraction grating. (b) The diffraction pattern resulting from a beam of electrons.

waves hit the edge of an object or pass through a narrow aperture, they bend around it and spread out, like the waves in a ripple tank. So, if electrons could behave like waves, it should be possible to diffract them, like light through a diffraction grating.

Three years after de Broglie's thesis appeared, G. P. Thomson provided experimental evidence which demonstrated that electrons can be diffracted in the same manner as X-rays. The conclusion was that electrons have both wave-like and particle-like properties. The realization that the electron shows **wave–particle duality** opened up a whole new theoretical method for describing the electronic structure of atoms based on the mathematics of waves.

The Bohr model

Bohr's model of the atom can explain the observed emission spectrum of hydrogen. However, it fails to explain the spectra of atoms more complex than hydrogen. If a model cannot explain observations fully, then that model is limited or is built on false assumptions. The Bohr model is built on the false assumption that the electron exists as a solid particle of matter in an atom.

Heisenberg's uncertainty principle

In 1927, Werner Heisenberg published details of his uncertainty principle. The **uncertainty principle** states that it is impossible to measure accurately both the position and the velocity of an electron at the same time. The principle can be justified as follows. You view the world about you with the help of radiation (light rays) which reflects from objects and enters your eyes. To see an object as small as an electron would require radiation of extremely low wavelength and hence extremely high frequency. The energy of this radiation would alter the position and velocity of the electron you were trying to observe. Once scientists accepted the uncertainty principle, they stopped trying to construct Bohr-style models of the atom that attempted to define the positions of electrons exactly.

The Schrödinger equation

Erwin Schrödinger, working separately at almost the same time as Heisenberg, and following de Broglie's ideas, founded the mathematical technique called **wave mechanics**. This technique produced a *mathematical* model of the atom. The model is described by the **Schrödinger equation**, which allows for the uncertainty principle and the wave-like properties of the electron. Solutions of the Schrödinger equation account for the quantization of electronic energy levels.

The solution of the Schrödinger equation leads to the idea that there are regions of space around the nucleus where there is a high probability (but not an absolute certainty) of finding an electron of a given energy. These regions are called **atomic orbitals**. Because it does not try to define the exact position and path of an electron, the Schrödinger equation is consistent with the uncertainty principle.

The Schrödinger equation explains the existence of both shells and subshells, as described in spread 4.9.

Five Nobel Prizes

The Nobel Prize in Physics was awarded in 1922 to Niels Bohr, in 1929 to Louis de Broglie, in 1932 to Werner Heisenberg, in 1933 to Erwin Schrödinger, and in 1937 to G.P. Thomson. Niels Bohr's prize was uniquely announced one year early (at the same time as the 1921 prize).

SUMMARY

- The electron shows wave–particle duality.
- The Bohr model ignores one of the fundamental attributes of matter at the atomic level – uncertainty.
- The Schrödinger equation is a model of the atom based on the mathematics of waves.
- Solutions of the Schrödinger equation define regions of space where an electron is likely to be found: these regions are called atomic orbitals.

PRACTICE

1 Summarize each of the following in no more than 30 words:

a Bohr model of the atom

b Wave–particle duality

c The uncertainty principle

d Atomic orbital.

4.5

OBJECTIVES

- Electron density
- The shapes of the s, p, and d orbitals
- Atomic orbitals visualized as 90% boundary surfaces

VISUALIZING ATOMIC ORBITALS

Atomic orbitals indicate the electron density for an electron of a given energy. When these electron densities are plotted in three dimensions they show the shapes that represent the various atomic orbitals. Atomic orbitals for the hydrogen atom have the same energies as the shells developed by Bohr. Atomic orbitals are labelled as s, p, or d, as they also naturally explain the subshells found from the atomic spectrum of sodium.

Electron density

Wave mechanics does not picture an electron as a point charge. Instead, the electron in an atomic orbital is imagined as being *smeared out*. The distribution of the electron is not uniform; rather, there is a high probability of finding the electron in some regions, and a low probability of finding it in others. This distribution of the probability of finding an electron at a certain position is the **electron density**. Regions of high probability have high electron density because the electron spends a greater proportion of its time in that region.

The 1s atomic orbital

An electron in the 1s atomic orbital has the lowest possible energy for an electron in that atom. The shape of the electron density in the 1s orbital is spherical. Look at the plot of the total electron density at r (distance from the nucleus) against r, shown below (c). It shows that the total electron density at the centre of the nucleus ($r = 0$) is zero, as might be expected. As the distance from the nucleus increases, so does the total electron density, until it reaches a maximum. The distance corresponding to the maximum total electron density is called the **Bohr radius** (a_0) and equals the radius of the orbit calculated by Bohr. Beyond the Bohr radius, the total electron density falls steadily, but does not reach zero. In theory, there is a finite (but *very* small) probability of the electron being found anywhere in the universe. Slicing the electron density in a plane through the nucleus produces circular contours. To get a picture of what is going on, we can find the 'shape' of the s orbital by defining a boundary surface that includes 90% of the electron density.

(a)

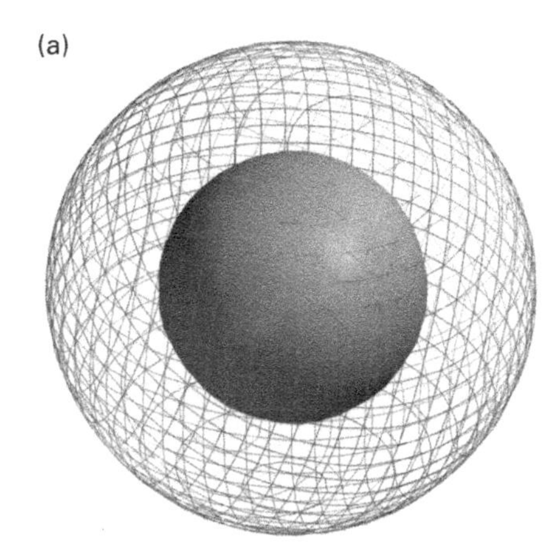

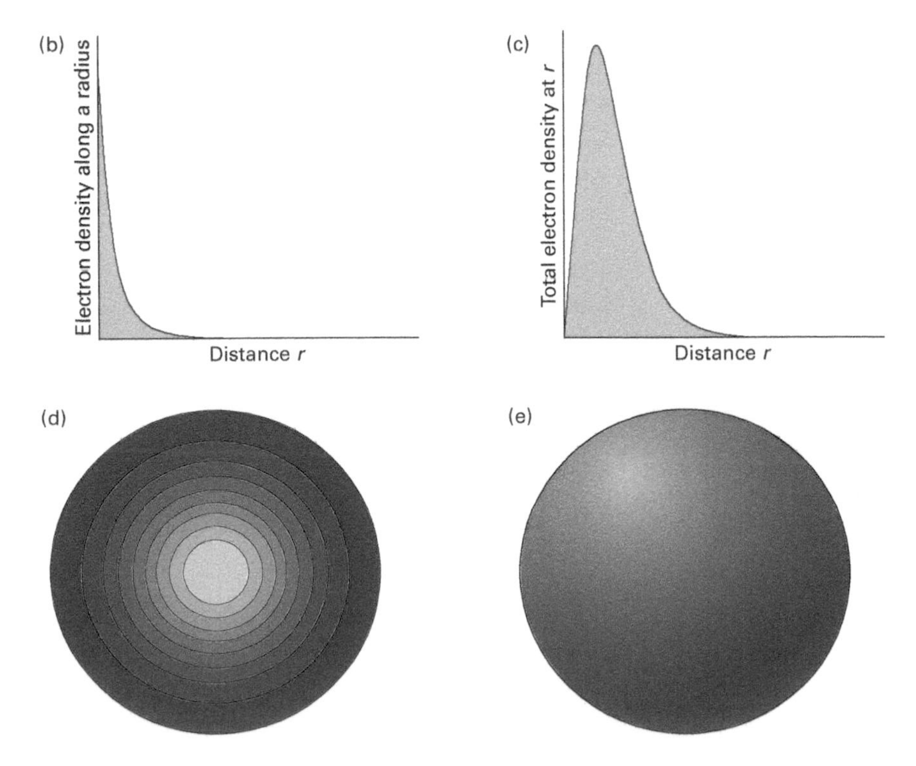

The electron density in the 1s orbital is spherical. (a) The solid figure joins points of equal electron density; the wire-frame joins points where the electron density is ten times lower.
(b) As you move out from the nucleus along any radius of the sphere that makes up the 1s orbital, the electron density falls steadily.
*(c) The total electron density at a particular distance from the centre of the sphere (rigorously called the **radial distribution function**) varies as shown here, because all points on the surface of the sphere (of area $4\pi r^2$) are at the same distance from the nucleus. (The maximum value corresponds to the Bohr radius. See appendix B.1.)*
(d) A cross-section through the electron density shows concentric circles, each successive circle including 10% more of the total electron density. The outer circle includes 90%.
(e) To sketch an s orbital in the rest of the book, we will draw the 90% contour.

The 2p atomic orbitals

There is just one 1s orbital and one 2s orbital, but there are three 2p orbitals of equal energy. The 2p orbitals are not spherically symmetrical about the nucleus. Electron density is more concentrated along one direction in space, so p orbitals must be drawn on three-dimensional axes to distinguish one orbital from another. Each orbital consists of two lobes with a region of zero electron density (a **node**) between them centred on the nucleus.

The three 2p orbitals are known as the $2p_x$, the $2p_y$, and the $2p_z$ orbitals. They are at right angles to each other.

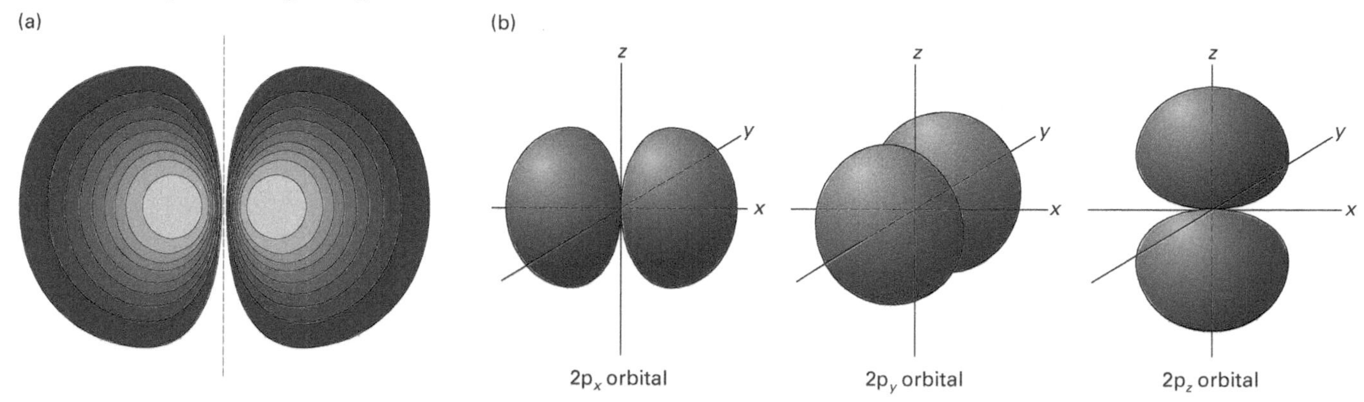

(a) Cross-section through the electron density of a 2p orbital: each successive contour includes 10% more of the total electron density. (b) The 90% boundary surfaces (i.e. 'the shapes') of the $2p_x$, the $2p_y$, and the $2p_z$ atomic orbitals.

The 3d atomic orbitals

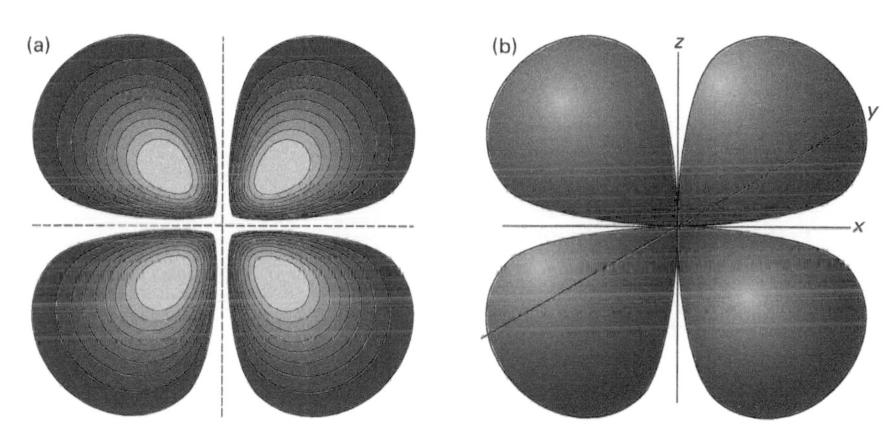

There are five 3d orbitals. This diagram shows (a) a cross-section through the electron density: each successive contour includes 10% more of the total electron density; (b) the shape of the $3d_{xz}$ orbital. One of the five d orbitals has a more unusual shape: see spread 19.11.

SUMMARY

- Solutions of the Schrödinger equation for atoms are called atomic orbitals.
- Atomic orbitals show the electron density for an electron of a given energy.
- The shape of an atomic orbital is visualized as a boundary surface which encloses 90% of the electron density.
- An s orbital is spherically symmetrical about the nucleus: it is sketched as a circle.
- A p orbital has two opposing lobes, one on each side of the nucleus.

PRACTICE

1 Explain why it is not possible to draw 100% boundary surfaces for orbitals.

4.6

OBJECTIVES

- Electron spin
- Listing atomic orbitals in order of increasing energy
- Electrons-in-boxes notation
- Electronic structures

Atoms with more than one electron

Bohr's model of the atom explains the wavelengths of the lines in the hydrogen emission spectrum. It treats the electron as a particle and describes the various energy levels that it can occupy. Current theory relies on atomic orbitals, which define regions of space around the nucleus where there is a high electron density for an electron of a given energy. Atomic orbitals for the hydrogen atom have the same energies as the orbits of Bohr's model. But what about atoms other than hydrogen? Two questions now arise: what are the energies of the electrons in atoms with more than one electron, and what sort of quantized energy level does each occupy?

Hydrogen-like orbitals

It seems reasonable to assume that the orbitals in any atom will be much the same as those in hydrogen. In its lowest energy state, the electron in a hydrogen atom occupies the 1s orbital. Elements other than hydrogen can be built up by filling the atomic orbitals found for the hydrogen atom, that is 1s, 2s, 2p, 3s, 3p, etc. The main difference is that the *energy* of each atomic orbital will be affected by the electrons occupying other atomic orbitals.

Electron spin

Some of the electron's properties must be interpreted in terms of a property that scientists have chosen to call 'spin'. Picturing an electron spinning like the Earth on its axis is helpful, but simplistic. The human mind always prefers to visualize the unseen in pictorial terms.

Split lines and electron spin

High-resolution spectrometers show that the yellow line at 589 nm in the emission spectrum of sodium is actually split into two closely spaced lines. The explanation is that an electron can exist in one of two states, called 'spin up' and 'spin down'. Two electrons with the same spin state are said to have **parallel spins**.

Finding electronic structures

The way in which an atom's electrons are arranged in its atomic orbitals is called its **electronic structure** or electronic configuration. The electronic structure can be worked out using three basic rules:

- The building-up (*Aufbau*) principle
- The Pauli exclusion principle
- Hund's rule

Simplified electronic structures

In earlier science courses, you probably saw electronic structures written without identifying subshells. So sodium, for example, was Na 2,8,1 or Na 2.8.1.

The building-up principle

The **building-up principle** states that electrons fill atomic orbitals in order of increasing energy, subject to the Pauli exclusion principle.

The Pauli exclusion principle

This principle, introduced by Wolfgang Pauli, allows *no more than two electrons* to occupy any orbital. For example, the 1s orbital may contain one or two electrons only. The three orbitals ($2p_x$, $2p_y$, and $2p_z$) in the 2p subshell (which we saw in the previous spread) may contain a total of up to six electrons. Electron spin explains why the Pauli exclusion principle arises: two electrons occupying the same orbital must have **paired** (opposite) **spins**.

Hund's rule

Hund's rule is applied where orbitals of equal energy are available, for example $2p_x$, $2p_y$, and $2p_z$. The orbitals will first fill with one electron each with parallel spins before a second electron is added with the paired (opposite) spin.

Filling atomic orbitals with electrons

The Pauli exclusion principle says that an orbital can hold no more than two electrons. The first eight subshells are listed to the right in order of increasing energy. *It is important to learn this order*. For elements other than hydrogen, the electrons interact with each other and alter the relative energies of the orbitals. Energy levels become closer to each other than in the hydrogen atom, with the result that the energy of the 4s orbital can be below that of the 3d orbital. This is the case for sodium, for example, see spread 4.3. Similar changes occur with higher orbitals.

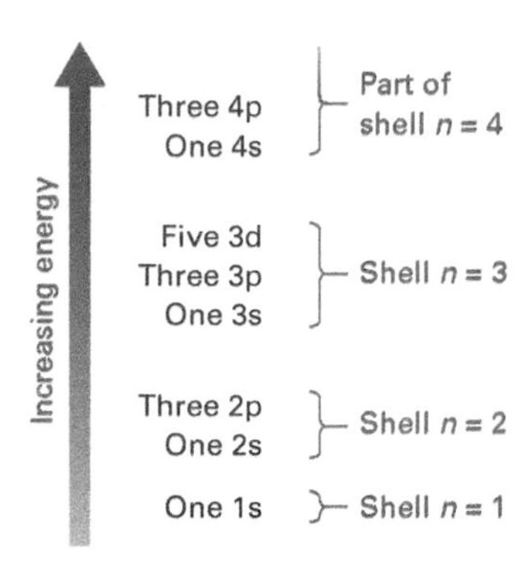

The relative energies of the 1s to 4p atomic orbitals for atoms with more than one electron. 4s orbitals are shown above 3d orbitals here, but they are actually of similar energy.

'Electrons-in-boxes' notation

A convenient way to represent electronic structures is a notation known as *'electrons-in-boxes'*. This notation shows each orbital as a box and the electrons as arrows in the boxes. The opposite spin of paired electrons is shown by arrows facing in opposite directions (usually up and down).

Although the 'electrons-in-boxes' notation is a useful visual aid, it is a tedious way to depict electronic structure routinely. We need an easier, shorter form. More usually, the occupied subshells are written in order of increasing energy with the number of electrons following as a superscript. For example, when there is only one electron in the 1s orbital, we write this as $1s^1$; similarly, $1s^2$ means two electrons are in the 1s orbital; and $2p^6$ means six electrons are in the 2p subshell (two each in the $2p_x$, $2p_y$, and $2p_z$ orbitals).

Element	Electrons-in-boxes: 1s	2s	2p: $2p_x$	$2p_y$	$2p_z$	Shorter form of electronic structure
H	↑					$1s^1$
He	↑↓					$1s^2$
Li	↑↓	↑				$1s^22s^1$
Be	↑↓	↑↓				$1s^22s^2$
B	↑↓	↑↓	↑			$1s^22s^22p^1$
C	↑↓	↑↓	↑	↑		$1s^22s^22p^2$
N	↑↓	↑↓	↑	↑	↑	$1s^22s^22p^3$
O	↑↓	↑↓	↑↓	↑	↑	$1s^22s^22p^4$
F	↑↓	↑↓	↑↓	↑↓	↑	$1s^22s^22p^5$
Ne	↑↓	↑↓	↑↓	↑↓	↑↓	$1s^22s^22p^6$

'Electrons-in-boxes' notation and corresponding electronic structures in shorter form for the elements hydrogen to neon.

Element	Atomic number	1s	2s	2p	3s	3p	3d	4s	4p
H	1	1							
He	2	2							
Li	3	2	1						
Be	4	2	2						
B	5	2	2	1					
C	6	2	2	2					
N	7	2	2	3					
O	8	2	2	4					
F	9	2	2	5					
Ne	10	2	2	6					
Na	11	2	2	6	1				
Mg	12	2	2	6	2				
Al	13	2	2	6	2	1			
Si	14	2	2	6	2	2			
P	15	2	2	6	2	3			
S	16	2	2	6	2	4			
Cl	17	2	2	6	2	5			
Ar	18	2	2	6	2	6			
K	19	2	2	6	2	6		1	
Ca	20	2	2	6	2	6		2	
Sc	21	2	2	6	2	6	1	2	
Ti	22	2	2	6	2	6	2	2	
V	23	2	2	6	2	6	3	2	
Cr	24	2	2	6	2	6	5	1	
Mn	25	2	2	6	2	6	5	2	
Fe	26	2	2	6	2	6	6	2	
Co	27	2	2	6	2	6	7	2	
Ni	28	2	2	6	2	6	8	2	
Cu	29	2	2	6	2	6	10	1	
Zn	30	2	2	6	2	6	10	2	
Ga	31	2	2	6	2	6	10	2	1
Ge	32	2	2	6	2	6	10	2	2
As	33	2	2	6	2	6	10	2	3
Se	34	2	2	6	2	6	10	2	4
Br	35	2	2	6	2	6	10	2	5
Kr	36	2	2	6	2	6	10	2	6

Electronic structures for the elements of atomic number $Z = 1–36$. The structures for chromium and copper are not as might be expected. This will be explained in chapter 20.
Transition metal ions have electronic structures of the following form:

Fe^{2+} $1s^22s^22p^63s^23p^63d^6$

Cu^{2+} $1s^22s^22p^63s^23p^63d^9$

Note that the 4s electrons are always the first to be lost.

Ions

An ion with a charge of 1+ has one *less* electron than its parent atom. Such a *positive* ion is formed by the *loss* of an electron from the highest-energy occupied atomic orbital. The electronic structures of ions are presented in the same way as those of neutral atoms. For example, for sodium: Na $1s^22s^22p^63s^1$; Na^+ $1s^22s^22p^6$. Negative ions are represented in a similar manner, remembering that a *negative* ion has *more* electrons than its parent atom. For example, for chlorine: Cl $1s^22s^22p^63s^23p^5$; Cl^- $1s^22s^22p^63s^23p^6$.

SUMMARY

- The building-up principle: electrons fill atomic orbitals in order of increasing energy.
- The Pauli exclusion principle: an atomic orbital can contain no more than two electrons. When two electrons occupy an atomic orbital, their spins must be paired.
- Hund's rule: when orbitals have the same energy, they first fill with one electron each (with parallel spins), before they start to pair up.

4.7

O B J E C T I V E S

- Patterns in electronic structures across periods and down groups

ELECTRONIC STRUCTURE AND THE PERIODIC TABLE

Mendeleyev constructed the first periodic table before anything was known about the structure of the atom. He organized it purely on the grounds of atomic mass and the properties of the elements. However, it is now known that chemical properties arise from the electronic structure of atoms and not from their relative masses. The shape of the periodic table may now be explained in terms of electronic structures.

Periods 1 and 2

The first shell, with principal quantum number $n = 1$, consists of just one s orbital. Period 1 therefore contains two elements: hydrogen H, with the electronic structure $1s^1$; and helium He, with the electronic structure $1s^2$. At the noble gas helium, the $n = 1$ shell is full.

The element lithium Li stands at the beginning of Period 2. The electronic structure of lithium is $1s^2 2s^1$. The second shell, with $n = 2$, has started to fill. This shell consists of one 2s orbital, which can contain two electrons, and three 2p orbitals, which can contain a total of six electrons. The $n = 2$ shell can therefore hold a maximum of eight electrons. Look at the outline periodic table below: the noble gas neon Ne marks the point at which the first shell ($n = 1$) and the second shell ($n = 2$) are both full.

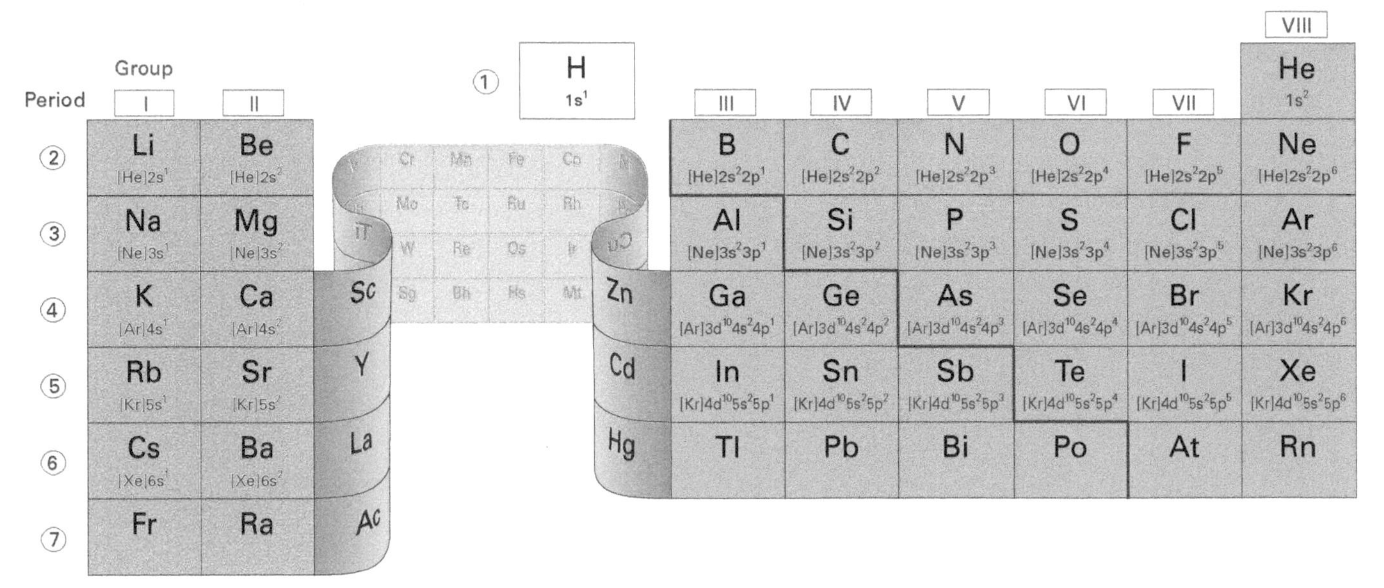

The periodic table with element symbols and short-form electronic structures up to barium Ba. Elements with electronic structures having full shells are the noble gases. As a short form, we use the symbols for the noble gases, enclosed in square brackets, to indicate electronic structures for full shells. For example, sodium Na $1s^22s^22p^63s^1$ can be written as Na $[Ne]3s^1$.

Groups

Notice how groups contain elements with similar electronic structures: Group I elements all end in ns^1; Group II elements all end in ns^2.

Period 3

Period 3 contains the sequence of elements from sodium Na to argon Ar, a total of eight elements. Period 3 fills the 3s and the 3p orbitals in the same way that Period 2 filled the 2s and the 2p orbitals. The third shell, with $n = 3$, consists of one 3s orbital, three 3p orbitals, and five 3d orbitals. You might expect the 3d to fill in the course of Period 3, but things are not that straightforward! The large number of electrons present causes the relative energies of the orbitals to change. The 4s orbital becomes lower in energy than 3d, and filling of the 3d orbitals is delayed until Period 4.

Period 4

Period 4 consists of a total of 18 elements, from potassium K to krypton Kr. The orbitals fill in the sequence 4s (potassium K, calcium Ca), 3d (scandium Sc to zinc Zn), and 4p (gallium Ga to krypton Kr).

Aluminium Al [Ne]$3s^23p^1$ is a metal. Silicon Si [Ne]$3s^23p^2$ is a metalloid. Phosphorus P [Ne]$3s^23p^3$ is a non-metal. These three elements, Al, Si, and P, have the sequential atomic numbers 13, 14, and 15, yet represent three different classes of element.

The s, p, d, and f blocks

The main body of the periodic table divides into three areas: the **s block**, the **p block**, and the **d block**.

- The s block, on the left, is where the s orbitals are filling.
- The p block, on the right, is where the p orbitals are filling.
- The d block, between the s and p blocks, is where the d orbitals of the previous shell are filling.

The two series of elements separated out at the bottom of the table make up the **f block**. There are 14 elements in each row (seven orbitals, two electrons each). The seven orbitals in each of the 4f and the 5f fill in the course of these series.

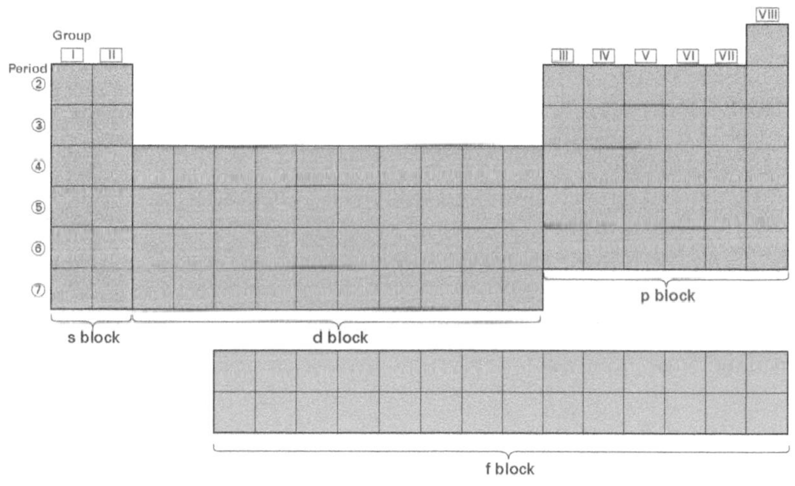

An outline of the periodic table showing the s, p, d, and f blocks.

SUMMARY

- The s block comprises Groups I and II (s orbitals filling – total two electrons).
- The p block comprises Groups III to VIII (p orbitals filling – total six electrons).
- The d block contains elements that have d orbitals filling (a total of 10 electrons for each of the 3d, 4d, and 5d subshells).

PRACTICE

1 In which block of the periodic table is each of the following elements?

a Strontium **b** Nickel

c Lithium **d** Arsenic

e Iron **f** Carbon.

2 Which orbitals are filling in the course of the following periods?

a Period 2 **b** Period 3

c Period 4.

4.8

OBJECTIVES

- First ionization energy
- Successive ionization energies

MORE EVIDENCE FOR SHELLS AND SUBSHELLS: IONIZATION ENERGIES

If you want to find out how something works, it is often a good idea to take it to pieces. Taking an atom to pieces involves removing the electrons one at a time, starting with the outermost. You already know that the process of removing electrons from an atom is called ionization. The energy needed to remove each successive electron from an atom is called the first, second, third, etc., ionization energy. The question is: can ionization energies confirm the arrangement of electrons in atoms?

Ionization energy

The **first ionization energy** is the minimum energy required to remove one electron from an isolated atom in the gas phase. In other words, it is the energy for the process:

$$E(g) \rightarrow E^+(g) + e^-(g)$$

where E represents an element. The value is usually quoted per mole of atoms. For example, the value for hydrogen is 1312 kJ mol^{-1}, and the value for sodium is 498 kJ mol^{-1}. All ionization energies are positive because it always requires energy to remove an electron. A plot of ionization energy against atomic number provides compelling evidence for the existence of shells and subshells.

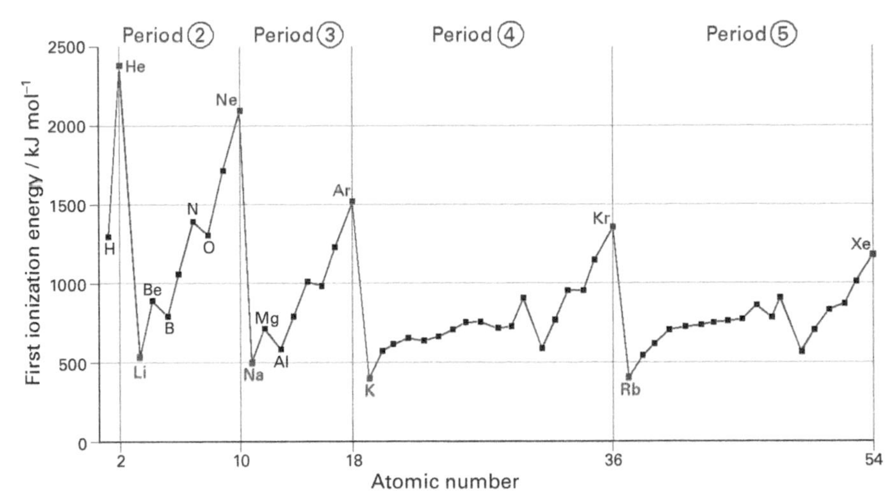

A plot of first ionization energy against atomic number for the elements hydrogen to xenon.

Evidence for shells

The noble gases helium to xenon stand at the peaks of the plot, where the values of first ionization energy are highest. These elements make up Group VIII: helium has one complete shell; neon has two complete shells; argon has completed the $n = 3$ shell (with the exception of the 3d subshell). Krypton and xenon follow a similar pattern.

Now turn your attention to the elements at the troughs in the plot, where the values of first ionization energy are lowest. These elements are all members of Group I. The outermost electrons of lithium, sodium, and potassium are all s electrons. From the graph we can see that the single s electron requires relatively little energy for removal compared with that needed to remove other electrons. This is because the s electrons are in 'new' shells further from the nucleus than the electrons in the preceding noble gases helium, neon, and argon. Note that the overall shape of the plot illustrates the periodic repetition of electronic structures that occurs in the periodic table.

Evidence for subshells

The plot of first ionization energies also provides evidence for subshells, although we need to look harder. See how the value *decreases* slightly between beryllium and boron, and between magnesium and aluminium. In both cases, the second element has its outermost electron in a new subshell (the p subshell) rather than an s subshell. As the p subshell is further from the nucleus, the attractive force is less and so the energy required to remove the electron is less.

There is also a decrease in ionization energy between nitrogen and oxygen. This decrease arises because nitrogen has three unpaired electrons in the 2p subshell, whereas oxygen has two unpaired electrons and two paired electrons. In the case of oxygen, the electron–electron repulsion between the two paired electrons in one p orbital makes one of the pair slightly easier to remove. The electronic structure of oxygen is therefore slightly less stable than the half-filled subshell in nitrogen, which contains no paired electrons. As a result, slightly less energy is required to remove an electron from oxygen than might otherwise be expected.

Some electronic structures

Beryllium	$_4Be$	$[He]2s^2$
Boron	$_5B$	$[He]2s^22p^1$
Nitrogen	$_7N$	$[He]2s^22p^3$
Oxygen	$_8O$	$[He]2s^22p^4$
Magnesium	$_{12}Mg$	$[Ne]3s^2$
Aluminium	$_{13}Al$	$[Ne]3s^23p^1$

Further evidence for shells

Further evidence for shells comes from the successive ionization energies needed to remove *all* the electrons from an atom. Look, for example, at removing all 11 electrons from a sodium atom. There is a significant increase in energy needed between removing the first and second electrons. There is another large increase between removing the ninth and tenth electrons. These observations suggest that the increases occur when the electron being removed has to come from a new shell, closer to the attractive charge of the nucleus. The pattern shows that the outermost $n = 3$ shell of sodium contains one electron, and the $n = 2$ shell has eight electrons, leaving two electrons in the innermost $n = 1$ shell.

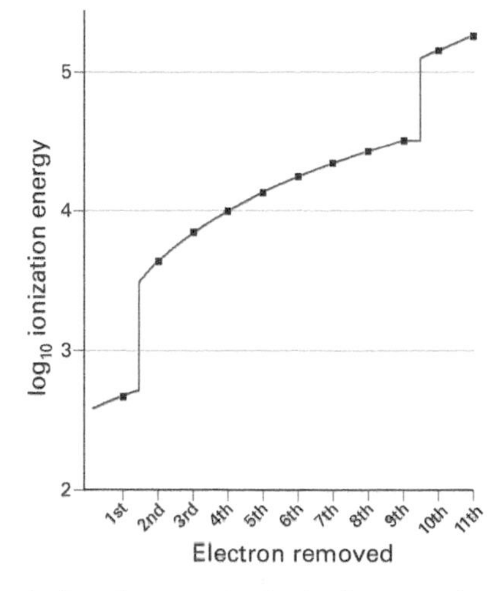

A plot of successive ionization energies for the sodium atom, $_{11}Na\ 1s^22s^22p^63s^1$. See spread 12.3 for the meaning of log_{10}.

Trends across a period

Each successive element in a period has one more proton than the previous one. As a result, the nuclear charge increases steadily. But the hold the nucleus has on the outer electrons does *not* change steadily. For successive elements across Period 3, electrons are added to the *same* outer shell. Since electrons in the same shell do not shield each other very well from the attraction of the nuclear charge, the *effective* nuclear charge experienced by the outermost electrons in each successive element increases, see spread 17.3. So the outermost electrons experience a greater attractive force towards the nucleus. The result is that the atomic radius *decreases* as the outermost electrons are pulled closer to the nucleus, as you go from sodium to argon, and the ionization energy *increases* significantly from sodium to argon.

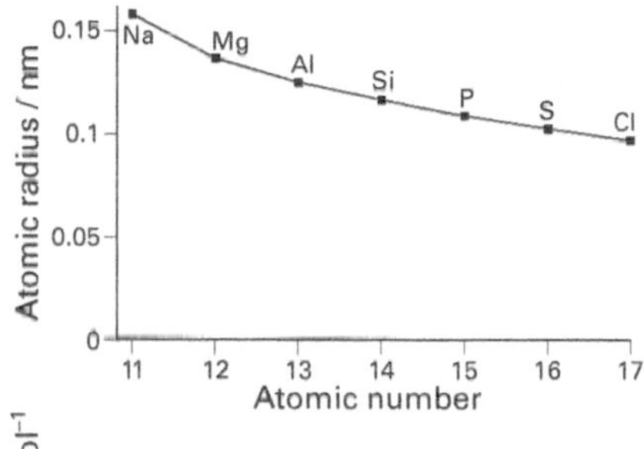

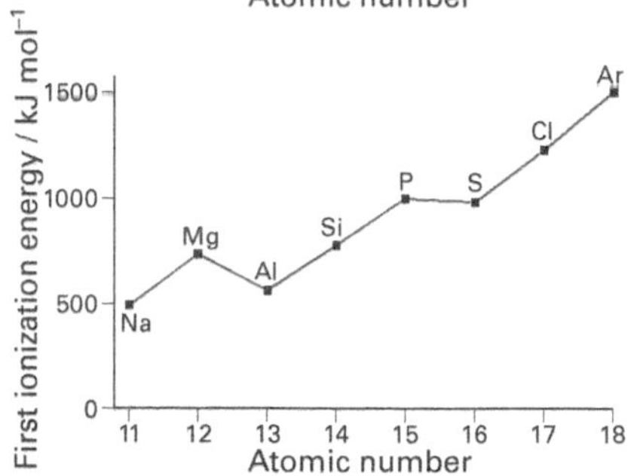

Plots of (a) atomic radius, and (b) first ionization energy against atomic number for Period 3.

SUMMARY

- Ionization energies provide evidence for shells and subshells.
- Atomic size decreases and first ionization energy generally increases across a period.

PRACTICE

1 Sketch a graph of the successive ionization energies of potassium. What does the shape of your graph tell you about the electronic structure of potassium?

2 The trend in ionization energies from lithium to neon is not smooth. Is the value for beryllium higher, or is the value for boron lower, than might be expected? Explain your answer.

3 Why do the ionization energies for the elements in the group helium to xenon show a downward trend?

4 The ionization energies of the d-block elements scandium to zinc are fairly similar. Refer to the electronic structures of these elements to explain this observation.

EXTENSION

4.9

OBJECTIVES

- The form of the Schrödinger equation
- Solutions of the Schrödinger equation
- Physical significance of the solutions

Principal quantum numbers

The Schrödinger equation for the hydrogen atom can only be solved analytically for certain values of E. The full set of solutions available can be labelled in terms of three quantum numbers. Remember that the Bohr model gave only one, the principal quantum number n. Solutions to the Schrödinger equation give values for n which concur with Bohr.

- n increases from 1 onwards in integer steps: 1, 2, 3, 4, etc.

Azimuthal quantum numbers

The second quantum number that arises from solutions of the Schrödinger equation is the **azimuthal quantum number** l. An s orbital has $l = 0$, a p orbital has $l = 1$, and a d orbital has $l = 2$.

- For any particular value of n, l can take all integer values from 0 to $n - 1$.

The first shell ($n = 1$) can therefore have only one subshell: 1s ($l = 0$). The second shell ($n = 2$) can have two subshells: 2s ($l = 0$) and 2p ($l = 1$). The third shell ($n = 3$) can have three subshells: 3s ($l = 0$), 3p ($l = 1$), and 3d ($l = 2$). The Schrödinger equation provides a natural explanation for the existence of subshells. (See spread 4.3.)

Magnetic quantum numbers

The third quantum number that arises from solutions of the Schrödinger equation is the **magnetic quantum number** m_l.

- For any particular value of l, m_l can take all integer values from $-l$ to l.

There is only **one s orbital** ($m_l = 0$) for any value of n. There are **three p orbitals** ($m_l = -1, 0, 1$) for $n \geq 2$, **five d orbitals** ($m_l = -2, -1, 0, 1, 2$) for $n \geq 3$, and **seven f orbitals** ($m_l = -3, -2, -1, 0, 1, 2, 3$) for $n \geq 4$.

The Schrödinger equation provides a natural explanation for the number of orbitals in each subshell.

Solving the Schrödinger equation

This spread looks at solutions of the Schrödinger equation, the mathematical model that gives us the shapes of atomic orbitals. **You are not expected to learn the contents of this spread**, but it gives more insight into the atomic orbital theory for students interested in mathematics. The Schrödinger equation gives a model of the atom based on the mathematics of waves, as we saw in spread 4.4 on wave mechanics. It is a *differential* equation. The two common techniques for solving such equations produce analytical solutions and numerical solutions. Early attempts to solve the equation focused on analytical solutions whereas numerical solutions are more common now, due to the advent of more powerful computers. In simple circumstances, the Schrödinger equation can even be solved on a graphical calculator, as shown below.

The Schrödinger equation

The form of the Schrödinger equation for the energy E of the 1s orbital in the hydrogen atom is as follows:

$$-\frac{h^2}{8\pi^2 m}\left[\frac{d^2\psi}{dr^2} + \left(\frac{2}{r}\right)\frac{d\psi}{dr}\right] - \frac{e^2}{4\pi\varepsilon_0 r}\psi = E\psi \qquad (1)$$

where h is the Planck constant, m is the mass of the electron, e is the charge on the electron, and ε_0 is a fundamental constant called the vacuum permittivity. The function ψ is called the **wavefunction.** The first (bracketed) term is the wave equation version of the kinetic energy of the electron. The second term is its potential energy due to electrostatic attraction to a single proton at a distance r.

Multiplying both sides of equation (1) by $(-8\pi^2 m/h^2)$ and collecting together fundamental constants using $a_0 = \varepsilon_0 h^2/\pi m e^2$, we find:

$$\frac{d^2\psi}{dr^2} + \left(\frac{2}{r}\right)\frac{d\psi}{dr} + \left(\frac{2}{a_0 r}\right)\psi = \left(\frac{-8\pi^2 mE}{h^2}\right)\psi \qquad (2)$$

The quantity a_0 has the units of length and is called the Bohr radius, the radius at which Bohr said the electron orbits the proton. The second and third terms cancel out if:

$$\frac{d\psi}{dr} = \frac{-\psi}{a_0}$$

This is the equation for exponential decay, whose solution is:

$$\psi = \psi_0 e^{-r/a_0}$$

The energy E for the 1s orbital is found by differentiating again, substituting into equation (2), and rearranging:

$$\frac{d^2\psi}{dr^2} = \frac{d}{dr}\left(\frac{d\psi}{dr}\right) = \frac{d}{dr}\left(\frac{-\psi}{a_0}\right) = \frac{\psi}{a_0^2}$$

$$\frac{1}{a_0^2} = \frac{-8\pi^2 mE}{h^2}$$

Thus,

$$E = \frac{-h^2}{8\pi^2 m a_0^2}$$

If you substitute the values $h = 6.626 \times 10^{-34}$ J s, $m = 9.109 \times 10^{-31}$ kg, $a_0 = 5.292 \times 10^{-11}$ m, and multiply by the Avogadro constant, 6.022×10^{23} mol^{-1}, to get the energy per mole, the numerical value is 1310 kJ mol^{-1} (to 3 sig. figs.), which is the ionization energy for the hydrogen atom, spread 4.8.

Using a graphical calculator

A graphical calculator can also be used to solve the Schrödinger equation numerically. With the substitutions:

$$x = \frac{r}{a_0}$$

and

$$E = -\frac{h^2}{8\pi^2 m a_0^2}$$

and using f in place of ψ for convenience, equation (2) (multiplied by a_0^2) simplifies to:

$$\frac{d^2f}{dx^2} + \left(\frac{2}{x}\right)\frac{df}{dx} + \left(\frac{2}{x}\right)f = f \qquad (3)$$

Equation (3) must be manipulated into a form suitable for use with a graphical calculator that can solve differential equations. Such a calculator usually needs a second-order differential equation (with a term $\frac{d^2f}{dx^2}$) to be re-written as a pair of first-order differential equations. This transformation is carried out as follows:

$$g = \frac{df}{dx}, \qquad (4)$$

$$\frac{dg}{dx} = \frac{d^2f}{dx^2}$$

$$= -\left(\frac{2}{x}\right)\frac{df}{dx} - \left(\frac{2}{x}\right)f + f$$

$$= -\left(\frac{2}{x}\right)g - \left(\frac{2}{x}\right)f + f \qquad (5)$$

This pair of equations (4 and 5) is now ready for solution once initial conditions are chosen. The following choice compares the value of ψ with its value at the origin (ψ_0):

$f = 1$ and $g = -1$ at $x = 0$

The solution is a *numerical* solution to the equation (which must avoid $x = 0$ itself, which leads to an infinite value). The result is shown to the right, a decaying exponential. See spread 4.5 for the related electron density (which is the square of the wavefunction).

If the pair of equations is solved again, with the term in front of the final f changed slightly (to simulate a value for the total energy *close to but different from* the true answer), the solution becomes unstable and does not correspond to physical reality. For example, for the equation:

$$\frac{dg}{dx} = -\left(\frac{2}{x}\right)g - \left(\frac{2}{x}\right)f + 1.01f$$

the numerical solution results in the lower figure shown to the right. The total energy is only *1% away from the true answer* and yet the solution is unacceptable as the value of the wavefunction rises to infinity (exceeding its initial value less than 12 Bohr radii from the nucleus).

The two plots shown provide graphic illustration of the quantization of energy. Quantization of energy arises naturally from the Schrödinger equation, whereas Bohr had to *assume* that energy is quantized.

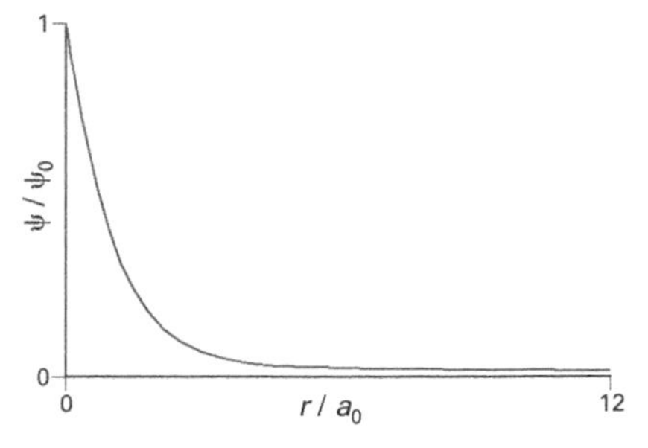

The numerical solution of the Schrödinger equation for the lowest-energy orbital (1s) is an exponential decay of the wavefunction ψ against distance from the nucleus.

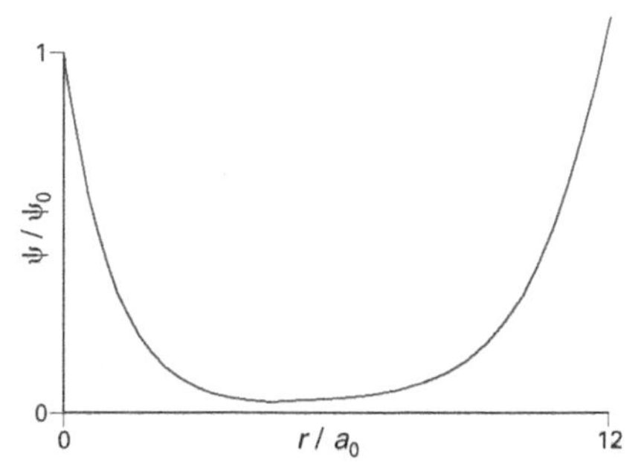

The plot of ψ against r, for a solution which does not correspond to physical reality.

Electron spin

The one failure of the Schrödinger equation is that it does not account for the existence of two electrons per orbital; it cannot explain electron spin or the Pauli exclusion principle.

PRACTICE EXAM QUESTIONS

1 The graph below shows the trend in first ionization energy from oxygen to magnesium.

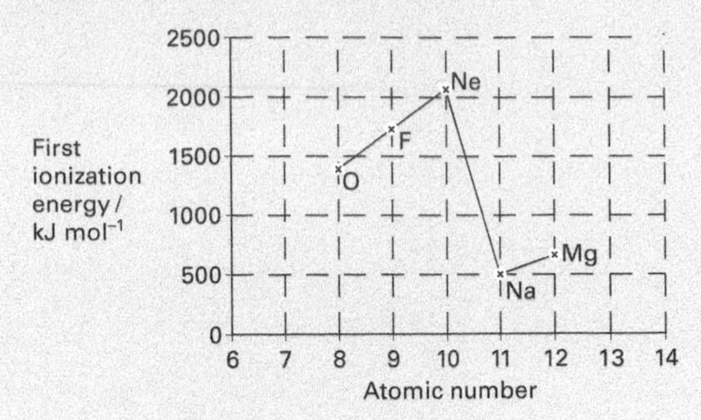

a Using crosses, mark on the graph the first ionization energies of nitrogen and of aluminium. Label each of your crosses with the symbol for the element. [2]

b Explain why the first ionization energy of neon is greater than that of sodium. [2]

c Of the elements neon, sodium, and magnesium, predict which one has the largest second ionization energy. Explain your answer. [3]

d Published values of electronegativity are available for oxygen, fluorine, sodium, and magnesium but not for neon.

i Explain why a value of electronegativity is not available for neon.

ii Of the elements oxygen, fluorine, sodium, and magnesium, predict which one has the smallest electronegativity value. [2]

e Explain, with reference to the bonding in sodium oxide, why this compound reacts with water to form a solution with a pH of 14. [3]

f What general type of oxide forms acidic solutions in water? Give the formula of one such oxide. [2]

(See also chapters 16 and 17.)

2 The diagram below shows the electronic structure of boron.

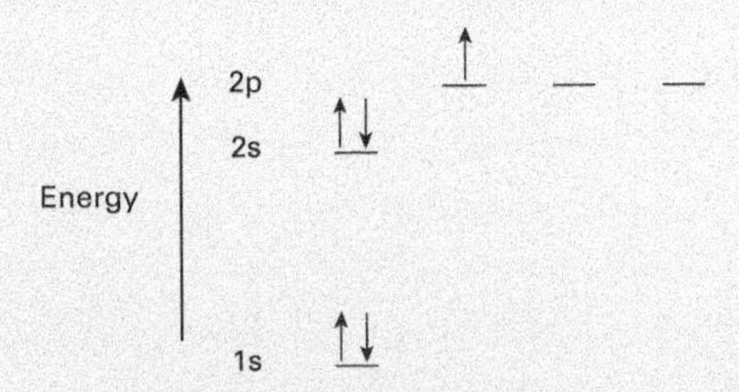

a The electrons are represented by arrows. What property of the electrons do these 'up' and 'down' arrows represent? [1]

b Suggest why electrons which occupy the 2p sub-levels have a higher energy than electrons in the 2s sub-level. [1]

c Complete the following energy level diagram to show the electronic structure of carbon. [2]

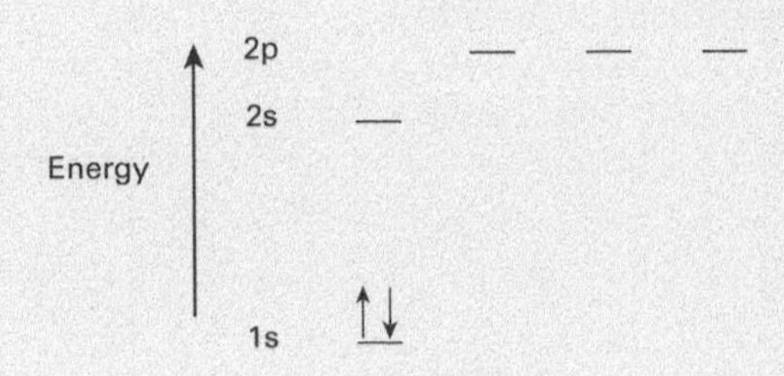

d Explain the meaning of the term *first ionization energy*. [2]

e Explain why oxygen has a lower first ionization energy than nitrogen. [2]

3 The following table contains ionization energy data.

Element	N	O	F	Ne	Na
First ionization energy/kJ mol^{-1}	1400	1310	1680	2080	494

a Explain the meaning of the term *first ionization energy* of an element. [2]

b Explain why neon has a higher first ionization energy than fluorine. [2]

c Explain why oxygen has a lower first ionization energy than nitrogen. [2]

d Explain why sodium has a lower first ionization energy than neon. [2]

e Predict an approximate value for the first ionization energy of carbon and explain your answer. [3]

4 a Explain why the first ionization energy of aluminium is less than the first ionization energy of magnesium. [3]

b Explain why the first ionization energy of aluminium is less than the first ionization energy of silicon. [2]

c Explain why the second ionization energy of aluminium is greater than the first ionization energy of aluminium. [2]

d Write an equation to illustrate the third ionization energy of aluminium. [1]

e Explain why the third ionization energy of aluminium is much less than the third ionization energy of magnesium. [2]

5 The Sun largely consists of a mixture of hydrogen and helium, the presence of each being detected by spectroscopy. The line emission spectrum of atomic hydrogen in the ultraviolet region of the electromagnetic spectrum is shown below.

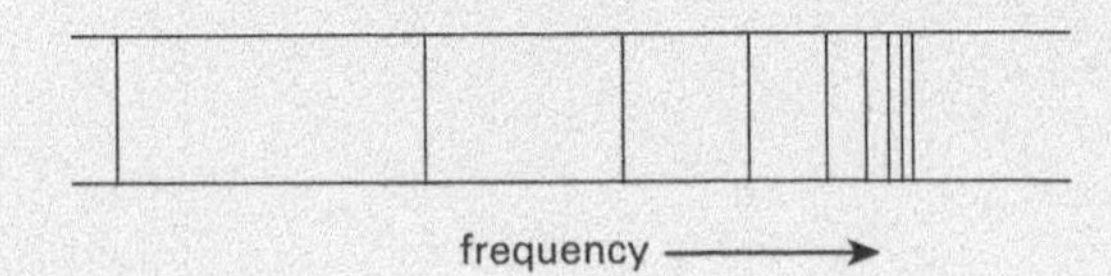

a Explain why this spectrum consists of lines which are converging.

(Up to 2 marks may be obtained for the quality of language in this part.) [7]

b Explain how the ionization energy of atomic hydrogen can be calculated from this spectrum. [2]

6 a Define the second ionization energy of fluorine. [2]

b Using axis as below, sketch a graph to show the successive ionization energies of fluorine. Give reasons for the shape of the line you draw.

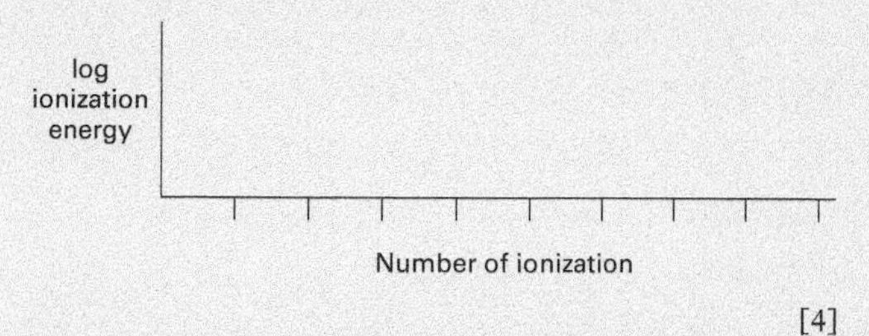

[4]

7 a i Give the equation which represents the first ionization energy of oxygen atoms. [1]

ii Why is the first ionization energy of helium the largest of all the atoms? [2]

iii Why is the first ionization energy of oxygen atoms less than that of nitrogen atoms? [2]

iv Why do the first ionization energies of the atoms in Group 1 decrease as the atomic number increases? [2]

b i Write equations which represent the first and second electron affinities of oxygen atoms. [2]

ii The first electron affinity of oxygen atoms is -141 kJ mol^{-1} and the second is $+798$ kJ mol^{-1}. Suggest why the first is exothermic and the second endothermic. [2]

c Magnesium burns brightly in oxygen to give magnesium oxide, which contains the ions Mg^{2+} and O^{2-}. The formation of both these ions from their elements is strongly endothermic. Why, therefore, should magnesium combine with oxygen? [2]

d The table below gives the successive ionization energies of sodium.

No. of ionization	1	2	3	4	5	6
Energy/kJ mol^{-1}	496	4563	6913	9644	13 352	16 611

No. of ionization	7	8	9	10	11
Energy/kJ mol^{-1}	20 115	25 491	28 934	141 367	159 079

What information about the electronic structure of sodium is provided by this data? [2]

(See also chapter 10.)

8 a Study the table of ionization energies below and answer the questions which follow.

Ionization energy/kJ mol^{-1}	1st	2nd	3rd	4th
Sodium	494	4560	6940	9540
Magnesium	736	1450	7740	10 500
Aluminium	577	1820	2740	11 600

Explain the relative magnitudes of the following:

i the 1st ionization energies of sodium and magnesium;

ii the 1st ionization energies of magnesium and aluminium;

iii the 2nd ionization energies of magnesium and aluminium;

iv the 3rd and 4th ionization energies of aluminium. [8]

b Consider the electron affinities for oxygen given below.

Electron affinity/kJ mol^{-1}	1st	2nd
	−142	+844

i Write equations representing the changes to which the 1st and 2nd electron affinities of oxygen relate.

ii Explain the relative magnitudes of the 1st and 2nd electron affinities of oxygen.

iii Given the endothermic nature of the 2nd electron affinity of oxygen, comment briefly on the thermodynamic stability of ionic metal oxides. [7]

9 The first ionization energies of the elements lithium to neon, in kJ mol^{-1}, are given below.

Li	Be	B	C	N	O	F	Ne
519	900	799	1090	1400	1310	1680	2080

a Write an equation representing the first ionization energy of oxygen. [1]

b i Explain why the ionization energies show an overall tendency to increase across the Period.

ii Explain the irregularities in this trend for B and O. [6]

c An element X has successive ionization energies as follows:

786; 1580; 3230; 4360; 16 000; 20 000; 23 600; 29 100 kJ mol^{-1}

i To which Group in the Periodic Table does **X** belong?

Explain your answer.

ii Write down the outer electronic configuration of an atom of **X**.

iii Suggest formulae for TWO chlorides of **X**. [5]

5 Chemical bonding

Excluding the 'exotic' elements produced in nuclear reactors, there are just 90 naturally occurring elements. These few elements combine to make over five million different compounds. A compound is any substance formed by the chemical combination of two or more elements in fixed proportions. The properties of a compound are usually completely different from the properties of its constituent elements. When atoms combine together to make compounds, there is a change in the arrangement of the electrons in the outermost shell of each atom. These electrons form links called **chemical bonds** between the atoms. The aim of this chapter is to investigate the nature of chemical bonds and to describe how they form.

Bonding revision

5.1

OBJECTIVES

- Ionic bonding
- Covalent bonding
- Bonding and properties

In your earlier studies you have met two main types of bonding:

- ionic bonding,
- covalent bonding.

Before studying the subject of bonding in greater depth, this spread will revise the two basic principles underlying the formation of bonds: electron transfer and electron sharing. This spread uses simple concepts which will probably be familiar to you.

Ionic bonding

Ionic bonds form between metals and non-metals. The name 'ion' is used to describe any species that has unequal numbers of electrons and protons and so carries an electric charge. Metals lose one or more electrons and become positively charged ions. Non-metals gain one or more electrons and become negatively charged ions. Both types of ion usually have full outer shells of electrons, corresponding to the stable electronic structures of the noble gases. The oppositely charged ions attract each other to form a rigid three-dimensional lattice. Each ion in the lattice is surrounded by others of opposite charge.

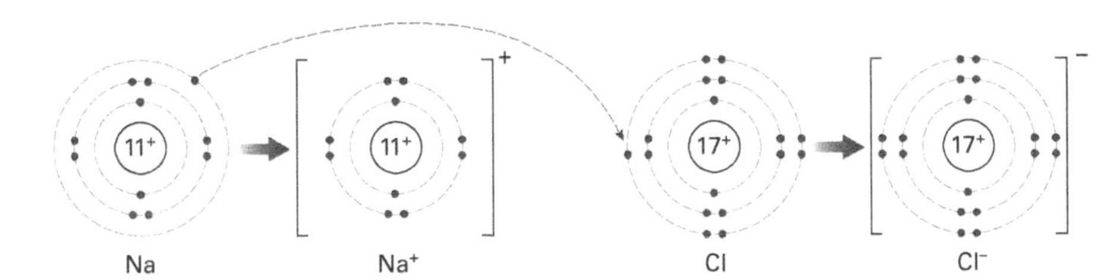

The formation of sodium and chloride ions from their atoms. The full inner shells of electrons are not usually shown.

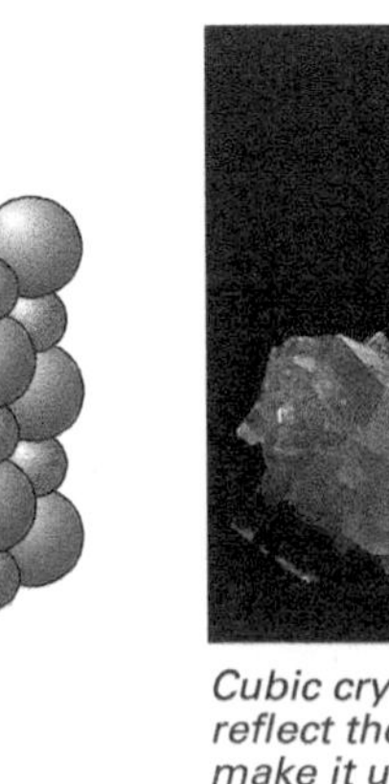

Cubic crystals of pure sodium chloride reflect the arrangement of the ions that make it up.

The sodium ion

A sodium *atom* Na has 11 protons in its nucleus. The nucleus is surrounded by 11 electrons. The atom is neutral overall, the total negative charge of the 11 electrons being exactly balanced by the total positive charge of the 11 protons, i.e. (+11) + (−11) = 0.

A sodium *ion* Na^+ forms when a sodium atom *loses* one electron. The ion has an overall charge of +1 because there are 11 protons in the nucleus with only 10 surrounding electrons, i.e. (+11) + (−10) = +1.

The chloride ion

A chlorine *atom* Cl has 17 protons in its nucleus. The nucleus is surrounded by 17 electrons.

A chloride *ion* Cl^- forms when a chlorine atom *gains* one electron. The ion has an overall charge of −1 because there are 17 protons in the nucleus with 18 surrounding electrons, i.e. (+17) + (−18) = −1.

In sodium chloride, Na^+ ions and Cl^- ions alternate in a three-dimensional lattice. The formula of sodium chloride is NaCl, which signifies that 1 mol of sodium ions is combined with 1 mol of chloride ions. It is incorrect *to speak of 'a molecule of sodium chloride'.*

Covalent bonding

Covalent bonds form between non-metals, which share electrons in order to attain full outer shells. Each shared pair of electrons is regarded as one covalent bond. Covalent bonding between atoms results in the formation of **molecules**. The simplest examples are **diatomic molecules** (molecules that contain two atoms), such as the elements hydrogen H_2 and chlorine Cl_2.

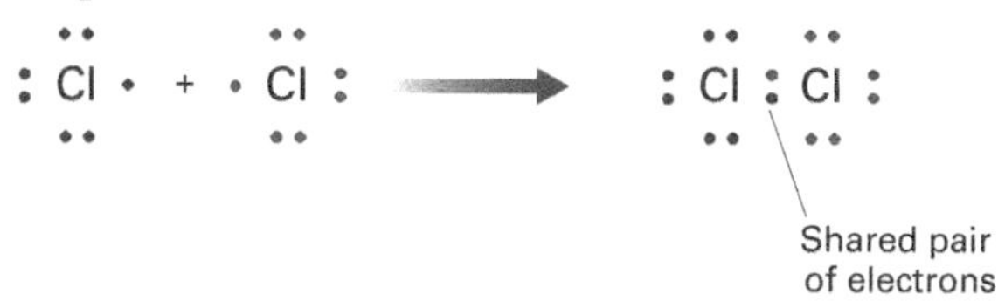

The formation of a chlorine molecule. Two electrons, one from each chlorine atom, form a shared pair. Including this shared pair, each atom now has a full outer shell of electrons, corresponding to the electronic structure of the noble gas argon.

The shared pair of electrons is concentrated in the region of space between the two atomic nuclei. The nuclei are each attracted towards this region of negative charge concentrated between them. Covalent bonds are therefore the result of electrostatic attraction.

Covalent bonds form in the same manner between atoms of different elements. For example, one hydrogen atom and one chlorine atom combine to form a hydrogen chloride molecule HCl; hydrogen and oxygen atoms form water H_2O.

Looking forward

The model for ionic bonding described above was introduced by Walther Kossel in 1916. In the same year, Gilbert Lewis presented the theory of covalent bonding based on the sharing of electron pairs between atoms. The following two spreads will use Lewis's ideas to describe the bonding in a number of substances. Looking at the electron pairs present in molecules will then allow us to make predictions about their shapes.

SUMMARY

- Ionic bonds are the electrostatic forces of attraction between oppositely charged ions; the ions are the result of electron transfer between atoms.
- Covalent bonds result from atoms sharing electron pairs; there is an attraction between these electrons and the nuclei of the atoms.
- Ionic compounds generally have high melting points and high boiling points.
- Covalent compounds composed of small molecules have low melting points and low boiling points.

Properties of ionic compounds

The electrostatic forces of attraction between oppositely charged ions are strong. Ionic compounds have high melting points (T_m) and high boiling points (T_b) (e.g. sodium chloride: $T_m = 801\,°C$, $T_b = 1413\,°C$). Solid ionic compounds do not conduct electricity. When solid ionic compounds are melted, they *do* conduct electricity because the ions are then free to move.

Properties of covalent elements and compounds

The covalent bonds between atoms within a molecule are strong, but the forces of attraction *between molecules* are weak. Covalent compounds made up of small molecules have low melting points and low boiling points (e.g. hydrogen chloride HCl: $T_m = -115\,°C$, $T_b = -85\,°C$). Solid, liquid, and gas states all consist of covalently bonded molecules. Covalently bonded substances do not conduct electricity in any of these states because there are no free charged species (electrons or ions) present.

Chlorine gas consists of covalently bonded diatomic molecules (i.e. molecules in which two chlorine atoms are held together by a covalent bond).

PRACTICE

1 Give the symbols of the stable ions that may be formed by the following atoms:
 a Lithium
 b Aluminium
 c Oxygen
 d Potassium
 e Fluorine
 f Strontium
 g Nitrogen.

2 Draw diagrams to show what happens to the electrons when the ionic compound potassium sulphide is formed from the elements potassium and sulphur.

3 Draw diagrams to show how the following elements bond together to form covalent molecules. Give the formula of each molecule.
 a Hydrogen and fluorine
 b Oxygen and hydrogen
 c Nitrogen and hydrogen
 d Oxygen and fluorine.

OBJECTIVES

- Drawing Lewis structures
- Why electrons pair
- Bonding pairs and lone pairs

LEWIS STRUCTURES FOR COVALENT MOLECULES – 1

You should already be familiar with drawing covalent molecules on paper, representing electrons as dots surrounding the symbols of each constituent element. These drawings are called Lewis structures. In the chlorine molecule Cl_2, one pair of electrons is shared between the two atoms. The sharing of a single electron pair constitutes a single covalent bond. This spread will look more closely at Lewis structures, and will show why electron pairs are central to the idea of covalent bonding.

(a)

: Cl ⁝ Cl ⁝

(b)

: Cl : Cl :

(a) The dot-and-cross diagram and (b) the Lewis structure illustrating the bonding in the chlorine molecule. Note that the bond contains one electron from each of the contributing atoms. From now on we will use only Lewis structures in this book.

Lewis structures

Lewis structures use dots to represent electrons; **dot-and-cross diagrams** are similar to Lewis structures, but the electrons are shown as dots or crosses. Dot-and-cross diagrams help by showing the sources of the electrons involved in covalent bonds. However, dot-and-cross diagrams can be misleading, as they suggest that the electrons are somehow different from each other. Therefore, Lewis structures will be used throughout this text, with the dots of different intensity to show the sources of the electrons.

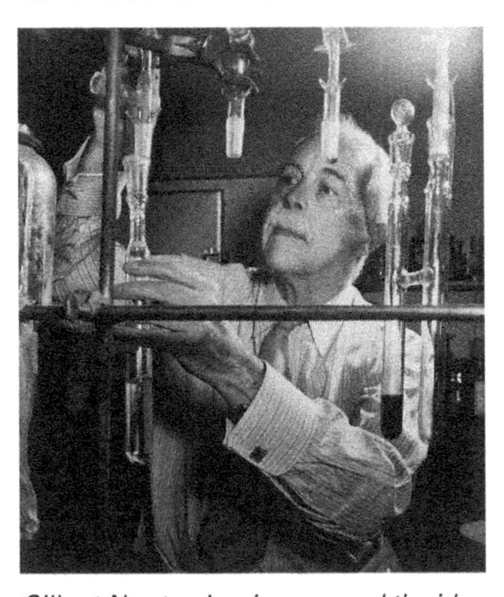

Gilbert Newton Lewis proposed the idea of covalent bonding and showed the importance of electron pairs in bonding. Lewis's treatment of covalent bonding was remarkable because the reasons for the formation of bonding pairs of electrons were not known at the time. Lewis first introduced his dot diagrams in 1916.

Electron pairs

A chlorine atom has 17 electrons; just two electrons are involved in the covalent bond between the atoms of a chlorine molecule. In this case it is clear that most of the electrons are not used for bonding. When two chlorine atoms form a covalent bond in Cl_2, each achieves the electronic structure of argon, a noble gas. The shared electrons involved in the formation of a covalent bond are called the **valence electrons**. The shell from which they originate is called the **valence shell** of the atom concerned. The valence shell is usually the *outermost* shell.

In the chlorine molecule Cl_2, the valence shell of each chlorine atom contains two electrons that are involved in bonding and six that are not. How are these electrons arranged? This question may be answered by considering the electronic structures of the atoms.

The atomic number of chlorine is 17. Its electronic structure in terms of occupied shells is therefore: Cl 2,8,7. This method of listing the electrons says little about the valence shell other than that it contains seven electrons. Stating the electronic structure in terms of atomic orbitals (as we did in the previous chapter) is more informative:

$$\mathrm{Cl}\ 1s^2 2s^2 2p^6 3s^2 3p^5 \quad \text{or} \quad \mathrm{Cl}\ [\mathrm{Ne}]3s^2 3p^5$$

This structure shows that the valence shell (n = 3) is occupied by two s electrons and five p electrons (a total of seven electrons). The actual occupancy of the 3s and the 3p orbitals is shown by the electrons-in-boxes notation (shown left).

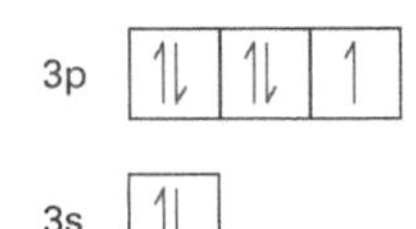

Electrons-in-boxes notation for 3s and 3p chlorine atomic orbitals. Note that there is no distinction between the three p orbitals. The unpaired electron could be in any one of them.

There is a pair of electrons in the 3s orbital. Two of the three equivalent 3p orbitals contain a pair of electrons each, while the remaining 3p orbital contains an unpaired electron. Look back to the Lewis structure for the chlorine atom and you will see that the electrons are grouped as three pairs with a single unpaired electron.

The bond in the chlorine molecule consists of a shared pair of electrons, which forms when two electrons (from atomic orbitals on two separate atoms) pair up and occupy a region of space between the two atoms. The electrons that form the bond are called a **bonding pair** of electrons. The bonding electrons must have opposite (paired) spins as a result of the Pauli exclusion principle. Each atom also has three pairs of electrons that are in the valence shell but do not take part in bonding. These electron pairs are called **lone pairs** (or non-bonding pairs).

More Lewis structures

The Lewis structures for the covalently bonded molecules hydrogen chloride HCl, hydrogen H_2, and water H_2O are shown below. Each electron pair constitutes a single covalent bond. The bonding in the three molecules may be represented respectively as H—Cl, H—H, and H—O—H. A single line conventionally represents a single shared pair of electrons (a single covalent bond).

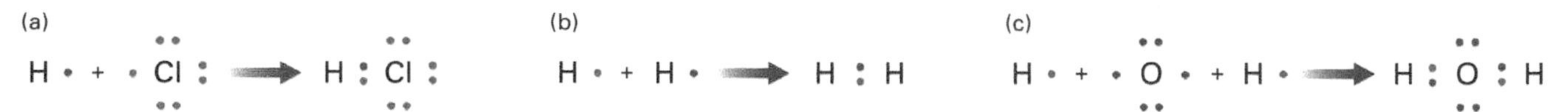

Lewis structures showing the formation of: (a) hydrogen chloride HCl; (b) hydrogen H_2; and (c) water H_2O.

The octet rule

When they combine to form molecules, many non-metal atoms share electrons in order to have eight electrons in their outermost (valence) shells. This electronic structure represents the particularly stable structure found in all the noble gases except helium. This pattern of electron sharing provides a useful rule of thumb, which helps to work out how atoms will bond together. Lewis called it the **octet rule** because atoms share electrons in order to achieve *eight* electrons in their valence shells. It is important to remember that hydrogen can only hold two electrons in its valence shell, so it can only share one pair of electrons (to gain the electronic structure of the noble gas helium).

For example, the ammonia molecule contains atoms of nitrogen and hydrogen. The electronic structures are N $1s^2 2s^2 2p^3$ and H $1s^1$. According to the octet rule, nitrogen must share *three* pairs of electrons with hydrogen atoms to bring the total population of its valence shell up from five electrons to eight.

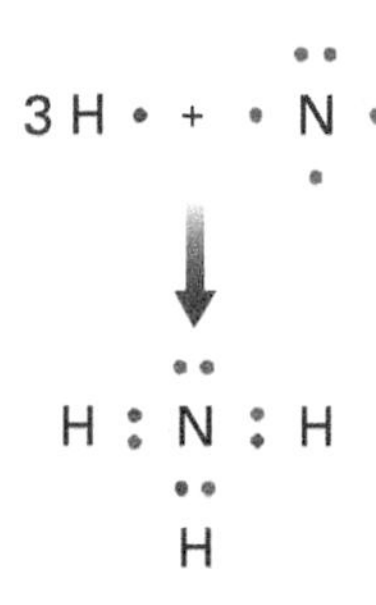

Lewis structure showing the formation of ammonia NH_3. A single covalent bond forms between each hydrogen atom and the nitrogen atom.

Double bonds

A carbon atom has four electrons in the valence shell and so needs to share a total of four electron pairs. An oxygen atom has six electrons in the valence shell and so needs to share a total of two electron pairs. So two oxygen atoms combine with one carbon atom to form the molecule carbon dioxide. The bonding may therefore be represented as O=C=O, indicating the presence of two *double* covalent bonds. Each double covalent bond is conventionally represented by a double line.

The Lewis structure for carbon dioxide CO_2.

Triple bonds

A *triple* covalent bond forms between two atoms when each contributes *three* electrons. A total of six electrons are shared. This occurs, for example, in nitrogen N_2, represented as N≡N. Each triple covalent bond is conventionally represented by a triple line.

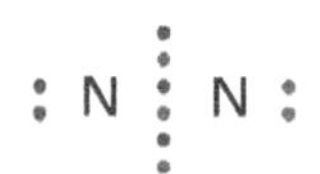

The Lewis structure for nitrogen N_2.

SUMMARY

- Lewis structures use dots to show the electrons involved in bonding pairs and lone pairs.
- The outer shells of atoms in molecules are usually full; i.e. the atoms have the electronic structures of the noble gases.
- A covalent bond forms when electrons from atomic orbitals on two separate atoms pair up and occupy the region of space between the two atoms.
- A single covalent bond is a shared electron pair.
- A double covalent bond consists of four electrons making up two shared pairs.
- A triple covalent bond consists of six electrons making up three shared pairs.

OBJECTIVES

- Failure of the octet rule
- Coordinate bonding

Lewis structures for covalent molecules – 2

A single covalent bond consists of a shared pair of electrons. Double and triple covalent bonds result from the sharing of two and three pairs of electrons respectively. Atoms combine to form molecules – the atoms share electrons so that there are eight in their valence shells, following the octet rule. Lewis structures help us to explain this, and show the numbers of bonding pairs and lone pairs of electrons. However, there is a wide range of compounds whose formulae do not follow the octet rule. Their valence shells do not contain eight electrons. This spread takes a first look at this problem.

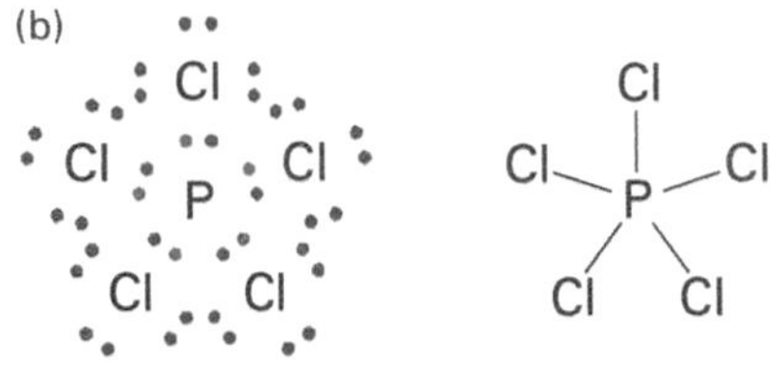

Lewis structures for (a) phosphorus trichloride (which follows the octet rule) and (b) phosphorus pentachloride (which does not follow the octet rule).

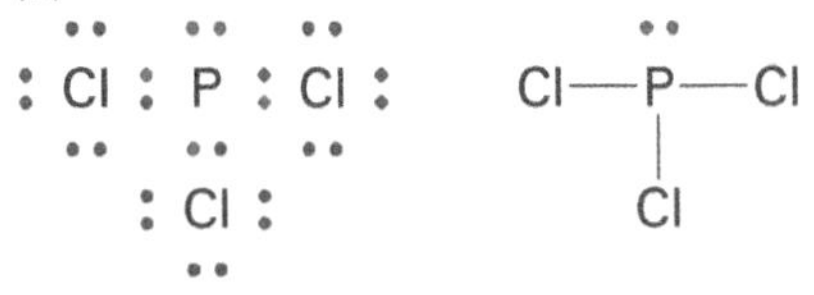

Expanding the octet

Phosphorus trichloride is a volatile liquid (T_b = 75 °C) which does not conduct electricity. This substance has the properties of a typical covalent compound, and the octet rule predicts correctly that phosphorus and chlorine atoms bond together to form molecules with the formula PCl_3.

Phosphorus trichloride reacts with chlorine to form phosphorus pentachloride. This substance contains phosphorus and chlorine in the mole ratio 1:5, and its properties suggest that it is covalently bonded. In the vapour state the molecules are completely separate from each other and have the formula PCl_5, although it has a more complicated structure in the solid state (but still with the same mole ratio).

The octet rule does not hold for the molecule PCl_5. Five P—Cl single covalent bonds involve five shared pairs of electrons, with the result that the valence shell of phosphorus contains a total of *ten* electrons. In this case, phosphorus is said to have 'expanded its octet', i.e. the valence shell contains more than eight electrons. You can still draw the Lewis structure of PCl_5, as shown to the left, but bear in mind that the electronic structure does not correspond to that of a noble gas and that it does not follow the octet rule.

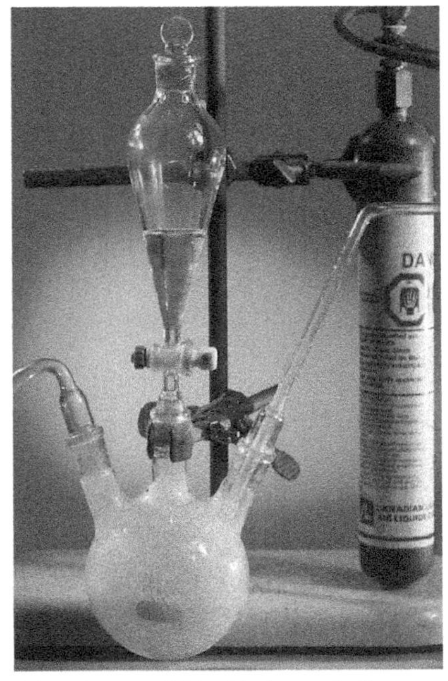

Phosphorus trichloride is a volatile liquid; it reacts with chlorine to form phosphorus pentachloride, which is a solid.

Hypervalent

A hypervalent species contains an element that has expanded its octet. Examples include PCl_5, SF_6 (see next spread), and the ion PCl_6^- present in solid phosphorus pentachloride, spread 17.7.

Coordinate bonds

Aluminium is in Group III of the periodic table. The electronic structure of aluminium is Al [Ne]$3s^2 3p^1$ and so each atom can only contribute to three pairs of electrons in a molecule. Drawing the Lewis structure of aluminium chloride results in a molecule of formula $AlCl_3$ with three chlorine atoms each joined by a single covalent bond to an aluminium atom.

However, the actual formula of solid aluminium chloride (as determined by experiment) is Al_2Cl_6. This molecule consists of two $AlCl_3$ molecules joined together. The illustration below shows how this arrangement is achieved. Each of the aluminium atoms accepts a share in a lone pair of electrons donated by a chlorine atom attached to the *other* aluminium atom. Such a bond is referred to as a **coordinate bond** (or coordinate covalent bond or dative covalent bond) to emphasize the fact that *both* of the electrons in the shared pair originated from the *same* atom. The bond may be represented as an arrow, which shows the direction of the electron pair donation.

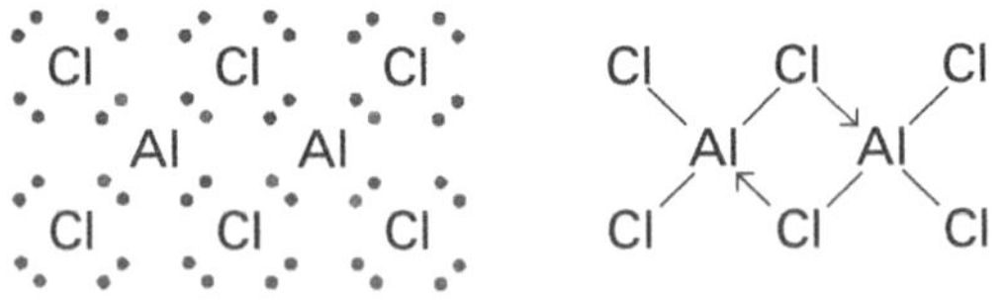

Lewis structure showing the electronic structure of Al_2Cl_6.

Compounds with an incomplete octet

Boron is also in Group III. Boron has the electronic structure B [He]$2s^2 2p^1$ and boron trichloride has the formula BCl_3. This substance has an incomplete octet because its valence shell contains fewer than eight electrons (BCl_3 does not follow the octet rule).

The octet of boron can only be completed if it accepts a share in a lone pair of electrons from another molecule, e.g. from an ammonia molecule. The bond that forms between boron and nitrogen is a coordinate bond because both the electrons have been donated by the nitrogen atom.

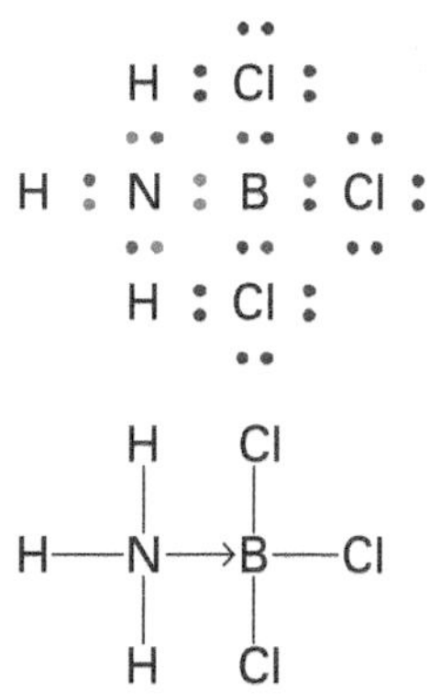

```
            ••
        H : Cl :
   ••   ••   ••
H : N : B : Cl :
   ••   ••   ••
        H : Cl :
            ••

     H    Cl
     |    |
H —— N ——→ B —— Cl
     |    |
     H    Cl
```

Lewis structure showing the formation of a coordinate bond between ammonia and boron trichloride, in which boron obtains an octet of electrons in its valence shell.

An adduct

The illustration to the left shows the Lewis structure of boron trichloride and its acceptance of the lone pair of the nitrogen atom of an ammonia molecule to form an *adduct.* This adduct may be written in a shortened representation as $H_3N{:}\rightarrow BCl_3$.

Ions containing more than one atom

The only type of ion considered so far is formed when a single atom loses or gains electrons. Ions can also consist of covalently bonded atoms that have an unequal number of protons and electrons. A good example is the ammonium ion NH_4^+.

The ammonium ion contains four identical N—H bonds. Three bonding pairs of electrons are formed by the sharing of one electron from the nitrogen atom and one electron each from three hydrogen atoms. The fourth bonding pair results from a hydrogen ion H^+ accepting a share in the nitrogen lone pair. The species is a positive ion because the total number of protons is one greater than the total number of electrons.

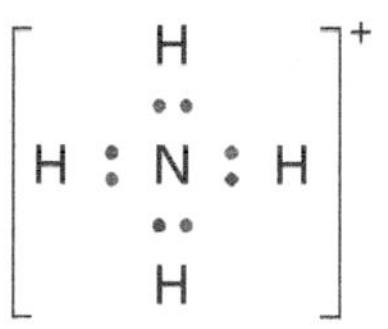

```
[     H     ]+
      ••
  H : N : H
      ••
      H     ]
```

Lewis structure showing the ammonium ion NH_4^+, which is formed from ammonia NH_3 and a hydrogen ion (proton) H^+.

SUMMARY

- A coordinate bond is a covalent bond in which one of the atoms donates both of the electrons in the shared pair.
- Expansion of the octet is possible in molecules such as phosphorus pentachloride PCl_5, in which the valence shell of the phosphorus atom contains ten electrons.
- Some compounds (such as boron trichloride BCl_3) have an atom with an incomplete octet.

PRACTICE

1 Draw Lewis structures for the following molecules, and write down the numbers of bonding pairs and lone pairs of electrons in the valence shells of the atoms concerned:

a Chlorine Cl_2

b Ammonia NH_3

c Hydrogen H_2

d Water H_2O

e Carbon dioxide CO_2

f Methane CH_4

g Oxygen O_2.

5.4

THE SHAPES OF MOLECULES – 1

OBJECTIVES

- Molecular shape from the VSEPR theory
- Five basic molecular shapes

Founders of the VSEPR theory

The VSEPR theory was introduced by Nevil Sidgwick and Herbert Powell, and was refined by Ronald Gillespie and Ronald Nyholm.

Understanding the shapes of molecules can help to explain many of their properties. This spread and the next describe how to predict the shapes of molecules by following a few simple rules.

The VSEPR theory

The best simple approach to determining the shape of a molecule is the **valence-shell electron-pair repulsion (VSEPR)** theory. This theory suggests that the electron pairs around an atom repel each other; bonding pairs and lone pairs arrange themselves to be as far apart as possible. The pairs take up this arrangement so that the potential energy due to their electrostatic repulsion is at a minimum.

In the case of molecules that have only bonding pairs and no lone pairs on the central atom, the total number of electron pairs equals the number of bonding pairs. It is quite straightforward to use the VSEPR theory to work out the shapes of these molecules, as follows:

1 Write a Lewis structure for the molecule.

2 Count the number of electron pairs around the central atom. (Treat multiple bonds in the same way as single bonds.)

3 Note the arrangement that this number of electron pairs will adopt:
two pairs = linear,
three pairs = trigonal planar,
four pairs = tetrahedral,
five pairs = trigonal bipyramidal,
six pairs = octahedral.

4 Write the symbol of the central atom and arrange the other atoms around it in the shape found in step 3.

The procedure is similar when there are one or more lone pairs on the central atom, as discussed in the next spread.

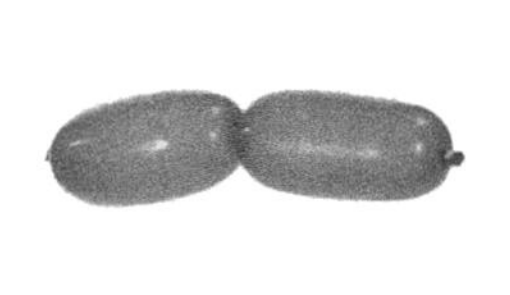
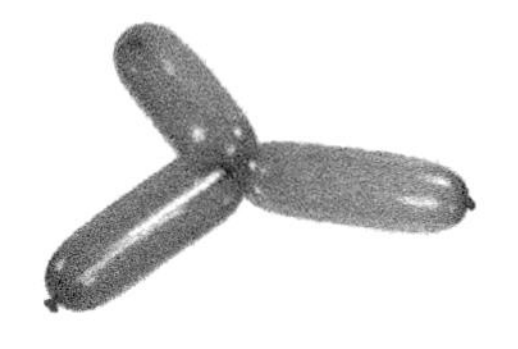

Balloons tied together will arrange themselves so that they are as far away from each other as possible. They form the same geometric arrangements as electron pairs in a molecule around a central atom. (a) Two balloons: linear; (b) three balloons: trigonal planar; (c) four balloons: tetrahedral; (d) five balloons: trigonal bipyramidal; (e) six balloons: octahedral.

Diatomic molecules

The simplest molecule of all is hydrogen H_2. It consists of two hydrogen atoms joined by a single covalent bond H—H. There are no electrons present in the molecule apart from the bonding pair. Hydrogen is a **linear** molecule because its two atoms are in a straight line (it is not possible to arrange them in any other way). All diatomic molecules are linear.

Linear molecules

Carbon dioxide CO_2 consists of two oxygen atoms each joined by a double covalent bond to one carbon atom, i.e. the arrangement of the atoms is O=C=O. There are two separate regions of high electron density, corresponding to the bonding electron pairs in the two C=O bonds. These two regions of high electron density repel each other and are therefore arranged so that they are as far apart as possible. The result is that CO_2 is a **linear** molecule. The arrangement of the electron pairs causes the three constituent atoms to be in a straight line. The bond angle O—C—O is 180°.

(a)

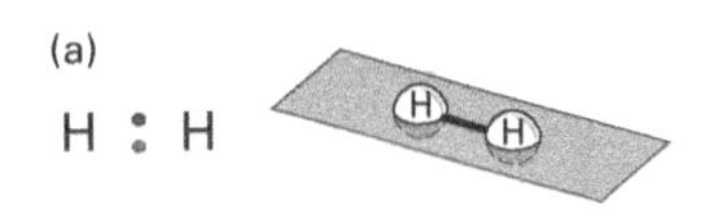

(b)

Linear

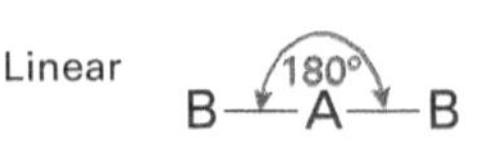

Two linear molecules: (a) hydrogen H_2 (diatomic molecules are all linear); and (b) carbon dioxide CO_2 (not all molecules with three atoms are linear). Beryllium chloride $BeCl_2$ (in the vapour) is another example of a linear molecule.

Trigonal planar molecules

Boron trichloride BCl_3 consists of three chlorine atoms each joined by a single covalent bond to one boron atom. There are three separate regions of high electron density, corresponding to the three B—Cl bonding pairs of electrons. These three bonding pairs repel each other equally, with the result that BCl_3 is a **planar** (flat) molecule. (Remember from your maths that a plane can always be drawn through any three points.) The arrangement of the electron pairs is **trigonal**: the three B—Cl bonds point towards the three corners of an equilateral triangle. The bond angles Cl—B—Cl are all equal at 120°. Boron trifluoride BF_3 has an identical structure.

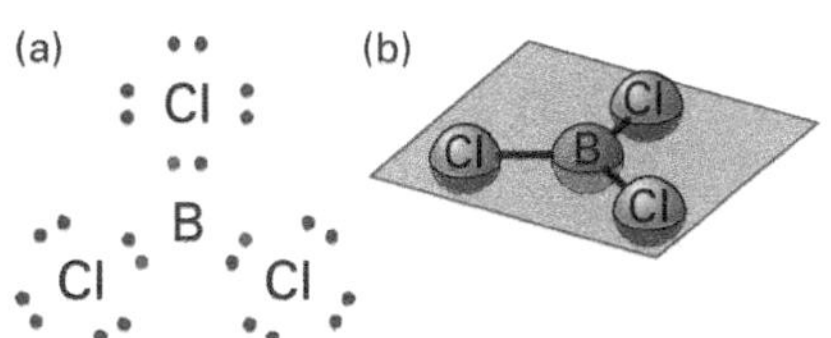

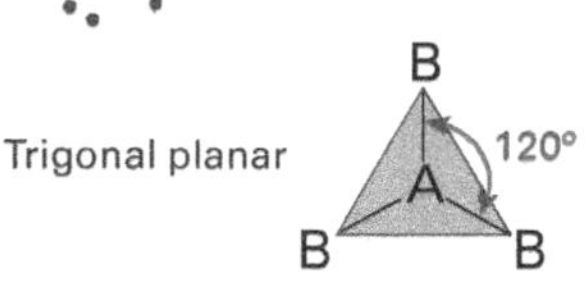

Boron trichloride BCl_3. (a) The Lewis structure shows that the central atom has three bonding pairs of electrons. (b) The arrangement of the electron pairs is trigonal planar.

Tetrahedral molecules

Methane CH_4 consists of four hydrogen atoms each joined by a single covalent bond to one carbon atom. There are four separate regions of high electron density, corresponding to the four C—H bonding electron pairs. These four bonding pairs repel each other equally. The arrangement of the electron pairs is **tetrahedral**: the four C—H bonds point towards the four corners of a regular tetrahedron. In other words, the four hydrogen atoms symmetrically surround the central carbon atom. The bond angles H—C—H are all equal at 109.5°. This angle is often referred to as the '**tetrahedral angle**', its precise value being 109° 28′.

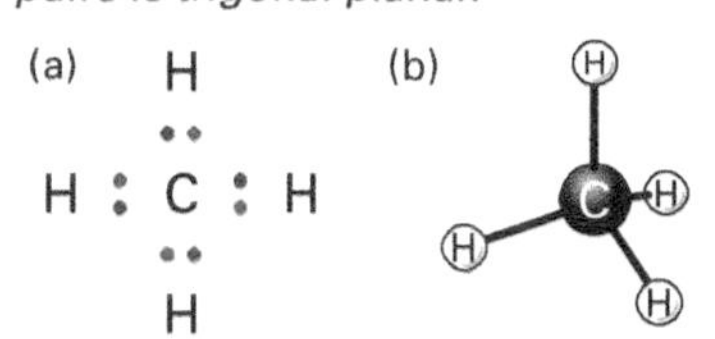

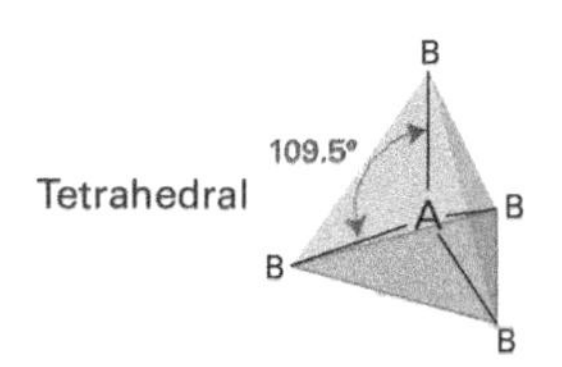

Methane CH_4. (a) The Lewis structure shows that the central atom has four bonding pairs of electrons. (b) The arrangement of the electron pairs is tetrahedral. (A tetrahedron is a three-dimensional shape with four triangular faces.)

Trigonal bipyramidal molecules

Phosphorus pentafluoride PF_5 consists of five fluorine atoms each joined by a single covalent bond to one phosphorus atom. There are five separate regions of high electron density, corresponding to the five P—F bonding electron pairs. The overall result is that the five P—F bonds point towards the five corners of a **trigonal bipyramid**. Phosphorus pentachloride PCl_5 has an identical structure.

The structure is not totally symmetrical: the bond angles F—P—F within the plane are 120°; the bonds above and below the plane are at 90°.

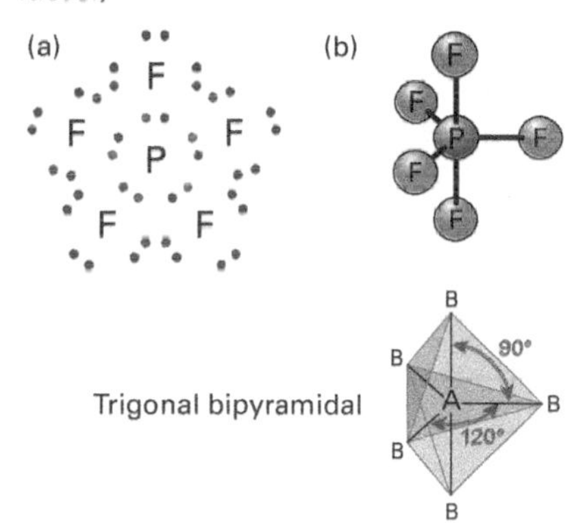

Phosphorus pentafluoride PF_5. (a) The Lewis structure shows that the central atom has five bonding pairs of electrons. (b) The arrangement of the electron pairs is trigonal bipyramidal. (A trigonal bipyramid consists of two pyramids with a common triangular base: bipyramid = two pyramids; trigonal = three-cornered.)

Octahedral molecules

Sulphur hexafluoride SF_6 consists of six fluorine atoms each joined by a single covalent bond to one sulphur atom. There are six separate regions of high electron density, corresponding to the six S—F bonding electron pairs. These six bonding pairs repel each other equally. The arrangement of the electron pairs is **octahedral**: the six S—F bonds point towards the six corners of a regular octahedron. In other words, the six fluorine atoms symmetrically surround the central sulphur atom. The bond angles F—S—F are all equal at 90°.

SUMMARY

- Molecular shapes can be predicted using the valence-shell electron-pair repulsion (VSEPR) theory.
- The number of electron pairs dictates the possible arrangements: linear (two electron pairs);
  trigonal planar (three electron pairs);
  tetrahedral (four electron pairs);
  trigonal bipyramidal (five electron pairs);
  and octahedral (six electron pairs).

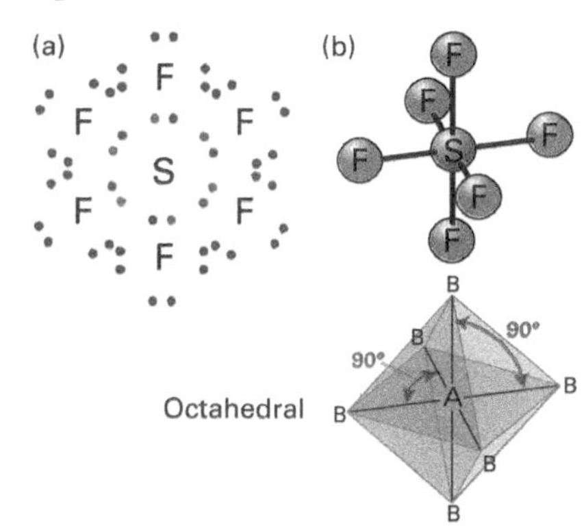

Sulphur hexafluoride SF_6. (a) The Lewis structure shows that the central atom has six bonding pairs of electrons. (b) The arrangement of the electron pairs is octahedral. (An octahedron is a three-dimensional shape with eight triangular faces.)

5.5

THE SHAPES OF MOLECULES – 2

OBJECTIVES

- Effect of lone pairs on molecular shape
- Repulsion between bonding pairs and lone pairs contrasted
- Multiple bonds

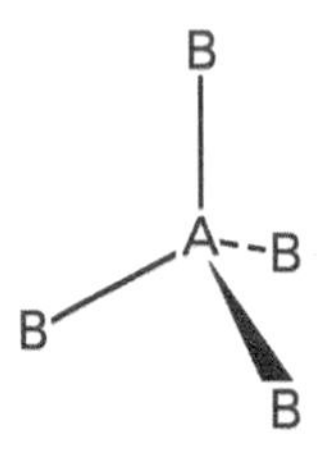

A representation of the tetrahedral arrangement. In this diagram, the solid lines each represent the axis of an electron pair (bonding pair or lone pair) in the plane of the paper; the wedge shape represents an axis coming out of the plane of the paper; and the dotted line represents an axis behind the plane of the paper.

The previous spread introduced the idea of determining molecular shape by using the valence-shell electron-pair repulsion (VSEPR) theory. However, that introduction was limited to considering molecules that have only bonding pairs of electrons on the central atom. Lone pairs repel bonding pairs, and so they must also be taken into account when determining molecular shape.

The ammonia molecule

The Lewis structure for the ammonia molecule NH_3 shows that the central nitrogen atom has three bonding pairs and one lone pair. These four electron pairs are arranged tetrahedrally. However, the H—N—H bond angle is not 109.5° as expected for a regular tetrahedral arrangement. The angle is measured at 107°. This observation can be explained by assuming that:

- Lone pairs repel bonding pairs slightly more than bonding pairs repel each other.

So lone pair / lone pair repulsion is greater than lone pair / bonding pair repulsion, which in turn is greater than bonding pair / bonding pair repulsion.

Look at the illustration below. The *electron pairs* in ammonia are arranged tetrahedrally, as shown in part (b). Now focus just on the *atoms*, as shown in part (c). The atoms are arranged in the shape of a pyramid, so the shape of the *molecule* is **pyramidal**.

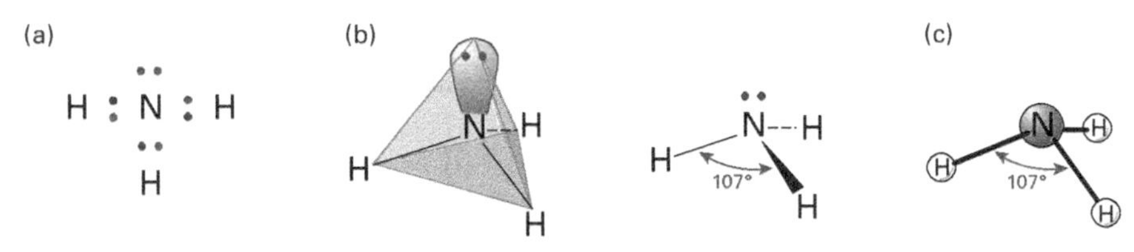

Ammonia NH_3. (a) The Lewis structure shows that the central atom has three bonding pairs and one lone pair. (b) The arrangement of the electron pairs is distorted tetrahedral. (c) The shape of the ammonia molecule (i.e. the arrangement of the atoms) is pyramidal.

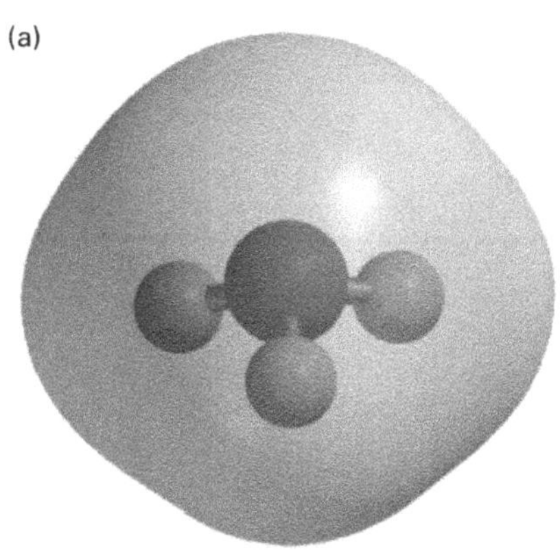

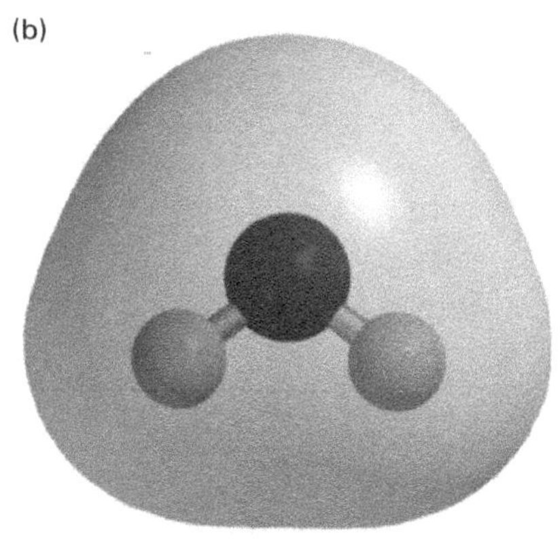

Electron density diagrams for (a) ammonia and (b) water. Each shows a contour of equal electron density and confirms the shapes suggested by the VSEPR theory. See following spread for details of the software used.

The water molecule

The Lewis structure for the water molecule H_2O shows two bonding pairs and two lone pairs around the oxygen atom. This gives a total of four electron pairs. The VSEPR theory indicates that four electron pairs will adopt a tetrahedral arrangement in space. The lone pairs take up two of the tetrahedral positions, with the bonding pairs taking up the other two positions.

If we focus on the electron pairs in the water molecule, the observed structure confirms the expected arrangement. However, the H—O—H bond angle is 104.5° instead of the tetrahedral angle of 109.5°. As in the ammonia molecule, the angle is smaller than expected because a lone pair repels more strongly than a bonding pair, and so the lone pairs distort

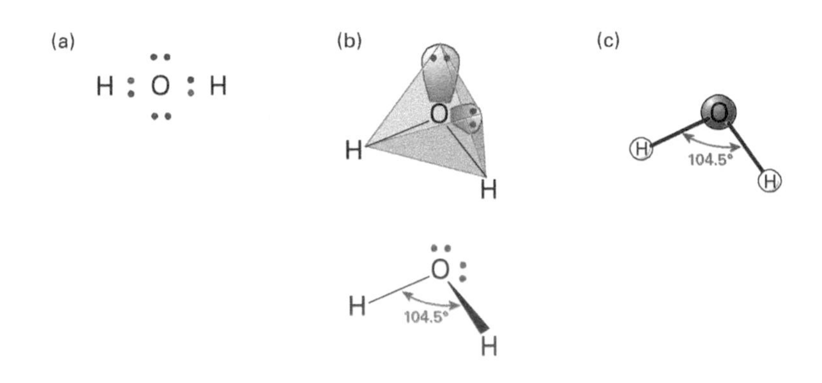

Water H_2O. (a) The Lewis structure shows that the central atom has two bonding pairs and two lone pairs. (b) The arrangement of the electron pairs is distorted tetrahedral. (c) The molecule is bent into a broad V-shape.

the shape by squeezing the angle between the O—H bonds. If we focus just on the atoms, we can consider the molecule to be bent into a broad **V-shape**.

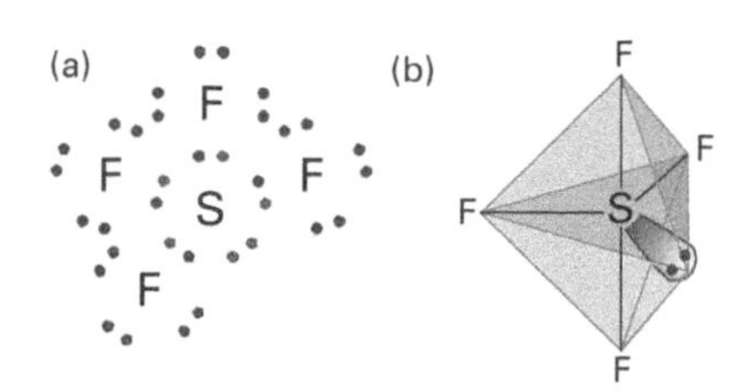

Sulphur tetrafluoride SF_4. (a) The Lewis structure. (b) The arrangement of the electron pairs. The three trigonal planar positions around the central atom are called the 'equatorial' positions. The two positions above and below the plane are called the 'axial' positions. The lone pair goes where the repulsions at the most acute angle (here 90°) are minimized. The shape is described as a see-saw.

The sulphur tetrafluoride and chlorine trifluoride molecules

The Lewis structure for the sulphur tetrafluoride molecule SF_4 shows four bonding pairs and one lone pair around the sulphur atom. These five electron pairs take up a trigonal bipyramidal arrangement, with the lone pair in an 'equatorial' position around the central sulphur atom, as shown in the diagram on the right. Chlorine trifluoride ClF_3 also has five electron pairs, *two* of which are lone pairs. The lone pairs take up equatorial positions and the resulting shape is described as T-shaped.

Multiple bonds

Double bonds involve two bonding pairs of electrons and triple bonds involve three pairs.

A double bond is a single region of high electron density. Ethene $CH_2{=}CH_2$ therefore contains five regions of high electron density: the C=C double bond and the four C—H single bonds. The arrangement is therefore trigonal planar around each carbon atom, and the molecule as a whole is planar. The bond angles H—C—H are slightly *less* than 120° because the two electron pairs of the double bond repel the electrons in the C—H bonds to a *greater* extent than a single pair.

A triple bond is also a single region of high electron density.

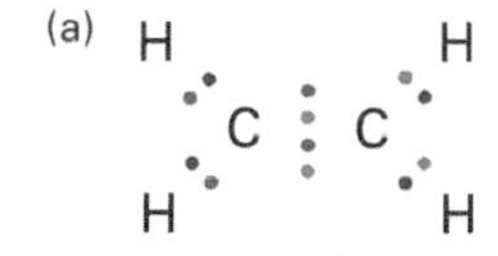

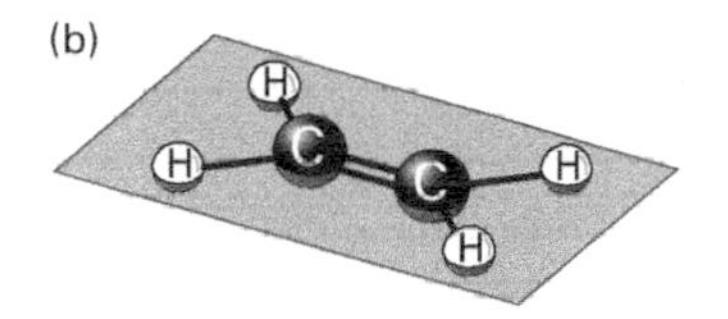

Ethene C_2H_4. (a) The Lewis structure. (b) The planar arrangement of the electron pairs.

VSEPR theory can be applied to ions

The VSEPR theory can be extended to the structure of ions as well as that of molecules. For example, the ammonium ion (NH_4^+) has exactly the same number of electrons as methane (CH_4): the two species are isoelectronic. The shape of the ammonium ion is exactly the same as that of the methane molecule, namely tetrahedral, because exactly the same number of bonding pairs and lone pairs are involved. As a second example, the oxonium ion H_3O^+, which will be centre stage in chapter 12 on acids and bases, is isoelectronic with ammonia, NH_3. Their shapes are the same, namely pyramidal. As a third example, the ion PCl_4^+ is isoelectronic with $SiCl_4$; both have four bonding pairs and no lone pairs, so both are tetrahedral.

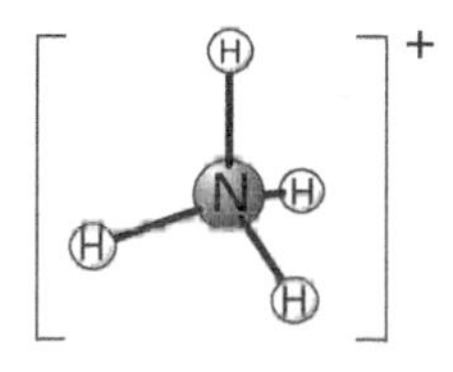

The ammonium ion is tetrahedral.

SUMMARY

- Lone pairs must be taken into account when determining molecular shape.
- Repulsion due to lone pairs is greater than repulsion due to bonding pairs.

PRACTICE

1 Draw Lewis structures for the following molecules. Using the VSEPR theory, sketch the arrangement of the electron pairs in space. Say what shape each molecule is.

  a Methane CH_4
  b Beryllium chloride $BeCl_2$
  c Tetrachloromethane CCl_4
  d Sulphur hexafluoride SF_6
  e Ammonium ion NH_4^+.

2 Draw Lewis structures for each of the following molecules. Draw the arrangement of the atoms and lone pairs. Say whether the bond angles are different from those expected for the regular geometrical shape.

  a Sulphur dioxide SO_2
  b Phosphine PH_3
  c Silicon tetrachloride $SiCl_4$
  d The ion PCl_4^+
  e Sulphur dichloride oxide SCl_2O.

5.6 Molecular orbital theory – 1

OBJECTIVES

- STM image shows interference
- Overlap of atomic orbitals
- Bonding and antibonding molecular orbitals

Wavefunction

John Pople shared the 1998 Nobel Prize in Chemistry for his work on molecular orbital theory. He cofounded with Warren Hehre the company Wavefunction Inc., who have kindly supplied a number of electron density diagrams from their SPARTAN software.

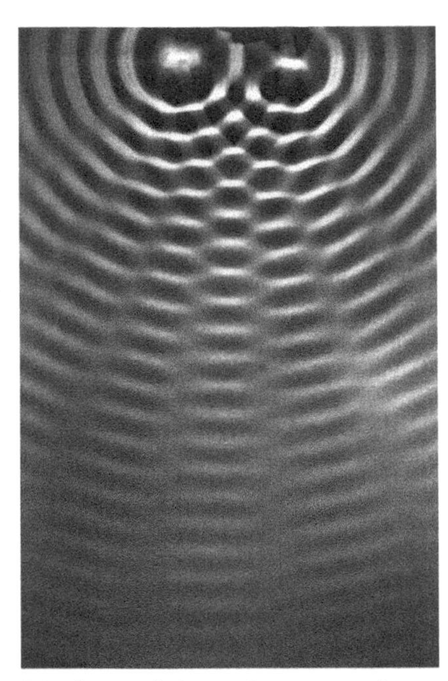

Interference between two sources in a ripple tank.

In chapter 4 you met the idea of atomic orbitals. Atomic orbitals define the regions of space around a nucleus where there is a high probability of finding an electron of a given energy (a high electron density). The electron density can be found from solutions of the Schrödinger equation for the atom concerned. Lewis structures represent covalent bonds as electron pairs shared between atomic centres. They do not explain *why* bonding electron pairs occupy these positions. A satisfactory explanation is provided by **molecular orbital theory** in which atomic orbitals overlap and combine to make molecular orbitals.

Visual 'proof'

Most of the electron density diagrams of atoms shown so far are pictures generated by a computer from solutions of the Schrödinger equation for that system of electrons and protons. They are visually convincing, with contours to show the magnitude of the electron density at each point. However, these pictures may seem a little contrived; they come from manipulating mathematical expressions. Can they really claim to answer the question 'What do atoms and molecules really look like?' The answer *yes they can!* is provided by some recent STM (scanning tunnelling microscope) images.

Look carefully at the STM image shown below. Until they understood what they were looking at, the workers who produced it thought their equipment had developed a fault. In fact, *the ripple effect on the surface layer of atoms in the crystal is the pattern resulting from the interference between electrons.* Interference is a phenomenon typical of waves.

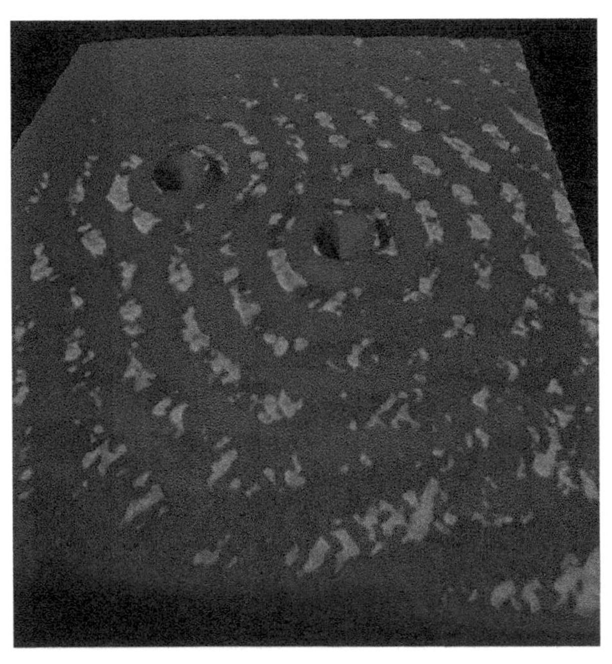

An STM image showing interference between electrons, confirming their wave nature. The two point defects (probably impurity atoms created in the preparation of the sample) scatter the surface state electrons, resulting in circular standing wave patterns.

Forming molecular orbitals

Imagine two hydrogen atoms moving towards each other. Each atom has one electron contained in a 1s orbital. The electron density in each separate atomic orbital is described by the Schrödinger equation as spherically symmetrical about the atomic nucleus.

Where the two atomic orbitals **overlap**, electron density is redistributed and new molecular orbitals are created. These define the regions of space around a *set* of nuclei where there is a high probability of finding an electron. **Molecular orbitals** are the solutions of the Schrödinger equation for molecules, in the same way that atomic orbitals are the solutions for atoms.

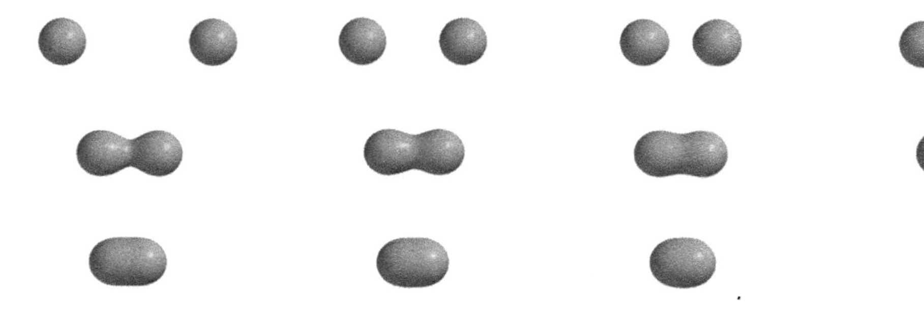

Electron density diagrams showing two hydrogen atoms approaching each other and forming a hydrogen molecule. Note the shift in the distribution of electron density as the nuclei get closer.

When the atoms are close together, the atomic orbitals interfere with each other. In a ripple tank, **constructive interference** results when waves of the same phase reinforce each other, and **destructive interference** results when waves of different phase cancel each other out. The same is true of the interference between atomic orbitals: the interference pattern is caused by orbitals of different phase interacting. In the diagrams, one phase is shown in red and the other phase in dark blue.

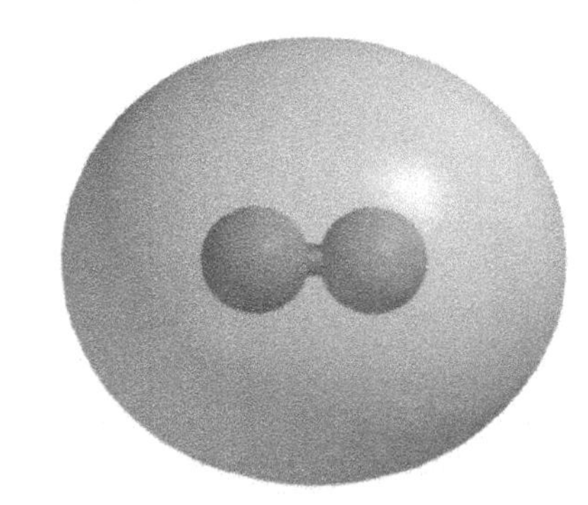

A contour of equal electron density for hydrogen H_2 at the equilibrium separation.

Bonding and antibonding molecular orbitals

Constructive interference between the two atomic orbitals (when they are in phase, shown by the same colour) *increases* the electron density between the two atoms and leads to a **bonding molecular orbital**. This orbital is of *lower* energy than that of either of the two individual atomic orbitals.

Destructive interference (when they are out of phase, shown by the different colours) *removes* electron density from between the nuclei and leads to an **antibonding molecular orbital**. This orbital is of *higher* energy than that of either of the two individual atomic orbitals.

The formation of these molecular orbitals may be represented on an energy level diagram.

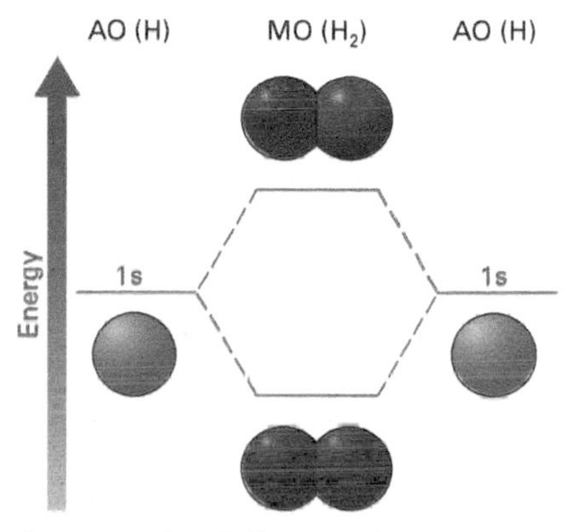

An energy level diagram showing two 1s atomic orbitals (AO) combining to form a lower-energy bonding molecular orbital (MO) and a higher-energy antibonding molecular orbital. The purple colour for the separate atoms signifies that the phase is unspecified.

The hydrogen molecule

In the case of the hydrogen molecule H_2, there is a total of two electrons (one electron per atom). Following the building-up principle, these electrons occupy the lowest energy orbital available, the bonding molecular orbital. The result is a single covalent bond containing two electrons; as required by the Pauli exclusion principle, the spins are paired. With two electrons in the bonding molecular orbital, the energy of the H_2 molecule is lower than that of the two separate atoms.

The non-existence of He_2

Molecular orbital theory explains very simply why helium atoms do not form a covalent bond with each other. A helium atom has two electrons contained in the 1s atomic orbital. The *hypothetical* helium molecule He_2 would therefore have a total of four electrons. These would need to be contained in molecular orbitals formed from the overlap of two 1s atomic orbitals. Referring to the energy level diagram, you can see that there would be two electrons in the bonding molecular orbital and two electrons in the antibonding molecular orbital. The energy of this arrangement is almost the same as the energy of two separate helium atoms. The energy of the system does not fall when two separate helium atoms approach each other closely. No covalent bond forms as it does in the case of hydrogen.

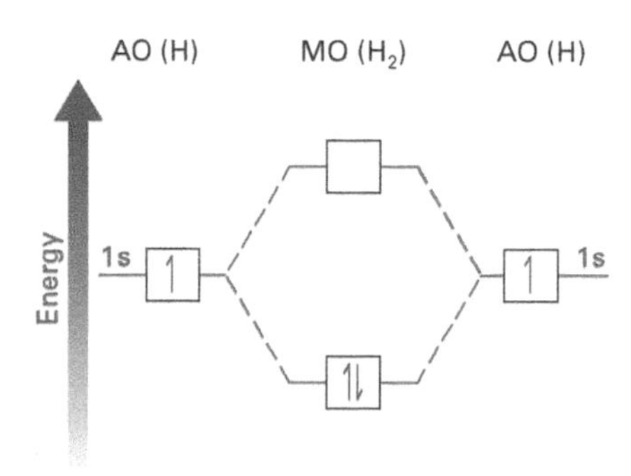

An energy level diagram for H_2 showing two electrons in the bonding molecular orbital of a hydrogen molecule. Note that the antibonding molecular orbital is empty. For the hypothetical molecule He_2, the bonding and the antibonding molecular orbitals would each contain two electrons; hence no bond forms between the two atoms.

SUMMARY

- Overlap of atomic orbitals leads to the formation of molecular orbitals.
- A bonding molecular orbital has a lower energy than either of the two atomic orbitals that overlap.
- An antibonding molecular orbital has a higher energy than either of the two atomic orbitals that overlap.

5.7 Molecular orbital theory – 2

OBJECTIVES
- σ and π molecular orbitals
- The O_2 molecule
- Electron density in covalent bonds

You should remember from the previous chapter that a p orbital has a very different shape from an s orbital. As a result, the overlap of two p orbitals to form molecular orbitals is significantly different from the overlap of two s orbitals. Overlap of two p orbitals forms two types of molecular orbital: sigma (σ) molecular orbitals and pi (π) molecular orbitals.

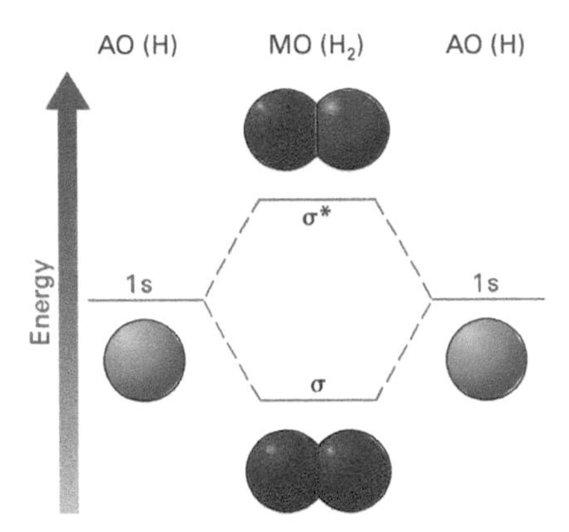

The overlap of two s orbitals produces σ and σ molecular orbitals. It is not possible to form π or π* molecular orbitals from s atomic orbitals.*

Sigma (σ) and pi (π) molecular orbitals

The overlap of two s orbitals produces a **sigma** (σ, the Greek letter corresponding to 's') bonding orbital when the two orbitals are in phase and a corresponding σ* antibonding orbital when they are out of phase.

Two electrons in a sigma bonding molecular orbital form a **sigma bond**.

There are *two* possible ways in which two p orbitals, the two lobes of which have opposite phase, can overlap. The first possibility is that they overlap end-on, giving rise to a sigma (σ) bonding molecular orbital when the overlapping lobes are in phase. (There will always be a corresponding antibonding orbital (σ*) when the overlapping lobes are out of phase.)

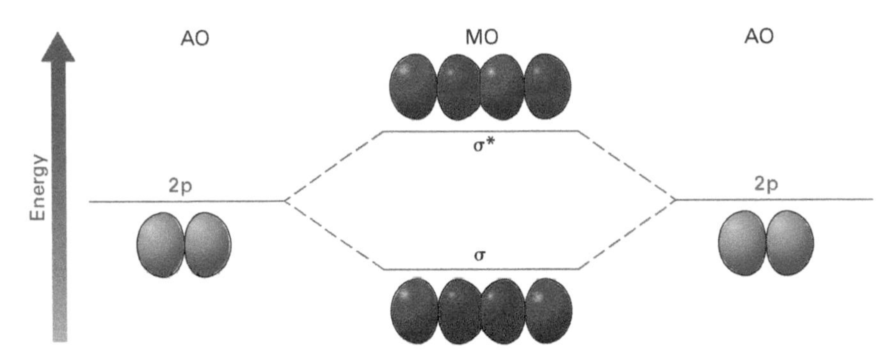

Overlap of two p orbitals end-on to form a sigma (σ) bonding molecular orbital and a sigma star (σ) antibonding molecular orbital.*

The second possibility is that the p orbitals at right angles to a sigma bond may overlap sideways. This arrangement gives rise to a **pi** (π) bonding molecular orbital when the orbitals are in phase. (Again, there is a corresponding antibonding orbital (π*) when the orbitals are out of phase.) Two electrons in a pi bonding molecular orbital form a **pi bond**. A pi bond is found in association with a sigma bond in molecules that contain a double covalent bond, such as ethene C_2H_4.

- A sigma molecular orbital is symmetrical about the internuclear axis (the imaginary line between the nuclei).
- A pi molecular orbital has a nodal plane (a plane in which the electron density is zero) along the internuclear axis.

The π bonding molecular orbital in ethene C_2H_4.

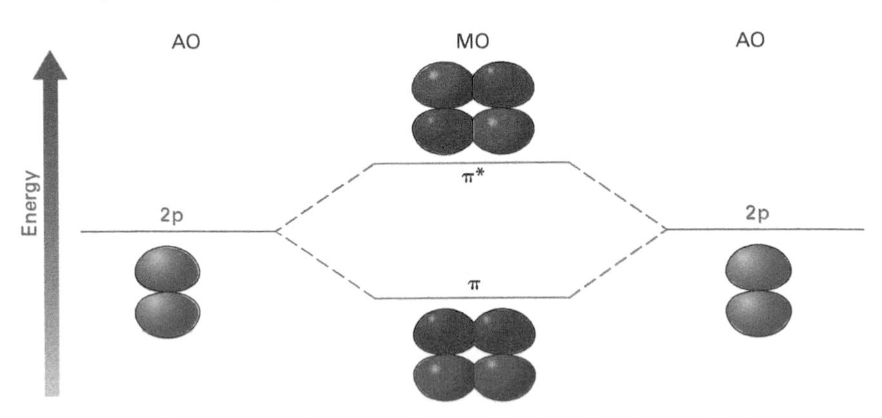

When two p orbitals overlap sideways, the resulting bonding is described as π bonding. When the orbitals overlap in phase (the lobes are the same colour), a π bonding molecular orbital is formed. When the orbitals overlap out of phase (the lobes of the orbitals are opposite colours), an antibonding (π) orbital is formed.*

Solving the oxygen problem

Molecular orbital theory has been especially successful in explaining the detailed electronic structure of the oxygen molecule. The Lewis structure for this molecule shows two bonding electron pairs between the oxygen atoms and two lone pairs located on each atom. All the electron spins are paired. However, the demonstration illustrated in the photograph below shows that liquid oxygen is paramagnetic. A **paramagnetic** substance is attracted into a magnetic field. Substances that are paramagnetic contain one or more unpaired electrons. The Lewis structure for oxygen cannot explain this phenomenon.

There is a powerful magnetic field between the jaws of this magnet. Liquid oxygen is attracted into this field. This phenomenon could not be explained before molecular orbital theory was introduced.

The energy level diagram for the atomic and molecular orbitals of oxygen is shown alongside. It can be seen that the electrons fill the orbitals following the building-up principle. The last two electrons have to fill the pair of antibonding π* molecular orbitals. The electrons go into the two orbitals of equal energy, with parallel spins (following Hund's rule, exactly as happens in atoms, spread 4.6). The two unpaired electrons in the antibonding orbitals cause the oxygen molecule to be paramagnetic.

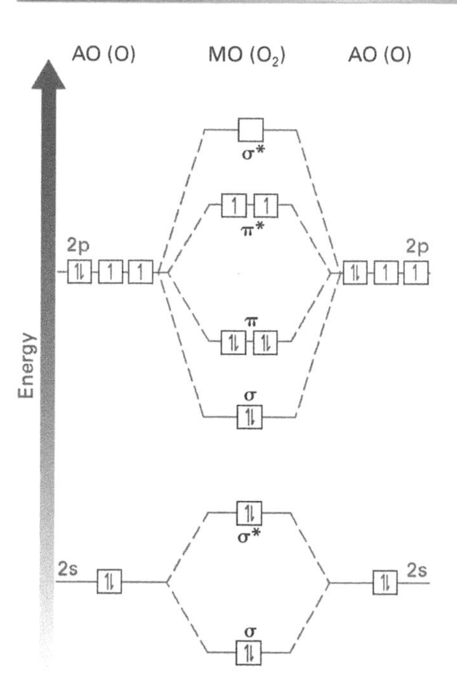

The energy level diagram for the formation of molecular orbitals from oxygen atomic orbitals. The molecular orbitals are filled according to the building-up principle and Hund's rule.

Bond order

The **bond order** is (the number of electrons in bonding orbitals minus the number of electrons in antibonding orbitals) divided by two. So oxygen has a bond order of $(8 - 4)/2 = 2$, as expected. It is especially important in the ongoing development of advanced catalysts involving transition metals (spreads 15.9 and 20.6).

The O_2^+ molecular ion

There is clear experimental evidence that antibonding molecular orbitals can physically exist.

The dioxygenyl ion O_2^+ (see spread 19.12) has a shorter and stronger bond than does the molecule O_2. This unusual situation occurs because the electron removed to form the ion comes from an antibonding molecular orbital. Removing the electron *strengthens* the overall bonding between the two atoms.

Electron density in covalent bonds

Molecular orbital theory can also describe the case where the original atomic orbitals that overlap have different energies. The resulting bonding molecular orbital more closely resembles the lower-energy atomic orbital and has a non-symmetrical distribution about the two atomic centres. This idea is central to the next spread.

(a)

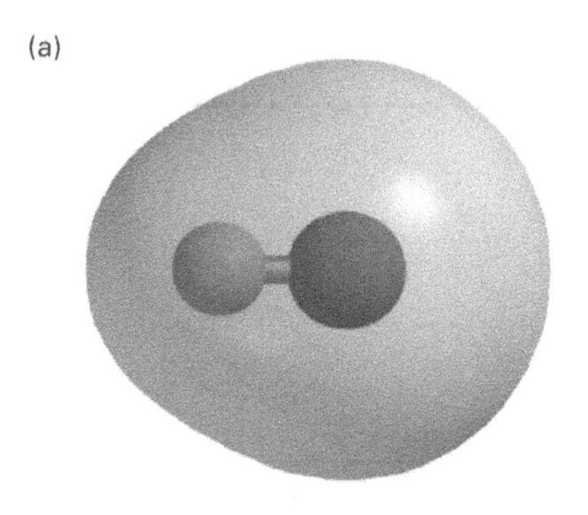

(b)

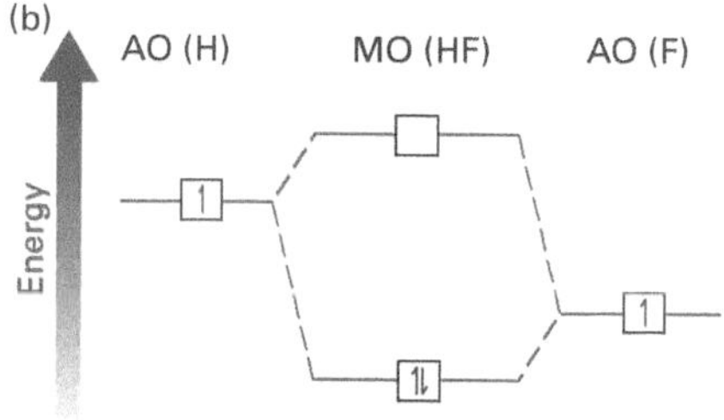

Hydrogen fluoride has a bonding molecular orbital formed from the overlap of the hydrogen 1s and one of the fluorine 2p atomic orbitals.
(a) Electron density diagram.
(b) Energy level diagram showing formation of the molecular orbitals and electron occupancy.

SUMMARY

- Overlap of two s orbitals produces a σ bond.
- Overlap of two p orbitals can produce a σ bond or a π bond.

PRACTICE

1 Draw the energy level diagrams for

a O_2^{2-} (peroxide ion)

b O_2^- (superoxide ion).

5.8

OBJECTIVES

- Electron density maps
- Polar covalent bonds
- Electronegativity

BOND POLARITY – 1

Lewis structures and the VSEPR theory can account for the shapes of a wide variety of molecules. The covalent bond is simply described as a pair of electrons shared between two atoms. In molecules where the atoms are of the same element, as in hydrogen H—H and chlorine Cl—Cl, equal sharing of the electron pair would be expected. However, where the atoms are of different elements, the sharing is *not* equal. A covalent bond in which the sharing is unequal is said to be a **polar covalent bond**. The unequal sharing of electron pairs is reflected in the properties of the compounds.

A reminder

Electron density describes how the charge of the electron is distributed. High electron density in a certain volume of space means that an electron has a high probability of being found there. Low electron density indicates that an electron has a low probability of being found there.

Electron density diagrams

Electrons in molecules inhabit particular regions of space. Bonding pairs are located in regions between the nuclei of the atoms that are bonded together. Lone pairs are located around a specific atom. Modern models of atomic and molecular structure concentrate on electron density. Electron density may be pictured in two main ways, as shown below.

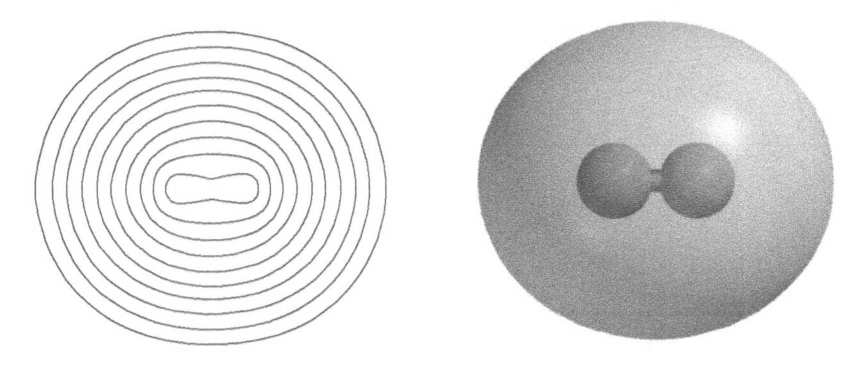

Two ways of picturing electron density in the hydrogen molecule H_2: (a) A contour map – showing lines of equal electron density. (b) An electron density diagram – a computer-generated plot using theoretically calculated values of electron density.

Evidence for unequal sharing

Compare the electron density diagrams shown below for the molecules hydrogen H_2, chlorine Cl_2, and hydrogen chloride HCl. The change is shown most clearly by calculating the electrostatic potential. An **electrostatic potential map** superimposes the colours representing the electrostatic potential onto a contour of equal electron density. A red colour indicates the most negative electrostatic potential and a blue colour indicates the most positive electrostatic potential. The potential increases in the order red < orange < yellow < green < blue.

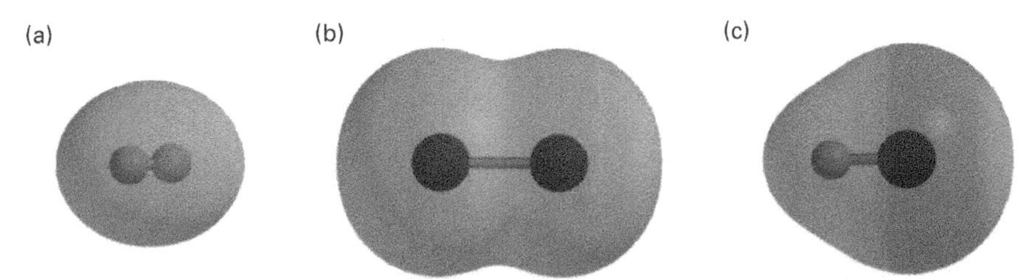

Electrostatic potential maps of (a) hydrogen, (b) chlorine, and (c) hydrogen chloride. The electron densities are symmetrical for hydrogen and chlorine but vary significantly from one end to the other in hydrogen chloride. The electrostatic potential shows that the hydrogen atom has a partial positive charge (blue).

You can see that, for hydrogen and chlorine, the electron density in both cases is symmetrically distributed between the two atoms. In the hydrogen chloride molecule, the electron density is *not* symmetrically distributed. The colour-coding shows there is greater electron density near the chlorine atom. The two bonding electrons that form the single covalent bond H—Cl are more likely to be found near the chlorine atom than near the hydrogen atom.

Hydrogen chloride exists as separate molecules of HCl. Gallium arsenide (a semiconductor used in the electronics industry) consists of a giant network of gallium and arsenic atoms covalently bonded to each other to form a solid lattice. The electron density diagram for the surface layer of a gallium arsenide crystal is shown below. It shows the high electron density between the two different atoms (gallium and arsenic) expected for a shared electron pair. However, careful inspection reveals that the electron density is slightly greater near the arsenic atom. This observation suggests that the electron pair is shared unequally. The arsenic atom has a greater share of the electron pair than the gallium atom.

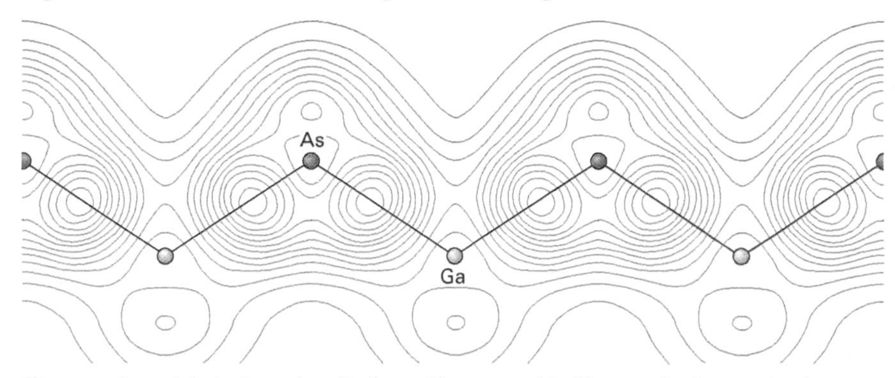

The experimental electron density for gallium arsenide. The smallest near-circular contour between the atoms lies closer to the arsenic atoms (coloured dark blue) than to the gallium atoms (coloured dark grey); arsenic is more electronegative than gallium.

Electronegativity

Where a covalent bond exists between two atoms that are not identical, one of the two will have a greater share of the electron pair. Unequal sharing happens because different atoms have different powers of attraction for electron pairs. The **electronegativity** of an atom gives a rough measure of its ability to withdraw electron density from a shared electron pair. Electronegativity is a relative quantity and is measured on the Pauling scale, ranging from 0.8 for caesium to 4.0 for fluorine. The atom with the greater electronegativity will have a larger share of the bonding pair of electrons. So the electron density will be skewed towards (i.e. higher near) the *more electronegative* atom. Thus in the example above, the smallest near-circular contour lies closer to arsenic (electronegativity 2.2) than to gallium (electronegativity 1.8).

Some electronegativity values

A list of electronegativity values on the Pauling scale is shown below for a selection of elements. Two successive elements from each of Groups I to VII are shown. The elements across Period 3 are shown in **bold** type.

Group	Element	Electronegativity
	Hydrogen	2.2
I	Lithium	1.0
	Sodium	0.9
II	**Magnesium**	1.3
	Calcium	1.0
III	**Aluminium**	1.6
	Gallium	1.8
IV	Carbon	2.5
	Silicon	1.9
V	Nitrogen	3.0
	Phosphorus	2.2
VI	Oxygen	3.4
	Sulphur	2.6
VII	Fluorine	4.0
	Chlorine	3.2

Polar covalent bonds

When the electron pair forming a covalent bond is shared unequally between two atoms, the atom with the greater share acquires a partial negative charge because electron density is displaced towards it and electrons are negatively charged. This partial negative charge is written as $\delta-$. The atom with the smaller share of the electron pair has a partial positive charge, written as $\delta+$.

For example, consider the bond between hydrogen and chlorine in the hydrogen chloride molecule. Chlorine is more electronegative than hydrogen, and so electron density is higher near the chlorine atom. As a result, chlorine has a partial negative charge and hydrogen a partial positive charge. The bond is a polar covalent bond, as is indicated by the formula $H^{\delta+}$—$Cl^{\delta-}$.

(a)

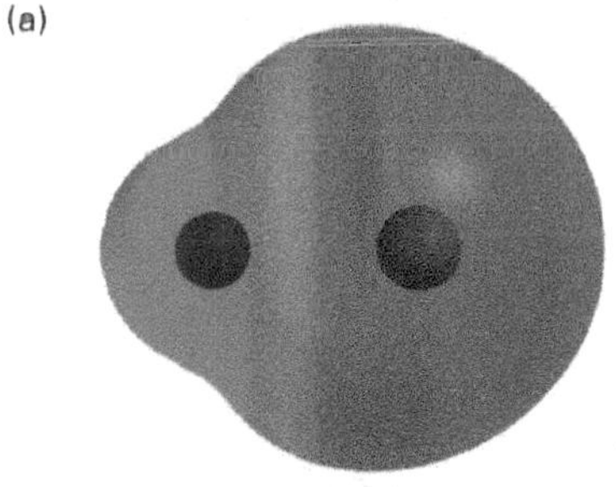

(b)

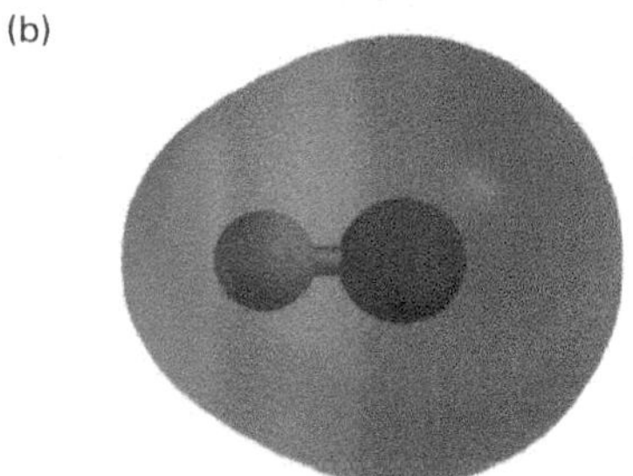

The electrostatic potential maps for the compounds LiH and HF. (a) Hydrogen is more electronegative than lithium and so the electrostatic potential shows that hydrogen has a partial negative charge (red). (b) Hydrogen is less electronegative than fluorine, and so the electrostatic potential shows that hydrogen has a partial positive charge (blue).

SUMMARY

- A covalent bond between two different atoms is a polar covalent bond.
- The electronegativity of an atom is a measure of its ability to withdraw electron density from a shared electron pair.
- Electron density in a covalent bond is skewed towards the atom of higher electronegativity.
- The atom of higher electronegativity has a partial negative charge $\delta-$.
- The atom of lower electronegativity has a partial positive charge $\delta+$.

OBJECTIVES

- Polar covalent bonds
- Polar molecules

Bond polarity – 2

Polar covalent bonds result from the unequal sharing of bonding pairs of electrons in covalent molecules. Electrons are shared unequally when there are differences in the electronegativity values of the two atoms concerned. This spread looks at the presence of polar covalent bonds in molecules and their influence on the polarity of the molecule *as a whole*.

Electronegativity trends

There is a clear trend in electronegativity in the periodic table. Non-metals are elements that gain electrons to complete their outer shells. They have greater electronegativity values than metals, which lose electrons to achieve full outer shells. Electronegativity values generally decrease as atomic number increases down a group. They increase as atomic number increases across a period.

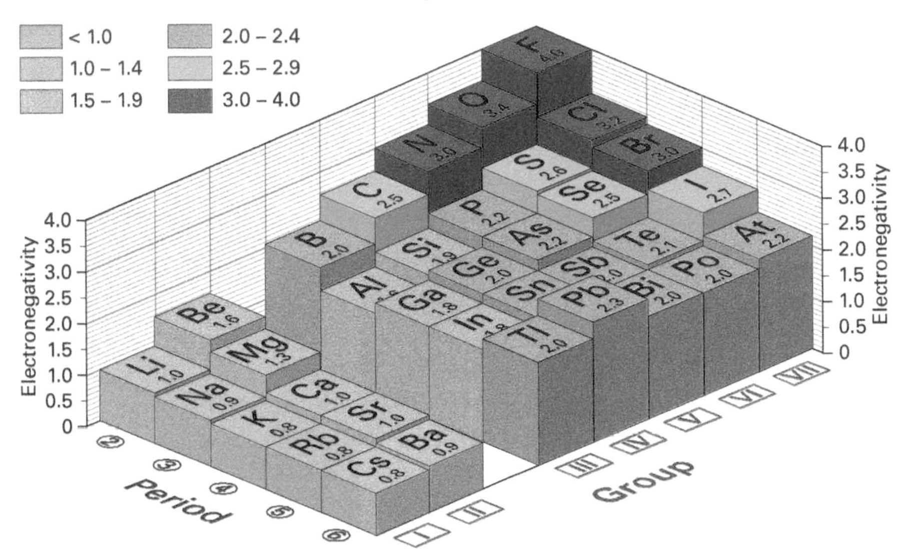

The elements and their electronegativity values arranged in a periodic table.

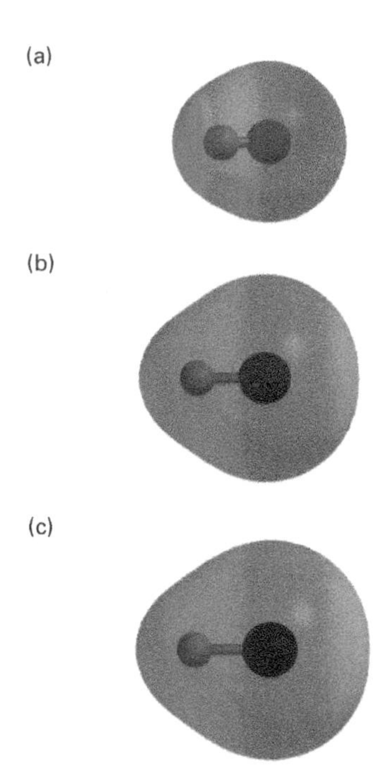

Electrostatic potential maps for (a) hydrogen fluoride HF; (b) hydrogen chloride HCl, and (c) hydrogen bromide HBr. Note that the colour-coding shows that the polarity decreases in the order HF > HCl > HBr.

The hydrogen halides

The electronegativity of the halogens (Group VII) decreases in the order

F > Cl > Br > I

All these elements have greater electronegativity values than hydrogen, with the result that the hydrogen halides HX are polarized $H^{\delta+}—X^{\delta-}$. Fluorine is the most electronegative of the halogens. As a result the H—F bond is polarized to a greater extent than in the other hydrogen halides. The order of decreasing bond polarity is

H—F > H—Cl > H—Br > H—I

Polar molecules

Hydrogen chloride, like all the hydrogen halides, has a polar covalent bond. It is said to be a **polar molecule**. The overall molecule has a **dipole**, that is, a pair of separated charges of opposite sign. Because the positive end of one molecule is attracted to the negative end of another, polar molecules interact with each other. These interactions affect the melting and boiling points of substances, and the types of reaction in which they take part. These effects will be discussed in depth in subsequent chapters.

(a) The hydrogen chloride molecule is a polar molecule. (b) A dipole can be represented as two opposite charges (+ and –) separated by a distance d.

Polar bonds but non-polar molecules

Carbon dioxide is a linear molecule. Both of the oxygen atoms are joined to the central carbon atom by a double covalent bond, so the molecule can be written as O=C=O. The electronegativity values for carbon and oxygen are 2.5 and 3.4 respectively. As a result, the C=O *bonds* are polarized, and can be written as $C^{\delta+}$=$O^{\delta-}$. However, in effect the carbon dioxide molecule O=C=O has *two* polarized bonds back-to-back, i.e. $O^{\delta-}$=$C^{\delta+}$ and $C^{\delta+}$=$O^{\delta-}$. These dipoles cancel out so that the molecule as a whole is **non-polar**.

Boron trifluoride has the formula BF_3. The arrangement of the electron pairs in the three B—F bonds is trigonal planar, spread 5.4. Although each B—F *bond* is polar ($B^{\delta+}$—$F^{\delta-}$), the *molecule* as a whole is non-polar: it has a zero dipole when considered *overall*.

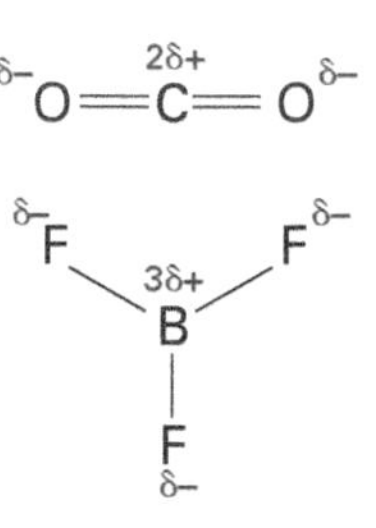

Although their molecules contain polar bonds, both carbon dioxide and boron trifluoride are non-polar molecules.

Water

Each of the hydrogen atoms in the water molecule H_2O is joined to the central oxygen atom by a single covalent bond. Writing the formula of water as H—O—H shows the two covalent bonds but does not take account of the two lone pairs of electrons on the oxygen atom. The arrangement of the four electron pairs (two bonding pairs and two lone pairs) is *distorted tetrahedral*. As a result, the bond angle H—O—H is 104.5°, as discussed in spread 5.5. If we focus just on the atoms, the water molecule is a broad V-shape. The O—H bonds are polar ($O^{\delta-}$—$H^{\delta+}$). The electrostatic potential map shows that the oxygen atom has a partial negative charge (indicated by the red colour). Because the two dipoles do not cancel out, the molecule as a whole is polar; that is, the molecule has a dipole.

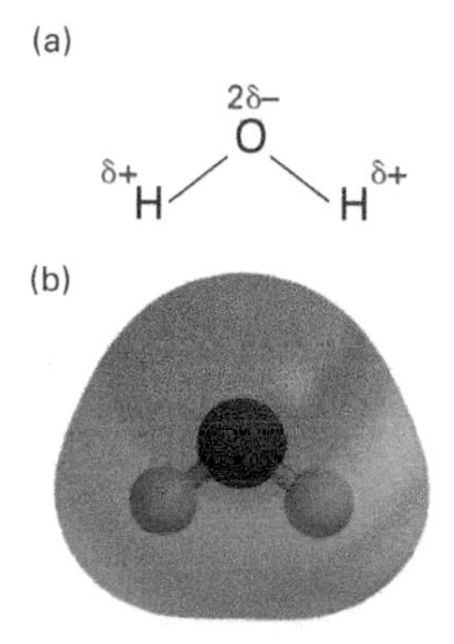

The water molecule: (a) Water has a dipole. (b) The electrostatic potential map, showing the polar nature of the overall molecule.

Ammonia

Each of the hydrogen atoms in the ammonia molecule NH_3 is joined to the central nitrogen atom by a single covalent bond. Writing the formula of ammonia as $:NH_3$ emphasizes the fact that the molecule has a lone pair on the nitrogen atom as well as three bonding pairs. The arrangement of the four electron pairs is *distorted tetrahedral* with H—N—H bond angles of 107°, spread 5.5. If we focus just on the atoms, the ammonia molecule is pyramidal. The N—H bonds are polar ($N^{\delta-}$—$H^{\delta+}$). The electrostatic potential map shows that the nitrogen atom has a partial negative charge (indicated by the red colour). Because the three dipoles do not cancel out, the overall molecule (like water) has a dipole. Notice how the hydrogens are less intensely blue than they are in water, because nitrogen is less electronegative than oxygen.

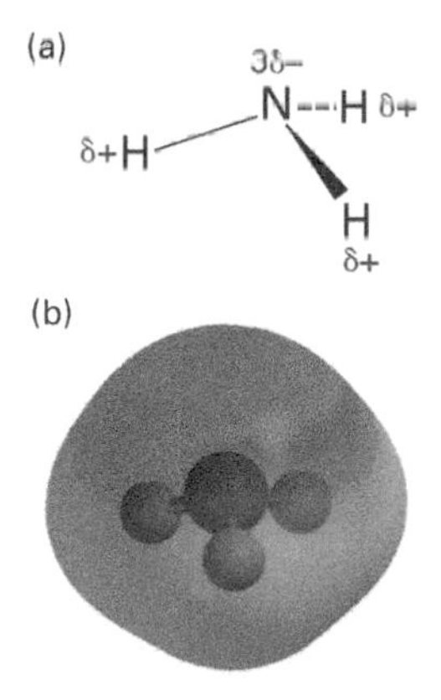

The ammonia molecule: (a) Ammonia has a dipole. (b) The electrostatic potential map, showing the polar nature of the overall molecule.

SUMMARY

- Polar covalent bonds result from the unequal sharing of bonding pairs of electrons in covalent molecules.
- Considered *overall*, a polar molecule has two oppositely charged regions: it acts as a dipole.
- The hydrogen halides, water, and ammonia are all polar molecules.
- CO_2 and BF_3 have polar bonds but are not polar molecules.

PRACTICE

1 For each of the following molecules, identify any polar bonds and label the constituent atoms appropriately δ+ or δ–. Say how you arrived at each decision.

a HBr **b** N_2 **c** ClF
d CCl_4 **e** CH_3Br **f** SO_2.

2 State whether each molecule in question 1 is polar or not, giving your reasons.

5.10 Ionic bonding revisited

OBJECTIVES

- The relationship between polar covalent bonds and ionic bonds
- Ion formation and electronegativity differences

As the electronegativity difference between two covalently bonded atoms increases, the partial charges (δ+ and δ–) developed on the atoms become larger. As the partial charges increase, the bond becomes more and more polar. If the electronegativity difference is more than about 1.8 on the Pauling scale, the bond is so polar that we can describe the electrons as having transferred essentially completely to the more electronegative atom. The bonding is now ionic because the transfer of electrons from one atom to another creates oppositely charged **ions**.

Ions – another reminder

When electrons are transferred from one atom to another, the atoms end up with unequal numbers of electrons and protons, and become ions. A **positive ion** has more protons than electrons, whereas a **negative ion** has more electrons than protons.

Metals and non-metals

As you can see from the periodic table in the previous spread, generally the metals (on the left-hand side) have low values of electronegativity and the non-metals (on the right) have high values. When a metal combines with a non-metal there is a large difference in electronegativity, and so ionic bonding results. For example, all the metallic elements combine with oxygen to form oxides, and most metallic oxides are ionic. All metals also react with chlorine to give chlorides, and most metallic chlorides are ionic. The illustration shows the spectacular reaction between sodium and chlorine to produce sodium chloride. There are many other examples of ionic compounds, such as potassium fluoride, calcium chloride, and magnesium oxide.

State symbols

The bracket (s) indicates a solid; the bracket (g) indicates a gas.

The metallic element sodium combines with the non-metallic element chlorine to form the ionic compound sodium chloride:

$$2Na(s) + Cl_2(g) \rightarrow 2NaCl(s)$$

Note how different sodium chloride looks from sodium and chlorine. You need to think carefully about what is happening to the arrangement of electrons in these species. Remember that bonding tends to result in an outer octet of electrons for each atom.

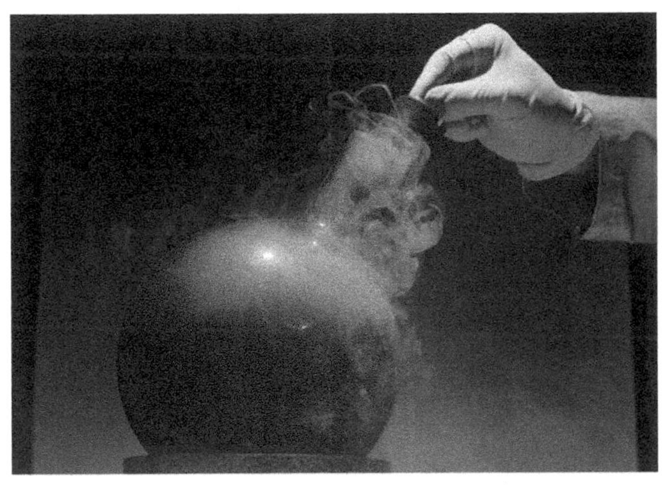

Sodium and chlorine react together vigorously. One electron transfers from the metal sodium to the non-metal chlorine.

Solid sodium metal, gaseous non-metallic chlorine, and the crystalline ionic compound sodium chloride.

Sodium (Na [Ne]$3s^1$) is in Group I of the periodic table. All the atoms of Group I have one electron in their valence shell. The ionization energy of sodium is quite low, so the metal readily loses this outermost electron. The nucleus is now surrounded by full shells only.

Chlorine is in Group VII of the periodic table. All the atoms of Group VII have seven electrons in their valence shell. Chlorine readily completes its octet of electrons by gaining an electron from another atom, such as sodium.

Sodium (Na 2,8,1) transfers the electron in its valence shell to the valence shell of chlorine (Cl 2,8,7), resulting in the sodium ion (Na^+ 2,8) and the chloride ion (Cl^- 2,8,8).

As a result of this transfer of one electron from sodium to chlorine, the sodium has lost one electron and become a sodium ion Na^+. The single positive charge indicates that the sodium ion has one more proton than it has electrons. The chlorine atom has gained one electron and become a chloride ion Cl^-. The single negative charge indicates that the chloride ion has one more electron than it has protons. The electrostatic attraction between Na^+ and Cl^- ions bonds them together.

Electrons transfer readily from Group I atoms to Group VII atoms, for the reasons given in the previous three paragraphs. The compounds that result are most accurately described by the ionic model of bonding. Lewis structures help to picture the transfer of an electron from one atom to another.

Ionic compounds – lattices

Ionic compounds in the solid state consist of oppositely charged ions. The attraction between positive and negative ions constitutes an **ionic bond**. The ions arrange themselves so that each ion is surrounded by ions of opposite charge. The result is a regular three-dimensional array of alternating charges called an **ionic lattice**. This model for the solid ionic state is confirmed by electron density measurements.

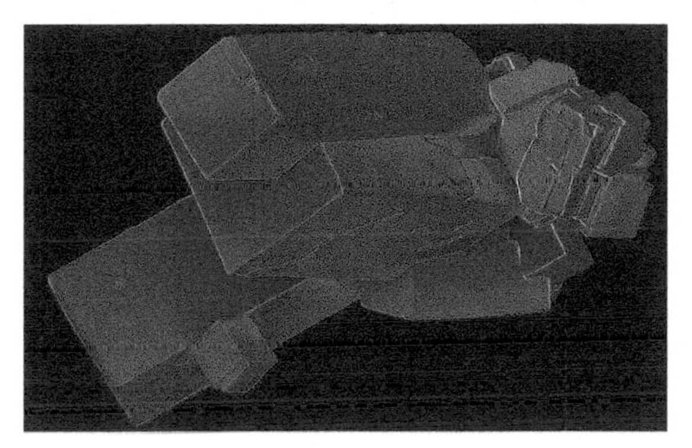

False-coloured scanning electron micrograph of sodium chloride.

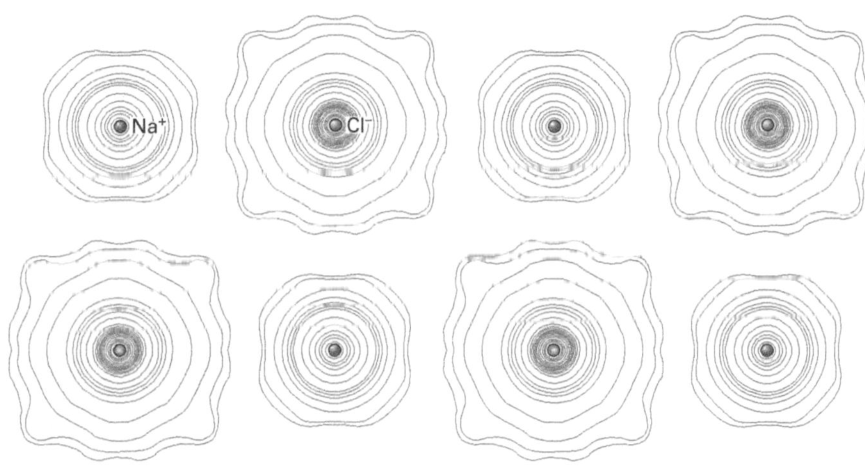

The experimental electron density for sodium chloride shows very little electron density between the sodium and the chloride ions. This picture supports the idea that electrons have been transferred from one atom to another to form ions. Diagrams like this are made by analysing the diffraction of X-ray beams by crystals of ionic compounds. Experimental measurement indicates 10.05 electrons per Na^+ ion: complete electron transfer would give a value of exactly 10.

Ionic bonding involves electron transfer. Covalent bonding involves electron sharing. The final spread of this chapter investigates compounds that exist between these two extremes.

SUMMARY

- Covalent bonds become more polar as the difference in the electronegativity values of the two bonded elements increases.
- An electronegativity difference greater than about 1.8 on the Pauling scale implies that the bond is sufficiently polar for the bonding to be ionic.
- Positive ions form when atoms lose electrons.
- Negative ions form when atoms gain electrons.
- Ions pack together in an array of alternating opposite charges to form a solid ionic lattice.

Magnesium and oxygen

The reaction between the Group II element magnesium and the Group VI element oxygen involves the transfer of two electrons from the magnesium atom to the oxygen atom. Magnesium has two valence electrons (Mg 2,8,2) and oxygen has six (O 2,6). The result is a magnesium ion with a double positive charge (Mg^{2+} 2,8) and an oxide ion with a double negative charge (O^{2-} 2,8).

Magnesium and chlorine

These two elements react together to form the ionic compound magnesium chloride $MgCl_2$. Each magnesium atom loses two electrons to form a magnesium ion:

$Mg \rightarrow Mg^{2+} + 2e^-$

These two electrons are accepted by a chlorine molecule to form two chloride ions:

$2e^- + Cl_2 \rightarrow 2Cl^-$

PRACTICE

1 Use Lewis structures to explain how the following ionic compounds form:
  a Potassium fluoride
  b Calcium chloride
  c Magnesium oxide.

2 Explain what you understand by each of the following terms:
  a Polarity
  b Ion
  c Ionic lattice.

5.11 Covalent or ionic bonds?

OBJECTIVES

- Ionic and covalent character
- Ionic and covalent bonding as extreme cases
- Polarizing power of ions
- Polarizability of ions

Molecular orbital theory

Molecular orbital theory can describe the electron density produced on overlap of atomic orbitals of different energies. See the diagram of the electron density and energy levels of HF in spread 5.7.

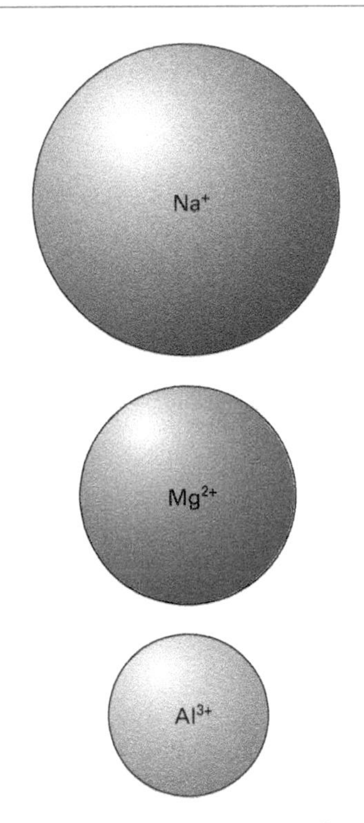

The relative sizes of the Na^+, Mg^{2+}, and Al^{3+} ions.

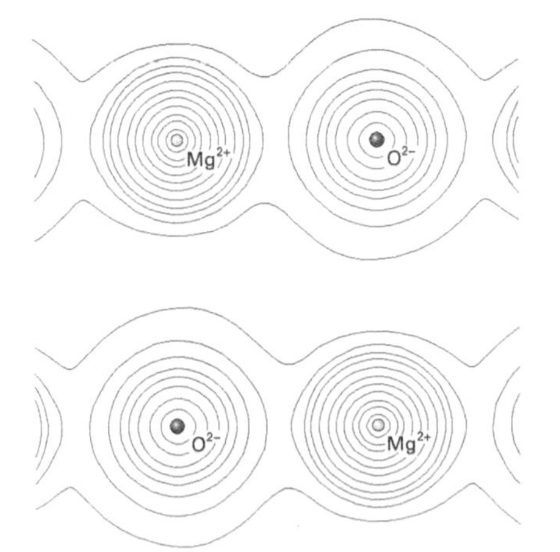

The experimental electron density for magnesium oxide. The minimum electron density between the ions is almost three times as large as that in sodium chloride, see previous spread.

In the simplest model of ionic bonding, electrons transfer completely from one atom to another. Ions of opposite charge result, and the ions pack together to form a solid ionic lattice. Conversely, the simplest model of covalent bonding has electron pairs being shared between atoms. These two situations are extremes, and the actual arrangement in reality often lies somewhere between them. You already understand that a polar covalent bond has some ionic character. It is now time to consider the converse: how an ionic bond can acquire covalent character.

Covalent character – polarizing power of the positive ion

An ionic bond will acquire a degree of covalent character if the positive ion can attract electron density from the negative ion back into the region between the nuclei. The covalent character arises because the valence electrons will then be partially shared. The extent to which a positive ion can do this is called its **polarizing power**. The polarizing power of a positive ion depends on two factors, its charge and its size:

- The larger the positive charge on the ion, the greater is its attraction for the valence electrons. Doubling the charge doubles the force that a positive ion can exert.
- The smaller the size of the positive ion, the closer it is to the valence electrons, and the larger the force it exerts on them.

These two factors together combine to make up the **charge density** on the positive ion:

- The greater the charge density on the positive ion, the greater its polarizing power and the greater the covalent character of the bond it forms with a given negative ion.

The effect of polarizing power is illustrated by the chlorides of the metals of Period 3: NaCl, $MgCl_2$, and $AlCl_3$. The charge on the ions increases across the period: $Na^+ < Mg^{2+} < Al^{3+}$. The size of the ions decreases across the period: $Na^+ > Mg^{2+} > Al^{3+}$. The magnesium ion has a higher charge density than the sodium ion, because it has a greater charge *and* is smaller. The aluminium ion has the highest charge density because it is the smallest ion *and* has the greatest charge. The bonding in sodium chloride and in magnesium chloride is described adequately by the simple ionic model. However, the intensely polarizing Al^{3+} ion places the bonding in $AlCl_3$ on the borderline between ionic and covalent bond character.

The effect of polarizing power is also illustrated by silicon tetrachloride $SiCl_4$. The formation of purely ionic bonds in this compound would involve forming the Si^{4+} ion. This ion is so polarizing that it would attract back the electrons it had lost, so it is very unlikely to form. The bonding in $SiCl_4$ is essentially covalent.

Covalent character – polarizability of the negative ion

The extent to which a positive ion can pull electrons back into the space between the nuclei depends not only on its own polarizing power but also on how easy it is to polarize the negative ion. The **polarizability** of a negative ion measures the ease with which its electron density can be distorted towards the positive ion.

Polarizability increases with the charge on the negative ion and also increases with the number of electrons the ion has. Large, highly charged, negative ions can be polarized easily, and form bonds with greater covalent character. Small, singly charged ions are not easy to polarize. There is a clear trend in the polarizability of the ions of the Group VII elements. Fluoride ions have fewer electrons and are smaller than chloride ions. Fluoride ions are therefore less polarizable than chloride ions; all fluorides have greater ionic character than the corresponding chlorides. For example, aluminium fluoride is a crystalline solid melting above 1000 °C. Anhydrous aluminium chloride (which is the dry form, with no water associated with it) shows much covalent character.

The iodide ion I^- is approximately 20% bigger than the chloride ion Cl^-. The iodide ion is therefore more polarizable than the chloride ion. The charge density of the Na^+ ions distorts the electron density around the iodide ions in the NaI lattice and draws electron density into the region between the oppositely charged ions. NaI therefore has a degree of covalent character. The Cl^- ion is less distorted because it is smaller in size.

The combined effect of polarizing power and polarizability is illustrated by considering a series of compounds that all have the same total number of electrons: sodium fluoride NaF, magnesium oxide MgO, aluminium nitride AlN, and silicon carbide SiC. Passing along this series, the polarizing power of the positive ion increases as the ion becomes smaller and more highly charged, and the negative ion becomes larger (the same number of electrons are under the control of fewer protons) and more polarizable as its charge increases. Bonding character changes from ionic to covalent across this series.

Bond character and chemical properties

Sodium chloride and magnesium chloride both exist as crystalline solids at room temperature, and have the typical properties of ionic compounds (e.g. high melting point, aqueous solution conducts electricity). These compounds both dissolve in water without reaction. The solid crystalline lattice breaks down and the ions are surrounded by the polar water molecules:

$$NaCl(s) \rightarrow Na^+(aq) + Cl^-(aq)$$

Anhydrous aluminium chloride is not crystalline, has the relatively low melting point of 180 °C, and reacts violently with water:

$$AlCl_3(s) + 3H_2O(l) \rightarrow Al(OH)_3(s) + 3HCl(aq)$$

Many other covalent chlorides (e.g. silicon tetrachloride) react similarly with water:

$$SiCl_4(l) + 4H_2O(l) \rightarrow Si(OH)_4(s) + 4HCl(aq)$$

This behaviour suggests that the bonding in anhydrous aluminium chloride is also essentially covalent. Reactions with water are referred to as **hydrolysis reactions**. The bracket (aq) indicates an aqueous solution; the bracket (l) indicates a liquid.

SUMMARY

- The covalent character of an ionic bond depends on the polarizing power of the positive ion and the polarizability of the negative ion.
- The polarizing power of a positive ion depends on its charge density, which increases with increasing charge and decreasing size.
- The polarizability of a negative ion increases with increasing charge and increasing number of electrons.
- Bond character is reflected in the chemical properties of a compound.

Van Arkel diagram

The gradual change from one type of bonding to another can be shown by a diagram introduced by Dutchman Anton van Arkel. Recognizing the three extreme types of bonding (ionic, covalent, and metallic), he used a triangular display. The line linking the ionic and covalent extremes contains binary compounds of elements with varying electronegativity differences. The line linking the covalent and metallic extremes contains elements with varying extents of delocalization.

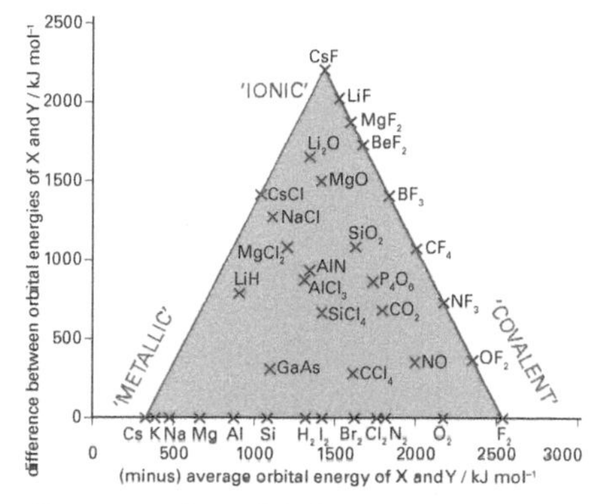

One type of van Arkel diagram.

The reaction between silicon tetrachloride and water is vigorous. The glass rod is moistened with concentrated aqueous ammonia, which produces white fumes with the hydrogen chloride produced on hydrolysis.

PRACTICE

1 Using one of the following headings:

- non-polar covalent molecules,
- highly polar covalent molecules,
- an essentially ionic lattice,
- an ionic lattice with a marked degree of covalent character,

describe the bonding in each of the substances listed alongside.

a Methane CH_4
b Tetrachloromethane CCl_4
c Rubidium iodide RbI
d Sodium chloride NaCl
e Magnesium bromide $MgBr_2$
f Strontium oxide SrO
g Phosphine PH_3

Explain your reasoning in each case.

PRACTICE EXAM QUESTIONS

1 **a** Copy and complete the following table by giving, in each case, the formula of a molecule or ion which has the bond angle shown. Use a different molecule or ion for each angle. [4]

Bond angle	Formula of molecule or ion with this bond angle
90°	
109° 28′	
120°	
180°	

b Draw a diagram of a water molecule and on your diagram indicate the value of the bond angle. [2]

c Explain why the value of the bond angle in part **b** is different from any of those values in part **a**. [2]

d Draw a diagram showing how two water molecules attract each other by hydrogen bonding. [3]

2 **a** The nitrogen and hydrogen atoms in an ammonia molecule are held together by covalent bonds. What is meant by the term covalent bond? [2]

b By referring to the formation of the ammonium ion from ammonia give the meaning of the term coordinate bond. [2]

c Suggest the difference in bond strength, if any, between the bond formed by co-ordination in the ammonium ion and the other N—H bonds. Explain your answer. [2]

d Give the bond angle in the ammonium ion and predict, with an explanation, an approximate value for the bond angle in the ammonia molecule. [3]

e Name the major force of attraction which exists between molecules in liquid ammonia and explain how this type of force arises. [4]

3 **a** State the type of bonding in a crystal of potassium bromide. Write an equation to show what happens when potassium bromide is dissolved in water and predict the pH of the resulting solution. [3]

b When iodine reacts directly with fluorine, a compound containing 57.2% by mass of iodine is formed.

i Determine the empirical formula of this compound.

ii The empirical formula of this compound is the same as the molecular formula. Write a balanced equation for the formation of this compound. [4]

c **i** Sketch a diagram to show the shape of a BrF_3 molecule. Show on your sketch any lone pairs of electrons in the outermost shell of bromine and name the shape.

ii BrF_3 reacts with an equimolar amount of potassium fluoride to form an ionic compound which contains potassium ions. Give the formula of the negative ion produced, sketch its shape, show any lone pair(s) and indicate the value of the bond angle. [6]

(See also chapter 9.)

4 **a** Explain what is meant by the terms **ionic bond, covalent bond, dative covalent bond.** [3]

b Copy the table below and indicate whether each of the following molecules has an overall polarity. [3]

	Yes/No
Tetrachloromethane	
Carbon dioxide	
Methane	
Ethanol	
Trichloromethane	
propanone	

c Below are given the atomic and ionic radii for a number of elements.

Atom	Radius/nm	Ion	Radius/nm
Na	0.157	Na^+	0.102
Mg	0.136	Mg^{2+}	0.072
Al	0.125	Al^{3+}	0.053
F	0.071	F^-	0.133
Cl	0.099	Cl^-	0.180
I	0.133	I^-	0.216

Give an explanation for each of the following:

i The magnesium atom is smaller than the sodium atom.

ii The sodium ion is smaller than the sodium atom.

iii Aluminium fluoride is ionic, aluminium iodide is covalent. [9]

5 **a** What is a polar covalent bond? [1]

b In what circumstances will a covalent bond be polar? [1]

c In what circumstances will an anion be polarized? [1]

d How does a polarized anion differ from an unpolarized anion? [1]

e **i** Draw diagrams to show the shapes of the following molecules and in each case show the value of the bond angle of the diagram.

I $BeCl_2$

II NCl_3

III SF_6

ii State which of the above molecules is most likely to form a coordinate bond with a hydrogen ion. Give a reason for your answer. [8]

6 This question is about geometry and molecular shapes.

a Distinguish between the terms *covalent bond* and *dative bond (coordinate bond)*. [1]

b i When ammonia and boron trifluoride (BF_3) are mixed, a reaction occurs and a compound of molecular formula NBH_3F_3 is formed. Using the usual symbols for covalent and dative bonds, represent the molecular bonding in a diagram.

What geometry would you expect around the nitrogen atom in NBH_3F_3?

ii Boron trifluoride is a planar molecule but nitrogen trifluoride (NF_3) is pyramidal. Use the electron-pair repulsion theory to explain the difference in shape. [5]

c i What shape is molecular phosphorus pentachloride?

ii Five electron pairs arrange themselves as shown in the diagram. **X** is the central atom. Any lone pairs in such systems go into the equatorial plane.

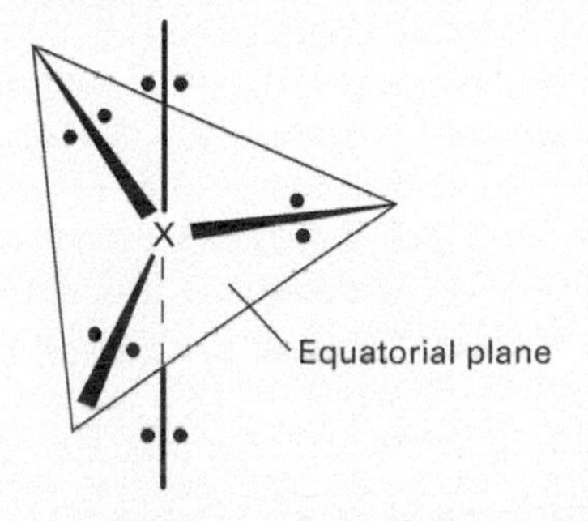

Use this fact to draw the shapes of the following molecules. You should indicate clearly the geometry about the central atom and show all lone pairs which are relevant to the shape.

Bromine trifluoride, BrF_3

Krypton difluoride, KrF_2 [5]

d A compound $Pt(NH_3)_2Cl_2$ is used in medicine as the modern drug cisplatin.

i Given that platinum(II) compounds are almost always square planar, draw the possible isomers of this compound and name the type of isomerism exhibited.

ii When $Pt(NH_3)_2Cl_2$ is synthesised in the laboratory the effectiveness of the resulting compound depends on the method of preparation. Suggest an explanation for this fact. [6]

(See also chapter 20.)

7 Ammonia, NH_3, is a polar covalent molecule.

a State the general rules which determine the shape of a covalent molecule. [3]

b Draw the shape of the ammonia molecule. [1]

c Why is the bond angle in ammonia 107° rather than 109° 28'? [1]

d Explain why the molecule of ammonia is polar. [1]

8 a This part of the question concerns the hydrides CH_4, NH_3, and H_2O.

i Draw the shapes of each molecule, suggesting values for the bond angles.

ii State the type of intermolecular forces present for each hydride.

iii Account for the variation in bond angles in these molecules.

iv In which hydride are the intermolecular forces strongest? Explain how you decided upon your answer. [10]

b i Draw 'dot-and-cross' diagrams to show the structures of the ammonia molecule and of the ammonium ion.

ii Using the ammonium ion as an example, explain what is meant by the term dative covalent bond.

iii The ammonium ion has a tetrahedral shape. Explain what this suggests about the four N—H bonds. [5]

(See also chapter 7.)

9 a Describe the motion of the ions in

i solid sodium chloride,

ii molten sodium chloride. [4]

b A lithium iodide crystal has some covalent character.

i Explain the meaning of the terms *covalent bond* and *some covalent character*.

ii Explain why lithium iodide has more covalent character than sodium iodide. [5]

10 The table below shows electronegativity values for some atoms.

H	N	O	F	Cl	Cs
2.1	3.0	3.5	4.0	3.0	0.7

a What do you understand by the term **electronegativity**? [1]

b The nature of the bonding in substances depends partly on the electronegativities of the atoms concerned. *Use the data* in the table above to suggest the nature of the bonding in each of the following substances.

i caesium fluoride; [1]

ii water; [1]

iii chlorine. [1]

Solids

All around you are examples of the three main states of matter: solids, liquids, and gases. Solids are distinguished from liquids and gases by having definite shapes, whereas liquids and gases take on the shape of their containers. A definite external shape implies a regular internal structure where atoms, ions, or molecules are held in a fixed array. The previous chapter discussed the nature of ionic bonding between metallic and non-metallic elements as well as covalent bonding between non-metallic elements. The aims of this chapter are to investigate the forces that maintain the shapes of solids and to describe and explain their internal structures. It opens with an investigation of metallic solids.

METALLIC SOLIDS: BONDING AND PROPERTIES

OBJECTIVES

- Lattice structure of metals
- Delocalized valence electrons
- Electrical conductivity
- Physical properties

Metallic elements all conduct electricity in the solid state. This observation cannot be explained by either an ionic or a covalent model of bonding. Ionic bonding requires the transfer of electrons to form oppositely charged ions; covalent bonding involves the sharing of electron pairs to form molecules. In neither model are there any species that are free to move around in the solid state. The bonding between metal atoms must be described by a different model.

Metallic bonding

When metal atoms are close to each other in the solid state, each atom loses control over one or more of its valence electrons. These electrons are no longer associated with a particular metal atom but are free to move throughout the solid piece of metal: the electrons are said to be **delocalized**. With valence electrons now delocalized, the metal atoms are effectively ionized. For example, the electronic structure of a sodium atom is Na $1s^2 2s^2 2p^6 3s^1$. The two inner shells are full. The 3s electron is delocalized, leaving each sodium atom as the sodium ion (electronic structure Na^+ $1s^2 2s^2 2p^6$). The bonding force is the attraction between the positive sodium ions and the delocalized electrons. The delocalized electrons are shared between all the sodium ions and act as a sort of 'glue' holding them together.

The whole system of ions and electrons has the lowest energy when the ions arrange themselves in a symmetrical pattern. In the solid state, **metallic bonding** therefore consists of regular arrangements of metal ions (called a **lattice**) surrounded by a 'sea' of delocalized electrons.

Metals consist of a regular array of metal ions surrounded by a 'sea' of delocalized valence electrons. Note that the overall structure remains electrically neutral.

Electrical conductivity

An electric current is a flow of charged particles. In the case of metals, the charged particles are the delocalized valence electrons, which are free to move through the three-dimensional lattice of metal ions. Connecting a source of electric potential difference (e.g. a battery) across a piece of metal causes the electrons to move. The negative terminal of a battery has a negative potential, which means there is a surplus of electrons. The positive terminal has a positive potential, which means there is a deficit of electrons. Electrons flow through the metal from the negative to the positive terminal.

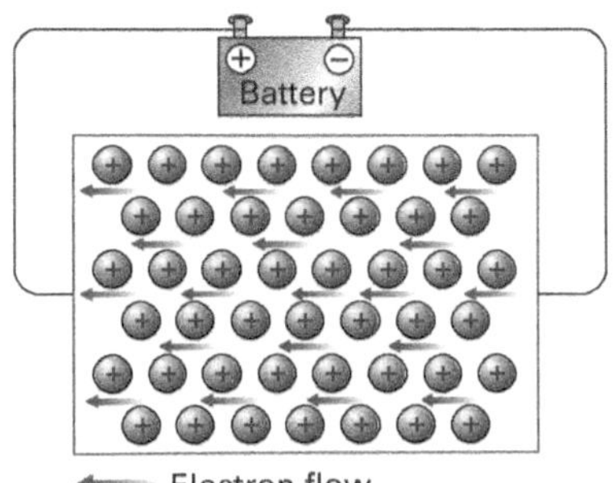

An electric current flowing through a metal is a flow of delocalized electrons. They flow from a region of negative electric potential to a region of positive potential.

Thermal conductivity

Thermal conductivity measures the rate of heat flow through a substance when one part of a sample is maintained at a higher temperature than the rest. The graph shows that the thermal conductivity of metals is linked closely to their electrical conductivities. This link indicates that electrons are the main conductors of heat through metals. In a metal, high temperature is represented by electrons with high kinetic energy and by metal ions with high vibrational energy. Energy is conducted through a piece of metal as the more energetic ('hotter') electrons collide with and speed up the slower, less energetic ('colder') electrons.

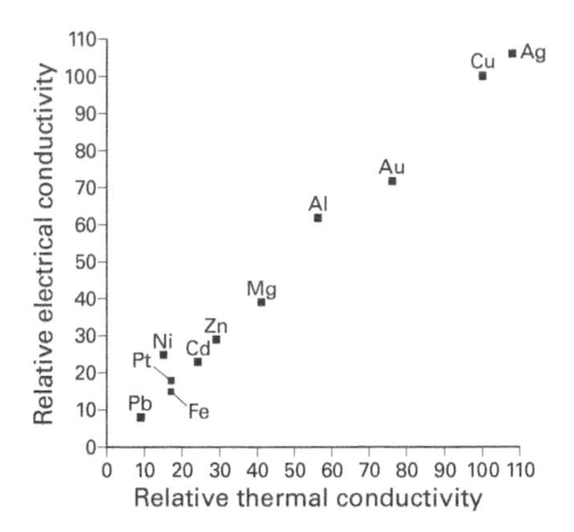

A scatter plot of relative thermal and electrical conductivities. The scales are relative to copper = 100.

Malleability and ductility

Metals are usually malleable: a **malleable** material may be beaten or pressed into new shapes. They are also usually ductile: a **ductile** material may be drawn out into a wire and made thinner by stretching. When metals are deformed, the metal ions in the solid structure slide past each other as the overall shape of the sample changes. The internal arrangement of ions and delocalized electrons allows this slippage to take place without significant disruption of the bonding forces.

When a metal is deformed, layers of metal ions can slip past one another without breaking up the whole structure.

Copper wire is made by drawing a thin cylinder of the metal through successively smaller holes. This process depends mainly on the property of ductility.

Car body panels are pressed from flat sheets of steel. This process takes advantage of both of the properties of malleability and ductility.

Metal crystals and alloys

A metal sample does not consist of a single regular lattice throughout its entire structure. The structure is broken up into tiny crystals, which may be seen through a microscope. The surface of a sample must first be carefully polished and then etched. When magnified, the surface appears to be divided into irregular areas. These areas are called grains; each **grain** represents a single crystal. The grains are separated by **grain boundaries**. We say that metals are polycrystalline.

Metals are mixed together to form **alloys**. The mechanical properties of alloys are controlled by the structures of the grains and by the composition of the grain boundaries.

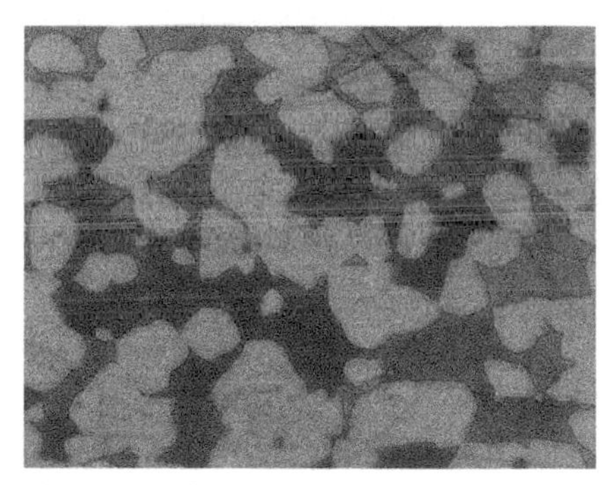

This magnified view of an etched and polished surface reveals the grains and grain boundaries in brass. ***Etching*** *is the removal of certain parts of a surface by means of a chemical reaction. It occurs selectively at the edges of the crystals, and so makes the edges much easier to see. The etching agent chosen depends on the type of metal: dilute nitric acid is used for iron or lead, and iron(III) chloride for copper.*

SUMMARY

- Metals consist of a regular array of metal ions surrounded by a 'sea' of delocalized valence electrons.
- The electrical and thermal conductivities of metals depend on the ability of the delocalized electrons to move freely.
- Metals are usually malleable and ductile because the metal ions can slip past each other without significantly disrupting the bonding forces.

PRACTICE

1 Explain why metals can conduct electricity in the solid state.

2 What do the terms *malleable* and *ductile* mean when applied to metals? Explain how pressing a sheet of steel to make a car body panel depends on both the malleability and the ductility of the steel.

3 Using a reference book or computer database, make a list of some alloys, along with their constituents, properties, and uses.

6.2

OBJECTIVES

- Sodium chloride crystal structure
- Electrical conductivity
- Physical properties

Ionic solids: bonding and properties

Sodium chloride is a good example of an ionically bonded compound. A crystal of sodium chloride consists of equal numbers of sodium ions Na^+ and chloride ions Cl^-. Each ion is surrounded by six others of opposite charge. The crystal structure is a regular three-dimensional pattern of sodium ions alternating with chloride ions. Such a regular arrangement is called an **ionic lattice**. The internal structure gives solid ionic compounds a regular and sharply defined external shape. For sodium chloride, crystals grown by evaporating a solution have a cubic shape.

Fluorite CaF_2 and dolomite $CaCO_3{\cdot}MgCO_3$ are both crystalline ionic compounds.

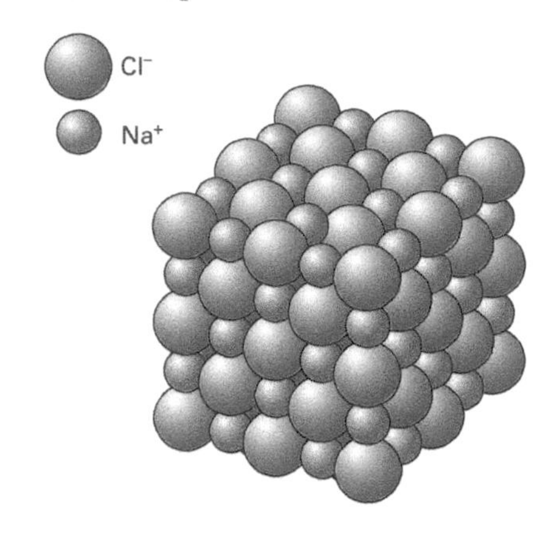

(a) A crystal of sodium chloride photographed in polarized light, (b) a space-filling model of the crystal lattice.

Melting ionic solids

Melting an ionic compound weakens the influence of the forces of attraction between the oppositely charged ions. The ions become free to move in the liquid state and are fairly independent of each other. The melting points of ionic compounds are generally high (e.g. T_m of sodium chloride = 801°C), indicating that the forces of attraction between the ions are strong. Because ionic compounds have high melting points, they are always found as solids at room temperature. If a substance is a liquid or a gas at room temperature, we can safely assume that it does not have ionic bonding. On the other hand, a high melting point does not necessarily mean that a compound is ionic.

Solubility in water

Like many other ionic compounds, sodium chloride dissolves easily in water. The solid crystals gradually disappear as their constituent ions disperse throughout the solution. We can ask the question: if melting and dissolving both involve overcoming the forces of attraction between ions, why does melting take place at 801°C while dissolving can take place at room temperature?

As we saw in spread 5.9, water is a V-shaped molecule with polar covalent bonds. The oxygen atom bears a partial negative charge and the hydrogen atoms each bear a partial positive charge. During solution, each positive sodium ion on the surface of the crystal attracts the partial negative charge on the oxygen atoms of water. Each negative chloride ion attracts the partial positive charges on the hydrogen atoms of the water molecules. The result is that the ions become surrounded by water molecules and are removed from their positions in the crystal lattice, as shown in spread 10.7.

> **Ion–dipole force**
>
> There is an attraction between a positive ion such as Na^+ and the negative end of the water dipole (the partial negative charge on the oxygen atom). This attractive force is called an **ion–dipole force**. The ion–dipole force between the ions and several water molecules almost exactly compensates for the attraction between the Na^+ and Cl^- ions.

If an ionic compound is soluble, the attractions between water molecules and the ions must be comparable with the energy required to separate the ions. This energy requirement is not always met and several ionic compounds, such as magnesium oxide and calcium carbonate, dissolve only to a very small extent in water.

Electrical conductivity

In general terms, an electric current will flow if charged particles are free to move when an electric potential difference is applied. Metals conduct electricity because they contain delocalized valence electrons. Ionic compounds conduct electricity when molten or when dissolved in water. In both cases, the electric current is carried by ions that are free to move. Solid ionic compounds do not conduct electricity because the ions cannot move freely.

Sodium chloride conducts electricity when in aqueous solution.

Ionic crystals – hard and yet brittle

Ionic crystals are not malleable or ductile. They cannot be squeezed or beaten into new shapes or stretched to make them thinner. A crystal of sodium chloride shatters if an attempt is made to cut it with a knife. The ions in the crystal are moved so that each sodium ion is immediately next to another sodium ion, and each chloride ion is next to another chloride ion. Instead of being strongly attracted by their neighbours, the ions of like charge are strongly repelled and the crystal flies apart. Contrast this behaviour with that of metallic crystals, which do not shatter when cut or even when hit.

A sharp blow along specific planes of this potassium bromide crystal with a knife edge causes the crystal to split.

SUMMARY

- Solid ionic compounds consist of ions packed into a regular lattice.
- In the lattice, ions of one charge are surrounded by ions of the opposite charge.
- Ionic compounds have high melting points because of the strong forces of attraction between the ions.
- Aqueous solutions of ionic compounds contain separate ions each surrounded by water molecules.
- Ionic compounds conduct electricity when molten or when in aqueous solution because they contain ions that are free to move.
- Ionic compounds are brittle; they shatter when hit.

PRACTICE

1 For the ionic solid sodium chloride, draw a diagram to show the arrangement of a sodium ion and its nearest neighbours.

2 Draw the structures of the sodium ion and the chloride ion as they exist dissolved in water. Explain what happens as the solution evaporates and crystals of solid sodium chloride grow.

3 Write equations with state symbols to show:

a Solid sodium chloride melting

b Solid sodium chloride dissolving in water.

6.3

O B J E C T I V E S

- Diamond
- Graphite
- Silica
- Glass
- Ceramics

GIANT COVALENT SOLIDS

Metallic and ionic solids are made up of ions held in place by electrostatic forces. In contrast, some covalent solids consist of networks of atoms, held in place by covalent bonds, that stretch throughout the whole structure. These substances are called **giant covalent solids**: they form the focus of this spread. Familiar examples include diamond, graphite, and silica. Giant covalent solids are sometimes also called **macromolecules** to reflect the immense size of the molecule involved.

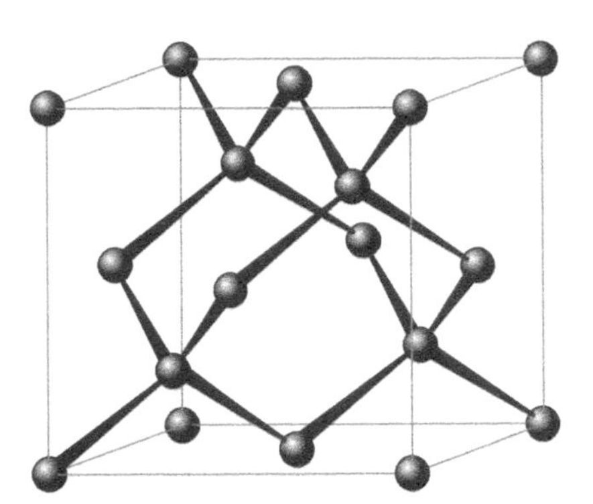

The three-dimensional structure of diamond.

Diamond

Diamond is a good example of a giant covalent solid. Each carbon atom in the structure is covalently bonded to four others. The bonding forces are uniform throughout the structure. As with ionic solids, giant covalent solids have very high melting points. The energy required to break the covalent bonds is very high. But, unlike most ionic solids, giant covalent solids do not dissolve in water because there are no ions to attract the water molecules.

The electronic structure of carbon is C $1s^2 2s^2 2p^2$. In diamond, the four valence electrons are all involved in bonding pairs located between the atoms of carbon. The other two electrons per atom are in a filled shell (n = 1). There are no free electrons and no ions present, so diamond does not conduct electricity. It is also the hardest substance known due to the strength of the carbon–carbon bonds and the geometrical rigidity of the structure. Many cutting tools are tipped with powdered diamond.

*The diamond merchant Harry Winston sold three of the eight largest uncut diamonds ever found (*The Star of Sierra Leone, Presidente Vargas, *and* Jonker*). When this photograph appeared in* National Geographic *during his lifetime, his face was blacked out at the insistence of his life insurance company.*

Graphite

Graphite is also a giant covalent solid, but its structure consists of two-dimensional layers of joined hexagonal rings rather than a three-dimensional network like diamond. Within a layer, each carbon atom is joined to three others by strong covalent bonds. The fourth valence electron from each carbon atom is delocalized within the layers. The delocalized electrons are free to move and so graphite can conduct electricity. The layers are not held tightly together and can slide over one another because of the weak forces of attraction between the layers (see the following chapter). This arrangement gives graphite a slippery feel and makes it a good high-temperature lubricant. Notice that the strong covalent bonds *within* the layers hold the carbon atoms close together; the weak forces *between* the layers result in a much wider separation. This wider separation explains why graphite's density (2.3 g cm^{-3}) is lower than diamond's (3.5 g cm^{-3}).

Boron nitride

The compound boron nitride, BN, is isoelectronic with carbon and also has two main forms, one with the graphite structure and one with the diamond structure. The form with the diamond structure, **Borazon**, is the second-hardest substance known.

Fullerite

See spread 19.4 for the structure of this third form of carbon.

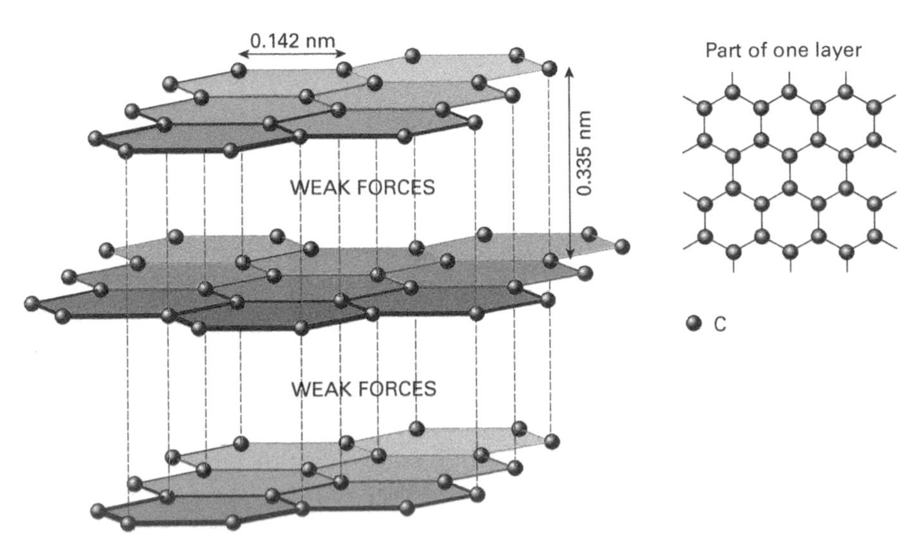

The structure of graphite. Compare the distance between the atoms within the layers with the distance between the layers.

Silica

Giant covalent solids can contain more than one sort of atom. Sand consists mostly of silicon dioxide, also called silica, whose formula is usually written as SiO_2. This formula gives the simplest ratio of the atoms concerned. Silica is a giant covalent solid with a structure similar to that of diamond. Because of its structure (i.e. strong covalent bonds, no ions, and no free electrons), silica is hard, has a high melting point, does not dissolve in water, and does not conduct electricity.

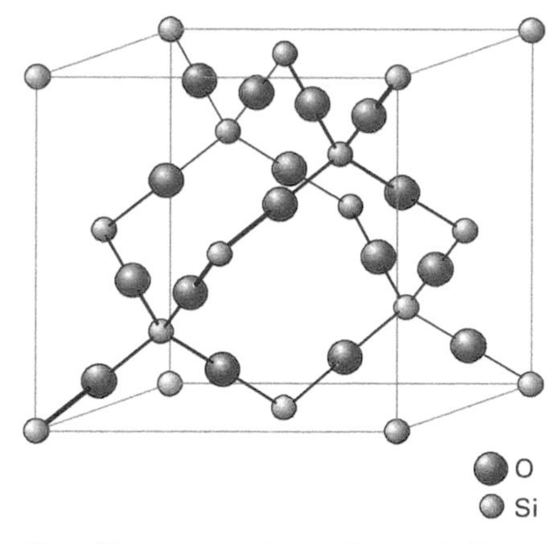

The silicon atoms in one form of silica adopt the diamond structure; there are oxygen atoms between each pair of silicon atoms.

Glass

All the giant covalent solids described above have regular structures. Glasses are made by melting silica with small amounts of other substances, but a glass does not have the regular structure of pure silica. The giant covalent structure is random, with ions from the other ingredients scattered throughout it. A glass does not have a distinct melting point.

Ceramics

Ceramics are manufactured solids that have a giant structure made up of two or more elements. The bonding can be either covalent, or ionic, or both. Familiar examples are bricks and porcelain, made by heating mixtures of sand and clays or feldspars to a high temperature (1500 °C) in a kiln. Clays and feldspars are aluminosilicates, spread 19.5, which contain aluminium, silicon, and oxygen, and often other metals such as magnesium. Because of their giant structure, ceramics typically show the following properties:

- They have high melting points.
- They are strong.
- They are good electrical and thermal insulators.

This glass building (Science Park at Futuroscope, Poitiers) echoes the shape of a quartz crystal, spread 17.4.

Superconductivity

Recent high-tech ceramics are of great interest because they exhibit **superconductivity**, which is the ability to conduct electricity without any resistance to current flow. Some metals develop this property, but only when cooled below 25 K by liquid helium (which is very expensive to make). Some ceramics become superconducting at much higher temperatures, above the boiling point of liquid nitrogen (77 K). Liquid nitrogen is cheaper than bottled water, which makes the technology affordable. These so-called 'high-temperature superconductors' offer the potential to store and transmit electricity without any loss of energy. A typical ceramic superconductor is an oxide of yttrium, barium, and copper, $YBa_2Cu_3O_7$.

Placing a magnet over this superconductor induced an electric current in it. The current circulates without loss, creating a magnetic field, which opposes that of the magnet above. The magnet 'levitates'.

SUMMARY

- Giant covalent solids have high melting points.
- Most giant covalent solids are electrical insulators because all of the electrons either are in filled shells or are involved in covalent bonds.
- Diamond has a tetrahedral structure in which each carbon atom is bonded to four others.
- Graphite has a layered structure; within each hexagonal layer, each carbon atom is bonded to three others.
- Graphite is an electrical conductor because one valence electron per atom is delocalized within the layers.

PRACTICE

1 Explain why giant covalent substances are:
   a usually hard,
   b electrical insulators,
   c stable (i.e. do not decompose) at high temperatures.

2 Explain how graphite conducts electricity.

3 What type of bonds are present in silica? What type of bonds are present in sodium chloride? How do the strengths of these bonds compare? Give some evidence to support your argument.

6.4 DELOCALIZATION

OBJECTIVES

- Delocalization in metals and graphite
- Delocalization in benzene
- Delocalization in ions

As you read in the spread on bonding in metals at the beginning of this chapter, valence electrons in a metal are not associated with a particular metal atom but are free to move throughout the sample: the electrons are said to be **delocalized**. These delocalized electrons constitute a sort of 'glue' that holds the resulting metal ions in a regular lattice structure. Graphite is a form of carbon, which is a non-metallic element. The atoms are covalently bonded together in layers of hexagonal rings joined together. Although a non-metal, graphite conducts electricity because one valence electron per atom is delocalized within these layers. As you will see in later chapters, delocalization is present in many other substances and is an important idea in helping to explain their properties.

Graphite – a closer look

The concept of delocalization might seem to be a cunning invention to explain away the ability of graphite, a covalently bonded non-metallic element, to conduct electricity. However, the distances between the atoms (the **bond lengths**) in graphite show the bonding to be unusual.

The C—C single bond in diamond has a length of 0.154 nm. The C=C double bond in ethene $CH_2{=}CH_2$ is shorter at 0.134 nm. This is because the *two* pairs of electrons in the double bond draw the atoms closer together. The bond length between adjacent carbon atoms within the layers in graphite is 0.142 nm, a value *intermediate* between those for the C—C bond and the C=C bond. It is therefore reasonable to assume that the number of bonds between each pair of adjacent carbon atoms in graphite is between 1 and 2. Considering a simpler molecule, benzene, will explain why and how the bonds have this intermediate nature.

Benzene

Benzene is a compound of carbon and hydrogen with the formula C_6H_6. X-ray diffraction techniques confirm its structure as a hexagon of carbon atoms. Each carbon atom is joined to two carbon atoms, and to a hydrogen atom by a single covalent bond. All six carbon–carbon bonds have the same length, 0.139 nm. This value too is intermediate between the lengths for C—C and C=C.

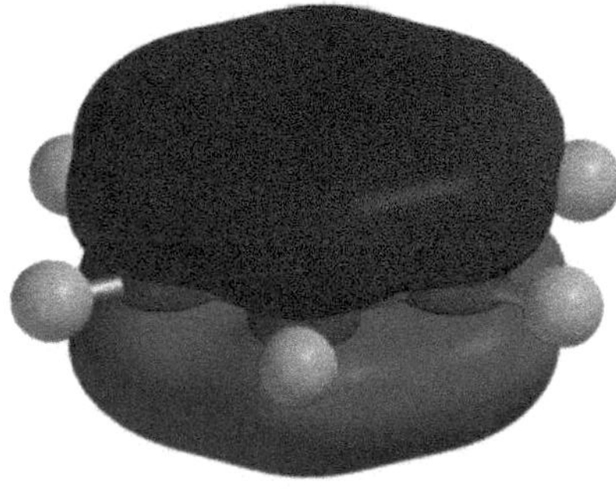

The p orbitals on all six carbon atoms of the benzene ring overlap in phase to make this lowest-energy delocalized molecular orbital. See spread 23.1.

The electronic structure of carbon is C $1s^2 2s^2 2p^2$. So each carbon atom, in joining to two carbon atoms and one hydrogen atom, has so far not used one p orbital (out of three) and one valence electron (out of four). To understand what this means and to progress further, we need to apply molecular orbital theory.

Delocalization in benzene

The one p orbital per atom not so far used in bonding projects above and below the plane of the hexagonal ring. The p orbitals on adjacent carbon atoms overlap to form molecular orbitals which are spread over all six carbon atoms. The lowest-energy molecular orbital is shaped like two hexagonal doughnuts positioned above and below the plane of the ring. The pair of electrons in the lowest-energy molecular orbital bonds *all six atoms* together. (There are also two other bonding molecular orbitals of more complicated shape each containing a pair of electrons.) As shown to the left, this model is confirmed by a computer-generated electron density diagram.

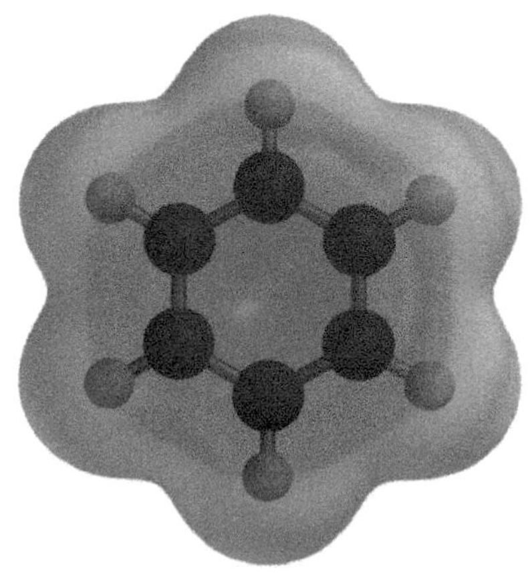

Electrostatic potential map for benzene: six bonding electrons enter the three delocalized orbitals to form a delocalized cloud of electrons. Note that the ball-and-stick model visible inside the electrostatic potential map shows atoms that are bonded; the number of bonds is intentionally not shown.

Graphite again

Benzene and graphite consist of hexagonal rings of carbon atoms. In benzene, each carbon atom is bonded to two other carbon atoms; a carbon–hydrogen bond extends out from each atom in the ring. In graphite, further carbon–carbon bonds extend out of each ring to link with other rings.

The p orbitals on each of the carbon atoms in graphite therefore overlap to form delocalized molecular orbitals, which cover the entire layer of joined hexagonal rings. Electrical conduction in graphite is therefore a flow of electrons within delocalized molecular orbitals.

Delocalization in ions

The electrostatic potential map for ethanoic acid CH_3COOH shows that one of the oxygen atoms is bonded to a hydrogen atom and that the other oxygen atom is in a different environment. The heavy concentration of red shows that the carbon-oxygen double bond is strongly polar. The other red patch is around the other oxygen atom.

When a hydrogen ion H^+ is lost from ethanoic acid to form the ethanoate ion, the electron density shows a significant change has occurred. A simplistic analysis would suggest the formula CH_3COO^-. However, the electron density shows that the two oxygen atoms are no longer different. It is impossible to tell either to which oxygen atom the hydrogen was attached or which was the doubly bonded one. The ion is much better represented by the formula $CH_3CO_2^-$, the two oxygens now being equivalent. This will have important consequences for the acidity of organic acids, as explained in spread 27.1. The nitrate ion NO_3^- and the carbonate ion CO_3^{2-} are both delocalized; their shapes are trigonal planar. The sulphate ion SO_4^{2-} is also delocalized; its shape is tetrahedral.

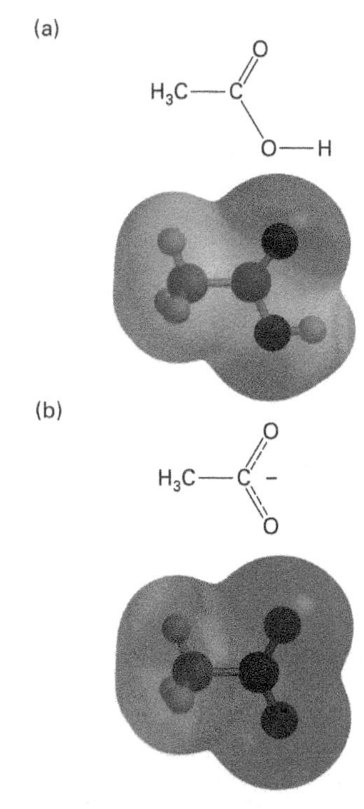

Electrostatic potential maps for (a) ethanoic acid and (b) the ethanoate ion. The two oxygen atoms are identical in the ion. Delocalization may be shown using dashed lines.

The H_3^+ ion

This ion is the simplest example of delocalization. Three protons positioned at the corners of an equilateral triangle are held together by just two electrons shared between all three. This ion was first observed by mass spectrometry by J. J. Thomson in 1912, before a model existed which could explain its stability. The presence of H_3^+ has recently been discovered in the atmosphere of Jupiter.

Important features about delocalization

- Delocalization stabilizes a molecule or ion.

The reason for this lies in the wave nature of electrons. The more spread out a wave is, the less severe its curvature becomes; and hence its kinetic energy becomes lower.

- Most molecular orbitals are delocalized. The only exceptions to this occur in molecules that have exactly two atoms (diatomic molecules). See the discussion of the bonding in methane in spread 22.12.
- The lowest-energy orbital of each symmetry (σ or π) is easy to visualize: each atom has all the relevant orbitals in phase. See for example the shape of the lowest-energy π orbital in benzene shown on the opposite page.
- If n electron pairs in delocalized molecular orbitals bond $n+1$ atoms together (in CH_4 four electron pairs bond five atoms), n *localized* bonds can be made which give the correct electron density, albeit the wrong energy; see spread 22.12. When fewer than n electron pairs are involved, as is the case for the π electron pairs in benzene (three electron pairs bond six atoms), the localized picture fails to get *either* the electron density *or* the energy correct; see spread 23.1.

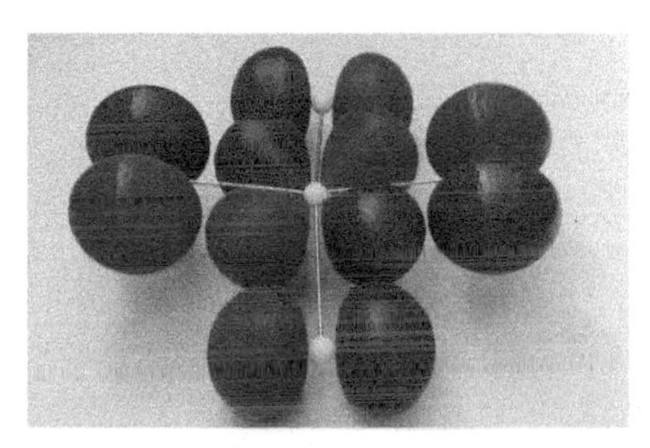

The two delocalized pi orbitals in sulphate ion. The unusually shaped d orbital (the one with a torus) is called d_{z^2}. The other d orbital has the usual four lobes (spread 4.5).

SUMMARY

- Delocalization is the extension of bonding electron density to cover more than two atomic centres.

PRACTICE

1 Look up spread 23.1 and then write an account of the delocalization in benzene.

σ / π delocalization

The concept of delocalization is well accepted for π orbitals. That σ **orbitals are also delocalized** is not well recognized, because of point 4 alongside. Except in spread 22.12 this idea will not be raised again.

6.5 Metallic solids: structures

OBJECTIVES

- Metallic crystal structures
- Simple cubic
- Body-centred cubic
- Hexagonal close packing
- Cubic close packing
- Coordination number

What are the actual structural arrangements in metals? What sorts of patterns do the ions adopt? Look at apples piled up on a market stall. A pile consists of flat layers in which the apples are as close as possible to each other. The apples in one layer fit into the hollows in the layer below. In this spread, we look at the four main structures that result when metal ions pack together.

Layers of metal ions

The simple model of a metallic crystal assumes that the ions are identical spheres which touch each other. It is helpful to think about the crystal lattice first in terms of the structure of each layer, and then in terms of how these layers fit together. There are two main ways in which spheres can be arranged in a single layer: a square pattern and a hexagonal pattern. The square pattern makes less efficient use of the space compared with the hexagonal pattern (i.e. there is a greater volume of empty space in the square pattern). Layers of spheres in either the square or the hexagonal pattern may then be stacked in different ways to make the four basic crystal structures: simple cubic, body-centred cubic, hexagonal close-packed, and cubic close-packed. To help us describe these structures, we shall give each layer a label, A, B, or C.

Packing of spheres in a layer: (a) square pattern where each sphere has four nearest neighbours; and (b) hexagonal pattern where each sphere has six nearest neighbours.

The unit cell

Crystals are made up from countless numbers of particles arranged in layers stacked together. The **unit cell** is the simplest pattern containing the appropriate particles that, when replicated, reproduces the overall arrangement of the lattice. Unit cells are usually represented by ball-and-stick models, which show the arrangement of the particles more clearly than space-filling models.

Simple cubic structure

This crystal structure results when square layers are placed exactly one on top of another. The layers repeat in the sequence AAA.... This means that the ions (spheres) in layers 2, 3, ... are in exactly the same positions as those in layer 1. This arrangement represents the least-efficient occupancy of space possible. The **coordination number** of an ion gives the number of its *nearest* neighbours. The coordination number for a simple cubic crystal structure is 6 because each ion has six nearest neighbours. The only metal with the simple cubic structure is polonium.

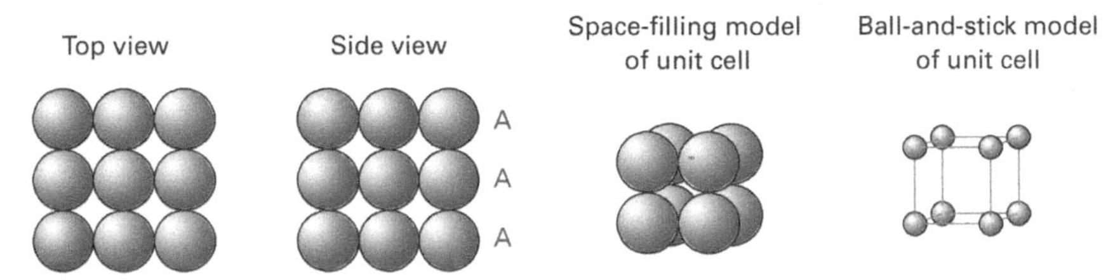

The simple cubic structure – showing the square layers stacked AAA....

Body-centred cubic structure

The ions in each square layer can be arranged to fit into the hollows between the ions in the layer immediately below. The square layers are then in the alternating sequence ABAB.... (This means that the ions in layers 1, 3, 5, ... are in exactly the same relative positions, but different from those in layers 2, 4, 6, ...) This is a more efficient use of space (68% of space is filled compared with only 52% for simple cubic). The coordination number is 8 and the unit cell is described as **body-centred cubic** (b.c.c.). Group I metals adopt the body-centred cubic structure, and their low densities reflect the relative openness of the b.c.c. structure, compared with the close packing described next.

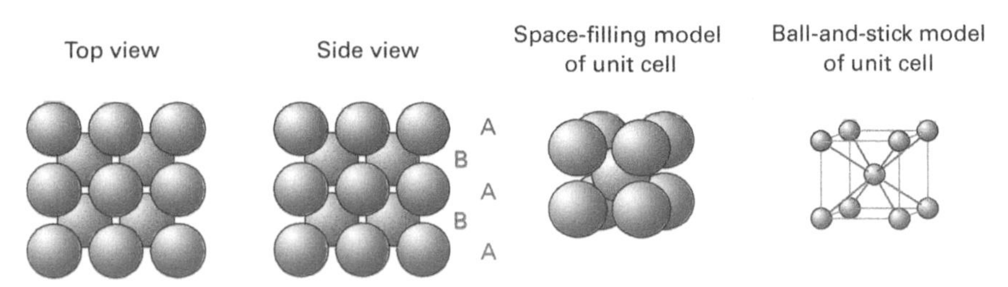

The body-centred cubic structure – showing the square layers stacked ABAB.... The layers have been coloured differently so that it is easier to see the arrangement, but all the spheres are identical.

Hexagonal close-packed structure

Hexagonal layers can stack one on another in the alternating sequence ABAB.... (Note that the labels A and B now refer to hexagonal layers.) The coordination number is 12, representing the most efficient use of space possible (74% of space is filled). The unit cell is called the **hexagonal close-packed** (h.c.p.) structure. Magnesium, zinc, and titanium are among the common metals to adopt the hexagonal close-packed structure in the solid state.

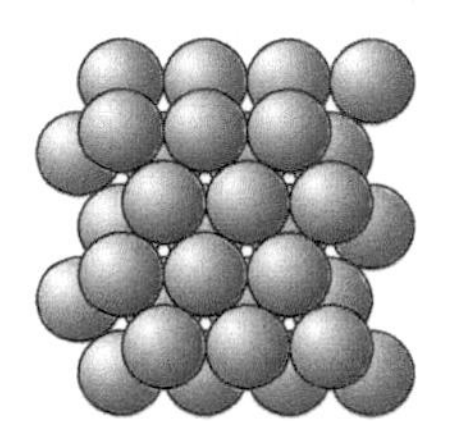
The ions in layer 2 lie in the hollows between the ions in layer 1

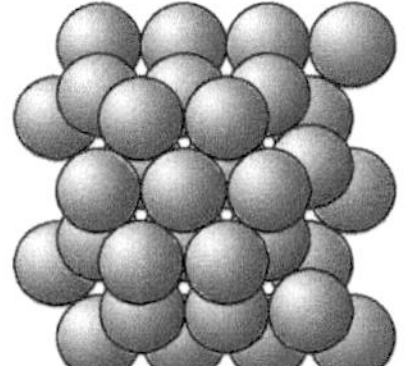
The ions in layer 3 lie directly above the ions in layer 1

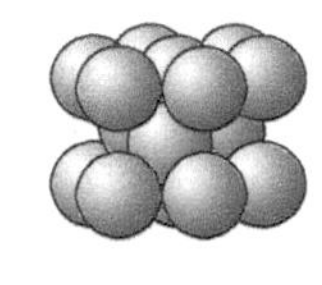
Space-filling model of unit cell

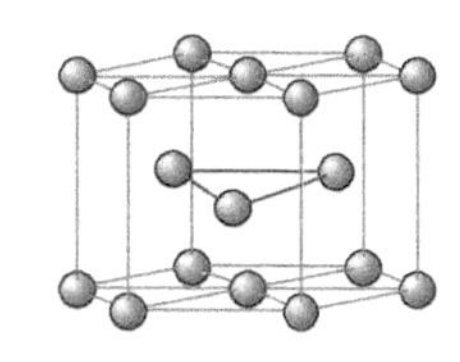
Ball-and-stick model of unit cell

The hexagonal close-packed structure – showing the hexagonal layers stacked ABAB.... The layers have been coloured differently so that it is easier to see the arrangement, but all the spheres are identical.

Cubic close-packed structure

There is another way in which hexagonal layers can stack together. Instead of layer 3 being the same as layer 1, it can fit over the hollows in *both* the first two layers. This structure results in the sequence ABCABC.... The occupancy of space is the same as in the hexagonal close-packed structure (74%) and the coordination number is also 12. This unit cell is called the **cubic close-packed** structure, but may also be described as **face-centred cubic** (f.c.c.). The easiest way to understand the alternative name is to view *at an angle* from the close-packed planes. There is a sphere at each corner of the cube and also spheres at the centre of each of the faces. Aluminium, copper, silver, and gold are among the common metals to adopt the cubic close-packed structure in the solid state.

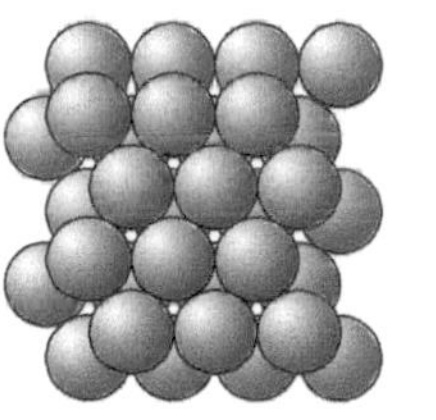
The ions in layer 2 lie in the hollows between the ions in layer 1

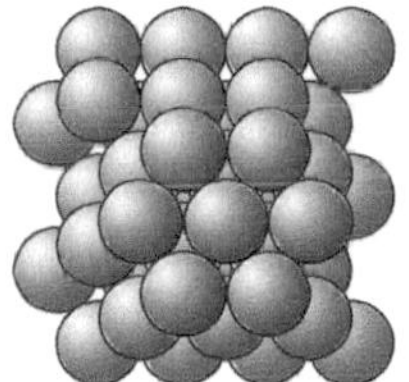
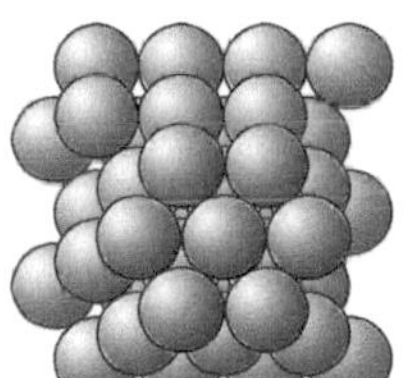
The ions in layer 3 lie above the hollows in layers 1 and 2

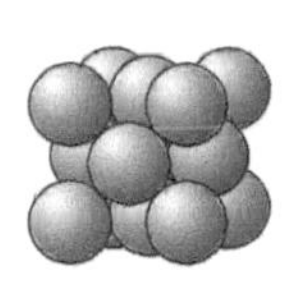
Space-filling model of unit cell

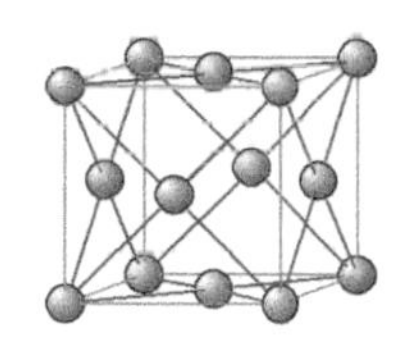
Ball-and-stick model of unit cell

The cubic close-packed structure – showing the hexagonal layers stacked ABCABC.... Again, the layers have been coloured differently so that it is easier to see the arrangement, but all the spheres are identical.

SUMMARY

- Square layers stack AAA... to give the simple cubic structure or ABAB... to give the body-centred cubic structure.
- Hexagonal layers stack ABAB... to give the hexagonal close-packed structure or ABCABC... to give the cubic close-packed structure.
- Cubic close packing may also be called face-centred cubic.

PRACTICE

1 Give the coordination number for each of the following lattice structures:
  - a Simple cubic
  - b Body-centred cubic
  - c Hexagonal close-packed
  - d Face-centred cubic.

2 a Which of the four structures in question 1 are close-packed?
  - b State what you understand by the term 'close packing'.
  - c Give examples of structures that are not close-packed and say what distinguishes them from those which are.

3 Draw the ball-and-stick unit cells for each of the structures in question 1.

6.6

OBJECTIVES

- Caesium chloride structure
- Rock-salt (sodium chloride) structure

IONIC SOLIDS: STRUCTURES

We have seen that metallic structures consist of *equal*-sized ions packed into crystalline lattices. There are four basic structures for arranging these metal ions: simple cubic, body-centred cubic, hexagonal close packing, and cubic close packing. So how do ionic substances, consisting of ions of generally *different* sizes, pack into regular lattices? As with metal ions in metallic structures, we can also think of the ions in ionic compounds as being spherical.

Ionic compounds

The crystals of ionic compounds are made up of ions that have opposite charges and generally different sizes. Packing these ions together presents a similar problem to packing oranges and grapefruit in the same box, with the additional complication that the positive and negative ions must be arranged to minimize repulsions and maximize attractions. The lattice adopted by ionic compounds depends largely on the relative numbers of ions and on their sizes. In this spread, we shall consider in detail two basic structures in which the ion ratio is 1:1.

The caesium chloride structure

The simplest structure found in ionic compounds is called the **caesium chloride structure**. In this structure, a simple cubic array of caesium ions interpenetrates a simple cubic array of chloride ions. It is not correct to refer to this structure as being body-centred cubic, because the ion at the centre of the unit cell has the *opposite* charge to the other ions. The arrangement of each constituent ion must be considered separately. There are eight caesium ions around each chloride ion, and eight chloride ions around each caesium ion. So the coordination number of each ion is 8.

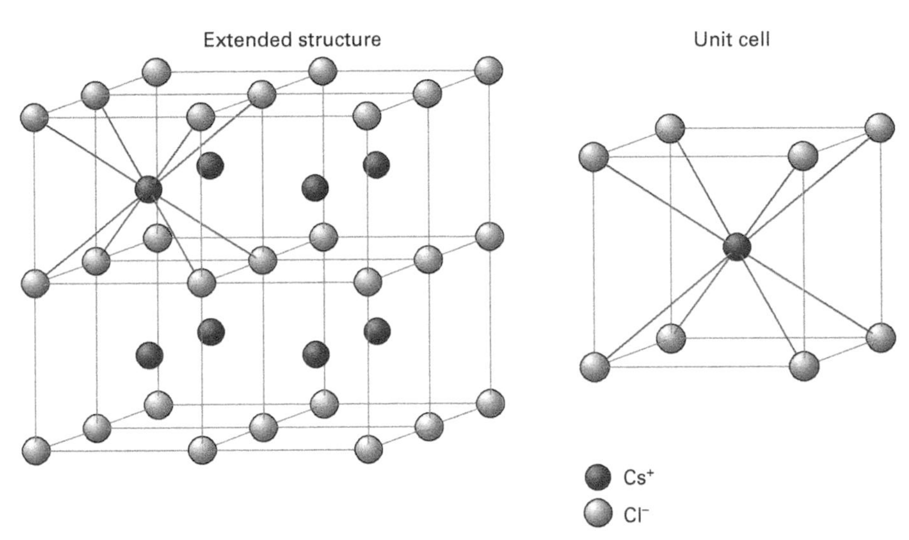

The caesium chloride structure.

Although the caesium chloride structure is the simplest way of packing ions of different sizes, it is not very common. It is found only in the chlorides, bromides, and iodides of caesium (Cs^+) and thallium(I) (Tl^+), where the positive and negative ions are of similar size. All the other alkali metal halides, including caesium fluoride, adopt the structure of sodium chloride, which we now consider.

The rock-salt (sodium chloride) structure

Sodium chloride is found naturally as rock salt. The crystal structure of sodium chloride is called the **rock-salt structure**. Each sodium ion is surrounded by six chloride ions. The coordination number of the sodium ion is 6. Similarly, each chloride ion is surrounded by six sodium ions. The coordination number of the chloride ion is also 6. Note that the sodium ion is significantly smaller than the chloride ion.

Look carefully at the diagram below. If we think just about the chloride ions in the rock-salt structure, we can see that they form a face-centred cubic array. The sodium ions similarly form a face-centred cubic array. The overall structure consists of two interpenetrating face-centred cubic arrays, one of sodium ions and one of chloride ions.

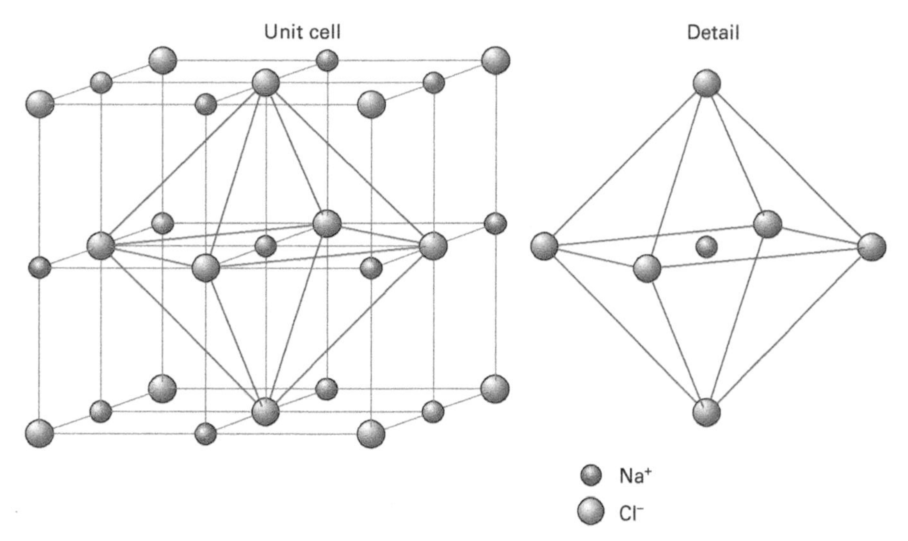

The rock-salt (sodium chloride) structure.

Compounds with the rock-salt (NaCl) structure

The rock-salt structure is by far the most common structure for compounds with a simple 1:1 ratio of ions. Examples include:

- seventeen Group I metal halides,
- silver chloride and silver bromide,
- magnesium oxide, calcium oxide, strontium oxide, and barium oxide,
- iron(II) oxide, cobalt(II) oxide, and nickel(II) oxide.

The structure is also adopted by compounds which show considerable covalent character, such as lead(II) sulphide PbS and titanium carbide TiC.

SUMMARY

- The lattice adopted by ionic compounds depends on the relative numbers of ions and on their sizes.
- Where the positive and negative ions are of approximately equal size, they may pack into the caesium chloride structure.
- The caesium chloride structure consists of two interpenetrating simple cubic arrays.
- The rock-salt (sodium chloride) structure consists of two interpenetrating face-centred cubic arrays.
- The rock-salt structure is by far the most important for compounds with a 1:1 ion ratio.

PRACTICE

1 Explain the meaning of both of the following sentences:

  a The coordination number of the sodium ion in solid sodium chloride is 6.

  b The coordination number of the chloride ion in the caesium chloride structure is 8.

2 Draw simple structures to represent the arrangement of the ions in the following:

  a Thallium bromide (TlBr)

  b Magnesium oxide (MgO).

OBJECTIVES

- Fluorite structure
- Zinc-blende structure
- Crystal systems

More about crystal structures

The caesium chloride and rock-salt structures described in the last spread are adopted by ionic compounds with a one-to-one ratio of ions. Compounds with a different ion ratio have different structures. Several examples related in simple ways to the rock-salt structure will be described in this spread.

The general structures met in any crystal will then be described. Every crystalline substance falls into one of the seven classes shown on the right-hand page.

The rock-salt structure revisited

The rock-salt structure described on the previous spread can be interpreted in another way. Imagine that the face-centred cubic (f.c.c.) array of chloride ions is expanded so that the chloride ions do not touch each other. The ions retain the geometrical arrangement of close packing, leaving holes in the structure. It can be seen that one type of hole has six spheres around it: this is called an **octahedral hole**. The rock-salt structure can therefore equally well be described as one in which *the chloride ions are in an f.c.c. array with all the octahedral holes filled with sodium ions.*

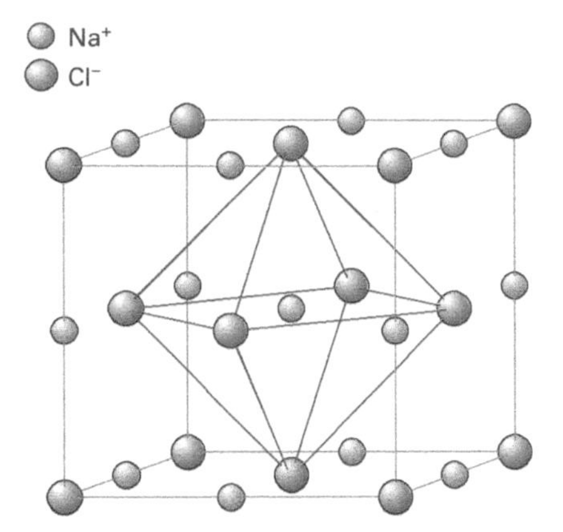

The rock-salt structure viewed as an f.c.c. array of chloride ions with all octahedral holes filled with sodium ions.

The fluorite structure

In addition to octahedral holes, there is one other type of hole: a **tetrahedral hole** has four spheres around it. There are twice as many tetrahedral holes as octahedral holes.

The most important structure adopted by compounds with a 2:1 ratio of ions is the **fluorite structure** (the mineral fluorite is calcium fluoride CaF_2, spread 6.2). In the fluorite structure, the positive (calcium) ions are in an f.c.c. array, with the negative (fluoride) ions in all the tetrahedral holes. The fluorite structure is adopted by, for example, the fluorides of calcium, strontium, and barium, and the oxide of uranium, UO_2. In the **anti-fluorite structure**, the *negative* ions are in the f.c.c. array, with the positive ions in all the tetrahedral holes. The anti-fluorite structure is adopted by, for example, the oxides of lithium, sodium, potassium, and rubidium.

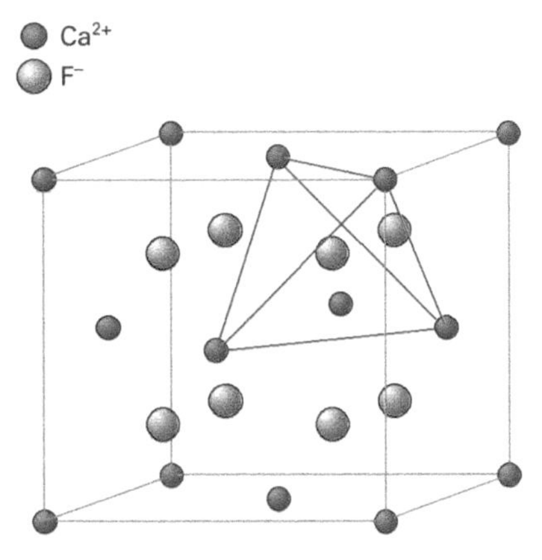

The fluorite structure.

The zinc-blende structure

Finally, another important structure can be created by filling only *half* of the tetrahedral holes. This gets us back to a 1:1 ion ratio. The mineral **zinc blende**, ZnS, gives its name to this structure. Note the close similarity with the diamond structure discussed in spread 6.3.

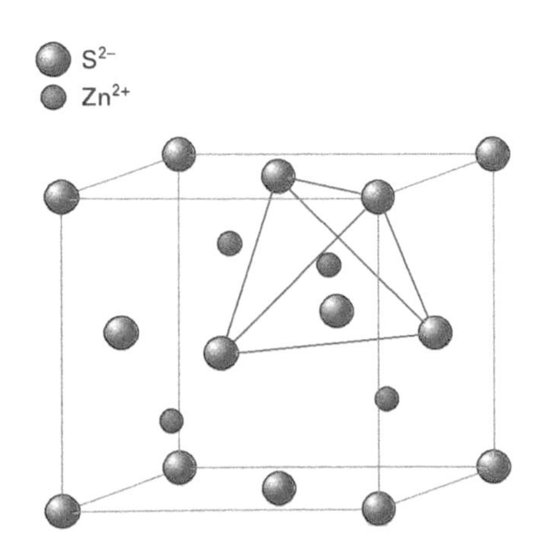

The zinc-blende structure.

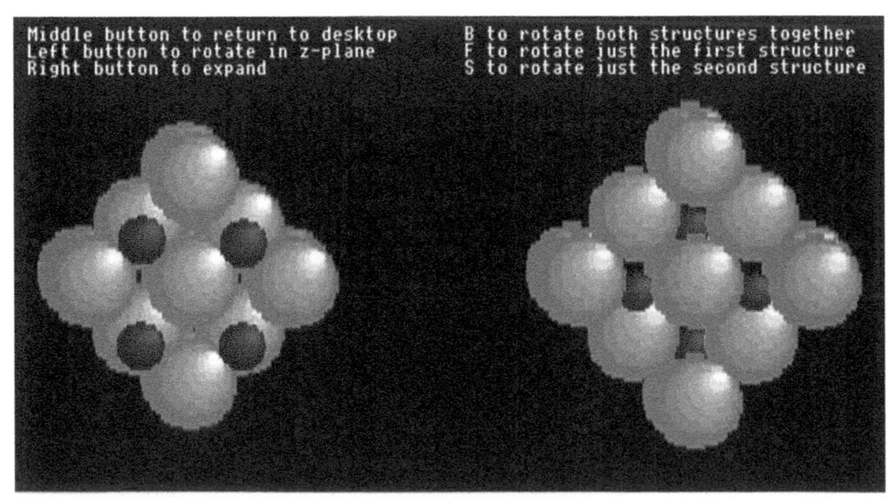

Computer programs such as ChemSoft are available which allow the user to display structures (sodium chloride (left) and zinc blende (right), for example), which can be expanded and rotated in three dimensions in real time. This is a great aid to understanding the structures better. See Appendix B.1 on the use of computers in chemistry.

The crystal systems

All substances crystallize in one of the following seven **crystal systems**. In the following three systems, the axes are at right angles:

- **cubic**, with three sides equal
- **tetragonal**, with two sides equal
- **orthorhombic**, with all sides unequal.

Tin is cubic at low temperatures ('grey tin'), transforming to a tetragonal structure ('white tin') at 13 °C; sulphur at room temperature is orthorhombic ('rhombic sulphur').

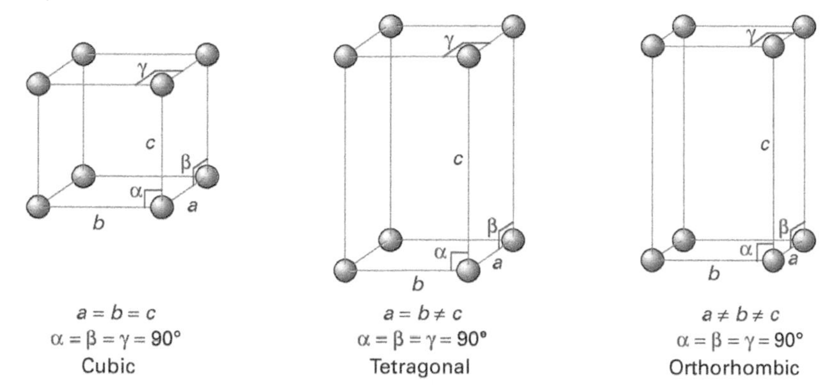

The cubic, tetragonal, and orthorhombic crystal systems have all axes at right angles.

The next set of three structures can be considered as distortions of each of the three structures above.

- **Rhombohedral** is formed from cubic by pulling it out along a diagonal line between opposite corners; the angles between the axes remain equal, but they are not right angles.
- **Hexagonal** is formed from tetragonal by pushing opposite corners of the square face until one angle is 60° and the other 120°.
- **Monoclinic** is formed from orthorhombic by pushing it over in one direction so one angle is no longer a right angle.

Calcite, a form of calcium carbonate ($CaCO_3$), is rhombohedral, as is obvious from the photograph in spread 9.3; quartz, a form of silica (SiO_2), is hexagonal; just below 100 °C, rhombic sulphur changes into monoclinic sulphur.

The least symmetrical structure of all is **triclinic**, which has no equal angles and no sides equal. The most common example of a substance with this structure is hydrated copper(II) sulphate $CuSO_4{\cdot}5H_2O$.

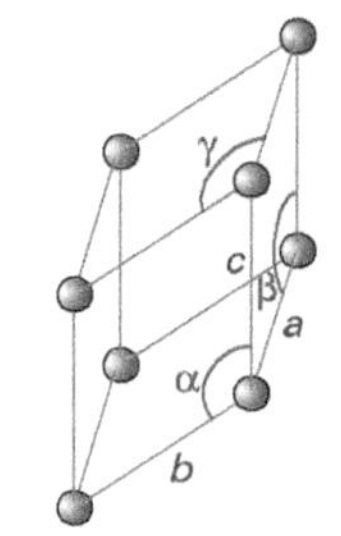

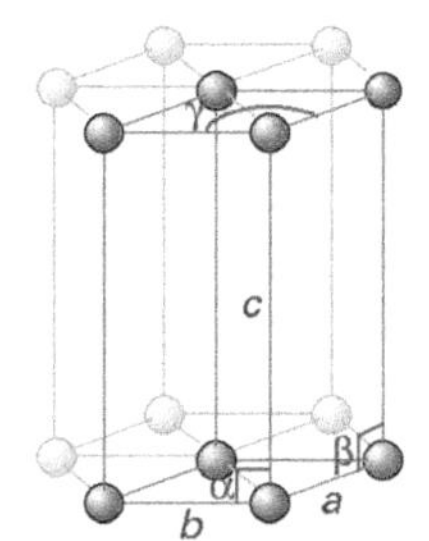

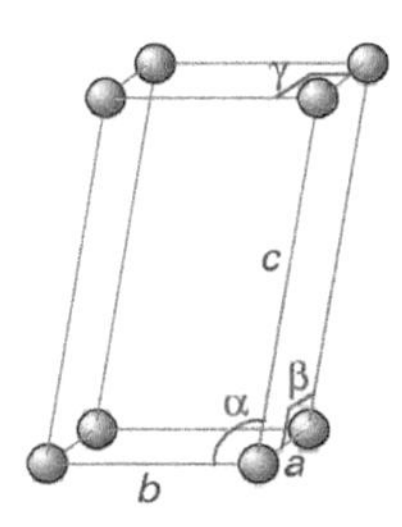

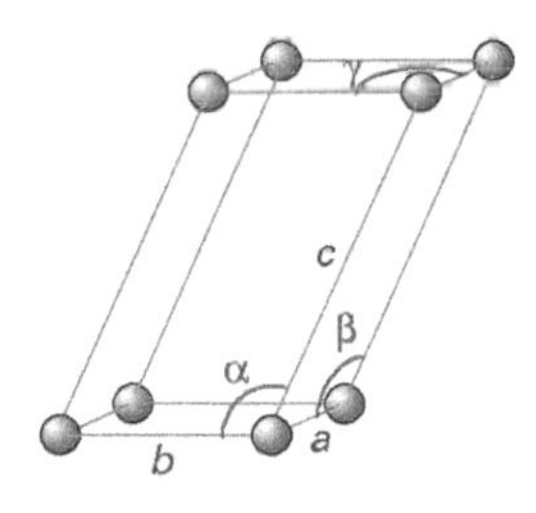

The rhombohedral, hexagonal, and monoclinic crystal systems have at least one angle that is not a right angle. The triclinic crystal system is the least symmetrical.

Centred structures

Some of the above crystal systems can be found in a few variants. For example the cubic system has three variants: simple cubic, body-centred cubic, and face-centred cubic.

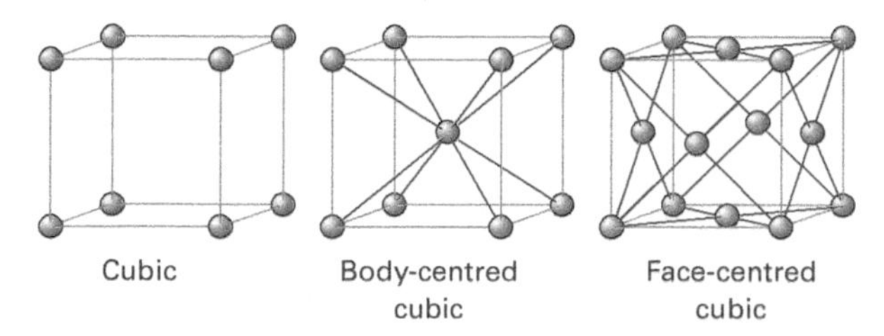

The cubic crystal system has three variants: simple cubic, body-centred cubic, and face-centred cubic.

Polonium is an example of an element which crystallizes as simple cubic, potassium as body-centred cubic, and platinum as face-centred cubic.

SUMMARY

- The most common structure for compounds with a 2:1 ion ratio is the fluorite (or anti-fluorite) structure.
- All crystals fall into one of the seven crystal systems.

Triclinic crystal of copper(II) sulphate.

PRACTICE EXAM QUESTIONS

1 a The table below gives some data for an element in the Periodic Table. Copy and complete the table for the other elements shown. [6]

Element	Electronic configuration	Block
Sodium	$1s^2\ 2s^22p^6\ 3s^1$	s
Copper		
Gallium		
Phosphorus		

b Using the axes in the figure below, sketch the graph obtained when all the electrons are successively removed from an aluminium atom. [3]

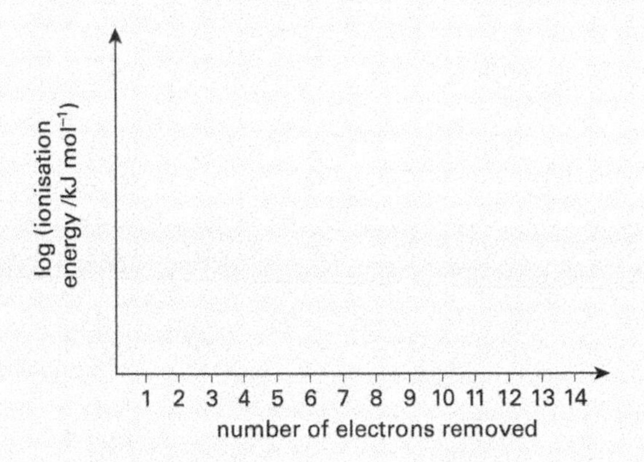

c The figure below shows the crystal structures of sodium and magnesium.

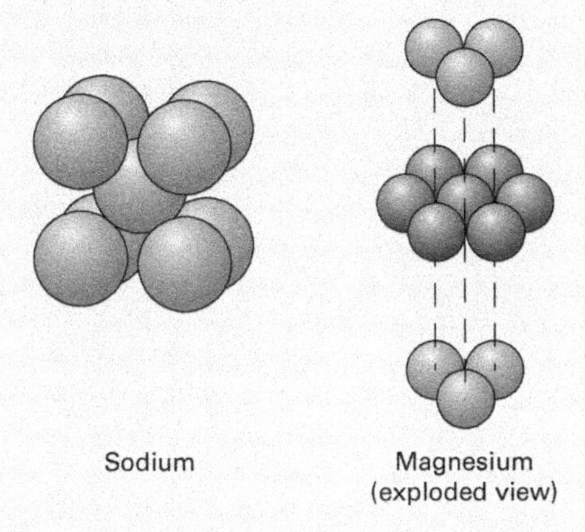

Give the name of the structure and state the coordination number of the atoms in each structure. [4]

2 a The diagram below represents part of a sodium chloride crystal with the position of one sodium ion shown by a plus (+) sign in a circle.

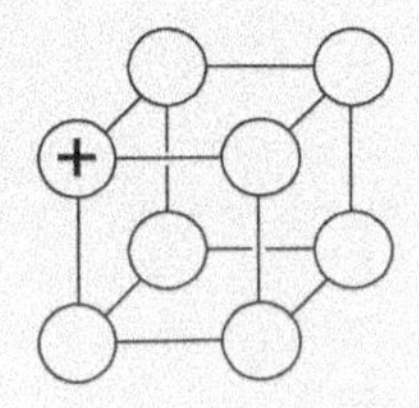

i Copy the diagram and mark with minus (–) signs all the circles in the above diagram which show the positions of chloride ions.

ii How many nearest sodium ions surround each chloride ion in a sodium chloride crystal? [2]

b Describe a simple test to show that sodium chloride is ionic. [2]

c A crystal of aluminium chloride vaporizes when heated to a relatively low temperature. In the gas phase, aluminium chloride exists as a mixture of $AlCl_3$ and Al_2Cl_6 molecules. The Al_2Cl_6 molecule is formed when $AlCl_3$ molecules are linked by two coordinate (dative covalent) bonds. The structure of Al_2Cl_6 is shown below with one of the coordinate bonds labelled.

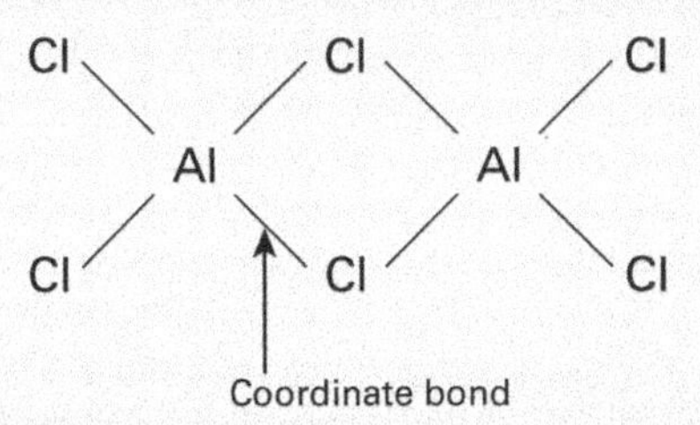

i Explain the meaning of the term *coordinate bond*.

ii Using an arrow, indicate on a copy of the diagram the other coordinate bond.

iii Explain briefly why solid aluminium chloride vaporizes at a relatively low temperature. [5]

3 a Sodium chloride is a crystalline solid, melting point 801 °C. It is soluble in water.

i State the type of bonding present in sodium chloride. [1]

ii Describe, in terms of the motion and arrangement of the particles, what happens when solid sodium chloride is steadily heated from room temperature until it melts. [2]

iii Account for the relatively high melting point of sodium chloride and explain why it dissolves in water at room temperature. [3]

b In carbon dioxide the carbon atom is joined to each oxygen atom by a double covalent bond, O=C=O. Each of the double bonds is made up of one σ bond and one π bond.

Explain, either in words or in clear diagrams, what is meant by a

i σ bond;

ii π bond. [2]

4 a Describe the bonding present in solid aluminium. Explain why aluminium is a conductor of electricity. [4]

b Aluminium combines readily with both dry fluorine and dry chlorine. Anhydrous aluminium chloride is a white solid which sublimes at about 200 °C; it reacts with water and dissolves in non-polar solvents.

Aluminium fluoride is a crystalline solid up to a temperature in excess of 1290 °C; it is insoluble in non-polar solvents.

i Suggest, using the information above, the name of the bond type present in:

anhydrous aluminium chloride;

anhydrous aluminium fluoride. [2]

ii Give an explanation for the difference in bond type present in the two anhydrous aluminium halides. [3]

iii Explain why the bonding in anhydrous aluminium fluoride leads to a high melting temperature. [1]

5 This question deals with bonding in molecules.

a Define the following terms, giving an example in each case:

i dative covalent bond.

ii ionic bond. [4]

b Both aluminium chloride and molecular iodine sublime.

i Define the term sublimation.

ii Explain why aluminium chloride dissolves readily in water, whereas molecular iodine does not. [3]

c Explain how metallic bonding accounts for the following characteristic properties of metals:

i high electrical conductivity,

ii high thermal conductivity. [2]

6 a Describe, with the aid of diagrams, the bonding and structure of:

i iodine; [3]

ii diamond [3]

b State and explain the effect, if any, of heat on separate samples of solid iodine and diamond. [4]

c Sodium chloride and caesium chloride have different crystal structures. Describe, with the aid of diagrams, the structure (stating the lattice type and coordination numbers) and bonding found in each solid. Explain why these salts have different crystal structures. [10]

d Copper metal has a high electrical conductivity. Using a diagram, show the bonding in copper metal and explain its high electrical conductivity. [5]

7 Electric cable is made of copper wire surrounded by poly(chloroethene) which is also called polyvinyl chloride, PVC. Another type of electric cable used in fire alarm systems has the copper wire surrounded by solid magnesium oxide, which acts as an insulator, the whole encased in a flexible copper tube, itself covered with PVC.

a Describe, with the aid of diagrams where appropriate,

i the bonding in copper, and hence explain how the metal conducts electricity;

ii the two types of bonding present in PVC, and hence explain why PVC acts as an insulator;

iii the bonding in magnesium oxide, giving the formulae of the particles, and hence explain why magnesium oxide does not conduct. [6]

b i Suggest **two** reasons why magnesium oxide is preferred to PVC as an insulator in fire alarm cabling.

ii Suggest why copper is used to encase the magnesium oxide. [2]

8 a Sodium chloride and caesium chloride have different ionic lattices.

i State what is meant by the term ionic lattice and briefly explain how such structures are held together. [3]

ii Draw diagrams to show the ionic arrangements found in caesium chloride and sodium chloride lattices; state the coordination number in each case. [5]

iii Use the information in the table below to explain why sodium chloride and caesium chloride have different crystal structures. [2]

Ion	Na^+	Cs^+	Cl^-
Ionic radius / nm	0.095	0.169	0.181

iv Give the name of a technique that can be used to determine crystal structure. [1]

b The reaction between phosphorus and fluorine produces a covalent compound of formula PF_5.

i Complete the electronic configuration of phosphorus. [1]

ii Draw a diagram to show the shape of the PF_5 molecule, state the bond angles present in the molecule, and give the name of this shape. [3]

iii Explain, in terms of the electron pairs present, why the molecule has this shape. [2]

7 Changes of state and intermolecular forces

The common states of matter are solid, liquid, and gas. The previous chapter discussed the structures of solids. This chapter will consider how structure affects the temperatures at which solids melt and liquids boil. **Melting** is the change of state from solid to liquid. **Boiling** is the change of state from liquid to gas. These changes happen for different substances over a wide range of temperatures. For example, ice cubes melt rapidly when taken from the freezer. Butter melts if left in the sun. On the other hand, have you ever seen table salt melt? Boiling water is familiar, but boiling salt is not. The reasons for these different behaviours lie in the forces between the particles that make up the solid.

States of matter

7.1

OBJECTIVES

- States of matter: solid, liquid, and gas
- Changes of state: melting, freezing, evaporation, boiling, condensation
- Arrangement of particles

Matter consists of particles, which are either atoms, molecules, or ions. For example, a sample of a noble gas (e.g. argon) consists of separate atoms; whereas carbon dioxide gas, liquid water, and solid sugar (sucrose) consist of molecules; and salt (sodium chloride) consists of ions. The forces between the particles that make up a substance determine whether it is a solid, a liquid, or a gas at a given temperature.

Solids

Solids are difficult to compress because the particles that make up a solid are very close to each other. Solids have fixed shapes because attractive forces bind the particles together in a fixed pattern. The particles vibrate about their fixed positions, and the magnitude of the vibrations becomes greater as the temperature increases.

(a)

(b)

(c)

(a) A solid has a fixed shape. (b) A liquid adopts the shape of the lowest part of its container. (c) A gas fills the whole of its container.

The temperature of this mixture of ice and water remains at 0 °C while both ice and water are still present.

Take ice out of the freezer and measure its temperature while it warms up. The temperature of the solid gradually increases as it takes in energy from its surroundings. Once the ice begins to melt, its temperature remains constant until melting is complete. This constant temperature is called the **melting point** (T_m). The energy the solid substance takes in during melting is needed to overcome the forces holding the particles together. It does not raise the temperature of the substance. During melting, the structure of the solid breaks down and the particles become free to move relative to each other. Once melting is complete, the energy supplied causes the temperature to rise again.

Liquids

Liquids can flow because the particles in a liquid can move past one another. However, the forces of attraction between the particles are strong enough to hold a liquid together in one place. A liquid takes up the shape of the vessel that holds it. It may also be poured from one container into another. The volume of a liquid is constant at a particular temperature, and its density usually decreases with increasing temperature. The particles are almost as close together as in the solid state, which makes liquids very difficult to compress.

Some US farmers spray crops with water during unseasonable frosts. While both liquid water and solid ice are present, the temperature of the mixture cannot fall below 0 °C, the freezing point of water. The crops do not suffer frost damage because the temperature of the ice- and water-coated plants does not fall below 0 °C.

As a liquid is heated, the total energy of the sample increases. This energy mostly takes the form of kinetic energy as the particles move from place to place. Random collisions cause the energy of each particle to change continually. As a result, some particles at the surface gain sufficient energy to overcome the forces holding them within the liquid. Some of the particles at the liquid surface enter the space above the liquid and become a gas. This change of state is called **evaporation**. Evaporation occurs at the surface of a liquid at any temperature, below the boiling point.

Gases

Gases can be compressed easily because the particles in a gas are widely separated. Gases have very low densities compared with liquids and solids. They expand to fill the space available uniformly. The total energy of the particles is sufficiently high to overcome completely the forces of attraction that hold them together in the liquid state.

The change of state from gas to liquid is called **condensation**. It occurs at temperatures below the boiling point when gas particles collide with insufficient energy to rebound from each other. Particles coalesce and droplets of liquid form.

SUMMARY

- Solids and liquids are extremely difficult to compress; gases can be compressed easily.
- The particles in solids and liquids are close to each other; those in gases are relatively far apart.
- Solids have fixed shapes; in crystalline solids, the particles are fixed in regular patterns.
- Liquids are mobile; they stay at the bottom of their container.
- Gases are mobile and uniformly fill their container.

Freezing

The opposite of melting is **freezing**, when a liquid changes into a solid. The temperature of a liquid stays constant as it freezes.

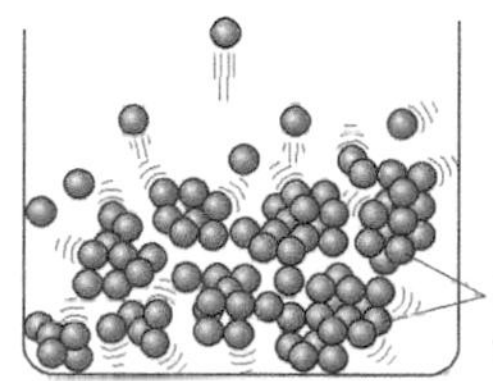

Evaporation occurs when particles in the liquid surface have energy greater than the forces of attraction between the particles.

Boiling

Boiling occurs when bubbles of vapour form *within the body of the liquid* (not just at the surface). A pure liquid boils at a fixed temperature called the **boiling point** (T_b). Energy supplied during boiling is needed to overcome the forces of attraction between the particles in the liquid state.

PRACTICE

1 Name each of the following state changes:
 a Solid to liquid
 b Liquid to solid
 c Liquid to gas, at any temperature
 d Liquid to gas, at a fixed, maximum temperature
 e Gas to liquid.

7.2

OBJECTIVES

- How structure affects melting and boiling points
- Atomic elements have very low melting and boiling points
- Metallic elements generally have high melting and boiling points
- Simple molecular substances have low melting and boiling points
- Giant covalent solids have very high melting and boiling points
- Ionic compounds have high melting and boiling points

Changes of state and the forces between particles

Elements in the solid state have one of four main structures: atomic, metallic, simple molecular, or giant covalent. Compounds in the solid state have one of three main structures: simple molecular, giant covalent, or ionic. When solids melt, these structures break down. The temperature at which melting takes place depends on the forces between the particles that make up the solid.

Atomic elements

The noble gases helium to radon in Group VIII are the only elements that exist in the solid state as separate atoms arranged in an ordered lattice. The force of attraction between the atoms is very weak. As a consequence, the melting points of these elements are extremely low (e.g. solid argon, $T_m = -189\,°C$).

(a)

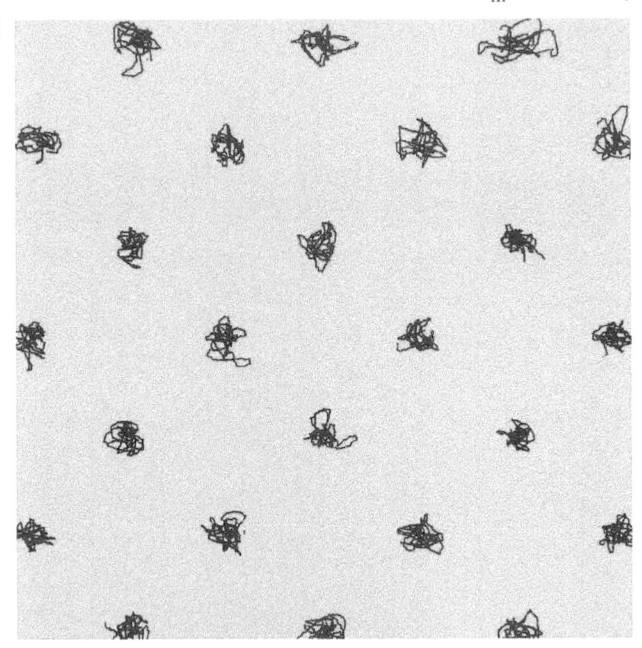

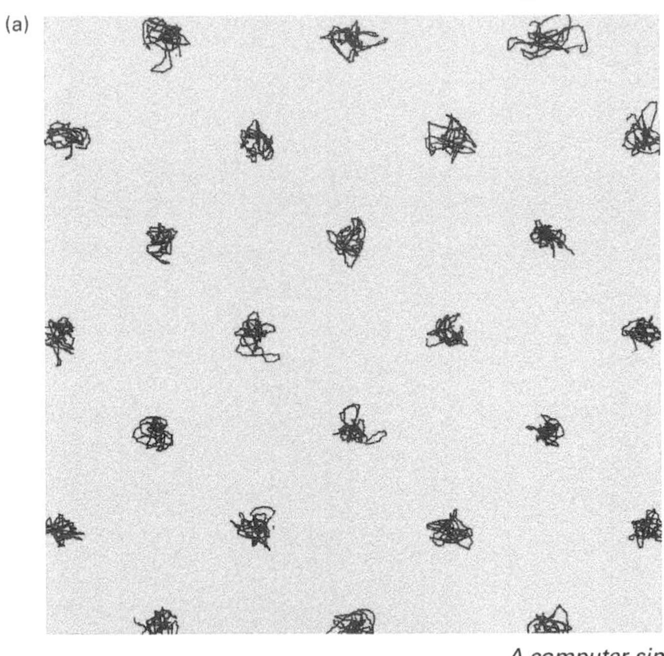

(b)

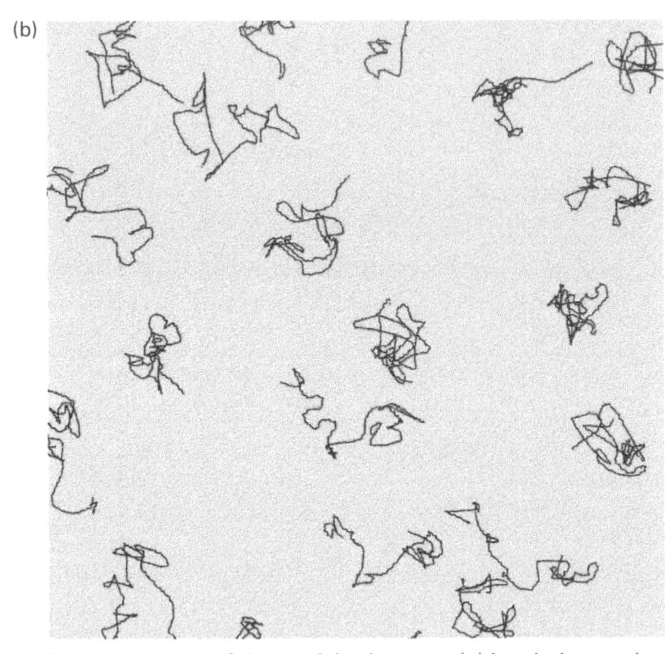

A computer simulation of the movement of the particles in argon (a) just below and (b) just above its melting point. Solid argon has the f.c.c. structure, spread 6.5.

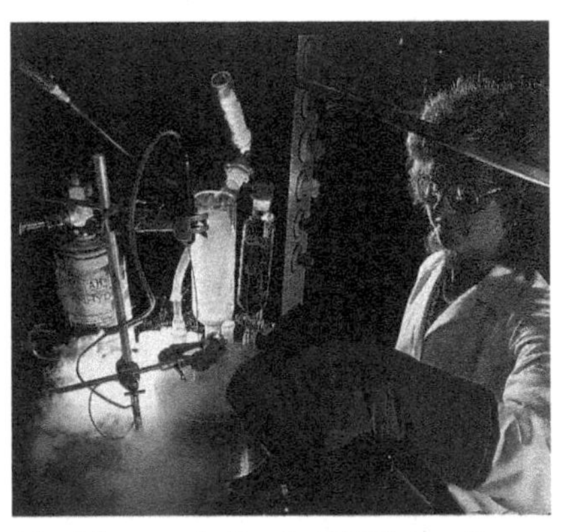

Liquid argon boils at −186 °C, less than 100 °C above absolute zero (−273 °C). It vaporizes rapidly at room temperature when outside its insulating flask.

The forces between atoms of the noble gases are also weak in the liquid state, with the result that their boiling points are less than 10 °C above the melting points (e.g. liquid argon, $T_b = -186\,°C$).

Metallic elements

Metallic elements consist of a regular lattice of positively charged metal ions bonded together by a 'sea' of delocalized valence electrons. The bonding forces are strong and their effects are felt uniformly throughout the lattice. The melting points of the first three metals in Period 3 increase as follows: sodium, $T_m = 98\,°C$; magnesium, $T_m = 649\,°C$; and aluminium, $T_m = 660\,°C$. In general, Group I metals melt below 200 °C and Group II and the p-block metals usually melt below 1000 °C. These differences are due mainly to:

- the number of electrons delocalized from each atom;
- the size of the ions and their charge.

The small, highly charged Al^{3+} ions are held together strongly by three delocalized electrons per ion. The larger Na^{+} ions are only held together by one electron per ion, so sodium melts at a lower temperature.

Metals often boil at very high temperatures (e.g. iron, $T_b = 2750\,°C$). A metal atom must regain control over its required number of valence electrons before it is able to move into the gas phase as an isolated and independent atom.

Simple molecular elements and compounds

Simple molecular solids consist of covalently bonded molecules held together by weak **intermolecular forces** (the forces *between* molecules). Simple molecular solids melt when sufficient energy is provided to overcome these weak forces. The covalent bonds between the atoms in the molecules remain intact. Simple molecular substances therefore have relatively low melting points (e.g. oxygen O_2, T_m = –219 °C; methane CH_4, T_m = –183 °C; iodine I_2, T_m = 114 °C; sulphur S_8, T_m = 115 °C).

Simple molecular liquids contain molecules whose structure is similar to that in the solid state. The same weak intermolecular forces operate but the molecules have sufficient energy to move past each other. Simple molecular liquids boil at relatively low temperatures (e.g. O_2, T_b = –183 °C; CH_4, T_b = –161 °C).

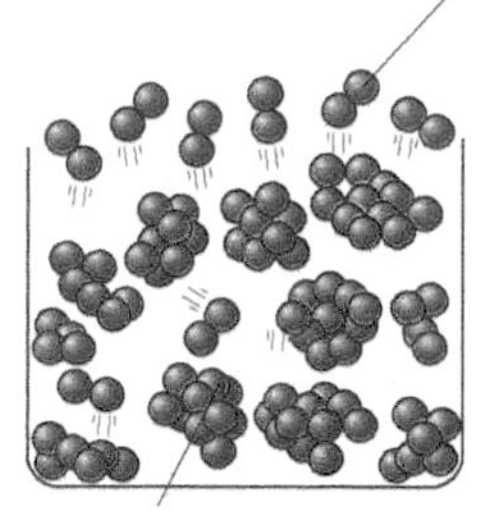

This diagram represents liquid oxygen O_2 boiling to form oxygen gas. Note that the covalent bonds between the oxygen atoms do not break.

Giant covalent elements and compounds

The structures of these substances consist of atoms covalently bonded to each other. The strong covalent bonds operate *throughout* the structure, making a sample of the substance effectively one enormous single molecule. Melting a giant covalent solid involves breaking these strong bonds. The liquid consists of separate free atoms. Melting points are therefore extremely high, e.g. diamond (elemental carbon) T_m = 3550 °C. Boron nitride BN (a compound with the diamond structure) breaks down above 3000 °C to liberate atoms in the gas phase.

Ionic compounds

Ionic compounds consist of oppositely charged ions held together by strong electrostatic forces between the ions. These strong forces operate throughout the structure. The process of melting an ionic solid requires sufficient energy to break the strong bonds holding each ion to its neighbours. The melting points of ionic compounds are therefore generally high (e.g. sodium chloride NaCl, T_m = 801 °C).

In the liquid state, ionic compounds consist of separate ions which are able to move past each other. Strong electrostatic forces operate between the ions, as in the solid state. The ions must achieve high energies before they can break free from each other and become a gas. Boiling points are therefore high (e.g. NaCl, T_b = 1413 °C).

Sublimation

The change of state directly from solid to gas is called **sublimation**. The substance does not pass through the liquid state during sublimation. Carbon dioxide is a simple molecular substance that sublimes when heated, spread 7.7.

The sublimation of carbon dioxide CO_2. Solid carbon dioxide changes directly from solid to gas on warming up; 'dry ice' is used to produce smoke in this display at EuroDisney featuring the caterpillar from Alice in Wonderland.

SUMMARY

- Substances boil when their constituent particles have sufficient energy to break free from the forces holding the particles together.
- Melting and boiling points generally follow the sequence: atomic elements < simple molecular substances << metallic elements ≈ ionic compounds < giant covalent substances.

PRACTICE

1 The melting point of iron is 1536 °C and its boiling point is 2750 °C. Draw diagrams and write a description to explain the changes that take place at an atomic level when iron is heated from room temperature to 1600 °C.

2 Repeat the exercise outlined in question 1 for the following substances and temperature ranges:

a Argon (T_m = –189 °C, T_b = –186 °C) from –188 °C to –184 °C.

b Silicon (diamond structure: T_m = 1410 °C, T_b = 2355 °C) from room temperature to 1500 °C.

c Sodium chloride (T_m = 801 °C, T_b = 1413 °C) from 1200 °C to 1500 °C.

7.3

O B J E C T I V E S

- Bond polarity and polar molecules
- The effects of shape
- Dipole-dipole forces

	Dipole/D		Dipole/D
He	0	Ar	0
H_2	0	N_2	0
O_2	0	CH_4	0
CCl_4	0	CO_2	0
HF	1.91	HCl	1.08
HBr	0.80	HI	0.42
H_2O	1.85	NH_3	1.47
CH_3Cl	1.87	CH_3OH	1.71

The dipoles of some atoms and molecules in debyes (D). (One debye is 3.34×10^{-30} coulomb metre.)

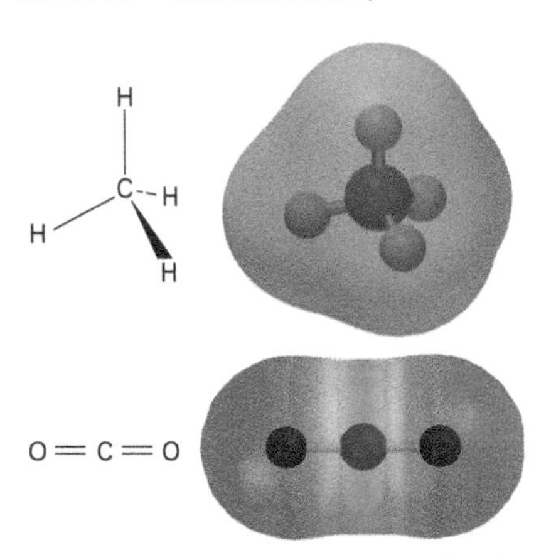

The individual bonds in these molecules are polar. However, the molecules are symmetrical so the overall effects of the bonds cancel out and the molecules are non-polar. They have a dipole of zero.

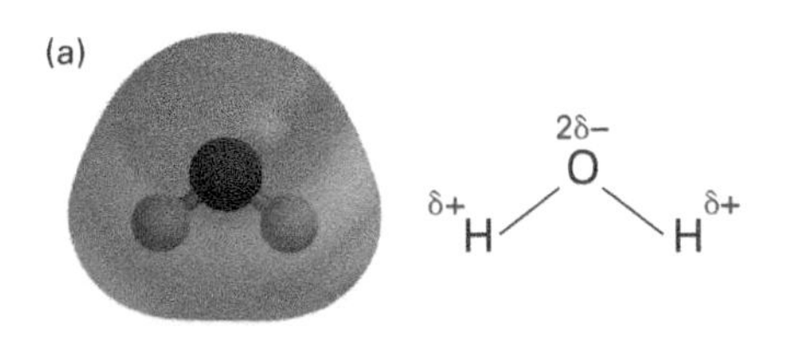

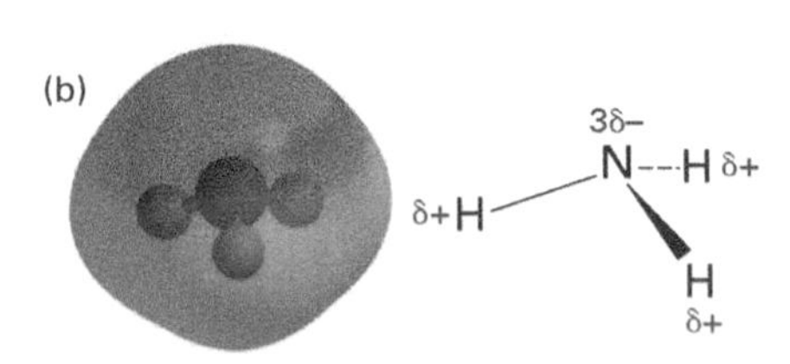

The individual bonds in these molecules are polar. The molecules are non-symmetrical and so they are both polar molecules.

Polar molecules and dipole–dipole forces

Covalent bonds are the result of atoms sharing electron pairs. When an electron pair is shared in a bond between atoms of *different* elements, the sharing will be *unequal*. The bond will be polar and the atoms concerned will bear partial charges. For a diatomic molecule X—Y, the polarity of the bond $X^{\delta+}$—$Y^{\delta-}$ makes the molecule as a whole polar. Forces of attraction exist between polar molecules because of the electrostatic forces of attraction between the partial charges δ+ and δ– on separate molecules. This spread will continue the discussion started in the previous two chapters by investigating molecules containing polar bonds.

Dipoles

A polar molecule has a *dipole*. This term refers to the separation of charge in a molecule, causing it to behave as a pair of point charges of opposite sign separated from each other, spread 5.9. A non-symmetrical molecule that has polar bonds has a **permanent dipole**. The intermolecular forces between molecules that have permanent dipoles are called **dipole–dipole forces**.

Symmetrical polyatomic molecules

Most **polyatomic molecules** (molecules containing three or more atoms) contain polar bonds. However, the polarity of a molecule *as a whole* depends on its shape. A molecule is said to be **polar** if its overall charge distribution is equivalent to a pair of separated opposite charges.

Tetrachloromethane CCl_4 contains four polar $C^{\delta+}$—$Cl^{\delta-}$ bonds. However, the bonds are arranged tetrahedrally and the molecule is symmetrical. The four partial negative charges of the chlorine atoms are the same distance from the partial positive charge of the carbon atom. Tetrachloromethane is a non-polar molecule. There are no intermolecular dipole–dipole forces as the molecule has zero dipole.

Similarly a linear carbon dioxide molecule CO_2 has no dipole–dipole forces, despite the strong polarity of each C=O bond.

Non-symmetrical polyatomic molecules

Water H_2O and ammonia NH_3 are two familiar examples of non-symmetrical polyatomic molecules. The molecules are non-symmetrical because of the presence of lone pairs, two in the water molecule and one in ammonia.

Each oxygen–hydrogen bond in water is polarized $O^{\delta-}$—$H^{\delta+}$. The two lone pairs and the two bonding pairs take on the overall shape of a distorted tetrahedron. The two O—H bonds are thus arranged in a V-shape. As a result, the negative region of charge due to the oxygen atom is concentrated at one end of the molecule (at the point of the 'V'), spread 5.9. The positive region of charge due to the two hydrogen atoms is concentrated at the other end of the molecule. So the water molecule is a polar molecule.

In ammonia, each nitrogen–hydrogen bond is polarized $N^{\delta-}$—$H^{\delta+}$. The single lone pair and the three bonding pairs take on the overall shape of a distorted tetrahedron. The three N—H bonds are thus arranged in a pyramidal shape. The negative region of charge due to the nitrogen atom is concentrated at the apex of the pyramid and the positive region of charge due to the three hydrogen atoms is concentrated at the base, spread 5.9. So the ammonia molecule is a polar molecule.

Polar molecules and boiling points

Boiling involves the change from liquid (l) to gas (g). The molecules in the liquid state become widely separated in the gaseous state as intermolecular forces are overcome.

The presence of dipole–dipole forces can have a significant effect on the boiling point of a substance. For example, compare the boiling points of hydrogen chloride (–85 °C) and argon (–186 °C). The HCl molecule has exactly the same number of electrons as the Ar atom. But in the HCl molecule the electron density has been 'stretched out' along the direction of the bond, creating a molecule with a dipole. When HCl boils, the dipole–dipole forces *between* the HCl molecules must be overcome. There is no dipole in the Ar atom, so no such forces exist in argon. As a result of the dipole–dipole forces, HCl has the higher boiling point.

Dipole
$H^{\delta+}$ $Cl^{\delta-}$
Dipole–dipole intermolecular force

Dipole–dipole intermolecular forces between hydrogen chloride molecules.

Hydrogen halides revisited

Intermolecular forces are not simply restricted to dipole–dipole forces. The boiling points of the hydrogen halides reveal that there are *two other* factors at work.

First of all, the boiling point of hydrogen fluoride is anomalously high. To fit in with the overall trend in the group, the value should be in the region of –100 °C. Instead of having the lowest value, it actually has the highest value. The anomalously high boiling point of hydrogen fluoride is the result of a special variety of dipole–dipole force called *hydrogen bonding*. This effect will form the focus of the next-but-one spread.

The second consideration refers directly to the magnitude of the dipoles. The dipoles of the hydrogen halides decrease in the order HF > HCl > HBr > HI, as shown in the table opposite. The size of the dipole depends partly on the electronegativities of the halogens, which decrease in the sequence F (4.0) > Cl (3.2) > Br (3.0) > I (2.7). Putting the anomalous hydrogen fluoride to one side for the moment, the magnitudes of the dipoles indicate that dipole–dipole forces should decrease in the order HCl > HBr > HI, with the result that the boiling points should decrease in the same order. However, the boiling points *increase* in this order, i.e. HCl < HBr < HI. There must be another, more important,intermolecular force to consider: this is the *dispersion force*, which is discussed in the next spread.

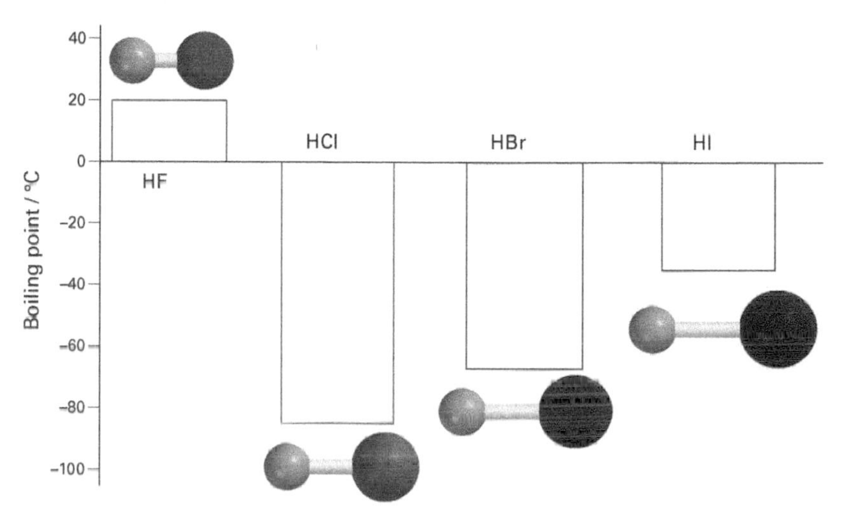

Hydrogen halide boiling points.

SUMMARY

- Polarity in molecules results from unequal sharing of electron pairs in covalent bonds.
- Unequal sharing of electron pairs results from differing electronegativity values.
- If dipole–dipole forces are the dominant intermolecular forces, boiling points increase with increasing polarity.

PRACTICE

1 Explain why tetrachloromethane CCl_4 is a non-polar molecule whereas trichloromethane $CHCl_3$ has a permanent dipole.

2 Explain why water H_2O is a more polar molecule than hydrogen sulphide H_2S (electronegativities: H, 2.2; O, 3.4; S, 2.6).

3 Explain the relationship between the terms 'polar bond' and 'polar molecule'.

7.4

Dispersion forces

OBJECTIVES

- Origin of dispersion forces
- Polarizability
- Significance in noble gases and simple molecular substances

Strengths of forces

The strength of a force or bond is expressed as the energy per mole required to break the bond. Strengths have the following typical values:

Single covalent bonds
150–550 kJ mol^{-1}

Dipole–dipole forces
0–5 kJ mol^{-1}

Dispersion forces
1–15 kJ mol^{-1}

Two specific examples (values in kJ mol^{-1}) are as follows:

	Dipole–dipole	Dispersion
HCl	0.2	2
NH_3	6	13

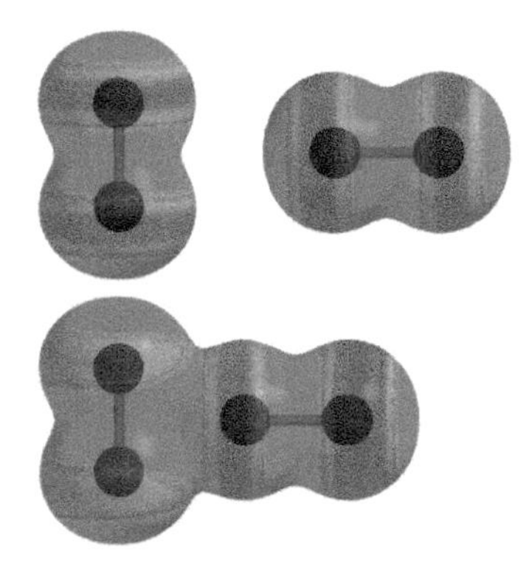

Electrostatic potential map of two chlorine molecules as they come closer to each other. The closer they approach the more polarized the molecule becomes.

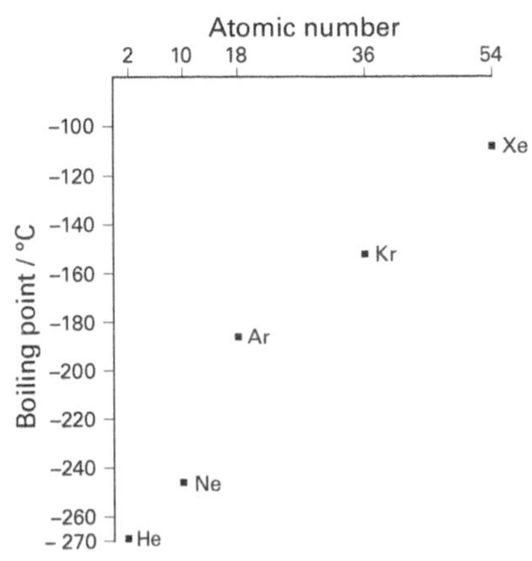

Plot of boiling point against atomic number for the noble gases. The forces between the atoms are dispersion forces, which increase as the number of electrons increase with increasing atomic number.

Dispersion forces are forces of attraction that operate between atoms and between molecules. They are weak, less than 5% of the strength of a covalent bond. Dispersion forces result from an instantaneous uneven distribution of electron density within atoms and molecules. The focus of this spread is to investigate the origin of dispersion forces, and to discuss their strengths relative to other bonding forces and their influence on the physical properties of various substances.

Origin of dispersion forces

On average, the electron density in a non-polar molecule (or an individual atom) is evenly distributed. At any one instant, however, the distribution may not be even and an **instantaneous dipole** will result. This *instantaneous* dipole will then cause the electrons in a neighbouring molecule to arrange themselves so that the force is attractive. An instantaneous dipole of this sort is also called an **induced dipole**. Dipoles flicker in and out of existence, inducing dipoles in other molecules in their vicinity. All these dipoles adjust to maintain an attractive force between the molecules by attracting or repelling electron density in the adjacent molecules.

The force between the instantaneous dipole of one molecule and that of another is a type of intermolecular force called the **dispersion force** (or **London force**, after Fritz London). Dispersion forces exist between molecules with or without permanent dipoles, between ions, and between the single atoms of noble gases. Their influence is generally ignored in metals, in ionically bonded compounds, and in giant covalent solids, because other forces of attraction between the particles are much stronger. But dispersion forces are a significant feature to consider in molecules where permanent dipoles result in dipole–dipole forces.

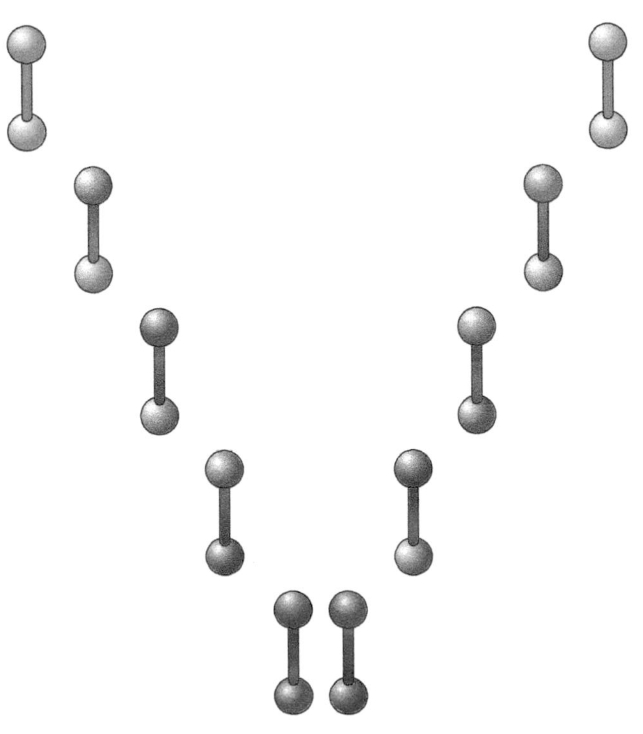

A schematic diagram explaining the origin of dispersion forces. When two non-polar molecules (such as two chlorine molecules) approach each other, instantaneous dipoles are produced that get into step. *The opposite partial charges on the molecules then cause attraction. As the molecules approach closer, the instantaneous dipoles become larger in magnitude (shown by the greater intensity of colouring). The force is always an attraction but the direction of the dipole on an individual molecule is random; hence the diagram shows the direction switching twice.*

Dispersion forces increase with (and are directly proportional to) increasing polarizability of the molecule concerned. **Polarizability** measures how easily the electron density is distorted when subjected to an external electric field. It depends on a number of factors; for example, polarizability generally increases with the surface area of the molecule and with the number of electrons present.

Dispersion forces between atoms

The elements of Group VIII are called the noble gases. They have full shells and exist in all states (solid, liquid, or gas) as separate individual atoms. Dispersion forces are the only forces of attraction which operate between these atoms. As a result, melting and boiling points are extremely low, reflecting the weakness of these forces. However, the values of the melting and boiling points increase with increasing atomic number (see opposite). The atomic number indicates the number of electrons present and is hence a rough measure of the polarizability of the atom concerned.

Dispersion forces between molecules

The elements of Group VII (the halogens) exist as diatomic molecules, denoted generally as X_2. These molecules do not have permanent dipoles because each molecule consists of two atoms of the same element. There is a clear trend of increasing melting and boiling points with increasing atomic number (and hence increasing numbers of electrons). Melting and boiling points are higher overall than those of the noble gases. For example, compare chlorine Cl_2 with the noble gas krypton Kr. They have almost equal numbers of electrons (Cl_2, 34; Kr, 36), but the melting and boiling points for chlorine are a good deal higher than those for krypton. The reason for this difference is that the chlorine molecule consists of two atoms with a covalent bond between them. The electron density in chlorine is more spread out than that in the krypton atom and chlorine is therefore more polarizable.

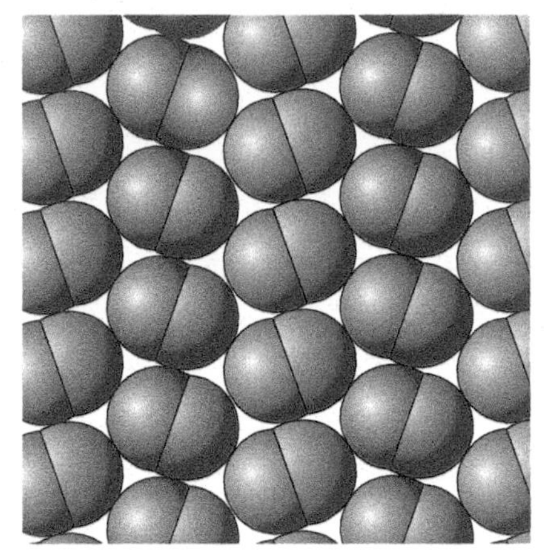

The structure of solid iodine. The molecules are held together by dispersion forces.

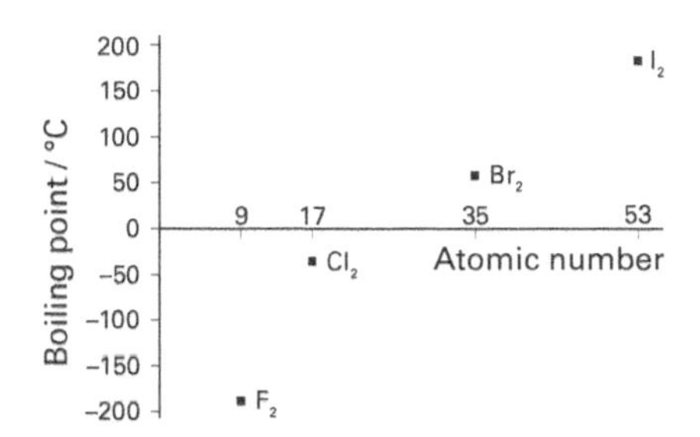

Plot of boiling point against atomic number for the halogens.

Dispersion forces and molecular size

You should remember from your earlier studies that the alkanes have the general formula C_nH_{2n+2} and consist of chains of n carbon atoms with hydrogen atoms attached along the sides. The illustration alongside shows the structure of the alkane pentane C_5H_{12}. Data for the boiling points of the alkanes with one to eight carbon atoms are also given. Notice that, as the number of carbon atoms increases, so does the total number of electrons present in the molecule. The C—H bonds have low polarity because the electronegativities of carbon (2.5) and hydrogen (2.2) are very similar. As a result, dispersion forces are the only significant intermolecular forces in alkanes. The total number of electrons increases with increasing numbers of carbon and hydrogen atoms; boiling points therefore also increase.

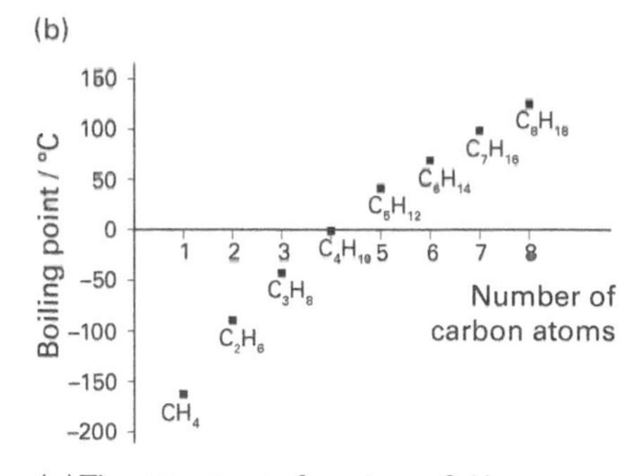

(a) The structure of pentane C_5H_{12}.
(b) Boiling point data for the alkanes CH_4 to C_8H_{18}. These differences in boiling point allow mixtures of alkanes present in crude oil to be separated by fractional distillation, see spread 22.1.

SUMMARY

- Dispersion forces are significant when considering the melting and boiling points of the noble gases and of both polar and non-polar molecules.
- Dispersion forces are weak, less than 5% of the strength of a covalent bond.
- The magnitude of dispersion forces increases with the polarizability of the molecule, which generally increases with the number of electrons.

PRACTICE

1 Describe the factors that influence the magnitude of dispersion forces.

2 Explain why dispersion forces are not taken into account when describing the bonding in a giant covalent solid such as diamond.

3 Explain why dispersion forces are a significant feature in maintaining the structure of graphite.

4 State the total number of electrons in each of bromine Br_2 and iodine monochloride ICl.

Which substance would you expect to have the higher boiling point? Explain your reasoning. [Electronegativities: chlorine Cl, 3.2; bromine Br, 3.0; iodine I, 2.7.]

7.5

HYDROGEN BONDING

OBJECTIVES

- Relationship to other intermolecular forces
- Origin
- Effects on structure and boiling points
- Group trends in boiling point

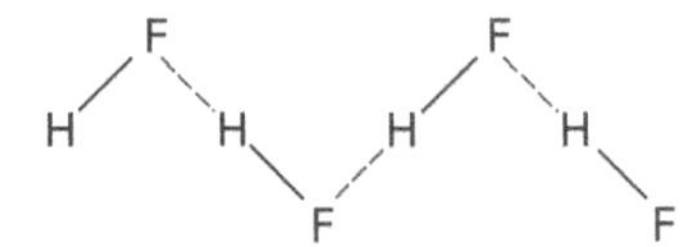

Hydrogen bonding in hydrogen fluoride $H^{\delta+}—F^{\delta-}$ causes the molecules to associate together in chains. Electronegativity values are: H, 2.2; F, 4.0. The hydrogen bonds are shown by orange dashed lines.

Hydrogen bonding is the strongest of the intermolecular forces. It is a force of attraction that can have about 5% of the strength of a covalent bond. **Hydrogen bonds** form between molecules that contain a hydrogen atom bonded to one of the *small, highly electronegative* elements fluorine, oxygen, or nitrogen. Water H_2O is a good example of a molecule that has hydrogen bonds. When present, hydrogen bonding has a marked influence on melting and boiling points and on the structures of solids.

A special case: δ+ hydrogen

Any atom will have a partial positive charge (δ+) when it is bonded to another atom of greater electronegativity. Electron density is withdrawn from the atom with the lower electronegativity.

A covalently bonded hydrogen atom has a share of one bonding pair of electrons and no other shells. As a result, the hydrogen atom is significantly smaller than other atoms. The δ+ charge of the bonded hydrogen atom is therefore spread over a smaller volume and so its polarizing power is unusually high. The highly polarizing δ+ hydrogen atom then attracts electron density (commonly a lone pair) from a small, highly electronegative atom in another molecule. The δ+ hydrogen becomes 'sandwiched' between two small, highly electronegative atoms; it is covalently bonded to one atom and hydrogen-bonded to the other.

A computer-enhanced image of a snow crystal.

Water and the structure of ice

In liquid water, hydrogen bonds form as δ+ hydrogen atoms attract the lone pairs on oxygen atoms of nearby molecules. Water molecules group together in clumps, which are constantly losing and gaining molecules as a result of random collisions.

In ice, hydrogen bonding is maximized because the water molecules line up in an ordered way to form a regular lattice. In this ordered structure the water molecules are further apart than they are in the (more random) liquid state. As a result, at 0 °C ice is *less* dense than liquid water. This is unusual: in general, substances have *higher* densities in the solid state than in the liquid state.

Experiments show that the maximum density of water is at 4 °C. In cold weather ice forms on the top of a lake, while water at 4 °C sinks to the bottom and allows aquatic life to survive under the ice.

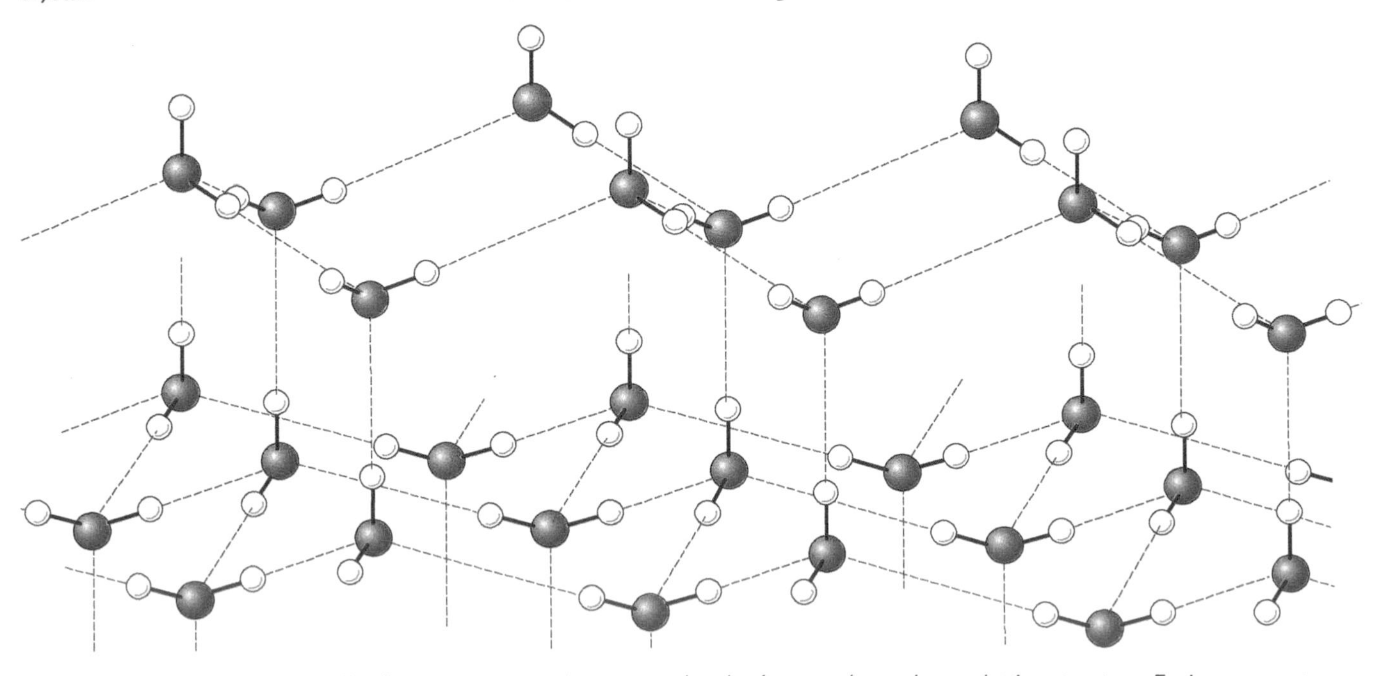

A three-dimensional network of hydrogen bonds holds water molecules in a regular and open lattice structure. Each oxygen atom forms two covalent bonds and two hydrogen bonds.

Hydrogen bonding and boiling points

When a liquid boils, energy is required to overcome all the forces of attraction between the molecules in the liquid state. Look at the trends in boiling points for the hydrides of Groups V and VII. The values for ammonia NH_3 and hydrogen fluoride HF are anomalously high because there is hydrogen bonding between the molecules. On the other hand, there is a steady increase in the boiling points of the hydrides of Group IV consistent with a steady increase in the strength of the dispersion forces. Methane CH_4 does not have an anomalously high boiling point because there is no hydrogen bonding between molecules of methane.

Biochemistry

Hydrogen bonding also plays a crucial role in the structure and function of large organic molecules such as DNA, proteins, and enzymes, which contain covalently bonded H, O, and N atoms. The role of hydrogen bonding in biochemistry is discussed in chapter 30.

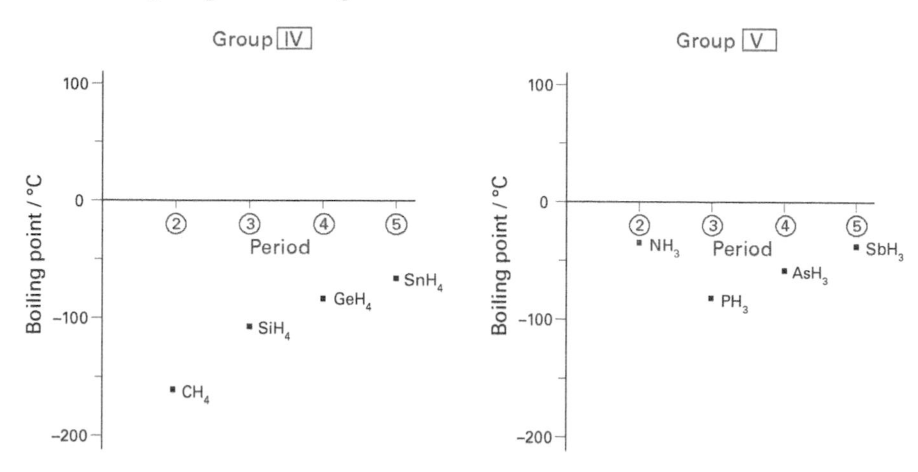

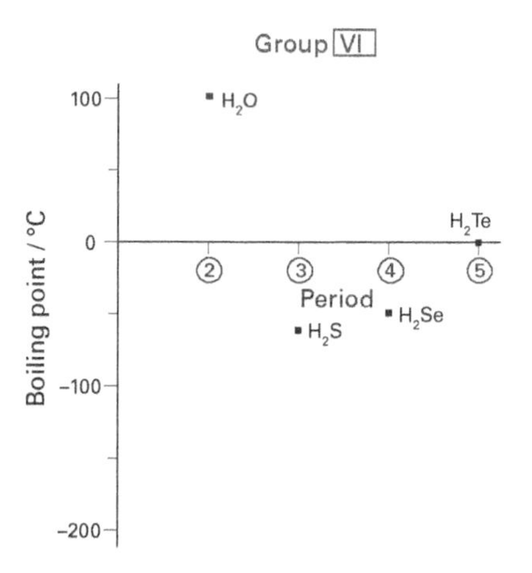

The boiling points for the hydrides of Groups IV, V, VI, and VII. Hydrogen bonding causes the hydrides NH_3, H_2O, and HF (highlighted in red) to have anomalously high boiling points.

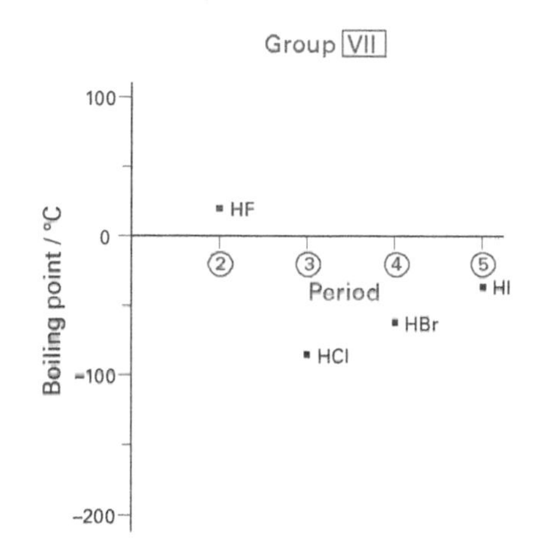

Water and the hydrides of Group VI

The most significant intermolecular force present in water is hydrogen bonding. The effect of the hydrogen bonds is for water H_2O to have a significantly higher boiling point than the other hydrides of Group VI (H_2S, H_2Se, and H_2Te). The steady upward trend in boiling points for the last three substances is due to increasing dispersion forces as the number of electrons present increases. Hydrogen bonds are not present in these compounds because the elements sulphur, selenium, and tellurium do not have sufficiently high electronegativity values.

SUMMARY

- Hydrogen bonds form between molecules that contain hydrogen bonded to fluorine, oxygen, or nitrogen atoms.
- Hydrogen bonds arise because a covalently bonded hydrogen atom has no electrons other than the bonding pair and is very small.
- Hydrogen bonding leads to unexpectedly high melting and boiling points, most importantly for water.
- Hydrogen bonding is especially important in biochemistry.

PRACTICE

1 Carbon, nitrogen, oxygen, and fluorine are the elements that stand at the heads of Groups IV, V, VI, and VII respectively. Explain why hydrogen bonds do not form between molecules of methane CH_4 whereas they do between the hydrides of the other named elements.

2 Water H_2O and ammonia NH_3 both show hydrogen bonding. Electronegativity values are: N, 3.0; O, 3.4; H, 2.2. Explain why the boiling point of ammonia is 133 °C *lower* than that of water: include reference to the structures of these molecules in your answer.

7.6

OBJECTIVES

- Changes in intermolecular forces on mixing
- Criteria for miscibility
- Viscosity
- Surface tension

Mixing and entropy

The magnitude of the intermolecular forces is not the only criterion that determines the likelihood of mixing. Another important factor is entropy. **Entropy** is a measure of the disorder of a system: entropy increases when a system of particles becomes more disordered. Mixing two liquids together causes an increase in entropy. However, entropy must be considered *together with* the strengths of the intermolecular forces in order to make accurate predictions. The topic of entropy is discussed at length in chapter 14.

Hydrogen bonding is the predominant intermolecular force between water and ethanol molecules in a mixture of the two substances.

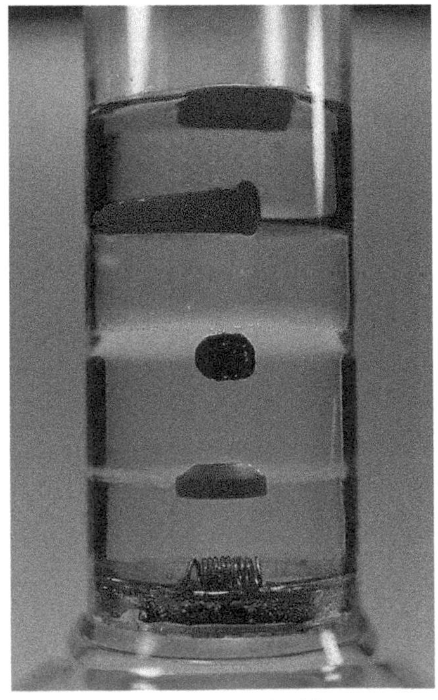

A variety of immiscible liquids; mercury is the densest and corn oil the least dense.

FURTHER EFFECTS OF INTERMOLECULAR FORCES

The intermolecular forces discussed so far in this chapter include those due to permanent dipoles (dipole–dipole forces), instantaneous dipoles (dispersion forces), and hydrogen bonds. These forces can help to explain the melting and boiling points of substances. This spread shows that knowledge of intermolecular forces can also help us to explain the miscibility of liquids, as well as viscosity and surface tension.

Mixing liquids

When two liquids mix, the molecules of one become surrounded by the molecules of the other. Changes in the intermolecular forces result from this process of mixing. For example, the separate liquids A and B have intermolecular forces between their molecules, which may be represented as A–A and B–B. Mixing the two liquids breaks down these two sets of intermolecular forces and establishes a new set A–B.

It is possible to predict whether liquids A and B will mix by comparing the intermolecular forces in the separate liquids (A–A and B–B) and in the mixture (A–B). The ability of liquids to mix, their **miscibility**, depends on the relative strengths of these three sets of intermolecular forces. If the intermolecular forces between molecules in the mixture are stronger than those between the molecules in the separate liquids, then mixing will occur.

Water and ethanol

Water H_2O and ethanol CH_3CH_2OH mix together in all proportions. A single, uniform, and homogeneous mixture results. The most significant intermolecular forces between water molecules are very strong hydrogen bonds. Ethanol possesses highly polar O—H bonds, which cause hydrogen bonds to form between its molecules. When the two liquids are mixed together, water forms hydrogen bonds with ethanol almost as readily as it does with itself.

Water and tetrachloromethane

Tetrachloromethane CCl_4 is a liquid organic solvent. It can dissolve a wide range of organic compounds such as greases and oils. Dispersion forces are the only intermolecular forces because the molecules have no dipole. Water and tetrachloromethane do not mix because the only forces of attraction between tetrachloromethane and water would be dispersion forces. These are very much weaker than the hydrogen bonds between the water molecules. The water molecules therefore remain firmly hydrogen-bonded to each other.

Miscibility

The hydrogen bonding between ethanol and water is strong enough to replace the hydrogen bonds in pure water and in pure ethanol. Water and ethanol are said to be totally **miscible** because they mix together to form a single liquid mixture.

Water and tetrachloromethane do not mix for the reasons just given. Water and tetrachloromethane are said to be **immiscible**: when shaken together, the two substances separate out into two distinct layers.

Viscosity

The viscosity of a fluid is a measure of its resistance to flowing when poured. For example, treacle and engine oil are more viscous than water, which in turn is more viscous than liquid nitrogen. Viscosity may be explained in terms of the intermolecular forces present in a liquid. Treacle is a highly concentrated solution of sucrose (sugar) in water. Each sucrose molecule $C_{12}H_{22}O_{11}$ has a total of eight polar O—H groups.

These groups form hydrogen bonds with water molecules, which themselves are hydrogen-bonded to each other and to other sucrose molecules. Hydrogen bonds must break and reform as treacle flows.

Molecules of engine oil consist of long alkane chains up to 30 carbon atoms long. The total number of electrons in each molecule is high, with the result that extensive dispersion forces hold oil molecules side by side in a tangled mass. These dispersion forces must break and reform as oil molecules move past each other in the flowing liquid.

Liquid nitrogen is far less viscous than water because the intermolecular forces in nitrogen are very weak dispersion forces. Note that liquid helium (T_b = –269 °C) has the lowest viscosity of all liquids.

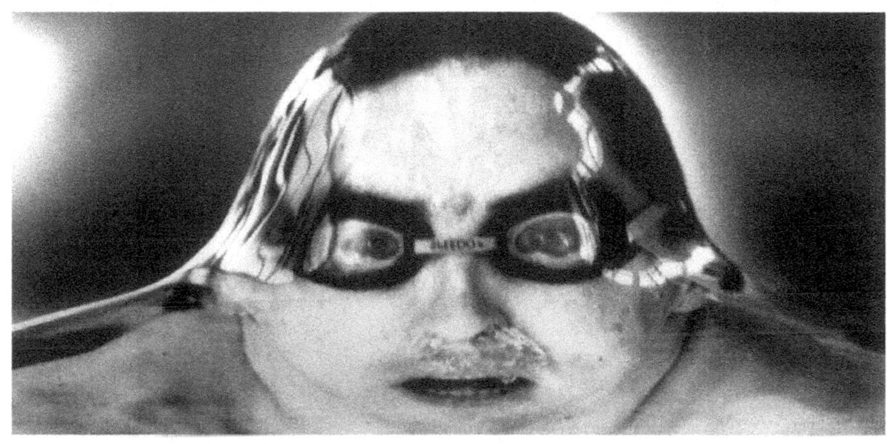

This remarkable prize-winning photo of the swimmer Matt Dunn taken by Tim Clayton was first published in the Sydney Morning Herald. *Dunn's head is surrounded by a thin film of water.*

Surface tension

Surface tension is a property of liquid surfaces that causes them to appear to be covered by a thin elastic 'skin' in a state of tension. Surface tension arises from the attractive forces between the molecules of the liquid. Surface tension causes falling drops of liquid and soap bubbles to have spherical shapes. Water has a high surface tension because there is strong hydrogen bonding between the molecules.

Producing bubbles of this size requires skill!

SUMMARY

- As an approximate rule, liquids mix when there are similar intermolecular forces between molecules in the mixture and between the molecules in the separate liquids.
- The magnitudes of viscosity and surface tension increase with increasing strength of intermolecular forces.

PRACTICE

1 Propane is a gas at room temperature; propan-2-ol and propanone are liquids.

```
     H   H   H
     |   |   |
 H — C — C — C — H        Propane
     |   |   |
     H   H   H

     H   OH  H
     |   |   |
 H — C — C — C — H        Propan-2-ol
     |   |   |
     H   H   H

         O
    H    ||    H
     \   C    /
      C / \  C             Propanone
  H /  \    / \ H
       H   H
```

a Referring to their structures, explain why propane is a gas and the other two substances are liquids.

b Which of the two liquids will have the higher boiling point? Explain your answer.

c Use diagrams to explain why propan-2-ol and propanone both dissolve in water.

2 Hexane is a liquid alkane with the formula C_6H_{14}. Explain with the aid of diagrams why it does not dissolve in water, but does dissolve in tetrachloromethane.

3 Explain why treacle is more viscous than engine oil.

EXTENSION

7.7

OBJECTIVES

- Phase diagram for water
- Phase diagram for carbon dioxide
- Two-component phase diagrams
- Eutectic

Phase diagrams

A '**phase**' is defined as a state of matter that is uniform throughout, not only in chemical composition but also in physical state. For a pure substance, the words 'phase' and 'state' have essentially the same meaning. Water's solid state could be described as the ice phase. When a *mixture* of two substances is considered, it is inappropriate to use the word 'state'. Two liquids that mix together, for example, form a single liquid phase, not a single liquid state.

The conditions of temperature and pressure under which different phases are stable can be shown on a phase diagram. A **phase diagram** describes which phase is most stable under particular conditions of temperature and pressure. We start first with the phase diagrams for two important pure substances and then consider the phase diagram for mixtures of two components.

Phase diagram for water

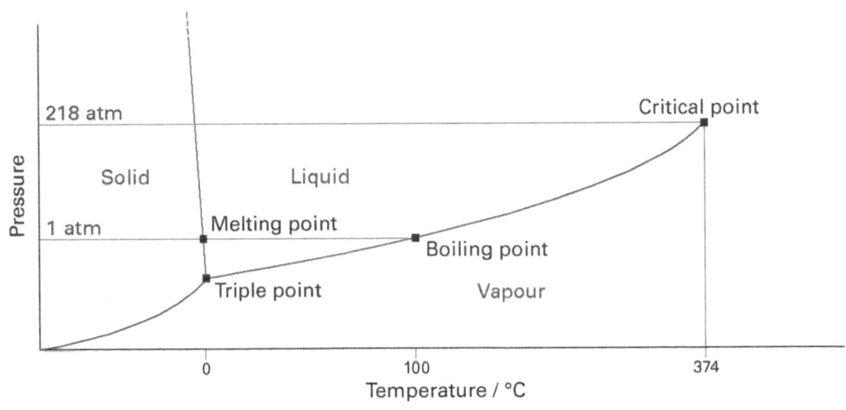

The phase diagram for water is shown above. At low temperatures under atmospheric pressure, the stable phase is the solid (ice). At 1 atm, the temperature at which ice and water are in equilibrium is called the **melting point** (0 °C, 273.15 K). This is shown by the line separating the solid and liquid phases. As this line is not vertical, the transition from solid to liquid takes place at different temperatures under different pressures. The line slopes backwards, which is unusual. The unusually open structure of ice is caused by hydrogen bonding, spread 7.5. Ice skating depends on this phenomenon. The pressure exerted by the skater on the narrow edge of the skates causes the melting point to lower and the ice melts locally (friction from the blade aids this process).

Why snowballs bind. When snow is squeezed in the hands, the pressure increase lowers the melting point: the snow melts. When the pressure is released, the water refreezes and holds the snowball together.

At temperatures between 0 °C and 100 °C under atmospheric pressure, the stable phase is liquid water. At 100 °C, the line representing the equilibrium between liquid and gas is reached and the temperature at which boiling takes place under atmospheric pressure is the **boiling point**. Again, the boiling point varies as the pressure varies. This phenomenon is well known to climbers because cooking takes much longer on the tops of mountains as water boils at a lower temperature. A 'three-minute' egg can require 30 minutes' cooking.

The unique point at which the three lines representing the solid/liquid, liquid/vapour, and solid/vapour equilibria meet is called the **triple point**, as it is the only condition of pressure and temperature at which all three phases are present. The triple point of water (at 611 Pa and 273.16 K) is chosen as the second fixed point on the Kelvin scale (the other being absolute zero).

At the other extreme of the liquid/vapour line is the **critical point**, above which it is not possible to liquefy a gas, however great the pressure. Indeed the term 'vapour' should only be used below the critical point, as it implies that it is possible to form a liquid. We will return to the idea of critical behaviour in the next chapter.

Phase diagram for carbon dioxide

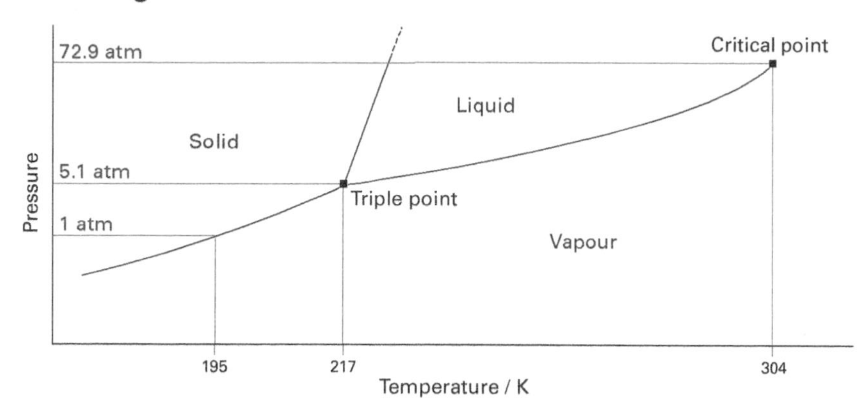

The phase diagram for carbon dioxide, shown above, is similar to that of water with two notable exceptions. First, the line joining solid and liquid slopes forwards, as is the common behaviour because the solid is more dense than the liquid. Second, the triple point lies *above* atmospheric pressure, and so carbon dioxide sublimes, spread 7.2. Ice too can sublime, but only if the pressure is below 611 Pa, the pressure of its triple point. When this happens on a cold winter's day, **hoarfrost** is formed. Frost, on the other hand, is just frozen dew.

Two-component phase diagrams

When two substances are present, such as two solids, the phase diagram usually consists of specifying the stable phase as temperature is changed at constant pressure. A typical example is shown to the right. Adding the other substance lowers the melting point of either component. So an alloy will typically melt at a lower temperature than the melting point of the lower-melting pure metal.

The lowest temperature that can be reached corresponds to the horizontal line. The point marked E corresponds to the composition that melts at the lowest temperature. This is called the **eutectic** (from the Greek for 'easily melted'). The eutectic can be distinguished from either pure A or pure B because adding a little of either component will increase its melting point. Microscopic examination shows that the solid with the eutectic composition consists of small individual crystals of the two components.

When a mixture of composition c_1 starts to cool, it remains solid until it reaches the line between the two phases (at c_2). At this temperature, pure lead will crystallize out. The composition of the remaining mixture gets more concentrated in tin and follows the line down to the point *e*, when all the rest of the mixture crystallizes out.

Eutectics have a number of applications. For example, tin (T_m = 232 °C) and lead (T_m = 327 °C) form a eutectic (63% Sn, 37% Pb) that melts at the conveniently low temperature of 183 °C; this is used as electrical solder.

Common salt (sodium chloride) added to water lowers its melting point. The eutectic mixture melts at –21 °C. This is the reason that salt is spread on roads to prevent ice forming. However, reaching the eutectic composition could only occur by chance.

The phase diagram for potassium iodide and water is shown to the right. The line to the right of the eutectic represents the solubility curve for the salt in water.

SUMMARY

- A phase diagram describes which phase is most stable under particular conditions of temperature and pressure.
- The eutectic melts at the lowest temperature.

Entropy again

The reason why the lowering of melting point takes place is that the entropy of the liquid is greater when there are two substances mixed together to form a solution.

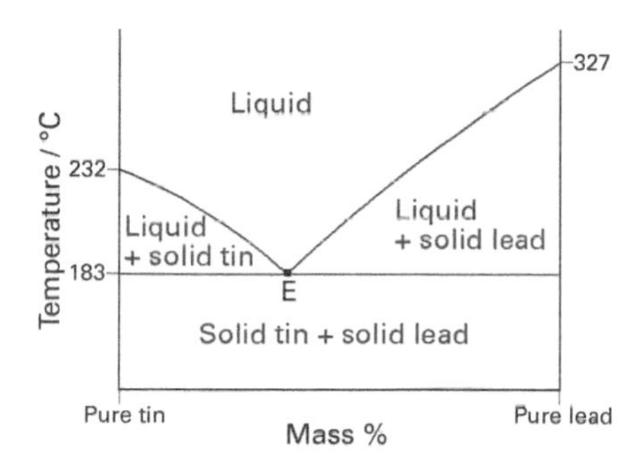

The phase diagram for mixtures of tin and lead; the point E marks the eutectic.

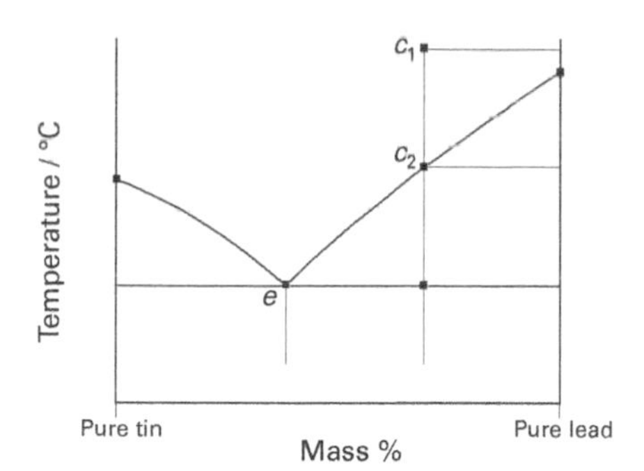

Cooling the mixture of composition c_1 first forms pure lead and then the eutectic.

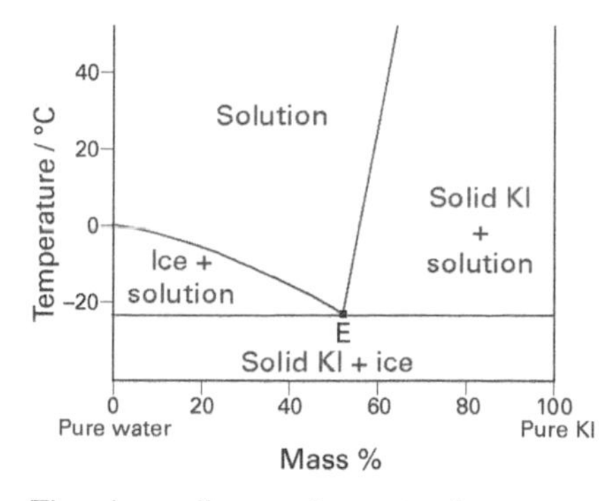

The phase diagram for potassium iodide and water.

PRACTICE EXAM QUESTIONS

1 a i Copy the diagram below and by using the symbols δ+ and δ–, indicate the polarity of the covalent bonds shown. [3]

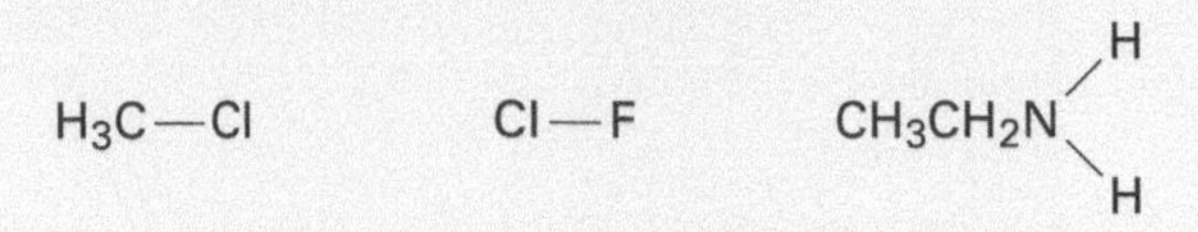

ii Explain the term *electronegativity*. [2]

b The diagram below shows the structure of ethanoic acid when dissolved in benzene.

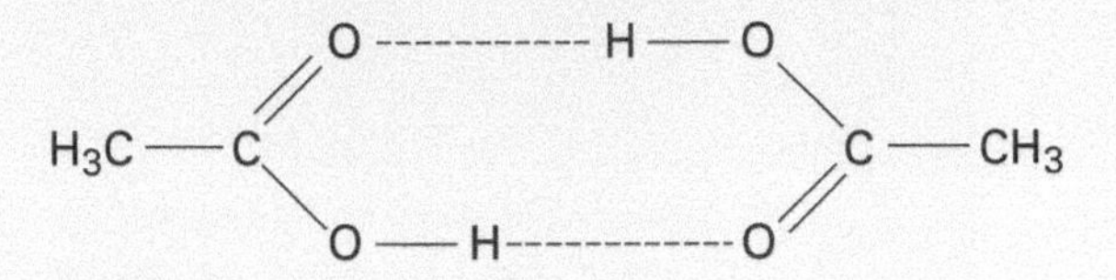

i Calculate the apparent relative molecular mass of ethanoic acid when dissolved in benzene. [2]

ii Give the name of the type of bonding shown by the dashed lines in the diagram above and explain how it arises. [4]

iii When ethanoic acid is dissolved in water the relative molecular mass is slightly less than half the value calculated in **b i**. Explain this observation by referring to the bonding shown in the diagram above. [3]

2 In order to counter the effects of accidental fires, many stores have fitted sprinkler devices. These are fitted with plugs, made of a mixture of two metals, which withstand the pressure of the water behind them. When the temperature rises above a particular value, water is released in a spray. The diagram below is a phase diagram for two such metals, A and B.

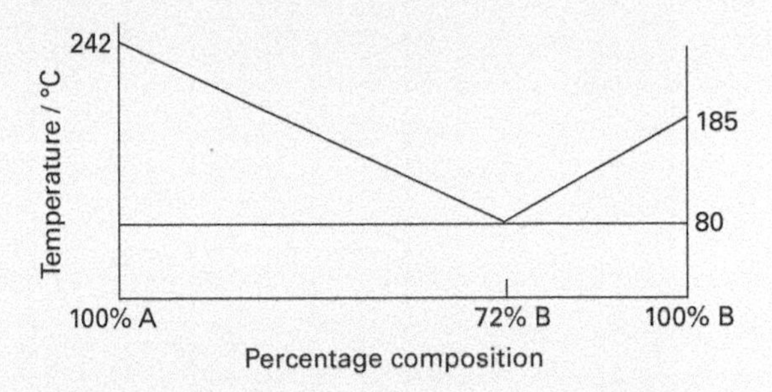

a i A fire starts in such a store and a plug with composition 60% B and 40% A is exposed to temperatures which rise from 20 °C to 300 °C over a short period of time. Explain, using a graph of temperature against time, what happens to the metallic mixture in the plug as the temperature rises. [4]

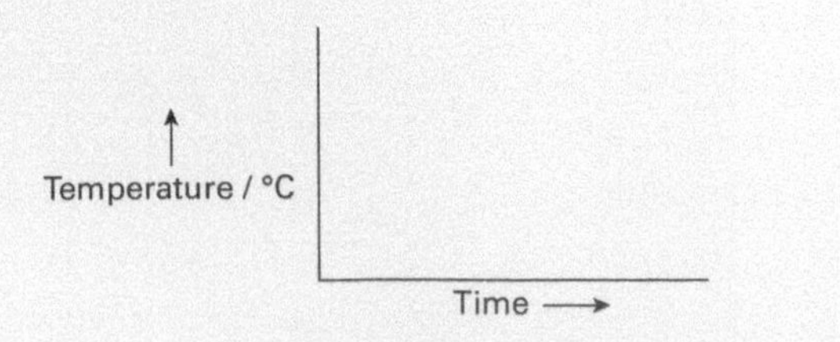

ii How would the graph differ if the plug had the composition of 72% B?

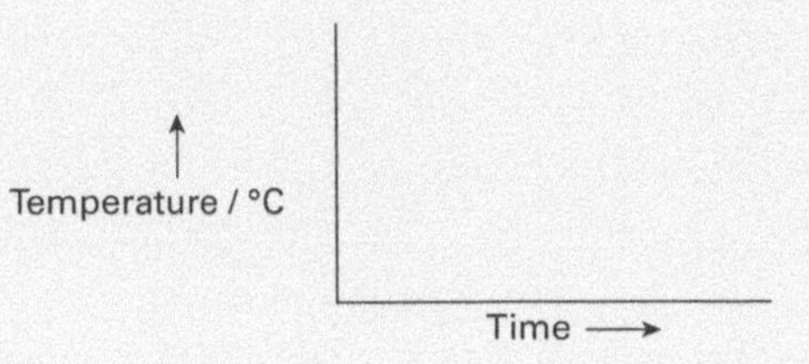

[3]

iii What is the name given to the mixture containing 72% B? [1]

b If the relative atomic masses of A and B are 127 and 181 respectively, calculate the mole fraction of A in this mixture. [2]

(See also chapter 11.)

3 a Define the term electronegativity. [2]

b The diagram below shows the trend in the boiling points of the hydrides of the Group VI elements, oxygen to tellurium.

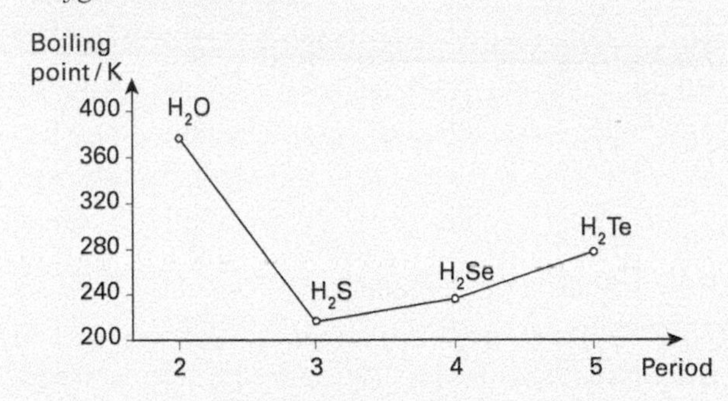

i Explain this rising trend in the boiling points of the compounds H_2S to H_2Te. [3]

ii Hydrogen bonding is said to account for the anomalously high boiling point of H_2O. With reference to the nature of the atoms involved, explain why intermolecular forces are so strong in H_2O. Draw a diagram, containing at least three molecules, to show hydrogen bonding in H_2O. [5]

c Protein molecules are composed of sequences of amino acid molecules joined together in long chains. The sequence of amino acids in a protein chain is illustrated below.

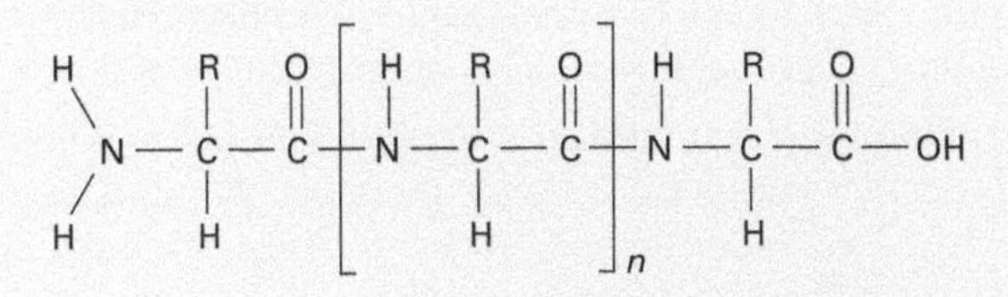

The protein chains are organized into complex three-dimensional structures. Briefly describe how the three-dimensional structure of a protein is held in place. [3]

(See also chapter 30.)

4 Typical surface temperatures of the planets Mars, Jupiter, and Saturn are:
Mars, –60 °C; Jupiter, –140 °C; Saturn, –180 °C.

The surfaces of these planets contain methane, ammonia, and water. The phase diagrams of these three substances can be used to predict their physical states on these three planets. The values of the triple points are:

	Triple point / °C	Approx. pressure / kPa	$\log_{10}$ (pressure / kPa)
Methane	–170	10^1	1
Ammonia	–78	10^2	2
Water	0	1	0

a On the same sheet of graph paper and using the same axes, sketch the phase diagrams of methane, ammonia, and water. Plot the values of $\log_{10}$ (pressure) on the *y*-axis. Label the areas of the three phases on your phase diagram for water. [6]

b Use your sketch to predict whether each of these three substances exists as a solid, a liquid, or a gas on the surface of each of the three planets. [You may find it helpful to mark the planetary temperatures on your sketch.]

Mention any necessary assumptions for making your predictions. [4]

5 a A sketch of the plot of the logarithms of the first seven successive molar ionization energies of silicon against the number of electrons removed is shown below.

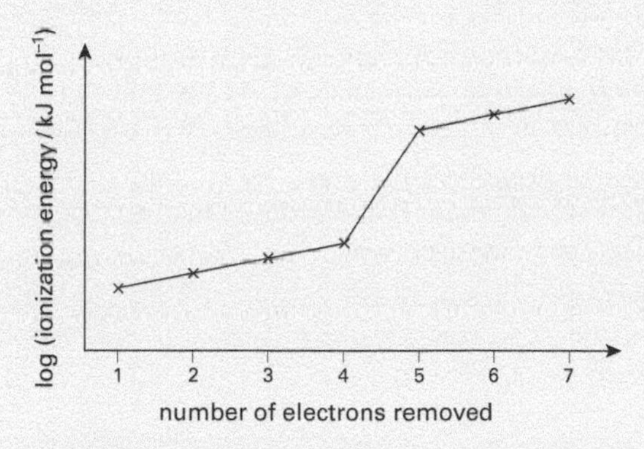

i Define the term *molar first ionization energy* of an element. [2]

ii Write an equation to represent the process whose energy change is equal to the molar second ionization energy of silicon. [2]

iii Explain the gradual increase in values as the second, third, and fourth electrons are removed. [2]

iv Explain why the molar fifth ionization energy is very much greater than the fourth. [2]

b The compounds silicon dioxide, SiO_2, and silicon tetrachloride, $SiCl_4$, both contain covalent bonding but their melting points are very different, being 1610 °C and –70 °C respectively.

i Give the name of the type of structure present in each of the compounds.

SiO_2 $SiCl_4$ [2]

ii Explain how the melting points of the two compounds are related to their structures. [4]

6 a Consider the information about halogens and hydrogen halides below, then answer the questions below.

	Fluorine	Chlorine	Bromine	Iodine
Electronegativity of halogen	4.0	3.0	2.8	2.5
Boiling point of hydrogen halide/K	293	188	206	238

i Define the term *electronegativity*. [2]

ii Briefly explain the steady increase in the boiling points of the hydrogen halides from HCl to HI.[2]

iii Explain why the boiling point of hydrogen fluoride is higher than that of any of the other hydrogen halides. [4]

b A carbonyl group (C═O) contains a double covalent bond between the carbon and oxygen atoms, which is made up of one σ bond and one π bond. Explain, in words or diagrams, what is meant by:

i a σ bond: [1]

ii a π bond. [1]

c i Draw a diagram to show the shape of the carbonate ion (CO_3^{2-}) and the bonding in it. [2]

ii State the similarity between the bonding in the carbonate ion and that present in the benzene molecule. [1]

7 The table below shows some boiling temperatures (T_b) at a pressure of 100 kPa.

Substance	H_2	CH_4	HCl
T_b/K	21	112	188

a In liquid hydrogen, the atoms are held together by covalent bonds.

i What is a covalent bond?

ii How are the hydrogen molecules held together in liquid hydrogen? [2]

b Explain why methane has a higher boiling temperature than hydrogen. [2]

c i Give the meaning of the term *electronegativity*.

ii The electronegatives of hydrogen, carbon and chlorine are 2.1, 2.5, and 3.0, respectively. Use these values to explain why the boiling temperature of hydrogen chloride is greater than that of methane. [6]

Gases

The properties of gases are very different from those of solids and liquids. Solids have fixed volumes and definite shapes. Liquids have a fixed volume but they flow and may be poured. As a result, liquids adopt the shape of the lower part of the container in which they are placed. In contrast, gases have neither definite shape nor fixed volume. They spread out to fill uniformly the whole of any space they enter. Solids and liquids expand when heated, but the extent of expansion depends on the substance present. When different gases are heated or compressed, they all behave in approximately the same manner, irrespective of the gas that is actually present. This chapter explores the distinctive properties of gases. The properties are explained in terms of the behaviour of the molecules that make up the gas.

Gases: three basic ideas

8.1

OBJECTIVES

- Boyle's law
- Charles's law
- Avogadro's principle

Gases may consist of separate atoms (e.g. argon Ar) or separate molecules (e.g. oxygen O_2, and carbon dioxide CO_2). At very high temperatures, gases may also consist of ions (e.g. $Na^+(g)$ and $Cl^-(g)$ from vaporized sodium chloride) or metal atoms (e.g. Na(g) from vaporized sodium). This chapter is mainly concerned with substances that are gases at room temperature and pressure. The vast majority of these gases are made up of simple molecules; so, for the sake of convenience, the term 'molecule' will be used to describe the particles in gases throughout this chapter. There are three important ideas describing the behaviour of gases: Boyle's law, Charles's law, and Avogadro's principle. They were discovered by these three scientists long ago in the history of modern science.

Applying Boyle's law

The **Heimlich manoeuvre** to relieve a person who is choking uses Boyle's law. Clasping the sufferer from behind and thrusting the fist forwards and upwards under the breastbone reduces the chest volume. The resulting pressure increase expels the blockage in the windpipe like a champagne cork.

Boyle's law

In 1662, Robert Boyle wrote down a law that summed up the experimental evidence he had gathered by measuring the volumes of air at different pressures. Expressed in modern language, **Boyle's law** is as follows:

- The volume of a fixed mass of gas (at constant temperature) is inversely proportional to its pressure.

In other words, if the pressure of a sample of gas is doubled, its volume is halved. Boyle's law may be expressed graphically, as shown below left, and can be written in symbols as $V \propto 1/p$.

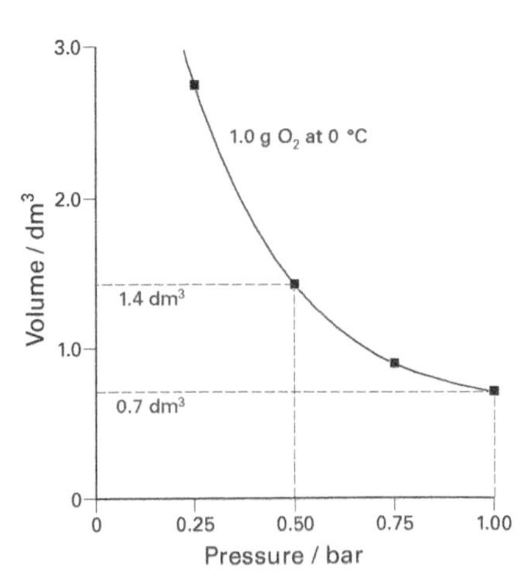

Boyle's law states that $V \propto 1/p$ or that pV = constant.

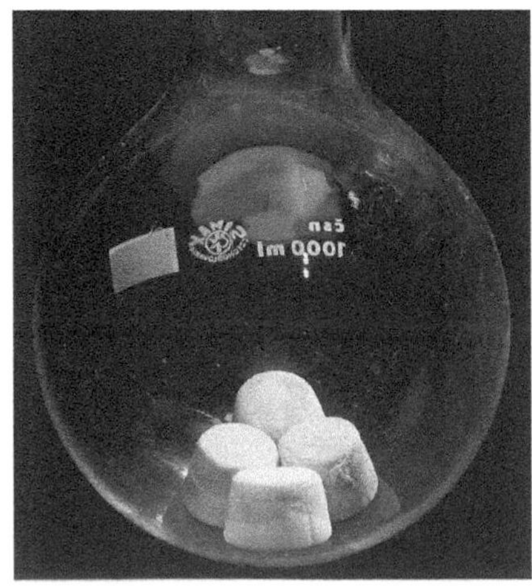

A vacuum pump reduces the pressure around marshmallows. Bubbles of air trapped inside them expand. The result is bigger marshmallows – until they are removed from the flask!

Charles's law

This law provides a quantitative version of the familiar 'rule of thumb' that heating a gas causes it to expand. **Charles's law** is as follows:

- The volume of a fixed mass of gas (at constant pressure) is directly proportional to the thermodynamic temperature.

Notice that the thermodynamic scale of temperature is used (see box to the right). Charles's law may be expressed graphically, as shown on the right, the plot of volume *V* against temperature *T* being a straight line, i.e. $V \propto T$. Extrapolating the graph to the point of zero volume suggests the idea of **absolute zero**, the lowest possible attainable temperature.

> **Jacques Charles**
>
> Jacques Charles stated his law in 1787, four years after he smashed the world altitude record. By using a hydrogen-filled balloon instead of a hot-air balloon, he reached 25 times the previous record height.

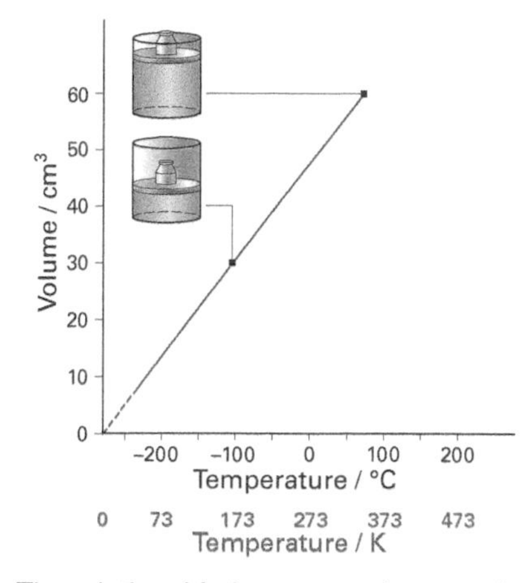

The relationship between volume and temperature is linear (the graph is a straight line), i.e. $V \propto T$.

The air inside a balloon shrinks to a small volume when placed in liquid nitrogen at −196 °C. The air expands again when warmed to room temperature.

> **Thermodynamic temperature**
>
> The work of Charles led to the idea of an absolute zero of temperature, now known to be −273 °C. This temperature is designated as the zero point (0 K) on the Kelvin temperature scale. The value of a temperature measured on the Kelvin scale (e.g. hydrogen boiling point = 20 K) is sometimes referred to as an 'absolute temperature' or, more correctly, as a 'thermodynamic temperature'; the latter will be used throughout this text. All temperatures measured on the Kelvin scale are thermodynamic temperatures.

Avogadro's principle

Amedeo Avogadro proposed a theoretical idea about gases in 1811. **Avogadro's principle** is as follows:

- Equal volumes of different gases at the same temperature and pressure contain the same number of molecules.

In other words, if there are *x* molecules of O_2 in 10 cm³ of oxygen gas, then there are *x* molecules of N_2 in 10 cm³ of nitrogen gas and there are *x* molecules of CO_2 in 10 cm³ of carbon dioxide gas at the same temperature and pressure. Another way of putting this is that the volume of a gas depends on the amount in moles, $V \propto n$.

Working almost 200 years ago, Avogadro had no direct experimental evidence to support this statement. He simply talked about gases in terms of 'molecules'. Later experimental evidence showed the significance of Avogadro's principle, especially in recognizing that atoms can combine to form molecules.

An Italian stamp depicting Avogadro's principle.

SUMMARY

- Unlike solids and liquids, gases have neither definite shape nor fixed volume.
- Boyle's law: at constant temperature, $V \propto 1/p$ or pV = constant.
- Charles's law: at constant pressure, $V \propto T$ (*T* in kelvin).
- Avogadro's principle: equal volumes of different gases at the same temperature and pressure contain the same number of molecules.

PRACTICE

1 Jane puts her finger over the hole at the end of a bike pump. She pushes the plunger half way in. What is the pressure inside the pump?

2 A balloon contains 500 cm³ of air at 300 K (27 °C). What is the volume of the balloon when it is placed in a freezer at 255 K (−18 °C)?

8.2 IDEAL GASES

OBJECTIVES

- The ideal gas equation
- The kinetic theory of gases

The greatest difference between gases and solids or liquids is that the molecules of a gas are very much further apart. Much of the chapter on solids was spent discussing the attractive forces that exist between the particles. When describing the behaviour of gases, it is assumed that there are no forces between the molecules. A gas in which there are no intermolecular forces is called an **ideal gas** (sometimes **perfect gas**). The behaviour of an ideal gas is described by the ideal gas equation. The molecular model of ideal gases is developed using the kinetic theory of gases. Real gases behave very much like ideal gases under normal conditions of temperature and pressure.

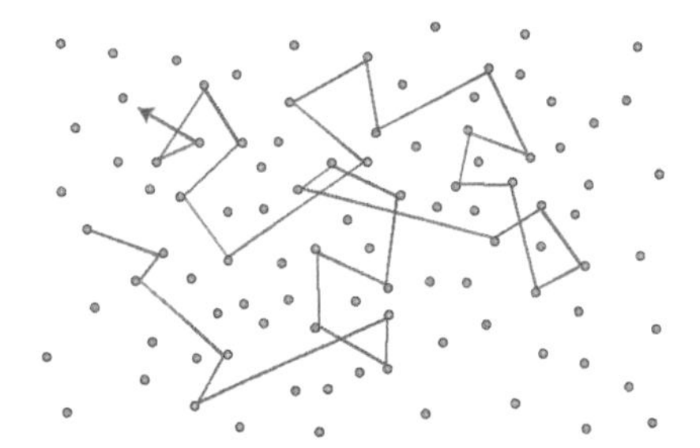

The path of a single molecule zig-zags randomly as the molecule repeatedly collides with other molecules.

The ideal gas equation

Boyle's law and Charles's law are respectively expressed as:

$V \propto 1/p$ (at constant temperature) and

$V \propto T$ (at constant pressure)

Combining these expressions gives:

$V \propto T/p$ or $pV \propto T$

Adding a constant of proportionality results in the relationship:

$$pV = kT$$

where k is a constant for a fixed mass of a particular gas.

Avogadro's principle shows that the volume of a gas depends on the number of molecules. So the volume is proportional to the amount of gas (in moles), i.e. $V \propto n$. The behaviour of an ideal gas can therefore be described by the **ideal gas equation**:

$$pV = nRT$$

where R is a constant for all gases called the **gas constant**. Its value in SI units is $R = 8.31\ \text{J K}^{-1}\text{mol}^{-1}$ or in other useful units $8.31\ \text{Pa m}^3\text{K}^{-1}\text{mol}^{-1}$.

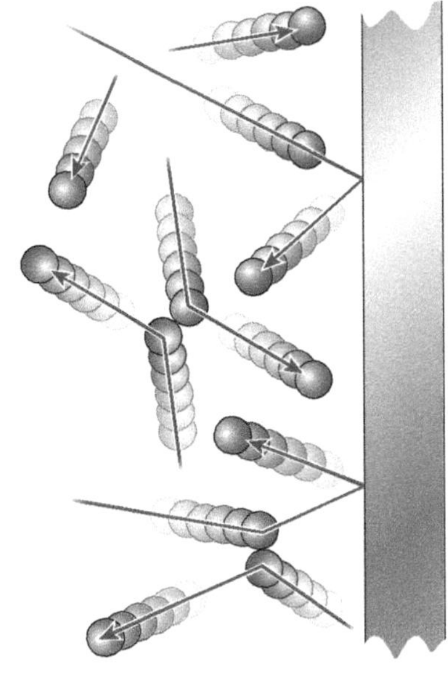

A gas exerts pressure as the molecules collide with and rebound from the walls of its container.

The kinetic theory of gases

An ideal gas follows the gas laws exactly. The **kinetic theory of gases** puts forward a model for gases that explains this behaviour. It makes the general assumption that an ideal gas is composed of independent molecules widely separated from each other, as described by the following four statements:

- The molecules in a gas are in continuous random motion.
- There are no intermolecular forces, so the only interactions between molecules are collisions.
- All collisions are perfectly 'elastic': the molecules bounce off each other without their total kinetic energy changing.
- The molecules themselves have no size, i.e. they occupy zero volume.

Pascals, bars, and atmospheres

Pressure is defined as force per unit area. The unit of pressure is the **pascal** (Pa): 1 Pa = 1 N m^{-2}. Atmospheric pressure at sea level is about 10^5 Pa (100 kPa); 1 **bar** is defined as 1×10^5 Pa. Weather forecasts give atmospheric pressures in millibars, mbar (1000 mbar = 1 bar). Finally, 1 **atmosphere** (atm) is equal to 101.325 kPa and is the pressure at which the boiling point of water is exactly 100 °C.

SI units and the ideal gas equation

SI units should be used consistently in calculations involving the ideal gas equation.

- The SI unit of pressure is the **pascal, Pa**, where 1 Pa = 1 N m^{-2}. Since the pascal is a very small unit, it is often more convenient to express pressures in bars or atmospheres (atm); 1bar is 1×10^5 Pa and 1 atm is 1.01×10^5 Pa, so they are almost the same size. When using the ideal gas equation, however, the pressure should be converted to pascals.
- The SI unit of volume is the **cubic metre, m³**, but the volume of a gas is most often expressed in cubic decimetres, dm^3; 1 m^3 = 1000 dm^3. Note that 1 dm^3 is the same volume as 1 litre. The diagram at the top of the opposite page shows how these units of volume are related. Again, when using the ideal gas equation, the volume should be converted to cubic metres.

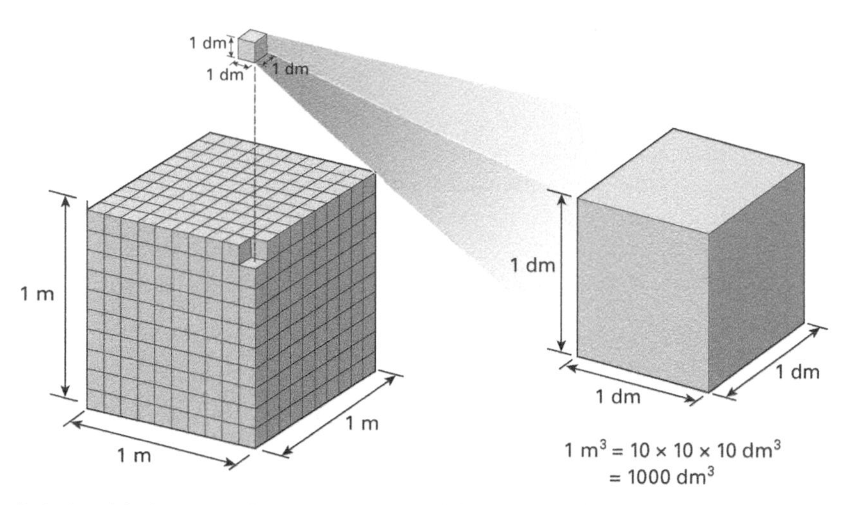

Relationship between the cubic metre and the cubic decimetre.

- The SI unit of temperature is the **kelvin, K**. Temperatures are most commonly measured in degrees celsius. Conversion to kelvins is very easy because 0 °C is equivalent to 273 K, so adding 273 to a temperature in degrees celsius converts it into a temperature (the *thermodynamic temperature*) in kelvins. When using the ideal gas equation, the temperature should be converted to kelvins.

Gas law summary

Boyle's law:
$V \propto 1/p$ (at constant n and T)
Charles's law:
$V \propto T$ (at constant n and p)
Avogadro's principle:
$V \propto n$ (at constant p and T)
The ideal gas equation:
$pV = nRT$

Kelvin scale

The thermodynamic temperature scale is often called the Kelvin scale (with a capital 'K'), after William Thomson, Lord Kelvin. The unit has a small letter when spelled out, kelvin, and is abbreviated to (capital) K. (Also note that it is **wrong** to say 'degrees Kelvin'.)

Worked example on calculating pressure

Question: Chloroethene is a chemical used to make the polymer PVC. A sample of the gas that contains 3.93 mol has a volume of 98.4 dm³ at 24 °C. What is the pressure of the gas?

Strategy: Use the ideal gas equation. Convert all quantities to SI units. Write the ideal gas equation in the form:

$$p = \frac{nRT}{V}$$

Answer:

$V = 98.4\,\text{dm}^3 = 98.4 \times 10^{-3}\,\text{m}^3$

$n = 3.93\,\text{mol}$

$T = (24 + 273)\,\text{K} = 297\,\text{K}$

So we get

$$p = \frac{nRT}{V}$$

$$= \frac{(3.93\,\text{mol}) \times (8.31\,\text{Pa m}^3\,\text{K}^{-1}\,\text{mol}^{-1}) \times (297\,\text{K})}{(98.4 \times 10^{-3}\,\text{m}^3)}$$

$$= 9.86 \times 10^4\,\text{Pa}$$

A raincoat and umbrella made of the plastic PVC (short for polyvinyl chloride). Vinyl chloride is the traditional name for chloroethene. PVC results when molecules of chloroethene join to make long chains, spread 22.10. PVC has many other uses.

SUMMARY

- The molecules of an ideal gas:
  are in continuous random motion,
  experience no intermolecular forces,
  have perfectly elastic collisions,
  occupy zero volume.
- The ideal gas equation is: $pV = nRT$.
- SI units should be used in calculations involving the ideal gas equation.

PRACTICE

1 Calculate the volume of 0.5 mol of gas at 1×10^5 Pa and 20 °C.

2 What will be the volume of the gas in question 1 when the temperature is raised to 100 °C?

3 Calculate the pressure inside the cathode-ray tube in a television set, assuming that it contains 3.6×10^{-7} mol of gas in a volume of 5.0 dm³ at 20 °C.

8.3

OBJECTIVES

- The molar volume of an ideal gas
- Molar masses by experiment

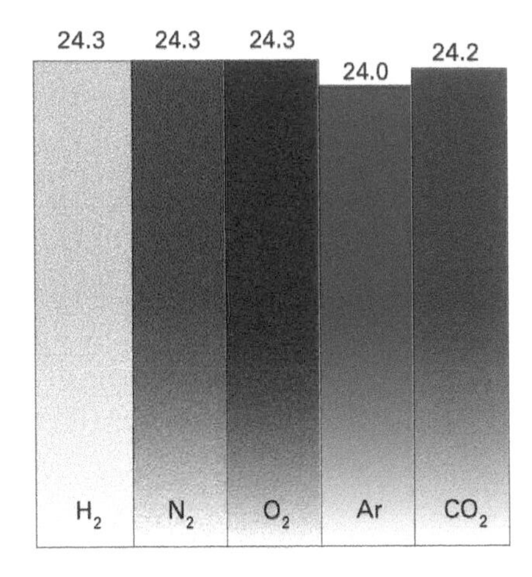

The volumes per mole of different gases are all very similar. Figures are given in $dm^3\ mol^{-1}$ at 20 °C and 1 bar.

Conditions: *T* and *p*

The numerical value of the molar volume depends on the temperature and pressure of a gas. The usual conditions chosen are 20 °C and 1 bar. You should not refer to these conditions as 'room temperature and pressure' (because the conditions of the room in question are not specified!). For calculations, the conditions of a gas must be specified numerically.

The volume of one mole of an ideal gas at 20 °C and 1 bar is 24 dm³.

Molar volume of an ideal gas

Avogadro's principle gives us some useful conclusions. As we saw in the previous two spreads, Avogadro's principle states that equal volumes of different gases contain the same number of molecules. Doubling the number of molecules present doubles the volume (at constant pressure). Also, according to the laws we met in the previous spread, doubling the number of molecules present will double the pressure (at constant volume).

The molar volume

Another way of stating Avogadro's principle is to say that:

- One mole of an ideal gas occupies the same volume under the same conditions of temperature and pressure.

This assertion leads to the idea of the **molar volume**, the volume per mole of any gas under stated conditions. You should find this idea easier to understand if you work through the following two examples.

Worked example on calculating the molar volume of a gas at room temperature and pressure

Question: What is the volume occupied by 1.00 mol of an ideal gas at 20 °C and 1.00 bar pressure?

Strategy: Use the ideal gas equation. Convert all quantities to SI units. Write the ideal gas equation in the form:

$$V = \frac{nRT}{p}$$

Answer:

$n = 1.00\,\text{mol}$

$T = (20 + 273)\text{K} = 293\,\text{K}$

$p = 1.00\,\text{bar} = 1.00 \times 10^5\,\text{Pa}$

So we get

$$V = \frac{nRT}{p} = \frac{(1.00\,\text{mol}) \times (8.31\,\text{Pa m}^3\,\text{K}^{-1}\,\text{mol}^{-1}) \times (293\,\text{K})}{(1.00 \times 10^5\,\text{Pa})}$$

$= 0.0243\,\text{m}^3 = 24\,\text{dm}^3$ (2 sig. figs)

Note: The units mol, Pa, and K cancel, leaving simply m³, the SI unit of volume. In the more usual units of dm³ (1 m³ = 1000 dm³) the equation predicts that the molar volume of an ideal gas at 20 °C and 1 bar (about room temperature and atmospheric pressure) is **24 dm³ mol⁻¹**. At standard temperature and pressure (273 K and 1.00 atm), abbreviated to STP, the molar volume of an ideal gas is $22.4\,\text{dm}^3\,\text{mol}^{-1}$. The volume is smaller because the temperature is lower.

Worked example on calculating the molar volume of a gas under alternative conditions

Question: On the surface of Venus the temperature is 470 °C and the pressure is 100 atm. What is the volume occupied by 1.00 mol of an ideal gas under these conditions?

Strategy: Use the ideal gas equation. Convert all quantities to SI units. Write the ideal gas equation in the form:

$$V = \frac{nRT}{p}$$

Answer:

$n = 1.00\,\text{mol}$

$T = (470 + 273)\,\text{K} = 743\,\text{K}$

$p = 100\,\text{atm} = 1.01 \times 10^7\,\text{Pa}$

So we get

$$V = \frac{nRT}{p} = \frac{(1.00\,\text{mol}) \times (8.31\,\text{Pa m}^3\,\text{K}^{-1}\,\text{mol}^{-1}) \times (743\,\text{K})}{(1.01 \times 10^7\,\text{Pa})}$$

$= 6.11 \times 10^{-4}\,\text{m}^3 = 0.61\,\text{dm}^3$ (2 sig. figs)

Note: The molar volume is much less than on Earth because the atmospheric pressure is much greater.

Molar masses by experiment

We can use the idea of the molar volume of a gas to find out the molar mass of a volatile liquid, because a volatile liquid may be made into a gas easily. To do this we vaporize a known mass of the liquid and measure its volume. Then we assume that the vapour behaves as an ideal gas to work out its molar mass.

The experiment is carried out using the apparatus shown below. The steps are as follows:

1 Weigh the hypodermic syringe containing the volatile liquid (m_1).

2 Inject a small quantity of liquid into the gas syringe. Reweigh the hypodermic syringe (m_2) and hence find the mass of the vapour ($m = m_1 - m_2$).

3 Measure the volume of vapour that results in the gas syringe (V), the temperature (T), and the atmospheric pressure (p).

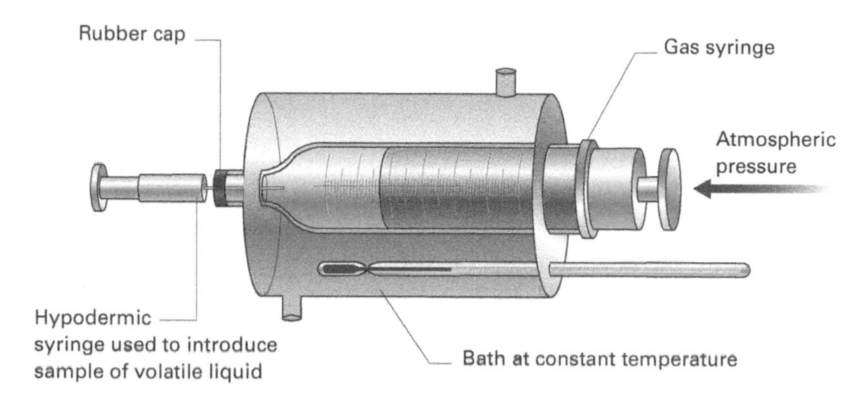

The apparatus for measuring the molar mass of a volatile liquid.

Reminder

Whenever starting a calculation involving gases, your first thought should be:

'The ideal gas equation is $pV = nRT$. How can I apply this equation to this problem?'

Calculating molar mass

Once we have carried out the experiment above, we are ready for the calculation. First we must ensure that all quantities are expressed in SI units. Then, from spread 3.3 on the mole, we know that the amount of gas in moles (n), the mass (m), and the molar mass (M) are related by the expression

$$n = \frac{m}{M} \qquad (1)$$

The ideal gas equation is

$$pV = nRT \qquad (2)$$

Substituting equation (1) into equation (2) gives us

$$pV = \frac{mRT}{M} \qquad (3)$$

Rearranging equation (3) to make M the subject of the equation gives

$$M = \frac{mRT}{pV} \qquad (4)$$

Now you can substitute your results from the experiment above into equation (4) and obtain a value for the molar mass M of the volatile liquid.

When fully pressurized at cruising altitude, a Jumbo Jet contains about one tonne of air. Knowledge of the pressure and temperature enables the calculation to be done.

SUMMARY

- The molar volume of an ideal gas at standard temperature and pressure (273 K, 1 atm) is 22.4 $dm^3 mol^{-1}$.
- The molar volume of an ideal gas at 20 °C and 1 bar is 24 $dm^3 mol^{-1}$.
- The molar mass of a volatile liquid may be measured using a gas syringe.

PRACTICE

1 On evaporation at 100 °C and 1.00 atm, 0.124 g of a volatile liquid hydrocarbon gave 45.3 cm^3 of vapour. Calculate the molar mass of the hydrocarbon.

2 The hydrocarbon in question 1 contained hydrogen and carbon in the mole ratio 2:1. What is the molecular formula of the hydrocarbon?

THE MOVEMENT OF GAS MOLECULES

OBJECTIVES

- Effusion
- Graham's law
- Diffusion
- The Maxwell–Boltzmann distribution

Bicycle tyres slowly deflate and balloons go limp and flabby. The walls of tyres and balloons are full of microscopic holes called pores. The molecules in a gas are in continuous random motion. If a molecule is by chance moving in the direction of a pore, it will enter it and eventually escape to the outside. This escaping tendency is called effusion. The rate of effusion of gases is linked to their molar masses. The model behind this leads to the idea of molecular kinetic energies and then to the Maxwell–Boltzmann distribution.

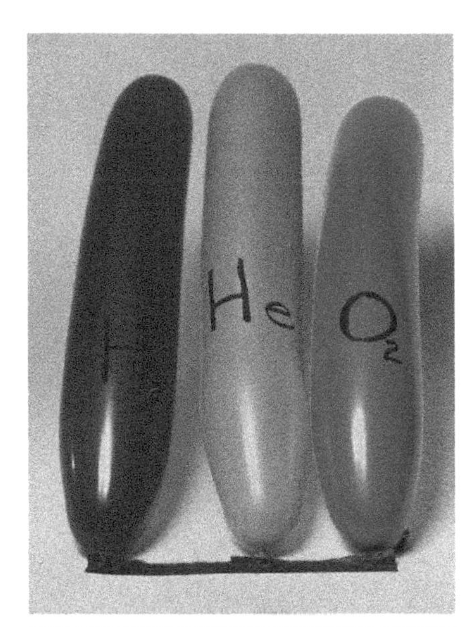

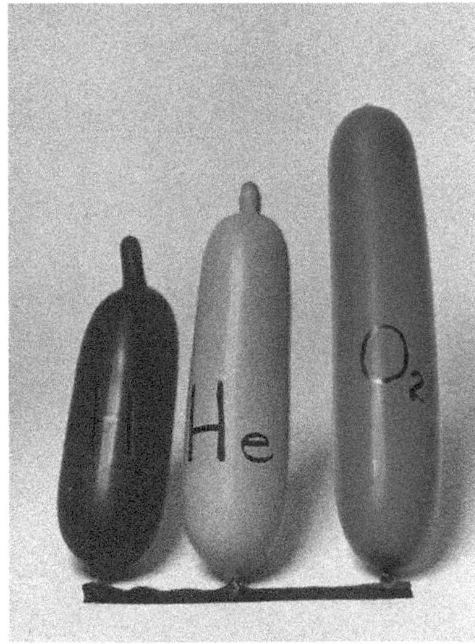

The rate of effusion is inversely proportional to the square root of the relative formula masses of the gases, which are: H_2, 2; He, 4; O_2, 32.

Rates of effusion and Graham's law

Effusion is the escape of gas molecules through a small hole. Experiments show that some gases effuse faster than others. This observation may be explained in terms of the kinetic energy of the molecules. Temperature is a measure of the average kinetic energy of the gas molecules. Average kinetic energy is proportional to the thermodynamic temperature.

The *average* kinetic energy of the molecules in the gas is constant at a given temperature.

You should remember the equation for kinetic energy from your earlier studies. So

$$\text{average KE} = \tfrac{1}{2}mv^2 = \text{constant}$$

$$mv^2 = \text{constant, which means} \quad v^2 = \frac{\text{constant}}{m} \quad \text{and } v \propto \frac{1}{\sqrt{m}}$$

The speed of the molecules in the gas and therefore the rate of effusion is inversely proportional to the square root of the mass of the molecule. This means that lighter gases effuse faster at a given temperature because the molecules move faster.

This was discovered experimentally by Thomas Graham in 1846 before the kinetic theory of gases had been developed. He summarized his observations in a statement now known as **Graham's law**:

- The rate of effusion of a gas is inversely proportional to the square root of its relative formula mass, i.e. rate $\propto 1/\sqrt{M_r}$.

Diffusion

Diffusion is the mixing of one type of molecule throughout a space containing another type of molecule. For example, a bad smell diffuses until it uniformly fills a room. Experiments show that some gases diffuse faster than others. The process of gas diffusion obeys Graham's law, as shown by the illustration below: NH_3 diffuses faster as it is lighter than HCl.

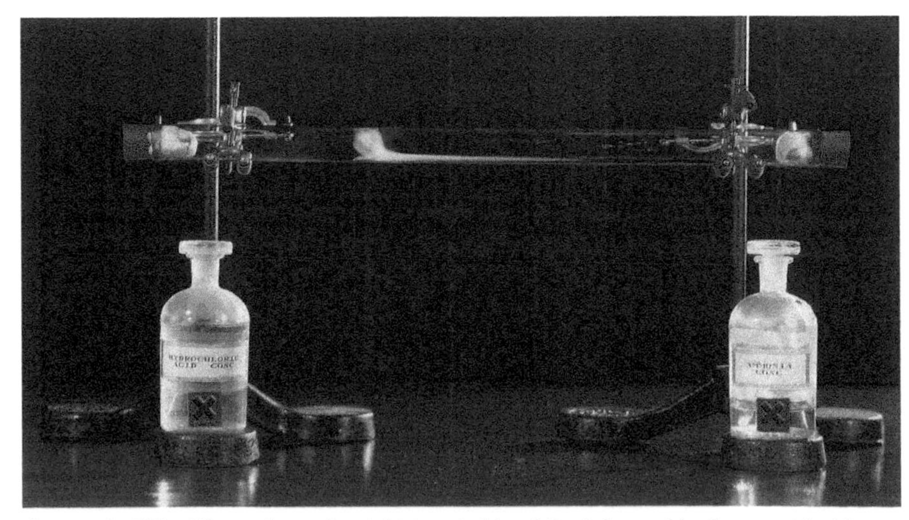

Ammonia NH_3 diffuses from the right-hand side of the tube and hydrogen chloride HCl from the left. The gases react to form solid ammonium chloride when they meet.

The Maxwell–Boltzmann distribution

Graham's law has been explained in terms of the *average* kinetic energy of the molecules. Since lighter gas molecules travel faster, they effuse faster. Many combinations of energy will, however, give the same average energy. For example, to produce an average of 4 kJ mol^{-1}, all the molecules could have that energy, or half of them could have 2 kJ mol^{-1} and the other half could have 6 kJ mol^{-1}. Alternatively, the molecules could have a wide spread of energies.

James Clerk Maxwell looked at the problem of how many molecules in a gas had a particular energy, an area of study later reconsidered by Ludwig Boltzmann. Maxwell succeeded in calculating the probability of a gas molecule having a particular energy. For each energy, the **Maxwell–Boltzmann distribution** shows the number of molecules that have that energy. The energies of individual molecules in a gas are continually changing due to random collisions, but the calculations give the total number of molecules with that energy at a particular instant.

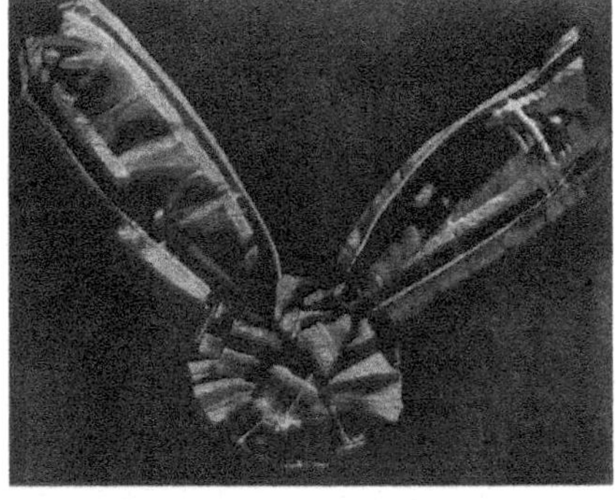

James Clerk Maxwell was the first person to suggest how to make a colour photograph. The photo above, of a tartan ribbon, is the result of superimposing three photographs taken with different coloured filters.

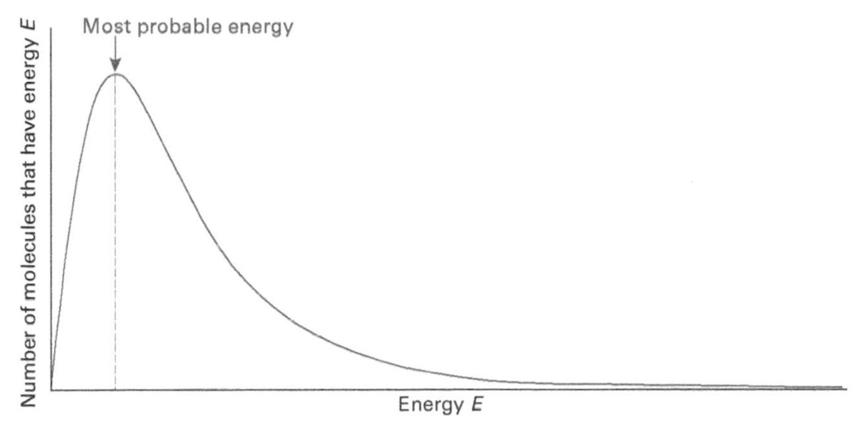

The Maxwell–Boltzmann distribution of molecular energies for a gas at a fixed temperature.

The highest point on the curve represents the most probable energy. The greatest number of molecules have this energy at a given instant. No molecules are stationary and a few molecules have exceptionally high energy. The area under the curve is *not symmetrical* about the most probable energy. There is a larger 'tail' at higher energy, and the graph drops off more sharply on the low-energy side.

At higher temperatures, the average energy of the molecules is greater. The range of the energies is also greater, with a higher chance of very high energy. The peak of the curve is broader and lower at higher temperature. The area under each of the curves represents the total number of molecules. This number is fixed, so the curve must have a lower peak at higher temperatures as more molecules have high energy.

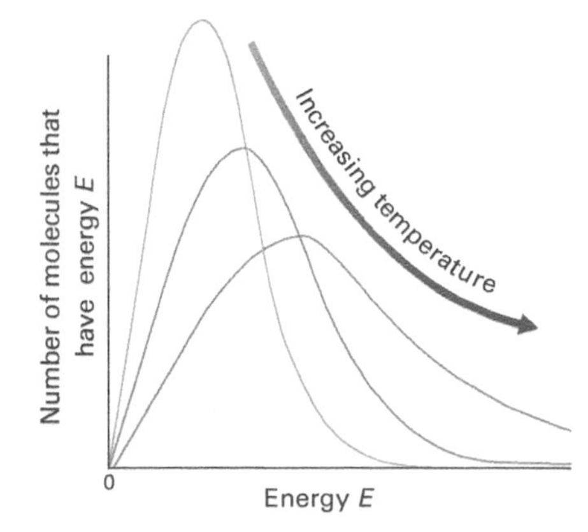

The Maxwell–Boltzmann distribution of molecular energies for a fixed mass of gas at three different temperatures. Spread 15.4 explains why the reaction rate varies with temperature.

SUMMARY

- Effusion is the escape of a gas through a small hole.
- The rate of effusion (and the rate of diffusion) of a gas is inversely proportional to the square root of its relative formula mass.
- Average molecular kinetic energy is proportional to the thermodynamic temperature of the gas.
- The number of molecules with a given energy is shown by the Maxwell–Boltzmann distribution. (Memorize the shape.)

PRACTICE

1 In an experiment, 100 cm^3 of a gas diffused out of a gas syringe in 14.1 s. The same volume of oxygen took 10 s. Calculate the relative formula mass of the unknown gas. Suggest an identity for the gas.

2 Arrange the following gases in order of increasing rate of effusion under the same conditions: ammonia NH_3; carbon dioxide CO_2; chlorine Cl_2; helium He; xenon Xe.

EXTENSION

8.5

OBJECTIVES

- Deviations of real gases from ideal gas behaviour

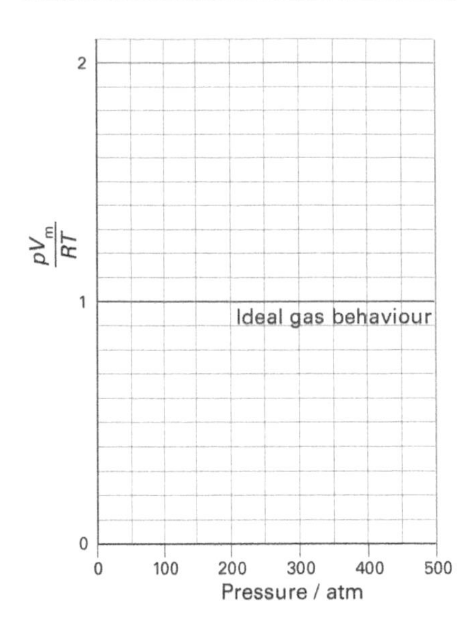

Isothermals

The plot of pV_m/RT against p shown above is called an **isothermal**. This term means that it is the result of measurements made at equal temperature.

Johannes Diderik van der Waals

The Dutch scientist Johannes van der Waals was the first person to take account of the forces between molecules in a gas. The general term 'van der Waals forces' is sometimes applied to intermolecular forces.

REAL GASES

The ideal gas equation is an empirical law, which means that it is based on observations. The kinetic theory of gases is an attempt to explain the law by providing a model for the behaviour of gas molecules. Gases obey the ideal gas equation $pV = nRT$ under almost all common conditions of temperature and pressure. But are there any conditions under which the ideal gas equation is not valid? Two assumptions made by the kinetic theory about ideal gases are that (i) there are no intermolecular forces and (ii) the molecules themselves occupy zero volume. Real gases behave as ideal gases under conditions where these assumptions hold true. At high pressure and low temperature they do not, and real gases depart fairly significantly from ideal gas behaviour.

Real gases and ideal behaviour

It is important to note that deviations from ideal gas behaviour only occur under conditions of high pressure or low temperature, and that the deviations are usually quite small. The ideal gas equation remains an excellent approximation to the behaviour of real gases under normal laboratory conditions. A check for ideal behaviour depends on the following reasoning.

The ideal gas equation is:

$$pV = nRT$$

Remembering that the molar volume $V_m = V/n$, this rearranges to

$$\frac{pV_m}{RT} = 1$$

This relationship should apply for all values of pressure and temperature. The first diagram on spread 8.3 shows that the molar volume for many gases is within 2% of the ideal gas value at 20 °C and 1 bar, indicating that these real gases are behaving as an ideal gas. The assumptions made by the kinetic theory of gases apply to these gases *under these conditions*.

Deviations at moderately high pressure

Increasing the pressure on a gas forces its molecules closer to each other. When the pressure is high, the molecules are sufficiently close together for their intermolecular forces to become significant. So, the assumption underlying the kinetic theory of gases that there are no forces between molecules can no longer be made. The gas deviates from the ideal gas equation under these conditions. At high pressure (typically hundreds of atmospheres), the *attractive* forces between the molecules *decrease* the volume occupied by the gas: a real gas is more compact than an ideal gas would be under the same conditions.

Deviations at very high pressure

Remember that the kinetic theory of gases also assumes that the molecules themselves occupy zero volume. At *very* high pressure, the size of the molecules themselves becomes significant relative to their distance apart. Molecular size can no longer be ignored and the gas deviates from ideal behaviour. At very high pressures, the *repulsive* forces between the molecules *increase* the volume occupied by the gas: a real gas takes up more space than an ideal gas would under the same conditions.

Deviations at low temperature

Gases deviate from the ideal gas equation when the temperature is very low. As temperature decreases, the energies of the molecules decrease. The energy with which the molecules collide with each other is therefore less. At low temperatures, intermolecular forces become more significant. The molecules clump together as the collisions between them are not completely elastic and become more 'sticky'.

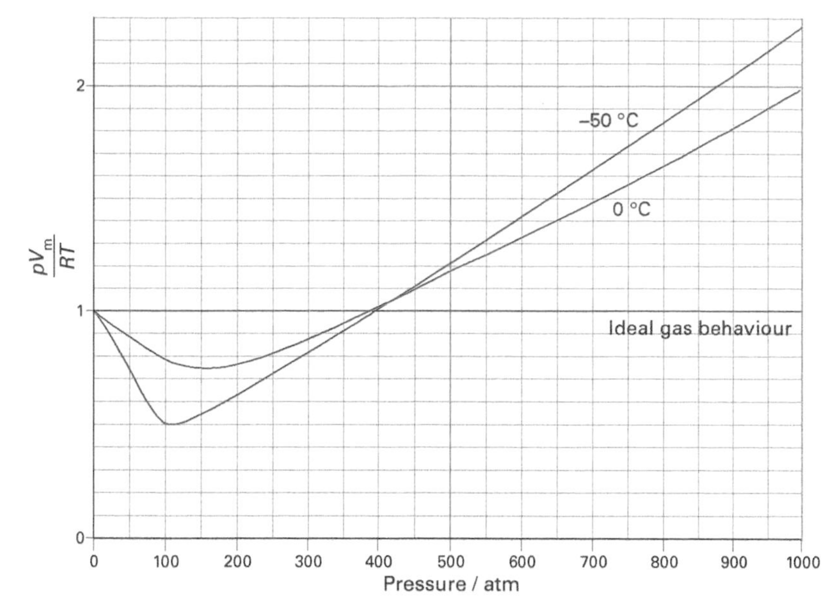

Plots of pV_m/RT against p for methane at 0 °C and at –50 °C at high pressures up to 1000 atm. The downward slopes of the plots at moderately high pressure result from the influence of attractive intermolecular forces. The subsequent upward slopes at very high pressures are due to repulsive forces operating between molecules where the size of the molecules themselves becomes significant.

Liquefaction of gases

A change of state occurs when a gas liquefies. At the point of liquefaction, at a sufficiently low temperature and high pressure, gas molecules collide and then stay close to each other. The collisions are no longer elastic and the gas condenses to form a liquid held together by attractive intermolecular forces. The temperature at which a gas liquefies depends on the pressure. There is a maximum temperature called the **critical temperature** above which a gas cannot be liquefied by the application of pressure alone. Values for ammonia, methane, and hydrogen are respectively 406 K, 191 K, and 33 K. As a result, liquid ammonia may be stored under pressure in steel containers at ambient temperature (20 °C). Liquid methane and hydrogen must be kept in refrigerated containers whose temperatures are below the relevant critical temperature.

Supercritical fluids

A modern method of making decaffeinated coffee depends on the properties of carbon dioxide above its critical temperature (31 °C). This supercritical fluid (SCF) can dissolve caffeine from raw coffee beans, without dissolving most of the substances responsible for coffee's taste and aroma. After pumping away the solution of caffeine in SCF carbon dioxide, the coffee beans are free of caffeine; any residual carbon dioxide disperses away. Older methods left a residue of organic solvent. SCF carbon dioxide is also a safer solvent for dry cleaning than the traditional chlorinated hydrocarbons, spread 24.4.

SUMMARY

- At high pressure, attractive forces between molecules become significant; the volume of a real gas is less than that of an ideal gas.
- At very high pressure, repulsive forces between molecules become significant; the volume of a real gas is greater than that of an ideal gas.
- At low temperatures, intermolecular forces become more significant.
- The critical temperature is the temperature above which a gas cannot be liquefied by the application of pressure alone.

PRACTICE

1 Which of the following pairs of gases would you expect to behave most like an ideal gas? Explain your answers.

  a Helium He and carbon dioxide CO_2

  b Helium He and xenon Xe

  c Carbon dioxide CO_2 and methane CH_4.

2 For both of the gases helium and carbon dioxide, sketch the plot of pV_m/RT against p you would expect for pressures ranging from 1 atm to 1000 atm. Compare and contrast the two plots and explain their shapes.

3 Explain the significance of critical temperature for the following observation:

'Camping gas' for small portable stoves is supplied as liquid butane C_4H_{10} in a disposable steel cartridge. The pressure of the gas above the liquid is about 3 atm.

PRACTICE EXAM QUESTIONS

1 **a** Discuss the bonding present in each of the following, indicating any bond polarity:

i hydrogen chloride molecule, HCl, [2]

ii ammonium chloride crystal, NH_4Cl, [2]

iii water molecule, H_2O. [2]

b Ice is a solid melting at 0 °C. Name the intermolecular bonding present in ice and explain the origin of this bonding. [2]

c Explain why solid iodine is much more volatile than solid sodium chloride. [2]

d Calculate the volume of 4.509 g of iodine vapour at 343 K and 1.01×10^5 Pa pressure, if the vapour contains only I_2 molecules. [3]

[The molar volume of a gas at 273 K and 1.01×10^5 Pa is $2.24 \times 10^4\ \text{cm}^3\ \text{mol}^{-1}$.]

2 The equation $pV = nRT$ may be used in the determination of the relative molecular mass of a volatile liquid.

The diagram below shows the apparatus that could be used in this determination. A known mass of a volatile liquid is injected through the self-sealing rubber cap into the gas syringe, which is then heated in a boiling water bath for several minutes before the volume of the vapour in the syringe is noted. [15]

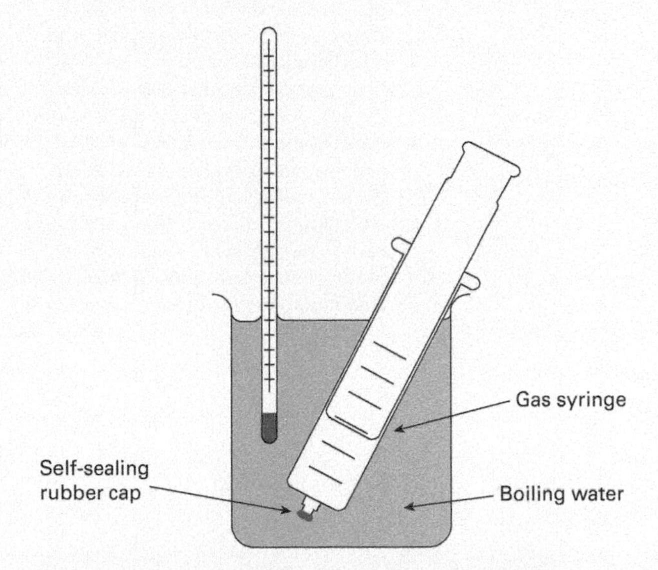

a **i** State what is meant by the term *volatile liquid*. Give the name of the equation $pV = nRT$, and state what the symbols n and R represent in this equation. [4]

ii Suggest why it is essential to leave the syringe in the boiling water bath for several minutes before reading the volume of the vapour. [2]

iii State **three** of the basic assumptions made in the kinetic theory of gases regarding the behaviour of gaseous molecules. [3]

b Using apparatus similar to that shown on the left, 0.167 g of ethanol, C_2H_5OH, was injected into a gas syringe and the syringe was then placed in a boiling water bath for several minutes. The atmospheric pressure was 101 300 Pa and the temperature of the bath was 100 °C. [5]

i Calculate the volume, in cm^3, of ethanol vapour that would have been produced under these conditions. [5]

$R = 8.314\ \text{J K}^{-1}\ \text{mol}^{-1}$

ii Explain why a gas syringe of 100 cm^3 capacity was found to be unsuitable. [1]

3 **a** With reference, where appropriate, to the kinetic theory of matter:

i Explain why a gas exerts a pressure; [2]

ii describe what happens as an ionic substance such as sodium chloride, melting point 801 °C, is steadily heated from room temperature until it melts; [2]

iii account for the relatively high melting points of ionic substances. [1]

b A volatile liquid, of mass 0.148 g, when vaporized occupies a volume of 63.0 cm^3 at a pressure of 1.01×10^5 Pa and a temperature of 100 °C. State the ideal gas equation and use it to calculate the relative molecular mass of the liquid. [5]

4 **a** **i** Copy the axis below and sketch a graph of V (the volume of an ideal gas) against $1/P$ (the reciprocal of its pressure) at a constant temperature. [1]

ii State **two** factors which cause real gases to deviate from ideal behaviour. [2]

b **i** Copy the axis below and sketch a graph to show the variation of the saturated vapour pressure of a liquid with temperature.

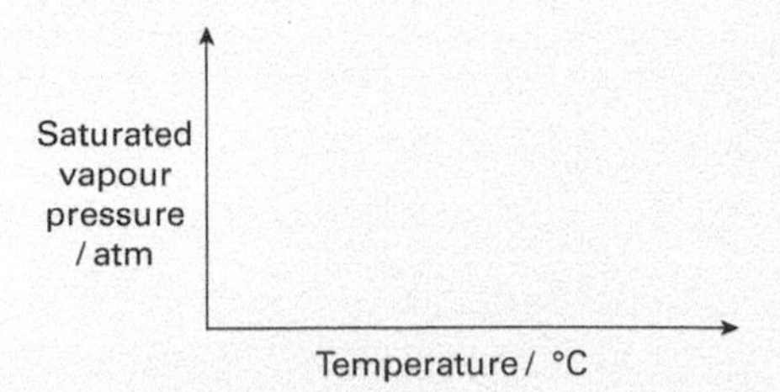

ii Explain why the boiling temperature of a liquid varies with the external pressure. [1]

iii Give **one** reason why some liquids are purefied by distillation under reduced pressure. [1]

c Calculate the molar mass of a volatile organic liquid, given that 0.597 g of the liquid when vaporized gives an ideal gas of volume 153 cm^3 at 100 °C and

1.01×10^5 Pa pressure (1 atmosphere).

[1 mole of an ideal gas occupies 2.24×10^4 cm^3 at 0 °C and 1.01×10^5 Pa pressure (1 atmosphere).] [3]

(See also chapter 22.)

5 The kinetic theory of gases explains the physical properties of a gas in terms of the behaviour of its particles. A gas which obeys the assumptions made in this theory is said to be an ideal gas.

a Give the ideal gas equation. [1]

b 10.0 g of carbon dioxide were placed in a vessel of volume 5.00 dm^3 and the temperature maintained at 0 °C. Calculate the pressure exerted by the gas.

[The value of the gas constant R is 8.31 J K^{-1} mol^{-1}.] [3]

c Explain why the actual pressure exerted was found to be slightly less than the calculated value. [1]

6 a i Draw a **labelled** diagram which shows the bonding in solid sodium. [2]

ii Use your diagram to explain two different physical properties which are **typical** of metals. [2]

b A mass spectrum of sodium vapour is shown in the following trace.

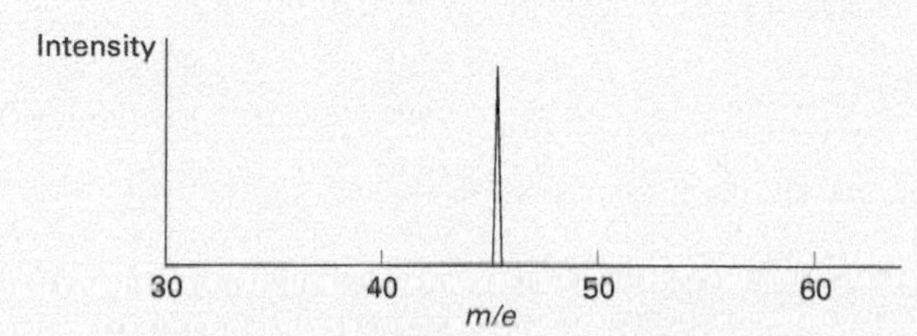

Explain the trace and draw a diagram showing the bonding in sodium under these conditions. [2]

c i A street lamp contains sodium vapour at a temperature of 20 °C and a pressure of 25 N m^{-2}. The volume of the tube containing the sodium is 50 cm^3. Calculate the mass of sodium in the lamp. [3]

ii What important assumption have you made in your calculation? [1]
[Gas constant R = 8.31 J mol^{-1} K^{-1}.]

d i Give the electron arrangement/configuration of the sodium **atom**, in terms of orbitals, in its ground state. [1]

ii Comment on the relative energies of the atomic orbitals given in **i**. [1]

iii Suggest the electron arrangement of a sodium atom in an **excited** state. [1]

7 a State the ideal gas equation. [1]

b The density of ethanoic acid, CH_3COOH, in the vapour state, is 2.74 g dm^{-3} at 400 K and 101 kPa.

i Calculate the apparent relative molecular mass of the acid under these conditions. [Gas constant R = 8.31 J K^{-1} mol^{-1}.] [3]

ii Suggest why the measured relative molecular mass differs from 60. [1]

c The figure below shows the variation in pV plotted against p for one mole of argon, at a constant temperature.

i Copy the figure and draw a line to show the behaviour of one mole of an ideal gas at the same temperature. [1]

ii Explain why argon does not behave as an ideal gas at pressures above zero. [2]

9 Reacting masses and volumes

You were introduced to the idea of using the mole as a unit of quantity in chemistry in spread 3.3. You should now understand the term 'molar mass' and be able to work out the mass per mole of any element or compound. This chapter extends these ideas and shows you how to calculate the yields of reactions, the formulae of compounds from analysis data, the volumes of gases evolved in reactions, and the concentrations and volumes of reacting solutions. These are all essential skills for practical chemists carrying out reactions in the laboratory or in industry.

Calculating masses involved in reactions

9.1

OBJECTIVES

- Mass conservation
- The relationship between amount in moles, mass, and molar mass
- Calculating masses in reactions

In a chemical reaction, reactants change into products. A chemical reaction may be represented by a balanced chemical equation, showing the formulae of the reactants and products, and the changes that take place. For example:

$$CaCO_3(s) + 2HCl(aq) \rightarrow CaCl_2(aq) + H_2O(l) + CO_2(g)$$

There are two important facts about any chemical equation: first, all the atoms present in the reactants are also present in the products; and secondly, a balanced chemical equation shows the *numbers* of species – molecules, ions, atoms, etc. – taking part in the reaction. These two facts allow the *masses* of substances involved in reactions to be calculated. By the end of this spread, you should be able to calculate the mass of product to be expected from a given mass of reactant, and the mass of reactant required to produce a given mass of product.

Mass conservation

The **law of mass conservation** states that:

- The total mass of a system remains constant during a chemical reaction.

In other words, for the reaction

$$A + B \rightarrow C + D$$

the total mass of the products (C + D) is equal to the total mass of the reactants (A + B). A balanced chemical equation obeys the law of mass conservation.

The law may be easily demonstrated by carrying out a chemical reaction on the pan of a balance, as shown to the left.

Mass remains constant during a chemical reaction. Before reaction, a flask with aqueous silver nitrate and a measuring cylinder with potassium chromate(VI) are weighed. Mixing the solutions forms a solid precipitate of silver chromate(VI) and aqueous potassium nitrate. There is no change in mass, as indicated by the balance.

Moles and mole ratios

Calculations to find the masses of substances in chemical reactions always start from the balanced chemical equation for the reaction. The equation shows the amount in moles of each substance involved in the reaction. For example, the reaction between hydrogen and oxygen to produce water is:

$$2H_2(g) + O_2(g) \rightarrow 2H_2O(l)$$

This equation translates into plain language as: '2 moles of hydrogen molecules react with 1 mole of oxygen molecules to give 2 moles of water molecules'.

The ratios between the amounts in moles of reactants and the amounts in moles of products are called the **mole ratios** (previously called *stoichiometric ratios*). In this example, the mole ratios H_2:O_2:H_2O are 2:1:2. In other words, the amount in moles of hydrogen is twice the amount in moles of oxygen required to react with it; the amount in moles of hydrogen is the same as the amount in moles of water; and the amount in moles of water is twice the amount in moles of oxygen. The concept of mole ratio is the central idea on which the following calculations depend.

Mass-to-mass calculation

In this type of calculation, you might be asked to work out, for example, the mass of reactant needed to produce a certain mass of product or the mass of product that results from using a certain mass of reactant.

A calculation of this sort requires the balanced chemical equation and data about the mass of one reactant or product (A). The question will then ask you to work out the mass of another reactant or product (B). The route for the calculation follows a set path, with the following steps:

1 Use data about the mass of A to calculate the amount in moles of A.

2 Use the balanced chemical equation to state the mole ratio B:A.

3 Calculate the amount in moles of B.

4 Use the amount in moles of B to calculate the mass of B.

A mass-to-mass calculation therefore follows the sequence:

$$\text{Mass of A} \rightarrow \text{Moles of A} \xrightarrow{\text{Mole ratio B:A}} \text{Moles of B} \rightarrow \text{Mass of B}$$

Worked example of a mass-to-mass calculation

Sodium is produced in industry by the electrolysis of molten sodium chloride, spread 16.1.

Question: What mass of sodium chloride is needed to produce 1.00 kg of sodium?

Strategy: The calculation follows the four steps outlined above.

Answer:

Step 1 Use data about the mass of sodium to calculate its amount in moles.

We have

$$\text{amount in moles of Na} = \frac{\text{mass of Na}}{\text{molar mass of Na}}$$

so

$$n(\text{Na}) = \frac{1000\,\text{g}}{23.0\,\text{g mol}^{-1}} = 43.5\,\text{mol}$$

Step 2 Use the balanced chemical equation to state the mole ratio of NaCl:Na.

The balanced chemical equation for the reaction is

$2NaCl(l) \rightarrow 2Na(l) + Cl_2(g)$

The mole ratio of NaCl:Na is therefore 2:2, i.e. 1:1.

Step 3 Calculate the amount in moles of sodium chloride.

The amount of Na is 43.5 mol. The mole ratio NaCl:Na is 1:1. Therefore the amount of NaCl is 43.5 mol.

Step 4 Use the amount in moles of NaCl to calculate the mass of sodium chloride.

We have

mass of NaCl = amount in moles of NaCl × molar mass of NaCl

The molar mass of NaCl is $(23.0\,\text{g mol}^{-1} + 35.5\,\text{g mol}^{-1}) = 58.5\,\text{g mol}^{-1}$. Therefore:

$$m(\text{NaCl}) = (43.5\,\text{mol}) \times (58.5\,\text{g mol}^{-1}) = 2540\,\text{g (3 sig. figs)}$$
$$= 2.54\,\text{kg}$$

That is, 2.54 kg of sodium chloride is needed to produce 1.00 kg of sodium.

SUMMARY

- The total mass of the products in a chemical reaction is equal to the total mass of the reactants.
- Amount in moles = mass/molar mass; in symbols, $n = m/M$.
- A mass-to-mass calculation follows the path: $\text{Mass of A} \rightarrow \text{Moles of A} \xrightarrow{\text{Mole ratio B:A}} \text{Moles of B} \rightarrow \text{Mass of B}$

Reminder – amount in moles, mass, and molar mass

The molar mass (symbol M, unit g mol^{-1}) of a substance is defined as the mass (symbol m, unit g) present divided by the amount in moles (symbol n, unit mol), i.e.

$$\text{molar mass} = \frac{\text{mass}}{\text{amount in moles}}$$

or

$$M = \frac{m}{n}$$

The numerical value of the molar mass of a species is the same as its relative formula mass, but the molar mass has units, g mol^{-1}, whereas the relative formula mass does not.

Multiplying both sides of the equation above by n shows that the mass is equal to the amount in moles multiplied by the molar mass, i.e.

mass = amount in moles × molar mass

or

$$m = nM$$

Dividing both sides of the equation by M shows that the amount in moles is equal to the mass divided by the molar mass, i.e.

$$\text{amount in moles} = \frac{\text{mass}}{\text{molar mass}}$$

or

$$n = \frac{m}{M}$$

% atom economy

When there is more than one reactant, it is useful to consider the following:

$$\textbf{\% atom economy} = \frac{\text{mass of desired product}}{\text{total mass of reactants}} \times 100$$

Reactions with a higher atom economy produce fewer waste materials. One example of a process with a high atom economy is an addition reaction such as the manufacture of ethanoic acid by methanol carbonylation, spread 27.1: methanol can itself be manufactured with a high atom economy, spread 25.1.

PRACTICE

1 Calculate the mass of sodium that results from the electrolysis of 50.0 g of molten sodium chloride.

2 Calculate the mass of chlorine that accompanies the production of 1.00 kg of sodium during the electrolysis of molten sodium chloride.

9.2

OBJECTIVES

- Aluminium production
- The 'top 10' chemicals
- A laboratory reaction
- Percentage yield

FURTHER MASS-TO-MASS CALCULATIONS

The concept of molar mass is the key to carrying out mass-to-mass calculations. These calculations provide answers to questions of the sort: 'What mass of this reactant is required to produce x grams of that product?' or ' What mass of this product results from y grams of that reactant?' This spread provides further worked examples of mass-to-mass calculations, in both industrial and laboratory-based settings.

The crash test result of the Jaguar XJ220, which has an aluminium frame. Unlike steel-framed cars there is no deformation of the area around the passengers and no bursting at the seams of the doors.

Bauxite and aluminium

Aluminium is produced from the electrolysis of molten aluminium oxide, which is prepared from the ore *bauxite*. You should be familiar with the worked example given in the previous spread involving the extraction of sodium from sodium chloride. Answering the question 'What mass of aluminium oxide is needed to produce 1 kg of aluminium?' involves a similar calculation.

Worked example on aluminium production

Question: What mass of aluminium oxide is needed to produce 1.00 kg of aluminium?

Strategy: The calculation is carried out in the same four steps as in the previous spread.

Answer:

Step 1 We have

$$\text{amount in moles of Al} = \frac{\text{mass of Al}}{\text{molar mass of Al}}$$

so

$$n(\text{Al}) = \frac{1000\,\text{g}}{27.0\,\text{g mol}^{-1}} = 37.0\,\text{mol}$$

Step 2 The balanced chemical equation for the reaction is

$$2Al_2O_3(l) \rightarrow 4Al(l) + 3O_2(g)$$

The mole ratio Al_2O_3:Al is 2:4, i.e. 1:2.

Step 3 The amount of Al is 37.0 mol. The mole ratio Al_2O_3:Al is 1:2. Therefore the amount of Al_2O_3 is 37.0/2 = 18.5 mol.

Step 4 We have

mass of Al_2O_3 = amount in moles of Al_2O_3 × molar mass of Al_2O_3

The molar mass of Al_2O_3 is $2 \times (27.0\,\text{g mol}^{-1}) + 3 \times (16.0\,\text{g mol}^{-1}) = 102\,\text{g mol}^{-1}$.

Therefore

$$m(Al_2O_3) = (18.5\,\text{mol}) \times (102\,\text{g mol}^{-1}) = 1890\,\text{g (3 sig. figs)}$$
$$= 1.89\,\text{kg}$$

That is, 1.89 kg of aluminium oxide is needed to produce 1.00 kg of aluminium.

The Hall–Héroult cell for the extraction of aluminium. Graphite anodes dip into molten aluminium oxide dissolved in cryolite. The steel cell is also lined with graphite, which acts as the cathode. Oxygen gas is evolved at the anodes and liquid aluminium is produced at the cathode. The cell contents are kept molten at around 950 °C by the passage of the electric current.

The top 10 chemicals manufactured

The top 10 chemicals (ignoring water, steel, and sodium chloride) manufactured in the USA between 2000 and 2009 are listed on the opposite page by mass in megatonnes. The top six are clear of the field; there is a tight bunch jostling for seventh place.

One megatonne is 10^6 tonnes, which is 10^9 kg or 10^{12} g. In terms of moles, the top three chemicals are nitrogen, ethene and oxygen, in that order. The *amount in moles* of nitrogen manufactured is

$$\frac{34 \times 10^{12}\,\text{g}}{28\,\text{g mol}^{-1}} = 1.2 \times 10^{12}\,\text{mol}$$

which is more than one thousand billion moles.

The rise of China

In 2004 both the USA and China produced 38 megatonnes of sulphuric acid. Five years later, the USA produced 30 megatonnes: China produced exactly twice as much.

Worked example on sulphuric acid production

Sulphuric acid, the chemical produced in highest mass, is produced in a series of reactions from elemental sulphur. The overall reaction may be represented as:

$2S + 2H_2O + 3O_2 \rightarrow 2H_2SO_4$

Question: What mass of sulphur was used to make the 43.3 megatonnes of sulphuric acid produced in the USA in 1995? Less is made now (see table).

Strategy: The calculation follows the same steps as before, but now we do not list all the lines.

Answer:

43.3 megatonnes = 43.3×10^{12} g

The molar mass of H_2SO_4 is

$2 \times (1.0\,g\,mol^{-1}) + 32.1\,g\,mol^{-1} + 4 \times (16.0\,g\,mol^{-1}) = 98.1\,g\,mol^{-1}$

We have

$$n(H_2SO_4) = \frac{43.3 \times 10^{12}\,g}{98.1\,g\,mol^{-1}} = 4.41 \times 10^{11}\,mol$$

The mole ratio of $S{:}H_2SO_4$ is 2:2, i.e. 1:1.

The amount in moles of sulphur required is the same as the amount in moles of sulphuric acid produced, i.e. 4.41×10^{11} mol. So

mass of sulphur = amount in moles of sulphur × molar mass of sulphur

$m(S) = (4.41 \times 10^{11}\,mol) \times (32.1\,g\,mol^{-1}) = 14.2 \times 10^{12}\,g$

That is, 43.3 megatonnes of sulphuric acid are produced from 14.2 megatonnes of sulphur.

Chemical	Mass produced /megatonnes (10^{12} g)	Molar mass /g mol^{-1}
1 Sulphuric acid	36	98
2 Nitrogen	34*	28
3 Oxygen	25*	32
4 Ethene	24	28
5 Calcium oxide (lime)	21*	56
6 Propene	15	44
7 Ammonia	11	17
8 Phosphoric acid	11	98
9 Chlorine	11	71
10 1,2-Dichloro-ethane	10	99

**Approximate values. The average mass of the top 10 chemicals produced annually in the USA between 2000 and 2009. The relative positions of ammonia, phosphoric acid, and chlorine varied from year to year.*

We can express the success of a particular reaction as a **percentage yield**:

$$\text{percentage yield} = \frac{\text{(actual mass of product)}}{\text{(theoretical mass of product)}} \times 100$$

How big is 14.2 megatonnes of sulphur? What size cube would it fill?

We know that

$$\text{density} = \frac{\text{mass}}{\text{volume}}$$

i.e.

$$\text{volume} = \frac{\text{mass}}{\text{density}}$$

The density of sulphur is 2.07 g cm^{-3}, which is equal to 2.07 tonnes m^{-3}. The mass of sulphur is 14.2×10^6 tonnes. The volume of this sulphur is

$$\text{volume} = \frac{14.2 \times 10^6\,\text{tonnes}}{2.07\,\text{tonnes}\,m^{-3}} = 6.86 \times 10^6\,m^3$$

This volume is equivalent to a cube of side 190 m. To put this value into perspective, the height of the Eiffel Tower in Paris is 300 m.

Worked example on calculating a percentage yield

Question: What mass of ethanol could form from fermentation of 90.0 g of glucose $C_6H_{12}O_6$? What is the percentage yield if 6.0 g of ethanol forms?

Strategy: The calculation follows the same steps. Then find the percentage yield.

Answer: As the molar mass of $C_6H_{12}O_6$ is

$6 \times (12.0\,g\,mol^{-1}) + 12 \times (1.0\,g\,mol^{-1}) + 6 \times (16.0\,g\,mol^{-1}) = 180\,g\,mol^{-1}$

$$n(C_6H_{12}O_6) = \frac{90.0\,g}{180\,g\,mol^{-1}} = 0.500\,mol$$

The chemical equation, spread 25.1, is:

$C_6H_{12}O_6(aq) \rightarrow 2CH_3CH_2OH(aq) + 2CO_2(g)$

The mole ratio of $CH_3CH_2OH{:}C_6H_{12}O_6$ is 2:1 so

$n(CH_3CH_2OH) = 2 \times (0.500\,mol) = 1.00\,mol$

The molar mass of CH_3CH_2OH is

$2 \times (12.0\,g\,mol^{-1}) + 6 \times (1.0\,g\,mol^{-1}) + 16.0\,g\,mol^{-1} = 46.0\,g\,mol^{-1}$

Hence the theoretical mass of ethanol = $(1.00\,mol) \times (46.0\,g\,mol^{-1}) = 46.0\,g$

$$\text{The percentage yield} = \frac{6.0\,g}{46.0\,g} \times 100 = 13\%$$

An industrial-scale pile of sulphur.

SUMMARY

Mass-to-mass calculations follow a set pathway:

- Use data about the mass to calculate the amount in moles of the substance whose mass is known.
- Use the balanced chemical equation to state the mole ratios of the substances in the question.
- Find the amount in moles of the substance whose mass is unknown.
- Use the amount in moles to find the unknown mass.

9.3

FINDING EMPIRICAL FORMULAE

OBJECTIVES

- Empirical and molecular formulae
- Calculating empirical formulae from mass composition data

We can use calculations involving the masses of the elements that combine together to find the formulae of compounds. The specific version of the formula that can be found is the simplest whole-number ratio of the atoms present.

Empirical and molecular formulae

The **empirical formula** of a compound gives the simplest whole-number ratio of the atoms of each element present. The **molecular formula** of a covalent compound that consists of small molecules gives the actual numbers of atoms of each element present in one molecule.

The molecular formula of a covalent compound that consists of small molecules is a whole-number multiple of the empirical formula. For example, the empirical formula of ethane is CH_3, as determined by various methods of analysis. Each molecule contains three times as many hydrogen atoms as there are carbon atoms. The relative formula mass corresponding to the empirical formula CH_3 is

$$12.0 + (3 \times 1.0) = 15.0$$

Experimental data show that the molar mass of ethane is $30.0\,g\,mol^{-1}$. The relative formula mass of ethane is therefore 30.0. This value, which corresponds to the molecular formula, is twice the value for the empirical formula. The molecular formula of ethane is therefore $(2 \times CH_3) = C_2H_6$. Each molecule of ethane contains two atoms of carbon and six atoms of hydrogen.

The formula of an ionic compound is an empirical formula. Ionic compounds do not exist as separate molecules but as giant lattices consisting of ions. The formula for common salt is NaCl. This empirical formula indicates that the ratio Na:Cl is 1:1. Empirical formulae may be calculated from data about the composition by mass.

Empirical formula calculations – the three steps

1 Find the amount in moles of each element present by dividing the mass of each element by its molar mass.

2 Find the ratio of the number of atoms of each element by dividing the amounts in moles by the smallest value found in step 1.

3 Convert these numbers into whole numbers, because atoms combine together in whole-number ratios.

The mineral calcite is a crystalline form of calcium carbonate $CaCO_3$. Calcite is birefringent, producing two refracted rays.

Worked example on the empirical formula of calcite

Data: The percentage by mass of the elements in calcite is 40.0% calcium, 12.0% carbon, and 48.0% oxygen.

Question: What is the empirical formula of calcite?

Strategy: Follow the three steps shown above.

Answer:

Step 1 Divide each mass by the molar mass of the element.

The data are given as percentages, and so this step will give the amount in moles of each element in 100 g of calcite. We get

$$\text{amount in moles of Ca} = \frac{40.0\,g}{40.1\,g\,mol^{-1}} = 0.998\,mol$$

$$\text{amount in moles of C} = \frac{12.0\,g}{12.0\,g\,mol^{-1}} = 1.00\,mol$$

$$\text{amount in moles of O} = \frac{48.0\,g}{16.0\,g\,mol^{-1}} = 3.00\,mol$$

Step 2 Divide by the smallest value.

The smallest value in this case (0.998 mol) is very close to 1 mol, so we can move on to step 3.

Step 3 Convert to whole numbers.

We get

Ca:C:O = 1:1:3

The empirical formula of calcite is $CaCO_3$.

Worked example on the empirical formula of aspirin

Data: A sample of aspirin contains 6.00 g of carbon, 0.44 g of hydrogen, and 3.56 g of oxygen.

Question: What is the empirical formula of aspirin?

Strategy: We follow the same three steps as before.

Answer:

Step 1

We have

$$\text{amount in moles of C} = \frac{6.00\,\text{g}}{12.0\,\text{g mol}^{-1}} = 0.500\,\text{mol}$$

$$\text{amount in moles of H} = \frac{0.44\,\text{g}}{1.0\,\text{g mol}^{-1}} = 0.44\,\text{mol}$$

$$\text{amount in moles of O} = \frac{3.56\,\text{g}}{16.0\,\text{g mol}^{-1}} = 0.223\,\text{mol}$$

Step 2

The smallest value is the amount in moles of O (0.223 mol). So

$$C = \frac{0.500\,\text{mol}}{0.223\,\text{mol}} = 2.24$$

$$H = \frac{0.44\,\text{mol}}{0.223\,\text{mol}} = 2.0$$

$$O = \frac{0.223\,\text{mol}}{0.223\,\text{mol}} = 1.0$$

Step 3

The ratio C:H:O is 2.24:2.0:1.00. It is necessary to multiply by 4 to obtain numbers that are all near to whole numbers, i.e. 8.96:8.0:4.00. So C:H:O = 9:8:4.

The empirical formula of aspirin is $C_9H_8O_4$.

Crystals of aspirin $C_9H_8O_4$ viewed in polarized light. The aspirin molecule (below) consists of a six-membered ring of carbon atoms with two side-groups attached. The synthesis of this compound will be discussed in spread 27.2.

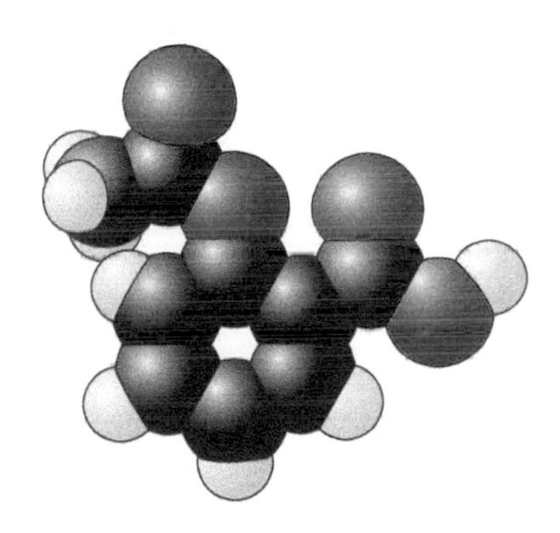

Aspirin

In the case of aspirin, the relative formula mass corresponding to the empirical formula $C_9H_8O_4$ is:

$(9 \times 12.0) + (8 \times 1.0) + (4 \times 16.0) = 180$

Experimental data show that the relative formula mass of aspirin is 180. The molecular formula of aspirin is therefore the same as the empirical formula, i.e. $C_9H_8O_4$.

SUMMARY

- The empirical formula gives the simplest whole-number ratio of the atoms of each element in a compound.
- The molecular formula gives the actual numbers of atoms of each element in one molecule of a covalent compound.
- Empirical formulae are found by calculating the mole ratio of the constituent elements from mass composition data.

PRACTICE

1 Calculate the empirical formula of the ore haematite. Its percentage composition is 70% iron and 30% oxygen.

2 Calculate the empirical formula of the ore iron pyrite (also known as 'fool's gold'). It was found that 1.00 g of the ore contains 0.47 g of iron and 0.53 g of sulphur.

9.4 Calculations involving gases

OBJECTIVES

- Mass-to-volume calculations
- Volume-to-volume calculations

You should now be familiar with calculating the masses of substances involved in reactions, given a balanced chemical equation for the reaction and the mass of one of the reactants or products. Many reactions involve gases, as reactants and/or as products. For example, hydrogen gas is evolved when zinc granules react with dilute hydrochloric acid; and the gases hydrogen and oxygen react together to form water. The aim of this spread is to develop the concepts underlying mass-to-volume and volume-to-volume calculations, so enabling you to calculate the volumes of gases involved in chemical reactions.

Mass-to-volume calculations – overall strategy

In the reaction between zinc and excess hydrochloric acid, the volume of hydrogen evolved depends on the mass of zinc that has reacted. A typical question might ask you to calculate the volume of hydrogen evolved when, for example, 5.00 g of zinc reacts with excess hydrochloric acid. The strategy for this type of calculation follows a set path, with the following steps:

1. Use data about the mass of the solid to calculate the amount in moles of solid.
2. Use the balanced chemical equation to state the mole ratio of the gas to the solid.
3. Calculate the amount in moles of the gas.
4. Use the ideal gas equation $pV = nRT$ to calculate the volume of the gas at the stated conditions of temperature and pressure.

A mass-to-volume calculation therefore follows the path:

$$\text{Mass of solid} \rightarrow \text{Moles of solid} \xrightarrow{\text{mole ratio gas:solid}} \text{Moles of gas} \xrightarrow{\text{ideal gas equation}} \text{Volume of gas}$$

You will be asked to carry out a calculation for the reaction between zinc and hydrochloric acid in the practice questions at the end of the spread. The following worked example concentrates on a more practical use of a mass-to-volume reaction.

Excess reactant and limiting reactant

The phrase 'excess hydrochloric acid' implies that there is more than enough acid present to react with all the zinc added to it. When the reaction is complete, all the zinc has reacted, and some acid remains unused.

The substance that is completely used up is called the **limiting reactant**; it limits the yield of product. In the example to the right, the zinc metal is the limiting reactant and the acid is the reactant present in excess.

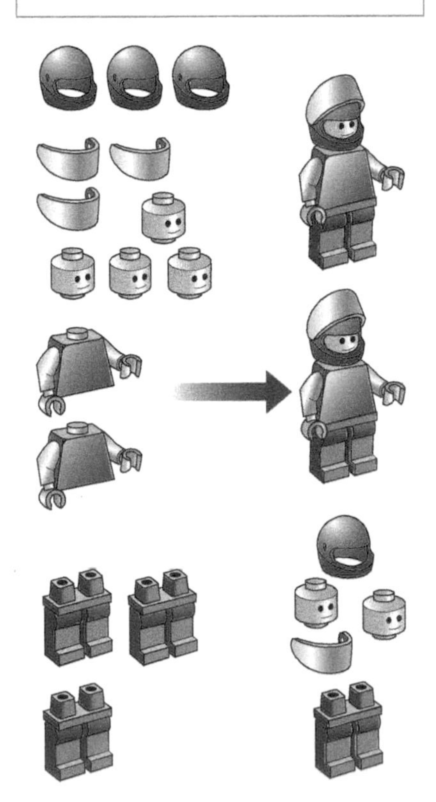

Only two figures can be made, because there are only two bodies. Bodies are limiting; the other components are present in excess.

The chemistry of an airbag

Airbags are fitted to many cars as safety equipment. An airbag has a sensor that can detect a severe collision. The airbag inflates about 20 milliseconds after the collision has been detected. Once fully inflated, it is designed to collapse gradually.

The most important chemical in most airbags is sodium azide NaN_3. When an impact has been detected, the NaN_3 is electrically ignited. It then decomposes very rapidly to produce sodium metal and nitrogen gas. It is the nitrogen gas that inflates the airbag. The equation for the decomposition is:

$$2NaN_3(s) \rightarrow 2Na(s) + 3N_2(g)$$

Worked example on an airbag

Data: An airbag contains 75 g of solid sodium azide. Assume that the temperature is 20 °C and the pressure is 1.00 bar.

Question: What is the volume of nitrogen gas produced?

Strategy: Follow the four steps shown above.

Answer:

Step 1 Calculate the amount in moles of solid.

The molar mass of NaN_3 is

$23.0\,g\,mol^{-1} + 3 \times (14.0\,g\,mol^{-1}) = 65.0\,g\,mol^{-1}$

So

amount in moles of $NaN_3 = \frac{75.0\,g}{65.0\,g\,mol^{-1}} = 1.15\,mol$

Step 2 State the mole ratio gas:solid.

The mole ratio $N_2:NaN_3 = 3:2$.

Step 3 Calculate the amount in moles of gas.

Therefore,

amount in moles of N_2 produced $= (\frac{3}{2}) \times 1.15\,mol = 1.73\,mol$

Step 4 Use the ideal gas equation to calculate the volume of gas.

The volume at 20 °C and 1.00 bar is obtained from the ideal gas equation $pV = nRT$. After dividing both sides by p, it can be written as

$$V = \frac{nRT}{p}$$

so for the airbag

$$V = \frac{(1.73\,mol) \times (8.31\,Pa\,m^3\,K^{-1}\,mol^{-1}) \times (293\,K)}{(1.00 \times 10^5\,Pa)}$$

$= 0.042\,m^3 = 42\,dm^3$ (2 sig. figs)

The decomposition can therefore provide 42 litres of gas very quickly indeed. Manufacturers determine the optimum volume of gas required in the inflated airbag and then vary the mass of sodium azide accordingly.

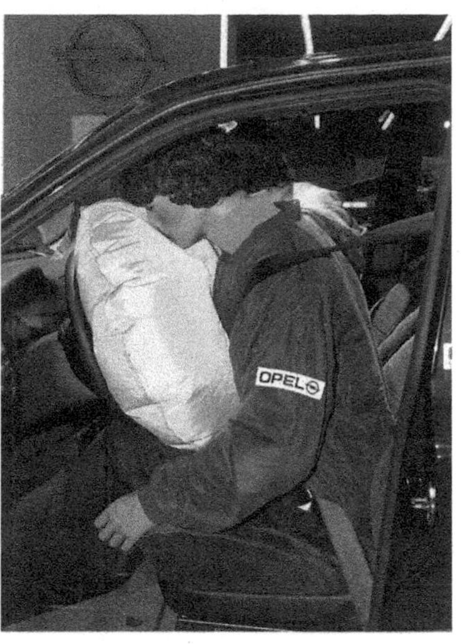

Sensors detect that a crash is happening and the airbag inflates within about 20 milliseconds. A canister of compressed gas would not operate with sufficient speed. The gas in the bag must be non-toxic and not flammable in case of leaks, so nitrogen is an ideal choice.

Volume-to-volume calculations

This type of calculation is extremely simple and straightforward, and may be carried out in a matter of seconds. For example, hydrogen and oxygen react to form water.

$$2H_2(g) + O_2(g) \rightarrow 2H_2O(l)$$

Avogadro's principle states that:

- Equal volumes of different gases contain the same number of molecules, under the same conditions of temperature and pressure.

A consequence of Avogadro's principle is that one mole of *any gas* has the same volume under stated conditions of temperature and pressure.

The balanced equation for the reaction between hydrogen and oxygen indicates that 2 mol of hydrogen reacts with 1 mol of oxygen. Thus, whatever the volume of hydrogen, half that volume of oxygen will be required for reaction, i.e. $50.0\,cm^3$ of hydrogen require $25.0\,cm^3$ of oxygen.

- The volumes of gases that react (and the volumes of the products if gaseous) are proportional to their mole ratio.

This is sometimes called **Gay-Lussac's law**, after Joseph Gay-Lussac.

Molar volume of a gas

In the calculations to the left, the ideal gas equation is used to convert the amount in moles of a gas to the volume of the gas. This equation includes the variables p and T and thus enables calculation of the volume under *any* conditions of temperature and pressure. However, a simpler expression may be used to convert the amount in moles of gas to gas volume:

$$\text{amount in moles} = \frac{\text{volume}}{\text{molar volume}}$$

This uses the molar volume of the gas. At 20 °C and 1 bar, you should remember, spread 8.3, that 1 mol of any gas occupies $24\,dm^3$, i.e.

molar volume of any gas is $24\,dm^3\,mol^{-1}$ (at 20 °C, 1 bar)

and this value can be used in the above expression under these conditions.

SUMMARY

- Mass-to-volume calculations follow the path:

$$\text{Mass of solid} \rightarrow \text{Moles of solid} \xrightarrow{\text{mole ratio gas:solid}} \text{Moles of gas} \xrightarrow{\text{ideal gas equation}} \text{Volume of gas}$$

- The limiting reactant is the substance that is completely consumed in the course of a reaction; its amount limits the amounts of products that may form.

PRACTICE

1 Zinc reacts with hydrochloric acid to produce hydrogen gas. Calculate:

a The volume of hydrogen produced at 293 K and 1.00 bar when 5.00 g of zinc reacts with excess dilute hydrochloric acid.

b The mass of zinc required to produce $1000\,cm^3$ of hydrogen at 293 K and 1.00 bar (excess acid present).

2 Calculate the volume of sulphur dioxide produced at 293 K and 1.00 bar when 5.00 g of sulphur burns in oxygen.

9.5 SOLUTION CONCENTRATIONS

OBJECTIVES

- Solutes, solvents, and solutions
- Definition of molar concentration
- Calculating molar concentration
- Calculating amount in moles from molar concentration

Warning

Do not use the words 'strong' or 'weak' to refer to solutions of high or low concentration. These technical terms are used to describe acids and bases (see chapter 12).

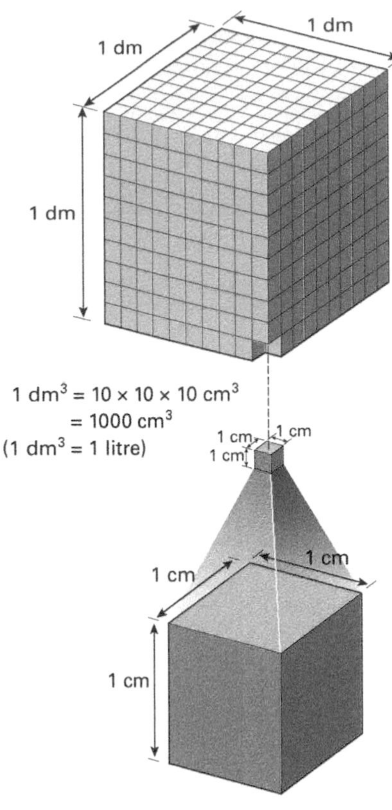

Volume interconversions. The SI unit of volume is the cubic metre (m³), spread 8.2. Of more practical use for laboratory solutions are the cubic decimetre (dm³) and the cubic centimetre (cm³).

A **solution** consists of a substance dissolved in a solvent. The dissolved substance is called the **solute** and may be a solid, a liquid, or a gas. The **solvent** is almost always a liquid, although solid solutions can occur (e.g. some metal alloys). The **concentration** of a solution measures how much of a dissolved substance is present per unit volume of a solution. A solution is described as 'concentrated' if it consists of a large quantity of solute in a small quantity of solvent. A small quantity of solute in a large quantity of solvent results in a 'dilute' solution.

Solution concentration

The common unit of volume for solutions is the cubic decimetre (dm^3), also known as the litre (L or l): 1 cubic decimetre (1 dm^3) is equivalent to 1000 cubic centimetres (1000 cm^3); 1 cm^3 is also known as 1 ml (millilitre). The concentration of a solution is usually defined as the quantity of solute per unit volume of the solution.

The **mass concentration** of a solution may be written, for example, as mass per cubic decimetre; for example, the concentrations of the ions present in bottled spring water are usually stated in milligrams per cubic decimetre (mg dm^{-3}).

Molar concentration

Chemists usually express the concentration of a solution in terms of its molar concentration (a traditional name for this was molarity). By definition, the **molar concentration** (c) of a solution is equal to the amount in moles of a solute (n) divided by the volume (V) of the solution, i.e.

$$\text{molar concentration} = \frac{\text{amount in moles of solute}}{\text{volume of solution}}$$

or

$$c = \frac{n}{V}$$

Multiplying both sides of this equation by V makes the amount in moles of solute the subject of the expression, i.e.

amount in moles of solute = volume of solution × molar concentration

or

$$n = Vc$$

The units of molar concentration are moles per cubic decimetre (mol dm^{-3}).

Often a 0.1 mol dm^{-3} solution is referred to as a 0.1 M solution.

Worked example 1

Question: The concentration of calcium ions in a sample of spring water is 100 mg dm^{-3}. Calculate the molar concentration of the calcium ions in mol dm^{-3}.

Answer: 1 dm^3 of water contains 100 mg of $Ca^{2+}(aq)$ ions. Dividing this mass by the molar mass of calcium (40.1 g mol^{-1}) gives the amount in moles of $Ca^{2+}(aq)$ ions, i.e.

$$\text{amount in moles of } Ca^{2+}(aq) = \frac{(100 \times 10^{-3}\,g)}{(40.1\,g\,mol^{-1})} = 2.5 \times 10^{-3}\,mol$$

This amount of calcium ions is present per dm^3 of solution. Therefore, the molar concentration of calcium ions is 2.5×10^{-3} mol dm^{-3}.

Worked example 2

Question: 250 cm³ of a solution contains 5.85 g of sodium chloride. Calculate the molar concentration of sodium chloride in $mol\,dm^{-3}$.

Answer: We have

$$\text{amount in moles of NaCl} = \frac{\text{mass of NaCl}}{\text{molar mass of NaCl}} = \frac{5.85\,g}{58.5\,g\,mol^{-1}} = 0.100\,mol$$

Also

$$\text{molar concentration} = \frac{\text{amount in moles of solute}}{\text{volume of solution}} \qquad c = \frac{n}{V}$$

First convert the volume in cm^3 to a volume in dm^3 by dividing by 1000:

$$250\,cm^3 = \frac{250}{1000}\,dm^3 = 0.250\,dm^3$$

Therefore:

$$\text{molar concentration} = \frac{0.100\,mol}{0.250\,dm^3} = 0.400\,mol\,dm^{-3}$$

The molar concentration of the solution is $0.400\,mol\,dm^{-3}$.

*The **volume concentration** of a solution is usually stated as a percentage. This wine is labelled '13% vol.', which means that it contains 13 cm³ of alcohol in 100 cm³ of wine. Health workers call 10 cm³ of alcohol '1 unit', so a 0.75 litre bottle of wine contains nearly 10 units of alcohol, sufficient in the UK to put you more than twice over the drink–drive limit.*

Worked example 3

Question: What volume of hydrogen gas (at 20 °C and 1 bar) can be produced from 250 cm³ of $2.0\,mol\,dm^{-3}$ hydrochloric acid and excess zinc?

Answer: We have

amount in moles of HCl = volume of HCl × molar concentration of HCl

First convert the volume in cm^3 to a volume in dm^3 by dividing by 1000:

$$250\,cm^3 = \frac{250}{1000}\,dm^3$$

So the amount in moles of HCl is

$$n(HCl) = V(HCl) \times c(HCl) = \left(\frac{250}{1000}\,dm^3\right) \times (2.0\,mol\,dm^{-3}) = 0.50\,mol$$

The chemical equation is:

$$Zn(s) + 2HCl(aq) \rightarrow ZnCl_2(aq) + H_2(g)$$

The mole ratio H_2:HCl is 1:2. So the amount in moles of H_2 produced is

$$n(H_2) = \tfrac{1}{2} \times n(HCl) = \tfrac{1}{2} \times 0.50\,mol = 0.25\,mol$$

The molar volume at 20 °C and 1 bar is $24\,dm^3\,mol^{-1}$. The volume of the 0.25 mol of hydrogen gas produced is therefore

$$(0.25\,mol) \times (24\,dm^3\,mol^{-1}) = 6.0\,dm^3 \text{ (2 sig. figs)}$$

Note: In the example in the previous spread, the volume of hydrogen produced was limited by the mass of zinc, because the *acid* was present in excess. Here, the *zinc* is in excess, and so the acid limits the volume of hydrogen.

Laboratory solutions

Laboratories are usually equipped with a selection of reagent bottles containing solutions of acids, alkalis, and other substances. Knowing the molar concentration of a reagent in solution allows you to calculate the amount of product to be expected from a reaction.

Zinc reacts with hydrochloric acid to produce hydrogen gas and aqueous zinc chloride. When excess zinc is present, the limiting reactant is the acid.

SUMMARY

- Molar concentration = $\dfrac{\text{amount in moles of solute}}{\text{volume of solution}}$

PRACTICE

1 Calculate the amount in moles of solute in 35.0 cm³ of a solution that has a concentration of $0.100\,mol\,dm^{-3}$.

2 Calculate the volume of a solution that has a concentration of $0.25\,mol\,dm^{-3}$ and contains 0.125 mol of solute.

9.6

OBJECTIVES

- Standard solutions
- Titration: burette, pipette, indicators

USING TITRATION TO MEASURE CONCENTRATION

Suppose you are in charge of the quality control department in a vinegar factory. How can you be sure that each batch of vinegar contains the right concentration of acid? The answer is to carry out a **titration**: add vinegar to a known volume of alkali, of known molar concentration, until the solution just becomes acidic. You know the molar concentration and volume of the alkali. You also know the volume of vinegar used. You can perform a calculation involving mole ratio, molar concentration, and volume to find the concentration of the acid in the vinegar. This spread gives details about the titration technique; the following spread discusses the necessary calculations.

Standard solutions

One of the solutions used in a titration must be a **standard solution**, a solution whose concentration is accurately known. A standard solution is made by adding a known mass of solute to the solvent and making up the solution to a known volume. The accuracy of a standard solution in the school laboratory depends partly on the accuracy with which the substances are weighed out or measured, and partly on the purity of the solute used. Chemical manufacturers supply many reagents of especially high purity for use in industry and in the laboratory.

Solid sodium hydroxide reacts with atmospheric carbon dioxide:

$$2NaOH(s) + CO_2(g) \rightarrow Na_2CO_3(s) + H_2O(l)$$

So, an accurate concentration cannot be given for aqueous sodium hydroxide made from an *ancient* sample of solid that may be only 95% pure. On the other hand, reagents such as solid sodium carbonate and ethanedioic acid $(COOH)_2$ are available to at least 99.9% purity. These reagents are used to make **primary standard solutions** whose concentrations may confidently be stated to high precision.

Making a standard solution

Suppose that you want to make a standard solution of sodium chloride. What mass of sodium chloride must you dissolve in 250 cm³ of solution to obtain a concentration of 0.250 mol dm^{-3}?

We know that

amount in moles of solute = volume of solution × molar concentration

Therefore,

$$n(\text{NaCl}) = \left(\frac{250}{1000}\text{dm}^3\right) \times (0.250\,\text{mol dm}^{-3})$$

$$= 6.25 \times 10^{-2}\,\text{mol}$$

The molar mass of NaCl is (23.0 + 35.5) g mol^{-1} = 58.5 g mol^{-1}. We also know that

mass = amount in moles × molar mass

Therefore,

$$m(\text{NaCl}) = (6.25 \times 10^{-2}\,\text{mol}) \times (58.5\,\text{g mol}^{-1}) = 3.66\,\text{g}$$

So you would need to dissolve 3.66 g of sodium chloride.

Performing a titration

A titration can determine the volume of one solution required to react exactly with a known volume of another solution. Titrations frequently involve the reaction of acids with bases. Titrations can also involve reactions other than acid–base reactions, such as redox reactions and reactions involving precipitations.

Whatever the type of reaction, all titrations follow the same overall method and involve the same major components, namely:

- A **burette** containing a solution of one of the reagents.
- A **flask** containing an accurately known volume of the other reagent solution, added using a **pipette**.
- An **indicator** (added to the contents of the flask) that gives a visual indication of when the reaction is complete.

One of the two solutions must be a standard solution and the other is of unknown concentration. The burette is used to add a measured volume of one reagent solution to the other reagent solution in the flask.

There are three main stages to any titration. For a titration in which reagent A runs from the burette into reagent B in the flask:

1 *Initial stage:* A and B react together. A is consumed and B remains in excess.

2 *Equivalence point:* Sufficient A has been added to B to consume all of B initially placed in the flask. The flask contains the products of the reaction only.

3 *Final stage:* Adding more A to the flask does not result in further reaction because B has already been completely consumed. A is now in excess in the flask.

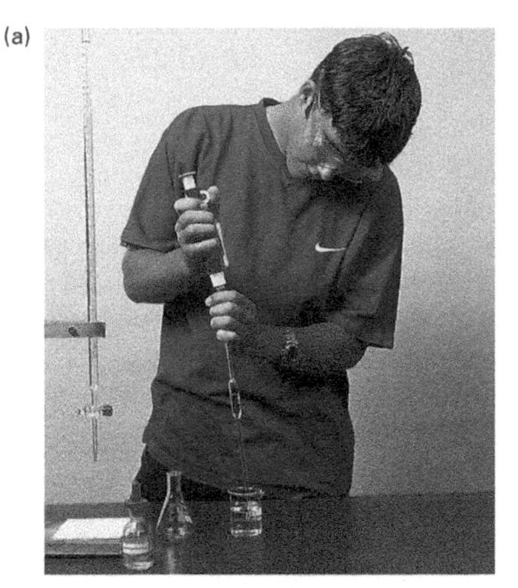
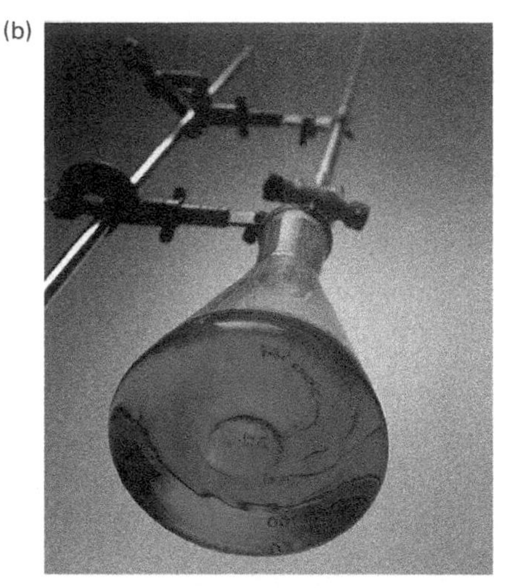

A titration involving dilute hydrochloric acid in the burette and aqueous sodium hydroxide in the flask. (a) Phenolphthalein indicator is then added to the aqueous sodium hydroxide, which turns pink (b). Hydrochloric acid is gradually added to the flask. Near the equivalence point, a small amount of base remains in the flask; the flask must be swirled after the addition of each drop of acid to ensure complete reaction. At the equivalence point, the addition of one more drop of acid will neutralize all remaining base: acid is now in excess and the indicator becomes colourless.

For almost all titrations, an indicator is added that has one colour when one of the two reagents is in excess and another colour when the second reagent is in excess.

End point

The change in colour of the indicator marks the **end point** of the titration. The **equivalence point** marks the stage at which the reaction is complete: neither reagent is in excess. Within the volume of one drop of reagent added from the burette, the end point is usually the same as the equivalence point, if the indicator is chosen carefully.

Indicators

The indicators chosen for acid–base titrations are usually different types of water-soluble dyes. The particular indicator chosen depends on the acid and base used in the titration. Examples of common acid–base indicators are given below, together with their colours in acidic and in basic solutions. The choice of indicators is discussed in spread 12.9.

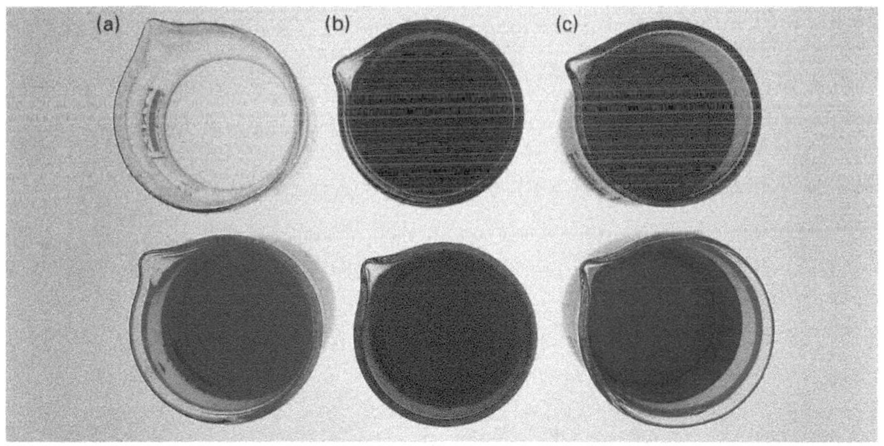

Common indicators and their colours in acidic (top) and in basic (bottom) solutions: (a) phenolphthalein, (b) methyl orange, and (c) bromothymol blue.

SUMMARY

- A titration can determine the volume of one solution required to react exactly with a known volume of another solution.
- The equivalence point occurs when neither reagent is in excess.
- An indicator is chosen that will change colour very close to the equivalence point of the reaction.

PRACTICE

1 Calculate the mass of sodium chloride that must be dissolved in 100 cm^3 of solution to give a molar concentration of 0.500 mol dm^{-3}.

2 Calculate the concentration of a standard solution that contains 1.70 g of silver nitrate dissolved in 250 cm^3 of solution.

9.7 ACID–BASE TITRATION CALCULATIONS

OBJECTIVES

- Calculating molar concentrations
- Industrial applications

Burette reading

A burette can be read to the nearest 0.05 cm^3, which corresponds to half-way between the markings.

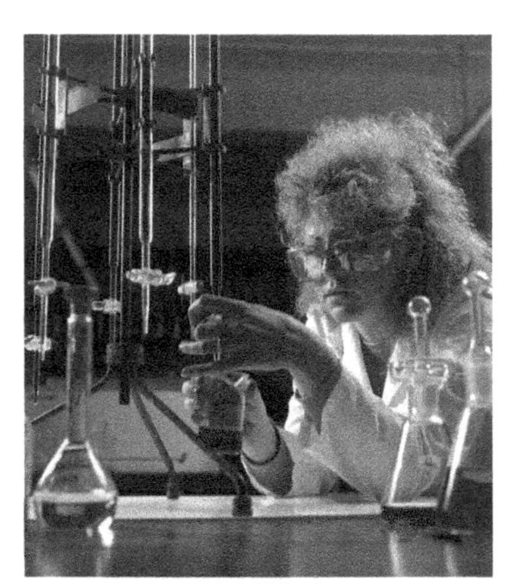

A quality control laboratory analyses a company's products to ensure they remain within specification.

Back titration

To find the molar concentration of ammonium ions, add a known excess of aqueous sodium hydroxide, then titrate the excess with standard hydrochloric acid, as shown alongside.

Suppose you need to determine the concentration of a solution of a base, for example, aqueous sodium hydroxide. You have carried out a titration between this solution and a standard solution of an acid, for example hydrochloric acid. Taking the average of a number of separate titrations, you are able to make a statement that summarizes your results. A typical statement of results might be:

25.0 cm^3 of aqueous sodium hydroxide (molar concentration unknown) reacts exactly with 35.0 cm^3 of hydrochloric acid (molar concentration 0.250 mol dm^{-3}).

This spread starts by showing you how to calculate the molar concentration of the aqueous sodium hydroxide, and goes on to discuss other related types of calculation.

Calculating the molar concentration

As in all mole calculations, this calculation starts by determining the amount in moles of the solution for which you know the molar concentration and have measured the volume used (in this case, the hydrochloric acid). The balanced chemical equation leads to the mole ratio between the acid and the base, which in turn leads to the amount in moles of the base (here, aqueous sodium hydroxide). Finally, the molar concentration of the base is calculated from the known volume and the amount in moles.

From the earlier spreads in this chapter we know that

amount in moles = volume of solution × molar concentration

Therefore, for the acid

$$n(\text{HCl}) = \left(\frac{35.0}{1000}\,\text{dm}^3\right) \times (0.250\,\text{mol dm}^{-3}) = 8.75 \times 10^{-3}\,\text{mol}$$

The balanced chemical equation is

$$\text{HCl(aq)} + \text{NaOH(aq)} \rightarrow \text{NaCl(aq)} + \text{H}_2\text{O(l)}$$

Therefore the mole ratio NaOH:HCl = 1:1. The amount in moles of NaOH present is 8.75×10^{-3} mol.

Now we use the equation

$$\text{molar concentration} = \frac{\text{amount in moles}}{\text{volume of solution}}$$

but this time for the base. First convert the volume of aqueous NaOH used to dm^3:

$$25.0\,\text{cm}^3 = \frac{25.0}{1000}\,\text{dm}^3 = 25.0 \times 10^{-3}\,\text{dm}^3$$

$$\text{Therefore, molar concentration of NaOH} = \frac{8.75 \times 10^{-3}\,\text{mol}}{25.0 \times 10^{-3}\,\text{dm}^3} = 0.350\,\text{mol dm}^{-3}$$

Worked example from industry

We now return to our vinegar factory, and consider the type of calculation that would be involved in the quality control laboratory.

Data: Vinegar must contain between 4% and 6% by mass of ethanoic acid (CH_3COOH) in water, i.e. the mass concentration of ethanoic acid is between 40 and 60 g dm^{-3}. Malt vinegar is made by extracting malt from barley and fermenting it with the aid of yeast. The result is a dilute solution of ethanol (an alcohol), which is then oxidized by atmospheric oxygen to give ethanoic acid. Batches of vinegar may be analysed by titration to ensure that the concentration of ethanoic acid falls within the prescribed limits. For example, suppose that 34.2 cm^3 of a sample of vinegar neutralized 25.0 cm^3 of aqueous sodium hydroxide (molar concentration = 0.500 mol dm^{-3}).

Questions: The relevant questions the quality controller asks are:

- What is the molar concentration of the vinegar in $mol\,dm^{-3}$?
- What is the mass concentration of the vinegar in $g\,dm^{-3}$?
- Is the batch from which the sample came fit for sale?

Answer: The calculation is similar to the one carried out above. We start with the equation

amount in moles = volume of solution × molar concentration

$$n(NaOH) = \left(\frac{25.0}{1000}\,dm^3\right) \times (0.500\,mol\,dm^{-3}) = 1.25 \times 10^{-2}\,mol$$

The equation for the reaction is:

$$CH_3COOH(aq) + NaOH(aq) \rightarrow Na^+CH_3CO_2^-(aq) + H_2O(l)$$

Therefore, the mole ratio CH_3COOH:NaOH = 1:1. The amount in moles of CH_3COOH present is 1.25×10^{-2} mol.

The volume of aqueous CH_3COOH used is

$$34.2\,cm^3 = \left(\frac{34.2}{1000}\,dm^3\right) = 34.2 \times 10^{-3}\,dm^3$$

Now we can use

$$\text{molar concentration} = \frac{\text{amount in moles}}{\text{volume of solution}}$$

$$\text{molar concentration of } CH_3COOH = \frac{1.25 \times 10^{-2}\,mol}{34.2 \times 10^{-3}\,dm^3} = 0.365\,mol\,dm^{-3}$$

Finally we use

mass concentration = molar concentration × molar mass

The molar mass of CH_3COOH is

$$2 \times (12.0\,g\,mol^{-1}) + 4 \times (1.0\,g\,mol^{-1}) + 2 \times (16.0\,g\,mol^{-1}) = 60.0\,g\,mol^{-1}$$

$$\text{mass concentration of } CH_3COOH = (0.365\,mol\,dm^{-3}) \times (60.0\,g\,mol^{-1})$$
$$= 22\,g\,dm^{-3} \text{ (2 sig. figs)}$$

This concentration falls outside the limits of 40 to 60 $g\,dm^{-3}$, so the batch is not fit for sale, being too dilute.

Another industrial application

Large nickel–cadmium batteries are used for stand-by lighting systems in trains. The cells in these batteries contain an electrolyte consisting of aqueous potassium hydroxide. The problem is that the electrolyte slowly absorbs carbon dioxide from the atmosphere, decreasing the concentration of the potassium hydroxide and forming potassium carbonate in the solution:

$$2KOH(aq) + CO_2(g) \rightarrow K_2CO_3(aq) + H_2O(l)$$

The electrical resistance of the cell increases, with the result that performance is reduced. Cells can be returned to full efficiency by renewing the electrolyte. Manufacturers of these nickel–cadmium cells suggest that the electrolyte is replaced when the mass concentration of potassium carbonate reaches 75 $g\,dm^{-3}$. They suggest that chemical analysis of the electrolyte is carried out every two to three years, more frequently when batteries are operated close to vehicle exhausts or at high temperatures. The analysis involves titrating a sample of the electrolyte against boric acid (in the burette). Phenolphthalein indicator shows an end point when all the potassium hydroxide in the sample has been neutralized. Methyl orange indicator is then added, which shows an end point when all the potassium carbonate has been neutralized. The *difference* between the two burette readings for the two end points is then read off against a chart that shows potassium carbonate mass concentration.

SUMMARY

Calculation route for determining concentration:

- Calculate the amount in moles of the solution for which the volume and molar concentration data are available.
- Find the mole ratio from the balanced chemical equation.
- Use the mole ratio to determine the amount in moles of the second solution used in the titration.
- Calculate the molar concentration of the second solution from its volume and the amount in moles.

Delocalization

See spread 6.4 for the reason why the formula of sodium ethanoate is written as $Na^+CH_3CO_2^-$.

PRACTICE

1 As a result of a series of titrations, it was found that 10.2 cm^3 of 0.100 $mol\,dm^{-3}$ sulphuric acid exactly neutralized 25.0 cm^3 of aqueous potassium hydroxide. Calculate the concentration of the aqueous potassium hydroxide in

a $mol\,dm^{-3}$ **b** $g\,dm^{-3}$.

2 Why do the nickel–cadmium cells mentioned in the box above deteriorate more rapidly when they are installed close to vehicle exhausts?

3 Imagine that you are the quality controller of our vinegar factory. To make as much profit as possible, you have to make the vinegar as weak as allowed. You carry out a titration using a vinegar sample with the minimum allowed concentration of ethanoic acid. What titration result would you expect for the volume of vinegar required to neutralize 25.0 cm^3 of the same aqueous sodium hydroxide?

PRACTICE EXAM QUESTIONS

1 50 kg of pure sulphuric acid were accidentally released into a lake when a storage vessel leaked. Two methods were proposed to neutralize it.

a The first proposal was to add a solution of 5 M (where M = mol dm^{-3}) NaOH to the lake. Sodium hydroxide reacts with sulphuric acid as follows.

$2NaOH(aq) + H_2SO_4(aq) \rightarrow Na_2SO_4(aq) + 2H_2O(l)$

Calculate the volume of 5 M NaOH required to neutralise the sulphuric acid by answering the following questions.

i How many moles of sulphuric acid are there in 50 kg of the acid?

ii How many moles of sodium hydroxide are required to neutralize this acid?

iii Calculate the volume, in dm^3, of 5 M NaOH which contains this number of moles. [4]

b The second proposal was to add powdered calcium carbonate which reacts as follows.

$CaCO_3(s) + H_2SO_4(aq) \rightarrow CaSO_4(s) + H_2O(l) + CO_2(g)$

Calculate the mass of calcium carbonate required to neutralize 50 kg of sulphuric acid. [3]

c Suggest **two** reasons why the addition of calcium carbonate to neutralize the acid is the preferred method in practice. [2]

2 a Sulphamic acid reacts with sodium hydroxide according to the following equation.

$NH_2SO_3H + NaOH \rightarrow NH_2SO_3Na + H_2O$

A standard solution of sulphamic acid was made by dissolving 5.210 g of the acid in water and making the volume up to exactly 250 cm^3 with more water.

i Calculate the number of moles of acid used and the molarity of the acid solution.

ii In a titration, 22.6 cm^3 of this acid solution were required to neutralize 25.0 cm^3 of sodium hydroxide solution. Calculate the molarity of the sodium hydroxide solution. [5]

b In a separate experiment sodium hydroxide was used to prepare a sample of Glauber's salt $Na_2SO_4.10H_2O$, by neutralisation with sulphuric acid followed by recrystallisation from aqueous solution.

$H_2SO_4 + 2NaOH \rightarrow Na_2SO_4 + 2H_2O$

Calculate the maximum mass of Glauber's salt which could be made from 5.0 g of sodium hydroxide. [4]

3 A tank contained 4 m^3 of waste hydrochloric acid. It was decided to neutralize the acid by adding slaked lime, $Ca(OH)_2$.

a The concentration of the acid was first determined by titration of a 25.0 cm^3 sample against 0.121 M sodium hydroxide of which 32.4 cm^3 were required.

i Calculate the molarity of the hydrochloric acid in the sample.

ii Calculate the total number of moles of HCl in the tank. [4]

b Calculate the mass, in kg, of slaked lime required to neutralize the acid. Slaked lime reacts with hydrochloric acid according to the equation shown below.

$Ca(OH)_2 + 2HCl \rightarrow CaCl_2 + 2H_2O$ [3]

c The slaked lime was manufactured by roasting limestone and then adding water.

$CaCO_3 \rightarrow CaO + CO_2$

$CaO + H_2O \rightarrow Ca(OH)_2$

Calculate the mass of limestone which is required to produce 1 kg of slaked lime. [2]

4 Elemental analysis of a salt **X** gave the following percentages by mass, the remainder being oxygen:

Silver: 71.05% Carbon: 7.89%

a Determine the empirical formula of the salt. [3]

b The relative molecular mass of **X** is 304. Use this to determine the molecular formula of **X**. [1]

c When heated, 5.00 g of **X** were decomposed completely to give a solid residue and 8.14×10^{-4} m^3 of carbon dioxide gas at 298 K and 100 kPa.

i Calculate the number of moles of **X** decomposed.

ii Use the volume of carbon dioxide gas to calculate the number of moles of carbon dioxide produced.

iii Using your answers to part **b** and parts **c i** and **ii**, identify the solid residue and write an equation for the thermal decomposition of **X**. [6]

5 a An aqueous solution of hydrogen peroxide, H_2O_2, decomposes in the presence of a catalyst according to the equation:

$2H_2O_2(aq) \rightarrow 2H_2O(l) + O_2(g)$

i Calculate the number of moles of H_2O_2 required to produce 10 dm^3 of oxygen gas measured at room temperature and pressure.

[1 mole of any gas occupies a volume of 24 dm^3 at room temperature and pressure.] [2]

ii The number of moles calculated in **a i** is present in 1 dm^3 of H_2O_2 solution.

Calculate the volume of this solution required to make 250 cm^3 of a 0.100 mol dm^{-3} solution, by dilution with water. [2]

b Calculate the mass of potassium manganate(VII), $KMnO_4$, (M_r = 158) required to make 200 cm^3 of a solution having a concentration of 0.020 mol dm^{-3}. [2]

c When $20.0\,cm^3$ of the $0.100\,mol\,dm^{-3}$ solution of H_2O_2 was acidified with dilute sulphuric acid and titrated against a $0.020\,mol\,dm^{-3}$ solution of potassium manganate(VII), $KMnO_4$, $40.0\,cm^3$ of the latter were required for complete reaction.

[No experience of the reaction in the laboratory is needed to answer this question.]

i Calculate the number of moles of $KMnO_4$ in $40.0\,cm^3$ of $0.020\,mol\,dm^{-3}$ solution. [1]

ii Calculate the number of moles of H_2O_2 in $20.0\,cm^3$ of $0.100\,mol\,dm^{-3}$ solution. [1]

iii Hence deduce the number of moles of H_2O_2 which react with 1 mole of $KMnO_4$. [1]

iv Use the result from **c iii** to balance the following equation for the reaction taking place in the titration.

$... KMnO_4 + ... H_2O_2 + ... H_2SO_4 \rightarrow$
$... MnSO_4 + ... K_2SO_4 + ... H_2O + ... O_2$ [2]

d The solution at the end of this reaction contains potassium sulphate and manganese(II) sulphate only.

i Write formulae for the cations present in this aqueous solution. [2]

ii Treatment of this solution with dilute sodium hydroxide gives a precipitate which does not dissolve in excess sodium hydroxide solution. Identify the precipitate by name or formula. [1]

e In addition to the normal oxide, sodium forms a peroxide, Na_2O_2. This reacts with carbon dioxide to form sodium carbonate and oxygen:

$2Na_2O_2 + 2CO_2 \rightarrow 2Na_2CO_3 + O_2$

i Calculate the volume of oxygen gas that would be formed by the reaction of 0.39 g of sodium peroxide with excess carbon dioxide.

[The molar volume of a gas at the temperature and pressure of the reaction should be taken as $24\,dm^3\,mol^{-1}$.] [3]

ii How many molecules of oxygen would be present in this volume of oxygen?

[The Avogadro constant, L, is $6.02 \times 10^{23}\,mol^{-1}$.] [1]

iii Use oxidation numbers to identify the type of process that occurs in the formation of oxygen from the peroxide ion. [2]

(See also chapter 13.)

6 A solution of a weak acid H_2X was made by dissolving 2.25 g of solid H_2X in water to give $500\,cm^3$ of solution. On titration, $25.0\,cm^3$ of this solution was completely neutralized by $25.0\,cm^3$ of sodium hydroxide solution containing $0.100\,mol\,dm^{-3}$.

a Write an equation for the reaction. [1]

b i Calculate the number of moles of NaOH in $25.0\,cm^3$ of $0.100\,mol\,dm^{-3}$ solution.

ii How many moles of H_2X would be required to react with this quantity of NaOH?

iii Calculate the relative molecular mass of H_2X.

iv A hydrated form of the acid also exists, $H_2X.yH_2O$. A solution containing $6.30\,g\,dm^{-3}$ of the hydrated acid has the same (molar) concentration as the solution of the anhydrous acid, H_2X, originally used. Using this information and your answer from **b**, calculate the value of y. [8]

c The presence of sulphur dioxide in the atmosphere is the main cause of acid rain. Outline a method which could be used to estimate quantitatively the concentration of sulphur dioxide in a sample of air. [5]

7 a Define what is meant by the *molecular formula* of a compound. [1]

b A 0.130 g sample of a liquid compound **A** was vaporized at 100 °C (373 K) at a pressure of 1 atm (101 kPa) and occupied a volume of $85.0\,cm^3$. Calculate the relative molecular mass of **A**. [2]

c Compound **A** has the following compsition by mass: C, 52.2%; H, 13.0%; O, 34.8%.

i Calculate the empirical formula of **A**.

ii Suggest an identity for **A** by giving a name or structural formula. [3]

8 a The table below gives the accurate masses of two atoms

	^{1}H	^{12}C
Mass / g	1.6734×10^{-24}	1.9925×10^{-23}

i Calculate accurate values for the mass of one mole of each atom.

[The Avogadro constant (L) = $6.0225 \times 10^{23}\,mol^{-1}$.]

ii Why is ^{12}C referred to when defining the relative atomic mass of an element? [3]

b i The carbon in a sample of pure graphite has a relative atomic mass of 12.011. Suggest why this value is different from the mass of one mole of ^{12}C which you have calculated in part **a i**.

ii This sample of graphite was burned completely in oxygen. The carbon dioxide produced occupied a volume of $1.85\,dm^3$ at 293 K and 98.0 kPa. Calculate the mass of carbon in the sample. [5]

c In a separate experiment, 1.54 g of carbon dioxide were produced and then absorbed in $50.0\,cm^3$ of a sodium hydroxide solution forming sodium carbonate (Na_2CO_3) in the solution. Calculate the molar concentration of the sodium carbonate in the solution. [3]

10 Thermochemistry

Thermochemistry is the study of the energy changes that accompany chemical reactions. In the course of a chemical reaction, bonds between atoms are broken and new bonds are made as the atoms regroup to form new substances. As we saw in chapter 5, these processes of breaking and making bonds involve changes in energy. Energy is needed to break bonds and is given out when new bonds form. This means that the reaction is accompanied by a change in energy, mainly in the form of heat. So, during the reaction, heat flows into or out of the reaction mixture and it returns to its original temperature. Thermochemistry investigates this flow and uses it to help explain the structures of substances and the bonding within them. Thermochemistry is a part of the larger subject of **thermodynamics**, which studies the laws that govern the conversion of energy from one form to another. The subject of thermodynamics is developed further in chapter 14 'Spontaneous change towards equilibrium'.

Chemical reactions and energy

OBJECTIVES

- Understanding basic terminology

The solid fuel which propelled the Space Shuttle was finely powdered aluminium mixed with ammonium perchlorate NH_4ClO_4. These substances were moulded with a small amount of iron catalyst into a solid plastic cylinder. When the fuel ignited, the ammonium perchlorate reacted with the aluminium to form white clouds of aluminium oxide. The temperature increased enormously and a mixture of high-pressure gases formed. These gases produced thrust as they expanded rapidly and roared out of the rocket engine at high speed, carrying the other reaction products with them.

General definitions

Energy A measure of a system's capacity to do work. The unit of energy is the joule (J).

Work An energy transfer that is the result of a force moving a body through a distance. The unit of work is the joule (J).

Heat An energy transfer that is the result of a temperature difference between a system and its surroundings. The unit of heat is the joule (J).

Temperature The property of a system that determines the direction of heat flow between the system and its surroundings. Heat flows from the hotter region to the colder. The unit of temperature is the kelvin (K).

Thermochemical terminology

There are two groups of key terms specifically used in thermochemistry. The first group contains the two linked terms 'system' and 'surroundings'. The second group contains words which are used casually in everyday life but which you must use with great care in the context of this subject. These four related terms are 'energy', 'work', 'heat', and 'temperature'. The six terms can be illustrated in the context of the Space Shuttle rocket engine, as follows.

The **system** consists of the collection of substances involved in the chemical reaction. In its initial state, the system consists of all the reactants in the rocket fuel. In its final state, the system consists of all the products of the reaction. The **surroundings** consist of everything else in the Universe with the exception of the system:

- The Universe = the system + the surroundings.

As a result of the reaction, the *temperature* of the exhaust gases is greater than the temperature of the surroundings. *Heat* flows from the system to the surroundings. After some time, the temperatures of the system and the surroundings become equal again, as they were at the start before take-off.

The reaction inside the rocket engine evolves gases. These gases do *work* as they push outwards against the atmospheric pressure of the surroundings. The rocket transfers energy to its surroundings by doing work on the surroundings and by heating them up.

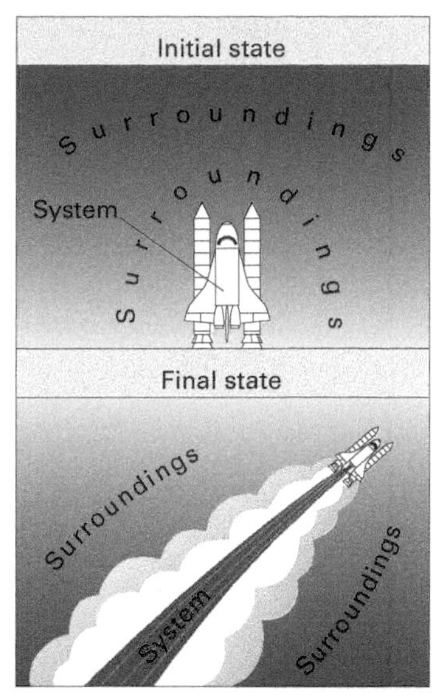

In the course of a chemical reaction, the system passes from its initial state into its final state. In the case of the rocket propellant, heat flows from the system to the surroundings; the system also does work on the surroundings.

Chemical reactions and internal energy

As you watch a rocket taking off, it seems obvious that the chemical reactants in the fuel contain more energy than the chemical products in the exhaust plume. The energy contained within a system is called its **internal energy** (symbol U). Internal energy cannot easily be measured because it is the sum of the kinetic energies of all the particles in the system and their potential energies.

Hours after the rocket has launched the Shuttle into space, the exhaust products have cooled down to the same temperature as the surroundings. Some heat has transferred from the system to the surroundings. Also, the exhaust gases have expanded so that they are at atmospheric pressure; and the Shuttle is in orbit. The expanding gases have done some work. If the internal energy of the fuel is $U_{reactants}$ and the internal energy of the exhaust products is $U_{products}$, then:

$U_{reactants}$ is greater than $U_{products}$

The difference between these two values is given the symbol ΔU (Δ is the Greek capital letter 'delta' and is used to indicate change):

$$\Delta U = U_{products} - U_{reactants}$$

The reaction in this case results in a *decrease* in internal energy, and the expression above gives a *negative* value for ΔU for the Space Shuttle example (because $U_{reactants}$ is greater than $U_{products}$), signifying that the internal energy has *decreased*.

- The *decrease* in internal energy of the system results in an *increase* in the energy of the surroundings.

Practical considerations

The object of thermochemistry is to use measurements taken during practical experiments to explain the structure and bonding of substances. Changes in bonding during reactions involve changes in internal energy. It is necessary to make measurements that reflect the changes in internal energy. It is fairly straightforward to measure the heat evolved during a reaction by measuring the temperature change in a calorimeter, for example (see spread 10.4), and carrying out a relatively simple calculation. It is not so easy to measure the work done. It is helpful to define a quantity which is easy to measure; this quantity is 'enthalpy' and it is introduced in the following spread.

A snowmaker uses the ideas of thermochemistry. It contains a mixture of compressed air and water vapour at about 20 atm pressure. Because of the large pressure difference to the outside atmosphere, when the mixture is sprayed into the atmosphere there is almost no heat exchanged with its surroundings. The large quantity of work done by the gas causes the mixture to cool, because of the law of energy conservation; snow forms because the mixture cools.

SUMMARY

- During a chemical reaction, energy is transferred between the system and the surroundings.
- Energy can be transferred as heat or as work.
- Heat flows as the result of a temperature difference between the system and the surroundings.

PRACTICE

1 Zinc and dilute hydrochloric acid react together according to the equation:

$Zn(s) + 2HCl(aq) \rightarrow ZnCl_2(aq) + H_2(g)$

The reaction takes place in an open test tube in a laboratory.

a What does the system consist of in its initial state?

b What does the system consist of in its final state?

c The reaction mixture becomes warmer. In which direction will heat flow?

d Does the system do work? Explain your answer.

e Does the internal energy of the system increase or decrease?

Enthalpy changes

OBJECTIVES

- Enthalpy changes ΔH
- Exothermic reactions
- Endothermic reactions

In a chemical reaction, existing bonds are broken and new bonds are made. Reactants are used up and products form. The internal energy of the system changes as it exchanges energy with the surroundings as heat or work. Changes in bonding are reflected in changes in internal energy. The problem is that internal energy cannot easily be measured. This problem is overcome by introducing and using a term called 'enthalpy', which is linked to internal energy but which can be measured more easily in practice.

Chemical changes and heat

A change in internal energy (ΔU) occurs when a system exchanges energy with its surroundings in the form of heat and/or work:

$$\text{internal energy change} = \text{heat added to the system} + \text{work done on the system}$$

Heat flows as the result of a temperature difference. If we know the mass of the system, its specific heat capacity (see spread 10.4), and use a thermometer to measure temperature difference, we can calculate the heat added to the system.

Work taking place during chemical reactions is more difficult to measure. In some chemical reactions work is obviously done, such as those that release a gas. The gas has to do work on the atmosphere to push the atmosphere out of the way and make room for itself. But it is inconvenient to collect the gas, and to measure its volume and the pressure of the surroundings, in order to calculate the work done. Some reactions take place at constant volume, such as precipitation reactions between solutions. In these cases, there is no change of volume and consequently no work is done. It is possible to study all chemical reactions in terms of changes in a quantity called 'enthalpy'.

Enthalpy

Enthalpy (symbol H) is a term that describes the heat content of a system. Just like internal energy, its actual value cannot be measured. However, an *enthalpy change* can easily be measured. An **enthalpy change** (ΔH, measured in joules, J) is defined as the heat added to the system under conditions of constant pressure:

enthalpy change = enthalpy of products – enthalpy of reactants

In symbols:

$$\Delta H = H_{\text{products}} - H_{\text{reactants}}$$

The typical practical procedure is to insulate the system from its surroundings, so that no heat can escape, and allow the temperature of the system to change during the reaction. You can then calculate the heat needed to bring the system back to its original temperature. This heat is the enthalpy change for the system:

ΔH = heat added to the system at constant pressure

For a reaction that evolves heat (i.e. heat flows *from* the system *to* the surroundings), the sign of ΔH is negative.

- A reaction with a negative value for ΔH is said to be **exothermic**.

For a reaction that absorbs heat (i.e. heat flows *from* the surroundings *to* the system), the sign of ΔH is positive.

- A reaction with a positive value for ΔH is said to be **endothermic**.

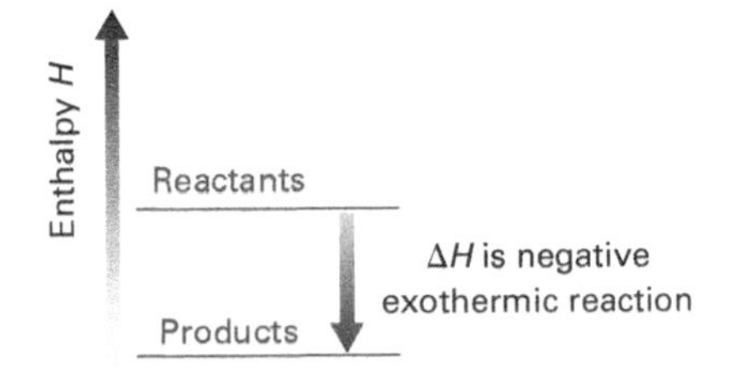

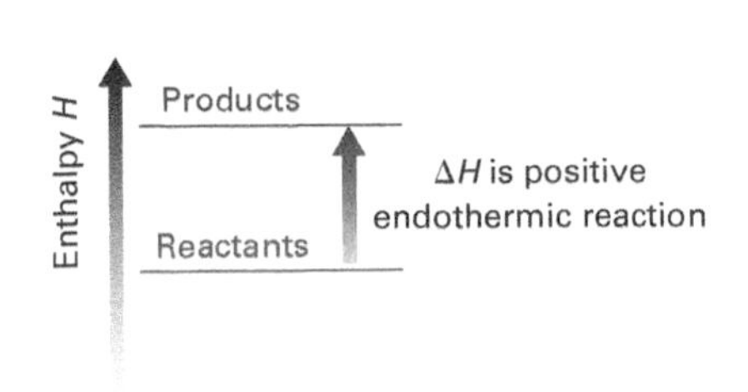

(a) For an exothermic reaction, the enthalpy change ΔH is negative. (b) For an endothermic reaction, the enthalpy change ΔH is positive.

Bond breaking and bond making

What actually happens when hydrogen burns in oxygen to form water vapour? The chemical equation is:

$$2H_2(g) + O_2(g) \rightarrow 2H_2O(g)$$

We will think about this reaction in terms of the bonds broken and made.

First of all, the molecules of hydrogen and oxygen must break apart to form separate atoms.

- Breaking the original bonds is an endothermic process.

Then the separate atoms combine in a new configuration to form molecules of water.

- Forming the new bonds is an exothermic process.

These changes may be represented on an **enthalpy profile diagram**, as shown to the right. Note that the reaction is exothermic overall.

Enthalpy H

$4H(g) + 2O(g)$

$2H_2(g) + O_2(g)$

Reactants

Products $2H_2O(g)$

ΔH is negative

Breaking the bonds in the oxygen and hydrogen molecules is an endothermic process. Making the bonds in water is an exothermic process. The reaction is exothermic overall with $\Delta H = -242$ kJ per mole of water formed (at 298 K and 1 bar pressure), which we write as:

$H_2(g) + \frac{1}{2}O_2(g) \rightarrow H_2O(g)$

Exothermic reactions

For an exothermic reaction, the enthalpy change ΔH is negative and heat flows from the system into the surroundings.

Endothermic reactions

For an endothermic reaction, the enthalpy change ΔH is positive and heat flows from the surroundings into the system.

Solid hydrated barium hydroxide and excess solid ammonium nitrate react:
$Ba(OH)_2{\cdot}8H_2O(s) + 2NH_4NO_3(s) \rightarrow Ba(NO_3)_2(s) + 2NH_3(g) + 10H_2O(l)$
The water formed dissolves the excess ammonium nitrate and this process is highly endothermic. If the flask is placed on a wet biscuit tin, the water freezes and sticks the flask to the tin.

The reaction between glycerol (see spread 30.1) and potassium manganate(VII) is highly exothermic.

SUMMARY

- For an endothermic reaction, ΔH is positive.
- For an exothermic reaction, ΔH is negative.

PRACTICE

1 State, with reasons, whether each of the following processes is exothermic or endothermic:

a A match burning

b An ice cube melting

c Photosynthesis.

2 A chemical reaction usually involves the endothermic process of breaking the bonds of the reactants followed by the exothermic process of making the bonds of the products. Compare the magnitudes of these two processes:

a in a reaction that is exothermic overall,

b in a reaction that is endothermic overall.

EXTENSION

10.3

OBJECTIVES

- A justification for using enthalpy
- Work done in expansion
- Comparing values of heat and work

ENTHALPY CHANGES EXAMINED IN DETAIL

You will have noticed in the previous spreads of this chapter that few equations are expressed in symbols. The text speaks of 'internal energy', 'work', 'enthalpy', and 'heat', but does not often express these quantities in terms of their symbols. The reason for this approach is that much care must be taken with the signs used. Reference must be made to the direction of heat flow and to the distinction between the system and the surroundings. When reading this spread, you must always note whether a sign is positive or negative and explain the distinction in terms of what is actually happening in the experiment.

A justification for using enthalpies

The total energy of a system, its internal energy, may be increased either by heating the system or by doing work on it. The change in the thermodynamic quantity called 'enthalpy' is defined in such a way as to make it equal to the heat added at constant pressure. The reasoning below explains why thermochemistry concentrates on enthalpy changes rather than on internal energy changes.

If the heat added *to* a system is q and the work done *on* this system is w, then the change in the internal energy U is given by the expression:

$$\Delta U = q + w \qquad (1)$$

There are many ways in which work can be done. By far the most usual is when a gas evolved by a reaction expands against the pressure of the atmosphere. Consider the gas behind a piston expanding just a little against the constant pressure of the external atmosphere. Calculating the work done then depends on the definition of work:

$$\text{work} = \text{force} \times \text{distance moved in the direction of the force} \qquad (2)$$

Pressure is defined as force per unit area, i.e.

$$\text{pressure} = \text{force/area}$$

Rearranging this, we have

$$\text{force} = \text{pressure} \times \text{area} \qquad (3)$$

Substituting equation (3) into equation (2) gives

$$\text{work} = \text{pressure} \times \text{area} \times \text{distance moved}$$

i.e.

$$\text{work} = \text{pressure} \times \text{change in volume}$$

In the case of an expanding gas moving a piston, the system is doing work on the surroundings:

$$w = -p\Delta V$$

Remember that the sign is *negative* when the system *does* work.

When heat q is added to a system from the surroundings and the system does expansion work $p\Delta V$ on the surroundings, the internal energy change ΔU is expressed by the relationship:

$$\Delta U = q - p\Delta V \qquad (4)$$

If a chemical reaction involves a volume change, the work done must in principle be measured in order to calculate the internal energy change. It is difficult to measure the work done. To avoid measuring it, a new quantity called **enthalpy** is defined, as follows:

$$H = U + pV \qquad (5)$$

For small changes, the enthalpy change is:

$$\Delta H = \Delta U + p\Delta V + V\Delta p \qquad (6)$$

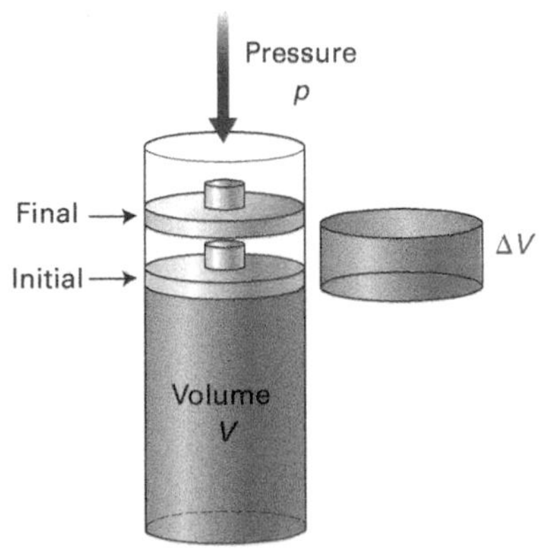

The pressure inside the cylinder is greater than the atmospheric pressure p outside. When the piston moves outwards, the volume inside the cylinder increases by ΔV. The gas inside the cylinder has done work.

Substituting the equation for the internal energy change (4) into equation (6), the end result of the derivation is that

$$\Delta H = q - p\Delta V + p\Delta V + V\Delta p$$

i.e.

$$\Delta H = q + V\Delta p \qquad (7)$$

If the change in pressure Δp is *zero*, then the enthalpy change is identical to the heat added to the system:

$$\Delta H = q \qquad \text{(at constant pressure)}$$

Chemical reactions almost always take place under conditions of constant pressure in an open laboratory because the atmospheric pressure is approximately constant.

In this way, the definition of enthalpy means that we can ignore the work term provided that the pressure is constant. The enthalpy change is the heat added to the system at constant pressure. Under conditions of constant pressure, a heat measurement *is* a measurement of an enthalpy change. The name 'enthalpy' appropriately comes from the Greek for 'inner warmth'.

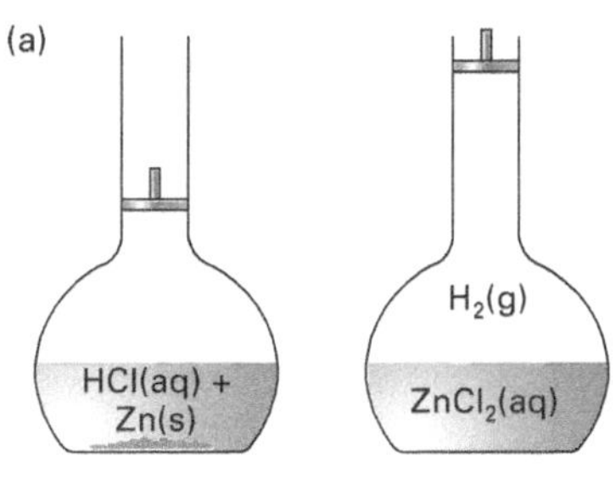

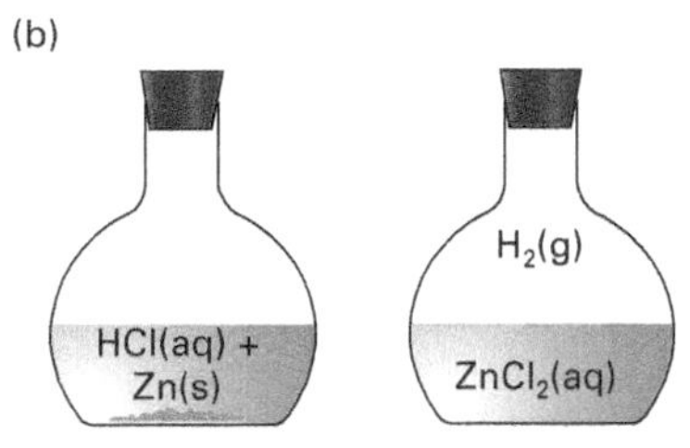

(a) At constant pressure, the reaction evolves gas, which expands and does work against the atmosphere. The enthalpy change = the heat added to the system. (b) If the same reaction takes place at constant volume, the reaction increases the pressure in the vessel and does more work on the bung. Now the heat added is not *equal to the enthalpy change.*

Worked example to measure work done in expansion

The enthalpy change per mole on vaporizing water at 100 °C and 1.0 bar is 40.7 kJ mol^{-1}. The molar volume of the liquid is 18 cm^3 mol^{-1} and the molar volume of the vapour is 24 dm^3 mol^{-1} (so the volume of the liquid is negligible). 1.0 bar is 1.0×10^5 Pa (N m^{-2}).

The work done on the surroundings is equal to $p\Delta V$. Therefore

$$\begin{aligned} \text{work done} &= (1.0 \times 10^5\ \text{N m}^{-2}) \times (24 \times 10^{-3}\ \text{m}^3\ \text{mol}^{-1}) \\ &= 2.4 \times 10^3\ \text{N m mol}^{-1} \\ &= 2.4\ \text{kJ mol}^{-1} \end{aligned}$$

This calculation shows that the work done is only about 6% of the enthalpy change.

SUMMARY

- The internal energy change $\Delta U = q + w$.
- When a system expands, the work done on the system equals $-p\Delta V$.
- The enthalpy change $\Delta H = q$ (at constant pressure).
- The enthalpy change is the heat added to the system at constant pressure.

PRACTICE

1 Zinc reacts with hydrochloric acid according to the chemical equation:

$Zn(s) + 2HCl(aq) \rightarrow ZnCl_2(aq) + H_2(g)$

1 mol of zinc liberates 24 dm^3 of gas at room temperature.

a Calculate the work done by the production of this volume of gas.

b Would the value of the enthalpy change be the same if this reaction were carried out at constant volume?

2 Explain why enthalpy H cannot easily be measured but an enthalpy change ΔH may be measured.

10.4

OBJECTIVES

- 'Coffee cup' calorimeter
- Flame calorimeter
- Bomb calorimeter

CALORIMETERS: MEASURING ENTHALPY CHANGES

Specific heat capacity

In this spread we need to use the equation

$q = mc\Delta T$

that is

heat = mass of solution × its specific heat capacity × temperature change

The **specific heat capacity** is the heat required to raise the temperature of 1 g of a substance by 1 K. The specific heat capacity of pure water is $4.18\ J\ g^{-1}\ K^{-1}$.

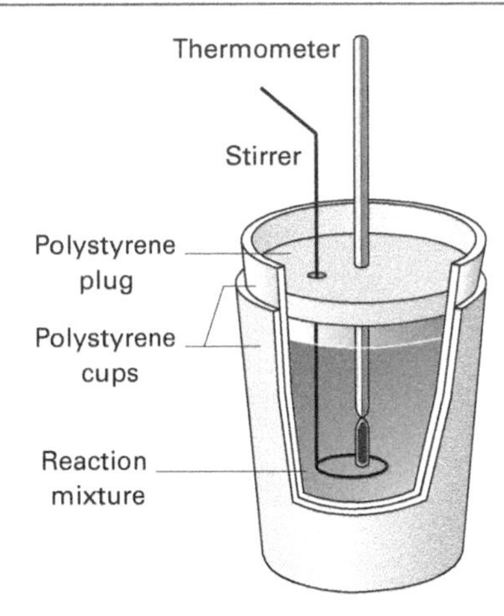

Two expanded polystyrene cups are placed one inside the other. An expanded polystyrene plug supports a thermometer and a wire stirrer. If the mass and specific heat capacity of the reaction mixture is large compared with that of the thermometer and stirrer, then the heat exchanged with the stirrer and thermometer is very much less than that exchanged with the reaction mixture, and we can make the assumptions mentioned to the right.

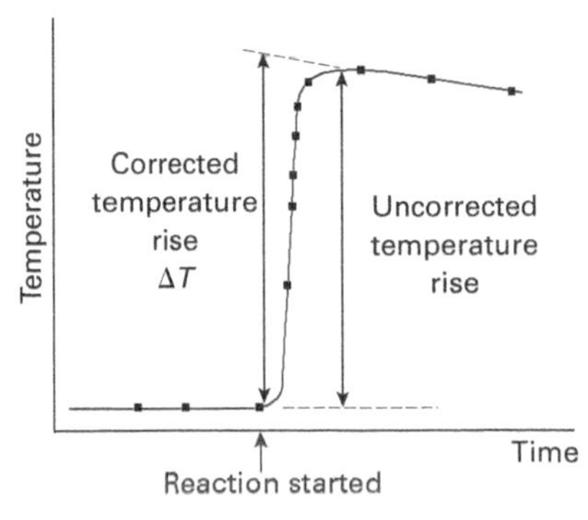

Cooling correction. Allowance may be made for heat lost from the calorimeter for an exothermic reaction by plotting a graph of temperature against time. Assuming that heat loss is constant during the experiment, a more accurate figure for the maximum temperature is obtained by extrapolating the graph back to the time of mixing.

A calorimeter is a piece of apparatus used to measure the enthalpy changes that accompany chemical reactions. Calorimeters are designed to insulate the reaction system thermally from its surroundings. The enthalpy change then causes a change in the temperature inside the calorimeter, which can be measured with a thermometer. The enthalpy change may then be calculated from a knowledge of the temperature change and the mass and specific heat capacity of the contents of the calorimeter. This spread describes three types of calorimeter.

The 'coffee cup' calorimeter

This simple type of calorimeter is suitable for changes that take place in aqueous solution. Examples include:

- neutralization reactions between an acid and a base;
- precipitation reactions where a solid forms on mixing two aqueous solutions;
- dissolving a solid to form an aqueous solution.

This calorimeter consists of an insulated cup containing an aqueous solution of the reacting substances. The chemical reaction that occurs is accompanied by an enthalpy change. This enthalpy change leads to a change in temperature of the calorimeter. The aqueous solution together with the cup, thermometer, and stirrer heat up or cool down. For the purposes of the calculation, we assume that all the heat is exchanged with the solution alone, and that the solution has the same specific heat capacity as pure water.

Worked example on measuring enthalpy of solution

The *enthalpy of solution* is the enthalpy change that accompanies the dissolution of 1.00 mol of a solid in a large excess of water. In this example, 8.00 g of ammonium nitrate (NH_4NO_3) was dissolved in 50.0 g of water in a simple expanded polystyrene calorimeter. The temperature fell by 10.1 °C.

Question: Calculate the enthalpy of solution for the process:

$NH_4NO_3(s) + water \rightarrow NH_4^+(aq) + NO_3^-(aq)$

Answer: The temperature fell as the solid dissolved: forming an aqueous solution of ammonium nitrate is an endothermic process, spread 10.2.

Step 1 Calculate the heat added from the change in temperature using

$q = mc\Delta T$

where q = heat, m = mass, c = specific heat capacity, and ΔT = temperature change. Substituting values we get

$q = (58.0\ g) \times (4.18\ J\ g^{-1}\ K^{-1}) \times (10.1\ K)$

$= 2.45 \times 10^3\ J = 2.45\ kJ$

Step 2 Calculate the amount in moles of ammonium nitrate dissolved.

Molar mass of $NH_4NO_3 = 2 \times (14.0\ g\ mol^{-1}) + 4 \times (1.0\ g\ mol^{-1}) + 3 \times (16.0\ g\ mol^{-1})$ $= 80.0\ g\ mol^{-1}$.

Amount in moles of NH_4NO_3 dissolved $= \dfrac{8.00\ g}{80.0\ g\ mol^{-1}} = 0.100\ mol$

Step 3 Calculate the enthalpy change per mole.

Solution of 0.100 mol of NH_4NO_3 required 2.45 kJ

Solution of 1.00 mol of NH_4NO_3 would require $\dfrac{2.45}{0.100}$ kJ = 24.5 kJ

The process is endothermic, so the accompanying enthalpy change is positive, i.e.

$\Delta H = +24.5\ kJ\ mol^{-1}$

Note that the accepted value is $+25.8\ kJ\ mol^{-1}$. The smaller value calculated in this example is the result of *heat losses* from the simple type of calorimeter used.

The flame calorimeter

A more sophisticated instrument than a simple polystyrene cup is the flame calorimeter, which is used for measuring the enthalpy changes during combustion. A measured mass of the substance being investigated is combusted in a small burner. The heat released warms the surrounding vessel and the temperature rise is recorded. Unlike the polystyrene cup calorimeter, the flame calorimeter has a significant specific heat capacity. The heat capacity of a calorimeter containing a specified quantity of water is called the **calorimeter constant** (units $J\,K^{-1}$). If we know the calorimeter constant, we can calculate the enthalpy change directly from a measured temperature change.

There are two main sources of error when using a flame calorimeter:

1 Combustion products do not completely transfer their heat to the spiral copper heat exchanger. Therefore, the temperature rise of the apparatus is smaller than it should be.

2 Combustion of the sample may be incomplete. For example, a compound containing only carbon and hydrogen (a hydrocarbon) should burn to produce carbon dioxide and water only. Incomplete combustion results in carbon monoxide and carbon also being produced, with the result that the enthalpy change (and the resultant temperature rise) is less than the theoretical maximum value.

The bomb calorimeter

A bomb calorimeter consists of a steel-walled pressure vessel in which a solid or liquid sample is burned in pure oxygen at 25 atm pressure. The main advantage of the bomb calorimeter is that the combustion is more likely to be complete due to the use of pure high-pressure oxygen.

Enthalpy changes are defined at constant *pressure*. The bomb calorimeter operates at constant *volume* and therefore measures the internal energy change for the reaction. Enthalpy changes calculated from bomb calorimeter data have to be corrected for work done as the result of a pressure change. The magnitude of the work done is typically a few per cent of the total energy change.

SUMMARY

- Calorimeters are used to measure temperature changes resulting from enthalpy changes accompanying chemical reactions and physical changes.
- The heat is calculated from the temperature change by using the relationship $q = mc\Delta T$.
- The calorimeter constant is the heat required to raise the temperature of a given calorimeter and its contents by 1 K.

PRACTICE

1 Calculate the heat required to raise the temperature of 150 g of water by 25.0 °C. [For pure water $c = 4.18\,J\,g^{-1}\,K^{-1}$.]

2 The calorimeter constant of a flame calorimeter is $5.83 \times 10^3\,J\,K^{-1}$. The complete combustion of 3.20 g of methanol CH_3OH raises the temperature by 12.3 °C. Calculate the enthalpy change for the combustion of 1.00 mol of methanol.

Calorimeter constant

The calorimeter constant (heat capacity) of a calorimeter may be calculated by measuring the temperature rise caused by a known quantity of electrical energy. This electrical energy is generated by an electric heater, the energy being given by

energy = voltage × current × time

Alternatively, the calorimeter may be calibrated by combusting a known mass of a substance whose enthalpy change on combustion is known.

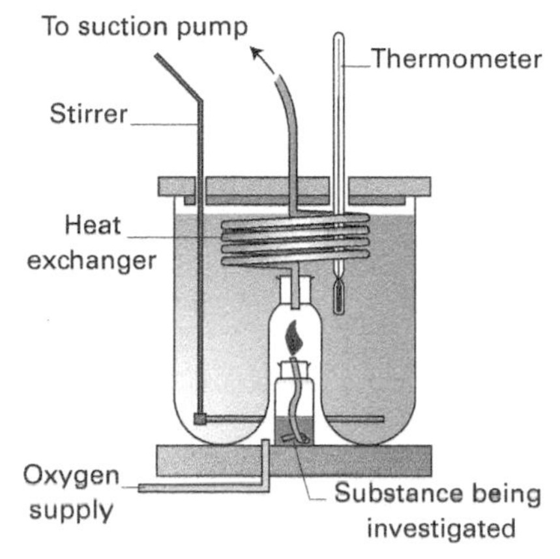

A flame calorimeter. A suction pump draws a steady stream of air through the apparatus so that hot combustion products flow through the copper heat exchanger coil.

Insulation

The flame calorimeter is not thermally insulated. The maximum temperature of the calorimeter is obtained by plotting a cooling correction graph and extrapolating as described opposite.

10.5

OBJECTIVES

- Standard enthalpy changes
- Standard enthalpy change of formation
- Standard enthalpy change of combustion
- Hess's law

Standard state

The **standard state** of a substance is the substance in its pure form at 1 bar and the stated temperature. Although values are commonly quoted at 298 K (25 °C), measurements can be made at other temperatures. For example, the processes happening in a blast furnace occur at around 2000 K, see chapter 14.

Reference state

Some elements can exist in more than one form. The form used should be the one that is most stable at 1 bar and the stated temperature; this form is the element's **reference state**. For example, the standard enthalpy change of formation of gaseous carbon dioxide at 25 °C is the standard enthalpy change when carbon dioxide gas is formed from solid graphite (not diamond, which is the less stable form of carbon) and oxygen gas:

$C(s) + O_2(g) \rightarrow CO_2(g)$;

$\Delta_f H^\ominus(298\,K) = -393.5\,kJ\,mol^{-1}$

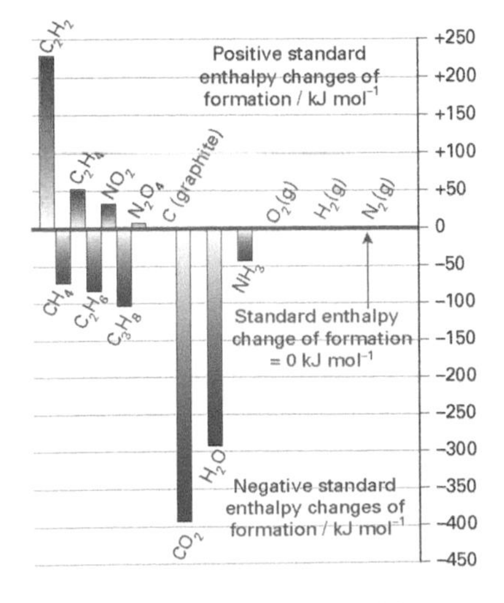

Standard enthalpy changes of formation are usually written in a table. Here they are compared diagrammatically.

Standard enthalpy changes and Hess's law

All chemical reactions have an associated enthalpy change. For a given chemical reaction, the value of the enthalpy change depends on three factors: the amounts of the substances involved, the temperature, and the pressure at which the reaction is carried out. So if you quote an enthalpy change for a reaction, you need to state the conditions under which the reaction takes place. This leads to the idea of 'standard enthalpy changes', which are measured under certain fixed conditions. If we measure standard enthalpy changes, we can compare the values for different reactions. We can use measured standard enthalpy changes to calculate the standard enthalpy changes for related chemical reactions, giving us a value for a reaction that is difficult to measure practically.

Standard enthalpy changes

Data books list 'standard enthalpy changes' or, occasionally, 'standard molar enthalpy changes' or 'standard reaction enthalpies'. Standard enthalpy changes are given the symbol $\Delta H^\ominus$. The superscript $^\ominus$ signifies that

- all substances are in their *standard states* (see Box),
- the *pressure is 1 bar*, and
- the enthalpy change is measured *per mole* of the specified substance.

The physical state of a substance and the numerical value of the enthalpy change both depend on the temperature at which the reaction takes place. The temperature is indicated in brackets after the $\Delta H^\ominus$ symbol. For example, if the temperature at which the reaction takes place is 298 K, the full symbol for the standard enthalpy change is $\Delta H^\ominus$ (298 K).

Standard enthalpy change of formation

This is defined as follows:

- The **standard enthalpy change of formation** $\Delta_f H^\ominus$ is the standard enthalpy change when a compound is formed from its *elements*. The symbol used to be written $\Delta H^\ominus_f$.

Note that the *standard enthalpy change of formation* is also sometimes called the 'standard enthalpy of formation' or the 'standard formation enthalpy'.

For example, the standard enthalpy change of formation of aluminium oxide is represented as follows:

$$2Al(s) + \tfrac{3}{2}O_2(g) \rightarrow Al_2O_3(s); \qquad \Delta_f H^\ominus\,(298\,K) = -1676\,kJ\,mol^{-1}$$

Here:

- the temperature must be stated (in this case, 298 K),
- all substances are in their standard states at the stated temperature,
- pressure is 1 bar,
- the enthalpy change is measured per mole of the compound formed.

Values of standard enthalpy changes of formation are given in the diagram to the left. Note that the standard enthalpy change of formation of an element in its reference state (see Box) is zero, by definition.

Standard enthalpy change of combustion

The standard enthalpy change of combustion is defined in a very similar way to the standard enthalpy change of formation:

- The **standard enthalpy change of combustion** $\Delta_c H^\ominus$ is the standard enthalpy change when a substance is fully combusted in oxygen.

For example, the standard enthalpy change of combustion of methane is represented as follows:

$CH_4(g) + 2O_2(g) \rightarrow CO_2(g) + 2H_2O(l)$;
$\Delta_c H^\ominus$ (298 K) = −890.7 kJ mol^{-1}

Here:

- the temperature must be stated (in this case, 298 K),
- all substances are in their standard states at the stated temperature,
- pressure is 1 bar,
- the enthalpy change is measured per mole of the substance combusted.

Combustion can be used for many tasks, such as propulsion. The first round-the-world flight by a hot-air balloon took place in 1999. Many fuels are organic and include methane and the other constituents of natural gas.

Standard enthalpy change of reaction

The term 'standard enthalpy change of reaction' (or 'standard enthalpy of reaction') may be used for any reaction, not just combustion or formation.

Hess's law – calculating standard enthalpy changes

Germain Hess was a Russian chemist born in Switzerland. In 1840 he developed a thermochemical version of the law of energy conservation, now known as **Hess's law**:

- The standard enthalpy change for a reaction is independent of the route taken from the reactants to the products.

Hess's law may be used to calculate the standard enthalpy change for any reaction from known standard enthalpy changes. See the next spread.

For any reaction, we could consider it as taking place in two halves, making the products from their elements after having turned the reactants back into their elements. The latter step corresponds to the reverse of the standard enthalpy change of formation of the reactants.

Hence the overall standard enthalpy change is equal to the standard enthalpy change of formation of the products minus the standard enthalpy change of formation of the reactants:

$$\Delta H^\ominus = \Delta_f H^\ominus(\text{products}) - \Delta_f H^\ominus(\text{reactants})$$

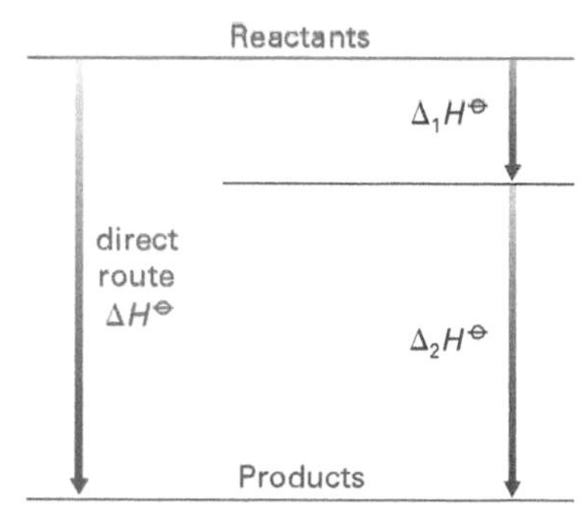

Hess's law summarized diagrammatically:
$\Delta H^\ominus = \Delta_1 H^\ominus + \Delta_2 H^\ominus$
where $\Delta H^\ominus$ refers to the direct route and $\Delta_1 H^\ominus$ and $\Delta_2 H^\ominus$ refer to the indirect route. Such a diagram is often called an enthalpy cycle.

SUMMARY

- The standard enthalpy change is the enthalpy change per mole for conversion of reactants in their standard states into products in their standard states, at a stated temperature.
- The standard state of a substance is the pure substance at 1 bar.
- The standard enthalpy change of formation is the standard enthalpy change when a compound is formed from its elements.
- The standard enthalpy change of combustion is the standard enthalpy change when a substance is fully combusted in oxygen.
- Hess's law states that the standard enthalpy change for a reaction is independent of the route taken from the reactants to the products.

10.6

OBJECTIVES

- Using Hess's law to do enthalpy calculations

Hess's law examples

Hess's law is very useful for performing calculations to find standard enthalpy changes for reactions, given appropriate data about other reactions. So long as the data given enable two routes to be taken from the same reactants to the same products, Hess's law guarantees that the standard enthalpy change calculated by either route must be the same. We will now do several examples to illustrate the law.

Data

The standard enthalpy changes of formation (in kJ mol^{-1}) are as follows:

$CH_4(g)$	−74.4
$CO_2(g)$	−393.5
$H_2O(l)$	−285.8

Worked example to calculate the standard enthalpy change of combustion of methane from standard enthalpy changes of formation

The chemical equation for the combustion of methane (the direct route) is:

$CH_4(g) + 2O_2(g) \rightarrow CO_2(g) + 2H_2O(l)$

Question: Calculate the standard enthalpy change of combustion of methane, using the values of the standard enthalpy change of formation for methane, carbon dioxide, and water shown in the margin.

Strategy: Use the equation
$\Delta H^{\ominus} = \Delta_f H^{\ominus}(\text{products}) - \Delta_f H^{\ominus}(\text{reactants})$

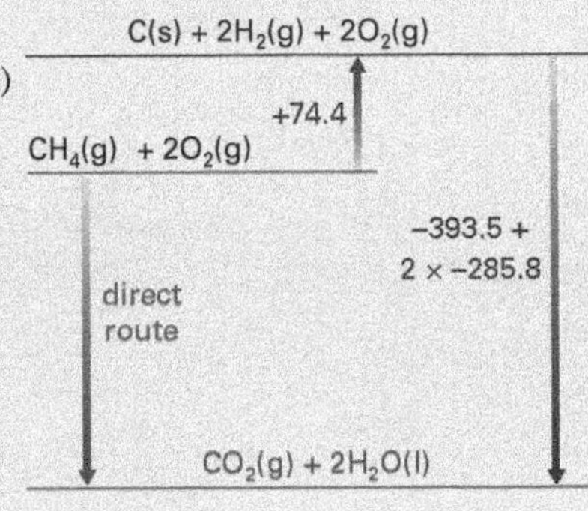

Answer:

Applying the principle explained above to this specific example, we get

$\Delta H^{\ominus} = \Delta_f H^{\ominus}(CO_2) + 2 \times \Delta_f H^{\ominus}(H_2O) - \Delta_f H^{\ominus}(CH_4) - 2 \times \Delta_f H^{\ominus}(O_2)$

where the numbers correspond to the mole ratios in the balanced equation.

Now substitute the values for the substances, remembering that the value for water is that for the liquid:

$\Delta H^{\ominus} = (-393.5\,\text{kJ mol}^{-1}) + 2 \times (-285.8\,\text{kJ mol}^{-1}) - (-74.4\,\text{kJ mol}^{-1}) - 2 \times (0\,\text{kJ mol}^{-1})$

i.e. we can write

$CH_4(g) + 2O_2(g) \rightarrow CO_2(g) + 2H_2O(l); \quad \Delta_c H^{\ominus}(298\,\text{K}) = -890.7\,\text{kJ mol}^{-1}$

Data

The standard enthalpy changes of combustion (in kJ mol^{-1}) are as follows:

C(s)	−393.5
$H_2(g)$	−285.8
$CH_4(g)$	−890.7

Worked example to calculate the standard enthalpy change of formation of methane from standard enthalpy changes of combustion

Question: Given the standard enthalpy changes of combustion shown in the margin, calculate the standard enthalpy change of formation of methane.

Strategy: Set up an enthalpy cycle that includes the data given and the unknown step.

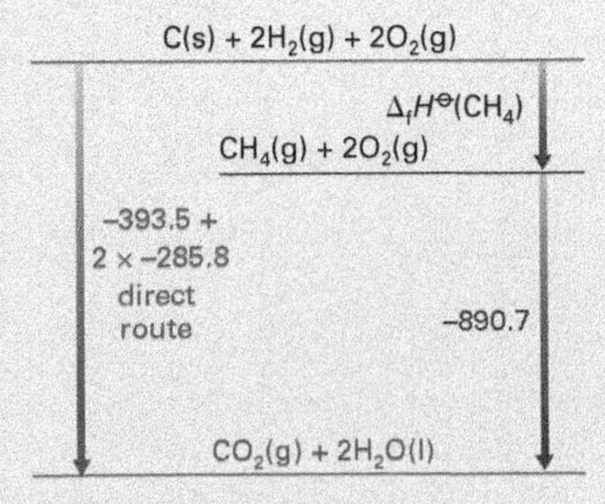

Answer: The enthalpy cycle shows that the direct route (the combustion of the elements) gives the same standard enthalpy change as the indirect route in which the elements are combined into methane and then methane is combusted:

$(-393.5\,\text{kJ mol}^{-1}) + 2 \times (-285.8\,\text{kJ mol}^{-1}) = \Delta_f H^{\ominus}(CH_4) + (-890.7\,\text{kJ mol}^{-1})$

Hence

$\Delta_f H^{\ominus}(CH_4) = (-393.5\,\text{kJ mol}^{-1}) + 2 \times (-285.8\,\text{kJ mol}^{-1}) - (-890.7\,\text{kJ mol}^{-1})$

$= -74.4\,\text{kJ mol}^{-1}$

Note: This calculation is essentially the reverse of the previous example.

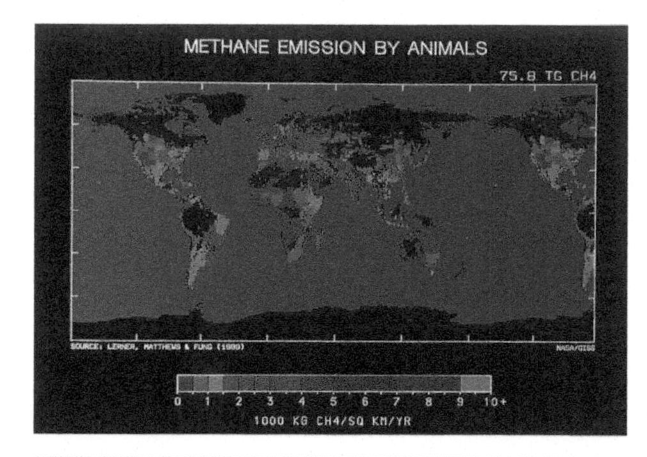

The intensity of methane emissions gives a clear view of the predominance of farming in different countries.

Worked example to calculate the standard enthalpy change of formation of nitromethane from standard enthalpy changes of combustion

Question: Given the standard enthalpy changes of combustion shown in the margin, calculate the standard enthalpy change of formation of nitromethane.

Strategy: Set up an enthalpy cycle that includes the data given and the unknown step.

Answer: The enthalpy cycle shows that the direct route (the combustion of the elements) gives the same standard enthalpy change as the indirect route in which the elements are combined into nitromethane and then nitromethane is combusted:

$C(s) + \frac{3}{2}H_2(g) + \frac{1}{2}N_2(g) + \frac{7}{4}O_2(g)$

$\Delta_f H^\ominus(CH_3NO_2)$

$CH_3NO_2(l) + \frac{3}{4}O_2(g)$

$-393.5 + (\frac{3}{2}) \times -285.8$ direct route

-709.2

$CO_2(g) + \frac{3}{2}H_2O(l) + \frac{1}{2}N_2(g)$

$(-393.5\ \text{kJ mol}^{-1}) + (\frac{3}{2}) \times (-285.8\ \text{kJ mol}^{-1})$

$= \Delta_f H^\ominus(CH_3NO_2) + (-709.2\ \text{kJ mol}^{-1})$

Hence

$\Delta_f H^\ominus(CH_3NO_2) = (-393.5\ \text{kJ mol}^{-1}) + (\frac{3}{2}) \times (-285.8\ \text{kJ mol}^{-1}) - (-709.2\ \text{kJ mol}^{-1})$

$= -113.0\ \text{kJ mol}^{-1}$

Data

The standard enthalpy changes of combustion (in kJ mol^{-1}) are as follows:

C(s)	–393.5
$H_2(g)$	–285.8
$CH_3NO_2(l)$	–709.2

The combustion of nitromethane being evaluated in a test rig.

Worked example to calculate the standard enthalpy change of reaction from standard enthalpy changes of formation

Question: Calculate the standard enthalpy change for the following reaction:

$AlCl_3(s) + 6H_2O(l) \rightarrow AlCl_3{\cdot}6H_2O(s)$

Strategy: Use the equation $\Delta H^\ominus = \Delta_f H^\ominus(\text{products}) - \Delta_f H^\ominus(\text{reactants})$

Answer:

$\Delta H^\ominus = \Delta_f H^\ominus(AlCl_3{\cdot}6H_2O(s)) - \Delta_f H^\ominus(AlCl_3(s) + 6H_2O(l))$

$= (-2680.0\ \text{kJ mol}^{-1}) - ((-704.2\ \text{kJ mol}^{-1}) + 6 \times (-285.8\ \text{kJ mol}^{-1}))$

$= -261\ \text{kJ mol}^{-1}$

Note: Remember to multiply by the mole ratio in the equation.

Data

The standard enthalpy changes of formation (in kJ mol^{-1}) are as follows:

$H_2O(l)$	–285.8
$AlCl_3(s)$	–704.2
$AlCl_3{\cdot}6H_2O(s)$	–2680.0

SUMMARY

- Hess's law can be used to calculate the standard enthalpy change during reactions.

PRACTICE

1 Calculate the standard enthalpy change for the reaction $CaO(s) + H_2O(l) \rightarrow Ca(OH)_2(s)$ given that the standard enthalpy changes of formation of calcium oxide, water, and calcium hydroxide are –635, –286, and –987 kJ mol^{-1} respectively.

2 Calculate the standard enthalpy change of formation of carbon monoxide given that the standard enthalpy changes of combustion of graphite and carbon monoxide are –393.5 and –283.0 kJ mol^{-1} respectively.

10.7 Some important enthalpy changes

OBJECTIVES
- Named standard enthalpy changes
- Breaking reactions into component steps

The heat needed to evaporate sweat is obtained from our hot bodies and so helps to keep us cool.

Lattice energy and ionization energy

Chapter 4 introduced the idea of removing electrons from atoms to form ions. The energy associated with that change was referred to as the 'ionization energy'. You will often find enthalpy changes being referred to colloquially as 'energies' rather than as 'enthalpies'. Similarly, 'lattice enthalpy' is often called 'lattice energy'.

Lattice formation enthalpy

Lattice enthalpy refers to the enthalpy change accompanying breaking of a solid lattice into separate ions in the gas phase. *Lattice formation enthalpy* refers to the opposite change, from gas-phase ions to form a solid. For example

$Na^+(g) + Cl^-(g) \rightarrow NaCl(s)$;

$\Delta_{latfor}H^\ominus(298\text{ K}) = -787\text{ kJ mol}^{-1}$

Compare this expression with the one for 'lattice enthalpy' alongside in the main text.

By now you should understand the meaning of the term 'standard enthalpy change'. You should also be able to carry out calculations involving Hess's law relating to standard enthalpy changes of combustion and standard enthalpy changes of formation. This spread introduces a further selection of important standard enthalpy changes. The spread concludes by showing you how to break reactions down and calculate the standard enthalpy changes for a series of steps.

Named standard enthalpy changes

It is conventional in all the following definitions to omit the word 'standard' preceding the name, despite the fact that they *are* standard enthalpy changes.

- The **enthalpy of fusion (melting)** is the standard enthalpy change accompanying the melting of a solid substance at its melting point. For example

  $Al(s) \rightarrow Al(l)$; $\Delta_m H^\ominus(933\text{ K}) = +10.7\text{ kJ mol}^{-1}$

- The **enthalpy of vaporization** is the standard enthalpy change accompanying the vaporization of a liquid substance at its boiling point. For example

  $H_2O(l) \rightarrow H_2O(g)$; $\Delta_v H^\ominus(373\text{ K}) = +40.7\text{ kJ mol}^{-1}$

- The **enthalpy of atomization** is the standard enthalpy change accompanying the production of separate gaseous atoms of an element. For example

  $\frac{1}{2}Cl_2(g) \rightarrow Cl(g)$; $\Delta_{at}H^\ominus(298\text{ K}) = +121\text{ kJ mol}^{-1}$

  $Na(s) \rightarrow Na(g)$; $\Delta_{at}H^\ominus(298\text{ K}) = +108\text{ kJ mol}^{-1}$

- The **first ionization enthalpy** is the standard enthalpy change accompanying the removal of one electron from an atom in the gas phase. For example

  $Na(g) \rightarrow Na^+(g) + e^-(g)$; $\Delta_{i.e.}H^\ominus(298\text{ K}) = +498\text{ kJ mol}^{-1}$

- The **second ionization enthalpy** is the standard enthalpy change accompanying the removal of a second electron from a singly positively charged ion in the gas phase. For example

  $Na^+(g) \rightarrow Na^{2+}(g) + e^-(g)$; $\Delta_{2nd\,i.e.}H^\ominus(298\text{ K}) = +4560\text{ kJ mol}^{-1}$

- The **electron-gain enthalpy** (previously called 'electron affinity') is the standard enthalpy change accompanying the addition of one electron to an atom in the gas phase. For example

  $Cl(g) + e^-(g) \rightarrow Cl^-(g)$; $\Delta_{e.g.}H^\ominus(298\text{ K}) = -351\text{ kJ mol}^{-1}$

  Note that the process of electron gain for chlorine is exothermic, so the enthalpy change has a negative sign.

- The **lattice enthalpy** is the standard enthalpy change accompanying breaking of a solid lattice into separate ions in the gas phase. For example

  $NaCl(s) \rightarrow Na^+(g) + Cl^-(g)$; $\Delta_{lat}H^\ominus(298\text{ K}) = +787\text{ kJ mol}^{-1}$

- The **enthalpy of hydration** is the standard enthalpy change accompanying the production of a hydrated ion from an ion in the gas phase. For example

  $Na^+(g) + \text{water} \rightarrow Na^+(aq)$; $\Delta_{hyd}H^\ominus(298\text{ K}) = -406\text{ kJ mol}^{-1}$

- The **enthalpy of solution** is the standard enthalpy change accompanying dissolving a solid in a large excess of water. For example

  $NaOH(s) + \text{water} \rightarrow NaOH(aq)$; $\Delta_{sol}H^\ominus(298\text{ K}) = -42.7\text{ kJ mol}^{-1}$

Lattice enthalpy

The magnitude of the lattice enthalpy depends on three main factors, which reflect how ions interact with each other. The three factors are:

1 *The charges on the ions*

The greater the charges on the ions, the greater the attraction between them, and the greater will be the lattice enthalpy. For example, sodium fluoride and magnesium oxide have similar structures. The halide is of the form M^+F^- and the oxide is $M^{2+}O^{2-}$. The attraction between the ions is proportional to the product of the charges. The lattice enthalpy of the oxide is expected to be *four times* that of the fluoride.

This prediction is quite close to the ratio of the observed values (1:4.2).

NaF $\Delta_{lat}H^{\ominus}$ (298 K) = +926 kJ mol^{-1}

MgO $\Delta_{lat}H^{\ominus}$ (298 K) = +3850 kJ mol^{-1}

2 *The distance between the ions*

The smaller the distance between the ions, the greater the attraction between them, and the greater will be the lattice enthalpy. For example, the sizes of the halide ions increase in the order $F^- < Cl^- < Br^- < I^-$. Lattice enthalpies for the sodium halides *decrease* in the same order (note that each substance has the same ionic charges).

The values for the sodium halides are:

NaF $\Delta_{lat}H^{\ominus}$ (298 K) = +926 kJ mol^{-1}

NaCl $\Delta_{lat}H^{\ominus}$ (298 K) = +787 kJ mol^{-1}

NaBr $\Delta_{lat}H^{\ominus}$ (298 K) = +752 kJ mol^{-1}

NaI $\Delta_{lat}H^{\ominus}$ (298 K) = +705 kJ mol^{-1}

The potassium ion is larger than the sodium ion, with the result that the values for the potassium halides are *smaller* than those for the corresponding sodium compounds.

The values for two potassium halides are:

KF $\Delta_{lat}H^{\ominus}$ (298 K) = +821 kJ mol^{-1}

KCl $\Delta_{lat}H^{\ominus}$ (298 K) = +717 kJ mol^{-1}

3 *The detailed crystal structure of the compound*

This last factor is usually the least important. The two factors discussed above have considered the interaction between only a single pair of ions. A real ionic crystal is far more complex, see spread 6.6.

Enthalpy of solution

Two processes take place when a solid dissolves in a solvent. First, the species in the solid become separated from each other. The standard enthalpy change for this endothermic process is the lattice enthalpy of the solid. Secondly, the separated species become surrounded by molecules of the solvent. This process is called **solvation**, or **hydration** when the solvent is water.

The illustration to the right shows water molecules clustering around the ions of dissolved sodium chloride. The attraction between the water molecules and the ions causes a total exothermic enthalpy of hydration of $\Delta_{hyd}H^{\ominus}$ (298 K) = −783 kJ mol^{-1} (for sodium chloride), i.e. the sum of

$$Na^+(g) + water \rightarrow Na^+(aq); \quad \Delta_{hyd}H^{\ominus}\ (298\ K) = -406\ kJ\ mol^{-1}$$

(as given opposite) plus

$$Cl^-(g) + water \rightarrow Cl^-(aq); \quad \Delta_{hyd}H^{\ominus}\ (298\ K) = -377\ kJ\ mol^{-1}$$

Water molecule

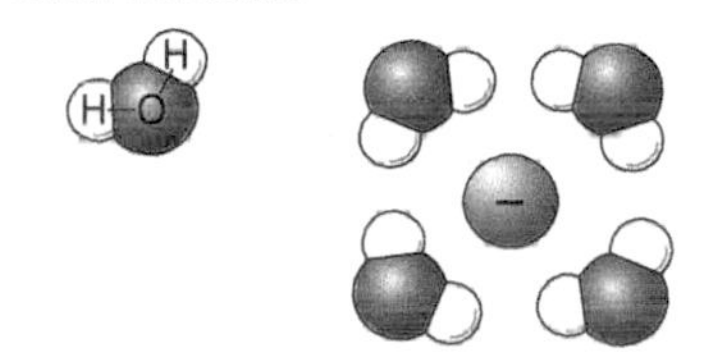

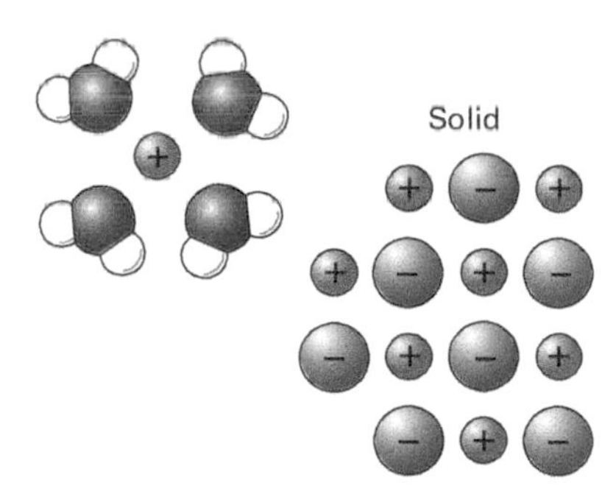

Solution takes place in two steps: the break-up of the lattice followed by hydration of the resulting ions.

The enthalpy of solution is the standard enthalpy change when a solid dissolves in a large excess of water. It is the sum of the lattice enthalpy and the total enthalpy of hydration of the ions. If a solid dissolves exothermically, the solution will become warm as it dissolves. If it dissolves endothermically, the solution will become cold. For example, when sodium chloride dissolves:

$$NaCl(s) + water \rightarrow Na^+(aq) + Cl^-(aq)$$

From above, the lattice enthalpy of sodium chloride is $\Delta_{lat}H^{\ominus}$ (298 K) = +787 kJ mol^{-1}. So, for sodium chloride, the enthalpy of solution is

$$\Delta_{sol}H^{\ominus}\ (298\ K) = \Delta_{lat}H^{\ominus}\ (298\ K) + \Delta_{hyd}H^{\ominus}\ (298\ K)$$
$$= (+787\ kJ\ mol^{-1}) + (-783\ kJ\ mol^{-1})$$
$$= +4\ kJ\ mol^{-1}$$

The enthalpy of solution is *just* endothermic.

SUMMARY

- Lattice enthalpy depends on the charges on the ions, the distance between the ions, and the detailed crystal structure of the compound.
- The enthalpy of solution of an ionic compound is the sum of the lattice enthalpy and the total enthalpy of hydration of the ions.

10.8

Born–Haber cycles

OBJECTIVES

- Born-Haber cycle for sodium chloride
- Enthalpy of formation and stability

Hess's law states that the value of the standard enthalpy change of a reaction is independent of the route taken. As a result, we can consider going from the reactants to the products by two different routes, in an enthalpy cycle. The values of the standard enthalpy changes for the direct and the indirect routes are equal, so we can calculate the value of a particular stage in the overall cycle. An enthalpy cycle used to calculate the standard enthalpy change of formation of ionic compounds is called a **Born–Haber cycle** after Max Born and Fritz Haber who developed it. A Born–Haber cycle pictures the formation of an ionic compound from its elements in a series of steps.

Cubic crystals of pure sodium chloride reflect the arrangement of the ions that make it up.

The Born–Haber cycle for sodium chloride

A diagram representing the formation of sodium chloride from its elements is shown below. The equation

$Na(s) + \frac{1}{2}Cl_2(g) \rightarrow NaCl(s)$

represents the *direct* route. The *indirect* route consists of five separate steps (temperature is 298 K throughout):

1 Atomize solid sodium

$Na(s) \rightarrow Na(g); \quad \Delta_{at}H^{\ominus} = +108\,kJ\,mol^{-1}$

2 Atomize chlorine gas

$\frac{1}{2}Cl_2(g) \rightarrow Cl(g); \quad \Delta_{at}H^{\ominus} = +121\,kJ\,mol^{-1}$

3 Form sodium ions from the sodium atoms

$Na(g) \rightarrow Na^+(g) + e^-(g); \quad \Delta_{i.e.}H^{\ominus} = +498\,kJ\,mol^{-1}$

4 Form chloride ions from the chlorine atoms

$Cl(g) + e^-(g) \rightarrow Cl^-(g); \quad \Delta_{e.g.}H^{\ominus} = -351\,kJ\,mol^{-1}$

5 Pack the sodium and chloride ions together to make solid sodium chloride

$Na^+(g) + Cl^-(g) \rightarrow NaCl(s); \quad \Delta_{latfor}H^{\ominus} = -787\,kJ\,mol^{-1}$

The sum of the standard enthalpy changes for this indirect route equals $-411\,kJ\,mol^{-1}$. This value is the same as the standard enthalpy change of formation of sodium chloride $\Delta_f H^{\ominus} = -411\,kJ\,mol^{-1}$. The standard enthalpy change corresponding to any one of the steps in a Born–Haber cycle may be calculated by applying Hess's law.

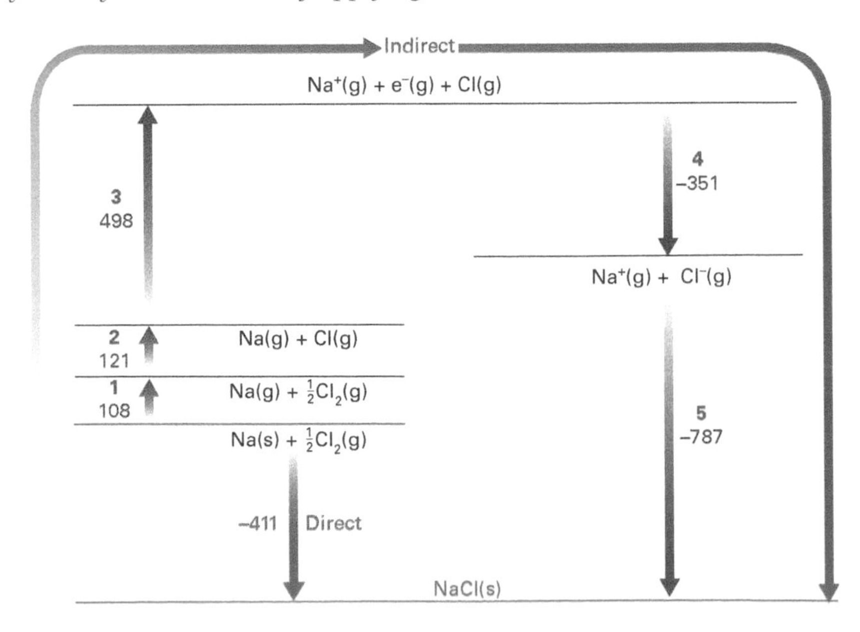

The direction of each arrow in this diagram signifies the sign of the enthalpy change: upwards represents an endothermic step; downwards an exothermic step. All figures are in kJ mol⁻¹.

Stability of ionic compounds

Overall the atomization and ionization processes (i.e. steps 1–4) in a Born–Haber cycle are endothermic (+376 kJ mol^{-1} for NaCl). When the ions in the gas phase pack together to form the solid compound (step 5), the process is exothermic (−787 kJ mol^{-1}). If a compound is to form, the exothermic lattice formation enthalpy must be large enough to compensate for the endothermic processes. As a general rule, the more negative the value of the standard enthalpy change of formation, the more stable the compound will be. (In this context, 'stability' refers to decomposition of the compound back to its elements on heating.) Conversely, an endothermic standard enthalpy change of formation indicates that the compound is likely to be thermally unstable or may not form at all.

Sodium reacts with oxygen to form sodium oxide Na_2O. The Born–Haber cycle for sodium oxide Na_2O shows that the exothermic lattice formation enthalpy is sufficiently large to compensate for the endothermic processes of atomization and ionization.

Note carefully that the *second* electron-gain enthalpy of oxygen is *endothermic*; once one electron has been gained, it repels further electrons.

Sodium does *not* react with oxygen to form the oxide NaO. The relevant Born–Haber cycle shows that the formation of the *hypothetical* ionic compound NaO would require the extremely endothermic step of forming Na^{2+} ions. The second electron would have to be removed from an inner shell closer to the nucleus, causing an enthalpy change of 4560 kJ mol^{-1}. Such a large endothermic process cannot be compensated for by the lattice formation enthalpy. The compound of formula NaO does not form because its standard enthalpy change of formation is too highly endothermic.

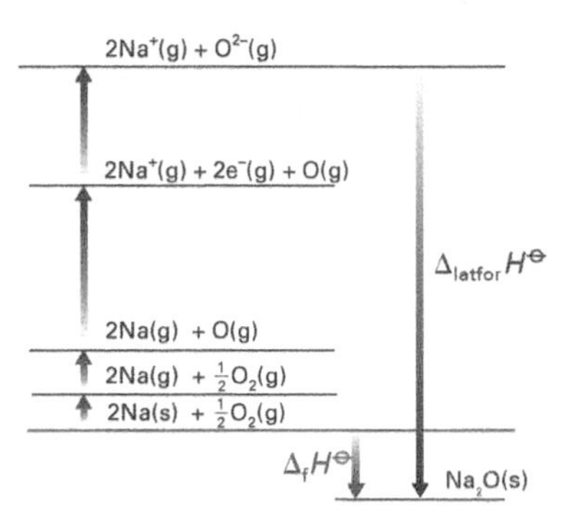

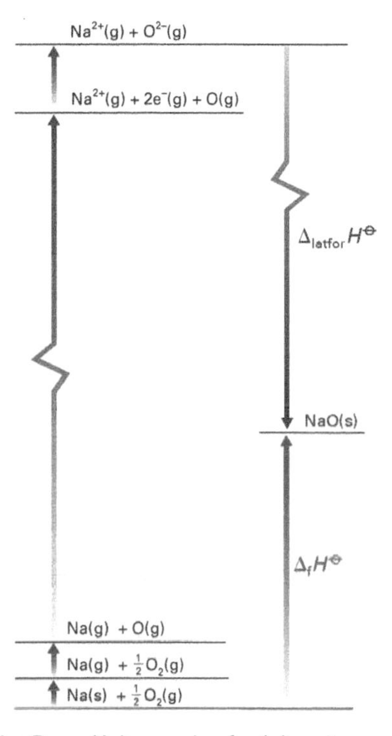

The Born–Haber cycles for (a) sodium oxide Na_2O and (b) the hypothetical compound NaO.

Born–Haber cycles and bond character

Lattice formation enthalpy terms refer to the change in which separated ions come together to form an ionic lattice. A model consisting of *separate uniformly charged spherical ions* represents a state of pure ionic bonding. Using this model, it is possible to calculate theoretical values for lattice formation enthalpies from a knowledge of the charges on the ions and the distance between the ions. Experimental values for lattice formation enthalpies are found from Born–Haber cycles in which the standard enthalpy changes of the various steps result from experimental data.

There is a discrepancy of just 2% between the theoretical and experimental values of lattice formation enthalpy for sodium chloride NaCl. The structure of sodium chloride closely resembles the model for pure ionic bonding. In the case of silver chloride AgCl, there is a discrepancy of 6%, indicating that the crystal lattice does not consist of separate spherical ions. Electron density concentrated *between* the two ions introduces a degree of covalent character to the bonding between the silver and chloride ions, so the lattice formation enthalpy differs from that predicted for a purely ionic compound.

Compound	Theoretical value /kJ mol^{-1}	Experimental value /kJ mol^{-1}
NaCl	−769	−787
NaBr	−732	−752
NaI	−682	−705
AgCl	−864	−915
AgBr	−830	−904
AgI	−808	−889

Theoretical and experimental values of lattice formation enthalpy. Notice that the discrepancy is more marked in AgBr and AgI than in AgCl. The bromide and iodide ions are larger than the chloride ion and are more polarizable (spread 5.11), causing a greater deviation from pure ionic bonding.

SUMMARY

- Born–Haber cycles are an application of Hess's law to ionic compounds.
- Compounds with large negative standard enthalpy changes of formation are likely to be thermally stable with respect to their elements.
- Born–Haber cycles may be used to calculate the theoretical standard enthalpy changes of formation of ionic compounds to see how likely they are to exist.

10.9

OBJECTIVES

- Bond enthalpy
- Average bond enthalpy
- Comparing calculation methods

Bonds

Bond breaking requires energy.
Bond making releases energy.

Bond enthalpies

Oxygen	O=O	+496 kJ mol^{-1}
Hydrogen	H–H	+436 kJ mol^{-1}
Water	O–H	+463 kJ mol^{-1}

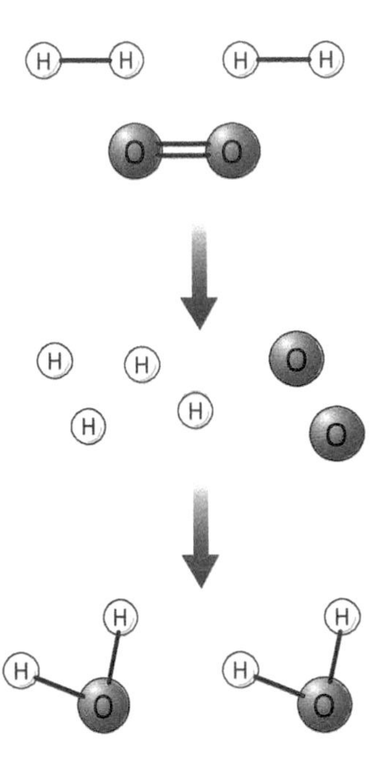

The reactants hydrogen H_2 (2 mol) and oxygen O_2 (1 mol) break into atoms, which combine to make 2 mol of the product, water H_2O.

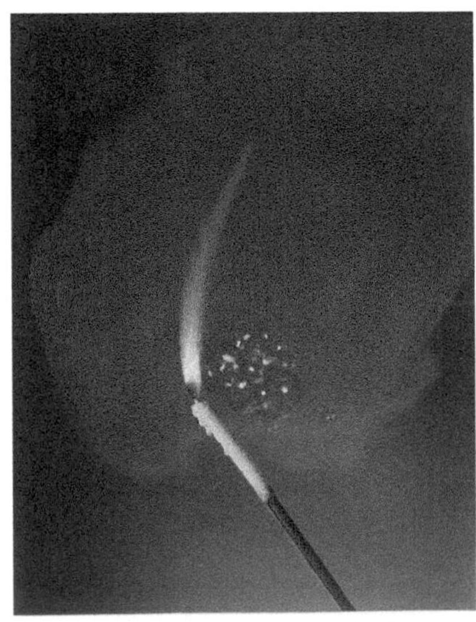

These bubbles are filled with hydrogen. They explode when ignited as the hydrogen meets the oxygen in the air.

BOND ENTHALPY

You are already familiar with using standard enthalpy change of formation data in conjunction with Hess's law to calculate standard enthalpy changes of combustion. Another way of calculating the enthalpy changes of reactions involving covalent compounds is to consider the enthalpy associated with each covalent bond. Calculating the standard enthalpy change is then a matter of comparing the standard enthalpy change involved in *breaking* the bonds in the reactants with that involved in *making* new bonds to form the products. Adding the standard enthalpy changes for the two processes leads to the standard enthalpy change for the overall reaction.

Making and breaking bonds

The **bond enthalpy** (which is often colloquially but inaccurately called the 'bond energy') is the standard enthalpy change associated with breaking A—B bonds into A atoms and B atoms, all species being in the gas phase:

$$A{-}B(g) \rightarrow A(g) + B(g)$$

For example, the H—Cl bond enthalpy is the standard enthalpy change associated with breaking H—Cl molecules into H atoms and Cl atoms. The values of bond enthalpies are always positive because *breaking* a bond always requires energy. The enthalpy change for *making* a bond is equal in magnitude, but opposite in sign.

Calculations involving bond enthalpies

The calculation of the overall standard enthalpy change for any reaction can be thought of as follows:

1. The first step involves calculating the standard enthalpy change accompanying breaking the bonds in the reactant molecules.
2. The second step involves calculating the standard enthalpy change accompanying making the bonds in the product molecules.
3. Then we add together these standard enthalpy changes to obtain the overall standard enthalpy change.

We now show this method for a particular reaction. The equation for the formation of water in the gas phase is:

$$H_2(g) + \tfrac{1}{2}O_2(g) \rightarrow H_2O(g)$$

The steps are as follows:

Step 1 Break into atoms 1 mol of H_2 and ½ mol of O_2.

The total standard enthalpy change for breaking the molecules apart is:

$$1 \times (+436\ kJ\ mol^{-1}) + \tfrac{1}{2} \times (+496\ kJ\ mol^{-1}) = +684\ kJ\ mol^{-1}$$

i.e.

$$H_2(g) + \tfrac{1}{2}O_2(g) \rightarrow 2H(g) + O(g); \qquad \Delta H^{\ominus}(298\ K) = +684\ kJ\ mol^{-1}$$

Step 2 Form 2 mol of O—H bonds per mole of water produced.

Each water molecule has two O—H bonds, so we need to form 2 mol O—H per 1 mol H_2O. The value of the standard enthalpy change is negative because bonds are being made. The standard enthalpy change is therefore:

$$2 \times (-463\ kJ\ mol^{-1}) = -926\ kJ\ mol^{-1}$$

i.e.

$$2H(g) + O(g) \rightarrow H_2O(g); \qquad \Delta H^{\ominus}(298\ K) = -926\ kJ\ mol^{-1}$$

Step 3 Sum the standard enthalpy changes to calculate the overall standard enthalpy change:

$$(+684\ kJ\ mol^{-1}) + (-926\ kJ\ mol^{-1}) = -242\ kJ\ mol^{-1}$$

The standard enthalpy change of formation of water in the gas phase is $-242\ kJ\ mol^{-1}$, spread 10.2.

Average bond enthalpies

The H—H bond enthalpy is quite straightforward because the H—H bond occurs only in the H_2 molecule. On the other hand, there are C—H bonds in many different compounds; for example, C—H bonds are present in every alkane. However, carbon atoms may be bonded together in various arrangements and to various other atoms. Because they may have these different environments, all C—H bonds do not have the same bond enthalpies. The bond enthalpies quoted in tables represent average (mean) bond enthalpies derived from the full range of molecules that contain a particular bond. The results of calculations using average bond enthalpies will therefore show discrepancies when compared with results from experiments with specific molecules.

Worked example on methane

Question: Using average bond enthalpy values, calculate the standard enthalpy change of formation of methane. $\Delta_{at}H^{\ominus}$ for carbon is +717 kJ mol^{-1}. How does the answer obtained compare with the experimental value of −75 kJ mol^{-1}?

Answer: The equation for the reaction is:

$C(s) + 2H_2(g) \rightarrow CH_4(g)$

Step 1 Atomize the reactants.

Break into atoms 1 mol of carbon atoms and 2 mol of hydrogen molecules. The standard enthalpy change is:

$1 \times (+717\ \text{kJ mol}^{-1}) + 2 \times (+436\ \text{kJ mol}^{-1}) = +1589\ \text{kJ mol}^{-1}$

Step 2 Form 4 mol of C—H bonds.

The standard enthalpy change is:

$4 \times (-413\ \text{kJ mol}^{-1}) = -1652\ \text{kJ mol}^{-1}$

Step 3 Sum the standard enthalpy changes to calculate the overall standard enthalpy change.

The standard enthalpy change of formation is:

$(+1589\ \text{kJ mol}^{-1}) + (-1652\ \text{kJ mol}^{-1}) = -63\ \text{kJ mol}^{-1}$

i.e.

$C(s) + 2H_2(g) \rightarrow CH_4(g); \quad \Delta_f H^{\ominus}(298\ \text{K}) = -63\ \text{kJ mol}^{-1}$

Note: This *calculation* therefore gives the standard enthalpy change of formation of methane as −63 kJ mol^{-1}. The calculation uses the *average* bond enthalpy for C—H bonds in the full range of compounds, not just in methane.

The value determined *from experiment* is based on the standard enthalpy changes of combustion of carbon, hydrogen, and methane. Using experimental enthalpy of combustion data, the standard enthalpy change of formation for methane is found to be −75 kJ mol^{-1} using Hess's law. This latter figure is accepted as being the correct value.

There is a difference between results of calculations based on *average* bond enthalpies and results based on *specific* experimental data. However, bond enthalpies are easy to understand and to manipulate, and usually give an accurate enough indication of the standard enthalpy change of a reaction involving covalent substances.

SUMMARY

- Bond enthalpy is the standard enthalpy change associated with breaking A—B bonds into A atoms and B atoms, all species being in the gas phase.
- Average bond enthalpies represent the average value of the standard enthalpy changes required to break a particular covalent bond in the full range of molecules in which that bond may be found.
- Calculations based on average bond enthalpies lead to only approximate results when compared with values obtained from specific experimental data.

Average (mean) bond enthalpies

Bond	Average bond enthalpy/kJ mol^{-1}
C—F	484
C—Cl	338
C—Br	276
C—I	238
C—H	413
C—C	348
C=C	612
C≡C	838
N—H	388
P—H	322
O—H	463
S—H	338
C=O	743

Bond enthalpies and molecular orbitals

The larger the difference in energy between the bonding molecular orbital (spreads 5.6 and 5.7) and the original atomic orbitals, the larger the bond enthalpy.

PRACTICE

1 Using the average bond enthalpy data given above:

a Calculate the standard enthalpy change of formation of ethane C_2H_6 and of propane C_3H_8

b Compare your two answers with one another and with the value of methane.

2 Calculate the standard enthalpy change of combustion for methane, given that the average bond enthalpy for the C=O bond is 743 kJ mol^{-1}.

10.10

OBJECTIVES

- Comparing calculation with experiment
- Insights into bonding

Bond enthalpies and bonding

Bond enthalpy values may be used to calculate the standard enthalpy change for almost any reaction involving covalent substances. The average bond enthalpy value for a given bond is the average for that bond in all its different common environments. Using these *average* values for a *specific* chemical reaction may give unusual or anomalous results when compared to those obtained from experimental data. Such *anomalies* imply that a specific bond does not behave like the 'average bond' represented by the average bond enthalpy value. Studying these anomalies can lead to deeper understanding of the nature of the chemical bonds involved.

Strategy

Step 1 Write down the balanced chemical equation.

Step 2 Identify the bonds that are broken; assign average bond enthalpy values to each bond; breaking bonds is an endothermic process.

Step 3 Identify the bonds that are made; assign average bond enthalpy values to each bond; making bonds is an exothermic process.

Step 4 Sum the exothermic and the endothermic processes to give the overall reaction.

Step 5 Remember that bond enthalpy values refer to the gas phase; water formed from the gas phase must be converted to its standard state (liquid). The enthalpy of vaporization of liquid water is $+44\,kJ\,mol^{-1}$ at 298 K.

The standard enthalpy change of combustion of methane

In the following example, we will calculate the standard enthalpy change of combustion of methane from bond enthalpy values. We follow the steps shown on the left.

Step 1 The balanced chemical equation for this reaction is:

$$CH_4(g) + 2O_2(g) \rightarrow CO_2(g) + 2H_2O(l)$$

Step 2 We must break four C—H bonds (in the methane molecule) and two O=O bonds (in the two oxygen molecules). The standard enthalpy change is

$$4 \times (+413\,kJ\,mol^{-1}) + 2 \times (+496\,kJ\,mol^{-1}) = +2644\,kJ\,mol^{-1}$$

Step 3 We must make two C=O bonds (in the carbon dioxide molecule) and four O—H bonds (in the two water molecules). The standard enthalpy change is

$$2 \times (-743\,kJ\,mol^{-1}) + 4 \times (-463\,kJ\,mol^{-1}) = -3338\,kJ\,mol^{-1}$$

Step 4 We calculate the standard enthalpy change of combustion in the gas phase by summing the values calculated above. This gives

$$(+2644\,kJ\,mol^{-1}) + (-3338\,kJ\,mol^{-1}) = -694\,kJ\,mol^{-1}$$

Step 5 We now convert the water formed in the gas phase to a liquid. When water in the gas phase condenses to a liquid, it releases $44\,kJ\,mol^{-1}$ at 298 K. In this reaction, 2 mol of water form per mole of methane combusted, so the standard enthalpy change is

$$2 \times (-44\,kJ\,mol^{-1}) = -88\,kJ\,mol^{-1}$$

The prediction for the value of $\Delta_c H^\ominus$ for the reaction with liquid water as a product is:

$$(-694\,kJ\,mol^{-1}) + (-88\,kJ\,mol^{-1}) = -782\,kJ\,mol^{-1}$$

i.e.

$$CH_4(g) + 2O_2(g) \rightarrow CO_2(g) + 2H_2O(l);$$
$$\Delta_c H^\ominus (298\,K) = -782\,kJ\,mol^{-1}$$

An average bond enthalpy term derives from a bond in a range of environments. A bond dissociation enthalpy refers to a specific bond in a specific molecule.

Bond broken	Bond dissociation enthalpy/ $kJ\,mol^{-1}$
CH_3CH_2-H	423
$(CH_3)_2CH-H$	413
$(CH_3)_3C-H$	403

The average bond enthalpy for the C—H bond is $413\,kJ\,mol^{-1}$

Bonding implications

The calculation in the example above gives the result $\Delta_c H^\ominus = -782\,kJ\,mol^{-1}$, which is far from the accepted standard enthalpy change of combustion ($-891\,kJ\,mol^{-1}$). The bond enthalpy value used for the C=O bonds in carbon dioxide assumes that each C=O bond is discrete (separate). In reality, electron density is not confined to the two C=O bonds separately. It is delocalized throughout the entire molecule, considerably strengthening each bond. As a result, the enthalpy required to break a C=O bond in carbon dioxide is significantly greater than that required to break the C=O bond in, for example, propanone CH_3COCH_3.

Spectroscopic measurements have determined the actual C=O bond enthalpy (called the **bond dissociation enthalpy**) in carbon dioxide to be $+804\,kJ\,mol^{-1}$. The value used in the calculation above ($+743\,kJ\,mol^{-1}$) is the *average* bond enthalpy for the C=O bond.

Repeating the calculation with –804 kJ mol^{-1} in place of –743 kJ mol^{-1} gives an answer of –904 kJ mol^{-1}, very close to the experimentally based value of –891 kJ mol^{-1}. The remaining error is due to the C—H and O—H values remaining as average values.

(a) The Lewis structure for carbon dioxide shows two electron pairs between each oxygen atom and the central carbon atom. (b) Overlap of the p orbitals in phase forms the molecular orbital (c). This is one of the two molecular orbitals, both of which are delocalized over both *oxygen atoms and the central carbon atom (the other orbital is in the perpendicular plane).*

Benzene

Average bond enthalpy values predict a standard enthalpy change of formation for the structure shown in (a) below of +215 kJ mol^{-1}. The experimentally based value is +49 kJ mol^{-1}. Benzene is thermodynamically more stable with respect to its elements than this structure suggests. Electron density diagrams show that there is equal electron density between each of the carbon atoms. There are six bonding electrons *delocalized* around the entire hexagonal ring of carbon atoms.

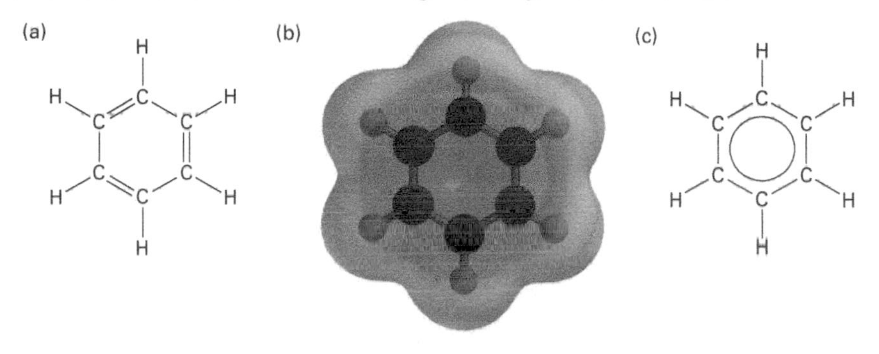

(a) The structure for benzene C_6H_6 suggested by Kekulé over 100 years ago. The structure shows carbon forming 4 covalent bonds and hydrogen forming 1. Six carbon atoms are held in a hexagonal ring structure by alternating C—C single and C=C double covalent bonds. (b) The electrostatic potential map for benzene showing that the carbon-carbon bonds are all the same length. (c) This modified structural formula indicates that the π electrons are delocalized.

Predictions using bond enthalpies

Despite discrepancies which are sometimes difficult to explain, average bond enthalpies are very useful for making general predictions. For example, the standard enthalpy changes of combustion of the alkanes are expected to increase in proportion to the number of carbon atoms present. On combustion, each additional CH_2 group will form one CO_2 molecule and one H_2O molecule. These predictions are borne out in practice and are a useful guide when choosing or designing fuels.

SUMMARY

- If results for $\Delta H^{\ominus}$ obtained from different calculations do not agree, this may be due to errors in assumptions about the nature of specific bonds.

PRACTICE

1 Butane and 2-methylpropane both have the molecular formula C_4H_{10}. Would you expect their standard enthalpy changes of combustion to be identical? Explain your answer.

2 Calculate the standard enthalpy change of combustion $\Delta_c H^{\ominus}$ (298 K) for ethane C_2H_6.

3 Estimate the standard enthalpy change of combustion of nonane C_9H_{20}.

4 The measured standard enthalpy change of formation of buta-1,3-diene $CH_2{=}CH{-}CH{=}CH_2$ is +112 kJ mol^{-1}.
Explain why the standard enthalpy change of formation calculated from average bond enthalpy terms is *more* endothermic than this value.

PRACTICE EXAM QUESTIONS

1 Below are some standard enthalpy changes including the standard enthalpy of combustion of nitroglycerine, $C_3H_5N_3O_9$:

$\frac{1}{2}N_2(g) + O_2(g) \rightarrow NO_2(g)$
$\Delta H^\ominus = +34\,kJ\,mol^{-1}$

$C(s) + O_2(g) \rightarrow CO_2(g)$
$\Delta H^\ominus = -394\,kJ\,mol^{-1}$

$H_2(g) + \frac{1}{2}O_2(g) \rightarrow H_2O(g)$
$\Delta H^\ominus = -242\,kJ\,mol^{-1}$

$C_3H_5N_3O_9(l) + \frac{11}{4}O_2(g) \rightarrow 3CO_2(g) + \frac{5}{2}H_2O(g) + 3NO_2(g)$
$\Delta H^\ominus = -1540\,kJ\,mol^{-1}$

a Standard enthalpy of formation is defined using the term *standard state*.

What does the term *standard state* mean? [2]

b Use the standard enthalpy changes given above to calculate the standard enthalpy of formation of nitroglycerine. [4]

c Calculate the enthalpy change for the following decomposition of nitroglycerine. [3]

$C_3H_5N_3O_9(l) \rightarrow 3CO_2(g) + \frac{5}{2}H_2O(g) + \frac{3}{2}N_2(g) + \frac{1}{4}O_2(g)$

d Suggest one reason why the reaction in part **c** occurs rather than combustion when a bomb containing nitroglycerine explodes on impact. [1]

e An alternative reaction for the combustion of hydrogen, leading to liquid water, is given below.

$H_2(g) + \frac{1}{2}O_2(g) \rightarrow H_2O(l)$ $\qquad \Delta H^\ominus = -286\,kJ\,mol^{-1}$

Calculate the enthalpy change for the process

$H_2O(l) \rightarrow H_2O(g)$ and explain the sign of $\Delta H^\ominus$ in your answer. [2]

2 The tables below contain data which are needed to answer the questions.

Name	Hydrazine	Ethane
Formula of compound	N_2H_4	C_2H_6
Boiling temperature / K	387	184

Formula and state of compound	$C_2H_6(g)$	$CO_2(g)$	$H_2O(l)$
Standard enthalpy of formation (at 298 K) / kJ mol^{-1}	−85	−394	−286

a Suggest why hydrazine has a much higher boiling temperature than ethane. [2]

b When liquid hydrazine burns in oxygen it forms nitrogen and water. The standard enthalpy change for this reaction when one mole of hydrazine forms water in the liquid state is $-624\,kJ\,mol^{-1}$.

i Write a balanced equation for the combustion of hydrazine in oxygen.

ii Calculate the standard enthalpy of formation of liquid hydrazine. [4]

c i Write an equation for the complete combustion of ethane.

ii Use the appropriate standard enthalpies of formation to calculate the standard enthalpy of combustion of ethane. [4]

d Suggest one reason why hydrazine is more suitable than ethane for use as a rocket fuel. [1]

3 a Define the term *standard molar enthalpy change of formation*. [3]

b State *Hess's law*. [1]

c The equation below shows the reaction between ammonia and fluorine.

$NH_3(g) + 3F_2(g) \rightarrow 3HF(g) + NF_3(g)$

i Use the standard molar enthalpy change of formation ($\Delta H_f^\ominus$) data in the table below to calculate the molar enthalpy change for this reaction.

Compound	NH_3	HF	NF_3
$\Delta H_f^\ominus$/ kJ mol^{-1}	−46	−269	−114

[4]

ii Use the average bond enthalpy data in the table below to calculate a value for the molar enthalpy change for the same reaction between ammonia and fluorine.

$NH_3(g) + 3F_2(g) \rightarrow 3HF(g) + NF_3(g)$

Bond	N–H	F–F	H–F	N–F
Average bond enthalpy / kJ mol^{-1}	388	158	562	272

[3]

d The answer you have calculated in **c i** is regarded as being the more reliable value. Suggest why this is so. [3]

4 The relationship between enthalpy of solvation, enthalpy of solution, and lattice enthalpy for sodium chloride and water may be represented by the diagram below.

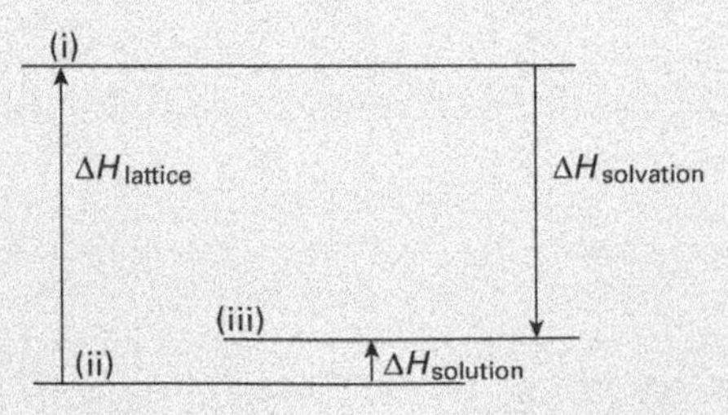

a Copy the diagram and on it, using symbols, indicate the species present at (i), (ii), and (iii). [6]

b Calculate the molar enthalpy of solution of sodium chloride if:

$\Delta H_{lattice} = +788\,kJ\,mol^{-1}$ and $\Delta H_{solvation} = -784\,kJ\,mol^{-1}$

$\Delta H_{solvation} =$ ______________ kJ mol^{-1} [2]

c State the tests you would carry out, giving experimental details, and observations you would make to correctly identify all three solutions labelled A, B, and C, suspected to be sodium chloride, potassium chloride, and lithium chloride. [4]
(See also chapter 16.)

5 a A Born–Haber cycle for the formation of calcium oxide is shown below.

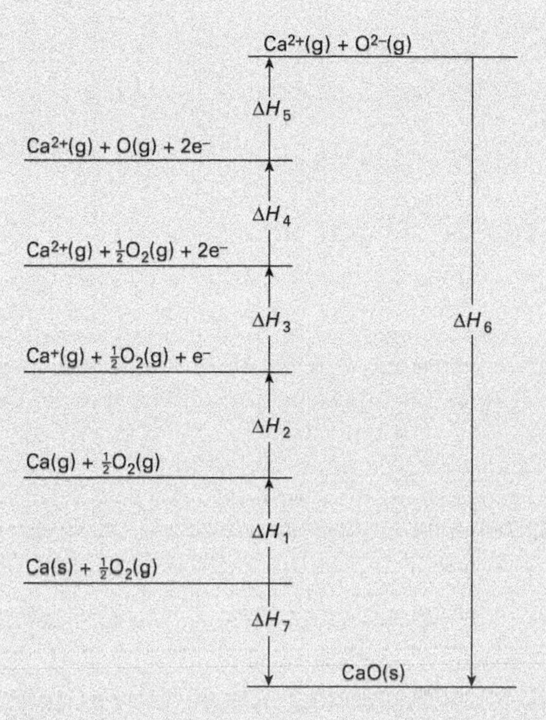

Data ΔH / kJ mol^{-1}:

$\Delta H_1 = +193$; $\Delta H_2 = +590$; $\Delta H_3 = +1150$; $\Delta H_4 = +248$; $\Delta H_6 = -3513$; $\Delta H_7 = -635$.

i Identify the change which represents the lattice enthalpy of CaO. [1]

ii Use the data above to calculate ΔH_5. [3]

iii Use this value of ΔH_5 to calculate the first electron affinity of oxygen, given that the second electron affinity of oxygen is +844 kJ mol^{-1}. [2]

b i What enthalpy change does the value of ΔH_2 represent? [1]

ii Would the value of ΔH_2 be larger or smaller for magnesium than it is for calcium? [1]

iii Explain your answer in **b ii**. [2]

c Given a sample of solid calcium chloride, contaminated with calcium carbonate, describe tests you would perform in order to confirm the presence of:

i calcium ions; [1]

ii chloride ions. [4]
(See also chapters 16 and 18.)

6 Sodium bromide is formed from its elements at 298 K according to the equation

$Na(s) + \frac{1}{2}Br_2(l) \rightarrow NaBr(s)$

The lattice dissociation enthalpy of sodium bromide refers to the enthalpy change for the process

$NaBr(s) \rightarrow Na^+(g) + Br^-(g)$

The electron addition enthalpy refers to the process

$Br(g) + e^- \rightarrow Br^-(g)$

Use the information and the data in the table below to answer the questions which follow.

Standard enthalpies		$\Delta H^\ominus$ / kJ mol^{-1}
$\Delta H_f^\ominus$	formation of NaBr(s)	−361
$\Delta H_{ea}^\ominus$	electron addition to Br(g)	−325
$\Delta H_{sub}^\ominus$	sublimation of Na(s)	+107
$\Delta H_{diss}^\ominus$	bond dissociation of Br_2(g)	+194
$\Delta H_{ion}^\ominus$	first ionization of Na(g)	+498
$\Delta H_L^\ominus$	lattice dissociation of NaBr(s)	+753

a Construct a Born–Haber cycle for sodium bromide. Label the steps in the cycle with symbols like those used above rather than numerical values. [6]

b Use the data above and the Born–Haber cycle in part **(a)** to calculate the enthalpy of vaporization, $\Delta H_{vap}^\ominus$, of liquid bromine. [3]

7 Beer was brewed by the ancient Egyptians and is thought to have been among the rations of the builders of the Pyramids.

The ethanol and glucose composition of a beer is given in the table.

Constituent	**Concentration / g dm^{-3}**
ethanol, C_2H_5OH	20
glucose, $C_6H_{12}O_6$	20

Ethanol can be regarded as a food as well as a drug and is a more concentrated energy source than carbohydrate.

a Write a balanced equation for the complete combustion of ethanol. [1]

b i Define the *term standard enthalpy change of combustion*.

ii The standard enthalpy change of combustion of ethanol and of glucose may be taken as

−1370 kJ mol^{-1} and −3000 kJ mol^{-1}, respectively.

Hence, calculate the enthalpy change per gram of ethanol and of glucose.

iii Calculate the total energy available in 1 dm^3 of the beer, as detailed in the table above. [5]

11 Chemical equilibrium

So far in this book, a chemical equation has implied that the reactants (the species on the left of the arrow) change completely to form the products (the species on the right of the arrow):

reactants $\rightarrow$ products

However, chemical reactions do not only move in the forward direction, from left to right in the chemical equation. This chapter explores the idea that chemical reactions also move in the backward direction, from right to left in the chemical equation. The relationship between the forward and the backward reactions and their effect on the overall yield of the reaction make up the study of *chemical equilibrium*.

The nature of dynamic equilibrium – 1

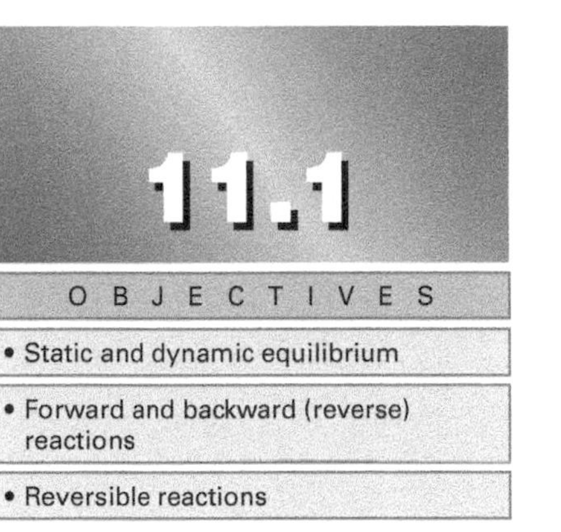

11.1

OBJECTIVES

- Static and dynamic equilibrium
- Forward and backward (reverse) reactions
- Reversible reactions

'Equilibrium' is a term used to denote balance. The two main types of equilibrium encountered in everyday life are static equilibrium and dynamic equilibrium. If you sit balanced against another person on a see-saw, you are in a state of *static* equilibrium. You do not have to move to maintain the state of balance. Dynamic equilibrium is very different. Imagine that you are on a downward-moving escalator and have set yourself the challenge: 'How can I remain at a fixed point between two floors?' You can achieve this objective only by climbing upwards at exactly the same speed as the escalator steps move downwards. In this way you establish a *dynamic* equilibrium, a balance between two objects actively moving in opposite directions.

Dynamic equilibrium for a physical process

Bromine is a dense poisonous fuming liquid. In a fume cupboard, pour some bromine into a large glass bottle and seal it with a stopper. Liquid bromine evaporates to form a vapour. To begin with, there is very little colour in the space above the liquid. But the red–brown colour of the vapour gradually darkens until the space is a uniformly dark shade. The colour change over time shows that the concentration of the bromine vapour increases until a maximum concentration is reached. After this initial change, the liquid and gaseous contents of the flask do not appear to alter their concentrations as time goes by (so long as the temperature remains constant). There will always be liquid bromine in the bottom of the flask with bromine vapour above. But, the equilibrium is not static.

Bromine vapour forms because the most energetic bromine molecules escape from the surface of the liquid: this process is called **vaporization** or **evaporation**. Also, the least energetic molecules in the vapour return to the liquid state: this process is called **condensation**. Vaporization and condensation are *both happening simultaneously* in the flask.

- When the concentration (colour) of the vapour remains steady, the rate of condensation equals the rate of vaporization.

Liquid bromine is in dynamic equilibrium with bromine vapour. This equilibrium process may be represented by a simple equation:

$$Br_2(l) \rightleftharpoons Br_2(g)$$

The symbol $\rightleftharpoons$, two half-arrows, means that it is an equilibrium process. Written in this form, the equation means that, at equilibrium, bromine liquid is changing into bromine vapour at the same rate as bromine vapour is changing into bromine liquid:

$Br_2(l) \rightarrow Br_2(g)$ is the forward change

$Br_2(g) \rightarrow Br_2(l)$ is the backward change

In a closed container, liquid bromine is in dynamic equilibrium with its vapour.

Chemical reactions and chemical equilibrium

Changes of state such as those detailed opposite are physical changes. A dynamic equilibrium will also develop during a chemical change. The balance of reactants and products at equilibrium is called the **equilibrium mixture** or equilibrium composition. All chemical reactions show a tendency to form an equilibrium mixture. The equilibrium mixture may consist mostly of reactants, which shows that very little reaction takes place overall. On the other hand, a reaction is commonly described as 'going to completion' if the equilibrium mixture consists mostly of products and hardly any reactants.

- **Chemical equilibrium** occurs when the concentrations (i.e. molar concentrations) of reactants and products remain constant.
- A **reversible reaction** describes the case where the equilibrium mixture contains significant amounts of the reactants.

Nitrogen, hydrogen, and ammonia

An example of a reversible reaction is the reaction between nitrogen and hydrogen at 450 °C to form ammonia:

$$N_2(g) + 3H_2(g) \rightleftharpoons 2NH_3(g)$$

Mixing nitrogen and hydrogen in a sealed reaction vessel under the right conditions will produce ammonia. The concentration of ammonia will reach a maximum value, but the concentrations of nitrogen and hydrogen do *not* fall to zero. Moreover, pure ammonia in a reaction vessel under the same conditions produces nitrogen and hydrogen. Whether starting from nitrogen and hydrogen or from ammonia, the system will always reach an equilibrium mixture of nitrogen, hydrogen, and ammonia. Once this state of chemical equilibrium is reached, the rates of the forward and backward reactions are the same and there is no further tendency for the composition to change.

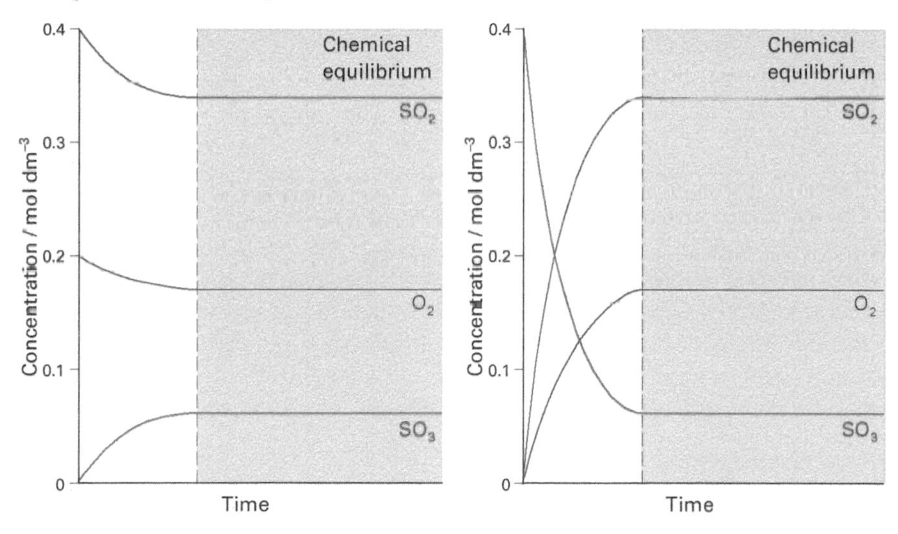

In the reaction $2SO_2(g) + O_2(g) \rightleftharpoons 2SO_3(g)$, the same concentrations are reached whether starting from pure reactants (SO_2 and O_2) or pure product (SO_3). After the time shown by the vertical dashed line, the composition does not change. This is the reversible reaction central to the manufacture of sulphuric acid (see alongside).

SUMMARY

- The forward reaction describes reactants changing into products.
- The backward reaction describes products changing back into reactants.
- Chemical equilibrium occurs when the concentrations of reactants and products remain constant.
- Chemical equilibrium is established when the rate of the forward reaction equals the rate of the backward reaction.

(a)

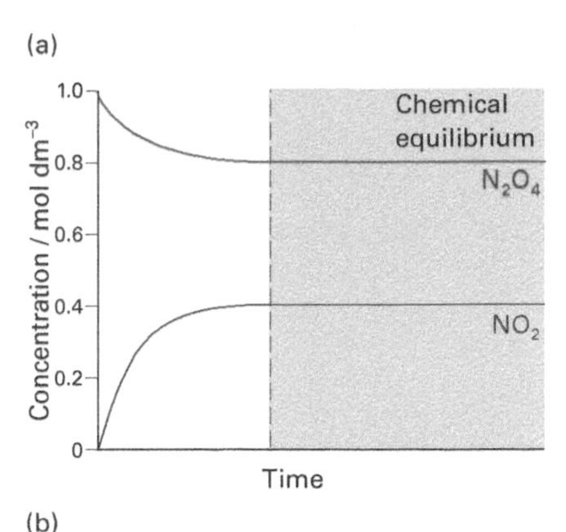

(b)

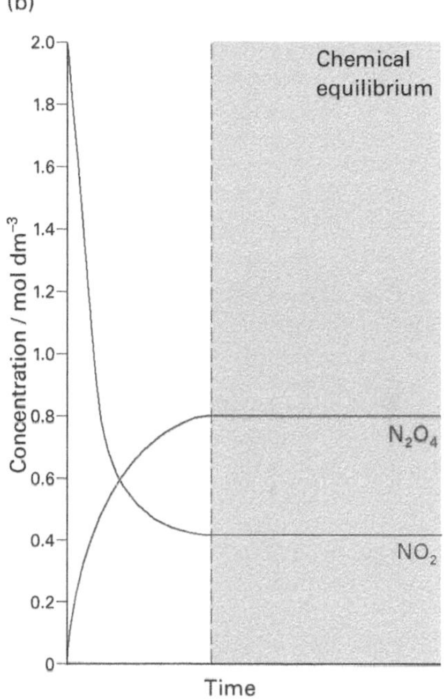

(a) Dinitrogen tetraoxide N_2O_4 decomposes into two molecules of nitrogen dioxide NO_2. (b) Two molecules of nitrogen dioxide combine to form dinitrogen tetraoxide. After the time shown by the vertical dashed line, the composition does not change. Whichever pure compound is chosen to start with, the same concentrations of the two compounds result once equilibrium is reached.

Large industrial plants, such as this one making sulphuric acid, often involve control of reversible reactions. Conditions of temperature and pressure are carefully controlled to produce the optimum yield.

11.2

OBJECTIVES

- Practical proof of the dynamic nature of chemical equilibrium
- Use of radioisotopes and 'heavy hydrogen'
- Ammonia, warfare, and food

THE NATURE OF DYNAMIC EQUILIBRIUM – 2

Nitrogen and hydrogen react to form ammonia; and ammonia decomposes to nitrogen and hydrogen. You are already aware that these two reactions make up a reversible reaction expressed as:

$$N_2(g) + 3H_2(g) \rightleftharpoons 2NH_3(g)$$

Once this system has reached chemical equilibrium, the equilibrium composition shows no further tendency to change. However, chemical reaction does *not* cease when equilibrium is reached. The reactants continue to react to form the products, and the products continue to react to form the reactants. Chemical equilibrium is *dynamic*; the forward and the backward reactions occur *at the same rate*.

Practical proof that chemical equilibrium is dynamic

Substances called radioisotopes (radioactive isotopes) may be used to show that chemical equilibrium is dynamic. Look back at chapter 2 'The nuclear atom' to remind yourself about isotopes. **Radioisotopes** are isotopes that have unstable nuclei. An example is iodine-131, whose nuclei contain 53 protons and 78 neutrons. This isotope has four extra neutrons compared with the stable isotope iodine-127. As a result, iodine-131 is unstable. The nuclei in a sample of iodine-131 decay randomly, emitting radiation, which may be detected with a device called a Geiger counter. The radioisotope iodine-131 may be used to demonstrate the dynamic nature of a chemical equilibrium, as follows.

When a saturated solution of silver iodide (one that contains as much dissolved silver iodide as possible) is in contact with solid silver iodide, an equilibrium is set up between the solid silver iodide and the silver ions and iodide ions in solution:

$$AgI(s) \rightleftharpoons Ag^+(aq) + I^-(aq)$$

Even though no more solid silver iodide can dissolve, exchange continues between the solid silver iodide and the ions in solution. At equilibrium, the solid silver iodide is changing into aqueous silver ions and aqueous iodide ions at the same rate as the ions form solid silver iodide.

This exchange may be demonstrated by using a sample of radioactive solid silver iodide-131 and non-radioactive saturated aqueous silver iodide-127. Remember that overall the solution is saturated and no more solid can dissolve. The radioactive solid is added to the solution and the mixture is left to stand for several hours. After this time, the filtered solution is found to be radioactive. This effect can only happen if the equilibrium is dynamic, and there is exchange between the solid and the solution.

Heavy hydrogen

A similar experiment may be carried out using a *heavy* isotope instead of a *radioactive* one. A **heavy isotope** is an isotope having an extra neutron or neutrons. For example, deuterium (symbol D) is an isotope of hydrogen in which the nucleus of each atom consists of one proton and one neutron. It is often referred to by the name 'heavy hydrogen'.

We now return to the ammonia equilibrium mixture:

$$N_2(g) + 3H_2(g) \rightleftharpoons 2NH_3(g)$$

Some of the hydrogen in the $N_2/H_2/NH_3$ equilibrium mixture is replaced by an equal amount of heavy hydrogen D_2. The D_2 isotope behaves chemically in the same way as H_2 and will take part in the above reaction. So some NH_2D, NHD_2, ND_3, and HD will be detected later (by the use of a mass spectrometer). As in the previous example, this effect can only happen if there is an exchange of atoms between the ammonia and the nitrogen, hydrogen, and deuterium.

Iodine-131

The full symbol for iodine-131 is $^{131}_{53}I$. During the decay of an iodine-131 nucleus, one neutron breaks down to form a proton and an electron (together with another sub-atomic particle called an antineutrino $\bar{\nu}_e$, which has zero charge and zero mass):

$$^{1}_{0}n \rightarrow {}^{1}_{1}p + {}^{0}_{-1}e + \bar{\nu}_e$$

The electron is ejected from the nucleus at high speed and may be detected as a beta particle (β) by a Geiger counter. This decay effectively adds a proton to the nucleus of the iodine-131, increasing its atomic number by one. The resulting nucleus contains 54 protons and 77 neutrons, which corresponds to the stable nuclide xenon-131. The overall change may therefore be represented by:

$$^{131}_{53}I \rightarrow {}^{0}_{-1}\beta + {}^{131}_{54}Xe$$

Uses of radioisotopes

Radiocarbon dating uses the convenient half-life (spread 15.7) of 5730 years for ^{14}C, radiocarbon, to date archaeological remains. Living organisms continuously ingest carbon compounds with carbon's natural $^{14}C/^{12}C$ ratio. When they die, they stop ingesting and the radiocarbon decays. Measuring the $^{14}C/^{12}C$ ratio using a mass spectrometer enables the time since death of the organism to be found. **Radiotherapy** uses the γ radiation from, for example, cobalt-60 to treat cancers; the cancer cells are targeted using multiple intersecting beams.

Smoke alarms commonly use about 0.3 μg of americium-241 (^{241}Am), an α emitter with a half-life of 432 years: α particles have low penetrating power and essentially all will be stopped by the plastic surround. The α particles produce ionization within the detector and hence a current between two charged plates; any reduction in the current due to the presence of smoke triggers the alarm. Because an α particle is a ^4He nucleus (spread 2.3), the daughter nuclide produced is neptunium-237 (^{237}Np).

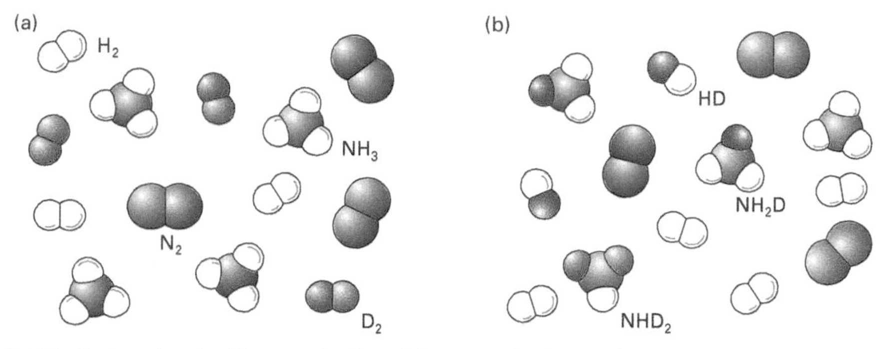

(a) Deuterium has just been substituted for some hydrogen in an equilibrium mixture of nitrogen/hydrogen/ammonia. (b) Some time later, the forward and the backward reactions have introduced deuterium atoms into ammonia and hydrogen molecules, producing HD, NH_2D, and NHD_2. ND_3 could also be found.

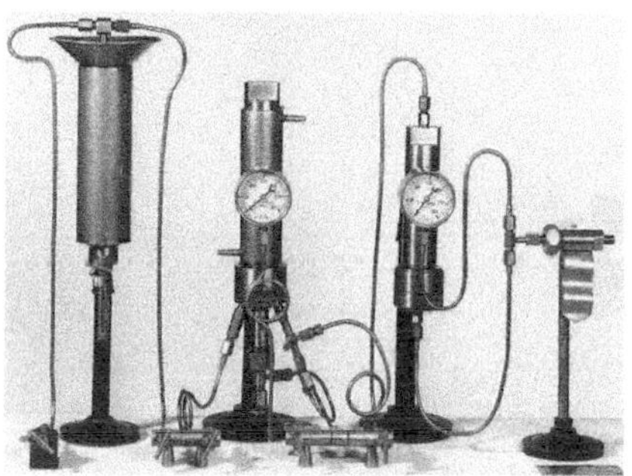

(a) Fritz Haber. (b) Haber's original laboratory apparatus for investigating the N_2/H_2 reaction at various temperatures and pressures.

Fritz Haber, Carl Bosch, and the ammonia synthesis

Fritz Haber was born in 1868 in Silesia (which is now Poland). In 1898 he became Professor of Physical Chemistry at the University of Dahlem, near Berlin. In 1908 he developed the synthesis of ammonia from nitrogen and hydrogen. Haber's chief discoveries were that high pressure was required and that iron catalyses the reaction.

Carl Bosch, who was born in Cologne in 1874, developed the process from the laboratory scale of Haber's experiments to the enormous industrial scale required. The first factory was built in 1913 at Ludwigshafen-Oppau by the Badische Anilin und Soda Fabrik (BASF).

At the start of the First World War, the British Navy blockaded all of Germany's ports. It was expected that the supply of nitrates for making explosives would dry up within a couple of years, as the only large-scale source of nitrates was in Chile. However, the Haber–Bosch process enabled the German chemical industry to make ammonia from nitrogen in the air and hydrogen synthesized from coke (carbon) and steam. Ammonia was then oxidized to make nitrates, which were then used to make explosives. Now independent of nitrate imports, Germany was able to fight on for four more years.

Ammonia is also used in the manufacture of ammonium and nitrate salts for use as plant fertilizers. The importance of ammonia for feeding the world's population was recognized by the award of the 1918 Nobel Prize to Haber. In some ways Haber appeared an extraordinary choice for the first post-war prize. He had prolonged the fighting and also supervised the production of chlorine, the first gas to be used in warfare. Thirteen years later, Bosch won the 1931 Nobel Prize for his contribution to solving the chemical engineering problems.

Human costs

Haber laboured to help his adopted country pay off the costs imposed on Germany by the armistice agreement signed at the end of the war. It should have been obvious to all that he was a patriot. However, in 1933 he faced an unexpected peril. Adolf Hitler had become Chancellor and Haber, of Jewish ancestry, was now in danger of imprisonment. He fled the country and died of a heart attack in Basle, Switzerland, just a few miles over the border from his beloved and ungrateful homeland. Haber's wife Clara, the first female Chemistry PhD from Dahlem, strongly disagreed with him about the use of chemical warfare. On the evening he was promoted for directing gas attacks, she committed suicide.

SUMMARY

- The concentrations of reactants and products do not change after they reach dynamic chemical equilibrium.
- Radioactive elements (radioisotopes) emit radiation, which can be detected by a Geiger counter.
- Radioisotopes and 'heavy' isotopes may be used to label substances and show how they react.
- The development of chemical processes can have profound social and political implications.

PRACTICE

1 In a closed container, liquid bromine is in dynamic equilibrium with its vapour.

a Explain the meaning of the term 'dynamic equilibrium'.

b Why would there no longer be an equilibrium if the stopper were removed from the bottle?

2 For the equilibrium

$N_2(g) + 3H_2(g) \rightleftharpoons 2NH_3(g)$

write down chemical equations for:

a the forward reaction,

b the backward reaction.

11.3

Le Chatelier's principle – 1

OBJECTIVES

- The response of an equilibrium to change
- The effect of concentration change
- The effect of pressure change

Equilibrium mixtures respond to external attempts to change the concentrations of their components. When a change in the concentration of one substance is made, the concentrations of all the other substances involved in the equilibrium will also change. The response of equilibrium systems to changes of concentration, pressure, and temperature was first summed up by Henri Le Chatelier in 1888. Le Chatelier's principle has important implications for many industrial processes, where conditions must be set to favour the production of a particular constituent in an equilibrium.

Statement of Le Chatelier's principle

The principle is a general rule-of-thumb for predicting the effect of changing conditions on the position of a dynamic equilibrium.
Le Chatelier's principle states that:

- If a system at equilibrium is subjected to a small change, the equilibrium tends to shift so as to *minimize* the effect of the change.

Le Chatelier's principle implies that changing the concentration of one substance causes a shift in the position of equilibrium which tends to minimize the change.

The effect of concentration change

The beaker on the left contains yellow CrO_4^{2-} ions. Adding sulphuric acid to a similar solution in the beaker on the right converts some CrO_4^{2-} to orange $Cr_2O_7^{2-}$ ions. The equilibrium has shifted to the right.

Sodium chromate(VI) is a yellow solid with the formula Na_2CrO_4. It dissolves in water to give a yellow solution containing the ions $Na^+(aq)$ and $CrO_4^{2-}(aq)$. Sodium dichromate(VI) is an orange solid with the formula $Na_2Cr_2O_7$. It dissolves in water to give an orange solution containing the ions $Na^+(aq)$ and $Cr_2O_7^{2-}(aq)$.

Adding acid $H^+(aq)$ to a solution of chromate(VI) ions establishes the equilibrium:

$$\underset{\text{YELLOW}}{2CrO_4^{2-}(aq)} + 2H^+(aq) \rightleftharpoons \underset{\text{ORANGE}}{Cr_2O_7^{2-}(aq)} + H_2O(l)$$

The concentrations of the various substances present at equilibrium are constant.

Adding *more* acid will increase the concentration of $H^+(aq)$ ions. This addition has the effect of disturbing the equilibrium. As a response, the balance between the various concentrations of the substances shifts to allow for the addition of H^+ ions. Some of the extra H^+ ions added react with the CrO_4^{2-} ions to form orange dichromate(VI) ions and so minimize the increase in concentration of $H^+(aq)$.

- An equilibrium will shift to the right when the concentration of a reactant (a species on the left of the equation) is increased.

Adding alkali (containing OH^- ions) will reduce the hydrogen ion concentration by neutralization:

$$H^+(aq) + OH^-(aq) \rightarrow H_2O(l)$$

The position of the equilibrium will shift to the left to minimize the decrease in $H^+(aq)$ ion concentration, forming yellow chromate(VI) ions.

The effect of pressure change – ammonia

The molecules in a gas are very far apart, and so gases can be compressed into a smaller volume by increasing the pressure. Under pressure in a smaller volume, the molecules of reacting gases are more likely to collide and to react together. In addition to such externally applied pressure, the reaction between gases might itself cause an overall change of volume (e.g. two gases reacting to form a solid or liquid) and thus pressure. Reactions involving gases are therefore sensitive to pressure change, and the position of equilibrium can be influenced by changes in pressure. Changing the equilibrium position changes the yield of the overall reaction.

Look again at the synthesis of ammonia from nitrogen and hydrogen:

$$N_2(g) + 3H_2(g) \rightleftharpoons 2NH_3(g)$$

There are 1 mol of N_2 and 3 mol of H_2 as reactants on the left of the equation (making 4 mol of gas altogether). On the right, there are only 2 mol of NH_3. Since 1 mol of any gas occupies the same volume, the forward reaction (from 4 mol to 2 mol) decreases the volume. If the pressure on the system is increased, Le Chatelier's principle says that the equilibrium will shift to minimize the pressure increase. The pressure will decrease if the equilibrium system contains fewer moles of gas. The equilibrium shifts to the right, increasing the concentration (and yield) of ammonia.

- In general, increasing the pressure shifts an equilibrium to whichever side of the equation has fewer gas-phase species.

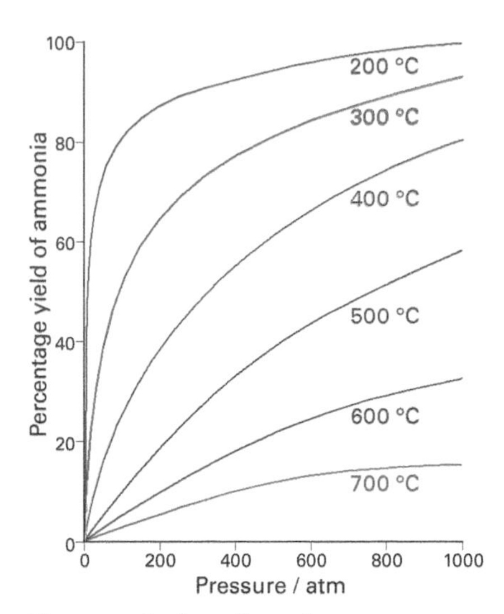

The quantitative effect of pressure on ammonia synthesis. At equilibrium, a larger percentage of ammonia is present at high pressure. In practice, there is a trade-off between initial costs and yield. UK plants operate at around 250 atm, but in France around 1000 atm is used. The higher pressure requires a greater initial investment (special chromium steel vessels are required), but the higher yield increases the profit in the long term.

The effect of pressure change – nitrogen dioxide

Another example of the effect of changing pressure is the dimerization of nitrogen dioxide NO_2. This red-brown gas is a major atmospheric pollutant from car exhausts. Two molecules of NO_2 tend to join together to form a colourless dimer called dinitrogen tetraoxide N_2O_4:

$$2NO_2(g) \rightleftharpoons N_2O_4(g)$$

In this equilibrium, there are more gas molecules on the left of the equation. Increasing pressure therefore shifts the equilibrium to the right, that is towards dimerization. This shift reduces the total number of molecules in the equilibrium system and so tends to minimize the pressure. Conversely, reducing the pressure shifts the equilibrium to the left, away from dimerization.

(a) Red–brown NO_2 and colourless N_2O_4 are at equilibrium in a gas syringe. (b) Pushing in the plunger increases the pressure. (c) After some time has passed, equilibrium has become re-established. The lighter colour indicates that the mixture contains a lower proportion of NO_2.

SUMMARY

- If a system at equilibrium is subjected to a small change, the equilibrium tends to shift so as to minimize the effect of the change.
- An equilibrium will shift to the right when the concentration of a reactant (a species on the left of the equation) is increased.
- An equilibrium will shift to the left when the concentration of a product (a species on the right of the equation) is increased.
- Increasing the pressure shifts an equilibrium to whichever side of the equation has fewer gas-phase species.

11.4

OBJECTIVES

- The effect of temperature change
- Industrial considerations
- Catalysts

Le Chatelier's principle – 2

The previous spread used Le Chatelier's principle to investigate the effects of changing concentration and pressure. This spread applies the principle to the effect of changing the temperature of an equilibrium system. It also considers the effect of catalysts on chemical equilibria.

The effect of temperature on an equilibrium between gases

You already know from the previous chapter that an endothermic reaction absorbs heat from the surroundings, whereas an exothermic reaction gives out heat to the surroundings. If an equilibrium is exothermic in the forward direction, then it is endothermic in the backward direction.

The dimerization of nitrogen dioxide at various temperatures. In hot water (left) red–brown NO_2 predominates; at 0 °C (middle) there is much less NO_2. In a freezing mixture (right) this equilibrium system appears almost colourless (nearly 100% N_2O_4).

For example, the dimerization of nitrogen dioxide

$$2NO_2(g) \rightleftharpoons N_2O_4(g)$$

is exothermic in the forward direction, because a bond is formed between the two nitrogen atoms and no other bonds are broken:

$$2NO_2(g) \rightarrow N_2O_4(g); \qquad \Delta H^{\ominus}(298\,K) = -24\,kJ\,mol^{-1}$$

The backward reaction, the decomposition of dinitrogen tetraoxide, is endothermic; the standard enthalpy change has the same magnitude as, but the opposite sign to, the forward reaction:

$$N_2O_4(g) \rightarrow 2NO_2(g); \qquad \Delta H^{\ominus}(298\,K) = +24\,kJ\,mol^{-1}$$

Le Chatelier's principle suggests that increasing the temperature shifts the equilibrium in the endothermic direction so that heat is absorbed and the temperature increase is minimized. That is, the concentration of NO_2 increases when the temperature is increased. Conversely, decreasing the temperature shifts the equilibrium in the direction that gives out heat. That is, the concentration of N_2O_4 increases when the temperature is decreased.

The pink $[Co(H_2O)_6]^{2+}$ ion predominates at low temperature and the blue $CoCl_4^{2-}$ ion at high temperature.

The effect of temperature on an equilibrium in solution

Dissolving cobalt(II) chloride in hydrochloric acid sets up the following equilibrium:

$$\underset{\text{BLUE}}{CoCl_4^{2-}(aq)} + 6H_2O(l) \rightleftharpoons \underset{\text{PINK}}{[Co(H_2O)_6]^{2+}(aq)} + 4Cl^-(aq)$$

The equilibrium as written is exothermic in the forward direction. If the temperature is increased, the mixture becomes blue. This colour change shows that the concentration of the chloro ion $CoCl_4^{2-}$ has increased. The equilibrium has shifted to the left (in the endothermic direction), thereby absorbing heat. If the temperature is decreased, more of the pink aqua ion $[Co(H_2O)_6]^{2+}$ forms as the equilibrium shifts to the right.

- In general, increasing the temperature shifts an equilibrium in the endothermic direction.

Test for water

You have probably already encountered anhydrous cobalt(II) chloride as a test for water. As the equilibrium equation shows, adding water to blue anhydrous cobalt(II) chloride produces a colour change to pink.

Temperature and the ammonia synthesis

The formation of ammonia from its elements is exothermic:

$$N_2(g) + 3H_2(g) \rightleftharpoons 2NH_3(g); \qquad \Delta H^{\ominus}(298\,K) = -92\,kJ\,mol^{-1}$$

As a result, Le Chatelier's principle predicts that the synthesis should be carried out at *low* temperature to obtain maximum yield. The yields at different temperatures are shown in the illustration on the right.

For industrial processes, the effect of temperature on the equilibrium yield is not the only consideration. The product must also be obtained as rapidly as possible, and so the rate of reaction is also important. As you might expect, reactions are slower at low temperatures and so a *compromise* has to be reached between yield and rate. The temperature used in the Haber–Bosch synthesis of ammonia is usually about 450 °C.

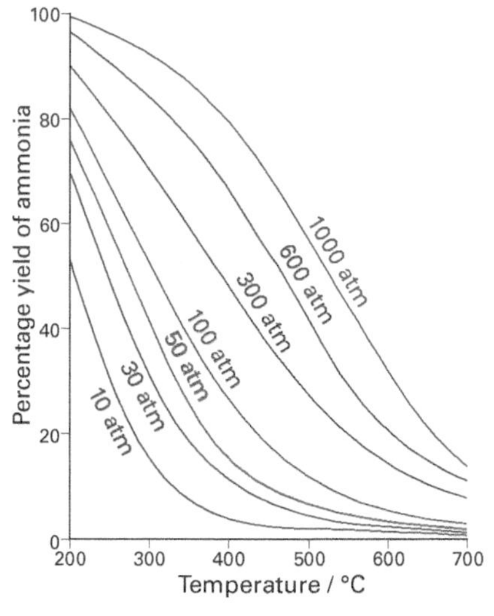

Graph of percentage yield against temperature at various pressures for the production of ammonia from nitrogen and hydrogen.

Catalysts

A **catalyst** is a substance that *speeds up* the rate of a chemical reaction. In the ammonia synthesis, a catalyst (iron) speeds up the rate at which equilibrium is reached. It does *not* affect the yield of ammonia. A catalyst increases the rates of the forward and the backward reactions *to the same extent*.

In some industrial processes, it may be uneconomical to keep the reaction going long enough for equilibrium to be reached. It may be necessary to compromise between the yield and the time taken to produce it.

SUMMARY

- If the forward reaction of an equilibrium is exothermic, then the backward reaction is endothermic; the standard enthalpy change has the same magnitude but opposite sign.
- An increase in temperature shifts an equilibrium in the endothermic direction; a decrease in temperature shifts an equilibrium in the exothermic direction.
- The temperature chosen for an industrial process is often a compromise between yield and rate.
- The equilibrium position is unaffected by the presence of a catalyst; equilibrium is reached faster in the presence of a catalyst.

PRACTICE

1 Consider the equilibrium:

$2CrO_4^{2-}(aq) + 2H^+(aq) \rightleftharpoons Cr_2O_7^{2-}(aq) + H_2O(l)$

a Give two ways of forcing the equilibrium to the right.

b Give two ways of forcing the equilibrium to the left.

c Describe the change in appearance of the solution when the equilibrium moves to the right.

2 What would be the effect of a decrease in pressure on each of the following equilibria?

a $N_2O_4(g) \rightleftharpoons 2NO_2(g)$

b $H_2(g) + I_2(g) \rightleftharpoons 2HI(g)$

c $CO(g) + 2H_2(g) \rightleftharpoons CH_3OH(g)$.

3 Ethanol CH_3CH_2OH is an important industrial compound made from the reaction between ethene C_2H_4 and steam. The three substances are involved in the equilibrium system:

$C_2H_4(g) + H_2O(g) \rightleftharpoons CH_3CH_2OH(g)$

How would you alter the pressure to increase the yield of ethanol?

4 An important step in the manufacture of sulphuric acid is the catalytic oxidation of sulphur dioxide SO_2 to sulphur trioxide SO_3:

$2SO_2(g) + O_2(g) \rightleftharpoons 2SO_3(g);$

$\Delta H^{\ominus}(298\,K) = -197\,kJ\,mol^{-1}$

For each of the following changes, say how the equilibrium will react to the change and give your reasoning:

a Increase in temperature.

b Decrease in pressure.

c Reduction of catalyst efficiency by 50%.

11.5

OBJECTIVES

- Equilibrium position
- The equilibrium constant K_c
- Calculating K_c

THE EQUILIBRIUM CONSTANT K_c

A chemical equilibrium is a reversible chemical reaction in which the overall concentrations of reactants and products are not changing with time. An equilibrium may be described as 'lying to the left', indicating that the reactants predominate; or as 'lying to the right', if products predominate. The actual position of the equilibrium ('to the left' or 'to the right') is a basic property of any particular equilibrium, under fixed conditions. However, as we saw in the previous two spreads, the equilibrium position *does* change when the temperature, pressure, and concentrations change. The equilibrium position may be described in precise terms by combining the equilibrium concentrations to give a value for K_c, the equilibrium constant.

The general expression for K_c

An equilibrium expression may be written for any reaction. Consider the general equilibrium equation:

$aA + bB \rightleftharpoons cC + dD$

The expression for the **equilibrium constant** in terms of concentrations is:

$$K_c = \frac{[C]_{eq}^c[D]_{eq}^d}{[A]_{eq}^a[B]_{eq}^b}$$

where, for example, the term $[C]_{eq}^c$ means the concentration ($mol\,dm^{-3}$) of C at equilibrium raised to the power c.

The equilibrium constant K_c

Nitrogen and hydrogen react together to form ammonia:

$$N_2(g) + 3H_2(g) \rightleftharpoons 2NH_3(g)$$

The equation indicates that 1 mol of N_2 would react with 3 mol of H_2 to form 2 mol of NH_3, *if the reaction went to completion*. In fact, significantly less NH_3 results than the equation predicts because the reaction is reversible and the three species $N_2(g)$, $H_2(g)$, and $NH_3(g)$ are *all* present at chemical equilibrium.

The table below shows some experimental results for this equilibrium. Experiment 1 starts with nitrogen and hydrogen; experiment 2 starts with ammonia; experiment 3 starts with the concentrations indicated. The expression $[H_2]_{eq}$ means 'the concentration of hydrogen gas at equilibrium, in moles of hydrogen molecules H_2 per cubic decimetre'. $[H_2]_0$ refers to the concentration at the start of the reaction.

Results of three experiments for the ammonia synthesis reaction: $N_2(g) + 3H_2(g) \rightleftharpoons 2NH_3(g)$ at 500 °C

Experiment	Initial concentrations /mol dm^{-3}	Equilibrium concentrations /mol dm^{-3}	Equilibrium constant $K_c = \frac{[NH_3]_{eq}^2}{[N_2]_{eq}[H_2]_{eq}^3}$ /dm^6 mol^{-2}
1	$[N_2]_0 = 1.00$ $[H_2]_0 = 1.00$ $[NH_3]_0 = 0$	$[N_2]_{eq} = 0.922$ $[H_2]_{eq} = 0.763$ $[NH_3]_{eq} = 0.157$	6.02×10^{-2}
2	$[N_2]_0 = 0$ $[H_2]_0 = 0$ $[NH_3]_0 = 1.00$	$[N_2]_{eq} = 0.399$ $[H_2]_{eq} = 1.197$ $[NH_3]_{eq} = 0.203$	6.02×10^{-2}
3	$[N_2]_0 = 2.00$ $[H_2]_0 = 1.00$ $[NH_3]_0 = 3.00$	$[N_2]_{eq} = 2.59$ $[H_2]_{eq} = 2.77$ $[NH_3]_{eq} = 1.82$	6.02×10^{-2}

Units

For the reaction:

$N_2(g) + 3H_2(g) \rightleftharpoons 2NH_3(g)$

the mathematical expression for K_c is:

$$K_c = \frac{[NH_3]_{eq}^2}{[N_2]_{eq}[H_2]_{eq}^3}$$

It contains concentrations expressed in units of $mol\,dm^{-3}$. The units for K_c for this reaction are therefore

$$\frac{(mol\,dm^{-3})^2}{(mol\,dm^{-3}) \times (mol\,dm^{-3})^3} = dm^6\,mol^{-2}$$

The units for K_c *for other reactions* may be different.

Look at the three main columns in the table. Each experiment starts with different initial concentrations. There is also a different set of concentrations for each of the equilibrium mixtures. However, substituting the equilibrium concentrations into the expression for K_c, gives a constant numerical result, $6.02 \times 10^{-2}\,dm^6\,mol^{-2}$.

Notice that the concentration of NH_3 is raised to the power of 2 and that of H_2 is raised to the power of 3. These powers are included because 2 mol of NH_3 and 3 mol of H_2 appear in the balanced equation.

- For K_c to have a constant value, the *concentrations* in an equilibrium expression must always be raised to powers corresponding to the mole ratios in the equation.
- No matter what concentrations of reactants or products are present at the start, a reaction always tends towards an equilibrium composition where the concentrations are related by K_c.
- The equilibrium constant is a constant for a particular reaction, at a constant temperature.

Esterification

The equilibrium discussed above takes place in the gas phase. An example of a *liquid-phase* equilibrium is the esterification reaction between an acid (ethanoic acid) and an alcohol (ethanol) to form an ester (ethyl ethanoate) and water, i.e.

$$\underset{\text{acid}}{CH_3COOH(l)} + \underset{\text{alcohol}}{CH_3CH_2OH(l)} \rightleftharpoons \underset{\text{ester}}{CH_3COOCH_2CH_3(l)} + \underset{\text{water}}{H_2O(l)}$$

The equilibrium concentrations of the four components may be determined by an acid–base titration, as shown in the next spread. The table below shows some experimental results that were obtained for this equilibrium.

Data for five experiments (at 373 K) showing initial concentrations of acid and alcohol, equilibrium concentrations of ester (and water), and associated K_c values.

Experiment	$[Acid]_0$ /mol dm^{-3}	$[Alcohol]_0$ /mol dm^{-3}	$[Ester]_{eq}$ /mol dm^{-3}	K_c
1	1.00	0.18	0.171	3.9
2	1.00	0.50	0.420	3.8
3	1.00	1.00	0.667	4.0
4	1.00	2.00	0.858	4.5
5	1.00	8.00	0.966	3.9

Taking experiment 3 as an example, the value for K_c is calculated as follows. The chemical equation indicates that 1 mol of ester requires the reaction of 1 mol of alcohol with 1 mol of acid. The equilibrium concentrations of acid and alcohol are

$(1.00 - 0.667)\ \text{mol dm}^{-3} = 0.333\ \text{mol dm}^{-3}$

Writing down the equation for K_c like that in the box opposite we get

$$K_c = \frac{[\text{ester}]_{eq}[\text{water}]_{eq}}{[\text{acid}]_{eq}[\text{alcohol}]_{eq}}$$

and substituting values

$$K_c = \frac{0.667 \times 0.667}{0.333 \times 0.333} = 4.0$$

Note that experiment 1 starts with an alcohol concentration more than five times smaller than experiment 3 and yet the values for K_c differ by just 2.5%. Experiment 5 starts with an alcohol concentration eight times larger than experiment 3 with the same resultant slight discrepancy.

- K_c remains essentially constant while the alcohol concentration changes by a factor of over 40.

SUMMARY

- K_c is the equilibrium constant for a chemical equilibrium expressed in terms of concentrations (mol dm^{-3}).
- A large numerical value of K_c indicates that the equilibrium lies to the right, i.e. products predominate in the equilibrium mixture.
- A small numerical value of K_c indicates that the equilibrium lies to the left, i.e. reactants predominate in the equilibrium mixture.
- The value of K_c is constant for a particular equilibrium reaction at a constant temperature.

Equilibrium constant and temperature

The value of the equilibrium constant is unaffected by changes in concentration or pressure. However, it *is* dependent on the temperature of the system. The response of an equilibrium to a change in temperature is to establish a new value for the equilibrium constant. Le Chatelier's principle cannot explain the *size* of this change. In chapter 14 'Spontaneous change towards equilibrium', we show how thermodynamics provides quantitative predictions and also explain why K_c changes with temperature.

Units

Note that, unlike the situation for the $N_2/H_2/NH_3$ mixture, in the expression for the equilibrium constant for this esterification reaction, the units 'cancel out'. In this case, the units for K_c are

$$\frac{(\text{mol dm}^{-3}) \times (\text{mol dm}^{-3})}{(\text{mol dm}^{-3}) \times (\text{mol dm}^{-3})} = 1$$

and the equilibrium constant is just a number, i.e. it has no units.

PRACTICE

1 Write down the equilibrium constant for the following equilibria:

a $2NO_2(g) \rightleftharpoons N_2O_4(g)$

b $2SO_2(g) + O_2(g) \rightleftharpoons 2SO_3(g)$

11.6

OBJECTIVES

- Esterification equilibrium
- Thermostatic control of temperature
- 'Freezing' equilibrium reactions
- Analysis of results

AN EXPERIMENTAL DETERMINATION OF K_c

The equilibrium constant K_c is calculated from the concentrations of the substances in an equilibrium mixture. The value of K_c is a constant for a given equilibrium *at a constant temperature*. Changing the temperature alters the value of K_c; changing pressure or concentration does not. Equilibrium constants are derived from experimental data, and this spread gives details of the analysis of an equilibrium mixture and shows how to calculate K_c from the resulting data.

The apparatus for performing an esterification.

Esterification

The equilibrium reaction examined here takes place in the liquid state. It involves water, an alcohol (ethanol), an organic acid (ethanoic acid), and an ester (ethyl ethanoate). We looked at it briefly in the previous spread. The reaction is called 'esterification' because an ester forms as one of the products. The equation for the reaction is

$$\underset{\text{ETHANOIC ACID}}{CH_3COOH(l)} + \underset{\text{ETHANOL}}{CH_3CH_2OH(l)} \rightleftharpoons \underset{\text{ETHYL ETHANOATE}}{CH_3COOCH_2CH_3(l)} + \underset{\text{WATER}}{H_2O(l)} \qquad (1)$$

Ethanoic acid + Ethanol ⇌ Ethyl ethanoate + Water

The esterification reaction.

Practical procedure

Mix 12.0 g (0.200 mol) of ethanoic acid in a conical flask with 9.20 g (0.200 mol) of ethanol and 18.0 g (1.00 mol) of water. Carefully add 4.90 g (0.050 mol) of concentrated sulphuric acid. This acid acts as a catalyst, which speeds up the attainment of equilibrium, but does not affect its position. Seal the flask with a bung and place it for a minimum of four days in a thermostatic water bath set at 25 °C. The total volume of the solution is 44.0 cm^3.

For the analysis of the mixture, set up a titration apparatus with aqueous sodium hydroxide (concentration 1.00 mol dm^{-3}) in the burette. Add base from the burette to the flask until the phenolphthalein indicator remains permanently pink. The base neutralizes the ethanoic acid remaining in the equilibrium mixture *plus* all the sulphuric acid added at the start as a catalyst.

For the purposes of the worked calculation below, assume that a total of 233 cm^3 of base was required.

Calculation

Step 1 Calculate the amount in moles of NaOH used in the titration.

We know from our previous work that

amount in moles = volume × concentration

so

$$\text{amount in moles of NaOH} = \left(\frac{233}{1000}\,dm^3\right) \times (1.00\,mol\,dm^{-3})$$

$$= 0.233\,mol$$

Step 2 Calculate the amount in moles of NaOH that reacts with the sulphuric acid.

Sulphuric acid reacts with sodium hydroxide:

$$H_2SO_4(aq) + 2NaOH(aq) \rightarrow Na_2SO_4(aq) + 2H_2O(l) \quad (2)$$

We know that 0.050 mol H_2SO_4 was added at the start. Therefore the amount in moles of NaOH that reacts with the sulphuric acid according to equation (2) is

$$2 \times 0.050 = 0.100 \text{ mol}$$

Step 3 Calculate the amount in moles of ethanoic acid.

From Step 1, the total amount in moles of NaOH is 0.233 mol. Of this, in Step 2 we found that 0.100 mol reacts with the sulphuric acid. Therefore

$$0.233 - 0.100 = 0.133 \text{ mol}$$

reacts with the ethanoic acid present at equilibrium.

Ethanoic acid reacts with sodium hydroxide:

$$CH_3COOH(aq) + NaOH(aq) \rightarrow Na^+CH_3CO_2^-(aq) + H_2O(l) \quad (3)$$

Therefore according to equation (3) there are 0.133 mol ethanoic acid present at equilibrium.

Step 4 Calculate the equilibrium constant K_c.

We are told that 0.200 mol of both ethanol and ethanoic acid were present at the start. We have worked out in Step 3 that 0.133 mol of ethanoic acid are present at equilibrium. Therefore according to equation (1) there is 0.133 mol of ethanol present at equilibrium. The increase in amount in moles of ester and water present at equilibrium is

$$0.200 - 0.133 = 0.067 \text{ mol}$$

A summary of the results is shown below:

	$CH_3COOH(l)$ +	$CH_3CH_2OH(l)$ $\rightleftharpoons$	$CH_3COOCH_2CH_3(l)$ +	$H_2O(l)$
Start/mol	0.200	0.200	0	1.00
Equilibrium/mol	0.133	0.133	0.067	1.067
Concentration /mol dm^{-3}	3.02	3.02	1.52	24.2

The values of the concentrations have been calculated from the expression:

$$\text{concentration} = \frac{\text{amount in moles}}{\text{volume}}$$

The equilibrium constant for reaction (1) is written in terms of concentrations as

$$K_c = \frac{[CH_3COOCH_2CH_3]_{eq}[H_2O]_{eq}}{[CH_3COOH]_{eq}[CH_3CH_2OH]_{eq}}$$

and substituting values from the last line of results above, we get

$$K_c = \frac{(1.52 \text{ mol dm}^{-3}) \times (24.2 \text{ mol dm}^{-3})}{(3.02 \text{ mol dm}^{-3}) \times (3.02 \text{ mol dm}^{-3})} = 4.0$$

Notice that the value of K_c for this equilibrium has no units – it is *dimensionless* – because the units in the numerator and denominator of the expression above cancel exactly.

SUMMARY

- A thermostatically controlled water bath is used to establish equilibria at temperatures above room temperature.
- An equilibrium involving an acid or a base may be analysed by acid–base titration.
- The equilibrium position is 'frozen' before titration by the addition of excess cold water.

'Freezing' a reaction

Before titrating, the equilibrium position is usually 'frozen' by the addition of a large volume of cold water (about 10 times the volume of the reaction mixture). This large dilution slows the rate at which the substances react together. Titration removes acid from the equilibrium and so alters the balance of the equilibrium, but the titration is completed before the reverse reaction replaces the ethanoic acid neutralized by the addition of base.

'Freezing' is particularly useful when the equilibrium is established at high temperature and the mixture is then titrated at room temperature.

Equilibrium concentrations

Remember that K_c is calculated from the equilibrium concentrations of the various substances (as indicated by the brackets [ ] in the equilibrium expression). Step 4 in the calculation converts amounts in moles into concentrations (mol dm^{-3}). The total volume of the mixture is 44.0 cm^3, equal to 0.044 dm^3.

Spreadsheets

The calculation of K_c follows a clearly defined route, irrespective of the actual amounts of the substances involved. A spreadsheet can be very useful in the calculation of K_c See spread B.1.

PRACTICE

1 Describe with full experimental detail how you would determine the equilibrium constant for the formation of ethyl ethanoate.

EXTENSION

11.7

OBJECTIVES

- Yield and equilibrium position
- Yield and the value of K_c
- Calculating yield

General strategy for calculating yield

Step 1 Write down the equation for the reaction, together with the expression for the equilibrium constant.

Step 2 Calculate the amounts in moles of acid and alcohol *at the start*, using the relationship:

$$\text{amount in moles} = \frac{\text{mass}}{\text{molar mass}}$$

Step 3 Find expressions for the amounts in moles *at equilibrium* on the basis that x mol of ester and x mol of water are formed.

Step 4 Substitute amounts in moles *at equilibrium* into the expression for the equilibrium constant. Arrange into quadratic form:

$ax^2 + bx + c = 0$

Step 5 Solve for x, using the quadratic formula:

$$x = \frac{-b \pm \sqrt{(b^2 - 4ac)}}{2a}$$

Step 6 Find the yield (in grams) from

mass =

amount in moles (x) × molar mass

The equilibrium constant and yield

How can you calculate the yield of a chemical reaction? This task is straightforward in the case of reactions that 'go to completion'. In such reactions, essentially all the reactants change into products. Yield may be calculated from the balanced chemical equation and a knowledge of the amounts of the reactants. In the case of equilibrium reactions, calculating the yield is more complicated. Mixing the reactants leads to chemical equilibrium, where steady concentrations of reactants and products are present together. Product yield depends on the extent to which the equilibrium 'lies to the right', which in turn is reflected in the value of the equilibrium constant.

Calculating the yield of ethyl ethanoate

The reaction between ethanoic acid CH_3COOH and ethanol CH_3CH_2OH produces ethyl ethanoate $CH_3COOCH_2CH_3$ and water as products:

$$CH_3COOH(l) + CH_3CH_2OH(l) \rightleftharpoons CH_3COOCH_2CH_3(l) + H_2O(l)$$

The equilibrium constant has a value of 4.00 at 100 °C. The reaction may be used to make the ester, but the yield will be less than the amount indicated by the chemical equation.

Worked example on calculating yield

We consider the above reaction between ethanoic acid and ethanol.

Question: What mass of ethyl ethanoate is formed at equilibrium if 90.1 g of ethanoic acid is added to 92.1 g of ethanol at 100 °C? K_c for the reaction is 4.00 at 100 °C.

Strategy: We follow the steps outlined in the box on the left.

Answer:

Step 1 The equation for the reaction is given above. The expression for the equilibrium constant for the reaction is:

$$K_c = \frac{[CH_3COOCH_2CH_3]_{eq}[H_2O]_{eq}}{[CH_3COOH]_{eq}[CH_3CH_2OH]_{eq}}$$

Step 2 Calculate the amounts in moles of acid and alcohol *at the start* using the molar masses:

$$n(CH_3COOH) = \frac{90.1\,\text{g}}{[(2 \times 12.0) + (4 \times 1.0) + (2 \times 16.0)]\,\text{g mol}^{-1}}$$

$$= 1.50\,\text{mol}$$

$$n(CH_3CH_2OH) = \frac{92.1\,\text{g}}{[(2 \times 12.0) + (6 \times 1.0) + (1 \times 16.0)]\,\text{g mol}^{-1}}$$

$$= 2.00\,\text{mol}$$

Step 3 If x mol of $CH_3COOCH_2CH_3$ is formed, the chemical equation requires that x mol of H_2O is also formed and that the amounts of CH_3COOH and CH_3CH_2OH will each have *decreased* by x mol. *At equilibrium*, therefore, the amounts in moles can be expressed as:

$CH_3COOH = (1.50 - x)$ mol

$CH_3CH_2OH = (2.00 - x)$ mol

$CH_3COOCH_2CH_3 = x$ mol

$H_2O = x$ mol

Step 4 For each substance, its concentration is its amount in moles divided by the volume V of the mixture. For example, for CH_3COOH, at equilibrium its concentration is $(1.50 - x)$ mol/V. Substituting the expressions for the concentrations of all four substances into the expression for the equilibrium constant, and noting that we are given $K_c = 4.00$ in the question, we obtain:

$$K_c = \frac{(x\,\text{mol}/V) \times (x\,\text{mol}/V)}{\{(1.50 - x)\text{mol}/V\} \times \{(2.00 - x)\text{mol}/V\}} = 4.00$$

In this case (but *not* in all cases), the volume (V) and the units (mol) cancel, leaving:

$$4.00 = \frac{x^2}{(1.50 - x)(2.00 - x)}$$

Multiplying out the brackets gives:

$$(1.50 - x)(2.00 - x) = 3.00 - 1.50x - 2.00x + (-x)^2$$

$$= 3.00 - 3.50x + x^2$$

So

$$4.00 = \frac{x^2}{3.00 - 3.50x + x^2}$$

Multiply both sides by $(3.00 - 3.50x + x^2)$ to get rid of the denominator:

$$12.00 - 14.00x + 4.00x^2 = x^2$$

Subtract x^2 from each side,

$$12.00 - 14.00x + 4.00x^2 - x^2 = x^2 - x^2$$

$$12.00 - 14.00x + 3.00x^2 = 0$$

Arrange into the form $ax^2 + bx + c = 0$:

$$3.00x^2 - 14.00x + 12.00 = 0$$

i.e. $a = 3.00$, $b = -14.00$ (don't forget the minus sign), and $c = 12.00$.

Step 5 Solve using the quadratic formula given in the box above. Hence

$$x = \frac{14.00 \pm \sqrt{\{(-14.00)^2 - (4 \times 3.00 \times 12.00)\}}}{(2 \times 3.00)}$$

$$= 1.13 \text{ or } 3.54$$

The second *mathematical* solution is *chemically* impossible; you cannot form 3.54 mol of $CH_3COOCH_2CH_3$ from 1.50 mol of CH_3COOH. The amount in moles of $CH_3COOCH_2CH_3$ is therefore 1.13 mol.

Step 6 The yield (mass m) of $CH_3COOCH_2CH_3$ can therefore be found as

$$m = (1.13\,\text{mol}) \times \{(4 \times 12.0) + (8 \times 1.0) + (2 \times 16.0)\}\,\text{g mol}^{-1}$$

$$= 99\,\text{g}$$

K_c and equilibrium position

The magnitude of the equilibrium constant describes the equilibrium composition. Look again at the general expression for the equilibrium constant:

$$K_c = \frac{[C]_{eq}^c[D]_{eq}^d}{[A]_{eq}^a[B]_{eq}^b}$$

If the equilibrium mixture contains mostly products, $[C]_{eq}^c[D]_{eq}^d$ is much larger than $[A]_{eq}^a[B]_{eq}^b$ and K_c is large. As a general guide, if K_c has a value above about 1000, the reaction essentially goes to completion. If K_c is below about 0.001, there is too little reaction to be useful. If K_c is around 1, the reaction can sometimes be made useful by careful choice of reaction conditions, guided for example by Le Chatelier's principle.

SUMMARY

- $K_c > 1000$: reaction 'goes to completion'.
- $K_c < 0.001$: negligible reaction.
- K_c around 1: useful amounts of product may be obtained if reaction conditions are carefully chosen.

PRACTICE

1 The equation for the decomposition of phosphorus pentachloride PCl_5 is:

$$PCl_5(g) \rightleftharpoons PCl_3(g) + Cl_2(g)$$

The equilibrium constant $K_c = 0.19\,\text{mol dm}^{-3}$ at 250 °C. At equilibrium the mixture contains 0.200 mol dm^{-3} PCl_5 and 0.010 mol dm^{-3} PCl_3. Calculate the equilibrium concentration of Cl_2 and state the units.

2 Hydrogen iodide (1.00 mol) is introduced into a sealed vessel of volume 12 dm^3 at 425 °C. The equilibrium

$$2HI(g) \rightleftharpoons H_2(g) + I_2(g)$$

is quickly established. The value of the equilibrium constant K_c at this temperature is 0.018. Calculate the amounts in moles of hydrogen and iodine present in the equilibrium mixture.

11.8

OBJECTIVES

- Partial pressure
- Mole fraction
- Calculating K_p

GAS MIXTURES AND THE EQUILIBRIUM CONSTANT K_p

The composition of a solution is usually described in terms of the concentrations of its components. As a result, the equilibrium constant K_c for a reaction in solution is expressed in terms of concentrations. It is difficult to describe the composition of a gas mixture in terms of concentrations; the volume of a solution is fixed whereas the volume of a gas is not. A more natural measure is the *pressure*. The quantity of each gas in an equilibrium mixture is described in terms of the pressure that it exerts – its *partial pressure*. The equilibrium constant can be expressed in terms of these individual partial pressures: it is given the symbol K_p.

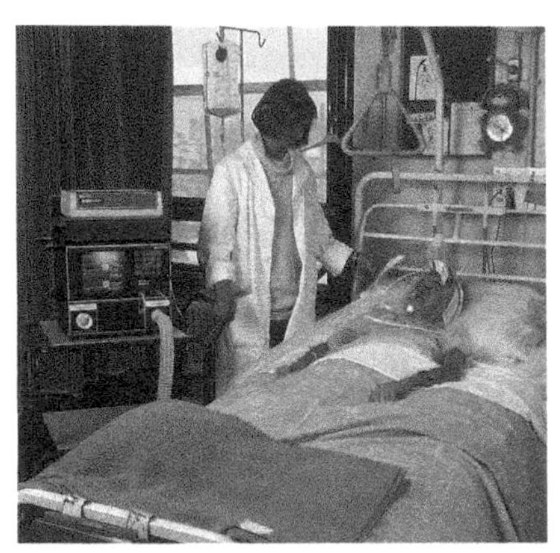

The measurement of the partial pressures of the gases carbon dioxide and oxygen enables a person's metabolic rate to be determined. The partial pressures of the exhaled and inhaled gases vary with the physical fitness, diet, and metabolism of each individual.

	Inhaled air	Exhaled air
$p(O_2)$/bar	0.21	0.16
$p(CO_2)$/bar	0.0003	0.04

Partial pressure

The **partial pressure** of a gas in a mixture of gases is the pressure it would exert if it alone occupied the total volume occupied by the gas mixture. **Dalton's law** of partial pressures states that:

- The total pressure exerted by a mixture of gases is the sum of the partial pressures of the gases.

Dalton's law can be rephrased in terms of the mole fractions of each of the components of the mixture of gases. The **mole fraction** x_A of a component gas A is its fraction of the total amount in moles. In symbols, for a mixture of two component gases A and B:

$$x_A = \frac{n_A}{(n_A + n_B)}$$

where n_A = the amount in moles of gas A, and n_B = the amount in moles of gas B. A mole fraction, of course, has no units because it is a ratio. The sum of the mole fractions of all the components equals one (see box).

In terms of the mole fractions, the partial pressure p_A of a gas A equals the mole fraction x_A of the gas multiplied by the total pressure p_{tot}:

$$p_A = x_A p_{tot}$$

The partial pressure of a gas in a mixture is proportional to its mole fraction in the mixture. An equilibrium constant K_p can therefore be defined in terms of *partial pressures*, in the same way that the equilibrium constant K_c is defined in terms of *concentrations*.

Mole fractions sum to one

$$x_A + x_B = \frac{n_A}{(n_A + n_B)} + \frac{n_B}{(n_A + n_B)} = \frac{(n_A + n_B)}{(n_A + n_B)} = 1$$

The equilibrium constant K_p

When an equilibrium involves gases, the equilibrium constant can be expressed in terms of the partial pressures of the gases. Consider the general equilibrium

$$aA(g) + bB(g) \rightleftharpoons cC(g) + dD(g)$$

The equilibrium constant expression in terms of *partial pressures* is

$$K_p = \frac{p(C)_{eq}^{\,c}\, p(D)_{eq}^{\,d}}{p(A)_{eq}^{\,a}\, p(B)_{eq}^{\,b}}$$

where the subscript 'eq' indicates that the pressures are those once equilibrium has been reached, and, for example, the term $p(C)_{eq}^{\,c}$ means the partial pressure (usually expressed in bar or atm) of C at equilibrium, raised to the power c. Just as K_c is constant if concentration changes, K_p has the same value whatever the starting partial pressures.

Equilibrium constant and temperature

The value of K_p is unaffected by changes in pressure but is dependent on the temperature of the system. Le Chatelier's principle can explain the qualitative effect (and this can be quantified by thermodynamics).

For example, hydrogen, iodine, and hydrogen iodide react together:

$$H_2(g) + I_2(g) \rightleftharpoons 2HI(g)$$

Since the reaction occurs in the gas phase, it is more convenient to write the equilibrium in terms of partial pressures and so calculate K_p:

$$K_p = \frac{p(HI)_{eq}^{\,2}}{p(H_2)_{eq}\, p(I_2)_{eq}}$$

Data for the reverse reaction at 450 K show that, starting with 1.000 mol HI, 0.778 mol HI remains at equilibrium. The amounts in moles of hydrogen and iodine present at equilibrium may be found from this figure: i.e. if 0.778 mol HI remains, then the amount in moles of HI that must have decomposed is

$$1.000 - 0.778 = 0.222 \text{ mol}$$

The chemical equation shows that, in the backward reaction, 2 mol HI decomposes into 1 mol H_2 and 1 mol I_2. Therefore, 0.222 mol HI decomposes into 0.111 mol H_2 and 0.111 mol I_2. These amounts in moles must be converted to partial pressures for the calculation of K_p.

The mole fraction of each substance has the same numerical value as its amount in moles, i.e.

$$\text{mole fraction of HI } (x_{HI}) = \frac{0.778 \text{ mol}}{1.000 \text{ mol}} = 0.778$$

The total amount in moles present at equilibrium is

(0.778 mol + 0.111 mol + 0.111 mol)

= 1.000 mol

The partial pressures are proportional to the mole fractions. For example, at equilibrium the partial pressure of HI is given by

$$p(\text{HI})_{eq} = 0.778 p_{tot}$$

where p_{tot} is the total pressure. The expression for the equilibrium constant becomes

$$K_p = \frac{(0.778 p_{tot})^2}{(0.111 p_{tot}) \times (0.111 p_{tot})} = \frac{0.778^2}{0.111^2} = 49$$

Units

In this case, we do not need to know the actual value of the total pressure, because the p_{tot} terms cancel in the equation, and K_p has no units because

$$\frac{(\text{bar})^2}{(\text{bar} \times \text{bar})}$$

also cancels.

Le Chatelier revisited

An earlier spread in this chapter investigated the effect of pressure change on the ammonia synthesis equilibrium:

$$N_2(g) + 3H_2(g) \rightleftharpoons 2NH_3(g)$$

Le Chatelier's principle indicates that, if the pressure of this system is increased, the equilibrium will shift to minimize the pressure increase and the yield of ammonia will increase.

Writing the equilibrium expression for the reaction in terms of partial pressures shows *quantitatively* the effect of changing pressure:

$$K_p = \frac{p(NH_3)_{eq}^{\ 2}}{p(N_2)_{eq} p(H_2)_{eq}^{\ 3}}$$

According to Dalton's law, each partial pressure may be rewritten in terms of the mole fraction of the gas at equilibrium, x:

$$K_p = \frac{x(NH_3)^2 p_{tot}^{\ 2}}{x(N_2) p_{tot} x(H_2)^3 p_{tot}^{\ 3}}$$

Dividing top and bottom by $p_{tot}^{\ 2}$:

$$K_p = \frac{x(NH_3)^2}{x(N_2) x(H_2)^3 p_{tot}^{\ 2}}$$

This expression shows that the equilibrium constant is proportional to a product of mole fractions divided by the square of the total pressure. As K_p is independent of pressure, K_p does not change if the pressure is changed. So if p_{tot} increases, the mole fraction term, $x(NH_3)^2/x(N_2)x(H_2)^3$, must also increase. But the sum of the mole fractions is always 1. So the mole fraction of ammonia, $x(NH_3)$, must increase while the mole fractions of nitrogen, $x(N_2)$, and hydrogen, $x(H_2)$, decrease.

Dinitrogen tetraoxide

For the dissociation of dinitrogen tetraoxide N_2O_4, the equilibrium constant K_p is

$$K_p = \frac{p(NO_2)_{eq}^{\ 2}}{p(N_2O_4)_{eq}}$$

Note that in this case K_p has units (e.g. bar).

Ethanol manufacture

The manufacture of ethanol by the direct hydration of ethene is a gas-phase reaction:

$$C_2H_4(g) + H_2O(g) \rightleftharpoons CH_3CH_2OH(g)$$

In a typical plant, 200 mol of ethene and 500 mol of steam are introduced into a reaction vessel at 300 °C and 70 atm pressure. Equilibrium is established rapidly in the presence of a phosphoric acid catalyst; the equilibrium mixture typically contains 180 mol of ethanol.

SUMMARY

- K_p is the equilibrium constant for gaseous equilibria; it is expressed in terms of the *partial pressures* of the gases involved in the equilibrium.
- The partial pressure of a gas in a mixture is equal to its mole fraction multiplied by the total pressure of the mixture.

11.9

HETEROGENEOUS EQUILIBRIA

OBJECTIVES

- Homogeneous and heterogeneous equilibria contrasted
- Heterogeneous equilibria and K_c, K_p
- Solubility products

All the chemical equilibria met so far have been **homogeneous equilibria**. The term 'homogeneous' means that all the component species are in the same physical state (phase). For example, the N_2/H_2/NH_3 system is a gas-phase equilibrium; and the CrO_4^{2-}/H^+/$Cr_2O_7^{2-}$ system is an aqueous equilibrium. **Heterogeneous equilibria** involve component species in more than one phase. For example, the ions $Ag^+(aq)$ and $Cl^-(aq)$ may be in equilibrium with solid AgCl(s):

$$AgCl(s) \rightleftharpoons Ag^+(aq) + Cl^-(aq)$$

This spread introduces some of the distinctive features of heterogeneous equilibria.

Heating limestone (calcium carbonate, $CaCO_3$) to a high temperature in a kiln causes decomposition to quicklime (calcium oxide, CaO).

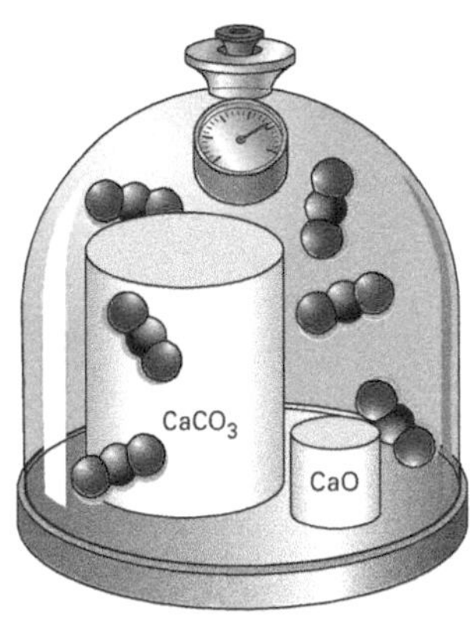

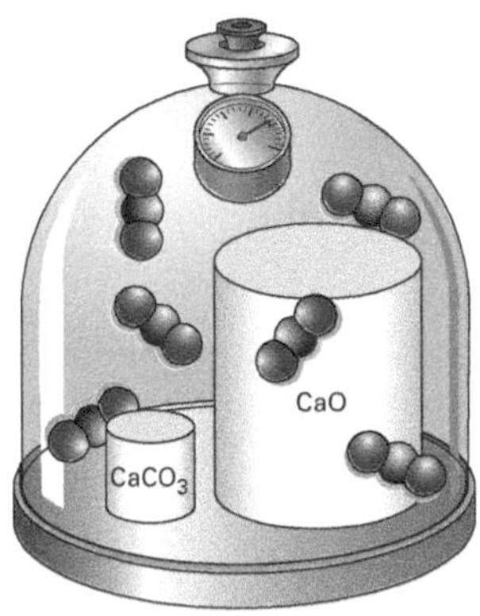

The concentration of a gas is proportional to its pressure. These two diagrams illustrate that the pressure of carbon dioxide is constant at a fixed temperature, regardless of the amounts of solid present.

K_c and the concentration of a solid

The concentration of a substance is defined as the amount in moles per unit volume. For a *solid*, this quantity is proportional to its density, which is its mass per unit volume. A large lump of material has the same density as a small lump. Unlike a solution or a gas, the concentration of a solid cannot change during a reaction. The solid may eventually all be used up; but, as long as any is present, its *concentration remains constant*.

When an equilibrium involves a solid, the concentration of the solid *is a constant*. Because the concentration of a solid cannot change during a reaction, it may be omitted from the expression for the equilibrium constant, K_c. For example, the decomposition of calcium carbonate is an important industrial process contributing to cement manufacture. It involves two solids and a gas:

$$CaCO_3(s) \rightleftharpoons CaO(s) + CO_2(g)$$

The concentration of each solid is a constant. So the expression for the equilibrium constant for this reaction, which is

$$K_c = \frac{[CaO]_{eq}[CO_2]_{eq}}{[CaCO_3]_{eq}} \quad (1)$$

can be reduced to

$$K_c' = [CO_2]_{eq} \quad (2)$$

The K_c in equation (1) is, however, a different number to K_c' in equation (2). The values of the concentrations of the solids have been included in the value of K_c'.

K_p and the concentration of a solid

K_p is the equilibrium constant for a reaction expressed in terms of the partial pressures of the constituent gases. Because calcium oxide CaO and calcium carbonate $CaCO_3$ are solids, the amount of each substance present has no effect on the position of the equilibrium, for the same reasons as given above in the discussion about K_c. The expression for the equilibrium constant in terms of partial pressures is therefore:

$$K_p = p(CO_2)_{eq}$$

Precipitates and equilibria

Aqueous solutions are examples of homogeneous systems: they consist of a solute dissolved in water. A precipitation reaction happens when two solutions are mixed and a solid forms. For example, mixing aqueous silver nitrate with aqueous sodium chloride gives a precipitate of silver chloride. The formation of the precipitate is described by the equilibrium:

$$Ag^+(aq) + Cl^-(aq) \rightleftharpoons AgCl(s)$$

which lies far to the right. This equilibrium is heterogeneous.

A **saturated solution** is a solution that contains as much dissolved solid as the solvent can dissolve at a particular temperature. In a saturated solution of silver chloride, silver ions and chloride ions are in equilibrium with solid silver chloride:

$$AgCl(s) \rightleftharpoons Ag^+(aq) + Cl^-(aq)$$

and the equilibrium constant may be written as:

$$K_c = \frac{[Ag^+]_{eq}[Cl^-]_{eq}}{[AgCl]_{eq}}$$

The concentration of the solid is a constant and may be left out of the equilibrium expression. The resulting equilibrium constant is called the **solubility product**:

$$K_{sp} = [Ag^+]_{eq}[Cl^-]_{eq}$$

The following spread will describe the uses of solubility products.

It is sometimes possible to make a solution that contains a greater amount of solute than a saturated solution. Such a solution is called a ***supersaturated solution*** *and is unstable. Adding one crystal of solid to a supersaturated solution will cause the crystal to grow rapidly until the solution is only saturated.*

When clear aqueous solutions of lead(II) nitrate $Pb(NO_3)_2$ and potassium iodide KI are mixed, a precipitate of solid lead(II) iodide PbI_2 forms.

Solubility product

Consider the general reaction:

$A_aB_b(s) \rightleftharpoons aA^{n+}(aq) + bB^{m-}(aq)$

The solubility product is

$K_{sp} = [A^{n+}]_{eq}^{a}[B^{m-}]_{eq}^{b}$

The equilibrium may be established by placing the solid in contact with water, or by forming a precipitate by mixing together solutions of the ions.

Calcium hydroxide

Calcium hydroxide is a *sparingly soluble* solid, one that dissolves only to a very small extent in water:

$Ca(OH)_2(s) \rightleftharpoons Ca^{2+}(aq) + 2OH^-(aq)$

The solubility product for calcium hydroxide is:

$K_{sp} = [Ca^{2+}]_{eq}[OH^-]_{eq}^{2}$

The concentration of the hydroxide ion has been raised to the power 2 because two hydroxide ions are formed in the balanced equation.

SUMMARY

- The concentration of a solid is constant, regardless of the amount present.
- Equilibrium expressions for K_c and K_p do not include terms relating to solids involved in the heterogeneous equilibrium mixture.
- A saturated solution is a solution that contains as much dissolved solid as the solvent can dissolve at a particular temperature.
- A sparingly soluble solute is a substance that has a very low solubility in the solvent.
- The solubility product is the equilibrium constant expressed in terms of concentrations of the ions produced from a sparingly soluble solid in contact with its saturated solution.

PRACTICE

1 State what you understand by each of the following terms, giving examples to illustrate your meaning:

a Equilibrium

b Sparingly soluble

c Solution

d Homogeneous

e Heterogeneous

f Amount in moles

g Molar concentration.

2 Write equilibrium expressions for K_c, K_p, or K_{sp} as indicated for each of the following equilibrium systems:

a The thermal decomposition of limestone to quicklime:
$CaCO_3(s) \rightleftharpoons CaO(s) + CO_2(g)$ (K_p)

b Solid yellow lead(II) iodide in contact with water:
$PbI_2(s) \rightleftharpoons Pb^{2+}(aq) + 2I^-(aq)$ (K_{sp})

c The dissociation of water:
$H_2O(l) \rightleftharpoons H^+(aq) + OH^-(aq)$ (K_c).

EXTENSION

11.10

OBJECTIVES

- Criteria for precipitation
- Qualitative analysis
- The common ion effect

SOLUBILITY PRODUCTS AND PRECIPITATES

If you mix two solutions containing ions, a solid precipitates if the ionic concentrations exceed those required for a saturated solution. For a sparingly soluble compound, the solubility product K_{sp} may be used to predict whether a precipitate will form. These predictions form the basis of much of qualitative analysis, which identifies substances by using specific reagents to form coloured precipitates.

Explaining precipitation

If you add dilute aqueous ammonia to a solution containing magnesium ions, magnesium hydroxide precipitates out from the solution. Repeat the experiment with a solution containing calcium ions, and no precipitate forms. These two practical results may be explained by considering the solubility products of calcium hydroxide $Ca(OH)_2$ and magnesium hydroxide $Mg(OH)_2$.

For both of the metal hydroxides, the expressions for the equilibrium reaction and the solubility product (using M to denote the metal) are:

$$M(OH)_2(s) \rightleftharpoons M^{2+}(aq) + 2OH^-(aq)$$

$$K_{sp} = [M^{2+}]_{eq}[OH^-]_{eq}^2$$

The equilibrium hydroxide ion concentration in dilute aqueous ammonia is typically $1.0 \times 10^{-3}\,mol\,dm^{-3}$, and hence $[OH^-]_{eq}^2 = 1.0 \times 10^{-6}\,mol^2\,dm^{-6}$. We now consider the solubility products of the two hydroxides.

The solubility product for *calcium hydroxide* is found from data tables to be $5.5 \times 10^{-6}\,mol^3\,dm^{-9}$, i.e.

$$K_{sp} = [Ca^{2+}]_{eq}[OH^-]_{eq}^2 = 5.5 \times 10^{-6}\,mol^3\,dm^{-9}$$

Substituting for $[OH^-]_{eq}^2$ from above, we get

$$[Ca^{2+}]_{eq} \times (1.0 \times 10^{-6}\,mol^2\,dm^{-6}) = 5.5 \times 10^{-6}\,mol^3\,dm^{-9}$$

Dividing both sides by $1.0 \times 10^{-6}\,mol^2\,dm^{-6}$

$$[Ca^{2+}]_{eq} = 5.5\,mol\,dm^{-3}$$

So a saturated solution of calcium hydroxide contains Ca^{2+} ions at a concentration of $5.5\,mol\,dm^{-3}$.

The solubility product for *magnesium hydroxide* is likewise found to be $1.1 \times 10^{-11}\,mol^3\,dm^{-9}$, i.e.

$$K_{sp} = [Mg^{2+}]_{eq}[OH^-]_{eq}^2 = 1.1 \times 10^{-11}\,mol^3\,dm^{-9}$$

Again, substituting for $[OH^-]_{eq}^2$ from above, we get

$$[Mg^{2+}]_{eq} \times (1.0 \times 10^{-6}\,mol^2\,dm^{-6}) = 1.1 \times 10^{-11}\,mol^3\,dm^{-9}$$

Dividing both sides by $1.0 \times 10^{-6}\,mol^2\,dm^{-6}$

$$[Mg^{2+}]_{eq} = 1.1 \times 10^{-5}\,mol\,dm^{-3}$$

So a saturated solution of magnesium hydroxide contains Mg^{2+} ions at a concentration of $1.1 \times 10^{-5}\,mol\,dm^{-3}$.

Most solutions of magnesium ions in the laboratory have concentrations greater than $1.1 \times 10^{-5}\,mol\,dm^{-3}$ and so $Mg(OH)_2(s)$ will precipitate if aqueous ammonia is added. It is unlikely that the calcium ion concentration in a solution will be as great as $5.5\,mol\,dm^{-3}$.

When clear aqueous solutions of iron(III) nitrate $Fe(NO_3)_3$ and potassium hydroxide KOH are mixed, a precipitate of solid iron(III) hydroxide $Fe(OH)_3$ forms.

Qualitative analysis

Precipitation reactions may be used in **qualitative analysis** to identify ions in solution. For example, adding dilute aqueous sodium hydroxide to an aqueous solution containing metal ions often causes precipitation. This happens because many metal hydroxides are only very slightly soluble (**sparingly soluble**) in water, i.e. they have very low values of solubility product. So a precipitate forms according to rule 3 in the box on the right. Many of the metal hydroxides precipitated have characteristic colours, which helps to identify the original aqueous metal ion, spread 20.11.

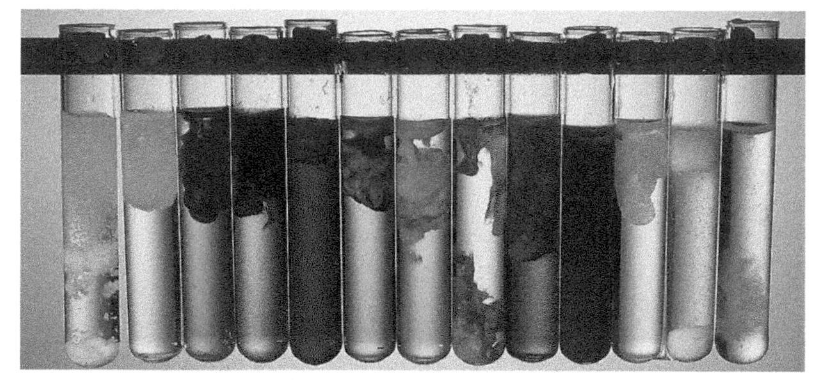
Precipitates of sparingly soluble hydroxides.

Aqueous halide ions are identified by their precipitation reactions with aqueous silver nitrate. The general equation is:

silver nitrate + metal halide ⇌ metal nitrate + silver halide
SOLUBLE SOLUBLE SOLUBLE INSOLUBLE PRECIPITATE

Silver *nitrate* is chosen because all nitrates have a high solubility in water. The nitrate ion in aqueous silver nitrate will not cause precipitation of metal nitrates from solutions to which it is added. The silver halides have characteristic colours, spread 18.6, so the halide ion can be identified.

The common ion effect

The solubility of a sparingly soluble compound is reduced further if another compound is added that contains an ion in common with it. For example, the expressions for the equilibrium reaction and solubility product for silver chloride are:

$$AgCl(s) \rightleftharpoons Ag^+(aq) + Cl^-(aq)$$

$$K_{sp} = [Ag^+]_{eq}[Cl^-]_{eq}$$

Adding sodium chloride (which is *very* soluble) to a saturated solution of silver chloride (which is only *sparingly* soluble) increases the chloride ion concentration. The value of K_{sp} for silver chloride must remain constant: but $[Cl^-]_{eq}$ has increased, so the equilibrium shifts to take up the extra Cl^- ions, and $[Ag^+]_{eq}$ must decrease. Thus the solubility of silver chloride is decreased; the production of a precipitate of silver chloride is correspondingly made more likely. The effect of the common Cl^- ion is to *reduce* the solubility of the silver chloride, so it precipitates.

The common ion effect is used in the Solvay process, the industrial preparation of sodium carbonate, spread 16.6. Carbon dioxide gas flows up a tower while brine (aqueous sodium chloride) saturated with ammonia flows down. The following equilibrium is established:

$$CO_2(g) + NH_3(aq) + NaCl(aq) + H_2O(l) \rightleftharpoons NH_4Cl(aq) + NaHCO_3(s)$$

The equilibrium shifts to the right because sodium hydrogencarbonate $NaHCO_3$ is almost insoluble in cold brine due to the common ion effect. The solid is filtered off and decomposed by heating to give the desired product (sodium carbonate).

SUMMARY

- A solid precipitates from solution when the product of the ionic concentrations in the solution exceeds the solubility product.
- Many metal ions form coloured hydroxide precipitates.
- The solubility of a sparingly soluble compound is reduced further on addition of another compound that contains a common ion.

Some basic rules

For an equilibrium such as

$AgCl(s) \rightleftharpoons Ag^+(aq) + Cl^-(aq)$

where a solid is in equilibrium with its ions in solution, the following rules hold:

1 If the product of the ionic concentrations on mixing two solutions is *exactly equal* to the solubility product, the mixed solution is saturated and no more solid will dissolve.

2 If the product of the ionic concentrations on mixing two solutions is *less than* the solubility product, the mixed solution is not saturated and a precipitate does not form.

3 If the product of the ionic concentrations on mixing two solutions is *greater than* the solubility product, the solubility of the solid is exceeded and a precipitate forms.

The brothers Solvay

Sodium carbonate is used in huge quantities to make glass. The Solvay process made this important raw material more cheaply than previous processes had done; as a result, the brothers Solvay made a large fortune. They funded the Solvay Congresses in the early twentieth century, which helped to solve the problem of the structure of the atom.

The Solvay Congress of 1927 with several of the individuals central to the solution of the structure of the atom highlighted. See spreads 4.1, 4.2, 4.4, and 14.4.

PRACTICE EXAM QUESTIONS

1 Each of the equations **A**, **B**, **C**, and **D** represents a dynamic equilibrium.

A $N_2(g) + O_2(g) \rightleftharpoons 2NO(g)$ $\Delta H^\ominus = +180\,kJ\,mol^{-1}$

B $N_2O_4(g) \rightleftharpoons 2NO_2(g)$ $\Delta H^\ominus = +58\,kJ\,mol^{-1}$

C $3H_2(g) + N_2(g) \rightleftharpoons 2NH_3(g)$ $\Delta H^\ominus = -92\,kJ\,mol^{-1}$

D $H_2(g) + I_2(g) \rightleftharpoons 2HI(g)$ $\Delta H^\ominus = -10\,kJ\,mol^{-1}$

a Explain what is meant by the term dynamic equilibrium. [1]

b Explain why a catalyst does not alter the position of any equilibrium reaction. [2]

c The units of the equilibrium constant, K_c, for one of the above reactions are $mol\,dm^{-3}$. Identify the reaction **A**, **B**, **C**, or **D** which has these units for K_c and write the expression for K_c for this reaction. [2]

d The graphs below show how the yield of product varies with pressure for three of the reactions **A**, **B**, **C**, and **D** given above.

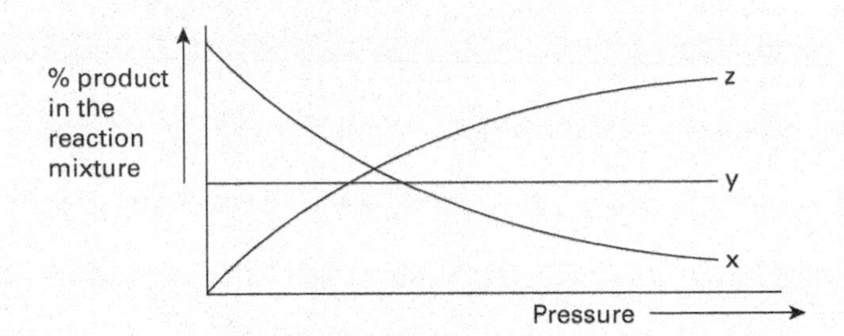

i Identify a reaction from **A**, **B**, **C**, and **D** which would have the relationship between yield and pressure shown in graphs x, y, and z.

ii Explain why an industrial chemist would not use a very low pressure for the reaction represented in graph x.

iii Explain why an industrial chemist may not use a very high pressure for the reaction represented in graph z.

iv Copy the above graphs and add a line to show how the product yield would vary with pressure if the reaction which follows curve z was carried out at a temperature higher than that of the original graph. [6]

2 At a temperature of 107 °C, the reaction

$CO(g) + 2H_2(g) \rightleftharpoons CH_3OH(g)$

reaches equilibrium under a pressure of 1.59 MPa with 0.122 mol of carbon monoxide and 0.298 mol of hydrogen present at equilibrium in a vessel of volume 1.04 dm^3.

Use these data to answer the questions that follow.

a Assuming ideal gas behaviour, determine the total number of moles of gas present. Hence calculate the number of moles of methanol in the equilibrium mixture. [3]

b Calculate the value of the equilibrium constant, K_c, for this reaction and state its units. [3]

c **i** Write an expression for the equilibrium constant K_p, for this equilibrium.

ii Calculate the mole fraction of each of the three gases present in the equilibrium mixture.

iii Calculate the partial pressure of hydrogen present in the equilibrium mixture.

iv Calculate the value of the equilibrium constant K_p, and state its units. [8]

3 In the Contact Process for the production of sulphuric acid, highly purified sulphur dioxide and oxygen react together to form sulphur trioxide. The process is carried out in the presence of vanadium(V) oxide at about 700 K and at a pressure of 120 kPa

$2SO_2(g) + O_2(g) \rightleftharpoons 2SO_3(g)$
$\Delta H^\ominus = -190\,kJ\,mol^{-1}$

Under these conditions, the partial pressures of sulphur dioxide and sulphur trioxide at equilibrium are 33 kPa and 39 kPa, respectively.

a What would be the effect on the yield of SO_3 of increasing the temperature? Explain your answer. [3]

b What would be the effect on the yield of SO_3 of increasing the total pressure? Explain your answer. [3]

c Determine the partial pressure and hence the mole fraction of oxygen in the equilibrium mixture. [3]

d **i** Write an expression for the equilibrium constant, K_p, for the reaction shown.

ii Calculate the value of the equilibrium constant, K_p, and state its units. [4]

4 Phosphorus(V) chloride dissociates at high temperatures according to the equation

$PCl_5(g) \rightleftharpoons PCl_3(g) + Cl_2(g)$

83.4 g of phosphorus(V) chloride are placed in a vessel of volume 9.23 dm^3. At equilibrium at a certain temperature, 11.1 g of chlorine are produced at a total pressure of 250 kPa.

Use these data, where relevant, to answer the questions that follow.

a Calculate the number of moles of each of the gases in the vessel at equilibrium. [3]

b **i** Write an expression for the equilibrium constant K_c, for the above equilibrium.

ii Calculate the value of the equilibrium constant, K_c, and state its units. [4]

c **i** Write an expression for the equilibrium constant, K_p, for the above equilibrium.

ii Calculate the mole fraction of chlorine present in the equilibrium mixture.

iii Calculate the partial pressure of PCl_5 present in the equilibrium mixture.

iv Calculate the value of the equilibrium constant, K_p, and state its units. [7]

5 Ammonia is manufactured by the Haber–Bosch process by mixing hydrogen and nitrogen at 700 K and 200 atmospheres pressure and using an iron catalyst with a potassium hydroxide promoter.

$N_2(g) + 3H_2(g) \rightleftharpoons 2NH_3(g)\ \Delta H^{\ominus} = -92\ kJ\ mol^{-1}$

a Explain the term **dynamic equilibrium**. [2]

b Explain why the use of high pressure favours ammonia formation. [2]

c Analysis of an equilibrium mixture, obtained by mixing nitrogen and hydrogen, showed 24.0 g of NH_3, 13.5 g of H_2, and 60.3 g of N_2 to be present. The total pressure of the system was 10 atmospheres.

i Copy this table and calculate the mole fraction of each of these substances present at equilibrium. [2]

N_2	H_2	NH_3
Mole fraction =	Mole fraction =	Mole fraction =

ii Calculate K_p and state its units. [4]

iii Explain the effect of increasing temperature on the value of K_p. [2]

d The reaction of ammonia with oxygen is exothermic:

$4NH_3 + 5O_2 \rightleftharpoons 4NO + 6H_2O$

However, mixtures of ammonia and oxygen do not catch fire spontaneously. Use this information to explain the difference between **kinetic** and **thermodynamic stability**. [4]

6 Consider the equilibrium

$N_2O_4(g) \rightleftharpoons 2NO_2(g)$

a **i** Write an expression for K_c, indicating the units. [2]

ii 1 mol of dinitrogen tetraoxide, N_2O_4, was introduced into a vessel of volume 10.0 dm^3 at a temperature of 70 °C. At equilibrium 50% had dissociated. Calculate K_c. [4]

	$\Delta H_f^{\ominus}$ / kJ mol^{-1}
N_2O_4	+9.70
NO_2	+33.9

iii Using the above data, calculate the enthalpy change for the forward reaction. [2]

iv If the same experiment is carried out at 100 °C, state qualitatively, giving your reasons, how the equilibrium composition will change. [2]

b Explain what you would do to increase the degree of dissociation of $N_2O_4(g)$ at constant temperature. [2]

c What is the effect of a catalyst on the following:

i the value of K_c; [1]

ii the equilibrium position; [1]

iii the rate of attainment of equilibrium? [1]

d Suggest why the reaction

$N_2 + 2O_2 \rightarrow 2NO_2$

is not a very useful method of making NO_2. [2]

7 This question concerns the reaction

$H_2(g) + I_2(g) \rightleftharpoons 2HI(g)$

which, even at high temperatures, is slow.

	H–H	H–I	I–I	Cl–Cl	H–Cl
Bond energy / kJ mol^{-1}	436	299	151	242	431

a **i** Calculate ΔH for the reaction between hydrogen and iodine. [2]

ii Sketch an energy level diagram for this reaction. [2]

b Indicate on your sketch in **a ii**:

i ΔH for the reaction;

ii the activation energy for the forward reaction ($E_{a(F)}$);

iii the activation energy for the reverse reaction ($E_{a(R)}$). [3]

c For the analogous reaction for the formation of hydrogen chloride

$H_2(g) + Cl_2(g) \rightarrow 2HCl(g)$

suggest how you would expect the activation energy of the forward reaction to compare with that shown for the formation of HI. Give a reason for your answer. [2]

d The reaction for the formation of hydrogen iodide does not go to completion but reaches an equilibrium.

i Write an expression for the equilibrium constant, K_c, for this reaction. [1]

ii A mixture of 1.9 mol of H_2 and 1.9 mol of I_2 was prepared and allowed to reach equilibrium in a closed vessel of 250 cm^3 capacity at 700 °C. The resulting equilibrium mixture was found to contain 3.0 mol of HI.

Calculate the value of K_c at this temperature. [3]

12 Acid–base equilibrium

What sort of substances are acids? Your earliest experience of an acid may well have been as a small child when you bit into a lemon. After this first encounter, you soon came to associate sour, sharp tastes with citrus fruits (citric acid), vinegar (ethanoic acid), and unripe apples (malic acid or 2-hydroxybutanedioic acid). In the late seventeenth century, Robert Boyle characterized acids by their sourness and attempted to explain this property by suggesting that the atoms of acids are coated with spikes. However, not all sour-tasting substances are acids, and very few acids are safe to taste to see if they are sour! (*Remember: Never taste laboratory reagents!*) Human tongues have sense receptors that respond specifically to sourness, but we have no specific sensitivity to bases. As a crude approximation, bases may be regarded as the opposites of acids: they are substances that react with acids. This chapter presents more precise, chemical definitions of acids and bases.

ACIDS AND BASES AND THEIR PROPERTIES

OBJECTIVES

- Acidic solutions and the H^+ ion
- Basic solutions and the OH^- ion
- Neutralization
- Brønsted–Lowry theory

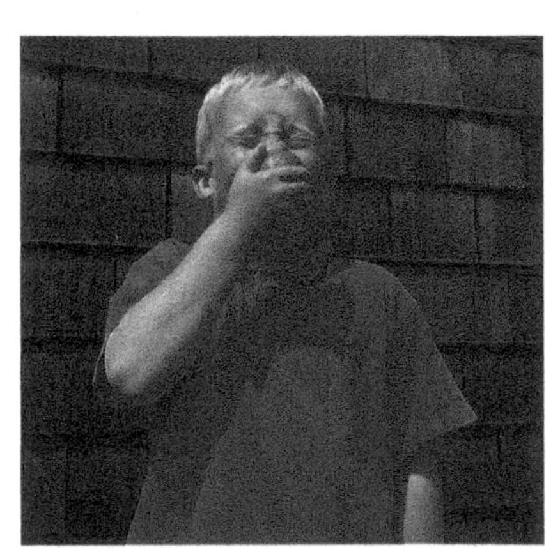

An early encounter with an acid – biting into a lemon.

Acids: a simple reminder

- Acids dissolve in water to give solutions of pH less than 7.
- Acids turn blue litmus red.
- Acids neutralize bases to give salts and water.
- Acids react with carbonates to give carbon dioxide gas.

The reaction between aqueous solutions of hydrochloric acid and sodium hydroxide is a typical acid–base reaction. Both solutions are good conductors of electricity; the electric current is carried by ions.

- Aqueous hydrochloric acid contains the ions $H^+(aq)$ and $Cl^-(aq)$.
- Aqueous sodium hydroxide contains the ions $Na^+(aq)$ and $OH^-(aq)$.

Acids and hydrogen ions

The ion common to all aqueous acids is the hydrogen ion or proton, H^+ (a proton results when a hydrogen atom loses its one electron). The simplest representation of the **aqueous hydrogen ion** is the symbol $H^+(aq)$. However, the hydrogen ion has a very large charge density, so each hydrogen ion attracts a lone pair of electrons on a neighbouring water molecule to form a single coordinate bond. The aqueous hydrogen ion is, therefore, more accurately described by the formula $H_3O^+(aq)$. The ion $H_3O^+(aq)$ is called the **oxonium ion** (or the **hydronium ion** or the **hydroxonium ion**).

The oxonium ion $H_3O^+(aq)$ forms when a hydrogen ion H^+ (a proton) from an acid bonds with a water molecule H_2O.

Bases and hydroxide ions

The ion common to all aqueous bases is the hydroxide ion, OH^-. This ion results when a soluble base dissolves in water. A base that is soluble in water is called an **alkali**. Dissolving an alkali in water gives an **alkaline** solution, which has a pH greater than 7. Some soluble bases (for example, sodium hydroxide) contain the hydroxide ion. The ion is released into solution as the substance dissolves:

$$NaOH(s) \rightarrow Na^+(aq) + OH^-(aq)$$

Some soluble bases generate hydroxide ions by reaction with water. For example, potassium oxide reacts as follows:

$$K_2O(s) + H_2O(l) \rightarrow 2K^+(aq) + 2OH^-(aq)$$

Neutralization reactions

Alkalis (soluble bases) and insoluble bases, such as copper(II) oxide CuO, *both* react with acids in **neutralization** reactions.

For example, aqueous hydrochloric acid and aqueous sodium hydroxide react according to the equation:

$$HCl(aq) + NaOH(aq) \rightarrow NaCl(aq) + H_2O(l)$$
an acid + an alkali → a salt + water

The neutralization reaction can be written in terms of the ions involved as:

$$H^+(aq) + Cl^-(aq) + Na^+(aq) + OH^-(aq) \rightarrow H_2O(l) + Cl^-(aq) + Na^+(aq)$$

The ions $Cl^-(aq)$ and $Na^+(aq)$ do not take part in the reaction. Ions that do not take part in a reaction are called **spectator ions**. All neutralization reactions between solutions involve the reaction between aqueous hydrogen ions $H^+(aq)$ and aqueous hydroxide ions $OH^-(aq)$:

$$H^+(aq) + OH^-(aq) \rightarrow H_2O(l)$$

This reaction is more accurately represented in terms of the oxonium ion:

$$H_3O^+(aq) + OH^-(aq) \rightarrow 2H_2O(l)$$

An insoluble base such as copper(II) oxide can also neutralize an acid. For example:

$$H_2SO_4(aq) + CuO(s) \rightarrow CuSO_4(aq) + H_2O(l)$$
an acid + a base → a salt + water

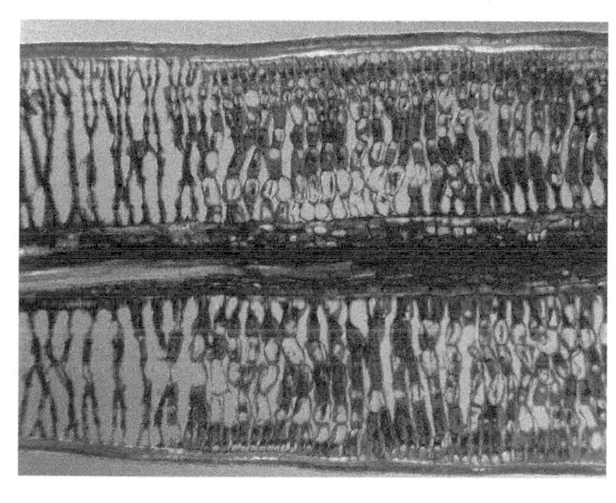

The spongy mesophyll of a leaf, with damage caused by acid rain visible on the left.

Brønsted–Lowry definitions

In aqueous solution, the underlying reaction in all acid–base neutralizations is

$$H_3O^+(aq) + OH^-(aq) \rightarrow 2H_2O(l)$$

Johannes Brønsted made a special study of acid–base reactions and in 1923 explained them in terms of **proton transfer**. Martin Lowry came to similar conclusions independently. According to the Brønsted–Lowry definition of acids and bases:

- A **Brønsted acid** is a proton donor.
- A **Brønsted base** is a proton acceptor.

During neutralization reactions, the oxonium ion acts as a Brønsted acid: it donates a proton to the hydroxide ion. The hydroxide ion acts as a Brønsted base: it accepts a proton from the oxonium ion.

SUMMARY

- Acidic solutions contain aqueous hydrogen ions $H_3O^+(aq)$.
- Basic solutions contain aqueous hydroxide ions $OH^-(aq)$.
- Acids neutralize bases.

Standard enthalpy changes of neutralization

The standard enthalpy changes of neutralization for the reactions between several different acids and bases are very similar. This observation supports the view that neutralization reactions involve the same ions in all cases, i.e.

$H^+(aq) + OH^-(aq) \rightarrow H_2O(l)$

Some typical values are given below:

Acid	Base	$\Delta H^\ominus$/kJ mol^{-1}
HCl	NaOH	–57.1
HCl	KOH	–57.2
HNO_3	NaOH	–57.3
HNO_3	KOH	–57.3

Acid rain

Rain water is naturally acidic because it contains dissolved carbon dioxide. The combustion of fossil fuels produces oxides of nitrogen and sulphur. These oxides dissolve in rain water and increase its acidity above natural levels. **Acid rain** is rain that is artificially more acidic than normal. The control of acidity is vital to living organisms, which have in-built systems for keeping the acidity of their tissues constant. The photograph alongside shows that trees are unable to cope with such dramatic increases in acidity.

Acid rain also leaches metals from soil and rocks into streams and lakes. These substances harm fish and other aquatic life-forms.

Monoprotic and diprotic acids

Many acids are called **monoprotic (monobasic) acids** because they can donate one mole of protons per mole of acid. Examples include hydrochloric and nitric acids, e.g.

$HCl(aq) + H_2O(l) \rightleftharpoons H_3O^+(aq) + Cl^-(aq)$

Some acids are called **diprotic (dibasic) acids** because they can donate two moles of protons per mole of acid. A common example is sulphuric acid H_2SO_4. When doing mole calculations with diprotic acids and sodium hydroxide, don't forget to take account of the mole ratio of 1:2.

A small number of acids are called **triprotic (tribasic) acids** because they can donate three moles of protons per mole of acid. A common example is phosphoric acid H_3PO_4.

12.2

O B J E C T I V E S

- More on Brønsted acids and bases
- Acid–base reactions involve proton transfer
- Conjugate acid–base pairs

ACIDS AND BASES AND PROTON TRANSFER

You should now understand that solutions of acids contain oxonium ions $H_3O^+(aq)$, and that solutions of bases contain hydroxide ions $OH^-(aq)$. The previous spread treated acid–base neutralization in terms of the reaction between these two ions. You will recall that acids are substances that dissolve in water, donating protons to water molecules to produce oxonium ions; and bases are substances that neutralize these ions to form water. According to the Brønsted–Lowry theory of acids and bases, a Brønsted acid is a proton donor and a Brønsted base is a proton acceptor.

Brønsted acids

For an acid to behave as a proton donor, a base must be present to accept protons from it. Hydrochloric acid is a solution of hydrogen chloride gas in water. You can imagine this acidic solution forming in two steps, the first of which is simply the dissolution of the gas in water:

$$HCl(g) \rightarrow HCl(aq)$$

The second step happens when a hydrogen chloride molecule donates a proton to a water molecule. The hydrogen chloride acts as a Brønsted acid; the water acts as a Brønsted base:

$$HCl(aq) + H_2O(l) \rightleftharpoons H_3O^+(aq) + Cl^-(aq)$$

You should recognize the ion $H_3O^+(aq)$ as a Brønsted acid. It is the **conjugate acid** of H_2O, because it is formed when H_2O accepts a proton. The ion $Cl^-(aq)$ is the **conjugate base** of HCl, because it is formed when HCl donates a proton.

The covalent H—Cl bond breaks as both of the shared pair of electrons go over to the Cl atom. The proton H⁺ released forms a new H—O bond with a lone pair of electrons on the H_2O molecule.

Non-aqueous solvents

Hydrogen chloride dissolves readily in benzene C_6H_6. Benzene is an example of a *non-aqueous* solvent. Hydrogen chloride does not behave as an acid when it is dissolved in benzene. There are no molecules present that can accept protons. For the same reason, dry hydrogen chloride gas is not acidic.

Brønsted bases

For a base to behave as a proton acceptor, an acid must be present to donate protons to it. Ammonia gas dissolves readily in water to produce a basic solution. You can imagine this basic solution forming in two steps, the first of which is simply the dissolution of the gas in water:

$$NH_3(g) \rightarrow NH_3(aq)$$

The second step happens when an ammonia molecule accepts a proton from a water molecule. The ammonia acts as a Brønsted base; the water acts as a Brønsted acid:

$$NH_3(aq) + H_2O(l) \rightleftharpoons NH_4^+(aq) + OH^-(aq)$$

You should recognize the ion $OH^-(aq)$ as a Brønsted base. It is the conjugate base of H_2O. The ion $NH_4^+(aq)$ is the conjugate acid of NH_3.

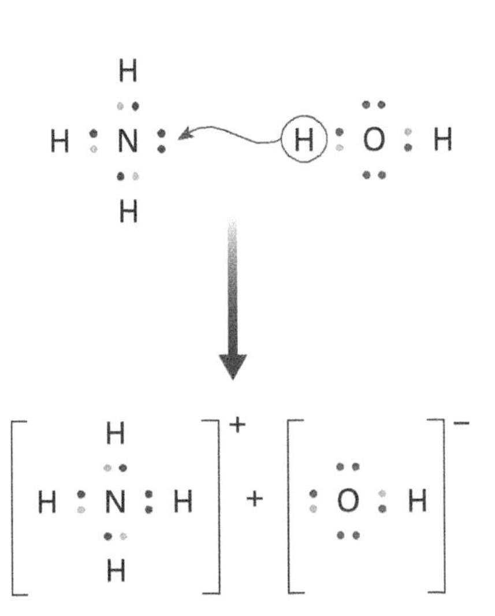

The Brønsted acid H_2O donating a proton to the Brønsted base NH_3. An H—O covalent bond breaks as both of the shared pair of electrons go over to the O atom. The proton H⁺ released forms a new bond with the lone pair of electrons on the NH_3 molecule.

Neutralization reactions revisited

All neutralization reactions may be regarded as equilibria involving proton transfer. For example, the reaction between ammonia and hydrochloric acid in aqueous solution can be described by the following equilibrium:

$$NH_3(aq) + H_3O^+(aq) \rightleftharpoons NH_4^+(aq) + H_2O(l)$$

In the forward reaction, the base NH_3 accepts a proton from the acid H_3O^+. In the reverse reaction, the base H_2O accepts a proton from the acid NH_4^+. The acid NH_4^+ and the base NH_3 are a **conjugate acid–base pair** because they are related by the transfer of a proton. Similarly, the acid H_3O^+ and the base H_2O are a conjugate acid–base pair. In the reverse reaction, water is acting as a base:

$$NH_3(aq) + H_3O^+(aq) \rightleftharpoons NH_4^+(aq) + H_2O(l)$$

BASE ACID CONJUGATE ACID CONJUGATE BASE

- The conjugate acid is the species that results when a base accepts a proton, e.g. the ammonium ion $NH_4^+(aq)$ is the conjugate acid of the base ammonia $NH_3(aq)$.
- The conjugate base is the species that results when an acid donates a proton, e.g. water $H_2O(l)$ is the conjugate base of the oxonium ion $H_3O^+(aq)$.

The hydroxide ion

The formula of the hydroxide ion is generally written as OH^-. However, it would be more correct to write the formula as HO^-. Remember that an oxygen atom has six electrons in its valence shell and a hydrogen atom has one electron. The Lewis structure for the hydroxide ion shows that the extra electron represented by the negative charge is used to complete *oxygen's* octet. It is therefore oxygen and not hydrogen that carries the negative charge.

(a) O (b) H (c) [O : H]⁻

The Lewis structures of (a) the oxygen atom, (b) the hydrogen atom, and (c) the hydroxide ion.

Sodium hydroxide should therefore have the formula NaHO, making the series H_2O, NaHO, Na_2O an obvious progression and similarly for H_3O^+, H_2O, HO^-. NaHO also lists the elements in the order of their electronegativities, following the general rule for formulae. Unfortunately, tradition dictates the illogical order NaOH.

SUMMARY

- Brønsted–Lowry theory explains acid–base reactions in terms of equilibria involving proton transfer.
- A Brønsted acid is a proton donor.
- A Brønsted base is a proton acceptor.
- Conjugate acid–base pairs are related by the transfer of a proton.

PRACTICE

1 State whether each of the following in aqueous solution can act as a Brønsted acid or a Brønsted base. Give reasons for your answers.
   a HCl
   b HI
   c NaOH
   d K_2O
   e $Ca(OH)_2$
   f NH_3.

2 For each of the following, give the formula of its conjugate base:
   a HCl(aq)
   b $H_3O^+(aq)$
   c $H_2O(l)$.

3 For each of the following, give the formula of its conjugate acid:
   a $NH_3(aq)$
   b $HO^-(aq)$
   c $H_2O(l)$.

4 Caves and potholes form in limestone, as carbon dioxide solution and rain water flow through and 'dissolve away' the rock:

   $$H_2O(l) + CO_2(aq) + CaCO_3(s) \rightleftharpoons Ca(HCO_3)_2(aq)$$

   Discuss this reaction in the light of the Brønsted-Lowry theory of acids and bases.

12.3

OBJECTIVES

- pH as a measure of acidity
- Definition of pH
- pH measurement

Definition and meaning of $\log_{10}$

For any positive number *n*, the common **logarithm** (symbol $\log_{10}$) of *n* is the power to which the **base** (in this case, 10) must be raised to make *n*. For example, for the number 2,

$\log_{10}2 = 0.3$ i.e. $2 = 10^{0.3}$

Logarithms to base 10 are often simply labelled 'LOG' or 'log' on a calculator.

Logarithms

The great advantage of logarithms is that they collapse very large numbers into more manageable ones. For example, the logarithm to base 10 of the Avogadro constant $6.0 \times 10^{23}\,mol^{-1}$ is 23.78. Similarly, very small numbers become more manageable. For example, the charge on the electron is 1.6×10^{-19} coulombs. The logarithm (to base 10) of this unwieldy number is –18.80. **Note that the units must be removed before a logarithm is taken.**

Antilogarithm

Finding the **antilogarithm** of a number is the reverse of the process of finding its logarithm. For example, the logarithm (to base 10) of the number 2 is 0.3. The antilogarithm (to base 10) of 0.3 is 2, i.e.

$\text{antilog}_{10}0.3 = 10^{0.3} = 2$

Aqueous hydrogen ion concentration and pH

Brønsted acids donate protons (hydrogen ions H^+) in aqueous solution. The **acidity** of a solution is a measure of the concentration of the aqueous hydrogen ion $H_3O^+(aq)$. The need to measure acidity quantitatively first arose at the end of the nineteenth century when brewers in Denmark were struggling to maintain the quality of their lager. They knew that the acidity of the lager was varying too much for the yeast to carry out optimum fermentation. But they could not control the acidity unless they could measure it. The Dane Søren Sørensen proposed the pH scale in 1909 for measuring acidity. The brewers were then able to provide the best conditions for their yeast. Chemists can now express the acidity of any solution as a simple number.

The pH scale

The name pH was chosen to conjure up the idea of the strength (*potenz* in German) of the hydrogen ions. It was known that aqueous hydrogen ion concentration, and therefore acidity, varies over a wide range. We shall see later in this chapter that solutions of concentration $1\,mol\,dm^{-3}$ (i.e. 1 mol of solute dissolved per dm^3 of solution) have aqueous hydrogen ion concentrations between 1 and $1 \times 10^{-14}\,mol\,dm^{-3}$. Because of this huge range of values, Sørensen adopted a *logarithmic* scale, where **pH** is defined as the negative logarithm, to base 10, of the aqueous hydrogen ion concentration $[H_3O^+]$ measured in $mol\,dm^{-3}$:

$$pH = -\log_{10}[H_3O^+]$$

For the sake of simplicity, this expression is sometimes written as:

$$pH = -\log_{10}[H^+]$$

The negative sign in the equation results in pH *increasing* as the aqueous hydrogen ion concentration *decreases*. For example, an aqueous hydrogen ion concentration of $1\,mol\,dm^{-3}$ gives a pH of:

$$pH = -\log_{10}1 = 0.0$$

so this solution is highly acidic.

- The pH scale typically ranges from 0 to 14, but these limits are not absolute.
- **Acidic** solutions have pH values less than 7.
- **Basic** solutions have pH values greater than 7.
- **Neutral** water has a pH of 7.

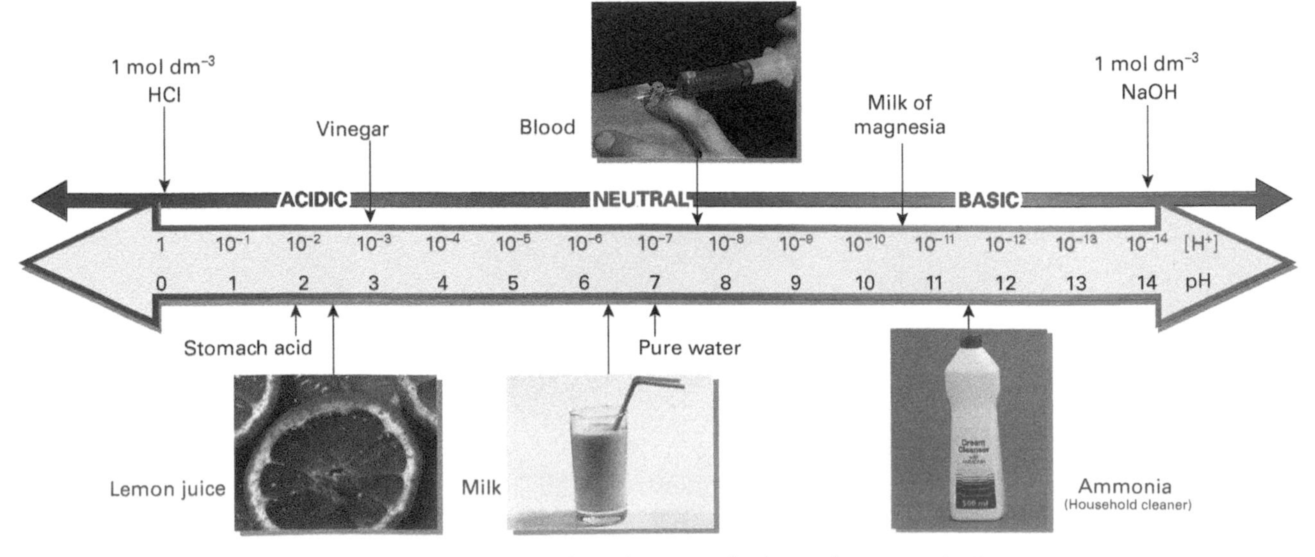

The pH scale and aqueous hydrogen ion concentration.

Measuring pH – indicators

Indicators are water-soluble dyes that have different colours in solutions of different pH. They may be used to indicate the approximate pH of a solution or reveal pH change.

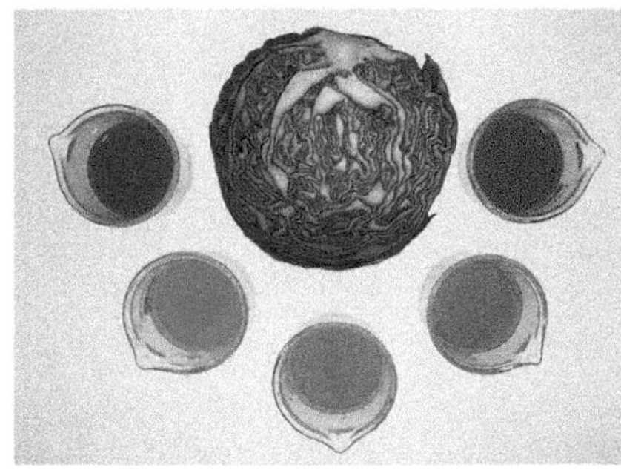

The juice of a red cabbage can act as an indicator. The solutions have (from left to right) pH values of 1, 4, 7, 10, and 13.

Measuring pH – pH meters

A pH meter consists of three main parts: a pair of electrodes that dip into the solution being measured; an electronic circuit; and a readout device. The electrode pair develops a small voltage, whose magnitude depends on the solution's aqueous hydrogen ion concentration. The electronic circuit amplifies this voltage sufficiently to drive the readout device. On a digital pH meter, the pH is displayed directly. On an analogue pH meter, the pH is shown by a pointer moving across a scale. pH meters must be calibrated periodically: the calibration is checked by immersing the electrode in solutions of known pH, spread 12.11.

Decimal places

The number of *decimal places* in a pH value is equal to the number of *significant figures* given in the concentration data.

pH values and aqueous hydrogen ion concentrations

Worked example on calculating pH from $[H_3O^+]$

Question: The aqueous hydrogen ion concentration in human blood is 4×10^{-8} mol dm^{-3}. What is the pH of human blood?

Strategy: Substitute the concentration into the definition of pH given opposite.

Answer: We have

$pH = -\log_{10}[H_3O^+]$

Therefore, for human blood

$pH = -\log_{10}(4 \times 10^{-8}) = 7.4$ (1 decimal place)

It is often useful to determine aqueous hydrogen ion concentrations from pH values, obtained, for example, from pH meter measurements.

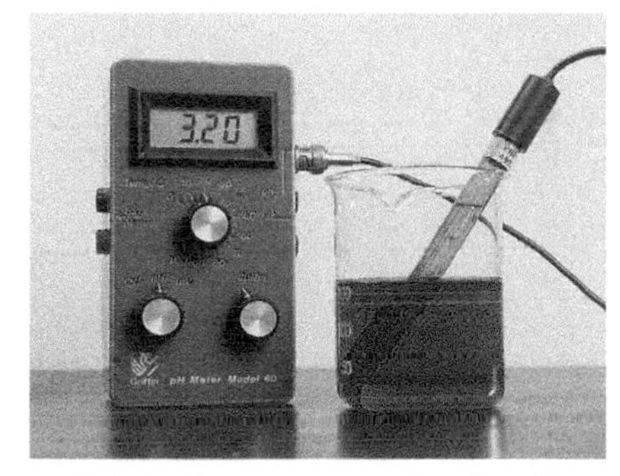

A digital pH meter measuring the pH of a solution containing aluminium ions. The solution is acidic because of the reaction between the metal ions and water molecules. The indicator, bromocresol green, turns yellow below pH 3.5; see the next but one spread.

Worked example on calculating $[H_3O^+]$ from pH

Question: A solution containing aqueous aluminium ions is quite acidic and has a pH of 3.2. What is the aqueous hydrogen ion concentration in the solution?

Answer: As before, pH is defined as:

$pH = -\log_{10}[H_3O^+]$

Multiply both sides of the equation by –1:

$\log_{10}[H_3O^+] = -pH$

Take the antilogarithm of both sides; add the units of mol dm^{-3}:

$[H_3O^+] = 10^{-pH}$ mol dm^{-3}

Substitute the pH value given

$[H_3O^+] = 10^{-3.2}$ mol dm^{-3}

i.e. the aqueous hydrogen ion concentration is 6×10^{-4} mol dm^{-3}

Note: We can check that the calculation has been performed correctly by evaluating $-\log_{10}(6 \times 10^{-4})$, whose value is 3.2.

PRACTICE

1 Calculate the pH of solutions with the following aqueous hydrogen ion concentration in mol dm^{-3}:
   a 0.1
   b 0.01
   c 0.05.

2 Calculate the aqueous hydrogen ion concentration corresponding to the following pH values:
   a 1.0
   b 2.5.

SUMMARY

- $pH = -\log_{10}[H_3O^+]$.
- The pH scale ranges typically from 0 to 14; neutral water has a pH of 7.
- Indicators and pH meters can be used to measure pH.

12.4

OBJECTIVES

- Strong and weak acids compared
- pH calculations for strong acids

STRONG ACIDS

The pH scale describes the aqueous hydrogen ion concentration of solutions. The lower the pH value, the greater the aqueous hydrogen ion concentration. Acidic solutions have pH values less than 7, and basic solutions have pH values greater than 7. As the pH decreases from 7, the acidity of the solution increases. The pH of an acidic solution depends on its concentration in mol dm^{-3}, and on the chemical properties of the acid itself – whether the acid is described as a *strong acid* or as a *weak acid*.

Strong and weak acids

The pH of 0.1 mol dm^{-3} aqueous hydrochloric acid is 1.0. The pH of 0.1 mol dm^{-3} aqueous ethanoic acid is 2.9 (see next spread). Both solutions have the same concentrations of acid yet different values of pH. The reason for this discrepancy is that hydrochloric acid is a strong acid and ethanoic acid is a weak acid.

Hydrochloric acid is described as a *strong* acid because the equilibrium

$$HCl(aq) + H_2O(l) \rightleftharpoons H_3O^+(aq) + Cl^-(aq)$$

lies essentially completely to the right. The acid is fully ionized in aqueous solution. Dissolving 1 mol of hydrogen chloride in water produces 1 mol of aqueous hydrogen ions. For example, a solution of hydrochloric acid of concentration 1 mol dm^{-3} has an aqueous hydrogen ion concentration $[H_3O^+] = 1$ mol dm^{-3}.

On the other hand, ethanoic acid is described as a *weak* acid because the equilibrium

$$CH_3COOH(aq) + H_2O(l) \rightleftharpoons H_3O^+(aq) + CH_3COO^-(aq)$$

lies to the left. Dissolving 1 mol of ethanoic acid in water does not result in 1 mol of aqueous hydrogen ions. For example, a solution of ethanoic acid of concentration 1 mol dm^{-3} has an aqueous hydrogen ion concentration $[H_3O^+] = 4 \times 10^{-3}$ mol dm^{-3}.

- A **strong acid** is *fully ionized* in aqueous solution. The aqueous hydrogen ion concentration is *equal* in magnitude to the concentration of the acid.
- A **weak acid** is *only partially ionized* in aqueous solution. The aqueous hydrogen ion concentration is *smaller* in magnitude than the concentration of the acid.

Ethanoate ion

The ion formed from ethanoic acid CH_3COOH, the ethanoate ion, is delocalized (see spread 6.4) and so should be written as $CO_3CO_2^-$.

To *simplify*, in this chapter we will use the formula CH_3COO^-.

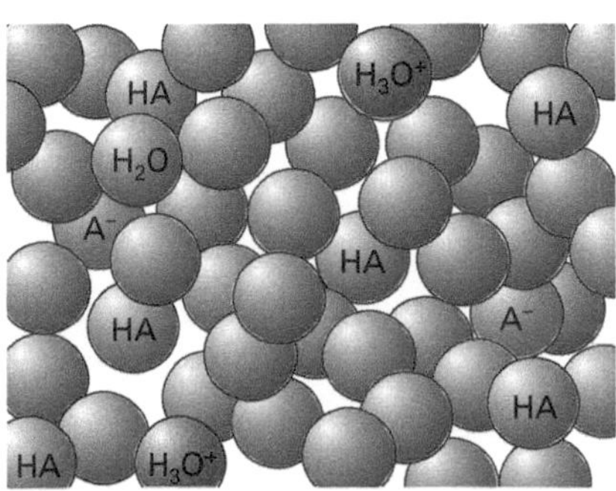

Ionization of a weak acid HA. The equilibrium $HA(aq) + H_2O(l) \rightleftharpoons H_3O^+(aq) + A^-(aq)$ lies far to the left. Very few acid molecules HA are ionized; the aqueous hydrogen ion concentration is therefore lower than for a strong acid of the same concentration.

Left: a solution of ethanoic acid of concentration 0.1 mol dm^{-3}. This acid is a weak acid and is only partially ionized in aqueous solution. The low concentration of ions makes the solution a poor electrical conductor. Right: A solution of hydrochloric acid of concentration 0.1 mol dm^{-3}. This acid is a strong acid and is fully ionized in aqueous solution. The high concentration of ions makes the solution a good electrical conductor.

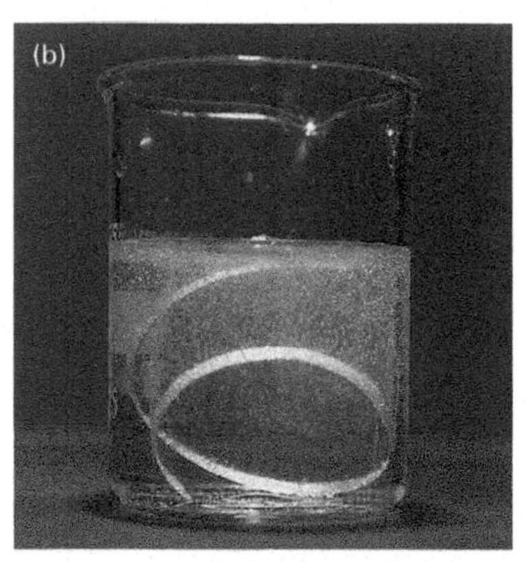

Identical samples of magnesium ribbon reacting with (a) 1.0 mol dm^{-3} hydrochloric acid, and (b) 1.0 mol dm^{-3} ethanoic acid. Hydrochloric acid is a strong acid (fully ionized in aqueous solution). The reaction is rapid because the solution contains aqueous hydrogen ions $H_3O^+(aq)$ at a concentration of 1 mol dm^{-3}, pH = 0. Ethanoic acid is a weak acid (only partially ionized in aqueous solution). The reaction is much slower because the aqueous hydrogen ion concentration is 4×10^{-3} mol dm^{-3}, pH = 2.4.

Calculating the pH of strong acids

The pH of a strong acid depends only on its concentration because a strong acid is fully ionized.

For example, nitric acid is a strong acid:

$$HNO_3(aq) + H_2O(l) \rightarrow H_3O^+(aq) + NO_3^-(aq)$$

As a result, nitric acid at a concentration of 1×10^{-3} mol dm^{-3} contains aqueous hydrogen ions at a concentration of 1×10^{-3} mol dm^{-3}. So the pH of this solution is:

$$\begin{aligned} pH &= -\log_{10}[H_3O^+] \\ &= -\log_{10}(1 \times 10^{-3}) \\ &= 3.0 \end{aligned}$$

Worked example on calculating the pH of a strong acid

Question: Calculate the pH of the following concentrations of hydrochloric acid:

(a) 1 mol dm^{-3}

(b) 0.1 mol dm^{-3}

(c) 0.05 mol dm^{-3}

Strategy: Hydrochloric acid is a strong acid and so the aqueous hydrogen ion concentration is the same as the acid concentration.

Answer:

(a) pH = $-\log_{10}1$ = 0.0

(b) pH = $-\log_{10}0.1$ = 1.0

(c) pH = $-\log_{10}0.05$ = 1.3

Negative pH values

A strong acid of concentration greater than 1 mol dm^{-3} will have a pH value *less than zero*. For example, 2 mol dm^{-3} aqueous hydrochloric acid has an aqueous hydrogen ion concentration of 2 mol dm^{-3}. The pH is therefore:

$$\begin{aligned} pH &= -\log_{10}[H_3O^+] \\ &= -\log_{10}2 \\ &= -0.3 \end{aligned}$$

pH and concentration

For a strong acid, each dilution of the concentration by a factor of 10 increases the pH by one unit. For example, 0.1 mol dm^{-3} aqueous hydrochloric acid has a pH of 1.0. Adding 1 cm^3 of this solution to 9 cm^3 of water gives 0.01 mol dm^{-3} aqueous hydrochloric acid and results in a solution of pH 2.0. This result is a consequence of using the logarithm to base 10 in the definition of pH.

SUMMARY

- Strong acids are fully ionized in aqueous solution.
- Weak acids are only partially ionized in aqueous solution.
- The aqueous hydrogen ion concentration of a strong acid is equal in magnitude to the concentration of the acid.
- The pH of a strong acid depends only on its concentration.

PRACTICE

1 Explain the difference between the terms *acid concentration* and *acid strength*.

2 Calculate the pH value for both of the following solutions:

a Hydrochloric acid HCl(aq), 0.05 mol dm^{-3}

b Nitric acid, 2×10^{-3} mol dm^{-3}.

12.5

OBJECTIVES

- Acid ionization constant K_a
- Definition of pK_a
- pH calculations for weak acids
- Indicators and pK_{in}

WEAK ACIDS

The *strength* of an acid must not be confused with its *concentration*. Strength does not depend on the amount in moles present but on the extent to which the acid ionizes in solution. For example, ethanoic acid CH_3COOH is a weak acid because it is *not* fully ionized in aqueous solution. Hydrochloric acid is a strong acid because it *is* fully ionized. A solution of either acid containing 5 mol dm^{-3} is concentrated, whereas a solution containing 5×10^{-3} mol dm^{-3} is relatively dilute. This spread introduces the terms used to describe weak acids, then explains how to calculate pH values for weak acids.

The acid ionization constant K_a

The equilibrium set up when a weak acid HA dissolves in water is

$$HA(aq) + H_2O(l) \rightleftharpoons H_3O^+(aq) + A^-(aq)$$

The position of equilibrium depends on the nature of the acid HA. The acid HA and the base A^- form a conjugate acid–base pair. In the forward reaction, the acid HA donates its proton to the base H_2O. In the reverse reaction, the acid H_3O^+ donates a proton to the base A^-. The expression for the equilibrium constant is:

$$K_c = \frac{[H_3O^+]_{eq}[A^-]_{eq}}{[HA]_{eq}[H_2O]_{eq}}$$

The concentration of water is essentially constant because it is so large compared to the concentrations of the other species present. It may therefore be omitted from the equilibrium expression because shifting the equilibrium changes its value insignificantly. The resulting equilibrium constant is given a special symbol K_a, and is called the **acid ionization constant** or the **acid dissociation constant**:

$$K_a = \frac{[H_3O^+][A^-]}{[HA]}$$

Note that this equilibrium is reached so quickly that the subscript 'eq' may be omitted.

To calculate K_a from pK_a

From the main text we have

$pK_a = -\log_{10}K_a$

Multiply both sides of the equation by −1:

$\log_{10}K_a = -pK_a$

Take the antilogarithm of both sides of the expression, remembering to add the units of mol dm^{-3}:

$K_a = 10^{-pK_a}$ mol dm^{-3}

For example, the pK_a for ethanoic acid is 4.8. Substitute this value into the expression above and check you obtain the answer $K_a = 2 \times 10^{-5}$ mol dm^{-3}.

pK_a

The values of K_a span a wide range, for example from 0.65 for trichloroethanoic acid CCl_3COOH to 1×10^{-10} for phenol C_6H_5OH, and even smaller for other substances. Such a wide range of values is better accommodated by using a logarithmic scale defined in a similar manner to pH, i.e.

$$pK_a = -\log_{10}K_a$$

The lower the value of pK_a, the larger the value of K_a and the greater the ionization of the acid in water. For example, at the same concentrations, aqueous hydrofluoric acid (pK_a 3.2) is ionized to a greater extent than aqueous ethanoic acid (pK_a 4.8). Some typical K_a and pK_a values are given in the table below.

Acid ionization constants of some weak acids in water at 25 °C.

Compound	Ionization equilibrium	Ionization constant K_a /mol dm^{-3}	pK_a
Hydrofluoric acid	$HF + H_2O \rightleftharpoons H_3O^+ + F^-$	6.3×10^{-4}	3.2
Ethanoic acid	$CH_3COOH + H_2O \rightleftharpoons H_3O^+ + CH_3COO^-$	1.6×10^{-5}	4.8
Benzoic acid	$C_6H_5COOH + H_2O \rightleftharpoons H_3O^+ + C_6H_5COO^-$	6.3×10^{-5}	4.2
Phenol	$C_6H_5OH + H_2O \rightleftharpoons H_3O^+ + C_6H_5O^-$	1.3×10^{-10}	9.9

Calculating the pH of a weak acid

The extent of ionization, and hence the pH, of a weak acid depends both on its concentration and on its pK_a value. These two variables must *both* be taken into account when calculating the pH of a weak acid. For example, the pH of 0.1 mol dm^{-3} aqueous ethanoic acid (K_a 1.6 × 10^{-5} mol dm^{-3}) is calculated as follows.

The equilibrium equation for aqueous ethanoic acid is:

$$CH_3COOH(aq) + H_2O(l) \rightleftharpoons H_3O^+(aq) + CH_3COO^-(aq)$$

Applying the general equilibrium expression for K_a to ethanoic acid, this expression becomes:

$$K_a = \frac{[H_3O^+][CH_3COO^-]}{[CH_3COOH]}$$

The proportion of molecules ionized is very small. The value of $[CH_3COOH]$, the concentration of the un-ionized acid, is effectively equal to the original concentration of the solution, and may therefore be taken as 0.1 mol dm^{-3}. Also, the values of $[H_3O^+]$ and $[CH_3COO^-]$ must be equal because 1 mol of acid ionizes to give 1 mol of each ion, hence:

$$1.6 \times 10^{-5}\,\text{mol dm}^{-3} = \frac{[H_3O^+]^2}{0.1\,\text{mol dm}^{-3}}$$

$$[H_3O^+]^2 = (0.1\,\text{mol dm}^{-3}) \times (1.6 \times 10^{-5}\,\text{mol dm}^{-3})$$
$$= 1.6 \times 10^{-6}\,\text{mol}^2\,\text{dm}^{-6}$$
$$[H_3O^+] = \sqrt{(1.6 \times 10^{-6}\,\text{mol}^2\,\text{dm}^{-6})}$$
$$= 1.3 \times 10^{-3}\,\text{mol dm}^{-3}$$

Substitute this value into the expression for pH:

$$\text{pH} = -\log_{10}(1.3 \times 10^{-3})$$
$$= 2.9 \quad (1\ \text{d.p.})$$

A short cut to this result is shown alongside.

A short cut

The pH, pK_a, and concentration A of a weak acid are related by the expression:

$\text{pH} = \frac{1}{2}pK_a - \frac{1}{2}\log_{10}A$

Its use is straightforward, as shown by the following example.

Suppose that we want to find the pH of a solution of ethanoic acid (pK_a 4.8) of concentration 0.1 mol dm^{-3}. We must substitute the data into the equation above, to obtain:

$\text{pH} = \frac{1}{2}pK_a - \frac{1}{2}\log_{10}A$
$= (\frac{1}{2} \times 4.8) - (\frac{1}{2}\log_{10}0.1)$
$= (\frac{1}{2} \times 4.8) - (\frac{1}{2} \times (-1.0))$
$= (2.4) - (-0.5)$
$= 2.9 \quad (1\ \text{d.p.})$

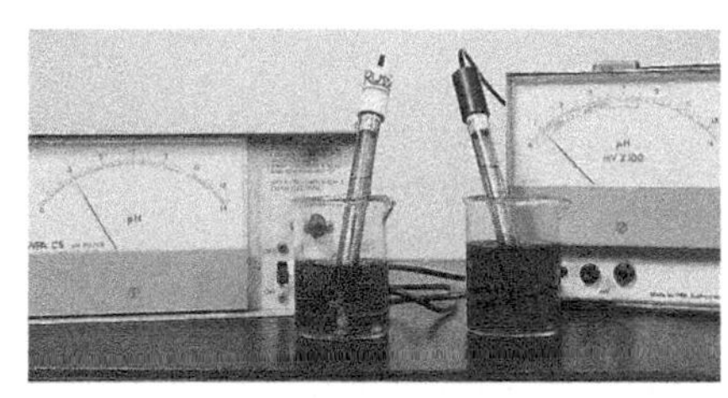

The pH meters show that 0.1 mol dm^{-3} aqueous ethanoic acid, a weak acid, has a pH of 2.9, whereas the same concentration of hydrochloric acid, a strong acid, has a pH of 1.0. The indicator thymol blue is orange at pH 1 and yellow at pH 2.9.

Indicators as weak acids

As you already know, indicators are water-soluble dyes used to indicate the pH of solutions. They are weak acids of the general form HIn, which exist in aqueous solution in equilibrium with their conjugate base In^-:

$$HIn(aq) + H_2O(l) \rightleftharpoons H_3O^+(aq) + In^-(aq)$$

$HIn(aq)$ and $In^-(aq)$ each have a different colour.

Changing the pH of a solution will bring about a change in the colour of an indicator added to it. The *range* of an indicator – the pH at which the colour change occurs – depends on the pK_a value of the indicator, designated pK_{in}. Typically the range of an indicator is one unit of pH either side of its pK_{in} value.

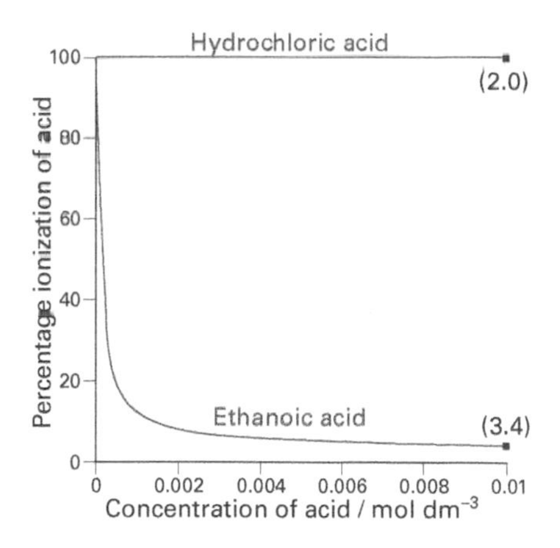

HCl is fully ionized at all concentrations. Ethanoic acid is just 4% ionized at 0.01 mol dm^{-3} concentration. Adding water to a solution of ethanoic acid forces the equilibrium $CH_3COOH(aq) + H_2O(l) \rightleftharpoons H_3O^+(aq) + CH_3COO^-(aq)$ to the right. The extent of ionization increases as the solution becomes more dilute. In extremely dilute solutions, the two acids have the same pH value. The figures in brackets are the pH values at 0.01 mol dm^{-3}.

The indicator bromocresol green is green at pH 4.5; below pH 3.5 it is yellow, above pH 5.5 it is blue.

SUMMARY

- Weak acids are only partially ionized.
- K_a is the acid ionization constant for a weak acid: $K_a = \frac{[H_3O^+][A^-]}{[HA]}$
- $pK_a = -\log_{10}K_a$

12.6

Strong bases

OBJECTIVES

- The ionic product of water
- Aqueous hydrogen ion concentrations of basic solutions
- Strong and weak bases compared
- pH calculations for strong bases

Acidic solutions have pH values less than 7; basic solutions have pH values greater than 7. But how are we able to apply the pH scale, which measures *hydrogen ion* concentration, to basic solutions, which contain the *hydroxide ion* $OH^-(aq)$? The key lies in the fact that pure water spontaneously ionizes to a small extent to produce both hydrogen ions and hydroxide ions.

The ionization of water

Water can act as an *acid* when reacting with ammonia,

$$H_2O(l) + NH_3(aq) \rightleftharpoons NH_4^+(aq) + OH^-(aq)$$

and as a *base* when reacting with hydrochloric acid,

$$H_2O(l) + HCl(aq) \rightleftharpoons H_3O^+(aq) + Cl^-(aq)$$

Now imagine a situation in which water acts *simultaneously* as an acid and a base. One molecule donates a proton (and so acts as an acid) to a second molecule that accepts the proton (and so acts as a base):

$$2H_2O(l) \rightleftharpoons H_3O^+(aq) + OH^-(aq)$$

This process is called **self-ionization** because two identical neutral molecules create ions. The equilibrium constant is:

$$K_c = \frac{[H_3O^+]_{eq}[OH^-]_{eq}}{[H_2O]_{eq}^2}$$

The concentration of water is essentially constant because it is so large compared with the concentrations of the other species present. It can therefore be omitted from the equilibrium expression. The resulting equilibrium constant is called the **ionic product of water**, K_w:

$$K_w = [H_3O^+][OH^-]$$

This equilibrium is reached so quickly that the 'eq' subscript may be omitted. The value of K_w varies with temperature, as does the value of any equilibrium constant. It has an experimentally measured value at 25 °C of $1.0 \times 10^{-14}\,mol^2\,dm^{-6}$. You have seen that we do not need to know about K_w to calculate the pH of a strong acid. However, K_w is central to calculating the pH of basic solutions.

The pH of neutral water

In *neutral* water, the concentrations of hydrogen ions and hydroxide ions must be equal, i.e.

$[H_3O^+] = [OH^-]$

In neutral water at 25 °C the aqueous hydrogen ion concentration is $1 \times 10^{-7}\,mol\,dm^{-3}$ (because the square of 1×10^{-7} is 1×10^{-14}, the value of K_w). The pH of neutral water at 25 °C is therefore:

$pH = -\log_{10}(1 \times 10^{-7})$

$= -(-7.0)$

$= 7.0$

The pH meters show that 0.1 mol dm⁻³ aqueous ammonia, a weak base, has a pH of 11.1, whereas the same concentration of aqueous sodium hydroxide, a strong base, has a pH of 13.0. The indicator is alizarin yellow, *see spread 12.9.*

Strong and weak bases

Sodium hydroxide is described as a strong base because it is essentially fully ionized in aqueous solution:

$$NaOH(aq) \rightarrow Na^+(aq) + OH^-(aq)$$

Dissolving 1 mol of solid sodium hydroxide in water results in 1 mol of aqueous hydroxide ions. For example, aqueous sodium hydroxide of concentration $1\,mol\,dm^{-3}$ has an aqueous hydroxide ion concentration $[OH^-] = 1\,mol\,dm^{-3}$.

Ammonia is described as a weak base because the equilibrium

$$NH_3(aq) + H_2O(l) \rightleftharpoons NH_4^+(aq) + OH^-(aq)$$

lies to the left. Dissolving 1 mol of ammonia gas in water does not result in 1 mol of aqueous hydroxide ions. For example, aqueous ammonia of concentration $1\,mol\,dm^{-3}$ has an aqueous hydroxide ion concentration $[OH^-] = 4 \times 10^{-3}\,mol\,dm^{-3}$.

- A **strong base** is *fully ionized* in aqueous solution. The aqueous hydroxide ion concentration is *equal* in magnitude to the concentration of the base.
- A **weak base** is *only partially ionized* in aqueous solution. The aqueous hydroxide ion concentration is *smaller* in magnitude than the concentration of the base.

Neutralization enthalpies

The standard enthalpy change of neutralization for $NH_3(aq)$ and $HCl(aq)$ is only $-52\,kJ\,mol^{-1}$, compared with $-57\,kJ\,mol^{-1}$ for $NaOH(aq)/HCl(aq)$ and $KOH(aq)/HCl(aq)$

The value for $NH_3(aq)$ and the weak acid $HCN(aq)$ is $-5\,kJ\,mol^{-1}$.

Calculating the pH of a strong base

The pH of a strong base depends only on its concentration because a strong base is fully ionized.

The calculation of pH for strong bases is of similar difficulty to that for strong acids. For example, aqueous sodium hydroxide at a concentration of 1 mol dm^{-3} has an aqueous hydroxide ion concentration of 1 mol dm^{-3}, i.e. $[OH^-]$ = 1 mol dm^{-3}.

The aqueous hydrogen ion concentration multiplied by the aqueous hydroxide ion concentration must equal the ionic product of water:

$$K_w = [H_3O^+][OH^-] = 1 \times 10^{-14}\,\text{mol}^2\,\text{dm}^{-6}$$

Divide both sides of the equation by the aqueous hydroxide ion concentration:

$$[H_3O^+] = \frac{K_w}{[OH^-]} = \frac{(1 \times 10^{-14}\,\text{mol}^2\,\text{dm}^{-6})}{(1\,\text{mol}\,\text{dm}^{-3})} = 1 \times 10^{-14}\,\text{mol}\,\text{dm}^{-3}$$

Substituting this value into the expression for pH:

$$\text{pH} = -\log_{10}[H_3O^+] = -(-14.0) = 14.0$$

This calculation shows that aqueous NaOH at a concentration of 1 mol dm^{-3} has a pH of 14. This is the typical upper limit for pH.

Worked example on calculating aqueous hydroxide ion concentration from pH

Question: The measured pH of an aqueous base is 12.0. Calculate the corresponding hydroxide ion concentration.

Answer: First we use

$\text{pH} = -\log_{10}[H_3O^+]$

i.e.

$12.0 = -\log_{10}[H_3O^+]$

Multiply both sides by –1:

$\log_{10}[H_3O^+] = -12.0$

Taking the antilogarithm of both sides of the expression, remembering to add the units, gives:

$[H_3O^+] = 1 \times 10^{-12}\,\text{mol}\,\text{dm}^{-3}$

Now we use

$K_w = [H_3O^+][OH^-] = 1 \times 10^{-14}\,\text{mol}^2\,\text{dm}^{-6}$

i.e.

$1 \times 10^{-14}\,\text{mol}^2\,\text{dm}^{-6} = 1 \times 10^{-12}\,\text{mol}\,\text{dm}^{-3} \times [OH^-]$

Rearranging gives:

$$[OH^-] = \frac{1 \times 10^{-14}\,\text{mol}^2\,\text{dm}^{-6}}{1 \times 10^{-12}\,\text{mol}\,\text{dm}^{-3}} = 1 \times 10^{-2}\,\text{mol}\,\text{dm}^{-3}$$

SUMMARY

- A strong base is fully ionized in aqueous solution; a weak base is only partially ionized.
- Water self-ionizes to a very small extent:
  $2H_2O(l) \rightleftharpoons H_3O^+(aq) + OH^-(aq)$
- The ionic product of water $K_w = [H_3O^+][OH^-]$.
- The value of K_w at 25 °C is 1.0×10^{-14} mol^2 dm^{-6}.

pOH and pK_w

Another way of doing pH calculations for bases involves introducing the quantity pOH, which is defined by analogy with pH as

pOH $= -\log_{10}[OH^-]$

From the equation shown alongside $K_w = [H_3O^+][OH^-]$

taking the negative logarithm of both sides gives

$\text{p}K_w = \text{pH} + \text{pOH}$

where **pK_w** $= -\log_{10}K_w$
$= -\log_{10}(1 \times 10^{-14})$
$= 14.0$ (at 25 °C)

For a strong base at a concentration of 1 mol dm^{-3}, pOH = 0.0

pH = 14.0 – pOH
= 14.0

pH and concentration

For a strong base, each dilution of the concentration by a factor of 10 decreases the pH by one unit. For example, 0.1 mol dm^{-3} aqueous sodium hydroxide has a pH of 13.0. This result is a consequence of using the logarithm to base 10 in the definition of pH.

PRACTICE

1 Calculate the pH of solutions with the following aqueous hydroxide ion concentrations in mol dm^{-3}:
  a 0.1
  b 0.01
  c 0.05.

2 Calculate the aqueous hydroxide ion concentration corresponding to the following pH values:
  a 13.0
  b 11.5.

EXTENSION

12.7

OBJECTIVES

- Base ionization constant K_b
- Definition of pK_b
- pH calculations for weak bases

WEAK BASES

As you might expect, we can do calculations on weak bases in a similar way to those on weak acids. Weak bases are only partially ionized in solution and the extent of the ionization is reflected in their respective values of K_b, the base ionization constant. Calculating the pH values of aqueous *strong* bases involves the ionic product of water K_w, which relates the concentrations of aqueous hydrogen ions and aqueous hydroxide ions. This relationship is also important in pH calculations involving weak bases.

The base ionization constant K_b

Sodium hydroxide is a solid which dissolves readily in water. It is a strong base because it is fully ionized in solution:

$$NaOH(s) \rightarrow Na^+(aq) + OH^-(aq)$$

Ammonia is a gas which dissolves readily in water:

$$NH_3(g) \rightarrow NH_3(aq)$$

It is a weak base because the equilibrium

$$NH_3(aq) + H_2O(l) \rightleftharpoons NH_4^+(aq) + OH^-(aq)$$

lies to the left.

In general, any base B acts as a base when it accepts a proton, the equilibrium in aqueous solution being:

$$B(aq) + H_2O(l) \rightleftharpoons BH^+(aq) + OH^-(aq)$$

The equilibrium constant K_c is given by the expression:

$$K_c = \frac{[BH^+]_{eq}[OH^-]_{eq}}{[B]_{eq}[H_2O]_{eq}}$$

Water is present in large excess and its concentration is essentially constant. The **base ionization constant** is therefore expressed as:

$$K_b = \frac{[BH^+][OH^-]}{[B]}$$

Note that the equilibrium establishes itself very quickly and so the 'eq' subscript is not needed.

pK_b

The pK_b value of a weak base is defined in a similar way to pH, i.e.

$$pK_b = -\log_{10}K_b$$

The lower the value of pK_b, the larger the value of K_b and the greater the ionization of the base in water. For example, at the same concentrations, aqueous ammonia $NH_3(aq)$ (pK_b 4.8) is ionized to a lesser extent than aqueous methylamine $CH_3NH_2(aq)$ (pK_b 3.4).

A comparison of K_a and K_b

For a weak acid HA:

$$HA(aq) + H_2O(l) \rightleftharpoons H_3O^+(aq) + A^-(aq)$$

ACID BASE CONJUGATE ACID CONJUGATE BASE

$$K_a = \frac{[H_3O^+][A^-]}{[HA]}$$

For a weak base B:

$$B(aq) + H_2O(l) \rightleftharpoons BH^+(aq) + OH^-(aq)$$

ACID BASE CONJUGATE ACID CONJUGATE BASE

$$K_b = \frac{[BH^+][OH^-]}{[B]}$$

To calculate K_b from pK_b

From the main text we have

$pK_b = -\log_{10}K_b$

Multiply both sides of the equation by −1:

$\log_{10}K_b = -pK_b$

Take the antilogarithm of both sides of the expression, remembering to add the units of $mol\,dm^{-3}$:

$K_b = 10^{-pK_b}\,mol\,dm^{-3}$

For example, the pK_b for ammonia is 4.8. Substitute this value into the expression above and check you obtain the answer

$K_b = 2 \times 10^{-5}\,mol\,dm^{-3}$.

Base ionization constants of some weak bases in water at 25 °C.

Compound	Ionization equilibrium	Ionization constant K_b /$mol\,dm^{-3}$	pK_b
Ammonia	$NH_3 + H_2O \rightleftharpoons NH_4^+ + OH^-$	1.6×10^{-5}	4.8
Methylamine	$CH_3NH_2 + H_2O \rightleftharpoons CH_3NH_3^+ + OH^-$	4.0×10^{-4}	3.4
Phenylamine	$C_6H_5NH_2 + H_2O \rightleftharpoons C_6H_5NH_3^+ + OH^-$	4.0×10^{-10}	9.4
Phenylmethylamine	$C_6H_5CH_2NH_2 + H_2O \rightleftharpoons C_6H_5CH_2NH_3^+ + OH^-$	2.5×10^{-5}	4.6

Calculating the pH of a weak base

The method of calculation follows a similar path to the one that uses K_a to find the pH of a weak acid.

As an example, we shall calculate the pH value of 0.1 mol dm^{-3} aqueous ammonia (K_b = 1.6 × 10^{-5} mol dm^{-3}).

The chemical equation for the equilibrium is:

$$NH_3(aq) + H_2O(l) \rightleftharpoons NH_4^+(aq) + OH^-(aq)$$

The expression for the base ionization constant is:

$$K_b = \frac{[NH_4^+][OH^-]}{[NH_3]}$$

Aqueous ammonia is only slightly ionized, so its concentration is effectively equal to the original concentration of the solution, i.e. $[NH_3]$ = 0.1 mol dm^{-3}. Substituting this value and the one for the base ionization constant K_b = 1.6 × 10^{-5} mol dm^{-3} gives:

$$1.6 \times 10^{-5}\ \text{mol dm}^{-3} = \frac{[NH_4^+][OH^-]}{0.1\ \text{mol dm}^{-3}}$$

The chemical equation indicates that the concentrations of ammonium ion and hydroxide ion are equal, i.e. $[NH_4^+] = [OH^-]$. So

$$1.6 \times 10^{-5}\ \text{mol dm}^{-3} = \frac{[OH^-]^2}{0.1\ \text{mol dm}^{-3}}$$

$$[OH^-]^2 = (0.1\ \text{mol dm}^{-3}) \times (1.6 \times 10^{-5}\ \text{mol dm}^{-3})$$

$$= 1.6 \times 10^{-6}\ \text{mol}^2\ \text{dm}^{-6}$$

$$[OH^-] = \surd(1.6 \times 10^{-6}\ \text{mol}^2\ \text{dm}^{-6})$$

$$= 1.3 \times 10^{-3}\ \text{mol dm}^{-3}$$

The ionic product of water requires that:

$$[H_3O^+][OH^-] = 1 \times 10^{-14}\ \text{mol}^2\ \text{dm}^{-6}$$

Substituting the value for $[OH^-]$ from above:

$$[H_3O^+] \times (1.3 \times 10^{-3}\ \text{mol dm}^{-3}) = 1 \times 10^{-14}\ \text{mol}^2\ \text{dm}^{-6}$$

$$[H_3O^+] = 7.7 \times 10^{-12}\ \text{mol dm}^{-3}$$

Substituting this value into the expression for pH:

$$\text{pH} = -\log_{10}[H_3O^+]$$

$$= -\log_{10}(7.7 \times 10^{-12})$$

$$= 11.1\ (1\ \text{d.p.})$$

Note: It is purely *coincidental* that pK_b for NH_3 and pK_a for CH_3COOH are both 4.8.

K_a values for weak bases

Aqueous ammonia is a weak base because it accepts protons from water:

$$NH_3(aq) + H_2O(l) \rightleftharpoons NH_4^+(aq) + OH^-(aq)$$

The base ionization constant K_b is given by the expression

$$K_b = \frac{[NH_4^+][OH^-]}{[NH_3]}$$

and has the value 1.6 × 10^{-5} mol dm^{-3}.

The conjugate acid of ammonia is the ammonium ion $NH_4^+(aq)$. It is a weak acid as shown by the equilibrium:

$$NH_4^+(aq) + H_2O(l) \rightleftharpoons H_3O^+(aq) + NH_3(aq)$$

The corresponding acid ionization constant K_a for the ammonium ion is

$$K_a = \frac{[H_3O^+][NH_3]}{[NH_4^+]}$$

and has the value 6.3 × 10^{-10} mol dm^{-3}.

Hence pK_a (NH_4^+) = 9.2

In this manner, the K_a values of weak acids and the conjugate acids of weak bases may be given as a continuous list. Note that

$K_a \times K_b = [H_3O^+][OH^-]$

= 1 × 10^{-14} mol^2 dm^{-6}

(K_w, the ionic product of water)

SUMMARY

- Weak bases are only partially ionized.
- K_b is the base ionization constant for a weak base:

$$K_b = \frac{[BH^+][OH^-]}{[B]}$$

- pK_b = $-\log_{10}K_b$

PRACTICE

1 Calculate the pH of a solution of ammonia, 0.2 mol dm^{-3}, K_b = 1.6 × 10^{-5} mol dm^{-3}.

2 Calculate the pH value of a solution of methylamine $CH_3NH_2(aq)$, 0.1 mol dm^{-3}, pK_b = 3.4.

3 Explain how the expression pK_a + pK_b = 14.0 may be applied to any aqueous acid–base system.

12.8

ACID–BASE TITRATIONS – 1

OBJECTIVES

- Practical technique
- End point and equivalence point
- Titration curve for strong acid–strong base
- Calculation from results

Acids and bases neutralize each other when mixed together in aqueous solution. An acid–base titration is a technique used to measure the volumes of acid and base required for neutralization. If the concentration of one solution is known, then the concentration of the other may be calculated. There are four possible combinations of strong and weak acids and bases that may be considered for a titration. This spread investigates the simplest example of the four, focusing on the neutralization reaction between a strong acid and a strong base.

A commercially available automatic titrator.

Titration technique – a reminder

The usual method followed in a titration is to place a known volume of aqueous base in a conical titration flask. A few drops of indicator are then added to show the pH change during the titration. Acid is then run into the flask from a burette until the indicator changes colour. The colour change usually happens on addition of just one final drop of acid. The **end point** occurs when the indicator changes colour.

The **equivalence point** of a titration occurs when there are equal amounts of $H_3O^+(aq)$ and $OH^-(aq)$ ions in the titration flask. At this point, the neutralization is complete:

$H_3O^+(aq) + OH^-(aq) \rightarrow 2H_2O(l)$

Titration of a strong acid with a strong base

A **titration curve** shows the pH changes that occur when an aqueous acid reacts with an aqueous base. It is a graph of pH against volume (usually volume of added acid). The titration of a strong acid with a strong base causes a large pH change on neutralization, as shown in the graph below. Both the acid and the base are fully ionized. Very small volumes of acid or base cause large changes in pH when the reaction mixture is very close to the point of neutralization.

Concentration

This titration curve, and the first two titration curves in spread 12.9, assume concentrations of 1 mol dm^{-3}.

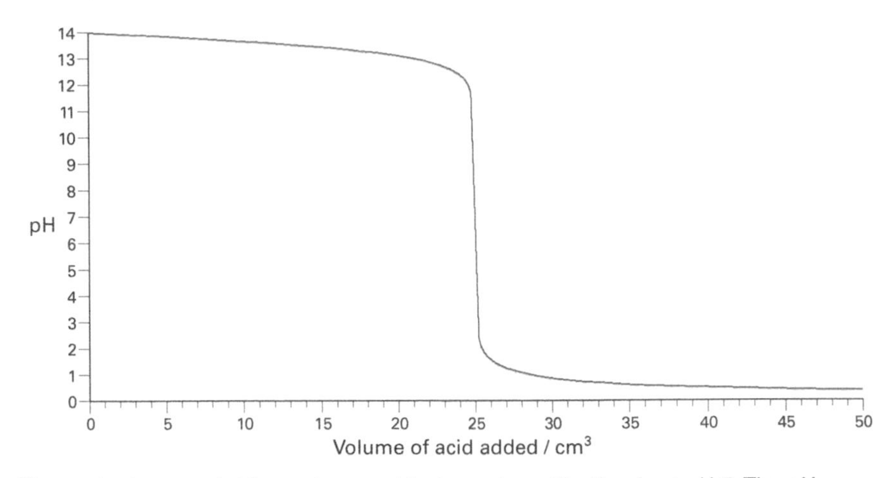

The equivalence point for a strong acid–strong base titration is at pH 7. The pH changes from 11 to 3 very rapidly so phenolphthalein (range pH 8–10), litmus (range pH 6–8), and methyl orange (range pH 3–5) would all be suitable indicators.

For example, imagine a titration of 25 cm^3 of 1 mol dm^{-3} aqueous sodium hydroxide with hydrochloric acid of the same concentration. The total volume of solution at neutralization will be exactly twice the initial volume, because the equation shows that 1 mol of NaOH reacts with 1 mol of HCl:

$$NaOH(aq) + HCl(aq) \rightarrow NaCl(aq) + H_2O(l)$$

The equivalence point of this titration occurs when the amounts of acid and base are exactly equal. Now imagine overshooting the equivalence point by *just one drop*. A drop has a volume of approximately 0.05 cm^3.

The excess acid is equivalent to 0.05 cm^3 in 50 cm^3, a 1000-fold dilution. This dilutes the acid from a concentration of 1 mol dm^{-3} (in the burette) to 1×10^{-3} mol dm^{-3} (in the flask) and produces a pH of 3. Just one drop (0.05 cm^3) *before* neutralization, the sodium hydroxide concentration was 1×10^{-3} mol dm^{-3} and the solution had a pH of 11.

In such a titration, adding just two drops of acid at the equivalence point causes a very large change in pH from about 11 to about 3. This change is characteristic of a strong acid–strong base titration and is easy to detect by using any indicator with a range between pH 3 and pH 11. At the start, the base is in large excess; there is only a small change in pH because any acid added is neutralized by the base. If excess acid is added after reaching the equivalence point, the pH remains very low.

The indicator must be chosen with care to ensure that the end point coincides with the equivalence point. The actual pH changes that happen during an acid–base titration depend on the strengths of the acid and base concerned (described in the following spread). During a strong acid–strong base titration, the pH swings from 11 to 3 on the addition of two drops of acid. All common indicators change colour within this range, and so the end point (when the colour of the indicator changes) coincides with the equivalence point.

Reading a burette

The volume on a burette should be read to the nearest 0.05 cm^3. The reading should be taken with your eye level with the bottom of the meniscus to avoid errors due to parallax.

Worked example on calculating concentration

Question: In a titration, 25.0 cm^3 of aqueous sodium hydroxide of concentration 1.00 mol dm^{-3} were exactly neutralized by 9.20 cm^3 of hydrochloric acid. Calculate the concentration of the acid.

Strategy: The calculation is carried out in four steps.

Answer: *Step 1* Write the balanced chemical equation for the reaction:
$HCl(aq) + NaOH(aq) \rightarrow NaCl(aq) + H_2O(l)$

Step 2 Calculate the amount in moles of sodium hydroxide:

$$n(NaOH) = \left(\frac{25.0}{1000} dm^3\right) \times (1.00 \text{ mol dm}^{-3}) = 0.0250 \text{ mol}$$

Step 3 Use the mole ratios indicated by the equation to calculate the amount in moles of HCl:

1 mol of HCl reacts with 1 mol of NaOH

Therefore the amount in moles of HCl is 0.0250 mol

Step 4 Calculate the concentration of the acid:

9.20 cm^3 (0.00920 dm^3) of the acid contains 0.0250 mol

$$\text{concentration} = \frac{\text{amount in moles}}{\text{volume}} = \frac{0.0250 \text{ mol}}{0.00920 \text{ dm}^3} = 2.72 \text{ mol dm}^{-3}$$

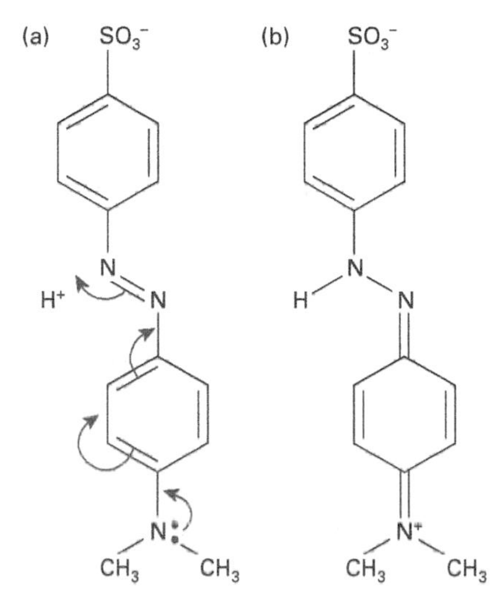

(a) Methyl orange in its base form. This structure predominates at pH 4.5 or greater and the solution colour is orange. (b) At pH 3 and below, the base form is protonated to give the conjugate acid, which is red.

Weak acids and bases

The titration curve for strong acid–strong base neutralization shows a large change in pH around the equivalence point. Choosing a suitable indicator is therefore a simple task. The following spread introduces neutralization reactions involving weak acids and bases. You will see that choosing an indicator is not so straightforward!

SUMMARY

- The end point of a titration occurs when the indicator changes colour.
- The equivalence point occurs when there are equal amounts of H_3O^+ and OH^- ions in the titration flask.
- Indicators with ranges between pH 3 and pH 11 are suitable for strong acid–strong base titrations.

12.9

ACID–BASE TITRATIONS – 2

OBJECTIVES

- Strong acid–weak base titrations
- Weak acid–strong base titrations
- Choice of indicator
- Weak acid–weak base conductometric titrations

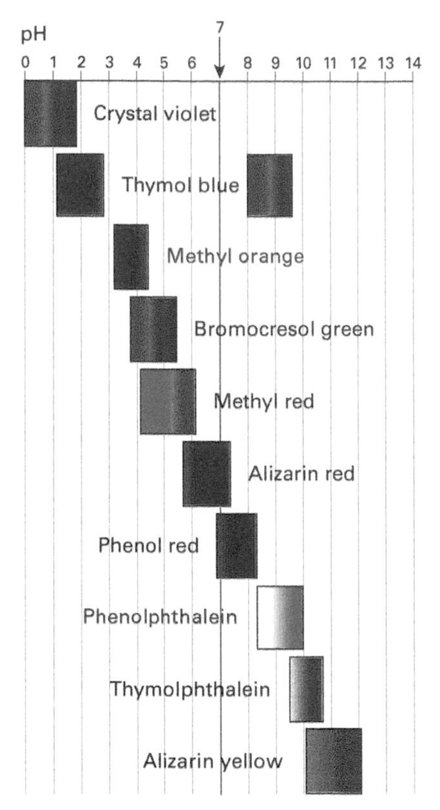

Common indicators and their ranges. The centre of the range is the value of pK_{in} where K_{in} is the indicator ionization constant.

The titration curve for a strong acid–strong base neutralization shows a large change in pH around the equivalence point. Any indicator with a range between pH 11 and pH 3 may be used. The titration curves for other combinations of strong/weak acid/base pairs offer a more limited change in pH. As a result, it is more difficult, if not impossible, to choose a suitable indicator.

Titration of a strong acid with a weak base

At the start of a strong acid–weak base titration, the weak base is in excess. Its pH is much lower (closer to 7) than that of a strong base because it is only partially ionized. When the weak base is just in excess (i.e. one drop before the equivalence point is reached), the pH will be only slightly greater than neutral (pH 7). (The exact pH value will depend on the K_b value of the weak base.) On adding excess strong acid (one drop after the equivalence point), a large change in pH value from 7 to about 3 occurs. A suitable indicator, such as methyl orange, will have a range between these two pH values: see spread 9.6.

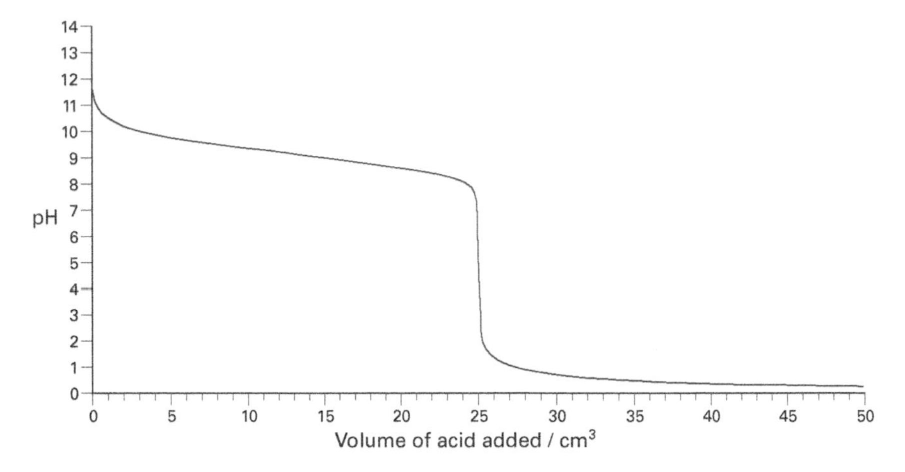

The titration curve of a strong acid (e.g. hydrochloric acid HCl) with a weak base (here ammonia NH_3).

Titration of a weak acid with a strong base

In this case, the strong base is in excess at the start of the titration and the weak acid is in excess at the end. You will see from the titration curve below that the large change in pH around the equivalence point occurs approximately between pH 11 and pH 7. A suitable indicator, such as phenolphthalein, will have a range between these two pH values.

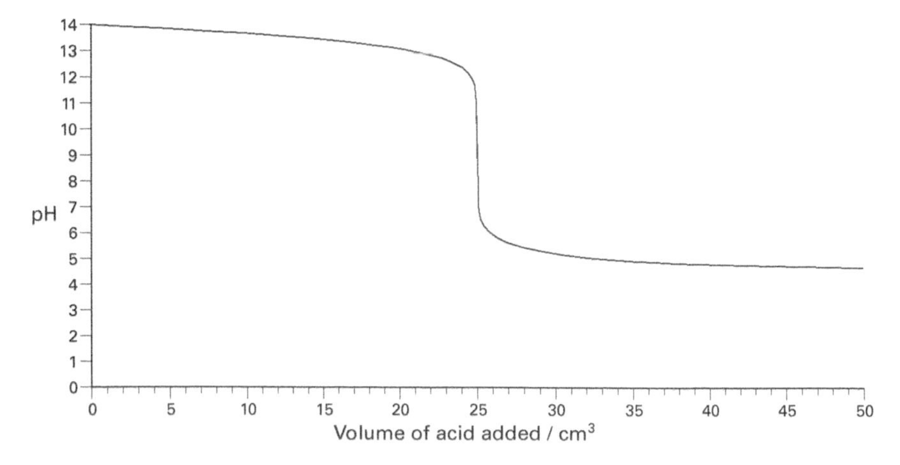

The titration curve of a weak acid (here ethanoic acid CH_3COOH) with a strong base (e.g. sodium hydroxide NaOH).

Choosing an indicator

The large changes of pH around the equivalence point can be summarized as follows:

- Strong acid–strong base: pH 11–3.
- Strong acid–weak base: pH 7–3.
- Weak acid–strong base: pH 11–7.

These regions of large pH change correspond to the near-vertical portions of the corresponding titration curves. In each case, a suitable indicator is one whose range falls within these pH limits. Note that the centre of the vertical region of a titration curve is called the **point of inflection**; it occurs at the point where the gradient of the curve changes direction.

It is possible to have two points of inflection in a titration, as shown in the figure on the right for the titration of sodium carbonate with hydrochloric acid. Two indicators are used in such a case, which is then called a **double indicator titration.** Two points of inflection also occur when a diprotic acid such as ethanedioic acid $(COOH)_2$ is titrated against sodium hydroxide.

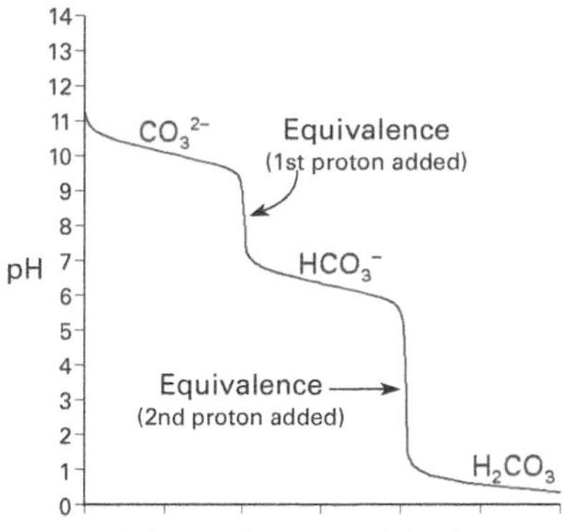

Two points of inflection. The titration curve for the weak base sodium carbonate Na_2CO_3 and the strong acid hydrochloric acid shows two points of inflection, i.e. two separate equivalence points. These two regions relate to the successive transfer of one and then two protons. The first corresponds to the conversion of carbonate to hydrogencarbonate:

$CO_3^{2-}(aq) + H_3O^+(aq) \rightleftharpoons HCO_3^-(aq) + H_2O(l)$

Phenolphthalein is a suitable indicator. The second equivalence point corresponds to the conversion of hydrogencarbonate to carbon dioxide:

$HCO_3^-(aq) + H_3O^+(aq) \rightleftharpoons H_2CO_3(aq) + H_2O(l) \rightleftharpoons CO_2(aq) + 2H_2O(l)$

Methyl orange is a suitable indicator.

Titration of a weak acid with a weak base

It is not possible to use an indicator to find the equivalence point of a titration involving a weak acid and a weak base. The variation of pH is too gradual for any indicator to show a distinct end point. Instead, a conductometric titration must be used.

At the start of the titration, the conductivity is not very great because the weak base in the titration flask is only partially ionized. Adding acid forms the salt, which is fully ionized. The conductivity rises steadily as more acid is added. Adding excess acid beyond the equivalence point barely changes the conductivity because the weak acid is also only partially ionized. The change in slope corresponding to the equivalence point is easy to detect.

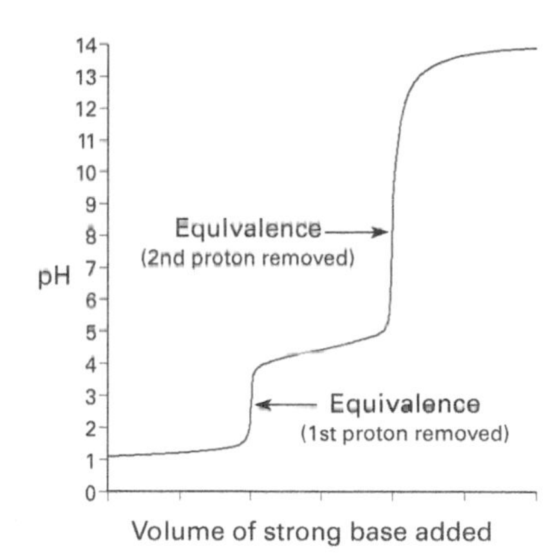

Ethanedioic acid / NaOH titration curve

SUMMARY

- The indicator range for a strong acid–weak base titration must fall between pH 7 and pH 3. Methyl orange is a useful choice.
- The indicator range for a weak acid–strong base titration must fall between pH 11 and pH 7. Phenolphthalein is a useful choice.
- Indicators cannot be used to find the equivalence point of a weak acid–weak base titration. A conductometric titration may be used.

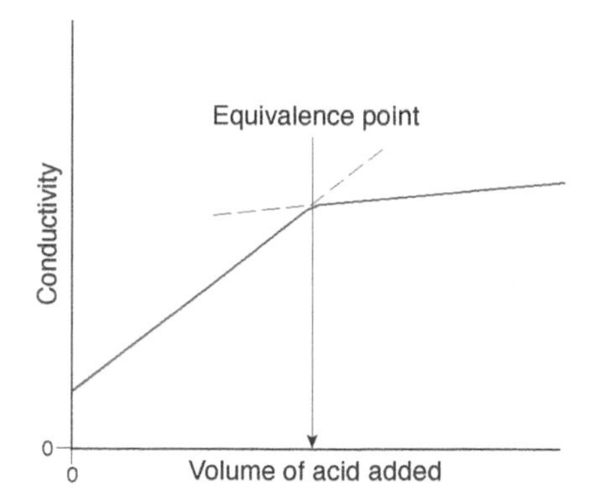

*The equivalence point in a weak acid–weak base neutralization may be determined by a **conductometric titration**, in which the electrical conductivity of the solution is measured.*

PRACTICE

1 Explain the terms *end point* and *equivalence point* with reference to the titration of ethanoic acid $CH_3COOH(aq)$ (a weak acid) against sodium hydroxide $NaOH(aq)$ (a strong base).

2 For which class(es) of titration (strong acid–strong base, etc.) would each of the following indicators be suitable?

a Alizarin yellow

b Thymolphthalein

c Bromocresol green.

3 In a titration, $36.2\,cm^3$ of $0.200\,mol\,dm^{-3}$ sulphuric acid $H_2SO_4(aq)$ neutralized $25.0\,cm^3$ of aqueous sodium hydroxide $NaOH(aq)$. Calculate the concentration of the aqueous sodium hydroxide.

12.10

OBJECTIVES

- Acidic buffer solutions
- Basic buffer solutions

Buffer solutions

Adding acid or base to a solution usually changes the value of its pH greatly. However, some mixtures, called **buffer solutions**, resist changes in pH despite small additions of acid or base. Buffer solutions may be made up to maintain specified acidic or basic pH values. This spread starts by identifying a region in a titration curve where buffer action is taking place, and then explains how to make buffer solutions.

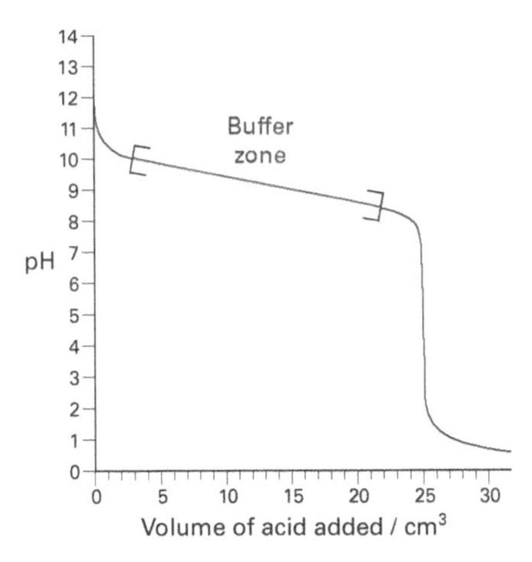

The titration curve resulting from the addition of the strong acid HCl(aq) to the weak base NH_3(aq). The buffer zone is highlighted in red.

Buffer action

Consider a titration in which the strong acid hydrochloric acid HCl(aq) is added to the weak base ammonia NH_3(aq). Initially the pH falls significantly; after this the titration curve shows only a moderate fall in pH, even though a strong acid is being added to the mixture. This portion of the titration curve, where buffer action is occurring, is called the **buffer zone**. The buffer present at this stage of the titration is a mixture of the weak base NH_3 and its salt NH_4^+, formed from the reaction between the acid and the base:

$$H_3O^+(aq) + NH_3(aq) \rightleftharpoons NH_4^+(aq) + H_2O(l)$$

This is an acid–base reaction, and the equilibrium lies far to the right.

Note that, in practice, buffer solutions are usually made by mixing suitable reactants in solution rather than by carrying out acid–base titrations.

Acidic and basic buffer solutions

Acidic buffer solutions maintain a nearly constant pH value less than 7. They consist of a solution of a weak acid and a salt that supplies the conjugate base of the weak acid. For example, an aqueous solution containing equal amounts in moles of ethanoic acid CH_3COOH and sodium ethanoate $Na^+CH_3COO^-$ constitutes an acidic buffer.

Basic buffer solutions maintain a nearly constant pH value greater than 7. They consist of a solution of a weak base and a salt that supplies the conjugate acid of the weak base. For example, an aqueous solution containing equal amounts in moles of ammonia NH_3 and ammonium chloride $NH_4^+Cl^-$ constitutes a basic buffer.

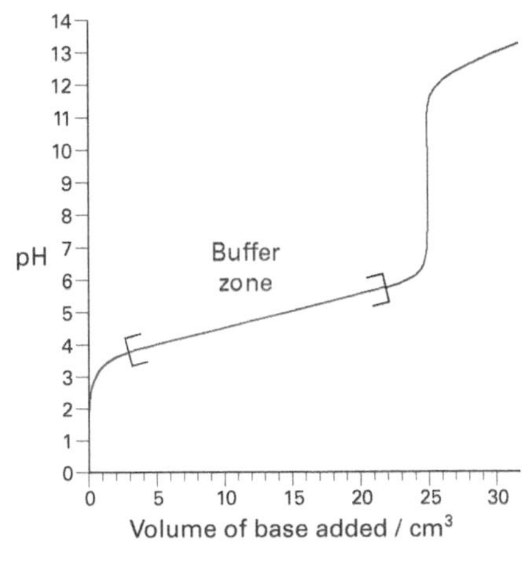

The titration curve resulting from the addition of the strong base NaOH(aq) to the weak acid CH_3COOH(aq). The buffer zone is highlighted in red. The centre of the buffer zone is at 4.8, the pK_a of CH_3COOH(aq), as explained in the next spread.

Acidic buffer action

First consider a solution of the weak acid CH_3COOH. The acid is partially ionized according to the equilibrium:

$$CH_3COOH(aq) + H_2O(l) \rightleftharpoons H_3O^+(aq) + CH_3COO^-(aq)$$

Le Chatelier's principle indicates that adding aqueous hydrogen ions H_3O^+ from a strong acid shifts the equilibrium (even further) to the left. Aqueous hydrogen ions are consumed, the concentration of CH_3COO^- ions decreases, and the concentration of CH_3COOH increases.

Conversely, adding aqueous hydroxide ions OH^- from a strong base will *remove* H_3O^+ ions and the equilibrium will shift to the right. The concentration of CH_3COO^- ions will increase and the concentration of CH_3COOH will decrease.

In the buffer, the salt $Na^+CH_3COO^-$ is also present. This salt is fully ionized in solution and supplies a large concentration of CH_3COO^- ions. So the buffer solution contains a large concentration of CH_3COO^- ions (from the salt) and a large concentration of CH_3COOH (the un-ionized acid).

The effect of adding acid to the buffer solution

When an acid is added, there is a large concentration of CH_3COO^- ions to react with the added H_3O^+ ions. The equilibrium

$$\underset{\text{FROM BUFFER}}{CH_3COO^-(aq)} + \underset{\text{ADDED}}{H_3O^+(aq)} \rightleftharpoons CH_3COOH(aq) + H_2O(l)$$

lies far to the right, and the pH remains almost unchanged.

The effect of adding base to the buffer solution

When OH^- ions from a base are added, they react with the high concentration of un-ionized CH_3COOH present. The equilibrium

$$\underset{\text{FROM BUFFER}}{CH_3COOH(aq)} + \underset{\text{ADDED}}{OH^-(aq)} \rightleftharpoons CH_3COO^-(aq) + H_2O(l)$$

lies far to the right, and the pH remains almost unchanged.

A natural (physiological) buffer system

Buffer solutions maintain optimum pH values for the biochemical processes taking place in living systems. The enzymes (biological catalysts) that enable biochemical reactions to take place can function only within a narrow range of pH. An important buffer system in human beings is the carbonate buffer consisting of carbonic acid H_2CO_3 and its conjugate base, the hydrogencarbonate ion HCO_3^-:

$$H_2CO_3(aq) + H_2O(l) \rightleftharpoons H_3O^+(aq) + HCO_3^-(aq)$$

Adding acid shifts the equilibrium to the left. Aqueous hydrogen ions are consumed (thus maintaining the pH value), reducing the concentration of HCO_3^- while increasing the concentration of H_2CO_3. Adding base shifts the equilibrium to the right. Aqueous hydrogen ions protonate the base and the concentration of HCO_3^- increases while the concentration of H_2CO_3 reduces.

Hyperventilation is needed to compensate for the low partial pressure of oxygen at high altitude. Climbers reaching the summit of K2 without additional oxygen have had their blood pH rise from 7.4 to 7.7.

SUMMARY

- Buffer solutions resist changes in pH despite small additions of acid or base.
- An acidic buffer consists of a weak acid and a salt that supplies its conjugate base.
- A basic buffer consists of a weak base and a salt that supplies its conjugate acid.
- Buffer solutions maintain optimum pH values for the biochemical processes taking place in living systems.

Basic buffer action

An example of a basic buffer pair is the salt ammonium chloride NH_4Cl dissolved in aqueous ammonia $NH_3(aq)$.

When aqueous hydrogen ions from an acid are added to this mixture, they react with the weak base NH_3. The equilibrium shifts to the right, taking up the added hydrogen ions:

$$NH_3(aq) + H_3O^+(aq) \rightleftharpoons NH_4^+(aq) + H_2O(l)$$

so the pH of the solution remains almost unchanged.

Adding base to this buffer solution introduces hydroxide $OH^-(aq)$ ions. The conjugate acid $NH_4^+(aq)$ donates protons to these hydroxide ions, shifting the following equilibrium to the right and resisting the change in pH:

$$NH_4^+(aq) + OH^-(aq) \rightleftharpoons NH_3(aq) + H_2O(l)$$

Buffers in industry

Many industrial processes take place successfully only within limited ranges of pH. Dyeing yarn and electroplating metals are two processes that are carried out in buffered solutions. Buffer solutions are also used to control the pH of shampoos.

PRACTICE

1 From the substances H_2CO_3, CH_3COOK, NH_3, CO_2, CH_3COOH and NH_4Cl, select:

a an acidic buffer pair

b a basic buffer pair.

2 Explain how a solution of a salt M^+A^- in a weak acid HA resists attempts to change its pH when aqueous hydrogen ions are added to it.

12.11 BUFFER SOLUTIONS: CALCULATIONS

OBJECTIVES

- The Henderson–Hasselbalch equation
- Preparing buffer solutions

(a)

(b)

(a) Both beakers contain solutions at pH 7 initially (indicated by the green colour), but the one on the left is a buffer solution.

(b) Dry ice (solid carbon dioxide) added to both beakers has changed the pH significantly only for the pure water, which has become acidic (indicated by the yellow colour).

Buffer pH on dilution

Because the Henderson–Hasselbalch equation involves a *ratio* of two concentrations, the volume of the solution cancels. So buffer solutions do not change pH on dilution.

Basic buffer solutions

The Henderson–Hasselbalch equation applies equally to basic buffer solutions. For example, the buffer solution containing aqueous ammonia NH_3 and ammonium chloride

NH_4Cl establishes the equilibrium:

$NH_4^+(aq) + H_2O(l) \rightleftharpoons H_3O^+(aq) + NH_3(aq)$

The base is NH_3 and the conjugate acid is NH_4^+. In this example, the Henderson–Hasselbalch equation becomes:

$$pH = pK_a(NH_4^+) + \log_{10}\left(\frac{[NH_3]}{[NH_4^+]}\right)$$

or

$$pH = pK_a + \log_{10}\left(\frac{[base]}{[conjugate\ acid]}\right)$$

Buffer solutions resist attempts to change their pH by the addition of acid or base. They have applications in general chemistry and in physiological systems. This spread shows you how to calculate the pH of a buffer solution and how to predict the pH of the reaction mixture during the course of a titration. It also describes how to prepare buffer solutions of known pH. The key to these calculations lies in the Henderson–Hasselbalch equation, which describes the pH of a buffer solution in terms of the concentrations of the acid–base pairs present.

Deriving the Henderson–Hasselbalch equation

In general, the equilibrium established in a solution of a weak acid is:

$$HA(aq) + H_2O(l) \rightleftharpoons H_3O^+(aq) + A^-(aq)$$

The acid ionization constant for this reaction is:

$$K_a = \frac{[H_3O^+][A^-]}{[HA]}$$

Making $[H_3O^+]$ the subject of this equation:

$$[H_3O^+] = \frac{K_a[HA]}{[A^-]}$$

Taking the negative logarithm of both sides, remembering that $pH = -\log_{10}[H_3O^+]$:

$$pH = -\log_{10}\left(\frac{K_a[HA]}{[A^-]}\right)$$

$$= -\left\{\log_{10}K_a + \log_{10}\left(\frac{[HA]}{[A^-]}\right)\right\}$$

$$= -\log_{10}K_a - \log_{10}\left(\frac{[HA]}{[A^-]}\right)$$

But $pK_a = -\log_{10}K_a$, so:

$$pH = pK_a - \log_{10}\left(\frac{[HA]}{[A^-]}\right)$$

This equation may be rewritten as:

$$pH = pK_a + \log_{10}\left(\frac{[A^-]}{[HA]}\right)$$

or

$$pH = pK_a + \log_{10}\left(\frac{[conjugate\ base]}{[acid]}\right)$$

This equation is known as the **Henderson–Hasselbalch equation.**

Preparing buffer solutions

It is often necessary to prepare buffer solutions of known pH, for example when calibrating a pH meter. There are two main steps involved in selecting suitable substances and deciding their concentrations. Look carefully at the Henderson–Hasselbalch equation as applied to acidic buffers. There are two terms on the right-hand side, which determine the final pH of the solution. The first term is pK_a, whose value is responsible for the 'coarse selection' of pH. The second term involves the ratio [conjugate base]/[acid] and provides 'fine tuning' to the final pH.

So the strategy for preparing an acidic buffer solution of known pH is to select an acid whose pK_a is within about one pH unit of the desired pH. The ratio of salt and acid concentrations is then adjusted to achieve the desired pH. This 'fine tuning' allows the pH to be varied by about ±1 unit either side of the pK_a value.

Worked example on calculating the pH of a buffer solution

Question: Consider an acidic buffer solution containing 0.1 mol dm^{-3} CH_3COONa and 0.1 mol dm^{-3} CH_3COOH, with pK_a 4.8. What is its pH?

Strategy: If the constituents of a buffer solution and their concentrations are known, then the pH can be calculated from the Henderson–Hasselbalch equation.

Answer: Ethanoic acid CH_3COOH is a weak acid and so is only partially ionized. Its concentration may be taken as 0.1 mol dm^{-3}. The conjugate base is the CH_3COO^- ion supplied by the salt CH_3COONa. This salt ionizes fully in solution, so the concentration of CH_3COO^- will equal the concentration of salt added, 0.1 mol dm^{-3}.

Substituting these values into the Henderson–Hasselbalch equation:

$$\text{pH} = \text{p}K_a + \log_{10}\left(\frac{[\text{conjugate base}]}{[\text{acid}]}\right)$$

$$= 4.8 + \log_{10}\left(\frac{0.1\ \text{mol dm}^{-3}}{0.1\ \text{mol dm}^{-3}}\right) = 4.8 + \log_{10}1.0 = 4.8$$

For example, suppose you wish to prepare an acidic buffer with a pH of 4.0. A suitable weak acid would be ethanoic acid CH_3COOH because its pK_a is 4.8. The conjugate base is the ethanoate ion CH_3COO^-, which can be provided by the salt sodium ethanoate CH_3COONa. Ethanoic acid is available as a laboratory bench reagent with a concentration of 1.0 mol dm^{-3}.

The question now becomes: 'What mass of sodium ethanoate must I add to the ethanoic acid?' The Henderson–Hasselbalch equation is:

$$\text{pH} = \text{p}K_a + \log_{10}\left(\frac{[\text{conjugate base}]}{[\text{acid}]}\right)$$

Substituting the known quantities:

$$4.0 = 4.8 + \log_{10}\left(\frac{[\text{conjugate base}]}{1.0}\right)$$

$$4.0 = 4.8 + \log_{10}[\text{conjugate base}]$$

Rearranging:

$$\log_{10}[\text{conjugate base}] = -0.8$$

Taking the antilog of both sides, adding the units:

$$[\text{conjugate base}] = 0.16\ \text{mol dm}^{-3}$$

i.e. the concentration of the conjugate base is 0.16 mol dm^{-3}.

The molar mass of sodium ethanoate is 82.0 g mol^{-1}. So 0.16 mol of sodium ethanoate has a mass given by

$$\text{mass} = \text{amount in moles} \times \text{molar mass}$$
$$= (0.16\ \text{mol}) \times (82.0\ \text{g mol}^{-1})$$
$$= 13\ \text{g}$$

Therefore, an acidic buffer solution of pH 4.0 will result when 13 g of sodium ethanoate is added to 1.0 dm^3 of 1.0 mol dm^{-3} ethanoic acid.

SUMMARY

- The pH of an acid–base buffer pair is found from the general equation:

$$\text{pH} = \text{p}K_a + \log_{10}\left(\frac{[\text{base}]}{[\text{acid}]}\right)$$

- When half the acid in a solution is neutralized by a base, the concentrations of acid and conjugate base are equal and pH = pK_a.

Acidic buffer pH and pK_a

A weak acid HA is partially ionized with an acid ionization constant K_a. The equilibrium lies well to the left:

$$HA(aq) + H_2O(l) \rightleftharpoons H_3O^+(aq) + A^-(aq)$$

$$K_a = \frac{[H_3O^+][A^-]}{[HA]}$$

A salt M^+A^- of the weak acid HA ionizes fully in solution:

$$M^+A^-(aq) \rightarrow M^+(aq) + A^-(aq)$$

If equal amounts in moles of the salt and the weak acid are added together in solution, then their concentrations are equal, i.e.

$[HA] = [A^-]$

The expression for K_a becomes:

$K_a = [H_3O^+]$

The pH at this point may be calculated as follows. By taking the negative logarithm of both sides of the above equation we get:

$-\log_{10}K_a = -\log_{10}[H_3O^+]$

But, by definition,

$-\log_{10}K_a = \text{p}K_a$

$-\log_{10}[H_3O^+] = \text{pH}$

and so we can see that

$\text{pH} = \text{p}K_a$

The pH of a solution consisting of equal amounts in moles of weak acid and one of its salts is equal to the pK_a value of the weak acid concerned.

This statement applies also to the central point of the buffer zone of a titration curve. In the first titration curve in spread 12.10, the centre of the buffer zone (at 12.5 cm^3 acid) is at pH = 9.2, the pK_a of ammonium ion, spread 12.7.

In the second titration curve on spread 12.10, the centre of the buffer zone (at 12.5 cm^3) is at pH = 4.8, the pK_a of ethanoic acid. Ten per cent of the way to neutralization (at 2.5 cm^3), [acid] = 10[conjugate base]

$$\text{So pH} = \text{p}K_a + \log_{10}\left(\frac{1}{10}\right)$$
$$= \text{p}K_a - \log_{10}10$$
$$= 4.8 - 1.0 = 3.8$$

pH may be measured accurately by using a pH meter, which incorporates a glass electrode similar to the one shown here. Buffer solutions of known pH are used to calibrate pH meters before use.

12.12

OBJECTIVES

- Lewis acids and bases
- Lewis and Brønsted–Lowry theories contrasted
- Coordinate bonds and complexes

LEWIS ACIDS AND BASES

The Brønsted–Lowry theory of acids and bases defines acids as proton donors and bases as proton acceptors. In the same year (1923), Gilbert Lewis, see spread 5.2, defined acids and bases in terms of the donation and acceptance of electron pairs. This definition broadens further the range of reactions described as 'acid–base' reactions. This spread applies the Brønsted–Lowry and Lewis theories to the reaction between hydrochloric acid and sodium hydroxide.

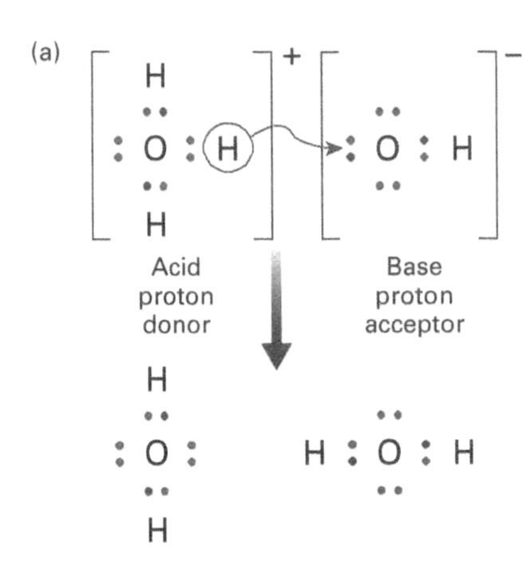

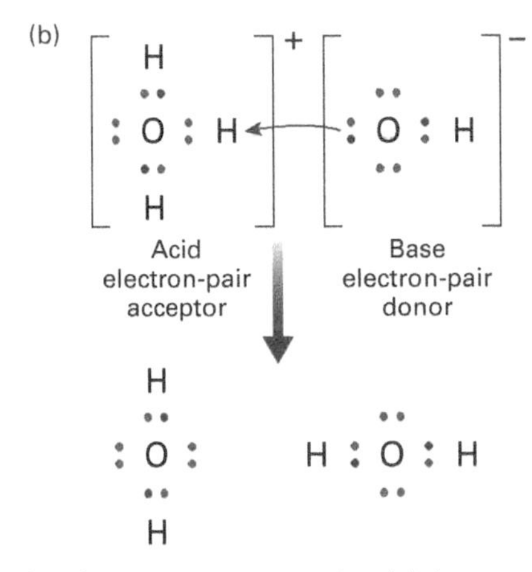

Lewis structures comparing (a) the Brønsted–Lowry and (b) the Lewis theories applied to the H_3O^+/OH^- acid–base neutralization reaction.

Brønsted–Lowry and Lewis theories

- A **Brønsted acid** is a proton donor.
- A **Brønsted base** is a proton acceptor.
- A **Lewis acid** is an electron-pair acceptor.
- A **Lewis base** is an electron-pair donor.

Hydrochloric acid and sodium hydroxide: a neutralization reaction

In aqueous solution, hydrochloric acid provides the aqueous hydrogen ion $H_3O^+(aq)$ and sodium hydroxide provides the aqueous hydroxide ion $OH^-(aq)$. These two ions react together, the $H_3O^+(aq)$ ion acting as an acid and the $OH^-(aq)$ ion as a base:

$$H_3O^+(aq) + OH^-(aq) \rightarrow 2H_2O(l)$$

As a Brønsted acid, H_3O^+ donates a proton, which is accepted by OH^- acting as a Brønsted base: see diagram (a) on the left.

Lewis theory provides an alternative, equally valid, view of the same reaction. As a *Lewis* base, OH^- donates a pair of electrons, which is accepted by a proton from H_3O^+ acting as a Lewis acid: see diagram (b) on the left.

In both cases, a pair of electrons forms a new bond between the proton and the hydroxide ion. Both of the theories describe the oxonium ion H_3O^+ as an acid and the hydroxide ion OH^- as a base.

The Lewis definition incorporates all Brønsted acids and bases. It is useful because it can also be used to describe reactions in which hydrogen ions are not involved, whilst applying the same terminology to more traditional acids. A good example is the reaction between ammonia and hydrogen chloride. In aqueous solution, these two substances neutralize each other on mixing:

$$NH_3(aq) + HCl(aq) \rightleftharpoons NH_4Cl(aq) + H_2O(l)$$

Ammonia and hydrogen chloride also react as covalent compounds in the gas phase to give white fumes of the ionic compound ammonium chloride:

$$NH_3(g) + HCl(g) \rightarrow NH_4Cl(s) \quad \text{[i.e. } NH_4^+Cl^-(s)]$$

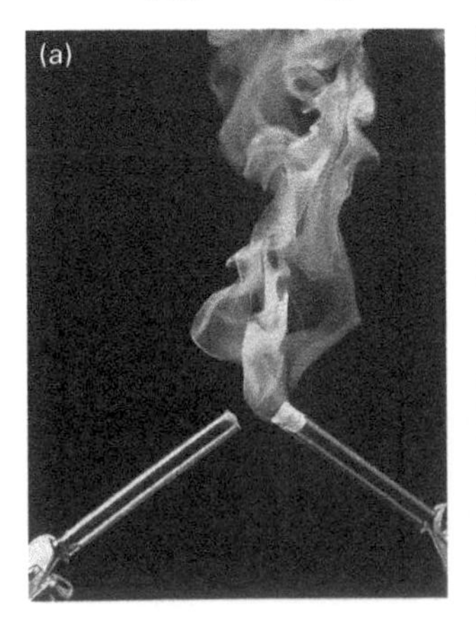

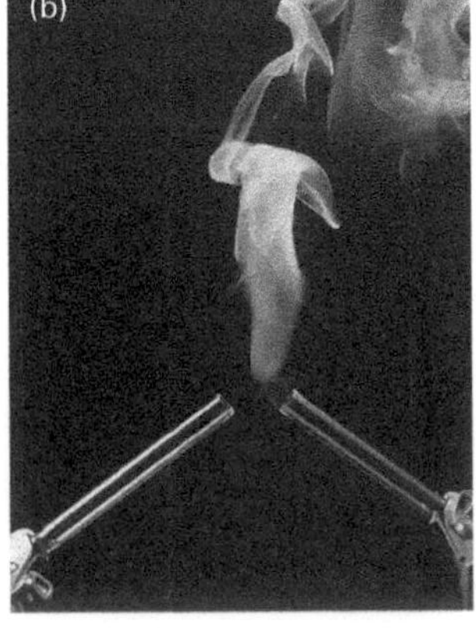

(a) Ammonia gas reacts with hydrogen chloride gas to produce clouds of solid ammonium chloride. The reaction is classified as an acid–base reaction by both Brønsted–Lowry theory and Lewis theory. (b) Ammonia reacts with boron trifluoride in a similar manner to hydrogen chloride. The reaction is classified as an acid–base reaction by Lewis theory only.

Magic Acid

A mixture of fluorosulphonic acid (FSO_2OH) and antimony pentafluoride (SbF_5) uses the combination of a Brønsted acid and a Lewis acid to protonate even weak bases. The name arose when a colleague working with George Olah added the remains of a Christmas candle to the acid, which promptly dissolved. The resulting solution gave a strong NMR signal (spread 32.5) characteristic of the $(CH_3)_3C^+$ carbocation (spread 22.8). His startled reaction 'Magic' stuck!

Brønsted–Lowry theory describes both of these reactions (aqueous and gas phase) as acid–base because they involve proton transfer. Ammonia may also be called a Lewis base because it donates a pair of electrons; hydrogen chloride is a Lewis acid because it accepts a pair of electrons.

The reaction between ammonia and the gas boron trifluoride also forms a solid product, as shown opposite:

$$NH_3(g) + BF_3(g) \rightarrow NH_3BF_3(s)$$

The product of the reaction is called an **adduct** because it is formed by the simple addition of one substance to another. Brønsted–Lowry theory does *not* recognize this reaction as an acid–base reaction because proton transfer does not take place. But from the standpoint of Lewis theory, the ammonia acts as a Lewis base because it donates an electron pair to form a bond. Boron trifluoride is a Lewis acid because it accepts a pair of electrons.

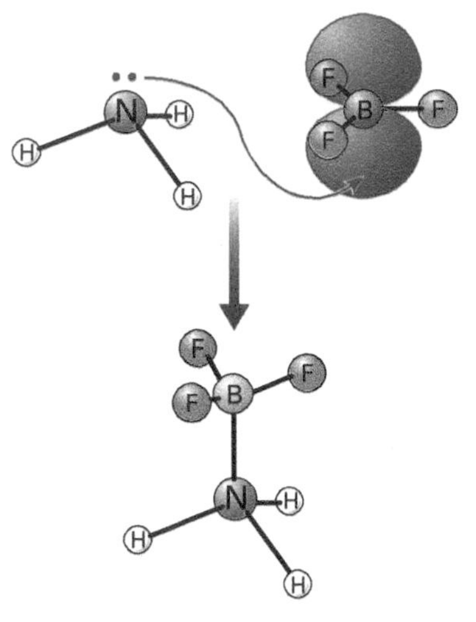

Boron trifluoride has an empty orbital in its valence shell. Ammonia donates a pair of electrons into this orbital to form a coordinate bond.

The formation of complex ions

Copper(II) sulphate dissolves in water to give a clear blue solution. A chemical equation representing this process is:

$$CuSO_4(s) + \text{water} \rightarrow Cu^{2+}(aq) + SO_4^{2-}(aq)$$

$Cu^{2+}(aq)$ is described as 'the aqueous copper(II) ion'. It behaves like all positively charged ions in aqueous solution, attracting the partial negative charge of the oxygen atoms of water molecules. However, the copper ion is small and bears a 2+ charge, with the result that it has a high charge density. The force of attraction is sufficient for oxygen atoms to donate a pair of electrons into vacant orbitals on the copper ion. In this way, the copper ion acts as a Lewis acid (an electron-pair acceptor) and the water molecules act as Lewis bases (electron-pair donors). Each copper ion forms a coordinate bond with each of six water molecules. The aqueous copper(II) ion $Cu^{2+}(aq)$ is therefore more accurately represented by the formula $[Cu(H_2O)_6]^{2+}$.

This structure is called a **complex ion** and is typical of the adducts formed between small, highly charged transition metal ions and substances that have lone pairs of electrons available to form coordinate bonds. Complex ions are Lewis acid–base adducts and are discussed in depth in chapter 20 'The transition metals'.

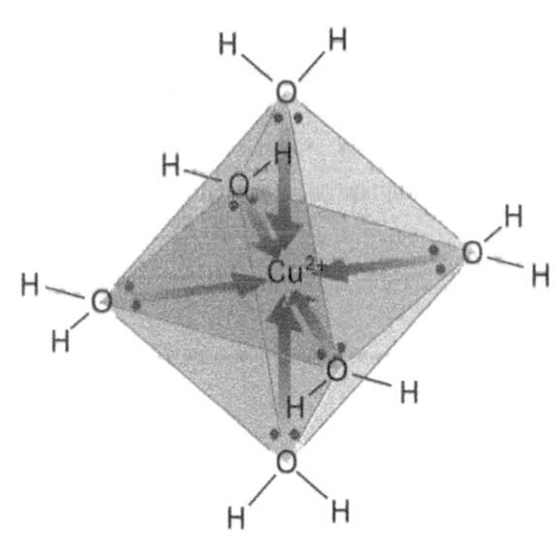

Blue aqueous copper(II) sulphate contains aqueous copper(II) ions $Cu^{2+}(aq)$. These ions are complex ions; their formula is better represented as $[Cu(H_2O)_6]^{2+}$ and their shape is octahedral.

SUMMARY

- A Lewis acid is an electron-pair acceptor.
- A Lewis base is an electron-pair donor.
- All Brønsted acids are Lewis acids; not all Lewis acids are Brønsted acids. (The same remarks apply to bases.)
- Many reactions cannot be described as acid–base reactions by Brønsted–Lowry theory, but may be described as acid–base reactions by Lewis theory. Such reactions include some that do not take place in aqueous solution.
- Complex ions result when Lewis bases form coordinate bonds with metal ions.

PRACTICE

1 Which of these reactions:
$NH_3(aq) + H_3O^+(aq) \rightarrow NH_4^+(aq) + H_2O(l)$
$2NH_3(aq) + Ag^+(aq) \rightleftharpoons [Ag(NH_3)_2]^+(aq)$
may be described as an acid–base reaction according to

a Brønsted–Lowry theory,

b Lewis theory?

Where appropriate, identify the substances acting as acids or bases, giving your reasoning in each case.

2 Liquid ammonia (T_b = –31°C) is slightly ionized in a similar manner to water:

$$2NH_3 \rightleftharpoons NH_4^+ + NH_2^-$$

AMMONIA — AMMONIUM ION — AMIDE ION

Identify the acid, the base, and the salt in the following reaction carried out in liquid ammonia ('am' means in solution in ammonia):
$NH_4Cl(am) + KNH_2(am) \rightarrow KCl(am) + 2NH_3(l)$
Explain your reasoning.

PRACTICE EXAM QUESTIONS

1 Propanoic acid occurs naturally in Swiss cheese in a concentration that can be as high as 1%. Its sodium and calcium salts are food additives used in processed cheeses to retard the formation of moulds.

a Propanoic acid is a weak acid with an acid dissociation constant of 1.22×10^{-5} mol dm^{-3}.

i Write an equation for the ionisation of propanoic acid. [2]

ii Calculate pK_a, for propanoic acid. [2]

b Sodium propanoate is made by the reaction of sodium hydroxide with propanoic acid.

i Write an equation for the reaction. [1]

ii If the reaction were carried out by titration using 0.1M solutions of the hydroxide and the acid, state the name of a suitable indicator. [1]

c A mixture of sodium propanoate and propanoic acid acts as a buffer solution.

i What is meant by the term **buffer solution**? [2]

ii Explain how the mixture of sodium propanoate and propanoic acid acts as a buffer solution.

(Up to 2 marks may be obtained for the quality of language in this part.) [5]

iii Calculate the pH of the solution formed by adding 15.0 cm^3 of 0.1M sodium hydroxide to 30.0 cm^3 of 0.1M propanoic acid. [3]

iv Name a buffer solution found in a biological system and explain its importance. [2]

2 a Explain the terms *acid* and *conjugate base* as used in the Brønsted–Lowry theory of acids and bases. [2]

b For each of the following reactions, give the name or formula of the acid and of its conjugate base.

i $NH_3 + HCl \rightarrow NH_4^+ + Cl^-$ [1]

ii $H_2SO_4 + HNO_3 \rightarrow HSO_4^- + H_2NO_3^+$ [1]

iii $NaNH_2 + NH_4Cl \rightarrow NaCl + 2NH_3$ [1]

c When ethanoic acid, CH_3COOH, dissolves in water, a solution of a weak acid is formed.

i Explain the difference between a *strong acid* and a *weak acid*. [2]

ii Write an equation to show the reaction of ethanoic acid with water. [1]

iii There are two conjugate acid–base pairs involved in the reaction in **c ii**. Copy the formulae, and use the symbols **A1** and **B1** to label one of the acid-base pairs, and the symbols **A2** and **B2** to label the other pair, by writing each symbol above the species to which it refers. [2]

d When dissolved in liquid ammonia, rather than in water, ethanoic acid becomes a strong Brønsted–Lowry acid.

i Write an equation to show the reaction between ethanoic acid and the ammonia solvent. [1]

ii Account for this increase in acid strength. [2]

3 The curves shown in **A**, **B**, and **C** below represent the variation of pH in three different acid–base titrations. In each case, 25.0 cm^3 of a solution **X** (0.1M) is titrated with a solution **Y** (also 0.1M).

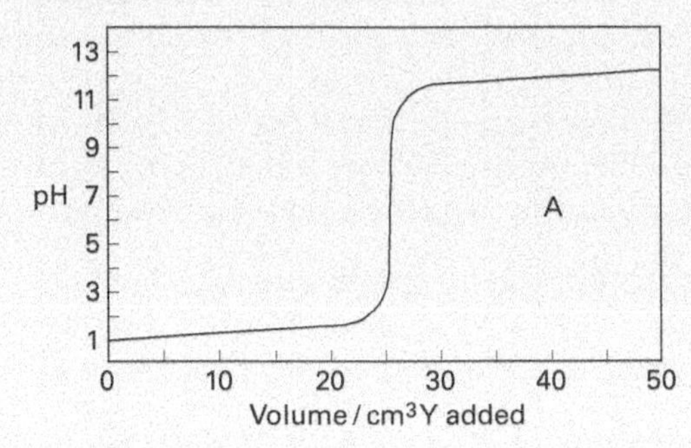

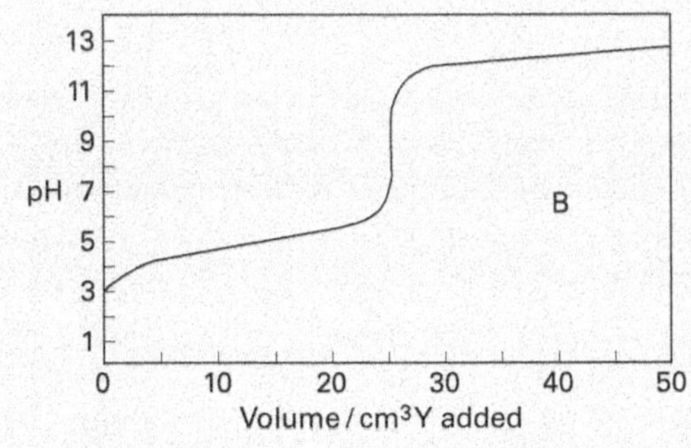

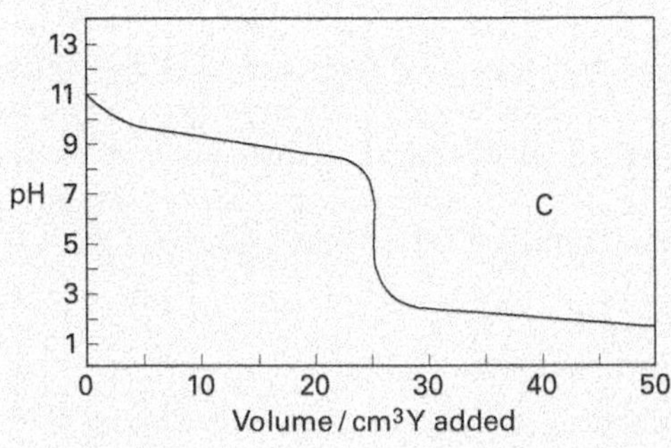

a State two possible components **X** and **Y** that would produce a titration curve similar to the one in **A** above. [2]

b Copy and complete the table below, by identifying the nature (i.e. strong acid/strong base/weak acid/weak base) of component **X** and component **Y** in curves **B** and **C** above. In **each** case give a reason to support your choice of the nature of both **X** and **Y**. [8]

Curve B	
Component X	**Component Y**
Nature of X Reason	Nature of Y Reason

Curve C	
Component X	**Component Y**
Nature of X Reason	Nature of Y Reason

c The pH of a 0.1M solution of HA is 2.30. Calculate the value of K_a for this acid. [3]

4 a i Aqueous ammonia is said to be a ***weak*** base. Explain what is meant by the term ***weak***. [1]

ii The expression for K_b for aqueous ammonia, $NH_3(aq)$, may be written as

$$K_b = \frac{[OH^-(aq)][NH_4^+(aq)]}{[NH_3(aq)]}$$

Write an expression for K_a for the aqueous ammonium ion, $NH_4^+(aq)$, given the equation

$NH_4^+(aq) \rightleftharpoons NH_3(aq) + H^+(aq)$. [1]

iii Hence write down an expression for the product $K_b \times K_a$. [1]

b i Write down an expression for the ionic product of water, K_w, and give its units. [2]

ii I. At 313 K, the numerical value of K_w is 2.91×10^{-14}. Calculate the pH of pure water at this temperature. [2]

II. The numerical value of K_w at 298 K is 1.01×10^{-14}.

Using Le Chatelier's principle, and the value of K_w at 313 K given in **b ii I.**, deduce the sign of $\Delta H^{\ominus}$ for the ionic dissociation of water. [2]

c In the reaction

$H_2O + H_2O \rightleftharpoons H_3O^+ + OH^-$

explain the role of the water molecules in terms of the Brønsted–Lowry theory of acids and bases. [2]

5 a Explain how an aqueous solution containing ethanoic acid and sodium ethanoate resists changes in pH when contaminated with small amounts of acid or alkali. [2]

b It can be shown that for solutions described in **a**

$$pH = pK_a + \log\frac{[\text{sodium ethanoate}]}{[\text{ethanoic acid}]}$$

where $pK_a = -\log K_a$.

i State the significance of the value of the pH when [sodium ethanoate] = [ethanoic acid]. [1]

ii Calculate the mass of sodium ethanoate which must be dissolved in 1 dm^3 of ethanoic acid of concentration 0.10 mol dm^{-3} to produce a solution with a pH value of 5.5 at 298 K. It can be assumed that there is no volume change on dissolving the salt.

[At 298 K the value of K_a for ethanoic acid is 1.8×10^{-5} mol dm^{-3}.] [3]

c i Name an indicator which would be suitable for the titration of aqueous ethanoic acid with aqueous sodium hydroxide, giving a reason for your choice. [1]

ii State whether the pH of an aqueous solution of pure ammonium chloride would be greater or less than 7 at 298 K, giving a reason for your answer. [1]

d The pH of human blood plasma is maintained at a value between 7.39 and 7.41 by the buffering action of dissolved carbon dioxide, hydrogencarbonate ions, and carbonic acid, H_2CO_3. Write equations to show what reactions may occur when the blood absorbs small amounts of acid or small amounts of alkali.

i Acid [1]

ii Alkali [1]

6 a The acid dissociation constant, K_a, of propanoic acid at room temperature is 1.35×10^{-5} mol dm^{-3}.

Calculate the pH of each of the following aqueous solutions at room temperature:

i Propanoic acid of concentration 0.05 mol dm^{-3}. [2]

ii A solution containing 0.05 mol dm^{-3} propanoic acid and 0.05 mol dm^{-3} sodium propanoate. [2]

b The experimentally determined graphs below show the changes in pH when aqueous sodium hydroxide of concentration 0.1 mol dm^{-3} is added **separately** to 25 cm^3 of aqueous hydrochloric acid and 25 cm^3 of aqueous propanoic acid both of concentration 0.1 mol dm^{-3}. The ranges of two indicators are also shown.

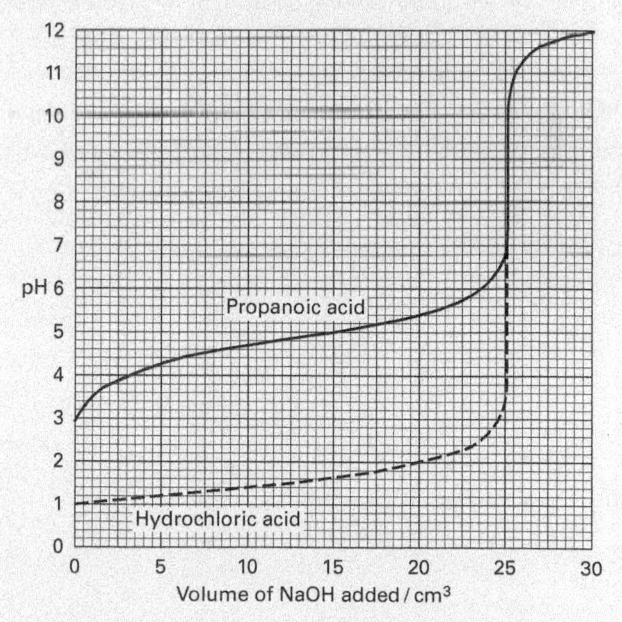

The pH ranges of two common indicators are

methyl orange 3.2 to 4.4

phenolphtalein 8.2 to 10

i State, giving a reason, which indicator or indicators would be suitable for the titration of each acid with aqueous sodium hydroxide. [2]

ii What is the value of the pH when the propanoic acid is half-neutralized? [1]

iii What is the significance of this value? [1]

13 Redox equilibrium

Redox reactions are reactions that involve the transfer of electrons between chemical species. Electron transfer from one place to another is a flow of electrons, and a flow of electrons is an electric current. The study of redox reactions is therefore the combined study of chemistry and electricity. In fact, an alternative title for this chapter could well be 'Electrochemistry'. We are all aware of a practical application of electrochemistry: the use of redox reactions inside batteries to generate an electric potential difference (voltage). A theoretical application uses knowledge about redox equilibria to predict whether chosen substances will take part in a redox reaction.

REDOX: OXIDATION AND REDUCTION

OBJECTIVES

- Oxygen and redox reactions
- Electron transfer and redox reactions
- Oxidants and reductants
- Half-equations

Magnesium Mg(s) burns in oxygen O_2(g) to form magnesium oxide MgO(s). Oxygen oxidizes magnesium; magnesium reduces oxygen. Magnesium loses electrons; oxygen gains electrons.

The term 'redox' is a contraction of the words *reduction* and *oxidation*. Oxidation and reduction take place together. One substance is reduced while the other is oxidized. In this introductory spread we look at an example of such a reaction and find the underlying process to be electron transfer.

The oxidation of magnesium

An example of a redox reaction is magnesium burning in oxygen to produce magnesium oxide:

$$2Mg(s) + O_2(g) \rightarrow 2MgO(s)$$

Magnesium is a metal, oxygen is a diatomic covalent gas, and magnesium oxide is an ionic compound containing Mg^{2+} and O^{2-} ions. A simple definition of redox reactions considers oxidation as the gain of oxygen atoms and reduction as the loss of oxygen atoms. Applying this definition to the burning of magnesium:

- magnesium is oxidized by oxygen;
- oxygen is reduced by magnesium.

The balanced chemical equation written above does not show electron transfer. Splitting the equation into two *half-equations* (below) reveals more clearly what is happening in the course of the reaction. One half-equation shows the loss of electrons and the other shows the gain of electrons.

- Magnesium loses electrons:

  $$2Mg(s) \rightarrow 2Mg^{2+}(s) + 4e^-$$

- Oxygen gains electrons:

  $$O_2(g) + 4e^- \rightarrow 2O^{2-}(s)$$

We can consider this reaction in terms of the underlying electron transfer:

- **Oxidation** is the loss of electrons (magnesium is oxidized).
- **Reduction** is the gain of electrons (oxygen is reduced).
- Oxidation is caused by the **oxidant** (or **oxidizing agent**): the oxidant (oxygen O_2) is itself reduced in the process (to O^{2-}).
- Reduction is caused by the **reductant** (or **reducing agent**): the reductant (magnesium Mg) is itself oxidized in the process (to Mg^{2+}).

The half-reaction described by the half-equation does *not necessarily* actually occur; the electrons may not become free in the way suggested. Half-equations are simply a useful way of thinking about the two component parts of a redox reaction. A half-equation must balance with respect to *charge* as well as with respect to the number of atoms. Notice how the sum of the charges on the left in the above equations equals the sum of the charges on the right.

Redox in the absence of oxygen

The example given above shows that the oxidation of magnesium may be interpreted in two ways: either as a gain of oxygen or as a loss of electrons. The idea of redox as electron transfer allows many reactions not involving oxygen to be classified as redox reactions. For example, the reaction between aqueous copper(II) sulphate and zinc metal is a redox reaction:

$$Zn(s) + CuSO_4(aq) \rightarrow ZnSO_4(aq) + Cu(s)$$

Remember that the formulae $CuSO_4(aq)$ and $ZnSO_4(aq)$ refer to ionic substances dissolved in water. $CuSO_4(aq)$ is therefore present as $Cu^{2+}(aq)$ and $SO_4^{2-}(aq)$ ions; $ZnSO_4(aq)$ is present as $Zn^{2+}(aq)$ and $SO_4^{2-}(aq)$ ions. The sulphate ion $SO_4^{2-}(aq)$ is a *spectator ion*; it does not take part in the chemical reaction and so we may disregard it.

The full balanced chemical equation is not very informative. However, the two half-equations reveal the redox nature of the reaction.

- The half-equation for zinc shows that the metal atoms are oxidized as the result of electron loss:

$$Zn(s) \rightarrow Zn^{2+}(aq) + 2e^-$$

- The half-equation for copper shows that the aqueous copper(II) ions are reduced as the result of electron gain:

$$Cu^{2+}(aq) + 2e^- \rightarrow Cu(s)$$

The balanced equation for the reduction of copper(II) ions by zinc results from adding together the two half-equations. In this example the combination is straightforward because both half-equations involve two electrons (note that we have not included the sulphate spectator ions):

$$Zn(s) + Cu^{2+}(aq) \rightarrow Zn^{2+}(aq) + Cu(s)$$

When a zinc strip is immersed in blue aqueous copper(II) sulphate, the result is copper metal and colourless aqueous zinc sulphate.

The outside of the Statue of Liberty is made from thin sheets of copper. Oxygen has oxidized the surface to copper(II) oxide. Further reaction with water and carbon dioxide has formed a blue-green coating of basic copper(II) carbonate of approximate formula $CuCO_3 \cdot Cu(OH)_2$.

OIL RIG

A 'mnemonic' for remembering the rule about redox and electron transfer is OIL RIG:

Oxidation	**R**eduction
Is	**I**s
Loss	**G**ain
(of electrons)	(of electrons)

Oxidants and reductants

The terms 'oxidant' and 'oxidizing agent' have the same meaning. Similarly 'reductant' and 'reducing agent' mean the same. In this text we will use the more modern terms 'oxidant' and 'reductant'.

- An **oxidant** is a substance that oxidizes another species by removing electrons from it.
- A **reductant** is a substance that reduces another species by donating electrons to it.

SUMMARY

- An oxidation reaction is always accompanied by a corresponding reduction reaction.
- Oxidation is loss of electrons.
- Reduction is gain of electrons.
- A redox reaction may be expressed by two half-equations, one representing oxidation and the other reduction.

PRACTICE

1 For each of the following redox reactions:

  a $2Ca(s) + O_2(g) \rightarrow 2CaO(s)$

  b $CuO(s) + H_2(g) \rightarrow Cu(s) + H_2O(g)$

  c $2AgNO_3(aq) + Cu(s) \rightarrow 2Ag(s) + Cu(NO_3)_2(aq)$

  identify (i) the oxidant, (ii) the reductant, (iii) which species is oxidized, and (iv) which species is reduced.

2 For each of the reactions in question 1, write half-equations to show (i) electron loss during oxidation and (ii) electron gain during reduction.

13.2

OXIDATION NUMBERS – 1

OBJECTIVES

- Oxidation number and oxidation state
- Rules for assigning oxidation number
- Identifying redox reactions

Chlorine oxidizes iron (and iron reduces chlorine) when they combine to form iron(III) chloride $FeCl_3$:

$2Fe(s) + 3Cl_2(g) \rightarrow 2FeCl_3(s)$

$2Fe \rightarrow 2Fe^{3+} + 6e^-$

$3Cl_2 + 6e^- \rightarrow 6Cl^-$

The compound $FeCl_3$ has a high degree of covalent character; however, we can still treat it as ionic for the purposes of determining oxidation numbers.

Fluorine and oxygen

Oxygen usually has an oxidation number of –2. However, in the compound OF_2 (oxygen difluoride), oxygen has an oxidation number of +2. Fluorine is the only element that is more electronegative than oxygen. So fluorine, rather than oxygen, gains the electrons. Oxygen loses electrons, i.e. is oxidized. Its oxidation number changes from 0 (in O_2) to +2 (in OF_2).

Redox reactions are concerned with electron transfer. Loss and gain of electrons during these reactions reflect changes in the way the elements concerned bond with each other. Up to now you have probably described the bonding ability of elements in terms of their valencies. The valency of an element is determined by the number of electrons it controls. A more precise term for valency is *oxidation number*. Elements gain and lose electrons during redox reactions, with the result that their oxidation numbers change. Oxidation numbers indicate where the electrons are, and so help us with chemical book-keeping. Oxidation numbers are assigned to elements on the basis of how readily they react in redox reactions.

Electron transfer and oxidation number

Each element in any chemical species may be assigned an oxidation number. The **oxidation number** of an element is the number of electrons that need to be added to the element to make a neutral atom. For example, the iron ion Fe^{2+} requires the addition of two electrons to make a neutral atom. The oxidation number for iron in the Fe^{2+} ion is therefore +2. The **oxidation state** of this ion is written as iron(II) or as Fe(II); it is expressed in Roman numerals and describes the extent of oxidation of the species. The Fe^{3+} ion has an oxidation number of +3. The oxidation state of this ion is written as iron(III) or as Fe(III), indicating that it is a more highly oxidized species than Fe^{2+}.

When an iron(II) ion (Fe^{2+}) reacts to form an iron(III) ion (Fe^{3+}), iron loses one electron and is therefore oxidized:

$$Fe^{2+}(aq) \rightarrow Fe^{3+}(aq) + e^-$$

When it is oxidized, the oxidation number change for iron is from +2 to +3; the oxidation state change for iron is from Fe(II) to Fe(III).

- Oxidation involves an *increase* in oxidation number.

In the reverse process, *reduction*, an iron(III) ion gains one electron to become an iron(II) ion:

$$Fe^{3+}(aq) + e^- \rightarrow Fe^{2+}(aq)$$

When it is reduced, the oxidation number change for iron is from +3 to +2; the oxidation state change for iron is from Fe(III) to Fe(II).

- Reduction involves a *decrease* in oxidation number.

Rules for assigning oxidation numbers.

Rule no.	Rule	Examples for each rule
1	The sum of the oxidation numbers of all the atoms in any species must equal the charge on the species. For an atom or a molecule, the sum is zero. For an ion, the sum is the charge on the ion.	Nitrate ion NO_3^-: the overall charge on the ion is –1, so the sum of the oxidation numbers must be –1. The oxidation number total for the oxygen atoms (see rule 4) is 3 × (–2) = –6. So the oxidation number of nitrogen is +5.
2	The oxidation number of an element in its elemental state is zero. So the oxidation number of magnesium in Mg(s) is zero, as is the oxidation number of oxygen in O_2(g).	Metallic iron: oxidation number is 0. Nitrogen gas N_2: oxidation number is 0.
3	The oxidation number of fluorine in its compounds is always –1. Fluorine is the most electronegative element and always gains electrons on bonding.	Sodium fluoride NaF: oxidation number of sodium is +1 and that of fluorine is –1.
4	The oxidation number of oxygen in its compounds is usually –2, except if fluorine is present (or in certain unusual compounds such as peroxides or superoxides).	Carbon dioxide CO_2: oxidation number of oxygen is –2 and that of carbon is +4. Water H_2O: oxidation number of oxygen is –2 and that of hydrogen is +1.
5	Each shared electron pair is assigned to the more electronegative element.	Hydrogen chloride HCl: chlorine is more electronegative than hydrogen; oxidation number of hydrogen is +1 and that of chlorine is –1.

Identifying redox reactions

How can you tell whether a reaction is a redox reaction? The strategy is to work out the oxidation number for each element in the reaction. If none of the elements changes its oxidation number, the reaction *is not* a redox reaction.

- If there is any change in oxidation number, this indicates that the reaction is a redox reaction.

  We will now apply this criterion to some reactions.

$$Fe(s) + S(s) \rightarrow FeS(s) \quad (1)$$

The reactants are elements and therefore both have oxidation number 0. FeS is the compound iron(II) sulphide: the electron pairs are assigned to sulphur, which has the higher electronegativity value. The oxidation number of iron is +2 and that of sulphur is –2. Reaction (1) is a redox reaction.

$$NaOH(aq) + HCl(aq) \rightarrow NaCl(aq) + H_2O(l) \quad (2)$$

There are no oxidation number changes, i.e. Na remains as +1, H as +1, Cl as –1, and O as –2. Reaction (2) is not a redox reaction; it is an acid–base reaction.

$$CuO(s) + H_2(g) \rightarrow Cu(s) + H_2O(g) \quad (3)$$

The oxidation number of copper is +2 in copper(II) oxide CuO and 0 in copper Cu, indicating reduction. The oxidation number of hydrogen increases from 0 in the element to +1 in water. Reaction (3) is a redox reaction.

$$MnO_2(s) + 4HCl(aq) \rightarrow MnCl_2(aq) + Cl_2(g) + 2H_2O(l) \quad (4)$$

Manganese is reduced from oxidation number +4 in MnO_2 to +2 in $MnCl_2$. Chlorine is oxidized from oxidation number –1 in HCl to 0 in the element Cl_2. The oxidation number of oxygen does not change. Reaction (4) is a redox reaction.

$$2H_2O_2(aq) \rightarrow 2H_2O(l) + O_2(g) \quad (5)$$

The oxidation number of oxygen in peroxides is –1, in water it is –2, and in the element 0. The oxidation number of hydrogen does not change. Oxygen is both oxidized and reduced in this reaction. Reaction (5) is a redox reaction.

$$Mg(s) + 2HCl(aq) \rightarrow MgCl_2(aq) + H_2(g) \quad (6)$$

The oxidation number of magnesium increases from 0 in the element to +2 in $MgCl_2$: magnesium is oxidized. The oxidation number of hydrogen decreases from +1 in HCl to 0 in the element: hydrogen is reduced. The oxidation number of chlorine remains at –1. Reaction (6) is a *redox* reaction of acids, unlike the acid-base reaction (2) above.

SUMMARY

- The oxidation number of an element is the number of electrons that need to be added to the element to make a neutral atom.
- A reaction is classed as a redox reaction if it involves a change in the oxidation number of any element taking part.
- Oxidation involves an increase in oxidation number.
- Reduction involves a decrease in oxidation number.

PRACTICE

1 In the reaction alongside, chlorine is produced by reaction of concentrated hydrochloric acid and potassium manganate(VII).

Is the change chloride ion to chlorine gas an oxidation or a reduction?

Expressing oxidation numbers and oxidation states

The name of the compound of formula $CuSO_4$ is usually given as copper(II) sulphate (pronounced 'copper-two sulphate'). In this context, it is correct to say that:

- The *oxidation number* of copper is +2.
- The *oxidation state* of copper is written as copper(II) or Cu(II).
- $CuSO_4$ is the chemical formula of copper(II) sulphate.

Comment on redox reactions

Look at reaction (1) on the left. All reactions between two elements are *necessarily* redox reactions. The oxidation number of an uncombined element is always zero; the oxidation number of an element in a compound is never zero. So there is a change in oxidation number, and such reactions are redox reactions.

Look at reaction (5) on the left. Oxygen in the peroxide ion O_2^{2-} has an unusual oxidation number (–1), because two oxygen atoms bond to form the ion $^-O{-}O^-$. In the reaction, the oxidation number of oxygen both *increases* (to form O_2) and *decreases* (to form H_2O). This is a redox reaction in which the *same* element is both oxidized *and* reduced. Such a reaction is also called a **disproportionation** reaction.

Gaseous chlorine oxidizes colourless aqueous bromide ions to red-brown bromine. This is an example of a redox reaction involving two non-metals:
$Cl_2(g) + 2Br^-(aq) \rightarrow 2Cl^-(aq) + Br_2(l)$

13.3

OXIDATION NUMBERS – 2

OBJECTIVES

- Systematic naming of compounds
- Traditional and systematic names

A redox titration. (a) Potassium manganate(VII) acts as its own indicator during titration with aqueous iron(II) ions. Purple $MnO_4^-(aq)$ ions are reduced to pale pink $Mn^{2+}(aq)$ ions, while pale green $Fe^{2+}(aq)$ ions are oxidized to pale yellow $Fe^{3+}(aq)$ ions. The oxidation state changes for manganese and iron are:

Mn(VII) to Mn(II) (reduction)
Fe(II) to Fe(III) (oxidation)

(b) The purple colour of $MnO_4^-(aq)$ can be seen when all the $Fe^{2+}(aq)$ has reacted.

The traditional names of many chemical substances date back to the nineteeth century. Many of these names are still in common use, especially those for acids (e.g. sulphuric acid H_2SO_4; sulphurous acid H_2SO_3) and their salts (e.g. sodium nitrate $NaNO_3$; potassium nitrite KNO_2). According to traditional nomenclature, an acid with the suffix '-ic' contains a larger amount of oxygen than one with the suffix '-ous'. (Look at the formulae of sulphuric and sulphurous acids.) In salts, the suffixes are '-ate' and '-ite', respectively representing greater and lesser amounts of oxygen. (Look at the formulae of sodium nitrate and potassium nitrite.) The modern systematic naming system uses the concept of oxidation number to help make the naming of chemical substances unambiguous.

Systematic names

Oxidation numbers can be used to name compounds unambiguously. No confusion is possible when an element exhibits only one common oxidation state. For example, Group I elements always have an oxidation number of +1 in their compounds. Sodium is present in the ionic compound sodium oxide as Na(I). From rule 4 in the previous spread, the oxidation number of oxygen is –2. Because oxygen is present as O(–II), the formula must therefore be Na_2O. The oxygen is present in the compound as the ion O^{2-}. The name 'sodium oxide' is adequate to describe this substance.

On the other hand, 'iron oxide' presents a problem because there are two possible oxidation states of iron, iron(II) and iron(III). Oxygen is present as O(–II). The compound FeO is therefore 'iron(II) oxide' and Fe_2O_3 is 'iron(III) oxide'. Similarly, MnO_2 is 'manganese(IV) oxide'.

When two non-metals are combined, the rules for naming are very different. The common name of the compound simply describes the number of atoms present. Thus CO_2 is 'carbon dioxide', CO is 'carbon monoxide', SO_2 is 'sulphur dioxide', SO_3 is 'sulphur trioxide', and N_2O_4 is 'dinitrogen tetraoxide'. This form of nomenclature is only used for simple covalent compounds containing non-metals.

Oxoanions and $KMnO_4$

Oxoanions are negatively charged ions that contain elements combined with oxygen. A common example is the ion MnO_4^- present in the substance $KMnO_4$, traditionally called 'potassium permanganate'. A solution contains the ions $K^+(aq)$ and $MnO_4^-(aq)$. The naming of MnO_4^- needs careful thought. You must first assign an oxidation number of –2 to oxygen. Now recognize that there are four oxygen atoms in the anion, and that the anion has an overall charge of –1. As a result, manganese is assigned an oxidation number of +7. The oxoanion MnO_4^- is therefore described as the manganate(VII) ion; $KMnO_4$ is called 'potassium manganate(VII)'. Note that an oxidation number of +7 does *not* imply that MnO_4^- contains the ion Mn^{7+}.

Traditional and systematic names

The following table will help you to move freely between the two naming systems. Only the *systematic name* should be used for acids and oxoanions containing metals. The *traditional names* are suitable where there is no chance of ambiguity. For example, the traditional name 'sulphuric acid' is less cumbersome for the substance H_2SO_4. Using this traditional name for the acid matches the use of the name 'sulphur trioxide' for SO_3 rather than the systematic name 'sulphur(VI) oxide'.

*Acids and oxoanions: traditional and systematic names (the names in **bold** are the ones that we shall use).*

Traditional name	Formula	Systematic name	
sulphuric acid	H_2SO_4	sulphuric(VI) acid	↑
sulphate ion	SO_4^{2-}	sulphate(VI) ion	
sulphurous acid	H_2SO_3	sulphuric(IV) acid	
sulphite ion	SO_3^{2-}	sulphate(IV) ion	
nitric acid	HNO_3	nitric(V) acid	
nitrate ion	NO_3^-	nitrate(V) ion	
nitrous acid	HNO_2	nitric(III) acid	
nitrite ion	NO_2^-	nitrate(III) ion	non-metals
permanganate ion	MnO_4^-	**manganate(VII) ion**	metals
chromate ion	CrO_4^{2-}	**chromate(VI) ion**	
dichromate ion	$Cr_2O_7^{2-}$	**dichromate(VI) ion**	↓

Orange dichromate(VI) ions poured into aqueous iron(II) ions result in a green solution. The ionic equation for the reaction (see next spread)

$$Cr_2O_7^{2-}(aq) + 6Fe^{2+}(aq) + 14H^+(aq) \rightarrow 2Cr^{3+}(aq) + 6Fe^{3+}(aq) + 7H_2O(l)$$

shows that dichromate(VI) ions are reduced to chromium(III) ions by iron(II) ions in acidified aqueous solution; the iron(II) ions are oxidized to iron(III).

SUMMARY

- Systematic names of substances include information about oxidation states.

Naming chlorine oxoanions

Chlorine forms four oxoanions. The systematic names for the ions ClO_4^-, ClO_3^-, ClO_2^-, and ClO^- are respectively chlorate(VII), chlorate(V), chlorate(III), and chlorate(I). The traditional names are perchlorate, chlorate, chlorite, and hypochlorite. See spread 10.1.

No ambiguity

The names actually only become unambiguous when the number of oxygen atoms is also specified. So the fully systematic name for MnO_4^- is tetraoxomanganate(VII) ion.

High oxidation states

High oxidation states are normally found either in oxides or oxoanions, such as dichromate(VI) or manganate(VII) ions, or in fluorides. Examples of fluorides with no corresponding chloride include platinum(VI) fluoride PtF_6 and xenon tetrafluoride XeF_4; see photographs of both in spread 19.12.

Reminder about oxidation number changes

- If the oxidation number of an element **increases** during a reaction, the element has been **oxidized**.
- If the oxidation number of an element **decreases** during a reaction, the element has been **reduced**.

PRACTICE

1 Give the oxidation numbers of each of the elements in the following species:
 a H_2O
 b K_2O
 c $AlCl_3$
 d NO_3^-
 e OF_2.

2 State whether each of the following may be classed as a redox reaction or not. Explain your reasoning by writing and discussing the full balanced chemical equations and half-equations:
 a Carbon dioxide gas being evolved from a mixture of calcium carbonate and dilute hydrochloric acid.
 b Chlorine gas being evolved from a heated mixture of manganese(IV) oxide and concentrated hydrochloric acid.

3 Name the following species:
 a H_2SO_4
 b H_2SO_3
 c NaClO
 d ClO_3^-
 e MnO_2
 f CuO.

13.4 Redox reactions and titrations

OBJECTIVES

- Manganate(VII) ion as oxidant
- Dichromate(VI) ion as oxidant
- Thiosulphate ion as reductant
- Calculations for a redox titration

In this dichromate(VI)/iron(II) titration, the intense orange colour of $Cr_2O_7^{2-}(aq)$ ions changes to green as they are reduced to $Cr^{3+}(aq)$.

Manganate(VII) ions and ethanedioate ions

Titrations between manganate(VII) ions and ethanedioate (oxalate) ions ($C_2O_4^{2-}$) show autocatalysis, as explained in spread 15.9.

Starch as an indicator

None of the ions present in the iodine/thiosulphate reaction is highly coloured. Starch, which forms a deep blue complex with iodine, can be used as an indicator to show a clear end point for the reaction. A small quantity of starch solution is added just before the end point when the iodine solution is a pale straw colour, i.e. when most of the iodine has reacted. The end point is when the intense blue colour of the starch–iodine complex finally disappears.

The previous chapter introduced you to the idea of using the method of titration to measure the concentrations of acids and bases in acid–base reactions. A **redox titration** uses the same technique but is applied to the reactants in a redox reaction. There are three reagents most commonly used in redox titrations: manganate(VII) ion, dichromate(VI) ion, and thiosulphate ion.

The manganate(VII) ion as an oxidant

Potassium manganate(VII) $KMnO_4$ can act as an oxidant in acidic solution. The half-equation for the reduction of manganate(VII) ion under these conditions is:

$$MnO_4^-(aq) + 8H^+(aq) + 5e^- \rightarrow Mn^{2+}(aq) + 4H_2O(l) \qquad (1)$$

Manganate(VII) ion is commonly used to determine the concentration of aqueous iron(II) ions $Fe^{2+}(aq)$, which it oxidizes to $Fe^{3+}(aq)$: see previous spread. The half-equation for the oxidation of iron(II) ions is:

$$Fe^{2+}(aq) \rightarrow Fe^{3+}(aq) + e^- \qquad (2)$$

This reaction supplies only one electron so the half-equation (2) must be multiplied by *five* before adding to the manganate(VII) half-equation (1). Doing this, we obtain the ionic equation for the redox reaction:

$$MnO_4^-(aq) + 5Fe^{2+}(aq) + 8H^+(aq) \rightarrow Mn^{2+}(aq) + 5Fe^{3+}(aq) + 4H_2O(l)$$

i.e. Mn(VII) is reduced to Mn(II); Fe(II) is oxidized to Fe(III).

- For titration calculations, the mole ratio is 5 mol Fe^{2+} to 1 mol MnO_4^-.

The dichromate(VI) ion as an oxidant

Potassium dichromate(VI) $K_2Cr_2O_7$ can act as an oxidant in acidic solution. The half-equation for the reduction of dichromate(VI) ion is:

$$Cr_2O_7^{2-}(aq) + 14H^+(aq) + 6e^- \rightarrow 2Cr^{3+}(aq) + 7H_2O(l) \qquad (3)$$

This reaction provides its own indicator because the colour changes as excess orange $Cr_2O_7^{2-}(aq)$ mixes with green $Cr^{3+}(aq)$ at the equivalence point. As in the example for manganate(VII) given above, iron(II) is a suitable source of electrons. But now we need to multiply the half-equation (2) by *six* before adding it to equation (3). Doing this, we get the ionic equation for the redox reaction:

$$Cr_2O_7^{2-}(aq) + 6Fe^{2+}(aq) + 14H^+(aq) \rightarrow 2Cr^{3+}(aq) + 6Fe^{3+}(aq) + 7H_2O(l)$$

i.e. Cr(VI) is reduced to Cr(III); Fe(II) is oxidized to Fe(III).

- For titration calculations, the mole ratio is 6 mol Fe^{2+} to 1 mol $Cr_2O_7^{2-}$.

The thiosulphate ion as a reductant

Sodium thiosulphate $Na_2S_2O_3$ can act as a reductant in acidic solution. The half-equation for the oxidation of thiosulphate ion $S_2O_3^{2-}$ to tetrathionate ion $S_4O_6^{2-}$ is:

$$2S_2O_3^{2-}(aq) \rightarrow S_4O_6^{2-}(aq) + 2e^- \qquad (4)$$

Iodine is a suitable oxidant to accept the electrons, itself being reduced to iodide ions:

$$I_2(aq) + 2e^- \rightarrow 2I^-(aq) \qquad (5)$$

We obtain the overall ionic equation for the redox reaction between iodine and thiosulphate ions by adding together half-equations (4) and (5):

$$I_2(aq) + 2S_2O_3^{2-}(aq) \rightarrow 2I^-(aq) + S_4O_6^{2-}(aq)$$

i.e. I(0) is reduced to I(–I); S(+2) is oxidized to S(+2.5). [You should note that the oxidation number of sulphur in tetrathionate is +2.5. This number cannot be written in Roman numerals, so the change is expressed here in terms of oxidation numbers, not oxidation states.]

- For titration calculations, the mole ratio is 1 mol I_2 to 2 mol $S_2O_3^{2-}$.

Worked example on a redox titration calculation

Data: Potassium iodate KIO_3 oxidizes potassium iodide KI to form iodine I_2 in acidic solution. In a laboratory reaction, 20.0 cm³ of aqueous potassium iodate containing 3.00 g KIO_3 per dm³ reacted with excess potassium iodide. The liberated iodine required 16.8 cm³ of 0.100 mol dm⁻³ aqueous sodium thiosulphate for complete reaction.

Question: Calculate the mole ratio between iodine and iodate ion and hence suggest an ionic equation for the reaction.

Answer:

Step 1 Calculate the amount in moles of thiosulphate ion used in the titration.

From earlier chapters we know that

amount in moles = volume of solution × molar concentration

So substituting the quantities given above for the thiosulphate ion $S_2O_3^{2-}$:

$$\text{amount in moles of } S_2O_3^{2-} = \left(\frac{16.8}{1000}\,\text{dm}^3\right) \times (0.100\,\text{mol dm}^{-3})$$
$$= 1.68 \times 10^{-3}\,\text{mol}$$

Step 2 Calculate the amount in moles of iodine liberated, using the ionic equation:

$I_2(aq) + 2S_2O_3^{2-}(aq) \rightarrow 2I^-(aq) + S_4O_6^{2-}(aq)$

The equation shows that 1 mol I_2 reacts with 2 mol $S_2O_3^{2-}$. So

$$\text{amount in moles of } I_2 = \frac{1.68 \times 10^{-3}\,\text{mol}}{2}$$
$$= 8.40 \times 10^{-4}\,\text{mol}$$

Step 3 Calculate the amount in moles of potassium iodate in the original solution.

The aqueous KIO_3 contained 3.00 g dm⁻³. The molar mass of KIO_3 is

$39.1\,\text{g mol}^{-1} + 126.9\,\text{g mol}^{-1} + 3 \times (16.0\,\text{g mol}^{-1}) = 214\,\text{g mol}^{-1}$

Therefore

$$\text{molar concentration of aqueous } KIO_3 = \frac{3.00\,\text{g dm}^{-3}}{214\,\text{g mol}^{-1}}$$
$$= 0.0140\,\text{mol dm}^{-3}$$

We now use

amount in moles = volume of solution × molar concentration

So substituting the quantities for the potassium iodate:

$$\text{amount in moles of } KIO_3 = \left(\frac{20.0}{1000}\,\text{dm}^3\right) \times (0.0140\,\text{mol dm}^{-3})$$
$$= 2.80 \times 10^{-4}\,\text{mol}$$

Step 4 Calculate the ratio of these two amounts, I_2/IO_3^-, and then convert it to a whole-number ratio.

The amounts in moles of I_2 and IO_3^- found in steps 2 and 3 are 8.40×10^{-4} mol and 2.80×10^{-4} mol respectively. The simplest ratio is

$$\frac{I_2}{IO_3^-} = \frac{8.40 \times 10^{-4}\,\text{mol}}{2.80 \times 10^{-4}\,\text{mol}} = \frac{8.40}{2.80} = \frac{3}{1}$$

Step 5 Suggest an ionic equation for the reaction.

According to this calculation, 3 mol of iodine I_2 results from the reaction of 1 mol of iodate IO_3^- with excess iodide. The equation for the reaction is:

$IO_3^-(aq) + 5I^-(aq) + 6H^+(aq) \rightarrow 3I_2(aq) + 3H_2O(l)$

I(V) is reduced to I(0); I(–I) is oxidized to I(0).

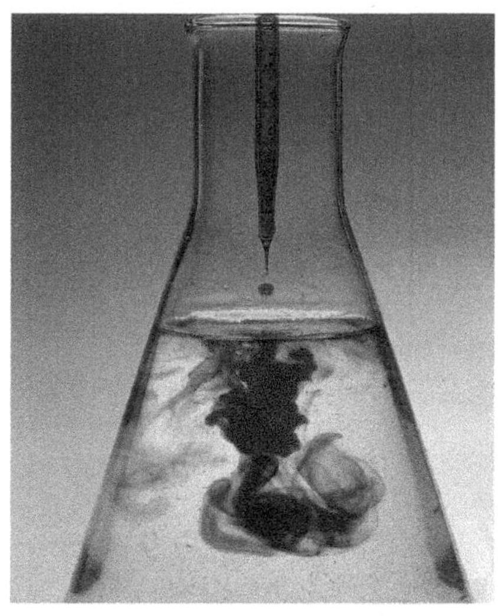

The intense blue starch–iodine complex consists of iodine molecules lying inside long spirals of starch molecules. Thiosulphate ions reduce this complex to colourless iodide ions.

Titrations and equations

Calculations such as this example were originally used to work out chemical equations. All required careful experimental evaluation of the amounts of one chemical needed to react with another. Although chemical reactions are rarely found in this manner from first principles, it is useful to see how one was actually determined.

SUMMARY

- $MnO_4^-(aq) + 8H^+(aq) + 5e^- \rightarrow Mn^{2+}(aq) + 4H_2O(l)$
- $Cr_2O_7^{2-}(aq) + 14H^+(aq) + 6e^- \rightarrow 2Cr^{3+}(aq) + 7H_2O(l)$
- $I_2(aq) + 2S_2O_3^{2-}(aq) \rightarrow 2I^-(aq) + S_4O_6^{2-}(aq)$

PRACTICE

1 Give two examples of solutions which can be used as oxidants in redox titrations. Describe the colour changes that occur.

2 Write down the equation for the reaction between iodine and sodium thiosulphate. How may the end point be detected?

13.5

OBJECTIVES

- Non-metal oxidants
- Metal reductants
- Displacement reactions and relative strengths

OXIDANTS AND REDUCTANTS

Several species may be conveniently classified as oxidants or reductants. If they are present in a reaction mixture, there is a strong likelihood that a redox reaction is taking place. You have already met many different examples of redox reactions in this chapter. This spread includes some further examples and introduces the idea of comparing the strengths of oxidants and of reductants.

Oxygen gas as an oxidant

Elemental (molecular) oxygen O_2 usually acts as an oxidant in reactions. It was oxygen that historically gave its name to the process of oxidation. The most obvious examples of redox reactions are those in which oxygen reacts with an element to form its oxide.

Oxygen acts as an oxidant because it has a very high electronegativity value: the rules for assigning oxidation numbers (which we met earlier in this chapter) assign to oxygen any electron pairs shared with *any* other element (except fluorine).

(a) 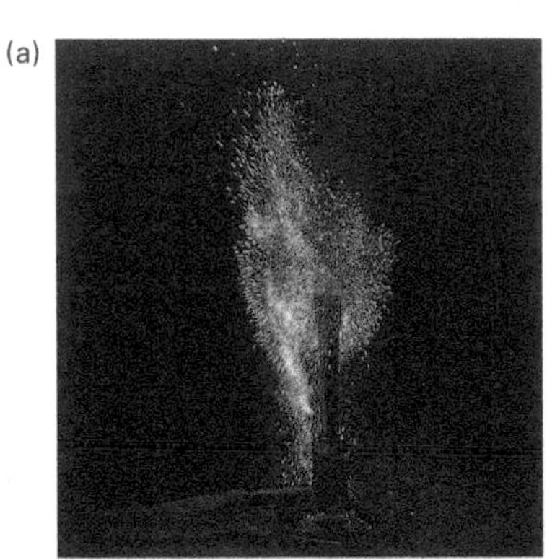

(b) 

(a) Oxidation of a metallic element: iron filings oxidize to a mixture of iron(II) oxide and iron(III) oxide when sprinkled into a non-luminous Bunsen flame:
$3Fe(s) + 2O_2(g) \rightarrow Fe_2O_3{\cdot}FeO(s)$
(b) Oxidation of a non-metallic element: heated molten sulphur burns in oxygen with a blue flame to produce sulphur dioxide gas:
$S(l) + O_2(g) \rightarrow SO_2(g)$

The halogen bromine reacts violently with phosphorus to form phosphorus tribromide PBr_3. Bromine is more electronegative than phosphorus so in forming PBr_3 the phosphorus has been oxidized. A later view of the same reaction is shown in spread 18.1.

Halogens as oxidants

Oxygen acts as an oxidant because it has a high electronegativity value (3.4) and so removes electrons from (oxidizes) most other species. The halogens fluorine and chlorine also have high electronegativity values (fluorine 4.0, chlorine 3.2). Fluorine is an extremely powerful oxidant and chlorine is comparable with oxygen.

Non-metal displacement reactions

Redox reactions may be considered as a competition between two species for control of electrons. For example, chlorine removes electrons from aqueous bromide ions, oxidizing them to bromine (spread 13.2):

$$Cl_2(g) + 2Br^-(aq) \rightarrow 2Cl^-(aq) + Br_2(l)$$

Chlorine is a more powerful oxidant than bromine, under these conditions, so the reverse reaction does not occur to any significant extent.

Gases as reductants

Hydrogen, carbon monoxide, and methane can all act as reductants. For example, each will reduce heated copper(II) oxide to copper:

$$H_2(g) + CuO(s) \rightarrow Cu(s) + H_2O(g)$$

$$CO(g) + CuO(s) \rightarrow Cu(s) + CO_2(g)$$

$$CH_4(g) + 4CuO(s) \rightarrow 4Cu(s) + CO_2(g) + 2H_2O(g)$$

These gases have important roles to play as reductants in industry. Amongst many other uses, hydrogen reduces vegetable oils in the manufacture of margarine. Carbon monoxide reduces iron ore in the blast furnace to make iron:

$$3CO(g) + Fe_2O_3(s) \rightarrow 2Fe(l) + 3CO_2(g)$$

Methane reduces steam in the presence of a nickel catalyst to produce hydrogen, used in the Haber–Bosch synthesis of ammonia:

$$CH_4(g) + H_2O(g) \rightarrow CO(g) + 3H_2(g)$$

It is important to remember, however, that it not possible to classify substances uniquely as *either* oxidants *or* reductants. The same substance may be an oxidant in one reaction and a reductant in another. Redox behaviour depends on *both* the reacting substances.

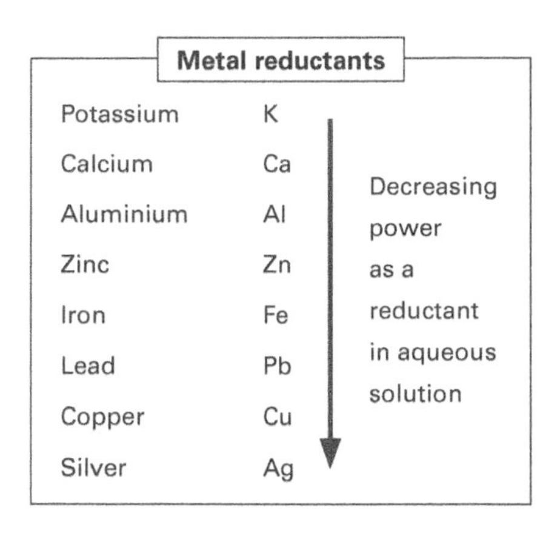

Metal reductants

Potassium	K	Decreasing power as a reductant in aqueous solution (↓)
Calcium	Ca	
Aluminium	Al	
Zinc	Zn	
Iron	Fe	
Lead	Pb	
Copper	Cu	
Silver	Ag	

Metals as reductants

Most metallic elements act as reductants, donating electrons to other species. For example, aluminium will reduce iron(III) oxide to metallic iron in the very vigorous 'Thermit' reaction, see spread 3.1:

$$Fe_2O_3(s) + 2Al(s) \rightarrow Al_2O_3(s) + 2Fe(l)$$

i.e. Fe(III) is reduced to Fe(0); Al(0) is oxidized to Al(III).

From your earlier work you may remember a list of metals called 'the reactivity series'. This list orders the metals according to their overall chemical reactivity. In general terms:

- A more reactive metal will reduce the oxide of a less reactive metal.

So you would expect zinc to reduce lead oxide, but you would not expect lead to reduce aluminium oxide.

- A more reactive metal will displace a less reactive metal from a solution containing ions of the less reactive metal.

So you would expect zinc to displace copper ions, spread 13.1, but you would not expect lead to displace zinc ions.

As another example, the reaction between copper metal and aqueous silver ions is

$$Cu(s) + 2Ag^{+}(aq) \rightarrow 2Ag(s) + Cu^{2+}(aq)$$

Copper is acting as the reductant, donating electrons to the silver ions. Copper metal is a more powerful reductant than silver, so the reverse reaction does not take place to any significant extent.

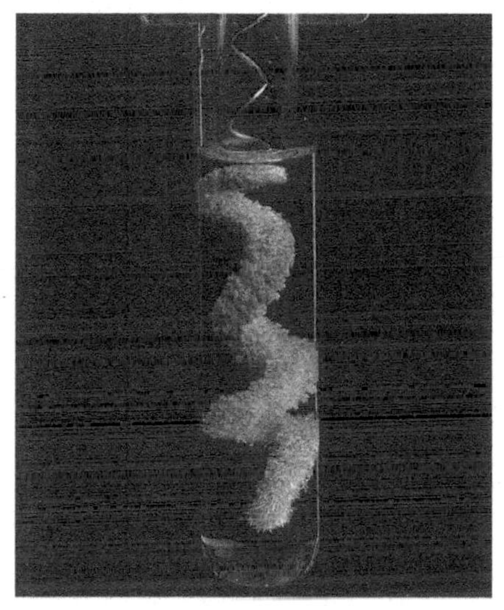

Copper metal reacts with aqueous silver ions to produce silver. Its nickname, the 'Christmas tree' reaction, is easy to understand.

The relative power of a metal as a reductant is therefore linked to its overall chemical reactivity, from potassium as the most powerful to silver as the weakest. This correlation is an approximate guideline only, and in the next few spreads we will introduce more precise ideas.

SUMMARY

- The non-metals fluorine, oxygen, chlorine, and bromine often act as oxidants.
- Metallic elements often act as reductants.
- The reactivity series of metals lists the elements in order of their strengths as reductants in aqueous solution.

PRACTICE

1 State the changes in oxidation number that happen to each of the elements involved in the following reactions:

a Methane burning in oxygen.

b Bromine displacing iodide ions from aqueous solution.

c Nickel displacing copper(II) ions from aqueous solution to form nickel(II) ions.

d The catalytic oxidation of methane to produce methanol

$$2CH_4(g) + O_2(g) \rightarrow 2CH_3OH(g)$$

2 For each of the reactions in question 1, identify (i) the oxidant and (ii) the reductant. Give reasons for your answers.

OBJECTIVES

- Half-equations and half-cells
- Electrochemical cells
- Potential difference and e.m.f.

REDOX AND ELECTROCHEMICAL CELLS

Some substances generally react as oxidants and others generally react as reductants. To predict whether a redox reaction will take place between two species, it is necessary to make measurements. The key to this problem is to separate the two species physically and to monitor the *direction* of electron flow between them. We need to remember that electrons flow *from* the reductant *to* the oxidant during a redox reaction.

Silvery-coloured nickel metal reacts with blue aqueous Cu^{2+} ions to form a red-brown deposit of copper and green aqueous Ni^{2+} ions. This is the result 1½ hours after mixing.

Metal displacement reactions

The previous spread introduced the idea of displacement reactions between a metal and the aqueous ions of another metal. We saw in spread 13.1 the example of immersing a strip of zinc metal in aqueous copper(II) ions, resulting in the following redox reaction:

$$Zn(s) + Cu^{2+}(aq) \rightarrow Zn^{2+}(aq) + Cu(s)$$

This chemical equation can be separated into two half-equations:

$$Zn(s) \rightarrow Zn^{2+}(aq) + 2e^-$$

i.e. zinc is the reductant; it loses electrons and is oxidized.

$$Cu^{2+}(aq) + 2e^- \rightarrow Cu(s)$$

i.e. copper(II) ion is the oxidant; it gains electrons and is reduced.

A displacement reaction also takes place between nickel metal and aqueous copper(II) ions:

$$Ni(s) + Cu^{2+}(aq) \rightarrow Ni^{2+}(aq) + Cu(s)$$

There are again two half-equations. One describes the oxidation of nickel:

$$Ni(s) \rightarrow Ni^{2+}(aq) + 2e^-$$

and the other describes the reduction of aqueous copper(II) ions:

$$Cu^{2+}(aq) + 2e^- \rightarrow Cu(s)$$

We might now ask an important question: 'Is it possible to measure the reducing powers of zinc and nickel and other metals to compare them?' The answer is 'yes', using measurements from half-cells.

Half-cells

Look again at the $Zn(s)/Cu^{2+}(aq)$ displacement reaction. It is possible to separate the reactions represented by the two half-equations given above. We can do so by carrying out each reaction in an electrochemical half-cell. A typical **electrochemical half-cell** consists of a metal in contact with an aqueous solution of its ions.

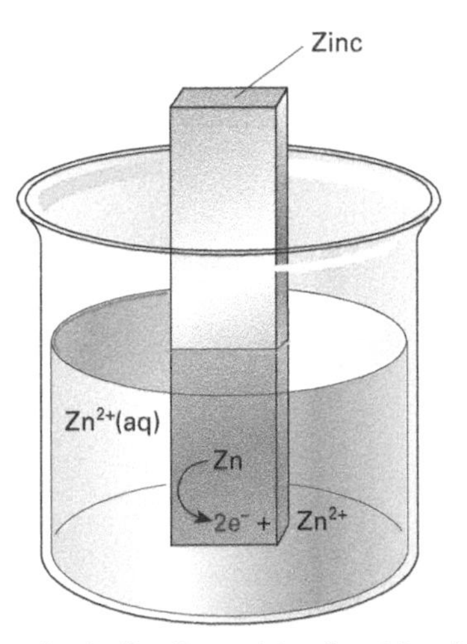

The zinc half-cell consists of a strip of zinc metal immersed in aqueous Zn^{2+} ions.

Imagine that a strip of zinc is placed in a beaker containing aqueous zinc ions (from zinc sulphate, for example). A dynamic equilibrium is quickly established between the surface of the metal and the aqueous ions:

$$Zn^{2+}(aq) + 2e^- \rightleftharpoons Zn(s)$$

The electrons liberated by the backward reaction are delocalized throughout the whole of the metal strip. In this context, the metal strip is called an **electrode** because electrons may enter or leave through it.

A similar equilibrium can be set up in a second beaker containing a copper electrode immersed in aqueous copper(II) ions (from copper(II) sulphate, for example):

$$Cu^{2+}(aq) + 2e^- \rightleftharpoons Cu(s)$$

Half-cells and electric potential

Any half-cell may be described by the corresponding redox half-equation, which we can write in general terms for a metal/metal ion half-cell using M for the metal and n for the number of charges:

$$M^{n+}(aq) + ne^- \rightleftharpoons M(s)$$

In terms of general chemical reactivity, zinc is more reactive than copper. You may therefore assume that the equilibrium in the zinc half-cell lies further to the left than does the equilibrium in the copper half-cell. As a result, there are more electrons on the zinc electrode than on the copper electrode.

As there are more electrons on the zinc electrode, its electric potential is more negative than that on the copper electrode. The zinc electrode is said to have a **negative potential** with respect to the copper electrode. (It is equally correct to say that the copper electrode has a **positive potential** with respect to the zinc electrode.)

Electrochemical cells

As shown in the diagram below, two half-cells may be connected together to make an **electrochemical cell**. The two electrodes are connected by wires to a voltmeter, which measures the **potential difference** (p.d.) between them. In this example, the voltmeter reading is 1.1 volts (symbol: V) when the solutions have concentrations of 1 mol dm^{-3} and the temperature is 298 K. This value is not affected by the size or shape of the electrodes.

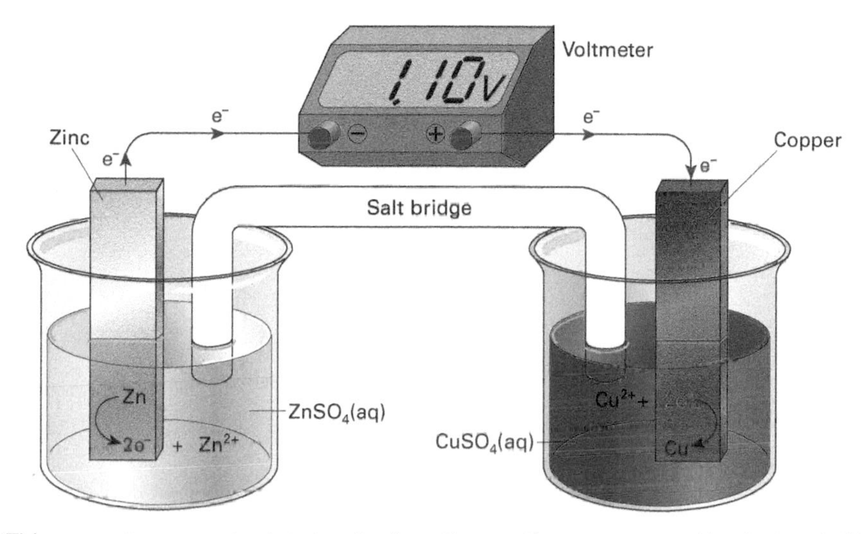

This apparatus separates into two beakers the reactions represented by the two half-equations for the zinc/copper(II) displacement reaction. See next spread.

The voltmeter indicates that the copper electrode is positive with respect to the zinc electrode. If the two electrodes are connected directly together by a wire, then electrons flow towards the positive (copper) electrode. The zinc electrode loses electrons and so the equilibrium

$$Zn^{2+}(aq) + 2e^- \rightleftharpoons Zn(s)$$

shifts to the left. Zinc metal reacts and the concentration of zinc ions increases. The copper electrode gains electrons and so the equilibrium

$$Cu^{2+}(aq) + 2e^- \rightleftharpoons Cu(s)$$

shifts to the right. Copper(II) ions accept electrons and a deposit of copper metal forms on the surface of the copper electrode.

The copper/zinc electrochemical cell has a measured e.m.f. of 1.1 V. The copper/nickel cell has a measured e.m.f. of 0.59 V. The zinc half-cell generates a more negative potential than does the nickel half-cell.

- Reductants are electron donors: therefore in aqueous solutions zinc must be a more powerful reductant than nickel.

The following spreads show how the direction of electron flow and the measured values of e.m.f. may be used to predict the outcome of redox reactions.

SUMMARY

- A half-cell contains the species that take part in a half-equation.
- The difference in electric potential between two half-cells is measured by a voltmeter. The unit of potential difference is the volt (symbol: V).

Electrons and ions

Note that there are no free electrons in the solution: they remain delocalized throughout the electrode. The reaction takes place on the surface of the electrode where the metal is in contact with the aqueous ions.

Measuring potential difference

A high-resistance digital voltmeter measures the potential difference (p.d.) between two half-cells. It allows little current to flow, but does indicate the direction in which a current *would* flow if the electrodes were directly connected together. The p.d. measured when negligible current flows is called the 'electromotive force' (**e.m.f.**) and is the maximum voltage that a cell can develop. If the voltmeter does allow an appreciable current to flow, then the p.d. measured is lower, because of the internal resistance of the cell.

Metals that react with water

For the most reactive reductants such as sodium, it is not possible to use a piece of the metal as an electrode immersed in an aqueous solution, because the metal would react with water. Instead we use an amalgam of the metal dissolved in liquid mercury as the electrode. Connection is then made to the external circuit and voltmeter via a platinum wire dipping into the amalgam.

The salt bridge

The **salt bridge** completes the electric circuit by allowing charged species to flow between the solutions. This flow of ions ensures that the contents of each half-cell remain electrically neutral while electrons flow from one electrode to the other. A simple salt bridge consists of a piece of filter paper soaked in saturated aqueous potassium chloride. A salt bridge which does not dry out so easily can be made using a glass tube filled with saturated aqueous potassium chloride in jelly form.

13.7

OBJECTIVES

- The standard hydrogen electrode
- Standard electrode potentials
- Cell diagrams

Sea level

Altitude in the U.K. is measured against an arbitrary zero, which is defined as the mean sea level at Newlyn, Cornwall. Assigning sea level requires an arbitrary decision about what zero means in much the same way as the standard electrode potential does.

Standard electrode potentials – 1

From previous work in this chapter, you should now understand that half-cells contain oxidized and reduced species, which are in equilibrium with each other. Two half-cells may be joined together to make an electrochemical cell, in which electrons flow from the electrode with the more negative potential to the electrode with the more positive potential. Whichever has the more negative potential is the stronger reductant. This spread shows that measuring each half-cell against a standard reference allows each to be assigned a 'standard electrode potential'.

The standard hydrogen electrode

It is only possible to measure a potential *difference*. It is not possible to measure the potential of an isolated electrode and assign an absolute value of potential to it. Chemists have found a way around this problem by agreeing the specification of a *standard half-cell*, which has a value of zero potential assigned to it. The potentials of other half-cells may then be measured against this arbitrary standard of zero potential. This standard half-cell is the **standard hydrogen electrode** shown below left.

The standard hydrogen electrode consists of a platinum electrode immersed in acid of pH 0 (i.e. where $[H^+(aq)] = 1\ mol\ dm^{-3}$) with hydrogen gas at standard pressure (1 bar) passing over it.

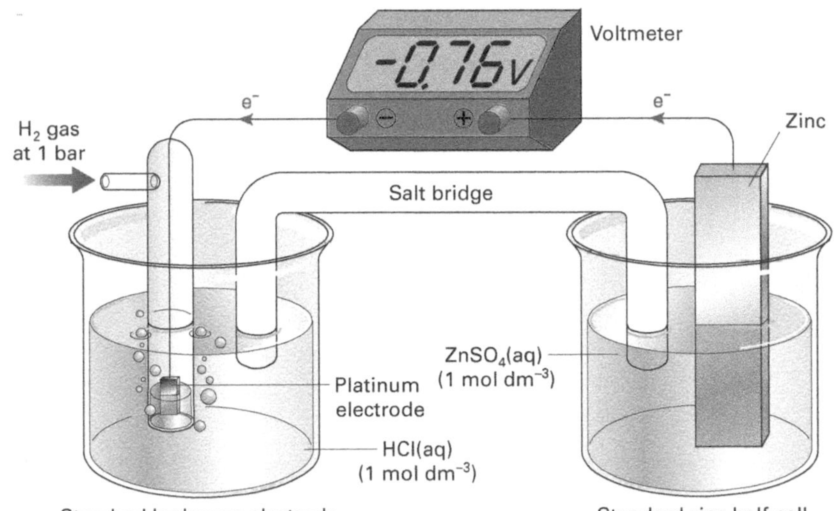

An electrochemical cell consisting of a standard hydrogen electrode and a standard zinc half-cell. The measured standard electrode potential for the redox equilibrium $Zn^{2+}(aq) + 2e^- \rightleftharpoons Zn(s)$ is –0.76 V.

The redox half-equation for the standard hydrogen electrode is:

$$2H^+(aq) + 2e^- \rightleftharpoons H_2(g) \qquad E^\ominus = 0\ V$$

The symbol $E^\ominus$ represents the **standard electrode potential** of a half-cell, which is the e.m.f. measured between a standard hydrogen electrode and a *standard* metal half-cell (set up using an aqueous solution of the metal ion at a concentration of $1\ mol\ dm^{-3}$) at 298 K.

By convention, standard electrode potentials are always recorded as standard *reduction* potentials. The reaction is written with the reduced species on the right, i.e.

oxidized species + ne^- $\rightleftharpoons$ reduced species

Some examples are given in the table to the left.

Standard electrode potentials for metal/metal ion half-cells.

Oxidized species	$\rightleftharpoons$	Reduced species	$E^\ominus$/V
$K^+(aq) + e^-$	$\rightleftharpoons$	K(s)	–2.92
$Ca^{2+}(aq) + 2e^-$	$\rightleftharpoons$	Ca(s)	–2.87
$Mg^{2+}(aq) + 2e^-$	$\rightleftharpoons$	Mg(s)	–2.37
$Al^{3+}(aq) + 3e^-$	$\rightleftharpoons$	Al(s)	–1.66
$Zn^{2+}(aq) + 2e^-$	$\rightleftharpoons$	Zn(s)	–0.76
$Fe^{2+}(aq) + 2e^-$	$\rightleftharpoons$	Fe(s)	–0.44
$2H^+(aq) + 2e^-$	$\rightleftharpoons$	$H_2(g)$	0
$Cu^{2+}(aq) + 2e^-$	$\rightleftharpoons$	Cu(s)	+0.34
$Ag^+(aq) + e^-$	$\rightleftharpoons$	Ag(s)	+0.80

The sign of the $E^\ominus$ value

The $E^\ominus$ value for the zinc half-cell is –0.76 V. The minus (–) sign signifies that the electric potential on the zinc electrode is more negative than the potential on the standard hydrogen electrode. The $E^\ominus$ value for the copper half-cell is +0.34 V. The plus (+) sign signifies that the standard hydrogen electrode has the more negative potential.

Cell diagrams

An electrochemical cell results when two half-cells are connected together. A **cell diagram** is an agreed way of depicting cells on paper. For example, combining hydrogen and zinc half-cells gives the overall cell diagram:

$Pt(s)|H_2(g)|H^+(aq)||Zn^{2+}(aq)|Zn(s)$

By convention:

- The two parallel lines represent the salt bridge.
- Each single line represents the change of phase, for example between aqueous ions and solid metal.
- The electrodes through which electrons flow are placed at the start and the finish of the cell diagram.
- The standard hydrogen electrode is placed on the left-hand side.

In the case of the zinc/copper electrochemical cell, the cell diagram is:

$Zn(s)|Zn^{2+}(aq)||Cu^{2+}(aq)|Cu(s)$

By convention, the cell diagram is written with:

- the half-cell undergoing oxidation on the left of the diagram
- the half-cell undergoing reduction on the right of the diagram.

So, when the electrodes are directly connected, electrons flow from the zinc half-cell to the copper half-cell. Reduction therefore takes place in the $Cu^{2+}(aq)|Cu(s)$ half-cell and oxidation in the $Zn(s)|Zn^{2+}(aq)$ half-cell.

The zinc half-cell is identified as the half-cell undergoing oxidation because its standard electrode potential is more negative than that of the copper half-cell. Look at the table of $E^\ominus$ values and you should understand that:

- Oxidation will take place in the half-cell with the more negative standard electrode potential.
- Reduction will take place in the half-cell with the less negative (more positive) standard electrode potential.

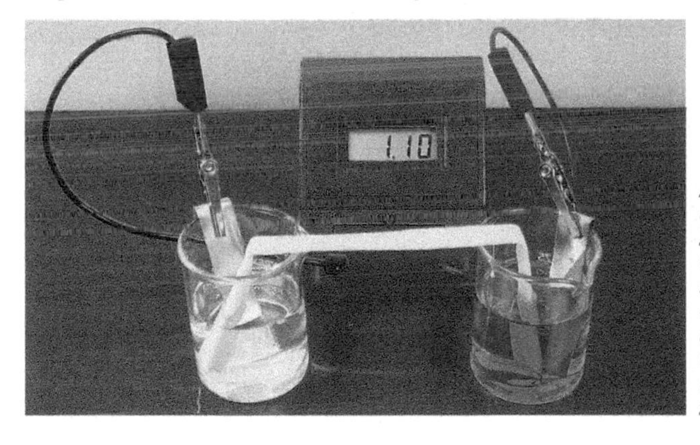

An electrochemical cell consisting of $Zn^{2+}(aq)/Zn(s)$ and $Cu^{2+}(aq)/Cu(s)$ half-cells. The standard cell e.m.f. is +1.10 V. If the electrodes are connected directly together, copper(II) ions are reduced and zinc metal is oxidized.

SUMMARY

- $E^\ominus$ is the standard electrode potential of a half-cell, measured against a standard hydrogen electrode at 298 K; solutions are at 1 mol dm^{-3}.
- The standard hydrogen electrode has a platinum electrode, H_2 gas at 1 bar, and acid at pH 0; it is assigned an arbitrary $E^\ominus$ of 0 V.
- Cell diagrams are written with the half-cell undergoing oxidation on the left and the half-cell undergoing reduction on the right, starting with one electrode and finishing with the other.
- The e.m.f. of a cell is the potential difference measured when negligible current flows.
- The standard cell e.m.f. is equal to the difference between the $E^\ominus$ values for the two half-cells concerned.

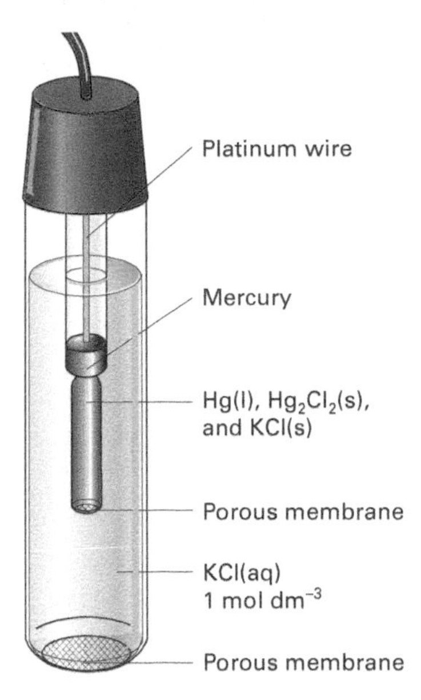

The 'standard calomel electrode' is easier to set up than a standard hydrogen electrode. ('Calomel' is the traditional name for mercury(I) chloride Hg_2Cl_2.) Its $E^\ominus$ value is +0.27 V and it is used as a secondary standard electrode against which to measure other electrode potentials. The redox reaction in the calomel half-cell is
$Hg_2Cl_2(s) + 2e^- \rightleftharpoons 2Hg(l) + 2Cl^-(aq)$

Oxidant/reductant power

The standard electrode potentials for the zinc and the copper half-cells are –0.76 V and +0.34 V respectively. So zinc metal (with the more negative $E^\ominus$ value) is a more powerful reductant than copper.

Standard cell e.m.f.

The **standard cell e.m.f.** is the difference in the standard electrode potentials of the two half-cells as calculated from the expression:

$$E^\ominus(\text{cell}) = E^\ominus(\text{right-hand electrode}) - E^\ominus(\text{left-hand electrode})$$

'Right' and 'left' signify the reactions as written in the cell diagram. For example, for the cell:

$Zn(s)|Zn^{2+}(aq)||Cu^{2+}(aq)|Cu(s)$

we have

$$E^\ominus(\text{cell}) = E^\ominus(Cu^{2+},Cu) - E^\ominus(Zn^{2+},Zn) = (+0.34\ V) - (-0.76\ V) = +1.10\ V$$

A high-resistance voltmeter connected to this electrochemical cell would indicate this value of e.m.f., as shown in the photo on the left.

Redox and $E^\ominus$ values: a reminder

The half-cell with the more negative ('lower') $E^\ominus$ will reduce the half-cell with the less negative $E^\ominus$.

OBJECTIVES

- Metal reactivity series and $E^{\ominus}$ values
- Other half-cells
- $E^{\ominus}$ values and predictions

STANDARD ELECTRODE POTENTIALS – 2

The *metal reactivity series* can be used to compare the relative reactivities of different metals. This spread starts by comparing the reactivity series with the relative reducing powers of metals according to their $E^{\ominus}$ values. There is an important distinction that we must draw between the *feasibility* of a chemical reaction and the *rate* at which it happens. The spread concludes with examples that use $E^{\ominus}$ values to explain which redox reactions happen.

Metal reactivity series and corresponding $E^{\ominus}$ values.

Metal reactivity series		Standard electrode potential/V	
K	↑ Increasing reactivity / Increasing reductant strength	K	−2.92
Na		Ca	−2.87
Ca		Na	−2.71
Mg		Mg	−2.37
Al		Al	−1.66
Zn		Zn	−0.76
Fe		Fe	−0.44
Pb		Pb	−0.13
		H_2	0
Cu		Cu	+0.34
Ag		Ag	+0.80

Metal reactivity and $E^{\ominus}$ values

The metal reactivity series lists metals in order of general chemical reactivity. Metals generally react by losing electrons to form positive ions. The more readily a metal loses electrons, the more reactive it is – and the greater its strength as a reductant. Metals higher up the series can reduce the ions of those lower down.

The standard electrode potential of a metal also indicates its strength as a reductant. The more negative the value of the standard electrode potential of a metal, the greater is its strength as a reductant. Hence, you might expect the metal reactivity series and standard electrode potentials to list metals in the *same* order. However, you must remember that the metal reactivity series is based on observing a *range* of reactions, such as displacement reactions between solid metals and solid metal oxides. Standard electrode potentials refer *specifically* to reactions taking place in aqueous solution.

The table (left) compares the metal reactivity series with the **electrochemical series**, which ranks metals according to their standard electrode potentials. The obvious discrepancy is the relative positions of sodium and calcium. Calcium is a stronger reductant than sodium according to $E^{\ominus}$ values, but the metal reactivity series suggests that calcium is *less* reactive than sodium. This discrepancy arises because calcium reacts at a much slower *rate*, in displacement reactions for example, which in turn happens because *two* electrons must be removed, not one as for sodium.

Note also that aluminium reacts readily with oxygen in the air, forming a layer of stable aluminium oxide on its surface. This impervious oxide coat often causes aluminium to exhibit lower reactivity than its position in the metal reactivity series indicates.

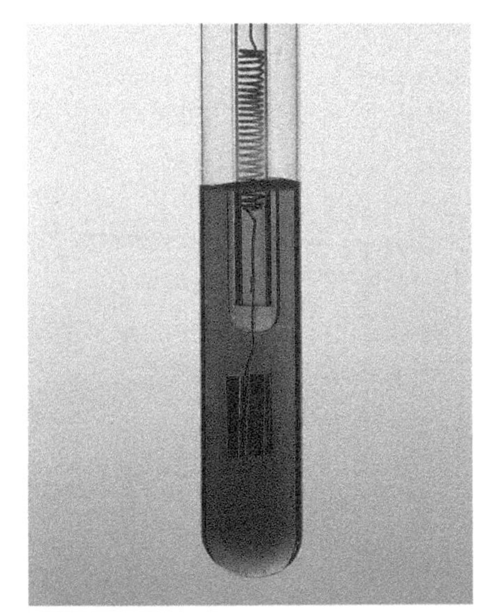

This half-cell consists of a platinum electrode dipping into an aqueous mixture of iron(II) and iron(III) ions.

Displacement reactions involving non-metals

With the exception of the standard hydrogen electrode, all the redox half-cells considered so far have consisted of a metal in contact with an aqueous solution of its ions. The photograph alongside shows a different type of half-cell involving iron ions. The platinum electrode is the point at which electrons enter or leave the system; the electrode does not take part in the redox reaction. Earlier in this chapter we introduced the concept of non-metal displacement reactions, including the reactions between halogens and aqueous solutions of halide ions. A typical example is the reaction between chlorine and bromide ions, spread 13.2:

$$Cl_2(g) + 2Br^-(aq) \rightarrow 2Cl^-(aq) + Br_2(l)$$

The relevant half-equations

$$Cl_2(g) + 2e^- \rightarrow 2Cl^-(aq)$$
$$2Br^-(aq) \rightarrow Br_2(l) + 2e^-$$

indicate that chlorine is reduced and bromide ions are oxidized, i.e. chlorine acts as an oxidant and bromide ion as a reductant. As with metal/metal ion half-cells, each of these reactions can be set up as a redox half-cell.

Halogen/halide standard electrode potentials.

Oxidized species	⇌	Reduced species	$E^{\ominus}$/V
$F_2(g) + 2e^-$	⇌	$2F^-(aq)$	+2.87
$Cl_2(g) + 2e^-$	⇌	$2Cl^-(aq)$	+1.36
$Br_2(l) + 2e^-$	⇌	$2Br^-(aq)$	+1.07
$I_2(s) + 2e^-$	⇌	$2I^-(aq)$	+0.54

Combining a standard halogen/halide half-cell with a hydrogen electrode allows the standard electrode potential to be measured. Values are shown in the table opposite. As mentioned in the previous spread, the equilibrium for a standard electrode potential is always written with the reduced species on the right:

oxidized species + $ne^- \rightleftharpoons$ reduced species

The oxidizing power of the halogens in aqueous solution decreases in the order $F_2 > Cl_2 > Br_2 > I_2$. The reducing power of the aqueous halide ions decreases in the order $I^- > Br^- > Cl^- > F^-$. See spreads 18.3 and 18.5.

Using standard electrode potentials to predict the outcome of redox reactions

Worked example 1

Question: Explain why zinc reacts with dilute hydrochloric acid to form hydrogen gas but copper cannot.

Answer: The balanced chemical equation for the reaction between zinc and hydrochloric acid is

$Zn(s) + 2HCl(aq) \rightarrow H_2(g) + ZnCl_2(aq)$

Because the chloride ions are spectator ions, the ionic equation is

$Zn(s) + 2H^+(aq) \rightarrow H_2(g) + Zn^{2+}(aq)$

The half-equations are

$Zn(s) \rightarrow Zn^{2+}(aq) + 2e^-$

$2H^+(aq) + 2e^- \rightarrow H_2(g)$

The relevant standard electrode potentials are

$Zn^{2+}(aq) + 2e^- \rightleftharpoons Zn(s)$; $E^{\ominus} = -0.76\ V$

$2H^+(aq) + 2e^- \rightleftharpoons H_2(g)$; $E^{\ominus} = 0\ V$

Electrons flow from the half-cell of more negative potential to the half-cell of more positive potential, i.e. from Zn^{2+}/Zn to H^+/H_2. Zinc metal is oxidized and hydrogen ions are reduced.

In the case of *copper*, the relevant standard electrode potential is

$Cu^{2+}(aq) + 2e^- \rightleftharpoons Cu(s)$; $E^{\ominus} = +0.34\ V$

In a copper/hydrogen electrochemical cell, electrons flow in the direction opposite to that required for copper metal to reduce aqueous hydrogen ions to hydrogen gas.

Worked example 2

Question: Copper *does* react with concentrated nitric acid, forming nitrogen dioxide (not hydrogen) according to the equation:

$Cu(s) + 4HNO_3(aq) \rightarrow Cu(NO_3)_2(aq) + 2NO_2(g) + 2H_2O(l)$

Explain why.

Answer: The relevant standard electrode potentials are

$Cu^{2+}(aq) + 2e^- \rightleftharpoons Cu(s)$; $E^{\ominus} = +0.34\ V$

$NO_3^-(aq) + 2H^+(aq) + e^- \rightleftharpoons NO_2(g) + H_2O(l)$; $E^{\ominus} = +0.80\ V$

Electrons flow from the more negative copper half-cell to the more positive nitrate half-cell. Copper is the reductant and *nitrate ion* is the oxidant.

SUMMARY

- Predictions using standard electrode potentials only apply to redox reactions taking place in aqueous solution.
- Standard electrode potentials may indicate that a reaction is feasible: however, the reaction rate may be so slow as to make the reaction appear not to happen.
- The concentrations of the ions concerned must be close to the standard concentration of $1\ mol\ dm^{-3}$ in order to apply $E^{\ominus}$ values.

Powerful oxidants and reductants

Powerful oxidants generally have standard electrode potentials more positive than about +1 V. For example, oxygen:

$O_2(g) + 4H^+(aq) + 4e^- \rightleftharpoons 2H_2O(l)$

$E^{\ominus} = +1.23\ V$

Fluorine is the strongest common oxidant:

$F_2(g) + 2e^- \rightleftharpoons 2F^-(aq)$; $E^{\ominus} = +2.87\ V$

Powerful reductants generally have standard electrode potentials more negative than about −1 V.

Potassium is the strongest common reductant:

$K^+(aq) + e^- \rightleftharpoons K(s)$; $E^{\ominus} = -2.92\ V$

The problem of reaction rate

Standard electrode potentials may be used to predict the outcome of redox reactions. However, it is important to note that these predictions do not take account of the *rate* of a reaction. A reaction that is predicted to occur may do so at a rate too slow to observe.

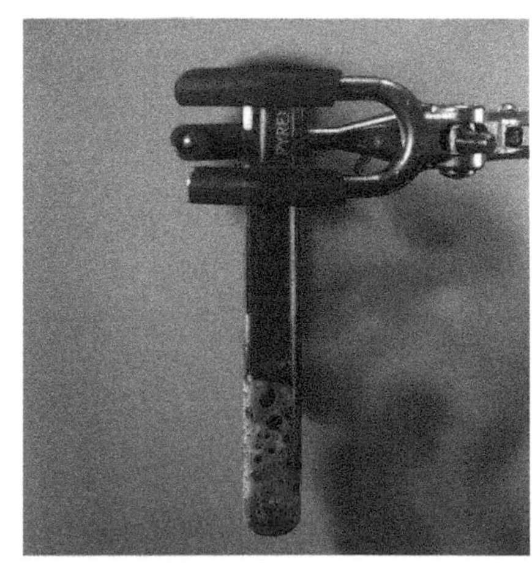

Copper metal reacts with concentrated nitric acid to give brown fumes of nitrogen dioxide.

EXTENSION

13.9

OBJECTIVES

- Changing concentration
- Nernst equation

NON-STANDARD CONDITIONS

All the redox reactions discussed so far have concerned aqueous solutions at 1 mol dm^{-3} concentration. Chemists are always looking for ways of controlling reactions and making use of them. This spread investigates the effect of non-standard conditions on redox equilibria, which may lead to a successful reaction.

Zinc and copper strips immersed in an electrolyte (lemon juice) can generate a potential difference despite the ionic concentrations being far away from 1 mol dm^{-3}.

Non-standard conditions

Standard electrode potentials may be used to predict whether a redox reaction will occur or not. If a reaction is feasible, then the greater the value of the standard cell e.m.f., the further to the right will be the chemical equilibrium for the overall reaction. When ion concentration is different from 1 mol dm^{-3}, the electrode potential for a cell may be greater or smaller than the standard electrode potential. Equilibrium position may be controlled by a suitable choice of concentrations.

Altering conditions to change the value of the cell e.m.f. will change the value of the equilibrium constant. For example, the power of an oxidant such as manganate(VII) ion is affected by the pH of its solution. The standard electrode potentials for manganate(VII) ion as oxidant are:

$$MnO_4^-(aq) + 8H^+(aq) + 5e^- \rightleftharpoons Mn^{2+}(aq) + 4H_2O(l) \qquad (1)$$

Mn(VII) to Mn(II) $\quad E^{\ominus} = +1.51$ V

$$MnO_4^-(aq) + 4H^+(aq) + 3e^- \rightleftharpoons MnO_2(s) + 2H_2O(l) \qquad (2)$$

Mn(VII) to Mn(IV) $\quad E^{\ominus} = +1.69$ V

Many reductants will reduce purple Mn(VII) to almost colourless Mn(II) via equilibrium (1) or to a brown suspension of manganese(IV) oxide via equilibrium (2).

Le Chatelier's principle indicates that, when $[H^+(aq)] < 1$ mol dm^{-3}, both equilibria will shift to the left. The values of their electrode potentials will become less positive and they will become less powerful as oxidants.

The Nernst equation

The potential E of a half-cell (measured against a standard hydrogen electrode) deviates from the standard electrode potential $E^{\ominus}$ according to the relationship:

$$E = E^{\ominus} + \frac{(0.059\,\text{V})}{z}\log_{10}\frac{[\text{oxidized species}]}{[\text{reduced species}]}$$

(z is the number of electrons transferred when the oxidized species changes into the reduced species). This relationship is called the **Nernst equation**, after Walther Nernst who developed the relationship. The Nernst equation provides a 'fine tuning' around $E^{\ominus}$ in a similar way to the Henderson-Hasselbalch equation for buffer solutions, spread 12.11.

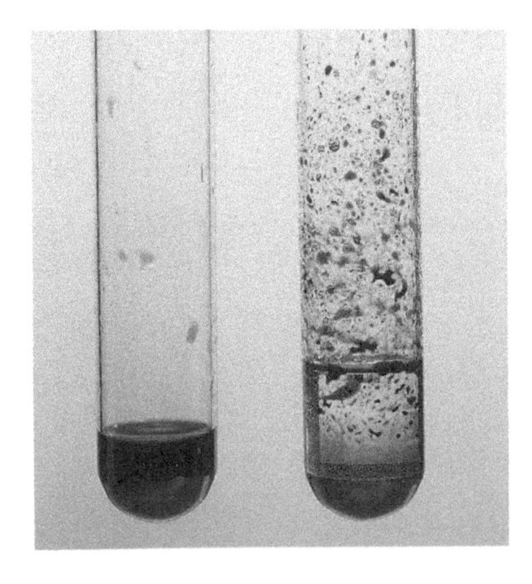

The alkene cyclohexene will reduce alkaline manganate(VII) ions to green aqueous manganate(VI) ions. The worked example opposite explains why this happens.

Worked example on using the Nernst equation

Question: What is the electrode potential *at pH 14* for these two reductions of manganate(VII) ion, MnO_4^-:

(a) $MnO_4^-(aq) + e^- \rightleftharpoons MnO_4^{2-}(aq)$

(b) $MnO_4^-(aq) + 8H^+(aq) + 5e^- \rightleftharpoons Mn^{2+}(aq) + 4H_2O(l)$

Strategy: Route (a) can be done by inspection: it has the same value as the standard electrode potential, +0.56 V, as no aqueous hydrogen ions are involved in the reduction. For route (b), the Nernst equation must be used as aqueous hydrogen ions are involved in the reduction and the pH is not 0.

Answer: The Nernst equation applied to route (b) gives:

$$E = E^{\ominus} + \frac{(0.059\,\text{V})}{5} \log_{10}[H^+]^8$$

where $E^{\ominus}$ is the standard electrode potential and five electrons are involved. This equation may be manipulated as follows, using the rules for logarithms (toolbox):

$$E = E^{\ominus} + (8/5)(0.059\,\text{V}) \log_{10}[H^+]$$

Recognizing $\text{pH} = -\log_{10}[H^+]$ leads to the equation

$$E = E^{\ominus} - (8/5)(0.059\,\text{V})\,\text{pH}$$

At pH 14, the electrode potential is $(+1.51\,\text{V}) - (8/5)(0.059\,\text{V}) \times 14 = +0.19\,\text{V}$.

Note: In *alkaline* solution, manganate(VII) ions can form the manganate(VI) ion, MnO_4^{2-}, which is a green solution. Alkaline manganate(VII) ions are used as a colour test for alkenes. The normal reduction to colourless manganese(II) ions, does *not* occur as it is now much more weakly oxidizing (has a lower electrode potential).

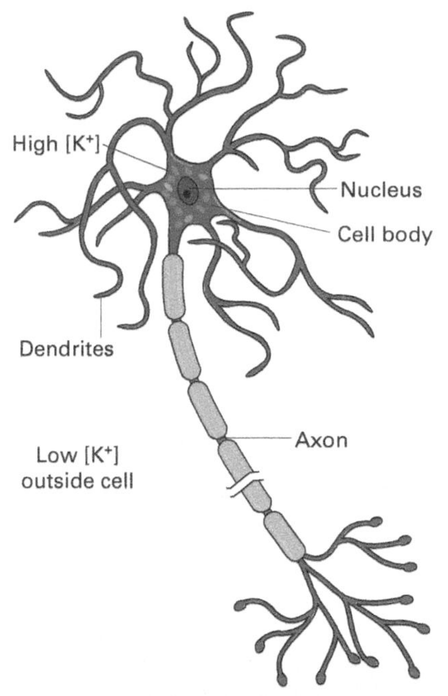

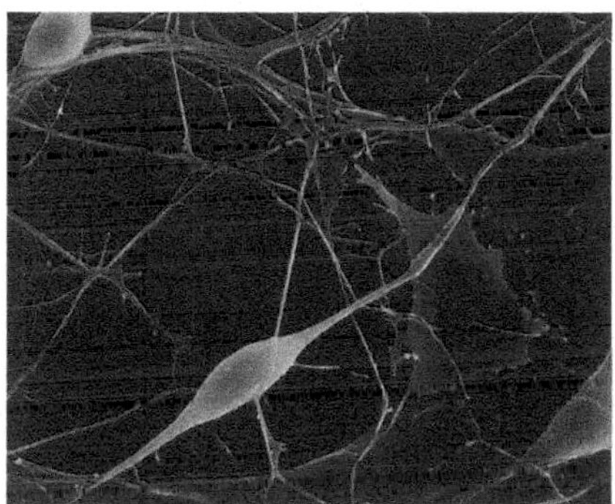

A ***concentration cell*** *generates a potential from two half-cells containing the same element, but with ions at different concentrations. A biochemical example occurs in nerve cells which have potassium ions at a concentration 25 times greater inside than outside the cell. The potential difference across a nerve cell membrane is about –70 mV. Electrical nerve impulses result from a change in this potential as sodium and potassium ions move across the membrane.*

The Nernst equation and nerve cells

One particularly important application of the Nernst equation arises when the two half-cells contain the same element, but with ions at different concentrations. Such a concentration cell can generate a potential. A biochemical example occurs in nerve cells. There are different concentrations of potassium ions inside and outside the cell. The potential across the nerve cell membrane can be approximated as:

$$E = -(0.059\,\text{V}) \log_{10}\left\{\frac{[K^+, \text{inside}]}{[K^+, \text{outside}]}\right\}$$

As the potassium concentration is about 25 times larger inside the cell, the membrane *resting potential* can be predicted to be

$$E = -(0.059\,\text{V}) \log_{10} 25 = -82\,\text{mV}$$

This is reasonably close to the actual value of –70 mV.

The Nernst equation and the effect of strong base

Although the concentration of *any* ion will affect the electrode potential, because $E^{\ominus}$ specifically refers to a concentration of 1 mol dm^{-3}, the only ion whose concentration can be altered easily over a wide range is the aqueous hydrogen ion. Its concentration can be changed by 14 orders of magnitude, simply by adding 1 mol dm^{-3} acid and then 1 mol dm^{-3} base.

In general, the presence of strong base causes the *higher* of two oxidation states to be preferentially stabilized, which is particularly important for the transition metals (see chapter 20). This is also relevant to the discussion of rusting in spread 13.12.

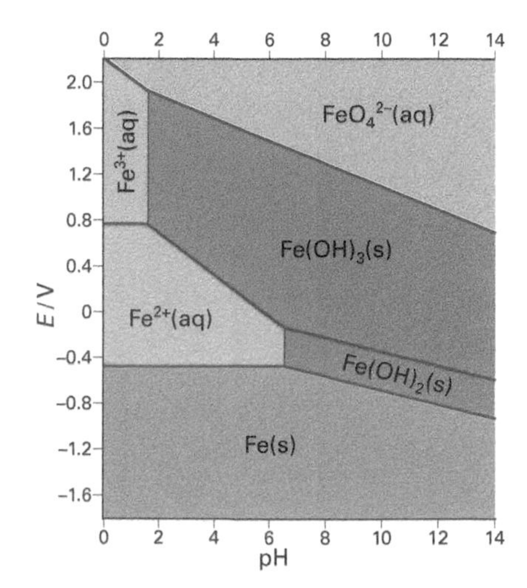

The variation of electrode potential E with changing pH. Note that iron(III) is preferentially stabilized at high pH.

SUMMARY

- Standard electrode potentials are measured at pH 0.
- In strongly basic solutions, electrode potentials may be very different from those at pH 0.
- Strong base preferentially stabilizes the higher of two oxidation states.

13.10

OBJECTIVES

- Primary and secondary cells
- Electrodes and electrolytes
- Anode and cathode reactions

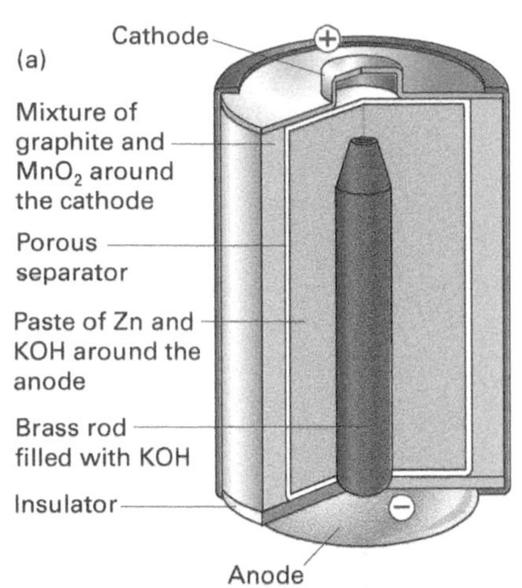

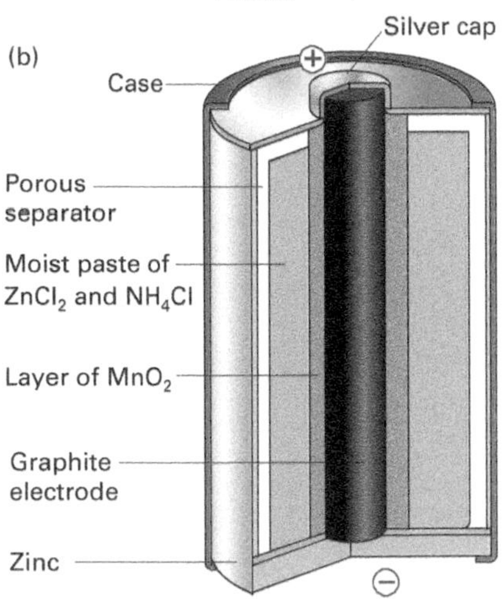

Alkaline dry cells (a) maintain operating voltage better at high current drain than the older type of carbon–zinc dry cells (b) (sometimes called Leclanché cells). Alkaline dry cells are marketed as 'high-power batteries' or as 'longer-lasting'.

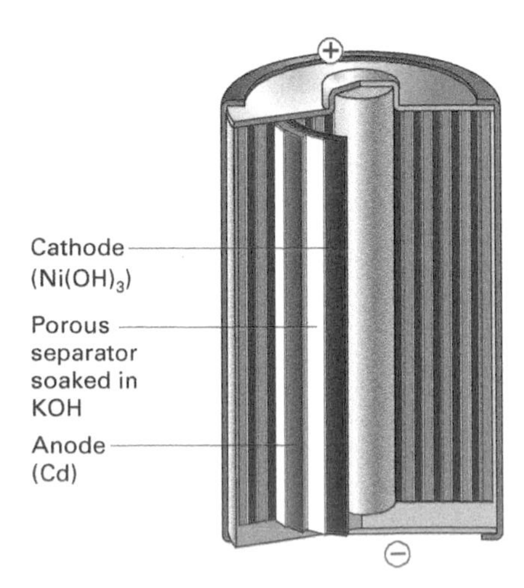

A rechargeable nickel–cadmium cell.

USING REDOX REACTIONS: GALVANIC CELLS

Switch on an electric torch or start a car engine and you are using electrical energy provided by redox reactions inside electrochemical cells. You already know that redox reactions involve a transfer of electrons and that moving electrons constitute an electric current. The everyday devices we usually call 'batteries' depend on these effects. They work by producing energy from chemical reactions inside a convenient portable package.

Galvanic cells

A **galvanic cell** (or **voltaic cell**) is an electrochemical cell used as a source of electric potential difference. Redox reactions inside the cell cause an electric current to flow in an external circuit, where useful work can be done. The first reliable galvanic cell, using the reaction between zinc and copper(II) sulphate, was invented in the early nineteenth century by John Daniell, though his cell was hardly portable. He developed the cell in response to the urgent need for an electricity supply to power the new technology of telegraphy.

Galvanic cells are classified as either primary or secondary cells. Both types of cell consist of two electrodes in contact with a liquid ('wet' cells) or moist paste ('dry' cells), called the electrolyte. Electrons flow in the electrodes and ions flow through the electrolyte. By definition:

- The **anode** is the site of oxidation, i.e. electrons are lost at the anode.
- The **cathode** is the site of reduction, i.e. electrons are gained at the cathode.

A **primary cell** (or **non-rechargeable cell**) may be used only once; it is thrown away when exhausted. In a chemical context, the term 'exhausted' means that current flow has ceased; the cell is discharged because the chemical contents have reached equilibrium. A common example of a primary cell is an alkaline dry cell used in torches and portable radios.

When a **secondary cell** (or **rechargeable cell**) is exhausted, it may be recharged by passing an electric current through it in the opposite direction to the current flow during discharge. In this manner, the discharge reactions are reversed and the original chemical state of the cell is restored.

The alkaline dry cell

The anode is a paste of powdered zinc with an electrolyte of potassium hydroxide, and the cathode is manganese(IV) oxide in the same electrolyte. The electrode reactions are

Anode: $Zn(s) + 2OH^-(aq) \rightarrow Zn(OH)_2(s) + 2e^-$

Cathode: $MnO_2(s) + 2H_2O(l) + e^- \rightarrow Mn(OH)_3(s) + OH^-(aq)$

This 'alkaline dry cell' produces a voltage of 1.5 V and is commonly known as a 'torch battery'.

The secondary nickel–cadmium cell

The **nickel–cadmium (nicad) cell** has an anode of cadmium and a cathode of nickel(III) hydroxide, of approximate formula $Ni(OH)_3$. The electrolyte is potassium hydroxide. The electrode reactions are

Anode: $Cd(s) + 2OH^-(aq) \rightarrow Cd(OH)_2(s) + 2e^-$

Cathode: $Ni(OH)_3(s) + e^- \rightarrow Ni(OH)_2(s) + OH^-(aq)$

This type of cell produces a voltage of 1.25 V and was widely used in portable equipment such as lap-top computers and mobile phones: they have been replaced by the similar nickel–metal hydride (NiMH) cell. Both may be recharged several hundred times during their life.

The secondary lead–acid cell

The **lead–acid cell** has an anode of spongy lead covering a lead–antimony alloy grid. The cathode is lead(IV) oxide in pockets on a lead–antimony alloy grid. The electrolyte is aqueous sulphuric acid. The electrode reactions are

Anode: $Pb(s) + SO_4^{2-}(aq) \rightarrow PbSO_4(s) + 2e^-$

Cathode: $PbO_2(s) + 4H^+(aq) + SO_4^{2-}(aq) + 2e^- \rightarrow PbSO_4(s) + 2H_2O(l)$

Typically, six cells are connected in series to produce a **battery** with a total voltage of 12 V. The most common use of these batteries is in cars.

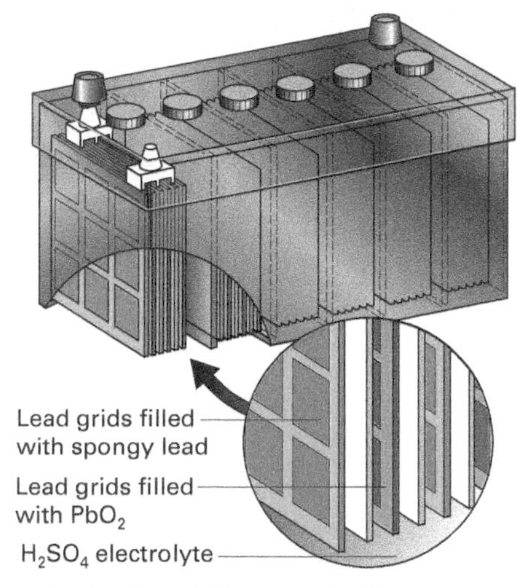

A lead–acid cell. These cells give off oxygen and hydrogen gas during recharging and must be periodically 'topped up' with pure water. 'Maintenance-free' batteries for cars use a calcium–lead alloy, which decomposes the water less rapidly.

The secondary lithium-ion cell

The **lithium-ion cell** uses a cathode which is a mixed metal oxide, such as lithium cobalt(III) oxide, $LiCoO_2$. During discharging the following cathode reaction involving the reduction of cobalt happens:

$CoO_2 + Li^+ + e^- \rightarrow LiCoO_2$

The anode is made of graphite. During discharging the anode reaction is

$LiC_6 \rightarrow Li^+ + e^- + 6C$

The structure of LiC_6 involves lithium inserted (**intercalated**) between the layers of graphite, spread 6.3. The lithium ion moves through the electrolyte, lithium perchlorate ($LiClO_4$) in an organic solvent.

Lithium-ion cells have the great advantage of being able to take repeated charge and discharge cycles without significant loss in performance.

Fuel cells

Fuel cells are galvanic cells that oxidize a fuel to produce an electric current. Fuel cells on-board manned space vehicles consume hydrogen and oxygen, producing electricity and drinking water. Fuel cells operate continuously as long as the reactants are supplied.

The electrode reactions for an alkaline fuel cell are

Anode: $H_2(g) + 2OH^-(aq) \rightarrow 2H_2O(l) + 2e^-$

Cathode: $O_2(g) + 2H_2O(l) + 4e^- \rightarrow 4OH^-(aq)$

A hydrogen–oxygen fuel cell used in the space program. It is about 70% efficient and is pollution-free.

Fuel cells can use other fuels too. The **direct methanol fuel cell** (DMFC) has the twin advantages over hydrogen fuel cells of easier storage of the fuel and a much higher energy density, spread 22.5. The cell reactions, using platinum as the catalyst, are:

Anode: $CH_3OH + H_2O \rightarrow CO_2 + 6H^+ + 6e^-$

Cathode: $3/2O_2 + 6H^+ + 6e^- \rightarrow 3H_2O$

Two concerns, however, are that methanol can permeate through the membranes used and that start-up at low temperatures is compromised by water being involved as a reactant. DMFCs have been used to power mobile phones and laptops.

Fuel cell vehicles

Current electric vehicles are typically hybrids, which use an electric motor at low speed and then a petrol engine at higher speed, which recharges the batteries. Several car manufacturers are working on fuel cell vehicles, typically using hydrogen as the fuel. The Toyota Mirai (Japanese for 'future') is the first hydrogen fuel-cell car on sale in the UK: it was voted World Green Car of the Year in 2016. The main problem with hydrogen fuel cells is the storage of the hydrogen: a **hydrogen economy** would require extensive work on the distribution network. In 2018, a journalist travelled from John O'Groats to Land's End in a Mirai, with the route dictated by stops at four of the nine refuelling stations available at the time. The longest run of 276 miles was from Beaconsfield to Land's End. Aberdeen, the first stop, is the busiest hydrogen fuel station in Europe.

A **methanol economy**, as advocated by 1994 Nobel Prize winner George Olah amongst others, would require less modification as methanol, like gasoline, is a liquid.

SUMMARY

- The anode is the site of oxidation.
- The cathode is the site of reduction.

PRACTICE

1 Write chemical equations to represent the anode and cathode reactions taking place in a Daniell cell when it provides an electric current.

2 Explain why it may not be practicable to power electric cars by:

a rechargeable nickel–cadmium cells,

b fuel cells.

3 Write chemical equations to represent the anode and cathode reactions during the recharging of a lead–acid cell.

4 Giving appropriate examples, explain and contrast the meanings of the terms: half-cell; cell; battery.

13.11

OBJECTIVES

- Electrolysis of molten substances
- Electrolysis of aqueous solutions
- Electroplating
- The Faraday constant

USING REDOX REACTIONS: ELECTROLYTIC CELLS

A galvanic cell uses chemical reactions to generate electricity, but an electrolytic cell does the opposite. An electrolytic cell uses electricity to cause chemical change. **Electrolysis** uses electrical energy to drive a chemical reaction in a direction opposite to that in which it naturally proceeds. As you will see, the uses of electrolytic cells include decomposing solutions and molten substances and the purification and plating of metals.

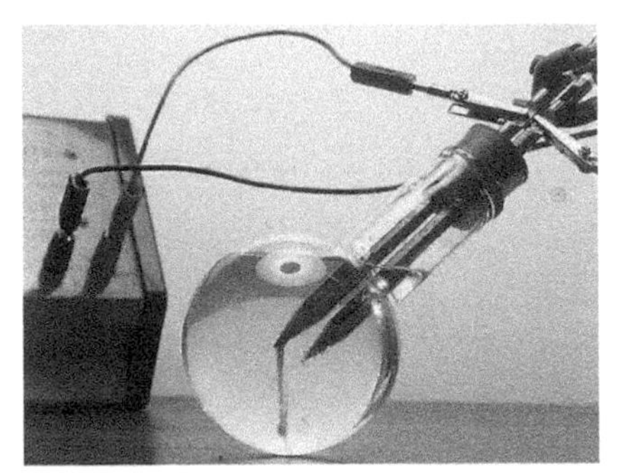

Electrolysis of aqueous potassium iodide produces iodine and hydrogen gas.

Cathodes

Note that the *sign* of the cathode is *opposite* in an electrolytic cell to that in a galvanic cell. It is the site of reduction in both cases.

Practical points – 1

The voltage applied to the cell must be sufficiently large to drive a current through it *against* the natural cell e.m.f. generated by the substances concerned. The reactions

$Na^+(l) + e^- \rightarrow Na(l)$

$2Cl^-(l) \rightarrow Cl_2(g) + 2e^-$

form the basis of the industrial extraction of sodium metal and chlorine gas from molten sodium chloride, spread 16.1.

Standard electrode potentials.

Oxidized species	⇌	Reduced species	$E^{\ominus}$/V
$2H^+(aq) + 2e^-$	⇌	$H_2(g)$	0
$Na^+(aq) + e^-$	⇌	$Na(s)$	−2.71
$O_2(g) + 4H^+(aq) + 4e^-$	⇌	$2H_2O(l)$	+1.23
$Cl_2(g) + 2e^-$	⇌	$2Cl^-(aq)$	+1.36

The principles of electrolytic cells

An **electrolytic cell** consists of two electrodes immersed in an electrolyte. The **electrolyte** may be either a molten ionic substance or a solution of an ionic substance in water. In either case, the electrolyte will contain positive and negative ions that are free to move. The electrodes are connected to a source of e.m.f., which removes electrons from one electrode and transfers them to the other electrode, via the external circuit. One electrode (the **anode**) bears a *positive* electric charge. The other electrode (the **cathode**) bears a *negative* electric charge.

Positively charged ions from the electrolyte are attracted towards the negatively charged cathode. Positive ions are therefore called **cations** because they are attracted to the cathode. Negatively charged ions (called **anions**) from the electrolyte are attracted towards the positively charged anode. Electron transfer reactions (that is, reduction at the cathode and oxidation at the anode) take place when these ions reach the electrode surfaces. In summary:

- Reduction occurs at the cathode where cations gain electrons from the cathode.
- Oxidation occurs at the anode where anions lose electrons to the anode.

The reactions that take place during electrolysis depend on a number of factors including:

- the state of the electrolyte;
- the concentration of the solution;
- the positions of the elements involved in the electrochemical series.

The electrolysis of a molten substance

When an ionic solid is melted, its ions are free to move in the liquid. We shall consider sodium chloride as a typical example. Molten sodium chloride produces *two* ionic species, $Na^+(l)$ and $Cl^-(l)$. In an electrolytic cell, the $Na^+(l)$ cations are reduced at the cathode to produce molten sodium metal:

$$Na^+(l) + e^- \rightarrow Na(l)$$

The $Cl^-(l)$ anions are oxidized at the anode to produce chlorine gas:

$$2Cl^-(l) \rightarrow Cl_2(g) + 2e^-$$

The electrolysis of an aqueous solution

When an ionic solid dissolves in water, its aqueous ions are free to move. At the same time, water ionizes partially to provide a small concentration of aqueous hydrogen and hydroxide ions:

$$H_2O(l) \rightleftharpoons H^+(aq) + OH^-(aq)$$

We shall continue with sodium chloride as our example. When sodium chloride dissolves in water, it produces $Na^+(aq)$ and $Cl^-(aq)$ ions. Water provides $H^+(aq)$ and $OH^-(aq)$ ions. Aqueous sodium chloride therefore contains a total of *four* ionic species. Which two of these will react (be **discharged**) to form products at the electrodes?

Considering the anions, chloride ion is discharged rather then hydroxide ion: chlorine gas, *not* oxygen, is usually evolved at the anode. The main reason for this result is that the concentration of chloride ions is far greater than usual. The industrial electrolysis of brine uses *saturated* aqueous sodium chloride. The high concentration of chloride ion causes the actual electrode potentials of Cl^- and OH^- to become very similar.

Now consider the cations. A list of standard electrode potentials shows that $H^+(aq)$ is a more powerful oxidant than $Na^+(aq)$, and so will be discharged (reduced) at the cathode as $H_2(g)$. In other words, hydrogen ions are more easily reduced than sodium ions, which remain in solution.

Practical points – 2

As electrolysis proceeds, aqueous hydrogen ions are discharged and hydroxide ions remain in solution. Water dissociates to maintain the value of K_w, the ionic product of water. The products of the electrolysis of aqueous sodium chloride are therefore hydrogen gas, chlorine gas, and aqueous sodium hydroxide, spread 18.2.

Electroplating and electrolytic purification

A metal that has a *positive* standard electrode potential is more readily reduced than hydrogen, and so will be discharged at the cathode in preference to hydrogen. One such metal is copper. In an electrolysis cell with an electrolyte of copper(II) ions, the cathode becomes plated with a layer of copper metal:

Cathode: $Cu^{2+}(aq) + 2e^- \rightarrow Cu(s)$

If a copper anode is used, it undergoes the following oxidation:

Anode: $Cu(s) \rightarrow Cu^{2+}(aq) + 2e^-$

(The anode is termed a *soluble anode*.) As a result, the concentration of copper(II) ions in the solution remains constant. This reaction is the basis of copper-plating and the industrial purification of copper. Using aqueous copper(II) sulphate as the electrolyte, an impure copper anode dissolves while pure copper plates onto the cathode. Impurities in the anode do not dissolve, but drop down to form a layer of 'anode sludge' at the bottom of the cell.

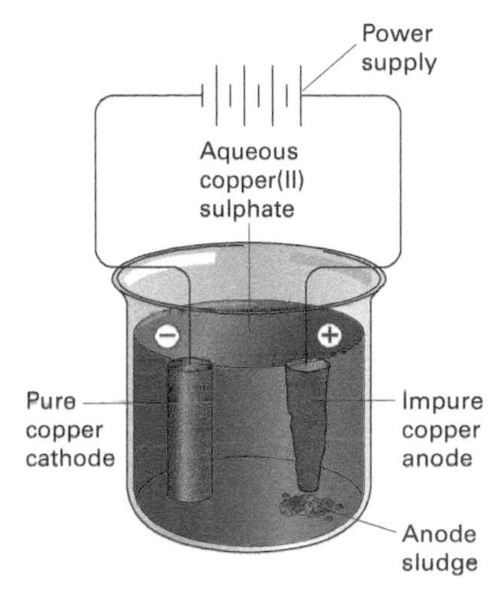

Extremely pure copper is required for electricity cables and water pipes. Precious metals such as gold and platinum are recovered from the 'anode sludge'.

The Faraday constant

The Faraday constant F is the magnitude of the charge per mole of electrons. It is expressed in coulombs per mole ($C\,mol^{-1}$), where one coulomb (1 C) is the total charge carried by one ampere (1 A) of electric current flowing for one second (1 s). It forms the link between the quantity of electricity and the amount in moles of substance discharged. The Faraday constant can be readily calculated by multiplying the charge e on a single electron by the number of electrons per mole L (the Avogadro constant):

$$F = eL = (1.602 \times 10^{-19}\,C) \times (6.022 \times 10^{23}\,mol^{-1})$$
$$= 96\,500\,C\,mol^{-1} \text{ (3 sig. figs)}$$

The Faraday constant may be used to calculate the charge required to deposit a particular mass of substance during electrolysis, as shown in the box (right).

How much metal?

We can calculate the mass of metal deposited during electrolysis by using the Faraday constant.

For example, 1 mol of silver Ag has a mass of 108 g. Silver ions $Ag^+(aq)$ discharge at the cathode during electrolysis:

$Ag^+(aq) + e^- \rightarrow Ag(s)$

So 96 500 C deposit 108 g of silver.

As a second example, 1 mol of copper Cu has a mass of 63.5 g. Copper(II) ions $Cu^{2+}(aq)$ discharge at the cathode during electrolysis:

$Cu^{2+}(aq) + 2e^- \rightarrow Cu(s)$

But now we need two moles of electrons per mole of Cu^{2+} ions. Therefore, we need 2 × 96 500 C to deposit 63.5 g of copper.

SUMMARY

The products obtained during electrolysis depend on a number of factors, including:

- the state of the electrolyte – molten or aqueous;
- the relative concentrations of the ions in an aqueous electrolyte;
- the relative values of the electrode potentials that apply to the conditions of the electrolysis cell.

PRACTICE

1 The anode and cathode in an electrolytic cell are made from copper; the electrolyte is aqueous copper(II) sulphate. What mass of copper will be transferred if a current of 0.5 A flows for 15 min?

13.12

Rusting

OBJECTIVES

- Cause of rusting
- Mechanism of rusting
- Prevention of rusting

One of the most important redox reactions is the corrosion or rusting of iron. It is estimated that this costs the UK in excess of £5 billion each year. Iron and steel objects have a natural tendency to rust. We first need to understand why rusting occurs before we can suggest strategies for reducing the rate of rusting.

The cause of rusting

Rusting can be predicted on the basis of standard electrode potentials. Oxygen has a standard electrode potential for the reaction

$$O_2(g) + 4H^+(aq) + 4e^- \rightarrow 2H_2O(l)$$

of $E^{\ominus} = +1.23$ V, spread 13.8; oxygen is a strong oxidant. Iron has a standard electrode potential for

$$Fe^{2+}(aq) + 2e^- \rightarrow Fe(s)$$

of $E^{\ominus} = -0.44$ V; iron is a moderately strong reductant. So oxygen should oxidize iron in aqueous solution.

It is well known that two of the requirements for rusting are the presence of water and of oxygen. An approximate equation describing rusting is:

$$4Fe(s) + 3O_2(g) + 2xH_2O(l) \rightarrow 2Fe_2O_3.xH_2O(s)$$

Notice how the exact number of water molecules present in rust is uncertain.

The hull of the Titanic has rusted since it sank in 1912.

The mechanism of rusting

We can examine the mechanism of rusting in more detail. A steel nail which has a drop of 'ferroxyl' indicator added shows two important aspects of the reaction.

A blue mass forms, indicating that iron(II) ions are present. This *anodic process* is the oxidation of iron:

$$Fe(s) \rightarrow Fe^{2+}(aq) + 2e^-$$

The pink colour indicates that hydroxyl ions are present. This *cathodic process* is the reduction of oxygen:

$$O_2(g) + 2H_2O(l) + 4e^- \rightarrow 4OH^-(aq)$$

Combining these two half-equations, multiplying the first by two, we find

$$2Fe(s) + O_2(g) + 2H_2O(l) \rightarrow 2Fe^{2+}(aq) + 4OH^-(aq)$$

The iron(II) hydroxide so formed is then rapidly oxidized under basic conditions, see spreads 13.9 and 20.5, to red-brown hydrated iron(III) oxide, best represented by the approximate formula $Fe_2O_3.xH_2O$. The rust flakes off the surface (partly because rust has a lower density than iron). This exposes more surface and rusting continues.

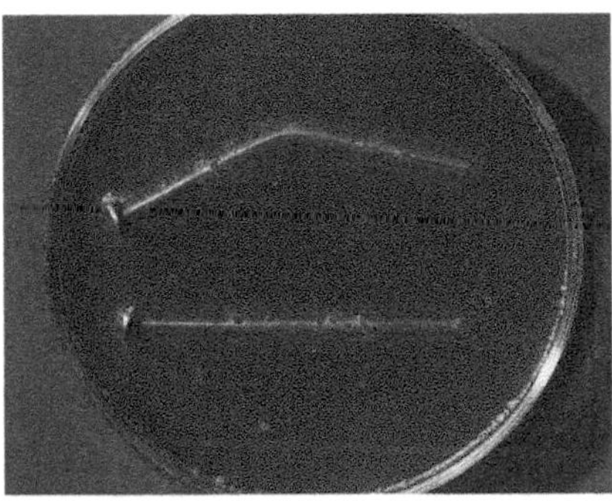

Nails placed in 'ferroxyl' indicator. 'Ferroxyl' indicator is a mixture of aqueous potassium hexacyanoferrate(III), which gives a blue colour with iron(II) ions, spread 20.11, and phenolphthalein, which indicates the presence of hydroxyl ions by turning pink in basic solution, spread 12.9.

Prevention of rusting

Iron and steel objects can be protected against rusting in a variety of ways. Painting would be perfectly adequate, as long as the surface is not damaged. In the case of a car, this is extremely unlikely to be the case once stone chips occur. Rusting can occur *anywhere* under the paint once the surface is damaged as the metal surface is a conductor and the electrons released by iron can travel through the rest of the car to the hole, where oxygen is then reduced. In the twentieth century, rust was a major cause of car scrappage.

A more intelligent solution results from consulting the table on spread 13.7. A metal with a *more negative* standard electrode potential will give up its electrons more readily than iron does. So if a more strongly reducing metal is attached to the iron, that protective metal will be oxidized in place of the iron. The protective metal is termed a **sacrificial anode**. To protect an object as large as a ship, pieces of magnesium ($E^{\ominus} = -2.37$ V) are attached to the hull. Replacement blocks are then substituted when the protective metal becomes depleted.

Bluebird

Donald Campbell was killed in his speedboat Bluebird K7 when he crashed at over 300 mph while trying to break his own world water speed record. (Campbell remains the only person to hold the world land and water speed records simultaneously.) The Bristol Siddeley Orpheus engine in Bluebird, made mainly of magnesium, acted as a sacrificial anode while the wreck of the boat lay under water for 34 years. Its decay effectively protected the less reactive aluminium and steel of the bodywork, as was discovered when Bluebird was raised from Coniston Water in 2001.

The same principle is applied to protecting cars using zinc ($E^{\ominus}$ = –0.76 V) in a process called **galvanizing**. Car manufacturers that use 100 per cent galvanized bodywork can offer longer warranties against body corrosion. Note that this slows the rate of corrosion, but does not prevent the iron from corroding once the zinc has been exhausted and so it does not protect indefinitely. Actually the situation is rather less simple than it appears at first sight. Zinc ions *do* go into solution in place of iron ions; aqueous zinc ions are colourless, as explained in spread 20.11, and escape visual detection: the car actually still disintegrates, only its decay is less easily seen!

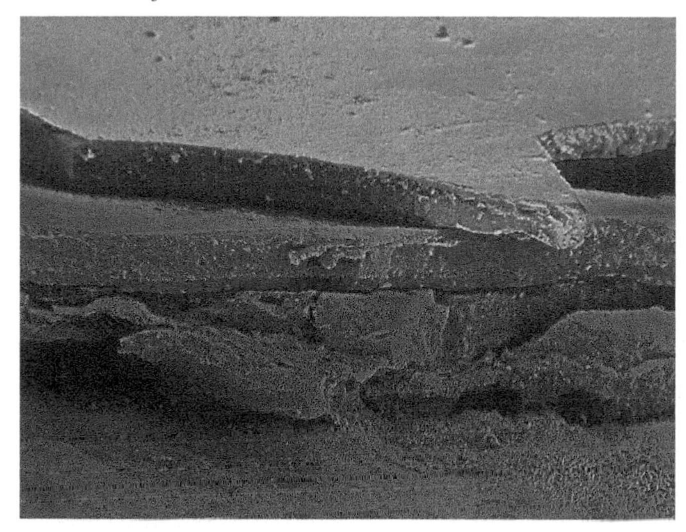

Scanning electron micrograph of resprayed paint (blue) poorly bonded to the original paint of a rusty car.

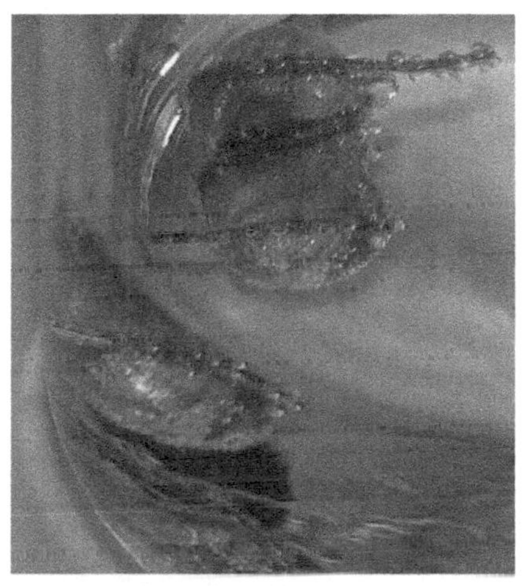

Acid reacting with a zinc granule wrapped in a copper wire. When in contact, electrons can flow from zinc to copper, where they can reduce hydrogen ions (from the acid) to form bubbles of hydrogen gas.

Making rusting more severe

Standard electrode potentials can also help us understand how rusting can be made more severe. The rate of rusting is accelerated if iron is in contact with a metal with a less negative standard electrode potential, such as tin ($E^{\ominus}$ = –0.14 V) or copper ($E^{\ominus}$ = +0.34 V). Tin-plated cans rust very quickly when opened, which explains why food should be removed rapidly (aluminium cans do not suffer in the same way). Gustav Eiffel, the designer of the Statue of Liberty in New York harbour, was aware that his support structure, built of iron, would rust if it contacted the copper cladding. He attempted to prevent this by placing asbestos pads between the metals. Unfortunately, contact anywhere along the structure can cause corrosion and Eiffel's original iron framework has had to be completely replaced.

A similar reaction can be demonstrated in the laboratory by wrapping a copper wire around a piece of zinc. When placed in acid, bubbles of hydrogen can be seen on the copper as well as on the zinc. It is well known that copper metal itself does not reduce hydrogen ions, spread 13.8. The electrons required for the reduction have come from the zinc; they are available at the copper as it is in electrical contact. The solution remains colourless when all the zinc has reacted, as aqueous zinc ions are colourless.

Railway trucks near a salt lake in Qinghai province, China: there is an impressive amount of rust, made worse by the high concentration of salt.

Rusting occurs faster when salt is present, which happens in winter as salt is spread on roads to try to keep them free from ice. The salt increases the conductivity of the water, which is the limiting factor in the rate of rusting as water's conductivity is much lower than that of the metal with which it is in contact.

SUMMARY

- Iron rusts in the presence of water and oxygen.
- The anodic process is the oxidation of iron: $Fe(s) \rightarrow Fe^{2+}(aq) + 2e^-$
- Iron can be protected using a sacrificial anode; zinc is used in galvanizing.
- Rusting is more severe with salt present.

PRACTICE

1. Describe the anodic and cathodic processes that occur during rusting.
2. How can you slow down the rate of rusting? Can you stop it permanently?

13.13 Principles of metal extraction

OBJECTIVES

- Carbon reduction
- Electrolysis
- Batch processes

The ideas introduced in this chapter can be used to understand the different extraction techniques used to win metals from their ores. The most widely applicable method is electrolysis; this is also the most expensive. Hence, this is the preferred option in industry only if it is not possible to extract the metal more cheaply by other methods. In this section we will examine the reasons why the various techniques can be used.

The least reactive metals

Some metals have been used since prehistory. They are the ones that are the least reactive (have the most positive standard electrode potentials). A famous example is gold. Because it is so unreactive, it exists in the environment 'native', which means uncombined with another element. In the previous two hundred years there have been a number of 'gold rushes', as for example in 1849 in California. If fortunate, a prospector could literally pick up a piece of gold. Hence no chemical techniques need to be employed to extract gold.

Gold has been prized for millennia and many of the most beautiful artefacts handed down from previous generations have been made of gold. One example is the saltcellar designed by Benvenuto Cellini.

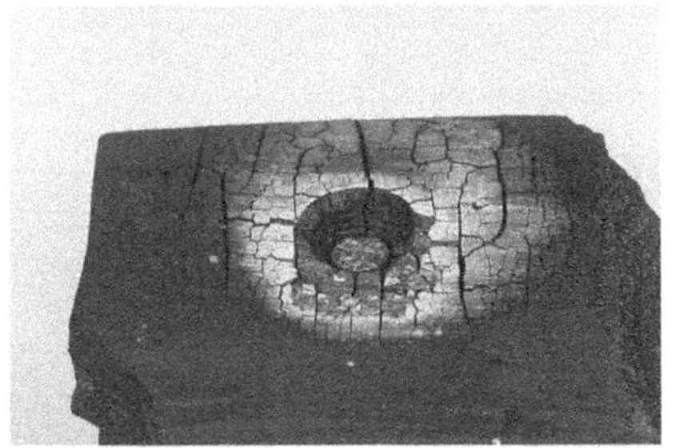

Shiny lead forms when lead(II) oxide is heated with carbon.

This magnificent saltcellar was made by Benvenuto Cellini in the sixteenth century for King Francis I of France.

The fairly reactive metals

When the metal is rather more reactive than gold, it will be found in the environment combined with other elements, most commonly with oxygen. The oxide will then need to be converted to the element. As has been explained in this chapter, conversion of an oxide to the element involves reduction. The choice of reductant needs to be carefully considered. Cost is a very significant factor and crucial for industrial success. The cheapest widely available reductant is the element carbon.

In the photograph in the margin above, lead(II) oxide is shown having been heated with carbon; shiny lead may be seen in places throughout the hole in the block. The equation for the reaction is

$$PbO(s) + C(s) \rightarrow Pb(s) + CO(g)$$

Iron has been extracted from its oxide ore haematite for over two thousand years by heating with carbon in a blast furnace. The temperature is crucial to the extraction process; the next chapter explains why the extraction requires high temperatures and also introduces the idea that both carbon and carbon monoxide (formed from carbon in the furnace) can be effective reductants. Iron is frequently converted into steel (spread 20.2).

Steel, 60% of which was recycled, was used in the construction of the ArcelorMittal Orbit, which was built to celebrate London 2012.

Andrew Carnegie became one of the richest people in the world at the turn of the twentieth century by gaining effective control of steel manufacture in the US. He built Carnegie Hall, New York, famous for its concerts and recitals.

Lakshmi Mittal

In the first decade of the twenty-first century, another steel magnate, Lakshmi Mittal, became one of the ten richest people in the world (and the richest in the UK). His London house was the most expensive house in the world when he bought it in 2003. History repeats itself!

The most reactive metals

Iron is the most familiar metal in everyday life for two reasons. The extraction technique is relatively inexpensive and the metal is very abundant in nature. However, there is one metal which is even more abundant, aluminium. Despite its abundance, elemental aluminium was not made until the nineteenth century because the temperature needed for extraction by carbon is uneconomically high, as explained in spread 14.6.

The extraction of aluminium, and that of other very reactive metals such as sodium and potassium, had to wait for a new technological advance. This was the discovery of electrolysis, which itself relied on the batteries first made around 1800.

Sodium and potassium proved easier to extract than aluminium. The principles of electrolysis require that the electrolyte is either molten or in solution (as discussed in a previous spread). The problem with aluminium is that its ore bauxite (hydrated aluminium oxide) has an enormously high melting point (over 2000 °C), thus preventing electrolysis at an economical temperature. Sodium chloride on the other hand melts at 801 °C, so its extraction is easier than that of aluminium.

The technological problems involved in the electrolysis of aluminium were overcome by Charles Hall, as described in spread 19.2. Like Carnegie, chemical manufacture made Hall extremely rich. As with iron manufacture the process is continuous, a typical Hall cell producing a little over a tonne of aluminium per day.

At the very top of the Washington Monument is an aluminium pyramid 22.6 cm high weighing 2.85 kg, which was at the time the largest single piece of cast aluminium. When the Monument was completed in 1884 it was the tallest man-made structure in the world (at 169 m). Aluminium was chosen for its rarity, conductivity, colour, and non-staining qualities.

Year	1852	1855	1858	1888	1895	1900	1950
Price of aluminium metal / $ per kg	1200	250	25	11.5	1.15	0.73	0.40
		→| Introduction of $Na/AlCl_3$ process		→| Introduction of Hall electrolysis			

The price of aluminium per kilogram fell significantly when the Hall process was introduced in the 1880s. The other major drop occurred in 1855 when it was first extracted in a batch process using sodium metal reducing aluminium chloride. Aluminium is now the second most widely used metal, after iron.

Batch processes

Metals required on a less extravagant scale than iron and aluminium do not need to be manufactured continuously. They can be made when needed in a **batch process**, the name being derived from the bread produced in a single baking in a baker's oven.

For example, aluminium can be used to extract chromium (spread 20.3); the driving force for the process is the extremely high lattice energy of aluminium oxide. The technique is very similar to that of the Thermit process shown in spread 3.1.

Similarly, magnesium can be used to extract titanium (spread 20.3). Titanium is the ninth most abundant element in the environment. Yet the fact that it is extracted in a batch process using a metal that itself must be extracted using the expensive option of electrolysis makes titanium a rare and prized metal. The monument commemorating the exploration of space erected in Moscow is made from titanium.

To celebrate their achievement of the first successful space flight by Yuri Gagarin and their subsequent exploration of space, the Russians built this monument out of titanium.

SUMMARY

- The least reactive metals, such as gold, are found 'native'.
- The fairly reactive metals, such as iron, are extracted by carbon reduction at high temperatures.
- The most reactive metals are extracted by electrolysis (as for aluminium) or by using an even more reactive metal, in a batch process (as for titanium).

PRACTICE

1 Explain the principles behind the extraction of metals.

PRACTICE EXAM QUESTIONS

1 a In terms of transfer of electrons, what is meant by the term *reduction*? [1]

b Name the electrode relative to which standard electrode potentials are measured, and give one reason why this electrode is not often used in experimental determinations of standard electrode potentials. [2]

c In both of the overall redox reactions given below, the reaction can be separated into two redox half-equations. In each case, identify the two redox half-equations, and state which species is being reduced.

i $2Ag^+(aq) + Cu(s) \rightarrow 2Ag(s) + Cu^{2+}(aq)$

ii $5S_2O_8^{2-}(aq) + 2Mn^{2+}(aq) + 8H_2O(l) \rightarrow$
$10SO_4^{2-}(aq) + 2MnO_4^-(aq) + 16H^+(aq)$ [6]

2 a Standard electrode potentials, $E^\ominus$, are measured relative to a standard reference electrode. What is the standard reference electrode and what is its potential? [2]

b State three conditions which must apply when values of $E^\ominus$ are being determined. [3]

c What is the function of a *salt bridge*, and what might it contain? [2]

d What is meant by the *Electrochemical Series*? [2]

e Consider the following standard electrode potentials.

$Fe^{2+}(aq) + 2e^- \rightarrow Fe(s)$ $E^\ominus/V = -0.44$

$Zn^{2+}(aq) + 2e^- \rightarrow Zn(s)$ $E^\ominus/V = -0.76$

State which species is reduced if these two half-cells are joined together in an electrochemical cell. Explain your answer. [3]

3 Use the standard potentials in the list below, as appropriate, to answer the questions that follow.

	$E^\ominus/V$
$MnO_4^-(aq) + 8H^+(aq) + 5e^- \rightarrow Mn^{2+}(aq) + 4H_2O(l)$	+1.51
$Cl_2(g) + 2e^- \rightarrow 2Cl^-(aq)$	+1.36
$Cr_2O_7^{2-}(aq) + 14H^+(aq) + 6e^- \rightarrow 2Cr^{3+}(aq) + 7H_2O(l)$	+1.33
$Fe^{3+}(aq) + e^- \rightarrow Fe^{2+}(aq)$	+0.78
$Fe^{3+}(aq) + 3e^- \rightarrow Fe(s)$	−0.04
$Fe^{2+}(aq) + 2e^- \rightarrow Fe(s)$	−0.44

a The two half-cells above which involve metallic iron are joined to produce an electrochemical cell. Using standard notation, give the conventional representation for this cell, calculate its standard potential, and write an equation for the spontaneous reaction that occurs. [6]

b Potassium manganate(VII) and potassium dichromate(VI) are both strong oxidizing agents. Which of the two is **not** used for the quantitative estimation of iron(II) ions in a solution of iron(II) chloride? Use data from the table to justify your choice. [3]

c Equimolar solutions of acidified potassium manganate(VII), manganese(II) sulphate, potassium dichromate(VI) and chromium(III) sulphate are mixed and allowed to come to equilibrium.

i State which ion is oxidised and which is reduced.

ii Use half-equations to construct the equation for the overall reaction that occurs. [5]

4 a Select **three** different general methods for the extraction of metals. For **each** method you select, state the starting materials, the conditions used and give one example of a metal extracted by this method. [9]

b i Indicate the essential chemistry involved in the removal of carbon from impure iron in the manufacture of steel.

ii Give **two** reasons why steel is less expensive to produce than titanium.

iii Give **one** reason why titanium is used for certain applications despite the extra cost of this metal as compared to steel. [5]

5 The apparatus below was used to measure the **standard** electrode potential of the Fe/Fe^{3+} electrode, copper being the positive electrode.

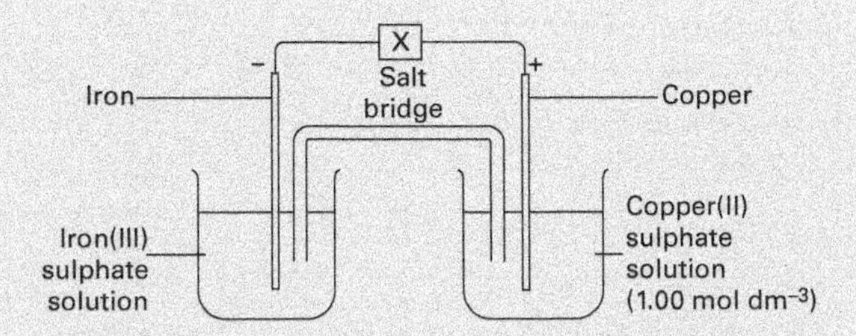

a i Name the instrument which could be used at **X** to measure the e.m.f. of the cell. Indicate its main characteristic.

ii What is the concentration of the Fe^{3+} ions in the iron(III) sulphate solution? [3]

b i What is the function of the salt bridge?

ii What might it contain?

iii Why is the salt bridge used and not a piece of wire? [3]

c i The e.m.f. of the cell is +0.38 V. Given that the standard electrode potential of the Cu^{2+}/Cu electrode is +0.34 V, calculate the standard electrode potential of the Fe^{3+}/Fe electrode and explain your reasoning.

ii Give the conventional representation of the cell.

iii Write an equation, with state symbols, to represent the cell reaction. [7]

d Use the data below to explain concisely why zinc is used in preference to tin for coating steel which is used to manufacture cars. [3]

	$E^{\ominus}$/V
$Sn^{2+}(aq) + 2e^- \rightleftharpoons Sn(s)$	−0.14
$Fe^{2+}(aq) + 2e^- \rightleftharpoons Fe(s)$	0.44
$Zn^{2+}(aq) + 2e^- \rightleftharpoons Zn(s)$	−0.76

6 A student set up 4 standard half-cells each containing one of the metals A, B, C and D. These half-cells were then used to make different electrochemical cells. The table below shows the standard cell potential, $E^{\ominus}_{cell}$, and the positive terminal of each electrochemical cell.

Cell	Metals used	$E^{\ominus}_{cell}$/V	Positive terminal
1	**A** and **B**	+1.10	**B**
2	**B** and **C**	+0.46	**C**
3	**B** and **D**	+0.47	**B**

a Draw a labelled diagram of cell 1. [3]

b Deduce the order of reactivity of the metals **A**, **B**, **C** and **D**. [3]

c Outline a method the student may have used to identify the positive terminal. [1]

d Calculate the standard cell potential of a cell made from the half-cells containing **A** and **D**. [1]

7 This question concerns the lead-acid battery.

The following data will be required.

	$E^{\ominus}$/V
$PbO_2(s) + 4H^+(aq) + SO_4^{2-}(aq) + 2e^- \rightleftharpoons PbSO_4(s) + 2H_2O(l)$	+1.69
$PbSO_4(s) + 2e^- \rightleftharpoons SO_4^{2-}(aq) + Pb(s)$	−0.36

a The lead-acid battery is one form of storage cell. What substance is used for:

i the negative pole; [1]

ii the positive pole; [1]

iii the electrolyte. [1]

b Give the equation for the overall cell reaction during discharge. [2]

c Calculate the e.m.f. of the cell. [2]

d A storage cell, as used in the lead-acid battery, is a simple cell in which the reactions are reversible i.e. once the chemicals have been used up they can be re-formed.

Write an equation for the chemical reaction which occurs on charging. [1]

e Give one disadvantage of such batteries for use in cars. [1]

f **i** State the essential requirement for the rusting of iron in water. [1]

ii Explain why corrosion of iron results in deep pitting of the metal surface. [1]

iii Explain why sheet iron which has been fabricated to a particular shape, sometimes under high pressure, is more likely to corrode than a single strip of pure iron. [2]

iv An underground iron pipe is less likely to corrode if bonded at intervals to magnesium stakes. Give a reason for this. Explain why aluminium would be a poor substitute for the magnesium. [2]

8 Aluminium/air electrochemical cells can be used to power golf trolleys and invalid carriages. One electrode is made of aluminium while the other is made by bubbling air through an inert porous material. The electrolyte is usually sodium hydroxide solution. The equations for the reactions taking place at the electrodes are:

I $Al(s) + 3OH^-(aq) \rightarrow Al(OH)_3(s) + 3e^-$

II $O_2(g) + 2H_2O(l) + 4e^- \rightarrow 4OH^-(aq)$

a Which electrode acts as the cathode of the cell? [1]

b Write an equation for the overall cell reaction. [1]

c Suggest one reason why the efficiency of the cell may be reduced over a period of time. [1]

9 **a** Explain why each of the following reactions can be classified as redox processes. Give a balanced equation for each reaction.

i chlorine reacting with aqueous potassium iodide;

ii chlorine reacting with hydrogen;

iii iron(II) ions reacting with acidified aqueous manganate(VII) ions. [5]

b **i** Explain why electrolysis is a redox process.

ii Describe, including the electrode reactions, the industrial extraction of aluminium from purified aluminium oxide. [7]

(See also chapters 18 and 19.)

Spontaneous change towards equilibrium

Stir a spoonful of sugar into a cup of coffee and the solid sugar *spontaneously* dissolves in the solution. Mix together aqueous silver ions and aqueous iodide ions and a yellow precipitate of silver iodide *spontaneously* forms. The Universe is full of spontaneous changes of this sort. Bicycles roll down hills, hot coffee cools down, our bodies age, and the steel bodies of motor cars rust slowly back to the ore from which the iron was originally extracted. **Spontaneous changes** are changes that have a natural tendency to occur. The question should now arise in your mind: 'Yes: but what makes spontaneous changes happen?' Developing an answer to this question is the aim of this chapter. At its end you should be able to complete the sentence: 'Spontaneous changes happen because ....'

SPONTANEITY AND SPREADING

14.1

OBJECTIVES

- System and surroundings
- Criterion for spontaneous change

The concept of *stability* is a useful point from which to start this topic. A stable substance, or mixture of substances, is one that does not change its structure or composition. Think about stirring sugar into your coffee. Kept in isolation, sugar and water are both *stable* substances. Each compound will remain unchanged for an indefinite period of time. Pouring sugar into water makes an *unstable* system. The state of the system spontaneously changes to make a solution. When the solution has formed, the system is more stable. **Spontaneous change**, change that has a natural tendency to occur, causes a system to move from a less stable state to a more stable state. So what *causes* spontaneous change?

When it is at the rim, the ball has minimum kinetic energy and maximum potential energy. When it is at the bottom of the bowl, the ball has maximum kinetic energy and minimum potential energy. In the absence of friction and air resistance, this system does not *tend towards the position of minimum potential energy.*

Spontaneous change and energy

The concept of energy change is often used to explain things that happen in the world. For example, think of a frictionless ball sliding down the inside edge of a large frictionless bowl. The ball travels down to the bottom of the bowl and up the opposite side. It is tempting to say that the ball is in an unstable state at the rim, and that it is more stable when it has fallen to the point of lowest potential energy, at the bottom of the bowl. However, there is no overall energy change happening in this theoretical system because *the total energy of the ball remains constant*. In the absence of friction, the ball will continue to oscillate and its total energy remains constant.

A real ball oscillating in a real bowl *does* spontaneously reach a final state

The direction of spontaneous change. A block of hot metal cools to the temperature of its surroundings. A block that is at the same temperature as its surroundings does not spontaneously become hotter.

System and surroundings: a reminder

Chemical changes and physical changes of state are accompanied by changes in energy. The substances undergoing the change are called 'the system'; everything else outside the system is called 'the surroundings'. An energy change in the system causes the temperature of the system to change. If the system and the surroundings are in contact, heat then flows between the system and the surroundings.

of minimum potential energy at the bottom of the bowl. However, the reason the ball settles at the bottom is not because that is the system's state of minimum potential energy. It is more accurate to say that friction causes *energy to spread out* from the ball. Kinetic and potential energy that were concentrated in the ball have *spread out* as heat into the ball, into the bowl, and into the surrounding air.

Spontaneous change and the state of matter

This change in our way of thinking about a situation helps to explain the physical change of sugar dissolving in a cup of coffee. Using ideas from thermochemistry, you might suggest that sugar dissolves in water because the change is *exothermic*. That is, the system gives out heat as the solution forms, because the energy of the separate hydrated sugar molecules is less than the energy of sugar molecules in a solid sugar crystal.

However, many substances have an *endothermic* enthalpy of solution. According to an argument based solely on enthalpy changes, these substances would not be expected to dissolve spontaneously in water. But, for example, ammonium nitrate *does* dissolve:

$$NH_4NO_3(s) \rightarrow NH_4^+(aq) + NO_3^-(aq); \quad \Delta_{sol}H^{\ominus}\,(298\,K) = +25.8\,kJ\,mol^{-1}$$

The positive value of the enthalpy change shows that heat has flowed from the surrounding water to increase the energies of the $NH_4^+(aq)$ and $NO_3^-(aq)$ ions (compared with their energies when in the solid state). While energy has not spread out during this spontaneous change, matter *has* spread out. As in all cases of dissolution, a solid ordered lattice has broken down and spread out into a more disordered state. The spreading out of matter – not the enthalpy change – is the common factor in the formation of solutions.

- Spontaneous changes are those that cause energy and/or matter to spread out.

You should now understand that the spreading out of energy and/or matter gives direction to spontaneous changes. Simply considering just decreases in energy does not lead to an adequate model that explains observations.

The next spread develops the idea of 'spreading out' by quantifying and measuring it. To proceed further, you must get to grips with a concept called 'entropy', and apply it to chemical as well as physical changes.

SUMMARY

- The structure and composition of a stable system do not change over time.
- Spontaneous change moves a system from a less stable state to a more stable state.
- Spontaneous change cannot be accounted for solely in terms of the system tending to a state of lower energy.
- Spontaneous changes are those that cause energy and/or matter to spread out.

Pendulums

A frictionless pendulum swinging in a vacuum is another example of a theoretical system in which the energy remains constant and does not spread out into the surroundings.

The direction of spontaneous change. A gas spontaneously diffuses to fill its container uniformly. For an ideal gas, there is no exchange of energy with the surroundings.

Spontaneity and rate

As in previous chapters, we must also consider the *rate* at which a reaction occurs. For example, hydrogen and oxygen are unstable with respect to water:

$H_2(g) + \frac{1}{2}O_2(g) \rightarrow H_2O(l);$

$\Delta H^{\ominus}(298\,K) = -286\,kJ\,mol^{-1}$

The reaction is highly exothermic. However, at room temperature (298 K), the rate of reaction is so slow as to be effectively zero. So, even though the reaction is spontaneous at room temperature, it occurs at a rate too slow to observe. Spontaneous does *not* mean fast.

PRACTICE

1 Describe each of the following changes in terms of the spreading out of matter or of energy. Where appropriate, describe changes in the structure of the system and changes in the energy distribution.

  a Melting ice.

  b The exothermic neutralization of aqueous hydrochloric acid by aqueous sodium hydroxide.

  c Gas diffusion.

  d The combustion of a liquid fuel.

2 Suggest why a spontaneous endothermic precipitation reaction is unlikely to occur.

14.2

OBJECTIVES

- Entropy and disorder
- Entropy and phase change
- Entropy changes in system and surroundings
- Second law of thermodynamics and spontaneous change

Boltzmann equation

Physicists commonly relate the entropy of a system to the number of ways in which the system's energy can be distributed amongst its particles. The more widely dispersed the energy, the greater the system's entropy. Quantitatively, the entropy S depends on W, the number of ways the energy can be spread, by the **Boltzmann equation** $S = k \ln W$, where k is the Boltzmann constant. The gas constant R is closely related to the Boltzmann constant: $R = Lk$ where L is the Avogadro constant.

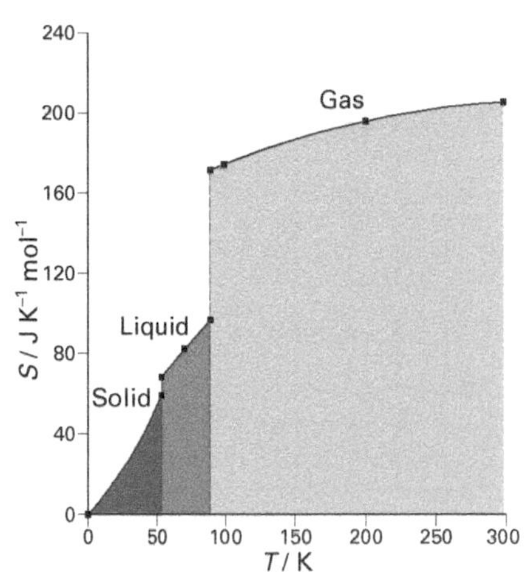

*The entropy **S** of a substance (in this case oxygen) increases as its temperature **T** increases. There is a sharp increase in entropy at the melting point as the structure changes from solid to the more random arrangement of particles in a liquid. There is an even greater increase in entropy at the boiling point because the particles are much more widely spread out.*

Comment

In both cases (a) and (b) in the worked example, heat has been added to the system (the water). This energy input has increased the internal energy of the water molecules.

At the molecular level, the increase in disorder is much greater in the case of vaporization than in the case of melting. This difference is reflected in the values of the entropy changes:

$\Delta S = +109\ \mathrm{J\ K^{-1}\ mol^{-1}}$ for vaporization,

$\Delta S = +22\ \mathrm{J\ K^{-1}\ mol^{-1}}$ for melting.

ENTROPY – 1

The previous spread revealed that spontaneous change involves an increase in disorder: either energy or matter or both must spread out. A large negative (exothermic) standard enthalpy change is *not* a sufficient criterion for spontaneous change. There are many endothermic changes which are spontaneous (such as the dissolution of ammonium nitrate). This spread introduces the concept of *entropy change* as a means of describing and quantifying disorder. The following spread uses calculations involving entropy to identify and predict spontaneous changes.

Disorder and entropy

Entropy is a measure of the disorder of a system. The entropy of a system increases when the matter or energy in the system spreads out or becomes more random in its arrangement. For example, entropy increases when a solid melts to produce a liquid, when a liquid vaporizes to produce a gas, and when substances dissolve or mix together. Entropy also increases when a substance is heated with no change of state. For example, heating a liquid increases the speed of the particles and the number of collisions per second; the entropy of the liquid increases as its particles become more disordered. At the same time, entropy increases as energy from the source of heat spreads out into the liquid.

Numerical values for entropy

Entropy is given the symbol S. Just as the enthalpy change ΔH was important, so it is the entropy *change* ΔS that is important. In general terms, the **entropy change** ΔS in a system is defined as:

$$\Delta S = q/T$$

where q is the heat added to the system from the surroundings and T is the thermodynamic temperature (in kelvin) at which the heat is transferred. To understand the significance of this expression, think about heating a system that is undergoing a change of state (e.g. ice melting). Heating the system does not increase the temperature of the system: it just makes the matter in the system more disordered.

Worked example on calculating entropy changes

Data: The enthalpy change during melting of ice $\Delta_{fus}H$ is $+6.01\ \mathrm{kJ\ mol^{-1}}$ at 273 K; the enthalpy change during vaporization of water $\Delta_{vap}H$ is $+40.7\ \mathrm{kJ\ mol^{-1}}$ at 373 K.

Question: Calculate the entropy change per mole when (a) ice melts at 0 °C and (b) water vaporizes at 100 °C. Remember that the enthalpy change equals the heat added to the system at constant pressure.

Answer:

(a) The equation is

$H_2O(s) \rightarrow H_2O(l)$; $\Delta_{fus}H(273\ \mathrm{K}) = +6.01\ \mathrm{kJ\ mol^{-1}}$

So

$$\Delta S = \frac{q}{T} = \frac{\Delta_{fus}H}{T} = \frac{+6010\ \mathrm{J\ mol^{-1}}}{273\ \mathrm{K}} = +22.0\ \mathrm{J\ K^{-1}\ mol^{-1}}$$

(b) The equation is

$H_2O(l) \rightarrow H_2O(g)$; $\Delta_{vap}H(373\ \mathrm{K}) = +40.7\ \mathrm{kJ\ mol^{-1}}$

So

$$\Delta S = \frac{q}{T} = \frac{\Delta_{vap}H}{T} = \frac{+40\,700\ \mathrm{J\ mol^{-1}}}{373\ \mathrm{K}} = +109\ \mathrm{J\ K^{-1}\ mol^{-1}}$$

System and surroundings

The total entropy change for any process can be broken down into the entropy change in the system, ΔS_{sys}, plus the entropy change in the surroundings, ΔS_{surr}. In an exothermic reaction, heat disperses from the system into the surroundings. Remember that, for an exothermic reaction, the enthalpy change for the system, ΔH, is negative; so the enthalpy change for the surroundings, $-\Delta H$, has a positive value.

In terms of entropy change, the significance of the heat released by an exothermic reaction into the surroundings depends on how hot the surroundings already are. The hotter they are, the less effect a given heat transfer will have. Quantitatively, the size of the entropy change in the surroundings is *inversely* proportional to the temperature:

$$\Delta S_{surr} = \frac{q}{T} \quad \text{or} \quad \Delta S_{surr} = \frac{-\Delta H}{T}$$

(The surroundings are always assumed to be so large that no change in temperature results.)

Combining the entropy changes in the system and the surroundings:

$$\Delta S_{tot} = \Delta S_{sys} + \Delta S_{surr}$$

which becomes

$$\Delta S_{tot} = \Delta S_{sys} - \frac{\Delta H}{T}$$

By convention, ΔS refers to the system, so the expression becomes:

$$\Delta S_{tot} = \Delta S - \frac{\Delta H}{T}$$

You should notice that the term $-\Delta H/T$, which is used to represent the entropy change in the *surroundings*, is expressed in terms that describe the *system* only.

The second law of thermodynamics

The system plus the surroundings taken together constitute the Universe. An observed fact is that, when any change takes place, the total entropy of the Universe ΔS_{tot} tends to increase. This observation is expressed as the **second law of thermodynamics**, simply stated as:

- The total entropy of the Universe always tends to increase; it never goes down.

As a result, ΔS_{tot} for any change is always greater than (or equal to) zero (the special case where it is zero is dealt with in spread 14.5). A negative total entropy change is not possible. For a spontaneous change to occur:

$$\Delta S_{tot} > 0$$

So substituting the equation above gives

$$\Delta S - \frac{\Delta H}{T} > 0 \quad \text{for a spontaneous change}$$

Now if we multiply through by T, the criterion for spontaneous change becomes:

$$T\Delta S - \Delta H > 0$$

Note that, as in any equation, the units are the same on both sides of this inequality, so the units of ΔH and of $T\Delta S$ are J mol^{-1}.

SUMMARY

- Entropy change $\Delta S = q/T$ (units J K^{-1} mol^{-1}).
- Entropy increases with the temperature of the system.
- For a spontaneous change, the total entropy tends to increase.
- For spontaneous change, $\Delta S_{tot} > 0$, or alternatively $T\Delta S - \Delta H > 0$

The first law of thermodynamics

The **first law of thermodynamics** states that:

- Energy cannot be created or destroyed.

The significance of this law to a system and its surroundings is that the enthalpy change in the surroundings is *exactly equal* in magnitude, and *opposite* in sign, to the enthalpy change in the system.

Spontaneous change

For a spontaneous change,

$\Delta S_{tot} > 0$ so $T\Delta S - \Delta H > 0$

Imagine a system consisting of solid ice and liquid water in equilibrium at 0 °C. Now imagine two situations:

1. If the surroundings have a temperature slightly greater than 0 °C (e.g. 10 °C, 283 K), heat flows *into* the system. Then from our data and calculations in the worked example opposite:

$H_2O(s) \rightarrow H_2O(l)$;

$\Delta H = +6.01$ kJ mol^{-1}

$\Delta S = +22.0$ J K^{-1} mol^{-1}

The value of $T\Delta S - \Delta H$ in this case is

[283 K × (+22.0 J K^{-1} mol^{-1})] − (+6010 J mol^{-1})

= +200 J mol^{-1} (1 sig. fig.)

This is greater than zero, so the ice melting is spontaneous.

2. If the surroundings have a temperature slightly below 0 °C (e.g. −10 °C, 263 K), heat flows *out of* the system. The value of $T\Delta S - \Delta H$ in this case is

[263 K × (+22.0 J K^{-1} mol^{-1})] − (+6010 J mol^{-1})

= −200 J mol^{-1} (1 sig. fig.)

This is less than zero, indicating that the total entropy would go down if the ice melted under these conditions – so melting is not spontaneous, but *freezing* is spontaneous.

PRACTICE

1 Explain whether entropy increases or decreases in the following changes:

 a melting of ice

 b boiling of water

 c decomposition of calcium carbonate.

14.3

ENTROPY – 2

OBJECTIVES

- Spontaneous change and $T\Delta S - \Delta H > 0$
- Standard entropy values
- Standard entropy change

A lime kiln in Iraq. The decomposition of calcium carbonate to produce quicklime (calcium oxide) is a highly endothermic process:
$CaCO_3(s) \rightarrow CO_2(g) + CaO(s)$;
$\Delta H^\ominus = +178\,kJ\,mol^{-1}$
There is no spontaneous decomposition at low temperature. At high temperature, the release of carbon dioxide gas increases the entropy of the system sufficiently to make the change spontaneous, i.e. to satisfy the requirement that $T\Delta S^\ominus > \Delta H^\ominus$. Limestone hills do not decompose to CO_2 and CaO until heated in a lime kiln!

Spontaneous endothermic changes

The entropy increase on dissolution explains why ammonium nitrate dissolves in water, spread 14.1. The endothermic reaction between solid hydrated barium hydroxide and excess solid ammonium nitrate shown in the photograph in spread 10.2 is driven by the production of ammonia gas and the fact that 3 reactants form 13 products, which together generate a positive entropy change: $T\Delta S$ at room temperature exceeds ΔH. Similarly the endothermic reaction between ethanoic acid and solid ammonium carbonate is driven by the production of carbon dioxide gas.

You should by now understand that the criterion for spontaneous chemical or physical change is that the total entropy change must be greater than zero, i.e. $\Delta S_{tot} > 0$. In other words, the total entropy of the Universe always tends to increase; it never goes down. The previous spread established the role of entropy and concentrated on physical changes. This spread introduces methods of calculation that can be applied to chemical reactions. However, our first task is to develop the point made in the previous spread about considering *both* enthalpy changes and entropy changes.

Two special cases

The general expression for spontaneous change described in the previous spread is

$$T\Delta S - \Delta H > 0$$

There are two special cases relating to this general expression.

First, if the enthalpy change ΔH is negligible, the processes that occur spontaneously are those that increase the entropy of the system so that $T\Delta S > 0$. Examples include the diffusion and mixing of gases. A similar example is dissolving a salt that has a small enthalpy of solution, such as sodium chloride (+4 kJ mol^{-1}). The salt dissolves because the particles can spread out into the solution rather than staying 'locked up' in the ordered crystal structure of the solid.

A second special case arises when the entropy change ΔS (and thus $T\Delta S$) is negligible. In this case the term $-\Delta H$ must be greater than zero, and this can only occur if the enthalpy change is negative. This conclusion explains why most reactions that occur spontaneously are exothermic. The fundamental reason for these *exothermic* reactions being spontaneous is that the entropy of the surroundings increases, as heat is released into them.

Both entropy change *and* enthalpy change must be considered in order to explain spontaneous endothermic processes. Endothermic processes have positive values for ΔH and can therefore only occur if the entropy change is positive and sufficiently large that:

$$T\Delta S > \Delta H$$

Spontaneous *endothermic* changes, such as the evolution of carbon dioxide when a dilute acid is added to a hydrogencarbonate, must be accompanied by a large increase in entropy in order that the total entropy of the Universe increases, i.e. to obey the second law of thermodynamics.

Standard entropies

The standard entropy of a substance at a specified temperature is the entropy change per mole that results from heating the substance from 0 K to the specified standard temperature. Here we look at how standard entropies are calculated.

At 0 K (–273 °C) the entropy of a perfect crystalline solid is zero. Heating a substance increases its temperature and increases its entropy. Remember that the heat q added to a system is related to the temperature change via the heat capacity C, which is equal to the mass multiplied by the specific heat capacity, spread 10.4:

$$q = C\Delta T \qquad (1)$$

The value of C itself depends on the temperature, so equation (1) applies only if the temperature change, ΔT, is small. Measuring the heat capacity of a substance down to very low temperatures allows the entropy at any temperature to be calculated, as follows.

Entropy change is defined as

$$\Delta S = \frac{q}{T} \qquad (2)$$

Combining equations (1) and (2) gives the entropy change for a small change in temperature,

$$\Delta S = \frac{C\Delta T}{T}$$

The **standard entropy** (symbol $S^{\ominus}$) of a substance is the sum of all the entropy changes resulting from heating the substance from 0 K to the specified temperature, i.e. written mathematically

$$S^{\ominus} = \sum_{T=0}^{T=\text{specified}} \frac{C\Delta T}{T} \quad \text{or} \quad \int_0^{\text{specified}} \frac{C}{T}\,dT$$

Data books usually list standard entropies of elements and compounds at 298 K, i.e. under the heading $S^{\ominus}$(298 K). (The standard pressure is 1 bar, as was the case for standard enthalpies.)

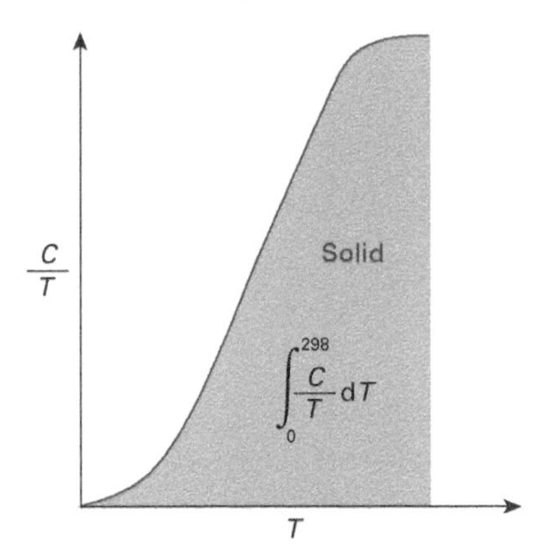

*The standard entropy of a solid at 298 K may be determined graphically by plotting **C/T** against **T** for values of **T** between 0 K and 298 K. The area under the curve is the standard entropy of the substance. Note that, if a phase change occurs below 298 K, then the corresponding $\Delta S = \Delta H/T$ must be included.*

Standard entropy change

You should remember, spread 10.5, that the *standard enthalpy change* is defined as the enthalpy change per mole for conversion of reactants in their standard states into products in their standard states, at a stated temperature. The *standard entropy change* $\Delta S^{\ominus}$ is defined in the same manner.

- The **standard entropy change** is the entropy change per mole for conversion of reactants in their standard states into products in their standard states, at a stated temperature.

In symbols, we can write this as

$$\Delta S^{\ominus} = S^{\ominus}(\text{products}) - S^{\ominus}(\text{reactants})$$

The following example shows how the standard entropy change may be calculated for the decomposition of ammonia:

$$2NH_3(g) \rightarrow N_2(g) + 3H_2(g)$$

The standard entropies of ammonia, nitrogen, and hydrogen are 192, 192, and 131 J K^{-1} mol^{-1} respectively. So

$$\Delta S^{\ominus} = S^{\ominus}(N_2) + 3S^{\ominus}(H_2) - 2S^{\ominus}(NH_3)$$
$$= (192\,J\,K^{-1}\,mol^{-1}) + 3 \times (131\,J\,K^{-1}\,mol^{-1}) - 2 \times (192\,J\,K^{-1}\,mol^{-1})$$
$$= +201\,J\,K^{-1}\,mol^{-1}$$

The standard entropy change (per mole of reaction as written) is +201 J K^{-1} mol^{-1}. So we can write

$$2NH_3(g) \rightarrow N_2(g) + 3H_2(g); \qquad \Delta S^{\ominus}(298\,K) = +201\,J\,K^{-1}\,mol^{-1}$$

The standard entropy change is positive, which signifies that there has been an increase in disorder. An inspection of the chemical equation confirms this: 2 mol of gas changes into 4 mol of gas. The following spreads show how to use the actual value of $\Delta S^{\ominus}$ to make predictions about spontaneity.

Substance	Formula	$S^{\ominus}$(298 K) /J K^{-1} mol^{-1}
Hydrogen	H_2	131
Nitrogen	N_2	192
Oxygen	O_2	205
Carbon dioxide	CO_2	214
Carbon monoxide	CO	198
Ammonia	NH_3	192
Water	H_2O	70
Ethanol	CH_3CH_2OH	161
Benzene	C_6H_6	173
Diamond	C	2.4
Graphite	C	5.7
Aluminium	Al	28
Lead	Pb	65
Calcium oxide	CaO	38
Calcium carbonate	$CaCO_3$	92

Standard entropies of substances at 298 K (25 °C).

Comparing standard entropies

Gases generally have larger standard entropies than liquids, which in turn have larger values than solids. Also, different substances in the same state can have different entropies. For example, diamond has a lower entropy than lead, partly because the carbon atoms in diamond are connected together by rigid bonds. Substances made up from heavy atoms generally have higher entropies than those made up from light atoms, e.g. N_2(g) has a higher entropy than H_2(g). Water H_2O has a lower entropy than ethanol CH_3CH_2OH.

- Exothermic reactions occur because the heat released increases the entropy of the surroundings.
- Endothermic reactions can only occur if the entropy change in the reaction is positive and large enough that $T\Delta S > \Delta H$.
- A value of standard entropy may be measured for any substance.
- The standard entropy change $\Delta S^{\ominus}$ is the entropy change per mole for conversion of reactants in their standard states into products in their standard states, at a stated temperature.

14.4

OBJECTIVES

- Standard Gibbs energy change $\Delta G^{\ominus}$
- Spontaneous change: $\Delta G^{\ominus} < 0$
- Calculating temperature for reaction

Standard Gibbs energy change

We have now clearly established the criterion for spontaneous change: the total entropy change (system plus surroundings) must be greater than zero ($\Delta S_{tot} > 0$). Simply expressed, spontaneous changes cause the entropy of the Universe to increase. However, as we saw in the previous spreads, deciding whether a reaction will be spontaneous requires knowledge of the entropy change ΔS, the enthalpy change ΔH, and the temperature T. It would be far more convenient to consider just one factor instead of several; that single factor is the standard Gibbs energy change.

Conditions

The entropy change in the surroundings is equal to $-\Delta H^{\ominus}/T$ when the change is carried out at:
(a) constant temperature T and
(b) 1 bar pressure (so that the heat added is equal to $\Delta H^{\ominus}$). The negative sign means that an exothermic reaction causes an *increase* in the entropy of the surroundings.

Defining standard Gibbs energy change

The criterion for spontaneous reaction is given in the previous two spreads. Now, at 1 bar, if we use the *standard* enthalpy change and the *standard* entropy change in this criterion to obtain a simple expression for spontaneity at the temperature T:

$$T\Delta S^{\ominus} - \Delta H^{\ominus} > 0$$

If the sign of both sides is changed, we arrive at an alternative expression for the criterion for spontaneous change:

$$\Delta H^{\ominus} - T\Delta S^{\ominus} < 0$$

The search throughout this chapter has been for a single quantity that will define spontaneity. The **standard Gibbs energy change** ('Gibbs energy' is sometimes called 'free energy') combines the three quantities above in one term and is defined at constant temperature as:

$$\Delta G^{\ominus} = \Delta H^{\ominus} - T\Delta S^{\ominus}$$

So, the Gibbs energy *decreases* in any spontaneous change.

- **The criterion for spontaneous change is that $\Delta G^{\ominus} < 0$.**

Both $\Delta H^{\ominus}$ and $T\Delta S^{\ominus}$ are measured in kJ mol^{-1}: $\Delta G^{\ominus}$ therefore also has units kJ mol^{-1}. A reaction with a negative value for $\Delta G^{\ominus}$ is said to be **exergonic**; this echoes the name 'exothermic' (spread 10.2).

- Exergonic reactions are spontaneous.

Josiah Willard Gibbs (1839–1903). When asked to name the most important scientist of his generation (after himself!), Albert Einstein immediately stated: Hendrik Lorentz (who discovered relativity at the same time as Einstein) but added 'I never met Willard Gibbs; perhaps, had I done so, I might have placed him alongside Lorentz.'

Standard Gibbs energy change

Standard Gibbs energy change is defined in the same way as the standard enthalpy change and the standard entropy change:

- The **standard Gibbs energy change** $\Delta G^{\ominus}$ is the Gibbs energy change per mole for conversion of reactants in their standard states into products in their standard states, at a stated temperature.

In simple terms, the criterion for spontaneous change is that $\Delta G^{\ominus} < 0$. When applied to a formation reaction, the standard Gibbs energy change of formation is related to the standard enthalpy and entropy changes of formation by

$$\Delta_f G^{\ominus} = \Delta_f H^{\ominus} - T\Delta_f S^{\ominus}$$

Data books provide values of standard Gibbs energy change of formation for a wide range of substances. For elements, $\Delta_f G^{\ominus}$ is by definition zero (as is the standard enthalpy change of formation).

Using $\Delta_f G^{\ominus}$ terms

Standard enthalpy changes may be calculated indirectly by using cycles based on Hess's law. Standard Gibbs energy changes may be manipulated in the same way. The standard Gibbs energy changes of formation of the reactants and of the products are obtained from data books. The standard Gibbs energy change $\Delta G^{\ominus}$ is given by

$\Delta G^{\ominus} = \Delta_f G^{\ominus}(\text{products}) - \Delta_f G^{\ominus}(\text{reactants})$

Worked example on methane

Question: Calculate the standard Gibbs energy change of combustion of methane from the standard Gibbs energy changes of formation of the substances involved in the reaction.

Answer: The first step is to write down the chemical equation, which is

$CH_4(g) + 2O_2(g) \rightarrow CO_2(g) + 2H_2O(l)$

With reference to the combustion of methane, the standard Gibbs energy change is given by

$\Delta_c G^\ominus = [\Delta_f G^\ominus(CO_2) + 2\Delta_f G^\ominus(H_2O)] - [\Delta_f G^\ominus(CH_4) + 2\Delta_f G^\ominus(O_2)]$

Substituting the relevant values:

$\Delta_c G^\ominus = [-394\,kJ\,mol^{-1} + 2 \times (-237\,kJ\,mol^{-1})] - [-50\,kJ\,mol^{-1} + 2 \times (0\,kJ\,mol^{-1})]$

$= -818\,kJ\,mol^{-1}$

So, the standard Gibbs energy change of combustion of methane is $-818\,kJ\,mol^{-1}$.

Comment – 1

The significance of the result for methane (negative sign, value in the order of hundreds of $kJ\,mol^{-1}$) will become clear in due course. For the present, you need simply to be familiar with this method of calculation.

Worked example on calcium carbonate

This example uses the expression for the standard Gibbs energy change to find the temperature at which $\Delta G^\ominus$ becomes negative. Remember, when $\Delta G^\ominus$ becomes negative, the reaction becomes spontaneous.

Question: The decomposition of calcium carbonate is an important process in the extraction of iron in a blast furnace (see the final spread in this chapter). Calculate the standard enthalpy change for the decomposition of calcium carbonate at 298 K. Also calculate the standard entropy change at 298 K. From these values, determine an approximate value for the decomposition temperature of the carbonate.

Answer:

Step 1 We first calculate the standard enthalpy change by combining the relevant standard enthalpy changes of formation in a Hess's law cycle. The equation for the decomposition is

$CaCO_3(s) \rightarrow CaO(s) + CO_2(g)$

So

$\Delta H^\ominus = \Delta_f H^\ominus(CaO) + \Delta_f H^\ominus(CO_2) - \Delta_f H^\ominus(CaCO_3)$

$= (-635\,kJ\,mol^{-1}) + (-394\,kJ\,mol^{-1}) - (-1207\,kJ\,mol^{-1})$

$= +178\,kJ\,mol^{-1}$

Step 2 We next calculate the standard entropy change in a similar manner. So

$\Delta S^\ominus = S^\ominus(CaO) + S^\ominus(CO_2) - S^\ominus(CaCO_3)$

$= (38\,J\,K^{-1}\,mol^{-1}) + (214\,J\,K^{-1}\,mol^{-1}) - (92\,J\,K^{-1}\,mol^{-1})$

$= +160\,J\,K^{-1}\,mol^{-1}$

Step 3 The reaction is spontaneous when $\Delta G^\ominus$ is negative, i.e. when

$\Delta H^\ominus - T\Delta S^\ominus < 0$

So, decomposition occurs when $T\Delta S^\ominus > \Delta H^\ominus$

i.e. when

$$T > \frac{\Delta H^\ominus}{\Delta S^\ominus}$$

Substituting the values calculated above:

$$T > \frac{+178\,000\,J\,mol^{-1}}{+160\,J\,K^{-1}\,mol^{-1}} \quad \text{i.e. } T > 1110\,K \quad \text{(3 sig. figs)}$$

The minimum temperature for the decomposition of calcium carbonate is 1110 K.

Comment – 2

In chapter 10 Thermochemistry we found that the standard *enthalpy* change of combustion of methane is $\Delta_c H^\ominus = -891\,kJ\,mol^{-1}$. The standard *Gibbs energy* change of combustion of methane worked out above is less negative ($\Delta_c G^\ominus = -818\,kJ\,mol^{-1}$) because in the combustion three moles of gas are consumed and only one mole of gas is formed: remember that the values are at 298 K so water is a liquid. This unfavourable entropy change causes the difference ($T\Delta S^\ominus$) between the two values.

Data – at 298 K or 1110 K?

The decomposition of calcium carbonate is strongly endothermic at room temperature (298 K); limestone hills do not spontaneously decompose. The predicted decomposition temperature is 1110 K, which is close to the actual value of 1170 K (900 °C). Using values for the standard enthalpy change and standard entropy change *at 1110 K* rather than 298 K would give even closer agreement with the observed decomposition temperature.

The concept of Gibbs energy helps to determine the feasibility of all reactions – from the functioning of an oil refinery to the biochemistry of the living cell.

- Standard Gibbs energy change is defined *at constant temperature* as $\Delta G^\ominus = \Delta H^\ominus - T\Delta S^\ominus$
- A spontaneous change is accompanied by a decrease in standard Gibbs energy, i.e. $\Delta G^\ominus < 0$.
- The standard Gibbs energy change may be calculated from standard Gibbs energy changes of formation $\Delta_f G^\ominus$.
- Standard Gibbs energy changes may be used to calculate the temperature at which a reaction becomes spontaneous.

$\Delta G^\ominus$ and $E^\ominus$

There is a direct proportionality between the standard Gibbs energy and the standard cell e.m.f.:

$\Delta G^\ominus = -zFE^\ominus$

where z is the number of electrons transferred and F is the Faraday constant, spread 13.11.

EXTENSION

14.5

OBJECTIVES

- $\Delta G^{\ominus} = 0$ as a special case
- Gibbs energy vs composition diagrams
- Standard Gibbs energy and equilibrium constant

Gibbs energy and chemical equilibrium

All the chemical equations written so far in this chapter have been represented as if they go to completion. The format has been: 'reactants' → 'products'. However, you should remember that all reactions form an *equilibrium mixture* of reactants and products, chapter 11. The position of the equilibrium is described by the equilibrium constant K. If the standard Gibbs energy change is the indicator for spontaneity of reaction, there should be a connection between $\Delta G^{\ominus}$ and K.

$\Delta G^{\ominus}$

The standard Gibbs energy change $\Delta G^{\ominus}$ is the difference between the standard Gibbs energy changes of formation of the pure products and the pure reactants. Under the conditions of the reaction, each substance is *not pure*; it is mixed with the other substances. The actual Gibbs energy change, established at a fixed composition of the reaction mixture, also contains a component due to mixing. The Gibbs energy of mixing is negative.

Gibbs energy of mixing

An important point to note is that standard Gibbs energy changes concern the change from *pure* reactants to *pure* products. Consider the imaginary reaction A + B → C + D for which $\Delta G^{\ominus} = 0\,\text{kJ mol}^{-1}$.

Does any reaction actually happen?

Entropy measures the disorder of a system. When two substances are mixed, there is an increase in disorder and so an increase in entropy. The **entropy of mixing** $\Delta_{mix}S$ is the entropy change due to the mixing of substances. The entropy of mixing is greatest when the two substances are present in equal proportions. There is an increase in entropy resulting from the mixing of reactants and products, which in turn means that the Gibbs energy *decreases* as the **Gibbs energy of mixing** is

$$\Delta_{mix}G = -T\Delta_{mix}S$$

In simple terms, the criterion for a spontaneous reaction is that $\Delta G^{\ominus}$ *is less than zero*. However, what would happen if $\Delta G^{\ominus}$ *equals* zero? Some reactants form products because the Gibbs energy of mixing favours a 50:50 equilibrium mixture over either pure reactants or pure products. The composition of the equilibrium mixture will be half-way between pure reactants and pure products. The **extent of reaction** ξ (the Greek letter xi) is defined such that pure reactants correspond to a value of 0 mol; pure products correspond to a value of 1 mol. For $\Delta G^{\ominus} = 0\,\text{kJ mol}^{-1}$, equilibrium occurs when $\xi = 0.5\,\text{mol}$.

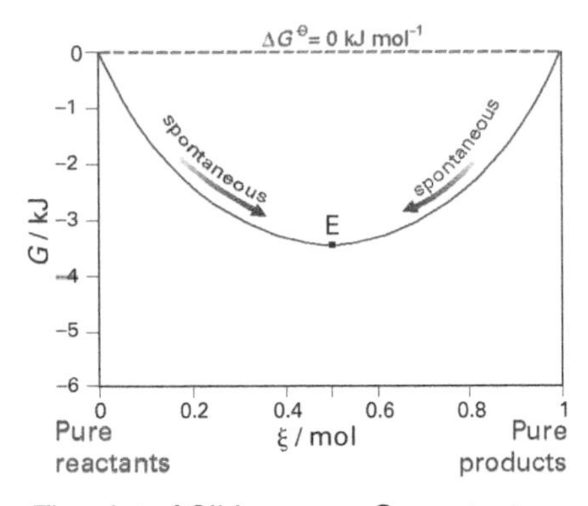

*The plot of Gibbs energy **G** vs extent of reaction for an imaginary reaction where $\Delta G^{\ominus} = 0$. In this example, the Gibbs energies of the reactants and the products are equal. Reaction spontaneously proceeds in either direction towards equilibrium E because the Gibbs energy of mixing of reactants and products is negative. Note that the Gibbs energy of mixing depends only on the number of species in the balanced chemical equation, and not on any energy terms.*

Equilibria favouring products

The example given on the left relates to an equilibrium reaction where $\Delta G^{\ominus} = 0$, i.e. the Gibbs energies of the reactants and products are equal. The diagram below left shows the plot of Gibbs energy vs extent of reaction when the Gibbs energy of the products is less than that of the reactants (i.e. $\Delta G^{\ominus} < 0$). The standard Gibbs energy change for the forward reaction is represented by the line joining pure reactants and pure products. If there were no mixing term, a *tiny* negative $\Delta G^{\ominus}$ would cause the composition to slide all the way down to products. A minimum in the Gibbs energy occurs because of the Gibbs energy of mixing.

Equilibrium occurs where Gibbs energy reaches a minimum

We can now state the exact criterion for any spontaneous reaction:

- Any composition of a reaction mixture will change spontaneously in the direction of decreasing Gibbs energy; no further change occurs when the Gibbs energy reaches its minimum value.
- In mathematical terms, the gradient is zero at equilibrium: $dG/d\xi = 0$.

The more negative the value for the standard Gibbs energy change $\Delta G^{\ominus}$, the steeper is the slope of the line from pure reactants to pure products, and the further towards the products the equilibrium lies.

Conversely, an equilibrium favours the reactants when the Gibbs energy of the products is greater than that of the reactants. Note that Gibbs energy of mixing is especially significant only where $\Delta G^{\ominus}$ lies between about +35 kJ mol^{-1} and −35 kJ mol^{-1}, as explained opposite.

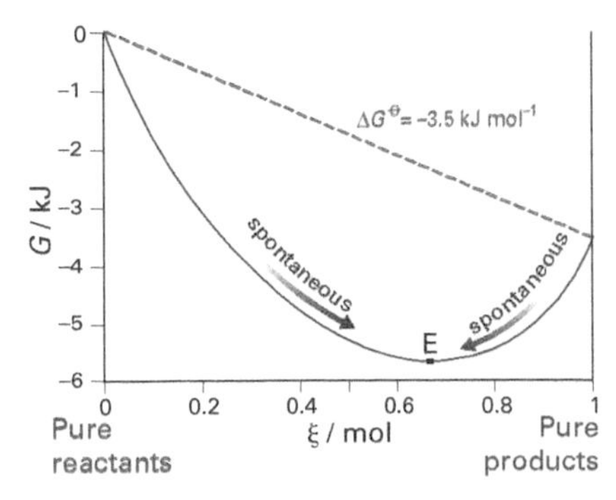

*The plot of Gibbs energy **G** vs extent of reaction for the formation of ethyl ethanoate. An extent of reaction of 0.67 mol corresponds to an equilibrium constant of*

$$\frac{(0.67\,\text{mol}) \times (0.67\,\text{mol})}{(0.33\,\text{mol}) \times (0.33\,\text{mol})} = 4$$

See spread 11.5.

Standard Gibbs energy change and the equilibrium constant

Standard Gibbs energy change and the equilibrium constant are linked together by the simple expression:

$$\Delta G^{\ominus} = -RT \ln K$$

where R is the gas constant (8.31 J K^{-1}mol^{-1}), T is the thermodynamic temperature, and $\ln K$ is the natural logarithm (see box) of the equilibrium constant. The significance of this relationship is shown below:

How the values of K and $\Delta G^{\ominus} = -RT\ln K$ are related to the equilibrium position.

Standard Gibbs energy change, $\Delta G^{\ominus}$/kJ mol^{-1}	Equilibrium constant K	Equilibrium position
More negative than −35	Greater than 10^6	'Reaction effectively complete'
Between −35 and 0	Between 10^6 and 1	Products predominate
Between 0 and +35	Between 1 and 10^{-6}	Reactants predominate
More positive than +35	Smaller than 10^{-6}	'Effectively no reaction'

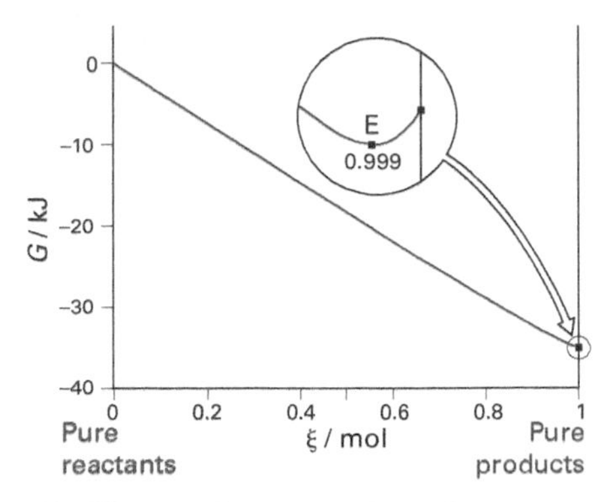

If $\Delta G^{\ominus}$ has a large negative value, Gibbs energy is at a minimum when the concentration of products in the equilibrium mixture E is much greater than the concentration of reactants. Then K >> 1, and the reaction goes almost to completion (i.e. the equilibrium lies very far to the right).

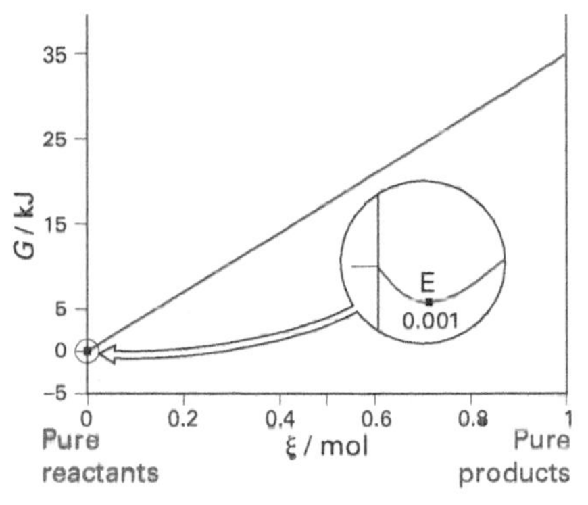

If $\Delta G^{\ominus}$ has a large positive value, Gibbs energy is at a minimum when the concentration of products in the equilibrium mixture E is much smaller than the concentration of reactants. Then K << 1, and the reaction 'does not go' (i.e. the equilibrium lies very far to the left).

Worked example on calculating an equilibrium constant

Question: Calculate the value of the equilibrium constant at 25 °C for the esterification reaction:

$$CH_3COOH(l) + CH_3CH_2OH(l) \rightleftharpoons CH_3COOCH_2CH_3(l) + H_2O(l)$$

Answer:

Step 1 We first use a data book to find the standard Gibbs energy changes of formation of the reactants and products. We find the values:

−389.9 kJ mol^{-1} (CH_3COOH)

−174.8 kJ mol^{-1} (CH_3CH_2OH)

−331.1 kJ mol^{-1} ($CH_3COOCH_2CH_3$)

−237.1 kJ mol^{-1} (H_2O)

Step 2 Now we calculate $\Delta G^{\ominus}$ for the forward reaction:

$$\Delta G^{\ominus} = [(-331.1) + (-237.1)]\,\text{kJ mol}^{-1} - [(-389.9) + (-174.8)]\,\text{kJ mol}^{-1}$$
$$= -3.5\,\text{kJ mol}^{-1}$$

The small negative value of the standard Gibbs energy change shows that an equilibrium slightly favouring the products will be established.

Step 3 Finally we insert values into the equation $\Delta G^{\ominus} = -RT\ln K$ and solve for K:

$$-3500\,\text{J mol}^{-1} = -(8.31\,\text{J K}^{-1}\,\text{mol}^{-1}) \times (298\,\text{K}) \times \ln K$$

$$\ln K = \frac{-3500\,\text{J mol}^{-1}}{-(8.31\,\text{J K}^{-1}\,\text{mol}^{-1}) \times (298\,\text{K})}$$
$$= 1.4$$

Now by reference to the box we find that

$K = e^{1.4} = 4$ (1 sig. fig.)

Note: The value for K may be used to calculate the composition of the equilibrium mixture (see chapter 11 Chemical equilibrium). The value calculated here from Gibbs energy changes matches the experimental value.

- Equilibrium occurs where the Gibbs energy reaches its minimum value.
- Equilibrium constant and standard Gibbs energy change are connected by the expression $\Delta G^{\ominus} = -RT\ln K$
- When $\Delta G^{\ominus}$ is large and positive, the reaction 'does not go'.
- When $\Delta G^{\ominus}$ is large and negative, the reaction 'goes to completion'.

Natural logarithms

These are like common logarithms, but to a different base, the number we call 'e'. For any positive number n, the **natural logarithm** (symbol ln or $\log_e$) of n is the power to which the base (in this case, e) must be raised to make n. For example, for the number 2,

$\ln 2 = 0.693$ i.e. $2 = e^{0.693}$

Natural logarithms are usually labelled 'ln' on a calculator.

For our purposes here, we need to do the reverse of finding the natural logarithm. If we have

$\ln K = n$ then $K = e^n$

so we need to use the button labelled 'e^x' on the calculator.

EXTENSION

14.6

OBJECTIVES

- Criterion for spontaneous decomposition of a metal oxide
- Carbon as a reductant
- Ellingham diagrams and predicting conditions

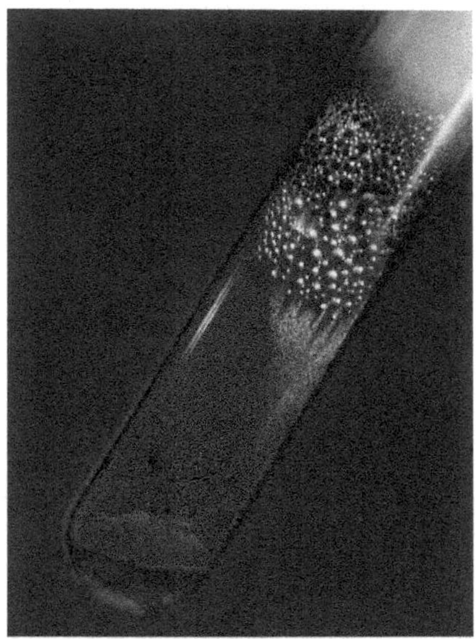

The decomposition of mercury(II) oxide to produce metallic mercury and oxygen gas. This was the method first used to make oxygen (with the energy supplied by the Sun via a magnifying glass).

K and $\Delta G^{\ominus}$

Remember that, for a reaction to favour products, the equilibrium constant K must be greater than 1. When $\Delta G^{\ominus} = 0$, $K = 1$, and reactants and products are present in equal concentrations.

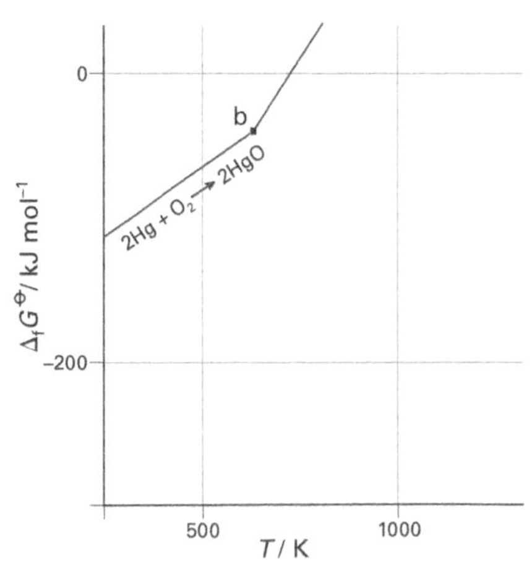

Gaseous oxygen reacts with mercury to form solid mercury(II) oxide. The standard entropy change of formation is therefore negative and the gradient of the line is positive. The gradient becomes steeper at b, the boiling point of the metal, because the solid oxide is now being formed from two gaseous species, oxygen and the metal.

GIBBS ENERGY AND METAL EXTRACTION

You are now well used to searching data books to enable you to calculate standard entropy, enthalpy, and Gibbs energy changes. However, you should bear in mind that the $^{\ominus}$ symbol in, for example, $\Delta G^{\ominus}$ refers to a pressure of 1 bar and a stated temperature that is *not necessarily 298 K*. The value of $\Delta G^{\ominus}$ for a reaction depends on the temperature. In this spread we look at how the temperature dependence of $\Delta G^{\ominus}$ determines the conditions under which metals may be extracted from their ores.

Changing metal oxides into metals

If you heat mercury(II) oxide alone at about 450 °C, it spontaneously decomposes into mercury and oxygen. But to extract zinc from its oxide ore, the ore must be reduced with carbon at about 1000 °C. The extraction of aluminium by carbon reduction would require a temperature greater than 2000 °C, too high for commercial use. The differences in behaviour between these metals may be explained by looking closely at the standard Gibbs energy changes for the reactions at different temperatures.

The dependence of $\Delta G^{\ominus}$ on temperature

When we are considering the extraction of metals from their oxide ores, the standard Gibbs energy change of particular interest is $\Delta_f G^{\ominus}$, the standard Gibbs energy change of formation of the oxide. The standard Gibbs energy change of formation is given by the equation:

$$\Delta_f G^{\ominus} = \Delta_f H^{\ominus} - T\Delta_f S^{\ominus}$$

The standard enthalpy change of formation $\Delta_f H^{\ominus}$ and the standard entropy change of formation $\Delta_f S^{\ominus}$ may be assumed to be roughly independent of temperature. Rewriting the relationship as

$$\Delta_f G^{\ominus} = -\Delta_f S^{\ominus} T + \Delta_f H^{\ominus}$$

gives the formula of a straight-line graph of the type

$$y = mx + c$$

That is, a plot of $\Delta_f G^{\ominus}$ against T gives a straight line with a gradient of $-\Delta_f S^{\ominus}$. The diagram to the left shows this plot for mercury(II) oxide. In the formation of the oxide, one mole of oxygen gas is destroyed, so $\Delta S^{\ominus}$ is strongly *negative* and the line generally slopes upwards.

The decomposition of mercury(II) oxide

As the temperature increases, the value of $\Delta_f G^{\ominus}$ for the reaction

$$2Hg + O_2(g) \rightarrow 2HgO(s)$$

becomes less negative, especially after mercury becomes a gas (at point b on the diagram). At approximately 750 K the value becomes zero; at this temperature, mercury, oxygen, and mercury(II) oxide exist together in an equilibrium with a K value of around 1. At higher temperatures, $\Delta_f G^{\ominus}$ becomes positive for this reaction, and therefore the standard Gibbs energy change for the *decomposition* of mercury(II) oxide

$$2HgO(s) \rightarrow 2Hg(g) + O_2(g)$$

becomes negative. Above 750 K, mercury(II) oxide spontaneously decomposes to its elements.

The majority of metal oxides remain stable up to extremely high temperatures. However, many metals can be extracted from their oxides with the assistance of a reductant, which is commonly carbon or carbon monoxide.

Ellingham diagrams

An **Ellingham diagram** consists of graphs of the standard Gibbs energy changes of formation for the metal oxides as a function of temperature. In an Ellingham diagram, the $\Delta_f G^{\ominus}$ values are quoted per mole of oxygen gas. As with the example of Hg/HgO given opposite, the gradients of the metal/metal oxide plots are generally positive. Many metals such as iron, lead, and zinc are produced by heating an oxide ore with carbon. Conditions in the metal smelting furnace may vary, so that the reductant is either carbon itself or carbon monoxide. Ellingham diagrams also contain the plots for the formation of the oxides of carbon:

$2C(s) + O_2(g) \rightarrow 2CO(g)$ (a)

$2CO(g) + O_2(g) \rightarrow 2CO_2(g)$ (b)

Notice that the line (a) for the conversion of carbon to carbon monoxide slopes *downwards*. The line slopes in this direction because 2 mol of carbon monoxide gas are produced for every 1 mol of oxygen gas used, and so the entropy change is positive. This downward-sloping line will eventually cross the upward-sloping line for the conversion of a metal to its oxide. Above the temperature at which the lines cross, carbon can reduce the metal oxide to the metal. Above 1000 K carbon can reduce iron(II) oxide. Zinc is also manufactured by carbon reduction at temperatures greater than 1200 K. However, the temperatures required for titanium, aluminium, and calcium are uneconomically high.

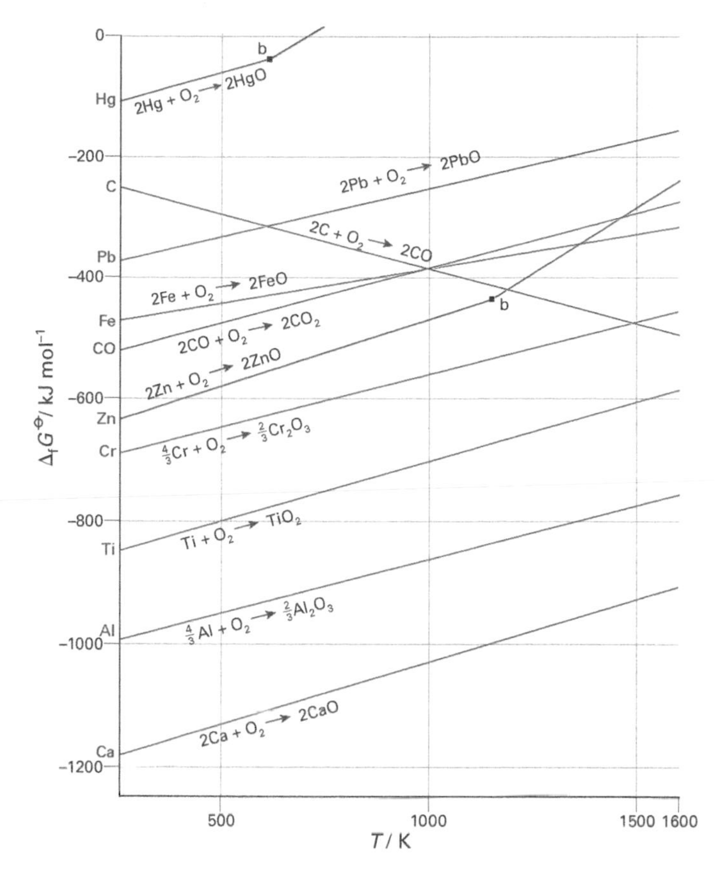

An Ellingham diagram showing the relationship between standard Gibbs energy change and temperature, for the oxides of carbon and various metal oxides. For zinc, the slope becomes significantly greater at the boiling point (b).

Generally, the reduction of the metal oxide may proceed by one of two possible pathways:

1. metal oxide + carbon → metal + carbon monoxide
2. metal oxide + carbon monoxide → metal + carbon dioxide

- A carbon–oxygen system will reduce a metal–oxygen system when the $\Delta_f G^{\ominus}$ value of the former is more negative than the $\Delta_f G^{\ominus}$ value of the latter.

Iron and the blast furnace

The extraction of iron in a blast furnace is a complex process that occurs in a number of steps. However, the final stage can be treated as the reduction of iron(II) oxide. Look at the Ellingham diagram plots for the formation of iron(II) oxide from its elements, and for the formation of carbon dioxide from carbon monoxide. Carbon monoxide will reduce iron(II) oxide at temperatures *up to* 1000 K – the point where the plot for $\Delta_f G^{\ominus}(CO_2$ from CO) cuts the line for $\Delta_f G^{\ominus}$(FeO from Fe):

$FeO(s) + CO(g) \rightarrow Fe(s) + CO_2(g)$

At temperatures *above* 1000 K, carbon acts as the reductant:

$FeO(s) + C(s) \rightarrow Fe(s) + CO(g)$

A blast furnace, showing the molten iron being run off.

- The standard Gibbs energy change is a function of temperature.
- Plots of $\Delta_f G^{\ominus}$ against T for metal oxides generally have positive gradients.
- Metal oxides spontaneously decompose into their elements at sufficiently high temperatures, when the value of $\Delta_f G^{\ominus}$ for the oxide becomes *positive*.
- Ellingham diagrams may be used to predict the temperature at which the reduction of a metal oxide by carbon or carbon monoxide becomes feasible.

PRACTICE EXAM QUESTIONS

1 The effect of temperature on chemical equilibrium is given by the equation

$\Delta G^{\ominus} = -RT \ln K_c$

Use this information, where relevant, in answering the questions that follow.

a To what does the symbol c refer in the equilibrium constant which is written K_c? [1]

b Derive an expression which shows the dependence of the logarithm of the equilibrium constant, $\ln K_c$, on temperature, T, standard enthalpy change, $\Delta H^{\ominus}$, and standard entropy change $\Delta S^{\ominus}$. [2]

c Assuming that $\Delta H^{\ominus}$ and $\Delta S^{\ominus}$ do not vary with temperature, use the expression you derived in **b** above to explain how the magnitude of $\ln K_c$ varies with *increase* in temperature for

i an exothermic reaction,

ii an endothermic reaction. [4]

2 The diagram below shows apparatus used to measure the standard e.m.f. of an electrochemical cell.

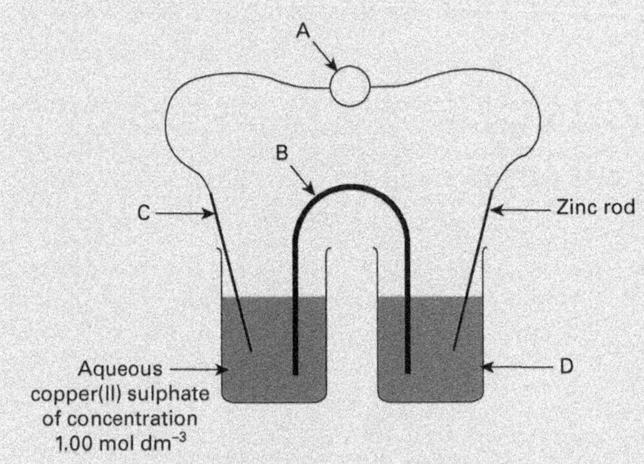

a i Give the name of each piece of apparatus **A** to **C**. [3]

ii Give the formula and concentration of the cation in solution **D**. [1]

iii State the condition required for this measurement. [1]

b The table below shows the standard electrode potentials for copper and zinc.

Electrode system	$E^{\ominus}$/V
$Zn^{2+}(aq) + 2e^- \rightleftharpoons Zn(s)$	−0.76
$Cu^{2+}(aq) + 2e^- \rightleftharpoons Cu(s)$	+0.34

i Write an equation to show the overall reaction in the cell. [1]

ii Calculate the e.m.f. of this cell. [1]

iii Calculate the free energy change, $\Delta G^{\ominus}$, for the reaction in **b i**.

[Faraday constant = 96 500 C mol^{-1}.] [3]

iv In which direction will the electrons move through **A** in the digram? [1]

3 The feasibility of a chemical reaction depends on the standard free energy change, $\Delta G^{\ominus}$, which is related to the standard enthalpy and entropy changes by the equation

$\Delta G^{\ominus} = \Delta H^{\ominus} - T\Delta S^{\ominus}$

Use this information, where relevant, in answering the questions that follow.

a Ice melts at atmospheric pressure only if the surrounding temperature rises about 0 °C.

i State the signs of the enthalpy and entropy changes during melting.

ii Explain why ice does not melt at temperatures below 0 °C. [4]

b When sodium hydrogencarbonate is added to dilute hydrochloric acid, the temperature of the reaction mixture drops. Despite this, the reaction is spontaneous. Explain how this can be. [3]

c Give the signs of enthalpy change, of the entropy change and of the free energy change for the combustion of propane. In **each** case give a reason for your answer. [6]

4 a The graph below shows the variation of ΔG with temperature for three reactions relevant to the extraction of iron from iron oxide.

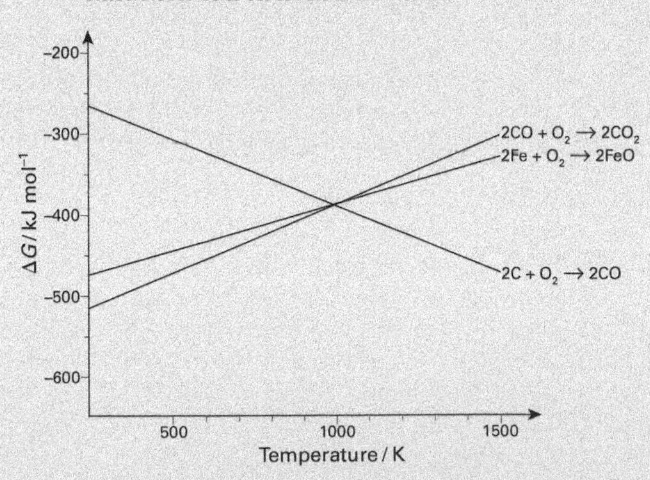

i Use this diagram to estimate values of ΔG at 800 K for the following reactions:

$2C + O_2 \rightarrow 2CO$

$2CO + O_2 \rightarrow 2CO_2$

$2Fe + O_2 \rightarrow 2FeO$ [3]

ii Calculate the values of ΔG for the following reactions at 800 K:

$FeO + C \rightarrow CO + Fe$

$FeO + CO \rightarrow CO_2 + Fe$ [3]

iii Use the values obtained in **a ii** to explain why carbon monoxide, rather than carbon, acts as the reducing agent at 800 K. [2]

iv Use **Figure 3** to estimate the minimum temperature at which carbon becomes capable of reducing iron oxide to iron. [1]

v Use the relevant equation given in **a ii** to explain why the reduction of iron oxide to iron using carbon as the reducing agent results in an increase in the entropy of the system. [2]

b The table below shows some values for standard electrode potentials.

Electrode system	$E^{\ominus}$/V
$Fe^{2+}(aq) + 2e^- \rightleftharpoons Fe(s)$	−0.44
$H^+(aq) + e^- \rightleftharpoons \frac{1}{2}H_2(s)$	0.00

i Give the emf of the cell. [1]

ii State which would be the positive electrode. [1]

iii Write an equation to show the overall reaction in the cell [1]

iv Write down the cell diagram that represents the overall reaction in the cell. [2]

v Give the name of an instrument that could be used to measure the emf of the cell. [1]

5 When zinc foil is placed in aqueous lead(II) nitrate solution, an exothermic reaction takes place. The enthalpy change can be measured using an electrical compensation calorimeter, and the following data were obtained in such an experiment in which an excess of zinc powder was added.

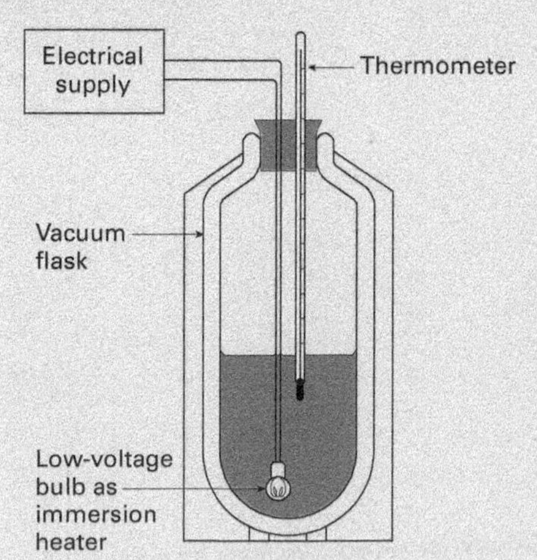

Output of immersion heater = 25.0 W (1 W = 1 J s^{-1})
Volume of 0.5 M lead(II) nitrate solution = 100 cm^3
Temperature rise in solution after reaction = 12.5 °C
Time taken for immersion heater to raise temperature by 12.5 °C = 305 s.

a Write two half-equations to represent the changes which take place when zinc reacts with lead(II) nitrate solution. [2]

b What are the two main advantages of the electrical compensation calorimeter over other simple methods for measuring enthalpy changes? [2]

c Calculate a value for the standard molar enthalpy change for this reaction from the data. [2]

d The same reaction can be used in an electrochemical cell. Under standard conditions this gives an e.m.f. of 0.63 V.

i Write down a conventional cell diagram for this cell.

ii Which is the positive electrode in this cell? Explain your reasoning.

iii State the names and concentrations of suitable solutions that could be used to set up the cell under standard conditions.

iv Draw a diagram to show how this cell could be set up in a laboratory, and its e.m.f. measured. You should clearly label the electrode materials and the solutions: the position of the salt bridge should be shown but details of its composition are not required. [6]

e i Calculate the entropy change at 298 K in the system during this reaction, using the relationships:

$\Delta G^{\ominus} = -zFE^{\ominus} = \Delta H^{\ominus} - T\Delta S^{\ominus}_{system}$
$[F = 9.65 \times 10^4\ C\ mol^{-1}]$

ii Calculate the entropy change at 298 K for the surroundings in this reaction ($\Delta S^{\ominus}_{surroundings}$).

iii Calculate the value of $\Delta S^{\ominus}_{system} + \Delta S^{\ominus}_{surroundings}$ for this reaction at 298 K, and explain clearly the significance of the result of this calculation. [5]

6 a Entropy is often linked with the idea of order and disorder. Write an equation for a chemical reaction in which the amount of *disorder* increases and state the sign of the entropy change which accompanies this reaction. [2]

b The combustion of graphite involves an increase in entropy from reactants to products of +3 J K^{-1} mol^{-1}. The standard entropy of oxygen is 205 J K^{-1} mol^{-1} and that of carbon dioxide is 214 J K^{-1} mol^{-1}. Calculate the standard entropy of graphite and suggest why it has a relatively low value. [4]

c i Write an equation that relates ΔG to ΔH and ΔS in a reaction.

ii Derive an equation which relates ΔS to ΔH when $\Delta G = 0$. [2]

d For certain reversible processes such as boiling and freezing, the Gibbs free energy change is zero. When water freezes, 6.0 kJ mol^{-1} of heat energy are evolved. Use the equation you derived in part **c ii** to calculate the entropy change in 54 g of water when it freezes at 0 °C. [4]

15 Chemical kinetics

Two questions are frequently asked about a chemical reaction. They are: 'Will it happen?' and 'How fast does it happen?' As we saw in earlier chapters, thermodynamics answers the first question by considering the implications of changes in enthalpy and especially Gibbs energy associated with chemical reactions. However, there are many reactions that, whilst thermodynamically feasible, proceed extremely slowly. Once we know that a reaction is feasible, knowledge about its rate (How fast?) is clearly of equal importance. This chapter is concerned with chemical kinetics – measuring the rates of reactions and trying to understand in detail how reactants change into products.

Reaction rate

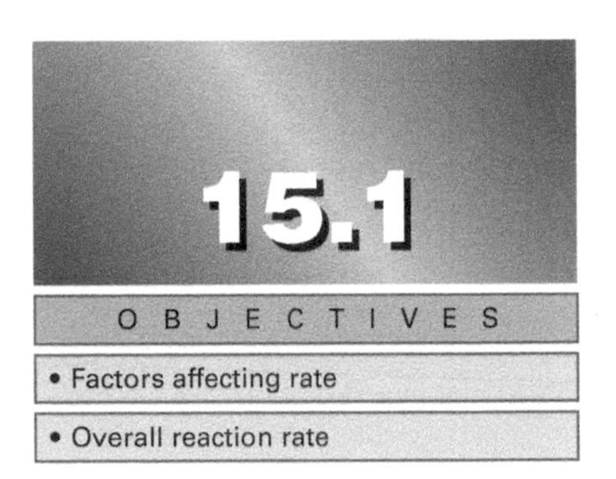

15.1

OBJECTIVES

- Factors affecting rate
- Overall reaction rate

Imagine a car travelling at a speed of 80 kilometres per hour ($km\,h^{-1}$) down the motorway. Each second, about 50 separate combustion cycles happen in the engine. Each cycle involves drawing a fuel/air mixture into the combustion chamber, compressing the mixture, burning it, and expelling the exhaust gases. The rate of this oxidation reaction is extremely rapid, taking less than five milliseconds. At the same time, a patch of rust hidden inside one of the doors is slowly growing. Rusting is the corrosion of iron (a reaction with oxygen and water). The rate of this oxidation reaction is extremely slow. It may take years before the first tell-tale brown spots appear under the gleaming paintwork.

A number of different factors affect the rates of reactions. We will summarize them now. Most of these factors are dealt with in this chapter, whereas others are studied elsewhere in this book.

Rate and collisions

In any reaction, the reactant species must collide with sufficient energy for reaction to happen. The reactant species could be molecules, atoms, or ions; for simplicity we will use the term 'molecules' from now on. Collisions break chemical bonds in the reactants; new bonds form between the fragments to create the products.

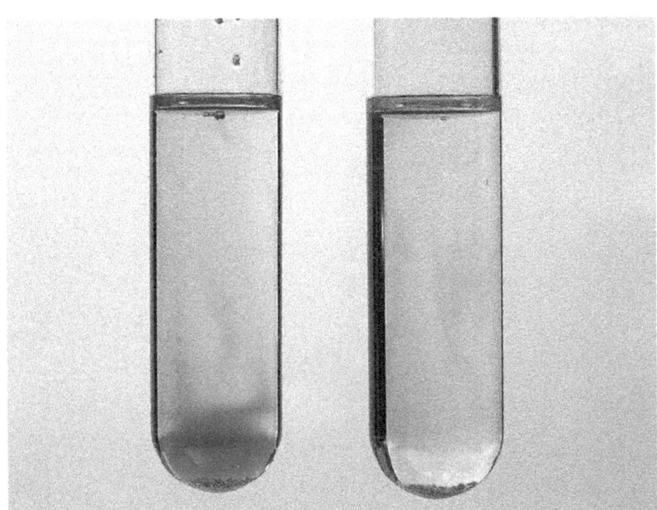

The effect of temperature on the reaction between magnesium and water (both beakers contain an indicator that changes from colourless to purple in the presence of the basic solution formed by the reaction): hot water at 80 °C (left) and cold water at 20 °C.

The effect of concentration on the reaction between magnesium and dilute sulphuric acid: $H_2SO_4(aq)$ of concentration $1.0\,mol\,dm^{-3}$ (left); and $H_2SO_4(aq)$ of concentration $0.2\,mol\,dm^{-3}$ (right).

Rate and temperature

Increasing the temperature of the reactants increases the rate of a reaction. At higher temperatures, the molecules are moving with greater average speed and so collide more frequently and with greater energy.

Rate and concentration

Increasing the concentration of the reactants increases the rate of a reaction. At greater concentration, there are more molecules in a given volume. Distances between these molecules are therefore reduced and there is an increased number of collisions per second.

Other factors

- *Pressure:* For reactions involving gases, increasing the pressure of a gas increases its concentration. A given volume contains a greater amount in moles of the gas.
- *Surface area:* In a reaction involving a solid, breaking the solid into smaller pieces increases its total surface area. There is more contact with the other reactant(s) and so the rate of reaction increases.

(a)

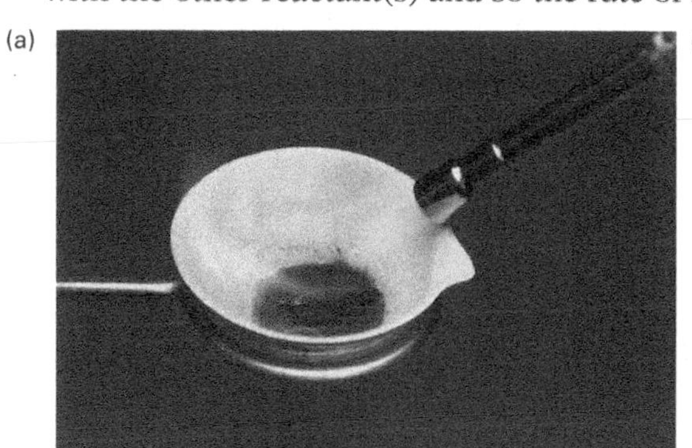

(b)

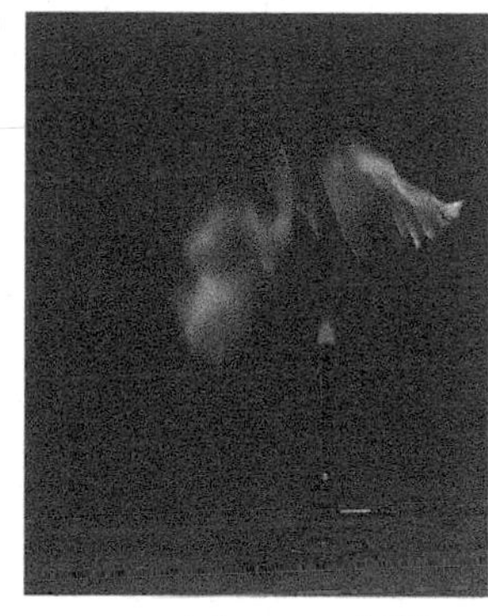

Lycopodium powder consists of the very fine spores of a common club moss. (a) The spores burn with difficulty when piled together in a dish. (b) When sprinkled into a flame, combustion is rapid because the spores are very well mixed with oxygen.

- *Light:* Some reactions are much more vigorous when carried out in bright light. Light is a form of electromagnetic radiation. If the frequency of the light is great enough, then the energy carried by each quantum of light may be sufficient to break a bond in one of the reactants and create a short-lived, reactive species (called a **reactive intermediate**) from which the products form.
- *Catalysts*: A catalyst is a substance that increases the rate of a chemical reaction without itself undergoing a permanent change.

Overall rate and instantaneous rate

You can measure the overall rate of a reaction by starting a clock as you mix the reactants and then stopping the clock when the reaction is complete. If you know the increase in concentration of a product, then the overall rate *with respect to that product* is given by

$$\text{rate of reaction} = \frac{\text{increase in concentration of product}}{\text{time}}$$

As in the box above, concentration is measured in moles per cubic decimetre ($mol\,dm^{-3}$) and time in seconds (s).

However, it may be obvious from watching a reaction that the *instantaneous* rate changes as the reaction proceeds. Rate usually decreases as time goes by because the reactant concentration is decreasing. (Returning to the car analogy, the *overall* speed for a journey may be $80\,km\,h^{-1}$. The *instantaneous* speed may vary between zero and $120\,km\,h^{-1}$.)

- The rate of a chemical reaction is the change in concentration per unit time.
- The rate of a reaction depends chiefly on temperature and on reactant concentration.
- Other factors that may affect reaction rate include pressure, light, the presence of a suitable catalyst, and the surface area of a solid reactant.

Car speed and reaction rate

From your science lessons in past years, you should remember that

$$\text{speed} = \frac{\text{change in distance}}{\text{time}}$$

So a car that travels 80 kilometres in one hour has an overall speed of

$$\frac{80\,km}{1\,h} = 80\,km\,h^{-1}$$

When we need to talk about how fast a reaction is, we use a similar expression, but using 'rate of reaction' and 'change in concentration':

$$\text{rate of reaction} = \frac{\text{change in concentration}}{\text{time}}$$

So a chemical reaction in which the concentration changes by $0.10\,mol\,dm^{-3}$ in 120 seconds has an overall rate of

$$\frac{0.10\,mol\,dm^{-3}}{120\,s} = 8.3 \times 10^{-4}\,mol\,dm^{-3}\,s^{-1}$$

(The units here are moles per cubic decimetre per second.)

Light

A mixture of the gases methane CH_4 and chlorine Cl_2 shows little sign of reaction. Fire a photographic flash gun close to the mixture and there will be an explosive reaction. Electromagnetic radiation forms highly reactive chlorine atoms, which trigger the reaction.

A sample of aqueous hydrogen peroxide $H_2O_2(aq)$ decomposes over a period of months into oxygen and water. If a platinum catalyst is immersed in the sample, the reaction is complete in a matter of minutes.

15.2

Some practical techniques

OBJECTIVES

- Titration
- Colorimetry
- Conductivity
- Flash photolysis
- Measuring gas volume

Chemical kinetics investigates the rates at which chemical reactions take place. The rate of a reaction is the rate of change in concentration of one of the products or reactants. The units of rate are usually moles per cubic decimetre per second ($mol\,dm^{-3}\,s^{-1}$). Some practical investigations directly monitor the concentration of one of the reactants or products as the reaction proceeds. Other investigations measure other variables such as mass or volume. All these measurements must be converted into changes in concentration over specified periods of time in order to calculate rate. This spread introduces the main methods of monitoring changes in the composition of reaction mixtures. Some are dealt with in more detail elsewhere in this book.

Titration

The reaction between aqueous hydrogen peroxide and acidified potassium iodide

$$H_2O_2(aq) + 2H^+(aq) + 2I^-(aq) \rightarrow I_2(aq) + 2H_2O(l)$$

may be investigated by monitoring the concentration of iodine. At measured (but *not* necessarily equal) time intervals, samples of the reaction mixture are extracted. The reaction in each sample taken is **quenched** (slowed down very significantly) by diluting in ice-cold water. Each sample is then titrated against standard aqueous sodium thiosulphate, and the concentration of the iodine is calculated. (We will look at this example in more detail in the next spread.)

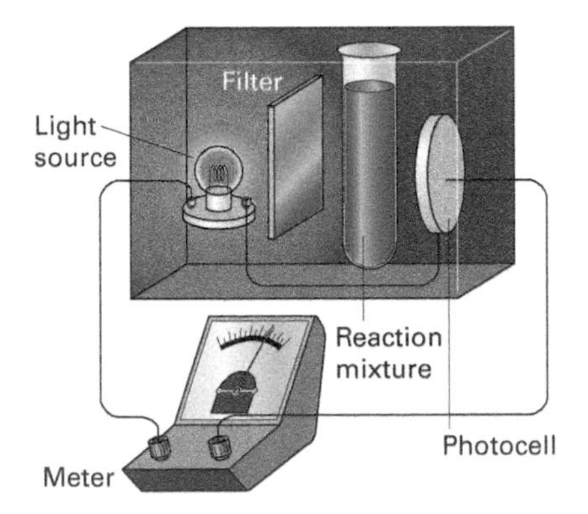

*A **colorimeter** indicates concentration by measuring the intensity of light shining through a coloured reaction mixture. The colorimeter produces an electrical signal, which can be analysed by computer.*

Colorimetry

Some reaction mixtures show a steady change of colour as the reaction proceeds. For example, colourless ethanedioate ions ($C_2O_4^{2-}$) reduce purple manganate(VII) ions (MnO_4^-) to colourless manganese(II) ions (Mn^{2+}) in acidic solution:

$$2MnO_4^-(aq) + 16H^+(aq) + 5C_2O_4^{2-}(aq) \rightarrow 2Mn^{2+}(aq) + 10CO_2(g) + 8H_2O(l)$$

The concentration of manganate(VII) ion may be monitored over time by using a colorimeter. Light of a fixed wavelength shines through the reaction mixture and onto a photocell. The photocell develops an e.m.f. proportional to the intensity of the light. If the photocell has been calibrated with solutions of known concentration, the e.m.f. can be converted to manganate(VII) ion concentration values.

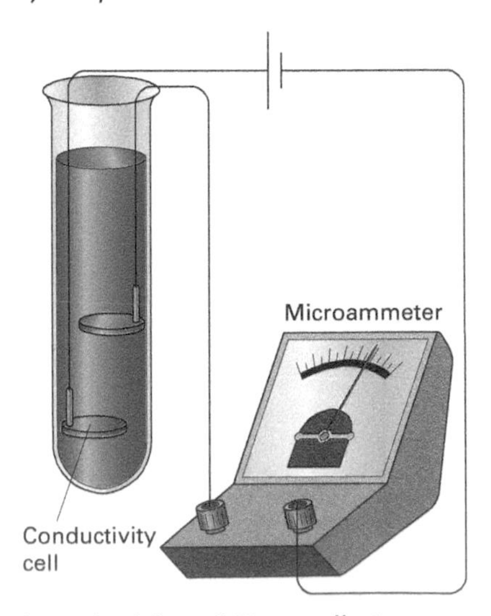

A conductivity cell. The smallest possible current is passed through the cell to limit electrolytic effects.

Conductivity

Look again at the above ionic equation for the reaction between manganate(VII) and ethanedioate ions. You should notice that the reactants include a total of 23 ions, whilst amongst the products there are just two ions. The electrical conductivity of a solution is proportional to the concentration of its ions and the charges they bear. The conductivity of the reaction mixture may be monitored over time by carrying out the reaction in a conductivity cell. As with the colorimeter, conductivity apparatus may be calibrated with solutions of known concentration. Readings of conductivity as the reaction proceeds may then be converted to concentrations of the ions present.

Laser flash photolysis and femtosecond spectroscopy

Some reactions are extremely rapid. The technique of **laser flash photolysis** uses a very short burst of light from a laser to start a reaction that is sensitive to light. Spectroscopic techniques, chapter 32, then monitor the concentration of the reactive intermediate formed. A particularly neat version of the technique uses the same laser to start the reaction and to measure the response of the system, as shown in the illustrations at the top of the opposite page.

A very short pulse of light of femtosecond duration (1 femtosecond = 1 fs = 10^{-15} s) starts out from a laser and is split into two beams. The direct beam causes the reaction to start; the indirect beam is delayed by being sent over a longer path to the reaction chamber. This probe pulse arrives a few femtoseconds later and causes the reaction mixture to absorb light, which is analysed to reveal the concentrations of the species present.

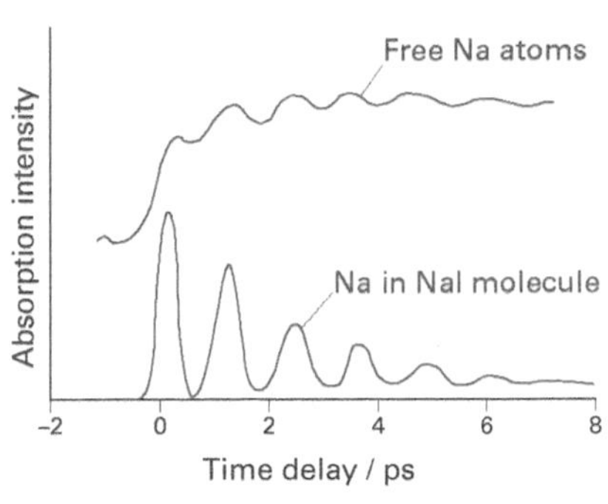

When excited by a laser pulse, the ion pair Na^+I^- becomes a covalently bonded pair NaI, which rapidly decomposes into separate atoms Na and I. The lower graph shows the intensity of the absorption by the molecule NaI. The upper graph shows the absorbance of free Na atoms (and hence their concentration).

Measuring the volume of gas produced

The decomposition of aqueous hydrogen peroxide occurs rapidly in the presence of a catalyst:

$$2H_2O_2(aq) \rightarrow 2H_2O(l) + O_2(g)$$

This reaction may be monitored by measuring the volume of oxygen gas evolved as a function of time elapsed after adding the catalyst.

The measured volume of gas may then be used to calculate the corresponding concentration of the reactant, hydrogen peroxide. The rate of the reaction is then expressed as

$$\text{rate of reaction} = \frac{\text{decrease in concentration of reactant}}{\text{time}}$$

Which product or reactant?

Look again at the reaction between aqueous hydrogen peroxide and acidified potassium iodide:

$$H_2O_2(aq) + 2H^+(aq) + 2I^-(aq) \rightarrow I_2(aq) + 2H_2O(l)$$

The reaction produces 2 mol of water for every 1 mol of iodine produced. Rate of reaction is defined as

$$\text{rate of reaction} = \frac{\text{change in concentration}}{\text{time}}$$

The rate of change in concentration with respect to water is therefore *twice* the rate of change in concentration with respect to iodine.

How this is dealt with in a formal way to give a unique rate for any reaction is explained in spread 15.7.

Temperature control

Reaction rates are obtained by monitoring the change in concentration of one substance with time. Rate is dependent on temperature, so reactions should be carried out in a thermostatically controlled water bath.

SUMMARY

- Concentration can be monitored using techniques such as titration, colorimetry, and conductivity.
- Ultra-fast reactions can be studied using laser flash photolysis.

PRACTICE

1 Suppose you mix acidified potassium iodide with aqueous hydrogen peroxide. You titrate the iodine in the sample against 0.20 mol dm^{-3} aqueous sodium thiosulphate and obtain the results shown (1 mol of iodine I_2 reacts with 2 mol of thiosulphate $S_2O_3^{2-}$). Calculate the rate of the reaction (include units) during:

a the first 15 minutes,

b the final 15 minutes.

Results:

t/min	Volume of $S_2O_3^{2-}$(aq)/cm^3
0	0
15	15.0
30	24.5
45	28.5
60	30.0
75	30.0

15.3

OBJECTIVES

- Rate changes during a reaction
- Plotting concentration–time graphs
- Finding rates from concentration–time graphs

Instantaneous reaction rate

The previous spread concluded by asking you a question about the reaction between acidified potassium iodide and aqueous hydrogen peroxide. Simply by looking at the experimental data (reproduced again below), it should be obvious that the reaction rate changes in the course of the reaction. The rate is greatest at the start of the reaction because the reactant concentration has its maximum value. The rate is zero at the end of the reaction because the concentration of one or more reactants is zero. How can you measure this changing rate and how can the results be expressed on paper? These are the problems tackled by this spread.

Results for the titration of 0.200 mol dm^{-3} aqueous sodium thiosulphate against the iodine liberated from the reaction between acidified potassium iodide and aqueous hydrogen peroxide.

t/min	Volume of $S_2O_3^{2-}$/cm^3
0	0
15	15.0
30	24.5
45	28.5
60	30.0
75	30.0

The practical procedure

The balanced chemical equation for the peroxide/iodide reaction is

$$H_2O_2(aq) + 2H^+(aq) + 2I^-(aq) \rightarrow I_2(aq) + 2H_2O(l)$$

The method of analysis is as follows:

1. Mix acidified potassium iodide with aqueous hydrogen peroxide; start the clock.
2. Every 15 minutes remove a 10.0 cm^3 sample from the reaction mixture and pour it into 100 cm^3 of water at 5 °C.
3. Titrate the iodine in the sample against 0.200 mol dm^{-3} aqueous sodium thiosulphate (1 mol of iodine I_2 reacts with 2 mol of thiosulphate $S_2O_3^{2-}$; see spread 13.4).

Typical results are shown in the table to the left. Note that the reaction is complete by 60 minutes: there is no increase in the volume of thiosulphate used and therefore no increase in the concentration of iodine in the reaction mixture.

Calculating concentrations

The rate of this reaction will be stated in terms of the concentration of the product, iodine. Iodine concentration is calculated from each of the titration results. The calculation for t = 15 min is as follows:

The volume of 0.200 mol dm^{-3} $S_2O_3^{2-}$ used is 15.0 cm^3.

Now we need to use

amount in moles = volume × concentration

So

$$\text{amount in moles of } S_2O_3^{2-} = \left(\frac{15.0}{1000}\,\text{dm}^3\right) \times (0.200\,\text{mol dm}^{-3})$$
$$= 3.0 \times 10^{-3}\,\text{mol}$$

The equation for the titration reaction is:

$$I_2(aq) + 2S_2O_3^{2-}(aq) \rightarrow 2I^-(aq) + S_4O_6^{2-}(aq)$$

From this, the mole ratio I_2:$S_2O_3^{2-}$ = 1:2. So

$$\text{amount in moles of } I_2 = \tfrac{1}{2} \times 3.00 \times 10^{-3}\,\text{mol}$$
$$= 1.50 \times 10^{-3}\,\text{mol}$$

This amount in moles was in a 10.0 cm^3 sample removed from the reaction mixture. Now we need to use

$$\text{concentration} = \frac{\text{amount in moles}}{\text{volume}}$$

So

$$\text{iodine concentration} = \frac{1.50 \times 10^{-3}\,\text{mol}}{10.0 \times 10^{-3}\,\text{dm}^3}$$
$$= 0.150\,\text{mol dm}^{-3}$$

Successive calculations along exactly the same lines as the above give the results shown in the table to the left.

Results showing the concentration of iodine at 15-minute intervals.

t/min	Concentration of I_2(aq) /mol dm^{-3}
0	0
15	0.150
30	0.245
45	0.285
60	0.300
75	0.300

Time and average rate

The results obtained in the course of the reaction may be plotted to give the graph shown on the right. The results in the graph show that the reaction produces a concentration of 0.300 mol dm^{-3} of iodine in 56 minutes, after which the reaction is complete. Rate is defined as:

$$\text{rate} = \frac{\text{change in concentration}}{\text{time}}$$

Therefore, the *overall* rate of the reaction is

$$\frac{0.300\ \text{mol dm}^{-3}}{56 \times 60\ \text{s}} = 8.9 \times 10^{-5}\ \text{mol dm}^{-3}\ \text{s}^{-1}$$

The rate *during the first 15 minutes* is

$$\frac{0.150\ \text{mol dm}^{-3}}{15 \times 60\ \text{s}} = 1.7 \times 10^{-4}\ \text{mol dm}^{-3}\ \text{s}^{-1}$$

These rates are average values, taken over a period of time. However, the requirement is usually for information about the magnitude of the rate *at a particular time.*

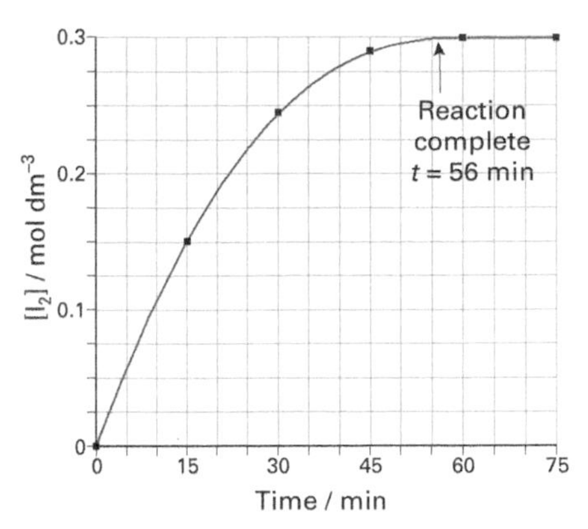

The graph of iodine concentration vs time.

Instantaneous rates

There are several important points to note about this graph:

- The rate of reaction is greatest at the start of the reaction (at t = 0 min).
- The graph becomes horizontal when the reaction is finished, i.e. when the gradient is zero.
- A time of just less than 60 minutes (about 56 minutes) elapsed from the start of the reaction to its completion. (Note that, from our results in the table, all we can say is that *the reaction was finished by the time we took our sample at 60 minutes*. On drawing the graph, we can say that *the reaction was finished just before we took our sample at 60 minutes.*)

To find the rate of reaction at a certain time, it is necessary to find the slope (gradient) of the tangent to the graph at that time.

- The instantaneous rate of reaction is equal to the slope (gradient) of the tangent to the concentration–time graph.

The technique is shown in the drawings alongside, using the graph obtained earlier from the experimental data and shown at the top of this page.

We calculate the instantaneous rate at t = 30 min as follows:

The line AB is the tangent to the graph at t = 30 min.

The rate of reaction at t = 30 min is equal to the slope of this line. So

$$\text{gradient AB} = \frac{\text{change in concentration}}{\text{time}}$$

$$= \frac{\text{MR}}{\text{RT}} = \frac{(y_2 - y_1)}{(x_2 - x_1)}$$

$$= \frac{(0.310 - 0.175)\ \text{mol dm}^{-3}}{60 \times (45.0 - 15.0)\ \text{s}}$$

$$= \frac{0.135\ \text{mol dm}^{-3}}{60 \times 30\ \text{s}}$$

$$= 7.5 \times 10^{-5}\ \text{mol dm}^{-3}\ \text{s}^{-1}$$

Remember that this value is for the rate of reaction at a given time – in the same way that the speedometer of a car will indicate 20 km h^{-1} as it is slowing down to a halt from an initial speed of 80 km h^{-1}.

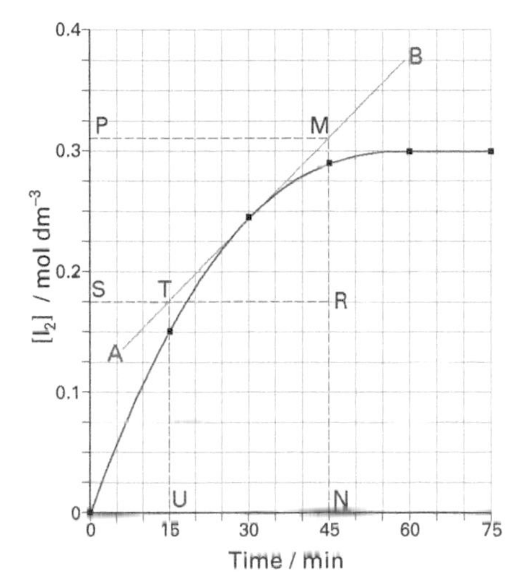

The construction lines that are necessary to find the rate at time t = 30 min:
1 Draw the tangent (AB) to the graph at t = 30 min.
2 At any convenient point M (x_2, y_2) on AB, draw the vertical line MRN (cuts the x-axis at x_2) and the horizontal line MP (cuts the y-axis at y_2).
3 At any other convenient point T (x_1, y_1) on AB, draw the vertical line TU (cuts the x-axis at x_1) and the horizontal line RTS (cuts the y-axis at y_1).

- Reaction rate is at a maximum at the start of the reaction; it is zero at the end of the reaction.
- Experimental results are used to plot concentration–time graphs.
- The rate at a given time is equal to the slope of the tangent to the concentration graph at that time.

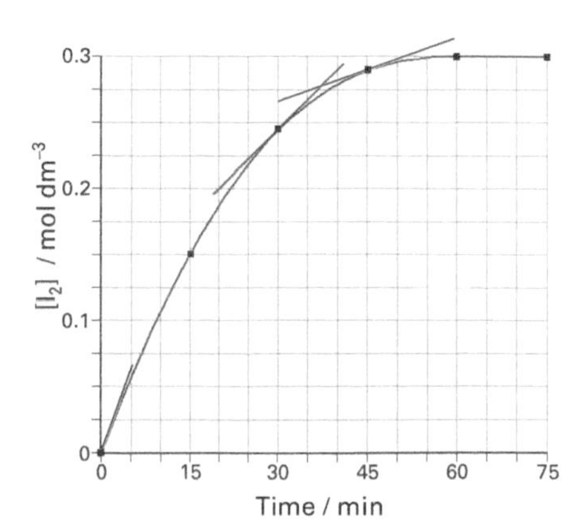

The slope is steepest at t = 0 min; it is less steep at t = 45 min than at t = 30 min. The reaction gets slower.

15.4

OBJECTIVES

- Temperature and kinetic energy
- Distribution of molecular energies
- Activation energy
- Reaction profiles

Reaction rate and collision theory

Reactant species: a reminder

The reactant species that collide in the course of chemical reactions may include molecules, atoms, or ions. For the sake of simplicity, we refer to reactant species simply as 'molecules'.

Temperature, energy, and speed

If we set T = thermodynamic temperature (in kelvin), E_k = kinetic energy, v = molecular speed, and m = molecular mass, then

$E_k \propto T$

and, by definition,

$E_k = \frac{1}{2}mv^2$

so,

$E_k \propto v^2$

and

$v^2 \propto T$

i.e.

$v \propto \sqrt{T}$

There are two assumptions about chemical reactions that it seems reasonable to make: molecules must physically meet before they can react; and molecules must collide with sufficient energy for reaction to take place. This spread examines these two assumptions and explores their influence on reaction rates. You will also see how they form the basis of 'collision theory' and the idea of 'activation energy'.

Temperature and average kinetic energy

Reactions occur faster at higher temperatures. For example, the reactions that cause milk and other food products to 'go off' occur more slowly at lower temperatures, which is why food is stored in a fridge. The even lower temperatures in a freezer keep food fresh for longer still.

The average kinetic energy of the molecules in a substance is proportional to its thermodynamic temperature. The kinetic energy of each of these molecules is proportional to the square of its speed. At higher temperature, the average speed of the molecules is greater. The greater average speed of molecules at higher temperatures has two consequences: the molecules collide more often, and each collision is more energetic.

Increasing the temperature from 25 °C to 35 °C causes the rate of the decomposition of dinitrogen pentaoxide N_2O_5:

$$2N_2O_5(g) \rightarrow 4NO_2(g) + O_2(g)$$

to increase by a factor of *four* (i.e. a 400% increase). You might conclude that the increase in reaction rate is due to the increase in molecular speed resulting in an increase in the collision frequency. However, increasing the temperature between these limits increases the average speed by just 2%. It seems the effect of temperature is far greater than can be explained simply by the molecules colliding more often.

The distribution of molecular energies

The molecules in a gas are continuously colliding with each other. The directions of the molecules are forever changing as they collide. An instantaneous 'snapshot' of a group of molecules would show a range of energies, from almost zero to very high values. This range may be represented by the Maxwell–Boltzmann distribution (see below and spread 8.4), which illustrates the distribution of molecular energies in a gas.

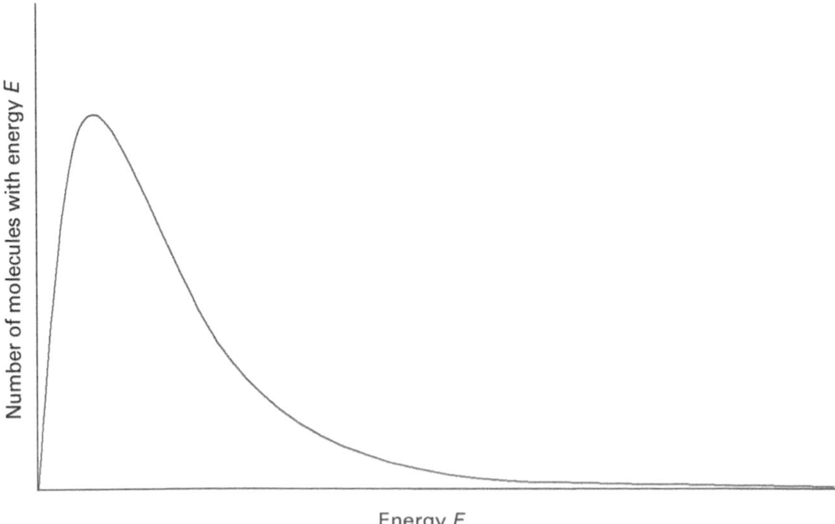

The Maxwell–Boltzmann distribution of molecular energies in a gas. You should note the following:
- *Only very small numbers of the molecules have either extremely high or extremely low energies.*
- *The curve is not symmetrical; the average energy is to the right of the peak of the curve.*

Activation energy

At the start of this spread, we made two assumptions: that molecules must come together to react; and also that they must collide with at least a certain minimum energy for reaction to occur. This minimum energy is called the **activation energy**, and it varies from one reaction to another. If the reactants collide with an energy at least equal to the activation energy, the collision is 'successful' and products will form. If the energy of the collision is less than the activation energy, the collision will be 'unsuccessful' and the molecules will simply rebound from each other unchanged.

The distribution of molecular energies shows how activation energy is the key to the dependence of reaction rate on temperature. The diagram below shows two superimposed molecular energy distributions at two temperatures, where T_2 is greater than T_1. At higher temperatures, there are many more energetic molecules, so the likelihood of a 'successful' collision increases rapidly with increase in temperature.

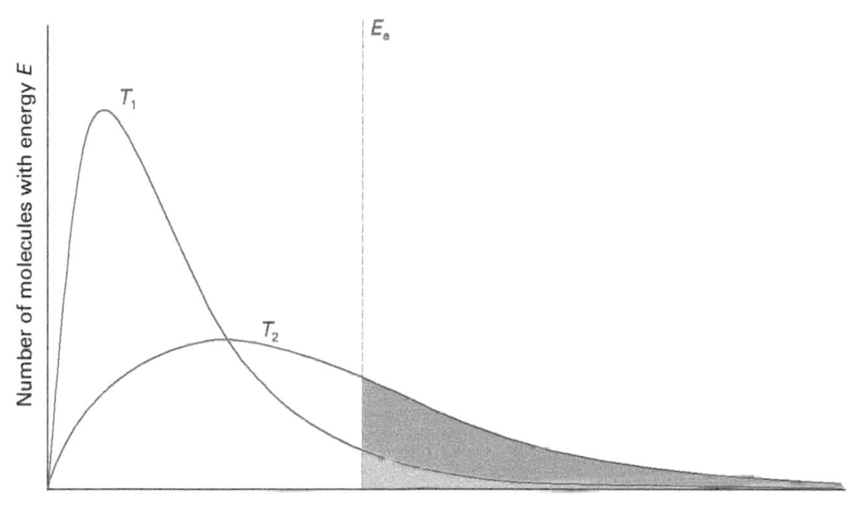

The distribution of molecular energies at two temperatures (T_2 is significantly higher than T_1). You should note the following:
•The area under each distribution curve represents the total number of molecules.
•There is a wider range of molecular energies at the higher temperature (T_2). The peak of the curve lies at higher energies at higher temperatures; the average energy also increases at higher temperatures.
•The activation energy is represented by the vertical line labelled E_a. The number of molecules that have at least this energy is much greater at the higher temperature than at the lower temperature (shown by the shading).

Reaction profiles

The activation energy may be shown on diagrams called 'reaction profiles'. These diagrams illustrate the role of activation energy as an energy barrier that must be overcome by reactants before they may form products. A question now needs to be answered: 'What state have the reactants reached at the point represented by the top of the reaction profile (the **transition state**)?' The answer will be discussed later in this chapter, as it is fundamental to understanding the action of catalysts.

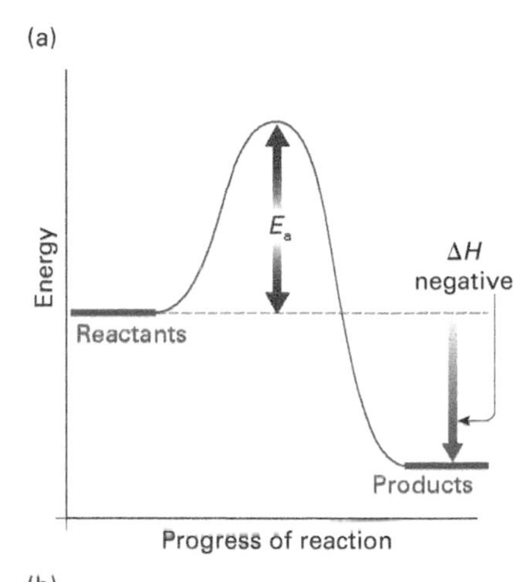

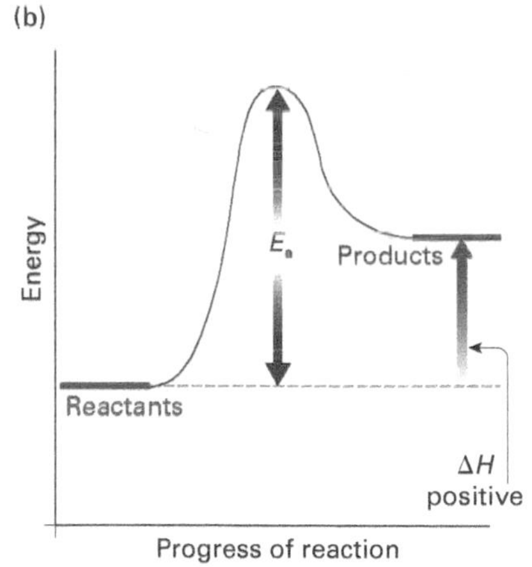

Reaction profiles for (a) exothermic and (b) endothermic reactions. E_a is the activation energy barrier that reactants must overcome before they can change into products. ΔH indicates the overall enthalpy change for the reaction.

SUMMARY

- The average energy of molecules is proportional to the thermodynamic temperature.
- The distribution of molecular energies is shown by the Maxwell–Boltzmann distribution.
- The energy of reactant molecules must be at least equal to the activation energy before reaction can take place.

PRACTICE

1 Explain each of the following terms or phrases, as used in the text above:
a Average energy **b** Distribution of energy **c** Energy barrier **d** 'Successful' collision **e** 'Unsuccessful' collision **f** Activation energy.

OBJECTIVES

- Rate and concentration
- Rate equations
- Order of reaction

Rate equations and order of reaction

Is there a relationship between the rate of a reaction and the concentration of the reactants? For a reaction to occur, the reactants have to collide. Higher concentration leads to a greater chance of the reactants colliding. So, at first sight, it seems reasonable to expect that the rate of a reaction *does* depend on the concentration of the reactants. Starting from this apparently common-sense point of view, two questions then arise: 'Can we express the rate of reaction *mathematically* in terms of reactant concentrations' and 'Can we then go on to *predict* the rate of a reaction?' The answer to both questions is 'Yes' – but we need to proceed with care!

The iodine clock reaction

The peroxodisulphate ion $S_2O_8^{2-}$ oxidizes iodide ions according to the reaction:

$$2I^-(aq) + S_2O_8^{2-}(aq) \rightarrow I_2(aq) + 2SO_4^{2-}(aq)$$

You can monitor the rate of this reaction in a convenient way by adding a fixed volume of aqueous thiosulphate ions to the reaction mixture together with a few drops of starch solution. The thiosulphate very rapidly reduces the iodine formed in the above reaction:

$$I_2(aq) + 2S_2O_3^{2-}(aq) \rightarrow 2I^-(aq) + S_4O_6^{2-}(aq)$$

When all the thiosulphate has reacted, free iodine rapidly forms a blue complex with the starch.

The iodine clock reaction.
(a) The flask contains the reactants, peroxodisulphate $S_2O_8^{2-}(aq)$ ions and iodide ions, together with thiosulphate $S_2O_3^{2-}(aq)$ ions and starch indicator. (b) As the thiosulphate is exhausted, the blue starch–iodine complex forms.

Experimental data for the peroxodisulphate/iodide reaction.

Experiment no.	Initial concentrations/mol dm^{-3} $[S_2O_8^{2-}]$	$[I^-]$	Initial rate of reaction /mol dm^{-3} s^{-1}
1	0.038	0.030	7.0×10^{-6}
2	0.076	0.030	14.0×10^{-6}
3	0.076	0.060	28.0×10^{-6}

Analysing the results

The table of results shows that doubling the concentration of peroxodisulphate (using the same concentration of I^-) *doubles* the rate of the reaction. For this reaction, reaction rate is proportional to peroxodisulphate concentration. In symbols:

$$\text{rate} \propto [S_2O_8^{2-}]$$

You will see from the table of results that the same relationship exists between rate and iodide concentration. In symbols:

$$\text{rate} \propto [I^-]$$

These two expressions may be combined with a constant of proportionality k to express the rate of the reaction in terms of the concentrations of both reactants:

$$\text{rate} = k[S_2O_8^{2-}][I^-]$$

The constant k is called the **rate constant**. (It may also be called the **rate coefficient**, because the value of k depends on temperature and so is not truly constant.)

The order of a reaction

In general, for a reaction between A and B, experimental results can be analysed to show that:

$$\text{rate} = k[A]^m[B]^n$$

Such an equation is called a **rate equation**. The power to which the concentration of A is raised in the experimental rate equation is called the **order** of reaction with respect to A. The order with respect to A is *m* and the order with respect to B is *n*. The overall order is the sum of the individual orders: $m + n$ in this case.

- Once the rate constant and order with respect to reactants A and B are known, the rate may be predicted at any concentrations of A and B.

The rate of the peroxodisulphate/iodide reaction above is directly proportional to the concentration of the peroxodisulphate ions; the rate is also directly proportional to the concentration of the iodide ions. This may be expressed as:

$$\text{rate} = k[S_2O_8^{2-}][I^-]$$

This rate equation indicates that the reaction is:

- first order with respect to peroxodisulphate ion;
- first order with respect to iodide ion;
- second order overall.

Note that orders must be determined *experimentally* and may not, as here, be equal to the mole ratios in the balanced chemical equation.

Determining order

One method of finding the order is to carry out a number of separate experiments with different initial concentrations and measure the *initial* rates of reaction, i.e. measure the time taken to produce a small fixed concentration of product.

- If *doubling* the concentration of A has *no effect* on the rate, the reaction is **zero order** with respect to A:

  $\text{rate} = k[A]^0$ i.e. $\text{rate} = k$

- If *doubling* the concentration of A *doubles* the rate, the rate and the concentration are directly proportional, and the reaction is **first order** with respect to A:

  $\text{rate} = k[A]^1$ i.e. $\text{rate} = k[A]$

- If *doubling* the concentration of A *increases* the rate by a *factor of four*, the reaction is **second order** with respect to A:

  $\text{rate} = k[A]^2$

SUMMARY

- The rate equation expresses the rate of a reaction in terms of the concentration of each reactant raised to a specific power.
- The power to which the concentration of a reactant is raised in the rate equation is the order of the reaction with respect to that reactant.
- The overall order of a reaction is the sum of the orders in the rate equation.

Units

In the rate equation

$\text{rate} = k[S_2O_8^{2-}][I^-]$

rate has units of $\text{mol dm}^{-3}\text{ s}^{-1}$ and the concentration of each reactant is expressed as mol dm^{-3}. So, for this reaction, *k* has units $\text{mol}^{-1}\text{ dm}^3\text{ s}^{-1}$. The units of *k* vary, depending on the overall order.

Predicting the rate

Dinitrogen pentaoxide gas N_2O_5 decomposes according to the chemical equation:

$2N_2O_5(g) \rightarrow 4NO_2(g) + O_2(g)$

This reaction is first order with respect to N_2O_5 as shown by the rate equation:

$\text{rate} = k[N_2O_5]$

At 298 K the value of *k* is $3.4 \times 10^{-5}\text{ s}^{-1}$. The instantaneous rate of this reaction may be calculated simply by substituting into the rate expression. For example, if the concentration of N_2O_5 is 1.0 mol dm^{-3}:

$\text{rate} = (3.4 \times 10^{-5}\text{ s}^{-1}) \times (1.0\text{ mol dm}^{-3})$

$= 3.4 \times 10^{-5}\text{ mol dm}^{-3}\text{ s}^{-1}$

Too many experiments

A disadvantage of this approach is that sufficient quantities of the reactants must be available to perform several experiments. These quantities are not always available, especially if the material is expensive or difficult to obtain. Later in this chapter we will look at a more elegant approach to determining order.

PRACTICE

1 The results shown refer to the oxidation of bromide ion by bromate ion in acidic solution. The equation for the reaction is:

$5Br^-(aq) + BrO_3^-(aq) + 6H^+(aq)$
$\rightarrow 3Br_2(aq) + 3H_2O(l)$

a Find the orders with respect to each of the three reactants.

b State the overall order of the reaction.

c Write the rate equation.

d Calculate the value of the rate constant and state its units.

e Calculate the rate when the concentrations of all three reactants are 0.2 mol dm^{-3}.

Results

$[Br^-]$ /mol dm^{-3}	$[BrO_3^-]$ /mol dm^{-3}	$[H^+]$ /mol dm^{-3}	Initial rate /mol dm^{-3} s^{-1}
0.10	0.10	0.10	8.0×10^{-4}
0.10	0.20	0.10	1.6×10^{-3}
0.20	0.20	0.10	3.2×10^{-3}
0.10	0.10	0.20	3.2×10^{-3}

15.6

OBJECTIVES

- Determining order from rate–concentration graphs
- Reaction mechanisms
- Order as evidence for mechanisms
- Rate-limiting step

Using order to find reaction mechanisms

The previous spread showed how the order of a reaction may be found by carrying out a series of separate reactions. If we know the order, we can calculate the rate constant, which in turn allows us to calculate the rate at given reactant concentrations. This spread shows how we may determine the order with respect to a reactant from a single reaction that is continuously monitored throughout its progress. Once we know the order for each reactant, we can suggest possible reaction mechanisms for the overall reaction.

(a)

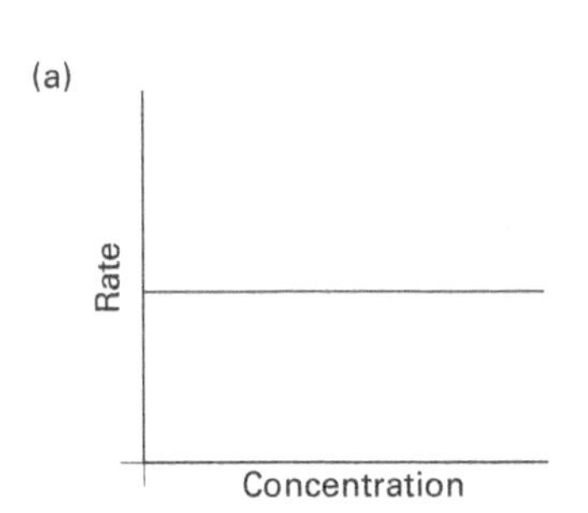

(b)

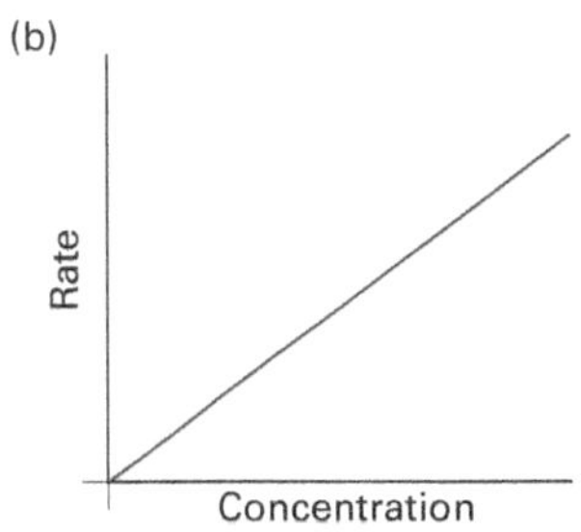

(c)

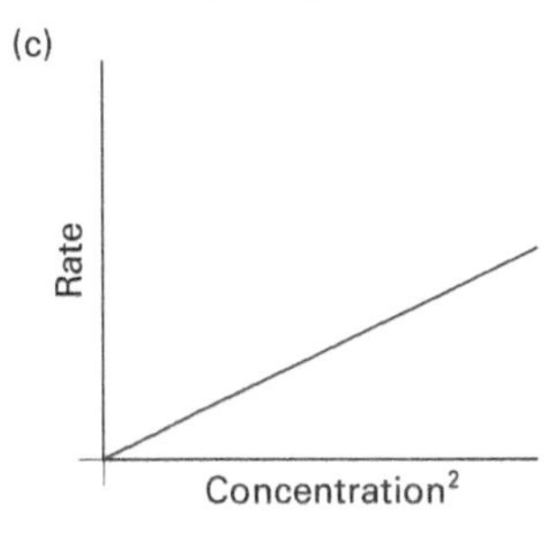

The three types of rate–concentration graphs used to indicate order with respect to a given reactant A.
(a) A zero-order reaction:
rate = $k[A]^0 = k$
(b) A first-order reaction:
rate = $k[A]^1 = k[A]$
(c) A second-order reaction:
rate = $k[A]^2$

Order from rate–concentration graphs

The concentration of a reactant falls as time elapses. If we monitor the concentration of a reactant during the course of a reaction, we will be able to plot a concentration–time graph. We can then draw a tangent at a point on this graph. The slope of the tangent gives the instantaneous rate of the reaction at that particular concentration. (The technique is detailed earlier in this chapter.)

We need to do this at several points on the concentration–time graph. Then we plot another graph. This time, the value of the instantaneous rate that we have found is plotted against concentration. The shape of our rate–concentration graph may then be used to determine order with respect to that reactant. There are three possible cases:

- For a zero-order reaction, the graph of rate as a function of concentration will be a horizontal straight line that cuts the vertical axis at k.
- For a first-order reaction, the graph of rate as a function of concentration will be a straight line of slope k passing through the origin.
- For a second-order reaction, the graph of rate as a function of the *square* of the concentration will be a straight line of slope k passing through the origin.

The iodination of propanone

The reaction between iodine and propanone CH_3COCH_3 is extremely slow, even at high temperatures:

$$I_2(aq) + CH_3COCH_3(aq) \rightarrow CH_3COCH_2I(aq) + HI(aq)$$

However, the addition of dilute acid provides $H^+(aq)$ ions, which catalyse the reaction and cause it to happen in a matter of minutes at 25 °C. The iodine concentration may be determined by titration. Details of a typical reaction mixture are given below together with expected results and their manipulation to find the order with respect to iodine.

(a) Initial concentrations

iodine	0.01 mol dm^{-3}
propanone	0.25 mol dm^{-3}
sulphuric acid	0.25 mol dm^{-3}

At $t = 0$, $[I_2] = 0.01$ mol dm^{-3}

(b)

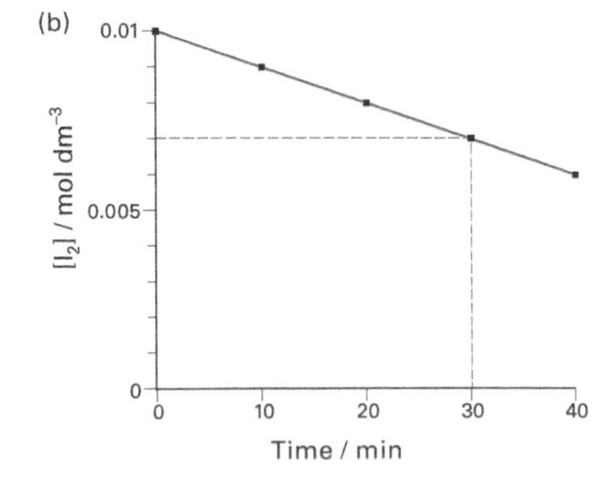

(c)

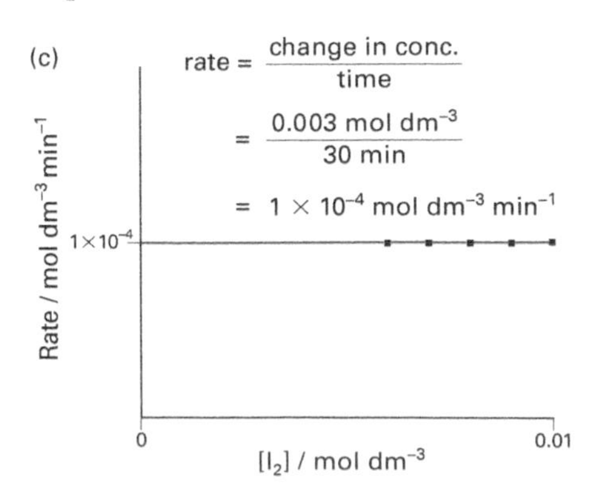

The experiment to determine the order of reaction with respect to iodine for the reaction between iodine and propanone. (a) Initial concentrations of the reaction mixture. (b) The results plotted as a graph (iodine concentration vs time). (c) The graph of rate vs iodine concentration.

The graph of rate against iodine concentration is a straight line parallel to the horizontal axis. This result shows that the reaction is zero order with respect to iodine, i.e. the rate is *independent* of the concentration of iodine. This is a strange result when you bear in mind that iodine is one of the reactants. Other kinetic studies show that the reaction is first order with respect to propanone and, perhaps more surprisingly, first order with respect to the catalyst (hydrogen ions from the dilute acid).

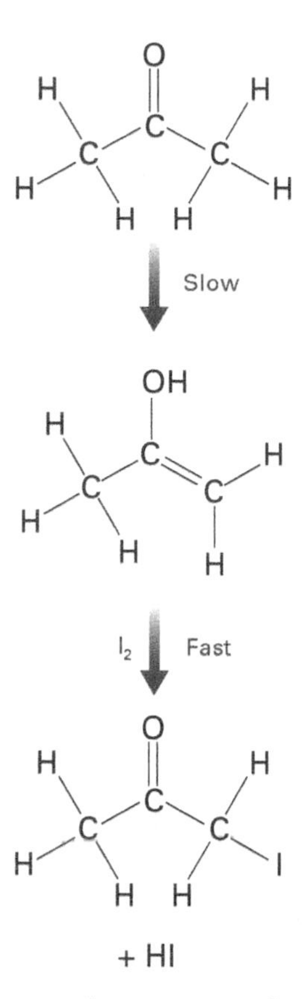

The suggested two-step reaction mechanism for the iodination of propanone.

Suggesting a reaction mechanism

A reaction **mechanism** is a detailed step-by-step account of how an overall reaction happens. It specifically states all intermediate stages and mentions all intermediate species formed, even though some do not appear as products. The kinetic observations suggest that the mechanism for the iodination of propanone occurs in at least two steps.

Rate is directly proportional to both propanone and aqueous hydrogen ion concentrations (both are first order), indicating that these two substances react together in a first step. In this first step, propanone is changed into a more reactive form (called the enol form). This is a *slow step* in comparison to the next one.

Rate is independent of iodine concentration (zero order). Because it does not affect the overall rate of the reaction, the second step must be a *fast step* in comparison to the first one. In this second step, the double bond between the two carbon atoms reacts rapidly with iodine to form the products of the reaction. The complete reaction mechanism is shown on the right.

Overall rate

The reaction above takes place in two steps. The first step is the slower of the two and is therefore called the **rate-limiting step** (sometimes called the *rate-determining step*). Increasing the rate of the second (faster) step will not increase the overall rate.

The enol form of propanone is referred to as the **reactive intermediate**. It reacts rapidly with the iodine and, in the process, regenerates the catalyst $H^+(aq)$.

Comment

This investigation shows that it is possible to find out important details about a reaction by interpreting information about orders of reaction derived from *experimental* data.

(a)

(b)

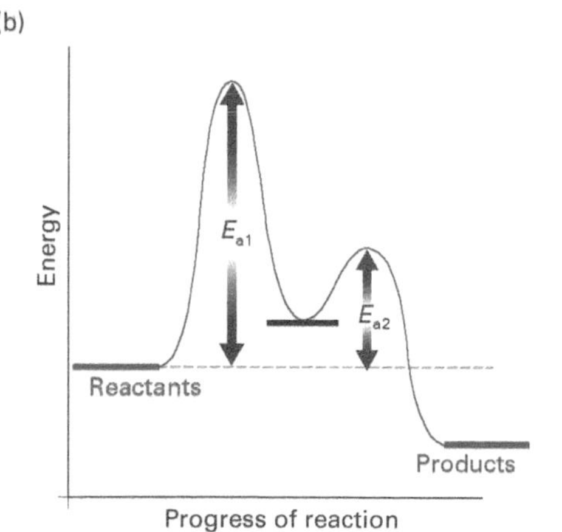

(a) This two-step reaction is like an assembly-line system. The overall rate of production is limited by the slowest step in the process. (b) The corresponding reaction profile includes two separate energy barriers E_{a1} and E_{a2}. Notice that E_{a2} is much smaller than E_{a1}, which explains the relative rates of the two steps.

SUMMARY

- A reaction mechanism describes in detail how reactants turn into products.
- The rate-limiting step is the slowest step in a multi-step reaction.

PRACTICE

1 Explain the meaning of the term *first-order* reaction.

15.7

Using calculus to find order

OBJECTIVES

- Using mathematics to find how rate depends on concentration
- Solving rate equations for zero- and first-order reactions
- First order and exponential decay
- Half-life

The substance being investigated in a rate experiment may not be available in unlimited quantities. In this case, the technique of doing multiple experiments with different initial concentrations and measuring the initial rates is too wasteful in material. It would be better to monitor the concentration as a function of time during a single experiment. This is the technique commonly used in research laboratories.

The disadvantage of this approach is that the analysis of the data becomes more complicated. The rate equations need to be solved mathematically and then appropriate graphs drawn. This spread introduces the mathematical solutions and explains how to analyse the experimental data.

Formal definition

The formal definition of the rate of a reaction is best presented in mathematical form. For a reaction $aA + bB \rightarrow cC + dD$ the rate can be defined uniquely by dividing the rate of change of any species by its mole ratio. That is rate =

$$\left(\frac{1}{c}\right)\frac{d[C]}{dt} = \left(\frac{1}{d}\right)\frac{d[D]}{dt} = -\left(\frac{1}{a}\right)\frac{d[A]}{dt} = -\left(\frac{1}{b}\right)\frac{d[B]}{dt}$$

Writing a rate equation mathematically

We can write a rate equation mathematically by recognizing that it corresponds to the time differential of a concentration c. If we are considering one of the *reactants*, the concentration of the reactant will decrease as the reaction proceeds. The rate of reaction can be written as

$$\text{rate} = -\frac{dc}{dt}$$

Zero order

For a zero-order reaction, the rate is proportional to the concentration raised to the power zero. As any number raised to the power zero is equal to one, the rate is a *constant*, k, and the rate equation for a zero-order reaction is

$$\text{rate} = kc^0 = k$$

$$-\frac{dc}{dt} = k$$

multiplying by –1 gives the equation

$$\frac{dc}{dt} = -k \qquad (1)$$

The solution of this equation is

$$c = c_0 - kt$$

where c_0 is the concentration at time zero. The solution can be differentiated to give equation 1; c_0 and k are constants.

The concentration in a zero-order reaction falls linearly with time, from the value c_0 at time zero. The slope of the line is $-k$. See, for example, the graph of the iodine concentration in the iodination of propanone shown alongside.

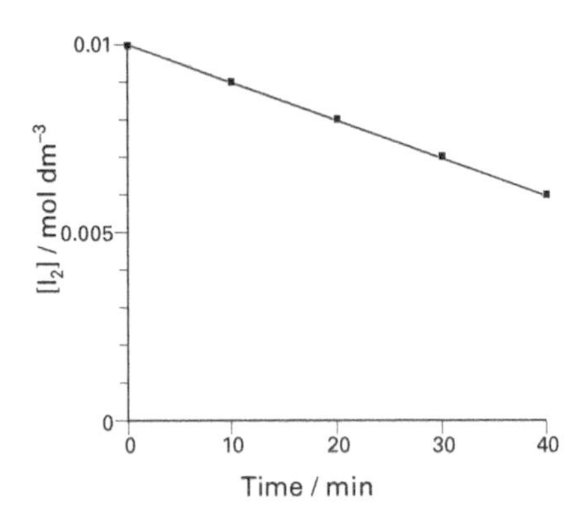

The graph of iodine concentration against time in the reaction between iodine and propanone is a straight line, so the reaction is zero order with respect to iodine.

Another zero-order reaction

The removal of alcohol from the human follows zero-order kinetics. Approximately one unit is removed each hour, independent of the amount consumed.

First order

For a first-order reaction, the rate is proportional to the concentration raised to the power one. The rate equation for a first-order reaction is

$$\text{rate} = kc$$

$$-\frac{dc}{dt} = kc$$

Multiplying by –1 gives the equation

$$\frac{dc}{dt} = -kc \qquad (2)$$

The solution to this equation involves an exponential, as this is the function whose differential is itself (see maths toolbox). Therefore the solution for the concentration as a function of time is as follows:

$$c = c_0 e^{-kt} \qquad (3)$$

Differentiating this equation produces equation 2:

$$\frac{dc}{dt} = c_0 \times (-k)e^{-kt} = -kc$$

Second order

The rate equation is $dc/dt = -kc^2$ whose solution is $1/c = 1/c_0 + kt$

Equation 3 is called an **exponential decay**: the concentration decays exponentially from the value c_0 at time zero. Taking natural logarithms (see maths toolbox) of this equation gives the following equation for the natural log of the concentration:

$$\ln c = \ln c_0 - kt$$

which is the equation of a straight line with a slope of $-k$.

So to prove that a reaction is first order it is necessary to plot the *natural log* of the concentration as a function of time. If the reaction is first order, this graph will be a straight line (see figure alongside).

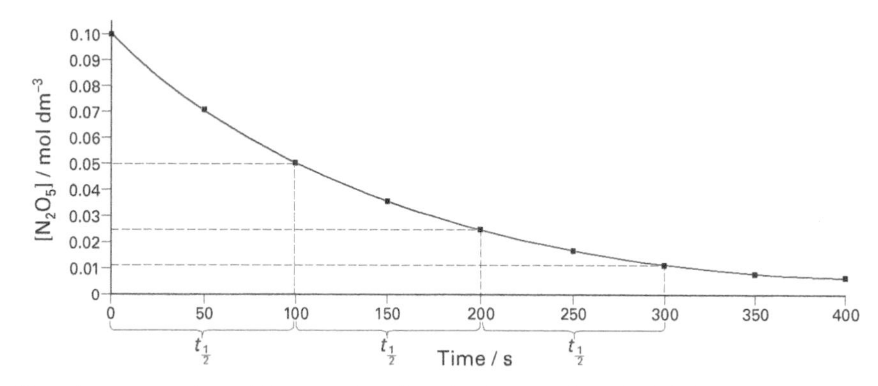

Plot of concentration of N_2O_5 against time for the reaction $2N_2O_5(g) \rightarrow 4NO_2(g) + O_2(g)$ at 70 °C. The graph shows that the concentration falls exponentially. The half-life $t_{½}$, discussed below, is constant for a first-order reaction.

Radioactive decay and half-life

An important example of a first-order reaction is radioactive decay. The emission of alpha or beta particles obeys a first-order rate equation (equation 2) because the probability of radioactive decay depends only on the number of radioactive atoms remaining.

The **half-life** of a first-order reaction is the time taken for the concentration to fall to half its original value. It is clearly related to the rate constant, the exact relationship being derived as follows.

$$c = c_0 e^{-kt}$$

Divide both sides by c_0 and take natural logs to find

$$\ln(c/c_0) = -kt$$

At the half-life $t_{½}$, the concentration c has become $½c_0$ and so

$$\ln(½) = -kt_{½}$$

Multiplying through by –1, using the rules of logarithms, gives the final answer that

$$\ln 2 = kt_{½}$$

Note that the half-life is independent of concentration.

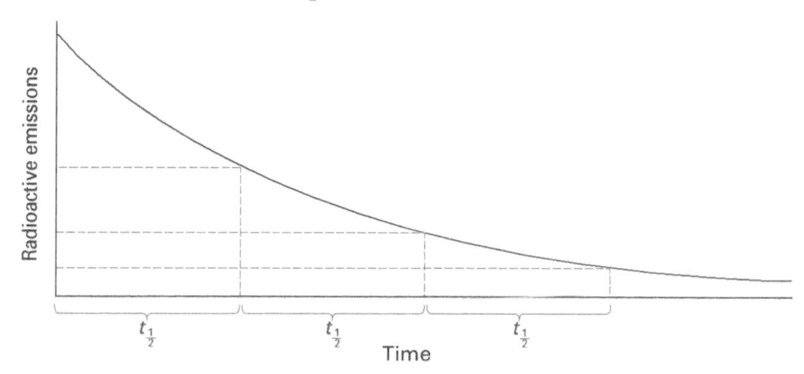

The half-life is the time taken for the concentration (the radioactive emissions in this instance) to fall to half its original value.

SUMMARY

- For a first-order reaction, the concentration decays exponentially.
- Radioactive decay is an example of a first-order reaction.

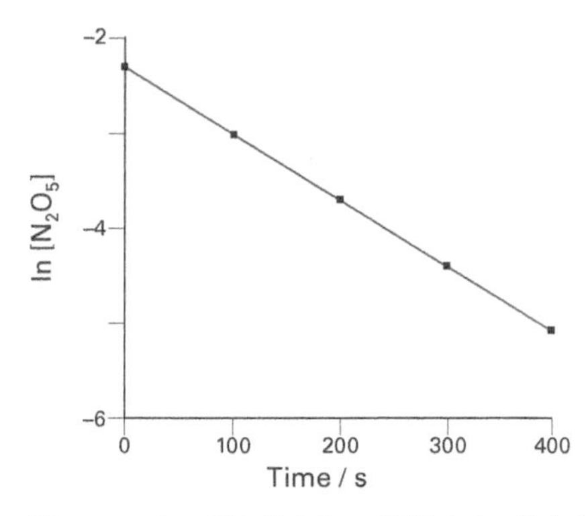

The reaction $2N_2O_5(g) \rightarrow 4NO_2(g) + O_2(g)$ is first order with respect to the reactant, as the natural log *graph is a straight line.*

In this experiment, the concentration is being displayed on the screen on the left. The exponential decay can be clearly seen.

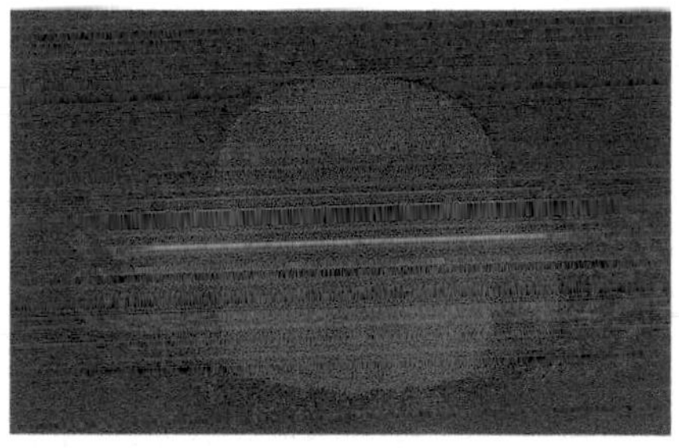

The radioactive element curium (named after Marie Curie), photographed in the dark, illuminated only by the radiation it emits itself.

Half-life and order

A half-life can be calculated for other orders of reaction; however, only first-order reactions have a *constant* half-life.

PRACTICE

1 Explain the meaning of the term *half-life*. For a first-order reaction, how many half-lives does it take for the concentration to reach one eighth of its original value?

15.8

OBJECTIVES

- Finding the activation energy for a reaction
- The Arrhenius equation
- Applications of the Arrhenius equation

A Cyalume™ light stick glows more brightly in warm water than in iced water.

The parameter A

The value of the **Arrhenius constant** A can be approximately interpreted as the rate at which the particles collide multiplied by a steric factor to account for the fact that not all collision directions lead to reaction.

T/°C	k/s^{-1}	ln k	1/(T/K)
25	3.4×10^{-5}	−10.29	3.36×10^{-3}
35	1.4×10^{-4}	−8.87	3.25×10^{-3}
45	5.0×10^{-4}	−7.60	3.14×10^{-3}
55	1.5×10^{-3}	−6.50	3.05×10^{-3}
65	4.9×10^{-3}	−5.32	2.96×10^{-3}

ACTIVATION ENERGY AND THE ARRHENIUS EQUATION

It may seem that the activation energy for a reaction is a concept that has been pulled like a rabbit from a hat. This is not the case; it is a measurable quantity. Detailed mathematical analysis allows the value of the activation energy to be extracted from experimental data. The Swede Svante Arrhenius suggested an empirical (based on observation and experiment) equation which could be used to measure the activation energy. It turned out to be consistent with the equations of thermodynamics.

The Arrhenius equation

Svante Arrhenius suggested how to calculate the activation energy from the rates of reaction at different temperatures. The Arrhenius equation predicts that the rate constant, k, defined in previous spreads, is related to the temperature by:

$$\ln k = \ln A - E_a/RT$$

The second term involves the activation energy E_a, the gas constant R (spread 8.2), and the thermodynamic temperature T (in kelvin). Each reaction has a particular value of A and E_a.

Although the Arrhenius equation seems complicated, it is actually the equation of a straight line ($y = mx + c$). To see this, think of $\ln k$ (the rate variable) as equivalent to y and $1/T$ (the temperature variable) as equivalent to x. The gradient, m, of the line is then the constant $-E_a/R$. The intercept, c, is the constant $\ln A$. A graph of $\ln k$ against $1/T$ is called an **Arrhenius plot.**

The figure below shows an Arrhenius plot of the data in the table on the effect of temperature on the decomposition of dinitrogen pentaoxide N_2O_5. It is clear that the graph is a straight line. The activation energy can be found from the gradient which equals $-E_a/R$ (where R is 8.31 J K^{-1} mol^{-1}). For this reaction, the activation energy is 105 kJ mol^{-1}.

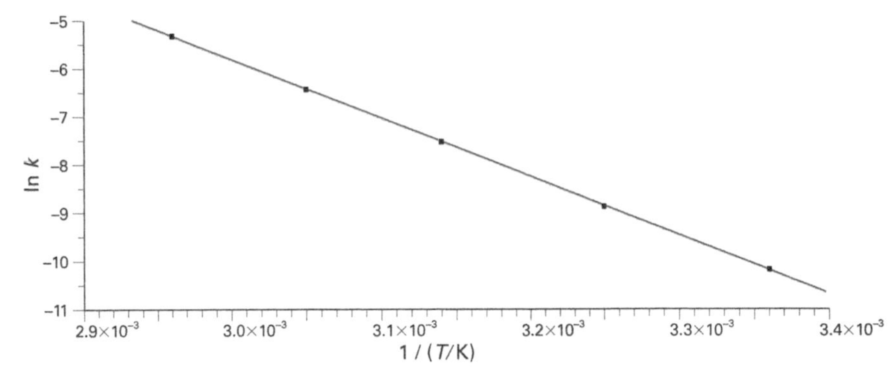

An Arrhenius plot for the decomposition of dinitrogen pentaoxide at different temperatures (shown in the table).

This sort of analysis can also be carried out for reactions in solution. Aqueous hydrochloric acid added to aqueous sodium thiosulphate causes a precipitate of sulphur to form: the time taken for the precipitate to obscure a card placed underneath the flask is measured. The third column in the table opposite on the sulphur precipitation ('sulphur clock') reaction, headed k, has been calculated as follows. The rate of reaction is the change in concentration divided by the time, t, taken for the change. Because the same concentration of sulphur is needed to hide the card, the relative rate is simply found by calculating $1/t$.

The table opposite shows how the relative rate varies as the temperature is increased from 25 °C to 65 °C. It is clear that this is not a linear change. The figure opposite shows an Arrhenius plot of the data in the table. This graph *is* a good straight line.

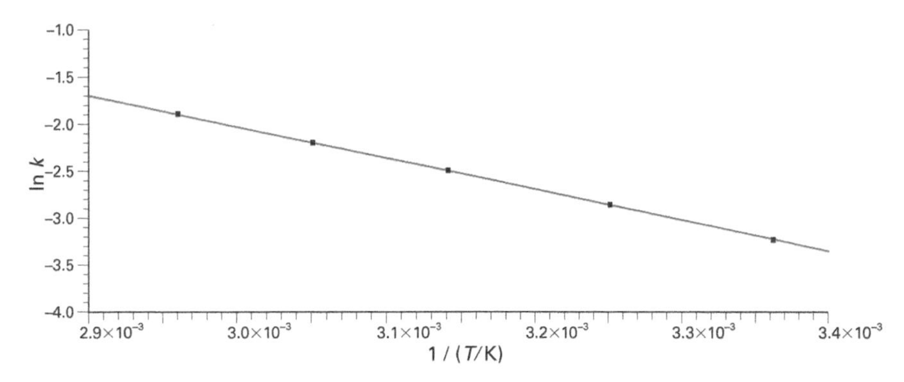

An Arrhenius plot for the sulphur precipitation reaction at different temperatures. See the previous table for 1/(T/K) values.

T/°C	t/s	k/s^{-1}	ln k
25	25.3	0.040	−3.22
35	17.9	0.056	−2.88
45	12.5	0.080	−2.53
55	9.0	0.111	−2.20
65	6.6	0.152	−1.88

The activation energy can be found from the gradient of the graph, which equals $-E_a/R$. For this reaction, the activation energy is 28 kJ mol^{-1}.

The Arrhenius plot is widely used to analyse the effect of temperature on reaction rate. The figure alongside shows data obtained by General Motors to monitor the effectiveness of their catalytic converters.

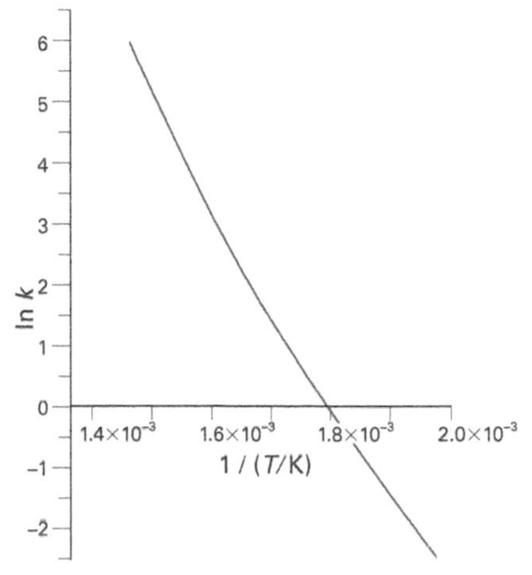

The effectiveness of a catalytic converter analysed using an Arrhenius plot. The reaction involved is discussed in the following spread.

The Arrhenius equation can be rearranged by taking exponentials of both sides, since natural logarithms ln and exponentials e^x are inverse operations. The resulting equation is

$$k = A\,e^{-E_a/RT}$$

The first factor on the right hand side of the equation, A, is called the **pre-exponential factor** or the **Arrhenius constant**. For gaseous reactants, it roughly takes account of the rate of collision. As only a certain small fraction of the collisions are successful, this needs to be multiplied by a second factor which depends on the activation energy to find the rate constant.

Worked example on using the Arrhenius equation

Question: The activation energy for a certain reaction is 50 kJ mol^{-1}. What is the effect on the rate constant of increasing the temperature by 10 K around room temperature (assumed to be 15 °C which is 288 K)?

Strategy: Substitute into the Arrhenius equation. Calculate the ratio of the rate constants at the two temperatures. The first factor in the Arrhenius equation is the pre-exponential factor and this is a constant for the reaction so it will cancel, leaving only the exponential term. So the effect of temperature will be explained by the second factor involving the activation energy.

Answer: The ratio of rate constants is

$$k(25\,°\text{C})/k(15\,°\text{C}) = \frac{\exp((-50\,000\ \text{J mol}^{-1})/(8.31\ \text{J K}^{-1}\ \text{mol}^{-1})*(298\ \text{K}))}{\exp((-50\,000\ \text{J mol}^{-1})/(8.31\ \text{J K}^{-1}\ \text{mol}^{-1})*(288\ \text{K}))}$$

$$= 2.0$$

Note: The rate doubles for a 10 °C rise in temperature if the activation energy is 50 kJ mol^{-1}.

Tropical fireflies flash faster on warm nights. The change in their flash rate produces an Arrhenius plot with an activation energy of about 50 kJ mol^{-1}.

SUMMARY

- The activation energy may be found using an Arrhenius plot.
- An Arrhenius plot is a graph of ln k against $1/T$ with T measured in kelvin: its gradient is $-E_a/R$.

15.9 Catalysis

OBJECTIVES

- General catalytic behaviour
- Alternative reaction routes
- Common features

Catalysts are substances that increase the rates of a wide variety of chemical reactions. They can be recovered at the end of the reaction, unchanged in mass or chemical composition. Industrial catalysts are used in the manufacture of an enormous variety of products, from margarine to plastics to fertilizers to cracking of crude oil; see spread 22.3. **Enzymes** are biological catalysts, which allow living systems to function. This spread looks at the general principles underlying catalyst function. The following two spreads deal with the two main categories of catalytic activity in greater depth.

Catalytic activity

Reaction profiles show that the activation energy acts as an energy barrier, which the reactants must overcome before they can change into products. Most chemical reactions proceed by a distinct route, which includes one or more intermediate steps. The reaction mechanism describes the steps by which a reaction takes place. When a reaction mixture includes a suitable catalyst, the reaction can occur by an *alternative route of lower activation energy* than the uncatalysed reaction. At the same temperature, a greater proportion of the reactant molecules will therefore have sufficient energy to overcome the (lower) energy barrier for the catalysed reaction, and so the rate is increased.

There are two important classes of catalysts: heterogeneous catalysts and homogeneous catalysts.

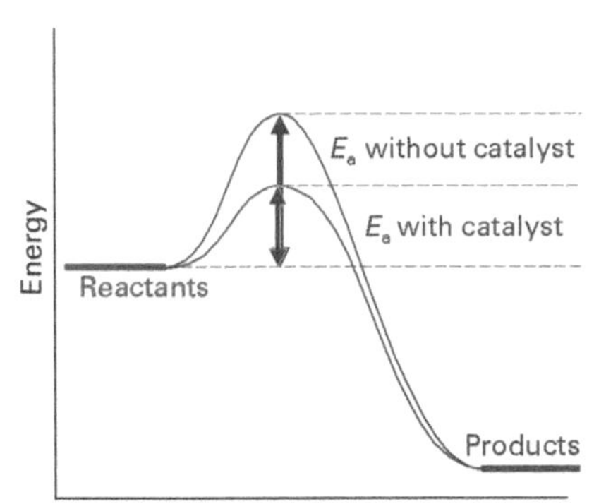

The reaction profiles for (a) an uncatalysed reaction and (b) the same reaction with a suitable catalyst present.

Autocatalysis

Autocatalysis occurs when a product of a reaction increases the rate of the reaction. For example, in the reduction of manganate(VII) ions by the similarly charged ethanedioate (oxalate) ions, $C_2O_4^{2-}$:

$$2MnO_4^-(aq) + 5C_2O_4^{2-}(aq) + 16H^+(aq) \rightarrow 2Mn^{2+}(aq) + 10CO_2(g) + 8H_2O(l)$$

the Mn^{2+} ions produced catalyse the reaction.

$$8Mn^{2+}(aq) + 2MnO_4^-(aq) + 16H^+(aq) \rightarrow 10Mn^{3+}(aq) + 8H_2O(l)$$

$$10Mn^{3+}(aq) + 5C_2O_4^{2-}(aq) \rightarrow 10Mn^{2+}(aq) + 10CO_2(g)$$

Heterogeneous catalysis

- A **heterogeneous catalyst** is in a different phase from the reactants.

An example of heterogeneous catalysis is the use of iron in the Haber–Bosch synthesis in which nitrogen and hydrogen react to form ammonia. The rate of this reaction is imperceptible at room temperature, and increasing the temperature and the pressure has relatively little effect because of the extremely high bond energy of the N≡N triple covalent bond. However, the equilibrium

$$N_2(g) + 3H_2(g) \rightleftharpoons 2NH_3(g)$$

is rapidly established in the presence of finely divided iron, which acts as a heterogeneous catalyst.

The catalysed reaction mechanism depends on the reactant molecules attaching themselves to the metal surface: the N≡N bond is broken as the molecules attach (this is called **dissociative adsorption**). Reaction occurs on the surface, and the ammonia molecules then detach (desorb) from the metal surface.

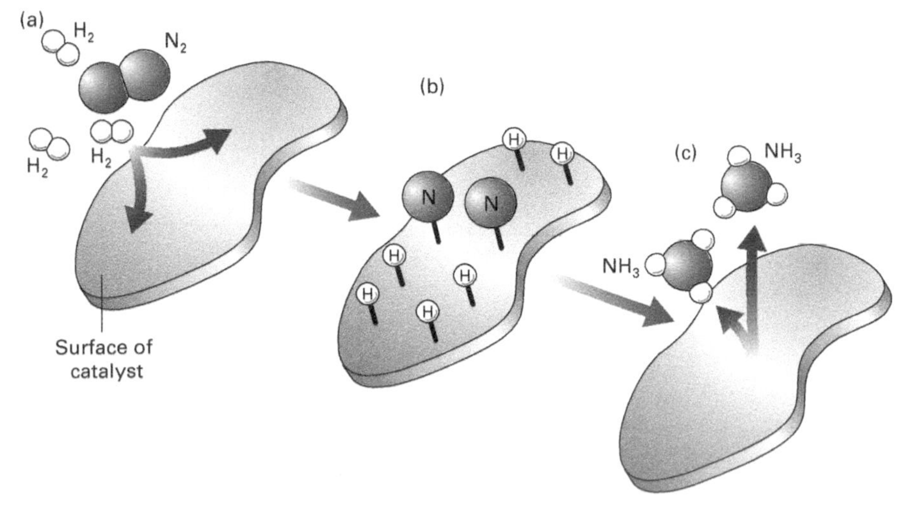

The reaction mechanism for the catalytic reduction of nitrogen by hydrogen on an iron surface. This diagram shows the three stages: (a) inward diffusion of reactants and (b) attachment to the catalyst surface; followed by reaction to form products; and (c) desorption and outward diffusion of products.

Margarine

Another important industrial application of heterogeneous catalysis is in the 'hardening' of vegetable oils in the production of margarine. Oil molecules contain large numbers of C=C double covalent bonds. Hydrogen reacts with these bonds in the presence of a nickel catalyst to produce molecules with C–C single bonds. The greater the proportion of these single bonds, the higher the melting point of the product. Liquid oils become solid fats.

Homogeneous catalysis

- A **homogeneous catalyst** is in the same phase as the reactants.

An example of homogeneous catalysis is the use of aqueous cobalt(II) chloride in the reaction between hydrogen peroxide and an aqueous solution of Rochelle salt (sodium potassium 2,3-dihydroxybutanedioate). When the catalyst is added, the reaction happens very much more quickly. During the reaction, the aqueous cobalt(II) chloride changes colour from pink to green before reverting to pink at the end, spread 20.6. This observation shows that the catalyst takes an *active* part in a reaction rather than being an inactive spectator.

Common features of catalysts

- *Catalysts may be very specific.*

A catalyst is generally specific to a single reaction or to a class of very similar reactions. For example, the enzyme urease catalyses the hydrolysis of urea, NH_2CONH_2, but not the hydrolysis of the molecule CH_3CONH_2. Platinum catalyses a number of hydrogenation reactions (addition of H_2) but not *all* addition reactions.

- *Catalysts do not affect the equilibrium position.*

Catalysts do not affect the equilibrium position, nor do they affect the value of the equilibrium constant. The position of equilibrium does not depend on the route taken by a reaction. The forward and backward reactions are speeded up *to the same extent*.

- *Small quantities of catalyst can usually achieve a huge increase in rate.*

Just 2 g of platinum can catalyse the decomposition of one million litres of hydrogen peroxide. One molecule of the enzyme triose-phosphate isomerase can catalyse the reaction of 400 000 molecules per second (the limiting factor is the rate of diffusion of reactant molecules to the enzyme).

- *The state of subdivision of the catalyst is significant.*

Solid catalysts are more effective when finely divided (very finely powdered), because they present a larger surface area to the reactants.

- *Catalysts may be poisoned by some other substances.*

Catalysts may be poisoned, and hence be rendered ineffective, by substances such as lead, arsenic, and the cyanide ion CN^-. Such substances bind strongly to the catalyst surface and block the adsorption of reactant molecules. These substances also often act as poisons to living organisms, binding to the active sites in enzymes and blocking their biological activity.

SUMMARY

- A catalyst speeds up a reaction by providing an alternative route of lower activation energy.
- A catalyst is specific to a single reaction or to a class of similar reactions.
- A heterogeneous catalyst is in a different phase from the reactants.
- A homogeneous catalyst is in the same phase as the reactants.
- A catalyst does not alter the equilibrium position.

The bombardier beetle stores hydrogen peroxide, water, and noxious substances in an abdominal sac. When threatened, it injects a catalyst into this mixture. The almost instantaneous exothermic decomposition of hydrogen peroxide generates steam, which ejects the contents of the sac as a hot and highly offensive spray.

Catalytic converters in the exhaust systems of modern cars show all the main features of a heterogeneous catalyst. The catalyst consists of about 2 g of finely divided platinum/rhodium, on a rigid ceramic support. The primary effect is to catalyse the conversion of the pollutants carbon monoxide and nitrogen monoxide to carbon dioxide and nitrogen:
$2CO(g) + 2NO(g) \rightarrow 2CO_2(g) + N_2(g)$
Note that leaded petrol will rapidly poison a catalytic converter.

Catalysts may be physically altered over time. The platinum/rhodium gauze used as a catalyst in the Ostwald process (the catalytic oxidation of ammonia; see spread 19.9) becomes crystallized over time. This roughens the surface and decreases the mechanical strength.

PRACTICE

1 Cover the text of this spread and define each of the following terms in one sentence each: reaction profile; reaction mechanism; homogeneous catalyst; heterogeneous catalyst; autocatalyst; poison; enzyme.

2 Explain what is meant by *catalytic activity*, including in your answer the terms 'activation energy', 'energy distribution', and 'reaction route'.

15.10

OBJECTIVES

- Surface adsorption
- Mechanism of heterogeneous catalysis
- Industrial considerations

(a)

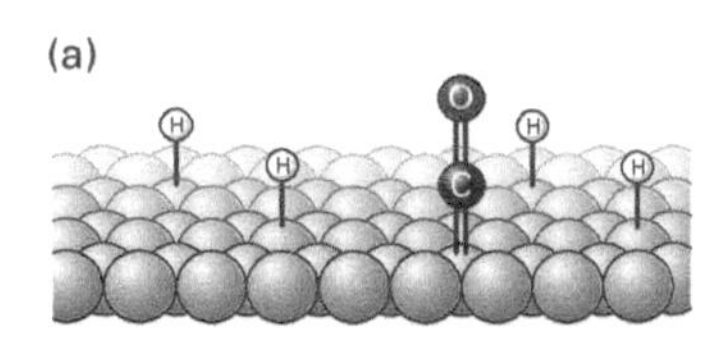

(b)

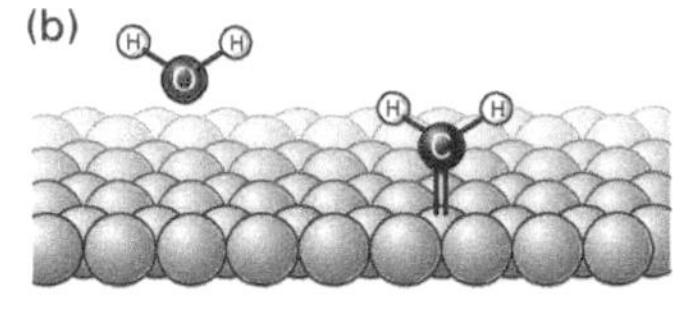

(c)

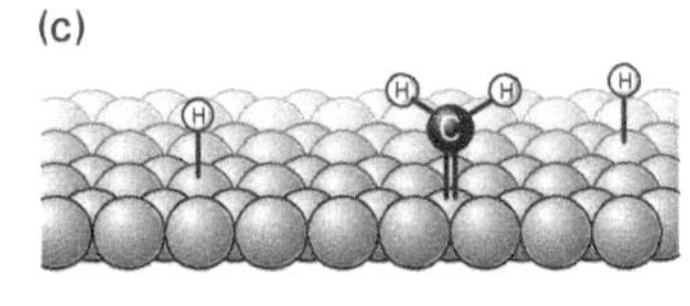

(d)

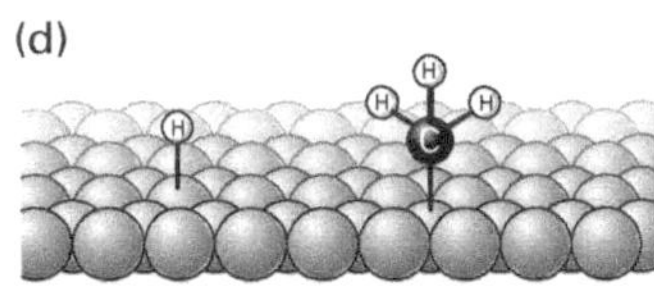

(e)

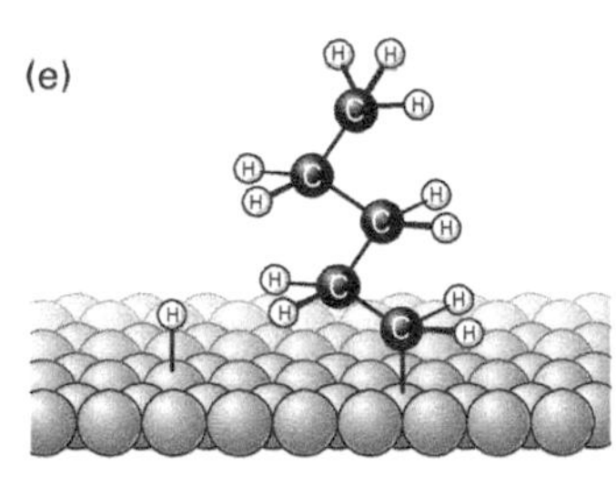

The mechanism that occurs over cobalt catalysts in the Fischer–Tropsch (FT) SASOL process:
(a) CO and H atoms on the Co surface.
(b) CH_2 on the surface and H_2O in the gas phase.
(c) An H atom on the surface reacts with $-CH_2$ on the surface to produce
(d) $-CH_3$, which then reacts further.
(e) The product (pentane here) forms on the surface by reaction with the H atom and then desorbs.
The FT reaction is likely to become even more important in the next couple of decades as it enables conversion of raw materials such as coal and natural gas into liquid fuels.

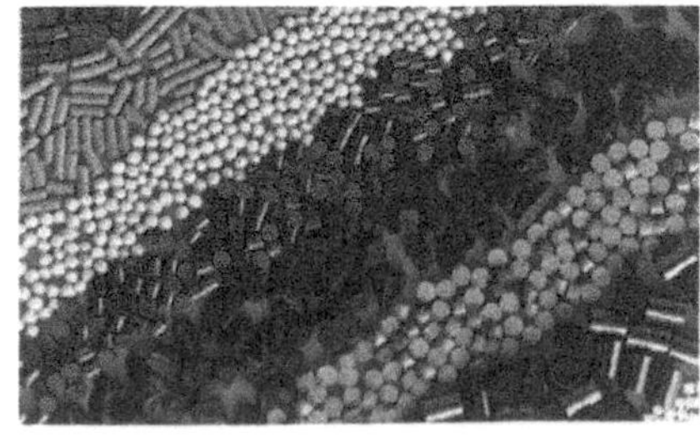

Various types of catalyst bead used in different stages of the Haber–Bosch synthesis of ammonia. Each bead is designed to present an optimum surface area to the reactants.

HETEROGENEOUS CATALYSIS

By definition, a heterogeneous catalyst is in a different phase from the reactants. The most common situation involves a solid catalyst in contact with liquid or gaseous reactants. For example, you should already be familiar with the laboratory preparation of oxygen by the catalytic decomposition of aqueous hydrogen peroxide by solid manganese(IV) oxide. This spread examines the mechanism of heterogeneous catalysis in detail and discusses the role of catalyst promoters and supports in industrial applications.

Adsorption

Adsorption describes the attachment of a species to the *surface* of a solid. A given catalyst surface has a finite number of **active sites** at which reactants may adsorb. When adsorption occurs by dispersion forces, it is described as **physisorption**; when it occurs by formation of a covalent bond, it is described as **chemisorption**. Chemisorption to transition metal catalysts often involves vacant d orbitals accepting electron density from the adsorbed atoms. Catalytic activity depends critically on the strength of the adsorption. The bond must be *strong* enough to bind the reactants, but *weak* enough to break again to release the product. So, for hydrogenation reactions, tungsten (W) forms too strong a bond, whereas silver (Ag) forms too weak a bond. Nickel (Ni) and platinum (Pt) are about right.

(a)

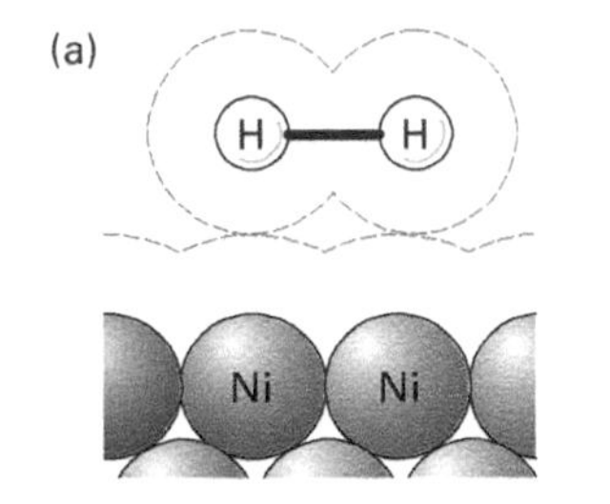

(b)

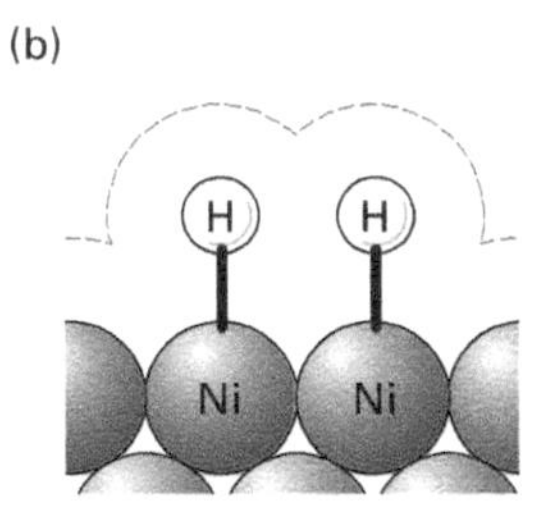

(a) Physisorption and (b) chemisorption of hydrogen on a nickel surface.

Adsorption and the Haber–Bosch synthesis

The reaction between nitrogen and hydrogen to form ammonia proceeds in a series of steps.

The first step is the adsorption of nitrogen and hydrogen onto the surface of the iron catalyst. The process of adsorption of nitrogen breaks the N≡N triple covalent bond and results in separate nitrogen atoms bound to the metal surface. Similarly, hydrogen molecules break up on attaching to the surface. The attachment to the surface is by chemisorption in both cases, involving the donation of electron density from nitrogen and hydrogen atoms into vacant d orbitals on the iron atoms.

The next step is the formation of three N—H bonds between atoms on the surface. The final step involves the desorption of the ammonia molecule from the surface. See previous spread.

The hydrogenation of ethene

Adsorption to metal surfaces is not limited to diatomic molecules. The hydrogenation of ethene to ethane is catalysed by nickel:

$$\underset{\text{ETHENE}}{CH_2{=}CH_2(g)} + H_2(g) \rightarrow \underset{\text{ETHANE}}{CH_3{-}CH_3(g)}$$

The first step involves the dissociative adsorption of hydrogen. An adsorbed hydrogen atom then approaches an ethene molecule and bonds to it ($H^\bullet$ represents a hydrogen atom and its associated electron):

$$CH_2{=}CH_2 + H^\bullet \rightarrow {}^\bullet CH_2{-}CH_3$$

This reaction forms a radical ${}^\bullet CH_2{-}CH_3$, which has an unpaired electron on a carbon atom. (A **radical** is an atom or molecule that contains one or more unpaired electrons.) The radical uses this unpaired electron to bond to the metal surface. Reaction with another adsorbed hydrogen atom follows:

$$^{\bullet}CH_2—CH_3 + H^{\bullet} \rightarrow CH_3—CH_3$$

The final step is the desorption of the product ethane.

Catalyst poisons

The function of a heterogeneous catalyst depends critically on the strength of adsorption of reactant and product molecules. Catalyst poisons are substances that bind strongly to the surface and do not readily desorb.

Hydrogen in the Haber–Bosch synthesis is derived from natural gas. Any sulphur impurities mixed with methane from a gas well must be removed. If they are not, the sulphur atoms will rapidly poison the iron catalyst by binding strongly to the active sites and blocking the adsorption of hydrogen and nitrogen.

In a similar manner, a vehicle's catalytic converter will be poisoned if leaded fuel is burned in the engine, because lead bonds strongly to the platinum/rhodium surface. Hence only unleaded fuel is sold in the UK.

Promoters

The effectiveness of a heterogeneous catalyst may be improved by the use of a promoter. A **promoter** is a substance that does not catalyse the reaction itself, but *does* further increase the rate when used with the catalyst. Promoters improve the surface area and also the electronic structure of the surface. The iron catalyst in the Haber–Bosch process for manufacturing ammonia is promoted by the oxides of potassium, calcium, and aluminium.

Catalyst supports

Heterogeneous catalysis of a reaction takes place where the reactant molecules meet the catalyst surface. Catalytic activity is affected by the surface area of the catalyst and not by its thickness. Many heterogeneous catalysts are expensive, e.g. platinum and rhodium. Efficient and cost-effective use can be made of these substances by depositing very thin layers of them on an inexpensive carrier material or **support**. Catalytic converters used in vehicle exhaust systems consist of a platinum/rhodium mixture on a ceramic support.

SUMMARY

- Heterogeneous catalysts are frequently transition metals or their compounds.
- Reaction takes place between reactant molecules adsorbed onto the catalyst surface.
- Catalyst poisons are substances that bind strongly to the surface and are not readily desorbed.
- Catalyst promoters improve the effectiveness of catalysts.
- Thin layers of expensive or mechanically weak catalyst materials can be deposited on the surfaces of strong, cheap supports.

PRACTICE

1 Draw Lewis structures of the hydrocarbon species to illustrate each of the stages in the catalytic hydrogenation of ethene to ethane.

2 Suggest a mechanism for the reaction $H_2(g) + I_2(g) \rightarrow 2HI(g)$ using a platinum catalyst. Explain why the activation energy of the catalysed reaction is approximately one-fifth that of the uncatalysed reaction.

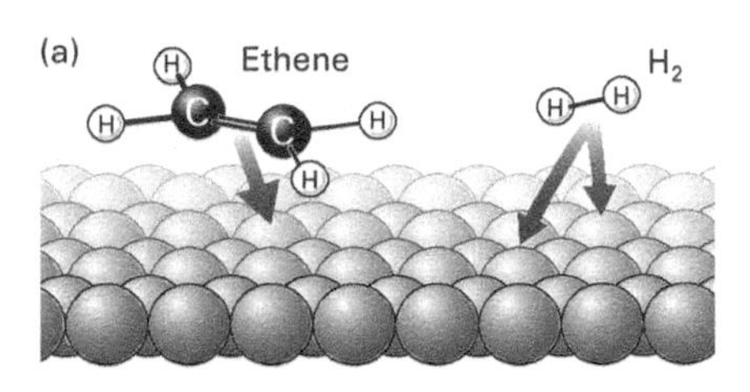

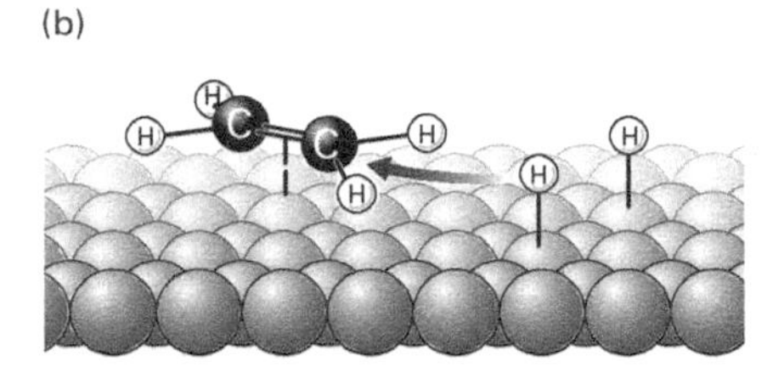

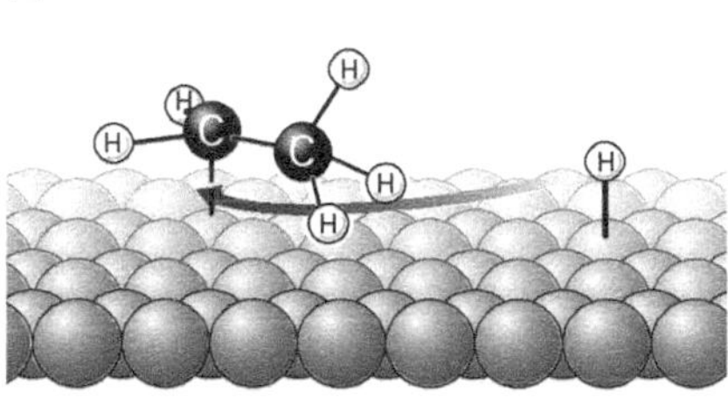

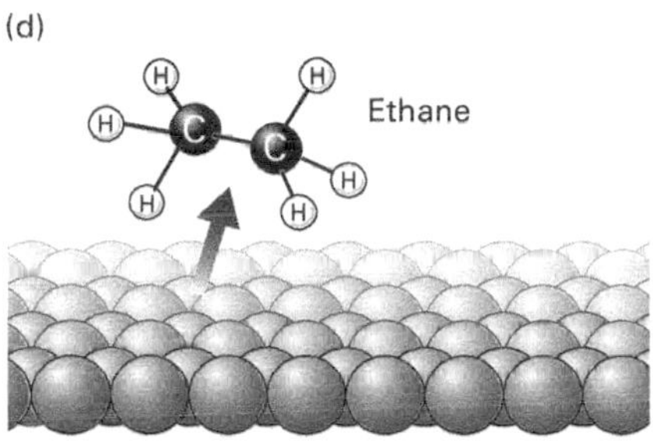

The four main steps in the catalytic hydrogenation of ethene: (a) Hydrogen atoms adsorb onto the surface; as does an ethene molecule. (b) Ethene reacts with a hydrogen atom; the radical $^{\bullet}CH_2—CH_3$ forms and attaches to the surface by means of the unpaired electron. (c) The radical reacts with another hydrogen atom to form $CH_3—CH_3$, which desorbs from the surface (d).

In a similar way, a catalytic converter adsorbs CO and NO molecules onto its surface, weakening their bonds – the reaction happens on the surface, and the product molecules (CO_2 and N_2, see previous spread) desorb from the surface.

Supports – another use

Some catalyst materials lack sufficient physical strength to be used on their own. A support stops the catalyst surface from collapsing. An example is the silica support for the phosphoric acid catalyst used in the direct hydration of ethene.

15.11

OBJECTIVES

- Mechanisms
- Transition metal ion catalysis
- Enzymes
- Acid catalysis

Inhibitors

The decomposition of hydrogen peroxide may be slowed down by the addition of ethanol. A substance that slows down a reaction is called an **inhibitor** (or, infrequently, a **negative catalyst**).

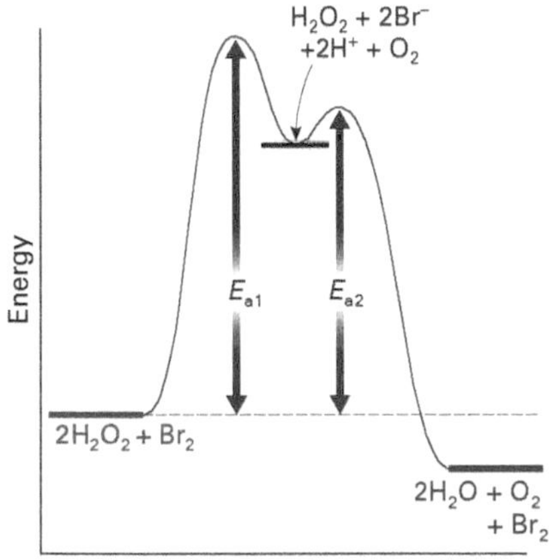

The reaction profile for the decomposition of hydrogen peroxide catalysed by bromine. The profile includes two separate energy barriers E_{a1} and E_{a2}. The peak of each 'hump' on the curve represents the transition state formed between (1) H_2O_2 and Br_2 and (2) H_2O_2 and H^+/Br^-.

Homogeneous catalysis

By definition, a homogeneous catalyst is in the same phase as the reactants. Whilst heterogeneous catalysis typically involves gases reacting on a solid catalytic surface, homogeneous catalysis typically involves liquid mixtures or substances in solution. All catalysts operate by making available reaction routes of lower activation energy; however, you will see that the mode of operation of homogeneous catalysts is distinctly different to that of heterogeneous catalysts.

Hydrogen peroxide (again)

Aqueous hydrogen peroxide $H_2O_2(aq)$ slowly decomposes to water and oxygen gas over a number of weeks. Adding a small quantity of bromine causes a similar sample to decompose in just a few minutes. The mechanism for the reaction involves two steps, as follows:

$$\mathbf{H_2O_2(aq)} + Br_2(aq) \rightarrow 2Br^-(aq) + 2H^+(aq) + \mathbf{O_2(g)} \quad (1)$$

$$\mathbf{H_2O_2(aq)} + 2Br^-(aq) + 2H^+(aq) \rightarrow \mathbf{2H_2O(l)} + Br_2(aq) \quad (2)$$

$$\text{overall } \mathbf{2H_2O_2(aq)} \rightarrow \mathbf{2H_2O(l) + O_2(g)}$$

You can see that bromine acts as a catalyst because it is involved in the reaction mechanism but emerges unchanged at the end of the two steps concerned.

Two-step reaction profiles

Reaction profiles imply that a transition state, spread 15.4, exists between the arrangement of atoms called 'the reactants' and the arrangement called 'the products'. The catalysed decomposition of hydrogen peroxide discussed above occurs in two steps. The reaction profile has two peaks with a trough between them. Each peak represents a transition state. Reactants achieve this transition state when they come together with energy at least equal to the activation energy for that step. At the transition state, an **activated complex** exists which is in equilibrium with the reactants and which may decompose to form the products. The groups of atoms (activated complex) corresponding to these transition states cannot be isolated. The trough on the reaction profile represents the **intermediate species**, in this case, $Br^-(aq)$ and $H^+(aq)$.

Transition metal ions

Many redox reactions may be catalysed by transition metal ions. These ions alternate between two oxidation states, transferring electrons between the oxidant and the reductant. For example, peroxodisulphate ions $S_2O_8^{2-}(aq)$ oxidize iodide ions to iodine (peroxodisulphate is reduced to sulphate):

$$S_2O_8^{2-}(aq) + 2I^-(aq) \rightarrow 2SO_4^{2-}(aq) + I_2(aq)$$

The uncatalysed reaction is slow, due largely to both the reactants bearing negative charges (which repel them from each other).

The two half-equations are:

Reduction: $S_2O_8^{2-}(aq) + 2e^- \rightarrow 2SO_4^{2-}(aq)$

Oxidation: $2I^-(aq) \rightarrow I_2(aq) + 2e^-$

Iron(II) ions catalyse the reaction by acting as an intermediate in the transfer of electrons from iodide to peroxodisulphate.

Step 1 Peroxodisulphate oxidizes Fe(II) to Fe(III):

$$S_2O_8^{2-}(aq) + 2Fe^{2+}(aq) \rightarrow 2SO_4^{2-}(aq) + 2Fe^{3+}(aq)$$

Step 2 Iron(III) ion oxidizes iodide ion to iodine:

$$2Fe^{3+}(aq) + 2I^-(aq) \rightarrow 2Fe^{2+}(aq) + I_2(aq)$$

Note that step 2 reduces iron(III) back to iron(II).

Enzymes

Enzymes act as homogeneous catalysts in living systems. They consist of complex protein chains coiled into specific shapes. Part of an enzyme, often containing a transition metal ion, is the active site where reaction takes place. Reactant molecules must fit the shape of the active site, rather like two jigsaw puzzle pieces fitting together, or a key fitting into a lock. Enzyme names often indicate the substances on which they act and end in *-ase*. For example, *peroxidase* is an enzyme found in mammalian livers and is responsible for decomposing harmful peroxides. It is interesting to note that the activation energy for the uncatalysed decomposition of hydrogen peroxide is 75 kJ mol^{-1}, whereas the corresponding values for the reaction catalysed by platinum and peroxidase are 49 kJ mol^{-1} and 23 kJ mol^{-1} respectively.

NB. Enzymes are discussed at greater length in spreads 30.6 and 30.7.

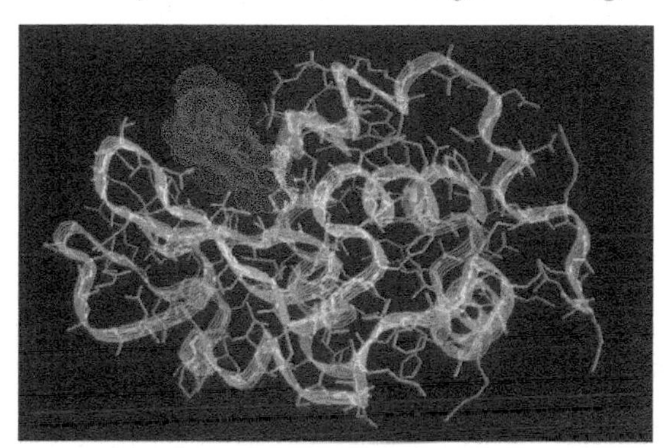

The lysozyme molecule (M_r 14100). This enzyme occurs in tears and nasal mucus and is responsible for destroying bacteria. The active site binds to specific chemical groupings on the bacterial cell wall and breaks them apart.

Depletion of the ozone layer

15–50 km above the Earth's surface, intense sunlight forms ozone (trioxygen O_3) from atmospheric oxygen (dioxygen O_2). The 'ozone layer' protects surface life by absorbing ultraviolet radiation from solar radiation. Chlorofluorocarbons (CFCs) found use as refrigerants and in plastics manufacture. A typical CFC is dichlorodifluoromethane CF_2Cl_2. Escaping into the ozone layer, it is broken down by UV radiation to give highly reactive radicals e.g.

$$CF_2Cl_2 \rightarrow {}^{\bullet}CF_2Cl + Cl^{\bullet}$$

The chlorine atom catalyses the decomposition of ozone to oxygen:

$$O_3 + Cl^{\bullet} \rightarrow O_2 + ClO^{\bullet}$$

$$ClO^{\bullet} + O_3 \rightarrow 2O_2 + Cl^{\bullet}$$

Other radicals, such as NO_x (see spread 19.7) formed for example from aircraft exhausts, are also capable of depleting ozone.

Acid catalysed esterification

Esterification is the reaction between a carboxylic acid and an alcohol to produce an ester, e.g. ethanoic acid CH_3COOH reacts with ethanol CH_3CH_2OH to form an equilibrium mixture, spread 11.5, also containing ethyl ethanoate $CH_3COOCH_2CH_3$ and water:

$$CH_3COOH(l) + CH_3CH_2OH(l) \rightleftharpoons CH_3COOCH_2CH_3(l) + H_2O(l)$$

The equilibrium constant K_c for the reaction is 4.0 at 80 °C. The uncatalysed reaction takes many weeks to reach equilibrium. However, the addition of concentrated sulphuric acid causes equilibrium to be established in a few hours. Note that the value of the equilibrium constant is identical for both catalysed and uncatalysed reactions (at the same temperature). The mechanism for this reaction, highlighting the role of the catalyst, is discussed in detail in spread 27.3.

SUMMARY

- A homogeneous catalyst is in the same phase as the reactants.
- Transition metal ions catalyse redox reactions by acting as intermediates in the electron-transfer process.
- Enzymes are highly-specific biological homogeneous catalysts.

PRACTICE

1 Explain the relationships between the three terms *intermediate species*, *transition state*, and *activated complex*.

2 The redox reaction between dichromate(VI) and iodide ions

$$14H^+(aq) + Cr_2O_7^{2-}(aq) + 6I^-(aq) \rightarrow 2Cr^{3+}(aq) + 3I_2(aq) + 7H_2O(l)$$

is catalysed by minute amounts of copper(II) ions.

a Write half-equations for the oxidation and the reduction reactions contained in the equation above.

b Outline an experimental procedure to show that the reaction is catalysed by copper(II) ions.

c Suggest a mechanism to explain the function of copper(II) as a catalyst in this reaction.

1 **a** A fixed mass of marble is reacted with dilute hydrochloric acid at a constant temperature. Explain why the rate of the reaction is increased if the lumps of marble are reduced in size. [2]

b The initial rate of the reaction between substances **A** and **B** was measured in a series of experiments and the following rate equation was deduced:

rate = k[**A**][**B**]^2

i Copy and complete the table of data below for the reaction between **A** and **B**.

Expt	Initial [A] /mol dm^{-3}	Initial [B] /mol dm^{-3}	Initial rate /mol dm^{-3} s^{-1}
1	0.020	0.020	1.2×10^{-4}
2	0.040	0.040	
3		0.040	2.4×10^{-4}
4	0.060	0.030	
5	0.040		7.2×10^{-4}

ii Using the data for Experiment 1, calculate a value for the rate constant, k, and state its units. [7]

2 **a** The diagram below shows the Maxwell–Boltzmann energy distribution curves for molecules of a gas under two sets of conditions **A** and **B**. The total area under curve **B** is the same as the total area under curve **A**.

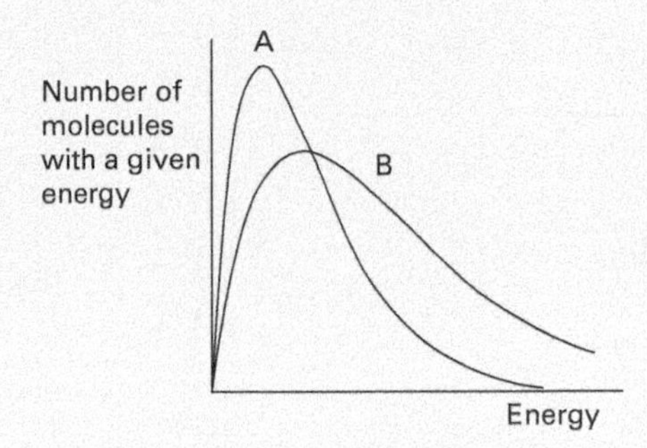

i What change of condition is needed to produce curve **B** from curve **A**?

ii What is represented by the total area under curve **A**?

iii Why is the total area under curve **B** the same as that under curve **A**? [3]

b **i** Explain the meaning of the term activation energy.

ii In a reaction involving gas molecules, if all other conditions are kept constant, state the effect, if any, on the value of the activation energy when a catalyst is added and the volume of the vessel is decreased. [3]

c Explain why reactions between solids usually occur very slowly, if at all. [2]

3 **a** State the difference between homogeneous and heterogeneous catalysis. [2]

b Explain the term activation energy. [2]

c A rate equation for a reaction between reagents A and B is

rate = $k[A]^n$

i State the meaning of all the terms other than rate which appear in this equation.

ii What can be deduced from the absence of reagent B from the rate equation? [4]

d In the examples below, decide whether the catalyst is homogeneous or heterogeneous, and explain how it provides an alternative route of lower activation energy in each case.

i The Contact Process

ii Enzyme catalysed reactions [6]

e Suggest **two** different measures that can be taken to maximize the efficiency and minimize the costs associated with a very expensive heterogeneous catalyst. [2]

(See also chapters 19 and 30.)

4 The following information refers to a procedure to determine the *order of reaction* with respect to iodide ions for the reaction represented by the equation

$2I^-(aq) + H_2O_2(aq) + 2H^+(aq) \rightarrow 2H_2O(l) + I_2(aq)$

Rate is measured by the time taken for the iodine produced to react with a small *fixed amount* of sodium thiosulphate added to the constant volume system.

The faster the iodine is produced, the shorter the time taken for the sodium thiosulphate to be used up.

The reciprocal of this time can be used as a measure of the initial rate of reaction. The results are given below.

Experiment	[KI(aq)]/mol dm^{-3}	Time (t)/s	Reciprocal of time (1/t)/s^{-1}
I	0.004	74	0.0135
II	0.006	49.4	0.0202
III	0.008	37	0.0270
IV	0.010	30	0.0333
V	0.012	25	0.0400

a **i** In the experiments the concentrations of acid and hydrogen peroxide were far more concentrated than that of potassium iodide. Explain why this was necessary. [1]

ii In each of the experiments the aqueous hydrogen peroxide was always added last. State why this was necessary. [1]

iii Explain why the volume of the system was kept constant for all the experiments. [1]

b **i** Copy showing the axes and plot, a graph of the initial rate ($\frac{1}{t}$) on the vertical axis against the concentration of potassium iodide used on the horizontal axis.

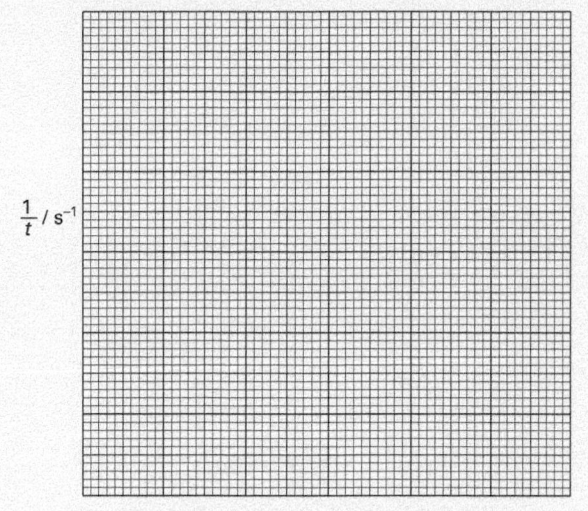

ii Use your graph to determine the order, ***n***, of the reaction with respect to iodide ions. Carefully state your reasoning. [3]

c Further studies show that the rate equation for the reaction is

Rate (mol dm^{-3} s^{-1}) = $k[I^-(aq)]^n[H_2O_2(aq)]$

i From this overall rate equation, state what can be deduced about the role of aqueous hydrogen ions. [1]

ii State the units of k in the above rate equation. [1]

5 **a** The Arrhenius equation, $k = Ae^{-E/RT}$, may be expressed in the following form:

$\ln k = \ln A - E/RT$

The decomposition of a gas, **Q**, was studied at a number of different temperatures (T), and the value of k was obtained at each temperature. Values were calculated for $1/T$ and for $\ln k$. A plot of $\ln k$ against $1/T$ is shown below.

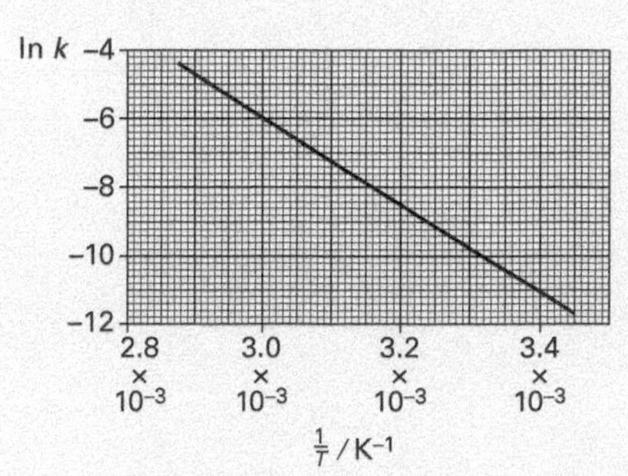

i What do the symbols k and A in the Arrhenius equation represent? [2]

ii Determine the gradient for the graph, $-E/R$. [2]

iii Using the gradient obtained in **a ii**, deduce the value of the activation energy, E, for the decomposition of **Q**. Include the units for E in your answer.

$R = 8.314\,J\,K^{-1}\,mol^{-1}$ [4]

b In separate reactions between sulphur dioxide and oxygen, an increase in the concentration of SO_2 from 0.180 mol dm^{-3} to 0.540 mol dm^{-3} was found to increase the initial rate of reaction by a factor of 9.

i State what is meant by the term *order of reaction*. [1]

ii Explain why the rate of the above reaction is increased by an increase in the concentration of SO_2. [2]

iii Deduce, using the information given above and showing your working, the order of the reaction with respect to SO_2. [2]

6 This question is about the reaction between bromomethane and aqueous hydroxide ions

$CH_3Br + OH^- \rightarrow CH_3OH + Br^-$

a An increase in temperature increases the rate of this reaction.

Explain this increase by referring to the collision frequency and the collision energy of the molecules. [3]

b By sketching the energy distribution of the molecules at a given temperature, T, show how the presence of a catalyst will increase the rate of the reaction. [3]

c Define the following terms used in reaction kinetics.

i Overall order of reaction [1]

ii Rate constant [1]

Time/min	0	10	20	30	40	50	60	70
$[CH_3Br]$/mol dm^{-3}	0.100	0.074	0.057	0.043	0.033	0.025	0.019	0

d In the reaction between bromomethane and aqueous hydroxide ions at constant temperature the concentration of bromomethane at various times is given in the table.

i Plot a graph to show that the reaction is first order with respect to bromomethane. [4]

ii If the concentration of the hydroxide ion doubles, all other factors remaining constant, the rate of the reaction doubles. What is the order of reaction with respect to the hydroxide ion? [1]

iii Hence write a rate equation for this reaction. [1]

iv Based on the kinetic information obtained above write the mechanism for the reaction between CH_3Br and aqueous OH^- ions. [3]

(See also chapter 24.)

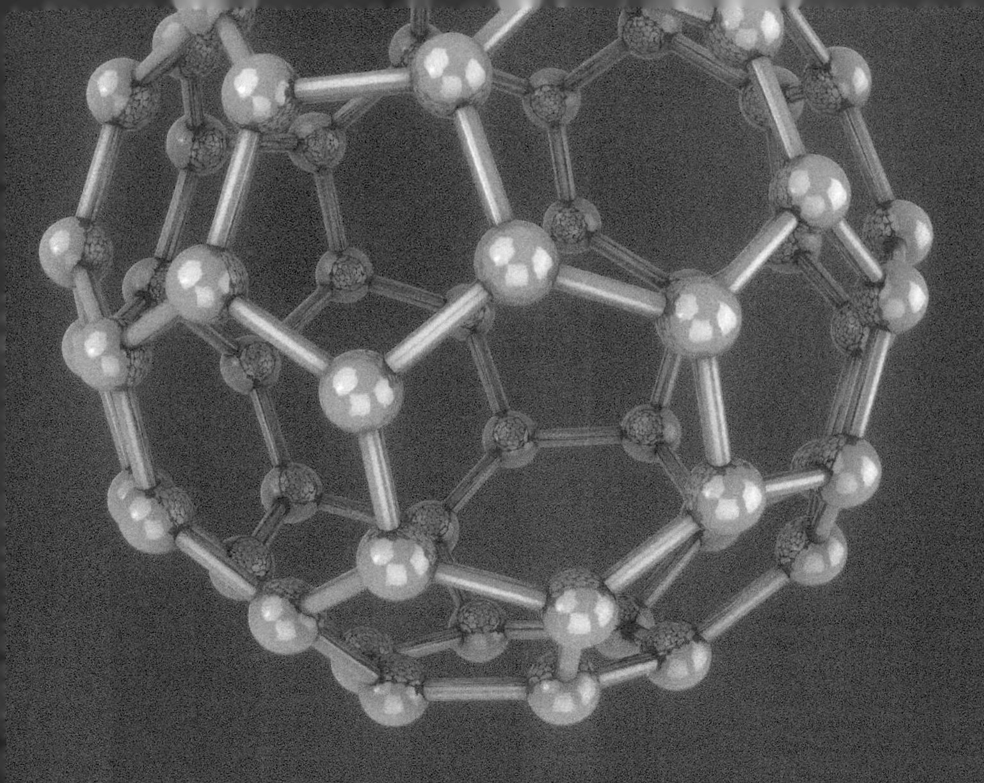

Inorganic **CHEMISTRY**

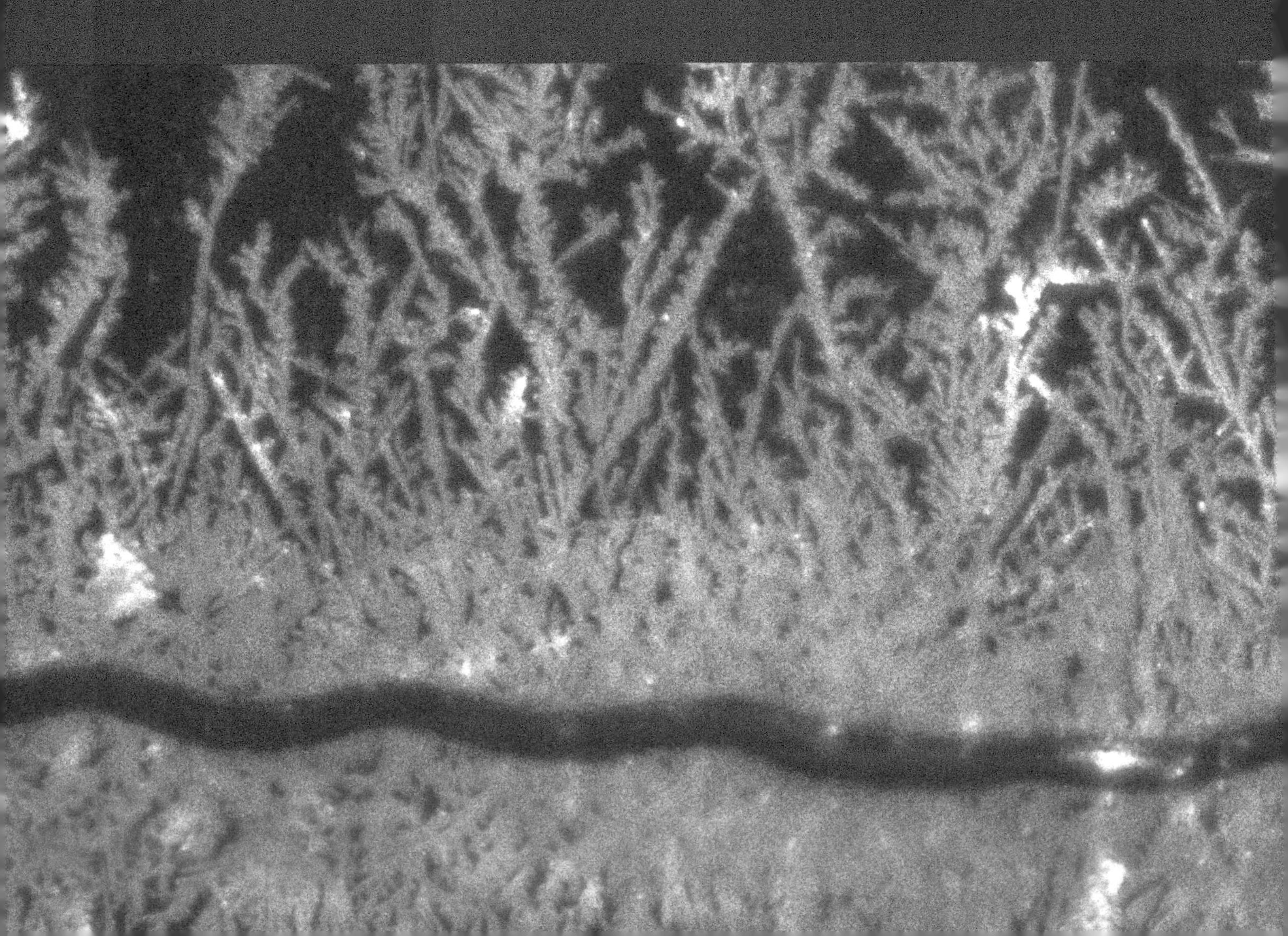

Inorganic chemistry studies the behaviour of the elements and their compounds, with the single exception of the element carbon (whose study is called organic chemistry). The most important unifying principle is the periodic table. (A simple introduction to the periodic table was given in chapter 1; further insight was provided in chapter 4, where the idea of the s, p, and d blocks was introduced.)

We study the periodic table in a particular order. The s block, which occurs on the far left of the periodic table, is looked at first (chapter 16). The general variation across the periodic table is considered (chapter 17), and then we turn to the halogens (chapter 18), the most important group of elements towards the right of the periodic table.

The halogens form a part of the p block. We have attempted to give a balanced approach to the most important elements in the p block in chapter 19.

The final area studied is the transition metals, which form the major part of the d block (chapter 20). They offer a visually interesting conclusion to the study of inorganic chemistry: much of the colour seen around us in nature is caused by transition metal compounds. Transition metals are also very important in biochemistry, being involved in the transport of electrons between species, as well as being central to the function of important molecules such as haemoglobin. Before studying the vital area of biochemistry, we first need to know about organic chemistry, the study of which begins in chapter 21.

A light micrograph of crystalline silver deposited on a copper wire. Copper wire is suspended in aqueous silver nitrate. The copper reacts with the silver ions, and silver is deposited as crystals on the wire. In time, all of the silver is displaced, leaving aqueous copper(II) nitrate. Magnification x65.

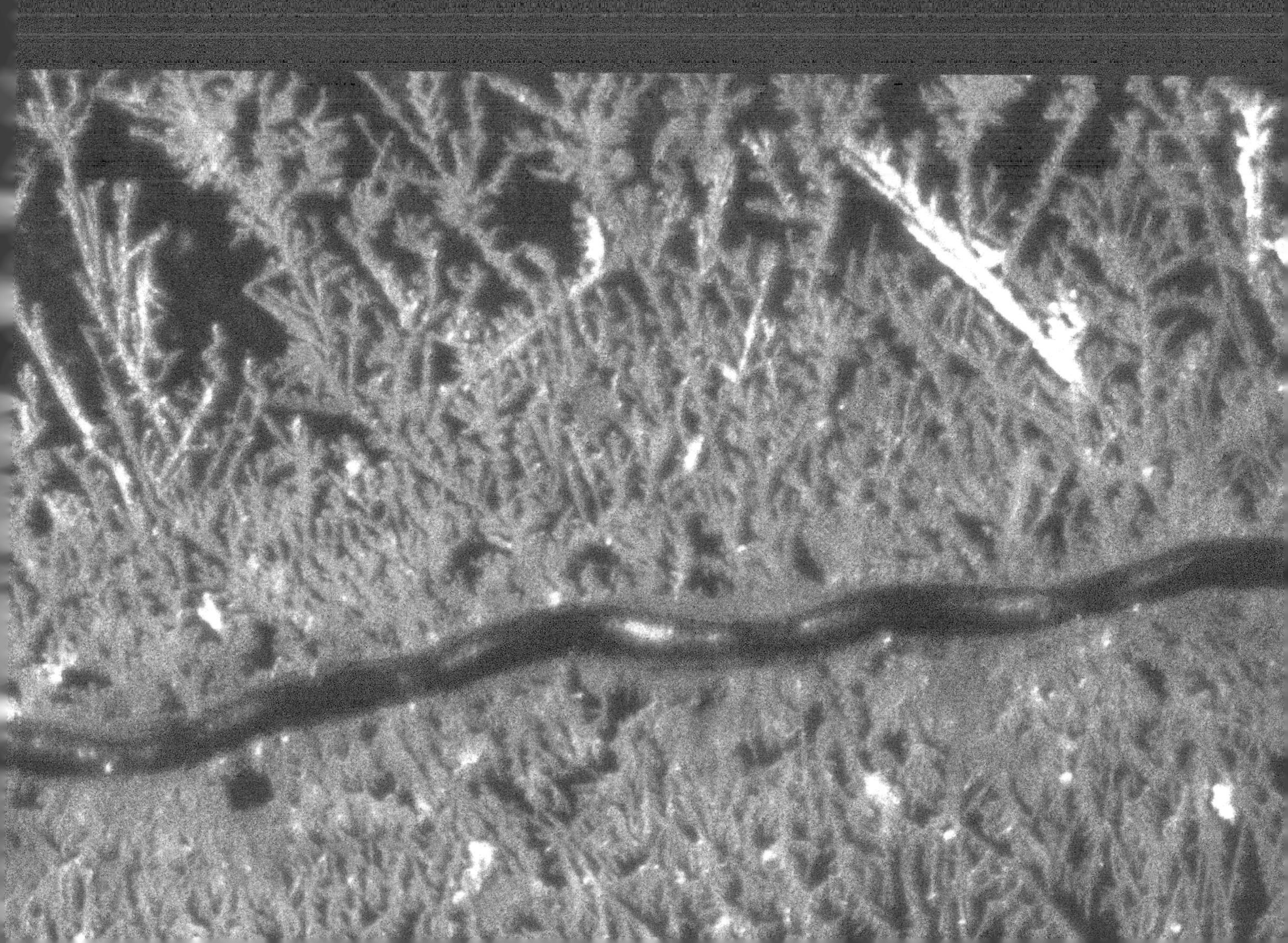

The s-block elements

The s-block elements consist of the metals contained in Groups I and II of the periodic table. These two groups are referred to as the s block because the elements in them have a valence-shell electronic structure of either ns^1 (Group I) or ns^2 (Group II). Group I contains the elements lithium to francium, and Group II contains the elements beryllium to radium. Compared to most other metals, the s-block metals have generally greater chemical reactivity. In this chapter, we shall look at the properties of these two groups of metals (with the exception of the radioactive elements francium and radium), and at the underlying reasons for their distinct identity.

GROUP I AND GROUP II METALS: AN OVERVIEW

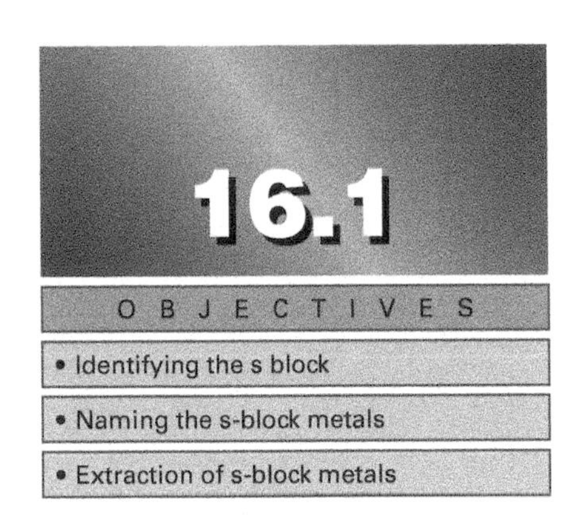

16.1

OBJECTIVES

- Identifying the s block
- Naming the s-block metals
- Extraction of s-block metals

The Group I elements show the typical properties of metals: a freshly cut surface is shiny, electrical and thermal conductivities are high, and the elements are ductile and malleable. The presence of just one electron in the valence shell leads to these elements forming compounds containing the M^+ ion.

The Group II elements likewise show typical properties of metals, although beryllium shows some anomalies (see later spread in this chapter). There are two electrons in the valence shell, so the elements magnesium to barium form compounds containing the M^{2+} ion.

For historical reasons, the Group I and Group II elements are sometimes known respectively as the **alkali metals** and the **alkaline earth metals**.

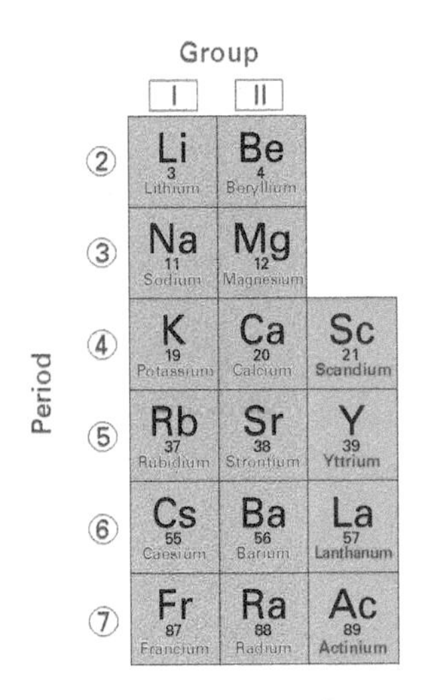

The s block comprises Groups I and II and occupies the extreme left-hand side of the periodic table.

Extraction of s-block metals

The s-block metals cannot be manufactured by the reduction of an oxide by carbon because the temperature required is too high to be economic. They are generally extracted by electrolysis of a molten salt, usually the chloride. For example, metallic sodium is manufactured in a Downs cell from molten sodium chloride at a temperature of about 600 °C.

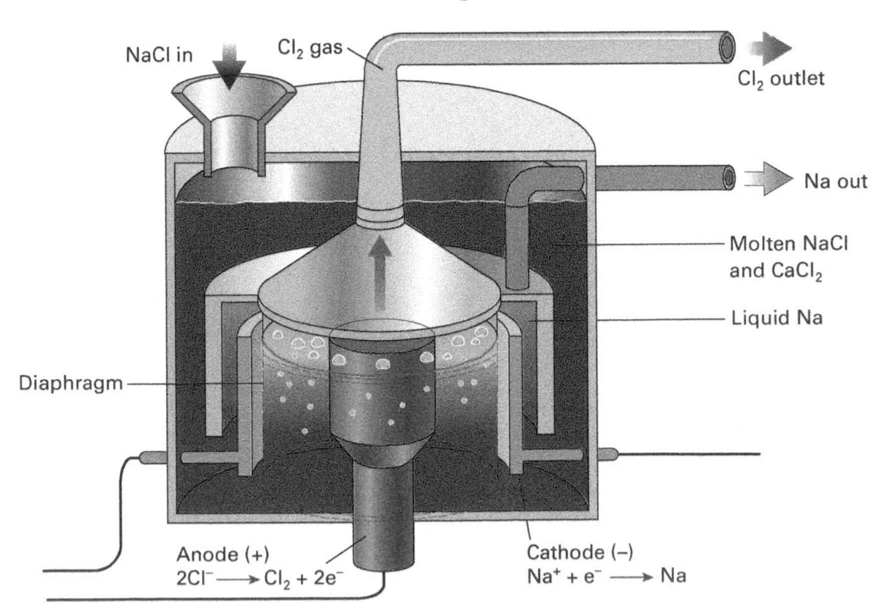

A Downs cell consists of a cylindrical graphite anode (which resists the attack of chlorine) and a steel cathode constructed in the shape of a ring. At the cathode, sodium ions are reduced to form sodium metal (which is molten at this temperature). At the anode, chloride ions are oxidized to form chlorine gas. The diaphragm keeps the molten sodium and gaseous chlorine apart. If the products were to come into contact, they would react, re-forming sodium chloride.

Downs cell

Electrolysis of aqueous sodium chloride reduces hydrogen ions in preference to the sodium ions; see spread 18.2. *Molten* sodium chloride contains sodium ion as the only positive ion. Some calcium chloride is added to the electrolyte because the mixture of chlorides melts at a lower temperature (about 600 °C) than the melting point of pure sodium chloride (801 °C), spread 7.7. Each Downs cell consumes about 200 kW at a current of about 25 000 A.

The s-block elements: table of data

For the purposes of this chapter, Group I includes the elements lithium to caesium, and Group II includes the elements beryllium to barium. The anomalous behaviour of lithium and beryllium will be covered separately in a later spread in this chapter. The radioactive elements francium and radium will not be covered here.

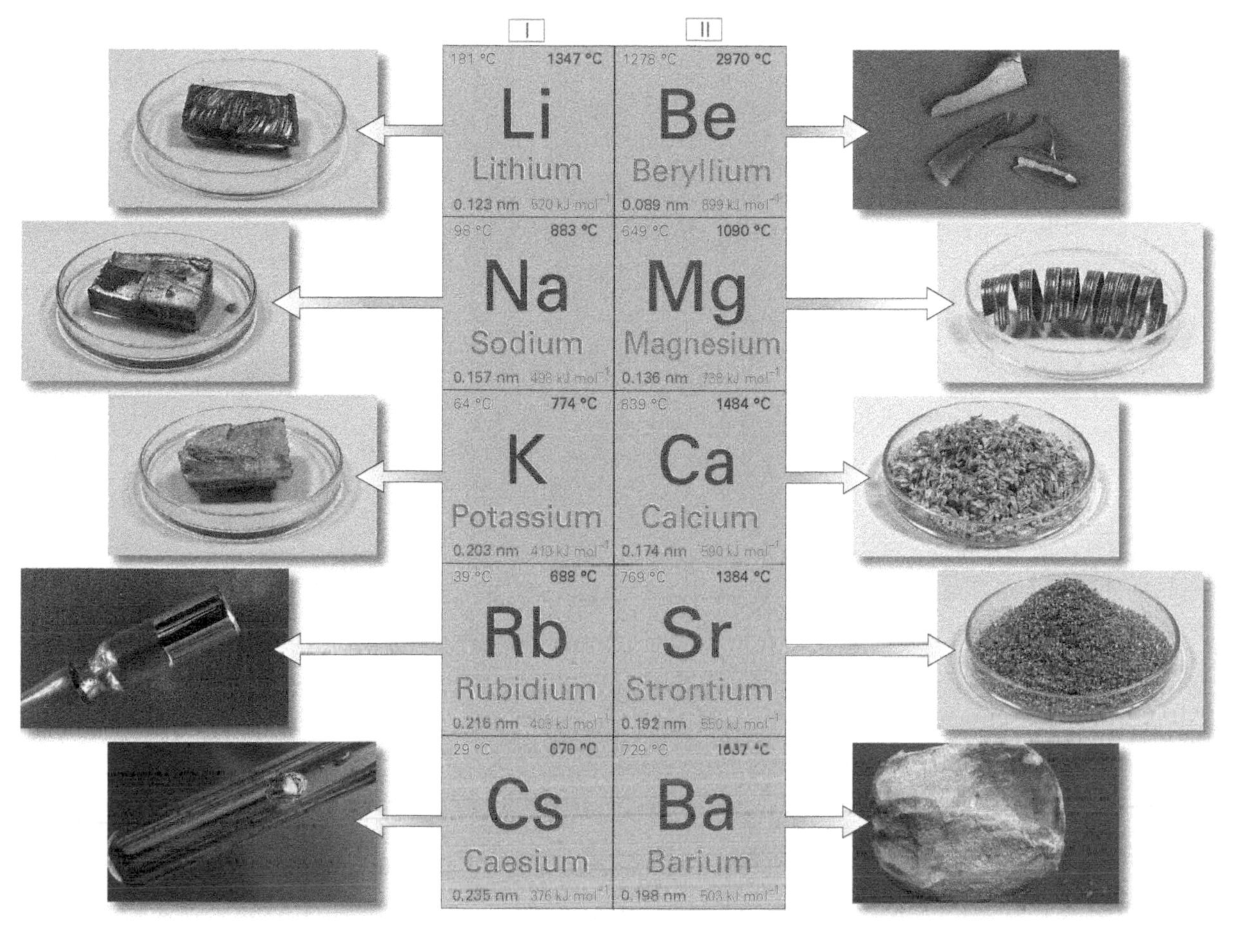

The major physical properties of the s-block elements: clockwise from top left = melting point, boiling point, first ionization energy, atomic radius. The Group I metals are very reactive: lithium, sodium, and potassium are coated with compounds that form on their surface. Rubidium and caesium are so reactive that they must be stored in sealed vessels.

Ionization energy: a reminder

In chapter 10 'Thermochemistry', where we were considering enthalpy changes, we associated the term 'ionization *enthalpy*' with the removal of electrons from atoms to form ions. In chapter 4 'Electrons in atoms', we associated the term 'ionization *energy*' with the same change. You will often find enthalpy changes being referred to colloquially as 'energies' and we will do so here.

SUMMARY

- The Group I metals studied are lithium, sodium, potassium, rubidium, and caesium.
- The Group II metals studied are beryllium, magnesium, calcium, strontium, and barium.
- The valence-shell electronic structures of the s-block metals are ns^1 (Group I) and ns^2 (Group II).
- The s-block metals are extracted by electrolysis of a molten salt.

PRACTICE

1 Plot separate graphs for Group I and Group II elements to illustrate the trends in each of the following properties:

a Melting point
b Boiling point
c Atomic radius
d First ionization energy.

When plotting your graphs, space each element equally along the x-axis in order of increasing atomic number.

2 For each of the plots that you produced in question 1, make a general comment about the trend in the property as the atomic number increases.

16.2

Some group trends

OBJECTIVES

- Trends in:
  atomic radius and ionic radius
  ionization energy
  melting point
  electronegativity
- Oxidation number

The s-block elements have distinctive characteristics that distinguish them from the metals in other regions of the periodic table. Within the s block, Group I and Group II are clearly distinguishable from each other, in particular because they have different oxidation numbers in their compounds. Within each group, properties such as atomic radius, ionization energy, and melting point show clear trends. This spread gives an overview of these properties of the s-block elements, and describes the trends within Group I and Group II.

Atomic radii

In chapter 4 'Electrons in atoms', we looked at the shapes and sizes of atomic orbitals. In terms of probability, we saw that the orbitals of an atom extend to infinity. So, describing the size of an atom (its radius) necessarily requires an arbitrary decision about where the boundary of an atom is.

The s-block elements have one or two valence electrons in orbitals outside filled shells of electrons. These filled shells are reasonably effective at shielding the valence electrons from the attraction of the nucleus. As a result, the atomic radii of the s-block metals are generally about 50% greater than those of other metals.

Comparisons

The average value for the atomic radii of the Group I metals is 0.187 nm; for the Group II metals it is 0.158 nm. For comparison, the average value for the d-block elements scandium to zinc is 0.122 nm. The atomic radius of the p-block metal aluminium is 0.125 nm.

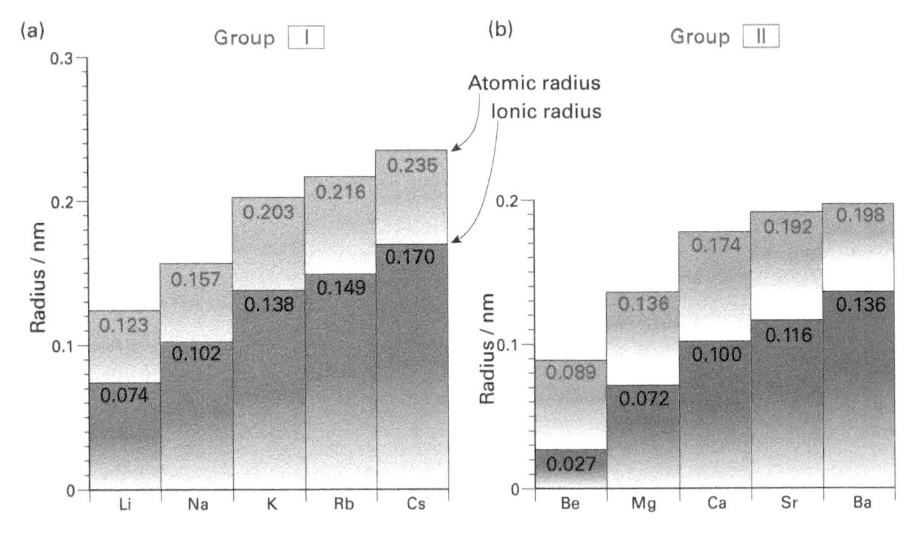

Atomic radius increases as atomic number increases from lithium to caesium (Group I) and from beryllium to barium (Group II). Each successive element in a group has an extra filled shell of electrons. Ionic radii increase similarly. All ions are smaller than their parent atoms as electrons have been lost.

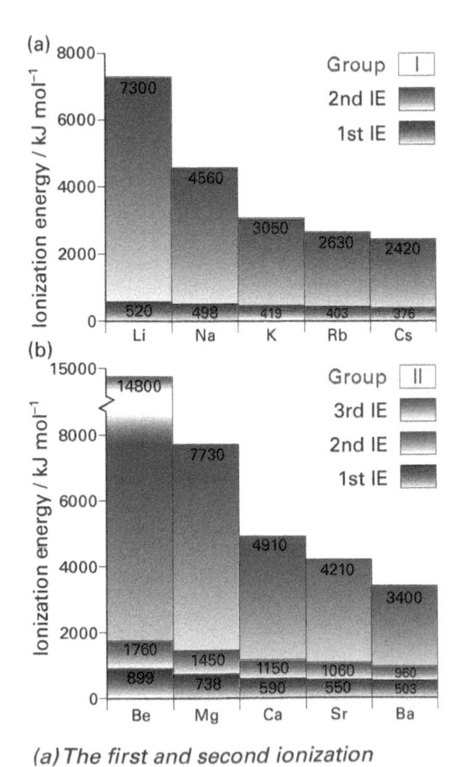

(a) The first and second ionization energies of the Group I elements.
(b) The first, second, and third ionization energies of the Group II elements.

See spread 18.1 for a more detailed explanation of the change in ionization energy down the group.

Ionization energies

The first ionization energy of an element involves removing an electron from its outermost orbital. The s-block elements have relatively large atomic radii and therefore have low values of ionization energy. For example

Group I: $Na(g) \rightarrow Na^+(g) + e^-(g)$; $\Delta H^\ominus = +498\,kJ\,mol^{-1}$

Group II: $Mg(g) \rightarrow Mg^+(g) + e^-(g)$; $\Delta H^\ominus = +738\,kJ\,mol^{-1}$

In each group, the atomic radius increases with increasing atomic number, and there is a corresponding decrease in the value of the ionization energy, as the electron being ionized is further from the nucleus. Within a given period, the Group II ion M^{2+} is smaller than the corresponding Group I ion M^+. This effect is due to the fact that the same number of electrons are under the control of more protons.

Group I elements have one electron in their valence shell. The energy required to remove that electron is relatively low (+498 kJ mol^{-1} for sodium), and so the elements readily form ions with a single positive charge. The second ionization energy is much greater because an electron needs to be removed from an inner shell much closer to the nucleus. For example, the second ionization energy of sodium is +4560 kJ mol^{-1}, about nine times greater than the first ionization energy. As a result, Group I elements do not form ions with a double positive charge.

Group II elements have two electrons in their valence shell. The first and second ionization energies are relatively low in value. The *third* ionization energies are large by comparison, corresponding to the removal of an electron from an inner shell much closer to the nucleus. So, Group II elements do form ions with a double positive charge.

Compared to other metals, the Group I elements are unusually soft. Sodium may be cut with a knife. Forces of attraction between the Na^+ ions and the delocalized electrons in the metallic lattice are relatively low. Because sodium is so reactive, it is stored under oil; some of that protective oil is visible here.

Melting points

Larger atoms tend to have smaller first ionization energies because the valence electron concerned is further from the nucleus. Larger positive ions have lower charge density, so they have a smaller attraction for the delocalized electrons within the metallic structure. As a result, melting points within a group generally decrease with increasing atomic number. Within a given period, the Group II metal has a higher melting point than its Group I counterpart because each atom delocalizes *two* electrons into the metallic lattice, increasing the forces of attraction between the metal ions and the delocalized electrons.

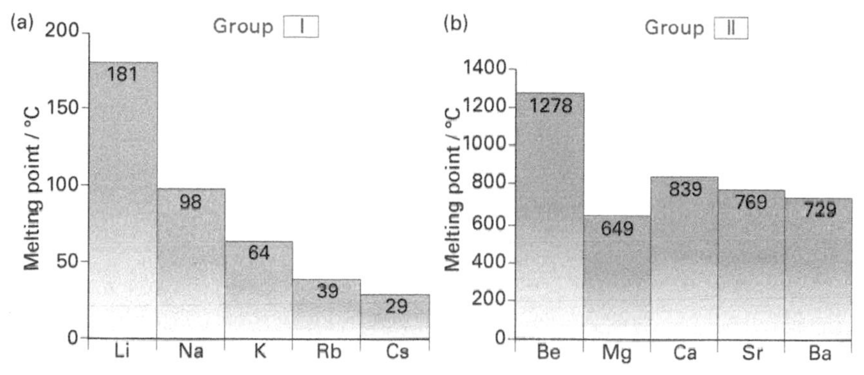

Melting points (T_m) of (a) Group I elements and (b) Group II elements; Be and Mg have a different crystal structure (h.c.p.) from Ca and Sr (f.c.c.); see spread 6.5.

Electronegativity

We can see from the values given on the right that, in a given group, metallic character increases with increasing atomic number. Within a given period, the Group I metal has greater metallic character than the corresponding Group II metal.

Electronegativity: a reminder

If you need to revise electronegativity, you should look back at chapter 5 'Chemical bonding'. Electronegativity is measured on the Pauling scale. Metals are elements which lose electrons when they form compounds. They have smaller electronegativities than non-metals. The smaller the electronegativity, the more metallic is the element.

Pauling electronegativity values for s-block elements.

Group I element	Pauling electronegativity
Li	1.0
Na	0.9
K	0.8
Rb	0.8
Cs	0.8
Group II element	**Pauling electronegativity**
Be	1.6
Mg	1.3
Ca	1.0
Sr	1.0
Ba	0.9

Oxidation numbers

Group I elements have an oxidation number of +1 in all their compounds, corresponding to the loss of the single ns^1 electron from the valence shell. Examples include the sodium ion Na^+ in sodium chloride NaCl, the potassium ion K^+ in potassium oxide K_2O, and the lithium ion Li^+ in lithium carbonate Li_2CO_3. The highly endothermic formation of the doubly charged ion prevents the element from reaching higher oxidation states.

Group II elements have an oxidation number of +2 in all their compounds, corresponding to the loss of both ns^2 electrons from the valence shell. Examples include the ions Mg^{2+} in magnesium oxide MgO, and Ca^{2+} in calcium carbonate $CaCO_3$. Both the ns^2 electrons can be removed during the formation of such compounds: the endothermic formation of the doubly charged ion M^{2+} is compensated for by the highly exothermic formation of the solid ionic lattice from its separate ions (see chapter 10 'Thermochemistry').

SUMMARY

- Within a group, the variations in atomic radius, first ionization energy, and melting point are due to the different number of electron shells in the atoms or ions.
- Group I compounds contain the M^+ ion. Group II compounds contain the M^{2+} ion.

PRACTICE

1 Describe and explain the trend in the ionization energies of the elements in Group II.

16.3 Reactions with water and oxygen

OBJECTIVES

- Trends in reactivity with water and oxygen
- Comparison between Groups I and II
- Trends in reactivity

The valence shells of Group I and Group II elements contain one and two electrons respectively. When the elements react to form compounds, these electrons are lost and positively charged ions form. The elements have relatively low ionization energies, which decrease with increasing atomic number. This spread gives details of the reactions of the elements with water and oxygen, and explains trends in reactivity in terms of trends in the values of ionization energies.

Reaction with water

The Group I metals are powerful reductants in aqueous solution. Their common name 'alkali metals' comes from their ability to reduce water to form alkalis (soluble bases) and hydrogen gas. For example, the reaction of sodium with water is:

$$2Na(s) + 2H_2O(l) \rightarrow 2NaOH(aq) + H_2(g)$$

The reaction of Group I elements with water increases in vigour with increasing atomic number. All the Group I metals are so reactive that they must be stored under oil to prevent reaction with water or air.

Redox reactions with water and acid

Reactions of metals with water and acids such as hydrochloric acid are redox reactions. See spread 13.2 for an explanation of the oxidation number changes in the reaction between Mg and HCl.

(a)

(b)

(c)

(d)

(a) Lithium reacts steadily with water in an unspectacular manner. (b) Sodium floats on water. The exothermic reaction melts the metal and the hydrogen released propels it across the surface. (c) The hydrogen that forms during the very exothermic reaction of potassium with water ignites. (d) Caesium reacts explosively, shattering its glass container.

An advanced explanation

The elements of Group II have two electrons that must be removed when they act as reductants. On average they react with water less quickly than the Group I elements. For example, calcium reacts about as vigorously with water as does sodium. You might expect calcium to be *more* reactive than sodium, because it has the more negative standard electrode potential. However, the *rate* of reaction of calcium is lower because *two* electrons must be removed when it reacts, so the activation energy for the reaction is greater than for sodium.

Group II metals react less vigorously than their Group I counterparts. Beryllium does not react with water, and magnesium reacts only very slowly. Reactivity increases down the group Ca–Sr–Ba, with calcium approaching the reactivity of sodium. As with the Group I elements, the hydroxide is formed and hydrogen is evolved:

$$Ca(s) + 2H_2O(l) \rightarrow Ca(OH)_2(aq) + H_2(g)$$

Magnesium reacting with steam: $Mg(s) + H_2O(g) \rightarrow MgO(s) + H_2(g)$ *Steam is introduced through the tube on the left. The flame visible at the small hole is caused by the ignited hydrogen.*

(a)
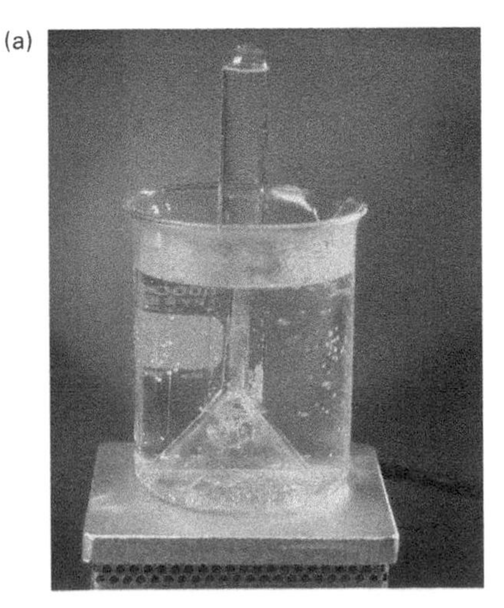
(b)

(a) The rate of reaction of magnesium with cold water is imperceptibly slow. It does react more rapidly with steam (see photo on the left). (b) Calcium reacts steadily with water. Calcium hydroxide has only limited solubility and gradually appears as a white cloudiness.

Reaction with oxygen

All Group I elements react with oxygen to form the oxide M_2O (where we use M to denote any Group I metal). For example, potassium reacts as follows:

$$4K(s) + O_2(g) \rightarrow 2K_2O(s)$$

All Group II elements react with oxygen to form the oxide MO (where now M denotes any Group II metal). For example, calcium reacts as follows:

$$2Ca(s) + O_2(g) \rightarrow 2CaO(s)$$

All Group II oxides are white ionic solids with extremely high melting points. Magnesium reacts very vigorously with oxygen, spread 13.1, to form magnesium oxide. Magnesium oxide is a particularly stable compound, which explains why magnesium reacts with carbon dioxide, as illustrated alongside.

Magnesium is a sufficiently powerful reductant to continue burning when placed between two pieces of dry ice (solid carbon dioxide). The reaction products are magnesium oxide and carbon:
$2Mg(s) + CO_2(g) \rightarrow 2MgO(s) + C(s)$

Peroxides and superoxides

In their reaction with oxygen, all the Group I metals from sodium to caesium and the Group II metals strontium and barium also form *peroxides*. A **peroxide** contains the ion O_2^{2-}, for example Na_2O_2 (Group I) and BaO_2 (Group II). The Group I metals from potassium to caesium also form *superoxides*. A **superoxide** contains the ion O_2^-, for example KO_2.

Trends in reactivity

Group I elements react more vigorously with water than their Group II counterparts. In each group, reactivity increases with increasing atomic number. The reason for these differences is that the elements form positive ions in the course of their reaction. Group I metals lose one electron and Group II metals lose two. The energy required for this change is the ionization energy.

In both groups, ionization energy decreases with increasing atomic number. The lower the ionization energy, the lower the activation energy for reaction, and so the faster the reaction. (Note that the word 'reactivity' here actually refers to the *rate* of a reaction.) So, the trends in reactivity *within* the two groups and the differences in reactivity *between* the two groups may be explained with reference to the ionization energies of the elements.

SUMMARY

- The reactivity of the Group I elements with water is greater than that of the Group II elements.
- Reactivity increases with increasing atomic number.
- The rate of reaction with water depends on the activation energy, which in turn depends on the ionization energy.

Underwater chemistry

Potassium superoxide is used to purify the air in submarines and in emergency breathing apparatus. It both removes the waste products of respiration and supplies fresh oxygen:

$4KO_2(s) + 4CO_2(g) + 2H_2O(g) \rightarrow 4KHCO_3(s) + 3O_2(g)$

Building with magnesium

The relatively slow rate of reaction between magnesium and water allows magnesium alloys to be used for construction. Magnesium has a low density (1.7 g cm^{-3}, compared with 2.7 g cm^{-3} for aluminium and 7.9 g cm^{-3} for iron), making it the least dense constructional metal. However, its strength as a reductant becomes important at high temperatures. In battle, disastrous fires have occurred in naval ships that used magnesium in their superstructures: the magnesium reacts violently once ignited.

PRACTICE

1 Write chemical equations for the reactions of rubidium and barium with water and with oxygen. For each of the four reactions, discuss the changes in oxidation state that take place.

2 Look at the illustrations on the left-hand page and make statements about the densities of the elements concerned. Are you able to comment on the densities of the s-block elements that are not illustrated?

3 Potassium is more dense than mineral oil and less dense than water. Describe how, with the aid of standard laboratory glassware, you would use these three substances to produce hydrogen *safely*. Give details of the *function* of your design – how it would work.

4 The text above states: 'The lower the ionization energy, the lower the activation energy for reaction, and so the faster the reaction.' Explain this statement with respect to the trend in the reactivity of the Group II metals.

16.4

OBJECTIVES

- Oxides and water
- Sodium hydroxide
- Trends in solubility of hydroxides
- The limewater test
- Base character

The s-block oxides and hydroxides

As noted in the previous spread, all the s-block metals form stable oxides. In this spread we shall investigate the properties of these oxides, particularly their solubility and their reaction with water to form hydroxides. You will see that both the oxides and hydroxides are ionic, and that they act principally as bases in their reactions with other substances. (Note that beryllium oxide has anomalous properties, which are specifically addressed in a later spread in this chapter.)

The solubility of the Group I and Group II hydroxides. Solubility data are given in grams of solid per 100 cm³ of water.

Group I hydroxide	Solubility
LiOH	13
NaOH	42
KOH	107
RbOH	180
CsOH	395
Group II hydroxide	**Solubility**
$Be(OH)_2$	ss*
$Mg(OH)_2$	0.0009
$Ca(OH)_2$	0.18
$Sr(OH)_2$	0.41
$Ba(OH)_2$	5.6

* ss = sparingly soluble.

Group I oxides

The Group I metal oxides do not simply dissolve in water, they *react* with it. For example, sodium oxide reacts with water to form aqueous sodium hydroxide:

$$Na_2O(s) + H_2O(l) \rightarrow 2NaOH(aq)$$

Group I metal oxides are ionic, and it is the oxide ion that reacts to form the hydroxide ion:

$$O^{2-}(s) + H_2O(l) \rightarrow 2OH^-(aq)$$

The resulting solutions are strongly basic because they contain a high concentration of the aqueous hydroxide ion.

Sodium hydroxide

Sodium hydroxide is typical of Group I hydroxides. Because it is cheap, it is the reagent of choice when a strongly basic solution is required. It is a white waxy solid usually supplied in the form of pellets. It is stable to heat, melting to a clear, colourless liquid at 318 °C. It dissolves in water exothermically to give a solution that is a strong base.

The three main classes of reaction involving sodium hydroxide are:

- Neutralization of acids (see spread 12.1)

  $$H_3O^+(aq) + OH^-(aq) \rightarrow 2H_2O(l)$$

- Precipitation of hydroxides (see spread 11.10)

  $$Zn^{2+}(aq) + 2OH^-(aq) \rightarrow Zn(OH)_2(s)$$

- Hydrolysis in organic chemistry

  $$CH_3CH_2Br(l) + OH^-(aq) \rightarrow CH_3CH_2OH(aq) + Br^-(aq)$$

Note that these reactions of aqueous sodium hydroxide (and of all Group I hydroxides) involve the aqueous hydroxide ion $OH^-(aq)$ acting as a Lewis base (see chapter 12 'Acid–base equilibrium'). (The Group I metal ion M^+ takes no part in the reactions and is a 'spectator' ion.)

Caustic soda

Caustic soda is the traditional name for sodium hydroxide. Sodium hydroxide is the most important alkali in industry: see spread 18.2 for its manufacture in the chlor -alkali industry. The name 'caustic soda' is still commonly seen labelling containers of solid sodium hydroxide pellets, which are sold for unblocking drains. **Caustic potash** is the traditional name for potassium hydroxide.

Pellets of sodium hydroxide. This substance is ***deliquescent****, which means that it absorbs water from the air to such an extent that it forms a concentrated solution. It must therefore be kept in an airtight container. This sample has been left for a short time and is already glistening.*

Group II oxides and hydroxides

There is a significant difference between the two groups of s-block elements in the solubility of their oxides. The oxides of the Group I metals react with water to form soluble hydroxides. Most of the oxides of the Group II metals are much less soluble in water. We can explain this effect by considering the energy changes that take place during dissolution.

The only Group II oxide to form a strong base in water is barium oxide:

$$BaO(s) + H_2O(l) \rightarrow Ba(OH)_2(aq)$$

Barium hydroxide is fully ionized in water. The other Group II hydroxides are far less soluble than those of Group I. Look at the table above and you will see that calcium hydroxide is only very slightly soluble and that magnesium hydroxide is even less so. Saturated solutions of these Group II hydroxides are only weakly basic, because the concentration of aqueous hydroxide ion is very low. In general, hydroxide solubility *increases* as the atomic number of the Group II metal increases. The main reason for this trend is that the lattice enthalpy decreases significantly as the ionic radius increases.

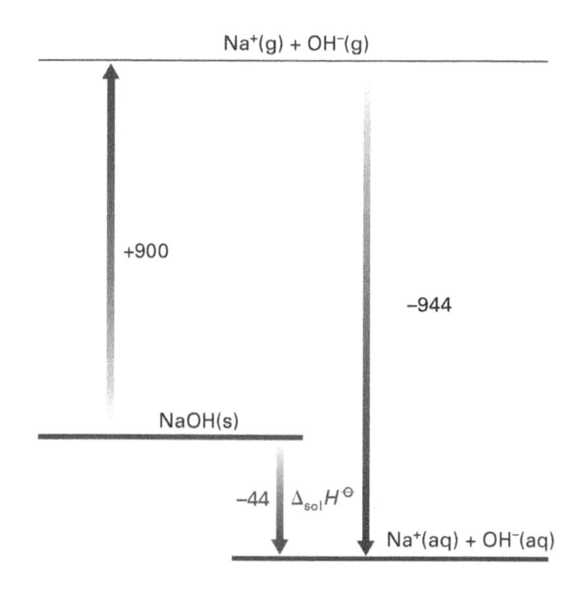

Enthalpy cycle for the dissolution of sodium hydroxide. The two most significant factors to consider about dissolving a metal hydroxide are the lattice enthalpy and the enthalpy of hydration; see spread 10.7.
(a) Breaking the ionic lattice into separate ions involves the highly endothermic lattice enthalpy $\Delta_{lat}H^\ominus$:
Group I: $NaOH(s) \rightarrow Na^+(g) + OH^-(g)$
Group II: $Mg(OH)_2(s) \rightarrow Mg^{2+}(g) + 2OH^-(g)$
(b) Hydrating the ions involves the highly exothermic *enthalpy of hydration $\Delta_{hyd}H^\ominus$:*
$OH^-(g) \rightarrow OH^-(aq)$
Group I: $Na^+(g) \rightarrow Na^+(aq)$
Group II: $Mg^{2+}(g) \rightarrow Mg^{2+}(aq)$
The lattice enthalpy for magnesium hydroxide is extremely large (3000 kJ mol^{-1}) because of the double charge on the Mg^{2+} ion. The magnitude of the enthalpies of hydration of the Mg^{2+} ion plus two OH^- ions is smaller than the magnitude of the lattice enthalpy. The reaction is therefore endothermic overall (+4 kJ mol^{-1}): $Mg(OH)_2$ is sparingly soluble. Sodium hydroxide has a much lower lattice enthalpy and the solution process is exothermic overall (−44 kJ mol^{-1}).

Magnesium, calcium, and strontium hydroxides are usually made by mixing aqueous sodium hydroxide with an aqueous salt of the metal. For example, for magnesium chloride:

$$MgCl_2(aq) + 2NaOH(aq) \rightarrow Mg(OH)_2(s) + 2NaCl(aq)$$

Magnesium hydroxide is only weakly basic, which is useful for its application as an antacid in indigestion tablets. The milky-white suspension is called 'milk of magnesia'.

Whilst they have limited reaction with water, all Group II oxides *do* react with an acid to form a salt and water. They may therefore be classified as *basic oxides*. For example, magnesium oxide reacts with hydrochloric acid to form magnesium chloride and water:

$$MgO(s) + 2HCl(aq) \rightarrow MgCl_2(aq) + H_2O(l)$$

The Group II hydroxides also react with an acid to form a salt and water, e.g.:

$$Mg(OH)_2(s) + 2HCl(aq) \rightarrow MgCl_2(aq) + 2H_2O(l)$$

$$Ba(OH)_2(aq) + 2HCl(aq) \rightarrow BaCl_2(aq) + 2H_2O(l)$$

SUMMARY

- All s-block oxides and hydroxides are basic; they can act as Brønsted bases.
- All Group I oxides react with water to form a solution of the hydroxide.
- All Group I hydroxides are stable to heat, but they dissolve readily in water.
- All Group II metal oxides and hydroxides (except those of barium) are weak bases with limited solubility in water. The solubility increases with increasing atomic number of the metal.

The limewater test for carbon dioxide

The test to identify carbon dioxide involves bubbling the gas through a solution of **limewater** (saturated aqueous calcium hydroxide), which has pH = 12. A milky precipitate of calcium carbonate forms:

$Ca(OH)_2(aq) + CO_2(g) \rightarrow CaCO_3(s) + H_2O(l)$

The precipitate reacts with excess carbon dioxide to form (a colourless solution of) aqueous calcium hydrogencarbonate:

$CaCO_3(s) + CO_2(g) + H_2O(l) \rightarrow Ca(HCO_3)_2(aq)$

Quicklime and slaked lime

Calcium oxide is called **quicklime**. It is produced in industry by roasting **limestone** (calcium carbonate). Quicklime may be used to neutralize acids and is especially useful for spreading on soils that are too acidic. Quicklime absorbs water to form calcium hydroxide, which is called **slaked lime**. Mixed with water and sand to make mortar (for bonding brick walls), it slowly sets by reaction with atmospheric carbon dioxide to form interlaced crystals of calcium carbonate.

PRACTICE

1 Write chemical equations for the following reactions (include state symbols):
  - **a** Potassium oxide and water
  - **b** Strontium oxide and water
  - **c** Magnesium oxide and nitric acid
  - **d** Magnesium nitrate and aqueous sodium hydroxide.

2 Which of the two solutions resulting in (a) and (b) in question 1 will be more strongly basic? Give your reasoning.

3 Write equations for the chemical changes discussed in the box above headed 'Quicklime and slaked lime':
  - **a** Production of quicklime
  - **b** Neutralizing soil acidity
  - **c** Formation of slaked lime.

16.5 THE S-BLOCK HALIDES

OBJECTIVES

- Bonding in the halides
- Trends in halide solubility
- Enthalpy changes and solubility

The s-block metals are highly reactive compared with other metals. The halogens are highly reactive non-metals. As you might expect, all the Group I and Group II metals react with all the halogens to form halides. In this spread we look at the methods of producing these halides, and we explore their bonding character and the reasons for their differing solubilities in water. The emphasis throughout is not so much on the properties themselves as on trends and the reasons for them.

Perhaps the most remarkable use of sodium chloride (and sugar) is in the rehydration tablets that help keep dehydrated patients alive. They are very effective and cheap.

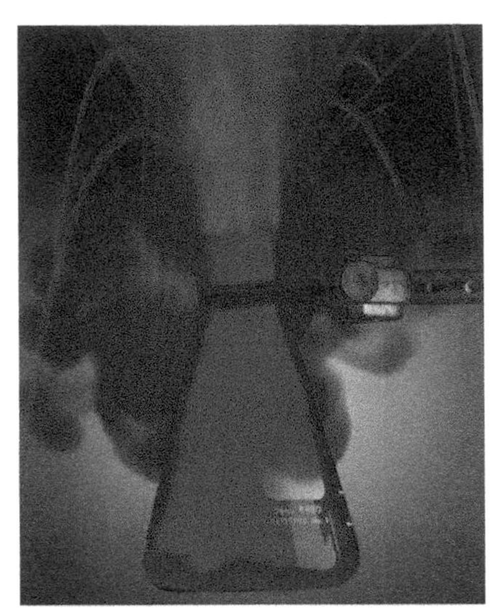

Sodium reacts very vigorously with bromine to form sodium bromide.

Group I halides

The halides of the Group I metals are all white crystalline solids. They are made either by directly combining the elements in a redox reaction, e.g.

$$2Na(s) + Br_2(l) \rightarrow 2NaBr(s)$$

or by a neutralization reaction, e.g.

$$HBr(aq) + NaOH(aq) \rightarrow NaBr(aq) + H_2O(l)$$

All Group I halides have the rock-salt structure (see chapter 6 'Solids'), with the exception of CsCl, CsBr, and CsI, which have the caesium chloride structure. All are ionic; they are soluble in water and their aqueous solutions conduct electricity.

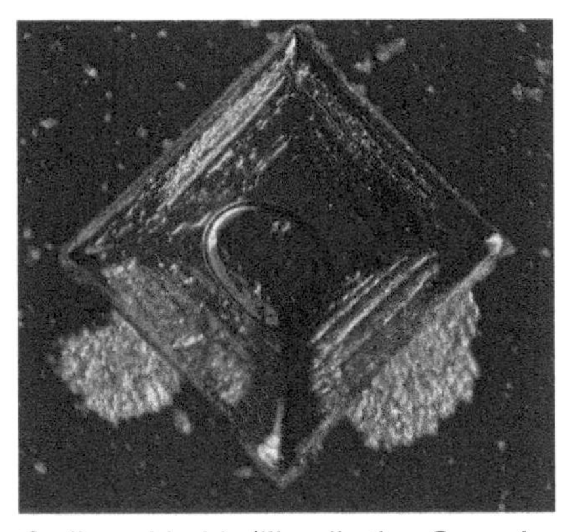

Sodium chloride (like all other Group I halides) consists of oppositely charged ions arranged in a regular crystalline lattice.

Group II halides

The halides of the Group II metals are off-white crystalline solids. CaF_2, SrF_2, BaF_2, $SrCl_2$, and $BaCl_2$ all have the fluorite structure (see chapter 6 'Solids'). Group II halides are predominantly ionic, with the exception of the compounds of beryllium. They may all be produced by direct combination of the elements. For example, magnesium and chlorine react:

$$Mg(s) + Cl_2(g) \rightarrow MgCl_2(s)$$

Reactivity decreases in the order $F_2 > Cl_2 > Br_2 > I_2$ and in the order

Ba > Sr > Ca > Mg > Be.

The melting points of the Group II metal chlorides.

Group II chloride	Melting point/°C
$BeCl_2$	415
$MgCl_2$	714
$CaCl_2$	772
$SrCl_2$	874
$BaCl_2$	962

Bonding character

The bonding in the chlorides is predominantly *ionic*. Ionic character and melting point increase with increasing atomic number of the metal. However, beryllium chloride has a relatively low melting point, which suggests that it has significant *covalent* character; see spread 16.8. Remember that covalent character increases where the cation has small size and high charge, that is, high charge density. For example, the small lithium ion has a greater charge density than the (larger) sodium ion, so its compounds have a relatively greater covalent character. Beryllium compounds have greater covalent character than the corresponding lithium compounds because the beryllium ion is *doubly* charged and so will have greater charge density.

Melting point and the nature of the anion

Bonding character also depends on the nature of the anion. For example, the melting points of the halides of magnesium show a decrease as halide ionic radius increases with atomic number: see table opposite. Covalent character also increases when the anion has high polarizability. Covalent character increases in the order fluoride < chloride < bromide < iodide.

The solubility of the s-block halides. Solubility data are given in grams of solid per 100 cm³ of water.

Cation	F^-	Cl^-	Br^-	I^-
Li^+	0.27	64	145	165
Na^+	4.2	36	116	184
K^+	92	34	53	128
Rb^+	131	77	98	152
Cs^+	367	162	124	44
Be^{2+}	reacts	reacts	s*	dec†
Mg^{2+}	0.008	54	102	148
Ca^{2+}	0.0016	75	142	209
Sr^{2+}	0.011	54	100	178
Ba^{2+}	0.12	38	104	205

* s = soluble (no figure available). † dec = decomposes.

Solubility

The solubilities of the s-block halides in water are summarized above.

Note the following points from the table:

- The covalent character of the beryllium halides results in their reacting with water (BeF_2, $BeCl_2$) or even decomposing (BeI_2).
- Solubility of the Group II halides generally increases in the order fluoride < chloride < bromide < iodide.

Remember that solubility is mainly affected by two opposing factors: the endothermic lattice enthalpy required to break the solid lattice into separate ions, and the exothermic enthalpy of hydration of the ions by water molecules. Both the lattice enthalpy and the enthalpy of hydration of the positive ions become smaller as the ions become larger. As an approximate guide, solubility increases with increasing enthalpy of solution; low solubility results where the lattice enthalpy exceeds the enthalpy of hydration, i.e. where the enthalpy of solution is endothermic overall.

SUMMARY

- The s-block halides may be made by direct combination of the elements.
- The bonding in the s-block halides is predominantly ionic.
- The bonding of beryllium halides and, to a lesser extent, lithium halides show significant covalent character.

The melting points of the magnesium halides.

Magnesium halide	Melting point/°C
MgF_2	1263
$MgCl_2$	714
$MgBr_2$	711
MgI_2	634

The enthalpies of solution ($\Delta_{sol}H^{\ominus}$) given by $\Delta_{sol}H^{\ominus} = \Delta_{lat}H^{\ominus} + \Delta_{hyd}H^{\ominus}(\text{metal}) + \Delta_{hyd}H^{\ominus}(F^-)$ and the solubilities of the Group I fluorides. Solubility data are given in grams of fluoride per 100 cm³ of water.

Group I fluoride	$\Delta_{sol}H^{\ominus}$ /kJ mol⁻¹	Solubility
LiF	+5	0.27
NaF	+1	4.2
KF	−18	92
RbF	−26	131
CsF	−37	367

Solubility and the entropy factor

Lattice enthalpies and the corresponding enthalpies of hydration are only an approximate guide to the solubility of a substance in water. Another factor is the entropy change (see chapter 14 'Spontaneous change ...') that takes place on dissolution. The ordered crystal lattice breaks down and the ions become randomized throughout the solution. This loss of order (increase in entropy) can be more significant than the enthalpy changes.

PRACTICE

1 Give the chemical equation for the preparation of the following:

a Rubidium bromide by a neutralization method

b Magnesium bromide by direct combination.

2 Arrange the following halides in order of increasing covalent character: magnesium iodide; potassium fluoride; potassium bromide. Give details of your reasoning.

3 Suggest reasons why the Group I fluorides show a much greater range of solubilities than the other Group I halides.

4 Look at the solubilities for the Group II fluorides. Suggest why the solubilities of the fluorides MgF_2 to BaF_2 are far smaller than those of all other s-block halides.

16.6

OBJECTIVES

- Reaction with acid
- Thermal decomposition of carbonates
- Carbonates in industry

Solution pH

Solutions of Group I carbonates are basic as the result of hydrolysis. For example, sodium carbonate dissolves to give aqueous ions:

$Na_2CO_3(s) \rightarrow 2Na^+(aq) + CO_3^{2-}(aq)$

Water acts as a Brønsted acid and protonates the carbonate ion. The pH becomes greater than 7 because the aqueous hydroxide ion $OH^-(aq)$ forms:

$CO_3^{2-}(aq) + H_2O(l) \rightleftharpoons OH^-(aq) + HCO_3^-(aq)$

The resulting hydrogencarbonate ion may itself then be protonated by water, further increasing the concentration of hydroxide ion:

$HCO_3^-(aq) + H_2O(l) \rightleftharpoons OH^-(aq) + H_2CO_3(aq)$

This equation indicates that solutions of hydrogencarbonates are also basic. However, these two equilibria lie to the left, so aqueous s-block carbonates have pH values about 11. Values for aqueous s-block hydrogencarbonates are lower (pH 8–9).

The stalactites that hang from the roofs of some caves form as follows. Rain water dissolves atmospheric carbon dioxide, producing (dilute) carbonic acid:

$H_2O(l) + CO_2(g) \rightarrow H_2CO_3(aq)$

This then percolates through carbonate rocks to form hydrogencarbonates:

$H_2CO_3(aq) + CaCO_3(s) \rightarrow Ca(HCO_3)_2(aq)$

The carbonate rock is slowly 'dissolved away' over millions of years to form caves. Water containing the dissolved calcium (or magnesium) hydrogencarbonates drips from the roof, and the water evaporates. The aqueous hydrogencarbonates decompose to form insoluble carbonates:

$Ca(HCO_3)_2(aq) \rightarrow CaCO_3(s) + H_2O(l) + CO_2(g)$

These carbonates are deposited to form stalactites (and build up the accompanying stalagmites on the floor below).

Carbonates and hydrogencarbonates

Both carbonates and hydrogencarbonates are salts of carbonic acid $H_2CO_3(aq)$, formed when carbon dioxide gas dissolves in water. Carbonates are compounds that contain the CO_3^{2-} ion, and hydrogencarbonates are compounds that contain the HCO_3^- ion. With the exception of beryllium, there is a complete series of s-block carbonates and hydrogencarbonates. A wide variety of s-block carbonate minerals exist, two of great commercial importance.

Solubility in water

The s-block carbonates are white solids. The Group I carbonates have the general formula M_2CO_3 and (with the exception of lithium carbonate) are soluble in water. The Group II carbonates have the general formula MCO_3 and are insoluble in water.

The Group I hydrogencarbonates have the general formulae $MHCO_3$; they exist as white solids which are soluble in water. Aqueous Group II hydrogencarbonates (except that of beryllium) may be obtained by bubbling carbon dioxide gas through an aqueous suspension of the carbonate, see spread 16.4:

$$CaCO_3(s) + CO_2(g) + H_2O(l) \rightarrow Ca(HCO_3)_2(aq)$$

If the solution is evaporated in an attempt to isolate the solid, the hydrogencarbonate decomposes by reversal of the above reaction.

Calcium and magnesium hydrogencarbonates are responsible for the temporary hardness of water.

Reactions with acid

Carbonates evolve carbon dioxide gas when added to acid:

$$MgCO_3(s) + 2HCl(aq) \rightarrow CO_2(g) + H_2O(l) + MgCl_2(aq)$$

$$CO_3^{2-}(s) + 2H^+(aq) \rightarrow CO_2(g) + H_2O(l)$$

Similarly, hydrogencarbonates evolve carbon dioxide gas when added to acid:

$$NaHCO_3(s) + HCl(aq) \rightarrow CO_2(g) + H_2O(l) + NaCl(aq)$$

$$HCO_3^-(s) + H^+(aq) \rightarrow CO_2(g) + H_2O(l)$$

Thermal decomposition

Of the Group I carbonates, only lithium carbonate decomposes on heating, giving carbon dioxide and lithium oxide:

$$Li_2CO_3(s) \rightarrow CO_2(g) + Li_2O(s)$$

All the others melt, with no significant decomposition. The hydrated crystalline forms of the Group I carbonates lose water on moderate heating. For example, sodium carbonate decahydrate:

$$Na_2CO_3{\cdot}10H_2O(s) \rightarrow Na_2CO_3(s) + 10H_2O(g)$$

Solid Group I hydrogencarbonates decompose when heated. For example:

$$2NaHCO_3(s) \rightarrow Na_2CO_3(s) + CO_2(g) + H_2O(g)$$

Group II carbonates decompose on heating:

$$CaCO_3(s) \rightarrow CaO(s) + CO_2(g)$$

Thermal stability

The carbonates of the Group II elements have different thermal stabilities, so they decompose at different temperatures when heated. The order of stability of the carbonates is $Mg < Ca < Sr < Ba$. To suggest reasons for this trend, consider calcium carbonate and barium carbonate. Calcium carbonate is stable up to about 900 °C; the corresponding figure for barium carbonate is about 1350 °C.

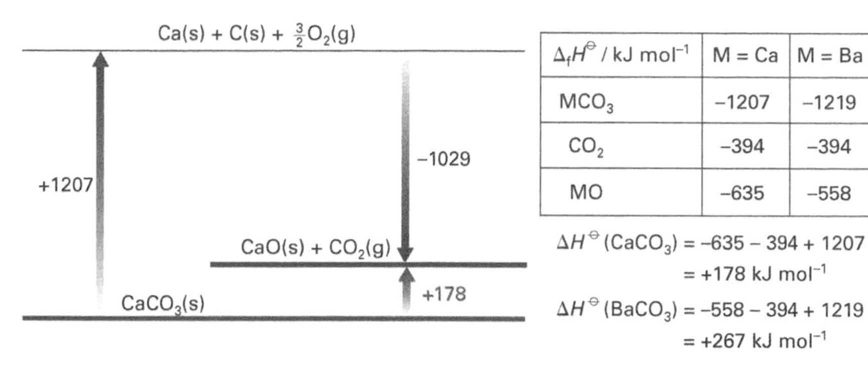

$\Delta_f H^\ominus$ / kJ mol^{-1}	M = Ca	M = Ba
MCO_3	−1207	−1219
CO_2	−394	−394
MO	−635	−558

$\Delta H^\ominus(CaCO_3) = -635 - 394 + 1207$
$= +178\ \text{kJ mol}^{-1}$

$\Delta H^\ominus(BaCO_3) = -558 - 394 + 1219$
$= +267\ \text{kJ mol}^{-1}$

Enthalpy cycle showing the decomposition of two Group II carbonates in terms of the standard enthalpy changes of formation of the constituents.

The difference in the standard enthalpy change for these two decompositions results from the difference in the values of the standard enthalpy change of formation of barium oxide compared with that of calcium oxide. Calcium oxide is more stable (has a more negative standard enthalpy change of formation) than barium oxide because its lattice enthalpy is larger, due to the calcium ion being smaller. So, calcium carbonate decomposes at a *lower* temperature than barium carbonate because calcium oxide is *more stable* than barium oxide.

By the same reasoning, magnesium carbonate decomposes more readily than calcium carbonate because magnesium oxide is more stable than calcium oxide, due to the magnesium ion being smaller than the calcium ion. Magnesium oxide has extremely high thermal stability and so is used as refractory (heat-resistant) bricks in furnaces.

Carbonates in industry

Calcium carbonate occurs as chalk, marble, and limestone. Finely ground chalk is incorporated in some toothpastes and cosmetics. Slabs of marble give a high-quality finish to buildings; and blocks may be carved to form sculptures. Millions of tonnes of crushed limestone are used each year to build roads. The most important chemical use of limestone is in the smelting of iron ore (see chapter 20 'The transition metals').

Sodium carbonate is by far the most commercially important Group I carbonate. It is manufactured by the Solvay process, as outlined below.

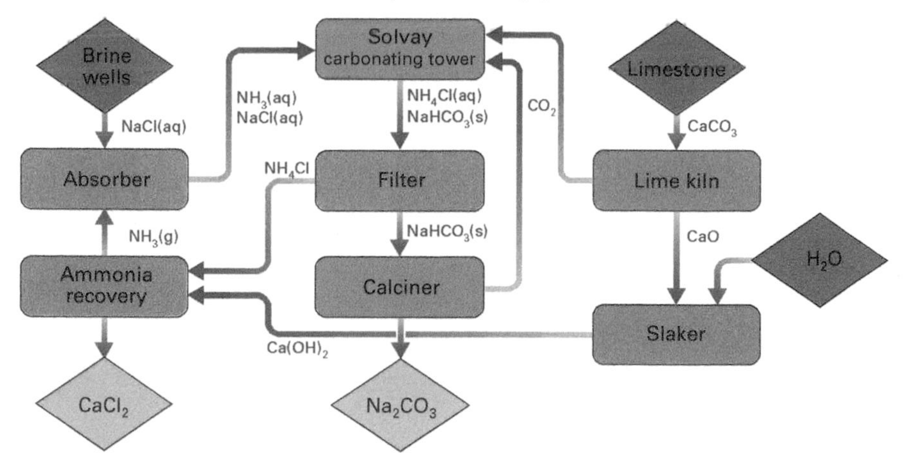

Flow scheme for the Solvay process. Carbon dioxide is passed up a carbonating tower (25 metres tall), down which flows an aqueous mixture of ammonia and concentrated sodium chloride:

$CO_2(g) + NH_3(aq) + NaCl(aq) + H_2O(l) \rightarrow NaHCO_3(s) + NH_4Cl(aq)$

The high concentration of sodium ion causes the solubility of sodium hydrogencarbonate $NaHCO_3$ to be exceeded through the common ion effect (see spread 11.10). Solid $NaHCO_3$ forms; it is filtered off and then decomposed by heating in rotating calciners (30 metres long) to obtain the carbonate:

$2NaHCO_3(s) \rightarrow Na_2CO_3(s) + H_2O(g) + CO_2(g)$

The carbon dioxide is recycled.

SUMMARY

- All s-block carbonates and hydrogencarbonates exist, with the exception of the beryllium compounds. They react with strong acids to give carbon dioxide gas.
- Group I carbonates are stable to heat, with the exception of lithium carbonate. All are soluble in water and form basic solutions.
- Group II carbonates are insoluble in water, and decompose when strongly heated.
- The thermal stabilities of the Group II carbonates increase with increasing atomic number.

Decomposition and entropy

Calcium carbonate does not decompose at 298 K because the reaction is strongly endothermic: limestone hills do *not* spontaneously decompose! However, reaction occurs above 900 °C because the entropy increase due to the release of carbon dioxide gas exceeds the entropy decrease in the surroundings due to the endothermic nature of the reaction (see chapter 14 'Spontaneous change...' and especially spread 14.4).

In the blast furnace

Limestone decomposes at about 900 °C:

$CaCO_3(s) \rightarrow CaO(s) + CO_2(g)$

The calcium oxide acts as a **flux** (a substance that reacts with unwanted impurities in the furnace to produce a waste product called a **slag**).

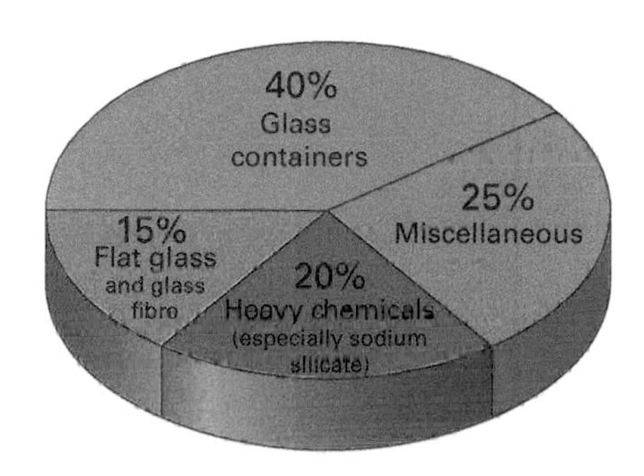

The major uses of sodium carbonate.

Trona

Although the Solvay process is still important world-wide, an alternative process has been gaining importance, especially in the USA. This involves mining the mineral *trona*, which is especially abundant in Wyoming. Trona has the approximate formula $Na_2CO_3 \cdot NaHCO_3 \cdot 2H_2O$. Heating the ore decomposes it and produces sodium carbonate.

16.7

OBJECTIVES

- Action of heat on nitrates
- Solubilities
- Barium sulphate

The s-block nitrates and sulphates

Nitrates and sulphates are salts containing respectively the ions NO_3^- and SO_4^{2-}. There is a full range of compounds consisting of s-block metal ions with either nitrate or sulphate ions. All these compounds exist as white crystalline solids or, when dehydrated, as white amorphous powders. As you might expect from reading the previous spreads in this chapter, the Group I and the Group II compounds show a distinct difference in behaviour, with lithium compounds being between the two.

Nitrates

Moderate heating decomposes Group I nitrates (except $LiNO_3$) to nitrites, which contain the ion NO_2^-:

$$2NaNO_3(s) \rightarrow 2NaNO_2(s) + O_2(g)$$

(More extreme temperatures form the oxide, nitrogen dioxide, and oxygen.)

Heating decomposes any Group II nitrate, and lithium nitrate, directly to the oxide. The difference in behaviour between Group I and Group II arises because the Group II oxides have more negative standard enthalpy changes of formation than the Group I oxides (see the previous spread for an explanation). Nitrogen dioxide (seen as brown fumes) and oxygen are evolved. For example:

$$2Ca(NO_3)_2(s) \rightarrow 2CaO(s) + 4NO_2(g) + O_2(g)$$

Sodium nitrate is the only nitrate to occur naturally in substantial amounts. Deposits of the sodium nitrate mineral *Chile saltpetre*, now largely mined out, are found in rainless areas of Chile. These deposits were the only source of nitrate before the introduction of the Haber–Bosch and Ostwald processes (see chapter 19 'The p-block elements'), which use atmospheric nitrogen to produce ammonia and then nitric acid. Much of the nitric acid now produced is used to make potassium and ammonium nitrate fertilizers. A minor use of potassium nitrate is in the gunpowder incorporated in fireworks.

Nitrate solubility

In common with all other nitrates, the s-block metal nitrates are soluble in water:

$Mg(NO_3)_2(s) \rightarrow Mg^{2+}(aq) + 2NO_3^-(aq)$

Potassium nitrate

Potassium nitrate is the only other nitrate ore. Major deposits are found in India.

Sulphates

The s-block sulphates are relatively stable to heat, decomposing only at high temperature. They have no significant chemical properties other than precipitating insoluble sulphates from solution. For example, for lead sulphate:

$$SO_4^{2-}(aq) + Pb^{2+}(aq) \rightarrow PbSO_4(s)$$

Of the Group I sulphates, sodium sulphate is a key chemical for making brown wrapping paper and corrugated cardboard. Potassium sulphate is used in glass-making and as a fertilizer ('sulphate of potash'). Group II sulphates occur naturally as the minerals *celestine* $SrSO_4$ and *barytes* $BaSO_4$. The main ores containing calcium are *gypsum* $CaSO_4{\cdot}2H_2O$ and *anhydrite* $CaSO_4$. The magnesium ore $MgSO_4{\cdot}7H_2O$ is called *Epsom salts*. In purified form it is used as a laxative and to make artificial snow on film sets.

Plaster of Paris is an insoluble form of calcium sulphate $2CaSO_4{\cdot}H_2O$. When mixed with water it forms $CaSO_4{\cdot}2H_2O$, expanding when it hydrates and hardening to a rigid, ceramic-like material. It is used for plaster-casts to immobilize broken bones, in dental work, and for gypsum wallboard ('plasterboard') in the construction industry.

This 2500 year old panel found in Pharaoh Rawer's tomb at Giza is made out of alabaster, a fine-grained form of gypsum $CaSO_4{\cdot}2H_2O$.

Sulphate solubility

In common with all other Group I salts, the sulphates are soluble in water. Group II sulphates show varying solubility: magnesium sulphate is soluble, calcium sulphate is sparingly soluble, and barium sulphate is very insoluble. Trends of this sort exist because solubility is a balance between the endothermic lattice enthalpy $\Delta_{lat}H^{\ominus}$ required to break the lattice into ions, and the exothermic enthalpy of hydration $\Delta_{hyd}H^{\ominus}$ accompanying the subsequent hydration of the ions. For example:

$$BaSO_4(s) \rightarrow Ba^{2+}(g) + SO_4^{2-}(g); \quad \Delta_{lat}H^{\ominus} = +2374\,kJ\,mol^{-1}$$

$$Ba^{2+}(g) + SO_4^{2-}(g) \rightarrow Ba^{2+}(aq) + SO_4^{2-}(aq); \quad \Delta_{hyd}H^{\ominus} = -2355\,kJ\,mol^{-1}$$

The lattice enthalpies of the three sulphates are quite similar because the sulphate ion is a very large ion; the changing size of the metal ion has little effect. On the other hand, the enthalpies of hydration become significantly smaller as metal ion size increases. The sum of the enthalpy of hydration and the lattice enthalpy therefore becomes correspondingly more endothermic (less favourable to solution) with increasing atomic number.

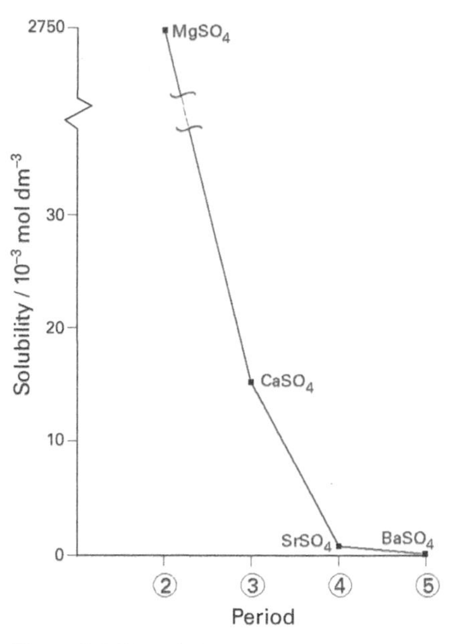

The solubility of Group II salts. The same decreasing trend as for sulphates is observed for other Group II salts with large, doubly charged negative ions such as carbonate CO_3^{2-}, ethanedioate $C_2O_4^{2-}$, and chromate(VI) CrO_4^{2-}.

Barium sulphate

Barium sulphate is extremely insoluble in water, and this feature is used both in the laboratory and in diagnostic medicine. In the laboratory, barium ions are used to identify aqueous sulphate ions. Addition of acidified barium nitrate (or barium chloride) to a solution containing sulphate ions causes a white precipitate of barium sulphate to form:

$$Ba^{2+}(aq) + SO_4^{2-}(aq) \rightarrow BaSO_4(s)$$

The acid is necessary to prevent the false detection of sulphite ion SO_3^{2-}. Barium sulphite $BaSO_3$ is also insoluble, and would form as a white precipitate. However, the acid reacts with sulphite ion: sulphur dioxide gas forms and the Ba^{2+} ions remain in solution. (Similarly, any carbonate ion present reacts to form carbon dioxide gas.)

Barium sulphate is used in medicine, as described in the caption to the photograph alongside.

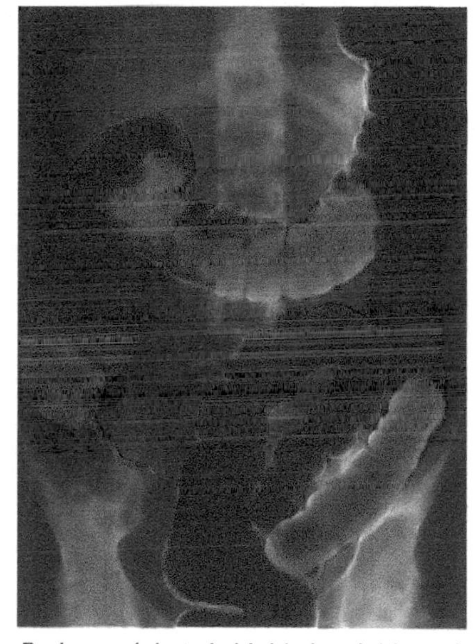

Barium sulphate is highly insoluble and is opaque to X-rays. This radiograph shows the human lower intestinal tract. The patient swallowed a suspension of barium sulphate about an hour before the X-ray was taken.

SUMMARY

- All s-block nitrates are soluble in water.
- All Group I sulphates are soluble in water. The solubility of the Group II sulphates decreases as atomic number increases.
- Group I nitrates (except $LiNO_3$) decompose to the nitrite and oxygen when heated.
- Lithium nitrate and Group II nitrates decompose to the oxide, nitrogen dioxide, and oxygen when heated.
- Acidified barium nitrate added to aqueous sulphate ions produces a white precipitate of barium sulphate.

PRACTICE

1 Barium ions $Ba^{2+}(aq)$ are highly poisonous. Explain why barium sulphate is used in preference to barium nitrate when taking X-ray pictures of the sort illustrated above.

2 a Why are minerals containing Group II sulphates so much more common than minerals containing Group I sulphates?

b Why are deposits of sodium nitrate found only in rainless regions of the world?

3 Suggest how 'plaster of Paris' is manufactured from gypsum.

EXTENSION

16.8

OBJECTIVES

- Increased covalent character
- Amphoteric beryllium hydroxide
- Diagonal relationships
- Thermal decomposition of lithium carbonate

The ionic radii of the s-block elements. Notice that each Group II ion is smaller than the corresponding Group I ion, because the same number of electrons is under the control of one more proton. The lithium ion in Group I is about the same size as the magnesium ion in Group II. The sodium ion is about 50% larger than the lithium ion.

Group I ion	Radius/nm
Li^+	0.074
Na^+	0.102
K^+	0.138
Rb^+	0.149
Cs^+	0.170

Group II ion	Radius/nm
Be^{2+}	0.027
Mg^{2+}	0.072
Ca^{2+}	0.100
Sr^{2+}	0.116
Ba^{2+}	0.136

Solid–phase structure

Cl—Be—Cl

Gas–phase structure

Solid anhydrous beryllium chloride consists of chains of beryllium and chlorine atoms bonded by covalent and coordinate bonds. In the gas phase, at temperatures around 900 °C, the compound consists of linear $BeCl_2$ molecules.

ANOMALOUS BEHAVIOUR OF LITHIUM AND BERYLLIUM

Lithium and beryllium stand at the head of Group I and Group II respectively. Many of the properties of these elements and their compounds are unlike those of the other members of their group. For example, lithium is the only Group I metal able to reduce nitrogen gas, this reaction being typical of Group II metals. Molten beryllium compounds are poor conductors of electricity, indicating that the liquid does not contain a substantial concentration of free ions. The main reason for these anomalous properties is the small size of each ion and hence the tendency towards covalent bonding.

Lithium: charge density

The very small lithium ion has a larger charge density than the sodium ion, so its compounds have a greater tendency towards covalency. The increased covalency is demonstrated by the lower solubility of lithium fluoride in water (spread 16.5), and by the fact that some lithium compounds are soluble in organic solvents. Organic solvents are generally non-polar when compared with the polar solvent water. Ionic solutes dissolve in polar solvents; covalent solutes dissolve in non-polar solvents. Lithium chloride has sufficient covalent character to be soluble in ethoxyethane; lithium iodide dissolves in both methanol and propanone.

The lithium ion is the smallest of the Group I metal ions, and so its ionic compounds have the largest lattice enthalpies. For example, the lattice enthalpy of lithium nitride is sufficient to compensate for the very strongly endothermic step of creating N^{3-} ions from a nitrogen N_2 molecule (containing a strong N≡N triple covalent bond). The reaction is therefore exothermic overall.

Beryllium: charge density

The compounds of beryllium tend to have significant covalent character because the beryllium ion is very small and doubly charged. It has the highest charge-to-radius ratio of all ions, with the exception of the hydrogen ion H^+, and its high charge density makes it highly polarizing. The Be^{2+} ion distorts the electron cloud around an adjacent negatively charged ion. Electron density becomes concentrated between the two ion centres, giving the bond substantial covalent character.

Beryllium: amphoteric character of hydroxide

Beryllium acts as a metal when it reacts with sulphuric acid:

$$Be(s) + H_2SO_4(aq) \rightarrow BeSO_4(aq) + H_2(g)$$

Adding aqueous sodium hydroxide to aqueous beryllium sulphate precipitates beryllium hydroxide:

$$BeSO_4(aq) + 2NaOH(aq) \rightarrow Na_2SO_4(aq) + Be(OH)_2(s)$$

Beryllium hydroxide acts as a base when it reacts with an acid:

$$Be(OH)_2(s) + 2HCl(aq) \rightarrow BeCl_2(aq) + 2H_2O(l)$$

All the behaviour mentioned so far is typical of a Group II metal such as calcium. However, beryllium hydroxide acts as an *acid* when it reacts with aqueous sodium hydroxide:

$$Be(OH)_2(s) + 2NaOH(aq) \rightarrow Na_2Be(OH)_4(aq)$$

Beryllium hydroxide shows both acidic and basic properties; it is therefore said to be **amphoteric**. In this respect it is similar to aluminium hydroxide, which also reacts with both acids and bases. Amphoteric behaviour is typical of the metallic elements closest in position in the periodic table to the non-metals.

Notice the similarities between the properties of lithium and those of magnesium, and between the properties of beryllium and those of aluminium. These *diagonal relationships* are due to the pairs of atoms concerned having similar electronegativities. Remember that electronegativity is a measure of the ability of an atom to attract electron density. Electronegativity increases as atomic number increases across a period in the periodic table; it decreases as atomic number increases down a group.

Because they have significant covalent character, beryllium salts (like lithium salts) are soluble in organic solvents. They also tend to hydrolyse when added to water. For example, beryllium chloride dissolves in water very exothermically. Strong coordinate bonds form between the beryllium ion and four water molecules, so that the beryllium ion exists in solution as a *complex ion* with the formula $[Be(H_2O)_4]^{2+}(aq)$.

Comparisons

- The solubilities of lithium salts often resemble those of magnesium salts rather than those of other Group I elements.
- Anhydrous beryllium chloride is similar to aluminium chloride. Both are deliquescent white solids, i.e. they absorb sufficient water from the atmosphere to form a concentrated solution.

Coordination number of beryllium

Solid beryllium chloride and beryllium sulphate have the chemical formulae $BeCl_2{\cdot}4H_2O$ and $BeSO_4{\cdot}4H_2O$ respectively. The beryllium ion exists as the complex ion $[Be(H_2O)_4]^{2+}$, in which the beryllium ion has a coordination number of 4. Note that all elements of Period 2 have this maximum coordination number (4). Higher coordination numbers involve d orbitals. For the elements of Period 2, the energies of the vacant 3d orbitals are too high compared with the occupied orbitals in the $n = 2$ shell, so no higher coordination numbers are possible.

Oxoanions, oxoacids, and oxosalts

Oxoanions are anions that contain oxygen; common examples are NO_3^- and SO_4^{2-}. **Oxoacids** are acids that form oxoanions; e.g. HNO_3, H_2SO_4 (so HCl and HF are *not* oxoacids). In an oxoacid, the hydrogen atoms are bonded directly to oxygen. **Oxosalts** are salts of oxoacids, so they contain oxoanions; e.g. Li_2CO_3, $CaSO_4$.

Thermal decomposition of lithium carbonate

Unlike the other Group I carbonates, lithium carbonate decomposes when heated. This behaviour is typical of Group II carbonates. The metal ions in the Group I oxides have a charge of 1+; those in the Group II oxides have a charge of 2+. As a result, the Group I oxides have lower lattice enthalpies than the Group II oxides. Group I carbonates are therefore more stable than Group II carbonates because the formation of the Group I oxides is less exothermic.

The lithium ion Li^+ is very small, so lithium oxide is more stable than other Group I oxides. Lithium carbonate is therefore relatively less stable when heated. For the same reasons, other lithium oxosalts (such as $LiNO_3$) decompose more readily than the oxosalts of other Group I metals.

Lithium and nerve impulses

The transmission of impulses along nerves depends on the movement of sodium and potassium ions across the nerve membrane (see spread 13.9). Lithium ions mimic these ions and modify the impulse transmission. Lithium carbonate may be administered to patients with psychiatric disorders (such as bipolar disorder) to help prevent swings of mood.

SUMMARY

- The properties of lithium and its compounds are often similar to those of magnesium (rather than to those of other Group I elements) – this is a 'diagonal relationship'.
- The beryllium ion has the second highest charge density of all ions, so it has great polarizing power. Its compounds with non-metals therefore have a high degree of covalent character.
- The properties of beryllium and its compounds are often similar to those of aluminium (rather than to those of other Group II elements) – this is another 'diagonal relationship'.

PRACTICE

1. Suggest reasons why the beryllium ion Be^{2+} is *so* small compared with the other Group II metal ions. (It is less than *half* the size of the Mg^{2+} ion.)
2. Lithium reacts with atmospheric nitrogen to form an ionic nitride Li_3N. Explain why you would also expect magnesium but not sodium to form a nitride.
3. The sum of the first and second ionization energies of beryllium is $+2660\,kJ\,mol^{-1}$ and of strontium is $+1610\,kJ\,mol^{-1}$. What implications do these figures have for the character of the bonding in the chlorides of these elements?
4. What do you understand by the term 'hydrolyse'? Explain why beryllium salts hydrolyse when added to water, but strontium salts do not.

16.9

OBJECTIVES

- The s-block elements and flame colours
- Biochemical roles of s-block elements

THE S-BLOCK ELEMENTS: FLAME TESTS AND LIVING SYSTEMS

All the common compounds of the alkali metals (the elements of Group I) are soluble. As a result, the metal ions cannot be identified by precipitation. The first part of this spread shows how flame tests may be used to identify these ions, as well as those of calcium, strontium, and barium. The spread concludes by introducing a completely different aspect of the s-block elements: their role in the biochemical reactions that support life. This discussion is continued at greater depth in chapter 30 'Biochemistry'.

Flame test colours

The characteristic colours for the Group I metals are:

- lithium: crimson-red
- sodium: yellow
- potassium: lilac
- rubidium: deep red
- caesium: sky blue

Some of the Group II metals also give flame colours:

- calcium: brick-red
- strontium: crimson
- barium: yellow–green (apple green)

Interference by sodium

The intensity of the colour is much greater for sodium than it is for potassium, so a small amount of sodium impurity can disguise the potassium colour. The presence of the potassium colour can be confirmed by looking at the flame through 'cobalt blue' glass. The blue glass absorbs the yellow sodium light, while allowing the lilac still to be seen.

Flame tests

An s-block element (with the exceptions of beryllium and magnesium) may be identified by a **flame test**. The unknown compound is vaporized in a flame. The electrons in its atoms are promoted into higher orbitals by the energy from the burning gas. The excited atoms then lose energy and undergo an electronic transition to a lower energy level, giving off the extra energy in the form of electromagnetic radiation. In the case of most s-block elements, the frequency of this radiation falls in the visible region. The electronic transitions are different for each element, so an s-block element in the unknown compound may be identified by the colour of the flame.

Neither beryllium nor magnesium colours a flame because their emissions lie outside the visible region of the electromagnetic spectrum.

The procedure for carrying out a flame test involves the following steps:

1. Clean a nichrome or platinum wire by dipping it in concentrated hydrochloric acid and placing it in a non-luminous Bunsen flame.
2. Continue this cleaning process until no colour at all is produced when the wire is in the flame.
3. Moisten the wire with concentrated hydrochloric acid, dip it in the unknown compound, and hold it in the flame again.
4. Check the colour observed against the list on the left or the photos below.

(a)

(b)
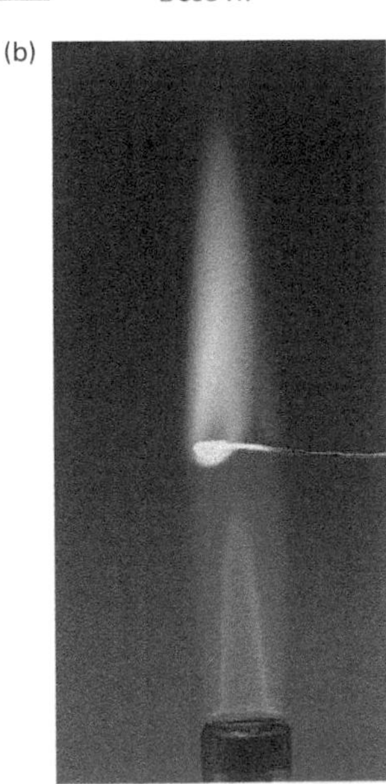
(c)
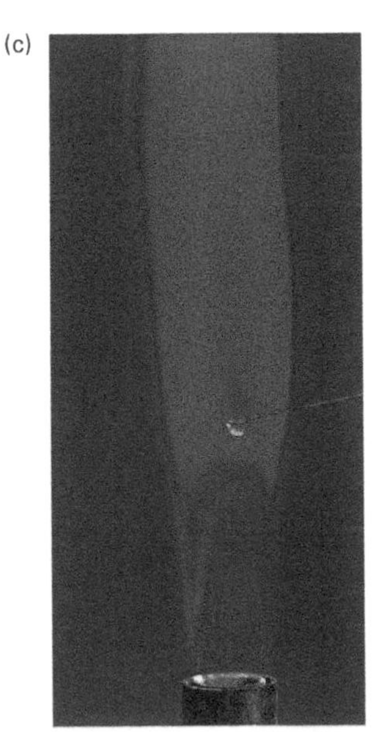
(d)
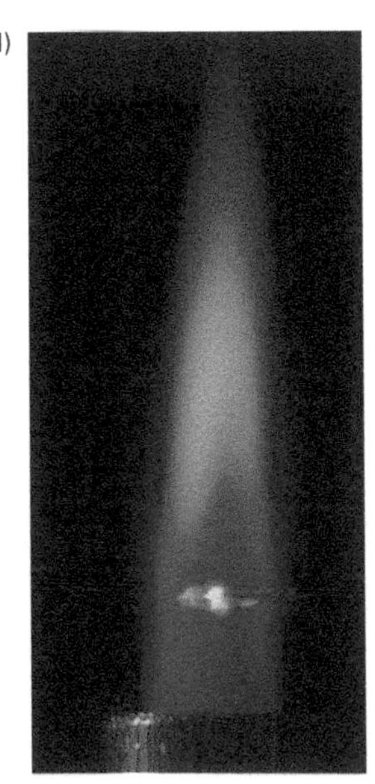

The flame colours for: (a) lithium, (b) potassium, (c) strontium, and (d) barium. Sodium is shown in spread 4.3.

The s-block elements in living systems

The s-block elements are important in biochemistry. In organisms, sodium is the principal cation in the fluid *outside* cells, and potassium is the principal cation in the fluid *inside* cells. The ions of sodium and potassium (together with chloride ion) are fundamental to many biochemical processes, e.g. in maintaining the balance between fluid pressures inside and outside the cells within an organism, in stabilizing certain species such as DNA, and in nerve action; see spread 13.9.

Magnesium is necessary for the stability of many biochemical anions. For example, adenosine triphosphate is the predominant supplier of energy in living cells. Its name is usually abbreviated to ATP, as if it were uncharged. In fact, it should be written as $ATP^{4-}{\cdot}Mg^{2+}$. The magnesium ion decreases the overall charge of the species, and so increases its stability.

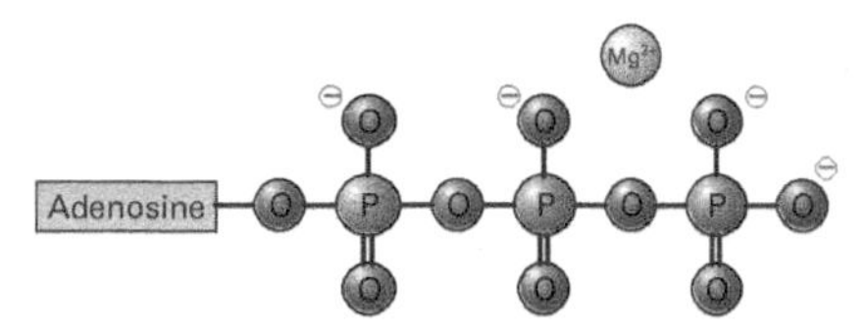

The complex $ATP^{4-}{\cdot}Mg^{2+}$.

Chlorophyll is an exceptionally important (and complex) compound of magnesium. This green pigment takes in energy from sunlight during photosynthesis. This fuels the growth of the plants. The green plants are at the bottom of many food chains, and so are very important sources of energy for the processes of life.

Calcium is also vital for a range of biochemical functions, which include: as a constituent of bones, shells, and teeth; in the transmission of impulses across the synapses where nerves meet; in the control of fertilization, permitting one sperm only to enter the egg; and in muscle contraction.

The concentration of calcium ion varies from 10^{-6} to 10^{-2} mol dm^{-3} in the environment, except in biological systems where it is rigorously controlled by hormones to be very close to 10^{-3} mol dm^{-3}. This concentration allows precipitation of carbonates (mollusc shells), phosphates (bones), and ethanedioates (plant structures). There is little room for error in the control of calcium ion concentration. In people, ageing is accompanied by loss of calcium ions from bone, which progressively loses its structural strength. The arteries may become calcified when the control systems no longer work properly. They become clogged with calcium salts and other debris, diminishing the blood flow.

SUMMARY

- A flame test can identify all s-block elements, with the exception of beryllium and magnesium.
- Electrons move to vacant outer orbitals when metal atoms are excited in the flame; visible light is emitted as the electrons move back to lower orbitals.
- The flow of an impulse along a nerve fibre depends on a change of sodium ion and potassium ion concentration inside the nerve fibre.
- Magnesium is fundamental to the action of chlorophyll and ATP.
- The control of calcium ion concentration is important for human health.

Anions and cations: a reminder

By definition, **anions** are ions that bear a negative charge; **cations** are ions that bear a positive charge.

Fireworks incorporate metal salts with gunpowder to make coloured effects. Which metals could be responsible for each of the colours in these bursting fireworks?

Toxic beryllium

The beryllium ion can replace the magnesium ion in biochemically active molecules. The beryllium ion mimics the magnesium ion, but it is smaller and so has a higher charge density. The ion is very toxic because it binds tightly to a site usually occupied by magnesium and so disrupts biochemical processes.

PRACTICE

1. Explain why the presence of magnesium in a compound cannot be established by a flame test.
2. Calcium ions are present in the cell sap of plants. Suggest why calcium ions do not bind to chlorophyll molecules.
3. Suggest why cadmium and mercury are toxic.

PRACTICE EXAM QUESTIONS

1 **a** The following table shows some physical properties of two s-block metals.

Metal	Hardness	Melting temperature/°C	Density/g cm^{-3}
Caesium	Very soft	28.7	1.9
Barium	Quite hard	714	3.51

i Suggest reasons for the differences in the physical properties of caesium and barium as shown in the table. The metals have the same crystal structure. [3]

ii Caesium gets its name from the blue colour it or its salts impart to a Bunsen flame. What process within the atom is responsible for the emission of this colour? [1]

iii If the light emitted from excited caesium atoms is passed through a spectrometer, what would you expect to see? [1]

b Sodium burns in excess oxygen to give a yellow solid, Y.

i Y contains 58.97% sodium. Find its empirical formula. [2]

ii The relative molecular mass of Y is 78. What is its molecular formula? [1]

iii If Y is reacted with ice-cold dilute sulphuric acid, a solution of Z is obtained which will react with potassium manganate(VII) solution. Describe the experimental procedure you would use to determine the mole ratio in which Z and potassium manganate(VII) react together. [3]

2 Sodium and sodium hydroxide are both manufactured by electrolytic processes.

a Name the electrolyte used in the manufacture of:

i sodium;

ii sodium hydroxide. [2]

b **i** What is produced at the anode during the manufacture of sodium hydroxide? Write an equation for its formation.

ii What other gaseous product might be given off at the anode under other conditions? Write an equation for its formation. [5]

c Suggest a reason why the product in **b i** is formed in the industrial process rather than that in **b ii**. [3]

d Describe what you would observe when dilute sodium hydroxide solution is added dropwise to a solution of aluminium sulphate until in excess. Give the formulae of the aluminium-containing species present in the original solution, and responsible for the observations you have described. [6]

(See also chapter 19.)

3 **a** **i** Complete the electronic configuration of a calcium atom: $1s^2$.... [1]

ii Describe the bonding present in solid calcium. [4]

b **i** Write an equation for the reaction of calcium with water. Identify the oxidation numbers of all the atoms involved by writing the numbers underneath each symbol in the equation. [2]

ii State which substance in **b i** has been oxidized and write a half-equation to show the oxidation process. [2]

iii What is the common name of the solution formed by the reaction of calcium with water? Suggest the likely pH of the solution. [2]

c What would be **observed** if aqueous sodium hydroxide were added dropwise, until in excess, to aqueous solutions of:

i magnesium chloride;

ii barium chloride? [2]

d Account for the observations in **c**, giving any relevant ionic equations. [3]

4 Magnesium oxide, MgO, is a white solid with a high melting temperature which is used as a furnace lining.

i State **two** ways in which magnesium oxide can be obtained giving a balanced equation in each case. [4]

ii Using **outer** electrons only, draw a dot-and-cross diagram showing the bonding in magnesium oxide. [2]

iii State why magnesium oxide is described as a basic oxide. [1]

iv One industrial method for obtaining magnesium from magnesium oxide involves heating magnesium oxide with silicon at a high temperature in the absence of air.

$$2MgO + Si \rightarrow 2Mg + SiO_2$$

I. State why the process must be carried out in the absence of air. [1]

II. Calculate the mass of silicon required to convert completely 500 kg of magnesium oxide into magnesium. [3]

5 **a** Magnesium occurs naturally as the mineral *carnallite*, $KCl{\cdot}MgCl_2{\cdot}6H_2O$.

i State what is **observed** and give a balanced equation for the reaction which occurs when a solution of carnallite is treated with sodium hydroxide solution. [2]

ii State how to test for the presence of chloride ions in the carnallite solution, giving details of the reagents added, **observation**, and an **ionic** equation for any precipitation reaction which may occur. [3]

b Both magnesium sulphate and barium sulphate occur naturally as minerals but only magnesium sulphate is soluble in water.

i Explain, in terms of hydration and lattice enthalpies, the reason why only one of the compounds is soluble in water. [2]

ii Name **two** features of magnesium sulphate which identify it as a **typical ionic compound**. [1]

c State why magnesium is an essential element for plant growth. [1]

(See also chapter 18.)

6 a i Write an equation for the reaction of barium with water. [1]

ii Would the reaction in **a i** occur more vigorously or less vigorously than the reaction of calcium with water? Identify one contributory factor and use it to justify your answer. [2]

iii Write an equation for the action of heat on solid barium carbonate. [1]

iv At a given high temperature which of the two carbonates, barium carbonate or calcium carbonate, would decompose more easily? [1]

v How would you distinguish between solutions of barium chloride and calcium chloride? State in each case what you would see as a result of the test on each solution. [2]

b 1.71 g of barium reacts with oxygen to form 2.11 g of an oxide **X**.

i Calculate the formula of **X**. [2]

ii Give the formula of the anion present in **X**. [1]

iii What is the oxidation number of oxygen in this anion? [1]

iv Sodium forms an oxide, **Y**, which contains this same anion. Give the formula of **Y**. [1]

c Treatment of either **X** or **Y** with dilute sulphuric acid leads to the formation of the sulphate of the metal, together with an aqueous solution of hydrogen peroxide, H_2O_2.

i Write an equation for the reaction of **Y** with dilute sulphuric acid. [1]

ii The hydrogen peroxide solution produced may be separated from the other reaction product. Explain briefly why this is easier to achieve if **X** is used as the initial reagent rather than **Y**. [2]

7 a i Using the data provided, construct a Born-Haber cycle for magnesium chloride, $MgCl_2$, and from it determine the electron affinity of chlorine. [5]

	ΔH /kJ mol^{-1}
Enthalpy of atomisation of chlorine	+122
Enthalpy of atomisation of magnesium	+148
First ionization energy of magnesium	+738
Second ionization energy of magnesium	+1451
Lattice enthalpy of magnesium chloride	−2526
Enthalpy of formation of magnesium chloride	−641

ii The theoretically calculated value for the lattice enthalpy of magnesium chloride is −2326 kJ mol^{-1}. Explain the difference between the theoretically calculated value and the experimental value given in data in **a i**, in terms of the bonding of magnesium chloride. [3]

b The table below gives some information about the sulphates of elements in Group 2.

Sulphate	Solubility /mol dm^{-3}	Lattice enthalpy /kJ mol^{-1}	Hydration enthalpy of M^{2+}/kJ mol^{-1}
$CaSO_4$	4.6×10^{-2}	−2480	−1650
$SrSO_4$	7.1×10^{-2}	−2484	−1480
$BaSO_4$	9.4×10^{-2}	−2374	−1360

i Suggest an explanation for the trend shown in the hydration enthalpies of the cations. [2]

ii Comment on the trend in the solubilities of these sulphates in relation to the lattice and hydration enthalpies given in the table. [4]

iii Barium sulphate, which is opaque to X-rays, is used for the 'barium meal' to enable X-ray pictures to be taken of the gut. Barium ions are very toxic; why is this not a problem here? [1]

iv Give the equation for the reaction of barium with cold water. [2]

v Suggest the practical procedure by which you might convert the solution of the product in the reaction in **iv** into a reasonably pure sample of barium sulphate. [3]

8 a i Describe what would be seen if dilute sodium hydroxide solution was added, until in excess, to aqueous solutions of magnesium nitrate and barium nitrate. [2]

ii Account for the observations given in **i** and write any relevant ionic equations. [3]

b i State, and explain, the trend in the thermal stability of Group 2 carbonates. [3]

ii Suggest, with a reason, how the thermal stability of sodium carbonate differs from that of magnesium carbonate. [2]

c One of the industrial processes for the manufacture of magnesium is similar in principle to that used for the manufacture of sodium.

i What type of process is used for the manufacture of sodium? [1]

ii Suggest which compound of magnesium could be used in its manufacture by a similar process and give its probable source. [2]

iii State the essential condition for this process, and write an ionic equation to represent the formation of magnesium. [2]

iv State the name of the other product of the process and give one of its uses. [2]

d i When LiCl is heated in a bunsen flame a characteristic red flame is seen. Explain why lithium compounds produce a coloured flame in these circumstances. [2]

ii Lithium chloride can be made by burning lithium in chlorine. Give a reason why rubidium chloride is not normally prepared in the laboratory in a similar way. [1]

17 Trends across a period

The periodic table is made up from elements arranged in vertical groups and horizontal periods. Look at successive elements as atomic number increases down a group. You will see that the number of valence electrons remains constant, while the number of filled inner shells increases. Now look at successive elements across a period. The filled inner shells of electrons remain constant, while the number of valence electrons steadily increases. The previous chapter considered the elements lithium to caesium and beryllium to barium arranged vertically in Groups I and II. In this chapter we consider the elements sodium to argon arranged horizontally in Period 3. We discuss their properties and the trends in those properties, explaining them in terms of the electronic structures of the elements concerned.

PERIOD 3: SODIUM TO ARGON

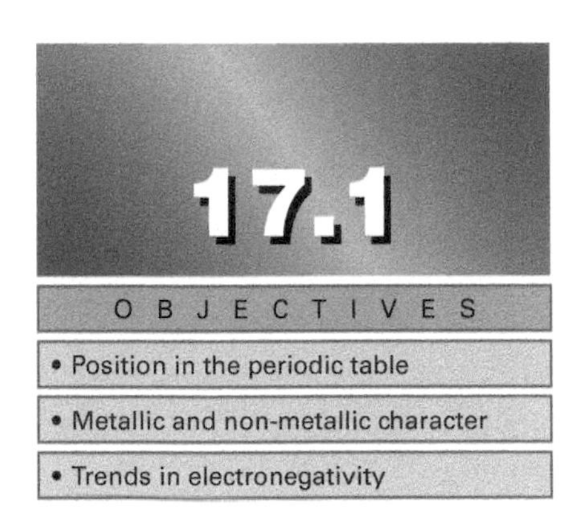

The elements sodium to argon are in Period 3 of the periodic table. They represent the most straightforward unbroken trend in properties across the periodic table. By comparison, Period 2 – the elements lithium to neon – includes elements that stand at the heads of their respective groups. As you will know from the previous chapter, these elements often have anomalous properties when compared with the other members of their group. Period 4 – the elements potassium to krypton – is interrupted by the d-block elements scandium to zinc. So in this spread we look at Period 3 in the context of the periodic table as a whole, and at the trend in the metallic/non-metallic nature of the elements.

Period 3 in the periodic table

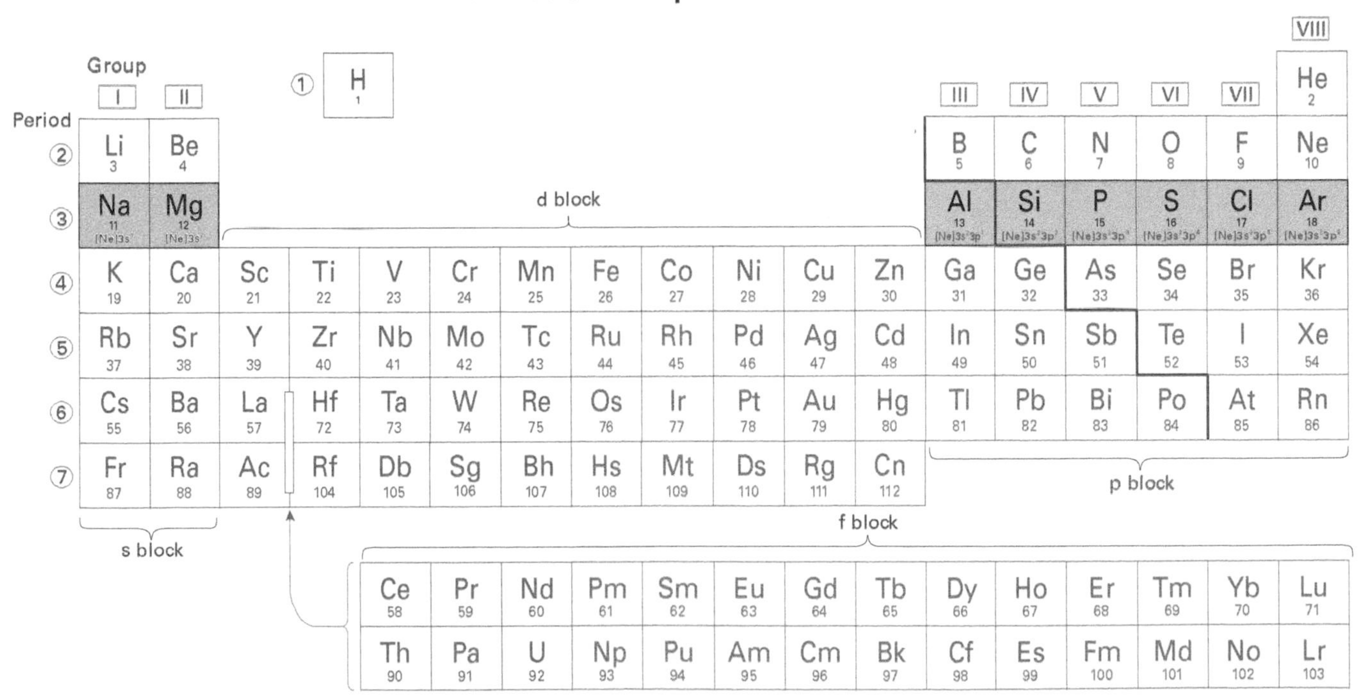

The elements of Period 3 in the periodic table. Note the successive filling of the 3s and 3p orbitals.

There are eight elements in Period 3, from sodium (atomic number $Z = 11$) to argon ($Z = 18$). All the elements have filled inner electron shells corresponding to the noble gas neon ([Ne] = $1s^2 2s^2 2p^6$). The two elements sodium and magnesium are members of the s block, with the electronic structures Na [Ne]$3s^1$ and Mg [Ne]$3s^2$ respectively. The six elements from aluminium to argon are members of the p block, where the three 3p orbitals fill, as shown above.

Metallic and non-metallic character

Metals are good electrical conductors and they are malleable and ductile. This behaviour is due to the presence of delocalized electrons in their solid structure. Metals tend to lose electrons during chemical reactions. Electronegativity is a measure of an element's ability to attract a shared electron pair to itself. Typical metals have electronegativities smaller than 1.7; typical non-metals have electronegativities greater than 2.4.

Electrical conductivity, electronegativity, and malleability data for the elements of Period 3.

Element	Electrical conductivity*	Electronegativity (Pauling scale)	Malleability
Sodium	37	0.9	good
Magnesium	38	1.3	good
Aluminium	64	1.6	good
Silicon	16	1.9	brittle
Phosphorus	10^{-16}	2.2	brittle
Sulphur	10^{-22}	2.6	brittle
Chlorine	negligible	3.2	–
Argon	negligible	–	–

* Relative scale based on Cu = 100.

The elements sodium (Na), magnesium (Mg), and aluminium (Al) are metals. They have conductivities at least one-third that of copper, are malleable, and have electronegativities between 0.9 and 1.6. The elements phosphorus (P), sulphur (S), chlorine (Cl), and argon (Ar) are non-metals. They have extremely low or negligible conductivities, the solid elements are brittle, and their electronegativities range from 2.2 to 3.2. The element silicon (Si) is a metalloid. It is a semiconductor with intermediate conductivity, its electronegativity is 1.9 (less than the values for the unreactive metals tin and lead), yet it is brittle, indicating a non-metallic structure.

The general trend is from metallic to non-metallic across the period. The following two spreads will look at trends in other physical properties of these elements.

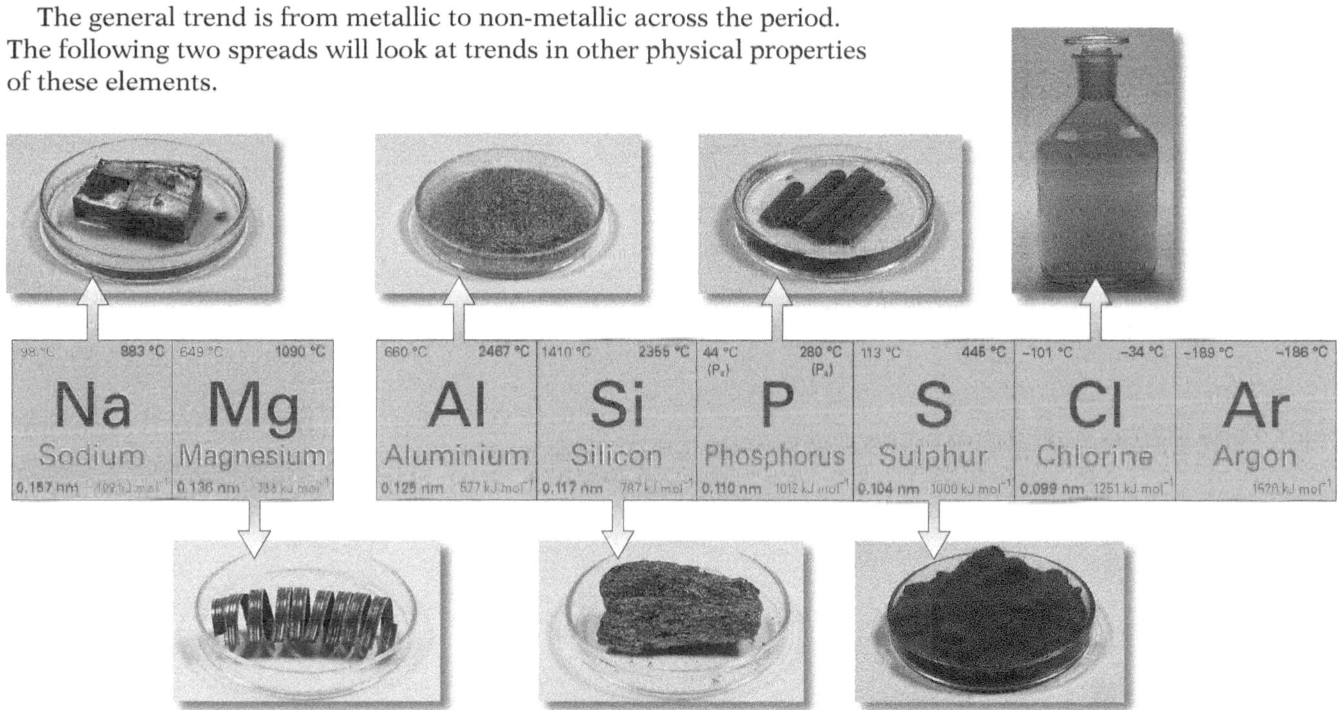

A summary of the major properties of the elements of Period 3 (the elements sodium to argon). The noble gas argon is a colourless gas.

SUMMARY

- As the atomic number of the elements increases across a period, the trend is from metallic to non-metallic character.
- Metals have electronegativities smaller than 1.7; non-metals have electronegativities greater than 2.4.

PRACTICE

1 Sketch an outline periodic table. Add a horizontal arrow above the table, pointing to the right. Add a vertical arrow to the left of the table, pointing downwards. Label each arrow twice to show overall trends in:

  a electronegativity,

  b metallic or non-metallic character.

2 Using information from this spread only (*don't* cheat!), suggest electronegativities for the elements caesium, barium, fluorine, and carbon. Comment on the metallic/non-metallic character of each.

17.2

MELTING POINTS, BONDING, AND STRUCTURE

OBJECTIVES

- The trend in melting point
- Bonding in the elements
- Some uses of the elements

Na, Mg, Al – facts and uses

Sodium

Liquid sodium is an ideal heat-exchange medium, because it has a high thermal conductivity and a high boiling point (883 °C). It is also little affected by radiation (it has a low neutron capture cross-section). For these reasons, liquid sodium is used to cool the small compact cores of 'fast breeder' nuclear reactors.

Magnesium

Low density alloys for the aircraft industry include mixtures of magnesium with zirconium and thorium. Magnesium is also used in the nuclear industry. Mixed with 0.8% aluminium and 0.5% beryllium to make the alloy 'Magnox', it is used to contain the uranium fuel in gas-cooled nuclear reactors.

Aluminium

The low density of aluminium makes it one of the metals most used for construction. Its malleability means that items with complex cross-sections may be made by extrusion: heated to about 200 °C, the metal is forced through a shaped die, just like toothpaste squeezed from a tube. Its low density and high strength mean that it can be used to construct aircraft and the overhead electricity cables slung from pylons.

Semiconductors

At temperatures above absolute zero, thermal energy causes some of the four valence electrons of silicon to become delocalized and move freely through the lattice. These electrons impart some conductivity to the silicon. So silicon (Group IV, Period 3) is a semiconductor: its electrical conductivity is intermediate between those of metals and those of non-metals. The conductivity can be altered and controlled by adding impurity atoms that contain three or five valence electrons. The 'silicon chips' so important in the electronic circuitry of computers, radios, and TVs rely on semiconduction for their function. Note that germanium (Group IV, Period 4) is also a semiconductor, as is gallium arsenide, which is composed of the two elements either side of germanium in Period 4.

The melting point of an element gives a direct indication of the forces between the particles that make up the solid element. A high melting point indicates metallic bonding or the presence of a giant covalent structure. A lower melting point is associated with covalently bonded molecules held in a solid lattice by weak dispersion forces. An extremely low melting point suggests a structure consisting of separate *atoms* held together by dispersion forces only. All these structural types are encountered in the elements of Period 3.

Melting points: sodium to argon

The graph of melting point (T_m) against increasing atomic number shows a distinct overall shape. There is a steady upward trend (sodium to silicon), followed by a sharp fall (silicon to phosphorus), and then an overall more gradual decline (phosphorus to argon). You should be able to identify the elements sodium, magnesium, and aluminium as metals; silicon is a metalloid; phosphorus, sulphur, chlorine, and argon are non-metals. Note also that the four non-metals fall into two groups: phosphorus and sulphur have melting points significantly greater than those of chlorine and argon (both of which are gases at 20 °C and 1 bar). These eight elements between them represent distinctly different types of structure.

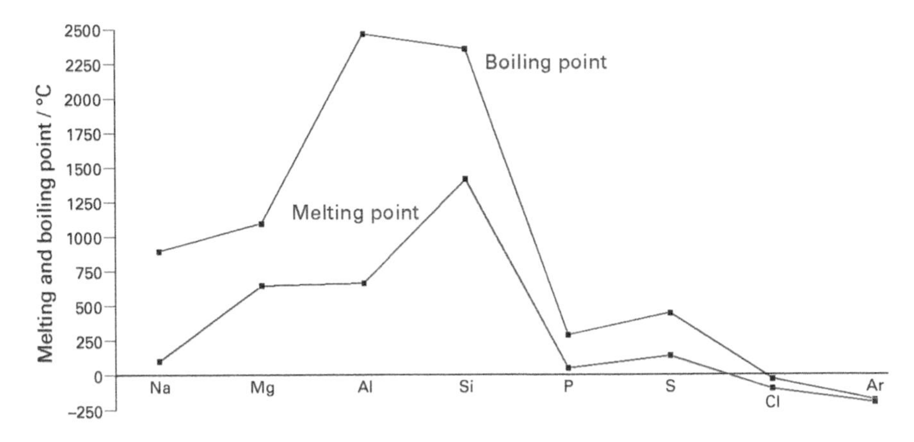

Trends in the melting and boiling points of the Period 3 elements.

The metals sodium, magnesium, and aluminium

The three elements sodium, magnesium, and aluminium have metallic structures consisting of a regular lattice of metal ions (Na^+, Mg^{2+}, and Al^{3+} respectively). These ions are surrounded by delocalized valence electrons, which come from the one, two, and three valence electrons per atom respectively.

The metalloid silicon

The atoms in solid silicon are bonded together covalently, with the atoms arranged in the diamond structure (a giant covalent structure – see chapter 6 'Solids'). Each silicon atom is bonded to four other silicon atoms; to melt silicon requires breaking all these covalent bonds. This needs a large amount of energy, so silicon has the highest melting point in Period 3. Silicon has a melting point of 1410 °C.

Phosphorus

Phosphorus exists in several crystalline forms (**allotropes**), the two most important of which are named according to their colour. Red phosphorus (T_m = 590 °C) consists of chains of phosphorus atoms. White phosphorus (T_m = 44 °C) consists of individual P_4 molecules. The bonds within these small tetrahedral molecules are very strained, which makes white phosphorus very reactive. White phosphorus must be stored under water because it ignites spontaneously in air above 35 °C.

(a)

(b)

(a) The structure of white phosphorus.
(b) White and red phosphorus.

Sulphur

There are two allotropes of sulphur: rhombic sulphur (T_m = 113 °C) and monoclinic sulphur (T_m = 119 °C). Both are composed of S_8 crown-shaped molecules and differ only in the arrangement in which the molecules are packed; see spread 6.7. At temperatures slightly above the melting point, liquid sulphur consists of separate S_8 crown-shaped ring molecules. The dispersion forces holding S_8 molecules together are stronger than those holding P_4 molecules together because there is virtually double the number of electrons. As a result, sulphur melts at a higher temperature than phosphorus. Sulphur occurs as the element in volcanic areas, including those found on Io, one of the moons of the planet Jupiter.

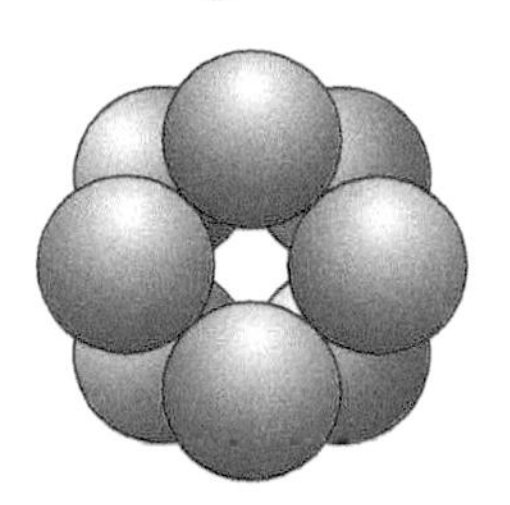

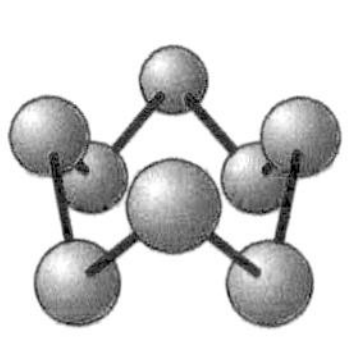

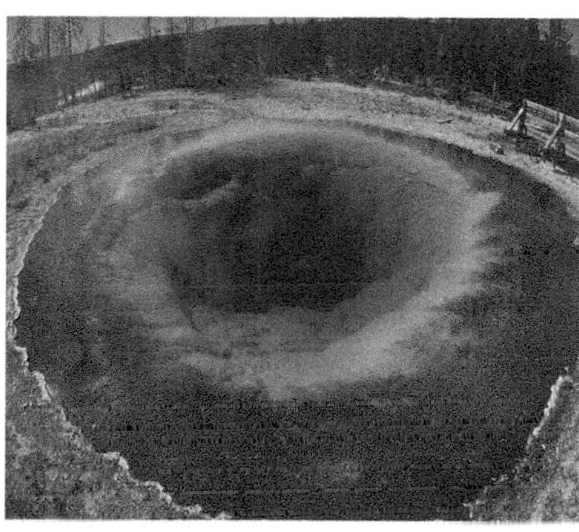

(a) The S_8 molecule. (b) Sulphur pool in Yellowstone National Park.

Chlorine

Solid chlorine consists of covalently bonded Cl_2 molecules held in a regular lattice by dispersion forces. The melting point is low because these dispersion forces are weak. Liquid and gaseous chlorine consist of separate Cl_2 molecules.

Argon

Solid argon consists of separate atoms held in a regular lattice by dispersion forces. These forces are extremely weak, so argon has a very low melting point.

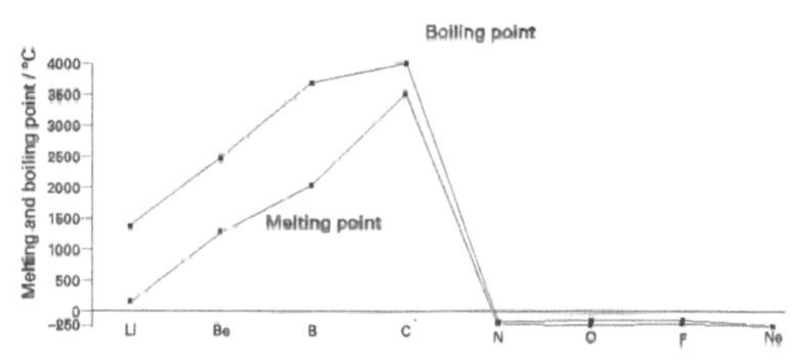

Trends in the melting points and boiling points of the Period 2 elements are very similar to the Period 3 trends, except that the values for nitrogen, oxygen, and fluorine are almost identical as they are all diatomic (N_2, O_2, F_2).

SUMMARY

- The melting points across Period 3 rise with increasing atomic number until silicon, after which they fall dramatically.
- Sodium, magnesium, and aluminium have metallic structures bonded by delocalized valence electrons.
- Silicon has a giant covalent structure.
- P_4, S_8, and Cl_2 are simple covalent molecules.
- The individual atoms in solid Ar are held together by dispersion forces only.

PRACTICE

1 Many of the ideas in this spread were covered in chapter 6 'Solids'. Refer to that chapter and make sketches that illustrate the structures of the following:
  a Magnesium
  b Silicon
  c Solid chlorine (include dispersion forces in your sketch)
  d Solid argon.

2 Explain why the bonds between the phosphorus atoms in the P_4 molecule are said to be 'strained'.

3 Explain the difference between the melting points of red and white phosphorus in terms of structure and bonding.

4 Suggest why the melting point of sodium is so much lower than those of magnesium and aluminium.

17.3

OBJECTIVES

- Trends in atomic radius
- Trends in ionization energy
- Trends in oxidation number

EFFECTS OF ATOMIC SIZE

Shielding

Each successive element across a period contains one more proton and electron. The extra electron might be expected to **shield** (cancel out the attraction of) the extra proton. This shielding is only *partially* successful; electron density is smeared out (as explained in chapter 4 'Electrons in atoms'), whereas the protons are definitely located in the nucleus.

This lack of perfect shielding means that the **effective nuclear charge** experienced by an electron increases across a period; the increasing nuclear charge outweighs the effect of an extra electron *in the same shell.*

Slater's rules

The effective nuclear charge experienced by an electron is given by $Z_{eff} = Z - S$, where Z is the number of protons in the atom and S is the **shielding constant**, which allows for the repulsive effect of all the other electrons. Rules introduced by John Slater allow approximate values for Z_{eff} to be found simply. Other electrons in the outer (n) shell contribute only 0.35 to S as they shield poorly. Electrons in the ($n - 1$) shell shield much more effectively (contributing 0.85), while electrons in shells even closer to the nucleus shield almost perfectly (contributing 1). Hence for Na $Z_{eff} = 11 - (8 \times 0.85) - 2 = 2.2$ and for Cl $Z_{eff} = 17 - (6 \times 0.35) - (8 \times 0.85) - 2 = 6.1$. This significant rise in the effective nuclear charge across the period, caused by the poor shielding of the electrons *in the same shell*, makes the ionization energy much larger for chlorine.

More exact values calculated by Enrico Clementi (and D. Raimondi) give 2.51 for Na and 6.12 for Cl, which confirms that the values from Slater's rules are reasonable.

First ionization energy

The first ionization energy is the minimum energy required to remove one electron from an isolated atom in the gas phase, represented for a general element E as:

$E(g) \rightarrow E^+(g) + e^-(g)$

(Values are usually quoted per mole.)

The value of the atomic radius of an element gives a measure of its size. The size of an atom has an influence on its ionization energy, which is the minimum energy required to remove one or more of the outermost electrons. In turn, the ionization energy influences the valency of an element expressed as the oxidation number. The values of these three interlinked attributes – atomic radius, ionization energy, and oxidation number – all show clear trends as the atomic numbers of the elements increase across a period.

Trends in atomic radius

In each period of the periodic table, the Group I metal is the element with the largest atomic radius. Each Group I metal has one valence electron outside filled inner shells which partially shield that electron from the nuclear charge. In Periods 2 and 3, the atomic radius decreases steadily as atomic number increases from Group I to Group VII. The steady increase in nuclear charge pulls all electrons closer to the nucleus. In Period 4, the trend seen in the previous two periods is less smooth because of the d-block elements, in which the *inner* 3d subshell is filling.

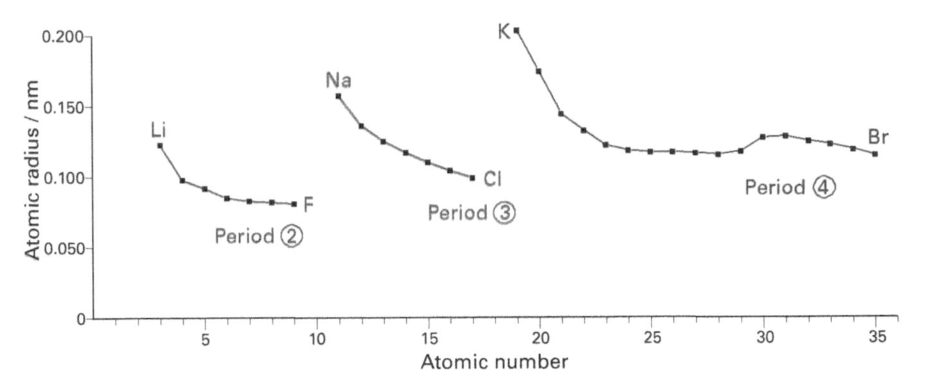

Plots of atomic radius against atomic number for Periods 2, 3, and 4.

First ionization energy

Atomic size decreases across a period as atomic number increases. It is therefore reasonable to expect the first ionization energy to increase as the valence electron becomes closer to the nucleus. While this is generally true, there are two other points of interest to note: 'dips' at Groups III and VI.

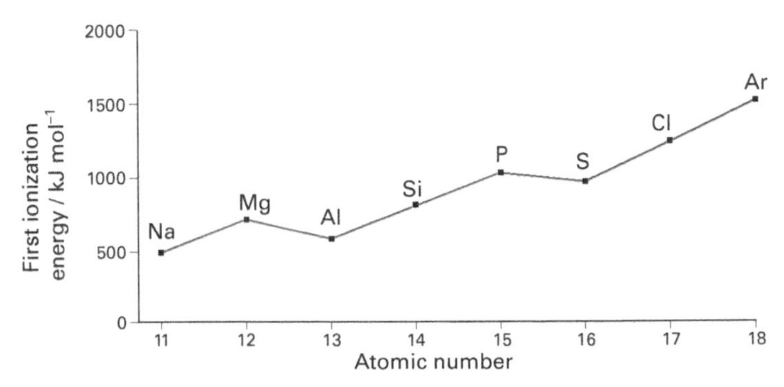

Plot of first ionization energy against atomic number for Z = 11 (Na) to Z = 18 (Ar). The plot shows that there is a periodic variation in ionization energy across Period 3.

The Group III/Group VI 'dips'

In Periods 2 and 3, there is a small dip in the plot at the Group III elements and at the Group VI elements. The dip at beryllium/boron (in Period 2) provides evidence for electron subshells (see chapter 4 'Electrons in atoms'). In Period 3, the first ionization energies of aluminium and sulphur are lower than expected. In the case of the pair of elements Mg/Al, the electron that is removed in Al is from the 3p subshell, which is further from the nucleus than the 3s electron that is removed from Mg. It therefore requires less energy to remove this electron during ionization. In the case of the pair of elements P/S, the 3p electrons in phosphorus all occupy separate orbitals. The fourth p electron in sulphur must enter one of these orbitals, resulting in increased electron–electron repulsion and a lower ionization energy than otherwise expected.

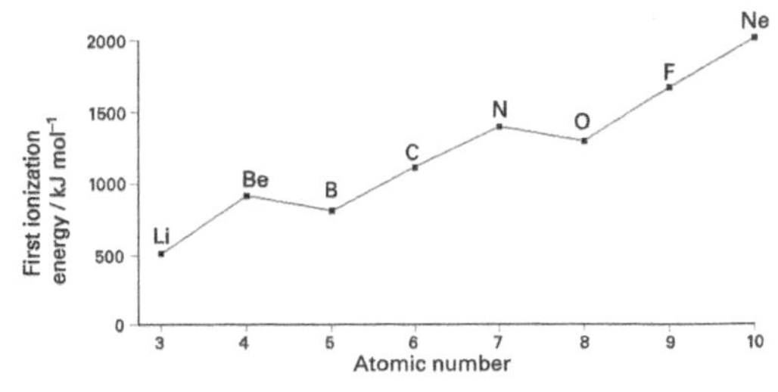

A very similar periodic variation in ionization energy is found across Period 2.

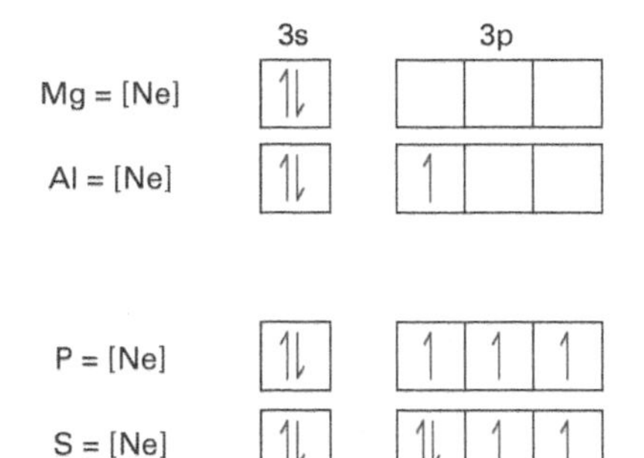

The electronic structures of Mg, Al, P, and S.

The d-block elements

There is a smaller overall change in ionization energy across the d-block elements than through the s-block and p-block elements. This is because d-block elements lose an outer s electron during ionization, rather than an electron from the inner incomplete d subshell. For example, the electronic structure of scandium is $[Ar]3d^1 4s^2$ and that of nickel is $[Ar]3d^8 4s^2$. Both elements lose a 4s electron of similar energy during ionization, so the ionization energies are similar.

Oxidation numbers

There is a general increase in first ionization energy as atomic number increases across Period 3. The relatively low values for the metals sodium, magnesium, and aluminium indicate that a positive oxidation number can be expected. In the s block, sodium (Group I) loses its single outer s electron and exhibits an oxidation number of +1 in all its compounds. Magnesium (Group II) has an oxidation number of +2 only, corresponding to the loss of both valence electrons. The p-block elements have a greater range of oxidation numbers.

The highest oxidation numbers are most likely to be found in fluorides, oxides, or oxoanions, because fluorine and oxygen are the most powerful oxidants. In Period 3, the highest oxidation number increases smoothly from +1 for sodium to +7 for chlorine, corresponding to the element concerned losing or sharing *all* of the electrons in its valence shell when forming a compound. However, ionization energy increases across the period, so the non-metals are much less likely to attain positive oxidation numbers than the s-block metals. In fact, the elements from silicon to chlorine exhibit *negative* oxidation numbers, which means that they *gain* rather than lose electrons to make up their octet. The trends in oxidation number exert a strong influence over the trends in chemical properties discussed later.

Filling d orbitals

Period 4 includes the elements scandium to zinc (10 elements) and Period 5 includes the elements yttrium to cadmium (10 elements). The five 3d orbitals fill in the course of Period 4 and the five 4d orbitals fill in Period 5.

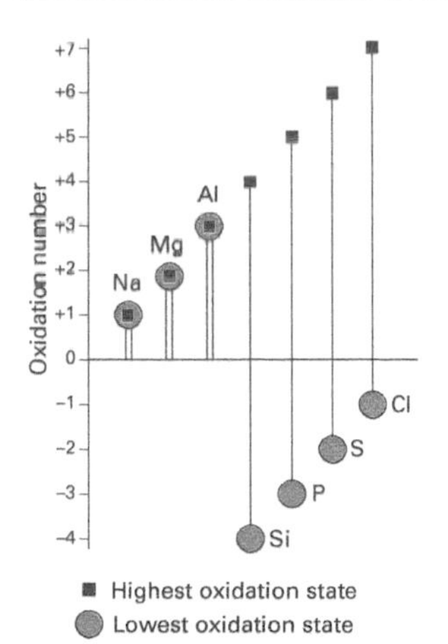

The oxidation numbers of the elements Na to Cl.

SUMMARY

- Within any period, atomic radii, ionization energies, and oxidation numbers all show a periodic change in their values.
- With increasing atomic number across Periods 2 and 3, atomic radii decrease, ionization energies increase (with small dips at Groups III and VI), and maximum oxidation number increases.
- The trends noted in Periods 2 and 3 (s- and p-block elements only) are repeated in later periods with the inclusion of a break corresponding to the d-block elements.
- The highest possible oxidation number of an element in its compounds is equal to the number of its valence electrons.

PRACTICE

1 Sketch the plots for atomic radius against atomic number for Periods 2, 3, and 4. Add an outline of the plot you would expect for Period 5. Label the regions occupied by s-, p-, and d-block elements.

2 The plot of first ionization energy against atomic number shows a series of peaks and troughs.

a Explain why Group I elements are found at the troughs and Group VIII elements at the peaks.

b Explain the trend in the ionization energies of the noble gases.

3 Write down the electronic structures of magnesium, aluminium, sulphur, and phosphorus. In each case, identify the electron lost during the process of ionization, represented in general as:

$E(g) \rightarrow E^+(g) + e^-(g)$

4 Construct an oxidation number diagram for the elements of Period 2, like the one given above for Period 3.

17.4

REACTIONS WITH OXYGEN

OBJECTIVES

- The oxides of Period 3
- Melting point and structure of oxides

White phosphorus spontaneously ignites in air at temperatures above 35 °C. Such substances are said to be ***pyrophoric.*** *A piece of white phosphorus also spontaneously ignites when placed in oxygen.*

Fighting fires

The pyrophoric property of white phosphorus means it has to be stored under water. Water is also the best way to fight a phosphorus fire (but completely disastrous on burning sodium, spread 16.3). You need to know your chemistry when fighting chemical fires!

Oxidation numbers

The formulae of the oxides show a smooth increase in *highest* oxidation number as atomic number increases across the period.

Na_2O	+1
MgO	+2
Al_2O_3	+3
SiO_2	+4
P_4O_{10}	+5
SO_3	+6
Cl_2O_7	+7

The *highest* oxidation number is equal to the number of valence electrons in the element, i.e. the maximum oxidation number is the same as the group number.

It was this clear trend that convinced Mendeleyev that his periodic table would prove useful.

All the Period 3 elements except silicon and chlorine react directly with oxygen to form oxides. There is a full range of oxides representing all the common oxidation states of the elements concerned. As you might expect, the properties of the oxides change steadily across the period as atomic number increases. This spread interprets the melting points of the oxides in terms of their bonding and structure; the next looks at their acid–base character.

The oxides of Period 3

The oxides of the elements of Period 3 have the following names and chemical formulae. Note that phosphorus, sulphur, and chlorine form more than one oxide, so more than one oxidation number is possible.

Oxide name	Formula	Oxidation number of the Period 3 element
Sodium oxide	Na_2O	+1
Magnesium oxide	MgO	+2
Aluminium oxide	Al_2O_3	+3
Silicon dioxide	SiO_2	+4
Phosphorus(III) oxide	P_4O_6	+3
Phosphorus(V) oxide	P_4O_{10}	+5
Sulphur dioxide	SO_2	+4
Sulphur trioxide	SO_3	+6
Dichlorine oxide	Cl_2O	+1
Chlorine dioxide	ClO_2	+4
Dichlorine heptaoxide	Cl_2O_7	+7

Direct combination of the elements

Sodium (Na), magnesium (Mg), aluminium (Al), phosphorus (P), and sulphur (S) all burn in oxygen to form an oxide. The balanced chemical equations are as follows:

$$4Na(s) + O_2(g) \rightarrow 2Na_2O(s)$$

$$2Mg(s) + O_2(g) \rightarrow 2MgO(s)$$

$$4Al(s) + 3O_2(g) \rightarrow 2Al_2O_3(s)$$

$$P_4(s) + 5O_2(g) \rightarrow P_4O_{10}(s)$$

$$S_8(s) + 8O_2(g) \rightarrow 8SO_2(g)$$

Melting points of the oxides

The plot of the melting point of the oxides of the elements Na to Cl shows a sharp initial rise followed by a steady decline; the values for P_4O_{10}, SO_3, and Cl_2O_7 are significantly lower. The overall shape of this plot indicates that there is a trend in the bonding and structure of the oxides. Remember that high melting points are associated with ionic or giant covalent structures, and low melting points with solids consisting of simple covalent molecules.

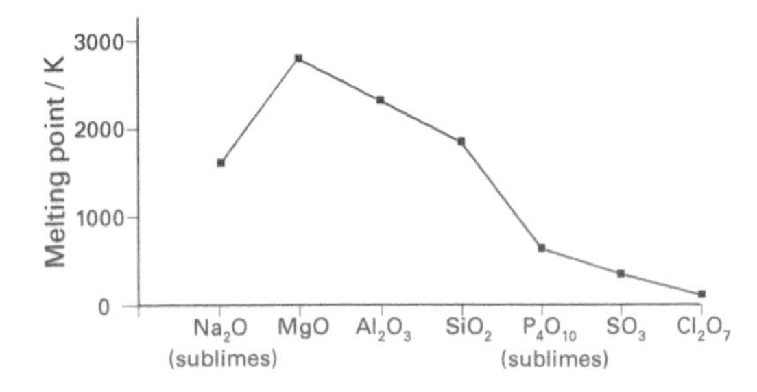

Plot of melting point for the oxides of the elements of Period 3. Note that, where an element has more than one oxide, the oxide corresponding to the highest oxidation number (as shown in the box on the left) is given.

Sodium, magnesium, and aluminium oxides

The oxides of metals are predominantly ionic, with the extent of covalent character increasing from sodium to aluminium, in line with the increasing charge density of the ions Na^+, Mg^{2+}, and Al^{3+}. The melting points are all relatively high. The very high value for MgO explains its use as a refractory lining.

Silicon dioxide (silica)

The oxide of the metalloid silicon is essentially covalent. Silicon dioxide SiO_2 is called *silica*, and it has a giant covalent structure, as illustrated in chapter 6 'Solids'. The formula SiO_2 gives the empirical formula of the substance, i.e. the simplest whole-number ratio of the atoms. Yellow sand is silica with impurities such as iron(III) oxide.

*Quartz is a mineral form of silica that crystallizes as molten rocks solidify underground. Quartz is **piezoelectric,** generating a potential difference when a stress is applied; this property explains its use in oscillators.*

The oxides of phosphorus

The oxides of phosphorus may be treated as molecules of formulae P_4O_6 and P_4O_{10}. Because the molecules are quite large, the dispersion forces are sufficiently strong for both oxides to exist as solids at room temperature. (Remember that dispersion forces are approximately proportional to the number of electrons in the molecule.)

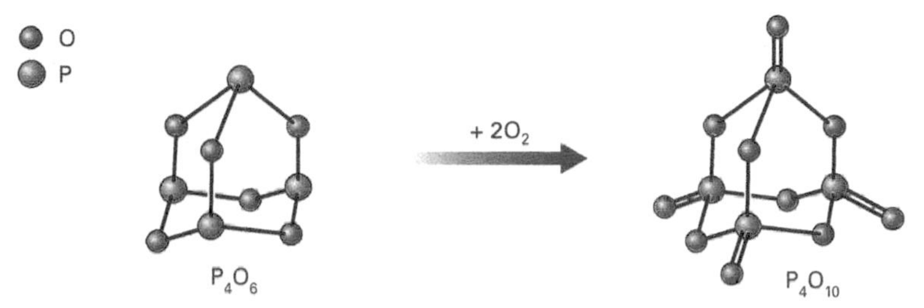

The structures of the molecules P_4O_6 and P_4O_{10}.

The oxides of sulphur

Sulphur dioxide SO_2 exists as individual molecules. The dipole–dipole and dispersion forces between SO_2 molecules are weak, so sulphur dioxide is a gas at room temperature. Sulphur trioxide SO_3 is composed mainly of rings of three molecules and chains whose length is not fixed. Below 17 °C, sulphur trioxide exists as a solid held together by the dispersion forces between the rings. A molecule of formula S_3O_9 would be expected to have a melting point between that of P_4O_{10} and that of Cl_2O_7: see graph opposite.

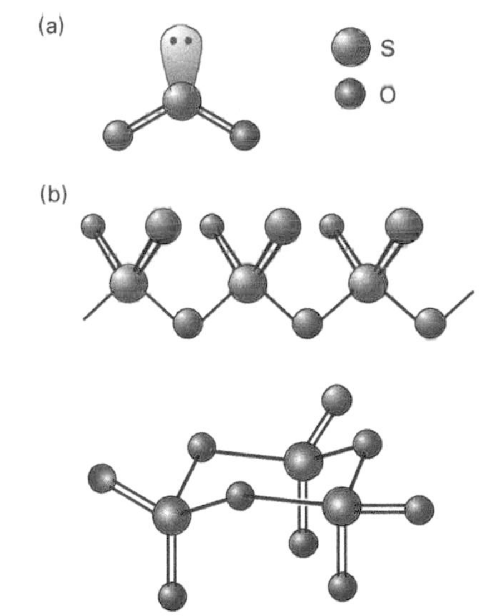

(a) The structure of an SO_2 molecule. (b) SO_3 molecules join together to make chains and rings. Sulphur uses its d orbitals to expand its valence shell.

The oxides of chlorine

The oxides of chlorine are highly reactive and unstable. Cl_2O and ClO_2 exist as separate molecules. These oxides therefore have low melting points and exist as gases at room temperature and pressure. The highest oxide of chlorine (Cl_2O_7) has greater dispersion forces and exists as an oily liquid under the same conditions.

SUMMARY

- As atomic number increases across Period 3, the oxide structure changes from ionic (Na_2O, MgO, Al_2O_3), to giant covalent (SiO_2), to simple molecular (P_4O_{10}, SO_2, Cl_2O_7).
- The melting points of the oxides follow the trend: Na_2O, MgO, Al_2O_3, SiO_2 – high; P_4O_{10}, SO_3, Cl_2O_7 – low.
- The melting point of SO_3 is greater than expected because it forms chains and three-membered rings.

PRACTICE

1 Write balanced chemical equations for the following reactions:
  a The combustion of phosphorus in a limited supply of oxygen to produce phosphorus(III) oxide.
  b The catalytic oxidation of sulphur dioxide by oxygen to produce sulphur trioxide.

2 Suggest why silicon reacts only very slowly with oxygen, even when at red-heat (temperature approximately 850 °C).

3 Outline the trends in melting points, structure, and bonding that you would expect for the oxides of Period 2.

17.5

OBJECTIVES

- Sodium and magnesium oxides are basic
- Aluminium oxide is amphoteric
- Non-metal oxides are acidic
- Trends in pH of aqueous solutions

ACID–BASE CHARACTER OF THE OXIDES

As the atomic number of the elements increases across Period 3 from sodium to chlorine, we shall see that there is a clear trend in the acid–base character of the oxides. The oxides of the metals on the left-hand side of the period are basic; the oxides of the non-metals on the right-hand side of the period are acidic. In this spread we discuss the nature of each oxide in turn. We conclude by summarizing these properties to show the overall trend across the period.

Solid sodium oxide $Na_2O(s)$ reacts with water exothermically. The product is aqueous sodium hydroxide $NaOH(aq)$, which turns the indicator blue.

Sodium and magnesium hydroxides

Sodium hydroxide NaOH is a strong base. It dissolves readily in water and ionizes fully to give a solution with pH ≈ 14:

$NaOH(s) \rightarrow Na^+(aq) + OH^-(aq)$

Magnesium hydroxide is only partially ionized in aqueous solution. The low hydroxide ion concentration results in a solution of pH ≈ 10:

$Mg(OH)_2(s) \rightleftharpoons Mg^{2+}(aq) + 2OH^-(aq)$

Hydrated aluminium oxide

It is much easier to demonstrate the chemical properties of aluminium oxide if the hydrated form is used. It forms when aqueous aluminium ions react with aqueous sodium hydroxide:

$Al^{3+}(aq) + 3OH^-(aq) \rightarrow Al(OH)_3(s)$

Although the formula of the precipitate has been shown as if it were a simple hydroxide, it is better represented as $Al_2O_3{\cdot}3H_2O$ (or, more realistically, as $Al_2O_3{\cdot}xH_2O$, where x can vary from 1 to 3).

Sodium oxide

Sodium oxide is a white solid. It reacts with water to give strongly basic aqueous sodium hydroxide with a pH of about 14:

$$Na_2O(s) + H_2O(l) \rightarrow 2NaOH(aq)$$

The oxide reacts vigorously with acids to produce an aqueous solution of a salt. For example

$$Na_2O(s) + H_2SO_4(aq) \rightarrow Na_2SO_4(aq) + H_2O(l)$$

These reactions show sodium oxide to be a *basic* oxide.

Magnesium oxide

Magnesium oxide is a white powder which is only slightly soluble in water. The aqueous solution is weakly basic with a pH of approximately 10:

$$MgO(s) + H_2O(l) \rightarrow Mg(OH)_2(aq)$$

The oxide reacts readily with acids to produce an aqueous solution of a salt. For example

$$MgO(s) + 2HCl(aq) \rightarrow MgCl_2(aq) + H_2O(l)$$

These reactions show magnesium oxide to be a *basic* oxide.

Aluminium oxide

Aluminium oxide Al_2O_3 is a white solid which is very insoluble in water. The hydrated oxide behaves as if it had the approximate formula $Al(OH)_3$. It acts as a base when it reacts with excess acid:

$$Al(OH)_3(s) + 3H^+(aq) \rightarrow Al^{3+}(aq) + 3H_2O(l)$$

The hydrated oxide acts as an acid when it reacts with excess aqueous sodium hydroxide to form a complex ion called the *tetrahydroxoaluminate* ion (sometimes abbreviated to aluminate ion):

$$Al(OH)_3(s) + OH^-(aq) \rightarrow [Al(OH)_4]^-(aq)$$

Aluminium oxide therefore has the properties of both a basic and an acidic oxide: it is an *amphoteric* oxide; see spread 16.8.

Silica

Silica SiO_2 has a giant covalent structure, and so is highly insoluble in water. It reacts only with highly concentrated alkalis or at high temperature. For example, it reacts with molten sodium hydroxide at 350 °C to form sodium silicate:

$$SiO_2(s) + 2NaOH(l) \rightarrow Na_2SiO_3(l) + H_2O(g)$$

The oxides of phosphorus

Phosphorus(V) oxide P_4O_{10} reacts with water to form aqueous phosphoric acid, which has a pH of about 1:

$$P_4O_{10}(s) + 6H_2O(l) \rightarrow 4H_3PO_4(aq)$$

Anhydrous phosphoric acid exists as deliquescent colourless crystals which melt at T = 42 °C. (Phosphorus(III) oxide P_4O_6 reacts with water to form aqueous phosphonic acid H_3PO_3.)

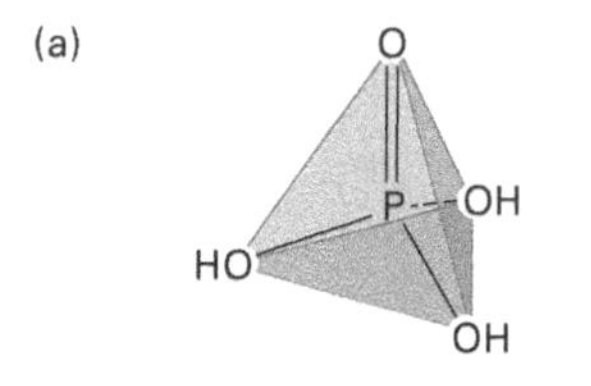

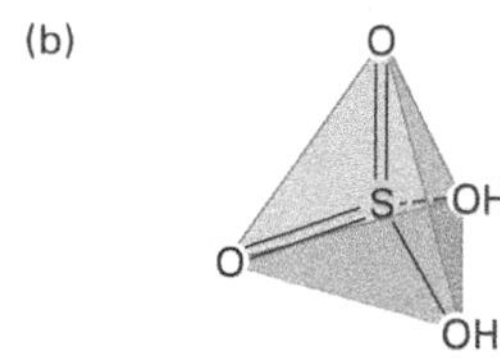

Oxoacid structures: (a) phosphoric acid; and (b) sulphuric acid.

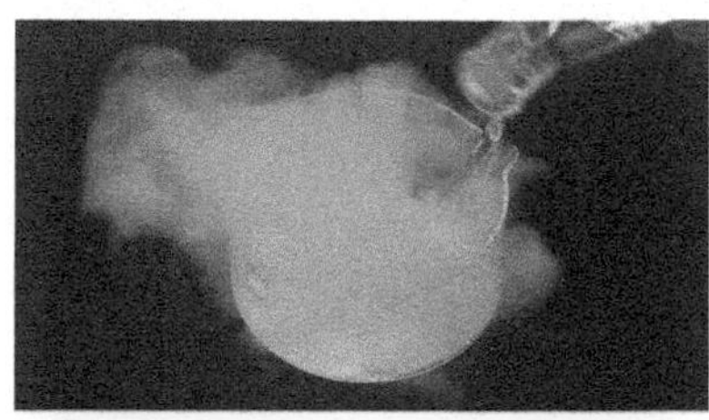

(a) Solid phosphorus(V) oxide $P_4O_{10}(s)$ reacts with water very exothermically. (b) The product is aqueous phosphoric acid $H_3PO_4(aq)$, which turns the indicator red.

The oxides of sulphur

Sulphur dioxide SO_2 is a gas with moderate solubility in water, forming an aqueous solution (pH ≈ 1) of sulphurous acid:

$$SO_2(g) + H_2O(l) \rightleftharpoons H_2SO_3(aq)$$

The acid cannot be isolated in the anhydrous state, as it decomposes back to the gas.

Sulphur dioxide reacts with sodium hydroxide to form sulphites (see spread 19.11):

$$SO_2(g) + 2NaOH(aq) \rightarrow Na_2SO_3(aq) + H_2O(l)$$

Sulphur trioxide SO_3 is a solid (T_m = 17 °C and T_b = 45 °C), which reacts with water to form aqueous sulphuric acid:

$$SO_3(s) + H_2O(l) \rightarrow H_2SO_4(aq)$$

Sulphuric acid is a strong acid and the aqueous solution has a pH close to 0. The anhydrous acid may be isolated as an oily liquid which boils at T_b = 338 °C.

Oxoacids and oxoanions: a reminder

Oxoacids are acids that contain oxygen. Oxoanions are anions that contain oxygen. Oxosalts are salts containing oxoanions.

The oxides of chlorine

The chemical properties of the chlorine oxides are dominated by their tendency to explode. However, about 100 000 tonnes of chlorine dioxide ClO_2 are manufactured each year. This explosive yellow gas is used to bleach wood pulp in the paper-making industry. It produces good whiteness without destroying the fibrous texture.

Dichlorine oxide Cl_2O reacts with water to give aqueous hypochlorous acid HClO(aq). Dichlorine heptaoxide Cl_2O_7 reacts with water to give aqueous perchloric acid $HClO_4(aq)$. The redox properties of perchloric acid and the other oxoacids of Period 3 will be dealt with in the following spread.

The properties of some oxides of the elements of Period 3. Note particularly the increasing acidity of the oxides and the corresponding decrease in the pH of their aqueous solutions.

Formula of oxide	Na_2O	MgO	Al_2O_3	SiO_2	P_4O_{10}	SO_2	SO_3
State of oxide at 25 °C	solid	solid	solid	solid	solid	gas	liquid
Bonding and structure in oxide	ionic lattice	ionic lattice	ionic lattice	giant covalent	covalent molecular	covalent molecular	covalent molecular
Effect of adding oxide to water	reacts to form strongly basic aqueous NaOH	reacts to form weakly basic aqueous $Mg(OH)_2$	does not react with water	does not react with water	reacts to form weakly acidic aqueous H_3PO_4	reacts to form weakly acidic aqueous H_2SO_3	reacts to form strongly acidic aqueous H_2SO_4
Aqueous solution pH	14	10	7	7	1	1	0
Nature of oxide	basic	basic	amphoteric	acidic	acidic	acidic	acidic

SUMMARY

- Sodium oxide and magnesium oxide are basic.
- Aluminium oxide is amphoteric.
- The non-metal oxides in Period 3 are acidic.
- The pH of an aqueous solution of the oxides decreases across the period: recall the actual values above.

PRACTICE

1 Describe the trend in the bonding and structure of the oxides of Period 3. Also describe the trend in the pH of their aqueous solutions.

17.6

The oxoacids

OBJECTIVES

- Non-metal oxoacids
- Trends in oxidizing power of oxoacids

Many of the oxides of the elements of Period 3 react with water to form aqueous solutions. The oxides of the non-metals phosphorus, sulphur, and chlorine form a range of covalent oxoacids such as phosphoric acid and sulphuric acid. Some of these oxoacids exhibit oxidizing properties during redox reactions with suitable reductants. A clear trend in oxidizing power emerges.

Some oxoacids of phosphorus, sulphur, and chlorine.

Name of acid	Simple formula	More exact formula
Phosphoric	H_3PO_4	$(HO)_3PO$
Sulphurous	H_2SO_3	$(HO)_2SO$
Sulphuric	H_2SO_4	$(HO)_2SO_2$
Hypochlorous	$HClO$	$HOCl$
Chlorous	$HClO_2$	$HOClO$
Chloric	$HClO_3$	$HOClO_2$
Perchloric	$HClO_4$	$HOClO_3$

The oxoacids of phosphorus, sulphur, and chlorine

There is a large number of oxoacids containing these three elements. Details of the acids covered by this text are listed on the left. The second, more cumbersome, formula makes clear their status as oxoacids, by showing that the hydrogen atoms are bonded directly to oxygen.

Phosphoric acid

Phosphoric acid H_3PO_4 is non-volatile and does not act as an oxidant. It is therefore the acid of choice when preparing hydrogen bromide or hydrogen iodide. For example, HBr can be prepared by heating solid sodium bromide with concentrated aqueous phosphoric acid:

$$H_3PO_4(aq) + NaBr(s) \rightarrow NaH_2PO_4(aq) + HBr(g)$$

(Concentrated sulphuric acid cannot be used as it would oxidize the bromide ion to bromine – see below.) H_3PO_4 ionizes in three stages to form $H_2PO_4^-$, HPO_4^{2-}, and PO_4^{3-} ions. The pK_a values for the successive loss of the three protons are $pK_{a1} = 2.2$, $pK_{a2} = 7.2$, and $pK_{a3} = 12.3$. The pK_{a2} value is very close to neutral; one of the two important buffer solutions in nature is the 'phosphate buffer' $H_2PO_4^-/HPO_4^{2-}$; see spread 12.10.

Notation: pK_{a1} and pK_{a2}

pK_{a1} refers to the loss of the first proton from a diprotic acid, and pK_{a2} refers to the loss of a second proton. For example:

$H_2SO_3(aq) + H_2O(l) \rightleftharpoons H_3O^+(aq) + HSO_3^-(aq)$; $pK_{a1} = 1.9$

$HSO_3^-(aq) + H_2O(l) \rightleftharpoons H_3O^+(aq) + SO_3^{2-}(aq)$; $pK_{a2} = 7.2$

Sulphurous acid

Sulphurous acid H_2SO_3 contains the element sulphur with oxidation number +4. It is a weak acid ($pK_{a1} = 1.9$, $pK_{a2} = 7.2$), forming salts containing the hydrogensulphite HSO_3^- and sulphite SO_3^{2-} ions. Sulphurous acid is an infamous constituent of acid rain. Sulphurous acid also acts as a reductant, and is oxidized to sulphate ion:

$$H_2SO_3(aq) + H_2O(l) \rightleftharpoons SO_4^{2-}(aq) + 4H^+(aq) + 2e^-$$

For example, sulphurous acid will reduce iron(III) ions to iron(II):

$$2Fe^{3+}(aq) + H_2SO_3(aq) + H_2O(l) \rightleftharpoons 2Fe^{2+}(aq) + SO_4^{2-}(aq) + 4H^+(aq)$$

Sulphuric acid

Sulphuric acid H_2SO_4 is a *strong* acid in aqueous solution. It forms *normal salts* such as sodium sulphate Na_2SO_4 and *acid salts* such as sodium hydrogensulphate $NaHSO_4$. Concentrated sulphuric acid (96% H_2SO_4, 4% H_2O) has three distinctive properties:

1 It is the *most involatile* acid.

That is, it is the acid with the highest boiling point, 338 °C. It can therefore be used to produce more volatile acids, displacing hydrogen chloride and nitric acid when heated with their salts:

$$H_2SO_4(aq,conc) + NaCl(s) \rightarrow NaHSO_4(aq) + HCl(g)$$

$$H_2SO_4(aq,conc) + NaNO_3(s) \rightarrow NaHSO_4(aq) + HNO_3(l)$$

2 It is a *powerful oxidant*.

For example, it oxidizes bromide ion to bromine. Concentrated sulphuric acid is reduced to sulphur dioxide (the oxidation number of sulphur goes from +6 to +4):

$$H_2SO_4(aq,conc) + NaBr(s) \rightarrow NaHSO_4(aq) + HBr(aq)$$

$$H_2SO_4(aq,conc) + 2HBr(aq) \rightarrow Br_2(aq) + 2H_2O(l) + SO_2(g)$$

Sulphuric acid

Sulphuric acid is crucial to a large number of industrial processes. More than 10 000 tonnes are manufactured in the UK each day, more than for any other oxoacid. To show explicitly that it is an oxoacid (that is, that the hydrogen atoms are bonded directly to oxygen), the formula may be written as $(HO)_2SO_2$ (see the previous spread for the detailed structure of the molecule).

Concentrated sulphuric acid similarly oxidizes iodide ion to iodine, and is itself reduced to SO_2. However, the iodide ion is a sufficiently powerful reductant to reduce the sulphuric acid to elemental sulphur (oxidation number 0) and further to H_2S (oxidation number −2); see the following chapter for further details. Concentrated sulphuric acid is not a sufficiently powerful oxidant to oxidize chloride ion to chlorine (as in 1 above).

3 It is a powerful *dehydrating agent*.

This means that it chemically removes the elements of water from a compound. It can for example dehydrate sugar (sucrose $C_{12}H_{22}O_{11}$), as shown alongside.

Concentrated sulphuric acid dehydrates sucrose by removing 11 molecules of water from each $C_{12}H_{22}O_{11}$ molecule. The result is carbon and steam:
$C_{12}H_{22}O_{11}(s) \rightarrow 12C(s) + 11H_2O(g)$

Oxoacids of chlorine

These acids are prepared in solution by adding a strong acid to a salt of the chlorine oxoacid. For example, aqueous hypochlorous acid HClO results when sulphuric acid is added to solid sodium hypochlorite:

$$NaClO(s) + H^+(aq) \rightarrow HClO(aq) + Na^+(aq)$$

The acid is only partially ionized ($pK_a = 7.4$) and is therefore a weak acid:

$$HClO(aq) + H_2O(l) \rightleftharpoons H_3O^+(aq) + ClO^-(aq)$$

Note that the oxygen atom bears the negative charge. Chlorine has oxidation number +1 in HClO. Its salts, such as sodium hypochlorite NaClO, are important bleaches in the home and in industry, acting by oxidation.

Oxidizing power increases in the order

$$HClO < HClO_2 < HClO_3 < HClO_4.$$

Perchloric acid $HClO_4$ is an extremely powerful oxidant and can react explosively with organic matter. Perchloric acid is also a strong acid.

Ionization of $H_2SO_4(aq)$

$H_2SO_4(aq)$ is a strong acid and so is fully ionized in water:

$H_2SO_4(aq) + H_2O(l) \rightarrow$
$H_3O^+(aq) + HSO_4^-(aq)$

The hydrogensulphate ion is, however, only partially ionized ($pK_a = 1.9$):

$HSO_4^-(aq) + H_2O(l) \rightleftharpoons$
$H_3O^+(aq) + SO_4^{2-}(aq)$

Trends in oxidizing power

Sulphuric acid H_2SO_4 is a stronger oxidant than phosphoric acid H_3PO_4; perchloric acid $HClO_4$ is stronger still. The oxoacids of chlorine with lower oxidation numbers are also oxidants, although less powerful than $HClO_4$.

Take care with sulphuric acid

The powerful oxidizing and dehydrating properties of concentrated sulphuric acid mean that it causes severe burns to living tissue.

SUMMARY

- Non-metal oxides react with water to form oxoacids, which contain hydrogen atoms bonded directly to oxygen.
- Oxidizing power of the oxoacids of phosphorus, sulphur, and chlorine increases with increasing oxidation state in the overall order P(V) < S(VI) < Cl(VII).
- Sulphuric acid is the most involatile acid and a powerful oxidant and dehydrating agent.
- Sulphuric acid and perchloric acid are strong acids.

PRACTICE

1 Arrange the oxoacids of chlorine in order of increasing acid strength.

2 a Draw three-dimensional representations of all the oxoacids described in this spread.

b For each of your sketches in (a), draw the Lewis structure and indicate the mechanism by which each acid dissociates.

3 Compare and discuss the pK_a data given in this spread.

17.7 THE CHLORIDES OF PERIOD 3

OBJECTIVES
- Formulae
- Formation
- Trends in melting point
- Bonding trends
- Reactions with water

All the elements of Period 3 from sodium to phosphorus react directly with chlorine to form chlorides. The metallic elements form ionic chlorides, and the non-metallic elements form covalent chlorides. As with the elements themselves and their oxides, there is a trend in properties of these chlorides as the atomic number of the element increases across the period. This spread deals with the formation of the chlorides and their reactions with water.

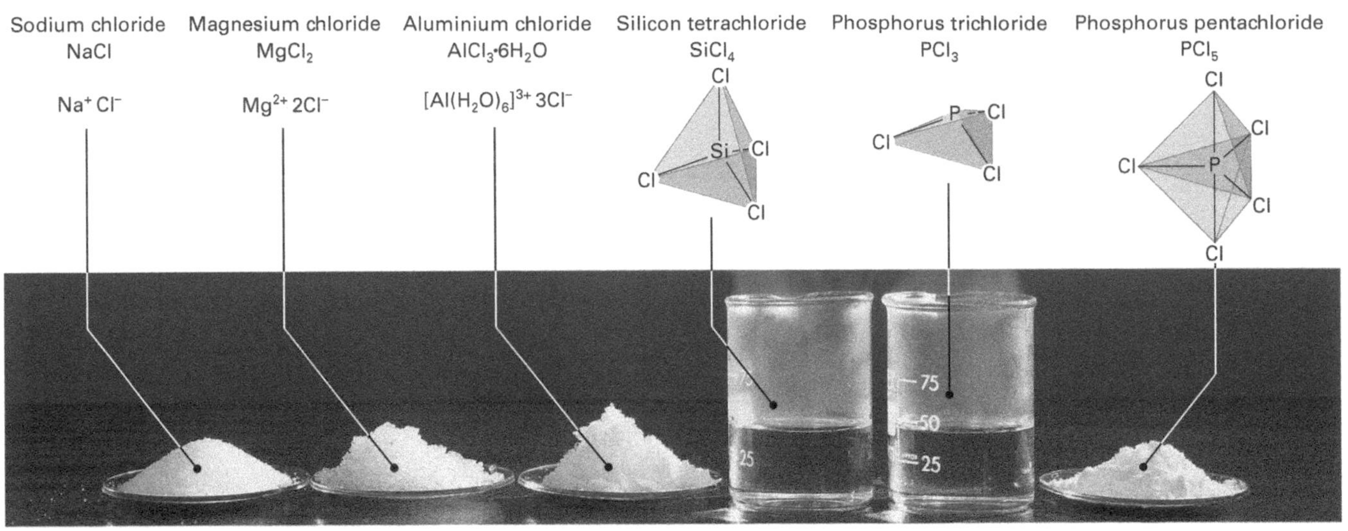

Physical state, formula, and bonding of the chlorides of the elements of Period 3.

The reaction of magnesium metal with chlorine gas is highly exothermic.

Oxidation numbers

Oxidation number increases from +1 to +5 in the sequence of chlorides from sodium chloride to phosphorus pentachloride: Na(I) in NaCl; Mg(II) in $MgCl_2$; Al(III) in $AlCl_3$; Si(IV) in $SiCl_4$; and P(V) in PCl_5. Note that phosphorus also forms a chloride of formula PCl_3. Sulphur forms a number of chlorides, such as S_2Cl_2, which are not important enough to describe in detail.

Sodium and magnesium chlorides

The chlorides of sodium and magnesium are formed by heating the metal in air until it burns, and then lowering it into a vessel containing chlorine. Sodium and magnesium both continue to burn in chlorine, with the white product coating the walls of the reaction vessel:

$$2Na(s) + Cl_2(g) \rightarrow 2NaCl(s)$$

$$Mg(s) + Cl_2(g) \rightarrow MgCl_2(s)$$

Aluminium chloride

Aluminium chloride reacts very readily with water, so it must be synthesized under anhydrous conditions. In the apparatus shown to the left, a stream of dry chlorine is passed over heated aluminium. The aluminium chloride product vaporizes and is carried through the apparatus by the flow of chlorine gas. It condenses as a white powder:

$$2Al(s) + 3Cl_2(g) \rightarrow Al_2Cl_6(s)$$

Similar apparatus may be used to synthesize the chlorides of the non-metals of Period 3.

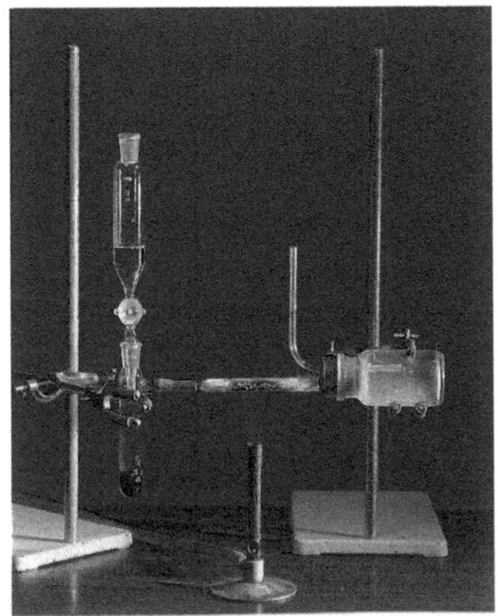

The reaction in the tube on the left supplies chlorine. Aluminium chloride condenses inside the flask seen on the right.

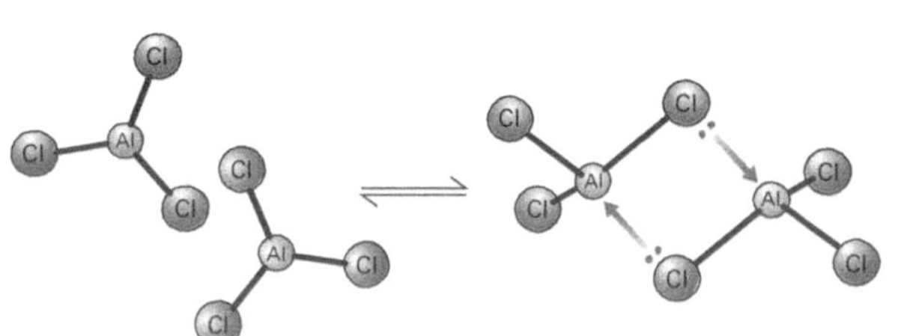

The structure of Al_2Cl_6. This anhydrous form results from $AlCl_3$ molecules forming dimers: chlorine atoms donate lone pairs of electrons into vacant orbitals in the valence shell of the aluminium atoms.

Silicon tetrachloride

The apparatus is similar to that used for the synthesis of aluminium chloride. However, $SiCl_4$ is a liquid at room temperature; the product is condensed in a water-cooled side-arm tube (which replaces the flask):

$$Si(s) + 2Cl_2(g) \rightarrow SiCl_4(l)$$

The chlorides of phosphorus

Heating phosphorus in excess chlorine forms the pentachloride PCl_5; excess phosphorus forms the trichloride PCl_3:

$$P_4(s) + 10Cl_2(g) \rightarrow 4PCl_5(s) \qquad \text{(excess } Cl_2\text{)}$$

$$P_4(s) + 6Cl_2(g) \rightarrow 4PCl_3(l) \qquad \text{(excess } P_4\text{)}$$

The pentachloride is a solid and the trichloride is a liquid at 25 °C and 1 bar. The hot product vapours must be condensed.

Phosphorus pentachloride

The structure of *solid* phosphorus pentachloride is unusual. It exists as the ion pair $PCl_4^+ PCl_6^-$. The PCl_4^+ ion has a tetrahedral shape; the PCl_6^- ion has an octahedral shape. See also the photo in spread 5.3.

Bonding in the chlorides and reaction with water

The chlorides of sodium and magnesium are predominantly ionic. They are white solids with high melting points, which dissolve in water to form a neutral solution (pH 7):

$$NaCl(s) \rightarrow Na^+(aq) + Cl^-(aq)$$

All the other chlorides of Period 3 have significant covalent character and *react* with water. These reactions are called **hydrolysis** reactions.

Aluminium forms two solid chlorides, one of which is anhydrous (Al_2Cl_6) and the other hydrated ($AlCl_3{\cdot}6H_2O$). The anhydrous chloride (see opposite) *reacts* with water, evolving hydrogen chloride gas:

$$Al_2Cl_6(s) + 6H_2O(l) \rightarrow 2Al(OH)_3(s) + 6HCl(g)$$

Some HCl dissolves in the water to give hydrochloric acid, a strong acid with pH ≈ 0. This behaviour is characteristic of covalent chlorides.

The hydrated chloride consists of a complex cation $[Al(H_2O)_6]^{3+}$ bonded ionically to three chloride ions. The formula of the solid is therefore more accurately expressed as $[Al(H_2O)_6]Cl_3$. It dissolves in water, as is typical of an ionic solid, to form the ions $[Al(H_2O)_6]^{3+}(aq)$ and $Cl^-(aq)$. The solution is slightly acidic (pH ≈ 3), for reasons discussed in chapter 19 'The p-block elements'.

The chlorides of silicon and phosphorus are predominantly covalent. They all react with water (are hydrolysed) rather than simply dissolving in water; $SiCl_4$ and PCl_3 fume in air (see opposite):

$$SiCl_4(l) + 2H_2O(l) \rightarrow 4HCl(g) + SiO_2(s)$$

$$PCl_5(s) + 4H_2O(l) \rightarrow 5HCl(g) + H_3PO_4(aq)$$

The hydrochloric acid formed turns the solution strongly acidic, with pH ≈ 0.

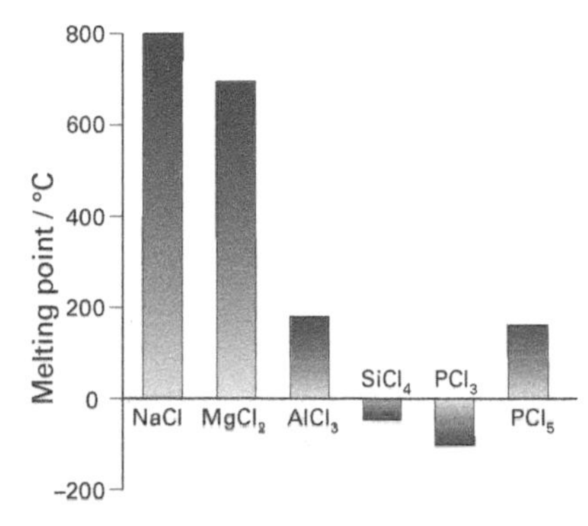

A bar chart of the melting points of the chlorides in Period 3.

Phosphorus trichloride is hydrolysed by water to aqueous phosphonic acid $H_3PO_3(aq)$ and hydrogen chloride: $PCl_3(l) + 3H_2O(l) \rightarrow H_3PO_3(aq) + 3HCl(aq)$ Some of the hydrogen chloride dissolves in the water to give an aqueous solution with a pH close to zero, which turns universal indicator red.

SUMMARY

- All the elements of Period 3 react directly with chlorine to form chlorides.
- The bonding in aluminium chloride is on the borderline between ionic and covalent.

Formula of chloride	NaCl	$MgCl_2$	$AlCl_3$	$SiCl_4$	PCl_5
State of chloride at 25 °C	solid	solid	solid	liquid	solid
Bonding and structure in chloride	ionic lattice	ionic lattice	see text	covalent molecular	covalent molecular
Effect of adding chloride to water	dissolves	dissolves	see text	reacts	reacts
Aqueous solution pH	7	7	3	0	0

PRACTICE

1 Describe the bonding and structure of the chlorides of sodium to phosphorus.

EXTENSION

17.8

OBJECTIVES

- Bonding character of the hydrides
- Reactions with water
- Trends in the hydrides

THE HYDRIDES: FROM IONIC TO COVALENT

Hydrides are compounds that contain hydrogen and one other element. There is a complete range of hydrides for the Period 3 elements, but the non-metal hydrides are much more important than the metal hydrides. The metal hydrides contain the hydride ion H^-, whereas the non-metal hydrides contain covalently bonded hydrogen atoms. The behaviour of the hydrides varies greatly, depending on the nature of the bonding within them. As you would expect by now, there is a trend in properties across the period from sodium hydride to hydrogen chloride.

The hydrides of Period 3

Hydride name	Formula
Sodium hydride	NaH
Magnesium hydride	MgH_2
Aluminium hydride	AlH_3
Silane	SiH_4
Phosphine	PH_3
Hydrogen sulphide	H_2S
Hydrogen chloride	HCl

Sodium hydride

Sodium hydride is a solid ionic hydride which reacts with water to form hydrogen gas and aqueous sodium hydroxide:

$$NaH(s) + H_2O(l) \rightarrow NaOH(aq) + H_2(g)$$

The hydride ion H^- (in which the oxidation number of hydrogen is −1) is unstable in water as shown in the photograph below left for the case of calcium hydride. Hydride ion reduces a hydrogen atom in the water molecule, forming hydroxide ion and hydrogen gas:

$$H^-(s) + H_2O(l) \rightarrow OH^-(aq) + H_2(g)$$

The hydride ion is a powerful reductant, and so ionic hydrides exist only where the metal has a low electronegativity.

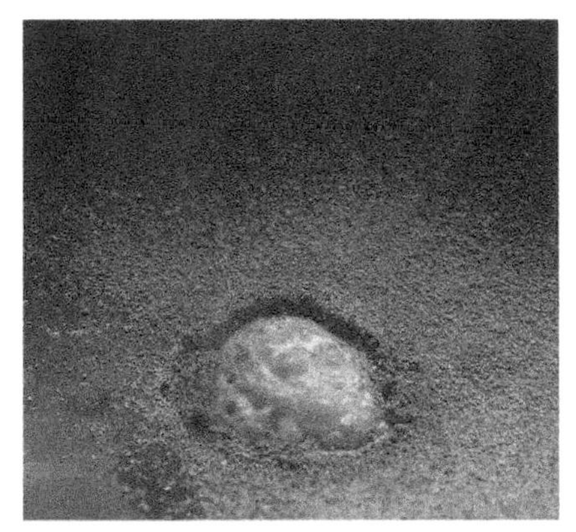

Calcium (Period 4) forms a solid hydride CaH_2. It reacts with water to form hydrogen, which is burning here with a yellow flame.

Aluminium hydride and the tetrahydridoaluminate ion

The hydride AlH_3 has little importance, but the closely related AlH_4^- ion is very important in organic chemistry. In the AlH_4^- ion, four hydrogen atoms are covalently bonded to a central aluminium atom. This ion is a powerful reductant and is stable only in the presence of metal ions with low electronegativity such as Li^+. Lithium tetrahydridoaluminate $LiAlH_4$ (also called lithium aluminium hydride) is a white solid which is used as a powerful reductant throughout organic chemistry, e.g. spread 26.2.

Silane

Silicon is in Group IV of the periodic table along with carbon. There is a series of silicon hydrides similar to the series of carbon hydrides (called *alkanes*, i.e. methane CH_4, ethane C_2H_6, propane C_3H_8, etc.). The first and most important of these is silane SiH_4, a gas which boils at T_b = −112 °C. Four hydrogen atoms are covalently bonded to a central silicon atom. Unlike methane, which is only mildly reducing, silane is a powerful reductant and spontaneously ignites in air:

$$SiH_4(g) + 2O_2(g) \rightarrow SiO_2(s) + 2H_2O(l)$$

Silane is hydrolysed by water (containing a trace of alkali) to hydrated silica and hydrogen gas:

$$SiH_4(g) + 2H_2O(l) \rightarrow SiO_2(s) + 4H_2(g)$$

Silane, produced by the reaction between magnesium silicide and acid, spontaneously inflames in air.

Phosphine

Phosphine PH_3 is a poisonous gas (T_b = −87 °C) which is almost insoluble in water. Its structure is similar to that of ammonia, the hydride of the Group V element nitrogen. Phosphine is a strong reductant, and is able to precipitate copper and silver from aqueous solutions of their salts. For example:

$$4Cu^{2+}(aq) + PH_3(g) + 4H_2O(l) \rightarrow 4Cu(s) + H_3PO_4(aq) + 8H^+(aq)$$

Hydrogen sulphide

Hydrogen sulphide H_2S is a gas (T_b = −61°C) which smells of 'bad eggs'. In solution, hydrogen sulphide is weakly acidic and is the parent acid of a series of salts called *sulphides*, e.g. sodium sulphide Na_2S.

Hydrogen sulphide is formed by adding a strong acid to a sulphide such as iron(II) sulphide:

$$FeS(s) + 2HCl(aq) \rightarrow FeCl_2(aq) + H_2S(g)$$

Hydrogen sulphide is a strong reductant, able to reduce sulphur dioxide:

$$2H_2S(g) + SO_2(g) \rightarrow 3S(s) + 2H_2O(l)$$

The reaction is catalysed by a trace of water.

Size

The covalent radius of the hydrogen atom H is 0.037 nm. The ionic radius of the hydride ion H^- is 0.208 nm. Compare these figures with those for the chlorine atom Cl (0.099 nm) and the chloride ion Cl^- (0.181nm).

Hydrogen chloride

Hydrogen chloride HCl is a gas (T_b = −85 °C) which dissolves very easily in water to form hydrochloric acid. Hydrochloric acid is a strong acid, being essentially fully ionized in solution:

$$HCl(g) + H_2O(l) \rightarrow H_3O^+(aq) + Cl^-(aq)$$

Trends in the hydrides

The trend in the bonding in the hydrides is similar to that in the chlorides. The elements on the extreme left of the period tend to form ionic hydrides, and the elements on the extreme right tend to form covalent hydrides.

The oxidation states of the Period 3 element in hydrides vary in the sequence: Na(I) in NaH; Mg(II) in MgH_2; Al(III) in AlH_4^-; Si(IV) in SiH_4; P(III) in PH_3; S(−II) in H_2S; and Cl(−I) in HCl. Note the change of oxidation number from positive on the left to negative on the right. Hydrogen has an electronegativity only *just* larger than that of phosphorus.

Where the hydrides dissolve in water or react with it, pH falls steadily in the sequence NaH (pH 14) to PH_3 (pH 7) to HCl (pH 0).

With the exception of hydrogen chloride, the hydrides are strong reductants.

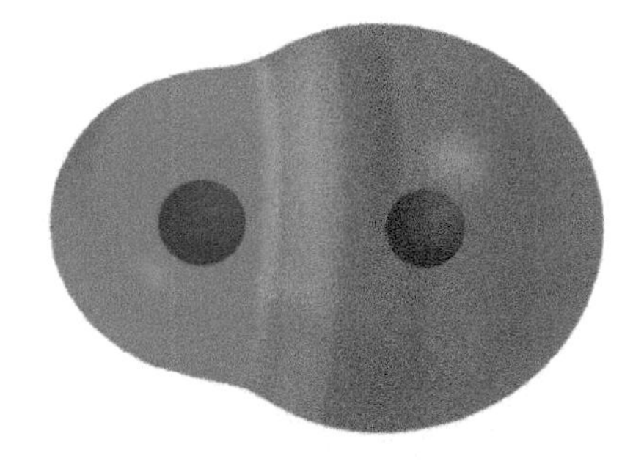

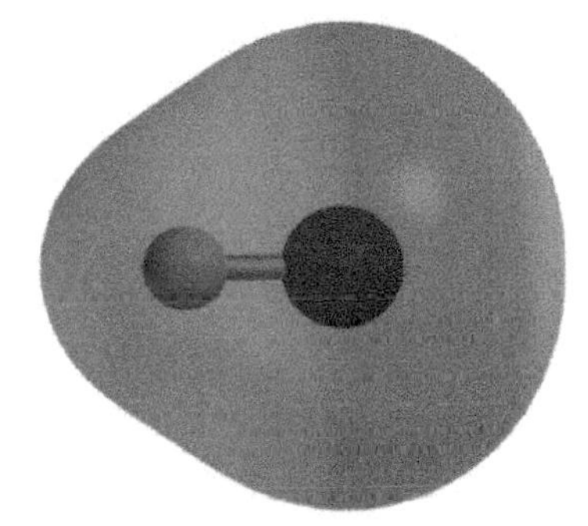

Electrostatic potential maps show that sodium hydride NaH has a large charge separation (is essentially ionic), with hydrogen partially negatively charged. Hydrogen chloride HCl is polar covalent with hydrogen partially positively charged. The other hydrides of the Period 3 elements show a steady change in bonding character between these two extremes.

SUMMARY

- The hydrides show trends across the period, from ionic to covalent compounds.
- Sodium hydride is an ionic solid which reacts with water to form an aqueous solution of pH ≈ 14.
- Hydrogen chloride is a covalent gas which reacts with water to form an aqueous solution of pH ≈ 0.

PRACTICE

1 Draw Lewis structures for the following species:
  a The hydride ion H^-
  b Silane SiH_4
  c Phosphine PH_3
  d Hydrogen sulphide H_2S
  e Hydrogen chloride HCl
  f The tetrahydridoaluminate ion AlH_4^-.

2 For each of the species in question 1(b) to (f), draw its shape showing the arrangement of its bonds, remembering to include lone pairs of electrons.

3 The text above states that: 'In solution, hydrogen sulphide is weakly acidic ...'. Write a chemical equation to illustrate this observation.

4 Explain why the hydride ion H^- is larger than most other monatomic (single-atom) ions.

5 Plot a graph of the boiling points of the hydrides SiH_4 to HCl. Comment on the shape of the plot. Explain any trend you can point out in terms of the intermolecular forces present.

PRACTICE EXAM QUESTIONS

1 a The graph below shows the melting points of the elements sodium to argon.

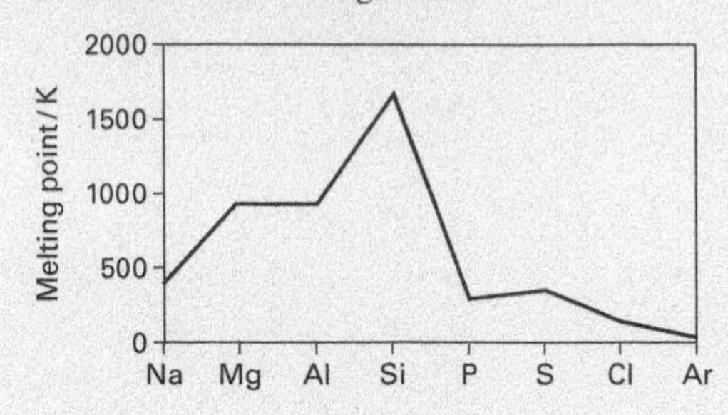

By reference to their structure and bonding, explain the melting points of the elements from sodium to argon. [5]

b The graphs below show the trends in atomic radius and molar first ionization energy for Group 1 elements.

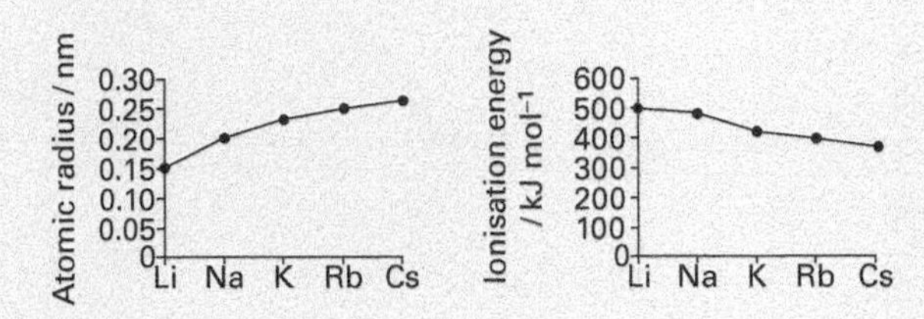

Explain why the atomic radius increases and the ionization energy decreases as the group is descended. [4]

c Write balanced equations to show the reaction of water with:

i sodium; [1]

ii sodium oxide; [1]

iii sodium hydride. [1]

d Suggest a pH value for the solution formed when sodium hydride reacts with water. [1]

2 The elements of the third period are as follows:

Na Mg Al Si P S Cl Ar

All of your answers below should relate to these elements.

a Which elements can exist:

i as diatomic molecules at room temperature;

ii as macromolecular structures? [2]

b Which pairs of elements combine to produce compounds with formulae of the type *XY*? [2]

c Two elements form chlorides with formula, of the type XCl_3. Draw displayed formulae for these two chlorides, and suggest values for the bond angles. [4]

d i One element combines with oxygen to form an oxide which reacts with water to give a strongly alkaline solution. Name the element, and write a balanced equation for the oxide reacting with water.

ii One element combines with oxygen to form an oxide of the type XO_2, which reacts with water to give an acidic solution. Name the element and write a balanced equation for the oxide reacting with water. [4]

3 a Explain the meaning of the term periodic trend when applied to trends in the Periodic Table. [2]

b Explain why atomic radius decreases across Period 2 from lithium to fluorine. [2]

c The table below shows the melting temperatures, T_m, of the Period 3 elements.

Element	Na	Mg	Al	Si	P	S	Cl	Ar
T_m/K	371	923	933	1680	317	392	172	84

Explain the following in terms of structure and bonding.

i Magnesium has a higher melting temperature than sodium.

ii Silicon has a very high melting temperature.

iii Sulphur has a higher melting temperature than phosphorus.

iv Argon has the lowest melting temperature in Period 3. [8]

4 a In the table below, give the formulae of the chlorides of the elements of Period 3, other than silicon.

Element	Na	Mg	Al	Si	P	S
Formula of chloride				$SiCl_4$		

[2]

b Calculate the percentage by mass of silicon in silicon tetrachloride. [2]

c i Draw a dot-and-cross diagram to show the bonding in silicon tetrachloride.

ii Draw the shape of this molecule.

Explain your answer in terms of the electron-pair repulsion theory.

iii State the shape of a molecule of $AlCl_3$ and explain why it is different from that of $SiCl_4$. [6]

d i Give an equation for the reaction of $SiCl_4$ with cold water.

ii How does the behaviour of carbon tetrachloride with cold water compare with this? Explain any differences. [4]

5 a Define the term **electronegativity**. [2]

b State and explain the trend in electronegativity across Period 3 from sodium to chlorine. [4]

c i What is the trend in bond type in the oxides of the Period 3 elements from sodium to sulphur?

ii Explain how this trend is related to the differences in electronegativity between the Period 3 element and oxygen. [3]

d Write an equation for the reaction of phosphorus(V) oxide, P_4O_{10}, with water. [2]

6 a Describe the nature of the attractive forces which hold the particles together in magnesium metal and in magnesium chloride. [4]

b Name the type of bond between aluminium and chlorine in aluminium chloride and explain why the bonding in aluminium chloride differs from that in magnesium chloride. [3]

c Write an equation, including state symbols, to show what happens when magnesium chloride dissolves in water. Explain, in terms of bonding, the nature of the interaction between water and magnesium in this solution. [3]

7 a Write equations to show what happens when the following oxides are added to water and predict approximate values for the pH of the resulting solutions.

i sodium oxide;

ii sulphur dioxide. [4]

b What is the general relationship between bond type in the oxides of the Period 3 elements and the pH of the solutions which result from addition of the oxides to water? [2]

c Write equations to show what happens when the following chlorides are added to water and predict approximate values for the pH of the resulting solutions:

i magnesium chloride;

ii silicon tetrachloride. [4]

8 The table below shows electronegativity values for some atoms.

Atom	H	N	O	F	Cl	Cs
	2.1	3.0	3.5	4.0	3.0	0.7

a What do you understand by the term electronegativity? [1]

b The nature of the bonding in substances depends partly on the electronegativities of the atoms concerned. Use the data in the table above to suggest the nature of the bonding in each of the following substances.

i caesium fluoride; [1]

ii water; [1]

iii chlorine. [1]

c Ammonia, NH_3, is a polar covalent molecule.

i State the general rules which determine the shape of a covalent molecule. [3]

ii Draw the shape of the ammonia molecule [1]

iii Why is the bond angle in ammonia 107° rather than 109° 28′? [1]

iv Explain why the molecule is polar [1]

d Each of the elements sodium to chlorine in Period 3 will react with oxygen given suitable conditions.

i Choose an element from this period which gives a basic oxide, and write equations both for its reaction with oxygen and to illustrate the basic nature of the oxide. [2]

ii Choose an element which forms an amphoteric oxide and write equations which illustrate this amphoteric nature. [2]

iii Carbon dioxide reacts readily with dilute aqueous sodium hydroxide whereas silicon dioxide does not. Explain this difference and suggest conditions under which silicon dioxide would react. [2]

18 The halogens

The halogens are the non-metallic elements fluorine (F), chlorine (Cl), bromine (Br), iodine (I), and astatine (At) that make up Group VII of the periodic table. They are p-block elements; the p orbitals are incompletely filled and the valence shell has the structure ns^2np^5. The halogens have generally greater chemical reactivity than other non-metals. They also have wide application in industrial processes and are often present in finished products. The properties of the halogens and their compounds (with the exception of the radioactive element astatine) form the subject of this chapter.

Group VII elements: an overview

18.1

O B J E C T I V E S

- The position of Group VII
- Naming the halogens
- Trends in boiling point
- Bonding in compounds

The halogens show the typical properties of non-metals: they have low melting and boiling points, when solid they are brittle, and when liquid they are extremely poor conductors of electricity. All halogens have seven electrons in the valence shell, so they all show similar chemical properties. The p orbitals are incompletely filled, being one electron short of the eight required for a full shell ns^2np^6. They form two main classes of compounds: ionic compounds containing the halide ion X^-, and covalent compounds where electrons are shared with other atoms.

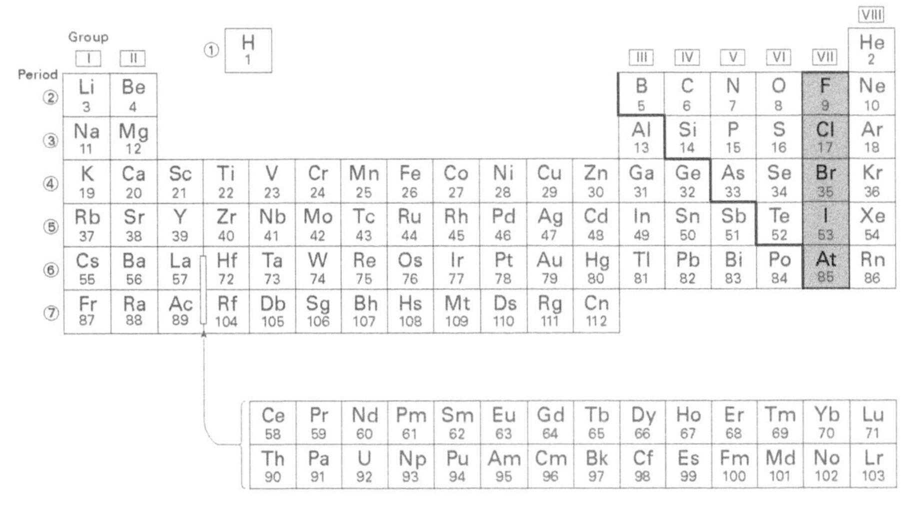

The halogens form Group VII of the periodic table. (Astatine is a radioactive element; its chemistry will not be dealt with in this text.)

Bromine reacts vigorously with red phosphorus. An earlier view of the same reaction is shown in spread 13.5.

The elements

The elements exist as diatomic molecules containing a single covalent bond X—X (where we use X to mean any halogen). The number of electrons increases with atomic number, resulting in an increase in the dispersion forces between the molecules (see spread 7.4). As a result, the melting and boiling points of the halogens increase as atomic number increases in the order $F_2 < Cl_2 < Br_2 < I_2$. Fluorine and chlorine are gases, bromine is a liquid, and iodine is a solid at 25 °C (298 K) and 1 bar.

: Cl : Cl :

Cl—Cl

Two halogen atoms share a pair of electrons to form a single covalent bond.

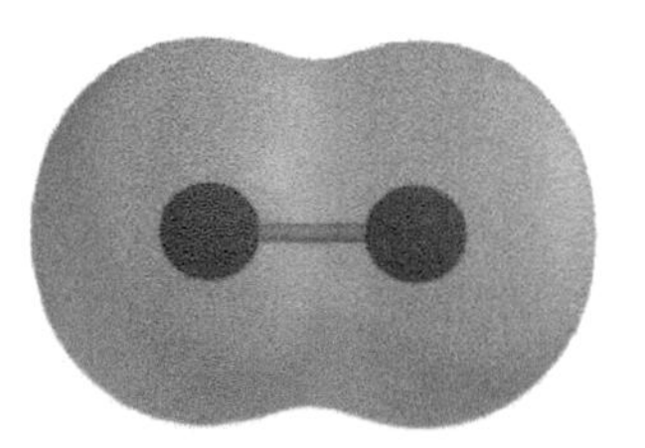

The electrostatic potential map of the chlorine molecule Cl_2. Note the concentration of electron density (forming the bond) between the two atomic centres. The symmetrical electron density shows that the molecule is non-polar.

Halogen compounds

The *binary compounds* (compounds in which just *two* elements are combined) containing halogens are called **halides**. Each of the four halogens shows similar formulae in their compounds. For example, the ionic sodium halides are sodium fluoride NaF, sodium chloride NaCl, sodium bromide NaBr, and sodium iodide NaI. All are white crystalline solids which are soluble in water. Halogens have an oxidation number of –1 in ionic compounds. Electronegativity decreases in the order F > Cl > Br > I, and so the covalent character of ionic compounds increases in the order fluoride < chloride < bromide < iodide. Aluminium fluoride melts above 1000 °C; aluminium chloride is more covalent and melts at 180 °C.

A similar pattern is noted in the covalent compounds. For example, the hydrogen halides are hydrogen fluoride HF, hydrogen chloride HCl, hydrogen bromide HBr, and hydrogen iodide HI. All are gases (at 25 °C and 1 bar) and are readily soluble in water, forming acidic solutions. The halogens have an oxidation number of –1 in the hydrogen halides. With the exception of fluorine, the halogens can also exhibit other oxidation numbers in some of their covalent compounds.

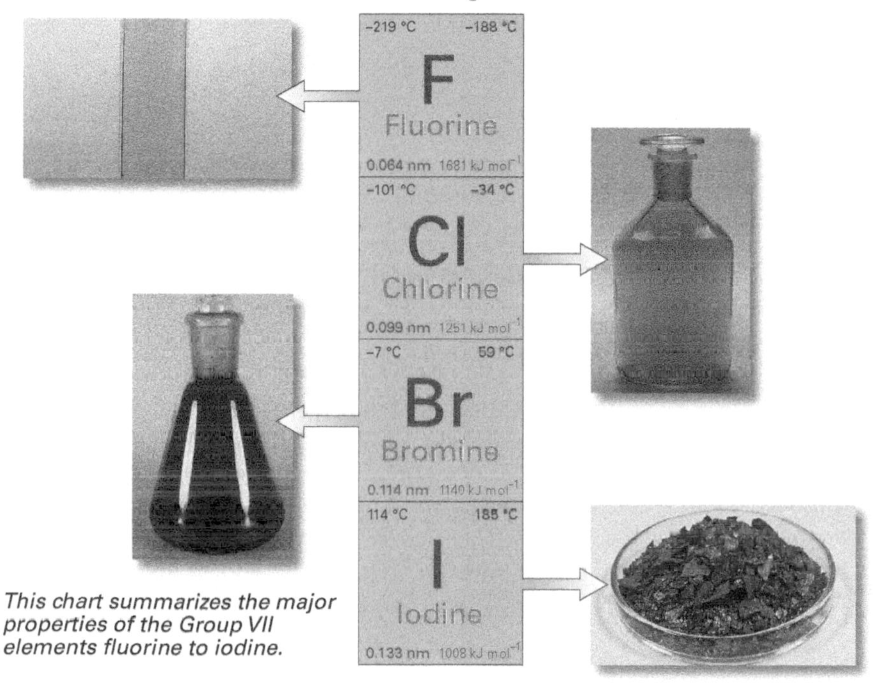

This chart summarizes the major properties of the Group VII elements fluorine to iodine.

SUMMARY

- The halogens are the elements in Group VII.
- They are non-metals with outer-shell electronic structures ns^2np^5.
- They form ionic compounds with metals, and covalent compounds with non-metals.
- As atomic number *increases* from fluorine to iodine: melting point, boiling point, and atomic radius *increase*; ionization energy and electronegativity *decrease*.

PRACTICE

1 Explain why the melting points of the halogens vary.

2 Draw Lewis structures to show the bonding in hydrogen bromide.

3 Draw diagrams to show how sodium and bromine bond together in sodium bromide.

4 Explain why sodium iodide melts at a lower temperature (661 °C) than sodium chloride (801 °C). Base your answer on a discussion of the character of the bonding, supported by data from the chart above.

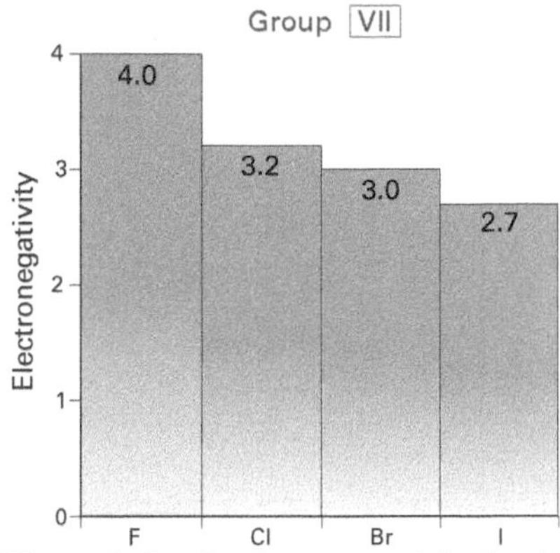

The variation in electronegativity for the halogens.

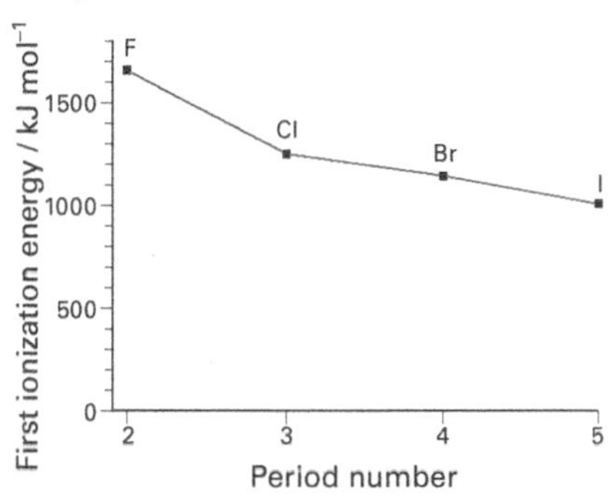

The ionization energy falls as the atoms get bigger because the electron being ionized is further from the nucleus.

Slater's rules (again)

The effective nuclear charge for Cl was found in spread 17.3 using Slater's rules to be 6.1. The value for F is 9 – 6 × 0.35 – 2 × 0.85 = 5.2. *Effective nuclear charge increases down a group.* This effect would make the electron *more* difficult to ionize so the distance effect described above is dominant. If, however, in spread 16.2 we had included the final elements in Groups I and II (Fr and Ra respectively) the ionization energy would have *increased* because the rise in effective nuclear charge is now the more important factor.

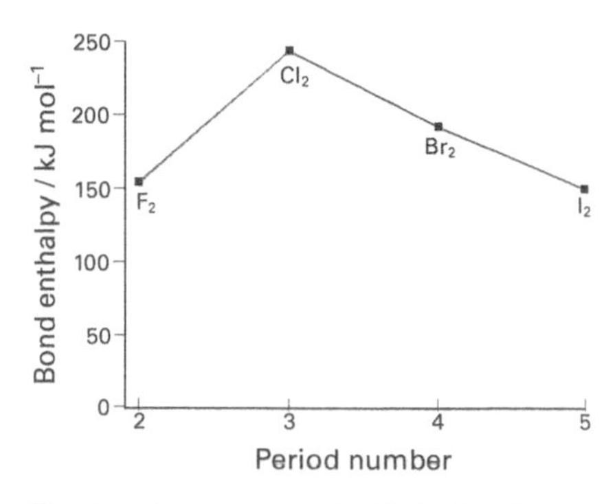

The bond enthalpies for Cl_2 to I_2 fall as the atoms get bigger: the bond is longer and hence weaker. The anomalously low value for F_2 is due to the repulsion between the three pairs of electrons on each fluorine atom as they get close to each other and is one reason for fluorine's high reactivity.

18.2

Manufacture and uses

OBJECTIVES

- Chlorine and fluorine: by electrolysis
- Bromine and iodine: by displacement
- Industrial roles

Uses of chlorine and its compounds

- Cl_2 – sterilizing water
- Cl_2 – recovery of tinplate from cans
- Cl_2 – making hydrochloric acid
- NaClO – bleaches for textiles and paper
- PVC – plastic
- Anaesthetics (e.g. Halothane), disinfectants (e.g. TCP)
- Chlorinated hydrocarbon solvents.

The greatest percentage use is PVC (27%), followed by water purification, bleaches, and then solvents.

At the anode

If the chloride ion concentration is low, oxygen is liberated at the anode from the oxidation of water, i.e.

self-ionization:
$H_2O(l) \rightleftharpoons H^+(aq) + OH^-(aq)$

at the anode:
$4OH^-(aq) \rightleftharpoons O_2(g) + 2H_2O(l) + 4e^-$

See spread 13.11.

'Mercury cells'

An earlier type of cell used a flowing mercury cathode. These 'mercury cells' have now been withdrawn from use because the inevitable loss of mercury caused environmental damage. The most notorious case occurred in the coastal town of Minamata in Japan during the 1950s. Contaminated waste from a new factory entered the food chain through plankton and fish, eventually causing babies to be born with severe abnormalities.

The halogens play an important role in an enormous number of modern manufacturing processes, and occur widely in many finished products. Chlorine is by far the most important halogen commercially. Millions of tonnes are manufactured each year from common salt, sodium chloride. In turn, chlorine plays a key role in the production of the halogens bromine and iodine. Together with fluorine, these elements are involved in the production of plastics, dyestuffs, pharmaceuticals and drugs, refrigerants, and photographic film.

Chlorine

The electrolysis of sodium chloride in the **chlor–alkali industry** produces chlorine, hydrogen, and sodium hydroxide. The two technologies currently used, the 'membrane cell' and the 'diaphragm cell', have several features in common. The electrolyte in both cells is saturated **brine** (aqueous sodium chloride) containing about 25% by mass of NaCl. The electrode reactions are the same in both cells.

At the anode, chloride ions lose electrons and are oxidized to chlorine gas, i.e.

at the anode: $2Cl^-(aq) \rightarrow Cl_2(g) + 2e^-$

The anode resists attack by chlorine because it is made from titanium with an inert coating of ruthenium(IV) oxide RuO_2. The high concentration of chloride ions in the solution results in chlorine being formed at the anode by oxidation of chloride ions. (If the chloride ion concentration were too low, oxygen would form from the oxidation of water.)

At the cathode (made from steel or nickel), water is reduced to hydrogen gas and hydroxide ions are formed, i.e.

at the cathode: $2H_2O(l) + 2e^- \rightarrow H_2(g) + 2OH^-(aq)$

The key problem in the manufacture is that the products of the electrolysis, i.e. Cl_2 and OH^-, react with each other. The **diaphragm cell** uses a porous asbestos diaphragm to keep the products separate. The **membrane cell** uses a polymer ion-exchange membrane that allows only cations to pass through it. The electrolyte is removed from the electrolysis cell and evaporated. The less-soluble sodium chloride crystallizes, leaving sodium hydroxide dissolved in the solution for later recovery.

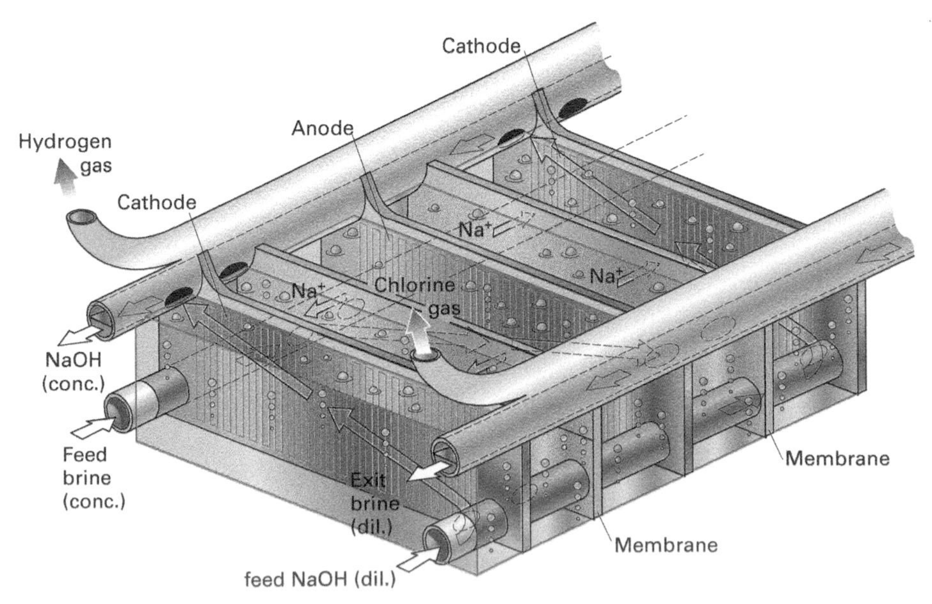

The membrane cell. The electrode processes are the same for the diaphragm cell; only the means of separating the products differs.

Fluorine

Fluorine is extracted by the electrolysis of a molten mixture of HF and KHF_2 at about 100 °C. Fluorine is released at the anode and hydrogen at the cathode. The extraction is extremely hazardous because fluorine will react violently with any moisture, oil, or grease present. The products combine explosively if allowed to mix, and so are kept apart by a diaphragm dipping below the electrolyte surface.

Bromine and iodine

The manufacture of these two halogens depends on the relative ease of oxidation of the halide ions:

$$2X^-(aq) \rightarrow X_2(aq) + 2e^-$$

The order is $I^- > Br^- > Cl^-$, which indicates that chlorine will displace both iodide and bromide ions, forming iodine and bromine respectively, as shown in the following spread.

Bromine may be produced from seawater, which contains about 0.006% by mass of bromide ion. Seawater is treated with chlorine and the free bromine is helped to vaporize by blowing air through the mixture:

$$2Br^-(aq) + Cl_2(aq) \rightarrow 2Cl^-(aq) + Br_2(aq)$$

The reaction is a redox reaction: each chlorine atom removes an electron from a bromide ion.

The red-brown element bromine is manufactured by the oxidation of bromide ions by chlorine.

Iodine is also present in seawater, at a concentration of less than one part per million. Some types of seaweed extract iodine during growth to the extent that their ash after combustion contains about 0.5% by mass of iodine. The ash is dissolved in dilute sulphuric acid to give aqueous iodide ion, and iodine is displaced from this solution by chlorine:

$$2I^-(aq) + Cl_2(aq) \rightarrow 2Cl^-(aq) + I_2(aq)$$

SUMMARY

- Chlorine (and sodium hydroxide) is manufactured by the electrolysis of saturated aqueous sodium chloride.
- Bromine and iodine are manufactured by using chlorine to displace the halogen from an aqueous solution of the halide.

PRACTICE

1 Explain why fluorine cannot be manufactured by the electrolysis of an aqueous fluoride compound.

Uses of fluorine and its compounds

- PTFE – 'non-stick' plastics and coatings
- UF_6 – uranium isotope separation
- BF_3 – petroleum industry catalysts
- Na_3AlF_6 – cryolite in aluminium smelting
- NaF – fluoridation of water.

Note that the previously very widely used chlorofluorocarbon (CFC) refrigerants and propellants (such as 'Freon' CCl_2F_2) are now banned in most countries, because of their adverse effect on the atmospheric ozone layer, spread 15.11.

Uses of bromine and its compounds

- $C_2H_4Br_2$ – petrol additive (dominant use in 1980s, now obsolete)
- AgBr – photographic film emulsion (now obsolescent)
- Medicines and drugs (2% of total use)
- Halons – fire extinguishers, e.g. $CBrClF_2$
- Synthesis of organic chemicals.

Note that the mass of bromine manufactured is one-hundredth that of chlorine.

Uses of iodine and its compounds

- Iodine ointment for skin ulcers
- Antiseptics (e.g. triiodomethane CHI_3)
- Antibacterial agents
- Polarizing light filters.

Iodine

The main source of the element is sodium iodate $NaIO_3$, which is present as an impurity in the mineral form of sodium nitrate, Chile saltpetre.

18.3

O B J E C T I V E S

- Halogens: relative oxidizing strength
- Halogen/halide displacement
- Halides: relative reducing strength
- Laboratory preparation of chlorine

Standard electrode potentials

The oxidizing ability of the halogens in aqueous solution is reflected in their standard electrode potentials. The more positive the value, the greater the oxidizing power of the halogen.

$F_2(g) + 2e^- \rightleftharpoons 2F^-(aq)$ $E^\ominus = +2.87\ V$

$Cl_2(g) + 2e^- \rightleftharpoons 2Cl^-(aq)$ $E^\ominus = +1.36\ V$

$Br_2(l) + 2e^- \rightleftharpoons 2Br^-(aq)$ $E^\ominus = +1.07\ V$

$I_2(s) + 2e^- \rightleftharpoons 2I^-(aq)$ $E^\ominus = +0.54\ V$

Recognition of halogens in organic solvents

The halogens exist as covalent diatomic molecules. Their solubility in the polar solvent water is limited, but they have greater solubility in organic solvents such as tetrachloromethane. Where water and this solvent are in contact, a halogen dissolves preferentially in the organic solvent. The solutions have distinctive colours: chlorine – pale yellow/green; bromine – red/brown; iodine – purple.

Iodine solutions

Iodine is brown in oxygen-containing solvents, such as water and ethanol. It is purple in solvents that do not contain oxygen, such as 1,1,1-trichloroethane and tetrachloromethane.

Fluorine as an oxidant

There are further examples of the great strength of fluorine as an oxidant. For example, it is the only element that is able to combine with the noble gas krypton. Sulphur forms a hexafluoride SF_6 (containing S(VI)) but there is no equivalent compound for the other halogens. Fluorine is so powerful an oxidant that it can oxidize water to molecular oxygen:

$2F_2(g) + 2H_2O(l) \rightarrow 4HF(aq) + O_2(g)$

HALOGENS AND HALIDES: REDOX BEHAVIOUR

In the manufacture of bromine and iodine, chlorine displaces the elements from their respective ions, bromide Br^- and iodide I^-. These displacement reactions follow the rule that a halogen in a *higher* position (of lower atomic number) in Group VII will oxidize a halide ion from *lower* in the group to the corresponding halogen. For example, chlorine added to aqueous iodide ion gives iodine and aqueous chloride ion; there is no reaction when bromine is added to aqueous chloride ion. These displacement reactions depend on the relative ability of the halogens to act as oxidants. Displacement reactions are redox reactions.

Halogen displacement: a reminder

The order of reactivity of the halogens can be demonstrated by reacting chlorine with aqueous bromide and iodide ions. In both cases, chlorine oxidizes the halide ion to the corresponding halogen and is itself reduced to chloride ion:

$$Cl_2(g) + 2Br^-(aq) \rightarrow Br_2(aq) + 2Cl^-(aq)$$

$$Cl_2(g) + 2I^-(aq) \rightarrow I_2(aq) + 2Cl^-(aq)$$

Chlorine is said to have 'displaced' bromine or iodine from solution.

Halogen/halide displacement reactions: upper layer aqueous; lower layer tetrachloromethane. Chlorine water added to fluoride: no reaction (left). Chlorine oxidizes bromide: red bromine in the lower layer (middle). Chlorine oxidizes iodide: purple iodine in the lower layer (right).

Oxidizing power

Fluorine is the strongest oxidant of all the elements. Chlorine is also a very strong oxidant. Bromine is a little weaker as an oxidant, and iodine weaker still. This trend is demonstrated very clearly by the compounds formed by iron when it combines directly with the halogens. Fluorine and chlorine form iron(III) fluoride and iron(III) chloride respectively; these compounds have iron in its higher oxidation state. Bromine forms both iron(III) bromide and iron(II) bromide.

Iodine is too weak an oxidant to form the higher oxidation state and so only forms iron(II) iodide. Indeed, iron(III) ions can oxidize iodide ions to form iodine and iron(II) ions:

$$2Fe^{3+}(aq) + 2I^-(aq) \rightarrow 2Fe^{2+}(aq) + I_2(s)$$

Reduction of iodine by thiosulphate

Fluorine, chlorine, and bromine can all oxidize thiosulphate ion $S_2O_3^{2-}$ to sulphate ion. Iodine is too weak an oxidant to achieve this. Instead it oxidizes thiosulphate ion to tetrathionate ion:

$$I_2(aq) + 2S_2O_3^{2-}(aq) \rightarrow 2I^-(aq) + S_4O_6^{2-}(aq)$$

The oxidation number of sulphur in thiosulphate is +2. Note that the oxidation number of sulphur in tetrathionate is +2.5, that is, not a whole number. As we saw in chapter 13 'Redox equilibrium', this very specific reaction for iodine is extremely useful in the quantitative analysis of oxidants, as follows.

The oxidant is first added to excess iodide ion to produce iodine, and then the liberated iodine is titrated against standard sodium thiosulphate. Sodium thiosulphate has the advantage that it is one of the few reductants that is not oxidized by air. The end point of the titration is easily seen using starch as an indicator added just as the end point is near and the iodine solution has become straw-coloured. Starch forms a deep blue complex in the presence of small quantities of iodine. The end point is when the last trace of blue is removed to leave a colourless solution.

Halide ions as reductants

A trend that follows directly from the fact that the *halogens* become weaker oxidants from fluorine to iodine is that the *halide ions* become stronger reductants from fluoride to iodide. That is, the strength of the halide ions as reductants is in the order $I^- > Br^- > Cl^- > F^-$.

Iodide ion is a common reductant in acidic solution. A useful illustration is the reaction of iodide ions with copper(II) ions, the result being a dull-looking precipitate of copper(I) iodide:

$$2Cu^{2+}(aq) + 4I^-(aq) \rightarrow 2CuI(s) + I_2(aq)$$

The dull colour is caused by the presence of iodine. This contaminant may be removed by adding aqueous sodium thiosulphate to the mixture. The iodine is reduced to iodide ion, revealing the white precipitate of copper(I) iodide.

Cu²⁺(aq) and Cl⁻(aq)

Unlike iodide ion, chloride ion does not reduce copper(II) ions; instead it forms a green complex $CuCl_4^{2-}$. No redox reaction has occurred: copper remains in oxidation state Cu(II).

Laboratory preparation of chlorine

Chlorine may be prepared by oxidizing concentrated hydrochloric acid using a suitably strong oxidant. A common oxidant which works at room temperature is potassium manganate(VII) $KMnO_4$:

$$2KMnO_4(s) + 16HCl(aq) \rightarrow 2KCl(aq) + 2MnCl_2(aq) + 5Cl_2(g) + 8H_2O(l)$$

Manganese(IV) oxide is also successful, although only when the reaction mixture is heated:

$$MnO_2(s) + 4HCl(aq) \rightarrow MnCl_2(aq) + Cl_2(g) + 2H_2O(l)$$

Note that fluorine cannot be obtained from the fluoride ion by chemical oxidation. This reaction requires a more powerful oxidant than fluorine – difficult if not impossible to obtain under normal laboratory circumstances. (Fluorine is manufactured by *electrolysis*, as discussed in the previous spread.)

Preparing chlorine. Concentrated hydrochloric acid is added from the funnel to solid manganese(IV) oxide in the flask. Warming the mixture releases chlorine. Chlorine is denser than air, so the gas is collected by downward delivery.

SUMMARY

- The oxidizing power of the halogens decreases in the order $F_2 > Cl_2 > Br_2 > I_2$.
- The general chemical reactivity of the halogens follows the order of their oxidizing power.
- Iodine reacts quantitatively with thiosulphate ion: $I_2(aq) + 2S_2O_3^{2-}(aq) \rightarrow 2I^-(aq) + S_4O_6^{2-}(aq)$
- A halogen will displace a halide that has a greater atomic number than itself.
- Halide ions may act as reductants; the order of reactivity is $I^- > Br^- > Cl^- > F^-$.
- Iodide ion is the most commonly used halide reductant.

Oxidation numbers

The reactions described above show that the halogens have oxidation number 0 as elements and oxidation number −1 as halide ions. You will encounter higher oxidation numbers in the following spreads.

PRACTICE

1 Decide whether a reaction takes place in each of the following mixtures, and write a balanced chemical equation where appropriate:

a Chlorine and aqueous bromide ions

b Bromine and aqueous iodide ions

c Bromine and aqueous chloride ions.

2 Chlorine gas is dissolved in 1,1,1-trichloroethane. The solution is shaken with an aqueous solution of sodium bromide and then allowed to stand. Describe what you would expect to see after the mixture had been standing for five minutes.

18.4

OBJECTIVES

- Halogen solubility in water
- Halogen reactivity with base
- Disproportionation reactions

Halogens in solution

The halogens dissolve in or react with water to varying extents and show a variety of behaviours. The nature of the behaviour depends largely on the strength of the halogen as an oxidant. For example, fluorine oxidizes water vigorously. The other halogens undergo a series of reactions in which higher halogen oxidation states are reached.

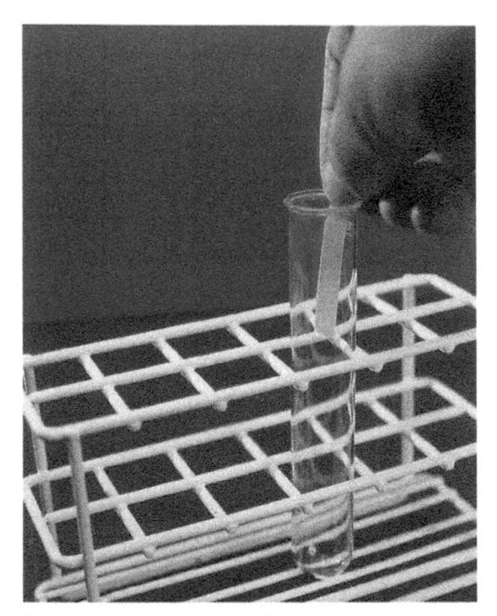

Testing for chlorine. Chlorine gas bleaches a piece of moist litmus paper held in it. The paper must be moist so that the chlorine may dissolve and react to form hypochlorous acid, which bleaches the paper. A red colour (indicating acidity) may be seen very briefly first.

Fluorine plus water

Unlike the other halogens, fluorine *oxidizes* water to oxygen, i.e.

$$2F_2(g) + 2H_2O(l) \rightarrow 4HF(aq) + O_2(g)$$

oxidation state changes: F(0) to F(–I); O(–II) to O(0)

Small amounts of hydrogen peroxide H_2O_2 and oxygen difluoride OF_2 are also formed. The solution contains essentially no dissolved fluorine gas, as it all reacts.

Chlorine plus water

Chlorine dissolves in water to give **chlorine water** $Cl_2(aq)$, and some chlorine then reacts with the water. Some oxidation of water occurs, especially in bright light, but the main reaction is **disproportionation**, which is a reaction in which a species is both oxidized and reduced:

$$Cl_2(aq) + H_2O(l) \rightleftharpoons HCl(aq) + HClO(aq)$$

oxidation state changes: Cl(0) to Cl(-I) and Cl(I)

The product HClO(aq) is called **hypochlorous acid**.

Other names

See spread 13.3 for alternative names used for the halogen oxoanions.

Hypochlorous acid and hypochlorite ion

Aqueous hypochlorous acid is a weak acid that ionizes to give the **hypochlorite ion** ClO^-:

$$HClO(aq) + H_2O(l) \rightleftharpoons H_3O^+(aq) + ClO^-(aq)$$

The hypochlorite ion ClO^- contains chlorine with an oxidation number of +1. It is an oxidant, especially in acidic solution. It will, for example, oxidize iodide ion to iodine:

$$2I^-(aq) + ClO^-(aq) + 2H^+(aq) \rightarrow I_2(aq) + Cl^-(aq) + H_2O(l)$$

The hypochlorite ion acts as a powerful disinfectant and bleach. Disinfectant action depends on its ability to oxidize organic material and thereby to disrupt the life processes of bacteria. Sodium hypochlorite is the active constituent in domestic bleaches and sterilizing fluids (e.g. *Milton*). 'Bleaching powder' contains calcium hypochlorite, and is used commercially for water treatment and for bleaching paper and textiles. Bleaching action depends on the ability of hypochlorite ions to break up the structures of organic colouring matter.

Pool water pH

The optimal pH range for bather comfort and efficiency of disinfection is 7.3 to 7.4. The pH can be controlled by adding sodium hydrogensulphate to reduce pH or sodium carbonate to increase it.

Hypochlorous acid is a good bactericide. 'Chlorine' is therefore used to kill bacteria in swimming pools and drinking water. As you might expect, it is important to use the optimum quantities in each case, with increased concentrations being required in swimming pools. The source of chlorine is often granular calcium hypochlorite $Ca(ClO)_2$ or tablets of 'Trichlor'.

Chlorine plus cold dilute aqueous sodium hydroxide

The equilibrium mixture of chlorine in water is acidic. If aqueous base is added, the equilibrium is shifted much further to the right. For example, sodium hypochlorite is produced when chlorine reacts with *cold* aqueous sodium hydroxide:

$$Cl_2(g) + 2NaOH(aq) \rightarrow NaCl(aq) + NaClO(aq) + H_2O(l)$$

Similarly, the reaction between chlorine and aqueous calcium hydroxide forms calcium hypochlorite:

$$2Cl_2(g) + 2Ca(OH)_2(aq) \rightarrow CaCl_2(aq) + Ca(ClO)_2(aq) + 2H_2O(l)$$

A solution of hypochlorite ion disproportionates at temperatures above about 75 °C to form chloride ion Cl^- and **chlorate ion** ClO_3^-. Note again the changes in oxidation state:

$$3ClO^-(aq) \rightarrow 2Cl^-(aq) + ClO_3^-(aq)$$

oxidation state changes: Cl(I) to Cl(-I) and Cl(V)

Bromine plus water or aqueous sodium hydroxide

Bromine undergoes a similar reaction with water to that of chlorine. The result is hydrobromic acid HBr(aq) and hypobromous acid HBrO(aq):

$$Br_2(aq) + H_2O(l) \rightleftharpoons HBr(aq) + HBrO(aq)$$

However, the equilibrium lies far to the left. Unless the solution is kept at around 0 °C, the **hypobromite ion** BrO^- disproportionates further to give bromide ion Br^- and **bromate ion** BrO_3^-:

$$3BrO^-(aq) \rightarrow 2Br^-(aq) + BrO_3^-(aq)$$

oxidation state changes: Br(I) to Br(-I) and Br(V)

Reacting aqueous bromine with aqueous sodium hydroxide encourages the disproportionation and increases the concentration of bromate ion.

Iodine plus water or aqueous sodium hydroxide

Iodine has very low solubility in water. However, it dissolves in water containing a high concentration of iodide ions, e.g. KI(aq). An equilibrium is established in which iodide ion reacts with iodine to form the **triiodide ion**:

$$I^-(aq) + I_2(s) \rightleftharpoons I_3^-(aq)$$

The solution behaves as if it is a solution of I_2(aq).

The concentration of HIO(aq) in this solution is negligible. Reacting iodine with aqueous sodium hydroxide does not increase the concentration of **hypoiodite ion** IO^-(aq). This species disproportionates to give the **iodate ion** IO_3^-:

$$3IO^-(aq) \rightarrow 2I^-(aq) + IO_3^-(aq)$$

oxidation state changes: I(I) to I(-I) and I(V)

SUMMARY

- Fluorine oxidizes water to give oxygen and hydrofluoric acid.
- Chlorine and bromine dissolve in water and disproportionate to give hypochlorous and hypobromous acids HXO.
- The concentrations of the hypohalite ions ClO^- and BrO^- may be increased by reacting the halogen with a base, e.g. $Cl_2(g) + 2NaOH(aq) \rightarrow NaCl(aq) + NaClO(aq) + H_2O(l)$
- The hypohalous acids are weak acids with stabilities in the order HClO > HBrO > HIO.
- Hypochlorite ion disproportionates to form chloride ion $3ClO^-(aq) \rightarrow 2Cl^-(aq) + ClO_3^-(aq)$

Decomposition

The hypochlorite ion slowly decomposes on standing as a result of two separate reactions. The first involves disproportionation to chloride and chlorate ClO_3^-:

$$3ClO^-(aq) \rightarrow 2Cl^-(aq) + ClO_3^-(aq)$$

oxidation state changes: Cl(I) to Cl(–I) and Cl(V)

The second reaction evolves oxygen gas, and is observed especially in sunlight:

$$2ClO^-(aq) \rightarrow 2Cl^-(aq) + O_2(g)$$

oxidation state changes: Cl(I) to Cl(–I); O(–II) to O(0)

So, in sunlight, chlorine gas reacts in a similar way to fluorine:

$$2Cl_2(g) + 2H_2O(l) \rightarrow 4HCl(aq) + O_2(g)$$

Sodium hypochlorite

This substance can be made commercially by the electrolysis of cooled brine. The electrolysis cell is arranged so that the chlorine produced at the anode mixes with the sodium and hydroxide ions in the solution.

Do not mix!

Some toilet cleaners contain acid to remove limescale. It is important not to mix these with bleach (which contains the hypochlorite ion) in an effort to remove staining. If these two cleaning chemicals are used together, they react according to the following chemical equation:

$$Cl^-(aq) + ClO^-(aq) + 2H^+(aq) \rightarrow Cl_2(aq) + H_2O(l)$$

Choking fumes of poisonous chlorine gas are given off! The acid reverses the equilibrium for the reaction of chlorine with water.

PRACTICE

1 Give the formulae of each of the following:

a Hypochlorous acid

b Hypochlorite ion

c Chlorate ion.

2 For each of the species in question 1, give the oxidation number of the halogen concerned.

18.5 The hydrogen halides

OBJECTIVES

- Physical properties
- Preparation
- Oxidation of hydrogen halides
- Acidic properties

The hydrogen halides are compounds containing a halogen and hydrogen only. They have the formula HX, where the halogen is bonded to hydrogen by a single covalent bond. They are all gases at 25 °C and 1 bar, and may be prepared by the action of a suitable concentrated acid on a halide salt. However, the redox properties of the hydrogen halides differ widely, so the acid used for preparation must be chosen with care.

Physical properties

The boiling points of the hydrogen halides increase from HCl to HI, because the dispersion forces increase as the number of electrons increases. The anomalously high boiling point of HF arises because there is hydrogen bonding between the highly polar molecules. This polarity results from the fact that fluorine is a small, highly electronegative element. See chapter 7 'Changes of state and ...'.

Hydrogen halide boiling points

Hydrogen halide	Boiling point/°C
HF	19
HCl	−85
HBr	−66
HI	−36

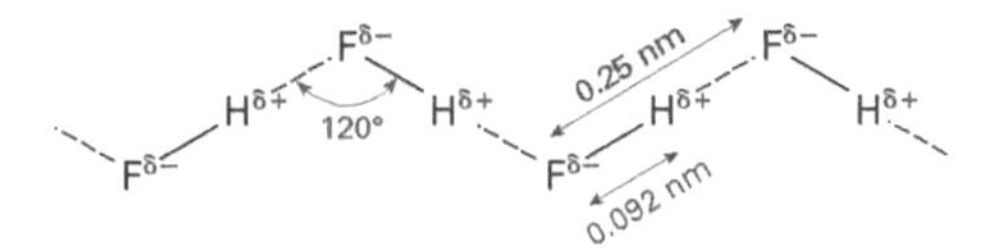

Hydrogen bonding occurs in hydrogen fluoride, due to the high electronegativity of fluorine and hence the extreme polarity of the $H^{\delta+}—F^{\delta-}$ bond.

Laboratory preparation

Hydrogen chloride can be made by heating a chloride salt with concentrated (96%) sulphuric acid:

$$NaCl(s) + H_2SO_4(aq, conc.) \rightarrow HCl(g) + NaHSO_4(aq)$$

The action of concentrated sulphuric acid with (a) NaCl, (b) NaBr, and (c) NaI. The products seen are: (b) some bromine, and (c) copious iodine vapour.

Concentrated sulphuric acid is a powerful oxidant. It may not be used to prepare HBr and HI because these substances are reductants and so are oxidized by concentrated sulphuric acid to bromine and iodine respectively. For example

$$2HBr(aq) + H_2SO_4(aq, conc.) \rightarrow 2H_2O(l) + SO_2(g) + Br_2(g)$$

Hydrogen iodide reduces concentrated sulphuric acid further, to sulphur and hydrogen sulphide. A summary of these reactions is given below.

Hydrogen bromide and hydrogen iodide are prepared by using phosphoric acid, an involatile acid that is *not* an oxidant. For example

$$NaI(s) + H_3PO_4(aq, conc.) \rightarrow HI(g) + NaH_2PO_4(aq)$$

Laboratory apparatus for the preparation of the hydrogen halides. A concentrated acid is run onto a halide salt. Heating the mixture drives off the volatile hydrogen halide, which is dried by concentrated sulphuric acid and collected by downward delivery. (HCl, HBr, and HI are all more dense than air.)

Summary table showing the products of reaction between sodium halides and concentrated sulphuric acid.

Halide	Observations	Products	Oxidation states
NaCl	Steamy fumes	HCl	Cl(−I)
NaBr	Steamy fumes	HBr	Br(−I)
	Brown fumes	Br_2	Br(0)
	Colourless gas	SO_2	S(+IV)
NaI	Steamy fumes	HI	I(−I)
	Purple fumes	I_2	I(0)
	Colourless gas	SO_2	S(+IV)
	Yellow solid	S	S(0)
	Smell of bad eggs	H_2S	S(−II)

Manufacture

Hydrogen chloride and hydrogen bromide are manufactured by direct combination of the elements. Hydrogen fluoride cannot be manufactured in this way because hydrogen reacts explosively with fluorine at temperatures down to –200 °C. Mixtures of hydrogen and chlorine also react explosively when exposed to a spark or to bright light, but a jet of hydrogen burns steadily in an atmosphere of chlorine. Hydrogen and bromine react more slowly, and commercial production requires a temperature of 300 °C and a platinum catalyst. This trend of decreasing ease of production of the hydrogen halides fits the pattern of the general chemical reactivity of the halogens and parallels the very large variation in standard enthalpy change of formation.

Hydrogen fluoride

This substance attacks glass. It is prepared by heating calcium fluoride with concentrated sulphuric acid:

$CaF_2(s) + H_2SO_4(aq, conc.) \rightarrow CaSO_4(aq) + 2HF(g)$

$\Delta_f H^\ominus$ (298 K) for HX(g)

Hydrogen halide	$\Delta_f H^\ominus$ (298 K) /kJ mol^{-1}
HF	–273
HCl	–92
HBr	–36
HI	+27

Thermal stability

Dipping a red-hot wire into hydrogen iodide gas produces copious violet clouds of iodine:

$$2HI(g) \rightarrow H_2(g) + I_2(g)$$

The order of ease of such decomposition of the hydrogen halides is the opposite of the order of reactivity between hydrogen and the halogens. The trend in ease of decomposition depends mainly on the strength of the hydrogen–halogen bond as measured by the bond enthalpy. The size of the atoms increases from fluorine to iodine; therefore the bond length increases and the bond enthalpy *decreases* in this order.

Bond lengths and enthalpies

Bond	Bond length /nm	Bond enthalpy /kJ mol^{-1}
H—F	0.092	562
H—Cl	0.127	431
H—Br	0.141	366
H—I	0.161	299

Hydrogen halides as acids

Dry, gaseous, hydrogen halides are not acidic; they do not affect dry indicator paper. However, their aqueous solutions *are* acidic. For example, hydrogen chloride dissolves readily in water to form hydrochloric acid:

$$HCl(g) + H_2O(l) \rightarrow H_3O^+(aq) + Cl^-(aq)$$

One surprise is that hydrofluoric acid is a *weak* acid (pK_a 3.45) whereas the other hydrogen halides all form strong acids (see chapter 12 'Acid–base equilibrium'). This is because the H—F bond is particularly strong and hydrogen bonding between hydrogen fluoride molecules and water molecules inhibits the ionization of HF. The other hydrogen halides do not show hydrogen bonding.

Weak but not safe

Although a weak acid, hydrofluoric acid causes burns, as do the other aqueous hydrogen halides. HF is particularly insidious because, especially in dilute solution, the skin burns do not usually cause pain until several hours later. Hence if HF penetrates protective clothing, catastrophic damage may occur before it is noticed.

SUMMARY

- Hydrogen halides are produced by the action of concentrated sulphuric acid or concentrated phosphoric acid on a metal halide salt.
- An aqueous solution of HF is weakly acidic; HCl, HBr, and HI are all strong acids in aqueous solution.
- Reducing power is in the order (powerfully reducing) HI > HBr > HCl > HF (non-reducing).
- Bond enthalpy (and hence thermal stability) follows the order HF > HCl > HBr > HI.

PRACTICE

1 Write balanced chemical equations for the laboratory preparation of HF, HCl, HBr, and HI.

2 Sketch the apparatus you would use to prepare a sample of dry hydrogen bromide. [Note that a suitable drying agent is solid anhydrous calcium bromide.]

3 Write balanced chemical equations for the following:

a The manufacture of HCl by burning hydrogen in an atmosphere of chlorine.

b The dissolution of HI gas in water to make an acidic solution.

c Liquid anhydrous HF protonating itself (thus accounting for the moderate electrical conductivity of the liquid).

4 Explain why hydrogen bonding occurs in hydrogen fluoride but not in the other hydrogen halides.

5 Explain why HI is a more powerful reductant than HCl.

18.6

OBJECTIVES

- Chemical tests for halide ions
- Solubility of silver halides

Some ionic halides

The physical properties of the halogens vary considerably, from gaseous fluorine and chlorine to liquid bromine and solid iodine. However, the ionic compounds that they form have similar formulae and are indistinguishable by appearance alone. For example, the compounds NaF, NaCl, NaBr, and NaI are all white crystalline solids which are soluble in water. While the physical properties are similar, the chemical properties are distinctive to each compound. These chemical properties form the basis of this spread. It ends with more details about the anomalous behaviour of fluorides.

Halogens

The name 'halogen' comes from the Greek for 'salt former'.

Oxidation number

Ionic halides contain the halide ions F^-, Cl^-, Br^-, and I^-. These ions form when a halogen accepts one electron to complete its valence shell of electrons. The oxidation number of –1 is found in all the ionic halides of metals. The formula of the Group I halides is MX; the formula of the Group II halides is MX_2.

A chemical test for halide ions

The silver halides are used to identify the chloride, bromide, and iodide ions. These halides may be identified in a solution by adding acidified silver nitrate and observing the colour of the precipitate formed. Silver fluoride is soluble and its solution is colourless, so the fluoride ion cannot be identified by the test. Silver chloride, silver bromide, and silver iodide, however, are insoluble in water and are precipitated. The colours of the precipitates are as follows:

- Silver chloride is white.
- Silver bromide is cream.
- Silver iodide is yellow.

Adding nitric acid

The solution to be tested for halide ions is acidified with nitric acid before adding the aqueous silver nitrate. The presence of this acid prevents the formation of other insoluble silver compounds, such as silver carbonate. The carbonate would react with the acid to form carbon dioxide gas.

The ionic equation for the precipitation of the chloride, for example, is

$$Ag^+(aq) + Cl^-(aq) \rightarrow AgCl(s)$$

Confirmation is obtained by observing the behaviour when aqueous ammonia is added to these precipitates:

- Silver chloride dissolves in dilute aqueous ammonia.
- Silver bromide dissolves in concentrated aqueous ammonia.
- Silver iodide remains as a precipitate in both dilute and concentrated aqueous ammonia.

Actually, the precipitate does not simply *dissolve* to give the aqueous ions $Ag^+(aq)$ and $X^-(aq)$. What happens is that a *complex ion* (see chapter 20 'The transition metals') is formed between the silver ion and ammonia:

$$AgCl(s) + 2NH_3(aq) \rightarrow [Ag(NH_3)_2]^+(aq) + Cl^-(aq)$$

Remember that bromide and iodide ions may also be identified from their reaction with concentrated sulphuric acid (see the previous spread).

Halide ionic radii (to scale).

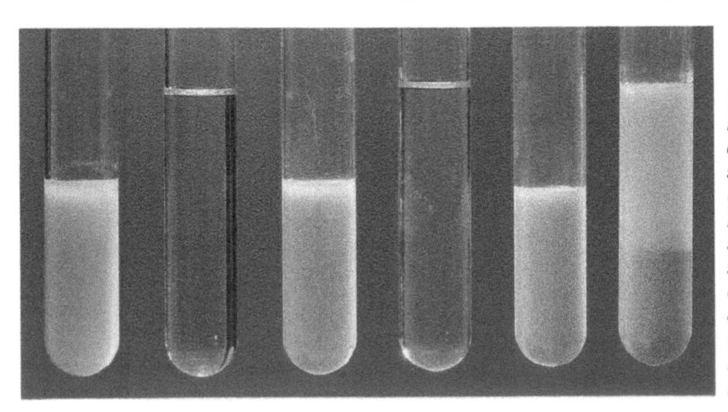
Tubes 1, 3, 5: Precipitates of silver chloride (white), silver bromide (cream), and silver iodide (yellow). The solids are light-sensitive and darken over time. Tubes 2, 4, 6: The effect on the precipitates of adding concentrated aqueous ammonia.

Trends in solubility

Why is silver chloride more soluble than silver iodide in aqueous ammonia? Solubility involves a balance between lattice enthalpy and enthalpy of hydration. Lattice enthalpy and enthalpy of hydration both become smaller when the anion is larger. The value of both of these enthalpy changes will therefore be smaller for silver iodide than for silver chloride because the iodide ion is larger than the chloride ion. The lower enthalpy of hydration for the iodide ion is mainly responsible for the lower solubility of silver iodide.

Quantitatively, the enthalpy of solution can be calculated by adding the (endothermic) lattice enthalpy and the (exothermic) enthalpies of hydration of both the silver and halide ions (see box). The enthalpy of solution of AgCl is +66 kJ mol^{-1}. For AgI the enthalpy of solution is +112 kJ mol^{-1}, i.e. AgI is less likely to dissolve than AgCl. These endothermic values should be compared with the enthalpy of solution of AgF, which is exothermic (–20 kJ mol^{-1}).

The enthalpy change on forming a complex ion with ammonia is exothermic. The value is *just* sufficient to take AgCl into solution, but is not large enough to give the same result with AgI.

Adding the enthalpies

Silver chloride	/kJ mol^{-1}
$\Delta_{lat}H^{\ominus}$	+915
$\Delta_{hyd}H^{\ominus}$ (Ag^+)	–472
$\Delta_{hyd}H^{\ominus}$ (Cl^-)	–377
$\Delta_{sol}H^{\ominus}$ (AgCl)	+66
Silver iodide	**/kJ mol^{-1}**
$\Delta_{lat}H^{\ominus}$	+889
$\Delta_{hyd}H^{\ominus}$ (Ag^+)	–472
$\Delta_{hyd}H^{\ominus}$ (I^-)	–305
$\Delta_{sol}H^{\ominus}$ (AgI)	+112
Silver fluoride	**/kJ mol^{-1}**
$\Delta_{lat}H^{\ominus}$	+967
$\Delta_{hyd}H^{\ominus}$ (Ag^+)	–472
$\Delta_{hyd}H^{\ominus}$ (F^-)	–515
$\Delta_{sol}H^{\ominus}$ (AgF)	–20

More about solubility

Ionic chlorides, bromides, and iodides of a given metal tend to have fairly similar solubilities in water, but the solubilities of the fluorides are often anomalous. You have already seen above that the silver halides are all insoluble with the exception of silver fluoride. On the other hand, calcium fluoride is *insoluble*, whereas the other halides of calcium are soluble.

Water of crystallization

The small size and large enthalpy of hydration of the fluoride ion mean that some fluorides retain water molecules as they crystallize. These water molecules are called **water of crystallization** (see spread 3.4). For example, crystalline potassium fluoride may have the formula $KF.2H_2O$. The other potassium halides crystallize with the formulae KCl, KBr, and KI, i.e. with no associated water of crystallization.

SUMMARY

- The halide ions all have an oxidation number of –1.
- A solution of silver nitrate acidified with dilute nitric acid is used to test for chloride, bromide, and iodide ions. The precipitates that form are white, cream, and yellow respectively. Aqueous ammonia is used to confirm the results.
- The enthalpy of solution for an ionic halide is the sum of the endothermic lattice enthalpy and the exothermic enthalpy of hydration.

PRACTICE

1 Why is nitric acid, and not sulphuric or hydrochloric acid, used to acidify solutions in the silver nitrate test?

2 Why would sodium carbonate not give a precipitate if you used it in the silver nitrate test?

3 Write the chemical formulae of the following substances:

a Strontium bromide

b Silver iodide

c Caesium fluoride

d Copper(II) bromide

e Chromium(III) chloride.

4 Use the following data to predict the relative enthalpies of solution of calcium fluoride and calcium chloride:

Calcium fluoride	/kJ mol^{-1}
$\Delta_{lat}H^{\ominus}$	+2630
$\Delta_{hyd}H^{\ominus}$ (Ca^{2+})	–1587
$\Delta_{hyd}H^{\ominus}$ (F^-)	–515
Calcium chloride	**/kJ mol^{-1}**
$\Delta_{lat}H^{\ominus}$	+2258
$\Delta_{hyd}H^{\ominus}$ (Ca^{2+})	–1587
$\Delta_{hyd}H^{\ominus}$ (Cl^-)	–377

PRACTICE EXAM QUESTIONS

1 **a** State and explain the trend in the electronegativity of the halogens down Group VII. [3]

b State and explain the trend in boiling temperatures of the halogens down Group VII. [3]

c The relative molecular masses of bromine, Br_2, and iodine monochloride, ICl, are almost the same, yet their boiling temperatures are quite different. Account for this difference in boiling temperature. [4]

2 Hydrogen iodide can be prepared by adding water to a mixture of red phosphorus and iodine, and then warming gently.

a Construct the following equations:

i phosphorus and iodine forming phosphorus tri-iodide;

ii phosphorus tri-iodide and water reacting to form hydrogen iodide and phosphoric(III) acid, H_3PO_3. [2]

b What would you expect to see when hydrogen iodide reacts with:

i aqueous silver nitrate, followed by aqueous ammonia;

ii warm concentrated sulphuric acid? [3]

3 **a** **i** How does concentrated sulphuric acid react with sodium chloride? Write an equation for the reaction. Suggest an appropriate temperature at which it might be carried out.

ii Sodium iodide does not react in this way. Give an equation for the reaction which occurs and explain the difference. [4]

b By reference to the structure and bonding in hydrogen fluoride explain why it is a much weaker acid than the other halogen hydrides. [3]

c Given samples of chloride and iodide salts, how would you distinguish them other than by using concentrated sulphuric acid? [3]

d **i** On the basis of the redox potentials

	$E^{\ominus}$/V
$Cl_2 + 2e^- \rightleftharpoons 2Cl^-$	+1.36
$Br_2 + 2e^- \rightleftharpoons 2Br^-$	+1.09

explain what occurs when chlorine is bubbled into a solution containing bromide ions.

ii What is the industrial significance of this reaction? [5]

4 **a** Phosphorus forms trihalides of formula PX_3 for X = F, Cl, Br and I, and pentahalides, PX_5, for X = F, Cl and Br.

i Give the name of the type of bonding present in PCl_3 and draw a dot-and-cross diagram to show it. (Only the outer electrons need be shown.) [2]

ii Draw a diagram to show the shape of the molecule of PCl_3 and briefly explain why the molecule has the shape given. [4]

iii Outline a suitable method for the preparation of PCl_5 in the laboratory. [4]

iv Give the formulae of the species present in solid phosphorus pentachloride. [2]

v Phosphorus pentafluoride is a gas at room temperature and pressure while phosphorus pentachloride is a solid. Suggest an explanation for this difference in physical states. [3]

vi Suggest why phosphorus pentaiodide is unknown. [1]

b PCl_5 can be used as a reagent in preparative organic chemistry as indicated in the scheme below:

$$\underset{\mathbf{R}}{C_3H_7OH} \longrightarrow \underset{\mathbf{S}}{C_2H_5COOH} \xrightarrow[\text{room temperature}]{PCl_5} \mathbf{T}$$

i Give the name and the graphical formula of compound **T**. [2]

ii Give **one** reason why care must be taken when phosphorus pentachloride is added to **S**, and one relevant observation. Write a balanced equation for the reaction. [3]

iii Draw the **two** graphical formulae for the isomeric alcohols of formula C_3H_7OH and state, with a reason, which isomer must be compound **R**. [4]

iv Give the names of the reagents required to convert **R** into **S** in the laboratory and identify the practical steps involved in obtaining a sample of **S**. [3]

(See also chapter 25.)

5 **a** Outline the process by which chlorine is manufactured from brine. [2]

b Describe how chlorine reacts with:

i hot, aqueous sodium hydroxide;

ii aqueous potassium bromide;

iii ethene.

In each case, describe what is seen, write an equation and identify the type of reaction occurring. [8]

6 **a** What oxidation numbers do the elements sodium to phosphorus show in their chlorides? Outline the reactions, if any, of these chlorides with water and relate these reactions to the bonding present. [5]

b Sulphur and chlorine can react together to form S_2Cl_2. When 1.00 g of this sulphur chloride reacted with water, 0.36 g of a yellow precipitate was formed, together with a solution containing a mixture of sulphurous acid, H_2SO_3, and hydrochloric acid.

i Use the above data to deduce the equation for the reaction between S_2Cl_2 and water. [3]

ii What volume of $1.00\ mol\ dm^{-3}$ sodium hydroxide would be required to neutralize the final solution? [2]

iii Write two ionic half-equations for this process which illustrate your definition of disproportionation. [4]

7 The major natural source of fluorine is the mineral fluorspar, which is mainly calcium fluoride, CaF_2.

a i Construct a Born–Haber cycle for the formation of CaF_2 from its elements.

ii Use the cycle to calculate the lattice energy of $CaF_2(s)$. Incorporate the following data as well as relevant data given in the *Data Booklet*.

$\Delta H^{\ominus}_{at}(Ca) = +178\ kJ\ mol^{-1}$

$F(g) \rightarrow F^{-}(g)$; $\Delta H^{\ominus} = -328\ kJ\ mol^{-1}$
(this is the electron affinity of fluorine)

$\Delta H^{\ominus}_{f}(CaF_2) = -1220\ kJ\ mol^{-1}$ [5]

b The first stage in liberating the fluorine from CaF_2 is to grind this compound up and react it with concentrated sulphuric acid. The products are hydrogen fluoride and calcium sulphate, $CaSO_4$.

i Write a balanced equation for this reaction.

ii Calculate the enthalpy change for this reaction, by using the following data in addition to those given above:

$\Delta H^{\ominus}_{f}(H_2SO_4) = -814\ kJ\ mol^{-1}$

$\Delta H^{\ominus}_{f}(HF) = -271\ kJ\ mol^{-1}$

$\Delta H^{\ominus}_{f}(CaSO_4) = -1434\ kJ\ mol^{-1}$

iii Should the reaction be heated or cooled? Give a reason for your answer. [5]

8 a This question concerns the essential features of the electrolytic process by which sodium hydroxide is manufactured in a diaphragm cell.

i Identify the electrolyte used. [1]

ii Give the equation for the reaction at the anode. [1]

iii Give the equation for the reaction at the cathode. [1]

iv Give the equation for the overall reaction occurring in the cell. [1]

b i How would the electrolytic process be modified in order to manufacture sodium chlorate(I)? [1]

ii Write an equation to show how sodium chlorate(I) is formed. [1]

iii Give one large scale use of sodium chlorate(I). [1]

c Chlorate(I) ions are capable of undergoing **disproportionation**.

i What is meant by the term disproportionation? [1]

ii Write an ionic equation for the disproportionation of sodium chlorate(I). Indicate the oxidation numbers of chlorine in each species in which it occurs. [2]

9 This question concerns the chemistry of halides.

a Describe and explain the trend in thermal stability of the hydrogen halides. [2]

b Halide ions react with concentrated sulphuric acid in different ways.

i Chloride ions react to give hydrogen chloride as the only gaseous product. Iodide ions react to give hydrogen iodide, HI, and other gaseous products. Identify two of these gaseous products.

ii Explain why these halide ions do not react in the same way with concentrated sulphuric acid. [3]

c An aqueous solution containing an unknown halide ion was acidified with nitric acid; aqueous silver nitrate was then added. A cream precipitate was obtained which dissolved in concentrated aqueous ammonia.

Identify the halide ion. [1]

d Aqueous KI was used to obtain I_2 by electrolysis. A current of 1.56 A was passed through the aqueous KI until 3.81 g of I_2 were collected at one electrode.

Calculate how many moles of I_2 were collected. [1]

10 Commercial bleaches contain sodium hypochlorite (sodium chlorate(I)), which acts as an oxidizing agent. The concentration of sodium hypochlorite in solution can be determined by reaction with acidified potassium iodide solution.

$NaOCl + 2KI + H_2SO_4 \rightarrow I_2 + H_2O + NaCl + K_2SO_4$

The liberated iodine is titrated with standard sodium thiosulphate solution:

a Which species is oxidised in this equation? [1]

b Explain the term **standard** sodium thiosulphate solution. [1]

c Name a suitable indicator for this titration, stating the expected colour change at the end point. [3]

d Write the equation for the reaction between iodine and sodium thiosulphate. [2]

e $15.0\ cm^3$ of a bleach sample was diluted to $250\ cm^3$ with de-ionized water. $25.0\ cm^3$ portions of the solution were treated with excess acidified potassium iodide solution and then titrated with 0.1 M sodium thiosulphate solution. The average titre was found to be $25.2\ cm^3$. Calculate the concentration of sodium hypochlorite in the original bleach sample. [4]

19 The p-block elements

The p block includes all the elements in Groups III to VIII inclusive in the periodic table. You have already encountered details of some p-block elements earlier in this book: chapter 17 'Trends across a period' included the elements aluminium, silicon, phosphorus, sulphur, and chlorine; and chapter 18 'The halogens' included the elements fluorine, chlorine, bromine, and iodine. The s-block elements are mostly highly reactive metals, while most d-block elements are relatively unreactive metals with high melting points. By comparison, the p-block elements show an enormously wide range of properties, from highly reactive non-metals, to mildly reactive metals, to chemically inert gases. However, there are patterns to be discovered amongst all this variety. As always, electronic structure and bonding dictate both the physical natures of the elements and the chemical properties they exhibit.

THE P-BLOCK ELEMENTS TO BE STUDIED

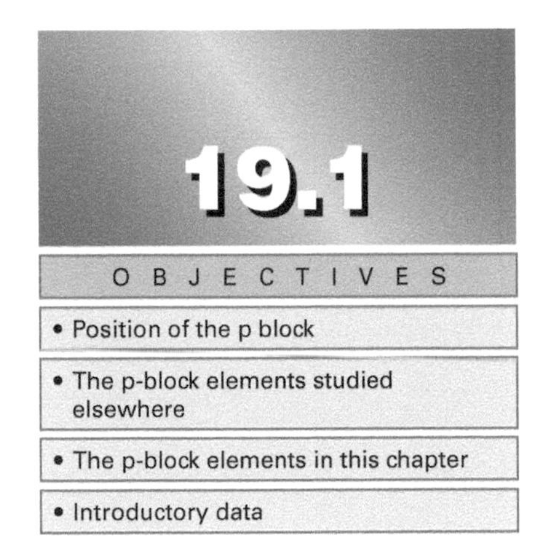

The position of the p block in the periodic table is shown below, with the elements studied in this chapter highlighted. You will notice that there are some elements, in addition to the halogens, that will *not* be discussed here in detail. Gallium, indium, and thallium of Group III, arsenic, antimony, and bismuth of Group V, and selenium, tellurium, and polonium of Group VI will be mentioned, but for comparison only. These elements fall outside the scope of this book because they are rare and have unusual properties and uses. Study of their detailed chemical properties is generally only carried out in universities and industrial research departments.

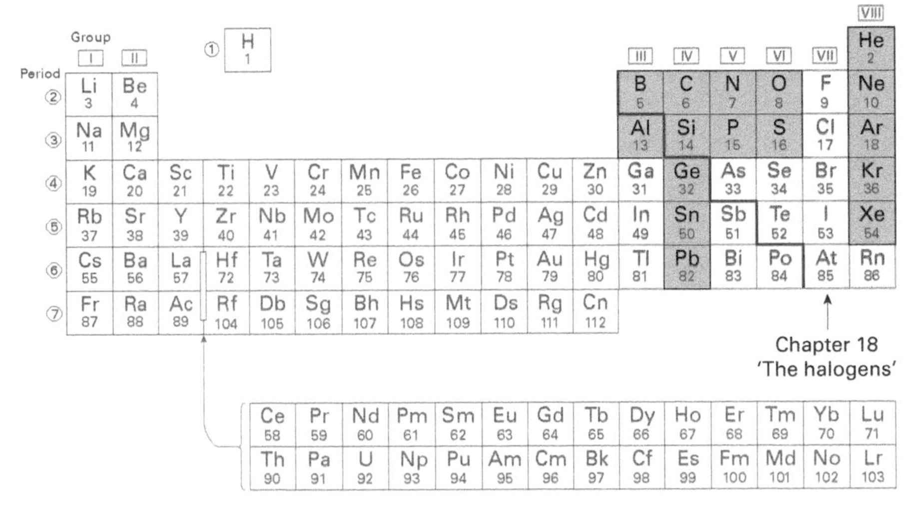

The position of the p block in the periodic table. Elements to be studied in this chapter are highlighted.

Electronic structures

The electronic structure of all p-block elements in Groups III to VII corresponds to the electronic structure of a noble gas together with a full outer s subshell and an incomplete p subshell. For example, the electronic structure of silicon Si (Period 3, Group IV) is $[Ne]3s^2 3p^2$. In addition, the p-block elements of Periods 4 and 5 have a complete inner d subshell, e.g. germanium Ge is $[Ar]3d^{10}4s^2 4p^2$. The p-block elements of Period 6 also have a complete f subshell, e.g. lead Pb is $[Xe]4f^{14}5d^{10}6s^2 6p^2$.

Two points that you should note are that the outermost orbitals are p orbitals, and that the total number of s and p electrons is equal to the group number. For example, the electronic structures of the three elements given above all end with ns^2np^2. Each of these elements is therefore a member of Group IV (i.e. 2 + 2 = 4). The element sulphur S has the electronic structure $[Ne]3s^23p^4$ and is therefore a member of Group VI.

Metal or non-metal?

Remember that, within a *group*, metallic character *increases* with increasing atomic number Z of the element. In Group IV, carbon ($Z = 6$) is a non-metal, germanium ($Z = 32$) is a metalloid, and lead ($Z = 82$) is a metal.

Within a *period*, metallic character *decreases* with increasing atomic number. In Period 3, aluminium ($Z = 13$) is a metal, silicon ($Z = 14$) is a metalloid, and phosphorus ($Z = 15$) is a non-metal.

The p-block elements for study in this chapter

- *Group III*: the non-metal boron and the metal aluminium.
- *Group IV*: the non-metal carbon, the metalloids silicon and germanium, and the metals tin and lead.
- *Group V*: the non-metals nitrogen (a gas) and phosphorus (a solid).
- *Group VI*: the non-metals oxygen (a gas) and sulphur (a solid).
- *Group VIII*: the noble gases helium, neon, argon, krypton, and xenon.

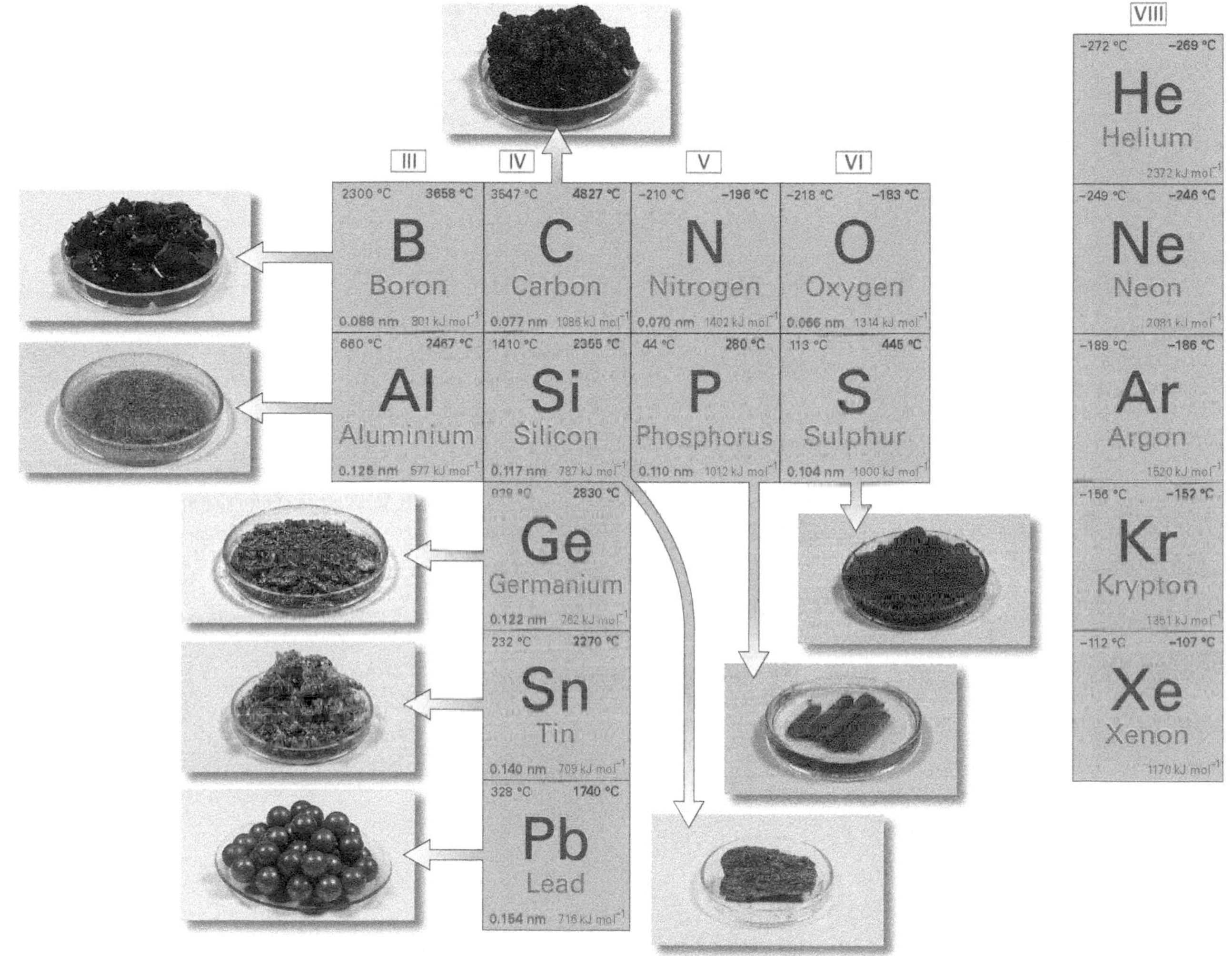

This chart summarizes the major properties of the p-block elements, with particular emphasis on those elements discussed in this chapter. The noble gases, nitrogen, and oxygen are all colourless gases.

SUMMARY

- The p-block elements in Groups III to VII inclusive have incompletely filled p orbitals.
- The elements of Group IV show the greatest change in character, from non-metallic (carbon) to metallic (lead).

PRACTICE

1 By inspecting the electronic structures only, identify the group number for each of the following p-block elements:

a $[He]2s^22p^2$ **b** $[Ne]3s^23p^6$

c $[Ar]3d^{10}4s^24p^4$

2 Sort the p-block elements to be studied under the three headings: 'Metal', 'Metalloid', and 'Non-metal'.

19.2 GROUP III: BORON AND ALUMINIUM

OBJECTIVES

- Boron and aluminium compared
- Manufacture of aluminium
- Uses of aluminium
- Boron and aluminium oxides compared

Hypothetical ionic boron compounds

The boron B^{3+} ion is extremely small and highly charged. The resulting high charge density would make the B^{3+} ion highly polarizing. In an ionic lattice, B^{3+} ions would withdraw electron density from neighbouring anions, concentrating it between the boron ions and the anions; the bonding becomes covalent.

Solid structures

Boron (melting point T_m = 2300 °C) has a unique covalent structure based on B_{12} molecules shaped as an icosahedron, a 20-sided polyhedron. Aluminium (T_m = 660 °C), on the other hand, has a close-packed metallic structure.

Charles Hall. His chemistry professor convinced him there was a fortune to be made if aluminium could be manufactured cheaply. The price of aluminium fell by a factor of 50 between 1855 and 1890. Earlier, no king nor emperor could afford to build a statue of himself out of aluminium: the first was that of Charles Hall, which stands in Oberlin College, Ohio.

The Group III elements are, in order of increasing atomic number, boron (B), aluminium (Al), gallium (Ga), indium (In), and thallium (Tl). All have a valence-shell electronic structure of ns^2np^1; all, except for boron, are metals. This spread and the following one concentrate mainly on the chemistry of aluminium.

Metallic and non-metallic character

Boron is a hard, brittle solid with poor electrical conductivity. Aluminium is a typical metal, being shiny, malleable, ductile, and with good electrical and thermal conductivities. Both boron and aluminium show an oxidation number of +3. The compounds of boron are all covalent; those of aluminium have considerable covalent character.

The reason that boron does not form ionic compounds is the extremely large value of the standard enthalpy change of formation of the B^{3+} ion ($\Delta H^\ominus$ = +6888 kJ mol^{-1}). To form an ionic solid, this endothermic step must be balanced by the exothermic lattice formation enthalpy. Compounds of boron are therefore essentially exclusively covalent.

The manufacture of aluminium

Aluminium is manufactured by the Hall process (or **Hall–Héroult process**). In 1886, Charles Hall in the USA and Paul Héroult in France independently developed the modern process for the extraction of aluminium by electrolysis. (Both were also 23 years old.) Electrolysis is very expensive but is necessary as carbon reduction is not possible, spread 13.13. The process uses an electrolyte of aluminium oxide, produced by purifying the ore *bauxite*, dissolved in molten cryolite (sodium hexafluoroaluminate Na_3AlF_6) at 950 °C. The melting point of aluminium oxide is 2070 °C, far too high for it to be used alone as an electrolyte. An electric current of about 100 000 A is passed through the liquid, heating it. Aluminium is formed by reduction at the cathode, spread 9.2, according to the electrode reaction:

$$Al^{3+}(l) + 3e^- \rightarrow Al(l)$$

Oxide ions are oxidized to oxygen at the anode: $2O^{2-}(l) \rightarrow O_2(g) + 4e^-$. The carbon anode is eroded as its surface is attacked by the oxygen.

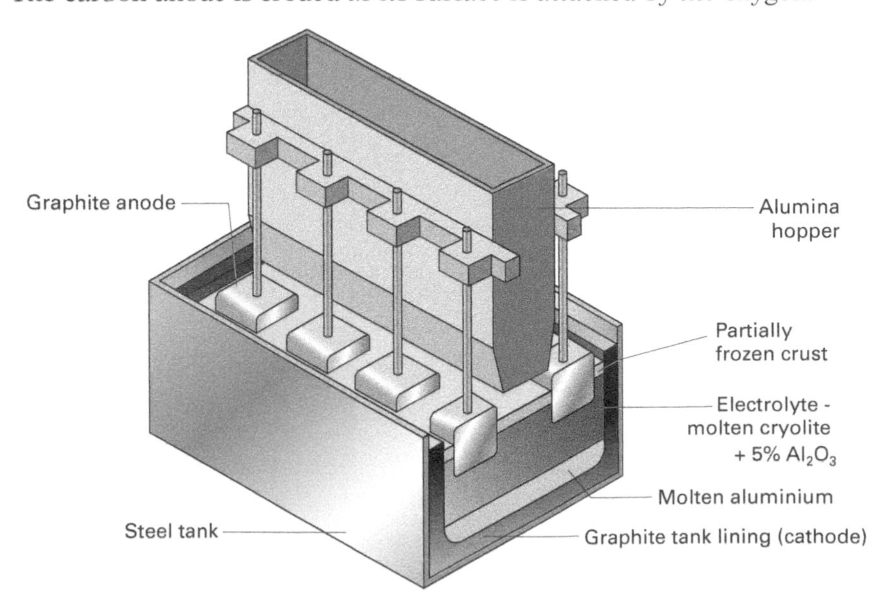

The Hall cell for the industrial extraction of aluminium.

Reactions of the elements

Boron is relatively inert, because of the short and strong B—B bonds. It ignites above 700 °C in air, forming the oxide B_2O_3 and the nitride BN. It does not react with dilute acids or aqueous bases, but combines at high temperatures with chlorine to form boron trichloride BCl_3.

The chemistry of aluminium is much more interesting.

- Aluminium burns when heated in oxygen

$$4Al(s) + 3O_2(g) \rightarrow 2Al_2O_3(s)$$

Aluminium is a strong reductant (standard electrode potential $E^{\ominus} = -1.66\,V$; compare with iron $E^{\ominus} = -0.44\,V$). However, its apparent reactivity is lower than that expected from these data. Aluminium does indeed react with air and water when exposed, but the reaction stops at the metal surface. The transparent aluminium oxide coat formed by this reaction is very thin indeed, about 10 nm (equivalent to a few atoms), but this layer protects the metal from further reaction. Note that an oxide layer does not always give protection. The rust that forms on iron is also an oxide, but rusting is a *corrosive* process, not a protective one.

- Aluminium reacts with acids and aqueous bases

Cleaned aluminium reacts with *acids* to give hydrogen and a solution of a salt:

$$2Al(s) + 6HCl(aq) \rightarrow 2AlCl_3(aq) + 3H_2(g)$$

It also reacts with concentrated aqueous *bases* to give an aqueous aluminate salt:

$$2Al(s) + 2NaOH(aq) + 6H_2O(l) \rightarrow 2NaAl(OH)_4(aq) + 3H_2(g)$$

- Aluminium combines directly with chlorine when heated (spread 17.7)

$$2Al(s) + 3Cl_2(g) \rightarrow 2AlCl_3(s)$$

Aluminium chloride has predominantly covalent character, and will be discussed further in the next spread.

Aluminium in the nineteenth century

In the nineteenth century, aluminium was rare and highly prized. At the Paris Exposition of 1855, it was exhibited next to the crown jewels. At state banquets, Emperor Napoleon III gave his favoured guests aluminium, rather than gold, cutlery. The Washington Monument was topped with aluminium in 1884, spread 13.13.

Aluminium powder reacts vigorously with oxygen.

Oxide character

Boron oxide B_2O_3 is *acidic*, as you would expect for an oxide of a non-metal. It reacts readily with water to form boric acid (a weak acid, $pK_{a1} = 9.2$):

$$B_2O_3(s) + 3H_2O(l) \rightarrow 2H_3BO_3(aq)$$

Aluminium oxide is *amphoteric*; see spread 17.5. It is insoluble in water but reacts with acids and bases:

in acids $\quad Al_2O_3(s) + 6H^+(aq) \rightarrow 2Al^{3+}(aq) + 3H_2O(l)$

in bases $\quad Al_2O_3(s) + 3H_2O(l) + 2OH^-(aq) \rightarrow 2[Al(OH)_4]^-(aq)$

The $[Al(OH)_4]^-$ ion is called the tetrahydroxoaluminate ion (usually abbreviated to 'aluminate ion').

Aluminium alloys (e.g. with copper, zinc, magnesium, and silicon) are strong and corrosion-resistant. Unlike iron, it is maintenance-free. Aluminium's density is about one-third that of iron, which explains its extensive use in the Apollo Lunar Excursion Module (LEM). Other uses include aircraft, window frames, overhead electricity cables, and soft drink cans.

SUMMARY

- All Group III elements have the valence-shell electronic structure ns^2np^1.
- All Group III elements, with the exception of boron, are metals.
- The compounds of boron are covalent; the compounds of aluminium are on the borderline between ionic and covalent, but AlF_3 and Al_2O_3 are predominantly ionic.
- Boron oxide is acidic; aluminium oxide is amphoteric.

PRACTICE

1 Write a balanced chemical equation for each of the following reactions:

 a The reduction of boron oxide by magnesium to form boron.

 b The reduction of boron tribromide by hydrogen to form boron.

 c Reacting aluminium metal with sulphuric acid.

 d Precipitating aluminium hydroxide from a solution of aluminium sulphate.

 e Heating aluminium hydroxide to form aluminium oxide.

 f Reacting aluminium oxide with sulphuric acid.

2 Explain why boron

 a is extremely hard, and

 b has a high melting point (above 2000 °C).

3 Explain why sodium chloride is ionic whereas aluminium chloride is on the borderline between ionic and covalent.

19.3

OBJECTIVES

- Acidic nature of $Al^{3+}(aq)$
- Reactions with anions
- Aluminium and boron halides as Lewis acids

Group III: the acidity of aluminium compounds

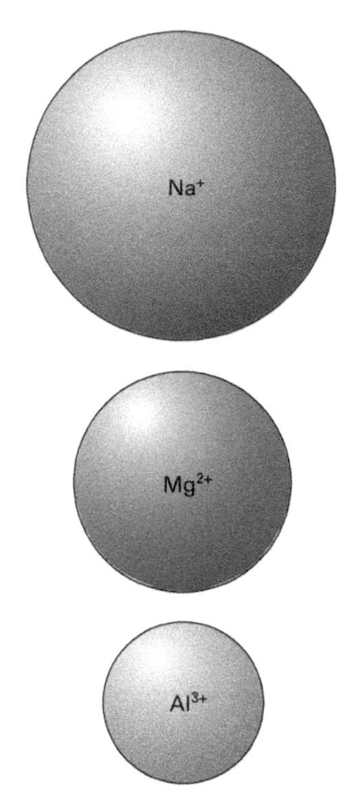

The relative sizes of the Na^+, Mg^{2+}, and Al^{3+} ions.

The aluminium ion Al^{3+} has a particularly high charge density due to its triple positive charge and small size. As a result, this ion has properties in solution which are different from those of 'well-behaved' ions such as Na^+ and Mg^{2+}. The symbol $Al^{3+}(aq)$ has a different meaning to the symbols $Na^+(aq)$ and $Mg^{2+}(aq)$. A further unusual property of aluminium (and boron) is electron deficiency in simple covalent compounds such as $AlCl_3$: the presence of just six electrons in the valence shell (rather than the usual eight) allows them to accept lone pairs from other atoms. Boron and aluminium halides are thus able to act as Lewis acids.

Aluminium chloride

The halides of aluminium may be prepared by direct combination of the elements on heating. They are borderline ionic/covalent with the exception of aluminium fluoride, which contains the Al^{3+} ion. Aluminium chloride is a solid that **sublimes** (changes directly from solid to gas) at 180 °C, to produce an equilibrium mixture of $AlCl_3$ and Al_2Cl_6 (for comparison, NaCl *melts* at T_m = 801 °C and $MgCl_2$ at T_m = 714 °C). The formation of the dimer Al_2Cl_6 is dealt with in spread 17.7.

With the exception of AlF_3, the anhydrous halides are hydrolysed by water:

$$Al_2Cl_6(s) + 6H_2O(l) \rightarrow 2Al(OH)_3(s) + 6HCl(g)$$

The solution equilibria of aluminium

The sodium ion Na^+, the magnesium ion Mg^{2+}, and the aluminium ion Al^{3+} all have 10 electrons. Each has the electronic structure $1s^2 2s^2 2p^6$. The aqueous sodium and magnesium ions, $Na^+(aq)$ and $Mg^{2+}(aq)$, consist of the metal ion surrounded by a number of water molecules making up a layer called the 'hydration sphere'. Water molecules are polar and the δ– oxygen atoms are attracted towards the positively charged metal ions.

An aluminium ion Al^{3+} has a much greater charge density than either Mg^{2+} or Na^+ because it is smaller and more highly charged. The high charge density makes the aluminium ion more strongly polarizing than sodium or magnesium ions. The aluminium ion attracts electron density from the lone pairs on the oxygen atoms of water molecules into its empty 3s, 3p, and 3d orbitals. Each aluminium ion becomes symmetrically surrounded by six water molecules, forming a complex ion (see the next chapter) with the formula $[Al(H_2O)_6]^{3+}$. Note that, although the water molecules are written as H_2O in this formula, bond formation is between the *oxygen* atom and the central aluminium ion.

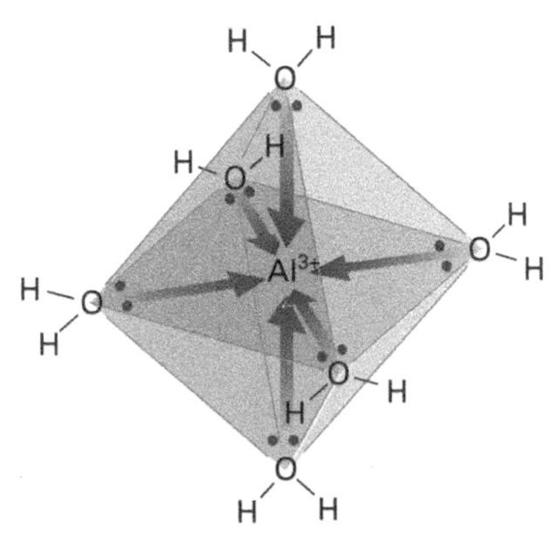

The structure of the $[Al(H_2O)_6]^{3+}$ complex ion. The six water molecules are referred to as ligands. They are evenly distributed about the central aluminium ion, giving the complex octahedral symmetry.

The acidic character of $Al^{3+}(aq)$

The high charge density of the central aluminium ion withdraws electron density from the O—H bonds in the water molecules, so weakening them. The hydrated complex ion is therefore likely to lose a proton and behave as an acid. The aluminium ion can act as a Brønsted acid (by donating a proton to a water molecule); for details see chapter 12 'Acid–base equilibrium'. As a result aqueous solutions of aluminium ions are about as acidic as aqueous solutions of ethanoic acid.

The first equilibrium established is:

$$[Al(H_2O)_6]^{3+}(aq) + H_2O(l) \rightleftharpoons H_3O^+(aq) + [Al(H_2O)_5OH]^{2+}(aq)$$

The complex ion on the right-hand side has a hydroxo ligand in addition to five water ligands. It can lose a further proton and form a complex ion with *two* hydroxo ligands:

$$[Al(H_2O)_5OH]^{2+}(aq) + H_2O(l) \rightleftharpoons H_3O^+(aq) + [Al(H_2O)_4(OH)_2]^+(aq)$$

This process occurs a third time; the difference is that the complex with three hydroxo ligands is uncharged and precipitates out of the solution:

$$[Al(H_2O)_4(OH)_2]^+(aq) + H_2O(l) \rightleftharpoons H_3O^+(aq) + Al(H_2O)_3(OH)_3(s)$$

All these equilibria can be shifted towards the products by adding a base to remove the aqueous hydrogen ions H_3O^+. Adding a small quantity of a base (e.g. aqueous sodium hydroxide) to aqueous aluminium ions therefore causes the precipitation of hydrated aluminium hydroxide $Al(OH)_3$. Further reaction can be achieved by adding more base. The solid hydroxide 'redissolves' (showing amphoteric behaviour):

$$Al(OH)_3(s) + OH^-(aq) \rightarrow [Al(OH)_4]^-(aq)$$

Note that this ion is *tetrahedral* in shape because four OH^- ligands surround the central Al^{3+} ion.

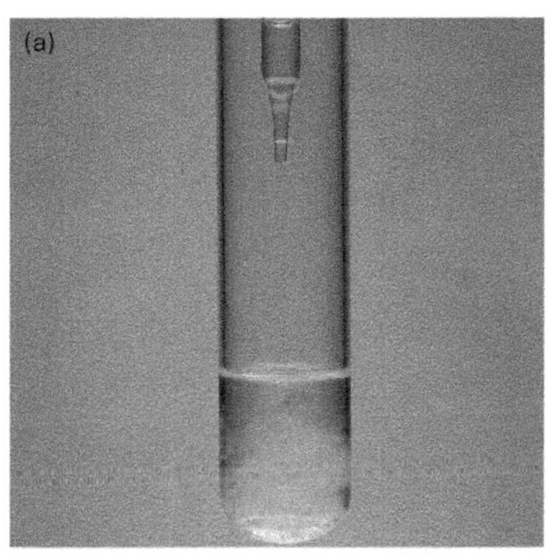

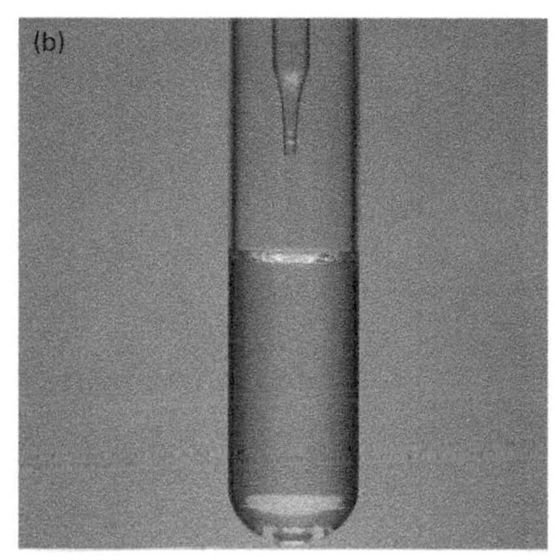

(a) Adding a small quantity of aqueous sodium hydroxide to aqueous aluminium ions causes a white precipitate of aluminium hydroxide $Al(OH)_3$. (b) Adding excess aqueous sodium hydroxide causes the precipitate to 'redissolve'.

Adding solid sodium carbonate to aqueous aluminium chloride causes the rapid release of carbon dioxide gas which might be rapid enough to push a cork out of a bottle. See spread 20.8 for more detail on the similar reaction of iron(III) ions and sodium carbonate. The process also occurs with anions from other weak acids, such as sulphide ion S^{2-} or sulphite ion SO_3^{2-}.

Further Lewis acid behaviour

The halides of boron and aluminium act as Lewis acids (spread 12.12) by accepting lone pairs. This behaviour is to be expected because the valence shells of the central atoms in $AlCl_3$ and BF_3, for example, contain just six electrons. During the reaction as a Lewis acid, the central atom gains a share in a further two electrons to complete the octet, as, for example, in the reaction between boron trifluoride BF_3 (Lewis acid) and ammonia NH_3 (Lewis base) to form the adduct $H_3N{:}\rightarrow BF_3$; see spread 12.12.

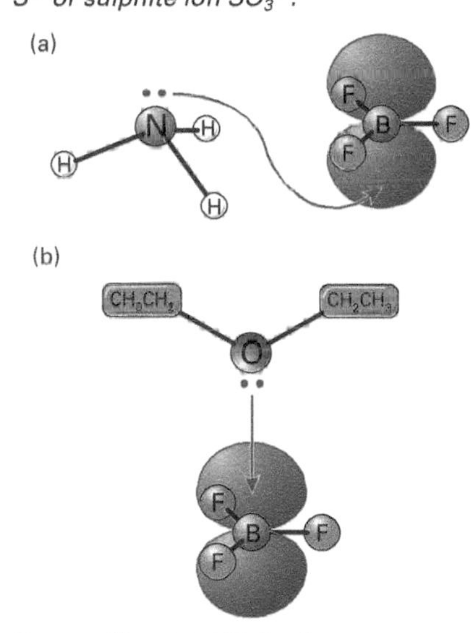

Boron trifluoride BF_3 is electron-deficient. It acts as a Lewis acid by accepting lone pairs from
(a) ammonia NH_3 or
(b) ethoxyethane $CH_3CH_2{-}O{-}CH_2CH_3$.

SUMMARY

- The Al^{3+} ion has a very high charge density; it is highly polarizing.
- The aqueous Al^{3+} ion, usually written as $Al^{3+}(aq)$, exists as a complex ion with the formula $[Al(H_2O)_6]^{3+}$.
- Aqueous aluminium salts are acidic; the ion $[Al(H_2O)_6]^{3+}$ protonates water molecules.
- Boron and aluminium halides have an incomplete octet. They can act as Lewis acids by accepting lone pairs from Lewis bases, thus forming adducts.

PRACTICE

1 Draw the structures of each of the following species:
 a $Al(OH)_3{\cdot}3H_2O$
 b $[Al(OH)_4]^-$
 c The adduct formed between BCl_3 and NH_3.

2 Describe what you would see when aqueous sodium hydroxide is slowly added with shaking to a solution of aluminium sulphate. In your answer, include balanced chemical equations and details of complex ion structure.

3 Draw Lewis structures to show the arrangement of valence electrons in boric acid and in the borate anion. Suggest shapes for these species.

4 Explain how the hydration sphere of water molecules around the aqueous sodium ion $Na^+(aq)$ differs from that in the aqueous aluminium ion $Al^{3+}(aq)$.

19.4 Group IV: carbon to lead

OBJECTIVES

- The elements
- Melting point and conductivity trends
- 'Buckyballs'
- Graphene

The Group IV elements are, in order of increasing atomic number, carbon (C), silicon (Si), germanium (Ge), tin (Sn), and lead (Pb). Carbon is a non-metal, silicon and germanium are metalloids, and tin and lead are metals. All have a valence-shell electronic structure of ns^2np^2. Of all the groups in the periodic table, Group IV shows the clearest trend from non-metallic to metallic character with increasing atomic number.

In November 2015 the second-largest diamond ever was found in Botswana: the stone has a mass of 1,111 carats.

Structures of the elements

The physical structures of the elements of Group IV are suggested by their melting points and electrical conductivities. The structures of the carbon allotropes, diamond and graphite, are described in chapter 6 'Solids'. Diamond has a giant covalent structure in which each atom is linked to four other atoms by single covalent bonds. Graphite has a layered giant covalent structure. Such structures are associated with very high melting points. The electrical conductivity of diamond is very low, confirming the lack of delocalized electrons. Graphite has an electrical conductivity 10^{15} times greater, owing to the presence of delocalized valence electrons; see spread 6.4.

Crystalline tin results when molten tin cools slowly. Tin is a typical metal, consisting of ions arranged in a metallic lattice bonded by delocalized valence electrons.

Silicon and germanium have similar structures to diamond. However, their electrical conductivities are 10^{14} (Si) and 10^{11} (Ge) times greater than that of diamond. These elements are *semiconductors*: they have conductivities intermediate between those of metals and those of non-metals, in keeping with their classification as metalloids. A small proportion of the localized bonding electrons break free from the covalent bonds and delocalize throughout the solid structure.

Tin and lead are metals, as confirmed by their high electrical conductivities.

Melting point and electrical conductivity data for the Group IV elements.

Element	Melting point /K	Electrical conductivity /S m^{-1}
C*	3800	7×10^4
C†	3820	1×10^{-11}
Si	1683	1×10^3
Ge	1211	2.2
Sn	505	9×10^6
Pb	601	5×10^6

* Carbon in the form of graphite.
† Carbon in the form of diamond.

Fullerenes – spherical carbon allotropes

During the 1970s, Harry Kroto, of Sussex University, became interested in the origin of black absorption bands in the visible spectrum of matter occurring between the stars. In the 1980s, Kroto suggested a collaboration with Bob Curl and Rick Smalley, of Rice University, USA, whose laboratory was equipped to bombard materials with very high energy lasers and examine, by mass spectrometry, any molecules produced. He was hopeful that they would find carbon species to explain the interstellar absorption bands. In 1985, they obtained a mass spectrum, shown on the left, with a very strong signal suggesting a stable molecule of formula C_{60}.

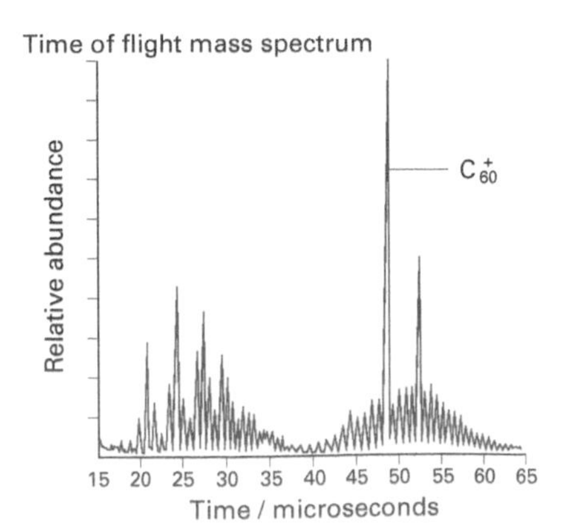

Mass spectrum showing the peak corresponding to C_{60}.

Speculating about possible structures for this molecule, Kroto kept thinking of the geodesic dome designed by the architect Buckminster Fuller for the *Expo '67* exhibition in Montreal. This structure was composed almost exclusively of hexagons with a few pentagons to close it into a spherical dome. They worked out from models how many carbon atoms arranged in hexagons and pentagons would be needed to form a sphere and to their delight found it was exactly 60! Kroto named C_{60} **buckminsterfullerene** in honour of the architect who provided the inspiration. The shape of C_{60} is more familiar as a football with its white (hexagonal) and black (pentagonal) patches. The molecule's spherical shape was confirmed by X-ray crystallography when Wolfgang Krätschmer, Donald Huffman, and their co-workers crystallized it from solution in benzene. The crystalline form was named **fullerite** (see photo opposite). The structure was only confirmed in 1990, when Kroto's team observed a single-line ^{13}C NMR spectrum which showed that all the carbon atoms are in identical environments in C_{60}. Since then a whole family of **fullerenes** has been discovered, informally referred to as '**buckyballs**'.

(a) The architectural form known as a geodesic dome, composed of interlocking hexagons and pentagons. (b) The structure of C_{60}: note the structural similarity to the geodesic dome. The model is held by Harry Kroto.

Graphene – the fundamental building block

The term **graphene** first appeared in 1987 to describe a carbon macromolecule consisting of hexagonally-arranged carbon atoms, in a sheet which is a single atom thick. It is essentially a single layer of the graphite structure (spread 6.3) and can be considered the basic structural element of graphite, carbon nanotubes, and fullerenes. Graphene's great importance became apparent in 2004 when Andre Geim and Konstantin Novoselov, working at Manchester University, managed to lift a single layer off a sample of graphite. Remarkably, they achieved this using ordinary sticky tape! Once isolated, it was possible to determine graphene's properties, revealing it as a material with huge potential in new technologies. It is as thin as any material can get, just one atom thick, and yet it is impermeable to gases and is the strongest two-dimensional material ever tested: about 50 times stronger in tension than steel. It is also transparent to light while being an excellent conductor of both heat and electricity.

Transparent conductors are fundamental to touch screens and flat screen televisions. Currently, these screens use indium tin oxide, requiring rapidly depleting sources of the metal indium. In 2010 the first sheets of graphene large enough to make a touch screen were produced.

Carbon cycle

The **carbon cycle** is a sequence of events in the environment by which carbon, in the form of carbon dioxide, is removed from the atmosphere and incorporated into carbon compounds during photosynthesis (see chapter 30 'Biochemistry'). Subsequently this carbon is returned to the atmosphere as a by-product of respiration. Note also that the oceans can dissolve carbon dioxide, as discussed in the next spread.

SUMMARY

- Carbon is a non-metal; silicon and germanium are metalloids; tin and lead are metals.
- Graphite and diamond are two allotropes of carbon; they have giant covalent structures.
- Carbon in the form of graphite is the only non-metallic element classed as an electrical conductor.
- Fullerite C_{60}, the solid form of buckminsterfullerene, is a third allotropic form of carbon.
- Graphene is essentially a single layer of the graphite structure.

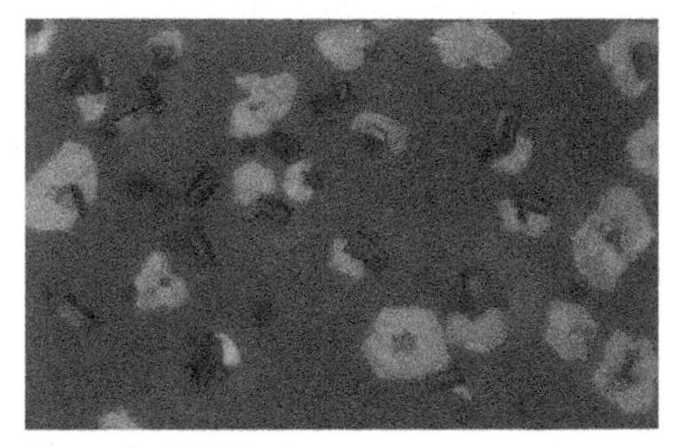

Crystals of fullerite.

Nanotechnology – engineering at the molecular scale

Carbon nanotubes were discovered as a spin-off from fullerene research, but are basically tubes of rolled up graphene typically about 1 nm in diameter. They can be produced in small quantities, as highly purified single-walled tubes, or as multi-walled tubes by the tonne. Their uses range from highly sophisticated drug-delivery systems to composite engineering materials with a higher strength-to-weight ratio than any so far known.

Nanotubes and drug delivery

Carbon nanotubes can also be imagined as constructed by splitting a buckyball into two and then adding a cylindrical section to join the two hemispheres together. Nanotubes tagged with a monoclonal antibody and enclosing a drug that can destroy malignant cells can attach to cancer cells. Irradiation of the body with near-IR radiation causes the nanotubes to fluoresce, showing the position of the target cells. A near-IR laser is then directed at the target areas; the heating causes the nanotubes to burst releasing the drug at exactly the right place.

The ultimate recognition

For their discovery of buckminsterfullerene, Kroto, Curl, and Smalley were awarded the 1996 Nobel Prize in Chemistry. Geim and Novoselov were awarded the 2010 Nobel Prize in Physics for their work in isolating graphene.

EXTENSION

19.5

GROUP IV: THE CHEMISTRY OF THE ELEMENTS – 1

OBJECTIVES

- Oxidation numbers
- Oxide character
- Silicates and carbonates

The oxidation numbers exhibited by Group IV elements in their compounds. The more stable oxidation number is in bold.

Element	Oxidation number
C	+2, **+4**
Si	**+4**
Ge	**+4**
Sn	+2, **+4**
Pb	**+2**, +4

The 'inert pair effect'

The increase in stability of the lower oxidation number with increasing atomic number is sometimes called the 'inert pair effect'.

Carbon monoxide

This gas is prepared in the laboratory by using warm concentrated sulphuric acid to dehydrate methanoic acid HCOOH:

$HCOOH(l) \rightarrow H_2O(l) + CO(g)$

Carbon monoxide burns with a pale blue flame to form carbon dioxide:

$2CO(g) + O_2(g) \rightarrow 2CO_2(g)$

Carbon monoxide's most notorious characteristic is its ability to bind tightly to haemoglobin in the blood. In this way it prevents haemoglobin from carrying oxygen around the body. Carbon monoxide gas forms as the result of incomplete combustion in petrol engines, and also in poorly ventilated gas fires, causing preventable deaths.

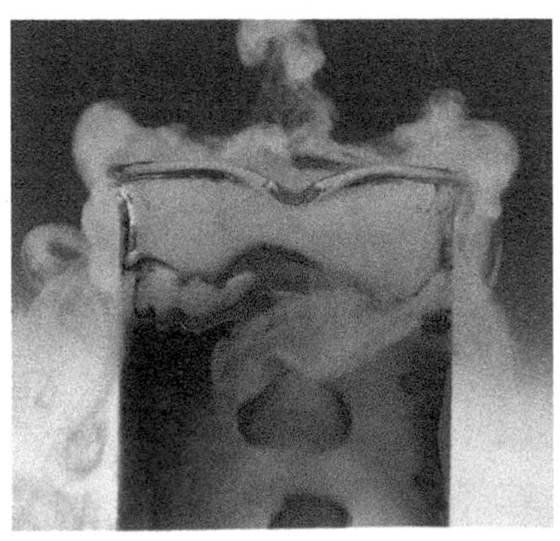

Solid carbon dioxide dissolves to produce some weak carbonic acid, $CO_2(g) + H_2O(l) \rightleftharpoons H_2CO_3(aq)$ Hence rainwater is naturally acidic.

The chemistry of the elements carbon to lead shows a clear trend from non-metallic to metallic properties. Consequently, the character of the bonding in the corresponding compounds shows a trend from covalent to ionic. Predominantly covalent bonds form when the element shares its four valence electrons ns^2np^2 to form four covalent bonds, as in silicon dioxide SiO_2. Ionic bonds form when the element loses the two outer p electrons to become a doubly charged ion M^{2+}, as, for example, in lead(II) nitrate $Pb(NO_3)_2$. As you might expect, the high values of ionization energy involved mean that ionic M^{4+} compounds are uncommon.

Oxidation numbers

The most common oxidation number in Group IV is +4, as in the compounds carbon dioxide CO_2, silicon tetrachloride $SiCl_4$, and lead(IV) oxide PbO_2. The state with an oxidation number of +2 becomes increasingly stable with increasing atomic number. For example, when the elements are heated with oxygen, they form the oxides CO_2, SiO_2, GeO_2, SnO_2, and PbO. Note that the elements have an oxidation number of +4 in all these oxides except lead(II) oxide PbO.

In carbon monoxide CO, carbon exhibits an oxidation number of +2 and thus stands outside the overall trend. This covalent compound is important as a reductant in industry, for example in the blast furnace where it reduces iron(III) oxide to iron (see next chapter). Tin and lead also form halides in which their oxidation number is +2. These compounds are predominantly ionic, containing the Sn^{2+} and Pb^{2+} ions.

Oxide character

As a general rule, the oxides of non-metals are acidic, the oxides of metalloids are usually amphoteric, and the oxides of most metals are basic, although some are amphoteric. As a result of the Group IV trend from non-metal to metal down the group, you might expect the oxides to become less acidic (i.e. more basic) as the atomic number of the element increases. Such a steady trend is *not* encountered in practice, because of the presence of more than one oxidation state for tin and lead.

Carbon dioxide CO_2 is an acidic gas. It dissolves in water to give an acidic solution (pH ≈ 5, as shown below left by the colour of universal indicator):

$$CO_2(g) + H_2O(l) \rightleftharpoons H_2CO_3(aq)$$

$$H_2CO_3(aq) + H_2O(l) \rightleftharpoons H_3O^+(aq) + HCO_3^-(aq)$$

It neutralizes bases to give carbonate salts:

$$CO_2(g) + 2NaOH(aq) \rightarrow Na_2CO_3(aq) + H_2O(l)$$

Silicon dioxide SiO_2 is also acidic, but it is a solid with a high melting point (1610 °C). Silicon dioxide reacts with molten bases to give silicates (in the same manner that carbon dioxide gives carbonates):

$$SiO_2(s) + 2NaOH(l) \rightarrow Na_2SiO_3(l) + H_2O(g)$$

Lead(II) oxide PbO is amphoteric; it reacts both with bases and with acids. Lead(II) oxide reacts with nitric acid as follows:

$$PbO(s) + 2HNO_3(aq) \rightarrow Pb(NO_3)_2(aq) + H_2O(l)$$

The oxides of lead

There are three oxides of lead: lead(II) oxide PbO,which is a yellow solid used to make lead glass which sparkles brilliantly, lead(IV) oxide PbO_2, which is a brown solid used in the lead–acid battery, and 'red lead' Pb_3O_4, which is named after its characteristic colour. Red lead is a mixed oxide, an array of oxide ions with two Pb^{2+} ions and one Pb^{4+} ion for each set of four O^{2-} ions. When red lead reacts with nitric acid, it turns to a brown suspension because PbO is more basic than PbO_2, hence leaving the brown solid behind.

The structures of CO_2 and SiO_2

The dramatically different physical properties of carbon dioxide and silicon dioxide are due to the different bonding in the two compounds. Carbon dioxide is a gas composed of individual small CO_2 molecules with carbon–oxygen double bonds. Silicon dioxide is a solid with a giant covalent structure similar to that of diamond; see spread 6.3.

Silicates and carbonates

Silicates account for over 90% of the rocks in the Earth's crust. The difference between the structures of silicates and carbonates mirrors the difference described above between the structure of silicon dioxide and carbon dioxide. Carbonates contain the carbonate ion, CO_3^{2-}, which has a double bond, delocalized between the three oxygen atoms and the carbon atom. The very important compound calcium carbonate was discussed in chapter 16 'The s-block elements'. Silicates are based instead on SiO_4 tetrahedra, with silicon–oxygen *single* bonds, linked together by sharing oxygen atoms at their vertices.

This sharing can result in rings, chains, double chains, sheets, and 3D networks. **Beryl** ($Be_3Al_2Si_6O_{18}$) is an example of a silicate containing a six-membered ring. Sodium silicate Na_2SiO_3 forms single-stranded chains; the notorious fibrous silicate **asbestos** has double chains. Sheet silicates are widely abundant and include kaolinite, micas, and talc. The most abundant of all minerals are **feldspars**, which are complex mixtures of 3D network **aluminosilicates** (silicates also containing aluminium); the alkali feldspars have the formula $Na_xK_{1-x}AlSi_3O_8$.

Clay minerals

Clays are layered aluminosilicates. Clays are the raw materials for some of our most ancient artefacts, such as pottery and bricks. An approximate equation describing the weathering of a potassium feldspar into a clay by the action of water and atmospheric carbon dioxide is:

$$2KAlSi_3O_8 + CO_2 + 2H_2O \rightarrow K_2CO_3 + 4SiO_2 + Al_2Si_2O_5(OH)_4$$

The structure of clay minerals is built up from two units: SiO_4 tetrahedra and AlO_6 octahedra. One layer of SiO_4 tetrahedra sharing corners with one layer of AlO_6 octahedra forms the 1:1 mineral **kaolinite** (china clay). The idealized composition of kaolinite is $Al_2Si_2O_5(OH)_4$.

Two layers of SiO_4 tetrahedra sharing corners with one layer of AlO_6 octahedra forms the 2:1 mineral **pyrophyllite**. The idealized composition of pyrophyllite is $Al_2Si_4O_{10}(OH)_2$. A closely related mineral in which two aluminiums are replaced by three magnesiums is talc $Mg_3Si_4O_{10}(OH)_2$. The layers are electrically neutral, and there are therefore only weak forces between the layers. Talc has the lowest value of 1 on the Mohs scale of hardness (diamond having the other extreme value of 10).

Replacement of one quarter of the silicons in pyrophyllite with aluminium creates negatively-charged layers, which can be balanced with positive ions (cations) such as potassium, to form muscovite (white mica) $KAl_3Si_3O_{10}(OH)_2$. This replacement, which depends ultimately on charge balance, is not specific for potassium ion, so such minerals allow cation exchange, where one cation takes the place of another. Plants get their cationic nutrients, such as iron(III), in a similar way from the soil.

SUMMARY

- The trend down the group is from non-metallic to metallic character.
- The most common oxidation number (except for lead) is +4.
- Tin and lead form ionic compounds containing Sn^{2+} and Pb^{2+} ions (i.e. oxidation number +2). Pb(II) is the stable oxidation state for lead.

PRACTICE

1 Explain why ionic compounds of Group IV containing M^{4+} ions are uncommon.

Carbon footprint

The term '**carbon footprint**' is a measure (of CO_2 emissions) used in assessing the impact human activities (at the individual or corporate level) have on the environment and how that contributes to climate change. The **primary footprint** measures direct emissions from burning of fossil fuels during domestic energy consumption or during transportation. The **secondary footprint** measures indirect emissions associated with the manufacture and distribution of the products we buy. An approximate calculation of the average annual carbon footprint in the UK is 12 tonnes of CO_2 per capita.

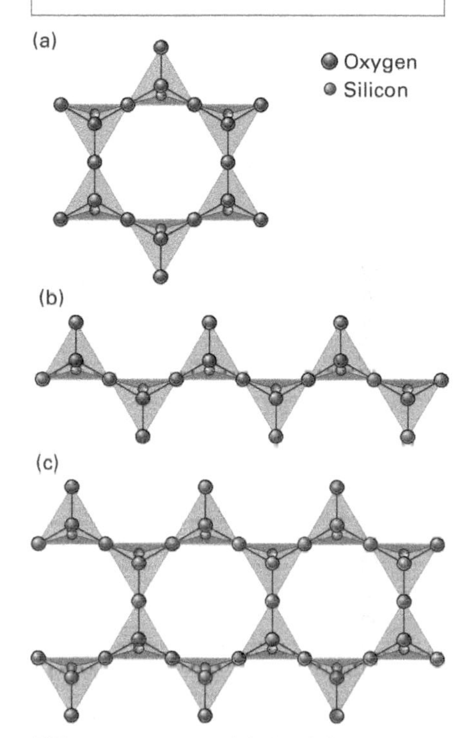

Silicate structures: (a) ring (b) single chain (c) double chain.

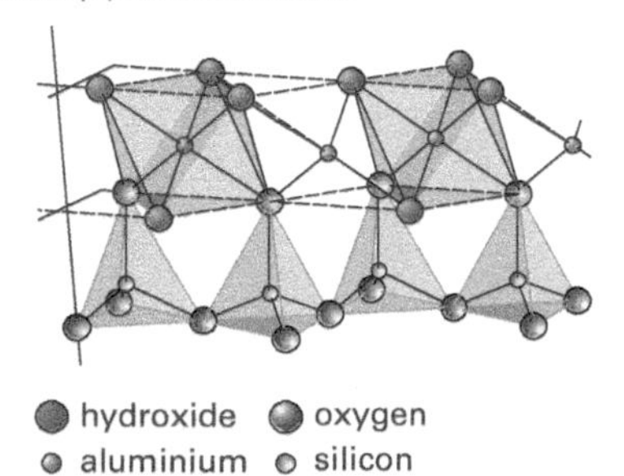

The structure of kaolinite.

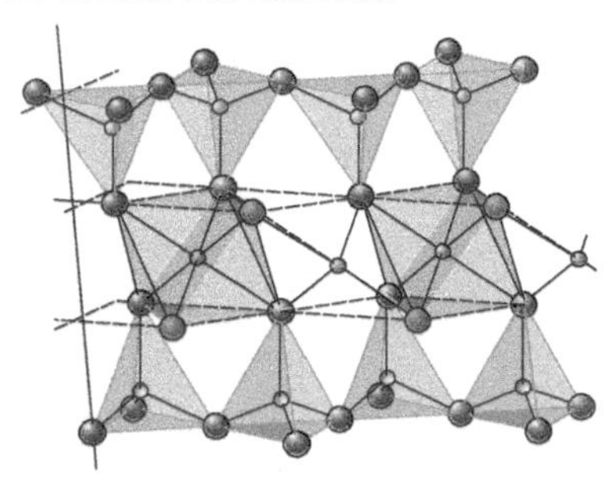

The structure of pyrophyllite.

19.6

Group IV: the chemistry of the elements – 2

OBJECTIVES

- Structure and bonding in the tetrahalides
- Hydrolysis of the tetrachlorides
- Environmental issues

The chemistry of the Group IV elements carbon to lead shows a clear trend from non-metallic to metallic properties down the group. Where oxidation numbers of +2 and +4 are available, compounds in which the element has oxidation number +2 tend to be predominantly ionic, and those in which the element has oxidation number +4 tend to be predominantly covalent in character. This spread concentrates particularly on the chemistry of the chlorides of the Group IV elements.

Structure and bonding in the tetrahalides

The predominant oxidation number for the elements carbon to tin is +4. However, remember that the stability of the state with oxidation number +2 increases down the group. Thermodynamically stable tin(II) compounds are available (but they are strong reductants), and the more stable state for lead is Pb(II). The tetrahalides all contain the Group IV element with oxidation number +4. Formulae, physical state, and boiling point data are summarized on the left.

The Group IV tetrachlorides.

Tetra-chloride	State at 25 °C and 1 bar	Boiling point/°C
CCl_4	liquid	77
$SiCl_4$	liquid	58
$GeCl_4$	liquid	87
$SnCl_4$	liquid	114
$PbCl_4$	liquid	decomp

The relatively low boiling points suggest that the bonding character of these compounds is predominantly covalent. This view is confirmed by their poor electrical conductivity in the liquid state and their ability to dissolve in non-polar (organic) solvents. Each of these tetrachlorides exists as discrete molecules of general formula ECl_4. The central Group IV element is attached to four chlorine atoms by four single covalent bonds. The shape of the molecule is tetrahedral.

The tetrahalides in general

All the tetrahalides of the Group IV elements are known, with the exception of the bromide and iodide of lead. Lead(IV) is an oxidant, and bromides and iodides are readily oxidized. Lead(IV) fluoride has appreciable ionic character, but lead(IV) chloride is predominantly covalent.

Hydrolysis of the tetrachlorides

Tetrachloromethane CCl_4 (previously called carbon tetrachloride) is immiscible with water and does not react with it. The C—Cl bond is strong (average bond enthalpy = 338 kJ mol^{-1}). However, the hydrolysis reaction

$$CCl_4(l) + 2H_2O(l) \rightarrow CO_2(g) + 4HCl(aq)$$

is favoured thermodynamically and will occur, given sufficient time (many years). This reaction proceeds at an extraordinarily slow rate.

The Si—Cl bond in silicon tetrachloride $SiCl_4$ is far more polar than the C—Cl bond in CCl_4; it is also longer and weaker. As a result, silicon tetrachloride gives off fumes in moist air and is hydrolysed rapidly by water:

$$SiCl_4(l) + 2H_2O(l) \rightarrow SiO_2(s) + 4HCl(aq)$$

The products of the reaction are hydrogen chloride fumes and a white precipitate of hydrated silicon dioxide.

The ease of hydrolysis decreases in the order Si > Ge > Sn >Pb as metallic character increases down the group.

(a) Tetrachloromethane reacts extraordinarily slowly with water. (b) Silicon tetrachloride is rapidly hydrolysed by water. The hydrogen chloride gas released forms white fumes with the top of a bottle of concentrated aqueous ammonia. Some HCl dissolves in water, turning the indicator yellow.

(a) $SiCl_4$ $\xrightarrow{+H_2O}$ Cl_4Si^{-}—$^{+}OH_2$ (b) Cl_4Si^{-}—^{+}OH—H $\xrightarrow[-H^+]{-Cl^-}$ Cl_3Si—OH

Because silicon has 3d orbitals close enough in energy to the occupied 3p orbitals, (a) the incoming water molecule can form a bond with the silicon before the existing bonds are broken. (b) Such a reaction mechanism is not possible for carbon because the 3d orbitals are too far away in energy from its 2p orbitals. Cyan highlights changes in an individual step.

Comparing tin and lead

The most stable oxidation number for tin is +4 and that for lead is +2. The increased stability of the lower oxidation state for lead is illustrated by lead(IV) chloride. It has a tendency to decompose near room temperature to form the more stable lower oxidation state:

$$PbCl_4(l) \rightarrow PbCl_2(s) + Cl_2(g)$$

This tendency is also illustrated when the 2+ ions of tin and lead react with acidified potassium dichromate(VI). Tin(II) ions are oxidized by the dichromate(VI) ions, and the colour changes from orange ($Cr_2O_7^{2-}$ ions) to green (Cr^{3+} ions). Lead(II) ions are not oxidized by dichromate(VI). Instead, a yellow precipitate of lead(II) chromate(VI) $PbCrO_4$ forms.

Heavy metals and the environment

Lead is called a 'heavy metal' because it has a high relative atomic mass. Other common heavy metals are cadmium and mercury. Lead in particular causes problems in the environment because it is a cumulative poison that is poorly excreted by humans. Small ingested quantities accumulate in the body over a period of years, ultimately causing damage to the central nervous system.

Because lead is malleable and ductile, it has been used for water pipes for centuries. In hard water areas, an impervious coating forms on the inside of the pipe. But soft water does not cause such a coating, and the water in the pipe becomes contaminated with lead. It has been suggested that the fall of the Roman Empire was due to richer administrators being able to afford piped water in their villas. The only suitable metal available at that time was lead. Neurological decay resulted in societal decay! Today, lead water pipes have mainly been replaced, with copper or plastic.

Until recently, vehicles used petrol that contained tetraethyllead as an 'anti-knocking' agent. Exhaust systems emitted tiny particles of lead compounds which contaminated the air and roadsides. Surveys revealed that children living near busy roads had elevated levels of lead in their blood and performed less well in cognitive and psychomotor tests. Modern cars now run on 'unleaded' petrol (that does not contain lead additives).

Franklin's naval expedition in 1847 to find the Northwest Passage (from the North Atlantic to the North Pacific, around the 'top' of Canada) ended with the deaths of all 147 men. The expedition was the first to use cooked canned food rather than salted and pickled preserves. The cans were made from sheet steel assembled with tin–lead solder. The lead leached into the food and slowly poisoned the men. Hair from bodies preserved in the ice was recently analysed and showed high lead content: the concentration increased the closer to the scalp the sample was taken.

SUMMARY

- Group IV compounds are predominantly covalent, with the oxidation number +4 being preferred.
- The stability of compounds of Group IV elements with oxidation number +2 increases with increasing atomic number.
- Lead(II) ionic compounds are more stable than lead(IV) covalent compounds.
- The tetrachlorides of Si, Sn, and Pb all hydrolyse to give HCl and hydrated oxides.
- Lead is a cumulative poison, particularly affecting the central nervous system.

PRACTICE

1 Draw Lewis structures of CCl_4 and $GeCl_4$. Sketch the shape of $SnCl_4$.

Electrode potentials

Tin(II) ions will reduce orange dichromate(VI) to green chromium(III). Lead(II) ions do not bring about this change. The different outcomes may be understood by looking at the standard electrode potentials:

$E^\ominus(Sn^{4+}/Sn^{2+})$	= +0.15 V
$E^\ominus(Cr_2O_7^{2-}/Cr^{3+})$	= +1.33 V
$E^\ominus(Cl_2/Cl^-)$	= +1.36 V
$E^\ominus(PbO_2/Pb^{2+})$	= +1.46 V
$E^\ominus(Pb^{4+}/Pb^{2+})$	= +1.69 V

Tin(II) ion reduces dichromate(VI), but lead(II) ion cannot reduce dichromate(VI).

Lead(IV) oxide is a strong oxidant, able to oxidize concentrated hydrochloric acid to chlorine:

$$PbO_2(s) + 4HCl(aq) \rightarrow Cl_2(g) + PbCl_2(aq) + 2H_2O(l)$$

Sheet lead is durable and easily beaten into shape to fit traditional roofs. People who work regularly with lead in the building industry have the level of lead in their blood measured twice yearly.

Lead: physiological and biological data

Lead is moderately toxic by ingestion and affects the gut and central nervous system. Lead compounds can be carcinogenic and teratogenic.

Biological role: none

Toxic intake: 50 mg

Levels in humans:

Blood/mg dm^{-3}: 0.2

Bone/p.p.m.: about 15

Liver/p.p.m.: about 10

Muscle/p.p.m.: about 1

19.7

OBJECTIVES

- Reactivity of the elements
- Oxidation numbers
- Oxides
- Biochemistry

Group V: nitrogen and phosphorus – 1

The Group V elements are, in order of increasing atomic number, nitrogen (N), phosphorus (P), arsenic (As), antimony (Sb), and bismuth (Bi). Metallic character increases with increasing atomic number. Nitrogen and phosphorus are non-metals, arsenic and antimony are metalloids, and bismuth is a metal. All have a valence-shell electronic structure of ns^2np^3. This spread and the following one will concentrate on the chemistry of nitrogen and phosphorus, with some reference to the remaining elements where they illustrate group trends.

Bismuth is the only member of Group V with all the attributes of a typical metal (although antimony has some metallic character). Bismuth is usually encountered in the laboratory as its oxoanion, the bismuthate(V) ion BiO_3^-. Bismuthate(V) is so strong an oxidant that it can oxidize manganese(II) ions (pale pink) to manganate(VII) ions (deep purple), hence providing a test for manganese.

Nitrogen and phosphorus: reactivity

The elements nitrogen and phosphorus have very different structures and therefore very different reactivities. Nitrogen is a gas at room temperature (melts at T_m = –210 °C, boils at T_b = –196 °C). A nitrogen molecule N_2 consists of two atoms bonded by a triple covalent bond (N≡N). This very strong bond (bond enthalpy +944 kJ mol^{-1}) makes nitrogen extremely stable and unreactive.

Phosphorus, on the other hand, is very reactive. It has several allotropes, none of which has the sort of strong bonding found in nitrogen. Of these allotropes, white phosphorus (T_m = 44 °C) is particularly reactive, since the P_4 molecule has a highly strained tetrahedral structure (see chapter 17 'Trends across a period').

Oxidation numbers

The valence-shell electronic structure of the Group V elements is ns^2np^3. All elements can show an oxidation number of +3 by forming covalent bonds, for example with chlorine. All elements except nitrogen readily show an oxidation number of +5 by using vacant d orbitals. Phosphorus forms two chlorides, PCl_3 and PCl_5. Nitrogen and phosphorus can also exhibit an oxidation number of –3.

For example, nitrides can be formed when nitrogen reacts with certain s-block elements. When heated with air, magnesium forms the nitride Mg_3N_2 as well as the oxide MgO. Similarly lithium reacts with nitrogen to form the compound Li_3N. Both Mg_3N_2 and Li_3N contain the nitride ion N^{3-}.

The oxidation number –3 is also shown in the important compounds ammonia and phosphine, which are discussed in the following spread.

Arsenic the poison

Arsenic, usually in the form of its oxide As_2O_3, is famous as a poison. It binds to the active sites in a variety of enzymes, inhibiting their biochemical function and ultimately causing death. Note that arsenic also poisons industrial catalysts such as nickel and platinum.

The oxides of nitrogen

The oxides of nitrogen and phosphorus are generally acidic, as is usual for oxides of non-metals. The three main oxides of nitrogen are the colourless gases nitrogen monoxide NO and dinitrogen oxide N_2O and the brown gas nitrogen dioxide NO_2. Remember that NO_2 exists in equilibrium with its dimer N_2O_4, the position of the equilibrium depending on the temperature and pressure (see chapter 11 'Chemical equilibrium'). The similar electronegativities of nitrogen and oxygen (3.0 and 3.4 respectively) imply that the N—O bonds will have only slight polarity. The boiling points are NO –152°C, N_2O –89 °C, and N_2O_4 +21 °C. These values fit the trend of increasing strength of dispersion forces as the number of electrons increases.

'Smog' was first used to describe a poisonous mixture of smoke, fog, and other chemicals that caused London air pollution; as late as 1952 4000 people died of respiratory diseases in a particularly severe week's smog. The Clean Air Act introduced in 1956 in the UK drastically improved the urban atmosphere. **Photochemical smog** (containing nitrogen oxides, ozone, together with organic compounds) has proved harder to avoid, especially in cities such as Los Angeles. Both NO and NO_2 have an unpaired electron (count the total number of electrons in the molecules).

Nitrogen monoxide reacts with oxygen

When nitrogen monoxide NO forms in the presence of oxygen, it immediately reacts to produce nitrogen dioxide NO_2:

$$2NO(g) + O_2(g) \rightarrow 2NO_2(g)$$

colourless → brown

Pain relief and anaesthesia

Dinitrogen oxide N_2O acts as an anaesthetic. It is a constituent of *Entonox*, the gas mixture that is sometimes given to ease pain in childbirth, and is also used as a short-lasting general anaesthetic in dental surgery.

A series of complex reactions involving radicals such as these, and involving absorption of radiation, create a number of dangerous chemicals. Humans, animals, plants, and even polymers all suffer damage.

Laboratory preparations

Nitrogen monoxide NO is prepared by pouring 50% nitric acid onto copper:

$$3Cu(s) + 8HNO_3(aq) \rightarrow 3Cu(NO_3)_2(aq) + 4H_2O(l) + 2NO(g)$$

Dinitrogen oxide N_2O is obtained by heating ammonium nitrate (produced in the reaction vessel from a mixture of ammonium chloride and sodium nitrate) very carefully:

$$NH_4NO_3(s) \rightarrow N_2O(g) + 2H_2O(g)$$

Nitrogen dioxide NO_2 is obtained when powdered lead(II) nitrate is heated:

$$2Pb(NO_3)_2(s) \rightarrow 2PbO(s) + 4NO_2(g) + O_2(g)$$

The lead(II) nitrate crackles as it produces brown fumes of nitrogen dioxide.

NO_x

High-temperature combustion of fuels, e.g. near the spark plug in a petrol engine, produces both NO and NO_2, as well as carbon dioxide and carbon monoxide. The term NO_x is often used to mean NO or NO_2 or a mixture of both. NO_x reacts with water and so contributes to the problem of acid rain. Most NO_x can be removed from car exhausts by catalytic converters (see chapter 15 'Chemical kinetics'). As the exhaust gases pass through the catalytic converter, NO_x reacts with carbon monoxide to form nitrogen and carbon dioxide; see spread 15.9.

The oxides of phosphorus

As we saw in chapter 17 'Trends across a period', the two oxides of phosphorus, phosphorus(III) oxide P_4O_6 and phosphorus(V) oxide P_4O_{10}, are molecular solids. Both structures bear a striking similarity to that of white phosphorus, as each is based on a tetrahedron of four phosphorus atoms. The oxide P_4O_6 forms when phosphorus burns in a limited supply of oxygen, and P_4O_{10} forms when it burns in a plentiful supply of oxygen. Both oxides dissolve in water to form acidic solutions.

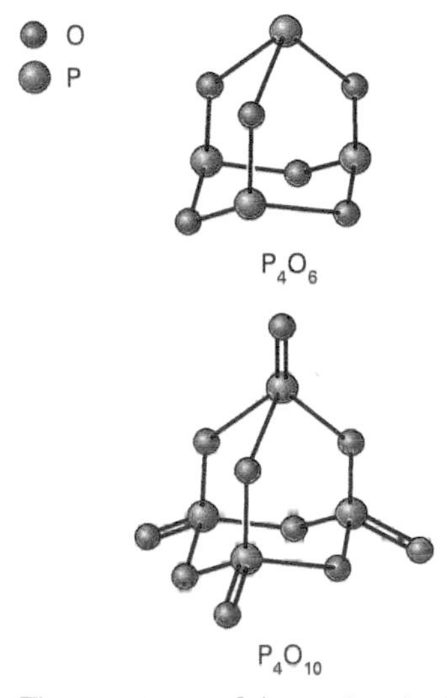

The structures of the molecules P_4O_6 and P_4O_{10}.

Nitrogen and phosphorus in biochemistry

Both nitrogen and phosphorus play vital roles in biochemistry. The total mass of nitrogen in an average 70 kg person is 1.8 kg; the corresponding value for phosphorus is 780 g.

Nitrogen is found in amino acids and the proteins that are built from them; see chapter 30 'Biochemistry'.

Phosphorus is an essential element in nucleic acids (DNA and RNA) and in adenosine triphosphate ATP, the carrier of chemical energy in all organisms. The availability of phosphorus in seawater is a limiting factor on the growth of plankton, and hence affects the biomass that occurs higher in marine food chains.

Hydroxyapatite $Ca_5(OH)(PO_4)_3$ is a phosphate mineral found in skeletons and teeth. Hydroxyapatite is piezoelectric, capable of producing electric charge in response to mechanical stress. This property allows bone to be the internal skeleton of multicellular organisms.

Nitrogen fixation

One of the most important processes in nature is nitrogen fixation, which enables leguminous plants (such as peas, beans, and clover) to make use of atmospheric nitrogen. Nitrogen fixation is carried out by bacteria such as *Rhizobium* in root nodules of the plants. Atmospheric nitrogen is converted into nitrate ion, which is then available for biochemical syntheses.

SUMMARY

- Oxidation numbers of nitrogen and phosphorus vary from –3 to +5.
- Nitrogen and phosphorus have important biochemical roles.

PRACTICE

1 Give the oxidation number of the Group V element in each of the following compounds:

a NH_3 **f** HNO_3
b NO **g** NO_2
c N_2O **h** BiO_3^-
d NO_2 **i** P_4O_{10}
e HNO_2 **j** BiOCl.

2 Elemental nitrogen exists as simple covalent molecules consisting of two triply bonded nitrogen atoms. Suggest reasons why phosphorus does not similarly form triply bonded P_2 molecules.

3 Correlate material from chapter 17 'Trends across a period' relating to phosphorus with the information in this spread.

19.8

OBJECTIVES

- Oxoacids
- Hydrides

Group V: nitrogen and phosphorus – 2

This spread focuses on the oxoacids and hydrides of nitrogen and phosphorus. The most important oxoacid of nitrogen is nitric acid HNO_3. The most important oxoacid of phosphorus is phosphoric acid H_3PO_4. As discussed in chapter 17 'Trends across a period', all these substances are formed by reacting oxides with water. Of the hydrides, ammonia NH_3 should be very familiar to you by now. It is often used as an example in connection with its synthesis, solubility, basic character (Brønsted–Lowry and Lewis), and range of salts. However, *all* the Group V elements form hydrides. There are some interesting differences to be noted between the properties of ammonia and those of phosphine PH_3, the hydride of phosphorus.

Oxoacids

When nitrogen dioxide reacts with water, it disproportionates to form **nitrous acid** HNO_2 (containing N(III)) and **nitric acid** HNO_3 (containing N(V)):

$$2NO_2(g) + H_2O(l) \rightarrow HNO_2(aq) + HNO_3(aq)$$

The salts of nitrous acid are called nitrites and contain the **nitrite ion** NO_2^-. The salts of nitric acid are called nitrates, and contain the **nitrate ion** NO_3^-. Nitric acid is an important reagent because it can act both as an acid and as an oxidant.

Nitric acid as a strong acid

Nitric acid is a strong acid. When added to water it becomes essentially fully ionized:

$$HNO_3(aq) + H_2O(l) \rightarrow H_3O^+(aq) + NO_3^-(aq)$$

To show that nitric acid is an oxoacid, its formula may be written as $HONO_2$.

Nitric acid as an oxidant

Nitric acid is a strong oxidant. For example, metallic copper does not react with hydrochloric acid, but *does* react with concentrated nitric acid. Copper metal is oxidized to blue aqueous copper(II) ions and a red–brown gas (nitrogen dioxide) is produced (see spread 13.8):

$$Cu(s) + 4HNO_3(aq, conc.) \rightarrow Cu(NO_3)_2(aq) + 2H_2O(l) + 2NO_2(g)$$

Copper and nitric acid

Note that, in the reaction between metallic copper and concentrated nitric acid, *nitrate ions* are reduced, not hydrogen ions. This result is unusual because acids usually react with metals to form hydrogen by reduction of hydrogen ions. See chapter 13 'Redox equilibrium'.

Ammonia and phosphine

All the Group V elements form hydrides of the general formula MH_3. They are named ammonia (NH_3), phosphine (PH_3), arsine (AsH_3), stibine (SbH_3), and bismuthine (BiH_3). Thermal stability decreases in the order $NH_3 > PH_3 > AsH_3 > SbH_3 > BiH_3$. Bismuthine is so unstable that it can only be obtained in very small amounts under controlled conditions. All the Group V hydrides are predominantly covalent compounds.

Ammonia is produced in the laboratory by heating an ammonium salt with a base. The usual mixture chosen is solid ammonium chloride and soda-lime (a mixture of solid sodium hydroxide and solid calcium hydroxide):

$$NH_4Cl(s) + NaOH(s) \rightarrow NH_3(g) + H_2O(g) + NaCl(s)$$

The gas is dried by passing it through calcium oxide; it is then collected by upward delivery (density of ammonia relative to air = 0.59).

Phosphine may be produced in the laboratory by heating calcium phosphide with water:

$$Ca_3P_2(s) + 6H_2O(l) \rightarrow 3Ca(OH)_2(aq) + 2PH_3(g)$$

The gas is collected over water.

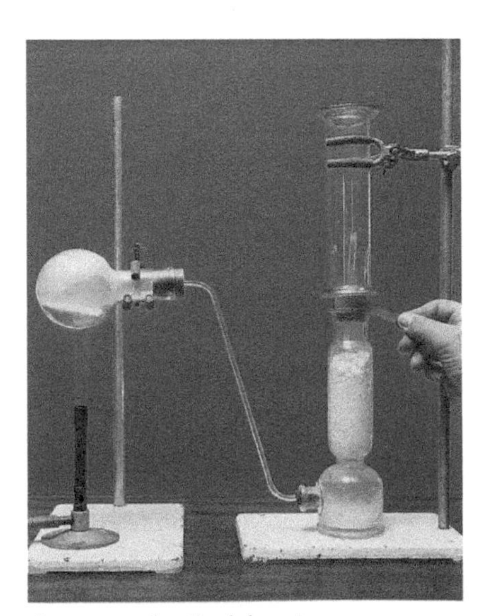

Apparatus for the laboratory preparation of ammonia. The damp litmus paper will turn blue showing that aqueous ammonia is basic.

Molecular shape

The molecules of both ammonia and phosphine are pyramidal (see spread 5.5). The H—N—H bond angle is 107° whereas the H—P—H bond angle is 93°. The P—H bonds are longer (0.142 nm) than the N—H bonds (0.101 nm), and so they have smaller electron density and therefore less mutual repulsion.

Boiling points

The boiling point of ammonia is –33 °C. If you consider dispersion forces alone, you would expect the boiling point of phosphine to be higher. However, phosphine boils at the *lower* temperature of –87 °C. The presence of hydrogen bonds (see chapter 7 'Changes of state and intermolecular forces') between ammonia molecules causes its boiling point to be higher than that expected on the basis of dispersion forces alone.

Water solubility

Because of the hydrogen bonds that form between ammonia and water molecules, ammonia is extremely soluble in water. The electronegativity of phosphorus (2.2) is equal to that of hydrogen (2.2), so the phosphine molecule is essentially non-polar and its solubility in water is very low.

Basic properties and salt formation

In solution, ammonia acts as a weak Brønsted base, accepting a proton from a water molecule:

$$NH_3(g) \rightarrow NH_3(aq)$$

$$NH_3(aq) + H_2O(l) \rightleftharpoons NH_4^+(aq) + OH^-(aq)$$

The resulting solution is basic: saturated aqueous ammonia has a pH of 11. Ammonia also acts as a Lewis base by donating a pair of electrons to a hydrogen atom on a water molecule. There are many examples of complex ions containing ammonia in the next chapter 'The transition metals'.

Ammonia readily forms solutions of salts with acids; and evaporation of the solution yields stable crystalline solids. For example:

$$NH_3(aq) + HNO_3(aq) \rightarrow NH_4NO_3(aq)$$

$$NH_3(aq) + H^+(aq) \rightarrow NH_4^+(aq)$$

Ammonium nitrate is a valuable fertilizer, see next spread. Phosphine is a much weaker base than ammonia. Phosphonium salts form only with difficulty and are unstable. For example, phosphonium iodide decomposes above 50 °C:

$$PH_4I(aq) \rightarrow PH_3(g) + HI(aq)$$

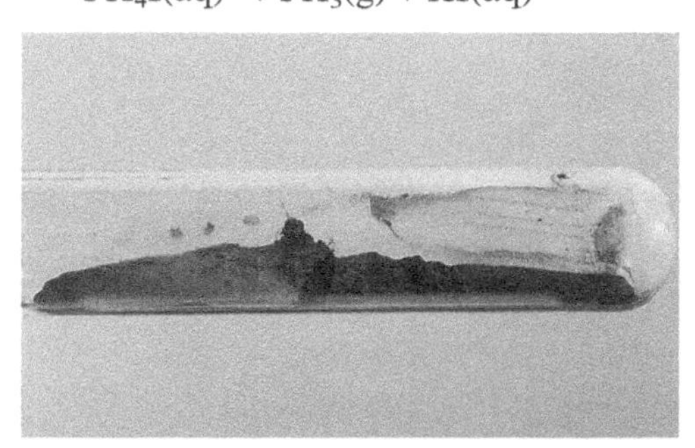

Redox properties. Ammonia acts as a reductant when passed over heated copper(II) oxide (a black solid):

$3CuO(s) + 2NH_3(g) \rightarrow 3Cu(s) + 3H_2O(g) + N_2(g)$

Oxidation state changes: Cu(II) to Cu(0); N(–III) to N(0)

Phosphine is a more powerful reductant than ammonia. It can precipitate copper and silver from their aqueous ions. Ammonia only forms a complex ion.

SUMMARY

- Nitric acid is an oxidizing acid; hydrochloric acid is not. (Reminder: sulphuric acid is oxidizing only when concentrated.)
- The boiling point of ammonia is greater than that of phosphine because there are hydrogen bonds between ammonia molecules.
- The solubility of ammonia in water is greater than that of phosphine because ammonia forms hydrogen bonds with water.
- Phosphine is a much weaker base than ammonia.

(a)

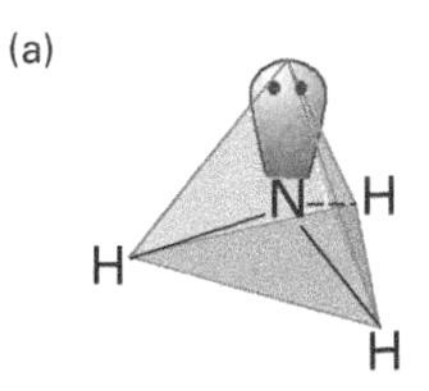

(b)

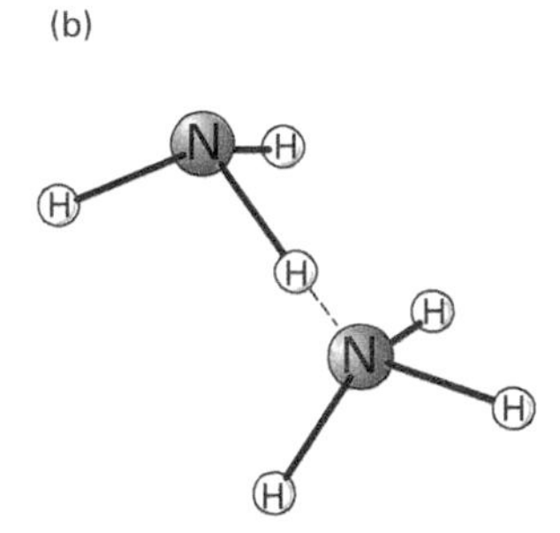

(a) The ammonia molecule. The tetrahedral angle is 109.5°. The 107° bond angles result because greater lone pair to bonding pair repulsions decrease the angle. (b) Hydrogen bonding in ammonia.

PRACTICE

1. Draw and label a diagram of the apparatus you would use to prepare and collect phosphine.
2. Draw labelled sketches to show how hydrogen bonding occurs between molecules of ammonia in the liquid phase.
3. Draw Lewis structures to show the electron arrangements in
   a PH_3
   b NH_4^+
4. Compare the reactions between:
   a Iron metal and nitric acid
   b Iron metal and hydrochloric acid.
5. Use the concept of electronegativity to explain why ammonia has a higher boiling point than phosphine. Sketch a graph of the boiling points of the Group V hydrides and explain its shape.

19.9

OBJECTIVES

- Ammonia: the Haber–Bosch process
- Nitric acid: the Ostwald process
- Sulphuric acid: the Contact process

THREE IMPORTANT INDUSTRIAL PROCESSES

The German chemist Baron Justus von Liebig commented in 1843 that the level of commercial prosperity of a country could be judged by the amount of sulphuric acid it consumed each year. This acid is one of a number of 'heavy chemicals' which are central to the manufacture of an enormous range of substances, from fertilizers and dyestuffs, to detergents, explosives, and artificial fibres. The production of the 'heavy chemicals' sodium hydroxide and chlorine from the electrolysis of brine was described in the previous chapter. This spread draws together the manufacture of ammonia, nitric acid, and sulphuric acid.

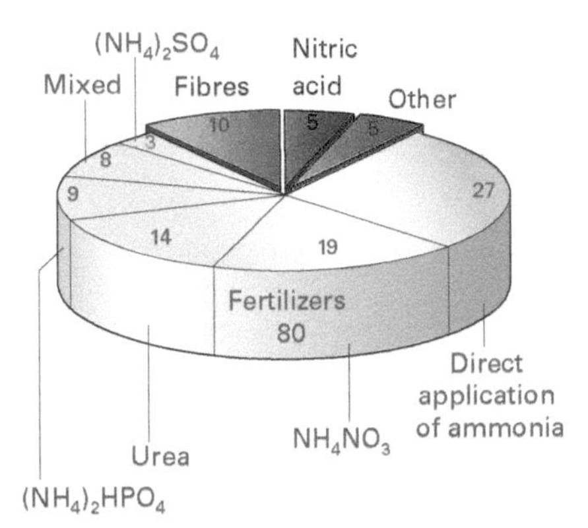

Ammonia uses: over 100 million tonnes of ammonia are manufactured each year world-wide.

Fertilizers and eutrophication

The demand for increased food production to keep the increasing world population alive resulted in much more efficient growth of crops during the twentieth century. Artificial fertilizers replace the nutrients taken out of the soil by plants, especially the three essential elements for plant growth, nitrogen N, phosphorus P, and potassium K. The main fertilizers are nitrates and phosphates. These inorganic fertilizers include potassium nitrate KNO_3, ammonium nitrate NH_4NO_3, and 'superphosphate' $Ca(H_2PO_4)_2$. Most commercial fertilizers are labelled with the amounts of nitrogen, phosphorus, and potassium that they contain. They are termed **NPK fertilizers** when they contain all three nutrients.

If more fertilizers are applied than is necessary, the excess may dissolve in rain water and run off into rivers and other watercourses. There, the fertilizers stimulate the growth of green and blue-green algae. This growth may become excessive and deprive the water of oxygen and light, and so cause other aquatic plants and animals to die. The watercourses become choked with decaying plant and animal matter. The excessive over-feeding and subsequent collapse of the water's ecosystem in this way is called **eutrophication**. Lake Balaton, the largest lake in Europe, turned visibly green in places in 1982 before the pollution problem was addressed.

Ammonia: the Haber–Bosch synthesis

Ammonia (NH_3) is manufactured from nitrogen and hydrogen by the Haber–Bosch synthesis. Hydrogen is produced in the 'primary reformer' from the reaction between methane and steam, using a nickel catalyst at about 750 °C and 30 atm:

$$CH_4(g) + H_2O(g) \rightleftharpoons CO(g) + 3H_2(g)$$

The methane must be desulphurized to avoid poisoning the catalyst.

Air is added to this gas mixture in the 'secondary reformer'. The oxygen is removed by reaction with some of the hydrogen. Carbon monoxide is removed in the 'shift reactor' by reaction with more steam in the presence of a freshly produced, finely divided Fe_3O_4 catalyst at 400 °C. The final mixture contains nitrogen and hydrogen in a ratio of 1:3.

In the synthesis reaction vessel, nitrogen, hydrogen, and ammonia exist together in the following equilibrium:

$$N_2(g) + 3H_2(g) \rightleftharpoons 2NH_3(g); \qquad \Delta H^{\ominus} = -92\ \text{kJ mol}^{-1}$$

Le Chatelier's principle (see chapter 11 'Chemical equilibrium') indicates that low temperature and high pressure increase the yield. In practice, a pressure of 250 atm and a temperature of 450 °C are used, giving a conversion of about 15%. The temperature is a compromise between yield and rate. A catalyst of iron is used, which is made more active by using the oxides of potassium, calcium, and aluminium as promoters.

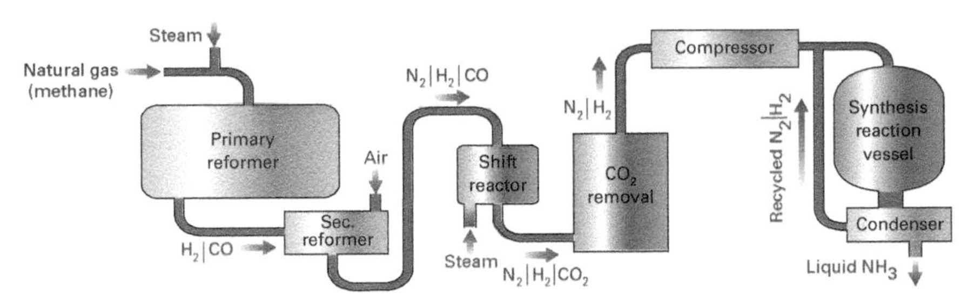

Flow diagram of the Haber–Bosch process for the synthesis of ammonia from its elements.

Nitric acid: the Ostwald process

The fertilizers (see box) that contain nitrogen are made from nitric acid, which is made from ammonia. The conversion of ammonia to nitric acid (HNO_3) is ingenious in its careful control of oxidation state change. Nitrogen has its lowest oxidation number of −3 in ammonia; it has its highest oxidation number of +5 in nitric acid.

The first stage of the process involves the catalytic oxidation of ammonia to nitrogen monoxide. A mixture of purified air and ammonia passes through a woven platinum–rhodium gauze at 850 °C:

$$4NH_3(g) + 5O_2(g) \rightarrow 4NO(g) + 6H_2O(g)$$

oxidation state changes: O(0) to O(-II); N(-III) to N(+II)

The conversion efficiency is about 96%.

The second stage of the manufacture involves cooling the nitrogen monoxide and mixing it with air; oxidation to nitrogen dioxide occurs:

$$2NO(g) + O_2(g) \rightarrow 2NO_2(g)$$

oxidation state changes: O(0) to O(–II); N(+II) to N(+IV)

The mixture of NO_2 and air then passes through an absorption tower containing water, in which the overall reaction is:

$$4NO_2(g) + O_2(g) + 2H_2O(l) \rightarrow 4HNO_3(aq)$$

oxidation state changes: O(0) to O(-II); N(+IV) to N(+V)

The acid that results is fairly concentrated (60%). This concentration is sufficient for most applications.

Nitric acid uses

Nitric acid is used in the manufacture of fertilizers (70% is used for NH_4NO_3 production), medicines, dyestuffs, and explosives. Its manufacture from ammonia is therefore of considerable commercial and social significance.

The platinum–rhodium gauze used in the Ostwald process.

Sulphuric acid: the Contact process

Sulphuric acid (H_2SO_4) is manufactured by the Contact process. The first step involves burning molten sulphur in air to produce sulphur dioxide:

$$S(l) + O_2(g) \rightarrow SO_2(g)$$

Sulphur dioxide contains sulphur with oxidation number +4; sulphuric acid contains sulphur with oxidation number +6. Further oxidation is therefore required to obtain sulphuric acid from sulphur dioxide.

This oxidation involves an equilibrium reaction, so the conditions must be carefully controlled to maximize the yield. The catalyst, spread 19.11, is vanadium(V) oxide V_2O_5, with potassium sulphate K_2SO_4 as a promoter, on a silica support. The equilibrium is:

$$2SO_2(g) + O_2(g) \rightleftharpoons 2SO_3(g); \qquad \Delta H^{\ominus} = -197\,\text{kJ mol}^{-1}$$

This reaction is exothermic: too high a temperature will cause the yield to be unacceptably low (equilibrium shifts to the left); too low a temperature results in an unacceptably low reaction rate. The best compromise temperature for this reaction is found to be 450 °C. Pressure is typically 2 atm. Under these conditions, conversion exceeds 99.5%.

The product from the reaction is sulphur trioxide, which is absorbed in concentrated sulphuric acid:

$$SO_3(g) + H_2SO_4(l) \rightarrow H_2S_2O_7(l)$$

The resulting liquid is called *oleum*. It is then diluted with water to form the oily concentrated acid, which is 96–98% pure H_2SO_4:

$$H_2S_2O_7(l) + H_2O(l) \rightarrow 2H_2SO_4(l)$$

This round-about method is used because if sulphur trioxide is added directly to water, it produces a stable mist, rather than dissolving in it.

Flue gas treatment

Some sulphur dioxide inevitably escapes during the production, the loss typically being below 0.5 per cent. The flue gases can be scrubbed with calcium oxide to remove sulphur dioxide from the gas stream by a neutralization reaction.

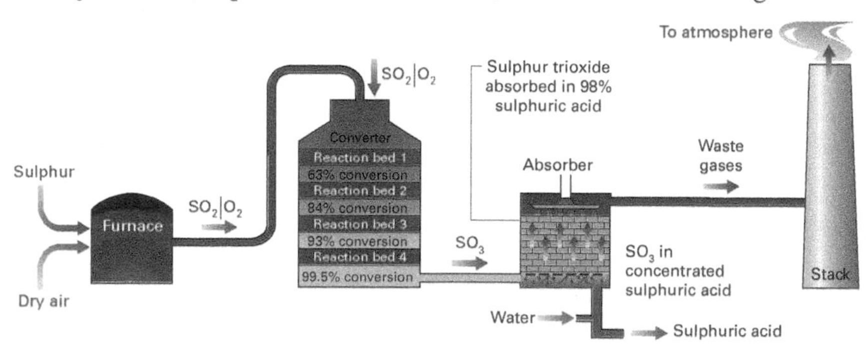

Flow diagram of the Contact process for the synthesis of sulphuric acid from water and the elements sulphur and oxygen. The absorber tower is packed with ceramic rings to ensure good contact between the acid and the gas. Most plants use two absorbers, the other after the reaction bed 3 which has been omitted for clarity.

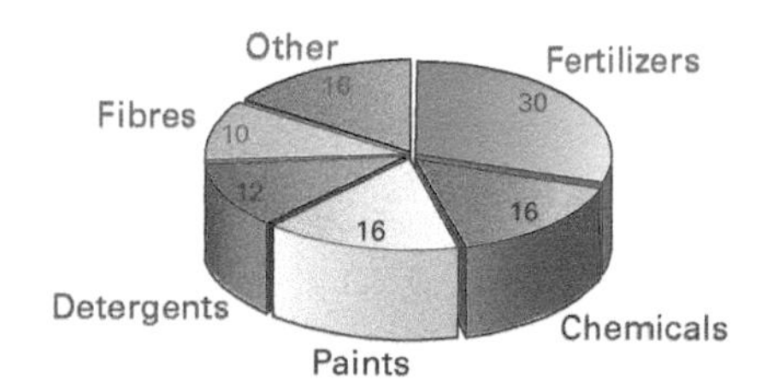

Sulphuric acid uses: over 150 million tonnes of sulphuric acid are manufactured every year throughout the world, a greater mass than for any other chemical.

SUMMARY

- Ammonia is manufactured from nitrogen (source: air) and hydrogen (source: methane). The catalyst is iron at 450 °C and 250 atm.
- Nitric acid is manufactured from ammonia, oxygen (source: air), and water. The catalyst is platinum–rhodium at 850 °C.
- Sulphuric acid is manufactured from sulphur (source: mining), oxygen (source: air), and water. The catalyst is vanadium(V) oxide at 450 °C.

19.10

Group VI: oxygen and sulphur – 1

OBJECTIVES

- Allotropes of oxygen
- Allotropes of sulphur
- Oxides and sulphides

The Group VI elements are, in order of increasing atomic number, oxygen (O), sulphur (S), selenium (Se), tellurium (Te), and polonium (Po). All have a valence-shell electronic structure of ns^2np^4. Oxygen and sulphur are non-metals, selenium and tellurium are metalloids, and polonium is a metal. Radioactive polonium has a simple cubic structure, which is unique among elements. You are already familiar with a good deal of the chemistry of oxygen and sulphur. This spread provides some more general background information about the elements and points to other chapters where more detailed information may be obtained.

Occurrence

Oxygen is the most common element in the Earth's crust, accounting for 46% of its solid constituents, 89% of water, and 21% of the atmosphere. It is obtained industrially from the fractional distillation of liquid air, itself produced by compressing and cooling air from the atmosphere. Oxygen is produced in the laboratory either (1) by adding hydrogen peroxide to manganese(IV) oxide or (2) by heating potassium chlorate with manganese(IV) oxide. In both cases the manganese(IV) oxide is acting as a catalyst, and the reactions are:

$$2H_2O_2(aq) \rightarrow 2H_2O(l) + O_2(g) \quad (1)$$

$$2KClO_3(s) \rightarrow 2KCl(s) + 3O_2(g) \quad (2)$$

Sulphur is found in the elemental state in volcanic areas. It is also present as hydrogen sulphide H_2S at concentrations of up to 6% in natural gas (methane) wells. The methane and hydrogen sulphide that make up the natural gas are both products of microbial action on organic matter. Sulphur is also extracted from crude oil (petroleum) during refining.

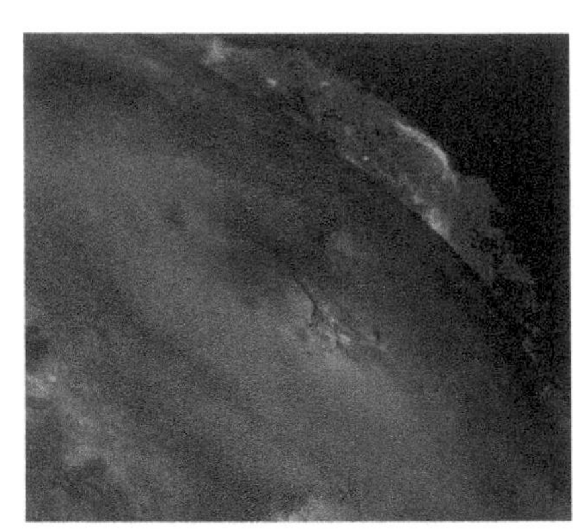

The active volcanoes on Io, one of the moons of the planet Jupiter, emit streams of sulphur.

Uses of oxygen

The chief industrial use of oxygen is steel-making, where it is blown into molten iron to oxidize and remove impurities. The gas is also used in some breathing apparatuses (e.g. for mountaineers and hospital patients). Mixed with oxygen, the gas ethyne (acetylene) burns at over 3000 °C, hot enough to cut and weld steel.

Uses of sulphur

The chief industrial use of sulphur is in the manufacture of sulphuric acid (described in the previous spread). Sulphur is used in its powdered elemental form to dust grape vines to prevent fungal growth. The sulphonamides (spread 23.6) are a group of antibiotics containing the $-SO_2NH_2$ group; they are used to treat infections of the gut and urinary system.

Allotropes of oxygen

The two allotropes of oxygen (O_2 and O_3) are both gases at 25 °C and 1 bar. The systematic names are dioxygen and trioxygen respectively. However, O_2 is usually referred to as 'oxygen' and O_3 as 'ozone'.

Ozone at ground level is hazardous as it is an extremely vigorous oxidant. It forms in the high-voltage discharge in sparking electrical equipment, where oxygen molecules split into oxygen atoms, which then combine with O_2 to form O_3. (Similar reactions initiated by solar radiation produce ozone in the atmosphere.) At low concentrations, ozone's strength as an oxidant is put to good use in the sterilization of water, for example in some swimming pools. Ozone has two advantages over chlorination: it kills micro-organisms more rapidly and there is no risk of forming potentially carcinogenic (cancer-inducing) chlorinated organic compounds.

Allotropes of sulphur

The two important solid allotropes of sulphur are rhombic sulphur and monoclinic sulphur, spread 6.7. Below 96 °C, rhombic sulphur is the stable allotrope; monoclinic sulphur is the stable allotrope above this temperature. Both are composed of ring-shaped S_8 molecules and differ only in the pattern in which these molecules pack (see chapter 17 'Trends across a period').

Oxides and sulphides

Oxides form when oxygen is heated with most elements; the oxygen acts as an oxidant by gaining electrons and forming the oxide ion O^{2-}. In the case of the s-block metals, the peroxide ion O_2^{2-} and the superoxide ion O_2^- may form (see chapter 16 'The s-block elements').

Sulphides form similarly when sulphur is heated with most elements. These compounds contain the sulphide ion S^{2-} in which sulphur has undergone reduction by gaining electrons.

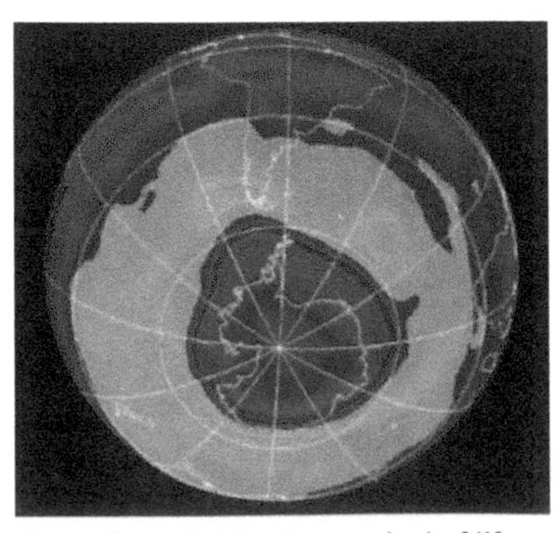

Ozone is crucial for the survival of life on Earth. The ozone layer in the upper atmosphere partially blocks harmful ultraviolet rays from the Sun. Depletion of ozone in the upper atmosphere, first discovered in the 1970s, has caused world-wide concern. See spread 24.4.

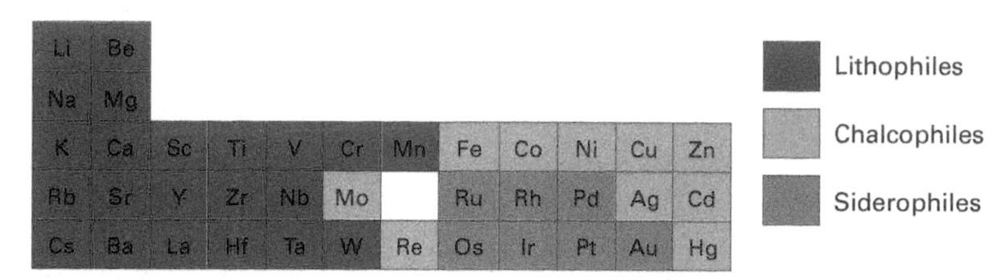

*The **lithophiles** tend to occur as oxides, silicates, sulphates, or carbonates. The **siderophiles** tend to occur native (in the elemental form). The **chalcophiles** tend to occur as sulphides. The element Tc (between Mo and Ru) is essentially unknown in nature.*

At first glance, the formation of the O^{2-} ion appears energetically highly unfavourable:

First electron-gain enthalpy of oxygen	$-148\,kJ\,mol^{-1}$
Second electron-gain enthalpy of oxygen	$+850\,kJ\,mol^{-1}$

However, the expenditure in producing the doubly charged ion is more than compensated by the lattice formation enthalpy of most oxides:

Lattice formation enthalpy of MgO	$-3800\,kJ\,mol^{-1}$
Lattice formation enthalpy of BaO	$-3050\,kJ\,mol^{-1}$

The ionization enthalpies of oxygen show that the formation of positive ions is highly unlikely, thus confirming its non-metallic nature:

First ionization enthalpy of oxygen	$+1314\,kJ\,mol^{-1}$
Second ionization enthalpy of oxygen	$+3390\,kJ\,mol^{-1}$

The non-metallic character is also confirmed by its electronegativity of 3.4.

Sulphur is a less powerful oxidant than oxygen. Sulphides also have greater covalent character than the corresponding oxides. This is because the electronegativity of sulphur (2.6) is lower than that of oxygen (3.4).

The majority of sulphides are extremely insoluble in water, with the exception of ammonium and Group I sulphides. The precipitates shown formed when hydrogen sulphide gas was bubbled into solutions of: $Fe^{2+}(aq)$ (left); $Cd^{2+}(aq)$ (middle); and $Sb^{3+}(aq)$ (right).

SUMMARY

- Oxygen and sulphur are both found as the elements: $O_2(g)$ and $S_8(s)$ respectively.
- Ozone O_3 is another allotrope of oxygen O_2.
- The two main allotropes of sulphur, monoclinic and rhombic, differ in the packing of the S_8 ring-shaped molecules.
- Sulphides are more covalent than the corresponding oxides because the electronegativity of sulphur is lower (O, 3.4; S, 2.6).
- The majority of metal sulphides are insoluble in water; the coloured precipitates may be used to identify metal cations.

Biochemistry

Oxygen is essential to most organisms for respiration – the oxidation of carbohydrates to carbon dioxide and water with the release of chemical energy. Death usually results after four minutes of oxygen starvation.

Sulphur is also an essential biochemical element. Links form between sulphur atoms at different points in a protein chain. These links are called **disulphide bridges** and they help to hold the protein chain in the correct shape for it to function correctly. Insulin is a hormone that controls the concentration of glucose in the blood: insulin deficiency causes diabetes. The shape of the insulin molecule (and hence its function) is partly controlled by disulphide bridges (spread 28.5).

Selenium

Selenium is important as a trace element in biological systems. In the late twentieth century, in northern central China, the local population suffered from a deficiency disease called Kashin–Beck syndrome, which caused muscular problems. The soil (and therefore the food crops grown in it) were found to have a very low selenium content. The soil is now treated with salts containing the selenate ion SeO_4^{2-} and the syndrome has effectively disappeared.

PRACTICE

1 Explain why the ozone layer is crucial for survival of life on Earth.

19.11

GROUP VI: OXYGEN AND SULPHUR – 2

OBJECTIVES

- Oxidation numbers
- Sulphur dioxide – sulphurous acid
- Sulphur trioxide – sulphuric acid

Oxygen and sulphur combine to make two oxides of sulphur, sulphur dioxide SO_2 and sulphur trioxide SO_3. These acidic oxides dissolve in water to form sulphurous acid H_2SO_3 and sulphuric acid H_2SO_4 respectively. These acids in turn give rise to a total of four series of stable, crystalline salts. Before looking more closely at these substances, this spread opens with an assessment of the oxidation numbers of oxygen and sulphur.

Oxidation numbers

By far the most common oxidation number for oxygen is –2; sulphur shows a wider range of oxidation numbers, including –2, +2, +4, and +6, because sulphur has d orbitals of suitable energy to involve in bonding, whereas oxygen does not. Particular use of d orbitals is made in compounds containing S(VI), e.g. SF_6.

In the definition of oxidation number, any shared electron pairs are assigned to the more electronegative atom. As oxygen is second only to fluorine in electronegativity, oxygen is always in a negative oxidation state except when combined with fluorine, e.g. OF_2 contains O(II) while H_2O contains O(–II).

Metal sulphides resemble the oxides in their chemical formulae and contain sulphur with oxidation number –2. While oxide ores are the most important sources for most elements, sulphide ores are especially important sources for the chalcophiles (see previous spread). Examples include zinc blende ZnS, copper pyrites $CuFeS_2$, and stibnite Sb_2S_3, used as eye-shadow since Biblical times.

The halides of sulphur also show a range of oxidation numbers; of the halogens, only fluorine can oxidize sulphur to oxidation number +6 in sulphur hexafluoride SF_6. Chlorine forms S_2Cl_2 and SCl_2, but not SCl_6. Sulphur is found with oxidation numbers +4 and +6 in its oxides and oxoacids. The oxides, SO_2 and SO_3, have been discussed in chapter 17 'Trends across a period'.

Oxidation state summary

Oxidation state	Found in
O(–II)	H_2O, CuO
O(II)	only OF_2
S(–II)	H_2S, CuS
S(II)	SCl_2
S(IV)	SO_2, SO_3^{2-}
S(VI)	SO_3, SO_4^{2-}

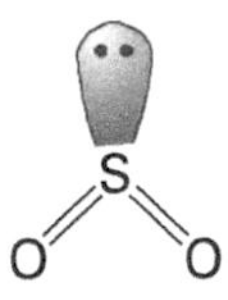

The Lewis structure for sulphur dioxide. Note that there are two S=O bonds and one lone pair. VSEPR theory (see chapter 5 'Chemical bonding') indicates that the molecule will be V-shaped; the O=S=O angle is 119°.

Sulphur dioxide and sulphurous acid

Sulphur dioxide is produced when sulphur burns in air:

$$S_8(s) + 8O_2(g) \rightarrow 8SO_2(g)$$

when a metal sulphide is heated in air, e.g.:

$$2CuS(s) + 3O_2(g) \rightarrow 2CuO(s) + 2SO_2(g)$$

and when an acidified solution of a sulphite is boiled, e.g.:

$$Na_2SO_3(aq) + H_2SO_4(aq) \rightarrow Na_2SO_4(aq) + SO_2(g) + H_2O(l)$$

The sulphur atom at the centre of the sulphur dioxide molecule has a lone pair because the two oxygen atoms form covalent bonds with only four of the six electrons on the sulphur atom. The shape of the molecule is therefore V-shaped.

Sulphur dioxide is an acidic gas. It dissolves in water to give **sulphurous acid**:

$$SO_2(g) \rightarrow SO_2(aq)$$

$$SO_2(aq) + H_2O(l) \rightleftharpoons H_2SO_3(aq)$$

Sulphurous acid is a weak dibasic acid:

$$H_2SO_3(aq) + H_2O(l) \rightleftharpoons H_3O^+(aq) + HSO_3^-(aq)$$

$$HSO_3^-(aq) + H_2O(l) \rightleftharpoons H_3O^+(aq) + SO_3^{2-}(aq)$$

Neutralization of sulphurous acid by base forms two series of salts, the **hydrogensulphites** (e.g. sodium hydrogensulphite $NaHSO_3$) and the **sulphites** (e.g. sodium sulphite Na_2SO_3):

$$H_2SO_3(aq) + 2NaOH(aq) \rightarrow Na_2SO_3(aq) + 2H_2O(l)$$

Food preservation

The reducing properties of sulphur dioxide make it useful as an antioxidant in food. Sulphite ions can also be used as antioxidants and are often more convenient than gaseous sulphur dioxide. In the presence of acids, they release free sulphur dioxide. A disadvantage is that SO_2 can cause an asthma attack in people who are sensitive to it.

Acid rain

Sulphur dioxide from the combustion of sulphur-contaminated fuels is a major contributor to acid rain. The main source of atmospheric SO_2 is coal-fired power stations. Sulphur dioxide dissolves in rain water to form sulphurous acid. Ultraviolet radiation in the upper atmosphere can initiate oxidation to SO_3, and sulphuric acid then forms.

Aqueous sulphites produce sulphur dioxide when acidified:

$$SO_3^{2-}(aq) + 2H^+(aq) \rightarrow SO_2(g) + H_2O(l)$$

Sulphur dioxide gas and sulphite ions are also reductants; they oxidize to sulphate ion in aqueous solution. The test for sulphur dioxide gas in the laboratory uses a piece of filter paper soaked in acidified potassium dichromate(VI). A change in colour from orange to green is a positive test for a reductant.

Acid ionization

See chapter 17 'Trends across a period' for the ionization of sulphurous and sulphuric acids.

Sulphur trioxide and sulphuric acid

Sulphur trioxide (T_m = 17 °C, T_b = 45 °C) is produced in the laboratory (and in industry) by passing dry sulphur dioxide and air over a heated vanadium(V) oxide catalyst (as shown on the right):

$$2SO_2(g) + O_2(g) \rightleftharpoons 2SO_3(g)$$

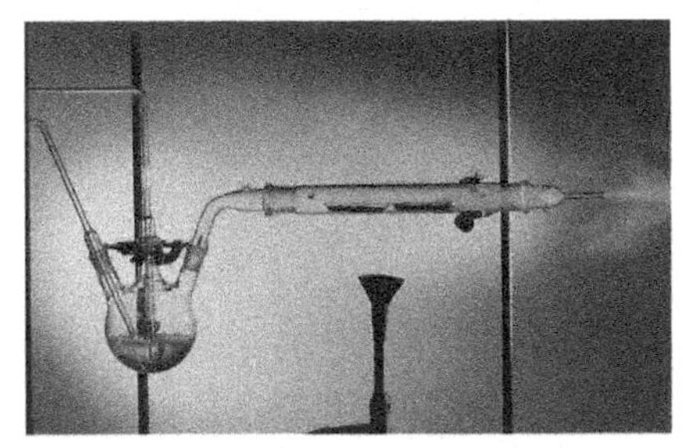

SO_2 and O_2 are dried by passing them through concentrated sulphuric acid. They combine on the surface of the V_2O_5 catalyst. SO_3 fumes in moist air. It can be condensed as a solid inside a cool dry flask.

Sulphur trioxide is acidic. It reacts violently with water to give **sulphuric acid** (often in the form of a stable acidic mist):

$$SO_3(g) + H_2O(l) \rightarrow H_2SO_4(aq)$$

Sulphuric acid is a dibasic acid; it forms two series of salts, the **hydrogensulphates** (e.g. sodium hydrogensulphate $NaHSO_4$) and the **sulphates** (e.g. sodium sulphate Na_2SO_4). Unlike aqueous sulphites, the aqueous sulphate ion is not affected by the presence of acid or base.

(a) H_2SO_4 (b) SO_4^{2-} (c)

(a) The sulphuric acid molecule has a distorted tetrahedral shape. (b) The sulphate ion has a regular tetrahedral shape. (c) The electrostatic potential map shows that electron density is delocalized equally throughout the four S—O bonds, proving that the simple structure shown is not adequate for the electronic structure in the ion.

The two delocalized pi orbitals in sulphate ion. The unusually shaped d orbital (the one with a torus) is called d_{z^2}. The other d orbital has the usual four lobes (spread 4.5).

SUMMARY

- Oxidation numbers: oxygen –2 (+2 with fluorine); sulphur -2, +2, +4, + 6.
- SO_2 dissolves in water to give sulphurous acid H_2SO_3.
- H_2SO_3 is the parent acid of the hydrogensulphite and sulphite salts.
- SO_2 and SO_3^{2-} are reductants.
- SO_3 reacts violently with water to give sulphuric acid H_2SO_4.
- H_2SO_4 is the parent acid of the hydrogensulphate and sulphate salts.

PRACTICE

1 Give the oxidation state of the Group VI element(s) in each of the following substances:
 a H_2SO_3
 b H_2SO_4
 c OF_2
 d Cl_2O
 e H_2Se
 f K_2SO_3
 g $Te(OH)_6$.

2 Write a balanced ionic equation for the reaction between aqueous sodium sulphite and dilute hydrochloric acid.

3 Write a balanced chemical equation for each of the following reactions:
 a Roasting solid iron(II) sulphide FeS
 b Heating solid potassium chlorate $KClO_3$
 c Bubbling hydrogen sulphide gas H_2S into a solution of copper(II) sulphate $CuSO_4$.

4 Explain why the following observations hold for the first and second electron-gain enthalpies ($\Delta H^{\ominus}_{e.g.}$) for oxygen and sulphur:
 a The first for oxygen is exothermic, but the second is highly endothermic.
 b The first for sulphur is *greater* than that for oxygen, whereas the second for sulphur is *smaller* than that for oxygen.

19.12

GROUP VIII: THE NOBLE GASES

OBJECTIVES

- Full valence shells
- Compounds of xenon and krypton
- 'Inert gases' or 'noble gases'?

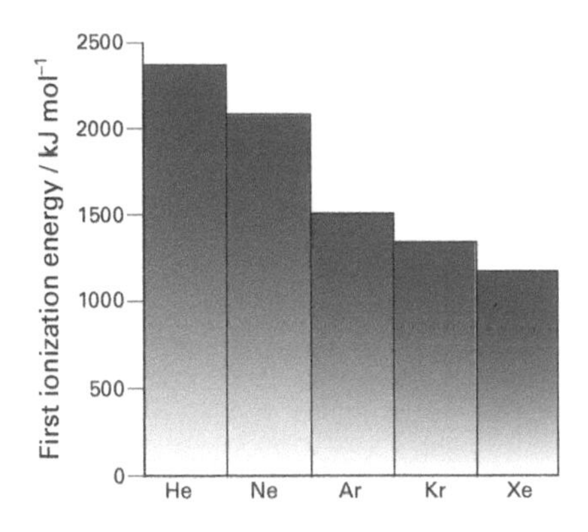

The first ionization energies of the noble gases decrease with increasing atomic number.

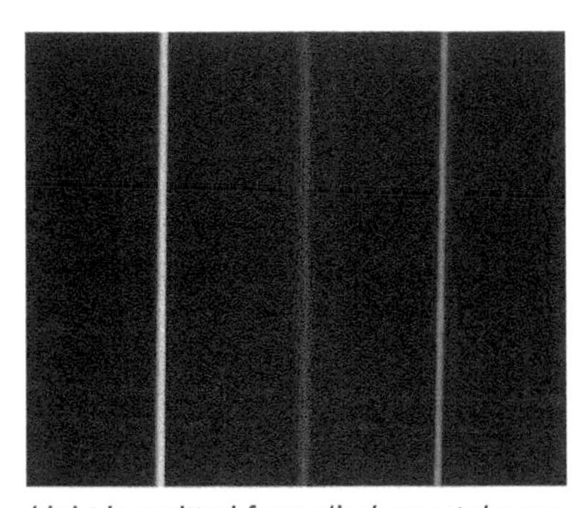

Light is emitted from discharge tubes as electrically excited electrons fall back to lower energy levels. The tubes contain helium (left), neon (centre), and argon (right).

He, Ne, and Ar *are* inert

The lack of any reactivity among the first three members of Group VIII can be understood in terms of their ionization energies. The values decrease with increasing atomic number because the electron being ionized is further from the attraction of the nucleus. Xenon forms compounds most readily because the ionization energy required is the lowest in the group.

The Group VIII elements are, in order of increasing atomic number, helium (He), neon (Ne), argon (Ar), krypton (Kr), and xenon (Xe). The sixth member of the group is radon (Ra). This element is radioactive and its study falls outside the scope of this book. All are colourless gases at room temperature. All the elements have full valence shells with an electronic structure ns^2np^6 (a complete octet), except He ($1s^2$). The members of this group were once regarded as being totally inert. However, compounds of xenon and krypton have been prepared.

Physical state

The noble gases have similar properties because the atoms all have a full valence shell. As a result, they show a great reluctance to combine with any other elements. The atoms do not interact to form molecules, as in the case of O_2, N_2, H_2, etc. They are therefore the only gases that exist as separate atoms.

The boiling points of the noble gases are extremely low. Helium boils within 5 K of absolute zero. The boiling points of the noble gases increase as their atomic number increases down Group VIII. The only forces between the atoms are dispersion forces (spread 7.4), which become larger as the size of the atoms increases. Larger atoms are more polarizable, because they have more electrons. Dispersion forces increase with increasing atomic number, as each atom has one more full shell of electrons compared with the previous noble gas.

Differences

Demonstrating the difference between the noble gases is fairly difficult, short of determining the values of a physical property such as relative atomic mass. They do, however, emit different colours of light when excited at low pressure in a discharge tube. The discharge excites the atoms, which causes them to produce their characteristic atomic emission spectrum. The spectrum varies from one element to another because the electronic energy levels vary. Atomic emission is used to distinguish between the noble gases as well as for the illuminated displays called 'neon lights'. Despite the name, these lights contain not just neon, but various noble gases to create the different colours.

Inert or noble?

The elements of Group VIII were called the 'inert gases' for 60 years after they were discovered just before the turn of the twentieth century. They were given this name because they appeared to react with no other element. They seemed to have no chemical properties (other than having no chemical properties!). While this observation remains true for the first three elements (helium, neon, and argon), the chemical research community was startled in 1962 by the discovery of the first compounds of xenon.

Xenon is now known to form a range of compounds with fluorine and oxygen, the two most electronegative elements. Examples include XeF_2, XeF_4, and XeF_6 together with XeO_3 and $XeOF_2$. There are fewer compounds of krypton because its ionization energy is higher. Again, the compounds consist mainly of fluorides such as the colourless solid KrF_2.

Once chemists knew that xenon reacted, the name for the group had to be changed. One suggestion was 'rare gases', but this was a poor choice, because argon is the third most abundant gas in the atmosphere (at about 1%). The name 'noble gases' has been chosen to replace 'inert gases'. It is reminiscent of the name 'noble metals' (for metals such as gold and platinum that do not react readily).

The preparation of xenon compounds

Louis Pasteur, who made important discoveries in chemistry, medicine, and microbiology, once said: 'In the field of observation, chance favours only the prepared mind.' The following discovery provides a good example of this general principle.

Neil Bartlett, a British-born Canadian chemistry professor, was working with the compound platinum(VI) fluoride PtF_6. One morning, instead of a red gas, his apparatus contained an orange solid. Rather than simply washing out the apparatus and starting again, Bartlett asked himself how the change had occurred. When he found a tiny hole in the apparatus, he suspected that air could have entered the apparatus overnight.

Analysis of the solid confirmed his guess. The solid had the formula O_2PtF_6. This unusual compound contained the ions O_2^+ (called dioxygenyl ion, spread 5.7) and PtF_6^-. This was an extraordinary discovery, as it meant that molecular oxygen, O_2, had been *oxidized* by loss of an electron. Platinum(VI) fluoride is such a powerful oxidant that it can oxidize molecular oxygen.

Following up the discovery, Bartlett wondered how much energy was needed to remove one electron from *molecular* oxygen. Data books showed the ionization energy to be $1175\,kJ\,mol^{-1}$. Bartlett then remembered that this value was very close indeed to the ionization energy of xenon, $1170\,kJ\,mol^{-1}$. He reasoned that, if this powerful oxidant could remove an electron from molecular oxygen, then maybe it could also remove one from xenon.

His colleagues were politely amused when Bartlett asked for some xenon 'so that I can try some reactions'. Their amusement turned to great excitement when he was able to prepare $XePtF_6$, the noble gas analogue of O_2PtF_6. Within a short space of time, xenon compounds exhibiting a wide range of oxidation states were prepared. The elements of Group VIII were no longer 'inert'; they were merely 'noble'.

(a)

(b)

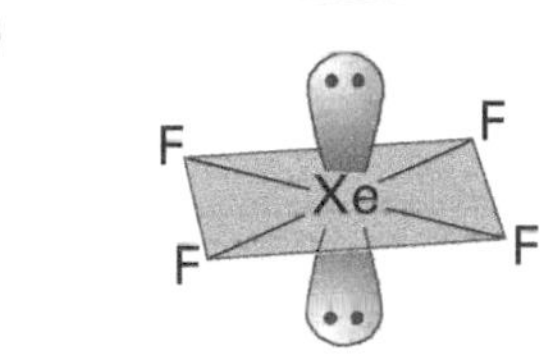

(a) Crystals of xenon tetrafluoride XeF_4.
(b) The structure of the molecule XeF_4.

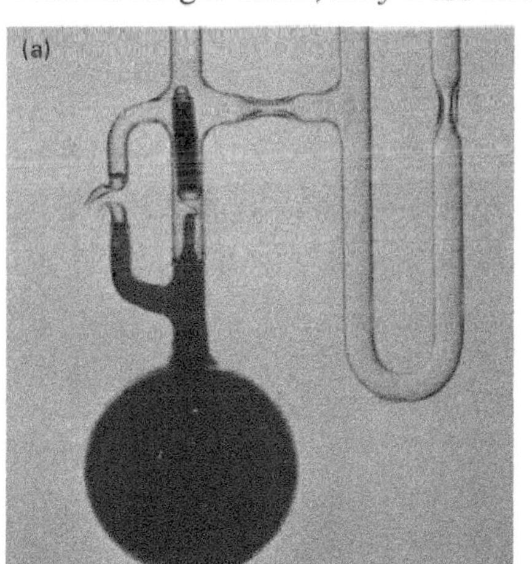

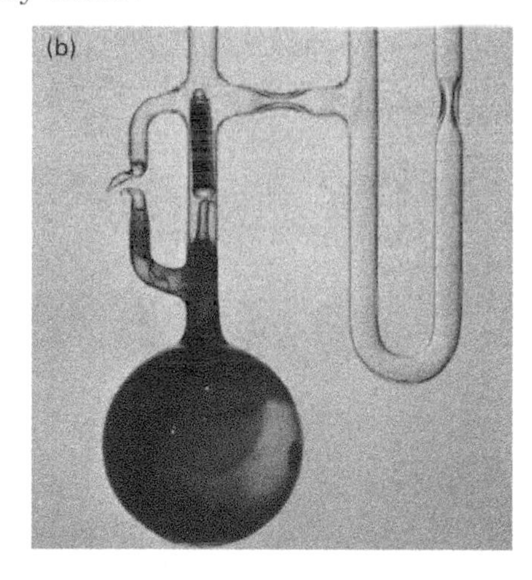

The preparation of the first xenon compound. (a) The apparatus contains the red gas PtF_6, and a seal to xenon gas. (b) After the seal is broken, it contains $XePtF_6$.

SUMMARY

- The Group VIII elements exist as separate atoms.
- The Group VIII elements have extremely low melting and boiling points; they are all gases at 25 °C and 1 bar.
- Helium, neon, and argon *are* inert; they do not form compounds.
- Xenon and, to a much lesser extent, krypton form compounds with the highly electronegative elements fluorine and oxygen.

PRACTICE

1. Describe the structure of and explain the bonding in solid argon.
2. Comment on the relative values of the boiling points of the noble gases.

PRACTICE EXAM QUESTIONS

1 Aluminium is produced commercially by the electrolysis of a 5% solution of aluminium oxide in molten cryolite. The cathode and anode can be made of carbon and the temperature of the electrolyte is maintained at around 1200 K.

a **i** Explain, with reference to economic considerations, why pure molten aluminium oxide is not used as the electrolyte. [2]

ii Write an equation for the reaction at the cathode and indicate the physical state of the aluminium as it is formed. [2]

iii Write an equation for the reaction at the anode and explain why the regular replacement of anodes is necessary. [3]

b The electrolysis uses large quantities of electricity. Identify the **two** main processes that have high electrical energy requirements. [2]

c Aluminium is readily recycled.

i Give **two** benefits of recycling rather than extracting aluminium from its ore. [2]

ii Identify **one** cost, other than electrical energy, in the process of recycling aluminium. [1]

2 **a** **i** Describe what you would observe when anhydrous aluminium chloride is added to an excess of water. Write an equation for the reaction.

ii Write an equation to show why an aqueous solution containing aluminium ions is acidic.

iii Describe what you would observe when solid sodium carbonate is added to a solution containing aluminium ions. Write an equation (or equations) for the reaction which occurs. [8]

b **i** Describe what is observed when dilute ammonia solution is added, dropwise until in excess, to a solution containing aluminium ions. Give the formula of the final aluminium-containing species.

ii How would your observations differ if dilute sodium hydroxide solution was used instead of dilute ammonia solution?
Give the formula of any different aluminium-containing species formed. [4]

3 **a** From the compounds of the Group IV elements C, Si, Ge, Sn and Pb choose an appropriate example for **each** of the following.

Type of compound	Name or formula of example
(i) A strongly reducing oxide	
(ii) A giant covalent oxide	
(iii) A strongly reducing chloride	
(iv) A covalent chloride which is **not** hydrolysed by water	

[4]

b **i** When metallic tin is placed in a solution of iodine in an organic solvent, a covalent compound of empirical formula, SnI_x, is formed which dissolves in the solvent.
When 0.4814 g of tin (an excess) was placed in a solution of iodine in the organic solvent, all the iodine reacted and formed SnI_x which dissolved in the solvent. Recovery of the covalent iodide yielded 1.987 g of product and 0.1048 g of unreacted tin. Determine the value of x. [2]

ii If the empirical formula and molecular formula of the covalent iodide are the same, state the shape of the molecule of SnI_x. [1]

c The element carbon exists as diamond and graphite which are covalent solids.

i State **two** ways in which the **structures** of these two forms of carbon differ. [2]

ii A third form of carbon C_{60}, called fullerene, is known, in which the carbon atoms form the shape of a football made up of five and six membered rings. State a physical method by which the existence of C_{60} particles could be verified. [1]

4 This question is concerned with elements in Group IV.

a Complete the following table:

Element	Formula of chloride of element in its highest oxidation state	Formula of oxide of element in its lowest oxidation state
Carbon		
Silicon		
Lead		

[2]

b A bottle of the chloride of silicon produces 'steamy' fumes when left open on the bench. A bottle of the chloride of carbon produces no such fumes.
Explain these observations, giving equations where appropriate. [5]

c Lead(IV) oxide oxidizes hydrochloric acid according to the following equation:

$$PbO_2 + 4HCl \rightarrow PbCl_2 + Cl_2 + 2H_2O$$

Write two ionic half-equations for this process. [2]

d Explain why carbon dioxide is a gas at room temperature whereas silicon dioxide is a solid with a melting point of 1710 °C. [3]

5 Ammonia, NH_3, and phosphine, PH_3, are the hydrides of the first two elements in Group V.

a **i** Draw a dot-and-cross diagram for the ammonia molecule. [1]

ii Sketch and explain the shape of the ammonia molecule. [3]

iii Ammonia molecules have σ-bonds. What do you understand by a **σ-bond**? [1]

iv Draw a charge-cloud (orbital) representation of the ammonia molecule. [1]

b Some physical properties of ammonia and phosphine are given in the following table:

	Boiling temperature/°C	Solubility in water/mol dm^{-3}
Ammonia	–33	31.1
Phosphine	–88	8.88×10^{-4}

i By reference to the nature of the intermolecular forces in both molecules, suggest reasons for the difference in boiling temperatures. [3]

ii Suggest why ammonia is much more soluble in water than phosphine. [2]

c Ammonium salts are widely used as fertilisers. One standard method for the analysis of ammonium salts (except the chloride) is to react them in a solution with methanal, HCHO. This forms a neutral organic compound together with an acid which can be titrated with standard alkali. For ammonium nitrate the equation for this reaction is:

$$4NH_4NO_3 + 6HCHO \rightarrow (CH_2)_6N_4 + 4HNO_3 + 6H_2O$$

15.0 g of a fertiliser containing ammonium nitrate as the only ammonium salt was dissolved in water and the solution made up to 1.00 dm^3 with pure water.

25.0 cm^3 portions of this solution were treated with saturated aqueous methanal solution and allowed to stand for a few minutes.

The liberated nitric acid was then titrated with 0.100 mol dm^{-3} NaOH solution. The volume of NaOH solution used was 22.3 cm^3. What percentage by mass of the fertiliser was ammonium nitrate? [4]

6 a Compare and contrast the properties of the Group IV chlorides by completing the table below.

	Tetrachloromethane	Silicon tetrachloride	Lead(II) chloride
physical state at room temperature			
electrical conductivity when liquid			
effect of adding water at room temperature			
type of bonding			

[5]

b i Write an equation for one of the Group IV oxides reacting with a base.

ii Write an equation for one of the Group IV oxides reacting with an acid. [2]

c i Which Group IV metal forms divalent ions that readily decolorize acidified, aqueous potassium manganate(VII)?

ii Use the redox half-equations in the *Data booklet* to write a balanced equation for the reaction in **c i**. [2]

7 a Sulphuric acid is produced in the UK by the Contact Process. The sulphur dioxide needed for this process is obtained by burning elemental sulphur, which is imported in the liquid state.

i State *one* advantage and *one* disadvantage of transporting sulphur in the liquid state. [2]

ii Write equations to show the conversion of sulphur to sulphur trioxide. [2]

b The figure below shows diagrammatically a sulphuric acid converter.

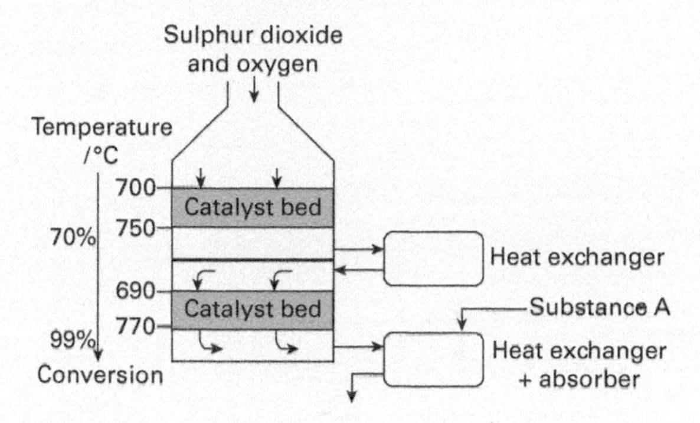

i Is the reaction between sulphur dioxide and oxygen endothermic or exothermic? [1]

ii Give the name of a catalyst used in this process. I

iii Give *two* reasons why heat exchangers are used in the Contact Process. [2]

iv Why is more than one catalyst bed used? [2]

v Give the name of substance A, shown in the diagram above which enters the absorber, and then write an equation for the process occurring in the absorber. [3]

c Give *two* reasons why a sulphuric acid plant should *not* be sited near a residential area. [2]

8 Copper and silicon(IV) oxide have properties that make them widely used materials.

a Copper is a good conductor of electricity whereas silicon(IV) oxide is not. Explain why this is so. [2]

b Suggest why silicon(IV) oxide has a higher melting point than copper. [2]

c Silicon(IV) oxide is known as a ceramic material. It is used to make hot plates for electric cookers. Suggest three reasons why it is used for this purpose. [3]

d Suggest one advantage and one disadvantage of using pure copper as the material of an axe-head. [2]

20 The transition metals

The transition metals have been of great importance throughout history. For example, gold jewellery and decorative artefacts were first made around 5000 BC. Two of the main stages of human development, namely the Bronze Age (3500–1100 BC) and the Iron Age (1100 BC – AD 100), are marked by the discovery and use of the two transition metals copper (bronze is an alloy of copper and tin) and iron. During the twentieth century, the catalytic activity of transition metals was exploited to make many key industrial processes possible. The transition metals are important in our modern industrialized society as catalysts and structural metals. When present in compounds, they show a range of distinctive properties.

The elements titanium to copper

20.1

OBJECTIVES

- Variable oxidation states and catalytic activity
- Coloured ions and complex formation
- Partially filled d subshell explains properties

Origin of the name

The name *transition* metals refers to the transition from the element calcium, in which the 3d orbitals are too high in energy to be used, to the element zinc, in which the 3d electrons are too strongly held to take part in chemical reactions.

The transition metals occupy the central region of the periodic table in Periods 4, 5, and 6. A set of five d orbitals fills with a total of 10 electrons in the course of each period. This chapter will concentrate on the chemistry of elements selected from the first row of the d block, the Period 4 elements scandium Sc to zinc Zn.

	d block										
Mg 12											Al 13
Ca 20	Sc 21	Ti 22 [Ar]$3d^24s^2$	V 23 [Ar]$3d^34s^2$	Cr 24 [Ar]$3d^54s^1$	Mn 25 [Ar]$3d^54s^2$	Fe 26 [Ar]$3d^64s^2$	Co 27 [Ar]$3d^74s^2$	Ni 28 [Ar]$3d^84s^2$	Cu 29 [Ar]$3d^{10}4s^1$	Zn 30	Ga 31
Sr 38	Y 39	Zr 40	Nb 41	Mo 42	Tc 43	Ru 44	Rh 45	Pd 46	Ag 47	Cd 48	In 49
Ba 56	La 57	Hf 72	Ta 73	W 74	Re 75	Os 76	Ir 77	Pt 78	Au 79	Hg 80	Tl 81
Ra 88	Ac 89	Rf 104	Db 105	Sg 106	Bh 107	Hs 108	Mt 109	Ds 110	Rg 111	Cn 112	

f block

Ce	Pr	Nd	Pm	Sm	Eu	Gd	Tb	Dy

The position of the d block in the periodic table.

The first row of the d block: scandium to zinc

Calcium is the metal immediately before scandium in Period 4. All the elements of the first row of the d block have greater first ionization energies than calcium. Notice that both the atomic radius and the ionization energy have similar values from titanium Ti to copper Cu. Atomic size decreases overall across Periods 2 and 3 (see chapter 17 'Trends across a period') because the increasing nuclear charge attracts the electrons in the outer orbitals more strongly. Throughout this first row of the d block, however, size changes very little. The electrons are being added to *inner* d orbitals, where they are much more effective at shielding the outer s electrons from the increasing nuclear charge. As a result, the ionization energy also increases very little across the period, because the outer 4s electrons are well shielded from the nuclear charge.

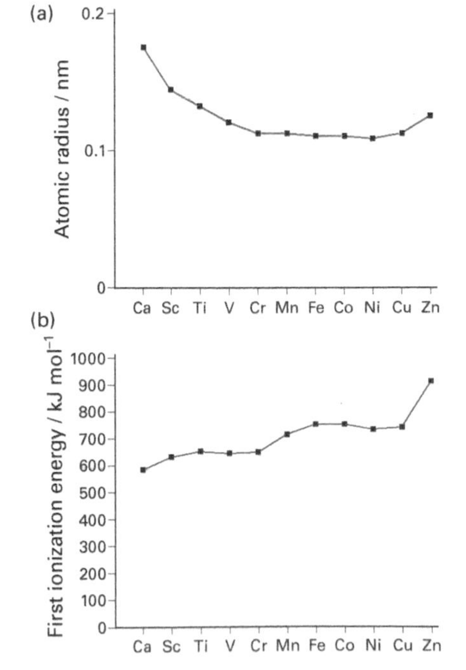

(a) The atomic radius decreases quite sharply from calcium to chromium, after which there is little change in the value. (b) The first ionization energy shows a rather gentle general increase across the d block, with a sharp increase from copper to zinc.

'Transition metals' or 'd-block elements'?

All the d-block elements of Period 4, with the exception of zinc, have much higher melting points and higher densities than calcium. Notice also that, with the exception of scandium and zinc, they show variable oxidation states in their compounds. Many d-block elements, again with the exception of scandium and zinc, have aqueous ions that are coloured. So, a transition metal is defined as follows:

- A **transition metal** (a transition element) is a d-block element forming at least one stable ion that has a *partially filled* d subshell.

Scandium and zinc are d-block elements but they are *not* transition metals. Each metal forms just one ion: there are *no* d electrons present in Sc^{3+}, and the d subshell is *full* in Zn^{2+}. This explains why scandium and zinc do not form coloured aqueous ions. Colour is due to electronic transitions between d orbitals (see the final spread in this chapter), so needs at least one d electron – and a vacant orbital in which to put it!

The major properties of the transition metals titanium to copper.

Electronic structures

The 3d orbitals are filling as atomic number increases from Ti to Cu. Look at the sequence of electronic structures opposite. You would expect the 3d subshell to fill with electrons smoothly from one to ten. However, there are two places where the filling does not follow this expected sequence. Chromium has the electronic structure $3d^54s^1$ and not $3d^44s^2$; and copper has the electronic structure $3d^{10}4s^1$ and not $3d^94s^2$. This is because the 3d and 4s orbitals are very close in energy; as electrons are added across the period, their *relative* energies change. At chromium, it is energetically favourable for all the orbitals to be half-full; in agreement with Hund's rule, the five electrons have unpaired spins (see chapter 4 'Electrons in atoms'). At copper, it is energetically favourable to fill the d orbitals completely, leaving the 4s orbital with an unpaired electron, as the 3d orbitals are now lower in energy than the 4s orbitals.

Ion formation

When the transition metals of Period 4 form ions, they lose the (outermost) 4s electrons *before* they lose the (inner) 3d electrons. For example, the Ti^{2+} ion is formed by losing the two 4s electrons rather than the two 3d electrons. The Ti^{2+} ion has the structure $[Ar]3d^2$ instead of $[Ar]4s^2$. Both Mn^{2+} and Fe^{3+} have the structure $[Ar]3d^5$.

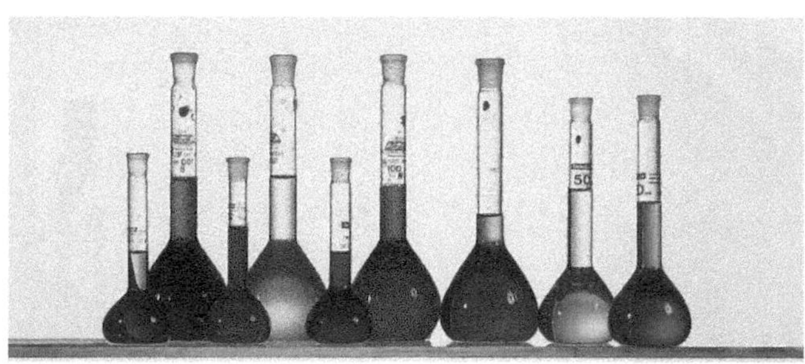

Common transition metal ions in aqueous solution (from left to right) Cr^{2+}, Cr^{3+}, $Cr_2O_7^{2-}$, Mn^{2+}, MnO_4^-, Fe^{3+}, Co^{2+}, Ni^{2+}, Cu^{2+}.

$$[H_3N{:} \longrightarrow Ag^+ \longleftarrow {:}NH_3]^+$$

Complex ions. Lone pairs are found on ions or molecules such as Cl^-, H_2O, NH_3, and CN^-. Transition metals can act as Lewis acids by accepting these lone pairs (they go into vacant d orbitals). This illustration shows the linear structure of the complex ion $[Ag(NH_3)_2]^+$ (see information about the 'silver nitrate test' in chapter 18 'The halogens').

SUMMARY

- Transition metal characteristics are:
  - variable oxidation states
  - catalytic activity
  - complex ion formation
  - formation of coloured ions.
- Transition metal characteristics result when a d-block element forms at least one stable ion that has a partially filled d subshell.
- The first row of transition metals comprises the eight elements titanium to copper.

PRACTICE

1 State the four common characteristics of transition metals.

20.2

OBJECTIVES

- The blast furnace
- Steel-making
- Specialist steels

IRON AND STEEL

Iron is the most widely used of all metals because it is abundant and relatively cheap to extract from its ores. Millions of tonnes are produced each year. As well as iron ore, the extraction uses coal and limestone mined from the ground together with oxygen from the air. Iron is used commercially in the form of steels, which consist of iron mixed with controlled amounts of other elements. Over 90% of all metallic objects are made out of steel. The chief drawback is that the cheaper types of steel corrode to produce a hydrated oxide (called rust; see spread 13.12), which is structurally weak.

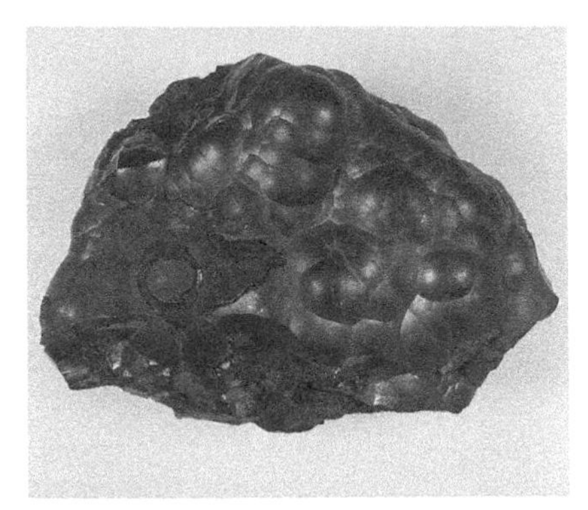

Kidney iron ore is a form of haematite Fe_2O_3 found as red-brown rounded nodules.

Iron ores and their treatment

The most important ore of iron is haematite, containing iron(III) oxide Fe_2O_3. High-grade haematite ores have iron concentrations over 60%, while low-grade ores may have an iron content of 20%. Other iron ores include magnetite Fe_3O_4 and siderite $FeCO_3$. The carbonate ore siderite is first roasted in air to convert it to iron(III) oxide:

$$4FeCO_3(s) + O_2(g) \rightarrow 2Fe_2O_3(s) + 4CO_2(g)$$

Sulphide ores on roasting produce some sulphur dioxide, which is a potential pollutant.

Iron smelting: the blast furnace

Oxides of iron can be reduced to the metal using carbon, as we saw in chapters 13 'Redox equilibrium' and 14 'Spontaneous change…'. The industrial process takes place in a **blast furnace**, shown below. The raw materials and the chemical reactions are described in the box on the left.

The chemistry of the blast furnace

Three solid raw materials are loaded into the top of the blast furnace:

- *Coke* (C) This is a form of carbon made by heating coal in the absence of air. It is the fuel and the source of the reductant.
- *Iron ore* (Fe_2O_3) This is the source of the iron. It is usually crushed haematite or roasted carbonate ore.
- *Limestone* ($CaCO_3$) This combines with high-melting-point impurities in the ore to form a liquid slag. Limestone is also a secondary source of the reductant.

A high-pressure blast of heated air enters the furnace. A series of chemical reactions take place:

1. Coke burns in the air:
   $C(s) + O_2(g) \rightarrow CO_2(g)$
2. Limestone decomposes:
   $CaCO_3(s) \rightarrow CaO(s) + CO_2(g)$
3. Coke reduces the carbon dioxide:
   $C(s) + CO_2(g) \rightarrow 2CO(g)$
4. Carbon monoxide reduces the iron oxide ore:
   $Fe_2O_3(s) + 3CO(g) \rightarrow 2Fe(l) + 3CO_2(g)$
5. Impurities in the ore (mostly silica) react with calcium oxide to form a liquid slag:
   $CaO(s) + SiO_2(s) \rightarrow CaSiO_3(l)$

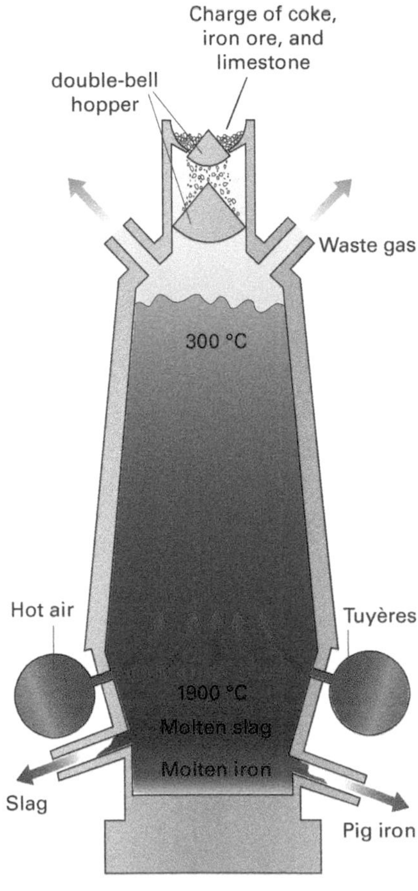

A modern blast furnace is a steel structure lined with refractory firebrick and is up to 30 metres high. The air blast is preheated to about 900 °C by heat exchangers (visible on the far left), which extract heat from burning the waste gases leaving the top of the furnace. This also greatly reduces levels of the pollutant carbon monoxide. A modern furnace produces around 10 000 tonnes of iron per day.

The liquid iron falls to the bottom of the furnace as it is formed, and the liquid slag floats on top of it. This protects the iron from oxidation by the air blast. Slag and iron are run off separately at intervals of about four hours. Slag is used to make building blocks and road foundations.

The manufacture of mild steel

Impure **pig iron**, the iron that comes out of a blast furnace, is very brittle. It contains about 4% carbon from the coke, together with nitrogen from the air blast, and silicon from impurities in the ore. The physical properties of iron improve dramatically when the carbon content is below 0.2%. Lowering the carbon content of iron and carefully controlling the amounts of other substances present is called steel-making. **Mild steel** is a tough steel that contains approximately 0.15% carbon. Pig iron is usually converted to mild steel by the **basic oxygen process**. Oxygen is blasted into molten iron through a water-cooled lance to oxidize the excess carbon to carbon dioxide. The 'basic' part of the name refers to the basic lining of the furnace and to the basic substances called fluxes that are added. These combine with the acidic impurities (oxides of other non-metals) to form a slag which can be removed from the iron.

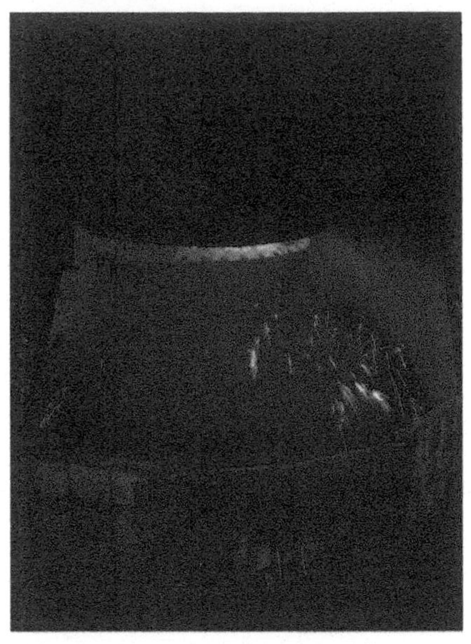

A basic oxygen converter contains up to 300 tonnes of molten pig iron mixed with up to 30% scrap steel. The oxygen lance oxidizes the chief non-metallic impurities, carbon, phosphorus, and nitrogen.

Alloy steels

Alloy steels have specific properties that suit their end-use. These properties are achieved by controlling the percentages of non-metals in the steel and by adding controlled quantities of other transition metals. For example, vanadium is used as an alloying agent in **ferrovanadium steel**, forming a carbide V_4C_3 that gives a fine grain to the steel and increases its resistance to wear. Ferrovanadium steel is used in springs.

High-tensile structural steels have a maximum of 0.3% carbon with 0.9% of both chromium and manganese, and 0.4% of copper. **Stainless steels** resist corrosion much better than iron itself. They typically contain 18% chromium and 8% nickel. Stainless steels for use above 450 °C also contain titanium. **Hadfield steel**, containing 13% manganese, is particularly frustrating for safe-crackers, as the metal becomes harder the more it is hit. **Armour plate** and some bicycle frames are made from steel containing 3% chromium and 0.5% molybdenum. **High-speed steels** include vanadium and cobalt as important constituents in a wide variety of wear-resistant alloy steels used to make cutting tools. **Alnico steel** contains iron, aluminium, nickel, and cobalt and is used to make permanent magnets.

Galvanizing involves immersing a steel or iron object in a tank of molten zinc; see spread 13.12. **Galvanized steel** or iron has a thin protective coating of zinc on its surface. The coating acts in two ways. It *physically* prevents water and oxygen reaching the iron below. But it continues to protect the iron or steel underneath in a more important *chemical* way, even if it is scratched or partly worn off. Zinc is more reactive than iron (it has a more negative standard electrode potential), so it will lose electrons more easily than iron. In contact with water and oxygen (from damp air, for example), it is therefore the zinc that is oxidized and preferentially corrodes away, rather than the iron rusting.

Removal of S

Sulphur is removed by blowing powdered magnesium into the melt; this reacts with sulphur to form magnesium sulphide, which is removed with the slag.

Recycling

Recycling scrap steel makes a helpful contribution to our care of the environment, as is the case with recycling aluminium cans.

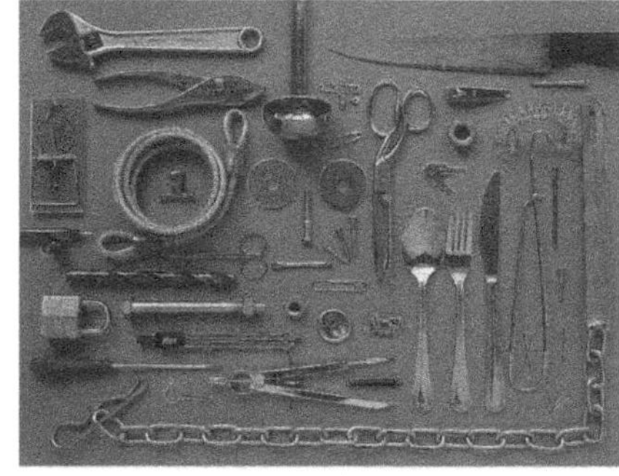

A selection of objects made from steel.

SUMMARY

- Pig iron is obtained by reduction of iron ore by carbon.
- Steel is made from pig iron by oxidizing most, but not all, of the carbon in the pig iron using a water-cooled oxygen lance.
- The properties of steels may be controlled by the addition of other transition metals.
- Stainless steels contain chromium and nickel alloyed with iron.
- Vanadium, manganese, and cobalt add toughness and hardness to steels.

20.3

Manufacture and uses of four transition metals

OBJECTIVES

- Manufacture and uses of chromium
- Manufacture and uses of titanium
- Manufacture and uses of copper
- Manufacture and uses of tungsten

Chromium, titanium, and copper are three transition metals with widely differing end-uses. The extraction method of each of these metals reflects its general reactivity and how readily it can be oxidized. Unlike iron, none of these three metals is extracted by carbon reduction in a blast furnace. Chromium would require a temperature 500 °C hotter than for iron for this process to be thermodynamically feasible. Titanium also is too strongly bonded to oxygen: furthermore, it forms a carbide with carbon. In the case of copper, the metal is so unreactive (i.e. the metal ions are so easily reduced) that a separate reductant is not required. Simply heating the ore causes decomposition to the metal. A further distinction is that iron, which is needed in huge quantities, is produced by a *continuous* process (the blast furnace), in which reactants are added and products are removed continuously over a long time. These three metals, which are required in smaller quantities, are each produced by a *batch* process; see spread 13.13. Batch processes are necessarily more costly than continuous processes.

Uses of chromium-plated steel. The layer of chromium plate is only a few hundredths of a millimetre thick. It protects the underlying steel and improves its appearance.

The manufacture of chromium

Chromium is a bright, shiny metal which forms a transparent and protective oxide layer on its surface. This makes chromium very useful for coating readily corroded metals such as iron.

The only commercially important chromium ore is chromite $FeCr_2O_4$ (i.e. $FeO{\cdot}Cr_2O_3$), 95% of the world's supplies of which are in South Africa. The element is extracted and mixed with nickel to make heat-resisting alloys, or with iron and nickel to make stainless steels. The presence of chromium also hardens steel. 'Chromium plate' is a thin protective and decorative layer of chromium electroplated onto steel.

If pure chromium is required (e.g. for electroplating), the ore is first concentrated and converted to chromium(III) oxide. The oxide is then reduced by the Thermit process, in which a powdered mixture of chromium(III) oxide and aluminium is ignited; see spread 3.1 for the equivalent process for iron(III) oxide:

$$Cr_2O_3(s) + 2Al(s) \rightarrow 2Cr(l) + Al_2O_3(s)$$

The result of this highly exothermic reaction is a pool of molten chromium at the bottom of the refractory container.

If a ferrochrome alloy for use in steel-making, rather than pure chromium, is required, then concentrated chromite ore is reduced by carbon in an electric arc furnace:

$$FeCr_2O_4(s) + 4C(s) \rightarrow Fe(l) + 2Cr(l) + 4CO(g)$$

The FFC Cambridge process

An alternative process for extracting titanium has been developed by Derek Fray, Tom Farthing, and George Chen, which uses molten salt electrolysis on a titanium(IV) oxide cathode (patent filed in 1998). The mechanism is currently not fully understood.

The manufacture of titanium: the Kroll process

Titanium is the ninth-commonest element in the Earth's crust. It has just over half the density of steel but comparable tensile strength (strength under tension, i.e. when pulled). Titanium is a reactive metal but it forms a protective oxide coating on its surface in the same way as aluminium does.

The main titanium ores are rutile TiO_2 and ilmenite $FeTiO_3$ (i.e. $FeO{\cdot}TiO_2$). Extraction is difficult because molten titanium reacts strongly with non-metals such as nitrogen, oxygen, and especially carbon: these impurities would render the metal brittle. The first stage is to produce titanium(IV) chloride (T_b = 136 °C) by heating the oxide with chlorine and carbon at about 900 °C:

$$TiO_2(s) + 2C(s) + 2Cl_2(g) \rightarrow TiCl_4(g) + 2CO(g)$$

The chloride is then reduced by magnesium at 1000 °C in an inert atmosphere of argon:

$$TiCl_4(g) + 2Mg(l) \rightarrow Ti(s) + 2MgCl_2(l)$$

Tungsten manufacture

Tungsten, like titanium, cannot be extracted by carbon reduction because of carbide formation. Its extraction uses a different reductant. Tungsten ore is first converted into the oxide WO_3, which is then reduced using hydrogen gas at 800 °C:

$$WO_3(s) + 3H_2(g) \rightarrow W(s) + 3H_2O(g)$$

Precautions have to be taken because of the flammable nature of hydrogen. Tungsten is used to make tungsten carbide drill bits and was used for the filaments in incandescent light bulbs.

In a Kroll reactor, the reduction of titanium(IV) chloride produces up to 3 tonnes of titanium, which must be bored or machined out of the cooled reactor. (This is a good example of a batch process.) Further processing is needed to remove magnesium and magnesium chloride impurities.

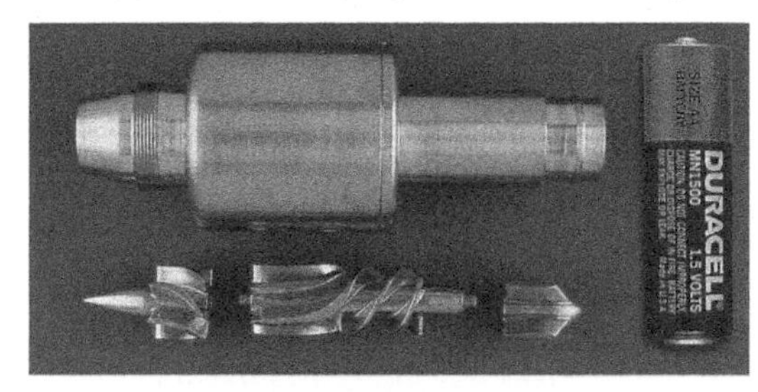

Titanium alloy is used to make the tiny turbine in this heart bypass pump. The shape of the blades required careful design to achieve a sufficient flow rate while avoiding cell damage. (The battery is shown for scale.)

Titanium cladding on the surface of the Guggenheim Museum in Bilbao.

The manufacture of copper

Copper is used as the pure metal, which resists corrosion and is an excellent conductor of heat and electricity. It polishes to a bright shine and so is decorative. **Bronze** is an alloy of about 90% copper with tin. **Brass** is an alloy of copper and zinc. **Coinage metals** contain about 75% copper together with nickel.

The chief copper ore is copper pyrite (chalcopyrite) $CuFeS_2$ (i.e. $CuS{\cdot}FeS$). Other commercial ores include copper glance Cu_2S, cuprite Cu_2O, azurite $2CuCO_3{\cdot}Cu(OH)_2$, and malachite $CuCO_3{\cdot}Cu(OH)_2$. Lumps of copper metal called 'native copper' are occasionally found, illustrating the low reactivity of copper compared to most other metals.

The ore malachite.

Most copper ores contain only a few per cent of the metal. The ore is concentrated by 'froth flotation', in which air is bubbled through a tank containing water, very finely powdered ore, and detergent. Copper-containing particles rise with the bubbles; waste (called 'gangue') sinks. The copper-containing froth can be skimmed off.

The next steps are most easily understood if we start from the ore Cu_2S. When roasted in air, copper(I) sulphide forms copper(I) oxide and sulphur dioxide:

$$2Cu_2S(l) + 3O_2(g) \rightarrow 2Cu_2O(l) + 2SO_2(g)$$

The oxide then reacts with unchanged ore to give the molten metal:

$$2Cu_2O(l) + Cu_2S(l) \rightarrow 6Cu(l) + SO_2(g)$$

The product contains 2–3% of impurities, mainly iron and sulphur. These impurities harden the metal and make it quite unsuitable for its major use in the manufacture of bendable wires and malleable copper piping. Electrolytic refining (see chapter 13 'Redox equilibrium') gives copper of almost 100% purity.

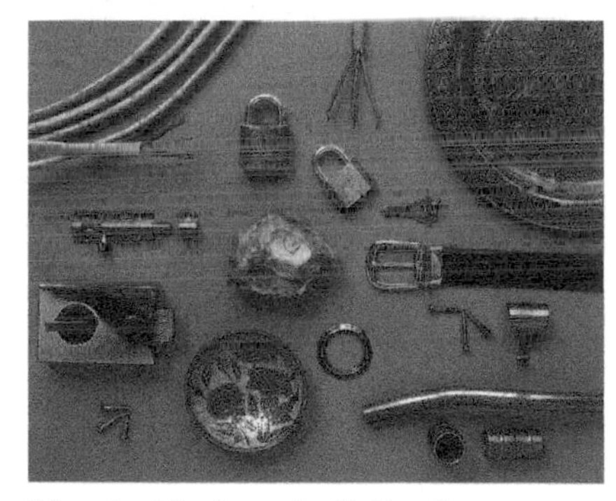
The electrical conductivity of pure copper is second only to that of silver. Its high ductility enables it to be drawn into wires. Good malleability means that wires and pipes may be bent when they are fitted into buildings.

Roasting of ores

In the roasting of sulphide ores in air, sulphur dioxide is formed. Release of this gas into the atmosphere causes acid rain, spread 19.11. It would in principle be possible to capture the gas and use it to manufacture sulphuric acid, spread 19.9.

SUMMARY

- Chromium is produced by the Thermit process: the reduction of chromium(III) oxide by aluminium.
- Uses of chromium include in stainless steels and as chromium plate.
- Titanium is produced by the reduction of gaseous titanium(IV) chloride by magnesium in an inert argon atmosphere.
- Titanium metal is strong and has a low density; titanium alloys perform well in high-temperature or corrosive environments.
- Copper is produced by roasting the ore in air.
- The use of copper in pipes and electrical wires depends on its chemical inertness, high ductility, malleability, and electrical conductivity.

PRACTICE

1 Write out the metal reactivity series, showing the relative reactivities of the metals calcium, iron, lead, tin, silver, sodium, and zinc. Insert copper, chromium, and titanium into this list. Justify your choice of position for each of the last three metals.

20.4

OBJECTIVES

- Range of oxidation states
- Bonding with 3d and 4s electrons
- Reduction of vanadium(V)
- Assigning oxidation states

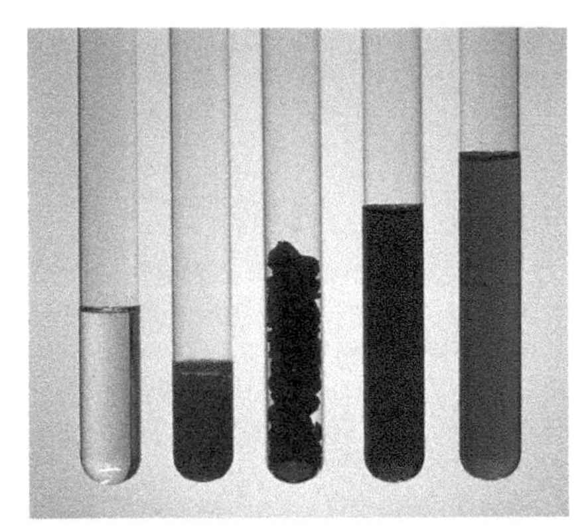

Compounds showing the common oxidation states of manganese. All are shown here as aqueous solutions, with the exception of manganese(III) hydroxide $Mn(OH)_3$ and manganese(IV) oxide MnO_2. (a) $Mn^{2+}(aq)$, (b) $Mn(OH)_3(s)$, (c) $MnO_2(s)$, (d) $MnO_4^{2-}(aq)$, and (e) $MnO_4^-(aq)$.

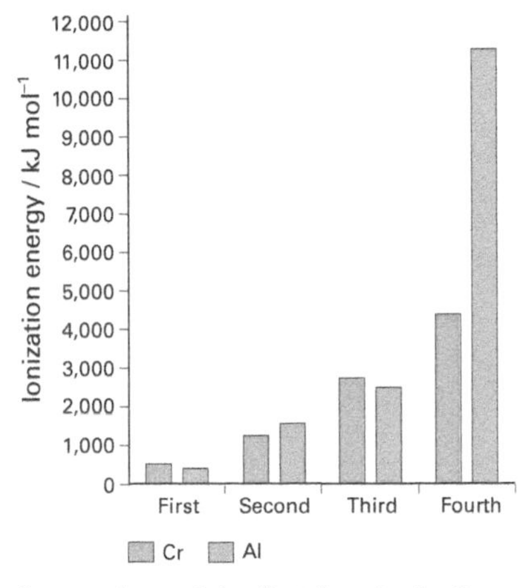

Comparison of the first four ionization energies of chromium and aluminium.

Titanium(IV) chloride reacts vigorously with water to form copious clouds of titanium(IV) oxide. This reaction is used to produce smoke trails during aerobatic displays.

VARIABLE OXIDATION STATES – 1

One of the characteristic properties of the transition metals is that they have variable oxidation states. This is because the five inner d orbitals are at a similar energy level as the single outer s orbital. In the transition metals, d electrons as well as s electrons are involved in bonding. Ionic bonds form when 4s and then 3d electrons are lost to produce positively charged ions. Covalent bonds form as unpaired electrons pair up with those on other atoms.

Electronic structures and oxidation states

You can see the wide range of oxidation states shown by the elements of the first row of transition metals below. Manganese, for example, shows most oxidation numbers from +2 to +7 inclusive. Notice that the *maximum* oxidation number is equal to the total number of 3d and 4s electrons for the elements titanium to manganese inclusive. As you might expect, compounds containing transition metals in higher oxidation states tend to be covalent, e.g. the manganate(VII) ion MnO_4^- is a covalently bonded oxoanion containing manganese with oxidation number +7. In this ion the two 4s electrons and all five 3d electrons are used for bonding. Lower oxidation states tend to involve ionic bonding, e.g. manganese(II) chloride contains the Mn^{2+} ion.

The figure alongside compares the first four ionization energies for aluminium and chromium, both of which are normally found in oxidation state III. The sum of the first three ionization energies differs by less than two per cent. However, the fourth ionization energy, from M^{3+} to M^{4+}, varies greatly. In aluminium, the fourth electron has to be taken from an *inner* (2p) orbital, much closer to the nucleus. Hence the huge rise in ionization energy and the fact that aluminium is never seen in oxidation state IV. Chromium's significantly lower fourth ionization energy can be compensated by the increased lattice energy caused by the higher charged ion. So chromium compounds in oxidation state IV exist, the most famous being chromium(IV) oxide, CrO_2, colloquially dubbed 'chrome dioxide' by the audio and video tape industry.

The main oxidation states (bold indicating the most important) and the electronic structures of the elements titanium to copper. Compounds containing a transition metal in its higher oxidation state tend to be oxidants. See spread 13.4 for redox titrations involving $Cr_2O_7^{2-}$ and MnO_4^-.

oxidation number								
+7				**+7**				
+6			**+6**	+6	+6			
+5		**+5**						
+4	**+4**	**+4**		**+4**				
+3	**+3**	+3	**+3**	+3	**+3**	**+3**	+3	
+2	+2	+2	+2	**+2**	**+2**	**+2**	**+2**	**+2**
+1								+1
Element	**Ti**	**V**	**Cr**	**Mn**	**Fe**	**Co**	**Ni**	**Cu**
Electronic structure	$[Ar]3d^24s^2$	$[Ar]3d^34s^2$	$[Ar]3d^54s^1$	$[Ar]3d^54s^2$	$[Ar]3d^64s^2$	$[Ar]3d^74s^2$	$[Ar]3d^84s^2$	$[Ar]3d^{10}4s^1$

Bond character: $TiCl_3$ and $TiCl_4$

The illustration to the left shows titanium(IV) chloride. Titanium(IV) chloride is a liquid at room temperature, confirming that it is covalent. In common with other covalent chlorides, it fumes in moist air as the result of hydrolysis. By contrast, titanium(III) chloride is a solid. It dissolves in water to form a solution containing the $Ti^{3+}(aq)$ ion, i.e. $[Ti(H_2O)_6]^{3+}$. This behaviour is typical of ionic chlorides. The reason for the difference in bonding is the higher charge density of Ti^{4+} compared with Ti^{3+}, making the bonds in titanium(IV) compounds more covalent.

Vanadium: V(V), V(IV), V(III), and V(II)

A striking illustration of variable oxidation states is shown by the chemistry of vanadium. The highest oxidation states occur when the element combines with the highly electronegative element oxygen. Vanadium shows its oxidation number +5 in the vanadate(V) ion. When zinc reduces ammonium vanadate(V) in concentrated hydrochloric acid, the solution changes from yellow to blue-green, and eventually violet as the oxidation number changes from +5 through to +2:

$$\underset{\substack{\text{V(V)}\\\text{yellow}}}{VO_2^+(aq)} \rightarrow \underset{\substack{\text{V(IV)}\\\text{blue}}}{VO^{2+}(aq)} \rightarrow \underset{\substack{\text{V(III)}\\\text{blue-green}}}{V^{3+}(aq)} \rightarrow \underset{\substack{\text{V(II)}\\\text{violet}}}{V^{2+}(aq)}$$

In a similar way, zinc can reduce dichromate(VI) ions $Cr_2O_7^{2-}$ in acidic solution via chromium(III) ions Cr^{3+} to chromium(II) ions Cr^{2+}. To prepare a transition metal in a low oxidation state, the general process is

1 add an acid
2 add a reductant

Zinc is the reductant in this case. See spread 20.10 for preparing a complex in a high oxidation state.

The oxidation states of vanadium may be demonstrated by carefully pouring a layer of aqueous potassium manganate(VII) over aqueous V^{2+} ions. After a few hours, coloured layers develop in the vanadium solution. The ion species are (from the lowest upwards): $V^{2+}(aq)$ – violet; $V^{3+}(aq)$ – blue-green; $VO^{2+}(aq)$ – blue; $VO_2^+(aq)$ – pale yellow. The green band in the middle is a mixture of the colours of VO^{2+} and VO_2^+. The brown and pink layers at the top contain $MnO_2(s)$ and $MnO_4^-(aq)$ respectively.

Worked example on assigning oxidation states

Question: Find the oxidation states of the transition metal(s) in each of the following reactions. Comment on whether the reaction is a redox reaction or not.

(a) $2Fe(s) + 3Cl_2(g) \rightarrow 2FeCl_3(s)$

(b) $Fe(s) + 2HCl(g) \rightarrow FeCl_2(s) + H_2(g)$

(c) $2CrO_4^{2-}(aq) + 2H^+(aq) \rightarrow Cr_2O_7^{2-}(aq) + H_2O(l)$

(d) $MnO_4^-(aq) + 5Fe^{2+}(aq) + 8H^+(aq) \rightarrow Mn^{2+}(aq) + 5Fe^{3+}(aq) + 4H_2O(l)$

Strategy: You may need to look back at the rules in chapter 13 'Redox equilibrium': see the box on the right.

Answer:

(a) The oxidation state of elemental iron is Fe(0). In $FeCl_3$, chlorine is the more electronegative element and has the oxidation state Cl(-I). Therefore the oxidation state of iron in $FeCl_3$ is Fe(III). Iron has been oxidized.

(b) The oxidation state of elemental iron is Fe(0). In $FeCl_2$, chlorine is the more electronegative element and has the oxidation state Cl(–I). Therefore the oxidation state of iron in $FeCl_2$ is Fe(II). Iron has been oxidized.

(c) Chromium in the chromate(VI) ion CrO_4^{2-} has the oxidation state Cr(VI). Chromium in the dichromate(VI) ion $Cr_2O_7^{2-}$ also has the oxidation state Cr(VI). The reaction is not a redox reaction.

(d) The oxidation number of the iron ion is equal to its charge. The change in oxidation state is therefore from Fe(II) to Fe(III). A redox reaction has occurred: iron has been oxidized. Manganese has the oxidation state Mn(VII) in the manganate(VII) ion MnO_4^-. In the ion Mn^{2+}, the oxidation number is equal to the charge on the ion, so the oxidation state is Mn(II). Manganese has been reduced from Mn(VII) to Mn(II).

Strategy for finding oxidation states

You find the oxidation states of each element by applying the rules listed in chapter 13 'Redox equilibrium'. Assign an oxidation number –2 to oxygen and remember that the sum of the oxidation numbers of the elements in an ion is equal to the charge on the ion. If the oxidation states for each element are the same on both sides of the chemical equation, then the reaction is not a redox reaction. If they are different, the reaction is a redox reaction.

Comment on reactions (a) and (b)

Notice how it takes the strong oxidant chlorine to force iron into the higher oxidation state Fe(III), whereas the weaker oxidant hydrogen chloride results only in oxidation state Fe(II).

Comment on reaction (d)

You should look back at chapter 13 'Redox equilibrium' for a discussion of redox titrations involving the iron(II) ion and the manganate(VII) ion.

Copper(I)

The only transition metal for which the oxidation number +1 is important is copper. Copper(I) oxide, for example, is formed by reduction of Fehling's solution using an aldehyde (see chapter 26 'Aldehydes and ketones'). Similarly, iodide ions can reduce aqueous copper(II) ions to a precipitate of copper(I) iodide; see spread 18.3.
As the iodine produced reacts quantitatively with thiosulphate ion, this can be used to estimate the copper content in alloys such as brass. Aqueous copper(I) solutions are not stable: see the next spread.

SUMMARY

- Variable oxidation states result from the outermost s and d orbitals having similar energies.

20.5

VARIABLE OXIDATION STATES – 2

OBJECTIVES

- Using standard electrode potentials
- Common oxidants and reductants
- Ions with oxidation numbers +2 and +3

All the transition metals titanium to copper can exist in two or more positive oxidation states. A species changing from a higher to a lower oxidation state is reduced and acts as an oxidant. A species changing from a lower to a higher oxidation state is oxidized and acts as a reductant. The likelihood of a given species being an oxidant or reductant is shown by its standard electrode potential; see chapter 13 'Redox equilibrium'.

What standard electrode potentials mean

- Oxidants tend to have large positive standard electrode potentials

For example, cobalt(III) ion tends to be oxidizing, removing electrons from other species and itself being reduced to cobalt(II) ion:

$$Co^{3+}(aq) + e^- \rightleftharpoons Co^{2+}(aq); \quad E^\ominus = +1.82\,V$$

- Reductants tend to have large negative standard electrode potentials

For example, chromium(II) ion tends to be reducing, donating electrons to other species and itself being oxidized to chromium(III) ion:

$$Cr^{3+}(aq) + e^- \rightleftharpoons Cr^{2+}(aq); \quad E^\ominus = -0.41\,V$$

- The half-cell with the more negative $E^\ominus$ will cause reduction; see spread 13.7.

For example, chromium(II) ion will reduce cobalt(III) ion in the redox reaction:

$$Cr^{2+}(aq) + Co^{3+}(aq) \rightarrow Cr^{3+}(aq) + Co^{2+}(aq)$$

Colours of some aqueous ions

Ion	Colour
$Ti^{3+}(aq)$	purple
$V^{2+}(aq)$	violet
$V^{3+}(aq)$	blue-green
$Cr^{2+}(aq)$	blue
$Cr^{3+}(aq)$	green
$Mn^{2+}(aq)$	pale pink
$Fe^{2+}(aq)$	green
$Fe^{3+}(aq)$	brown
$Co^{2+}(aq)$	pink
$Ni^{2+}(aq)$	green
$Cu^{2+}(aq)$	blue

See spreads 20.1, 20.4, 20.6, and 20.8 for photographs of these ions; spreads 20.8 and 20.11 explain the origin of the colours.

Reductants

Transition metals change from reducing to oxidizing as their oxidation state increases. In their *lower* oxidation states, the ions tend to be reductants. Metallic chromium and the Cr^{2+} ion are reductants.

- Chromium displaces hydrogen from acids to produce the blue $Cr^{2+}(aq)$ ion.

The relevant standard electrode potentials are:

$$Cr^{2+}(aq) + 2e^- \rightleftharpoons Cr(s); \quad E^\ominus = -0.91\,V$$
$$2H^+(aq) + 2e^- \rightleftharpoons H_2(g); \quad E^\ominus = 0\,V$$

Electrons flow from chromium to hydrogen, thus bringing about the overall change:

$$Cr(s) + 2H^+(aq) \rightarrow Cr^{2+}(aq) + H_2(g)$$

- Chromium(II) ion rapidly reduces oxygen in air to form the green chromium(III) ion.

The overall change is:

$$4Cr^{2+}(aq) + O_2(g) + 4H^+(aq) \rightarrow 4Cr^{3+}(aq) + 2H_2O(l)$$

Again, this change may be predicted from standard electrode potentials:

$$Cr^{3+}(aq) + e^- \rightleftharpoons Cr^{2+}(aq); \quad E^\ominus = -0.41\,V$$
$$O_2(g) + 4H^+(aq) + 4e^- \rightleftharpoons 2H_2O(l); \quad E^\ominus = +1.23\,V$$

- Iron(II) ion is a weak reductant that reduces strong oxidants like chlorine.

The overall change:

$$2Fe^{2+}(aq) + Cl_2(aq) \rightarrow 2Fe^{3+}(aq) + 2Cl^-(aq)$$

is confirmed by standard electrode potential data:

$$Fe^{3+}(aq) + e^- \rightleftharpoons Fe^{2+}(aq); \quad E^\ominus = +0.77\,V$$
$$Cl_2(aq) + 2e^- \rightleftharpoons 2Cl^-(aq); \quad E^\ominus = +1.36\,V$$

Green iron(II) hydroxide forms when aqueous base is added to aqueous iron(II) ions. The precipitate oxidizes on contact with air to red-brown iron(III) hydroxide. See spread 13.9.

Oxidants

Transition metal ions in *higher* oxidation states tend to be oxidants.

- Iron(III) is a moderately strong oxidant, reacting with copper to produce $Cu^{2+}(aq)$ ions.

The overall change:

$$2Fe^{3+}(aq) + Cu(s) \rightarrow 2Fe^{2+}(aq) + Cu^{2+}(aq)$$

may be predicted from standard electrode potentials:

$$Fe^{3+}(aq) + e^- \rightleftharpoons Fe^{2+}(aq); \quad E^\ominus = +0.77\,V$$

$$Cu^{2+}(aq) + 2e^- \rightleftharpoons Cu(s); \quad E^\ominus = +0.34\,V$$

(This reaction is used to etch copper in the manufacture of printed circuit boards in the electronics industry.)

- The purple manganate(VII) ion $MnO_4^-(aq)$ is a commonly used oxidant in acidic solution.

$$MnO_4^-(aq) + 8H^+(aq) + 5e^- \rightarrow Mn^{2+}(aq) + 4H_2O(l); \quad E^\ominus = +1.51\,V$$

In basic solution its final reduction product is manganese(IV) oxide:

$$MnO_4^-(aq) + 2H_2O(l) + 3e^- \rightarrow MnO_2(s) + 4OH^-(aq); \quad E = +0.59\,V$$

Note that here we have E not $E^\ominus$, because the reaction refers to non-standard conditions – there is an excess of base (see spread 13.9 in chapter 13 'Redox equilibrium').

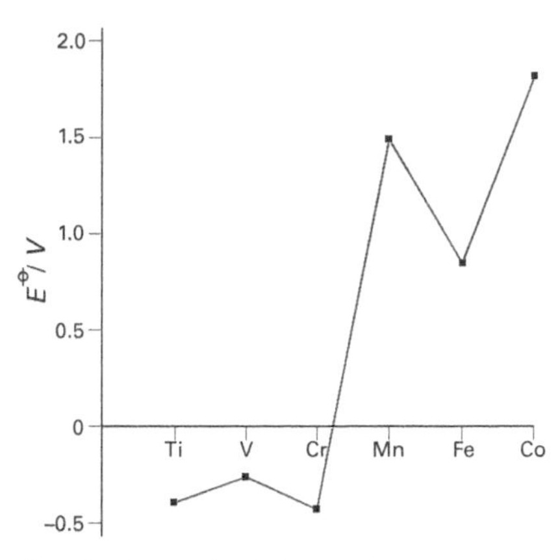

Graph of standard electrode potentials for M(III)/M(II) for transition metals Ti to Co. (In aqueous solution, the M(III) oxidation state is too unstable for either Ni or Cu to have a value.)

The M(II) and M(III) oxidation states

Both oxidation numbers +2 and +3 are possible for each of the elements from titanium to cobalt, and it is not obvious which is the more stable ion in aqueous solution. The standard electrode potential for the M^{3+}/M^{2+} half-cell (see the graph to the right) tells us which is the more stable ion for each transition metal M. The negative values for titanium, vanadium, and chromium indicate that the higher oxidation state is preferred and that a strong reductant such as metallic zinc ($E^\ominus$ = −0.76 V) must be used to reduce them to oxidation state II.

- A useful reductant to reduce Cr^{3+} to Cr^{2+} or V^{3+} to V^{2+} is zinc (see the previous spread).

The high positive standard electrode potential for Mn(III)/Mn(II) indicates that the more stable state for manganese is Mn^{2+}. For iron, *both* oxidation states can exist under normal conditions, as its standard electrode potential is much less positive. The drop from manganese to iron is due to the electronic structures of the ions concerned. Mn^{3+} has a $3d^4$ electronic structure, while the electronic structure of Mn^{2+} is $3d^5$. Fe^{3+} and Fe^{2+} are $3d^5$ and $3d^6$ respectively. The extra stability associated with a half-filled d subshell (see chapter 4 'Electrons in atoms') makes the change from Mn^{3+} to Mn^{2+} very favourable; the change from Fe^{3+} to Fe^{2+} is less favourable (i.e. the electrode potential is less positive) than expected. For cobalt, as for manganese, the more stable state is normally Co^{2+}; however

- Aqueous base makes the higher oxidation state relatively more stable, as explained in spread 20.10.

Disproportionation reactions can occur in compounds in which a transition metal has an intermediate oxidation state.

Zinc amalgam has reduced acidified vanadate(V) ions to vanadium(II) ions, V^{2+}, which are violet coloured. The reaction takes about a week.

$Cu^+(aq)$

Copper(I) ion disproportionates in aqueous solution. The relevant standard electrode potentials are:

$Cu^{2+}(aq) + e^- \rightleftharpoons Cu^+(aq)$ $\quad E^\ominus = +0.15\,V$

$Cu^+(aq) + e^- \rightleftharpoons Cu(s)$ $\quad E^\ominus = +0.52\,V$

The disproportionation reaction is

$2Cu^+(aq) \rightarrow Cu(s) + Cu^{2+}(aq)$

for which $E^\ominus = (+0.52\,V) - (+0.15\,V)$

$= +0.37\,V$

SUMMARY

- For a given transition metal, higher oxidation states are often oxidants, lower ones are often reductants.
- Higher oxidation states usually exist as covalently bonded (charged or neutral) molecules containing the transition metal with electronegative elements such as oxygen.
- The feasibility of a redox reaction is indicated by the standard electrode potentials (under *standard* conditions).

PRACTICE

1 Give examples of species in their common oxidation states for

a chromium

b manganese

c iron.

20.6

OBJECTIVES

- Homogeneous catalysis – role of variable oxidation states
- Transition metals in enzymes
- Heterogeneous transition metal catalysts

Catalytic activity

Catalysts speed up chemical reactions by providing an alternative route of lower activation energy (see chapter 15 'Chemical kinetics'). Transition metals show catalytic activity. Homogeneous catalysis involves a catalyst that is in the *same* phase as the reactants. Heterogeneous catalysis involves a catalyst that is *not* in the same phase as the reactants. Heterogeneous catalysts are generally used as the metals or as solid compounds; liquid- or gas-phase reactions take place at their surfaces.

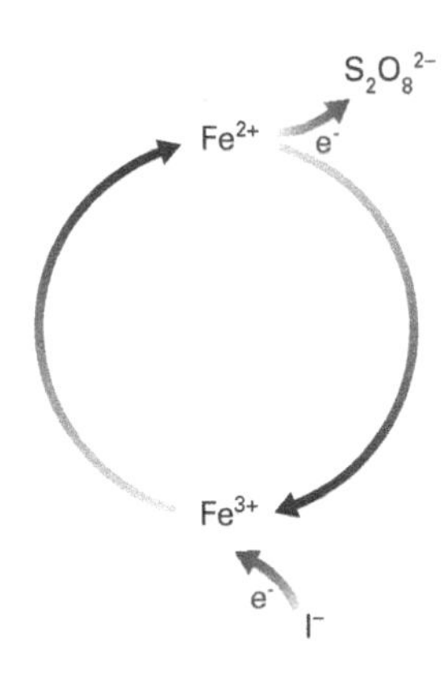

Summary diagram showing iron alternating between the states with oxidation numbers +2 and +3 as $S_2O_8^{2-}$ removes electrons and I^- donates them.

Homogeneous transition metal catalysis

Many transition metal compounds can act as homogeneous catalysts because they readily interconvert between oxidation states. In this way, they act as intermediaries for the exchange of electrons between the reactants. A good example is the reaction (outlined in spread 15.11) between peroxodisulphate ions $S_2O_8^{2-}$ and iodide ions I^-:

$$2I^-(aq) + S_2O_8^{2-}(aq) \rightarrow I_2(aq) + 2SO_4^{2-}(aq)$$

This redox reaction is much faster when Fe^{2+} ions are present.

The standard electrode potential of the Fe^{3+}/Fe^{2+} system is:

$$Fe^{3+}(aq) + e^- \rightleftharpoons Fe^{2+}(aq); \qquad E^\ominus = +0.77\,V$$

This value lies *between* the values of the relevant half-equations:

$$S_2O_8^{2-}(aq) + 2e^- \rightleftharpoons 2SO_4^{2-}(aq); \qquad E^\ominus = +2.01\,V$$

$$I_2(aq) + 2e^- \rightleftharpoons 2I^-(aq); \qquad E^\ominus = +0.54\,V$$

When Fe^{2+} ions are also present in the solution, peroxodisulphate ions oxidize Fe^{2+} ions *faster* than they oxidize I^- ions (which repel them):

$$2Fe^{2+}(aq) + S_2O_8^{2-}(aq) \rightarrow 2SO_4^{2-}(aq) + 2Fe^{3+}(aq)$$

The Fe^{3+} ions can oxidize I^- ions faster than $S_2O_8^{2-}$ ions can:

$$2I^-(aq) + 2Fe^{3+}(aq) \rightarrow I_2(aq) + 2Fe^{2+}(aq)$$

The result of the reaction is the same, whether catalysed or uncatalysed. In the catalysed reaction, the route has a lower activation energy and so the reaction is faster. The Fe^{2+} ions are regenerated.

Seeing is believing! In the redox reaction between Rochelle salt (sodium potassium 2,3-dihydroxybutanedioate) and hydrogen peroxide, the change in oxidation state of the catalyst can actually be seen. The catalyst, which consists of cobalt(II) ions, changes from pink to green – the colour of cobalt(III) – at the start of the effervescent reaction, and reverts back to pink again at the end of the reaction.

Transition metals in enzymes

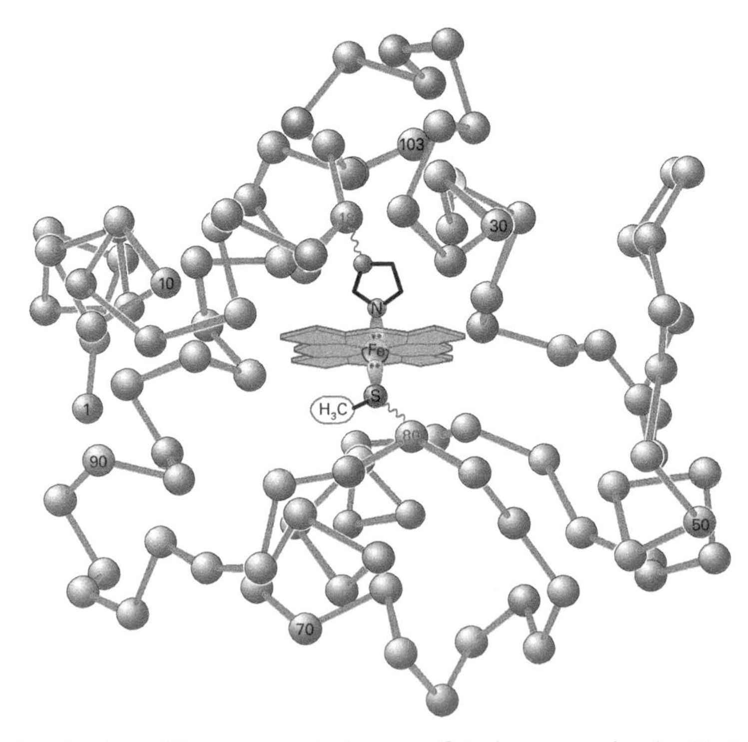

An outline structure of the enzyme cytochrome c. Cytochromes are involved in the biochemical electron-transport chain, i.e. respiration in mitochondria and photosynthesis in chloroplasts. The iron alternates between the Fe(II) and the Fe(III) oxidation states.

Heterogeneous transition metal catalysis

The mechanism of heterogeneous catalysis is discussed at the end of chapter 15 'Chemical kinetics'. With few exceptions, the catalyst is a solid and the reactants are usually gases. The activation energy is lowered because the catalyst adsorbs the reactants onto its surface and holds them close together in an orientation that favours reaction. Transition metals are very good at this because partially filled d orbitals can accept electron density from the adsorbed molecules. There are many important examples of heterogeneous catalysis using transition metals, some of which are mentioned below:

- Titanium(IV) chloride is used in the Zeigler–Natta polymerization (see chapter 22 'Alkanes and alkenes') of ethene to produce poly(ethene).
- Vanadium(V) oxide is used in the Contact process for making sulphuric acid (see the previous chapter). In this reaction, sulphur dioxide is oxidized by vanadium(V) oxide, which is reduced to vanadium(IV) oxide. The latter is then reoxidized by oxygen to vanadium(V) oxide.

  $SO_2(g) + V_2O_5(s) \rightarrow SO_3(g) + 2VO_2(s)$

  $4VO_2(s) + O_2(g) \rightarrow 2V_2O_5(s)$
- Manganese(IV) oxide catalyses the decomposition of hydrogen peroxide (see chapter 15 'Chemical kinetics').
- Metallic iron is the catalyst in the Haber–Bosch synthesis of ammonia (see the previous chapter).
- Nickel, palladium, and platinum are used as catalysts in making margarine from vegetable oils by hydrogenation (see chapter 22 'Alkanes and alkenes').
- Platinum and rhodium are used in the Ostwald process for making nitric acid (see the previous chapter).
- Platinum–rhodium alloy is used in vehicle exhaust system catalytic converters (see chapter 15 'Chemical kinetics').

Enzymes

Enzymes catalyse the biochemical reactions that support life. Many enzymes contain transition metals bonded within the structure of a protein. The transition metal alternates between oxidation states, and so allows other electron-transfer (redox) reactions to take place at a suitable rate. The role of enzymes is discussed in greater detail in chapter 30 'Biochemistry'.

Crystals of palladium (a transition metal of Period 5), a very good catalyst for hydrogenation.

SUMMARY

- A transition metal catalyst transfers electrons from the reductant to the oxidant. In doing so, it alternates between two oxidation states.
- A homogeneous transition metal catalyst system has a standard electrode potential *intermediate* between those of the redox systems being oxidized and reduced.
- Many enzymes consist of a transition metal bonded within the structure of a protein; such enzymes catalyse the biochemical electron-transport chain.
- On the surfaces of heterogeneous transition metal catalysts, incoming species donate electron density into vacant d orbitals.

PRACTICE

1 Explain why Fe(II)/Fe(III) would *not* catalyse the peroxodisulphate/iodide reaction if its standard electrode potential were *minus* 0.77 V.

2 Draw a labelled reaction profile diagram (see chapter 15 'Chemical kinetics') to illustrate the uncatalysed and the catalysed reactions between peroxodisulphate ion and iodide ion in solution.

3 The redox reaction between 2,3-dihydroxybutanedioate ions (found in Rochelle salt) and hydrogen peroxide is represented by the following chemical equations:

$(CHOHCO_2^-)_2(aq) + 2H_2O(l) \rightarrow 4CO_2(g) + 8H^+(aq) + 10e^-$

$2H^+(aq) + H_2O_2(aq) + 2e^- \rightarrow 2H_2O(l)$

a What substance is being reduced and what substance is being oxidized?

b Write an *overall* equation for this reaction.

c Suggest the mechanism by which $Co^{2+}(aq)$ ions catalyse this reaction.

d Suggest why the solution is pink at the start and finish of the reaction, yet green during it.

20.7

O B J E C T I V E S

- Ligands and bonding
- Coordination number
- Shapes

Lewis acids and bases

In a complex, the metal ion acts as a Lewis acid (see chapter 12 'Acid–base equilibrium') because it accepts electron pairs. Each donating species (ligand) acts as a Lewis base.

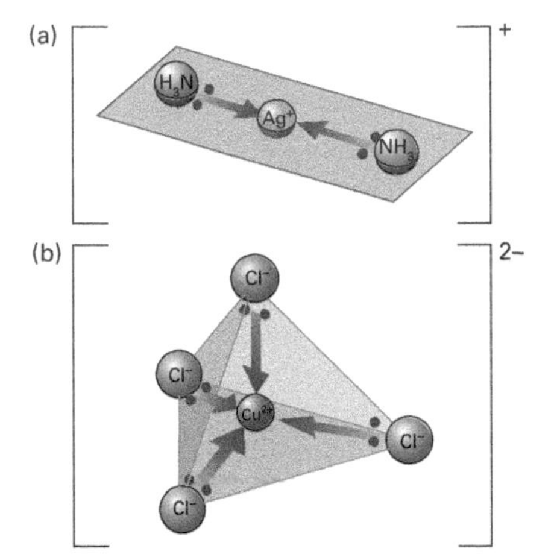

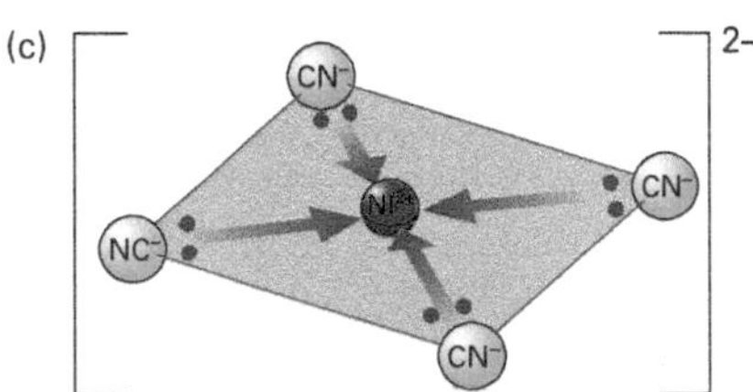

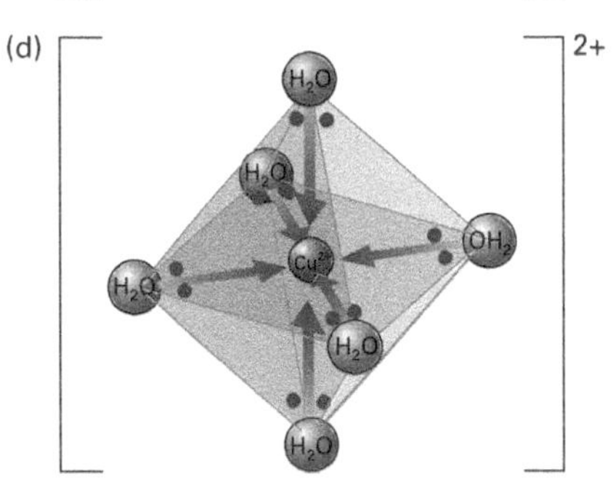

The four main shapes of complexes: (a) Linear – coordination number 2, e.g. $[Ag(NH_3)_2]^+$. (b) Tetrahedral – coordination number 4, e.g. $CuCl_4^{2-}$. $FeCl_4^{2-}$ and $CoCl_4^{2-}$ are also tetrahedral. (c) Square planar – coordination number 4 (rare, tends to occur in complexes where the metal ion has d^8 electronic structure), e.g. Ni^{2+} in $[Ni(CN)_4]^{2-}$. (d) Octahedral – coordination number 6, e.g. $[Cu(H_2O)_6]^{2+}$. $[V(H_2O)_6]^{3+}$, $[Cr(H_2O)_6]^{3+}$, $[Mn(H_2O)_6]^{2+}$, and $[Fe(H_2O)_5SCN]^{2+}$ are also octahedral.

TRANSITION METAL COMPLEXES: KEY FACTS

Complexes consist of a central metal ion surrounded by molecules or ions that form coordinate bonds (see chapter 5 'Chemical bonding') with it. Transition metals form a larger range of complexes than elements from other regions of the periodic table. This is partly because transition metals have d orbitals available for the formation of coordinate bonds. The metal ions also often have high positive charges (e.g. Cr^{3+}, Fe^{3+}), and the resulting high charge density strongly attracts the lone pairs on other species. This spread looks at these topics, and also introduces the terminology relating to complexes and describes their various shapes.

Some definitions

The species that donate electron pairs to the central metal ion are called **ligands**. The **coordination number** is the number of atoms donating electron pairs to the central metal ion and is most commonly 2, 4, or 6. The **shape** of a complex is its geometrical arrangement in space.

Ligands

Ligands donate at least one pair of electrons. Each electron pair forms a single coordinate bond with the central metal ion. Some ligands are negative ions, including cyanide CN^-, halide such as Cl^-, and hydroxide OH^- ions. Some ligands are uncharged molecules that have lone pairs to donate, such as water $:OH_2$ and ammonia $:NH_3$. These ligands are referred to as **unidentate** (or **monodentate**) because each ligand donates just one electron pair. Some other ligands are *bidentate*, because each ligand donates two electron pairs. Bidentate and multidentate ligands will be discussed in later spreads in this chapter.

Shape and coordination number

The shapes of complex ions can be predicted by using rules similar to those of VSEPR theory (see chapter 5 'Chemical bonding') for predicting the shapes of molecules. As a general rule (for unidentate ligands):

- Two ligands give a *linear* shape.
- Four ligands usually give a *tetrahedral* shape, although occasionally the shape may be *square planar*. Tetrahedral complexes are common with larger, negatively-charged ligands such as Cl^-.
- Six ligands give an *octahedral* shape. Octahedral complexes are common with small uncharged ligands such as H_2O and NH_3.

Naming complexes

There are four parts to the name of a complex. The first two parts identify the number and nature of the ligands present. The final two parts give the metal ion present and its oxidation state. With electrically neutral ligands, the charge on the complex simply equals the oxidation state of the metal ion. If the complex is negatively charged, the ending is changed to *-ate* and sometimes the Latin name for the metal is chosen. The four examples alongside are called respectively:

(a) diamminesilver(I)
(b) tetrachlorocuprate(II)
(c) tetracyanonickelate(II)
(d) hexaaquacopper(II)

The charge on the central ion

The *overall* charge on the complex depends on the charge on the central ion, and the *total* charge due to the ligands. Remember that ligands are usually either neutral (H_2O, NH_3) or negatively charged (F^-, CN^-), as shown by the following two examples.

Worked example on finding the charge on the central ion

Question: Find the charge on the central cobalt ion in (a) $[Co(NH_3)_6]Cl_3$ and (b) $CoCl_4^{2-}$.

Answer: (a) This formula corresponds to solid hexaamminecobalt(III) chloride. When dissolved in water, it ionizes to give $Cl^-(aq)$ ions and the aqueous complex ion $[Co(NH_3)_6]^{3+}(aq)$. The ammonia ligands are neutral; therefore the charge on the central cobalt ion is 3+.

(b) Each chloride ligand bears a charge of 1–. The total charge due to the ligands is therefore 4–. The overall charge on the complex is 2–; therefore the charge on the central cobalt ion is 2+.

Note: A **cationic complex** has a positive charge; an **anionic complex** has a negative charge.

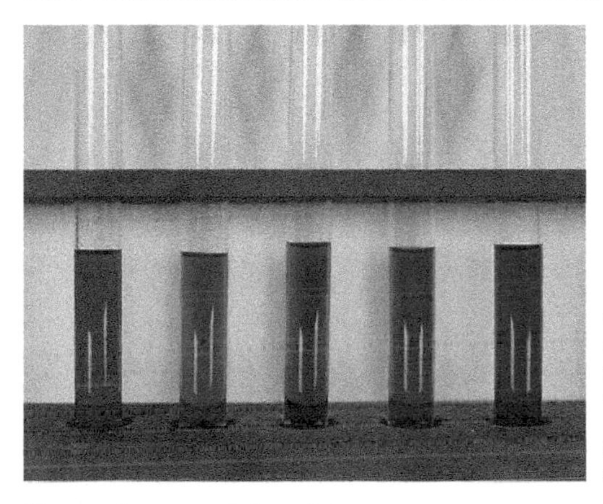

Copper(II) complexes change colour depending on the ligand. The tubes on the left have concentrated and dilute hydrochloric acid added. The tubes on the right have dilute and concentrated aqueous ammonia added.

Complexes and colour

Complex formation is often accompanied by a change in colour. Perhaps the most familiar is the formation of a deep blue solution when aqueous ammonia is added to aqueous copper(II) sulphate. The aqueous $Cu^{2+}(aq)$ ion exists in solution as the hexaaquacopper(II) complex, $[Cu(H_2O)_6]^{2+}$. Addition of aqueous ammonia displaces four of the water ligands, forming the tetraamminediaquacopper(II) complex, $[Cu(NH_3)_4(H_2O)_2]^{2+}$. Notice that ammonia does not displace all six of the water ligands, in the case of *copper*; two waters remain. Aqua ions can also exist in the solid state, as in $CoCl_2{\cdot}6H_2O$.

SUMMARY

- Complex ions consist of a central metal ion surrounded by ligands.
- Ligands donate electron pairs into vacant orbitals on the central metal ion. The bonding type is thus coordinate bonding.
- The number of atoms donating electron pairs to the central metal ion is its coordination number.
- Coordination number 2: linear shape.
- Coordination number 4: usually tetrahedral shape.
- Coordination number 6: octahedral shape.

Use of square brackets

When a complex ion contains a ligand that has *more than one atom*, such as water, the formula for the complex needs to have two types of brackets, round ones surrounding the ligand and square ones surrounding the whole complex. An example is $[Cu(H_2O)_6]^{2+}$. If the ligand is a simple ion, no brackets are needed: although square brackets are sometimes included for chloride complexes such as $CuCl_4^{2-}$; they are very rarely used for the logically equivalent oxide complexes such as CrO_4^{2-}.

Silver complexes

Three important linear complexes of silver are

$[Ag(NH_3)_2]^+$, used in Tollens' reagent (spread 26.4)

$[Ag(S_2O_3)_2]^{3-}$, used in photographic fixing

$[Ag(CN)_2]^-$, used in electroplating

Some more names

- Hexaaquairon(III)
Here *hexa* = six, and the ligands are *aqua* = water, i.e. six H_2O ligands. The central ion is Fe^{3+}, i.e. iron has oxidation number +3. The formula is $[Fe(H_2O)_6]^{3+}$.
- Hexaamminenickel(II)
Here *ammine* = ammonia. The formula is $[Ni(NH_3)_6]^{2+}$.
- Hexacyanoferrate(III)
Here *cyano* = cyanide ion CN^-, and *ferr* = iron (Latin *ferrum*, the central ion being Fe^{3+}). The formula is $[Fe(CN)_6]^{3-}$.

Changes of shape

Hydrated cobalt(II) chloride $CoCl_2{\cdot}6H_2O$ contains the hexaaqua complex ion $[Co(H_2O)_6]^{2+}$. When heated, the pink solid turns blue as the water is driven off and the tetrachloro complex $CoCl_4^{2-}$ is formed. The shape of the complex changes from octahedral for the hexaaqua complex to tetrahedral for the tetrachloro complex. The negatively charged chloride ions repel each other more than the uncharged water molecules repel one another.

PRACTICE

1 For each of the following complexes, give (i) the charge on the central transition metal ion and (ii) its coordination number:

a $[Co(NH_3)_6]^{3+}$

b $[Cu(NH_3)_4(H_2O)_2]^{2+}$

c $[Fe(CN)_6]^{4-}$

d $[Co(NH_3)_6]Cl_3$

e $CuCl_4^{2-}$

f $K_4[Fe(CN)_6]$

g Na_2CoCl_4

h $[Cu(CN)_2]^-$.

2 Sketch the shape of each of the complex ions in question 1.

3 Define each of the following terms, using the complex $[Ni(CN)_4]^{2-}$ to illustrate your answer:

a Ligand

b Coordination number

c Shape.

20.8

OBJECTIVES

- Ligand substitution reactions
- Types of ligand
- Applications

TRANSITION METAL COMPLEXES: SOME REACTIONS

Complexes may undergo reactions in which incoming ligands substitute for existing ligands. These **ligand substitution reactions** (or ligand exchange reactions) are equilibrium reactions, and have associated equilibrium constants. The extent to which one species of ligand will substitute for another therefore depends on the relative concentrations of the ligands and the enthalpy/entropy changes associated with the reaction. Some ligands are able to donate more than one electron pair per ligand. These ligands usually donate 2, 4, or 6 electron pairs. They have wide applications to analytical chemistry as well as to the medical and industrial fields.

A ligand substitution reaction in which one, two, and three en ligands (see opposite) substitute for two water molecules each around a nickel(II) ion: $[Ni(H_2O)_6]^{2+}$ is green, $[Ni(en)_3]^{2+}$ is pink.

Nickel(II) – a substitution reaction

Solid nickel(II) chloride exists as yellow crystals. It dissolves in water to give a green solution containing the complex $[Ni(H_2O)_6]^{2+}(aq)$. Pouring an excess of aqueous ammonia into this solution results in a colour change from green to violet. Ammonia ligands NH_3 replace all six water ligands:

$$[Ni(H_2O)_6]^{2+}(aq) + 6NH_3(aq) \rightleftharpoons [Ni(NH_3)_6]^{2+}(aq) + 6H_2O(l)$$

The acidic character of $M^{3+}(aq)$

The acidity of the aqueous aluminium ion $Al^{3+}(aq)$ was covered in the previous chapter. In a similar way, transition metal aqua ions with a charge of 3+ are notably acidic; their acidity is much greater than that of aqua ions with a charge of 2+ because of their higher charge density. For example, a piece of litmus placed in an aqueous solution containing iron(III) ions turns red. The following equilibrium is set up:

$$[Fe(H_2O)_6]^{3+}(aq) + H_2O(l) \rightleftharpoons H_3O^+(aq) + [Fe(H_2O)_5OH]^{2+}(aq)$$

The acidity of the solution means that when aqueous sodium carbonate is added to aqueous iron(III) ions, carbon dioxide gas is given off and a red-brown precipitate of iron(III) hydroxide $Fe(OH)_3$ forms. Similarly, when aqueous sodium carbonate is added to aqueous chromium(III) ions, carbon dioxide gas is given off and a green precipitate of chromium(III) hydroxide $Cr(OH)_3$ forms. Note that *no* metal(III) carbonate is formed.

The hydroxo complex, although a minor component of the mixture, absorbs light much more strongly than the hexaaqua ion does and so dominates the perceived colour. The colour of the unionized hexaaqua ion is seen in the solid state, where hydrated ammonium iron(III) sulphate (colloquially called 'ferric alum') is coloured pale purple. The aqueous iron(III) ions, $Fe^{3+}(aq)$, appear yellow-brown. A similar effect occurs for other ions such as chromium(III), for which the hexaaqua ion is violet but the hydroxo complex, formed when it is dissolved in water, is green.

The stability constant

A solution of an iron(II) salt such as iron(II) sulphate $FeSO_4$ contains the hexaaquairon(II) complex $[Fe(H_2O)_6]^{2+}(aq)$. Addition of cyanide ions displaces water ligands to produce the hexacyanoferrate(II) complex $[Fe(CN)_6]^{4-}(aq)$. Each step in the displacement of water by cyanide ion constitutes an equilibrium reaction and has an associated equilibrium constant.

$$[Fe(H_2O)_6]^{2+}(aq) + 6CN^-(aq) \rightleftharpoons [Fe(CN)_6]^{4-}(aq) + 6H_2O(l)$$

The equilibrium constant for the overall reaction is called the **stability constant** K_{stab} (note that the *outer* square brackets indicate the concentration, whereas the inner ones are part of the formula of the complex):

$$K_{stab} = \frac{[[Fe(CN)_6]^{4-}]}{[[Fe(H_2O)_6]^{2+}][CN^-]^6}$$

Stability constants for Cu^{2+} complexes

Ligand	$\log_{10} K_{stab}$
Cl^-	5.6
NH_3	13.3
EDTA	18.8
CN^-	27.3

EDTA will displace NH_3 and Cl^-.
CN^- will displace all other ligands.

Toxic cyanide

Cyanide ion has a high value of stability constant with respect to all transition metals. Many enzymes – particularly those responsible for respiration – contain a transition metal complexed with amino acids or other biological groups. Cyanide ion displaces these ligands and thus renders enzymes inoperative. Death rapidly follows as the energy-producing pathways in the organism cease functioning.

Changes of shape

Replacing small uncharged ligands with charged ligands may cause an octahedral complex to change its shape to tetrahedral. For example, adding concentrated hydrochloric acid to octahedral (6-coordinate) $[Cu(H_2O)_6]^{2+}(aq)$ forms tetrahedral (4-coordinate) $CuCl_4^{2-}(aq)$:

$$[Cu(H_2O)_6]^{2+} + 4Cl^- \rightleftharpoons CuCl_4^{2-} + 6H_2O$$

The colour change is from blue to green to yellow when the acid is in excess; see photo in previous spread.

The stability constant of a complex is a measure of its stability with respect to its constituent species. The larger the value, the more stable the complex. Stability constants have a wide range of values. They are usually expressed as their logarithm $\log_{10}K_{stab}$ in order to make the numbers more easily manageable.

- Comparing stability constants indicates the likelihood of one ligand species substituting for another.

For example, the K_{stab} values at 298 K for $[Cu(NH_3)_4(H_2O)_2]^{2+}$ and $[Cu(CN)_4]^{2-}$ are 2×10^{13} and 2×10^{27} respectively. Cyanide ligands CN^- will therefore displace ammonia ligands NH_3 from the complex $[Cu(NH_3)_4(H_2O)_2]^{2+}$. Ammonia ligands will not displace cyanide ligands from the complex $[Cu(CN)_4]^{2-}$ to any significant extent. (Note the common tetrahedral shape of a complex with four *anionic* ligands.)

Multidentate ligands

Ligands such as CN^- and NH_3 are referred to as unidentate ligands because they possess only one lone pair of electrons that may be used to form a ligand-metal bond. A ligand that donates two electron pairs is called a **bidentate ligand**. For example, the molecule 1,2-diaminoethane $NH_2CH_2CH_2NH_2$ (often known by its older name ethylenediamine and abbreviated to en) has two nitrogen atoms that can both donate their lone pair. Another example of a bidentate ligand is ethanedioate (oxalate) ion $C_2O_4^{2-}$, as in the complex $[Fe(C_2O_4)_3]^{3-}$. **Multidentate** (or **polydentate) ligands** donate more than two electron pairs. An example of a multidentate ligand is EDTA.

The porphyrin ligand in the haem group in haemoglobin is also multidentate.

The fact that multidentate ligands generally form very stable complexes is called the **chelate effect**: EDTA forms a much more stable complex than ammonia does, despite the fact that two of the atoms complexing in EDTA are also nitrogen atoms. The explanation for the chelate effect is that a single EDTA ligand liberates *six* water molecules, a process that is very favourable entropically (see chapter 14 'Spontaneous change...').

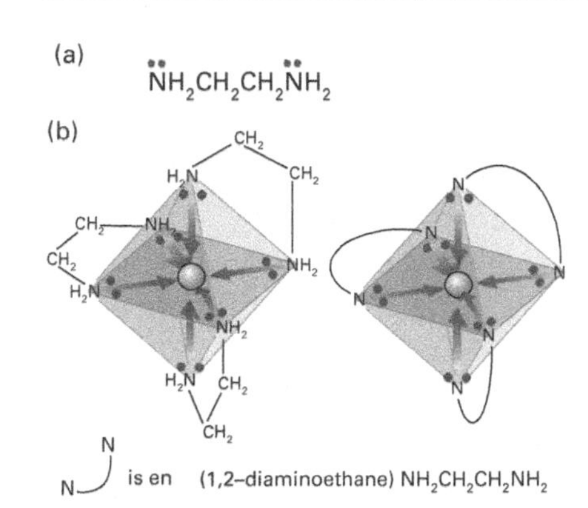

(a) The 1,2-diaminoethane ligand (en) with two available lone pairs. This bidentate ligand is neutral. (b) Structure of a metal-en complex.

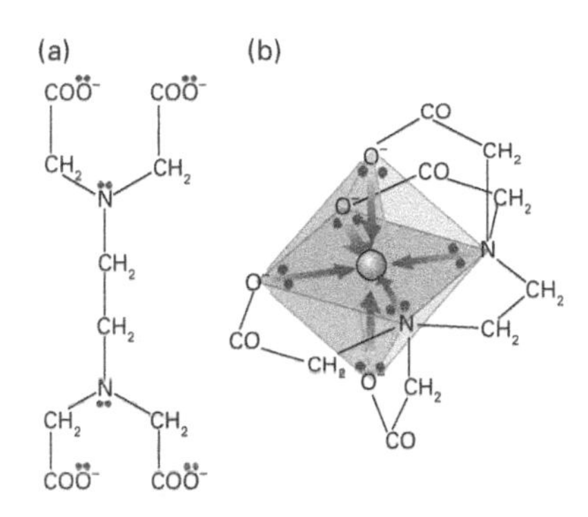

(a) Ethylenediaminetetraacetate EDTA: a hexadentate ligand. Each EDTA bears a total charge of 4⁻. (b) The structure of a metal-EDTA complex. The complex consists of a metal ion and a single EDTA and has an octahedral shape.

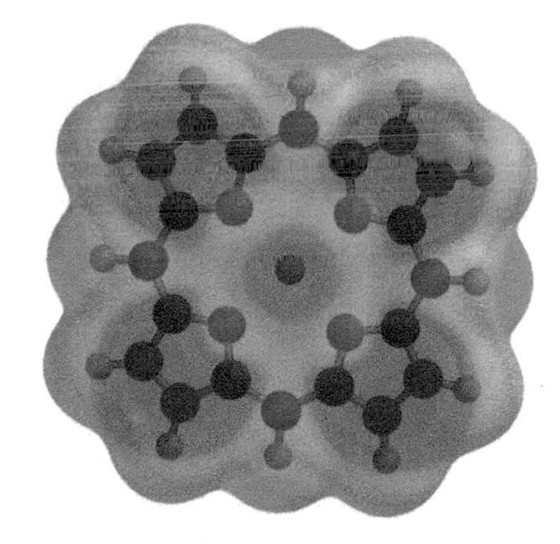

Haem, part of the haemoglobin molecule, has an Fe^{2+} complexed by a tetradentate porphyrin ligand. The electrostatic potential map shown here corresponds to the simplest porphyrin. Note that the porphyrin ring is delocalized; this cannot be shown accurately in the next but one spread where the actual haem group is drawn.

Applications

Unlike cyanide ion, EDTA is non-toxic. It has medical uses in extracting aqueous ions from the body fluids of patients suffering from poisoning by the aqueous ions of toxic metals such as lead and cadmium. EDTA acts as a *sequestering agent*, forming a complex with the toxic ion and preventing its physiological chemical effect. Other sequestering agents are present in soap powders and detergents and soften water by removing the $Ca^{2+}(aq)$ and $Mg^{2+}(aq)$ ions.

Standard solutions of EDTA may be used in complexometric titrations to estimate the concentrations of aqueous metal ions, especially magnesium and calcium. One mole of EDTA effectively removes one mole of aqueous metal ions, the end point of the reaction being indicated by a dyestuff that itself complexes with the metal ion. The dyestuff has one colour when complexed with the metal and another colour when displaced free into the solution by EDTA.

SUMMARY

- Complexes may undergo ligand substitution reactions in which incoming ligands displace existing ligands.
- Metal(III) aqua ions are acidic.
- Ligands of greater value of stability constant substitute for those of lower value.
- Bidentate ligands donate two electron pairs.
- Multidentate ligands donate more than two electron pairs.

PRACTICE

1 Draw the structure of the complex $[Cu(en)_2(H_2O)_2]^{2+}$. Suggest its geometry.

20.9

OBJECTIVES

- *Cis–trans* isomerism
- Optical isomerism

Square planar complexes

In general, if M is a metal ion and X and Y are unidentate ligands, square planar complexes with the formula MX_2Y_2 show *cis–trans* isomerism.

Cisplatin mechanism

Cisplatin (and related molecules such as carboplatin) is often highly effective at killing cancer cells. It acts by crosslinking DNA, most frequently by binding to two neighbouring guanine bases on a single strand of the double helix, spreads 30.8 and 30.9. This binding occurs by ligand substitution reactions, see previous spread, in which bonds are formed between the platinum and a nitrogen atom on each guanine. This crosslinking interferes with cell division. The damaged DNA activates DNA repair mechanisms. When repair turns out to be impossible, **apoptosis** (cell death) results.

Because any rapidly dividing cells are especially vulnerable, side-effects include hair loss and vomiting. Kidney damage and hearing loss are also concerns when cisplatin is administered.

Octahedral complexes

In general, if M is a metal ion and X and Y are unidentate ligands, octahedral complexes with the formula MX_2Y_4 show *cis–trans* isomerism.

ISOMERISM IN TRANSITION METAL COMPLEXES

Isomers are compounds that have the same molecular formula but different structures. Some transition metal complexes show isomerism. The form of isomerism most common in transition metal complexes is stereoisomerism, in which ligands are arranged differently in space around the central metal ion. Stereoisomerism can take two forms, *cis–trans* isomerism and optical isomerism. Each form has its own distinctive properties.

Cis–trans isomerism

***Cis–trans* isomers** of complexes have a different arrangement of ligands around the central ion. For example, there are two isomers of the square planar complex $[Pt(NH_3)_2Cl_2]$, which are called *cis-* and *trans-*diamminedichloroplatinum(II). The prefix *cis-* indicates that the two ammine ligands are next to each other in the complex ion structure. The prefix *trans-* indicates that the two ammine groups are on opposite sides of the structure. The isomer *cis*-$[Pt(NH_3)_2Cl_2]$ is called 'cisplatin' and is a highly effective anti-cancer agent. The *trans* isomer does not have anti-cancer properties. *Cis–trans* isomerism is sometimes known as **geometric isomerism.**

(a)

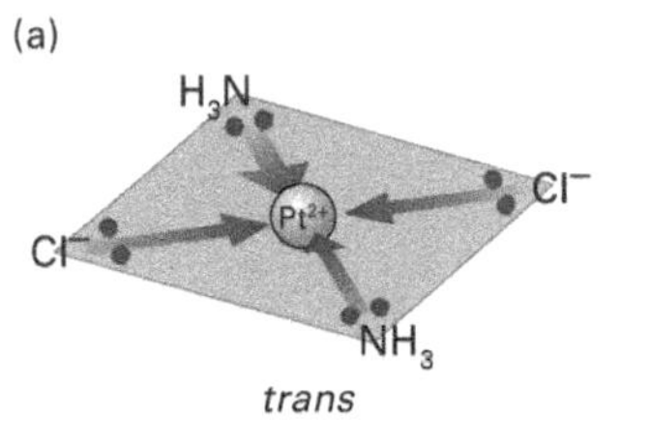

trans

(b)

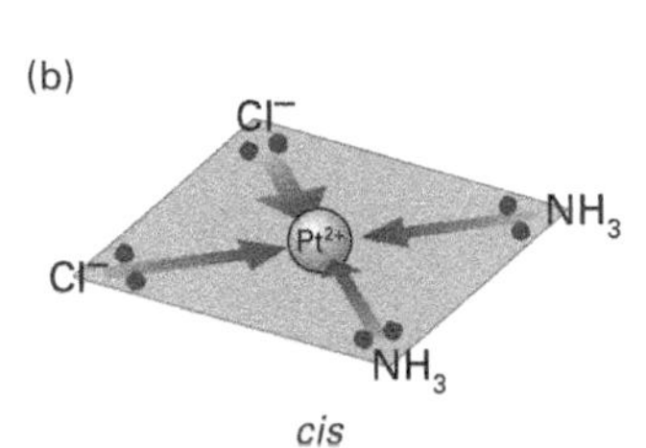

cis

(c)

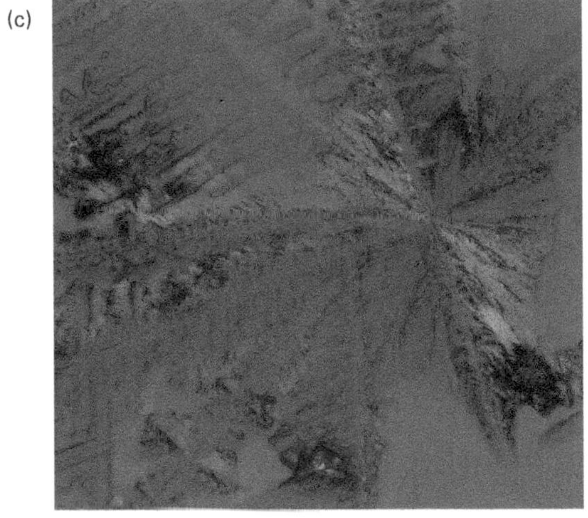

(a) The trans *isomer of $[Pt(NH_3)_2Cl_2]$. (b) The* cis *isomer of $[Pt(NH_3)_2Cl_2]$.*
(c) Crystalline cisplatin.

The octahedral tetraamminediaquacopper(II) complex ion $[Cu(NH_3)_4(H_2O)_2]^{2+}$ exists only as the *trans* isomer, in which the two water ligands are on opposite sides of the structure. The *cis* isomer does not exist, because of bonding and orbital energy constraints. However, there are *two* isomers of the tetraamminedichlorocobalt(III) ion $[Co(NH_3)_4Cl_2]^+$, as illustrated below.

(a)

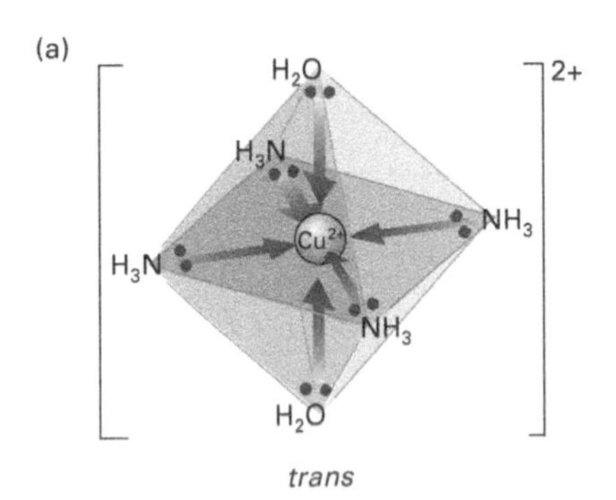

trans

(b)

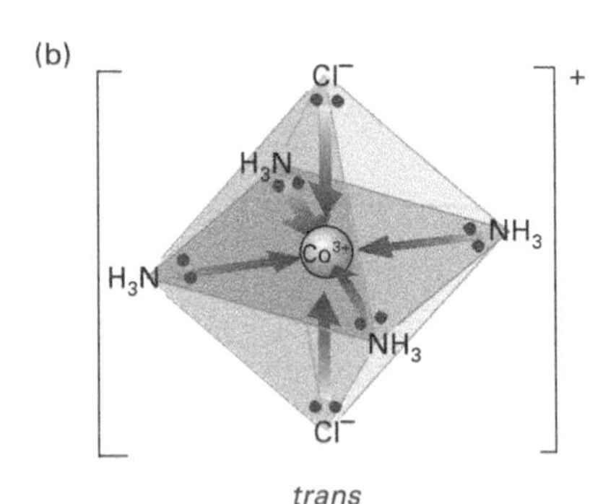

trans

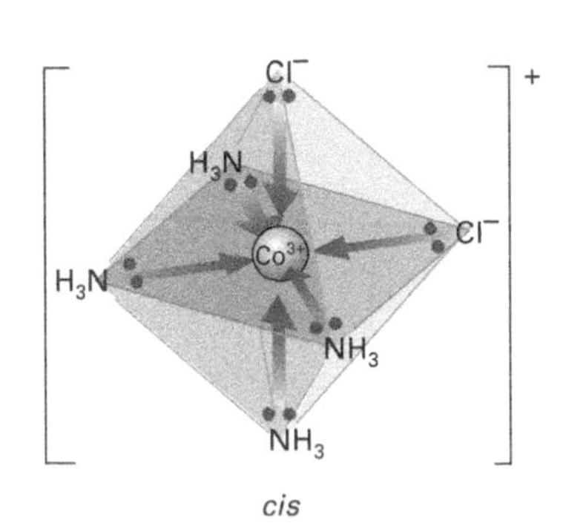

cis

(a) The single trans *isomer of $[Cu(NH_3)_4(H_2O)_2]^{2+}$. (b) The* trans *and* cis *isomers of $[Co(NH_3)_4Cl_2]^+$.*

Optical isomerism

Look again at the structure of the *cis*-tetraamminedichlorocobalt(III) ion $[Co(NH_3)_4Cl_2]^+$. Imagine a ligand substitution reaction in which two pairs of ammonia molecules are replaced by two molecules of the bidentate ligand 1,2-diaminoethane ('en'). There are two possible structures for the *cis* isomer of the bis(en)dichlorocobalt(III) ion, *cis*-$[Co(en)_2Cl_2]^+$. (Note that here we use the prefix 'bis'. This denotes 'two' and is used instead of 'di' with bidentate ligands.)

(a)

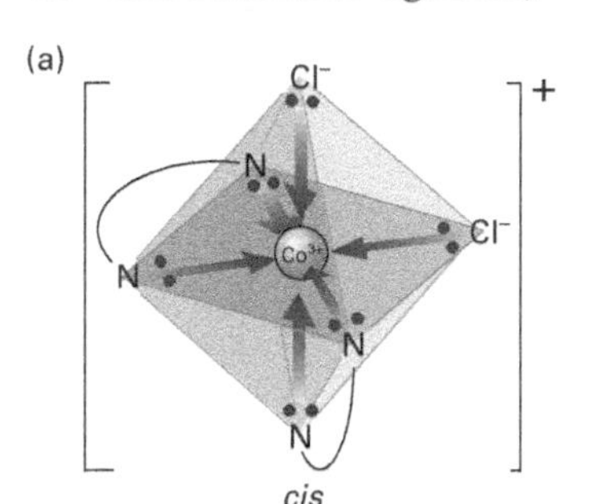

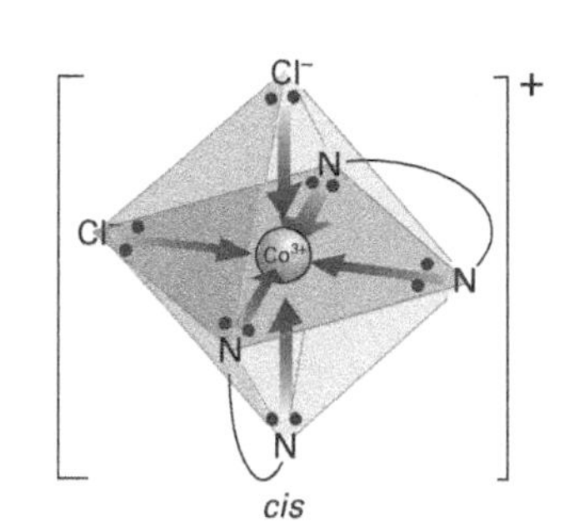

(b)

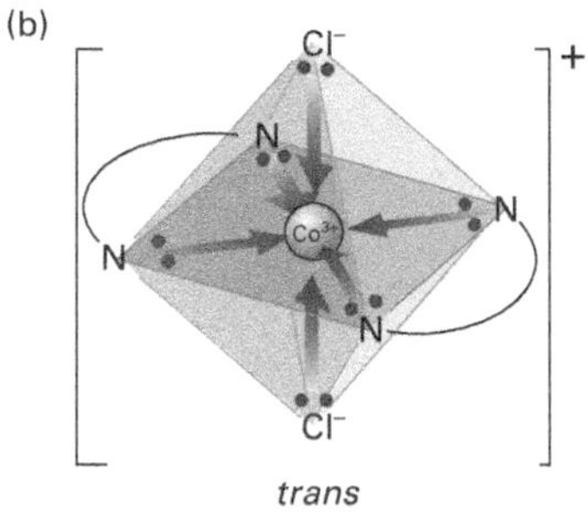

N
N is en
(1,2-diaminoethane)
$NH_2CH_2CH_2NH_2$

(a) The cis *isomer of $[Co(en)_2Cl_2]^+$ and its mirror image are not superimposable. They form a pair of optical isomers. (b) The* trans *isomer has a plane of symmetry. It is superimposable on its mirror image.*

The two *cis* isomers are called **chiral** because they exist as mirror images of, and are not superimposable on, each other, in the same way that your left hand is a mirror image of your right hand. These chiral isomers are also called **optical isomers** because their solutions have the ability to rotate the plane of plane-polarized light. Optical isomers are said to be optically active; one isomer will rotate the plane of plane-polarized light in one direction, the second isomer in the opposite direction. (Optical isomerism is described in more detail in chapter 28 'Amines and amino acids'.)

There is only one isomer of *trans*-bis(en)dichlorocobalt(III) ion. The *trans* isomer and its mirror image are superimposable because the *trans* isomer has a plane of symmetry.

Optical isomers also occur when there are three bidentate ligands. Examples include $[Fe(C_2O_4)_3]^{3-}$ and $[Ni(en)_3]^{2+}$.

Symmetry

The number 8 has two planes of symmetry (imagine slicing it in half with either a vertical cut, or a horizontal cut). The number 3 has one plane of symmetry (slice it horizontally), but the number 9 has no plane of symmetry.

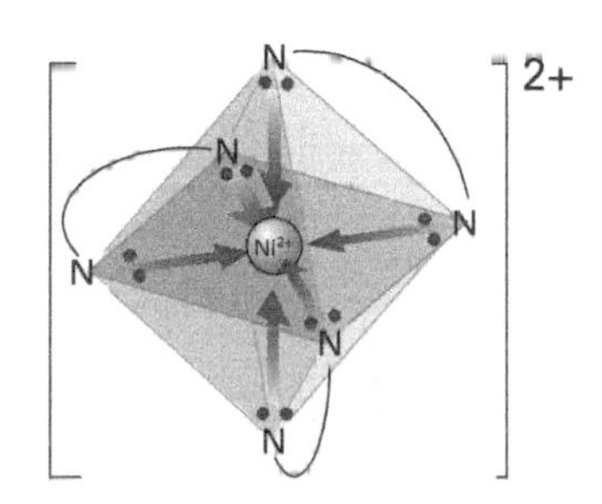

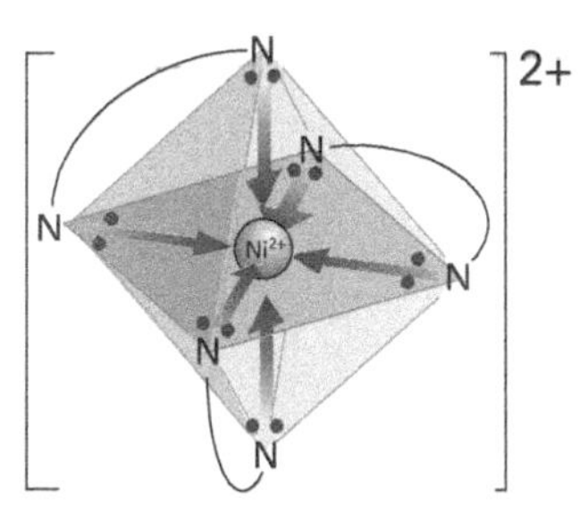

$[Ni(en)_3]^{2+}$ forms a pair of optical isomers. A solution containing $[Ni(en)_3]^{2+}$ is pink (see the photograph in the previous spread).

SUMMARY

- Stereoisomers have the same ligands, but they are arranged differently in space.
- There are two kinds of stereoisomerism – *cis–trans* isomerism and optical isomerism.
- Optical isomers are mirror images of each other.
- Chiral species are not superimposable on their mirror images.

PRACTICE

1 Explain why the tetraaquadichlorocobalt(III) complexes show *cis–trans* isomerism.

2 For the isomers of the complex $[Cr(en)_2Cl_2]^+$, sketch the pairs of structures that are:

a *cis–trans* isomers,

b optical isomers.

3 Explain why the *trans* isomer of $[Co(en)_2Cl_2]^+$ is not optically active.

4 Sketch the possible isomers of the complex $[Co(en)_3]^{3+}$. Select a pair that show optical isomerism.

20.10

OTHER REACTIONS OF COMPLEX IONS

OBJECTIVES

- Formation of an ammine complex of cobalt
- Redox reactions of transition metals in biochemistry

The presence of base stabilizes the higher oxidation states in transition metal complexes (see spread 13.9). A particularly famous application of this principle was the preparation by ligand substitution reaction of the ammine complexes of cobalt. Studying the chemistry of these substances helped Alfred Werner in 1893 to formulate his 'theory of coordination compounds', in which the concepts of oxidation state and coordination number were introduced for the first time. Ligand substitution reactions are also of great importance in the field of biochemistry. This spread looks at these topics and concludes by outlining how haemoglobin transports molecular oxygen in mammals.

Higher oxidation states

To prepare a metal complex that contains the metal in a high oxidation state, the general process is

1 add a base

2 add an oxidant.

In the example alongside, the base is ammonia, and the oxidant is air. An alternative oxidant is aqueous hydrogen peroxide. So for example addition of hydrogen peroxide to aqueous chromium(III) ions Cr^{3+} in sodium hydroxide results in yellow aqueous chromate(VI) ions CrO_4^{2-}.

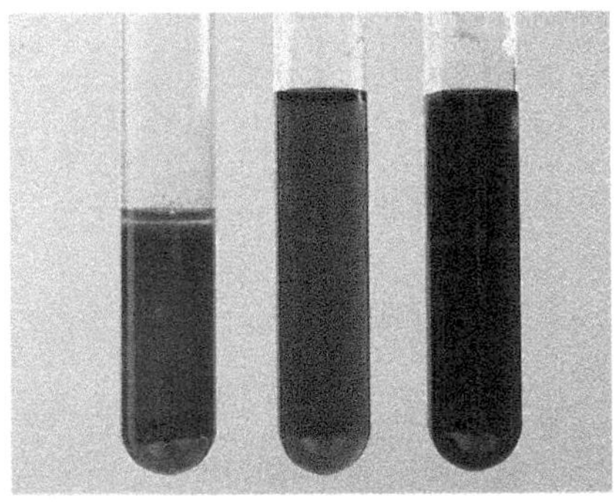

Stages in the preparation of a hexaamminecobalt(III) solution.

Preparing an ammine complex of cobalt

• The outline procedure

Werner found that the addition of aqueous ammonia to aqueous cobalt(II) chloride gave a blue-green precipitate of cobalt(II) hydroxide. Treating the precipitate with concentrated aqueous ammonia in the absence of air formed a pale brown (straw-coloured) solution of the hexaamminecobalt(II) complex. This solution rapidly oxidized in the presence of air to give a dark brown mixture containing the hexaamminecobalt(III) ion, $[Co(NH_3)_6]^{3+}$.

• The detailed steps

In acidic solution, the Co(II) oxidation state is more stable than the Co(III) state, as indicated by the standard electrode potential:

$$Co^{3+}(aq) + e^- \rightleftharpoons Co^{2+}(aq); \qquad E^\ominus = +1.82\,V$$

The value of this electrode potential becomes much less positive in basic solution (provided by excess aqueous ammonia), showing that the Co(III) oxidation state is now much *more* stable than it is under acidic conditions:

$$[Co(NH_3)_6]^{3+}(aq) + e^- \rightleftharpoons [Co(NH_3)_6]^{2+}(aq); \qquad E = +0.11\,V$$

The steps followed in Werner's preparation involve the following changes:

1 Dissolving cobalt(II) chloride in water gives pink aqueous $Co^{2+}(aq)$ ions present as the hexaaqua complex $[Co(H_2O)_6]^{2+}$.

$$CoCl_2(s) \rightarrow Co^{2+}(aq) + 2Cl^-(aq)$$

2 Adding concentrated aqueous ammonia forms a blue-green precipitate of cobalt(II) hydroxide $Co(OH)_2(s)$.

Aqueous ammonia forms the following equilibrium in water:

$$NH_3(aq) + H_2O(l) \rightleftharpoons NH_4^+(aq) + OH^-(aq)$$

The aqueous hydroxide ion causes the precipitation of cobalt(II) hydroxide (see spread 11.10):

$$Co^{2+}(aq) + 2OH^-(aq) \rightarrow Co(OH)_2(s)$$

3 Adding excess concentrated aqueous ammonia forms a straw-coloured solution containing the hexaamminecobalt(II) ion, $[Co(NH_3)_6]^{2+}(aq)$.

$$Co(OH)_2(s) + 6NH_3(aq) \rightarrow [Co(NH_3)_6]^{2+}(aq) + 2OH^-(aq)$$

4 Warming the solution and bubbling air through it oxidizes Co(II) to Co(III). A dark brown solution containing the hexaamminecobalt(III) complex $[Co(NH_3)_6]^{3+}(aq)$ together with others such as the pentaamminechlorocobalt(III) ion $[Co(NH_3)_5Cl]^{2+}(aq)$ results.

A biological substitution reaction

An earlier spread in this chapter discussed the role of transition metal ions in homogeneous catalysis. Simple *aqueous* species catalyse reactions by alternating between two oxidation states. The ion in an upper oxidation state accepts an electron from a reductant and is reduced to a lower state. Passing the electron to the oxidant, the catalyst then reverts to the upper oxidation state. In this manner, transition metal ions mediate in redox reactions by passing electrons from reductant to oxidant.

Many biological systems use *complexed* transition metal ions to mediate in biochemical reactions. For example, energy is generated in higher animals by the enzyme-mediated oxidation of sugars, particularly glucose. The oxidant is oxygen from the air. In the lungs, oxygen from the air is taken up into haemoglobin in the blood. Mammalian haemoglobin is an iron-containing protein complex with a relative formula mass of 64 450. It is haemoglobin that is the red pigment in red blood cells. The iron is present as the Fe^{2+} ion complexed by a tetradentate haem ligand; see spread 20.8.

The square planar shape is particularly prone to further coordination. A fifth position is coordinated by the amino acid histidine, which itself is part of the surrounding protein chain. The sixth position can be coordinated by the π-electron density in multiply bonded substances such as O_2, CO, and CN^-. Haemoglobin therefore takes up molecular oxygen in the lungs, forming oxyhaemoglobin. This is transported around the body in the bloodstream, and the oxygen is released in regions of low oxygen concentration. Note that the oxidation number of the complexed iron is +2 in *both* haemoglobin and oxyhaemoglobin.

Much of the detailed mechanism of the action of haemoglobin is now understood, through the pioneering work of Max Perutz. Deoxyhaemoglobin, lacking oxygen, contains a *'high-spin'* iron(II) ion which is converted into a *'low-spin'* iron(II) ion in the oxygenated state. The terms 'high-spin' and 'low-spin' depend on recognizing that the energy levels of the d orbitals are split, which will be explained in the next spread. The low-spin ion is smaller and as a result the iron ion can now slip into the plane of the haem ring.

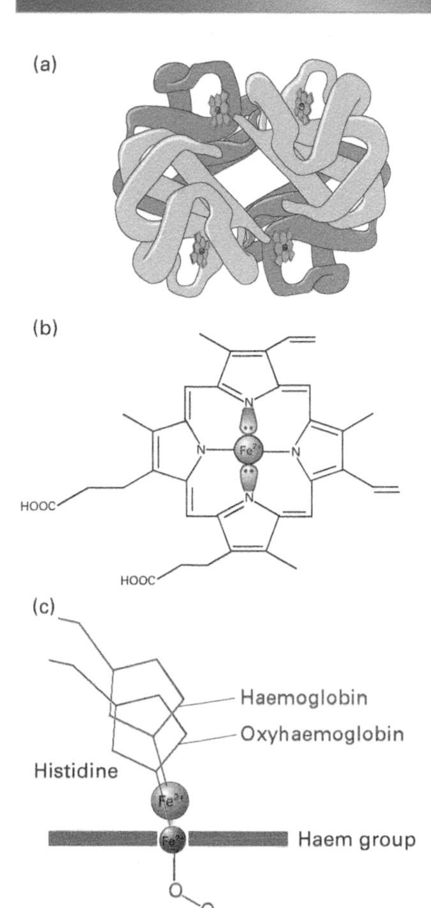

(a) Haemoglobin: the Fe^{2+} ion is located within a protein structure consisting of folded chains of amino acids. (b) The Fe^{2+} ion forms bonds with each of four nitrogen atoms on the haem ligand. Histidine, present in another section of the protein, forms a fifth coordinate bond to the Fe^{2+} ion. (c) Oxygen binds reversibly to the Fe^{2+} ion in haemoglobin to form oxyhaemoglobin. Note that CO and CN^- bind more strongly than O_2 and are strong poisons.

SUMMARY

- Higher oxidation states are stabilized by the addition of base.
- The presence of the base aqueous ammonia allows a cobalt(II) complex to be oxidized to cobalt(III) by the action of atmospheric oxygen alone.
- Transition metal complexes in living systems mediate in biochemical reactions. The metal ion is usually surrounded by and complexed by a protein of very high relative formula mass.

PRACTICE

1 Explain the action of cyanide ions as a poison.

2 Use the equilibrium expression

$$Hb + O_2 \rightleftharpoons HbO_2$$

where Hb refers to the haemoglobin molecule, to explain why haemoglobin complexes with oxygen in the lungs and releases oxygen elsewhere in the body tissues.

3 Lack of iron in the diet or poor uptake of iron from the diet can lead to a complaint called *anaemia*, in which the haemoglobin content of the blood is lowered. Suggest the major symptoms experienced by a person suffering from anaemia. Explain their origins.

4 There are three hydrates of the salt chromium(III) chloride $CrCl_3{\cdot}6H_2O(s)$: they are violet $[Cr(H_2O)_6]Cl_3$, green $[Cr(H_2O)_5Cl]Cl_2{\cdot}H_2O$, and green $[Cr(H_2O)_4Cl_2]Cl{\cdot}2H_2O$. Explain what you would expect to see when solutions of equal concentration of these compounds are each treated with an equal excess of aqueous silver nitrate.

20.11

Coloured ions

OBJECTIVES

- Complementary colours
- d orbitals are split by a ligand field
- d-to-d transitions
- Charge transfer transitions
- Qualitative analysis

(a)

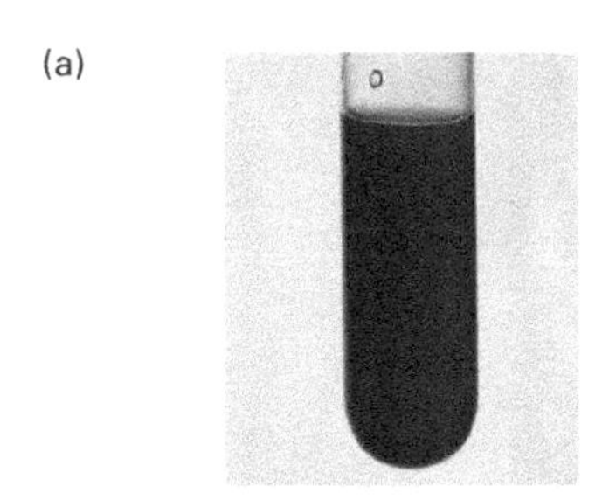

(b)

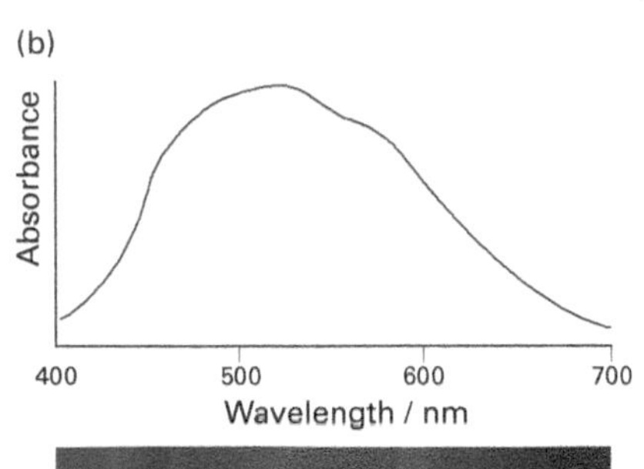

(a) A solution containing the complex ion $[Ti(H_2O)_6]^{3+}(aq)$. (b) The absorption spectrum of $[Ti(H_2O)_6]^{3+}(aq)$.

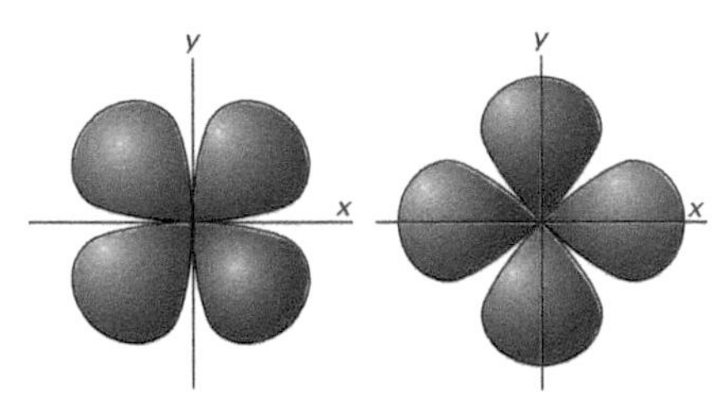

Three of the five d orbitals point between the axes; two point along the axes.

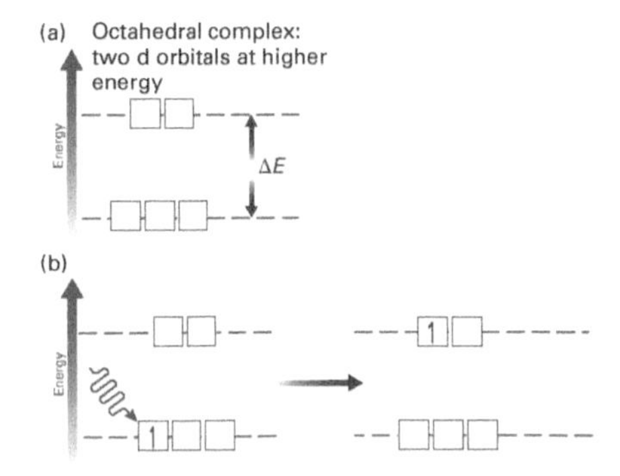

(a) The splitting of d orbitals by an octahedral ligand field. (b) An electron absorbs radiation (of frequency f such that $\Delta E = hf$) and is promoted to a higher-energy orbital. Note that the electron usually returns to its original energy level by an alternative route that does not emit visible radiation.

Look at aqueous $Ti^{3+}(aq)$ ions, shown on the left, and you see a purple colour. The solution absorbs green light from the continuous spectrum of white light. When white light passes through the solution, the colour you perceive is white minus green, which leaves purple. The absorption of light takes place by electronic transitions between different energy levels (see chapter 4 'Electrons in atoms'). In many transition metal ions, the differences between the energy levels correspond to energies within the visible spectrum. There are two ways in which this can come about.

d-to-d transitions

In an isolated atom or ion, the five d orbitals have the same energy. When ligands are present, some d orbitals are closer than others to the ligands, because of their distribution in space. The ligands may be roughly regarded as clouds of negative charge, which push the orbitals closest to them to slightly higher energy levels. In an octahedral complex, two of the d orbitals point along the axes *directly at* the ligands; these orbitals are at a higher energy level than the other three d orbitals. This is called **ligand field splitting**. The energy difference between the split d orbitals corresponds to frequencies within the visible spectrum.

The opportunity now exists for an electron to be excited from a lower-energy to a higher-energy d orbital, *provided* that there is at least one electron present to be excited *and* that there is space for it in one of the orbitals of higher energy. The condition for this **d-to-d transition** to occur and hence for an ion to be coloured is:

- the ion must have a partially filled d subshell.

(This confirms why neither $Zn^{2+}(aq)$ nor $Sc^{3+}(aq)$ are coloured.)

The frequency of the light absorbed, and so the colour seen, depends on the energy difference between the two levels. The energy difference between the split d orbitals (and hence the colour) depends on:

- the nature and oxidation state of the metal ion;
- the nature and number of the ligands.

Changing the ligands around a given metal ion changes the colour perceived. For example, the hexaaquacopper(II) ion $[Cu(H_2O)_6]^{2+}$ is pale blue; the tetrachlorocuprate(II) complex $CuCl_4^{2-}(aq)$ is yellow–green (spread 20.8); the complex $[Cu(NH_3)_4(H_2O)_2]^{2+}$ is deep blue. All three complexes are coloured as a result of d-to-d transitions.

A further consequence of ligand field splitting is that ions with between four and seven electrons can have them arranged in two ways. For example Fe^{2+} has six electrons. One possible electronic structure has all six in the lower energy level, in which case their spins must be paired and the ion is described as **low-spin**. Alternatively, four can be in the lower level and two in the higher; following Hund's rule, there are four unpaired electrons and the ion is described as **high-spin** (see previous spread).

Charge transfer transitions

An electron may be transferred from the *ligand* to the central metal ion in a **charge transfer transition**. When such a transfer occurs, it causes a much more intense absorption, as in the deep purple of manganate(VII) ions or the bright orange of dichromate(VI) ions.

The ease with which an electron transfers depends on the oxidizing power of the central ion. As the oxidizing power increases, the ion accepts the electron more readily and so the transition energy falls. For the vanadate(V) ion VO_4^{3-} the energy is high and the absorption lies in the ultraviolet; for the strongly oxidizing chromate(VI) ion CrO_4^{2-} the absorption extends into the blue end of the visible and the solution appears yellow. The very strongly oxidizing manganate(VII) ion MnO_4^- absorbs in the green and thus appears purple.

Colour and qualitative analysis

Addition of limited quantities of aqueous hydroxide ions to an aqua complex causes the metal hydroxide to precipitate. The hydroxides of transition metals are characteristically coloured, and this may be used to identify the metal ion. Those hydroxides that are amphoteric (e.g. chromium(III) hydroxide) 'dissolve' in excess aqueous sodium hydroxide. The addition of aqueous ammonia may have the same effect as adding aqueous sodium hydroxide because it also is a base. However, adding an excess of aqueous ammonia can dissolve the hydroxide precipitate by forming an *ammine* complex. These reactions are summarized in the following chart:

Tests for identifying some transition metal ions. The nature of the precipitates in oxidation state +3 are complex and they have been approximated as $M(OH)_3$.

Ion	Effect of adding dilute NaOH(aq)	Effect of adding dilute NH_3(aq)
Cr^{3+}(aq)	Green precipitate of $Cr(OH)_3$ 'redisssolves' in excess as it is amphoteric, forming $[Cr(OH)_6]^{3-}$	Green precipitate of $Cr(OH)_3$ 'redissolves' in excess, forming $[Cr(NH_3)_6]^{3+}$
Mn^{2+}(aq)	White precipitate of $Mn(OH)_2$	As for NaOH(aq)
Fe^{3+}(aq)	Red-brown precipitate of $Fe(OH)_3$	As for NaOH(aq)
Fe^{2+}(aq)	Green precipitate of $Fe(OH)_2$ turning brown on standing because of oxidation by air	As for NaOH(aq)
Co^{2+}(aq)	Blue-green precipitate of $Co(OH)_2$	Blue-green precipitate of $Co(OH)_2$; in excess gives a dark brown solution, as cobalt(III) ammine complex forms (spread 20.10)
Ni^{2+}(aq)	Green precipitate of $Ni(OH)_2$	Green precipitate of $Ni(OH)_2$; in excess gives a violet-purple solution as nickel(II) ammine complex forms
Cu^{2+}(aq)	Blue precipitate of $Cu(OH)_2$	Blue precipitate of $Cu(OH)_2$; in excess gives a dark blue solution as copper(II) ammine complex forms

Testing for Fe^{2+}(aq) and Fe^{3+}(aq)

Adding aqueous thiocyanate ions SCN^-(aq) to iron(III) ions Fe^{3+}(aq) forms the deep blood-red complex ion $[Fe(H_2O)_5SCN]^{2+}$. Iron(II) ions Fe^{2+}(aq) do not form this complex. But the colour formed with iron(III) is so intense that care is needed with this test. If iron(II) has been partially oxidized by air and there is even a small amount of iron(III) present in it, the red colour will still appear in this test. The result needs to be compared with a similar test on a known solution containing iron(III), and the intensity of the colours compared to be sure of a positive result.

A second test for Fe^{2+}(aq) and Fe^{3+}(aq) relies on adding hexacyanoferrate(II) ions $[Fe(CN)_6]^{4-}$ to iron(III) ions or hexacyanoferrate(III) ions $[Fe(CN)_6]^{3-}$ to iron(II) ions. The result is an extremely deep blue colour known as **Prussian blue** (and used traditionally for architects' 'blueprints'). This insoluble compound has the formula $Fe(III)_4[Fe(II)(CN)_6]_3$. Notice that *both* oxidation states have to be present to give the deep colour, which is due to a charge transfer transition.

SUMMARY

- Some transition metal compounds are coloured because they absorb certain colours from incident white light.
- An octahedral ligand field causes the d orbitals to split into two separate groups of differing energy.
- Coloured ions have fewer than 10 electrons and at least one partially filled d orbital.
- Transition metal ions may be identified by the formation of specific coloured species.

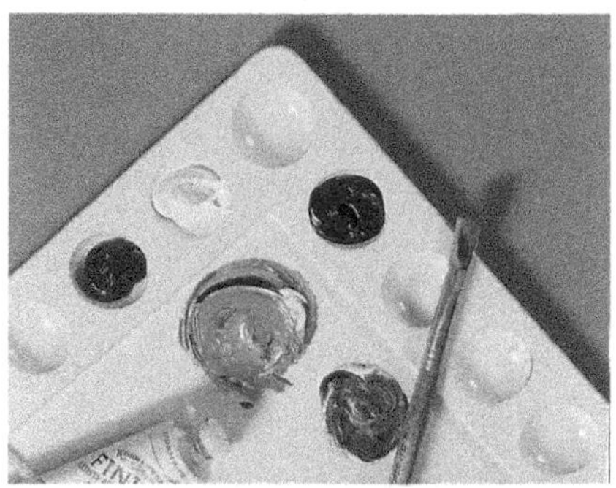

Many transition metal compounds are used as pigments, including those of titanium, iron, and copper: Monastral blue is a porphyrin complex of copper. Note that titanium(IV) oxide TiO_2, which is much used in paints, is white because Ti^{4+} contains no d electrons.

Zinc

Zinc forms a *white* precipitate of $Zn(OH)_2$ on addition of small quantities of either dilute aqueous sodium hydroxide or dilute aqueous ammonia. The zinc ion Zn^{2+} has a full d subshell, so no d-to-d transition can occur. The precipitate 'redissolves' in excess of either reagent, forming $[Zn(OH)_4]^{2-}$ or $[Zn(NH_3)_4]^{2+}$ respectively.

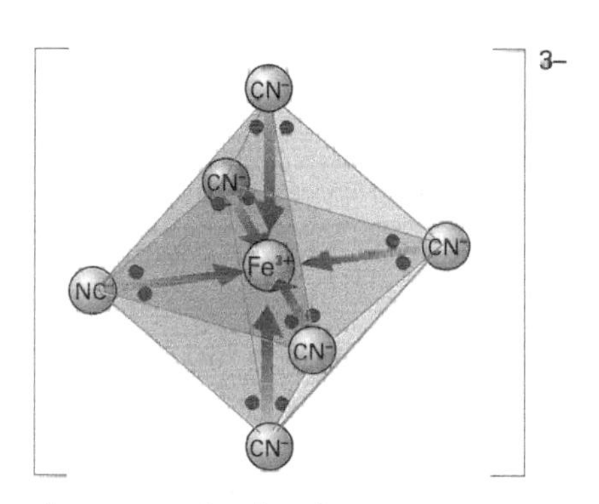

Aqueous potassium hexacyanoferrate(III) $K_3[Fe(CN)_6]$ contains the complex ion $[Fe(CN)_6]^{3-}$ (aq).

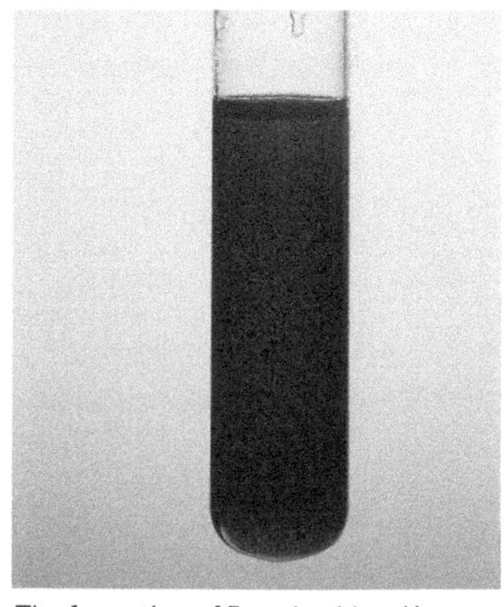

The formation of Prussian blue. Here, aqueous hexacyanoferrate(III) ions $[Fe(CN)_6]^{3-}$ are being added to aqueous iron(II) sulphate.

PRACTICE EXAM QUESTIONS

1 a Iron can be obtained from its oxide, Fe_2O_3, by reaction with carbon, aluminium, or hydrogen. Write an equation for the reaction in each case. [3]

b State, with an explanation, which of the reducing agents used above is likely to lead to:

i the cheapest iron;

ii the purest iron. [4]

c State two compounds formed during the extraction of iron in the blast furnace which lead to environmental pollution. In each case identify the type of pollution caused. [4]

d Outline the essential chemistry of the process for converting crude iron into a carbon steel. [2]

2 a Give two physical properties of iron metal which show it to be a typical d-block element. [2]

b i Complete the following ground-state electronic configurations, where [Ar] represents the ground-state configuration of the argon atom:
Fe [Ar] Fe^{2+} [Ar] [2]

ii Explain why the Fe^{2+} ion might be expected to be less stable than the Fe^{3+} ion. [2]

c i State the colour of aqueous solutions of iron(II) salts. Give the formula and shape of the complex ion responsible for the colour. [3]

ii What colour change slowly occurs if a solution containing this ion is allowed to stand? Explain your answer. [3]

d i What would be observed if aqueous potassium iodide were added to an aqueous solution containing Fe^{3+} ions? Explain any observations in terms of the reaction taking place. [3]

ii Write an ionic equation for the reaction. [1]

e Explain why compounds of d-block elements are often coloured. [3]

3 a i Explain the meaning of the term ligand.

ii Explain the meaning of the term bidentate as applied to a ligand. [2]

b i Anhydrous cobalt(II) chloride is a Lewis acid. Explain the meaning of the term Lewis acid.

ii Give the formula and shape of each of the complex ions formed when anhydrous cobalt(II) chloride is added separately to water and to concentrated hydrochloric acid.

iii Give a reason for the difference in shape of the complex ions formed in part **b ii** above.

iv Give a reason for the difference in colour of the complex ions formed in part **b ii** above. [7]

c 1,2–Diaminoethane, $NH_2CH_2CH_2NH_2$, acts as a bidentate ligand. Deduce the formula of the complex ion formed when cobalt(II) chloride is treated with an excess of 1,2–diaminoethane in the absence of air. (You may write 'en' as the formula of the ligand.) [2]

4 a Copy the boxes below and give the electronic structure of the vanadium atom and the V^{2+} ion:

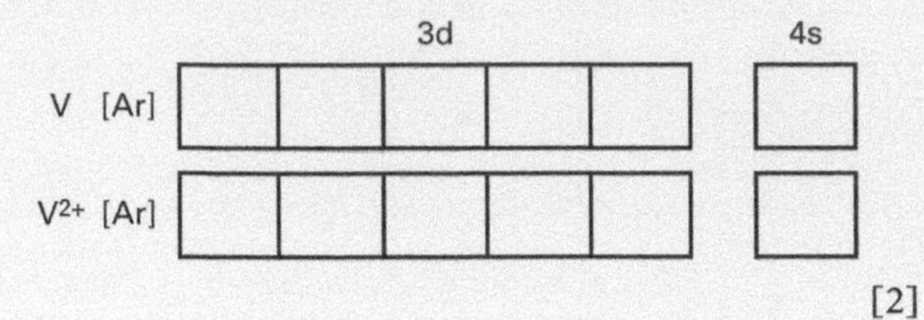

[2]

b i Suggest why the hydrated ion $[V(H_2O)_6]^{2+}$ is coloured.

ii Name the types of bonding within ions of this type. [3]

c Ammonium vanadate, NH_4VO_3, dissolves in aqueous sodium hydroxide with the evolution of a colourless gas. The solution becomes yellow after acidification. The gas has a pungent odour and produces a pale blue precipitate with copper(II) sulphate solution. The precipitate dissolves as more gas is passed in, to give a deep blue solution.

i Write an ionic equation for the reaction of the cation in NH_4VO_3 with alkali.

ii Name the pale blue precipitate.

iii Give the formula of the ion responsible for the colour of the deep blue solution.

iv Ammonium vanadate, on treatment with sulphuric acid, gives a yellow colour due to the $[VO_2]^+$ ion. Addition of zinc to the solution causes the solution colour to change to blue, then green, then violet. Give the oxidation number of vanadium in the vanadium-containing ions in each coloured solution. [5]

d The industrial production of sulphur trioxide from sulphur dioxide and oxygen is catalysed by vanadium(V) oxide. It has been proposed that the first stage of the reaction is

$$SO_2 + V_2O_5 \rightarrow SO_3 + 2VO_2$$

Write an equation for the second stage, thus showing the behaviour of vanadium(V) oxide as a catalyst. [1]

e Give the systematic name for each of these ions:

i $[VO_2]^+$

ii $[Cr(NH_3)_4Cl_2]^+$. [2]

f Draw and describe the shape of the ion in **e ii**. [2]

5 a Transition metals and their compounds can act as heterogeneous catalysts. Explain what is meant by both *heterogeneous* and *catalyst*. [2]

b State one feature of transition metals which makes them able to act as catalysts. [1]

c Write an equation for a reaction which is heterogeneously catalysed by a transition metal or one of its compounds. State the catalyst used. [2]

d When an acidified solution of ethanedioate ions is titrated with a solution of potassium manganate(VII), the colour of the manganate(VII) ions disappears slowly at first, but then more rapidly as the end point is reached. Explain this observation. [3]

6 a Give **three** changes which can result in a change of colour during the reaction of a transition metal ion. [3]

b For **each** of the observations described below, state which change or changes you have given in part **a** is responsible for the change in colour.

i A pale blue aqueous solution of copper(II) sulphate turns deep blue when added to an excess of aqueous ammonia.

ii A pink aqueous solution of cobalt(II) chloride turns blue when added to an excess of concentrated hydrochloric acid. [3]

c i Explain what is meant by the term *homogeneous catalyst*.

ii Which property of the transition metals makes their compounds particularly useful as catalysts?

iii The conversion of sulphur dioxide into sulphur trioxide by reaction with oxygen is catalysed industrially by vanadium(V) oxide. Write equations to show the catalytic role of vanadium(V) oxide in this reaction. [5]

7 a List **three chemical** characteristics of the transition elements. [3]

b Give **one** example, of your own choice, of the ***chemical*** use of a named transition metal of the 3d series or one of its compounds in a major industrial process. State the chemical property on which the use depends. [1]

c Copy and complete the boxes below by inserting arrows to show the ground state electronic configuration of:

i a chromium atom;

Argon core	3d					4s

ii a Cr^{2+} ion.

Argon core	3d					4s

[2]

d In acidic aqueous solution the dichromate(VI) ion, $Cr_2O_7^{2-}$, is a powerful oxidizing agent. The oxidation of iron(II) ions by dichromate(VI) ions may be represented by

$$Cr_2O_7^{2-}(aq) + 14H^+(aq) + 6Fe^{2+}(aq) \rightarrow 2Cr^{3+}(aq) + 6Fe^{3+}(aq) + 7H_2O(l)$$

i Deduce the change in the oxidation state of chromium in this reaction. [1]

ii Calculate the number of moles of $Fe^{2+}(aq)$ in 25.00 cm^3 of acidic aqueous iron(II) sulphate containing 12.15 g dm^{-3} of iron(II) sulphate, $FeSO_4$, (M_r =151.91). [2]

iii Calculate the volume of aqueous potassium dichromate(VI) of concentration 0.0200 mol dm^{-3} that will completely oxidize the number of moles of $Fe^{2+}(aq)$ in **d ii**. [2]

e State what you would expect to see if aqueous sodium hydroxide was added slowly to a solution of aqueous chromium(III) sulphate until the alkali was in excess. [2]

8 a When concentrated hydrochloric acid is added to aqueous copper(II) sulphate until in excess, a yellow solution containing the complex ion **P** is formed.

Give the formula of the complex ion **P**. [1]

b The addition of a slight excess of iron filings to aqueous copper(II) sulphate produces a solution of an ion **Q** and a solid which is then removed by filtration.

i Write the ionic equation for this reaction. [1]

ii Give two observations that could be made. [1]

iii In aqueous solution, **Q** exists as a complex ion. Give the formula of this complex ion. [1]

c Describe the charge and shape of each of the ions **P** and **Q**, using the appropriate words from the following list: anionic, cationic, neutral, octahedral, planar, and tetrahedral. [2]

d The addition of aqueous sodium hydroxide to the solution containing **Q** produces a precipitate of iron(II) hydroxide which, when filtered, slowly turns into a brown solid, **R**, on the filter paper.

i State the colour of the iron(II) hydroxide. [1]

ii Give the formula of **R**. [1]

iii With what does the iron(II) hydroxide react to produce **R**? [1]

iv What type of reaction occurs in the formation of both **Q** and **R**? [1]

Organic CHEMISTRY

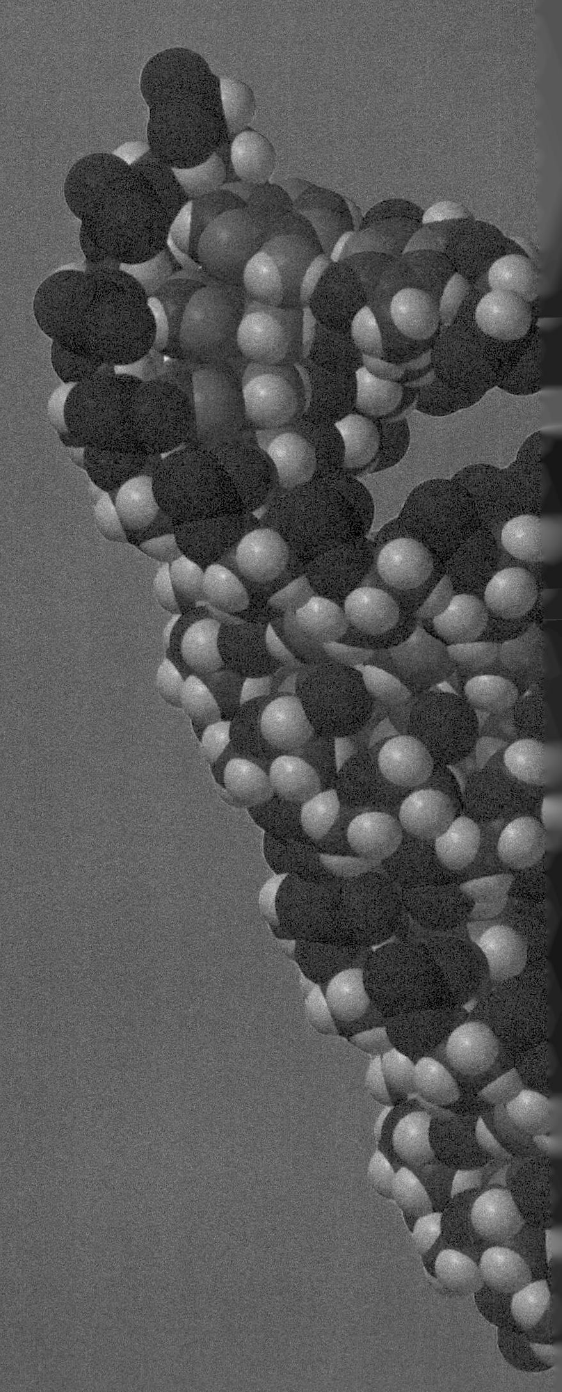

Organic chemistry involves the study of the compounds formed by carbon. There are over five million known, so their investigation is greatly simplified by considering the behaviour of a few significant classes of molecules (chapter 21). The most important classes are covered in turn, starting with the simplest compounds: those that contain only carbon and hydrogen atoms (chapters 22 and 23).

The next pair of chapters introduce atoms that create significant polarity in the bonding within the organic compounds, which has very significant effects on the types of reaction the molecules undergo. Halogen atoms are present in the halogenoalkanes (chapter 24); oxygen atoms forming single bonds are present in the alcohols (chapter 25).

Oxygen atoms are also capable of forming double bonds, and so the consequences of such a bond are discussed in the next two chapters, on aldehydes and ketones (chapter 26) and carboxylic acids (chapter 27). The final important element in organic chemistry is nitrogen, and chapter 28 surveys some of the molecules that contain nitrogen atoms. Chapter 29 seeks to draw together some of the ideas presented earlier and includes a discussion of the efforts organic chemists make to synthesize chemicals of significant value, especially to medicine, such as ibuprofen.

Chapter 30 tries to provide an insight into the exceptionally important area of biochemistry, arguably the most significant application of chemistry. In particular, the illustrations provided should prove helpful.

The final two chapters in the book, while not falling specifically under the heading 'organic chemistry', revolve around the separation and identification of chemicals. Powerful and widely used techniques such as chromatography (chapter 31) and spectroscopy (chapter 32) were introduced in the 20th century and transformed our ability to identify the structure of molecules.

Since 1950, advances have come most quickly in the area of biochemistry. In the second half of the 20th century, the detailed mechanism by which genetic information is passed from one generation to another via the replication, transcription, and translation of DNA was revealed. With little doubt, this is the most profound idea science has discovered during that period of time. At the heart of molecular biology is the double helix structure of the DNA molecule, shown here. You can read an account of this fascinating discovery in spreads 30.8–10.

There are still big questions to be answered. For example, it is unclear how (in full molecular detail) a single cell in the body differentiates to produce the appropriate limbs and organs. Maybe you will be the person who provides one of the vital links in this continuing quest.

Two computer-generated images of a space-filling model of DNA. The beauty of this molecule is even more apparent on the computer display, since the image can be rotated to show the three-dimensional structure to better effect. Here we have selected two views: one showing the grooves in the molecule and one looking down the inside of the double helix.

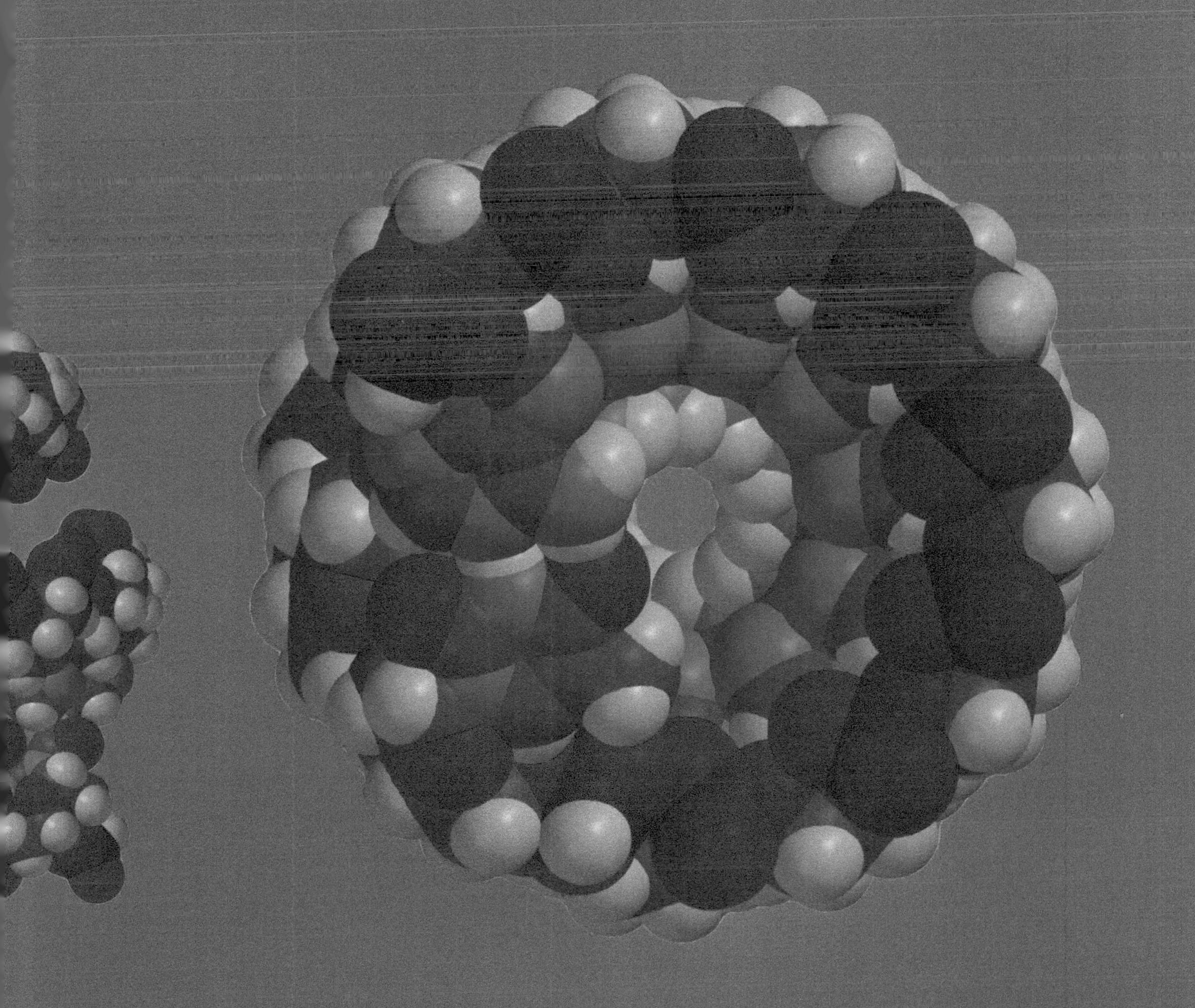

21 Introduction to organic chemistry

Organic chemistry is the study of the compounds of carbon. Carbon compounds are far more numerous than those of other elements because carbon atoms are able to bond together to form a wide range of chains and rings. The subject is called *organic chemistry* because living organisms are composed mainly of carbon compounds. Organic chemistry is the chemistry of life. It also includes the chemistry of an enormous range of substances, including food, fuels, textiles, plastics, drugs, dyes, explosives, pesticides, and paint. Most of the world's energy comes from burning carbon-based fuels, and the organic chemical industry is essential to most national economies. Major international crises have occurred over the supply and trade of carbon-containing substances such as crude oil, natural gas, coal, and their products. This chapter explores the basic principles of organic chemistry while giving you some feel for its scope. With *over five million* separate compounds, it would seem to be a vast subject. However, you will learn to group this huge array of organic compounds into separate families (called homologous series), each with its own distinct set of properties. This grouping together of compounds that react in a similar way makes studying them much simpler.

Organic chemistry and the carbon atom

21.1

OBJECTIVES

- Fuels and materials
- Basis of life
- Chains and rings of carbon atoms

Catenation

Elements other than carbon, especially other members of Group IV, are able to catenate. Silicon atoms bond to each other, but molecules containing more than eight silicon atoms are unstable. Sulphur atoms can form *very* long chains, especially in the liquid at a temperature of 170 °C, but sulphur cannot form complex molecules because each atom can only form two bonds, so chains cannot be branched. Carbon is unique in its ability to form molecules of almost any size and shape.

The basic structure of any organic compound – be it plastic, protein, medicine, fuel, or fibre – consists of a skeleton of carbon atoms joined together in chains and rings. The ability of an element to form chains of atoms bonded together is called **catenation**. This spread gives examples of the types of carbon skeleton possible. Later spreads will discuss the actual shapes of organic molecules and the conventions used to name them.

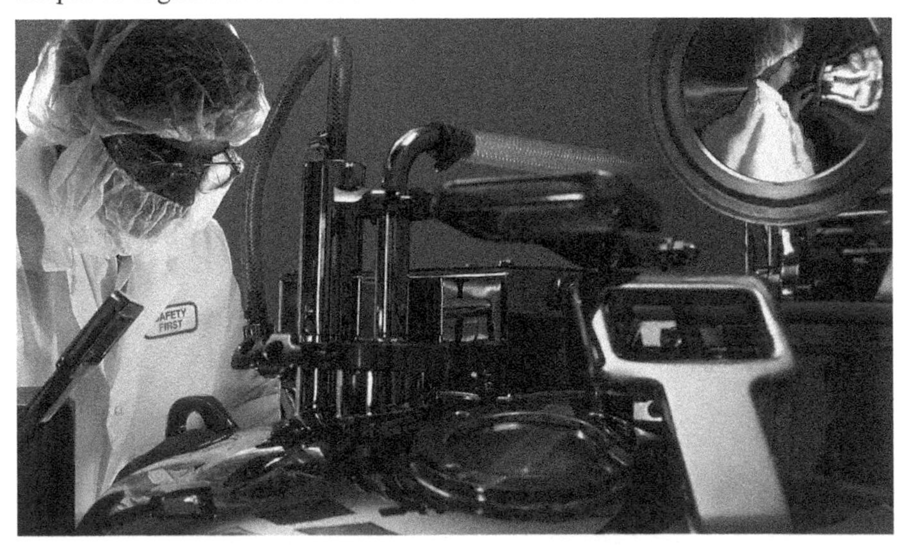

The economies of industrialized societies are underpinned by their chemical industries. Chemical companies generate some of the highest turnovers and the largest profits in British industry. Industrialized societies spend typically 10–25% of their income on fuel for transport, cooking, and heating. These fuels are provided by the organic chemical industry, whose function is to make useful compounds from readily available raw materials. The organic sector uses fossil fuels (coal, crude oil, and natural gas) and resources from living organisms as its raw materials.

The carbon atom

The carbon atom has four electrons in its valence shell. It can share each of these electrons with another atom to form *four* covalent bonds. The simplest organic compound is methane CH_4, which consists of a central carbon atom bonded to four separate atoms of hydrogen. A molecule of methane may be described in a number of different ways, as shown at the top of the opposite page.

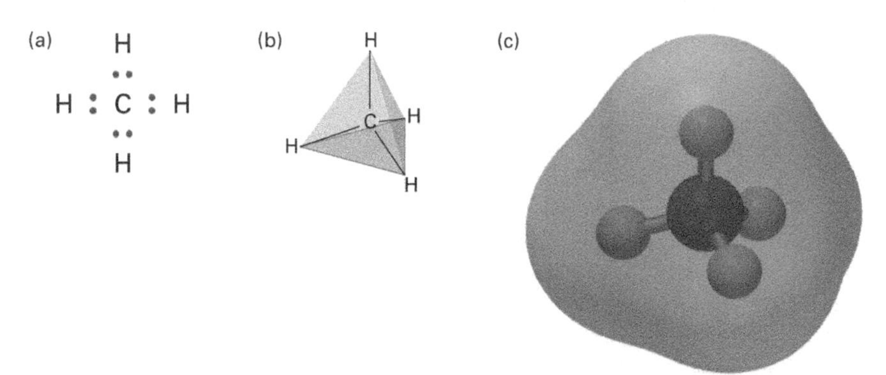

Various representations of the methane molecule CH_4. (a) The Lewis structure shows the arrangement of the bonding electron pairs. (b) VSEPR theory indicates that the molecule has a tetrahedral shape. (c) An electrostatic potential map confirms both the shape and the electron distribution suggested in (a) and (b). Note that there is very little polarity in the molecule.

Chains and rings of carbon atoms

Carbon is unique in that its atoms can bond together to make chains and rings of almost unlimited size and complexity. At the same time, bonds remain free to join with other atoms. Most of the diagrams below show only the carbon skeleton of some chains and rings; each line represents a bonding electron pair. Most of the diagrams do *not* show three-dimensional structure: they are *projections* onto the flat two-dimensional page. Carbon skeletons may also consist of branched chains or of two or more rings fused together:

(a) A single carbon atom and the four bonds it can form. (b) Two carbon atoms joined by a single covalent bond; six further bonds can be formed. (c) Five carbon atoms joined together to make a single chain. (d) and (e) Five and six carbon atoms joined to make ring-shaped structures.

A branched-chain carbon skeleton.

Kinetic and thermodynamic stability

Another reason why carbon forms so many compounds is that its compounds are *kinetically* very stable. The activation energy required to break a carbon–carbon bond is high. At normal temperatures compounds with carbon–carbon bonds are very stable. Once the activation energy is overcome at high temperatures, carbon compounds combust readily in the presence of oxygen. The reactions are highly exothermic because carbon compounds are thermodynamically unstable relative to their combustion products. It is this combination of a high activation energy and exothermic combustion that makes many organic compounds very useful as fuels.

SUMMARY

- Carbon compounds form the basis of living organisms. They form an enormous range of materials and fuels central to the function of any modern industrialized society.
- Catenation is the ability of an element to form chains with itself.
- Carbon catenates to form molecules of almost any size or shape.

Inorganic carbon chemistry

Organic chemistry studies carbon compounds, but a few carbon compounds such as carbon monoxide, carbon dioxide, metal carbonates, and carbon disulphide traditionally come under the umbrella of *inorganic* chemistry.

Hydrocarbons

Hydrocarbons are compounds that contain carbon and hydrogen only. Examples include the commercially important substances methane CH_4, benzene C_6H_6, and ethene C_2H_4.

Bond strengths

Bond	Bond enthalpy /kJ mol^{-1}
C–C	348
C=C	612
C≡C	838
C–H	413
C–O	360

Standard enthalpy changes of combustion

Compound	$\Delta_c H^{\ominus}$ /kJ mol^{-1}
Hydrogen H_2	–286
Methane CH_4	–891
Octane C_8H_{18} (in petrol)	–5512
Sucrose $C_{12}H_{22}O_{11}$ (sugar)	–5644

21.2

OBJECTIVES

- Five types of formulae
- Saturated and unsaturated molecules
- Homologous series

Empirical formula

The empirical formula, spread 9.3, shows the simplest whole-number *ratio* of the numbers of each type of atom present. The molecular formula shows the *actual* numbers of each type of atom present. The empirical formula of lactic acid (2-hydroxypropanoic acid) is CH_2O; its molecular formula is $C_3H_6O_3$.

(a) H–C(H)(H)–C(H)(H)–O–H CH_3CH_2OH

(b) H–C(H)(H)–O–C(H)(H)–H CH_3OCH_3

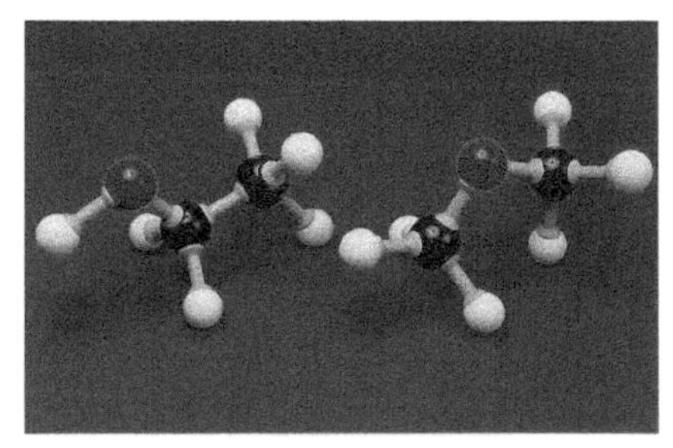

Displayed and condensed structural formulae: (a) ethanol; and (b) methoxymethane. Skeletal formulae (see below and page opposite) have also been added.

(a) OH OH

(b) O

Skeletal formulae: (a) ethanol (two versions); and (b) methoxymethane.

ORGANIC MOLECULES

The properties of organic molecules frequently depend on their shape as well as on the atoms they contain. As a result, chemists have developed various ways of representing the three-dimensional shapes of molecules, especially organic molecules, on two-dimensional sheets of paper. This spread begins with a discussion of empirical and molecular formulae, with which you are already familiar. The shortcomings of these formulae lead to the development of two types of formulae that represent molecular shape. The spread ends by listing the 10 basic carbon skeletons on which many common organic compounds are based.

Writing molecular formulae

It is possible to represent organic molecules in a number of different ways. The **molecular formula** (see chapter 9 'Reacting masses and volumes') of a compound shows the actual numbers of each type of atom present. For example, the molecular formula of ethanol is C_2H_6O. A problem immediately arises because the molecular formula of the compound methoxymethane is also C_2H_6O. The molecular formula does *not* distinguish between these two compounds.

Condensed and displayed structural formulae

The **structural formula** provides more information because it specifies exactly which atoms are bonded together. The **condensed** (**shortened**) **structural formulae** of ethanol and methoxymethane are CH_3CH_2OH and CH_3OCH_3 respectively. (The CH_3CH_2 group is sometimes written as C_2H_5, ethanol being represented as C_2H_5OH; this description is less clear and should be avoided.)

Structural formulae can also be *drawn* to show the bonds between atoms or groups of atoms. Such a **displayed formula** (or **graphic formula**) projects the three-dimensional structure of a molecule onto a flat two-dimensional page. As a result, bond angles are distorted and appear generally to be 90° or 180°.

Drawing stereochemical formulae

The **stereochemical formula** represents bond angles more accurately. It shows each atom separately and all the bonds present; it also attempts to represent the shape of the molecule. It is agreed among chemists that:

- A bond in the plane of the paper is shown as a solid line.
- A bond going behind the paper is shown as a dashed line (or as a diminishing wedge or as a 'striped' wedge).
- A bond coming in front of the paper is shown as an enlarging wedge.

Applying these rules, the tetrahedral shape of methane appears as shown below, together with the stereochemical formulae of ethanol and methoxymethane. Stereochemical formulae are only rarely used (but see spreads 22.8 and 28.6).

(a) H, C, H, H, H (b) H, H, C, H, O, H, C, H, H (c) H, O, H, C, C, H, H, H, H

Stereochemical formulae: (a) methane; (b) ethanol; and (c) methoxymethane.

Shorthand methods

The structural formulae of larger molecules containing rings and long chains of carbon atoms are tedious to draw out in full. A shorthand form of representation is often used, which makes the following assumptions:

- Each single line represents a single covalent bond between two carbon atoms; the symbols for the carbon atoms are not included.
- All other elements except for hydrogen are represented by their chemical symbols.

- The ends of all bonds are assumed to be occupied by hydrogen, unless otherwise indicated by a different chemical symbol.

The examples on the right show the **skeletal formulae** for cyclohexane C_6H_{12} and hexane C_6H_{14}.

(a)

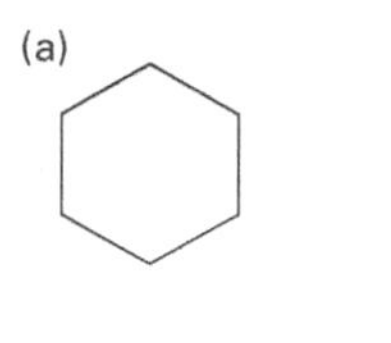

(b)

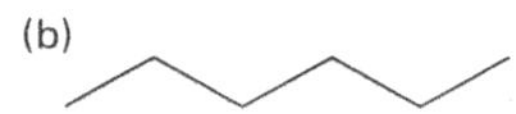

Skeletal formulae: (a) cyclohexane; and (b) hexane.

Saturated and unsaturated molecules

A molecule in which *all* the carbon atoms have four single covalent bonds is called a **saturated** molecule. A molecule containing one or more multiple bonds (i.e. double or triple covalent bonds) is said to be **unsaturated**.

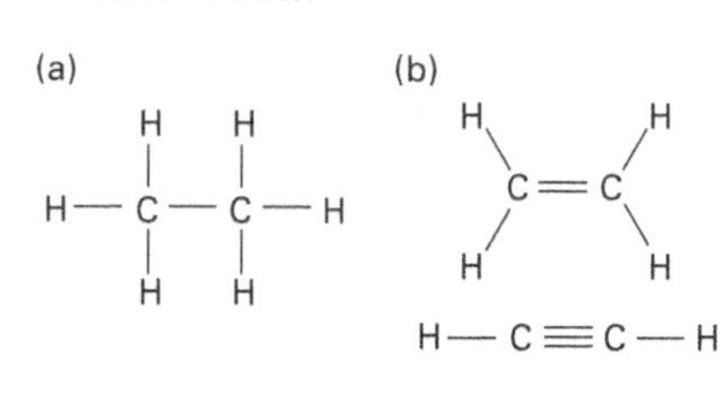

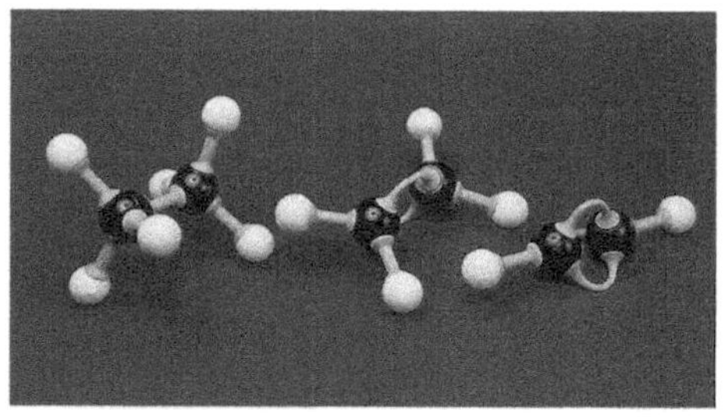

(a) A saturated molecule. (b) Unsaturated molecules.

Homologous series

The arrangement of carbon atoms may be thought of as the **skeleton** of an organic molecule. All the atoms of other elements are attached to that skeleton. The simplest skeleton is a single carbon atom; the next simplest consists of two carbon atoms bonded together. Then there can be chains of three, four, five, or more carbon atoms. A series of carbon compounds differing from each other only by the addition of more —CH_2— groups to increase the length of the carbon chain is called a **homologous series**. Compounds are named according to internationally agreed rules published by the International Union of Pure and Applied Chemistry (**IUPAC**) as explained in the next spread.

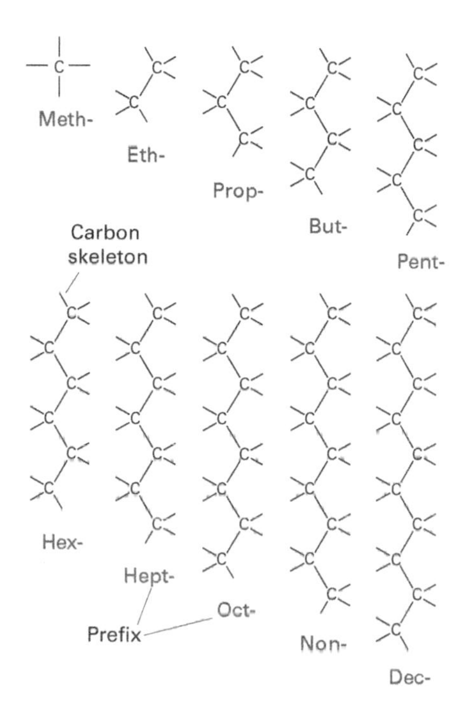

Carbon skeletons. Specific series of compounds (homologous series) are formed by attaching particular types or groups of atoms to each of these carbon chains. The name of each compound begins with the prefix that refers to the length of the carbon chain. For example, attaching hydrogen atoms only to these carbon chains results in a homologous series called the alkanes. The simplest member CH_4 is called methane; $C_{10}H_{22}$ is called decane. Notice that many of the names are similar to those used to describe polygons (e.g. pentagon, octagon, etc.).

SUMMARY

- The empirical formula shows the simplest whole-number ratio of the numbers of each type of atom present in a molecule.
- The molecular formula shows the actual numbers of each type of atom present in a molecule.
- The structural formula shows exactly which atoms are bonded together. It may be condensed or displayed.
- The stereochemical formula represents the individual atoms in a molecule, the bonds between the atoms, and the angles between the bonds.
- A saturated molecule contains single covalent bonds only.
- An unsaturated molecule contains one or more multiple (double or triple) covalent bonds.
- The members of a homologous series differ from each other only in the length of the carbon chain.

PRACTICE

1 Copy the 10 carbon skeletons shown above. To each, add the bonds and hydrogen atoms necessary to produce the alkane homologous series from methane to decane.

2 Why are the compounds in question 1 referred to as being *saturated?*

3 Give the following for the compound propane C_3H_8:

a Molecular formula

b Condensed structural formula

c Displayed formula.

4 Draw the structural formulae corresponding to each of the following skeletal formulae:

a **b**

21.3

OBJECTIVES

- Naming alkanes
- Structural isomers
- Chain isomers

STRUCTURAL ISOMERISM

The first organic compounds which we shall study are the hydrocarbons, which consist of a carbon skeleton to which only hydrogen atoms are attached. The simplest skeletons take the form of a single chain or a branched chain. The name of an organic compound gives information about its structure, and understanding how to name hydrocarbons prepares you for the study of more complex organic molecules containing elements other than carbon and hydrogen.

```
              H
              |
           H—C—H
   H   H      |      H   H
   |   |      |      |   |
H—C———C———C———C———C—H
   |   |      |      |   |
   H   H      H      H   H
```

3-Methylpentane.

```
         H
         |
      H—C—H
   H     |      H   H   H
   |     |      |   |   |
H—C————C————C———C———C—H
   |     |      |   |   |
   H     H      H   H   H
```

2-Methylpentane.

```
         H             H
         |             |
      H—C—H       H—C—H
   H     |      H      |       H
   |     |      |      |       |
H—C————C————C————C————C—H
   |     |      |      |       |
   H     |      H      H       H
      H—C—H
         |
         H
```

Name this hydrocarbon.

Straight-chain and branched-chain hydrocarbons

An organic compound with a straight-chain or branched-chain carbon skeleton that does *not* contain a benzene ring, spread 23.1, is called an **aliphatic** compound. The simplest aliphatic compounds are the **alkanes**, which contain carbon–carbon single bonds only. To find the IUPAC name for an alkane, you must carry out the following steps:

1. Identify the *longest* continuous carbon chain (called the **main chain**), count the number of carbon atoms in it, and name appropriately as meth-, eth-, prop-, but-, pent-, etc. (see the previous spread).
2. Identify any shorter branches (called **side chains**) attached to this main chain as methyl (CH_3—), ethyl (CH_3CH_2—), propyl ($CH_3CH_2CH_2$—) groups, etc.
3. Number the carbon atoms at which the branches are attached by counting the carbon atoms of the main chain from the end that will give the lower number at the first point of difference.

For example, the first compound to the left is 3-methylpentane. The name consists of:

- *-ane* because the compound only contains carbon–carbon single bonds and so is an alkane;
- *pent-* because there are five carbons in the main chain;
- *methyl-* because the branch contains one carbon atom;
- 3- because the branch is attached at the third carbon atom, no matter from which end of the main chain you count.

The second compound to the left would be called 2-methylpentane counting from one end of the main chain or 4-methylpentane counting from the other end. The correct name is 2-methylpentane because it contains the *lower* number.

Worked example on naming an alkane

Question: Name the compound shown to the left.

Strategy: Follow the steps mentioned in the text above.

Answer:

Step 1 Count the carbon atoms in the longest unbroken chain and name it.

The longest unbroken chain is five carbon atoms long and so the substance is a *substituted* pentane. The term *substituted* generally means that one or more hydrocarbon groups (e.g. methyl CH_3—) or atoms other than hydrogen (e.g. Cl—) are attached to the main chain.

Step 2 Count the carbon atoms in each side chain and name them.

The side chains all contain a single carbon atom and so they are all methyl groups. There are three methyl groups, so the substance is a trimethylpentane.

Step 3 Number the carbon atoms to which the side chains are attached, keeping the numbers as low as possible.

Numbering from the left gives two methyl groups on carbon 2 and one on carbon 4, i.e. 2,2,4-.... Numbering from the right gives one methyl group on carbon 2 and two on carbon 4, i.e. 2,4,4-.... Numbering from the left gives the lower number at the first point of difference. So the IUPAC name is 2,2,4-trimethylpentane.

The flame front of fuel burning in the cylinder of a petrol engine. A major component of petrol is octane C_8H_{18}. The straight-chain isomer (which occurs naturally in crude oil) gives explosive combustion that can result in engine 'knocking'. Combustion is smoother when the branched-chain isomers (such as the one in the worked example) are also used.

Structural isomerism

Alkanes with four or more carbon atoms can show different arrangements of the carbon atoms. For example, there are two structures (and two different substances) that have the molecular formula C_4H_{10}, namely butane and 2-methylpropane.

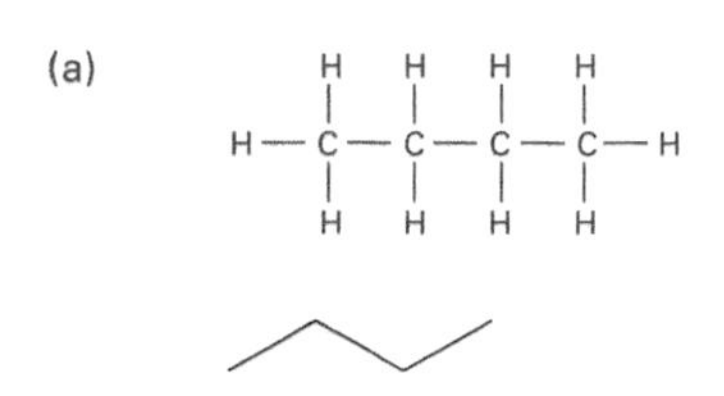

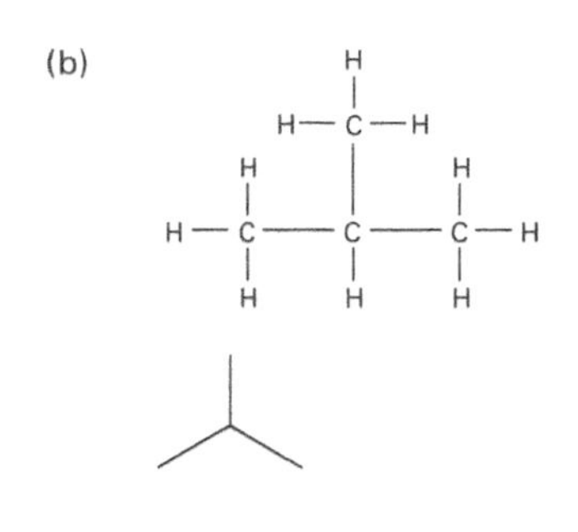

Molecules that share the same molecular formula but have different structural formulae are called **structural isomers**. There are three structures for the hydrocarbon C_5H_{12}, corresponding to the molecules pentane, 2-methylbutane, and 2,2-dimethylpropane.

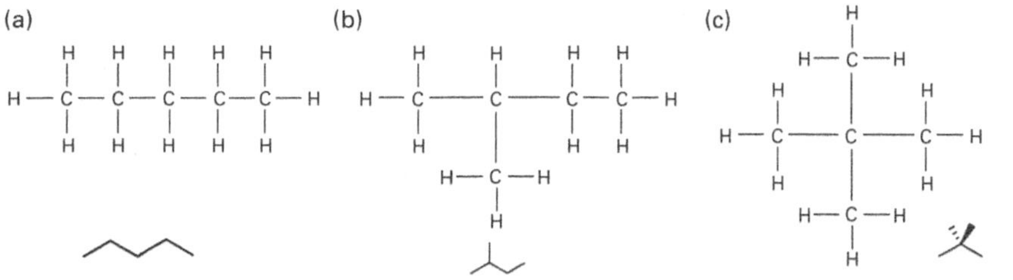

The three isomers with molecular formula C_5H_{12}: (a) pentane; (b) 2-methylbutane; and (c) 2,2-dimethylpropane.

Space-filling models for the two isomers with molecular formula C_4H_{10}: (a) butane; and (b) 2-methylpropane.

Structural isomers with different arrangements of the carbon skeleton are often called **chain isomers**. Chain isomers may have different *physical* properties because the different shapes will alter the strength of the dispersion forces. For example, pentane boils at 36 °C, 2-methylbutane at 28 °C, and 2,2-dimethylpropane at 10 °C. Their chemical properties, however, are virtually the same because they are all alkanes.

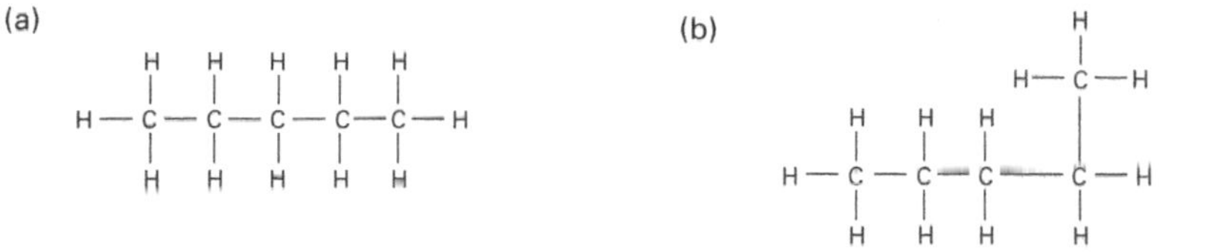

Twisting the end of a molecule (by rotating a single carbon–carbon bond) does not *create an isomer. For example, a bent chain is not an isomer of a straight chain, as shown in this example of pentane C_5H_{12}. Molecules are only isomeric if changing from one isomer to another involves breaking bonds and reassembling the molecule in a different way. Bond angles are represented as 90° in these structural formulae, which can be misleading. In reality, bond angles in alkanes are very close to the tetrahedral angle (109° 28′).*

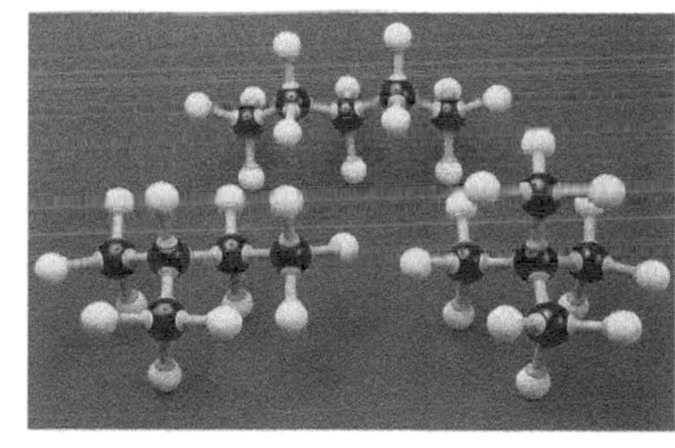

The three isomers with molecular formula C_5H_{12}.

SUMMARY

- The name of an organic compound is based on the number of atoms in the longest unbroken carbon chain.
- The carbon atoms in the main chain are numbered so that the positions of side chains are given the lower number at the first point of difference.
- Structural isomers have the same molecular formula but different molecular structures, i.e. the atoms are bonded together in different sequences.
- Chain isomers have different arrangements of their carbon skeletons.
- Chain isomers have similar chemical properties but different physical properties.

PRACTICE

1 Draw all the possible isomers of hexane C_6H_{14} and give the name of each.

2 Draw structural formulae for the following alkanes:
a 2,2-Dimethylbutane **b** 2-Methyl-4-ethylhexane
c Cyclohexane C_6H_{12}.

3 A student gave the name of a hydrocarbon as 2-methyl-2-ethylbutane. Give the *correct* name.

4 a Draw the structural formula of the most highly branched isomer of octane.
b Suggest why this isomer combusts more smoothly than the straight-chain isomer.
c Which of these two isomers has the higher boiling point? Give reasons why this isomer gives rise to the higher value.

21.4

OBJECTIVES

- Functional groups
- Positional isomers
- Functional group isomers

FUNCTIONAL GROUPS

The simplest organic molecules are the hydrocarbons, which consist of hydrogen atoms attached to a skeleton of carbon atoms. More complex substances result when hydrogen atoms are replaced by atoms of other elements. These atoms make up functional groups, which have significant effects on the properties of the substances. We will not look at the properties of the various functional groups in this introductory chapter. At this stage you simply need to recognize the groups, name them, and name molecules that contain them.

IUPAC nomenclature

When a functional group is present, the longest continuous chain *containing the functional group* is chosen as the main chain (see previous spread) and the numbering is adjusted so that the functional group has the lower number possible, so $CH_3CH_2CH_2CH(OH)CH_3$ is pentan-2-ol rather than pentan-4-ol.

Functional groups

Atoms other than carbon or hydrogen or groups of such atoms attached to the main carbon chain are called **functional groups**. As with hydrocarbons, the systematic IUPAC name of a substance that has a functional group allows you to determine its structural formula. For example, look at the structural formula of ethanol shown to the left. The first part of the name *ethan-* shows the length of the carbon chain (two carbon atoms in a saturated molecule CH_3CH_2—). The second part of the name (*-ol*) indicates the presence of the alcohol functional group —OH. There is a complete homologous series of alcohols (i.e. methanol, ethanol, propanol, butanol, etc.) that are based on the alkane homologous series. In each case, a hydrogen atom in an alkane molecule is replaced by a **hydroxyl group** —OH.

(a) H—C(H)(H)—C(H)(H)—O—H

(b)
CH_3OH
CH_3CH_2OH
$CH_3CH_2CH_2OH$
$CH_3CH_2CH_2CH_2OH$

(a) Ethanol. (b) The simplest members of the alcohol homologous series.

There are various different series of organic compounds, each with its own functional group. Examples include the ketones, aldehydes, carboxylic acids, ethers, and halogenoalkanes. The formulae and structures of these compounds are shown below. Note that, when writing a general formula for a homologous series containing a functional group, the symbol **R** ('residue') may be used to denote an alkyl group such as CH_3— or CH_3CH_2—.

R—OH	R—C(=O)—R′	R—C(=O)—H	R—C(=O)—O—H	R—O—R′	R—X (X = F, Cl, Br, I)
Alcohols	Ketones	Aldehydes	Carboxylic acids	Ethers	Halogeno-alkanes

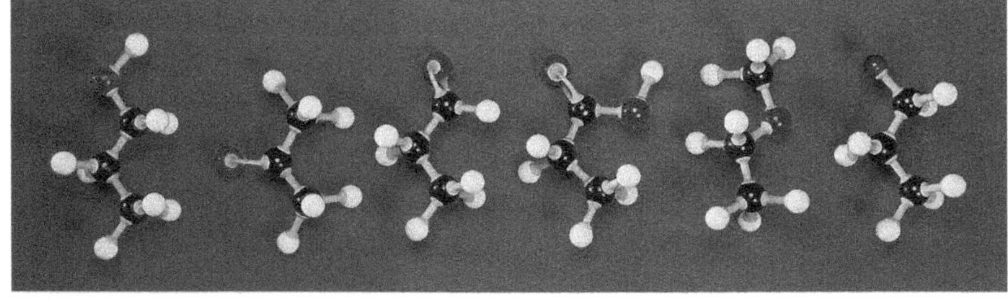

The structural formulae of some important homologous series containing functional groups: (a) alcohols, (b) ketones, (c) aldehydes, (d) carboxylic acids, (e) ethers, and (f) halogenoalkanes.

Chemical and physical properties

Functional groups determine the chemical reactions that a molecule can undergo, so *all members of a homologous series have very similar chemical properties*. The length of the carbon chain does not affect chemical properties because the C—C and C—H bonds have large bond enthalpies and so do not react easily. The chain length mainly affects physical properties such as melting point and boiling point. As additional —CH_2— units are added, the molecules have larger dispersion forces, and so the melting and boiling points increase.

Naming molecules with different functional groups

The four compounds shown below have names including 'but-', because the longest carbon chain consists of four carbon atoms (corresponding to the alkane butane). The other part of each name denotes the functional group present. The position of the functional group must be given in the name if there is any chance of ambiguity. For example, in (b), the hydroxyl group is attached to carbon atom 1 (butan-1-ol). It could also be attached to carbon atom 2 (butan-2-ol). These two substances both have the molecular formula $C_4H_{10}O$ but different structures. Functional groups give rise to *isomerism*.

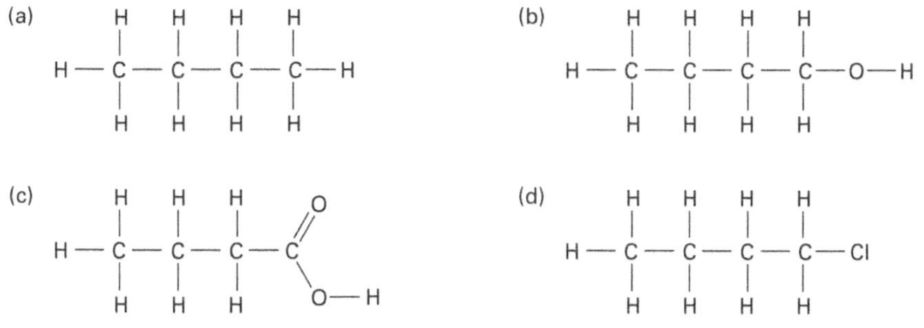

Naming molecules with functional groups: (a) butane; (b) butan-1-ol; (c) butanoic acid; and (d) 1-chlorobutane.

Isomers

Structural isomers can arise from placing a *functional group in a different position* on the chain. For example, C_3H_8O is the molecular formula for both propan-1-ol and propan-2-ol, but the functional group —OH occurs at position 1 or 2, giving different structural formulae.

These structural isomers are sometimes called **positional isomers.** The *physical* properties of positional isomers may vary slightly; for example, propan-1-ol boils at 97 °C whereas propan-2-ol boils at 82 °C. Their *chemical* properties may also vary because of the different position of the functional group on the carbon skeleton.

Structural isomers may also have *different functional groups*. The molecular formula C_3H_6O may correspond to propanal CH_3CH_2CHO (an aldehyde) or to propanone CH_3COCH_3 (a ketone) – see illustrations to the right. These structural isomers are sometimes called **functional group isomers**. Functional group isomers not only have different physical properties but they also have *different chemical properties* because the functional group is different.

Methoxymethane CH_3OCH_3 (see spread 21.2) is an ether; it is a functional group isomer of ethanol CH_3CH_2OH. Hydrogen bonding between ethanol molecules results in a much higher boiling point for ethanol (78 °C) than for methoxymethane (–25 °C). Ethanol reduces aqueous dichromate(VI) ions, turning the solution from orange to green. Methoxymethane does not react with aqueous dichromate(VI) ions.

Hydrocarbons revisited

The simplest hydrocarbons are saturated and are the homologous series called the **alkanes**. If a double bond is present, then the result is the homologous series of **alkenes**. The first few members of this series are shown to the right. The double bond in alkenes is called a functional group because it gives the series its distinctive chemical properties.

SUMMARY

- A functional group is an atom other than carbon or hydrogen (e.g. —Cl), a group of atoms (e.g. —OH), or a carbon–carbon double bond.
- The length of the carbon chain has little effect on the properties of a homologous series containing a functional group.
- Structural isomers of compounds that have functional groups may be positional isomers or functional group isomers.

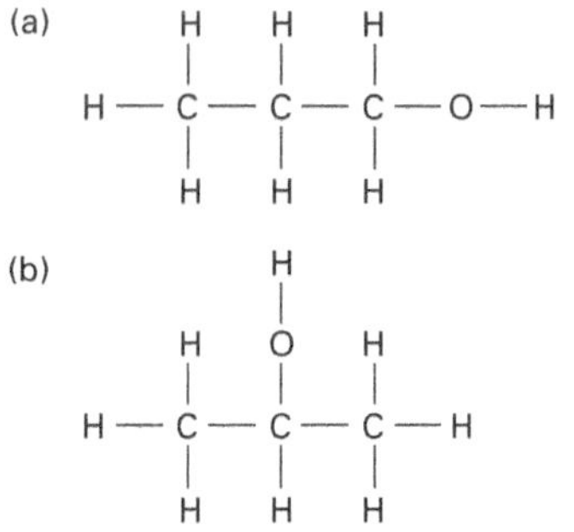

Positional isomers of C_3H_7OH: (a) propan-1-ol $CH_3CH_2CH_2OH$; and (b) propan-2-ol $CH_3CH(OH)CH_3$.

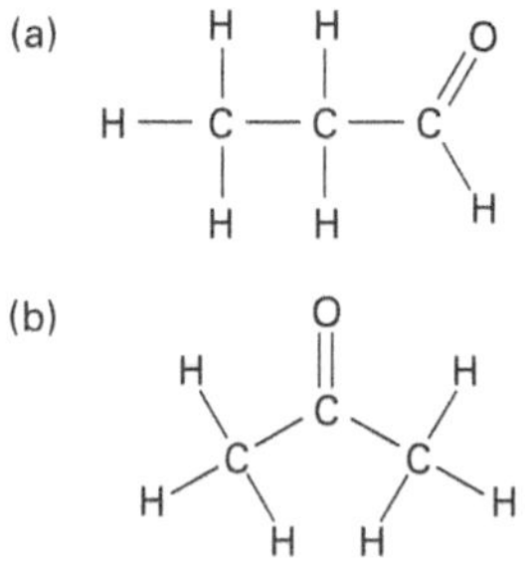

Functional group isomers with molecular formula C_3H_6O: (a) propanal CH_3CH_2CHO; and (b) propanone CH_3COCH_3.

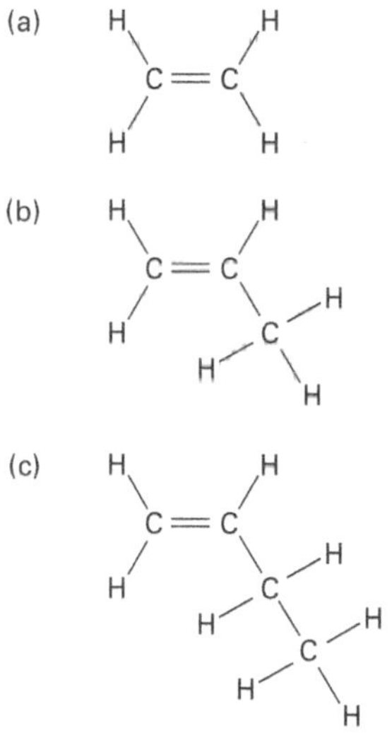

Straight-chain alkenes: (a) ethene; (b) propene; (c) but-1-ene. Note that (c) has an isomer but-2-ene in which the double bond is between carbon atoms 2 and 3. The lower number is used to describe the position of the double bond.

PRACTICE

1 Draw the structural formulae of all possible isomers corresponding to each of the following molecular formulae:

a C_4H_{10} **c** C_4H_9Cl

b C_5H_{12} **d** $C_4H_{10}O$.

21.5

OBJECTIVES

- Reactive sites
- Nucleophiles and electrophiles
- Substitution and addition

THE REACTIONS OF ORGANIC COMPOUNDS – 1

The reactions of organic compounds fall into a small range of distinct types. Apart from combustion, in which the molecular structure is completely destroyed, reactions tend to involve only part of the molecule, usually the functional groups. These act as reactive sites, reacting with other chemicals. You will meet the specific reactions of functional groups in the following chapters. The aim of the last two spreads in this chapter is to introduce the *types of reaction* undergone by organic molecules.

The site of reaction

Alkanes are very unreactive. The carbon–carbon and carbon–hydrogen bonds do not break easily because they have high average bond enthalpies (C—C, 348 kJ mol^{-1}; C—H, 413 kJ mol^{-1}). The elements have similar electronegativities (C, 2.5; H, 2.2) and so the bonds have little polarity.

Molecules with functional groups are generally more reactive. For example, the halogenoalkane bromoethane reacts to form ethanol when heated with aqueous base:

$$CH_3CH_2Br(aq) + OH^-(aq) \rightarrow CH_3CH_2OH(aq) + Br^-(aq)$$

Comparing bromoalkanes with alkanes, we can see that:

1 The C—Br bond is weaker than the C—C and C—H bonds (average C—Br bond enthalpy 276 kJ mol^{-1}).

2 The electronegativity of bromine is 3.0, so the C—Br bond is polarized $C^{\delta+}$—$Br^{\delta-}$.

H—C(H)(H)—C(H)(H)—Br + :Ö—H → H—C(H)(H)—C(H)(H)—O—H + :Br$^-$

Because the C—Br bond is polarized, the 'attacking' aqueous hydroxide ion OH$^-$(aq) is attracted towards the δ+ charge on the carbon atom. A C—OH bond forms (average bond enthalpy 360 kJ mol^{-1}) as the weaker C—Br bond breaks. Cyan highlights the parts of the species that are changing.

Successful reactions

Bond enthalpies and electronegativities are not the only factors that determine whether a reaction is successful. The nature of the products of the reaction, the conditions of temperature and pressure, and entropy changes (see chapter 14 'Spontaneous change ...') all play a part in deciding the likelihood of a reaction taking place.

Reactive sites

Functional groups in molecules contain reactive sites that are approached by incoming species. A **reactive site** is a region of higher or lower electron density.

Electron-deficient sites include:

- atoms with a partial δ+ charge because they are bonded to a more electronegative atom;
- positive ions.

Electron-rich sites include:

- atoms with a partial δ– charge because they are bonded to a less electronegative atom;
- negative ions
- double bonds between carbon atoms;
- lone pairs.

(a)

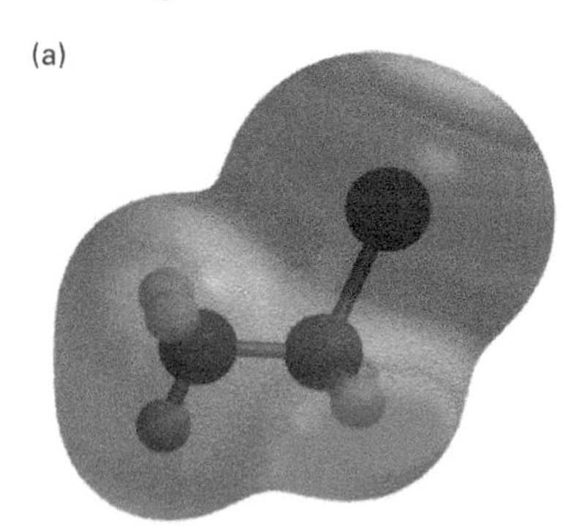

(b)

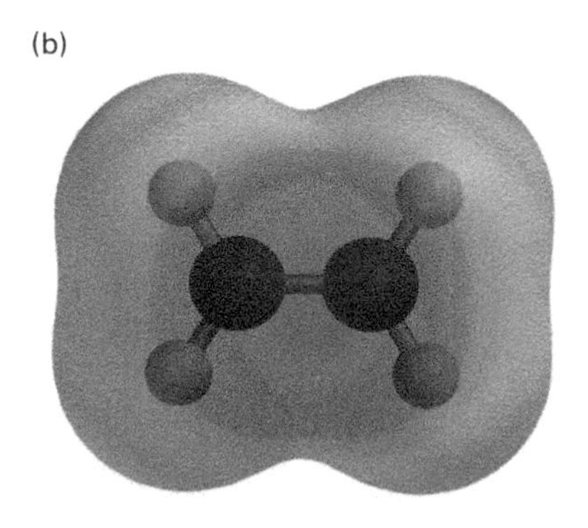

Reactive sites: electrostatic potential maps (see spread 5.8) of functional groups. (a) An electron-deficient site ($C^{\delta+}$) and an electron-rich site ($Br^{\delta-}$) in bromoethane. (b) An electron-rich site resulting from the increased electron density at the carbon–carbon double bond in ethene, indicated by the red colour.

The approaching species

Two important classes of reagent are used in organic reactions: these classes are called nucleophiles and electrophiles.

- **Nucleophiles** are species that are electron-pair donors.

Nucleophiles include negatively charged ions such as CN^-, OH^-, or Cl^-, as well as molecules bearing lone pairs of electrons like H_2O or NH_3. Nucleophiles attack the partially positively charged (δ+) atoms of polar covalent bonds such as the $C^{\delta+}$ in halogenoalkanes, aldehydes, or ketones. Nucleophiles donate electron pairs to form new bonds.

- **Electrophiles** are species that are electron-pair acceptors.

Electrophiles are typically positive ions like H^+ or NO_2^+. Electrophiles are attracted to the partially negatively charged (δ–) atoms of polar covalent bonds. They also often react with the regions of high electron density in the double bond in alkenes and other double-bonded structures. Electrophiles accept electron pairs to form new bonds.

Types of reactions

When an electrophile or a nucleophile approaches an organic molecule, several different reactions can follow. Two common types of reaction are substitution reactions and addition reactions.

- In a **substitution** reaction, an atom or group (X) in a molecule is *replaced* by another (Y). Substitution reactions are typical of saturated compounds.

Look again at the reaction between bromoethane and aqueous hydroxide ions described at the start of this spread. This reaction is classed as a substitution reaction because the hydroxyl group —OH substitutes for the bromine atom —Br.

- In an **addition** reaction, two molecules *add* together to form one larger one, as in the addition of a molecule X—Y across a double bond. Addition reactions are typical of unsaturated compounds.

Not all compounds undergo all types of reaction; for example, alkanes take part only in substitution reactions. To become a good synthetic organic chemist, you need to understand the properties of particular chemical bonds, and how and why they react. You need to know how to introduce different atoms into a given molecule. Planning a chemical synthesis involves considering the nature of the partial charges (if any) present in the molecule. Having located these 'points of weakness', you must then choose a reactant that will introduce the group you want into the molecule. You must consider whether there are other groups of atoms close by that will obstruct reaction with your chosen reagent. This apparently complex task is made much easier if you have a sure grasp of the terms introduced in this spread and in the one that follows.

SUMMARY

- Reagents react with regions of low or high electron density in organic molecules.
- Regions of low electron density (electron deficiency) result when highly electronegative atoms bonded to carbon atoms withdraw electron density from the bonds that join them.
- Regions of high electron density result around highly electronegative atoms or around double bonds.
- Nucleophiles are species that are electron-pair donors and attack regions of low electron density.
- Electrophiles are species that are electron-pair acceptors and react with regions of high electron density.

(a)

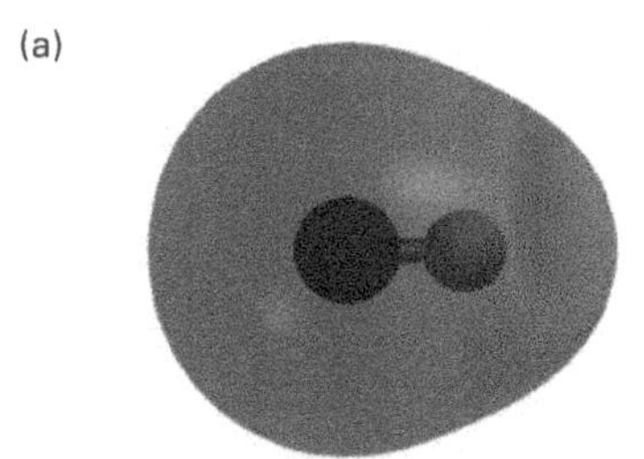

(b)

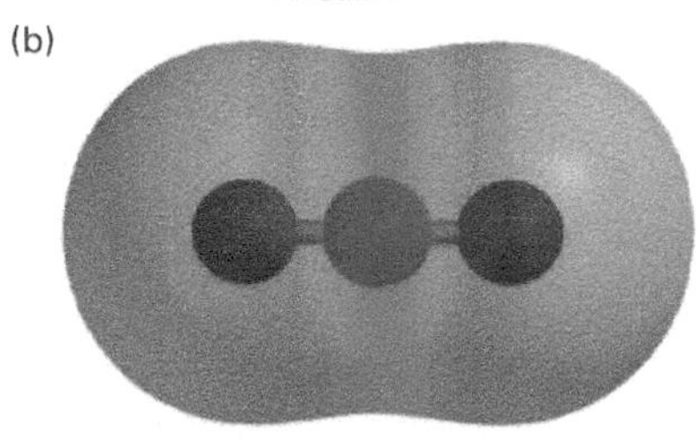

Electrostatic potential maps of a nucleophile and an electrophile. (a) The hydroxide ion OH^- is a nucleophile: note the high electron density due to the negative charge. (b) The nitronium ion NO_2^+ is an electrophile: note the low electron density due to the positive charge.

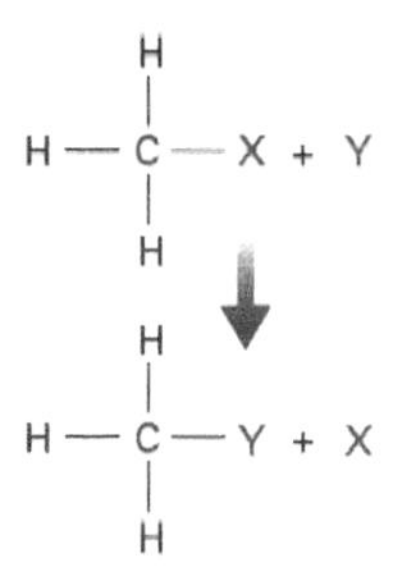

Substitution of X by Y.

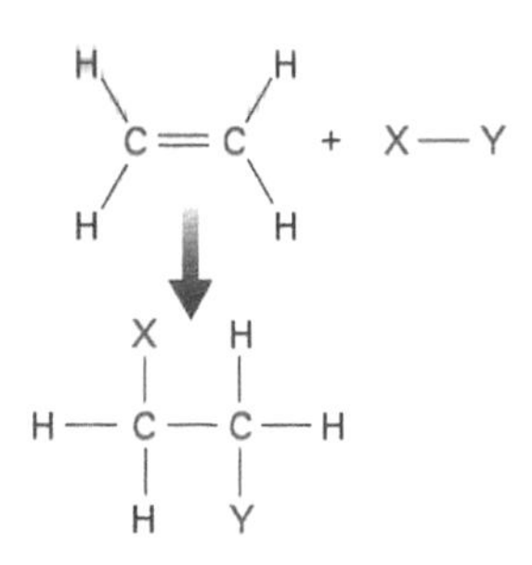

Addition of an incoming molecule X—Y across a carbon–carbon double bond.

Saturated and unsaturated compounds

You should remember that:

- *Saturated compounds* have only single covalent bonds between their atoms; for example, the alkanes are saturated hydrocarbons.
- *Unsaturated compounds* have one or more double bonds between their atoms; for example, carbon–carbon C=C double bonds in alkenes; and carbon–oxygen C=O double bonds in aldehydes and ketones.

21.6

OBJECTIVES

- Heterolytic fission
- Curly arrows
- Homolytic fission
- Radicals
- Oxidation

THE REACTIONS OF ORGANIC COMPOUNDS – 2

Organic compounds may contain regions of low electron density, which may be attacked by reagents that are nucleophiles, and regions of high electron density, which react with reagents that are electrophiles. Chemical reactions are concerned with the breaking and making of bonds. This spread starts by looking at the *ways* in which bonds can break, and ends with an introduction to the oxidation of organic compounds.

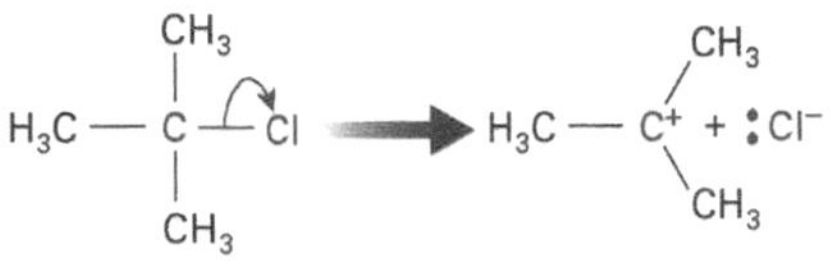

Breaking bonds: heterolytic fission

The breaking of bonds is often described as **bond fission**. When a covalent bond breaks, there are two ways in which the electron pair can be redistributed. During the process known as **heterolytic fission** (or **heterolysis**), *both* the electrons in the bond transfer to *one* of the two atoms:

$$A—B \rightarrow A^+ + B^-$$

$(CH_3)_3C—Cl \rightarrow (CH_3)_3C^+ + :Cl^-$

Heterolytic fission. The electron density in the C—Cl bond is displaced towards the chlorine atom. The bond breaks when the electron pair transfers entirely to the chlorine atom. Heterolytic bond fission should be shown by one double-headed curly arrow.

An ion pair results. Heterolytic fission produces *ions* because one atom gains an extra electron and the other loses one. The illustration to the left shows the heterolytic fission of a C—Cl bond. The species that result are highly reactive. They may form a new bond with an approaching reagent, or if not they may simply recombine. Polar covalent bonds often break heterolytically because the electron pair is already displaced towards one of the atoms.

The movement of an electron pair is shown by using a double-headed **curly arrow**. The arrow *starts* at the origin of the electron pair (in the case shown above left, the covalent bond between C and Cl). The arrow *ends* at the new position of the electron pair (in the case shown, on chlorine).

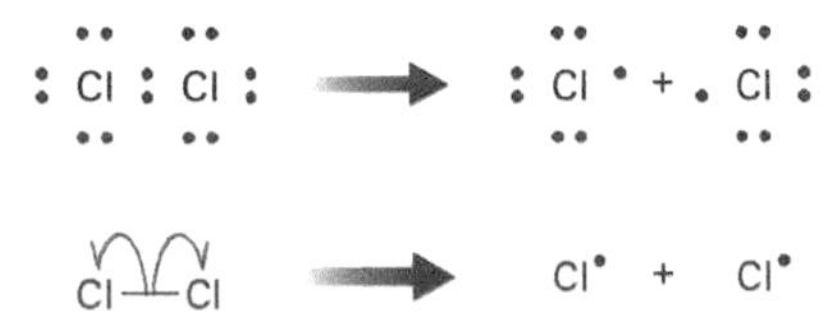

Breaking bonds: homolytic fission

Alternatively, during the process known as **homolytic fission** (or **homolysis**), the *two* atoms keep *one* electron each:

$$A—B \rightarrow A^\bullet + B^\bullet$$

$:Cl:Cl: \rightarrow :Cl^\bullet + {}^\bullet Cl:$

$Cl—Cl \rightarrow Cl^\bullet + Cl^\bullet$

Homolytic fission. Each chlorine atom receives one of the pair of electrons making up the single covalent bond. Homolytic bond fission may be shown by two single-headed curly arrows.

The electron pair that formed the single covalent bond is shared equally between the two atoms. Homolytic fission produces atoms or groups of atoms containing an *unpaired electron*. A species with an unpaired electron is called a **radical** (or sometimes a **free radical**). The unpaired electron is represented as a raised dot in diagrams.

A radical is a high-energy species. It is *very reactive* because forming a covalent bond by pairing with another electron is a highly exothermic process, releasing about 150–400 kJ mol^{-1}. Homolytic fission is typical of non-polar bonds such as the Cl—Cl bond in the Cl_2 molecule.

Fish hooks

In bond fission, electrons move from bonding orbitals positioned between atoms to orbitals located on individual atoms. A double-headed curly arrow shows the movement of an electron pair, as in heterolytic fission.

A single-headed curly arrow (sometimes called a 'fish hook') shows the movement of a single electron, as in homolytic fission.

Some chemists do not use 'fish hooks' to avoid confusion with double-headed curly arrows, whose use is both universal and exceptionally important. We colour 'fish hooks' differently to avoid confusion.

Causes of bond fission

Molecules are constantly colliding with each other. Bond fission occurs when they collide with sufficient energy and at the correct angle to each other. Non-polar bonds are more likely to undergo homolytic fission, particularly in the presence of ultraviolet light (see the following chapter). If a bond is already polarized (e.g. C—Cl), then the fission is likely to be heterolytic; the atom with the higher electronegativity gains a full negative charge.

Heterolytic bond fission can also be caused by the proximity of an approaching species. For example, a nucleophile has a region of high electron density that will repel electron density in a functional group. As the nucleophile approaches closer and closer, electron density transfers more and more to the most electronegative atom within the group. Eventually the electron density is completely displaced onto the most electronegative atom, and this atom is no longer bonded to the group.

Oxidation of organic molecules

The term 'oxidation' originally referred to the gain of oxygen (see chapter 13 'Redox equilibrium') or to the loss of hydrogen. This definition can quite often enable you to identify oxidation in equations representing organic reactions. Oxidation in organic chemistry is frequently carried out using potassium dichromate(VI) $K_2Cr_2O_7$ in acidic solution as the oxidant. For example, ethanol can be oxidized in two stages. During the first stage, hydrogen is removed from the molecule; during the second stage oxygen is added to the molecule (as shown below).

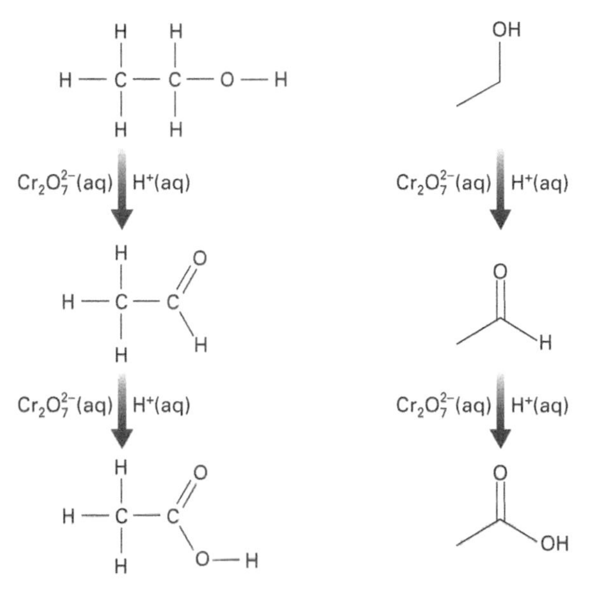

The oxidation of ethanol (CH_3CH_2OH) to ethanoic acid (CH_3COOH).

You are now familiar with the basic principles that underpin the reactions of organic molecules. We can now apply these principles to the reactions of specific functional groups and their homologous series.

SUMMARY

- In heterolytic fission, both electrons in a bond move to one of the two bonded atoms.
- Heterolytic fission tends to occur in polar bonds; the result is a pair of oppositely charged ions.
- In homolytic fission, the electron pair in a bond is split equally between the two bonded atoms.
- Homolytic fission tends to occur in non-polar bonds; the result is two radicals.
- A molecule is oxidized when it gains oxygen or loses hydrogen.
- A molecule is reduced when it loses oxygen or gains hydrogen.

Reduction

The process of reduction is the opposite of oxidation. An organic molecule is reduced when it loses oxygen or gains hydrogen. Ethanoic acid may be converted into ethanol by reduction, spread 27.8. Specific reduction reactions are discussed in later chapters.

Functional group level

It is possible to rationalize the redox reactions of organic molecules by considering the **functional group level**, which specifies the number of bonds to atoms that are more electronegative than carbon that a particular carbon atom makes.

The carbon atom singly bonded to an oxygen in an alcohol (chapter 25) is at functional group level 1 (FGL 1). The carbon atom doubly bonded to an oxygen in an aldehyde or ketone (chapter 26) is at FGL 2. The carbon atom doubly bonded to one oxygen and singly bonded to another in a carboxylic acid (chapter 27) is at FGL 3. The compound CO_2 has a carbon atom doubly bonded to two oxygen atoms and is at FGL 4.

To turn a carbon atom at a higher FGL to a lower FGL requires a reductant; to turn a carbon atom at a lower FGL to a higher FGL requires an oxidant. In the reactions shown on the left, the carbon atom coloured blue is oxidized from FGL 1 to FGL 2 and finally to FGL 3.

When there is more than one functional group present, the name gives priority to the carbon atom with the higher FGL.

Hence $CH_3CH(OH)CH_2CHO$ is 3-hydroxybutanal and $HOCH_2CH_2COOH$ is 3-hydroxypropanoic acid.

PRACTICE

1 a Name two electrophiles and two nucleophiles.
 b In a molecule of ethanol CH_3CH_2OH, with which atom would an electrophile react? Which atom would a nucleophile attack?

2 State whether each of the following reactions represents substitution or addition. Give reasons for your answer.
 a $CH_3CHBrCH_3 + OH^- \rightarrow CH_3CH(OH)CH_3 + Br^-$
 b $CH_3CH{=}CH_2 + H_2 \rightarrow CH_3CH_2CH_3$
 [Hint: start by drawing structural formulae for each of the compounds in these reactions.]

3 For each of the reactions in question 2:
 a Which part of the organic molecule reacts with the incoming reagent?
 b Which part of the reagent species reacts with the organic molecule?
 c Which bonds are broken and which bonds are made?

4 Using the molecule $(CH_3)_3CBr$ as an example, describe how the C—Br bond may undergo:
 a homolytic fission;
 b heterolytic fission.

5 Draw structural formulae for the ketone CH_3COCH_3 and the alcohol $CH_3CH(OH)CH_3$. Explain why conversion of the alcohol to the ketone is described as an oxidation reaction.

22 Alkanes and alkenes

The organic chemical industry produces millions of tonnes of products each year. These products may be fuels or materials such as plastics and fibres. They may also be the chemicals required for manufacturing processes, such as isoprene (2-methylbuta-1,3-diene) used to make the synthetic rubber for car tyres. Whatever the end product, the major raw materials are alkanes together with the alkenes derived from them. Alkanes are saturated hydrocarbons. They have few chemical properties because they have no functional groups. Alkenes are unsaturated hydrocarbons – they contain carbon–carbon double bonds. Substituting other elements for hydrogen atoms in alkanes and adding atoms to alkenes are the first steps to introducing the functional groups that allow the synthesis of almost any conceivable end product.

The oil industry: fractionation

22.1

OBJECTIVES

- Raw materials for the chemical industry
- Fractional distillation
- How fractionating columns work
- Fractions

The alkanes and the alkenes are two homologous series of hydrocarbons, their molecules consisting of hydrogen and carbon only. One of the major sources of alkanes is crude oil (also known as petroleum). Another major source is natural gas, which is found either dissolved in crude oil or else in underground reservoirs containing gases alone. Crude oil is a mixture of hundreds of different hydrocarbons, most of which are alkanes. This introductory spread shows how crude oil is separated by industrial-scale fractional distillation. It also discusses how fractional distillation is carried out in the laboratory and introduces you to ideas about vapour pressure that explain this process.

Fractional distillation of crude oil

The different components of a mixture have *different boiling points*, and fractional distillation separates the components as a result of this property. Fractional distillation of crude oil results in a series of **fractions**, each consisting of a mixture of hydrocarbons boiling within a given temperature range. The substances within a fraction may be separated by further distillation. The boiling point of a fraction increases with the number of carbon atoms, because of stronger dispersion forces, spread 7.4.

The actual composition of crude oil depends on where it was found. This sample comes from Ogoniland, Nigeria.

An oil refinery converts crude oil into a wide range of products. Some of the products are ready for use, e.g. gasoline (C_5–C_8) and kerosene/paraffin (C_{10}–C_{14}). Some of the products are used as chemical feedstocks for the manufacture of organic chemicals, e.g. gases (C_1–C_4) and naphtha (C_8–C_{10}).

Natural gas

Crude oil is the major source of alkanes. The second most important source of alkanes is natural gas, which is a mixture consisting mostly of methane CH_4 (at least 85%), ethane (up to 10%), propane (about 3%), and butane. Trace amounts of carbon dioxide, nitrogen, oxygen, and sometimes helium may also be present.

Theory and practice of fractional distillation

At a fixed temperature, the total vapour pressure above a liquid mixture depends on the vapour pressure of each pure component liquid and on the proportion of each liquid in the mixture. The total vapour pressure of a mixture of two liquids varies linearly with the composition of the mixture. The liquid boils when the total vapour pressure equals the external atmospheric pressure. See following spread for details.

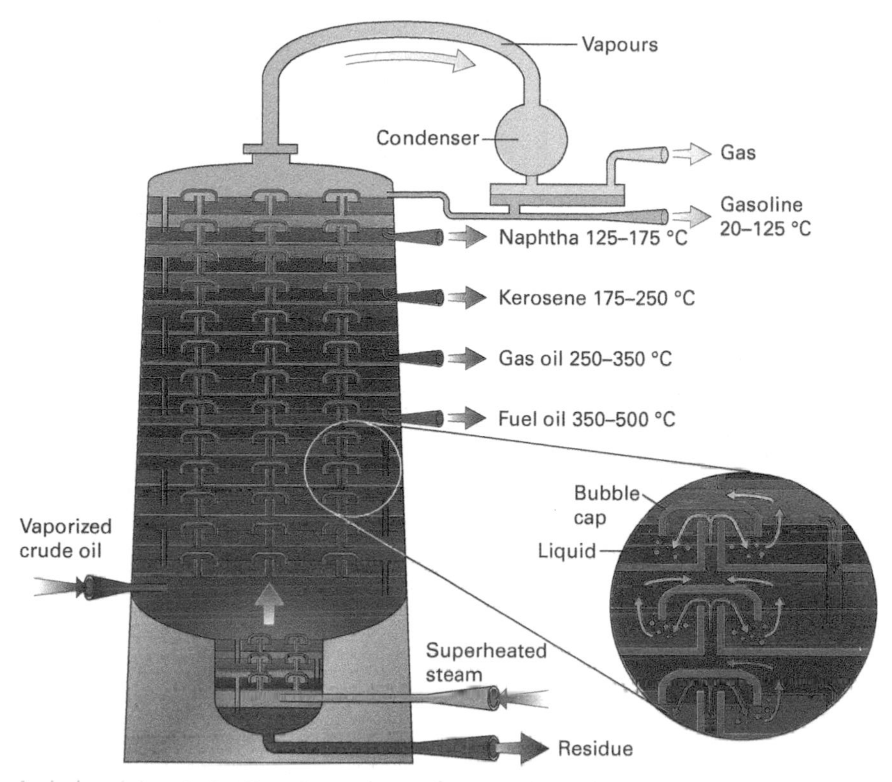

An industrial-scale fractionating column. Crude oil is pre-heated to about 450 °C in a furnace. It vaporizes on entering the fractionating column. Superheated steam maintains the temperature in the lower regions and ensures that the vapours rise through the column. Bubble caps cause rising vapour to mix intimately with the falling liquid that has condensed at each level. Liquid is either drawn off or is re-routed to a lower level. The boiling point in a fraction increases with the number of carbon atoms.

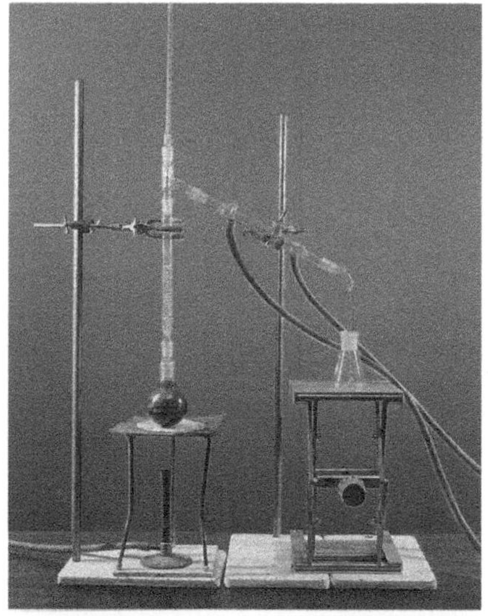

Laboratory-scale fractional distillation apparatus. When separating two liquids, the more volatile component distils over at a constant temperature as read by the thermometer. During this time, the less volatile component condenses and falls as a liquid down the column.

Imagine a liquid mixture consisting of 1 mol of each of two substances. As the temperature increases, the vapour pressures of each of the two liquids increase. However, the vapour pressure of the more volatile component increases to a greater extent. The vapour above the boiling liquid mixture becomes richer in the more volatile component. Condensing this vapour produces a *liquid* richer in the more volatile component. Repeated evaporation and condensation stages can completely separate the liquid mixture into its two components. This is what happens in a fractionating column.

In the laboratory-scale fractional distillation apparatus shown at top right, rising vapour and falling liquid meet and mix on the surface of the glass beads that fill the fractionating column. In an industrial-scale fractionating column, **bubble caps** allow the falling liquid to equilibrate fully with the rising vapour. Note that both columns have a temperature gradient; they are cooler at the top and hotter at the bottom.

The viscosity of the different fractions increases with increasing dispersion forces, which in turn increase with the number of carbon atoms in a hydrocarbon molecule.
(a) The liquid fraction C_5 to C_8.
(b) The liquid fraction C_{20} to C_{30}.

SUMMARY

- Crude oil and natural gas are the major raw materials of the organic chemical industry. They contain a high proportion of alkanes.
- Fractional distillation separates crude oil into fractions – mixtures of hydrocarbons boiling within a given temperature range.
- The boiling point of a hydrocarbon depends largely on the number of carbon atoms in its skeleton.
- A fractionating column allows intimate mixing between ascending vapour and descending liquid.

22.2

OBJECTIVES

- Raoult's law
- Fractional distillation explained
- Deviations from Raoult's law
- Azeotropes

FRACTIONAL DISTILLATION AND RAOULT'S LAW

It is possible to separate the fractions of crude oil using fractional distillation. This spread concentrates on explaining in detail how it is possible to separate a mixture of volatile liquids, such as crude oil. The principle depends on a law called Raoult's law. Fractional distillation will successfully separate a mixture of liquids that follows Raoult's law. At the end of the spread, we consider what happens when a liquid mixture deviates from Raoult's law.

Raoult's law: vapour pressure/composition varies linearly

Raoult's law states that, at a constant temperature, the total vapour pressure above a mixture of two liquids varies linearly with the composition of the mixture expressed as mole fractions, see spread 11.8. The vapour pressure of each component of the mixture varies as follows:

$$p_A = x_A p^*_A$$

where p_A is the vapour pressure of component A, p^*_A is its vapour pressure when pure, and x_A is the mole fraction of A in the liquid mixture. This law was first introduced by François Raoult, who made painstaking measurements of the vapour pressures of various mixtures. Raoult's law applies very well when the liquids in the mixture have similar intermolecular forces, which is the case for hydrocarbons with similar numbers of carbon atoms.

- A mixture that follows Raoult's law is called an **ideal solution**.

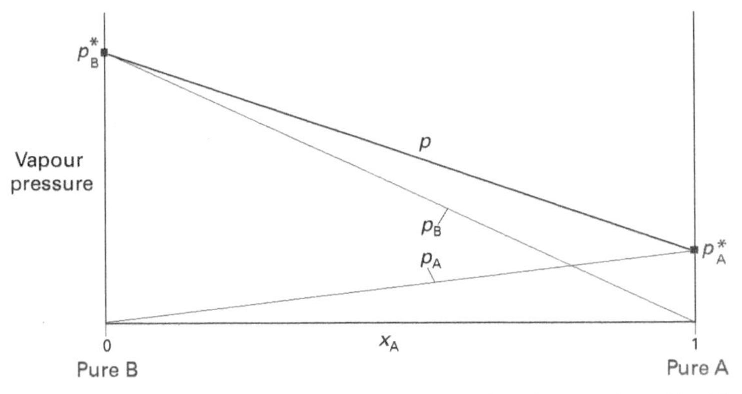

The vapour pressure at constant temperature of a mixture of two liquids varies linearly with the composition of the mixture expressed as mole fractions. This is Raoult's law.

Boiling-point curves

As the temperature increases, the vapour pressures of both the liquids in the mixture increase. However, the vapour pressure of the *more volatile* liquid increases more significantly, so a graph of the temperature at which the mixture boils when plotted against composition is a curve. Look at the diagrams alongside to see how this happens. This shows the conversion of a vapour pressure diagram into a boiling point curve, for a mixture of methanol and water.

The vapour above the boiling liquid mixture is richer in the more volatile component, so the composition of the vapour lies on the *other* side of the straight line joining the boiling points. This creates a cigar-shaped boiling-point/composition graph: see opposite.

To understand how to read this graph, consider heating a liquid of composition c_1 (which is richer in water, which has a higher boiling point). The vapour in equilibrium with the boiling liquid (at the same temperature) has the composition c_2. When this condenses, the liquid becomes richer in the more volatile component. If this process of boiling and condensing continues, the diagram shows that the mixture can be separated into its two components: pure methanol boils off first. When all the methanol has boiled off, water then starts to boil off.

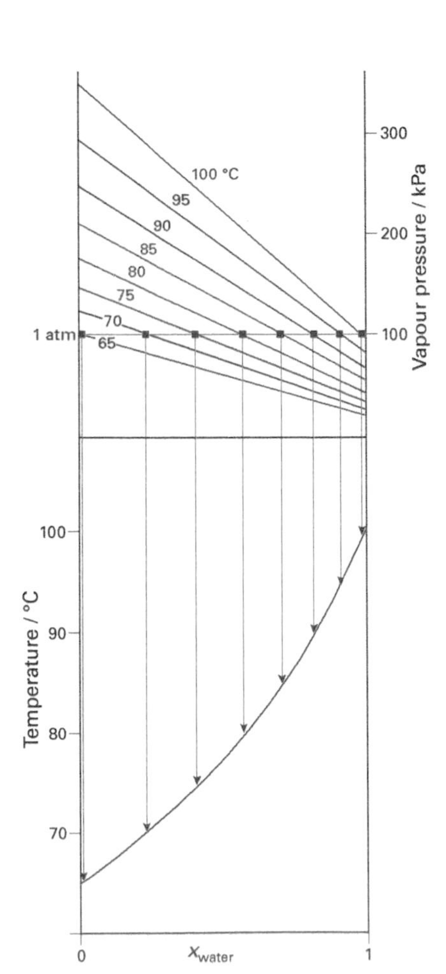

As the vapour pressure of the more volatile component (methanol) rises faster as the temperature is raised, the straight-line vapour pressure graph becomes a curved boiling-point graph.

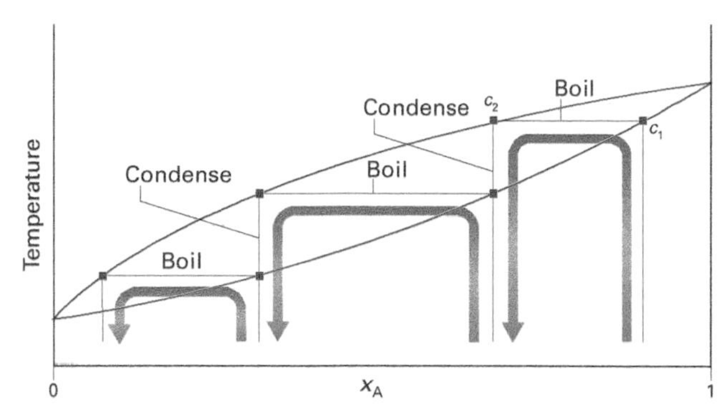

The vapour above the boiling liquid mixture is richer in the more volatile component. When the mixture of composition c_1 boils and then condenses, a liquid of composition c_2 forms.

The bubble caps in the industrial-scale apparatus, see previous spread, are the main means of establishing equilibrium at different temperatures between the falling liquid and the rising vapour. As long as the column is long enough and the number of bubble caps sufficient, it is possible to separate an ideal solution into its pure components.

- It is possible to separate an ideal solution into its pure components by fractional distillation.

Deviations from Raoult's law: azeotropes

Some liquid mixtures have boiling-point/composition graphs that do not look like the one shown above. Such mixtures are said to deviate from Raoult's law. For example, ethanol and water shows a graph which differs in a very important fashion from the methanol/water example discussed immediately above, despite the fact that ethanol is the next compound in the homologous series after methanol. There is a point on the graph, close to pure ethanol, where the vapour and liquid compositions become identical. Boiling this particular composition causes no further separation to occur: this particular mixture boils without change in composition and is called an **azeotrope**. Formation of an azeotrope occurs because the intermolecular forces between the components in the mixture vary too greatly.

Boiling an equal mixture (in terms of mole fraction) of ethanol and water causes a similar enrichment in the more volatile ethanol in a similar manner to that of an ideal mixture, until the azeotropic composition is reached, when no further separation occurs. Boiling a mixture of ethanol and water will produce the azeotropic mixture of 96% ethanol (by mass) and water.

- It is impossible to separate fully a mixture that forms an azeotrope by fractional distillation.

Further separation of the azeotropic mixture of ethanol and water requires a different strategy, such as using chemical drying by standing over calcium oxide.

Other mixtures form a maximum boiling azeotrope rather than a minimum boiling one, as was the case for ethanol/water. An example of a mixture forming a maximum boiling azeotrope is nitric acid/water. In this case, the deviation is more severe than for ethanol and the azeotrope occurs at 68 per cent nitric acid (by mass). So separation of an equal mixture of nitric acid and water by fractional distillation cannot concentrate the acid more than the azeotropic composition. This azeotropic mixture is typically labelled as concentrated nitric acid.

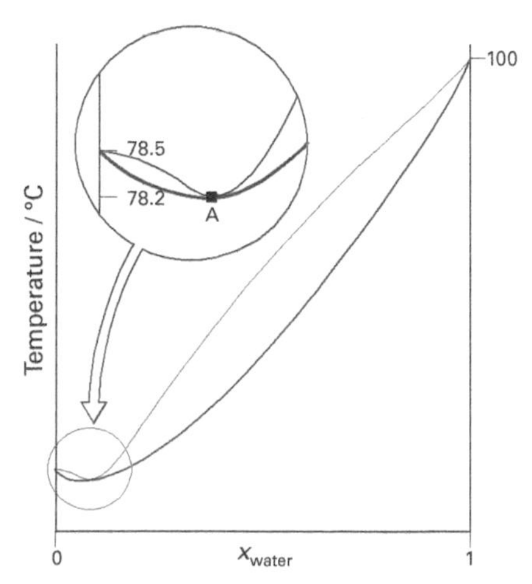

The boiling point / composition graph for ethanol and water. A marks the azeotrope.

SUMMARY

- Raoult's law: the vapour pressure of a liquid mixture varies linearly with its composition (as mole fractions).
- A mixture that follows Raoult's law is called an ideal solution.
- An azeotropic mixture boils without change in composition.

22.3

OBJECTIVES

- Alkanes are very unreactive
- Radical reactions
- Fractions and their uses
- Cracking

General formulae

The general formula for an alkane is C_nH_{2n+2}, where n is the number of carbon atoms. The general formula for an alkene is C_nH_{2n}.

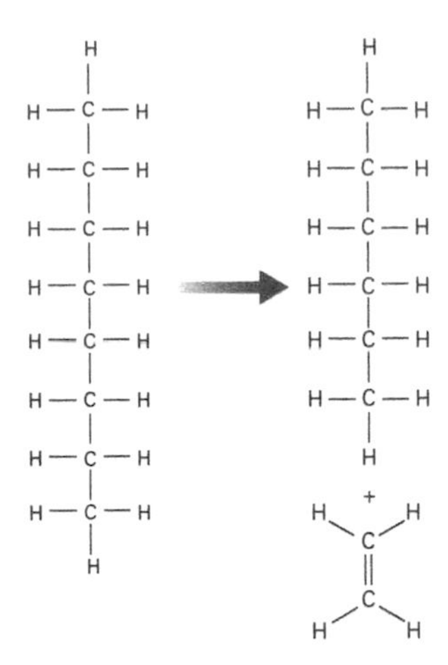

Structural formulae show how a molecule of octane splits to form hexane (an alkane) and ethene (an alkene).

Changing molecules

There are three main processes used to match the raw material, the fractions from crude oil, to the requirements of consumers.

- **Cracking** breaks large molecules into smaller molecules.
- **Reforming** rearranges the atoms in a molecule to make a new structure.
- **Polymerization** joins small molecules to make very large ones.

The oil industry: cracking

The alkanes in crude oil and natural gas are currently the most abundant and relatively cheap source of organic compounds. The alkanes, however, are very unreactive molecules. They contain C—C and C—H bonds only, and do not have any regions of high electron density because there are no double bonds. The bonds have low polarity, because carbon and hydrogen have similar electronegativities (C 2.5; H 2.2), so there are no centres of partial charge to attract electrophiles or nucleophiles. So, radicals are the only species reactive enough to overcome the high activation energy required to break the strong C—C and C—H bonds. The three most important radical reactions that alkanes undergo are:

- Thermal cracking
- Substitution with the halogens chlorine and bromine
- Combustion

This spread covers the topic of cracking; the next two spreads will deal with substitution and combustion.

Cracking

Carbon–carbon bonds are strong, so the skeleton of an organic molecule usually remains intact when reactions happen at functional groups. However, carbon–carbon bonds *can* be broken by cracking.

The table below shows that the percentage composition of crude oil does not match the market demand. One of the most important industrial processes at a refinery is **cracking** because it breaks down large molecules into smaller ones. When the original molecule is an alkane, the product molecules are a smaller alkane and an alkene. For example, octane could break into hexane and ethene:

$$C_8H_{18} \rightarrow C_6H_{14} + C_2H_4$$

or into pentane and propene:

$$C_8H_{18} \rightarrow C_5H_{12} + C_3H_6$$

Ethene and propene are particularly useful for making the polymers poly(ethene) and poly(propene), traditionally known as polythene and polypropylene respectively, spread 22.10.

There are two main cracking processes, which occur under different reaction conditions and proceed by different reaction mechanisms.

The composition of North Sea crude oil. The market demands a greater proportion of alkanes with smaller numbers of carbon atoms.

Fraction	Number of carbon atoms in molecules	Percentage of fraction from distillation	Approximate demand for fraction (%)	Typical use for fraction
Gas	1–4	2	4	refinery fuel
Gasoline	5–8	12	22	petrol
Naphtha	8–10	12	5	cracking
Kerosene	10–14	12	8	jet fuel
Gas oil	14–19	19	23	diesel
Fuel oil	19–35	43	38	power stations

Thermal cracking

In **thermal cracking** the alkane is heated to between 800 and 900 °C, for about half a second, sometimes in the presence of superheated steam. These high temperatures are sufficient to cause the strong carbon–carbon bonds to break. The reaction depends on the favourable entropy increase (see chapter 14 'Spontaneous change...') caused by forming two species in place of one. At high temperatures, this increase in entropy compensates for the unfavourable endothermic enthalpy change.

The species initially formed are radicals. A carbon–carbon bond undergoes homolytic fission, leaving each carbon with a single unpaired electron. For example, octane could break down into a hexyl radical and an ethyl radical:

$$C_8H_{18} \rightarrow C_6H_{13}^{\bullet} + C_2H_5^{\bullet}$$

The raised dot indicates an unpaired electron. These radicals can then undergo a number of reactions. For example, they can *lose a hydrogen atom* to form a stable molecule. The ethyl radical becomes ethene in such a reaction:

$$C_2H_5^{\bullet} \rightarrow C_2H_4 + H^{\bullet}$$

The hydrogen atoms, which are also radicals because they have an unpaired electron, may combine together to form hydrogen gas:

$$H^{\bullet} + H^{\bullet} \rightarrow H_2$$

Hydrogen gas is a useful by-product of thermal cracking.

The radicals formed may also *remove a hydrogen atom* from another molecule. For example, the ethyl radical can remove a hydrogen atom from octane to give ethane and an octyl radical:

$$C_2H_5^{\bullet} + C_8H_{18} \rightarrow C_2H_6 + C_8H_{17}^{\bullet}$$

There is always a mixture of products from cracking, including a high percentage of alkanes. However, the choice of reaction conditions does to some extent control the product mixture. The mixture is cooled and then separated by fractional distillation.

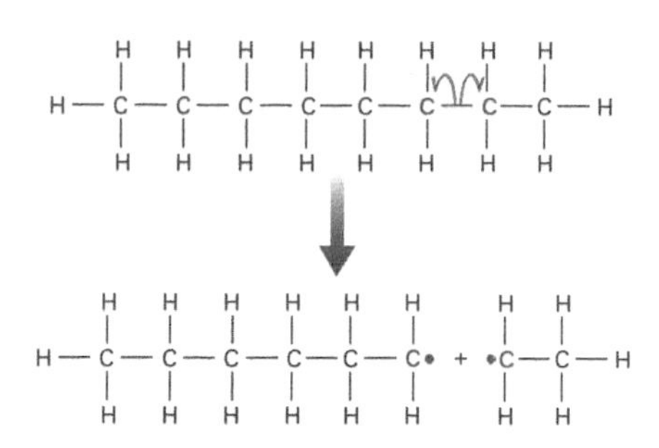

Structural formulae show how a molecule of octane splits to form two radicals – the hexyl and the ethyl radicals.

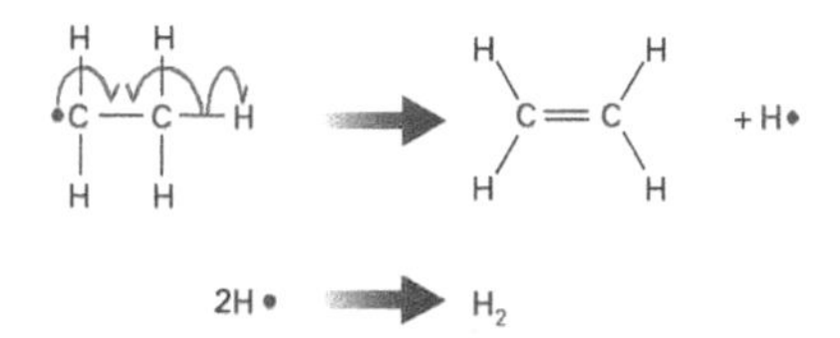

The ethyl radical loses a hydrogen atom to become ethene. Two hydrogen atoms then combine to form a molecule of hydrogen gas.

Catalytic cracking

Catalytic cracking (often abbreviated to 'cat-cracking') does not require such high temperatures as thermal cracking but does require a catalyst. A temperature of 500 °C is common, and the catalyst is often a finely divided mixture of silica SiO_2 and aluminium oxide Al_2O_3. The mechanism involves ions (carbocations, see spread 22.8) rather than radicals. Cracking dodecane $C_{12}H_{26}$ with a zeolite (a sodium aluminosilicate) catalyst gives a mixture in which just over half of the hydrocarbons contain four or five carbon atoms.

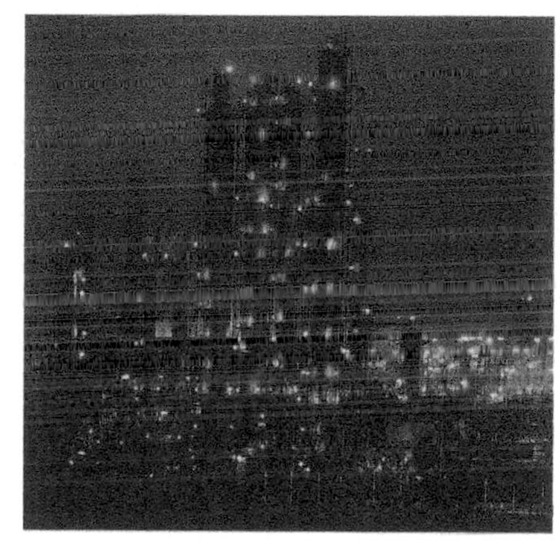

An industrial catalytic cracker operates at a temperature of 500 °C. The catalyst is often a mixture of SiO_2 and Al_2O_3.

Comparing thermal and catalytic cracking

The greatest difference between the products from the two types of cracking is that the carbon skeleton undergoes rearrangement to a greater extent in catalytic cracking. In a **rearrangement** reaction, carbon–carbon bonds are broken and reformed at different positions in the carbon skeleton, usually making it more branched. This result is put to good use in a related industrial process called reforming. The main purpose of **reforming** is to cause molecules to rearrange themselves to give branched-chain alkanes. These alkanes burn more steadily than their straight-chain isomers in internal combustion engines. Reforming also produces a significant quantity of aromatic hydrocarbons, such as benzene (see next chapter).

SUMMARY

- Alkane molecules are chemically unreactive because they do not have functional groups and the C—C and C—H bonds have high bond enthalpies.
- Cracking, reforming, and polymerization are used to match crude oil composition to market demand.
- Crude oil generally contains a higher proportion of relatively large molecules than the market requires.
- Cracking breaks large alkane molecules into a mixture of alkenes and smaller alkanes.
- Thermal cracking uses high temperatures alone; catalytic cracking uses more moderate temperatures and a catalyst.

PRACTICE

1 Use structural formulae to show how a molecule of octane forms pentane and propene.

22.4

THE ALKANES

OBJECTIVES

- Substitution by chlorine and bromine
- Combustion

Boiling points

The boiling points of the alkanes were explained in terms of dispersion forces in spread 7.4.

The alkanes have the general formula C_nH_{2n+2}. They lack functional groups, and so their chemistry is restricted to combustion and substitution reactions. Radicals are the only species reactive enough to overcome the high activation energy required to break the strong C—C and C—H bonds. The combustion of alkanes provides many of our energy requirements, from petrol burning in car engines to methane burning in domestic cookers. While the *overall* equations are fairly straightforward, combustion involves the formation and reaction of *intermediate* radicals in a series of extremely rapid reactions, which are complex and difficult to investigate experimentally. In this spread we will concentrate on the mechanism of the radical substitution of chlorine for hydrogen in an alkane, and only deal briefly with combustion, which will be considered in detail in the next spread.

Nomenclature

The overall reaction between an alkane and chlorine involves substitution of one or more chlorine atoms for hydrogen atoms. The reactive intermediates are radicals; the process is therefore called **radical substitution**. The reaction is termed 'dirty' as there is a mixture of products in which no one substance dominates.

Bromination

Bromination follows exactly the same mechanism as chlorination. In the case of iodination, substitution to form CH_3I is *endothermic* by 59 kJ mol^{-1} (for chlorination it is exothermic by 99 kJ mol^{-1}). The reaction is therefore energetically unfavourable; iodination does not occur under these conditions.

Substitution by the halogens chlorine and bromine

Halogen radicals such as $Cl^•$ or $Br^•$ may be produced from halogen molecules by shining ultraviolet light on the halogen (or by heating to high temperatures). Radicals form when ultraviolet light provides the energy needed to cause homolytic fission of the halogen–halogen covalent bond. Replacing hydrogen atoms in an alkane molecule with chlorine or bromine atoms is called **chlorination** or **bromination** respectively. For example, exposing a mixture of methane and chlorine to ultraviolet light results in the formation of chloromethane CH_3Cl:

$$CH_4(g) + Cl_2(g) \xrightarrow{\text{UV light}} CH_3Cl(g) + HCl(g)$$

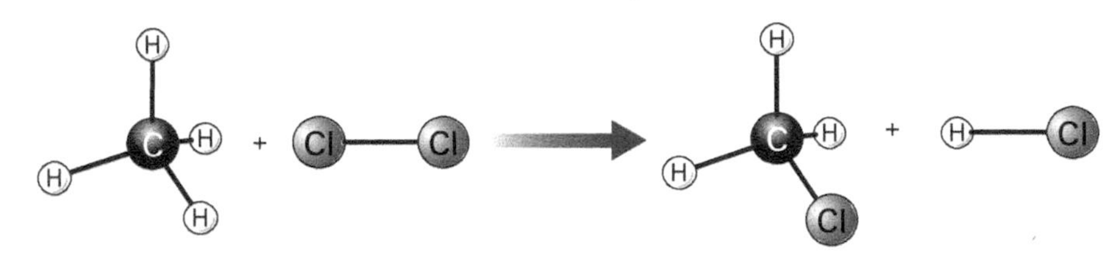

The radical substitution of chlorine in methane.

The mechanism of the chlorination of alkanes (chain reaction)

The reaction between an alkane and chlorine is a chain reaction: a **chain reaction** is a series of reactions in which a product of one reaction starts the next reaction in the chain.

The chain reaction is started when a reactive species is formed. The first step, called **chain initiation**, is the homolytic fission of a chlorine–chlorine bond to form two chlorine radicals, caused by absorption of ultraviolet light:

$$Cl_2 \xrightarrow{\text{UV}} 2Cl^• \qquad (1)$$

The chlorine radical is highly reactive and is likely to react with any radical or molecule that it meets.

The chlorine radical removes one of the hydrogen atoms from methane to form the stable molecule hydrogen chloride. The remaining species is the **methyl radical**, which has an unpaired electron resulting from homolytic fission of the bond:

$$Cl^• + CH_4 \rightarrow HCl + {}^•CH_3 \qquad (2)$$

The methyl radical is also highly reactive. It reacts with chlorine as follows:

$${}^•CH_3 + Cl_2 \rightarrow CH_3Cl + Cl^• \qquad (3)$$

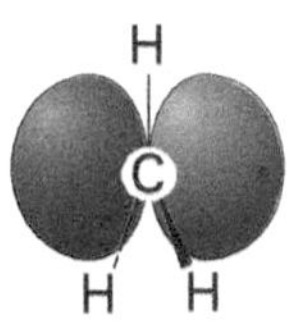

The methyl radical. Note that, whereas methane CH_4 is tetrahedral, the methyl radical ${}^•CH_3$ is planar.

Chain propagation and termination

The *pair* of reactions (2) and (3) is called **chain propagation** because the chlorine radical destroyed in reaction (2) is regenerated in reaction (3). The regenerated chlorine radical may then react with more methane. The reactions 'keep the chain going'; they *propagate* the chain. One CH_4 molecule and one Cl_2 molecule are consumed, generating one CH_3Cl molecule and one HCl molecule, as given by the overall equation above this box.

The alternating destruction and creation of chlorine radicals continues until one of several reactions ends, or terminates, the chain reaction. The **chain termination** reactions always involve *two* radicals reacting together to make a relatively unreactive molecule, e.g.

$^{\bullet}CH_3 + Cl^{\bullet} \rightarrow CH_3Cl$

$Cl^{\bullet} + Cl^{\bullet} \rightarrow Cl_2$

$^{\bullet}CH_3 + {}^{\bullet}CH_3 \rightarrow CH_3CH_3$

The presence of ethane in the product mixture provides strong evidence in support of this proposed mechanism.

Further substitution

Substitution reactions do not stop when just one hydrogen has been replaced. They continue until all the hydrogen atoms have been replaced. After some time, CH_3Cl molecules will be present in significant amounts. A chlorine radical colliding with CH_3Cl can produce HCl and a $^{\bullet}CH_2Cl$ radical, which can go on to form CH_2Cl_2. Further substitutions will result in a mixture of substitution products: CH_3Cl, CH_2Cl_2, $CHCl_3$, and CCl_4.

Combustion

As with the reaction of alkanes with chlorine, combustion is a radical chain reaction. The complication for combustion is that, as well as chain initiation, propagation, and termination, a further chain process called chain branching is possible.

Chain branching occurs when one reactant radical produces two product radicals. A common example is the reaction between a hydrogen molecule and an oxygen atom (which is a *biradical* $O^{\bullet\bullet}$ containing two unpaired electrons, spread 4.6) to produce a hydrogen atom (which is a radical $H^{\bullet}$) and the hydroxyl radical ($HO^{\bullet}$).

$$H_2 + O^{\bullet\bullet} \rightarrow H^{\bullet} + HO^{\bullet}$$

Chain branching can lead to an enormous increase in the rate of the reaction, because the number of radicals *doubles* after each reaction step. If the conditions during combustion favour chain branching, then an explosion occurs.

Cutting using a torch fuelled by propane.

SUMMARY

- Alkanes contain only C—C and C—H single bonds, which have high bond enthalpies.
- Only radicals (possessing unpaired electrons) are sufficiently reactive to break these bonds and force alkanes to undergo reaction.
- Chlorination and bromination substitute halogen atoms for hydrogen atoms in alkanes.
- Chlorination and bromination are radical chain reactions, involving the radicals $Cl^{\bullet}$ and $Br^{\bullet}$.
- Chain reactions proceed by means of chain initiation, chain propagation, and chain termination. (The exact steps for chlorination should be noted carefully.)
- Combustion is a complex radical chain reaction involving the oxygen biradical $O^{\bullet\bullet}$.

PRACTICE

1 Describe the mechanism of methane bromination.

22.5

Pluses and minuses of combustion

OBJECTIVES

- Liquid and gaseous fuels
- The 'greenhouse effect'
- Energy density of a fuel
- Biofuels
- Waste disposal

We saw in the last spread that the combustion reactions of alkanes provides many of our energy needs. However, we need to be aware of the problems that combustion can bring as well. This spread considers the advantages and disadvantages of combustion and the continuing search for the best alternatives.

Fuels

A good fuel must obviously have a high exothermic standard enthalpy change of combustion so that it is an efficient provider of heat. But that is not the only consideration; a fuel must also be safe, cheap, and practical to transport. Ideally then, the combustion reaction of a fuel must have a high activation energy, so that it does not easily catch fire when it is being transported, but it also needs to have a high energy yield for the mass or volume carried. Ideally too, a fuel should not contribute to pollution by producing toxic combustion products or those which will contribute to the 'greenhouse effect'.

Liquid fuels are relatively easy to transport and store compared to gases. The alkane octane C_8H_{18} and the alcohol ethanol CH_3CH_2OH are both liquids which can be used as fuels. However, their combustion reactions both produce CO_2 which is a greenhouse gas. Hydrogen gas H_2 is also a good fuel, but it is awkward to transport and store and forms an explosive mixture with air. Hydrogen has the added advantage that it burns to form water which does not pollute. The alkanes methane CH_4 and butane C_4H_{10} are gases used as fuels. Methane is the chief constituent of the natural gas which is piped to our houses. Methane is cheap to produce because it occurs in large amounts naturally. Butane is bottled under pressure so that it liquefies, and is familiar to many as camping gas. However, methane and butane both produce CO_2 when you burn them and are greenhouse gases in their own right.

Waste disposal

There are two basic methods available for the disposal of waste, landfill and incineration.

In **landfill**, the waste is compacted into natural or man-made cavities in the landscape. The most obvious objection is the eye-sore that this represents and the encouragement it gives to rats and other vermin. Microbiological action in organic landfill waste leads to the production of methane which is an explosion and fire hazard. Methane is also a greenhouse gas and, to avoid it being released to the atmosphere, action is being taken to collect it from landfill sites and either flame it off or use it as fuel. Microbiological action can also lead to toxic products such as hydrogen sulphide gas. Liquid leaching from landfill sites and polluting the ground water is yet another problem. This danger can be lessened by lining the site with impermeable material.

Incinerators use combustion reactions to dispose of waste.

Calculating the energy density of a fuel

The **energy density** of a fuel can be calculated as the energy generated per unit mass or per unit volume of the fuel. For example, if 0.200 g of fuel yields 15.7 kJ of heat, the energy density is given by:

energy density $= 15.7 \times 1000/0.200$

$= 78\,500\ \text{kJ kg}^{-1}$

Equally, if 0.200 cm^3 of fuel yields 15.7 kJ of heat, the heat from 1 dm^3 of fuel will be given by the same calculation except the units will be kJ dm^{-3}.

Carbon capture and storage

There is a need in the immediate future to reduce the amount of carbon dioxide released into the atmosphere, especially as the result of burning fossil fuels. One contribution to the reduction in CO_2 levels would be to capture the carbon dioxide emitted by point sources, such as power stations, and then store it permanently without returning it to the atmosphere. In September 2008 the first power station incorporating **carbon capture and storage** (CCS) technology started operation in the Lausitz region of Germany. It is a coal-fired power station with a modest output of 30 megawatt which compresses the CO_2 produced by a factor of 500 and then pumps it into a depleted gas field. As of 2008, coal power stations still produce around 50% of the UK's energy needs, so the ability to reduce the emitted CO_2 levels drastically is currently being very carefully investigated.

In 2014 two possible CCS projects were being investigated in the UK. In one, an existing gas-fired power station in Peterhead, Aberdeenshire, Scotland proposed to capture its emissions and transport the CO_2 100 km offshore to be buried in the depleted Goldeneye gas reservoir: CO_2 emissions should reduce by over 90%. The other proposal concerned the new-build White Rose project, co-fired by coal and sustainable biomass, in Drax, Selby, North Yorkshire. Both projects floundered at the end of 2015 after the withdrawal of a promise of government funding.

Landfill is an increasingly expensive way of disposing of wastes, due to the shortage of suitable sites and the cost of meeting higher environmental standards. So increasing amounts of waste are being burnt in incinerators. Incineration sterilizes the material for final disposal and reduces it in volume by about 90 per cent. Also, most new incinerators are designed to use the heat from burning rubbish to generate electricity.

However, it is the products of combustion that go up the incinerator flue into the atmosphere which must concern us. A large proportion of domestic waste is organic in origin and produces CO_2 (and H_2O) on combustion and so contributes to the greenhouse effect. Toxic gases such as carbon monoxide together with soot are produced by incomplete combustion. Sulphur-containing compounds produce sulphur dioxide and nitrogen-containing compounds produce NO_2; both can cause acid rain, spread 19.7. But by far the greatest concern is that an insufficiently high temperature can lead to chlorinated compounds producing dioxins.

Dioxins is the name given to a whole range of highly toxic compounds that consist of two benzene rings connected via two oxygen atoms.

dibenzodioxin

TCDD, tetrachlorodibenzodioxin

TCDD, tetrachlorodibenzodioxin, is a particularly dangerous product of incomplete combustion of the chlorinated polymer PVC.

In 1977 it was discovered that burning wood produced some dioxins. Burning anything that has a combination of carbon, oxygen, hydrogen, and chlorine, such as the polymer PVC, has the potential of producing dioxins. The largest source of dioxins in cities has been incinerators for burning domestic waste. However, incinerators have been greatly improved in recent years by passing the flue gases through carbon filters. This reduces the amount of dioxin released to trillionths of a gram of dioxin per cubic metre of gas emitted.

This negative effect must be balanced against the positive effects that a modern incinerator burns about 750 000 tonnes of waste a year, and generates electricity at the same time. The energy it provides saves the burning of 300 000 tonnes of fossil fuel. Added to this, an incinerator may in fact destroy more dioxins in the waste than it emits to the atmosphere.

SUMMARY

- A good fuel must have a highly exothermic standard enthalpy change of combustion.
- A good fuel must be cheap and easy to transport safely.
- Biofuels may have net zero effect on atmospheric CO_2.
- Incineration has advantages over landfill in waste disposal.

PRACTICE

1 Discuss the advantages and disadvantages of liquid and gaseous fuels.

2 Discuss the merits of incineration over landfill as a method for disposing of domestic waste.

A reminder about the greenhouse effect

Just as the glass of a greenhouse prevents the energy entering the greenhouse from the Sun being re-radiated out, so that the greenhouse warms up, certain gases in the atmosphere keep the Earth warm by the same effect – called the 'greenhouse effect'. However, there is concern that build up of certain gases, particularly CO_2, is causing an *enhanced* greenhouse effect which leads to 'global warming'.

Biofuels

These are fuels derived from biological sources. For example, a fuel known as 'biodiesel' (spread 27.4) is made from the oil extracted from rape seeds. Another biofuel is wood taken from sustainable forests. These fuels release CO_2 when they burn, but because the plants they come from also fix CO_2 from the atmosphere while they grow and photosynthesize, their use can have a net zero impact on atmospheric CO_2 levels and so does not contribute to the greenhouse effect. A **carbon-neutral biofuel** (see spread 25.4 for more details) would cause no net annual carbon emissions to the atmosphere.

A reminder about pollution from car engines

Because petrol and diesel consist mostly of alkanes, burning them releases CO_2, as we have seen. The poisonous gas carbon monoxide, CO, is produced when alkanes are burned in a limited supply of oxygen (incomplete combustion) as can happen in an internal combustion engine. Internal combustion engines also pollute the atmosphere with oxides of nitrogen (collectively known as NO_x) and unburned hydrocarbons. Most of these pollutants can be removed by fitting a catalytic converter to the exhaust system of the engine, spread 15.9.

22.6

OBJECTIVES

- Structure from VSEPR theory
- σ and π bonds
- *Cis–trans* isomers

Bonding in alkenes

Lewis structures give a reasonable model for the structures of methane and ethane; and the **VSEPR** theory predicts that methane has tetrahedral geometry. However, the final spread in the chapter shows that a fuller account of the bonding is given by considering the overlap of atomic orbitals to give molecular orbitals. Bonds form between atoms when these molecular orbitals are occupied by pairs of electrons. In ethane, a C—C single bond forms between the two carbon atoms. However, ethene has a C=C double bond. This spread investigates the bonding in ethene, first by the Lewis/VSEPR theory and then by using the idea of molecular orbitals.

Lewis structures and electron-pair repulsions

Ethene has the molecular formula C_2H_4. A Lewis structure suggests that each carbon atom can complete its octet of electrons by forming one C=C double bond and two C—H single bonds. The H—C—H angle at each end of the molecule is about 118° rather than the 120° expected from a regular trigonal planar shape. This is because the C=C double bond repels the C—H single bonds to a slightly greater extent than would a C—C single bond (as a result of the increased electron density in the C=C double bond).

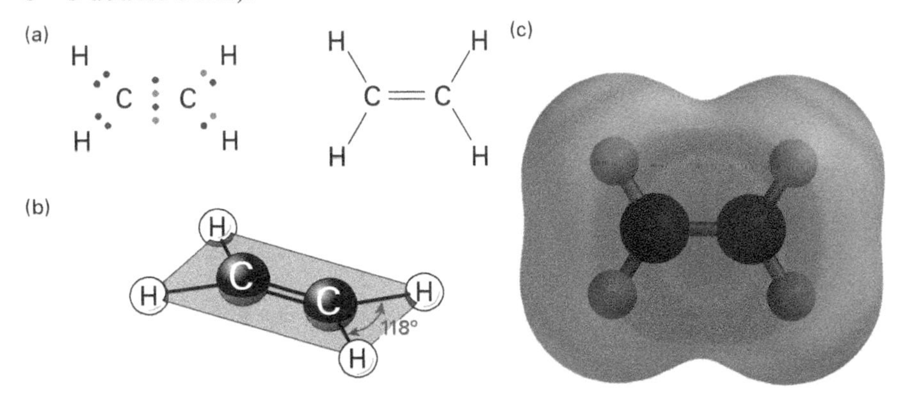

a) The Lewis structure of ethene shows that each carbon atom donates two electrons to form one C=C bond and two more electrons to form two C—H bonds. (b) The ethene molecule has trigonal planar geometry. (c) The electrostatic potential map for ethene. Note the red colour indicating the high electron density in the double bond.

Bonding electrons

Ethene consists of two carbon atoms and four hydrogen atoms. Four electrons ($2s^22p^2$) are available for bonding from each carbon atom; each hydrogen atom has the $1s^1$ electron available. The total number of bonding electrons to be contained in the molecular orbitals is therefore $2 \times 4 = 8$ from the carbon atoms and $4 \times 1 = 4$ from the hydrogen atoms, making a total of 12 electrons. They form six bonds: a C=C double bond and 4 C—H single bonds.

Ethene: overlapping atomic orbitals

As in spread 5.7, imagine two carbon atoms approaching each other. The first atomic orbitals to overlap are the 2p orbitals that point along the line of approach. A **sigma** (σ) **molecular orbital** results, which is symmetrical about the internuclear axis. This molecular orbital contains two electrons that bond the carbon atoms to each other, forming a single **sigma** (σ) **bond**.

As the carbon atoms approach even closer, the two 2p orbitals at right angles to those forming the sigma bond overlap *sideways* above and below the sigma bond. This overlap gives rise to another sort of molecular orbital called a **pi** (π) **molecular orbital**. It contains a pair of electrons that constitute a **pi** (π) **bond**. The π orbital is at a higher energy level than the σ orbital. The two carbon atoms now have a double covalent bond between them, consisting of one sigma bond and one pi bond; they are sharing two pairs of electrons. The carbon–hydrogen bonds form in a way similar to those in methane, see the final spread in the chapter.

The π molecular orbital of ethene.

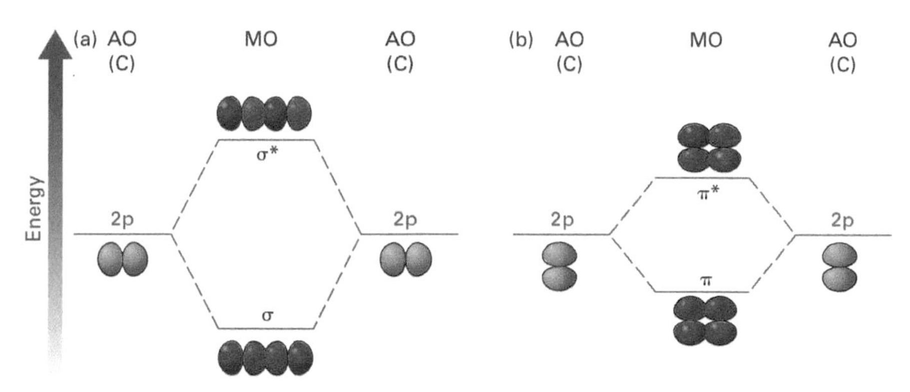

The formation of bonding (and antibonding) molecular orbitals in ethene: (a) the C—C sigma (σ) bond; (b) the C—C pi (π) bond. The C—H bonds form as described for methane in the spread 'Bonding in methane' at the end of the chapter.

trans-but-2-ene *cis*-but-2-ene

Cis–trans *isomers of but-2-ene C_4H_8. The chemical reactions of the double bond in these two isomers are the same. But their physical properties are slightly different:* cis-but-2-ene *boils at 4 °C whereas* trans-but-2-ene *boils at 1 °C because the molecules pack together differently, with resulting different values of dispersion force. There is recent strong evidence that consumption of* trans *unsaturated fatty acids, spread 30.1, can increase total cholesterol levels, with a consequent increased risk of coronary heart disease.*

The C=C bond and isomerism

Rotation about a C—C single bond is possible. For example, the two $—CH_3$ groups in ethane may alter their relationship to each other as they 'spin' like propellers at each end of the molecule. But twisting a π bond would weaken it, as the sideways overlap would be reduced. So, the π bond is rigid, and rotation about a C=C double bond is not possible. This *restricted rotation* about a double bond means that the two $=CH_2$ groups in ethene are fixed with respect to each other. As a result, there is a type of isomerism common in alkenes.

Compounds that contain double bonds can show *cis–trans* isomerism (formerly known as geometric isomerism). The groups around a double bond are fixed in space relative to each other because the double bond is a rigid structure and the carbon atoms at each end cannot move relative to each other. For example, the but-2-ene molecule has two alternative arrangements in space: *cis*-but-2-ene has the two methyl groups on the *same* side of the double bond, whereas *trans*-but-2-ene has the two methyl groups on *opposite* sides of the double bond.

E–Z nomenclature

An alternative notation uses *E* (German *entgegen*) instead of *trans*. *Z* (German *zusammen*) is used instead of *cis* (a mnemonic is that the groups are on **z**e **z**ame **z**ide). The *E–Z* notation is considered to be superior as it can be extended to deal with double bonds attached to three or four different groups, by assigning the groups different priorities (based on the Cahn–Ingold–Prelog priority rules, spread 28.7). The *Z* isomer has the two highest ranked groups on the same side.

Retinol

The chemical name for vitamin A is retinol. Green plants contain compounds, particularly the red/orange carotenes found in carrots and tomatoes, that the body can change into the vitamin. Retinol is converted into the aldehyde retinal in the liver. Retinal is a constituent of rhodopsin, the light-sensitive pigment that occurs in the rod cells in the retina of the eye. Part of the structure of retinal consists of four alternating single and double carbon–carbon bonds. A key reaction in the process that enables us to see is the conversion of a *cis-* arrangement of one of these double bonds to the *trans-* arrangement.

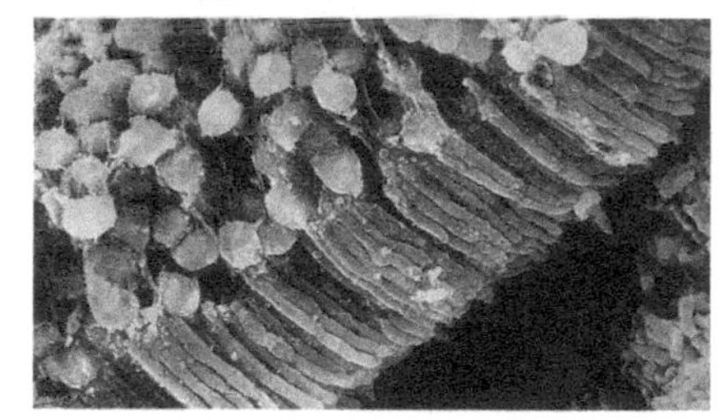

*Rod cells in the human retina. This type of light-sensitive cell contains rhodopsin and is responsible for vision in dim light. When light reaches rhodopsin, it converts 11-*cis*-retinal to all-*trans*-retinal and the rhodopsin molecule breaks up. This causes electrical changes in the rod cell, which can set up a nerve impulse. This is the first part of the complex process of sight in all animal eyes. Maximum sensitivity of these cells occurs when the light input to the retina is 10 000 quanta per second, equivalent to one quantum per rod cell every three minutes!*

SUMMARY

- A sigma (σ) molecular orbital results from s–s, s–p, or p–p atomic orbital overlap. It is symmetrical about the axis between the two atomic nuclei.
- A pi (π) molecular orbital results from sideways p–p atomic orbital overlap. It is not symmetrical about the axis between the two atomic nuclei.
- Rotation about a C—C single bond is possible, but rotation about a C=C double bond is restricted.
- Many alkenes show *cis–trans* isomerism as a result of the restricted rotation about the C=C double bond.

PRACTICE

1. Draw a Lewis structure showing the valence-shell electronic structure for propene.
2. Sketch the shape of the propene molecule indicated by VSEPR theory, showing the bond angles.
3. Explain why there are no geometrical isomers of but-1-ene.

22.7

ALKENES AND ADDITION – 1

OBJECTIVES

- Radical addition
- Electrophilic addition
- Bromine water test

Alkenes have the general formula C_nH_{2n} compared with C_nH_{2n+2} for the alkanes. Alkenes have greater overall reactivity than alkanes. The carbon–carbon double bond in alkenes is a centre of high electron density, which makes this homologous series more reactive. Like alkanes, alkenes can react by slow radical substitution (as discussed earlier in this chapter). However, alkenes undergo more rapid reactions involving the double bond between the carbon atoms. This functional group allows alkenes to react by *addition*, in which an incoming group splits and joins with the carbon atoms at each end of the double bond.

Addition of hydrogen to alkenes

Mixtures of hydrogen and an alkene are stable at room temperature. However, in the presence of a catalyst of finely divided nickel at about 150 °C, hydrogen adds across the double bond to form an alkane. For example, ethene is *hydrogenated* to ethane:

$$CH_2{=}CH_2(g) + H_2(g) \xrightarrow{Ni} CH_3CH_3(g)$$

Alternative catalysts for this **hydrogenation reaction** are palladium and platinum, which can be used at room temperature.

The mechanism of hydrogenation involves radicals, and the reaction between ethene and hydrogen is classed as **radical addition**. Ethene adsorbs on to the surface of the catalyst. Hydrogen molecules also adsorb, breaking into hydrogen atoms, which are radicals. One hydrogen atom combines with ethene to form an ethyl radical, which remains adsorbed on the catalyst. When a second hydrogen atom combines, the resulting ethane desorbs from the catalyst.

The addition of hydrogen to unsaturated vegetable oils is used industrially to make margarine. Hydrogenation lowers the number of double bonds in the polyunsaturated vegetable oils, 'hardening' the substance to make it a solid at room temperature.

Cycloalkenes

Cyclohexene, a six-membered ring containing a double bond, can be hydrogenated to cyclohexane, see spread 23.1.

Polyunsaturated

Polyunsaturated oils contain two or more double bonds. The prefix 'poly' usually means 'many'!

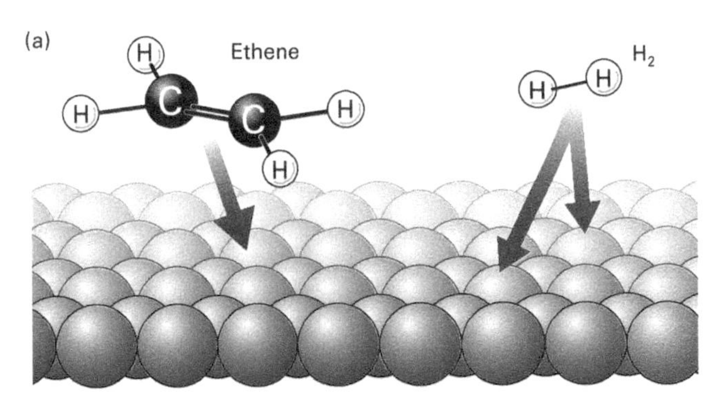

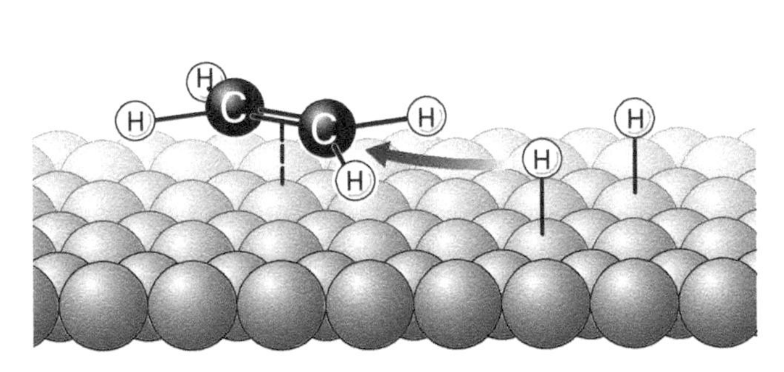

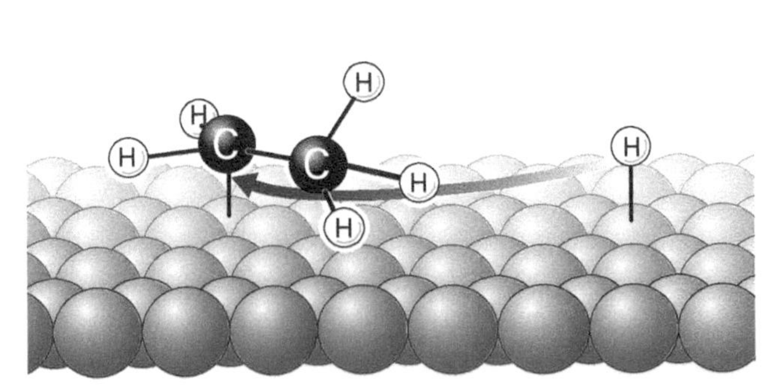

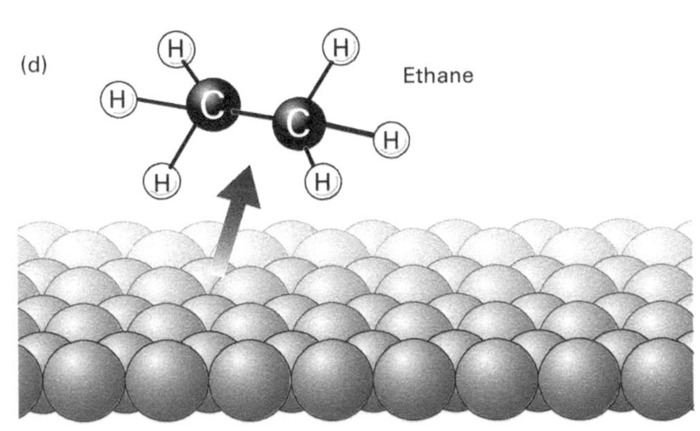

(a) Hydrogen and ethene adsorb on to the metal surface by donating electron density into vacant d orbitals on the catalyst metal atoms. (b) and (c) Hydrogen atoms (radicals) interact with the electron density in the ethene double bond, and bond with the carbon atoms. (d) The product ethane desorbs from the surface of the catalyst.

Addition of hydrogen halides or halogens to alkenes

Slow substitution between an alkane and a halogen results in a halogenoalkane. The results are unpredictable because substitution can replace more than one hydrogen atom on the alkane molecule, producing a mixture of products. Reacting an alkene with a hydrogen halide is a much more rapid reaction with a more predictable outcome. For example, ethene and hydrogen bromide react in aqueous solution:

$$CH_2{=}CH_2 + HBr \rightarrow CH_3CH_2Br$$

Hydrogen bromide ionizes in aqueous solution

$$HBr(aq) \rightarrow H^+(aq) + Br^-(aq)$$

and the reaction between ethene and hydrogen bromide is classed as **electrophilic addition** because the approaching species is the H^+ ion, which is attracted to the high electron density in the alkene C=C double bond. The full mechanism of the reaction is given in the following spread.

A good yield of product is obtained by bubbling ethene gas through concentrated hydrobromic acid and then distilling the mixture to isolate bromoethane. This reaction is straightforward to perform in industry because ethene is readily available from the cracking of the naphtha fraction of crude oil. The reaction takes place at room temperature and pressure.

Reacting a halogen (rather than a hydrogen halide) with an alkene also results in addition. For example, bromine and ethene react to form 1,2-dibromoethane:

$$CH_2{=}CH_2 + Br_2 \rightarrow BrCH_2CH_2Br$$

This reaction is also classed as an electrophilic addition, as will be explained in the next spread.

Molecules of sunflower oil (right) contain greater numbers of C=C double bonds than are found in peanut oil: sunflower oil is more polyunsaturated. As a result, sunflower oil decolorizes a dilute solution of iodine in ethanol more rapidly than peanut oil does.

The bromine water test

Bromine dissolves in water to form a red/brown solution, which is referred to in the laboratory as 'bromine water'. This solution is used in the '**bromine water test**' to distinguish between saturated and unsaturated hydrocarbons, which is carried out as follows:

1 Place the test substance in a test tube: if a gas, fill and stopper the tube; if a liquid, use a few drops.

2 Add a 1 cm depth of bromine water: shake, and allow the contents of the tube to settle.

3 If the test substance contains carbon–carbon double bonds, the liquid in the tube will be colourless. *An alkene decolorizes bromine water.* If the test substance is saturated, the colour of the bromine water will be unaffected.

A similar test with an alcoholic solution of iodine is used to detect double bonds in unsaturated oils (see top right).

Ethene decolorizes bromine water.

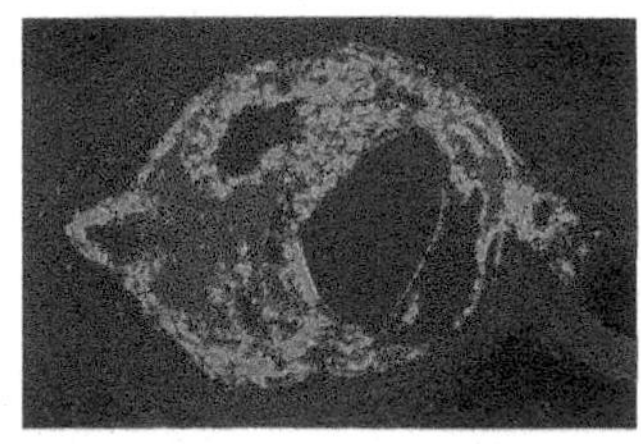

The presence of saturated fats in the diet can cause excess cholesterol in the body. Raised levels of blood cholesterol are associated with atherosclerosis, a condition in which fatty substances are deposited on the insides of artery walls, slowing and eventually stopping the flow of blood through the artery.

SUMMARY

- When mixed with a halogen such as bromine, alkanes undergo slow substitution reactions whereas alkenes undergo rapid addition reactions.
- Hydrogen halides add to alkenes to make the halogenoalkane; halogens can also add to alkenes.
- An alkene will rapidly decolorize bromine water.

PRACTICE

1 Explain why the catalytic addition of hydrogen to ethene is described as 'radical addition', whereas the addition of hydrogen bromide is described as 'electrophilic addition'.

2 Describe what you would expect to *see* when propene gas is bubbled into an aqueous solution of bromine.

22.8

ALKENES AND ADDITION – 2

OBJECTIVES

- Mechanism: HBr + alkene
- Mechanism: Br_2 + alkene
- Markovnikov's rule

Remember that a double-headed curly arrow indicates the movement of an electron pair.

Alkanes undergo substitution by means of radical chain reactions involving species with unpaired electrons. Alkenes undergo addition reactions. The previous spread gave details of several alkene reactions, including the addition of hydrogen bromide and bromine to ethene. You were also introduced to the bromine water test for unsaturation. This spread explains the mechanisms underlying these reactions. Addition reactions also proceed by a series of steps. However, the species involved are not radicals but *ions*.

Addition of hydrogen bromide to ethene

Step (1): Electron density in the ethene π bond attracts the H^+ ion (an electrophile).

When gaseous HBr is dissolved in a solvent such as dichloromethane, a similar reaction happens with the δ+ hydrogen atom of the HBr molecule.

Step (2): The final product forms when the carbocation reacts with the bromide ion (a nucleophile).

Hydrobromic acid is a strong acid, spread 12.4, and its solution therefore contains a high concentration of the ions $H^+(aq)$ and $Br^-(aq)$. The hydrogen ion is an electrophile, spread 21.5, because it has a positive charge. When ethene is bubbled through aqueous hydrobromic acid, the first step is the reaction between ethene and the H^+ electrophile:

$$CH_2{=}CH_2 + H^+ \rightarrow CH_3CH_2^+ \quad (1)$$

The resulting species contains a carbon atom bearing a positive charge. Because the charge is positive, the species is called a **carbocation** (sometimes less accurately termed a carbonium ion). The carbon–carbon linkage in $CH_3CH_2^+$ has a single σ bond in place of the double bond present in $CH_2{=}CH_2$. The electron pair that formed the π bond between the two carbon atoms now forms a carbon–hydrogen bond.

The carbocation can be attacked by a nucleophile, which provides an electron pair to form the new bond. The bromide ion in the solution is a nucleophile, and in the second step of the reaction it attacks the carbocation to form bromoethane:

$$CH_3CH_2^+ + Br^- \rightarrow CH_3CH_2Br \quad (2)$$

The overall result of these two steps is the addition of hydrogen and bromine across the double bond. The reaction is classed as an **electrophilic addition**, because the *first* step involves adding an electrophile (H^+).

Addition of bromine to ethene

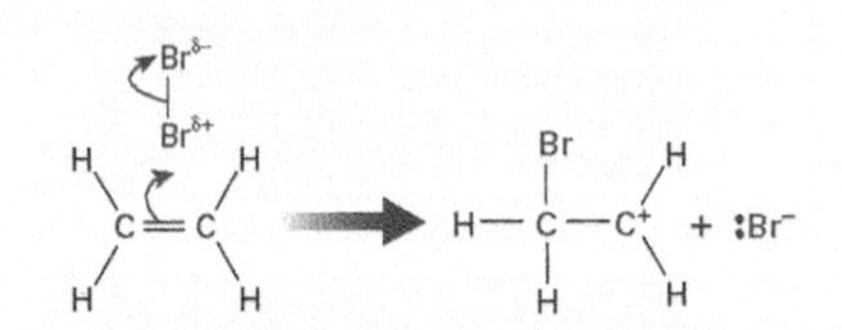

Step (1): The bromine molecule becomes polar as it approaches the region of high electron density in the ethene π bond. Ultimately, the π electrons form a bond with the δ+ bromine atom; the electron pair in the Br–Br bond migrates to the δ– bromine atom, giving it a full negative charge and breaking the bond.

Step (2): The final product forms when the carbocation reacts with the bromide ion (a nucleophile). This step is virtually identical to that shown in step (2) in the previous mechanism box.

The addition of bromine to ethene is another example of electrophilic addition. The bromine–bromine bond is not polar, so bromine is not apparently an electrophile. However, as the bromine molecule approaches the high electron density of the double bond, the electrons of the bromine–bromine bond are repelled: the approaching end of the molecule becomes partially positively charged. This end of the molecule acts as an electrophile and is attracted closer to the electron density in the double bond. The closer the bromine molecule approaches to the alkene π bond, the more polar it becomes. The bromine–bromine bond ultimately undergoes heterolytic fission; at the same time, a bond forms between one of the carbon atoms and the nearer ($Br^{\delta+}$) bromine atom, using the electrons of the double bond. The result is a carbocation and a bromide ion:

$$CH_2{=}CH_2 + Br_2 \rightarrow BrCH_2CH_2^+ + Br^- \quad (1)$$

The bromide ion now attacks the carbocation. The final result is the addition of two bromine atoms across the double bond:

$$BrCH_2CH_2^+ + Br^- \rightarrow BrCH_2CH_2Br \quad (2)$$

In the bromine *water* test, water (a nucleophile) is present in such huge excess that it attacks the carbocation instead of the bromide ion in an alternative second step. After loss of a proton (H^+), 2-bromoethanol forms:

$$BrCH_2CH_2^+ + H_2O \rightarrow BrCH_2CH_2OH_2^+ \rightarrow BrCH_2CH_2OH + H^+ \quad (2')$$

Markovnikov's rule

There is only one possible product in the electrophilic addition of HBr to ethene because the ethene molecule is a 'symmetrical' alkene – you cannot distinguish the two ends of the molecule either side of the double bond. If, however, the alkene molecule is *not* symmetrical about its double bond, two different products could result.

Where there is a choice of product in an electrophilic addition, the product that will be dominant can be predicted by **Markovnikov's rule**:

- The dominant product of electrophilic addition of HX to an unsymmetrical alkene has the hydrogen atom attached to the carbon atom that had more hydrogen atoms at the start.

Markovnikov's rule

Vladimir Markovnikov introduced his rule in the same year (1869) as Dmitri Mendeleyev introduced the periodic table.

Explanation for Markovnikov's rule

We shall consider the case of adding hydrogen bromide HBr to propene $CH_3CH{=}CH_2$. Does the Br atom add to the CH= carbon atom or to the $=CH_2$ carbon atom? In practice, the addition of HBr to propene results in a much greater quantity of 2-bromopropane than 1-bromopropane:

$CH_3CH{=}CH_2 + HBr \rightarrow CH_3CHBrCH_3$ (almost exclusively)
$CH_3CH{=}CH_2 + HBr \rightarrow CH_3CH_2CH_2Br$ (very little)

The explanation for this lies in the effect of an alkyl group (such as a methyl group) on the electron density in a molecule, compared with the effect of a hydrogen atom. A hydrogen atom has no electrons other than those in its bond. A methyl group has electron pairs in the three carbon–hydrogen bonds. In a carbocation, these electrons can be attracted towards the positive charge. Electron density feeds towards the charge, lowering the overall value of the positive charge on carbon and stabilizing the ion. A methyl group is described as **electron-donating** (or sometimes as having a **positive inductive effect**).

In the first step of the addition of HBr to propene, two possible carbocations may be formed:

$$CH_3CH{=}CH_2 + H^+ \rightarrow CH_3CH^+CH_3 \quad (1)$$
$$CH_3CH{=}CH_2 + H^+ \rightarrow CH_3CH_2CH_2^+ \quad (2)$$

The dominant carbocation (1) identified by Markovnikov's rule has *two* methyl groups attached to the charged carbon, which can stabilize the charge. The second carbocation (2) has only one alkyl group (an ethyl group) and two hydrogen atoms attached to the charged carbon. The carbocation is less stable because the charge is stabilized less. The **more stable carbocation** goes on to be attacked by a bromide ion to give the dominant product:

$$CH_3CH^+CH_3 + Br^- \rightarrow CH_3CHBrCH_3 \quad (3)$$

(a) A hydrogen ion could bond to a terminal carbon atom to form this carbocation. (The charged atom is attached to two alkyl groups.)

(b) A hydrogen ion could bond to the central carbon atom to form this carbocation. (The charged atom is attached to one alkyl group.)

(c) A bromide ion is attracted to the more stable carbocation and forms the product.

SUMMARY

- The first stage of an alkene electrophilic addition reaction involves the C=C π bond reacting with an electrophile; the result is a positively charged carbocation.
- The second stage of an alkene electrophilic addition reaction involves attack on the charged carbocation by a nucleophile.
- Markovnikov's rule: The dominant product of electrophilic addition of HX to an unsymmetrical alkene forms from the more stable carbocation.

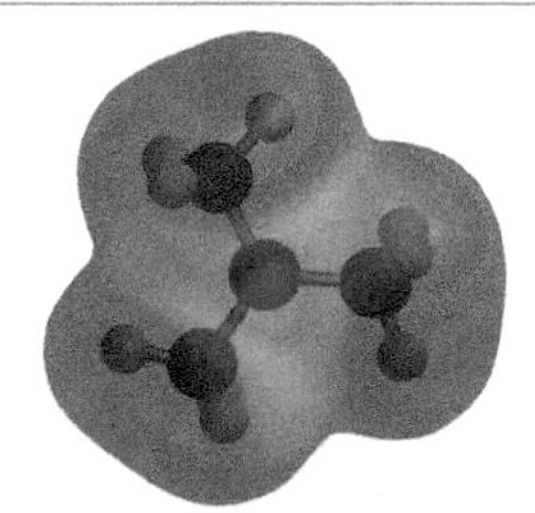

The electrostatic potential map for the carbocation $(CH_3)_3C^+$, which is very stable because of three electron-donating methyl groups.

PRACTICE

1 Explain why carbocations are attacked by nucleophiles and not by electrophiles.

2 Suggest reasons why alkene addition reactions are generally faster than alkane substitution reactions.

3 Write chemical equations involving structural formulae to explain the mechanisms in the reactions between the following:

a Propene and bromine

b Propene and hydrogen bromide

c But-1-ene and hydrogen bromide.

22.9

OBJECTIVES

- Ethanol from ethene
- Diols

Converting alkenes to alcohols

The addition of water to an alkene produces an alcohol: the reaction is called **hydration**. There is an abundant supply of alkenes in industry, and alcohols are important solvents. This reaction is therefore used in industry to produce millions of tonnes of various alcohols every year. The reaction may also be carried out in the laboratory under somewhat different conditions, the products of the reaction depending on the actual conditions chosen. (The mechanisms for some reactions are enclosed in boxes, which you may ignore if you do not require this level of detail.)

Ethanol: the industrial process

Ethanol is important in industry because it is widely used as a solvent. The main method of making ethanol is the **direct hydration of ethene**.

Hydration of ethene

An H^+ ion from the catalyst (phosphoric acid) reacts with the π bond of ethene. The ethene is protonated as it forms a bond with the hydrogen ion. As the proton H^+ does not have any electrons to bring to a bond, both of the bonding electrons come from the carbon–carbon π bond. As a result, the carbon atom that does not bond to the incoming hydrogen ion effectively loses an electron and becomes positively charged:

$$CH_2{=}CH_2 + H^+ \rightarrow CH_3CH_2^+$$

Water is a nucleophile and so is attracted to the carbocation:

$$CH_3CH_2^+ + H_2O \rightarrow CH_3CH_2OH_2^+$$

Finally, a proton is lost, regenerating the H^+ ion and forming ethanol:

$$CH_3CH_2OH_2^+ \rightarrow CH_3CH_2OH + H^+$$

This is very similar to the final step in the mechanism for the bromine water test (in the previous spread).

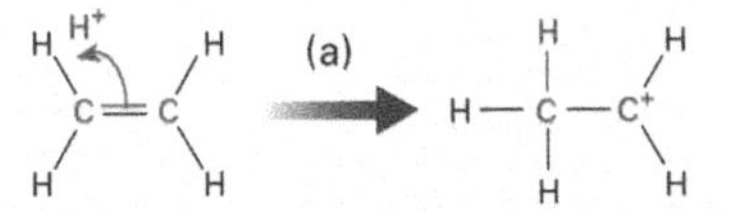

(a) An H^+ ion protonates ethene to form a carbocation.

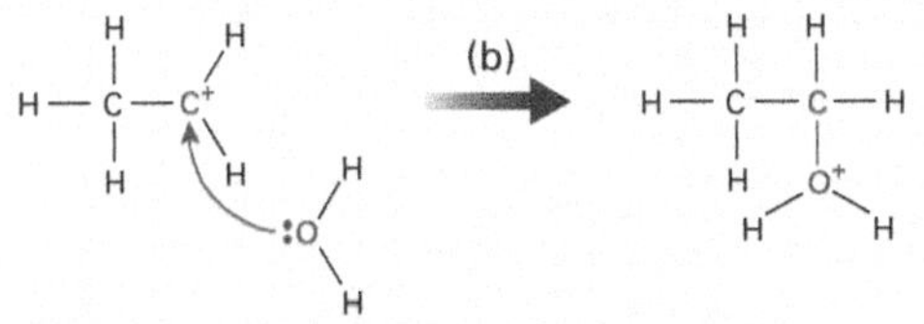

(b) Water acts as a nucleophile: oxygen donates a lone pair to the carbocation.

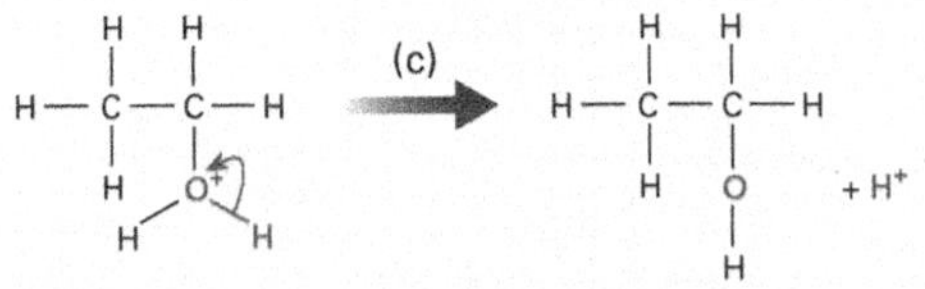

(c) Ethanol forms as the ion loses a proton (deprotonates).

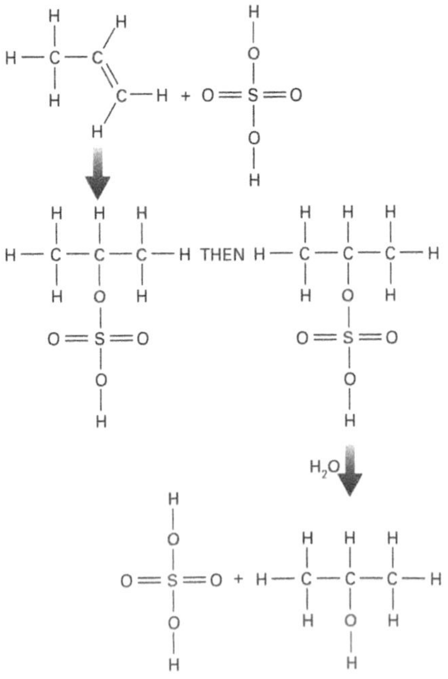

Ethene and steam react together at a temperature of 300 °C and a pressure of about 70 atm in the presence of a phosphoric acid catalyst held on a Celite (silica) support:

$$CH_2{=}CH_2 + H_2O \rightarrow CH_3CH_2OH$$

The phosphoric acid supplies H^+ ions that act as a catalyst: they take part in the reaction but are regenerated at the end. The overall reaction is reversible. The forward reaction is exothermic and so is favoured by low temperature. A compromise has to be reached when choosing the temperature because low temperatures lower the rate of reaction. Since one mole of gaseous product is produced from two moles of gaseous reactants, the forward reaction is favoured by high pressure.

Propene reacts with water to form propan-2-ol, following Markovnikov's rule (see previous spread).

Ethanol: the laboratory reaction

Concentrated sulphuric acid is used as the catalyst in the laboratory reaction to add water to ethene. Concentrated sulphuric acid H_2SO_4 may be looked upon as H—OSO_2OH (i.e. H—X), see spread 19.11, which adds across the C=C double bond in the usual manner:

$$CH_2{=}CH_2 + H{-}OSO_2OH \rightarrow CH_3CH_2OSO_2OH$$

The addition compound ethyl hydrogensulphate $CH_3CH_2OSO_2OH$ reacts with water to produce ethanol and reform the catalyst H_2SO_4:

$$CH_3CH_2OSO_2OH + H_2O \rightarrow CH_3CH_2OH + H_2SO_4$$

Adding *two* —OH groups across a double bond

Alcohols with two hydroxyl groups per molecule are called **diols**. Ethane-1,2-diol (common name *ethylene glycol*) is an important industrial chemical. It is mixed into the water in vehicle cooling systems during the winter as an 'antifreeze'. Ethane-1,2-diol is also the raw material for making polyester fibres (e.g. Terylene), spread 27.6.

Ethane-1, 2-diol in industry

The industrial production of ethane-1,2-diol is carried out in two stages, the first of which involves the oxidation of ethene by air using a silver catalyst at 300 °C and 15 atm. The product is a cyclic compound (a cyclic ether; considered in more detail in spread 25.6) epoxyethane:

$$CH_2{=}CH_2(g) + \tfrac{1}{2}O_2(g) \xrightarrow{Ag} CH_2(O)CH_2(g)$$

Epoxyethane is then hydrolysed by dilute acid:

$$CH_2(O)CH_2(g) \xrightarrow{H^+(aq),\ H_2O(l)} HOCH_2CH_2OH$$

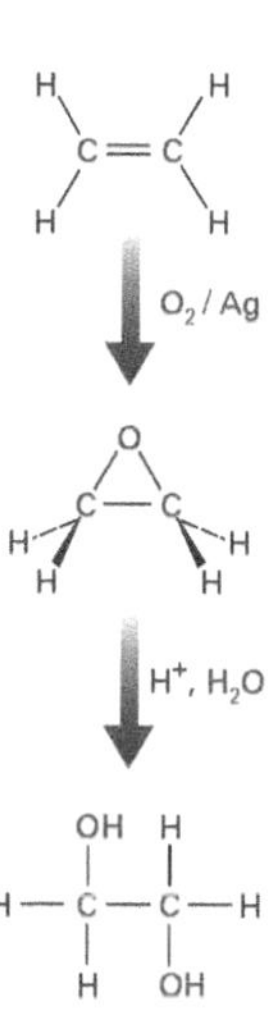

The manufacture of ethane-1,2-diol from ethene. Note the structure of the intermediate epoxyethane.

The laboratory reaction to add two —OH groups requires the use of a mild oxidant. Cold alkaline potassium manganate(VII) adds an —OH group to each side of an alkene double bond to give a diol. For example, ethene reacts to form ethane-1,2-diol:

$$CH_2{=}CH_2 \xrightarrow{MnO_4^-(aq)/\ OH^-(aq)} HOCH_2CH_2OH$$

This reaction may be used as a test for alkenes because the purple colour of the manganate(VII) ion changes to green aqueous manganate(VI) ion, spread 13.9. However, ethane-1,2-diol can be further oxidized under more strongly oxidizing conditions (see spread 26.5).

Ethane-1,2-diol (ethylene glycol) is used as antifreeze to prevent the water that cools an engine from freezing and damaging the engine as it expands. This is the engine from the Vauxhall Lotus Carlton, one of the fastest production saloon cars ever made.

SUMMARY

- Adding water across an alkene C=C double bond involves protonation by an acid catalyst followed by attack by water acting as a nucleophile.
- The industrial manufacture of ethanol adds steam to ethene with the aid of a phosphoric acid catalyst at 300 °C and 70 atm.
- Oxidation by alkaline manganate(VII) ion adds an —OH group to each of the carbon atoms in an alkene C=C double bond; the result is a diol.
- The change in colour by purple alkaline manganate(VII) ions acts as a test for the C=C double bond.

PRACTICE

1 Draw a flow diagram representing the industrial manufacture of ethanol from ethene. The conversion process operates at 70% efficiency, so you will need to separate unused reactants from products and recycle them. [Boiling points: water, 100 °C; ethanol, 78 °C; ethene, −104 °C]

2 Draw structural formulae to represent the following reactions:

a The industrial manufacture of ethanol from ethene.

b The action of cold dilute aqueous potassium manganate(VII) on propene.

c The protonation of propene to form a carbocation.

22.10

The polymerization of ethene

OBJECTIVES

- Discovery of polythene
- Addition polymerization
- Other addition polymers

IUPAC names

The IUPAC name for an **addition polymer** places the name of the monomer in brackets after the word 'poly', hence poly(ethene), poly(chloroethene), poly(propene), poly(phenylethene), and so on. These four have more common names: polythene, PVC, polypropylene, and polystyrene; see photos opposite.

The ethene molecule contains a C=C double bond, which can undergo addition reactions. In the polymerization of ethene, the molecule *adds* to *itself* to produce a saturated hydrocarbon chain that consists of repeating —CH_2CH_2— units. The resulting solid substance is called **polythene** (also poly(ethene) or polyethylene) and has many uses. Polythene is an example of a **polymer**, a large molecule made up of many identical repeating sub-units called **monomers**. Physical properties such as hardness depend on the number of carbon atoms (typically 40 000 to 800 000) in the polythene chain. Polythene was the first synthetic hydrocarbon polymer to be produced. We then discuss four other important polymers based on alkenes.

'Poly bags' and polythene food wrapping films have molecular chains with around 40 000 carbon atoms each. As the number increases, the polymer material becomes harder and more rigid. Milk containers have around 60 000 carbon atoms per molecule and bleach bottles around 80 000. When the number reaches 800 000, the material can be used in artificial ice rinks.

The accidental discovery of polythene

Serendipity

The discovery of polythene is an example of serendipity – a useful and unexpected discovery made by accident.

Starting in 1932, research was carried out at the ICI chemical works in Cheshire into the effects of high pressure on chemical reactions. Over 50 reactions were investigated but none appeared interesting. One of the 'failures' (in March 1933) was the reaction between ethene and benzaldehyde (C_6H_5CHO) at 1400 atm and 170 °C. A white waxy solid resulted, which was thought to be a polymer of ethene. Repeating the experiment with ethene alone caused a massive explosion.

In December 1935, the ethene experiment was repeated, with stronger equipment. As became clear later, the ethene was contaminated with traces of oxygen. Pumping ethene into the reaction vessel caused the pressure to increase: it then suddenly dropped as oxygen initiated the polymerization. The resulting polymer proved to be chemically inert and an excellent insulator. It was initially tested in the manufacture of cables. Commercial production started in 1939, one of the first wartime uses being the insulation of airborne radar aerials. Polythene later became the first polymer whose production exceeded 1 million tonnes per year.

Low-density poly(ethene), LDPE

Originally made by the high-pressure polymerization of ethene, LDPE has a density of around 0.92 g cm^{-3} and a melting point of around 110 °C. The polymer has considerable chain branching, leading to an open structure. LDPE is used as an electrical insulator and in packaging.

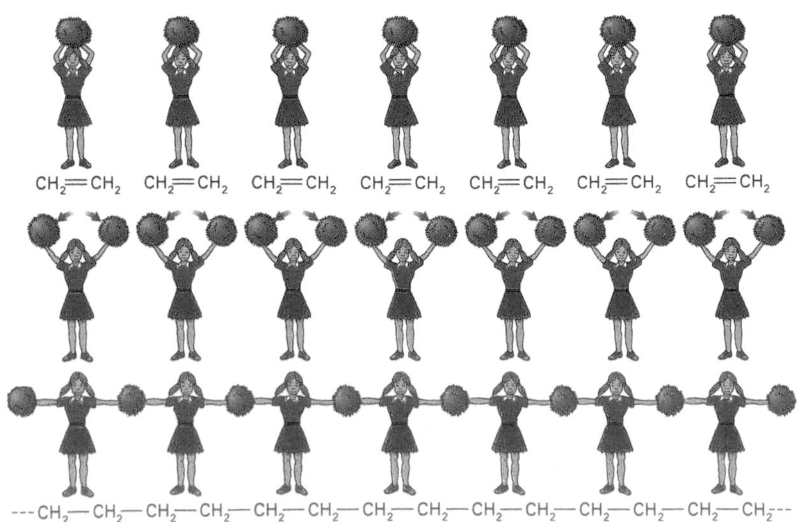

Polythene is formed by the addition polymerization *of ethene – the ethene monomers take part in an addition reaction across the double bond. The process can be pictured as the molecules using an internal linkage (the π bond) to create external linkages between each other.*

The variety of addition polymers

Ethene polymerizes to give a polymer of general formula $-[CH_2CH_2]_n-$. The general case is that the substituted alkene $CH_2{=}CHG$, where G is a side group, will polymerize to give the polymer $-[CH_2CHG]_n-$. Some common examples are shown below. PTFE, poly(tetrafluoroethene) is unusual in that four hydrogen atoms are substituted: the polymer is $-[CF_2CF_2]_n-$.

Uses of PVC

The unplasticized polymer (called **uPVC**) is a hard material used for buildings, especially window frames. Addition of the ester dioctyl phthalate as a **plasticizer** makes a more flexible material, used, for example, in waterproof clothing (see photo in spread 8.2).

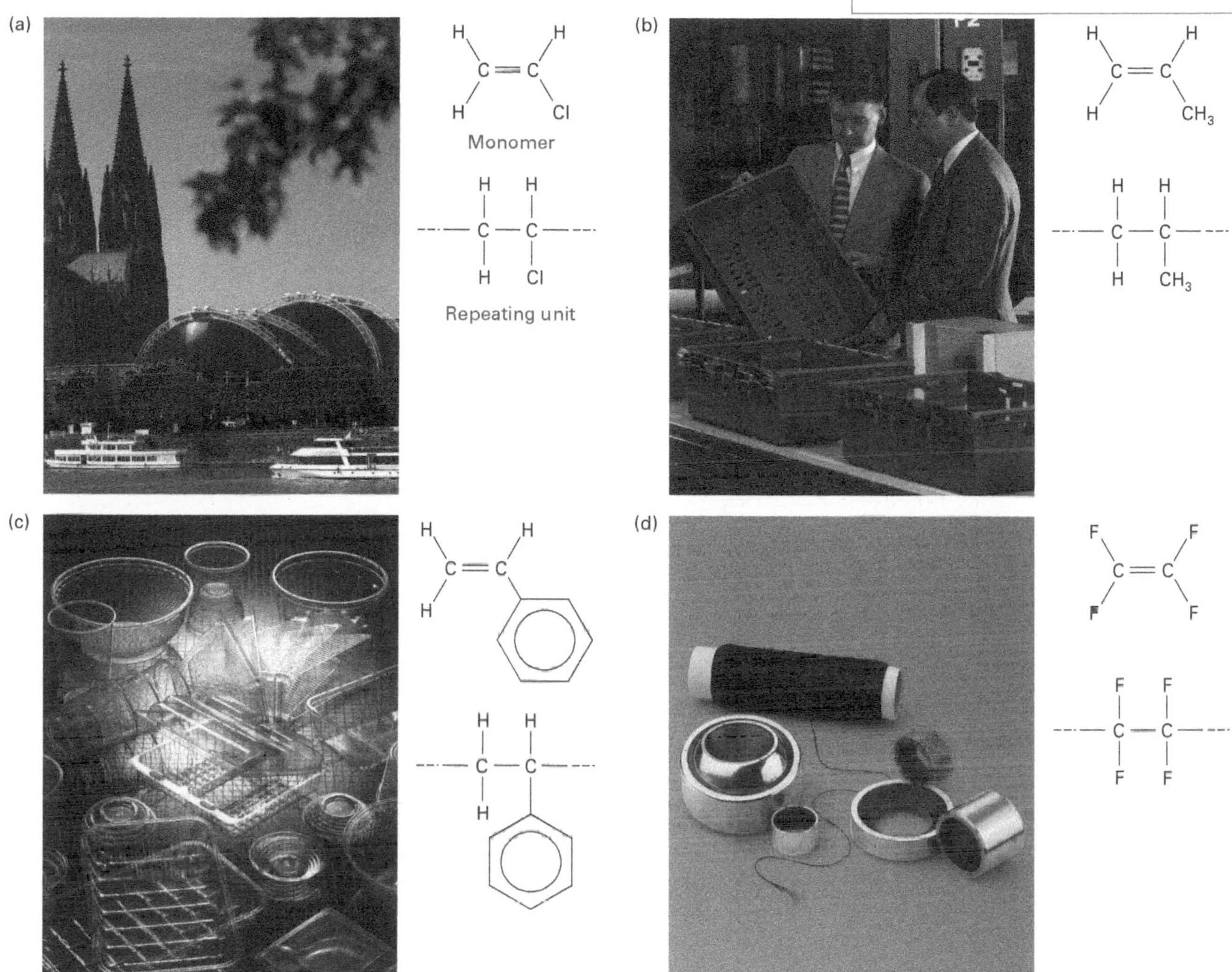

Some common addition polymers based on ethene derivatives together with their monomer and repeating unit. (a) PVC, (b) polypropylene (PP), (c) polystyrene (PS), and (d) PTFE.

SUMMARY

- Ethene undergoes addition to itself to form poly(ethene).
- Poly(ethene) is a polymer made up of ethene monomers. Its molecules are long chains consisting of repeating $-CH_2CH_2-$ units. Chains are between 10^4 and 10^6 carbon atoms long.
- Poly(ethene) is chemically inert as it consists of a saturated hydrocarbon chain.
- Increasing chain length leads to increased density, hardness, and melting point.
- For a given number of $-CH_2CH_2-$ units, increasing the degree of chain branching decreases the density, hardness, and melting point.
- Low-density poly(ethene) (LDPE) results from high-pressure polymerization (1400 atm and 170 °C).

Poly(ethenol)

The molecule poly(ethenol) or polyvinyl alcohol has a repeating unit with an OH in place of the Cl in PVC. This polymer has the convenient property (caused by hydrogen bonding) of dissolving in water: hospital disposal bags made from poly(ethenol) avoid human contact with the soiled contents but allow the contents to be washed subsequently. The polymer is made by transesterification, spread 27.4, using polyvinyl acetate PVA and methanol.

PRACTICE

1 List two uses each for LDPE, PP, PS, PVC, and PTFE.

22.11

THE MECHANISM OF POLYMERIZATION

OBJECTIVES

- Ziegler-Natta catalysis
- Polymer chains and properties
- Polymer geometry

Ethene $CH_2{=}CH_2$ undergoes addition polymerization by using the C=C π bond to form bonds between its molecules. The resulting polymer chain consists of thousands of $—CH_2CH_2—$ units joined together like beads on a necklace. The hydrogen atoms of ethene do not play a part in the polymerization reaction. Replacing one or more of the ethene hydrogen atoms by other atoms, or groups of atoms, results in polymer chains with repeating side groups. Side groups may be chosen to give polymers with specific desired properties. You will also see that the *arrangement* of the side groups along the polymer chain has an influence on the properties of the resulting polymer. This arrangement is affected by the mechanism of polymerization.

Mechanism of high-pressure polymerization

(a) RO—OR → 2RO•

(b) •OR + $CH_2{=}CH_2$ → $ROCH_2CH_2$• ; $ROCH_2CH_2$• + $CH_2{=}CH_2$ → $ROCH_2CH_2CH_2CH_2$• etc.

(c) $RO(CH_2CH_2)_n CH_2CH_2$• + •$CH_2CH_2(CH_2CH_2)_n OR$ → $RO(CH_2CH_2)_n CH_2CH_2CH_2CH_2(CH_2CH_2)_n OR$

The mechanism of commercial high-pressure addition polymerization (1400 atm and 170 °C) to form LDPE. (a) Initiation: radicals are generated from the decomposition of an organic peroxide. (b) Propagation: ethene adds to the radical; the unpaired electron is situated on the end carbon atom. (c) Termination: two radicals combine.

High-density poly(ethene), HDPE

This material is produced by Ziegler–Natta catalysis and has a very small degree of chain branching. As a result, the polymer chains are able to pack more closely than in LDPE, which has branched chains. HDPE is therefore harder and more rigid than LDPE, has a density of around 0.96 g cm^{-3}, and a melting point of around 135 °C. It is used to make containers.

Mechanism of polymerization with Ziegler–Natta catalysts

During the 1950s, Karl Ziegler and Giulio Natta discovered how to polymerize ethene at just 2 atm pressure and 70 °C. The key was a catalyst mixture consisting of titanium(IV) chloride $TiCl_4$ with triethylaluminium $Al(CH_2CH_3)_3$. Polymerization proceeds steadily as ethene passes into the catalyst mixture, allowing for a continuous manufacturing process. The equipment for the high-pressure production process is far more costly and this can only be carried out as a batch process. The mechanism of the Ziegler–Natta polymerization is complex and still only partially understood. The main points are summarized below.

$Cl_2Ti(R)$ + $CH_2{=}CH_2$ → $Cl_2Ti(R)\cdots CH_2{=}CH_2$ → $Cl_2Ti–CH_2CH_2–R$ → $Cl_2Ti(CH_2CH_2R)\cdots CH_2{=}CH_2$ → $Cl_2Ti–CH_2CH_2CH_2CH_2–R$

A summary of the Ziegler–Natta polymerization of ethene to form HDPE. The mechanism is called coordination polymerization *because coordination compounds (complexes) form as electron density from the ethene π bond is donated into d orbitals on an atom of titanium (the catalyst). One end of the growing hydrocarbon chain is attached to a titanium atom, while incoming ethene molecules form a complex with the same titanium atom. Step 2 is an example of an* insertion *reaction.*

Chain structure and polymer properties

The previous spread showed that chain length and chain branching in polythene affect the physical properties of the polymer material. Properties such as stiffness are particularly influenced by the manner in which the polymer chains interact with each other. The inclusion of side groups (e.g. —Cl in PVC, $—CH_3$ in polypropylene, $—C_6H_5$ in polystyrene) introduces extra factors that influence the interaction between chains.

First, the side groups may cause the polymer chains to align, which makes the material more **crystalline** (i.e. having a regular and repeated arrangement of atoms). For example, polythene is a substance that is fairly **amorphous** (i.e. lacking regular, crystalline structure), with the polymer chains in a random arrangement; whereas in polystyrene, interactions between the side groups lead to a more ordered and crystalline structure with increased hardness.

Secondly, different arrangements of the side groups along the polymer chain are possible. Polymerization of phenylethene (styrene) $CH_2{=}CHC_6H_5$ by means of a Ziegler–Natta catalyst results in an **isotactic polymer** with all the phenyl $—C_6H_5$ groups situated on the *same* side of the chain. Catalysts may also be tailored to produce the **syndiotactic polymer**, in which the phenyl side groups regularly *alternate* from one side of the chain to the other. Isotactic and syndiotactic polymers are called **stereoregular polymers.** Polymerization by radicals is a much more random process and results in the formation of the **atactic polymer**, in which the phenyl groups have a *random* arrangement along opposite sides of the polymer chain.

The three classes usually have significantly different properties. Physical properties depend on the interaction between the chains, which in turn depends on the type and extent of interaction between side groups. The positioning of the side groups with respect to each other has a marked effect on the opportunities for interaction between one chain and another. For example, atactic polystyrene softens and can be moulded at a much lower temperature than the isotactic form. Isotactic polypropylene can be used to make jug kettles.

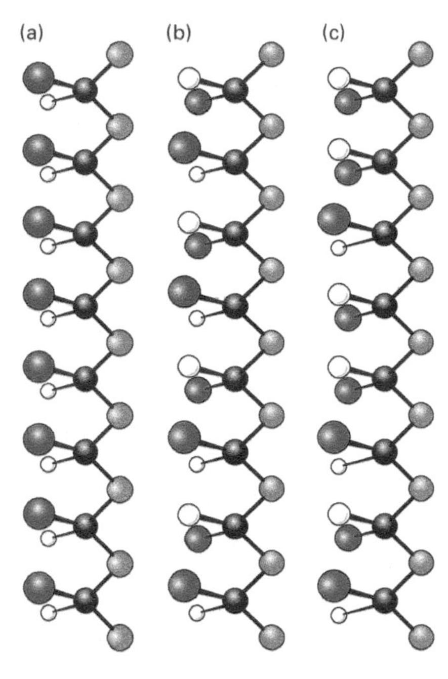

Polymer geometry: (a) isotactic (all groups on the same side of the hydrocarbon chain); (b) syndiotactic (alternating sides); and (c) atactic (random).

Isotactic PVC. The regular arrangement of the chloro side groups in the isotactic form allows the chains to approach each other more closely.

Recycling of plastics

Some plastics are being recycled by many local authorities. Codes for the different plastics were introduced in 1988: HDPE has code 2, PVC code 3, LDPE code 4, PP code 5 and PS code 6 (code 1 is the label for PET, spread 27.6). It is common for PET, HDPE, and PVC to be accepted for recycling; recently some garages have been able to repair rather than replace PP bumpers. Recycling promotes the sustainable use of materials. In 2008 the Closed Loop Recycling plant opened at a previously disused site on the outskirts of Dagenham: it aims to recycle 35 000 tonnes of PET and HDPE into food-grade plastic. See also spread 29.4.

Australian bank notes are made from two thin layers of poly(propene) joined under heat and pressure. They are more durable than paper notes and survive being washed in a pocket. The Bank of England introduced a plastic £5 note in 2016.

SUMMARY

- Ethene $CH_2{=}CH_2$ undergoes addition polymerization to give a polymer (polythene) of general formula $\text{+}CH_2CH_2\text{+}_n$.
- A substituted alkene $CH_2{=}CHG$ undergoes addition polymerization to give a polymer of general formula $\text{+}CH_2CHG\text{+}_n$.
- The side groups of isotactic polymers are all situated on the same side of the chain; in syndiotactic polymers, they regularly alternate; the arrangement in atactic polymers is random.
- Polymers produced by Ziegler–Natta catalysts are isotactic or syndiotactic; polymers produced by radical polymerization are atactic.
- Ziegler–Natta catalysis (based on titanium(IV) chloride and triethylaluminium) produces high-density poly(ethene) (HDPE) at low pressure (2 atm and 70 °C).

EXTENSION

22.12

OBJECTIVES

- Structures from VSEPR theory
- Bonding in methane
- Electron density models
- Delocalization

Bonding in methane

Alkanes contain carbon–carbon single bonds only. You should already be familiar with drawing Lewis structures to show the electronic structure in these types of molecules. However, it is now necessary to consider the bonding in these molecules more carefully. This spread starts by looking at two very different representations of the methane molecule. One is the Lewis structure, first put forward about 100 years ago. The other is the structure derived from overlapping atomic orbitals to form molecular orbitals; the quantitative calculations on which this model relies have only been possible since the early 1990s.

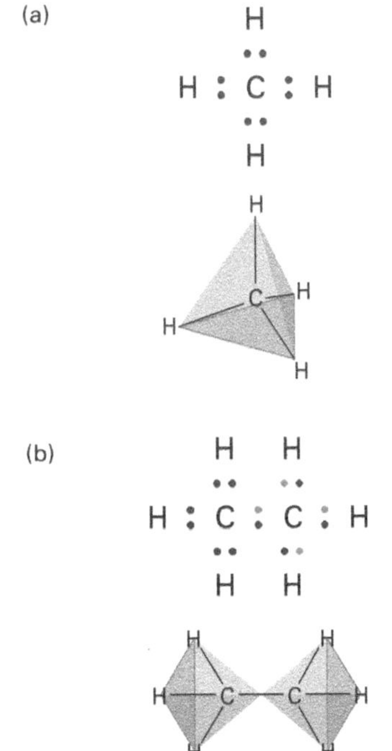

(a) Methane CH_4 has the shape of a regular tetrahedron. Four equivalent C—H bonds repel each other equally, giving a bond angle of 109.5°.
(b) The shape of the ethane molecule is equivalent to two methane molecules joined together.

Lewis structures and VSEPR

Lewis structures show the electronic structures of the valence shells in a covalently bonded molecule. VSEPR (valence-shell electron-pair repulsion) theory suggests three-dimensional shapes for molecules by considering the repulsion between electron pairs in bonds and lone pairs (when present). The shapes obtained for methane and ethane are shown alongside.

Methane: overlapping atomic orbitals

The electronic structures of carbon and hydrogen are $1s^2 2s^2 2p^2$ and $1s^1$ respectively. The valence shell in the carbon atom consists of the single 2s atomic orbital together with the three 2p atomic orbitals. The valence shell in the hydrogen atom consists of the single 1s atomic orbital. Now consider what happens as four hydrogen atoms approach a carbon atom and then bond together to form methane.

As the atoms approach, the atomic orbitals on the carbon atom overlap with those on the hydrogen atoms. Two sorts of overlap occur in methane. In the first, the 1s orbital on each of the four hydrogen atoms overlaps with the 2s orbital on the carbon atom. The combined volume of space that they now share is called a **molecular orbital**. This single molecular orbital is occupied by a pair of electrons *that bonds all five atoms together*.

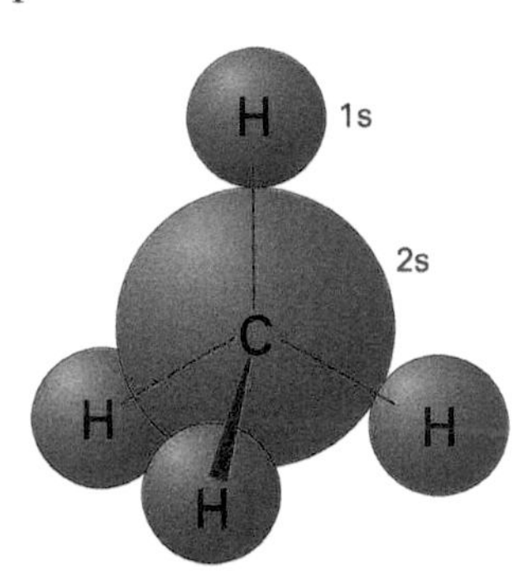

The formation of a bonding molecular orbital by the overlap of the carbon 2s orbital with the 1s orbitals of the four hydrogen atoms. The resulting molecular orbital is delocalized over all five atoms.

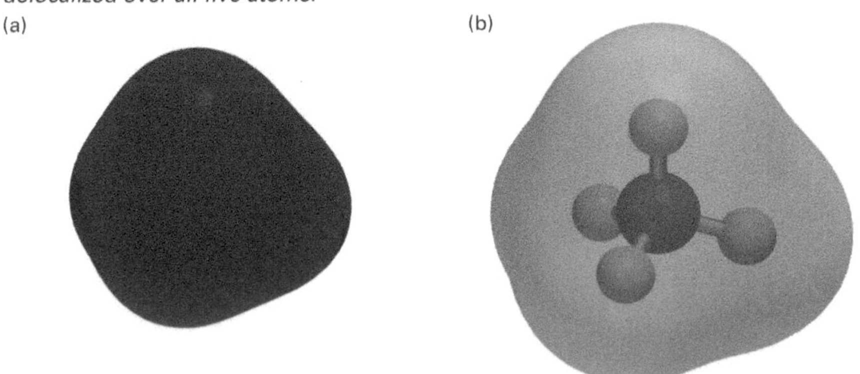

(a) The lowest-energy bonding molecular orbital formed by the overlap shown above is delocalized over all five atoms. The shape of this one orbital resembles the shape of the whole molecule, shown in the electrostatic potential map (b).

The second sort of overlap occurs between the four hydrogen 1s orbitals and the three carbon 2p orbitals. The result is three molecular orbitals containing six electrons that bond all five atoms together. These three orbitals have identical energies, but they are significantly higher in energy than the orbital arising from the overlap of the hydrogen 1s and carbon 2s orbitals. As a result, there are two sets of molecular orbitals in methane representing different energy levels.

There is a general result (see spread 6.4) that when 4 molecular orbitals are delocalized over 5 atoms, a set of 4 *localized* orbitals may be generated mathematically, each lying between a bonded pair of atoms; this reconstitutes the simple picture for methane of four equal bonds arranged tetrahedrally. Such a localization scheme provides a useful link with the Lewis structure: the electron density is accurately represented.

On the other hand, the *energies* of the molecular orbitals are actually very significantly different: the most stable delocalized molecular orbital involving overlap with the carbon 2s is *more than twice* as stable as the three molecular orbitals formed from the overlap with 2p. This feature does not appear in the localization scheme. The delocalized approach is confirmed both by computer calculation and by experimental measurement: there are two *different* ionization energies for the methane molecule. (This measurement was made by a spectroscopic technique called photoelectron spectroscopy.)

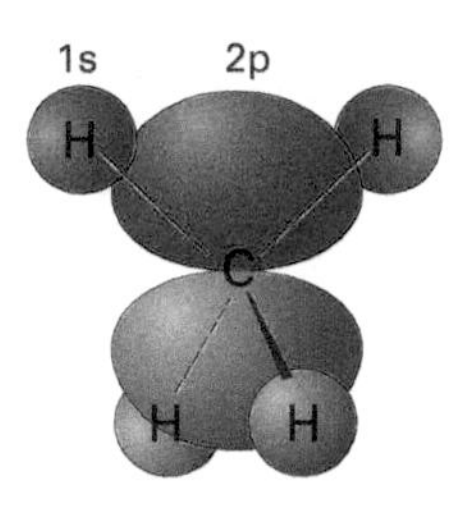

The formation of a bonding molecular orbital by the overlap of a carbon 2p orbital with the 1s orbitals of the four hydrogen atoms. The other two carbon 2p orbitals together bond with all four hydrogen atoms.

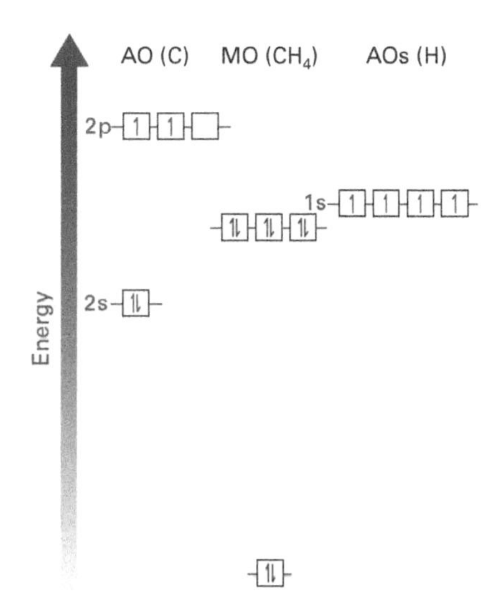

The bonding orbital formed by overlap of the carbon 2s is much more stable than the three equivalent bonding orbitals formed by overlap of the carbon 2p. (The figure is drawn to scale.)

SUMMARY

- Molecular orbitals result from the overlap of atomic orbitals on separate atoms.
- Each molecular orbital can hold a maximum of two electrons.
- Computer-generated electron density models show the distribution in space of the electrons in a molecule and illustrate its overall shape.
- The lowest-energy bonding molecular orbital in methane results from overlap of the carbon 2s and the four hydrogen 1s orbitals.
- The lowest-energy bonding molecular orbital in methane is delocalized over all five atoms.

PRACTICE

1 Draw a Lewis structure showing the valence-shell electronic structure for propane.

2 Sketch the shape of the propane molecule indicated by VSEPR theory, showing the bond angles.

3 Sketch the overlap of orbitals that forms the lowest-energy bonding molecular orbital in ethane.

Chapter 22 Reactions summary

- Cracking:
  alkane → smaller alkane + alkene
  $C_8H_{18} \rightarrow C_6H_{14} + C_2H_4$
- Alkane substitution:
  $CH_4 + Cl_2 \xrightarrow{UV} CH_3Cl + HCl$
  Mechanism: radical chain reaction
- Hydrocarbon combustion:
  $CH_4 + 2O_2 \rightarrow CO_2 + 2H_2O$
- Addition of hydrogen to an alkene:
  $CH_2{=}CH_2 + H_2 \xrightarrow{Ni} CH_3CH_3$
- Addition of hydrogen halide to an alkene:
  $CH_2{=}CH_2 + HBr \rightarrow CH_3CH_2Br$
  Mechanism: electrophilic addition
- Markovnikov addition to an unsymmetrical C=C bond:
  $CH_3CH{=}CH_2 + HBr \rightarrow CH_3CHBrCH_3$ (predominantly)
- Addition of bromine to an alkene:
  $CH_2{=}CH_2 + Br_2 \rightarrow BrCH_2CH_2Br$
- Action of bromine water on an alkene:
  $CH_2{=}CH_2 + Br_2(aq) \rightarrow BrCH_2CH_2OH$
  Bromine water is decolorized
- Addition of water to an alkene (acid catalyst):
  $CH_2{=}CH_2 + H_2O \rightarrow CH_3CH_2OH$
- Alkenes may be polymerized: examples include polythene, polypropylene, polystyrene, PVC, and PTFE.

PRACTICE EXAM QUESTIONS

1 The diagram below represents the industrial fractional distillation of crude oil.

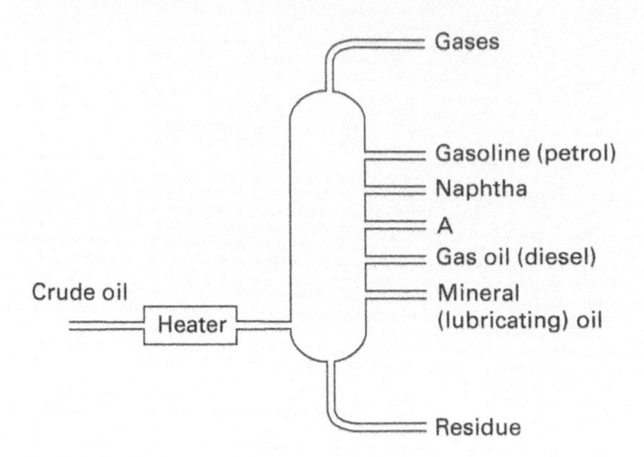

a Identify fraction **A**. [1]

b What property of the fractions allows them to be separated in the column? [1]

c Give **one** use each for the naphtha fraction and the gas oil fraction other than in diesel engines. [2]

d The gases include butane which is used as bottled fuel by campers.

i Write an equation for the complete combustion of butane.

ii Suggest a reason why gas oil would be unsuitable fuel for campers. [3]

e Why cannot the residue be further separated into paraffin waxes and tar by strong heating? [1]

2 a i State Raoult's law for an ideal mixture of two liquids. [2]

ii Benzene and methylbenzene may be separated by fractional distillation. Sketch the general form of the boiling-point/composition diagram for such a mixture and use it to explain the basis on which fractional distillation rests.

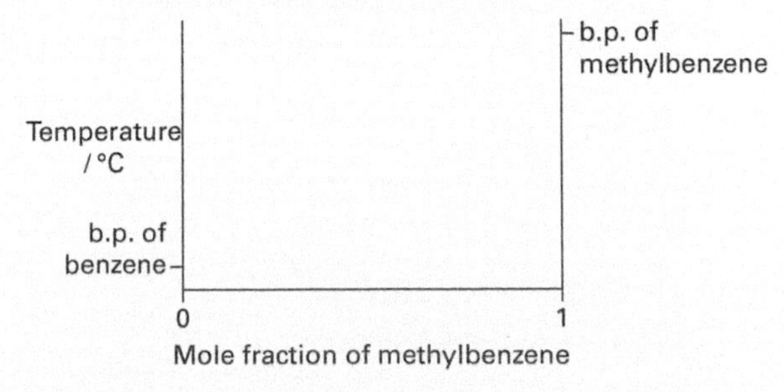

iii The first stage in the refining of crude oil (petroleum) is fractional distillation. State two ways in which the commercial fractional distillation of crude oil differs from the fractional distillation of a simple ideal mixture of two liquids. [6]

b i All lighter (more volatile) fractions from petroleum distillation are useful as fuels. Suggest two reasons why the liquid fractions with 8 to 12 carbon atoms per molecule are used as motor fuels, rather than the gaseous ones containing from one to four carbon atoms. [2]

ii Benzene is added to unleaded petrol to compensate for the absence of tetraethyllead. Both compounds are hazardous; which hazard is associated with benzene, other than its flammability? [1]

iii Tetraethyllead or benzene is added to petrol to prevent **pre-ignition**. What is pre-ignition, and why is it a problem? [2]

iv Suggest two reasons why unleaded fuel has been promoted by government and the petroleum industry. [2]

3 When irradiated with ultraviolet light, methane reacts with chlorine to form a mixture containing several organic products.

a Name the type of mechanism involved. [1]

b Explain why ultraviolet light is needed. [2]

c Write an equation to show the formation of chloromethane from chlorine and methane. [1]

d Another product has a relative molecular mass of 85. Identify this product and write that part of the mechanism which shows its formation from chloromethane. [3]

e Explain how a small amount of ethane is formed in the process and name the mechanistic step which leads to its formation. [2]

f Give the major product formed when methane reacts with a large excess of chlorine. [1]

4 Alkenes such as but-2-ene are used by the petrochemical industry to produce many useful materials. Some reactions of but-2-ene are shown below.

a Draw structures to represent possible compounds **A–D** in the reactions of but-2-ene below.

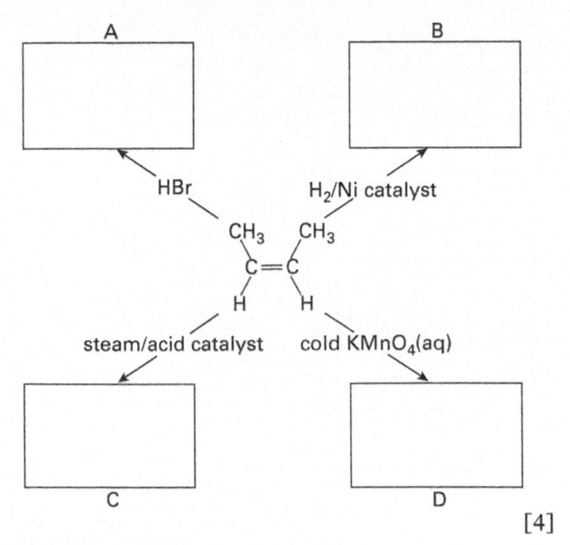

[4]

b Bromine reacts with but-2-ene in an addition reaction.

i What type of reagent is bromine in this reaction?

ii Complete the mechanism below for this reaction.

$(CH_3)HC{=}CH(CH_3)$ →   → $H-C(CH_3)(Br)-C(CH_3)(Br)-H$

$Br-Br$

c But-2-ene can be converted into buta-1,3-diene by a process called dehydrogenation. Buta-1,3-diene is used to make synthetic rubber.

i Suggest the structure of buta-1,3-diene.

ii Construct an equation for the dehydrogenation of but-2-ene. [2]

5 Low density poly(ethene) is used for packaging and plastic bags. The exothermic reaction by which poly(ethene) is made is shown by the following equation:

$nC_2H_4 \rightarrow (C_2H_4)_n$

a Write the structural formulae, showing all the bonds, of

i ethene;

ii poly(ethene), showing three repeating units. [2]

b Explain why typical conditions used in the process are a high pressure of 2000 atmospheres and a relatively low temperature of 200 °C. [4]

c The reaction proceeds via a free radical mechanism. The production of poly(ethene) may be initiated by the reaction.

$R\cdot + CH_2{=}CH_2 \longrightarrow R-CH_2-CH_2\cdot$

Suggest an equation to show how

i a subsequent stage occurs;

ii the polymerisation might terminate. [4]

d High density poly(ethene), used for articles such as buckets and crates, is made under other conditions, using a catalyst.

i Suggest why this form of poly(ethene) has a higher density.

ii Other than density, suggest ONE physical property which would be different for high density poly(ethene). [2]

e Draw a representative length of the molecule of poly(2-methylpropene), showing three repeating units. [1]

f i Write the structural formula, showing all covalent bonds, for the product obtained by reacting 2-methylpropene with bromine.

ii Write the structural formula of the compound formed by the reaction of aqueous sodium hydroxide with the product of the reaction in **f i**. [2]

6 Poly(chloroethene), PVC, is manufactured on a large scale from ethene. The process is essentially in three stages.

Stage 1 The formation of 1,2-dichloroethane from ethene.

Stage 2 The formation of chloroethene from 1,2-dichloroethene.

Stage 3 The polymerization of chloroethene.

a **Stage 1**, in some processes, may be carried out by the reaction below

$CH_2{=}CH_2 + \frac{1}{2}O_2 + 2HCl \rightarrow CH_2Cl-CH_2Cl + H_2O$; $\Delta H^{\ominus} = -242\ \text{kJ mol}^{-1}$

at a pressure of 5 atmospheres and a temperature of 570 K. A copper(II) chloride catalyst is used and the reaction vessel cooled.

i State one reason why the reaction vessel is cooled. [1]

ii The gases are washed to remove any unreacted hydrogen chloride. Give the name of a common aqueous solution which could be used. [1]

iii State what influence the presence of a catalyst has on the operating temperature and the economics of **Stage 1**. [2]

b **Stage 2** is brought about by thermal cracking. The reaction mechanism involves three steps.

$CH_2Cl-CH_2Cl \rightarrow CH_2Cl-CH_2\cdot + Cl\cdot$

$CH_2Cl-CH_2Cl + Cl\cdot \rightarrow HCl + CH_2Cl-CHCl\cdot$

$CH_2Cl-CHCl\cdot \rightarrow CH_2{=}CHCl + Cl\cdot$

i Write down the initiation step for **Stage 2**. [1]

ii Write down an equation for the overall reaction in **Stage 2**. [1]

iii The product, chloroethene (boiling temperature −13 °C), has to be separated from unchanged liquid 1,2-dichloroethane (boiling temperature 84 °C). State how this may be achieved. [1]

iv State what happens to the 1,2-dichloroethane recovered in **b iii**. [1]

c In **Stage 3** chloroethene undergoes polymerization.

i Draw the repeating unit in the polymer.

ii Give **one** everyday domestic use of the poly(chloroethene). [1]

iii Apart from the necessity to deal carefully with chloroethene, which is highly toxic, state **two** other safety considerations which would be necessary at a plant manufacturing PVC. [2]

23 Arenes

The **arenes** are an enormous group of compounds based on the benzene ring C_6H_6. This ring gives distinctive properties to any molecule that contains it. By comparison, aliphatic organic compounds are based on chains of carbon atoms. **Alicyclic** organic compounds contain a ring structure other than benzene; their properties are similar to those of aliphatic compounds. The chemistry of the arenes is so distinctive that it is usually studied separately in its own right. This chapter concentrates on the reactions of benzene and of arenes containing functional groups attached to a single benzene ring. Later chapters will deal with the properties of these functional groups in aliphatic compounds.

THE UNIQUE CHARACTER OF BENZENE

23.1

OBJECTIVES

- Alicyclics and arenes compared
- Delocalized electrons
- Delocalized π cloud

The benzene molecule consists of six carbon atoms joined to form a ring. Benzene is a hydrocarbon in which each carbon atom is joined to two other carbon atoms and to a single hydrogen atom. Molecules of cyclohexane and cyclohexene also consist of six carbon atoms joined in a ring. However, the properties of these compounds are very different from those of benzene. The reasons for these differences lie in the character of the bonds between the carbon atoms in each of the different types of ring. Cyclohexane, cyclohexene, and benzene all have six-membered rings of carbon atoms: so how do they differ?

Alicyclic compounds: cycloalkanes

The carbon atoms in cycloalkanes are joined to form a ring structure. These compounds are named by prefixing the appropriate alkane name with 'cyclo'. For example, cyclohexane has the formula C_6H_{12}. It is saturated and reacts by substitution only, in the same manner as aliphatic (straight-chain and branched-chain) alkanes.

In general, cycloalkanes are as unreactive as aliphatic alkanes, undergoing just the same types of radical processes. The two exceptions are cyclopropane and cyclobutane. Their bond angles are 60° and 90° respectively, compared with the normal tetrahedral angle of 109°. The ring structure is **strained** and causes these two compounds to be more reactive than their straight-chain analogues propane and butane.

Alicyclic compounds may consist of hydrocarbon rings with functional groups attached. For example, attaching an —OH group to cyclohexane C_6H_{12} results in the compound cyclohexanol $C_6H_{11}OH$ (in the same manner that attaching this group to ethane CH_3CH_3 results in the aliphatic compound ethanol CH_3CH_2OH). Cyclohexanol has almost identical chemical properties to ethanol.

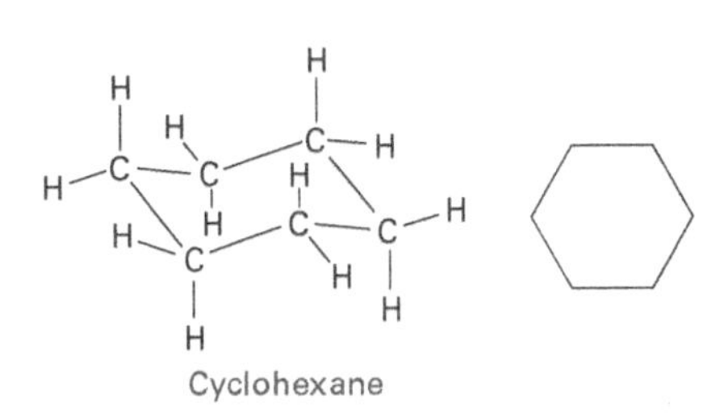

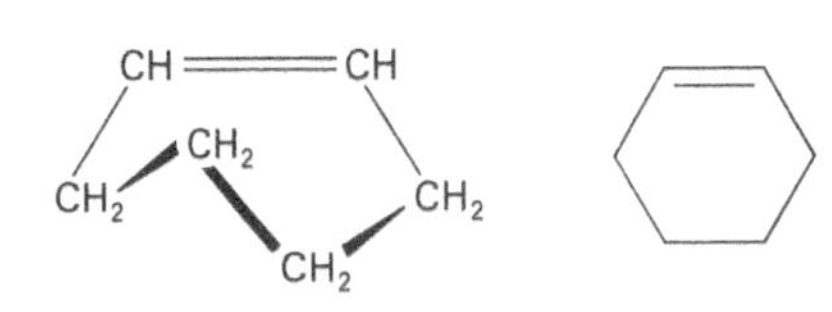

Formulae for two alicyclic compounds. Note that neither of these molecules is flat – the carbon atoms are not all in the same plane.

Alicyclic compounds: cycloalkenes

These compounds are similar to cycloalkanes in their overall structure, but contain one or more C=C double bonds. For example, cyclohexene has the formula C_6H_{10} and contains one C=C double bond and five C—C single bonds. It is unsaturated and reacts mainly by addition, in the same manner as aliphatic alkenes. The structures of cyclopropene and cyclobutene are strained for the same reasons as are the corresponding cycloalkanes. These two cycloalkenes therefore have greater reactivity than propene and butene.

Aromatic arenes

Benzene C_6H_6 is the parent molecule of all arene compounds. Arenes were earlier called 'aromatic' compounds because they have characteristic (strongly 'aromatic') smells.

Arenes: benzene

Benzene is a liquid (T_b = 80 °C) available from the fractional distillation of crude oil and also made from gasoline by catalytic reforming, spread 22.3. The molecular formula of benzene is C_6H_6. Using the standard form of single and double covalent bonds, Friedrich August Kekulé proposed the structure shown to the right in 1865. However, the following points show the Kekulé structure to be incorrect:

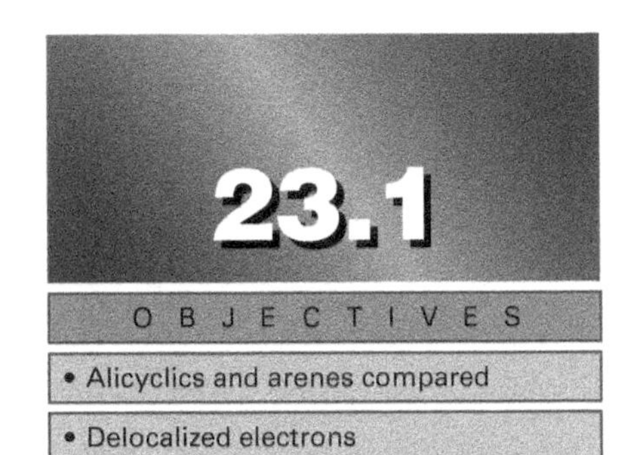

- The Kekulé structure contains three double bonds. Molecules with several double bonds are usually quite reactive and take part in addition reactions, spread 22.7. Benzene is relatively *unreactive*, slow *substitution* reactions being most common.
- The Kekulé structure suggests alternating double and single bonds. Double bonds are shorter than single bonds, but X-ray analysis of solid benzene shows a regular hexagon with all the bonds *the same length*, intermediate between a typical single and double bond.
- The Kekulé structure would be correctly described as cyclohexa-1,3,5-triene (triene because it has three double bonds). On this basis, the standard enthalpy change (see chapter 10 'Thermochemistry') associated with hydrogenating the three unsaturated C=C bonds would be expected to be three times that for hydrogenating the one double bond of cyclohexene (–120 kJ mol^{-1}). However, the experimental value of –208 kJ mol^{-1} is much smaller than the predicted value of –360 kJ mol^{-1}.

The enthalpies of hydrogenation show that benzene actually has a *more stable* structure than the Kekulé structure (by about 150 kJ mol^{-1}). The explanation is that the p orbitals at right angles to the plane of the carbon ring do not just overlap in pairs to form bonds, but all six overlap together.

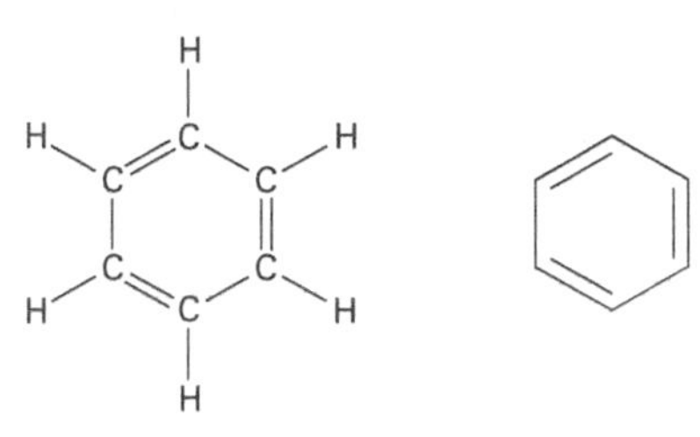

The Kekulé structure (displayed formula) for benzene with its skeletal version. Note that the carbon atoms are in one plane, i.e. benzene is 'flat'. The hydrogen atoms are also coplanar with the carbon atoms.

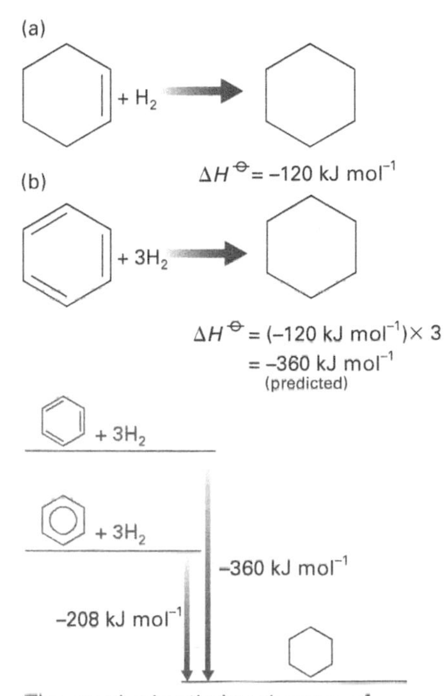

The standard enthalpy changes of hydrogenation for cyclohexa-1,3,5-triene and benzene.

Delocalization of electrons in the benzene ring

A molecule of benzene contains a total of 30 valence electrons (four for each of the six carbon atoms, and one for each of the six hydrogen atoms). Of these electrons, 12 are involved in six C—H σ bonds, and another 12 form six C—C σ bonds. These bonds establish the basic planar hexagonal **sigma framework** of the molecule. The remaining six electrons are not associated with any particular pair of carbon atoms. They are **delocalized** in ring-shaped molecular orbitals above and below the main hydrocarbon skeleton.

Each carbon atom uses its 2s and two of its 2p atomic orbitals to form the σ framework of the benzene molecule. The remaining 2p atomic orbitals all combine together to form a set of delocalized π molecular orbitals. The six remaining valence electrons fill three bonding molecular orbitals, each of which is delocalized (together constituting a delocalized π cloud). The lowest-energy orbital arises from overlap of all six orbitals in phase, spread 6.4. Thus the total number of π bonds is correctly described by the Kekulé structure, but the detailed electron density is not.

Delocalization, spread 6.4, lowers the energy of the structure by an amount often described as the **delocalization enthalpy** (or delocalization energy) (about 150 kJ mol^{-1}). This delocalized electronic structure has a dominant influence on the properties of molecules that contain benzene rings. The delocalization is now emphasized using the following symbol for benzene:

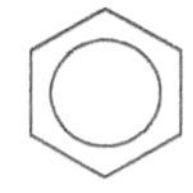

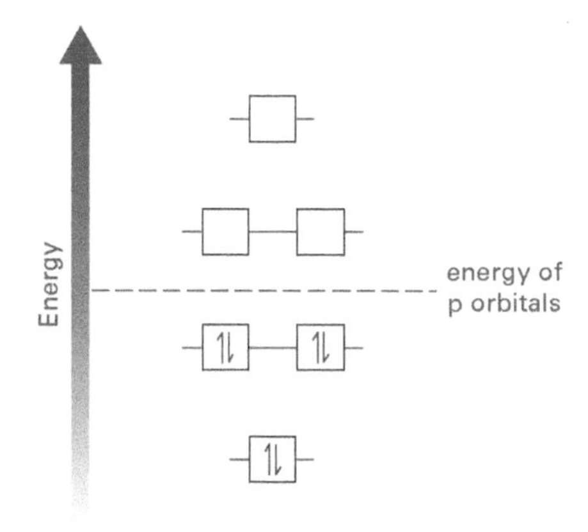

The three bonding orbitals are each occupied by two electrons: the shape of the lowest energy orbital is shown above left. The other two bonding π orbitals (lying below the energy of the p orbitals) have more complicated shapes; the average electron density resembles that of the lowest-energy orbital. (The three antibonding orbitals, of even more convoluted shape, are not occupied.)

SUMMARY

- Arenes contain the benzene ring C_6H_6, which is unsaturated but does not contain localized C=C carbon–carbon double bonds.
- Three bonding electron pairs in benzene are delocalized in molecular orbitals above and below the plane of the ring.

23.2

OBJECTIVES

- Naming arenes
- Electrophilic substitution
- Attack by radicals

Naming arenes

Naming arenes is generally simple: most are named as substituted benzenes. Examples include methylbenzene, chlorobenzene, and nitrobenzene. Most exceptions switch to using the name 'phenyl' for the group C_6H_5- Hence the molecule $C_6H_5CH{=}CH_2$ is phenylethene; the molecule C_6H_5OH is phenol.

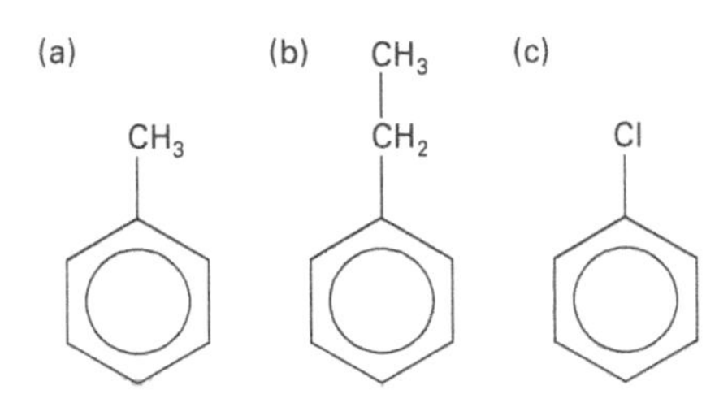

Structural formulae of substituted arenes: (a) methylbenzene (toluene); (b) ethylbenzene; (c) chlorobenzene.

HALOGENATION OF ARENES

Simple arenes generally consist of a functional group or a carbon chain attached to a benzene ring. Despite the unsaturated nature of benzene, it does not readily undergo addition reactions. Most functional groups and carbon chains are introduced into the benzene molecule by means of *substitution* reactions. As you will see, this type of reaction allows the delocalized π cloud of electrons to remain intact.

Addition of halogens to benzene is difficult

The unsaturated benzene molecule does not readily undergo electrophilic addition reactions (which are common for alkenes), because saturating the carbon–carbon bonds destroys the delocalized π cloud. Remember that this delocalized structure is particularly stable. Although benzene is unsaturated, it does not decolorize bromine water (the usual test for unsaturation).

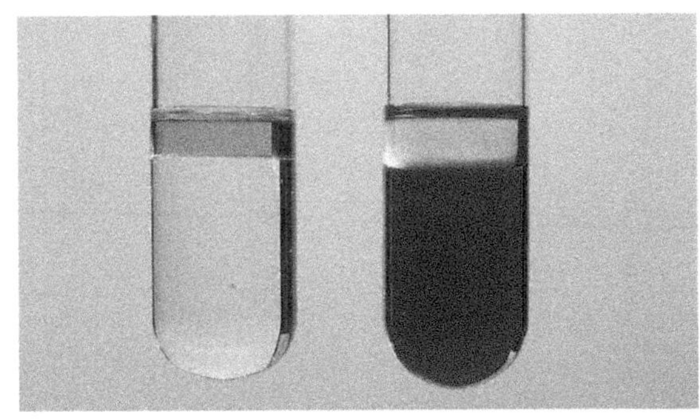

Bromine water is decolorized by cyclohexene, which contains a bonding pair of electrons localized between *two adjacent carbon atoms in a C=C double bond (left). Benzene does not decolorize bromine water: benzene contains three pairs of bonding π electrons that are* delocalized *over six carbon atoms (right).*

Electrophilic substitution

Benzene usually undergoes electrophilic *substitution* reactions in which the delocalized π cloud is preserved. The product of such a substitution reaction with a halogen is a **halogenoarene** such as chlorobenzene. To *substitute* a halogen for hydrogen in benzene, a catalyst called a **'halogen carrier'** is needed as well as the halogen itself, as the π cloud is more stable than the simple π bond in an alkene.

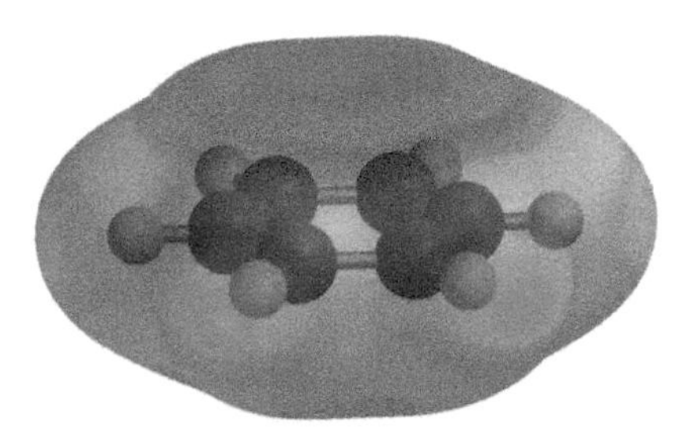

The electrostatic potential map for benzene: note the high electron density above (and below) the ring.

The six-membered carbon skeleton in the benzene molecule is sandwiched between the two halves of the delocalized π electron cloud. The π cloud is a region of high electron density, coloured red in the image alongside. Benzene therefore reacts with electrophiles; nucleophiles would be repelled. Halogen carriers make the halogen strongly electrophilic. They cause heterolytic fission of the Cl—Cl bond in the chlorine molecule (or polarize it to form $Cl^{\delta+}$—$Cl^{\delta-}$). The electrophilic Cl^+ ion then reacts with the ring. Typical halogen carrier catalysts are Lewis acids such as aluminium chloride $AlCl_3$ or iron(III) chloride $FeCl_3$. Metallic iron is also often used because it reacts with chlorine to form iron(III) chloride within the reaction mixture. Equivalent bromine-carrying catalysts, $FeBr_3$ and $AlBr_3$, are used with bromine to make bromoarenes.

- Electrophilic substitution is the characteristic reaction of all arenes.

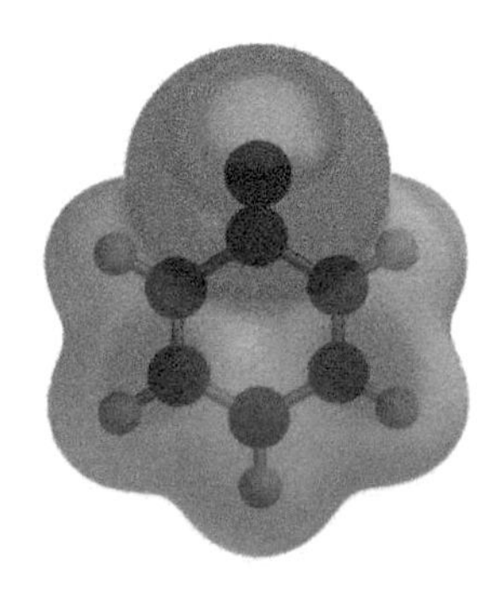

The electrophilic Cl^+ ion approaches benzene from on top, where the electron density is high, to form this Wheland intermediate.

The mechanism of electrophilic substitution starts with the delocalized electrons in the benzene ring reacting with the electrophilic Cl^+ ion. As the electrostatic potential map alongside shows, the high electron density is above (and below) the plane, hence this is the direction of approach. An intermediate called a **Wheland intermediate** forms, in which the positive charge is shared by the five other carbon atoms in the ring (see spread 23.4). This intermediate loses a proton to form chlorobenzene. Overall:

$$C_6H_6 + Cl^+ \rightarrow C_6H_5Cl + H^+$$

The mechanism of halogen substitution. (Note that, although chlorine is shown here, the same mechanism applies to bromination.) (a) Chlorine donates electron density into a vacant orbital on the aluminium atom in aluminium chloride. (b) The benzene π cloud attracts the Cl^+ ion. The Cl^+ ion bonds to carbon; four π electrons are shared by five carbon atoms (c). (d) The molecule loses a proton (deprotonates) *to form the product. This regains the delocalization energy of the benzene ring.*

Attack by radicals on side chains

Radicals normally attack alkanes indiscriminately. For example, the radical reaction between chlorine and an alkane in the presence of ultraviolet light gives a wide range of products (see previous chapter). In the case of an alkylbenzene, radical attack occurs preferentially at the carbon atom nearest the ring. For example, the radical chlorination (Cl_2, UV) of ethylbenzene $C_6H_5CH_2CH_3$ results in one dominant product, $C_6H_5CHClCH_3$.

Chlorine substitutes at this position because the radical formed during the attack is much more stable than alternative radicals. The unpaired electron is delocalized with the π cloud.

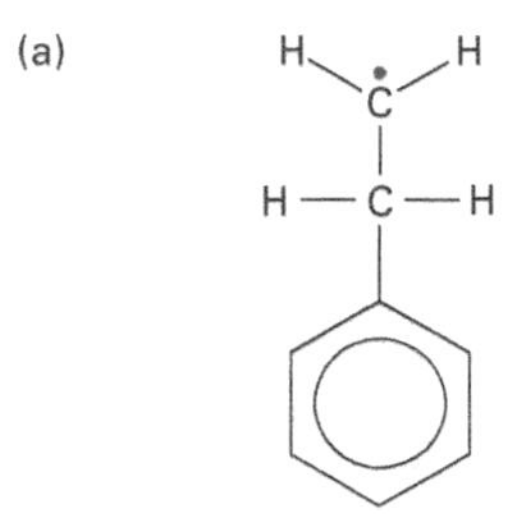

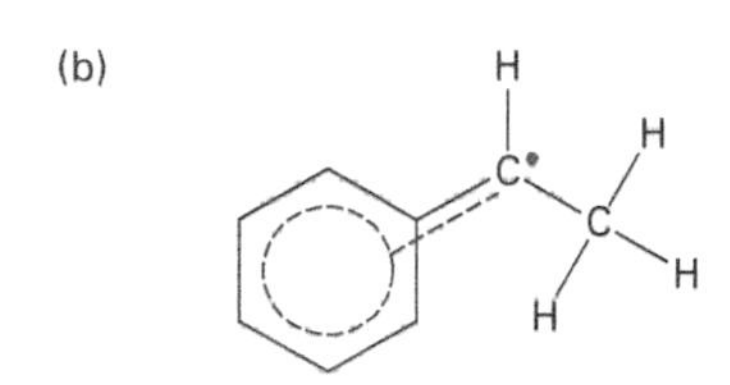

(a) Less stable radical. (b) More stable radical with its electron delocalized with the ring. Being more stable, it will stand a better chance of reacting to form the product.

Chlorine radical or chlorine ion

Remember that, if the chlorine species approaching the arene is the electrophilic Cl^+ ion, reaction will not occur at the alkyl side chain. Instead the electrophile will react with the benzene ring, and substitute a chlorine for a hydrogen in the ring. The products of chlorination are therefore strongly dependent on the exact nature of the species that approaches and hence on the reaction conditions.

To *add* chlorine to benzene itself, it is necessary to use the same vigorous conditions as those required to *substitute* chlorine for hydrogen in an alkane. In the presence of ultraviolet light, chlorine *adds* to benzene in a radical chain reaction to form the saturated halogeno compound 1,2,3,4,5,6-hexachlorocyclohexane, which has eight structural isomers, one of which is the 'organochlorine' insecticide *Lindane*:

$$C_6H_6(l) + 3Cl_2(g) \rightarrow C_6H_6Cl_6(l)$$

Adding chlorine to benzene destroys the delocalized π cloud of electrons. The product molecule is saturated.

SUMMARY

- Arenes most commonly react by electrophilic substitution.
- Halogen carriers polarize halogens so they become electrophiles capable of attracting electron density from the benzene ring π cloud.
- Halogen carriers include $FeCl_3$ and $AlCl_3$ for chlorination and $FeBr_3$ and $AlBr_3$ for bromination.

PRACTICE

1 In the chlorination of benzene by electrophilic substitution, chlorine and the halogen carrier $AlCl_3$ form Cl^+ and $AlCl_4^-$.

a Why is $AlCl_3$ described as being 'electron-deficient'?

b Why are Lewis acids used as halogen carriers?

c How many electrons are delocalized around the benzene ring (i) in the reaction intermediate $[C_6H_6Cl]^+$ and (ii) in the product C_6H_5Cl?

2 Draw structural formulae to show the reactant, reaction intermediate, and product in the radical reaction between propylbenzene and chlorine in the presence of ultraviolet light.

23.3

OBJECTIVES

- Nitrating mixture
- Nitrating benzene
- Nitrating nitrobenzene
- TNT

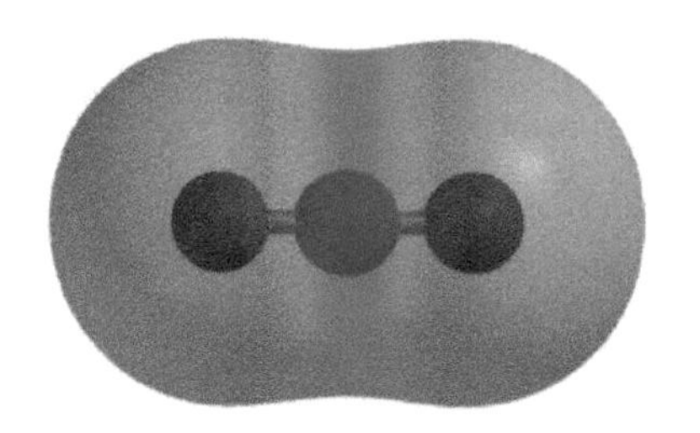

The nitrogen atom of the nitronium ion NO_2^+ carries most of the positive charge (indicated by the blue colour).

NITRATION AND NITROBENZENE

The nitro group $—NO_2$ is a very important functional group when attached to a benzene ring. Nitrobenzene is a yellow oily liquid (T_m = 6 °C; T_b = 211 °C) which smells of bitter almonds. It is far more important commercially than is chlorobenzene (discussed in the previous spread). Nitrobenzene is the starting material for the production of a wide variety of substances, including dyes and explosives, see spreads 28.3 and 28.4 for details. The colours of the clothes you are wearing are probably courtesy of a few teaspoonsful of nitrobenzene – as are the continuing tragic effects of the land mines hidden in abandoned battlefields around the world.

The nitrating mixture

Substituting a nitro group into a benzene ring is called **nitration**. The reagent that brings about the reaction is called a **nitrating mixture**. When nitrating benzene itself, the nitrating mixture consists of concentrated nitric acid and concentrated sulphuric acid, at a temperature of about 50 °C. These two acids react together to produce a powerful electrophile called the **nitronium ion** NO_2^+.

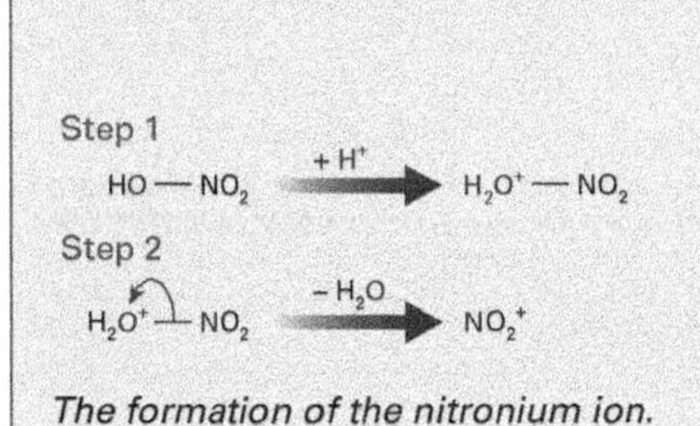

The formation of the nitronium ion.

The overall reaction is:

$HNO_3 + 2H_2SO_4 \rightarrow NO_2^+ + H_3O^+ + 2HSO_4^-$

The nitronium ion can be thought of as forming in two steps.

Step 1 Sulphuric acid protonates nitric acid HNO_3 (more correctly written as $HONO_2$, spread 19.8):

$HONO_2 + H_2SO_4 \rightarrow H_2O^+—NO_2 + HSO_4^-$

Step 2 The protonated nitric acid then loses water:

$H_2O^+—NO_2 \rightarrow NO_2^+ + H_2O$

The ion NO_2^+ is a powerful electrophile because it has a full positive charge.

Nitrating benzene

The overall reaction for the preparation of nitrobenzene may be represented by the chemical equation:

$$C_6H_6 + NO_2^+ \rightarrow C_6H_5NO_2 + H^+$$

The actual substitution reaction happens in the following three distinct steps.

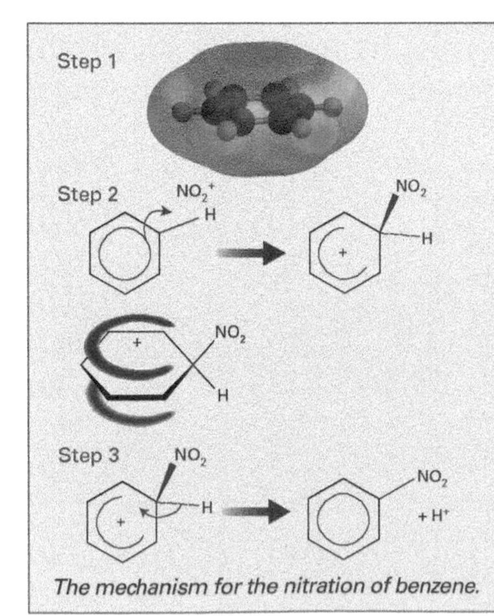

The mechanism for the nitration of benzene.

Step 1 The NO_2^+ ion is attracted to the high electron density in the π cloud of delocalized electrons above and below the plane of the benzene ring.

Step 2 An electron pair from the ring forms a bond between one of the carbon atoms and the approaching electrophile. This carbon atom has now formed four single bonds and so *no longer contributes to the delocalized system*. The intermediate species is positively charged because the positively charged electrophile did not bring any bonding electrons with it; *both* the electrons of the new C—N bond came from the ring.

Step 3 The pair of electrons in the C—H bond returns to the delocalized orbitals on the benzene ring. The hydrogen atom no longer possesses its valence electron and so is ejected as the ion H^+.

Note that, in step 2, the benzene ring exists as a positively charged ion. This situation is equivalent to the carbocation formed when an electrophile reacts with an alkene (see previous chapter). In that case a nucleophile attacks the carbocation, resulting in an *addition* reaction. In the case of the benzene ring, such a reaction would mean a significant loss of delocalization, which is energetically unfavourable. Instead, the intermediate deprotonates, which restores the delocalized system over all six carbon atoms and results in overall *substitution*.

Apparatus for the nitration of benzene. The nitrating mixture is prepared in the flask. Benzene is slowly run in from the funnel while the flask is cooled in water. Then the flask is heated to 50 °C by a water bath to convert the benzene to nitrobenzene.

Dinitrobenzene

The nitration of benzene takes place at 50 °C. If the temperature is increased to 100 °C, a second nitro group will substitute. When two or more groups are attached to a ring, their relative positions are indicated by numbering the six carbon atoms that make up the ring. In this *disubstitution* of benzene by nitro groups, 1,3-dinitrobenzene (T_m = 90 °C) will result. The reason for the production of this particular isomer will be explained in the following spread.

Nitration of substituted benzene rings

The nitrating mixture will also nitrate other substituted benzene rings. For example, phenol (C_6H_5OH) forms 2,4,6-trinitrophenol (formerly known as picric acid). Methylbenzene forms 2,4,6-trinitromethylbenzene (formerly known as trinitrotoluene, **TNT**). Notice that both these reactions happen readily in comparison to the nitration of nitrobenzene, which requires higher temperatures. The influence of functional groups on the rates of further substitution reactions will be dealt with in the following spread.

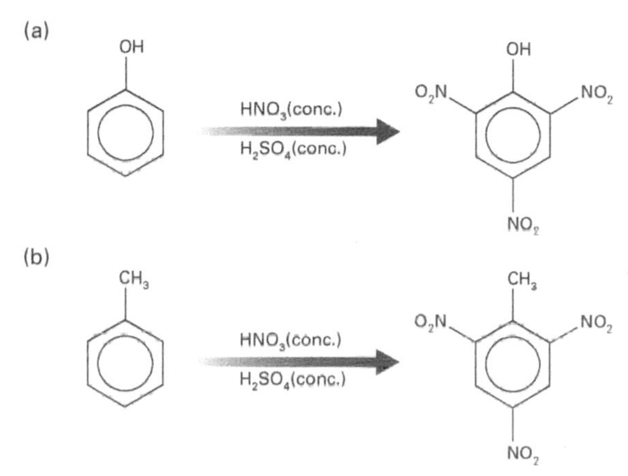

(a) Phenol consists of a hydroxyl (—OH) group attached to a benzene ring. The nitrating mixture readily converts it to bright yellow 2,4,6-trinitrophenol. This substance was used as a high explosive, particularly during the First World War. (b) Methylbenzene converts eventually and with difficulty to 2,4,6-trinitromethylbenzene.

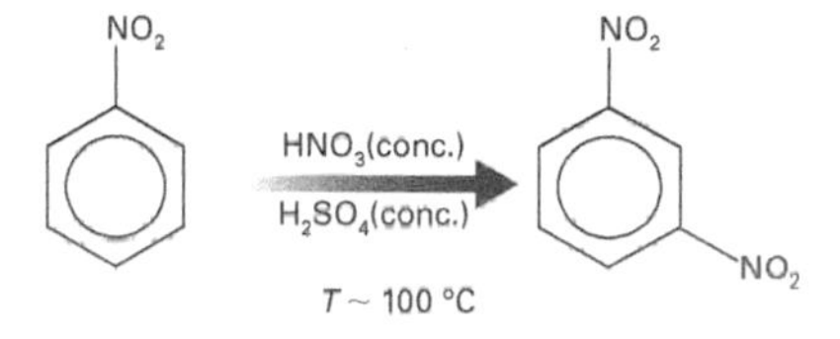

The reaction of nitrobenzene to form 1,3-dinitrobenzene.

SUMMARY

- The formula of the nitro group is $—NO_2$.
- The nitro group is substituted into the benzene ring by the nitrating mixture, which consists of concentrated nitric acid and concentrated sulphuric acid.
- The nitrating mixture produces the powerfully electrophilic nitronium ion NO_2^+.
- Further nitro groups may be substituted into nitrobenzene, at higher temperatures.

A space-filling model of 2,4,6-trinitro-methylbenzene (top). This is the high explosive trinitrotoluene (TNT).

PRACTICE

1 Draw Lewis structures to show the electronic structures of nitric acid and sulphuric acid. Show how these molecules interact to produce the nitronium ion NO_2^+.

2 With reference to the nitration of benzene to form nitrobenzene:

a Why is the NO_2^+ ion described as an electrophile?

b Why does the nitro group substitute into the benzene ring and not add to it?

c Each product molecule contains one nitro group only and does not go on to further substitution. Explain why.

3 Methylbenzene nitrates to form 2,4,6-trinitromethylbenzene. Look at the photograph of the space-filling model above and suggest why the formation of 2,3,4,5,6-pentanitromethylbenzene is highly unlikely.

23.4

OBJECTIVES

- The 2-/4-directing and 3-directing groups
- Activating and deactivating groups
- Wheland intermediate

THE EFFECT OF AN EXISTING FUNCTIONAL GROUP

Arenes most commonly react by undergoing electrophilic substitution of hydrogen atoms attached to the benzene ring. When the arene already has a functional group attached to the ring, this group may have two distinct effects on subsequent substitution reactions. First, the existing group may activate (or deactivate) the ring, making the ring respectively more (or less) likely than benzene to substitute further hydrogen atoms. Secondly, an existing group may direct incoming groups to specific positions on the ring relative to itself. As you will see, the key to this behaviour is the charged reaction intermediate – the so-called 'Wheland intermediate'.

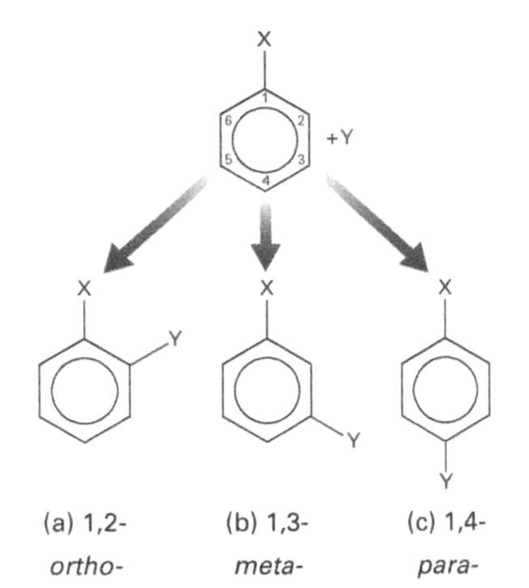

The three possible isomers of a disubstituted benzene compound.

Directing incoming groups

When one group is already attached to a benzene ring, a second incoming group may attach itself at three possible positions. The nature of the first group 'directs' (controls) the position where the second substitutes. For example, nitration is a reaction that substitutes a $—NO_2$ functional group for a hydrogen atom on the benzene ring. Nitration of nitrobenzene forms 1,3-dinitrobenzene. On the other hand, nitration of methylbenzene forms a mixture of 2- and 4-nitromethylbenzene. (Positions 2 and 6 are identical owing to the symmetry of the ring, as are positions 3 and 5.) Extreme conditions cause three nitro groups to be attached to form 2,4,6-trinitromethylbenzene (TNT). Note that two of the hydrogen atoms remain; they are *not* substituted by nitro groups.

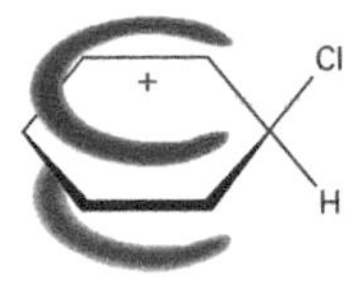

Electrophilic substitution of benzene – a reminder. The electrophilic substitution of benzene proceeds via a Wheland intermediate. *This has a positive charge; the incoming group and the departing hydrogen atom are both bonded to one of the carbon atoms of the benzene ring.*

The 2-/4-directing methyl group

In the Wheland intermediate, electrons are delocalized around five carbon atoms. However, it is most important to note that the positive charge is *not* spread *evenly* around these five carbon atoms, as shown by the colour coding in its electrostatic potential map alongside. Consider the nitration of methylbenzene. There are three possible Wheland intermediates corresponding to the final products 2-, 3-, and 4-nitromethylbenzene (see the diagrams below). In the intermediate, the positive charge is concentrated on the carbon atoms in positions 2, 4, and 6 *relative to the incoming electrophile*. If an electron-donating group (see the previous chapter) is already present in any of these positions, it will donate electron density, lessening the positive charge on the ring, and so stabilize the Wheland intermediate. The Wheland intermediate leading to the 1,2- and 1,4- isomers will have lower energy than the Wheland intermediate leading to the 1,3- isomer.

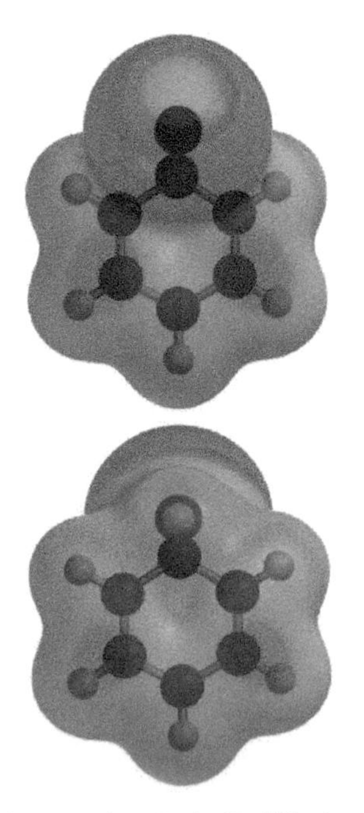

The electron density in the Wheland intermediate is not evenly distributed as is made clear by this electrostatic potential map; the lower view is from underneath (compared with the image on spread 23.2).

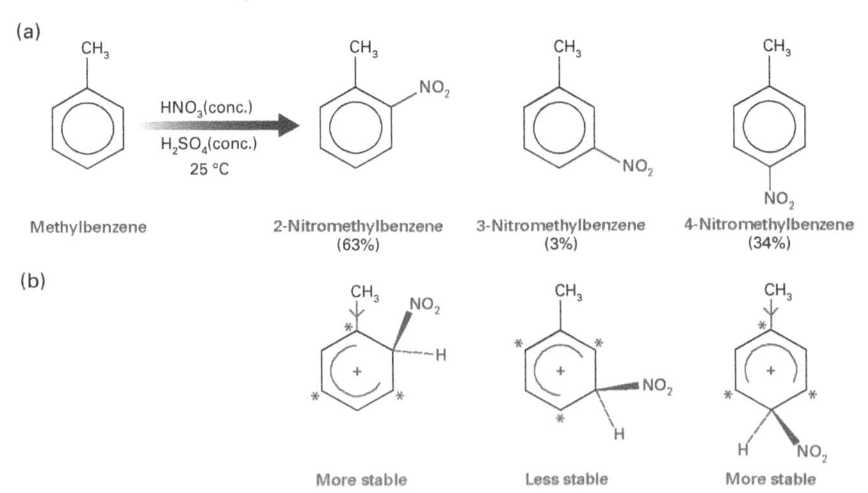

(a) The nitration of methylbenzene produces more of the 2- and 4- isomers than of the 3- isomer. (b) The Wheland intermediates leading to 2-nitromethylbenzene, 3-nitromethylbenzene, and 4-nitromethylbenzene. The positions at which the positive charge is concentrated are labelled with a star (positions 2, 4, and 6 relative to the incoming $—NO_2$ group). The $—CH_3$ group donates electron density to the ring and therefore stabilizes the two intermediates with an adjacent positive charge.

The ring-activating methyl group

The nitration of methylbenzene requires a temperature some 25 °C lower than that required for the nitration of benzene. The methyl group therefore has the effect of activating the ring to further substitution. By stabilizing the 1,2- and 1,4- Wheland intermediates, the methyl group lowers the activation energy and increases the rate of the reaction. In general:

- An *electron-donating group* attached to a benzene ring will *increase* the rate of further substitution relative to benzene; it will also direct the incoming group to the 2 or 4 position.

The 3-directing deactivating groups

Electron-withdrawing groups, such as the nitro —NO_2 group, deactivate the ring relative to benzene. As before, the positive charge on the Wheland intermediate is concentrated on the carbon atoms in positions 2, 4, and 6 *relative to the incoming electrophile*. The presence of an electron-withdrawing group in any of these positions *destabilizes* the Wheland intermediate by increasing the positive charge on the ring. This effect is reduced if the second group attaches to the 3 (or the equivalent 5) position relative to the existing nitro group. More severe conditions are needed for the nitration of nitrobenzene than for the nitration of benzene; the nitro group deactivates the ring to further substitution.

- An *electron-withdrawing group* attached to a benzene ring will *decrease* the rate of further substitution relative to benzene; it will also direct the incoming group to the 3 position.

2-/4- directing
— OH (and — NH_2) strongly activating
— CH_3 } activating
— R (other alkyl groups) } activating

3- directing
— NO_2 strongly deactivating
—C(=O)CH_3 deactivating

The influence of existing groups.

Phenol

The —OH (and —NH_2) group stabilizes the Wheland intermediate by another mechanism, namely extending the delocalization out onto the oxygen atom. This is possible only in the 2-and 4- positions and so phenol is also 2-/4-directing.

(a) NO_2 → HNO_3(conc.), H_2SO_4(conc.), $T \sim 100$ °C → NO_2, NO_2

(b) Less stable — More stable — Less stable

(a) The nitration of nitrobenzene produces 1,3-dinitrobenzene and very little of the 1,2- and 1,4- isomers. (b) The Wheland intermediates leading to 1,2-dinitrobenzene, 1,3-dinitrobenzene, and 1,4-dinitrobenzene. The positions at which the positive charge is concentrated are labelled with a star (positions 2, 4, and 6 relative to the incoming —NO_2 group). The existing —NO_2 group withdraws electron density from the ring and therefore destabilizes the two intermediates with an adjacent positive charge.

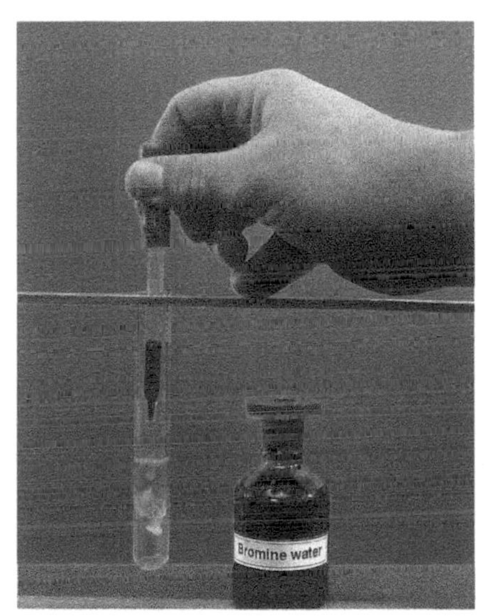

The —OH group in phenol C_6H_5OH activates the ring to substitution. Phenol decolorizes bromine water (remember that benzene does not); the product is an immediate white precipitate of 2,4,6-tribromophenol.

2,4,6-tribromophenol.

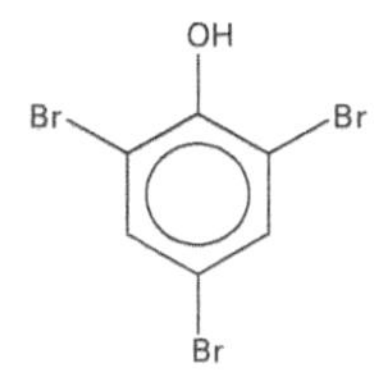

The equivalent chlorine molecule, 2,4,6-trichlorophenol, is known as TCP: TCP is a well-known antiseptic.

SUMMARY

- Disubstituted arenes have three isomers; the 1,2-, the 1,3-, and the 1,4- isomers.
- Electrophilic substitution reactions involve a positively charged Wheland intermediate. The charge is delocalized around five of the carbon atoms of the benzene ring.
- The positive charge of the Wheland intermediate is concentrated at positions 2, 4, and 6 relative to the position of the incoming electrophile.
- Electron-donating groups activate the ring and direct incoming groups to positions 2 and 4 relative to the existing group.
- Electron-withdrawing groups deactivate the ring and direct incoming groups to position 3 relative to the existing group.

PRACTICE

1 Give the formula(e) of the major disubstituted product(s) of each of the following reactions. State also whether the ring will be activated or deactivated by the existing functional group.

a Nitration of ethylbenzene
b Nitration of nitrobenzene.

23.5

THE FRIEDEL–CRAFTS REACTION

OBJECTIVES

- The Friedel–Crafts reaction
- Alkylation
- Acylation
- Oxidation of alkyl groups

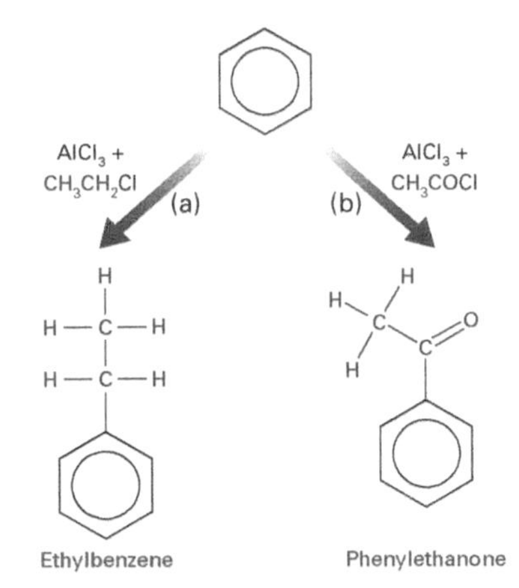

(a) Alkylation and (b) acylation reactions.

Friedel and Crafts

The Friedel–Crafts reaction arose from an unexpected laboratory accident in 1877. James Crafts, a native Bostonian, was collaborating in Paris with Charles Friedel, whom he had met sixteen years earlier on a previous trip to Europe. Crafts had actually returned to Paris mainly because of ill health. Because of the change of climate (along with their discovery!), his health improved dramatically. Crafts did eventually return to the USA and in 1897 was elected President of the Massachusetts Institute of Technology (MIT).

Aeroplanes of the Second World War era were propelled by high-performance piston engines that required high-octane fuel. The yield of this fuel from refining techniques at that time was very low. Friedel–Crafts alkylation of benzene and its hydrocarbon derivatives produced suitable fuel, which made great contributions to the war effort of the Allies.

Synthesis in organic chemistry involves making complex molecules from simpler starting materials. You are familiar with substituting chlorine, bromine, and nitro groups into benzene and its derivatives. These groups are described as *functional groups*; they can undergo further changes in subsequent reactions. Many arenes contain relatively inert alkyl groups attached to the benzene ring. Attaching alkyl groups involves a number of problems.

The Friedel–Crafts reaction

The **Friedel–Crafts reaction** attaches extra carbon atoms to a benzene ring. In this reaction, the arene reacts with a halogen-containing organic molecule. The whole of the molecule attached to the chlorine atom substitutes into the benzene ring. There are two variants of the Friedel–Crafts reaction: **alkylation** attaches an alkyl group (—R); and **acylation** attaches an acyl group (—COR).

In both cases the C—Cl bond in the approaching molecule is polarized, $C^{\delta+}—Cl^{\delta-}$ (because carbon has a lower electronegativity than chlorine). However, the halogeno compound alone is not a sufficiently powerful electrophile to react with the benzene ring. Adding a Lewis acid such as aluminium chloride breaks the C—Cl bond producing an electrophile, which can successfully react with the benzene ring.

Friedel–Crafts alkylation

If the halogen reagent in a Friedel–Crafts reaction is a halogenoalkane, the attached group is an alkyl group and the reaction is called a Friedel–Crafts alkylation. For example, chloroethane CH_3CH_2Cl (in industry C_2H_4 and HCl are used in place of CH_3CH_2Cl) reacts with benzene in the presence of an aluminium chloride $AlCl_3$ catalyst (under *anhydrous* conditions, as $AlCl_3$ hydrolyses) to give ethylbenzene $C_6H_5CH_2CH_3$, which is an intermediate in the manufacture of polystyrene:

$$CH_3CH_2Cl + C_6H_6 \xrightarrow{AlCl_3} C_6H_5CH_2CH_3 + HCl$$

The reaction takes place in a series of steps, as follows:

Step 1 The chloroethane donates electron density into a vacant valence orbital on electron-deficient aluminium chloride (the $AlCl_3$ is thus acting as a Lewis acid):

$$CH_3CH_2Cl + AlCl_3 \rightarrow CH_3CH_2^+ + AlCl_4^-$$

giving rise to the ethyl carbocation, which is a powerful electrophile.

Step 2 The electrophile reacts with the benzene ring, forming the Wheland intermediate (see opposite page):

$$C_6H_6 + CH_3CH_2^+ \rightarrow [C_6H_6CH_2CH_3]^+$$

Step 3 The Wheland intermediate deprotonates, forming the product (ethylbenzene in this example) by substitution, and thus reforming the delocalized π cloud of the benzene ring, and the catalyst is regenerated:

$$[C_6H_6CH_2CH_3]^+ \rightarrow C_6H_5CH_2CH_3 + H^+$$
$$AlCl_4^- + H^+ \rightarrow AlCl_3 + HCl$$

The main problem with Friedel–Crafts *alkylation* is that the alkyl group is electron-donating and *activates* the ring (see previous spread), making it more susceptible to further reaction. The initial product is *more* reactive than the original benzene, and there is a good chance that further alkyl groups will substitute as well, giving a mixture of products. The required product then has to be separated out from the mixture.

Friedel–Crafts acylation

This reaction is similar to the alkylation reaction above, except that the acyl group —COR becomes attached to the benzene ring.

> For example, ethanoyl chloride CH_3COCl reacts with the halogen carrier $AlCl_3$ as follows:
>
> $CH_3COCl + AlCl_3 \rightarrow [CH_3CO]^+ + AlCl_4^-$
>
> The ethanoyl ion $[CH_3CO]^+$ is a powerful electrophile. It approaches the benzene ring, which undergoes an electrophilic substitution reaction. The chemical equation for the overall reaction is:
>
> $CH_3COCl + C_6H_6 \xrightarrow{AlCl_3} C_6H_5COCH_3 + HCl$
>
> The compound formed in a Friedel–Crafts acylation is a *ketone*, phenylethanone $C_6H_5COCH_3$ in this example.

Friedel–Crafts *acylation* produces a product that is *less* reactive than the original benzene. The acyl group withdraws electron density and so deactivates the ring (see previous spread), making it less susceptible to further electrophilic reaction. As a result, the initial product is unlikely to react with a second acyl group. Unlike the case of Friedel–Crafts alkylation, acylation gives one predominant product.

Oxidizing an alkyl group

Alkanes are almost completely unaffected by oxidants. Alkenes split their carbon chains when treated with powerful oxidants such as hot concentrated acidified manganate(VII) ions (see chapter 29 'Organic synthesis'). In complete contrast, alkyl groups attached to a benzene ring undergo oxidation relatively readily. The whole alkyl group is oxidized to a carboxyl group —COOH.

> For example, alkaline potassium manganate(VII) oxidizes methylbenzene to benzoic acid (after neutralization):
>
> $C_6H_5CH_3 \xrightarrow{[O]} C_6H_5COOH + H_2O$
>
> (here the symbol [O] indicates oxidation). The product always has a carboxyl group attached directly to the benzene ring. Methylbenzene, ethylbenzene, or any other benzene derivative with an alkyl side chain (however long) will produce the *same* product, *benzoic acid*. The rest of the side chain is destroyed, forming carbon dioxide and water. For example,
>
> $C_6H_5CH_2CH_3 \xrightarrow{[O]} C_6H_5COOH + CO_2 + 2H_2O$
>
> The resulting benzoic acid is stabilized by delocalization of the carboxyl group with the benzene ring.

SUMMARY

- Friedel–Crafts alkylation attaches an alkyl group (e.g. the ethyl group —CH_2CH_3) to a benzene ring.
- Friedel–Crafts acylation attaches an acyl group (e.g. the ethanoyl group —$COCH_3$) to a benzene ring.
- Friedel–Crafts reactions use a Lewis acid catalyst such as aluminium chloride.
- For alkylation, use a halogenoalkane, e.g. chloroethane CH_3CH_2Cl.
- For acylation, use an acyl chloride, e.g. ethanoyl chloride CH_3COCl.
- Alkylbenzene compounds all oxidize to benzoic acid C_6H_5COOH, regardless of the length of the alkyl group attached to the benzene ring.

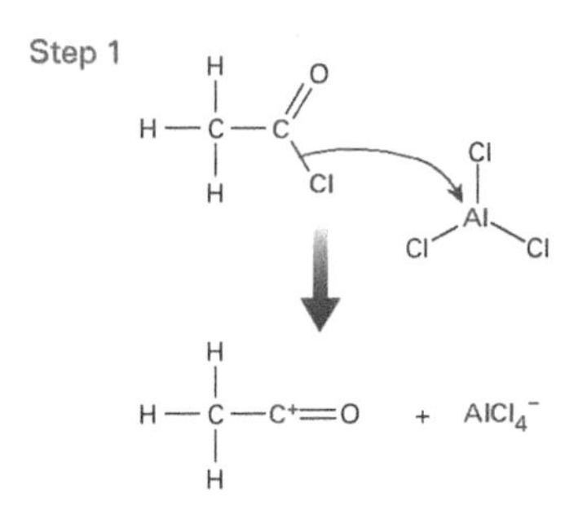

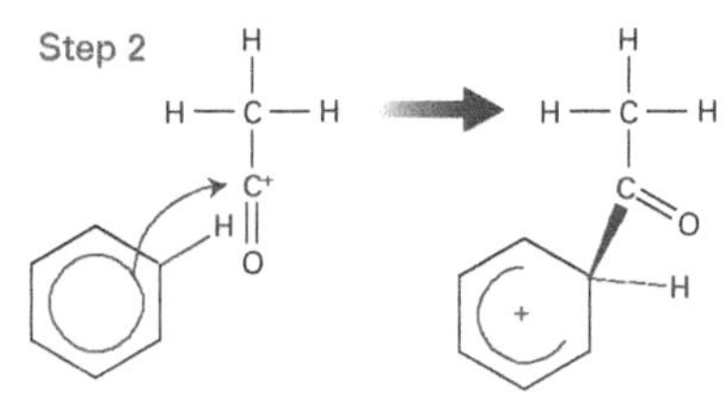

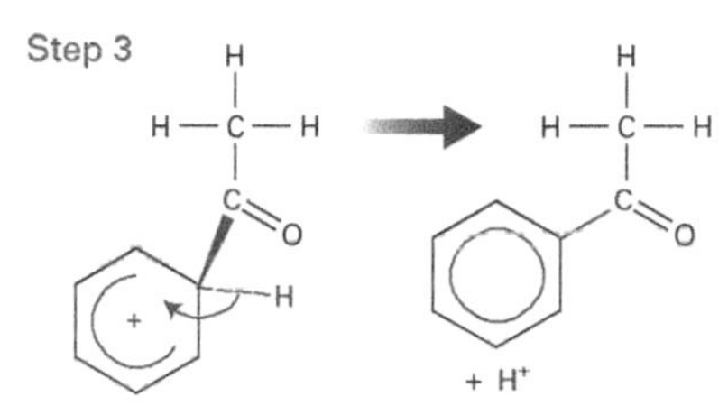

Friedel–Crafts acylation.
Step 1: Formation of the electrophile.
Step 2: Formation of Wheland intermediate.
Step 3: Deprotonation.

$CH_3CH_2CH_2CH_2CH_2CH_2CH_2CH_2CH_2CH(CH_3)–C_6H_4–SO_3^-Na^+$

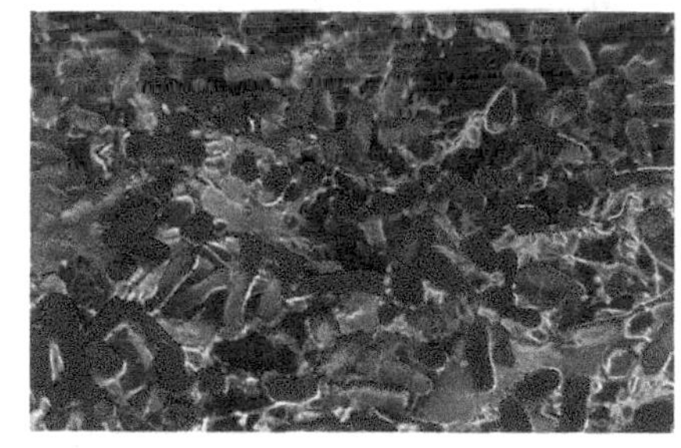

The Friedel–Crafts reaction is extremely useful in industry. Synthetic rubber, plastics, detergents, and high-octane unleaded petrol all depend on this one reaction. For example, the common biodegradable synthetic detergent sodium dodecylbenzenesulphonate has the 12-carbon side chain attached by a Friedel–Crafts reaction. (The next spread describes how the sulphonate group can be attached.) Bacteria, such as Escherichia coli *(shown magnified 42 000 times) can degrade this detergent.*

PRACTICE

1 Starting with benzene, outline the reaction(s) you would use to produce:
 a Propylbenzene
 b Diphenylmethanone ($C_6H_5COC_6H_5$)
 c Benzoic acid.

2 Draw structural formulae to describe the mechanism of the Friedel–Crafts reaction being used to attach an ethyl group to a benzene ring.

23.6

Sulphonation of benzene

OBJECTIVES

- Benzenesulphonic acid
- Preparation
- Uses

Benzene undergoes reactions and forms compounds which are unlike those of aliphatic compounds. For example, phenol C_6H_5OH is acidic, whereas hexan-1-ol $C_6H_{13}OH$ is neutral; ethylbenzene $C_6H_5CH_2CH_3$ forms an acid on oxidation, whereas octane C_8H_{18} does not react. The final spread of this chapter introduces the sulphonate group —SO_2OH. Joining this group to the benzene ring by a carbon–sulphur bond forms benzenesulphonic acid. This compound is of great industrial importance in the manufacture of detergents, dyestuffs, and drugs. The alkyl analogue is unstable in the presence of water and is of no commercial interest. As before, the delocalized electrons around the benzene ring are the key to the stability of the arene compound.

Sulphonation of benzene

Benzene is sulphonated by refluxing it with concentrated sulphuric acid. The formula of this acid is usually written as H_2SO_4. However, the form $(HO)_2SO_2$ (see chapter 17 'Trends across a period') reminds you that it is an oxoacid, with two —OH groups and two oxygen atoms attached to a central sulphur atom.

The actual electrophile is sulphur trioxide SO_3 and the overall reaction may be represented as:

$$C_6H_6 + SO_3 \rightarrow C_6H_5SO_2OH$$

The resulting product is benzenesulphonic acid, which is a colourless oily liquid. The reaction proceeds in a series of steps.

Step 1

$$(HO)_2SO_2 \rightleftharpoons SO_3 + H_2O$$

Step 2

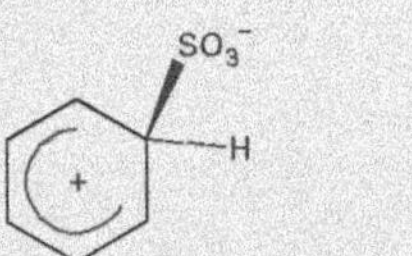

Step 3

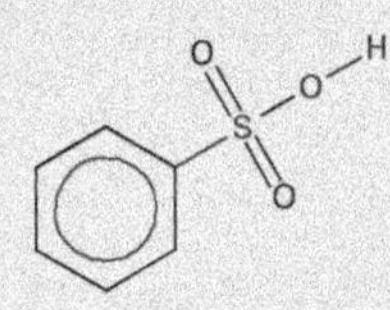

The mechanism of the sulphonation of benzene.

Step 1 Concentrated sulphuric acid contains a proportion of free sulphur trioxide as a result of the equilibrium:

$$H_2SO_4 \rightleftharpoons SO_3 + H_2O$$

Although it is a neutral molecule, sulphur trioxide is a powerful electrophile because oxygen has a greater electronegativity than sulphur, and so the sulphur atom has a significant partial positive charge.

Step 2 The sulphur trioxide electrophile is attracted to the delocalized π cloud of the benzene ring and forms a Wheland intermediate by bonding with a carbon atom on the ring:

$$SO_3 + C_6H_6 \rightarrow [C_6H_6]^+SO_3^-$$

Note that the intermediate bears a negative charge on the —SO_3^- group and a positive charge delocalized around the benzene ring.

Step 3 The benzene ring deprotonates from the carbon atom attached to the —SO_3^- group. Another proton then adds on to the sulphonate group:

$$[C_6H_6]^+SO_3^- \rightarrow C_6H_5SO_2OH$$

Acid character

Benzenesulphonic acid is a weak acid:

$$C_6H_5SO_2OH + H_2O \rightleftharpoons H_3O^+ + C_6H_5SO_3^-$$

It forms salts on reaction with base. For example,

$$C_6H_5SO_2OH + NaOH \rightarrow Na^+C_6H_5SO_3^- + H_2O$$

The salt $Na^+C_6H_5SO_3^-$ is called sodium benzenesulphonate.

The sulphonic acid group in benzenesulphonic acid is stabilized by delocalization of its electron density around the benzene ring. The three oxygen atoms in the ion $C_6H_5SO_3^-$ are all equivalent, as shown in the electrostatic potential map alongside.

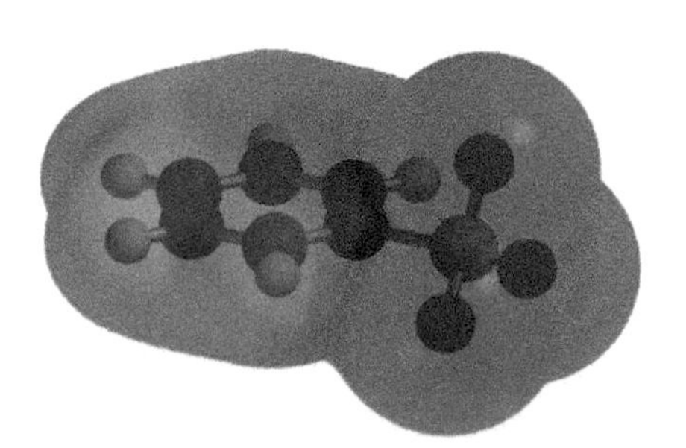

Delocalization of electron density within the benzenesulphonate ion.

Applications

Benzenesulphonates are widely used as a feedstock from which detergents are made, see previous spread. The sulphonate group introduces a polar region into an otherwise non-polar molecule, improving its overall solubility in water. Sulphonic acid groups are also present in some dyes. Their acidic groups attach the dye molecules to basic amino groups in wool or silk. In the case of cotton, sulphonic acid groups form ester linkages with the hydroxyl groups on the cellulose fibres of the cotton. A family of antibiotics called the *sulphonamides* have very similar molecular structures to sulphonates (see spread 28.7).

The structure of the sulphonamide antibiotic sulphathiazole, effective against Staphylococci.

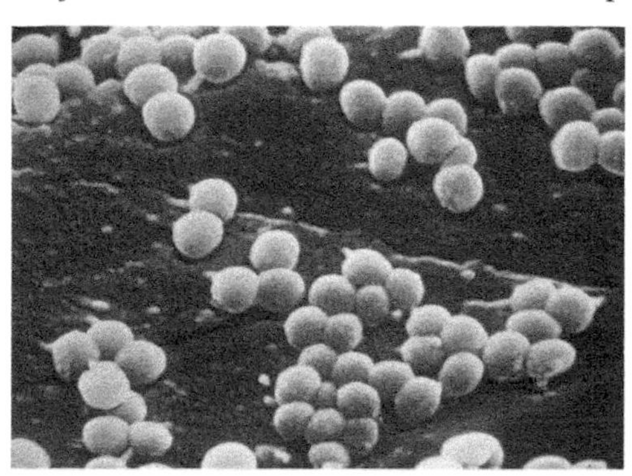

The bacteria Staphylococci *form clusters resembling grapes.*

SUMMARY

- The sulphonic acid group has the formula $—SO_2OH$.
- Benzenesulphonic acid is a weak acid in solution, forming salts such as $Na^+C_6H_5SO_3^-$.
- The sulphonating agent is concentrated sulphuric acid; it provides the electrophile sulphur trioxide SO_3.
- The sulphonic acid and sulphonate groups are highly polar, increasing the water solubility of molecules to which they are attached.
- Derivatives of benzenesulphonic acid include detergents, dyestuffs, and drugs.

PRACTICE

1 Draw structural formulae and Lewis structures for sulphuric acid and sulphur trioxide.

2 Use 'curly arrows' to show how electron pairs move to break and make bonds when:

a sulphuric acid protonates water;

b benzenesulphonic acid protonates water.

3 Explain why it is difficult to substitute a second sulphonic acid group into the benzenesulphonic acid molecule.

4 What would you expect to see when benzenesulphonic acid is added to aqueous potassium carbonate? Write a *balanced* chemical equation for the reaction.

Chapter 23 Reactions summary

- Chlorine substitution using a Lewis acid as a halogen carrier:
  $C_6H_6 \xrightarrow{Cl_2/AlCl_3} C_6H_5Cl$
- Bromine substitution using a Lewis acid as a halogen carrier:
  $C_6H_6 \xrightarrow{Br_2/AlBr_3} C_6H_5Br$
- The nitrating mixture producing the nitronium ion:
  $HNO_3 + 2H_2SO_4 \rightarrow NO_2^+ + H_3O^+ + 2HSO_4^-$
- Preparation of nitrobenzene:
  $C_6H_6 + NO_2^+ \xrightarrow{50\,°C} C_6H_5NO_2 + H^+$
  Further nitration forms 1,3-dinitrobenzene.
- Friedel–Crafts alkylation:
  $C_6H_6 + CH_3CH_2Cl \xrightarrow{AlCl_3} C_6H_5CH_2CH_3 + HCl$
- Friedel–Crafts acylation:
  $C_6H_6 + CH_3COCl \xrightarrow{AlCl_3} C_6H_5COCH_3 + HCl$
- Oxidation of alkyl side chain (however long) on benzene:
  $C_6H_5CH_2CH_3 \xrightarrow{[O]} C_6H_5COOH + CO_2 + 2H_2O$
- Sulphonation of benzene by concentrated sulphuric acid:
  $C_6H_6 + SO_3 \rightarrow C_6H_5SO_2OH$

PRACTICE EXAM QUESTIONS

1 Alkylation of an aromatic compound can be carried out using the Friedel–Crafts reaction. For example, benzene reacts with chloromethane in the presence of aluminium chloride to give methylbenzene.

$$C_6H_6 + CH_3Cl \xrightarrow{AlCl_3} C_6H_5CH_3 + HCl$$

a Hydrogen chloride is produced as the reaction proceeds. Describe how you could show when the reaction had finished. [2]

b The chloromethane reacts with aluminium chloride to give $CH_3^+[AlCl_4]^-$.

$CH_3Cl + AlCl_3 \rightarrow CH_3^+[AlCl_4]^-$

i The carbonium ion, CH_3^+, is an **electrophile**. Explain this term.

ii The mechanism for the reaction of CH_3^+ with a benzene ring is similar to that of the nitronium ion with benzene. Using a flow scheme suggest a mechanism for the reaction of CH_3^+ with benzene. [5]

c A problem with alkylation is that the reaction does not stop at methylbenzene; further alkylation of the benzene ring takes place to form dimethylbenzenes.

i Draw the structures of the three dimethylbenzenes and give the systematic name of **one** of them.

ii State a technique by which you could separate a mixture of the dimethylbenzenes. [5]

d One molecule of benzene reacts with one molecule of a chloroalkane to form an alkylbenzene:

$$RCl + C_6H_6 \xrightarrow{AlCl_3} C_6H_5R + HCl$$

The percentage composition by mass of the alkylbenzene is 90.0% carbon and 10.0% hydrogen.

i Calculate the empirical formula of the alkylbenzene.

ii Determine the molecular formula of the alkylbenzene. Explain your method. [4]

e Acylation of the benzene ring can occur with an acyl chloride.

$$C_6H_6 + CH_3COCl \xrightarrow{AlCl_3} X + HCl$$

Draw the structure of the product X. [2]

2 The following is a modified account of a method for the preparation of phenylethanone, $C_6H_5COCH_3$.

- Place 3 g of finely powdered anhydrous aluminium chloride and 7.5 cm^3 (6.6 g) of dry benzene in a 50 cm^3 round-bottomed flask.
- Fit the flask with a reflux condenser and place in a cold water bath in a fume cupboard.
- Slowly add, down the condenser, 2 cm^3 (2.2 g) of ethanoyl chloride.
- Heat the flask in a water bath at 50 °C for 30 minutes, or until no further hydrogen chloride is evolved.
- Pour the reaction mixture into a 100 cm^3 flask containing 20 cm^3 of water, and shake vigorously.
- Transfer to a separating funnel and discard the lower aqueous layer.
- Add a dilute solution of sodium hydroxide to the separating funnel. Shake the mixture, and again discard the aqueous layer.
- Dry the organic layer with anhydrous calcium chloride and fractionally distil, collecting the fraction boiling between 195 °C and 205 °C.
- The yield of phenylethanone is 2 g.

a **i** Give the equation for the reaction between ethanoyl chloride and benzene, using structural formulae. [2]

ii Suggest a mechanism for this reaction. [3]

b Suggest reasons for each of the following steps in the preparation:

i finely powdering the aluminium chloride; [1]

ii drying the benzene; [1]

iii adding the ethanoyl chloride slowly; [1]

iv using a fume cupboard; [1]

v shaking the product with sodium hydroxide solution. [1]

c The benzene is in excess in this preparation, being used as the solvent also.

i Calculate the mass of phenylethanone which could theoretically be obtained from this preparation based on the mass of ethanoyl chloride used. [The molar mass of ethanoyl chloride is 78.5 g mol^{-1} and that of phenylethanone is 120 g mol^{-1}.] [3]

ii What is the percentage yield in this preparation? [1]

iii Suggest reasons why the actual yield for organic reactions generally is significantly less than the theoretical yield. [2]

3 The following passage gives a method for the preparation of nitrobenzene. Place 35 cm^3 of concentrated nitric acid in a 500 cm^3 flask, and add slowly 40 cm^3 of concentrated sulphuric acid, keeping the mixture cool during the addition by immersing the flask in cold water. Place a thermometer in the nitrating mixture, and add very slowly 29 cm^3 of benzene. The benzene should be added about 3 cm^3 at a time, and the contents of the flask thoroughly mixed after each addition; the temperature of the mixture must not be allowed to rise above 40 °C, and the flask must be cooled in cold water if necessary.

When all the benzene has been added, fit a reflux condenser to the flask, and heat the latter in a water bath at 60 °C for 45 minutes. During this period the flask

should be removed from the water bath from time to time and vigorously shaken.
After this heating period, pour the contents of the flask into a large excess of cold water (about 300 cm^3), and stir the mixture vigorously. Decant off as much of the upper aqueous layer as possible, and transfer the residue to a separating funnel. Run off the lower nitrobenzene layer into a beaker, and reject the aqueous layer. Then wash the nitrobenzene successively first with an equal volume of cold water and then with dilute sodium carbonate solution.
Transfer the nitrobenzene to a small flask, add some granular calcium chloride, and leave until the liquid is quite clear. Filter the nitrobenzene into a small, dry flask, and distil, collecting the fraction which boils between 207 °C and 211 °C.

(Adapted from Mann F.G. & Saunders, B.C., *Practical Organic Chemistry*, 4th edn: Longman, 1960.)

a Suggest reasons for each of the following:

i the use of a mixture of concentrated nitric and sulphuric acids; [2]

ii the slow addition of the benzene to the nitrating mixture; [1]

iii the need to keep the temperature below 40 °C during the addition; [1]

iv the need to keep the reaction temperature at 60 °C; [1]

v the necessity to shake the mixture from time to time; [2]

vi the addition of the reaction mixture to an excess of cold water. [1]

vii the need to wash with water and then with sodium carbonate solution. [1]

viii the distillation between 207 °C and 211 °C. [1]

b Nitrobenzene can be nitrated further to give 1,3-dinitrobenzene:

NO_2 / NO_2

By analogy with the reduction of nitrobenzene to phenylamine, suggest reagents and conditions by which 1,3-dinitrobenzene can be converted to 1,3-diaminobenzene. [3]

c Nomex is a du Pont fibre, used for flame-retardant clothing, which can resist temperatures of 1000 °C for 12 seconds, enough to have enabled the Benetton Formula 1 team to have survived a serious fire during the 1994 Grand Prix season. Nomex is a polymer which could in principle be made from 1,3-diaminobenzene and benzene-1,3-dicarboxylic acid:

NH_2 / NH_2 COOH / COOH

i Draw a representative length of the Nomex polymer chain. [2]

ii What type of polymer is Nomex? [1]

iii Dicarboxylic acids are not usually used in making this type of polymer. They are generally made from acid chlorides. Suggest why this is so. [2]

iv Suggest how benzene-1,3-dicarboxylic acid could be converted to its diacid chloride, and draw its structure. [2]

4 a Under different reaction conditions, methylbenzene reacts with chlorine by different reaction mechanisms. One product of each reaction is shown in the figure below.

CH_2Cl (chloromethyl)benzene CH_3, Cl 2-chloromethylbenzene

In each reaction other organic products are also formed.

i For the reaction leading to the formation of (chloromethyl)benzene give:

the reaction conditions;

the formula of another organic product of the reaction. [2]

ii For the reaction leading to the formation of 2-chloromethylbenzene give:

the reaction conditions;

the formula of another organic product which is isomeric with 2-chloromethylbenzene. [2]

b The reaction giving (chloromethyl)benzene as a product is a free radical substitution. Write equations to show the following steps in the reaction mechanism:

i the initiation step; [1]

ii two propagation steps; [2]

iii two termination steps. [2]

c Give the name of the type of mechanism involved in the reaction which gives 2-chloromethylbenzene as a product. [1]

d When the compound with the formula shown in the figure below reacts with aqueous sodium hydroxide, one of the chlorine atoms is replaced.

CH_2Cl, Cl

i Draw the structure of the product of the reaction. [1]

ii Explain the difference in the reactivities of the two chlorine atoms. [4]

24 Organic halogeno compounds

There are two main reasons for making an organic compound: it may be useful in its own right, or it may be a route to some other molecule. Halogeno compounds consist of a halogen atom attached to a carbon skeleton. Their immediate uses are restricted because some are toxic and have damaging effects on the environment. However, they are extremely useful as intermediates for making other types of molecule. The carbon–halogen bond is polar and polarizable, and so reagents may attack it in order to substitute other functional groups. Inserting a halogen into a carbon skeleton often represents the first stage in a multi-step synthetic pathway.

POLAR CARBON–HALOGEN BONDS

OBJECTIVES

- Chloro-, bromo-, and iodoalkanes
- Preparation outlines
- Halogenoalkanes react by nucleophilic substitution
- Halogenoalkanes and halogenoarenes

The halogeno compounds of most interest are the chlorides, bromides, and iodides. Halogen atoms may be attached to alkyl, alicyclic, or aryl carbon skeletons. As you might expect, the alkyl halides and alicyclic halides have similar properties, which are distinctly different from those of the aryl halides (see previous chapter). This introductory spread is concerned with naming the various types of organic halides, revising the reactions by which they are prepared, and outlining the main type of reaction that halogenoalkanes undergo.

Naming halogeno compounds

The names of **halogenoalkanes** (also called **haloalkanes**) result from prefixing the name of the parent alkane with chloro-, bromo-, or iodo-, and indicating, where necessary, the number of the carbon atom to which the halogen is attached. The number of halogens in each molecule is indicated by the prefix di-, tri-, etc. The products of the reaction between chlorine and methane (see chapter 22 'Alkanes and alkenes') contain only one carbon atom, so their names are chloromethane (CH_3Cl), dichloromethane (CH_2Cl_2), trichloromethane ($CHCl_3$), and tetrachloromethane (CCl_4). In a similar manner, the names of halogenoarenes result from adding the appropriate halogeno prefix to the name of the arene, e.g. bromobenzene.

Structural formulae of some halogeno compounds: (a) trichloromethane; (b) 1-bromobutane; (c) 2-chloropropane; (d) 1,1,1-trichloroethane; and (e) bromobenzene.

Electronegativities: a reminder

Atom	Electronegativity
C	2.5
H	2.2
Cl	3.2
Br	3.0
I	2.7

Bond polarity and polarizability

Halogens are electronegative atoms. Attaching a halogen atom to a carbon atom creates a polar covalent bond. Therefore, the carbon–halogen bond provides a δ+ carbon atom for attack by nucleophiles. Furthermore, carbon–halogen bonds, especially C—I bonds, are polarizable by approaching nucleophiles. For this reason, introducing halogen atoms is a useful way of starting the synthesis of complex substances from the hydrocarbons obtained from refining and cracking crude oil.

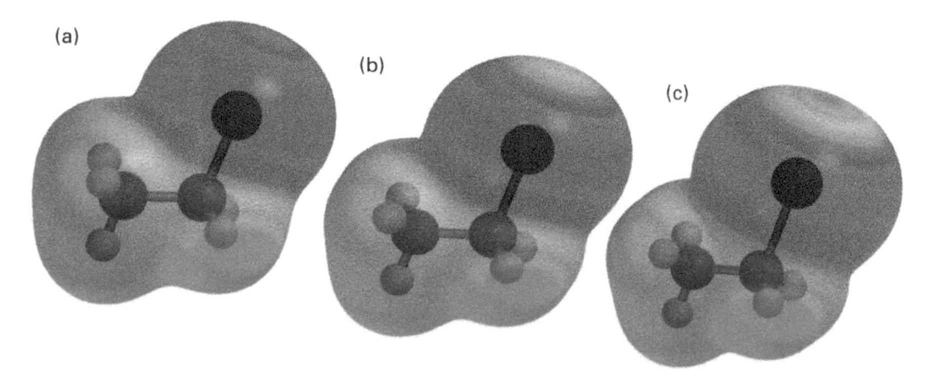

Electrostatic potential maps of the halogenoethanes. Note that the atomic radii are: Cl, 0.099 nm; Br, 0.114 nm; I, 0.133 nm. (a) The C—Cl bond (length 0.177 nm). (b) The C—Br bond (length 0.193 nm). (c) The C—I bond (length 0.214 nm).

Making halogenoalkanes

There are three main methods of attaching halogen atoms (X) to a carbon skeleton, as described in earlier chapters:

- Radical substitution of an alkane

  $CH_3CH_3 + Cl_2 \xrightarrow{UV} CH_3CH_2Cl + HCl$
  (and all other possible products up to CCl_3CCl_3)

- Electrophilic addition of HX or X_2 to an alkene

  $CH_2{=}CH_2 + HBr \rightarrow CH_3CH_2Br$

  $CH_2{=}CH_2 + Br_2 \rightarrow BrCH_2CH_2Br$

Reaction by nucleophilic substitution

The carbon–halogen bond in halogenoalkanes usually reacts by *nucleophilic substitution*. Nucleophiles are electron-pair donors and are attracted to the δ+ carbon atom of the carbon–halogen bond. Furthermore, the approaching nucleophile repels electron density towards the polarizable halogen atom, thus *inducing* a larger dipole in the bond. The carbon–halogen bond eventually breaks and the incoming nucleophile takes its place.

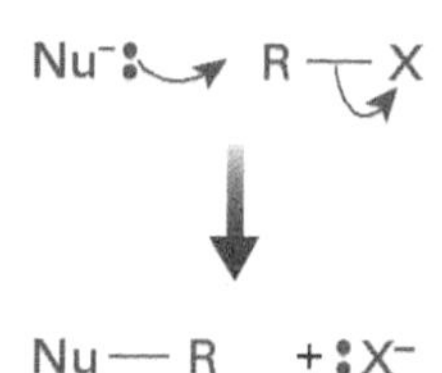

The approach and attack of a nucleophile on a halogenoalkane, causing a substitution reaction.

Reaction rates: halogenoarenes and halogenoalkanes

In halogenoarenes, a lone pair of the halogen atom is delocalized with the aromatic ring. This strengthens the carbon–halogen bond and makes it much harder to break: the C—X bond enthalpy is larger for a halogenoarene than for the corresponding halogenoalkane. (Note that the aryl C—Cl bond length is 0.169 nm compared with the alkyl C—Cl bond length of 0.177 nm.) For this reason, the chlorine atom in chlorobenzene C_6H_5Cl is *not* substituted by aqueous sodium hydroxide to give phenol C_6H_5OH, whereas chloroethane CH_3CH_2Cl forms ethanol CH_3CH_2OH. We shall therefore now concentrate on halogenoalkanes.

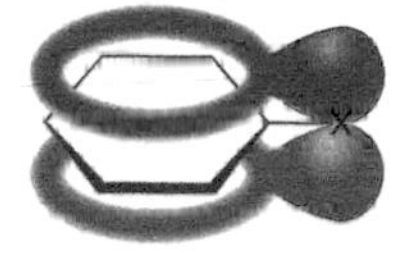

A lone pair from the halogen atom feeds into the delocalized π cloud. The electronegative halogen atom withdraws electron density from the delocalized π cloud strengthening the C—X bond.

SUMMARY

- Attaching a halogen to an alkane provides a means of substituting other functional groups.
- The carbon–halogen bond reacts because it is polarized and polarizable.
- The carbon–halogen bonds decrease in strength (and so increase in reactivity) in the order Cl > Br > I and increase in length in the order Cl < Br < I.
- Halogenoarenes are generally very much less reactive than halogenoalkanes.
- Halogenoalkanes react by nucleophilic attack on the carbon atom of the carbon–halogen bond.

PRACTICE

1 Draw structural formulae for the following substances:
- a Chloroethane
- b Bromobutane
- c 2-Bromopropane
- d 2-Iodobutane
- e 2,2-Dichloropropane
- f 1,4-Dibromobenzene.

2 a Describe the relationship between the length of a bond and its strength.
- b Compare the C—Cl bond lengths in chlorobenzene and chloroethane.
- c Use your answers to (a) and (b) and any other information to explain why chlorobenzene is much less reactive than chloroethane.

3 With reference to carbon–halogen bond strengths (see next spread), suggest an order of reactivity for the halogenoalkanes: chloroethane, bromoethane, and iodoethane.

4 Write a balanced equation for the reaction between chloroethane and aqueous sodium hydroxide. Identify the attacking nucleophile and the atom on the chloroethane molecule that it attacks.

24.2

OBJECTIVES

- Hydroxide ion as a nucleophile
- Cyanide ion as a nucleophile
- Ammonia as a nucleophile
- Relative reaction rates Cl/Br/I
- Primary, secondary, and tertiary halogenoalkanes

REACTIONS WITH HYDROXIDE, CYANIDE, AND AMMONIA

The major reason for introducing a halogen, and therefore bond polarity, into an organic molecule is to make nucleophilic substitution reactions possible. The carbon–fluorine bond is very strong and therefore difficult to break. However, nucleophilic substitution readily converts chloro-, bromo-, and iodoalkanes into other useful molecules. A nucleophile is attracted to the δ+ carbon atom bonded to the halogen. Typical attacking nucleophiles are hydroxide ion OH^-, cyanide ion CN^-, and ammonia NH_3. The result in all cases is the removal of a halide ion and its substitution by the incoming nucleophile. These reactions are easy to carry out and give a good yield of a relatively pure product.

Heating a reaction mixture in a flask with a vertical condenser is called 'heating under reflux' or 'refluxing'. Organic reactions are often slow and reactants may need heating together for several hours. The reflux condenser ensures that volatile substances are not lost during prolonged boiling; rising vapours condense and fall back into the flask.

Hydroxide ion as a nucleophile

- Reaction of a halogenoalkane RX with hydroxide ion OH^- produces an *alcohol* ROH, thus replacing a halogen atom with an *oxygen* atom:

$$CH_3CH_2Br(aq) + OH^-(aq) \rightarrow CH_3CH_2OH(aq) + Br^-(aq)$$

This reaction is described as **hydrolysis**. It is carried out by refluxing the halogenoalkane with dilute aqueous potassium (or sodium) hydroxide. After the hydrolysis is complete, the alcohol may be isolated by fractional distillation of the reaction mixture.

Water

Water H_2O is a weaker nucleophile than hydroxide ion OH^-, but water can hydrolyse halogenoalkanes to the same products as aqueous hydroxide ion:

$$CH_3CH_2Br(aq) + H_2O(l) \rightarrow CH_3CH_2OH(aq) + Br^-(aq) + H^+(aq)$$

However, the rate of reaction is very much slower.

Cyanide ion as a nucleophile

- Reaction of a halogenoalkane RX with cyanide ion CN^- produces a *nitrile* RCN, thus adding a *carbon* atom to the carbon skeleton:

$$CH_3CH_2Br + CN^- \rightarrow CH_3CH_2CN + Br^-$$

The halogenoalkane is heated under reflux with potassium cyanide dissolved in ethanol. Great care must be exercised as potassium cyanide is toxic (spread 20.10). This reaction is very useful in synthesis. The cyanide (nitrile) group converts readily to other useful functional groups, especially the —COOH group (see chapter 29 'Organic synthesis').

Ammonia as a nucleophile

- Reaction of a halogenoalkane RX with ammonia NH_3 produces an *amine* RNH_2, thus replacing a halogen atom with a *nitrogen* atom:

$$CH_3CH_2Br + NH_3 \rightarrow CH_3CH_2NH_3^+ + Br^-$$

The halogenoalkane is usually heated with an excess of ammonia dissolved in ethanol. Adding base then liberates the amine:

$$CH_3CH_2NH_3^+ + OH^- \rightarrow CH_3CH_2NH_2 + H_2O$$

But the amine produced is also a nucleophile, and so will attack and further substitute the halogenoalkane:

$$CH_3CH_2Br + CH_3CH_2NH_2 \xrightarrow{OH^-} (CH_3CH_2)_2NH + H_2O + Br^-$$

The reaction of ammonia with a halogenoalkane has a low yield, and the mixture of products obtained because of the further substitution results in this reaction being little used.

Bond strengths

Bond	Average bond enthalpy/kJ mol^{-1}
C—C	348
C—H	413
C—F	484
C—Cl	338
C—Br	276
C—I	238

The strength of the C—F bond results in the fluoroalkanes being unreactive. They are of little use in organic synthesis compared with other halogenoalkanes.

Halogen and rate of reaction

The rate of nucleophilic substitution depends on which halogen is present. Carbon–halogen bonds have different strengths, as seen in the box alongside. The C—Cl bond is stronger than the C—Br bond, which is stronger than the C—I bond. It is the carbon–halogen bond that breaks during a nucleophilic substitution reaction. As a result, iodoalkanes substitute faster than bromoalkanes, and bromoalkanes substitute faster than chloroalkanes.

Hydrolysing halogenoalkanes and then adding acidified silver nitrate shows how the rate of reaction varies with the halogen. Covalently bonded halogenoalkanes do not react with Ag^+. However, hydrolysis of the halogenoalkane releases halide ions into solution, as described above. The halide ions immediately precipitate out as insoluble silver halide:

$$RX(aq) + OH^-(aq) \rightarrow ROH(aq) + X^-(aq)$$

$$Ag^+(aq) + X^-(aq) \rightarrow AgX(s)$$

When the three halogenoethanes CH_3CH_2Cl, CH_3CH_2Br, and CH_3CH_2I react with silver nitrate in ethanol solution, the yellow colour of silver iodide forms faster than the cream colour of silver bromide, which itself forms faster than the white colour of silver chloride.

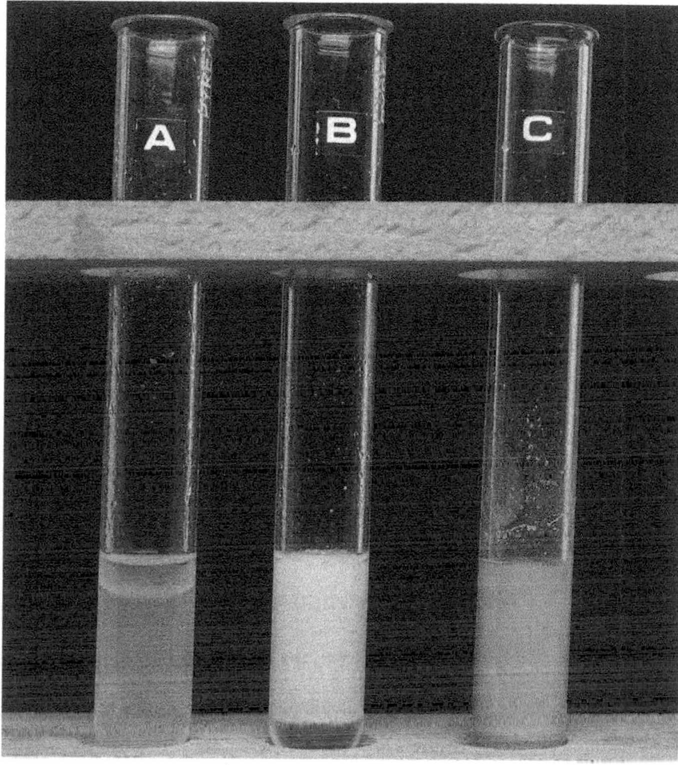

(a) CH_3CH_2Cl: faintest of white precipitates.
(b) CH_3CH_2Br: significant cream precipitate.
(c) CH_3CH_2I: distinct yellow precipitate.

SUMMARY

- The halogen of a halogenoalkane may be substituted by an attacking nucleophile.
- Hydroxide ion hydrolyses halogenoalkanes to alcohols.
- Halogenoalkanes show varying rates of hydrolysis in the order RI > RBr > RCl.
- Cyanide ion adds a carbon atom to form a nitrile RCN.
- Ammonia forms amines RNH_2; but further substitution occurs and the reaction is of little use.

Grignard reaction

Halogenoalkanes also undergo one other very important reaction, the Grignard reaction (see spread 29.2).

(a) CH_3-CH_2-Br

Br

Bromoethane,
a primary halogenoalkane

(b) $CH_3-CH(CH_3)-Br$

Br

2-Bromopropane,
a secondary halogenoalkane

(c) $CH_3-C(CH_3)_2-Br$

Br

2-Bromo-2-methylpropane,
a tertiary halogenoalkane

The relative reaction rates also depend on the structure of the halogenoalkane, as explained in the next spread.

PRACTICE

1 Write structural formulae to show the reaction between each of the following:

a Bromoethane and hydroxide ion

b Chloroethane and cyanide ion

c Iodoethane and water.

2 Heating iodoethane with aqueous ammonia in a sealed tube forms ethylamine $CH_3CH_2NH_2$. The reaction effectively substitutes an ethyl group for a hydrogen atom on the ammonia molecule. Further products form as ethylamine reacts with iodoethane. Give the structural formulae of two further products.

3 Halogenoalkanes are almost insoluble in water, but they dissolve readily in ethanol. Explain why ethanol is added to the mixture of aqueous silver ion and halogenoalkane described in the illustration above.

24.3

OBJECTIVES

- The S_N2 mechanism
- The S_N1 mechanism

Rates and reaction mechanisms

Hydroxide ions hydrolyse halogenoalkanes to alcohols. The previous spread showed that the reaction rates for halogenoalkanes increase in the sequence chloride < bromide < iodide. The reaction rate also depends on the detailed structure of the halogenoalkane. This spread considers the *order* of each reaction (see chapter 15 'Chemical kinetics'), and then goes on to explain these orders in terms of the different reaction mechanisms involved.

Primary and tertiary halogenoalkane hydrolysis

Halogenoalkanes are classified depending on the number of alkyl groups attached to the carbon atom bonded to the halogen (see previous spread):

- **primary** if there is *one* alkyl group
- **secondary** if there are *two* alkyl groups
- **tertiary** if there are *three* alkyl groups

1-Bromobutane is a primary halogenoalkane, and 2-bromo-2-methylpropane is a tertiary halogenoalkane. Both have the molecular formula C_4H_9Br; each is hydrolysed by hydroxide ion to the corresponding alcohol. Investigating the chemical kinetics of these hydrolysis reactions reveals the following information:

- 1-Bromobutane hydrolysis is *first* order with respect to 1-bromobutane and *first* order with respect to hydroxide ion.
- 2-Bromo-2-methylpropane hydrolysis is *first* order with respect to 2-bromo-2-methylpropane and *zero* order with respect to hydroxide ion.

These data indicate that the two reactions proceed by different reaction mechanisms.

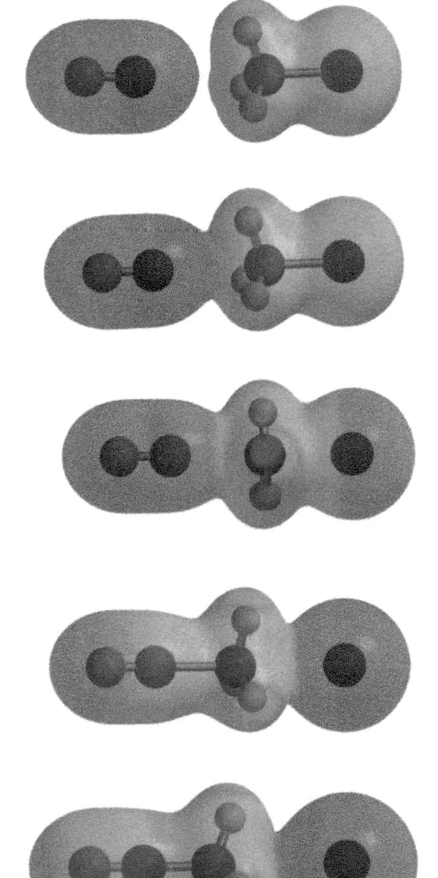

Five stages in the S_N2 reaction between cyanide ion and a halogenoalkane.

S_N2 inversion

The tetrahedral arrangement of the molecule around the δ+ carbon atom has flipped, like an umbrella turning inside out. This is called **Walden inversion.** Its experimental observation confirms that the molecule is attacked from the opposite side to the halogen.

Bromoethane and the S_N2 reaction mechanism

For a *primary* halogenoalkane reacting with dilute aqueous sodium hydroxide, the kinetics shows that the reaction is second order overall. The rate depends on the concentrations of *both* the halogenoalkane *and* the hydroxide ion (the nucleophile):

$$\text{rate} = k[\text{RX}][\text{OH}^-]$$

This shows that both the halogenoalkane and the OH^- ion must be involved in the rate-limiting step (see chapter 15 'Chemical kinetics'). The following mechanism fits the observed kinetics.

The OH^- ion is attracted to the δ+ carbon atom of the C—X bond. It approaches the molecule from the opposite side to the halogen, where its attack is not impeded by the bulky halogen with its δ– charge.

A transition state forms in which the halogen atom and the oxygen atom are both *partially* bonded to the carbon atom. Note that the transition state involves a p orbital on the carbon atom attached to the halogen, and so forces a 'reverse-side' attack by the incoming nucleophile.

The halide ion leaves the transition state and the product is formed.

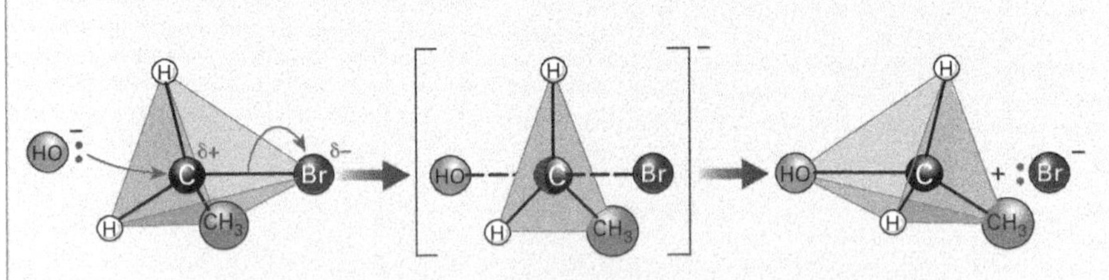

The mechanism of the S_N2 reaction of hydroxide ion with bromoethane.

Primary halogenoalkanes are said to react by an $\mathbf{S_N2}$ mechanism – S for substitution, N for nucleophilic, 2 for bimolecular. (The term 'bimolecular' means that there are two species involved in the rate-limiting step.) According to this mechanism, bond breaking and bond making happen *simultaneously*.

(a)

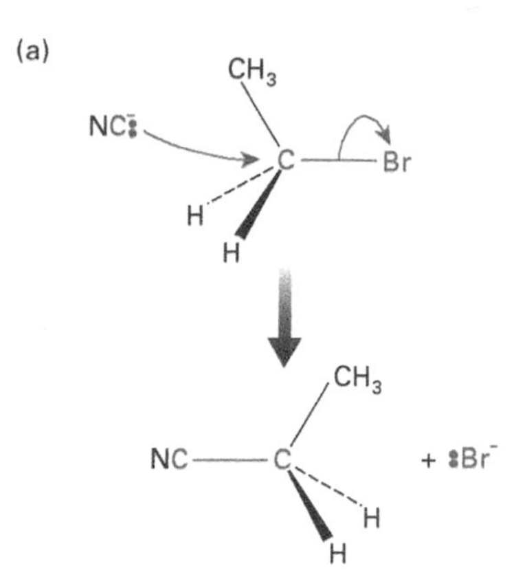

(b)

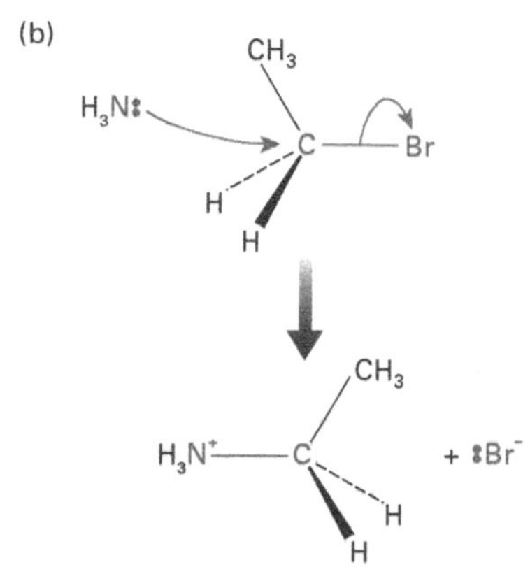

The S_N2 mechanism for reaction between (a) a cyanide ion and (b) ammonia with bromoethane.

2-Bromo-2-methylpropane and the S_N1 reaction mechanism

For a *tertiary* halogenoalkane reacting with dilute aqueous sodium hydroxide, the kinetics are different: the reaction is *first order* overall. The rate depends *only* on the concentration of the halogenoalkane:

$$\text{rate} = k[\text{RX}]$$

The concentration of the hydroxide ion does not have any effect on the rate of reaction, indicating that the OH^- ion is not involved in the rate-limiting step. The mechanism is discussed below.

Comparing the structure of a primary and a tertiary halogenoalkane shows that the reaction mechanisms are likely to be different. The $\delta+$ carbon atom in a tertiary halogenoalkane is surrounded by large alkyl groups, which obstruct the attack of the nucleophile. This effect is called **steric hindrance**. The hydrolysis of a tertiary halogenoalkane proceeds by the following steps.

Step 1 In the course of random collisions, typically with solvent molecules, the halide ion ionizes by heterolytic fission. The result of this rate-limiting step is a carbocation and a halide ion:

$$(CH_3)_3C\text{—}Br \rightarrow (CH_3)_3C^+ + Br^-$$

Step 2 The carbocation attracts a hydroxide ion OH^- and bonds to it, forming the product. This step happens much faster than step 1, so the overall rate does not depend on the hydroxide ion concentration. Notice also that the S_N1 mechanism does not necessarily involve an inversion because an OH^- ion can attack from either side of the *planar* carbocation.

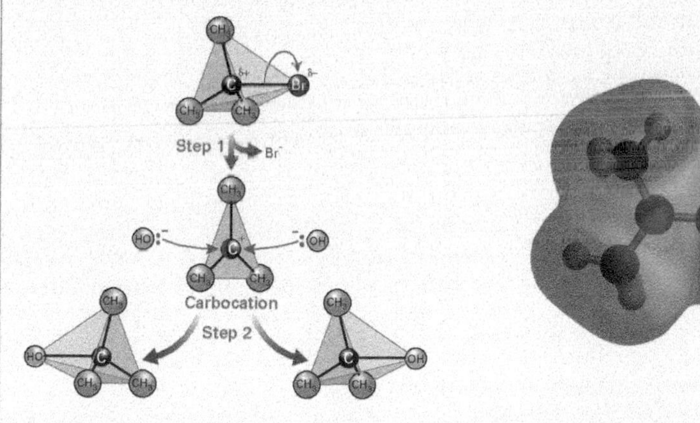

The mechanism of the S_N1 reaction.

The electrostatic potential map for the carbocation $(CH_3)_3C^+$.

Tertiary halogenoalkanes are said to react by an $\mathbf{S_N1}$ mechanism – S for substitution, N for nucleophilic, 1 for unimolecular. According to this mechanism, bond making happens *after* bond breaking.

Intermediate carbocations

The intermediate tertiary carbocation is particularly stable due to the electron-donating effect of the *three* alkyl groups attached to the positive carbon atom (see chapter **22** 'Alkanes and alkenes'). The hydrolysis of a primary halogenoalkane does not follow this mechanism because the primary carbocation is far less stable, having just one alkyl group attached to the positive carbon atom.

Leaving groups

The reaction rate in nucleophilic substitution reactions depends to a certain extent on the nature of the leaving group. The **leaving group** is the species that becomes displaced from the carbon atom and leaves the molecule. In the example alongside, bromide ion is the leaving group.

Secondary halogenoalkanes

Kinetic data are more complicated in the case of secondary halogenoalkanes because the S_N1 and S_N2 mechanisms are approximately equally likely.

SUMMARY

- Primary, secondary, and tertiary halogenoalkanes have respectively one, two, and three alkyl groups on the carbon atom bonded to the halogen.
- Primary halogenoalkanes substitute by the S_N2 reaction mechanism. The C—X bond breaks as the bond with the incoming nucleophile forms.
- Tertiary halogenoalkanes substitute by the S_N1 reaction mechanism. The C—X bond breaks before any reaction with the incoming nucleophile occurs. The reaction proceeds via a carbocation.

24.4

OBJECTIVES

- Elimination vs. substitution
- Direct uses and drawbacks
- The ozone layer
- Alternatives

Substitute or eliminate?

2-Bromopropane $CH_3CHBrCH_3$ can react with potassium hydroxide by *substitution* to give $CH_3CH(OH)CH_3$ or by *elimination* of HBr to give $H_2C{=}CHCH_3$. The actual outcome depends on the reaction conditions chosen.

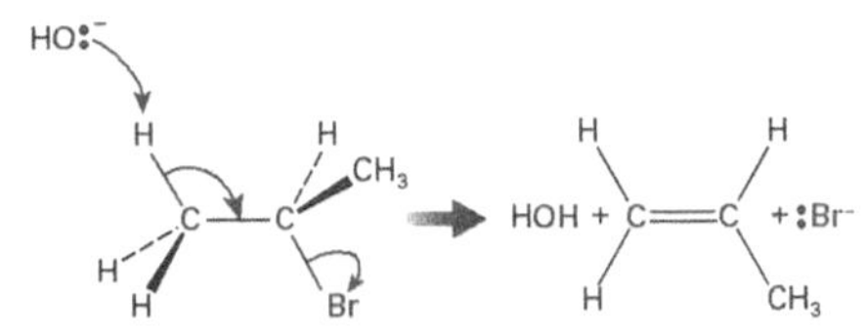

Elimination of 2-bromopropane using sodium hydroxide in ethanol produces propene. This ***E2 mechanism*** *involves two species in the rate-limiting step whereas the* ***E1 mechanism*** *shown below involves only one species.*

The carbocation $(CH_3)_3C^+$ formed during the S_N1 mechanism (see previous spread) may lose a proton instead of reacting with a nucleophile. The result is an elimination reaction: Br^- is lost in the first step and H^+ in the second.

Sevoflurane is a very effective general anaesthetic.

Health and safety

Some halogenoalkanes such as tetrachloromethane CCl_4 are toxic. They are no longer permitted for use as solvents at all. Legislation to control substances hazardous to health (in the UK these are called the 'COSHH Regulations') requires companies to seek safer alternative solvents.

Elimination reactions and uses of halogeno compounds

Halogeno compounds are extremely useful as *intermediates* in the synthesis of more complex molecules. The carbon–halogen bond can be substituted; halogenoalkanes can also eliminate, as discussed below. Halogeno compounds also have many direct uses. They are relatively straightforward to produce industrially, and may be used as anaesthetics, aerosol propellants, foaming agents, refrigerants, insecticides, herbicides, and fire extinguishers. However, we now know that some of them cause thinning of the ozone layer, thus endangering all life on Earth.

Elimination reactions

We shall now compare substitution and elimination reactions. As we have seen, dilute aqueous potassium (or sodium) hydroxide will *substitute* hydroxide ion for the halide ion:

$$CH_3CH_2Br + OH^- \xrightarrow[\text{dilute}]{\text{(aq)}} CH_3CH_2OH + Br^-$$

However, hot concentrated sodium hydroxide in ethanol ('alcoholic sodium hydroxide') will *eliminate* hydrogen bromide, leaving an alkene:

$$(CH_3)_3CBr + OH^- \xrightarrow[\text{hot concentrated}]{\text{(ethanol)}} (CH_3)_2C{=}CH_2 + H_2O + Br^-$$

In the substitution reaction, the hydroxide ion acts as a nucleophile and attacks the δ+ carbon atom. In the elimination reaction, the hydroxide ion acts as a *base* and removes a hydrogen ion. To some extent, substitution and elimination are always in competition because the attacking reagent OH^- is both a nucleophile and a base. In general:

- Substitution is fastest with primary halogenoalkanes.
- Elimination is fastest with tertiary halogenoalkanes.

Anaesthetics

In 1956 a team at ICI developed 'halothane', $CF_3CHBrCl$. It was the first effective and non-explosive general anaesthetic. Currently, the two most common general anaesthetics are isoflurane $CF_3CHClOCHF_2$ and sevoflurane $(CF_3)_2CHOCH_2F$, which, like halothane, contain the trifluoromethyl group.

BCF

Bromochlorodifluoromethane ('BCF') $CBrClF_2$ makes a good fire extinguisher, blanketing a fire in a dense, non-flammable, oxygen-excluding vapour. Combustion is a radical reaction. At high temperatures BCF molecules produce bromine radicals, which combine with the radicals present in the flame and inhibit the combustion process.

Solvents

Simple chloroalkanes are very good solvents. Some, such as 1,1,1-trichloroethane CCl_3CH_3 and tetrachloromethane CCl_4, were once familiar 'organic' solvents used to dissolve non-polar solutes. (They are now mostly considered too toxic for common use.) They dissolve grease well and were used for 'dry' cleaning.

CFCs and ozone layer depletion

Compounds in which some or all of the hydrogen atoms of an alkane have been replaced by chlorine or fluorine are called **chlorofluorocarbons**, abbreviated to **CFCs**. CFCs are even less reactive than alkanes because the C—F bond is extremely strong, spread 24.2. They are also non-toxic, volatile liquids which do not easily catch fire. These properties made them very useful as aerosol propellants, refrigerants, and as 'blowing agents' for making expanded ('foamed') plastics such as those used in heat-insulating packaging. CFCs (and BCF) have one very unfortunate property. When released during use, they diffuse unchanged into the upper atmosphere. Under the influence of intense ultraviolet radiation in the **stratosphere** (altitude of 11–50 km; the layer below 11 km is called the **troposphere**), they decompose to form radicals. These radicals react with the ozone formed at this altitude and break it down, producing the 'hole in the ozone layer' (see spread 19.10). As ozone shields the surface of the Earth from harmful ultraviolet radiation, its depletion is causing ultraviolet levels at the Earth's surface to increase. The average depletion at the latitude of London is about 0.35% per year.

The main culprit is the chlorine radical $Cl^•$. The carbon–chlorine bond is the weakest in the molecule, so it is the easiest to break. The depletion of the ozone occurs by reaction of the chlorine radical with ozone (see spread 15.11):

$$Cl^• + O_3 \rightarrow ClO^• + O_2$$

Furthermore, by initiating a *chain* reaction, one chlorine radical can destroy thousands of ozone molecules. Searches for replacements have focused on molecules that do not contain chlorine. Useful, but more expensive, alternatives are **hydrofluorocarbons** (HFCs) such as CH_2FCF_3. Manufacture started in 1990 and increased 10-fold over the next two years. HFCs produce almost zero ozone depletion, and are non-toxic and non-flammable.

Unfortunately, though HFCs provide an effective alternative to CFCs they are also powerful greenhouse gases (thousands of times more potent than CO_2, as were CFCs) with long atmospheric lifetimes. The 2016 Kigali Amendment to the Montreal Protocol (see margin), which aims to reduce the use of HFCs, has been described by the UN Environment Programme as potentially the single largest contribution to limiting global warming.

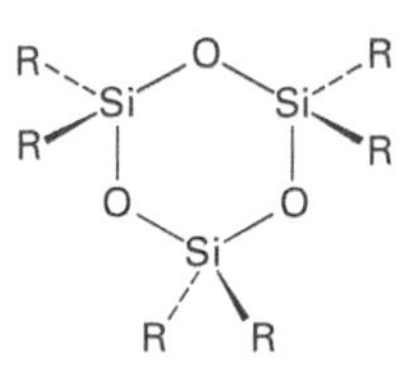

The international agreement for environmental protection called the Montreal Protocol (1989) effectively banned the use of chlorinated solvents in industrialized countries. The new generation of replacement solvents are based on cyclic siloxanes.

F. Sherwood Rowland (left) predicted that the ozone layer would be depleted by the use of CFCs before the problem was first detected. Here he is receiving the Nobel Prize in Chemistry from the King of Sweden.

The rise and fall of DDT

The insecticide DDT is a complex halogenoarene and an extremely effective insecticide. Developed in 1939, it acts as a nerve poison, which causes convulsions, paralysis, and death of insects. Furthermore, no documented human death has been attributed to DDT. Malaria, typhus, and sleeping sickness are diseases that have long been established in Africa and in Southern Europe. These diseases are caused by parasites spread by biting insects. The death toll in Italy from malaria was 400 000 in 1945. By using DDT to kill the mosquitoes that carry the parasites, this figure was brought down to 40 in 1968. The World Health Organization suggests that five million human lives were saved in the first eight years of the use of DDT; one billion people (a *seventh* of the Earth's population) have had their lives improved. However, DDT is so unreactive that it passes unchanged along food chains. As a consequence, there is now a world-wide ban on the use of DDT.

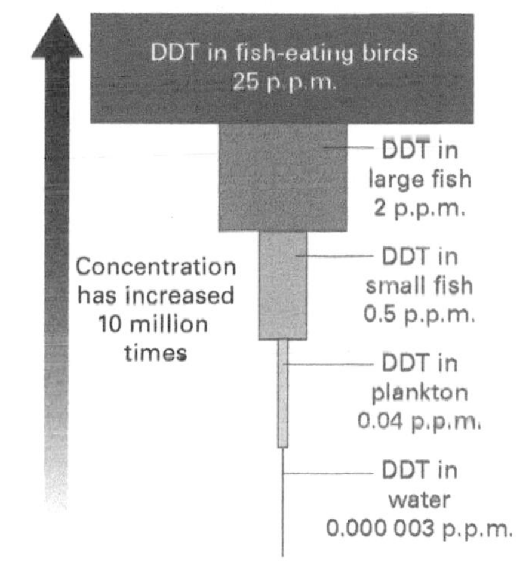

Because DDT is very unreactive, it remains unchanged in the bodies of organisms. Organisms high up in the food chain accumulate high concentrations of DDT and suffer toxic effects.

SUMMARY

- Halogenoalkanes undergo elimination to form alkenes when heated with concentrated sodium hydroxide in alcohol.
- Halogenoalkanes are used as solvents, anaesthetics, and fire extinguishers.

PRACTICE

1 Explain why the chemical properties of hydrofluorocarbons such as CH_2FCF_3 lead to them being described as 'ozone-friendly'.

2 Ultraviolet radiation increases the incidence of skin cancers (especially in pale-skinned people) and causes cataracts to form in unprotected eyes. Suggest why developing countries wish the industrialized countries to phase out CFCs before they do so themselves.

Chapter 24 Reactions summary

- Radical substitution of an alkane:
  $CH_3CH_3 + Cl_2 \xrightarrow{UV} CH_3CH_2Cl + HCl$
  and all other possible products up to CCl_3CCl_3.
- Electrophilic addition of HX or X_2 to an alkene:
  $CH_2{=}CH_2 + HBr \rightarrow CH_3CH_2Br$
  $CH_2{=}CH_2 + Br_2 \rightarrow BrCH_2CH_2Br$
- Reaction with hydroxide ion to give an alcohol:
  $CH_3CH_2Br + OH^- \rightarrow CH_3CH_2OH + Br^-$
- Reaction with cyanide ion to give a nitrile:
  $CH_3CH_2Br + CN^- \rightarrow CH_3CH_2CN + Br^-$
- Reaction with ammonia to give an amine (after neutralization):
  $CH_3CH_2Br + NH_3 \rightarrow CH_3CH_2NH_3^+ + Br^-$
  $CH_3CH_2NH_3^+ + OH^- \rightarrow CH_3CH_2NH_2 + H_2O$
- Elimination to yield an alkene:
  $(CH_3)_3CBr \xrightarrow[\text{hot concentrated}]{OH^-\text{(ethanol)}} (CH_3)_2C{=}CH_2 + H_2O + Br^-$
- The S_N2 mechanism (spread 24.3) should be noted.

PRACTICE EXAM QUESTIONS

1 **a** Explain what is meant by the term nucleophilic substitution. [2]

b Explain why 1-bromopropane reacts with nucleophiles but propane itself does not. [2]

c Write an equation and a mechanism for the reaction of 1-bromopropane with an excess of ammonia. [4]

d Give the starting materials and state the conditions for the preparation of CH_3CH_2CN by nucleophilic substitution. [3]

2 **a** Give the structural formula of 2-bromo-3-methylbutane. [1]

b Write an equation for the reaction between 2-bromo-3-methylbutane and dilute aqueous sodium hydroxide. Name the type of reaction taking place and outline a mechanism. [4]

c Two isomeric alkenes are formed when 2-bromo-3-methylbutane reacts with ethanolic potassium hydroxide. Name the type of reaction occurring and state the role of the reagent. Give the structural formulae of the two alkenes. [4]

3 Under appropriate reaction conditions, 2-bromo-3-methylbutane, $(CH_3)_2CHCHBrCH_3$, can be converted into an alcohol or into two isomeric alkenes.

a Name the type of reaction taking place and give the role of the reagent when 2-bromo-3-methylbutane reacts with **aqueous** potassium hydroxide. Give the structure of the alcohol and outline a mechanism for this reaction. [5]

b **i** Name the type of reaction taking place and give the role of the reagent when 2-bromo-3-methylbutane reacts with **ethanolic** potassium hydroxide.

ii One of the reaction products is 2-methylbut-2-ene. Outline a mechanism for the formation of this compound and give the structure of the second alkene which is also formed. [7]

4 **a** There are four **structural** isomers of molecular formula C_4H_9Br. The formulae of two of these isomers are given.

```
   H  H  H  H              H  CH3 H
   |  |  |  |              |  |   |
H—C—C—C—C—H          H—C—C—C—H
   |  |  |  |              |  |   |
   H  H  Br H              H  Br  H
```

Isomer 1 **Isomer 2**

i Draw the remaining two structural isomers. [2]

ii Give the name of **Isomer 2**. [1]

b All four structural isomers of C_4H_9Br undergo similar reactions with ammonia.

i Give the name of the mechanism involved in these reactions. [1]

ii Draw the structural formula of the product formed by the reaction of **Isomer 1** with ammonia. [1]

iii Select the isomer of molecular formula C_4H_9Br that would be the most reactive with ammonia. State the structural feature of your chosen isomer that makes it the most reactive of the four isomers. [2]

c The elimination of HBr from **Isomer 1** produces two structural isomers, compounds **A** and **B**.

i Give the reagent and conditions required for this elimination reaction. [3]

ii Give the structural formulae of the two isomers, **A** and **B**, formed by elimination of HBr from **Isomer 1**. [2]

d Ethene, C_2H_4, reacts with bromine to give 1,2-dibromoethane.

i Give the name of the mechanism involved. [1]

ii Show the mechanism for this reaction. [3]

5 **a** Outline a mechanism for the reaction of 1-bromopropane with an excess of ammonia. [4]

b Explain how you could distinguish between samples of 1-bromopropane and 2-bromopropane using

i infrared spectroscopy

ii low-resolution proton NMR spectroscopy. [4]

(See also chapters 31 and 32.)

6 **a** **i** One product of the reaction of ethene with aqueous bromine in the presence of sodium chloride is 1,2-dibromoethane. Give the mechanism for the formation of 1,2-dibromoethane.

ii Give the structural formula of ONE other compound formed.

iii Explain mechanistically why the following compound is not formed.

```
    H  H
    |  |
Cl—C—C—Cl
    |  |
    H  H
```

[6]

b **i** Give a reagent, and the conditions under which it might be used, to convert 1,2-dibromoethane to ethane-1,2-diol.

ii How would the rate of this reaction differ if 1,2-diiodoethane was used instead of 1,2-dibromoethane at the same temperature?

Explain your answer. [4]

7 a Myrcene has the molecular formula $C_{10}H_{16}$ (M_r = 136) and its structure is shown as **A** below. This compound is produced by pine trees in North America. Unfortunately, it attracts females of the beetle *Dendroctonus brevicomins*, which lays eggs in the tree. The drilling of the tree injects pathogenic fungus, causing the tree's death.

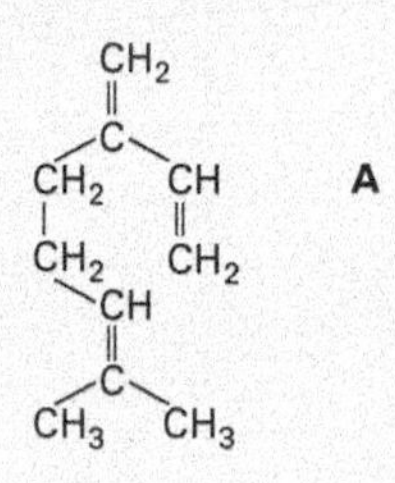

A

i State what you would see when bromine dissolved in an inert organic solvent is added to myrcene. [1]

ii A sample of Myrcene weighing 2.72 g is reacted with a 1.00 mol dm^{-3} solution of bromine; 60.0 cm^3 was required. Show that this is consistent with the structure of myrcene. [3]

b Myrcene reacts with hydrogen bromide to give **B**:

```
CH3   Br
   \ /
    C
   / \
CH2   HC—Br     B
 |     |
CH2   CH3
   \
   CH2
    |
CH3—C—Br
    |
   CH3
```

The double bonds can be regenerated by an elimination reaction; one elimination product among many is **C**:

```
    CH3
     |
     C
   // \
CH     CH       C
 |     ||
CH2    CH2
   \
   CH
   ||
    C
   / \
CH3   CH3
```

i State the reagents and conditions required to form **C** from **B**. [2]

ii Explain why products other than Myrcene itself are produced in this elimination reaction. [2]

iii What type of isomerism is shown by the bromo compound **B** which is not shown by Myrcene itself? Give a reason. [2]

8 a When 3-bromo-2-methylpentane, $(CH_3)_2CHCHBrCH_2CH_3$, reacts with aqueous potassium hydroxide, an alcohol is formed.

i Name the type of reaction taking place and give the role of the reagent.

ii Give the structure of the alcohol formed and outline a mechanism for its formation. [5]

b i Give the structural formula of the product obtained when the alcohol formed in part **a** is treated with acidified potassium dichromate(VI).

ii This product shows dominant peaks at *m/z* values of 57 and 71 in its mass spectrum. Give the structures of the species responsible for these two peaks.

Peak at *m/z* = 57.
Peak at *m/z* = 71. [3]

c When 3-bromo-2-methylpentane reacts with ethanoic potassium hydroxide, two isomeric alkenes are formed.

i Name the type of reaction taking place and give the role of the reagent.

ii One of the reaction products is 2-methylpent-2-ene. Give the structure of this alkene and outline a mechanism for its formation.

iii Give the structure of the second alkene which is also formed in this reaction. What type of stereoisomerism is shown by this compound? [8]

9 a i Give the structure of 3-bromopentane.

ii Name and outline the mechanism for the reaction taking place when 3-bromopentane is converted into pent-2-ene in the presence of a strong base.

iii What type of stereoisomerism is shown by pent-2-ene? [7]

b i Give the structures of the two carbocation intermediates formed when pent-2-ene reacts with hydrogen bromide.

ii Name the two isomeric organic products which result from this reaction.

iii Indicate why these two products are obtained in approximately equal amounts. [5]

25 Alcohols

There are two main categories of oxygen-containing organic compounds. In the first category, an oxygen atom has a *single* bond to two other atoms. The most familiar compounds of this kind are the **alcohols**, which contain the —OH group attached to an alkyl group. If the —OH group is directly attached to a benzene ring, the compounds are called **phenols**. If the oxygen atom is sandwiched between two alkyl groups to give an R—O—R' type structure, the result is an **ether**. This chapter discusses the properties of all three of these classes of molecule, with the main emphasis on alcohols. In the second category of oxygen-containing compounds, an oxygen has a *double* bond to a carbon atom. In the following two chapters you will meet the **carbonyl group** ⊃C=O, present in **aldehydes** and **ketones**, and, at the highest level of oxidation, the **carboxylic acids** (that contain the —COOH group), and their derivatives.

An introduction to alcohols

25.1

OBJECTIVES

- Structure of alcohols
- Manufacture of ethanol
- Manufacture of methanol
- Uses of alcohols

Carbonyl group symbol

Typographically we have used the symbol ⊃C=O, to represent the carbonyl group

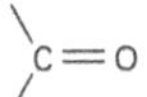

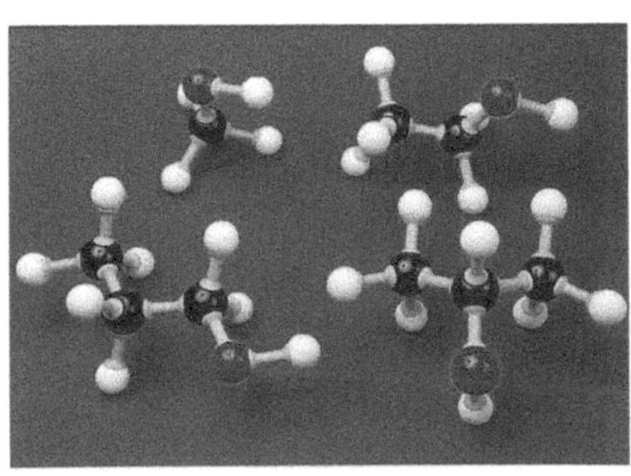

The structures of some alcohols: methanol, ethanol, and the two propanol isomers, propan-1-ol and propan-2-ol; see spread 21.4.

IR spectra

The presence of an —OH group in an alcohol can be detected by infrared spectroscopy, chapter 32.

Alcohols consist of a covalently bonded hydroxyl group —OH attached to a carbon skeleton. In the case of common aliphatic alcohols, the carbon skeleton is an alkyl group. While there are many different alcohols, the most widely used is ethanol CH_3CH_2OH. This substance is so well known that its everyday name is simply 'alcohol'.

The manufacture of ethanol

Two main processes are currently used to make ethanol. The process chosen depends on economic factors and on the end use of the product.

The first process is **fermentation**, which has been used for thousands of years to make ethanol. The illustration below shows an industrial-scale fermentation reaction in a modern brewery. The process uses a sugar such as sucrose (from sugar cane or sugar beet) or glucose (from the digestion of starch). In fermentation, the sugars are slowly decomposed by enzymes produced by organisms called yeasts. The process is **anaerobic** – it takes place in the absence of oxygen. For example, glucose is fermented by a series of enzymes to form ethanol and carbon dioxide:

$$C_6H_{12}O_6 \rightarrow 2CH_3CH_2OH + 2CO_2$$

When the concentration of ethanol reaches about 13%, the yeasts and their enzymes no longer function and fermentation stops. Distillation is used to achieve higher concentrations of ethanol in the end product.

The second process used to produce ethanol is the **direct hydration of ethene**, which uses ethene as a chemical feedstock. Ethene is currently available in huge quantities from the cracking of crude oil (see chapter 22 'Alkanes and alkenes'). Ethene and steam react together in the presence of a phosphoric acid catalyst. This catalyst is chosen because ethene is susceptible to reaction with electrophiles; the hydrogen ion H^+, which is characteristic of all acids, is an electrophile. Phosphoric acid, H_3PO_4, is the second least volatile acid after sulphuric acid, and so it may be used

Modern brewing takes place on an industrial scale. Most beers contain about 3–6% ethanol; distilled spirits contain about 40% ethanol. This fermentation vat is in Jamaica.

at moderately high temperatures. The actual conditions are 300 °C and 70 atm pressure. The phosphoric acid catalyst is supported on a silica-based material called Celite. The overall reaction is represented by the following chemical equation:

$$CH_2{=}CH_2(g) + H_2O(g) \rightarrow CH_3CH_2OH(g)$$

The manufacture of methanol

Methanol cannot be manufactured by fermentation. One major process uses the reaction between carbon monoxide and hydrogen (synthesis gas, spread 15.10):

$$CO(g) + 2H_2(g) \rightarrow CH_3OH(g)$$

Because three moles of gas form one mole, a high pressure is favoured (spread 11.3): the current process uses around 50 atm. The temperature of 250 °C is a compromise between rate and yield. The catalyst was traditionally a mixture of zinc oxide and chromium oxide; since 1966 a mixture of zinc oxide, copper, and aluminium oxide has been preferred. Production exceeds 20 million tonnes a year. One major use is in the manufacture of ethanoic acid, spread 27.1.

Some uses of alcohols

- The lower alcohols – methanol, ethanol, and propanol – are mainly used as solvents. Ethanol is the major component in methylated spirits.
- Large quantities of methanol produced by the catalytic oxidation of methane are oxidized to methanal HCHO, which is used as a chemical feedstock to make so-called 'formaldehyde' resins such as melamine.
- Methanol is used as a petrol additive to improve combustion. It was also the fuel in the motor racing IndyCar series in the US until the introduction of fuel-grade ethanol in 2007.
- Ethanol has a wide range of applications, including the extraction of essences from fruits and spices, as a dispersant for dyes in lacquers, and as a solvent for fragrances in perfumes and after-shave lotions.
- Ethanol is the active constituent of alcoholic drinks. In humans, it suppresses inhibitions and thereby brings more enjoyment, and misery, to the world than all other drugs put together.
- Propan-2-ol is the solvent in ink-jet printer ink, many cosmetics, and certain food flavourings. It is also present in a wide range of cleaning fluids for items such as compact discs.

SUMMARY

- Alcohols contain the hydroxyl group —OH bonded to an alkyl group.
- Ethanol is produced by (i) the fermentation of sugars by yeasts or (ii) the direct hydration of ethene.
- The lower alcohols are widely used as solvents.
- Methanol and propan-2-ol are raw materials for the plastics industry.

Fermentation and direct hydration of ethene compared

Direct hydration links the price of ethanol to the price of ethene which is linked to the price of crude oil. On the other hand, fermentation of sugars is seen as a means of using agricultural surplus; ethanol production by this means is subsidized by the EU, bringing the price down. Although the actual production cost of ethanol by direct hydration is less, the price is dependent on price stability in the oil markets. Environmentally, there is no waste from the catalytic hydration of ethene, but fermentation produces carbon dioxide, which contributes to the enhanced greenhouse effect.

Perfumes frequently contain ethanol as solvent.

Biochemistry

The —OH group is present in many important molecules, including retinol, menthol, cholesterol, vitamin C, adrenaline, and the anabolic steroid stanozolol abused by Ben Johnson, the disqualified 'winner' of the 1988 Olympic 100 m Gold medal.

PRACTICE

1. Explain each of the following terms/phrases in the context of this spread:
   - **a** Alkyl group
   - **b** Crude oil fractions
   - **c** Anaerobic decomposition
   - **d** Ethene is susceptible to reaction with electrophiles.
2. Explain why ethanol from the direct hydration of ethene is rarely used in the manufacture of alcoholic drinks.
3. Phosphoric acid is the catalyst used in the direct hydration of ethene. Suggest reasons for each of the following:
   - **a** Concentrated sulphuric acid is not used as the catalyst.
   - **b** Concentrated hydrochloric acid is not used as the catalyst.
   - **c** High pressure (70 atm) is used.

25.2

OBJECTIVES

- Effects of hydrogen bonding
- Nucleophilic substitution: halogenoalkanes
- Chlorinating agents
- Reaction with sodium

Polarity in alcohols

Alcohols contain the functional group —OH attached to an alkyl group. Oxygen is more electronegative than either carbon or hydrogen. As a result, the oxygen atom is charged δ– and the adjacent hydrogen and carbon atoms are charged δ+. Because of the polarity of the —OH group, alcohol molecules hydrogen-bond with each other and with water when in aqueous solution (see chapter 7 'Changes of state and intermolecular forces'). Also, as in the case of the halogenoalkanes (see previous chapter), nucleophiles are attracted to the δ+ carbon atom in the molecule. Nucleophilic substitution follows, in which the hydroxyl group —OH is replaced by the incoming nucleophile.

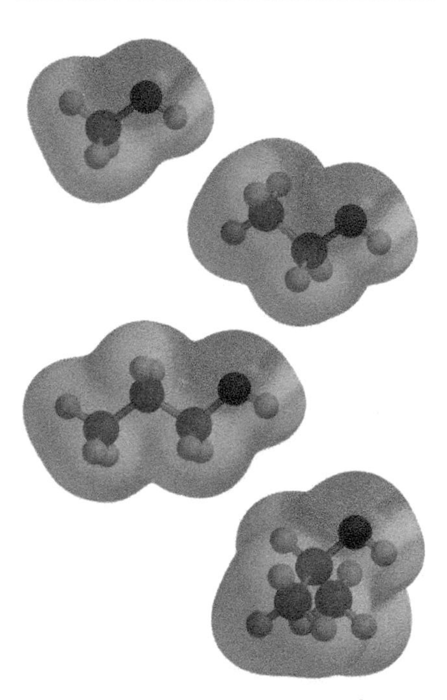

The electrostatic potential maps for methanol, ethanol, propan-1-ol, and propan-2-ol show the polarity of the C—O—H bonds.

Hydrogen bonding

Considering dispersion forces alone would suggest that the boiling point of an alcohol should be about the same as that of the alkane with the same number of electrons (see chapter 7 'Changes of state and intermolecular forces'). But alcohols have much *higher* boiling points than these corresponding alkanes: e.g. ethanol CH_3CH_2OH (26 electrons), T_b = 78 °C; propane $CH_3CH_2CH_3$ (26 electrons), T_b = –42 °C. This is because hydrogen bonding between the molecules of ethanol raises the boiling point of ethanol relative to propane.

When an alcohol is mixed with water, hydrogen bonds form between the alcohol molecules and the water molecules. As a result, the lower alcohols are miscible with water. As the length of the hydrocarbon chain increases, solubility in water decreases.

Nucleophilic substitution by HBr

Hydrogen bromide is a polar molecule in which the hydrogen and bromine atoms bear partial charges $H^{\delta+}$—$Br^{\delta-}$. In aqueous solution, HBr exists as the ions H^+ and Br^-. The Br^- ion is a nucleophile because of its negative charge. Aqueous hydrogen bromide (hydrobromic acid) is often produced *in situ* by adding potassium bromide (or sodium bromide) to concentrated sulphuric acid.

Heating an alcohol together with potassium bromide in concentrated sulphuric acid produces a halogenoalkane:

$$ROH + HBr \xrightarrow[\text{KBr}]{\text{conc. } H_2SO_4} RBr + H_2O$$

Concentrated sulphuric acid is run in to a flask containing the alcohol from a dropping funnel, while the mixture is cooled. Solid potassium bromide is then added, and the bromoalkane product is distilled into a cooled receiver.

Iodoalkanes may also be produced by this method. However, concentrated *phosphoric acid* must be used, to avoid the problem of concentrated sulphuric acid oxidizing iodide ions to iodine (see chapter 18 'The halogens').

Mechanism for the nucleophilic substitution reaction between ethanol and potassium bromide in concentrated sulphuric acid. (a) Protonation of the alcohol's oxygen atom (which donates a lone pair to form a single covalent bond). (b) Reverse-side attack by the Br^- nucleophile. (c) The C—Br bond forms and the C—O bond breaks: water is liberated.

> In a reaction of this sort, substitution occurs and OH^- 'leaves' the molecule. The OH^- group is said to be a 'poor leaving group' because the carbon–oxygen bond is short and consequently very strong. Because of this, the hydroxide ion leaves only with great difficulty. To make the reaction proceed more readily, a reactive intermediate is created that has a better leaving group.
>
> The first step is the protonation of the alcohol by the concentrated acid:
>
> $CH_3CH_2OH + H^+ \rightarrow CH_3CH_2OH_2^+$
>
> The leaving group is now water H_2O, which is a good leaving group.
> The nucleophilic bromide ion attacks the protonated alcohol to form the bromoalkane:
>
> $CH_3CH_2OH_2^+ + Br^- \rightarrow CH_3CH_2Br + H_2O$

Chlorinating agents

A **chlorinating agent** is a substance that replaces a hydroxyl group —OH with a chlorine atom —Cl.

The most important chlorinating agents are:

- Phosphorus trichloride PCl_3
- Phosphorus pentachloride PCl_5
- Sulphur dichloride oxide (thionyl chloride) $SOCl_2$

PCl_3 and $SOCl_2$ are liquids, and the reaction is carried out by heating the alcohol and the chlorinating agent under reflux. The equation for the reaction of sulphur dichloride oxide with ethanol is:

$$CH_3CH_2OH + SOCl_2 \rightarrow CH_3CH_2Cl + SO_2 + HCl$$

Reaction of an alcohol with solid PCl_5 occurs at room temperature:

$$CH_3CH_2OH + PCl_5 \rightarrow CH_3CH_2Cl + PCl_3O + HCl$$

The production of HCl gas on adding phosphorus pentachloride is a useful test for an alcohol.

Phosphorus pentachloride

Solid PCl_5 contains the ions PCl_4^+ and PCl_6^-. The PCl_4^+ ion performs a similar role to the H^+ ion in the HBr example given opposite, weakening the carbon–oxygen bond and assisting the nucleophilic substitution reaction. The PCl_4^+ ion and the H^+ ion both act as *Lewis* acids (see chapter 12 'Acid–base equilibrium'), accepting a lone pair from the oxygen atom in the hydroxyl group on the alcohol molecule.

Iodination

The hydroxyl group can be replaced by an iodine atom by using red phosphorus and iodine.

The reaction between alcohols and sodium

Nucleophilic substitution reactions break the polar C—OH bond. The polar O—H bond is broken when an alcohol is reduced by sodium. Hydrogen gas is slowly released, forming a solution of the product **alkoxide ion** RO^- in ethanol (et):

$$2ROH(l) + 2Na(s) \rightarrow 2Na^+RO^-(et) + H_2(g)$$

The reaction is very similar to the reaction of sodium with water:

$$2HOH(l) + 2Na(s) \rightarrow 2Na^+HO^-(aq) + H_2(g)$$

If the mixture resulting from the reaction between sodium and ethanol is carefully evaporated, the solid product is sodium ethoxide:

$$2CH_3CH_2OH(l) + 2Na(s) \rightarrow 2Na^+CH_3CH_2O^-(s) + H_2(g)$$

The alkoxides are useful nucleophiles, as you will see in a later spread.

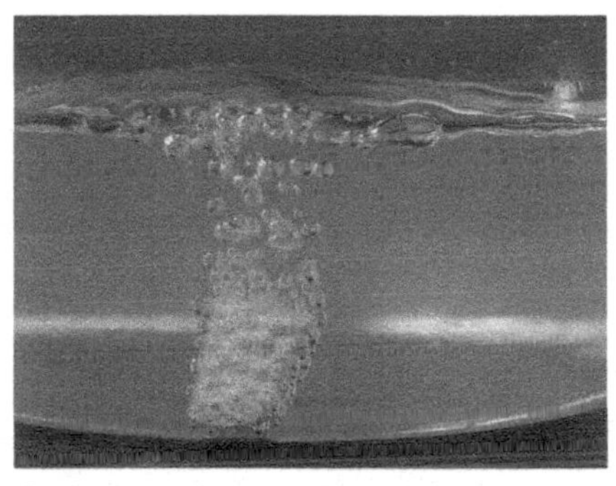

Sodium reacts less vigorously with ethanol than it does with water. Hydrogen is steadily evolved and a solution containing the ethoxide ion forms.

SUMMARY

- There is extensive hydrogen bonding between alcohol molecules, and between alcohol molecules and water molecules.
- Hydrogen bonding results in the alcohols having relatively high boiling points and good miscibility with water.
- The C—OH bond is polarized $C^{\delta+}—O^{\delta-}$; it reacts when nucleophiles attack the δ+ carbon atom.
- Hydrogen bromide (produced *in situ* from potassium bromide and concentrated sulphuric acid) causes nucleophilic substitution of the —OH group in alcohols.
- Chlorinating agents (PCl_3, PCl_5, and $SOCl_2$) replace the —OH group with a chlorine atom.
- The production of HCl gas on adding phosphorus pentachloride is a useful test for an alcohol.
- Alcohols react with sodium to produce an alkoxide and hydrogen gas.

PRACTICE

1 Explain why decan-1-ol $CH_3(CH_2)_8CH_2OH$ has only limited solubility in water, whereas ethanol CH_3CH_2OH is miscible with water in all proportions.

2 Give details of the reagents and the practical procedures required to prepare each of the following substances from a named alcohol:

a 1-Chloropropane

b 2-Bromopropane.

3 Explain why sodium metal reacts more slowly with ethanol CH_3CH_2OH than with water HOH.

4 Draw Lewis structures to show the formation of the ethoxide ion by the action of sodium on ethanol.

25.3

OBJECTIVES

- Elimination vs. substitution
- Dehydration

(a)

(b)

(c)

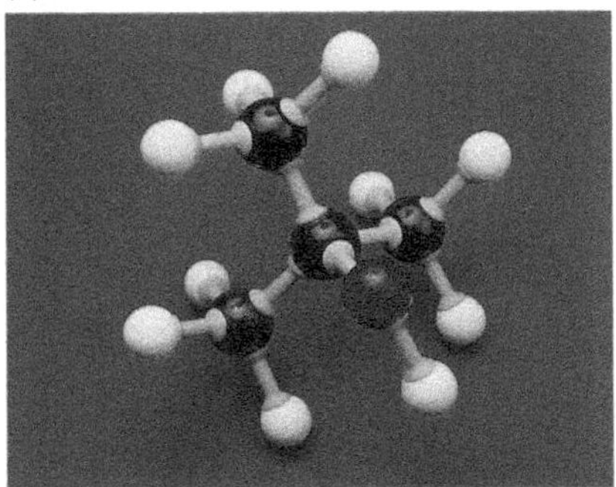

Three alcohols of molecular formula C_4H_9OH: (a) butan-1-ol is a primary alcohol; (b) butan-2-ol is a secondary alcohol; and (c) 2-methylpropan-2-ol is a tertiary alcohol.

(a)

(b)

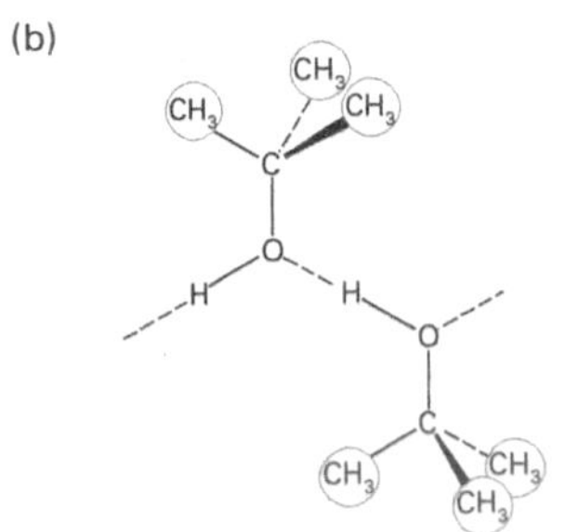

Hydrogen bonding between adjacent molecules of (a) a primary alcohol and (b) a tertiary alcohol. Note that steric hindrance by three alkyl groups inhibits close approach between tertiary alcohol molecules.

PRIMARY, SECONDARY, AND TERTIARY ALCOHOLS AND DEHYDRATION

You know that alcohols contain the functional group —OH attached to an alkyl group. The most typical reactions of alcohols involve nucleophilic substitution reactions in which the —OH group is replaced by an incoming group such as a halogen. However, substitution reactions generally take place in competition with elimination reactions (see previous chapter). In the case of alcohols, elimination reactions involve the formation of a carbon–carbon double bond as the —OH group and a hydrogen atom are removed from across a carbon–carbon single bond. Substitution and elimination reactions both involve the —OH functional group. Which of the two reactions predominates depends on the nature of the hydrocarbon chain attached to the functional group.

The three classes of alcohols

There are three main classes of alcohols: primary, secondary, and tertiary alcohols. As for the halogenoalkanes, spread 24.3, this classification is based on the number of alkyl groups (R) attached to the carbon atom bearing the —OH group:

- a **primary alcohol** has one R group and two hydrogen atoms;
- a **secondary alcohol** has two R groups and one hydrogen atom;
- a **tertiary alcohol** has three R groups.

Comparing physical properties

The electronegativity of oxygen is 3.4 and that of hydrogen is 2.2. As a result, the O—H bond is polarized $O^{\delta-}$—$H^{\delta+}$. This bond polarity has an effect on the intermolecular forces in alcohols. The structures of three alcohols of molecular formula C_4H_9OH are given on the left. Now look at their boiling points:

- butan-1-ol (a primary alcohol) 117 °C
- butan-2-ol (a secondary alcohol) 99 °C
- 2-methylpropan-2-ol (a tertiary alcohol) 82 °C

These alcohols have the same molecular formula. Weak dispersion forces are proportional to the number of electrons, so are comparable in these molecules. The most significant intermolecular force is hydrogen bonding, which depends both on the polarity of the O—H bond and on the structure of the molecule. The strength of the hydrogen bonding in an alcohol increases in the order tertiary < secondary < primary.

The illustration to the left shows that the —OH group in the tertiary alcohol is surrounded by methyl groups. These groups prevent the close approach of other molecules and inhibit the formation of hydrogen bonds. Such hindering of a chemical reaction (or, in this case, hydrogen bonding) by the arrangement of the atoms in space is called **steric hindrance**. The primary alcohol allows closer approach by adjacent —OH groups and increases the extent of hydrogen bonding.

Elimination of water

Alcohols are manufactured by adding water to an alkene, using phosphoric acid as a catalyst (see earlier in this chapter). This reaction may be reversed to eliminate water from an alcohol and produce an alkene. The elimination process involves **dehydration**, i.e. the removal of a molecule of water. Alcohols are dehydrated by warming with concentrated sulphuric acid at about 180 °C:

$$CH_3CH_2OH \rightarrow CH_2{=}CH_2 + H_2O$$

As with the hydration reaction, a strong acid acts as a catalyst. Remember that a catalyst will always catalyse the reverse reaction as well as the forward reaction. The acid first protonates the oxygen atom of the alcohol, to form $[CH_3CH_2OH_2]^+$, then water is lost, forming a carbocation; finally the catalytic hydrogen ion is regenerated.

Dehydration is most likely for tertiary alcohols because the intermediate carbocation is stabilized by the three alkyl groups attached to it. Remember that the stability of carbocations decreases in the order tertiary > secondary > primary, spread 22.8. Of the isomers of C_4H_9OH, the one that dehydrates most easily is the *tertiary* alcohol 2-methylpropan-2-ol:

$$CH_3-\underset{CH_3}{\overset{CH_3}{\underset{|}{\overset{|}{C}}}}-OH \rightarrow (CH_3)_2C{=}CH_2 + H_2O$$

The full mechanism is given below.

Another method of dehydration

An alcohol is dehydrated when its vapour passes over heated powdered aluminium oxide. The powder offers a large surface area on which the decomposition takes place. The chemical equation for the dehydration of ethanol under these conditions is:

$CH_3CH_2OH \rightarrow CH_2{=}CH_2 + H_2O$

(a) The oxygen atom of the tertiary alcohol, with its two lone pairs, provides a site for protonation by the catalyst, concentrated sulphuric acid.

(b) The protonated alcohol loses water; water, as mentioned in the previous spread, is a good leaving group. This loss creates a tertiary carbocation, which is stabilized by three electron-donating methyl groups.

(c) The tertiary carbocation loses a proton, regenerating the catalyst (H^+).

The product of dehydration is an alkene.

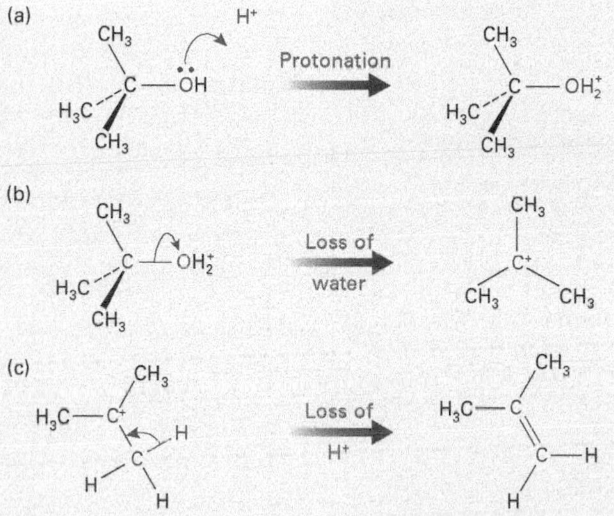

Mechanism for the production of 2-methylpropene by elimination of water from 2-methylpropan-2-ol.

SUMMARY

- Elimination reactions in alcohols bring about dehydration and the formation of an alkene.
- Ease of dehydration increases in the order primary < secondary < tertiary alcohols.
- Ethanol is dehydrated to ethene using concentrated sulphuric acid at 180 °C.

Dehydration conditions

Reaction conditions must be controlled to favour dehydration rather than the reverse reaction. Dehydration is favoured by high temperature (about 180 °C) because the reaction is endothermic. The entropy change is in favour of dehydration because one molecule decomposes to make two molecules.

PRACTICE

1 Draw structural formulae for each of the following substances and state whether each is a primary, a secondary, or a tertiary alcohol:

  a Butan-2-ol

  b 2-Methylbutan-2-ol

  c Cyclohexanol.

2 Arrange the alcohols in question 1 in order of increasing boiling point. Give reasons for your answer.

3 Give details of reagents and the practical procedures required to prepare each of the following substances from a named alcohol:

  a Propene

  b But-2-ene.

4 Suggest a reaction mechanism (complete with structural formulae) to explain the dehydration of ethanol using concentrated sulphuric acid as a catalyst to produce ethene. Include the following steps: (i) protonation of the alcohol; and (ii) loss of water.

25.4

OXIDATION OF ALCOHOLS

OBJECTIVES

- Combustion
- Oxidation of primary, secondary, and tertiary alcohols

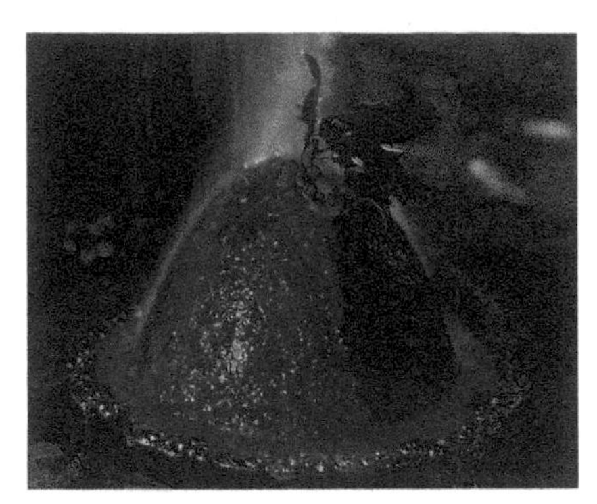

Brandy contains up to 40% ethanol. It ignites readily and burns with a clear non-sooty flame.

Carbon-neutral biofuels?

Methanol is manufactured from carbon monoxide, spread 25.1. Hence manufacture and then combustion of methanol would effectively replace CO by CO_2. Manufacture of ethanol from sugar cane (spread 25.1) and then combustion would generate the same levels of CO_2 as were there initially when the sugar was formed during photosynthesis, spread 30.6.

In 2008 a Royal Society report on sustainable biofuels explained that in order to decide which biofuels could be classed as carbon-neutral the environmental and economic aspects of the *whole* cycle needed to be assessed. The whole cycle involves growth of the plant and its transport to the refinery, the refining process including the waste materials produced, the distribution of the resultant fuel to consumers, and finally the combustion process itself including possible pollutants created. Such a detailed analysis has not so far been completed. Furthermore, widespread use of biofuels would have profound implications for land use (unintended consequences may outweigh any benefit).

Second-generation biofuels seek to use residual non-edible portions of the crop, such as stems and husks.

$Cr_2O_7^{2-}/H^+$

$Cr_2O_7^{2-}/H^+$

The oxidation of ethanol to ethanal.

Oxidation may take place by the process of combustion in oxygen or by chemical reaction with oxidants such as acidified potassium dichromate(VI). Like most organic compounds, alcohols combust readily. Complete combustion of primary, secondary, and tertiary alcohols always produces water and carbon dioxide. The products of chemical oxidation by potassium dichromate(VI) are more specific to the class of alcohol: primary alcohols yield aldehydes and carboxylic acids; secondary alcohols yield ketones; tertiary alcohols do not react.

Oxidation: combustion

Alcohols burn in air to form carbon dioxide and water. The molecular structure is completely destroyed and the constituent atoms oxidize to carbon dioxide and water. An example is the combustion of ethanol:

$$CH_3CH_2OH(l) + 3O_2(g) \rightarrow 2CO_2(g) + 3H_2O(l)$$

This reaction is strongly exothermic, and ethanol is used as a fuel in areas where it can be produced cheaply, e.g. in Brazil where conditions are suitable for growing sugar cane. In the UK, ethanol is sold as a fuel in the form of 'methylated spirits', which contains added methanol (about 5%) and a blue dye to make it unsuitable for drinking.

Oxidation and oxidants

It is also possible to oxidize an alcohol in ways that keep the carbon skeleton intact. Two oxidants are commonly used for this: potassium dichromate(VI), which contains the $Cr_2O_7^{2-}$ ion; and potassium manganate(VII), which contains the MnO_4^- ion. Used in acidified solution, both these oxidants change colour when reduced. This change is a useful sign that the organic compound has been oxidized. Use of sodium dichromate(VI) in sulphuric acid and propanone is called **Jones oxidation**.

Oxidation of the hydroxyl group —OH of an alcohol involves the removal of a hydrogen atom from the carbon atom bearing the hydroxyl group. The reaction product depends on the position of the —OH group in the alcohol, and on the reaction conditions used. Whereas all three classes of alcohol react in the same way during nucleophilic substitution, they differ greatly in their behaviour on oxidation.

Oxidation: primary alcohols

The structure of a *primary* alcohol is RCH_2OH. The —OH group is attached to a carbon atom bearing *two* hydrogen atoms. Oxidation (indicated in outline by the symbol [O]) removes one of these hydrogen atoms together with the hydrogen atom in the —OH group. The result is an aldehyde RCHO:

$$RCH_2OH \xrightarrow{[O]} RCHO$$

For example, ethanol is converted to ethanal CH_3CHO:

$$CH_3CH_2OH \xrightarrow{[O]} CH_3CHO$$

The oxygen atom is doubly bonded to the carbon atom, forming a carbonyl group $\supset C{=}O$. The illustration on the left gives structural formulae for these molecules.

Further oxidation: carboxylic acids

The aldehyde functional group —CHO consists of a hydrogen atom attached to the carbonyl $\supset C{=}O$ carbon atom. This structure may be oxidized further. The result is a carboxylic acid RCOOH:

$$RCHO \xrightarrow{[O]} RCOOH$$

Overall, the primary alcohol ethanol oxidizes in two stages: first to the aldehyde (ethanal) and then to the carboxylic acid (ethanoic acid):

$$CH_3CH_2OH \xrightarrow{[O]} CH_3CHO \xrightarrow{[O]} CH_3COOH$$

Oxidants are generally very unspecific in their reaction. Oxidizing a primary alcohol produces an aldehyde. Leaving the aldehyde product in contact with the oxidant will allow further oxidation to the carboxylic acid. This may be avoided by distilling the aldehyde away from the reaction mixture as it forms. The aldehyde lacks hydrogen bonding (hydrogen is bonded to *carbon*, not oxygen) and so has a lower boiling point than the corresponding alcohol.

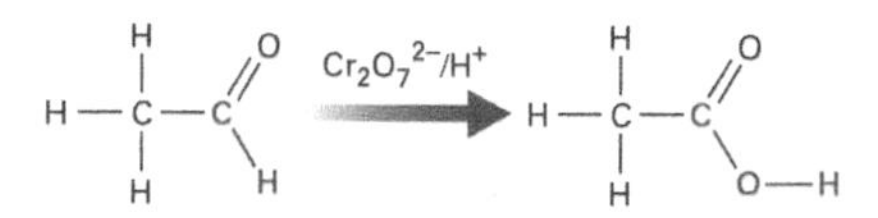

The oxidation of ethanal to ethanoic acid.

Aldehydes and ketones

See the following chapter for more details on aldehydes and ketones, such as their reduction to alcohols (26.2) and how to distinguish between them (26.4).

Breathalyser

The original **breathalyser** used the colour change in the reaction of dichromate(VI) as a test. More recently, ethanol levels have been measured by infrared spectroscopy, spreads 32.3 and 32.4, or an ethanol fuel cell (similar to a methanol fuel cell, spread 13.10).

Colour changes on heating alcohols with acidified potassium dichromate(VI): primary and secondary alcohols reduce orange $Cr_2O_7^{2-}(aq)$ to blue-green $Cr^{3+}(aq)$. Tertiary alcohols (right) have no effect.

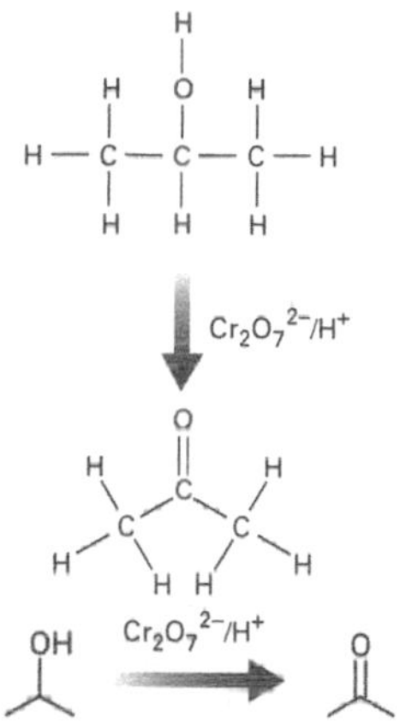

Oxidation: secondary alcohols

The structure of a *secondary* alcohol is RR'CHOH. The —OH group is attached to a carbon atom bearing *one* hydrogen atom. As in the case of primary alcohols, oxidation removes this hydrogen atom together with the hydrogen atom in the —OH group. The result is a ketone RR'C=O. Note that ketones and aldehydes both contain the carbonyl group ⊃C=O. There is no hydrogen atom attached to the carbon atom of the carbonyl group in ketones, just two alkyl groups. In general terms:

$$RR'CHOH \xrightarrow{[O]} RR'CO$$

Ketones resist further oxidation.

The oxidation of propan-2-ol to propanone.

Tertiary alcohols are not easily oxidized

The structure of a *tertiary* alcohol is RR'R"COH. In this case, there are no hydrogen atoms attached to the carbon atom bearing the hydroxyl group. As a result, tertiary alcohols *cannot* be oxidized (except under extreme conditions, when carbon–carbon bonds break).

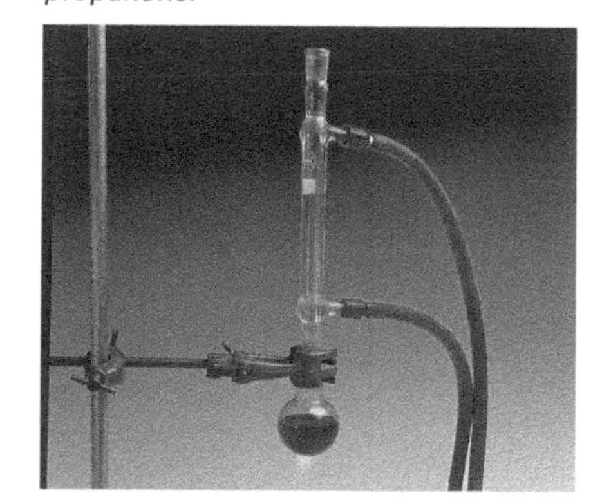

Refluxing a secondary alcohol with acidified potassium dichromate(VI) produces a ketone.

SUMMARY

- Primary alcohols are oxidized to aldehydes and then to carboxylic acids.
- Secondary alcohols are oxidized to ketones.
- Primary and secondary alcohols are oxidized by acidified potassium dichromate(VI) or acidified potassium manganate(VII).
- Tertiary alcohols are not readily oxidized.

PRACTICE

1 Methoxymethane and ethanol both have the molecular formula C_2H_6O. Explain in terms of molecular structure why ethanol (T_b = 78 °C) has a much higher boiling point than methoxymethane (T_b = −28 °C).

2 Acidified potassium dichromate(VI) was added to each of the following substances:

a Propanone

b 2-Methylpropan-2-ol

c Propan-2-ol

Write the structural formula for the product (if any).

25.5

Phenols and ethers

OBJECTIVES

- Phenols and alcohols contrasted
- Ether synthesis
- Ethers are inert

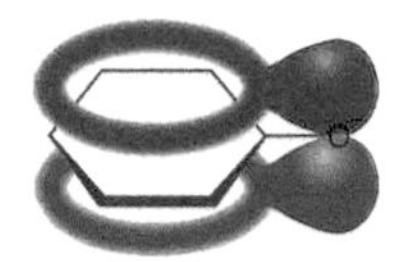

Aqueous ethanol has a pH of 7. The alkoxide ion is formed only in the presence of a powerful reductant. Sodium is capable of reducing alcohols to hydrogen gas. In contrast, aqueous phenol has a pH of about 4 (phenol is partially miscible with water). The —OH group attached to a benzene ring shows acidic properties. The negative ion is stabilized by delocalization of the negative charge, as shown in the electrostatic potential map below.

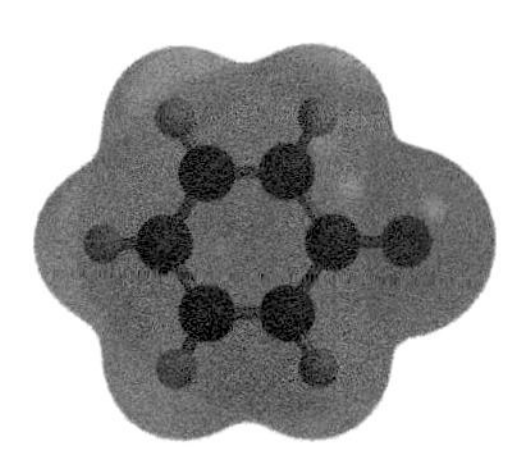

Phenylmethanol

This compound has the formula $C_6H_5CH_2OH$. The —OH group is not attached directly to the benzene ring, and so its properties are the same as those of an *alcohol*.

Phenol acidity

Phenol is not a strong enough acid to liberate carbon dioxide from aqueous sodium carbonate. Sodium phenoxide $Na^+C_6H_5O^-$ is the salt of a *weak* acid. The acid is regenerated on addition of *strong* acid:

$Na^+C_6H_5O^- + HCl \rightarrow C_6H_5OH + NaCl$

Test for phenol

A characteristic test for a phenol is the formation of a purple colour with aqueous iron(III) chloride.

The *alcohols* are a class of compounds in which a hydroxyl group —OH is attached directly to an alkyl group. By comparison, the *phenols* are a class of compounds in which the —OH group is attached directly to a benzene ring. The properties of alcohols and phenols are distinctly different. Different properties are also shown by the *ethers*, which consist of an oxygen atom joined to two separate alkyl groups. The lack of a hydroxyl group means that ethers are chemically inert.

The polar O—H bond in alcohols and phenols

The O—H bond is polarized $O^{\delta-}$—$H^{\delta+}$. Complete charge separation would cause the bond to break, resulting in the loss of a proton H^+ and a negative charge on the oxygen atom. However, aqueous solutions of alcohols are neutral. The terminal hydrogen atom on the —OH group is not acidic and does not protonate water to form the aqueous hydrogen ion (oxonium ion) $H_3O^+(aq)$. Moreover, alcohols do not act as acids in the presence of aqueous hydroxide ion $OH^-(aq)$ from bases such as aqueous sodium hydroxide.

In contrast, a solution of phenol is acidic (≈ pH 4):

$$C_6H_5OH + H_2O \rightleftharpoons H_3O^+ + C_6H_5O^-$$

Loss of a proton is possible in phenol because the negative charge in the **phenoxide ion** is delocalized around the benzene ring. Phenol reacts with strong bases to form salts:

$$C_6H_5OH + NaOH \rightarrow Na^+C_6H_5O^- + H_2O$$

Alcohols do not show this behaviour in solution or with aqueous bases because the negative charge remains localized on the oxygen atom.

A comparison of the properties of phenol and ethanol.

Property or reaction	Phenol C_6H_5OH	Ethanol CH_3CH_2OH
Solubility in water at 20 °C	Partially miscible	Miscible in all proportions
pH of solution	approx. pH 4	pH 7
Action of CH_3COOH	No reaction	Forms ester* $CH_3CH_2OCOCH_3$
Action of CH_3COCl	Rapid reaction, good yield of $C_6H_5OCOCH_3$	Rapid reaction, good yield of $CH_3CH_2OCOCH_3$
Action of bromine water	Immediate reaction to give 2,4,6-tribromophenol	No reaction
Action of $Cr_2O_7^{2-}/H^+$	Gives a complex mixture of products	Forms ethanal, then ethanoic acid
Action of sodium	Evolves $H_2(g)$	Evolves $H_2(g)$

* See chapter 27 'Carboxylic acids and their derivatives'.

Phenols can be used to manufacture plastics, such as the thermosetting (spread 29.4) polymer Bakelite, antiseptics such as TCP (spread 23.4), disinfectants, and resins for paints.

Ethers

The alkoxide ion RO^- is negatively charged and is therefore a nucleophile. Alkoxides dissolved in ethanol react with halogenoalkanes. The alkoxide ion substitutes for the halide ion to form an ether. An **ether** consists of an oxygen atom with two single covalent bonds each joined to a separate alkyl group.

The most important ether is ethoxyethane $CH_3CH_2OCH_2CH_3$ (diethylether, or in everyday use just 'ether').

The reaction in which an alkoxide ion substitutes for the halide ion is called the **Williamson synthesis** of an ether. The diagram shows the mechanism for this reaction.

For example, sodium ethoxide when refluxed with bromoethane forms ethoxyethane:

$CH_3CH_2O^- + CH_3CH_2Br \rightarrow CH_3CH_2OCH_2CH_3 + Br^-$

The alkoxide ion RO^- is a nucleophile, in a similar way to the hydroxide ion HO^-. Halogenoalkanes react with nucleophiles; the reaction follows the typical nucleophilic substitution mechanism of halogenoalkanes (see previous chapter). One advantage of the Williamson synthesis is that the alkoxide can have a different number of carbon atoms from the halogenoalkane and unsymmetrical ethers can be made this way: sodium methoxide, $Na^+CH_3O^-$, will react with bromoethane to give methoxyethane $CH_3OCH_2CH_3$.

Alexander William Williamson had to overcome considerable adversity in his desire to be a chemist. He had lost an arm and the use of an eye in childhood. His inquiring mind remained thankfully intact. Despite being essentially an organic chemist, he made major contributions to physical chemistry too, as he was for example the first person to formulate the idea of chemical equilibrium.

The reaction between ethoxide ion and bromoethane.

Ethers are generally unreactive and are mostly used as solvents. Note that ethers have lower boiling points than the corresponding alcohols. For example, the ether methoxybutane $CH_3—O—CH_2CH_2CH_2CH_3$ and the alcohol pentan-1-ol both have the molecular formula $C_5H_{12}O$. As a result, the dispersion forces between the molecules of each compound should be similar. However, methoxybutane boils at 70 °C and pentan-1-ol at 138 °C, confirming the absence of hydrogen bonds in the ether.

Ethers and petrol

In response to the need to remove lead from petrol, alternative petrol additives have been developed to prevent engine damage through 'knocking'. One such additive is an ether derived from methanol, methyl tertiary butyl ether (MTBE, $CH_3OC(CH_3)_3$). Petrol may contain 15% MTBE. Production of this chemical has increased from virtually zero in the 1970s to several million tonnes per year.

Ethoxyethane is being used here to extract chromium peroxide $CrO(O_2)_2$ from the reaction between hydrogen peroxide and acidified potassium dichromate(VI). A deep blue solution results.

SUMMARY

- The —OH group is acidic when *directly* attached to a benzene ring in a phenol.
- Alcohols and phenols have distinctly different properties.
- Alcohols react at the O—H or the C—O bond; phenols react at the O—H bond or undergo substitution reactions involving hydrogen atoms on the benzene ring.

PRACTICE

1. Explain why phenol is acidic in solution, whereas phenylmethanol is neutral.
2. Draw structural formulae to illustrate the reactions listed in the phenol/ethanol comparison table opposite.
3. Give details of a possible laboratory synthesis of MTBE.

25.6

OBJECTIVES

- Epoxyethane
- Ethane-1,2-diol
- Propane-1,2,3-triol

Alcohols with more than one —OH group

All of the alcohols met so far have been **monohydric** alcohols: each alcohol molecule (whether primary, secondary, or tertiary) has contained just *one* hydroxyl group —OH. Alcohols with *two* hydroxyl groups are called **dihydric** alcohols; those with *three* hydroxyl groups are called **trihydric** alcohols. The term **polyhydric** describes a molecule that contains a large number of hydroxyl groups. Common polyhydric alcohols include the sugars, which will be described in chapter 30 'Biochemistry'. This spread concentrates on di- and trihydric alcohols, two of which have great commercial importance.

Epoxyethane and ethane-1,2-diol

Epoxyethane is one of the most useful compounds made industrially from ethene. The molecule has one oxygen atom that is attached to two carbon atoms to form a triangular ring. The compound is therefore a *cyclic ether*. This arrangement is an example of a **heterocyclic** compound, where the ring contains another atom in place of carbon. (In contrast, in alicyclic compounds all the atoms in the ring are carbon.)

Epoxyethane is manufactured from ethene obtained from the cracking of crude oil fractions. The reaction is carried out by passing a mixture of ethene and oxygen over a silver catalyst at 300 °C and 15 atm. Below its boiling point of 11 °C, the epoxyethane produced is a colourless volatile liquid.

Considerable care must be taken in the formation and handling of epoxyethane, as it is both flammable and explosive. Furthermore, epoxyethane may irritate the respiratory system and may cause neurological disorders.

The bond angles in the epoxyethane ring are about 60°, so the structure is extremely strained. The molecule is very responsive to reactions that involve opening the ring, thereby relieving the strain. Ethers such as ethoxyethane $CH_3CH_2OCH_2CH_3$ do not have this strained ring structure and are generally unreactive. As discussed earlier in this chapter, the reactions of alcohols often involve the protonation of oxygen before the main reaction can occur. In the case of epoxyethane, dilute sulphuric acid acts as a catalyst and causes the molecule to react with water, forming ethane-1,2-diol $HOCH_2CH_2OH$ (a dihydric alcohol, traditionally called **ethylene glycol**):

$$CH_2(O)CH_2 + H_2O \rightarrow HOCH_2CH_2OH$$

Ethane-1,2-diol is used as antifreeze in car cooling systems (see spread 22.9) and in the manufacture of polyester fabrics, including the synthetic fibre Terylene (see chapter 27 'Carboxylic acids and their derivatives').

Trihydric alcohols

A common trihydric alcohol is propane-1,2,3-triol (traditionally called *glycerol* or *glycerine*: the latter name should be avoided). This trihydric alcohol is an extremely viscous liquid due to the high degree of hydrogen bonding between its molecules. It occurs widely in Nature as a component of fats and oils (see spread 30.1 'Lipids'). Heating propane-1,2,3-triol with a nitrating mixture of concentrated sulphuric and nitric acids results in **nitroglycerine**. This substance is the 'active ingredient' in dynamite, used for blasting rock out of quarries and when building structures such as roads.

Epoxides

Epoxides contain oxygen atoms in their molecules as part of a three-membered ring. Epoxides are cyclic ethers. They react with compounds such as phenols to form polymeric epoxy resins.

$H_2C{=}CH_2$ → (O_2 / Ag / 300 °C) → epoxyethane → (H_2O / H^+) → $HOCH_2CH_2OH$

The industrial conversion of ethene to epoxyethane and then to ethane-1,2-diol. Note that epoxyethane is hazardous, being both flammable and explosive.

Alkoxy alcohols

Epoxyethane reacts with an alcohol in the presence of an acid catalyst to yield a compound that is both an ether and an alcohol.

$ROCH_2CH_2OH$

These alkoxy alcohols are useful as solvents, and in the production of plasticisers and non-ionic detergents. For example, an alkoxy alcohol is used as a solvent for cellulose ethanoate in the manufacture of films, lacquers, and varnishes.

H H H
H—C—C—C—H
ONO_2 ONO_2 ONO_2

Pure nitroglycerine is a shock-sensitive liquid: it will explode if dropped. Working at the end of the nineteenth century, Alfred Nobel absorbed nitroglycerine into kieselguhr (a type of clay). The result was dynamite*, which explodes only when detonated. He hoped his invention would alleviate the dangers experienced by miners and other labourers. However, Nobel was greatly disappointed by the use of dynamite in warfare. He sold the explosives factories that had made him a rich man, and set up a fund to award annual prizes for outstanding work in the fields of science, writing, and the advancement of world peace. The first Nobel Prizes were awarded in 1901: they have their origins in a trihydric alcohol.*

SUMMARY

- Oxidation of ethene over a silver catalyst at 300 °C and 15 atm forms epoxyethane, a cyclic ether.
- Acidic hydrolysis of epoxyethane produces ethane-1,2-diol.
- Dihydric alcohols contain two hydroxyl groups, e.g. ethane-1,2-diol ('ethylene glycol'), used as an antifreeze additive.
- Trihydric alcohols contain three hydroxyl groups, e.g. propane-1,2,3-triol ('glycerol'), used to make nitroglycerine.

PRACTICE

1 Explain why, compared to ethanol, ethane-1,2-diol:

a is more viscous;

b has much higher melting and boiling points.

2 Explain why the structure of epoxyethane is described as 'strained'. What effect do you think this factor will have on the general reactivity of the molecule?

3 To what extent do you think propane-1,2,3-triol is soluble in water? Give reasons for your answer.

Chapter 25 Reactions summary

- Fermentation of glucose by yeast:
  $C_6H_{12}O_6 \rightarrow 2CH_3CH_2OH + 2CO_2$
- Direct hydration of ethene (phosphoric acid catalyst, 300 °C, and 70 atm):
  $CH_2{=}CH_2(g) + H_2O(g) \rightarrow CH_3CH_2OH(g)$
- Nucleophilic substitution of the —OH group by a halide:
  $CH_3CH_2OH + HBr \xrightarrow[KBr]{conc.\ H_2SO_4} CH_3CH_2Br + H_2O$
- Chlorination of ethanol by $SOCl_2$ (or PCl_3, PCl_5):
  $CH_3CH_2OH + SOCl_2 \rightarrow CH_3CH_2Cl + SO_2 + HCl$
- Elimination of water (dehydration) from ethanol (conc. sulphuric acid catalyst, 180 °C):
  $CH_3CH_2OH \rightarrow CH_2{=}CH_2 + H_2O$
- Combustion of ethanol:
  $CH_3CH_2OH(l) + 3O_2(g) \rightarrow 2CO_2(g) + 3H_2O(l)$
- Oxidation of a primary alcohol to an aldehyde and then to a carboxylic acid:
  $CH_3CH_2OH \xrightarrow{[O]} CH_3CHO \xrightarrow{[O]} CH_3COOH$
- Oxidation of a secondary alcohol to a ketone:
  $RR'CHOH \xrightarrow{[O]} RR'C{=}O$
- Tertiary alcohols do not oxidize.
- Synthesis of epoxyethane and subsequent hydrolysis:
  $CH_2{=}CH_2 \xrightarrow[300\ °C]{O_2,\ Ag} H_2C{-}CH_2$ (bridged by O) $\xrightarrow[H^+]{+H_2O} HOCH_2CH_2OH$

PRACTICE EXAM QUESTIONS

1 Butan-1-ol can be oxidized by acidified potassium dichromate(VI) using two different methods.

a In the first method, butan-1-ol is added dropwise to acidified potassium dichromate(VI) and the product is distilled off immediately.

i Using the symbol [O] for the oxidizing agent, write an equation for this oxidation of butan-1-ol, showing clearly the structure of the product. State what colour change you would observe.

ii Butan-1-ol and butan-2-ol give different products on oxidation by this first method. By stating a reagent and the observation with each compound, give a simple test to distinguish between these two oxidation products. [6]

b In a second method, the mixture of butan-1-ol and acidified potassium dichromate(VI) is heated under reflux. Identify the product which is obtained by this reaction. [1]

c Give the structures and names of two branched chain alcohols which are both isomers of butan-1-ol. Only isomer 1 is oxidised when warmed with acidified potassium dichromate(VI). [4]

2 The infrared spectrum of ethanol is shown below.

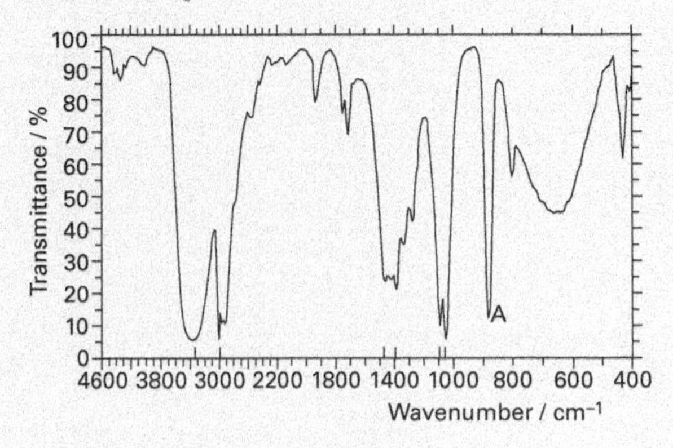

Table of infrared absorption data

Bond	Wavenumber / cm^{-1}
C–H	2840–3300
C–C	750–1100
C=C	1610–1680
C=O	1680–1750
C–O	1000–1300
O–H	3230–3550

a Using this table of data, identify a bond that could be responsible for the absorption labelled A on the spectrum. [1]

b i With the aid of the spectrum, suggest an approximate range for the fingerprint region.

ii How does the fingerprint region enable a compound to be identified? [3]

c Modern roadside breathalysers measure the absorption due to ethanol at 2950 cm^{-1}. Propanone, which is often found in the breath of diabetics, also absorbs at 2950 cm^{-1} but the breathalyser is able to distinguish between ethanol and propanone and to eliminate any signal from the latter.

i Suggest why the absorption at 3340 cm^{-1} is not used to analyse the amount of alcohol in a person's breath.

ii At what wavenumber approximately would a breathalyser indicate strong absorption of propanone but almost none for ethanol?

iii Describe a simple chemical test which would enable you to distinguish between ethanol and propanone. Give the reagent(s) used and the observation with each compound. [6]

(See also chapter 32.)

3 a Why is it necessary, in the direct synthesis of epoxyethane from ethene and air, to have efficient removal of the heat generated? [1]

b i Explain briefly why epoxyethane is highly reactive and write an equation for the reaction between one mole of epoxyethane and one mole of ethanol.

ii Give the repeating unit of the polymer formed when one mole of ethanol reacts with an excess of epoxyethane. [5]

c Predict, by means of an equation, how one mole of epoxyethane reacts with one mole of ammonia. [1]

4 Three different reactions of propan-2-ol are shown below.

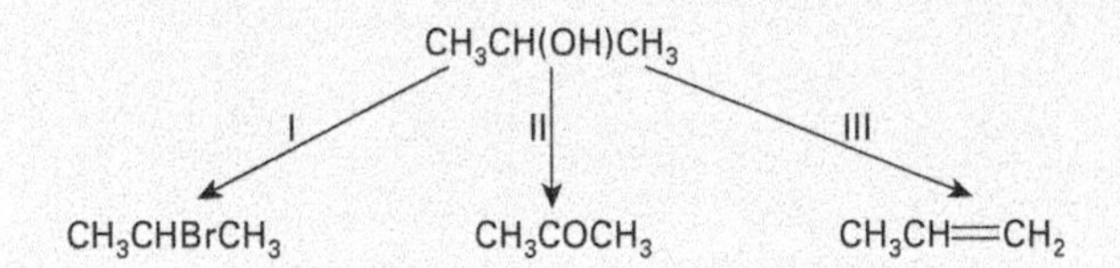

a For each of the reactions I, II, and III, give suitable reagents and conditions. [6]

b If 2-methylpropan-2-ol, $(CH_3)_3COH$, was used as the starting material in **a** instead of propan-2-ol, identify the organic products, if any, of reactions I, II and III. You should indicate if no reaction occurs. [3]

5 There are four alcohols of molecular formula $C_4H_{10}O$ which are structural isomers.

Three of these alcohols are given below.

$CH_3CH_2CH_2CH_2OH$ butan-1-ol

$CH_3CH(OH)CH_2CH_3$ butan-2-ol

$(CH_3)_3COH$ 2-methylpropan-2-ol

a Give the structural formula of the fourth alcohol that is isomeric with those above. [1]

b On heating with concentrated sulphuric acid, butan-2-ol is converted into a mixture of alkenes.

i Give the name of the type of reaction taking place. [1]

ii Give the structural formula of one of the alkenes formed. [1]

c i Give the name or structural formula of the organic compound produced when butan-1-ol is heated with acidified potassium dichromate(VI) and the product is removed by distillation as it forms. [1]

ii Give the name or structural formula of the organic compound produced when butan-1-ol is heated under reflux for 20 minutes with acidified potassium dichromate(VI). [1]

iii Give the structural formula of the organic product formed when butan-2-ol is heated under reflux for 20 minutes with acidified potassium dichromate(VI). [1]

iv State the type of reaction occurring in **c iii**. [1]

d When 2-methylpropan-2-ol is heated with a carboxylic acid in the presence of a catalyst, an ester, $C_6H_{12}O_2$, is formed.

i Give the structural formula of this ester of molecular formula $C_6H_{12}O_2$. [1]

ii Give the name of the carboxylic acid needed to form the ester in **d i**. [1]

iii Suggest a suitable catalyst for the reaction. [1]

e Lucas' test may be used to distinguish between the three alcohols butan-1-ol, butan-2-ol and 2-methylpropan-2-ol. A mixture of anhydrous zinc chloride and hydrochloric acid is warmed with each of the three alcohols under identical conditions.

i Describe how the results of Lucas' test lead to the identification of the three alcohols. [2]

ii Give the name and the structural formula of the organic compound formed by butan-2-ol in Lucas' test. [2]

(See also chapter 27.)

6 Alcohols and ethers have the same general formula, $C_nH_{2n+2}O$. Ethers contain a C—O—C linkage as, for example, in methoxypropane, $H_3C—O—CH_2CH_2CH_3$.

a i Give the number of peaks in the low-resolution proton NMR spectrum of methoxypropane and the ratio of the areas under the peaks in the spectrum.

ii Draw the structure of an ether which is an isomer of methoxypropane and which produces only 2 peaks in its low-resolution proton NMR spectrum. [3]

b i Alcohols can be prepared from haloalkanes by reaction with hydroxide ions. Name and outline the mechanism for the preparation of propan-1-ol from bromopropane.

ii Ethers can be prepared from haloalkanes by a similar reaction. Complete the following equation which shows the formation of methoxypropane.

$$\_\_\_\_\_ + BrCH_2CH_2CH_3 \rightarrow CH_3OCH_2CH_2CH_3 + Br^-$$

[5]

c Ethers are not oxidized by acidified potassium dichromate(VI). Name the type of alcohol which is also not oxidized by acidified potassium dichromate(VI) and draw the structure of an alcohol of this type which is an isomer of methoxypropane. [2]

d Write an equation for the complete combustion of methoxypropane in an excess of oxygen. [1]

e Ethers and alcohols can be distinguished by studying their infrared spectra. Using the table of data given below, state where, other than in the fingerprint region, their infrared spectra will be different and explain what causes this difference. [2]

Table of infrared absorption data

Bond	Wavenumber / cm^{-1}
C—H	2850–3300
C—C	750–1100
C=C	1620–1680
C=O	1680–1750
C—O	1000–1300
O—H in alcohols	3230–3550
O—H in acids	2500–3000

(See also chapter 32.)

26 Aldehydes and ketones

Aldehydes and ketones are two classes of compound that both contain the carbonyl group. The **carbonyl group** consists of an oxygen atom attached to a carbon atom by a double covalent bond, usually represented by the formula ⊃C=O. The carbon atom is referred to as the **carbonyl carbon atom**. **Aldehydes** have the structure RCHO where R is an alkyl or aryl group (methanal has the formula HCHO). **Ketones** have the structure RR′CO where R and R′ are alkyl or aryl groups. Both these types of **carbonyl compound** show the reactions of the ⊃C=O carbonyl group. In aldehydes, the hydrogen atom attached to the carbonyl group leads to different behaviour on oxidation compared to that seen with ketones.

Introduction to the carbonyl group

26.1

OBJECTIVES

- Aldehydes
- Ketones
- Nucleophilic addition reactions
- Addition of HCN

Carbonyl group symbol

Typographically we have used the symbol ⊃C=O, to represent the carbonyl group

>C=O

Solubility

The oxygen of the carbonyl group can form hydrogen bonds with water, which improves the water solubility of lower aldehydes and ketones (especially methanal, ethanal, and propanone).

Rate of reaction

The rate of nucleophilic addition to the carbonyl group increases with the size of the positive charge on the carbon atom. Groups that release electron density towards the carbonyl group will decrease the value of the charge. The order of reactivity is therefore:

H(H)C=O methanal > R(H)C=O aldehyde > R(R′)C=O ketone

Remember that the geometry of the R groups ('steric hinderance', see previous chapter) will also influence reaction rates.

Aldehydes and ketones are of immense industrial importance. Propanone CH_3COCH_3 (traditionally called *acetone*) is a ketone widely used as a solvent. Its derivative methyl isobutyl ketone (MIBK) is an important solvent used in the manufacture of plastics. Propanone is also used to form methacrylates, which are the chemical feedstocks for making plastics such as Perspex and acrylic resins. Propanone reacts with phenol to form bisphenol A, which is used to make epoxy resins. Methanal is an industrially important aldehyde used to make plastics such as 'Melamine', spread 29.4. All these substances and their various uses stem from the unique properties of the carbonyl group ⊃C=O and the opportunities it offers for chemical synthesis.

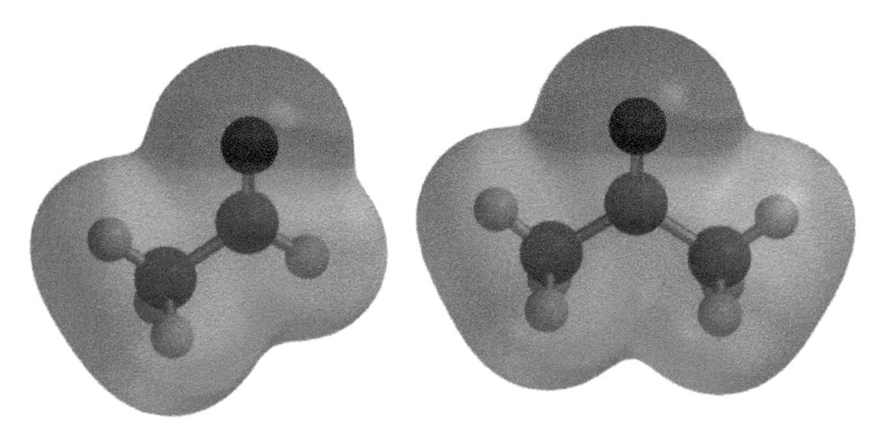

Electrostatic potential maps of (a) ethanal and (b) propanone show the polar nature of the carbonyl group ⊃C=O. The oxygen atom is δ– (shown by red colour); the carbon atom is δ+. Remember that the Spartan software intentionally does not show the number of bonds; see spread 6.4.

The carbon–oxygen double bond

The structure of the C=O bond makes it a useful starting point for the synthesis of other molecules. At first sight, the C=O bond looks similar to the C=C bond, and indeed both react to saturate the double bond by addition. However, the oxygen atom is much more electronegative than carbon and will attract electron density in the double bond towards itself. The bond is polarized $C^{\delta+}{=}O^{\delta-}$, thus offering *two* main centres of reactivity, the $C^{\delta+}$ atom and the $O^{\delta-}$ atom. Remember that the C=C bond is non-polar.

- The C=C bond reacts with electrophiles. The characteristic reaction of alkenes is *electrophilic addition*.
- The $C^{\delta+}$ atom in the ⊃C=O bond is attacked by nucleophiles. There are two possible outcomes: *nucleophilic addition* reactions and *condensation* reactions, which involve nucleophilic addition followed by elimination.

Nucleophilic addition of HCN

Nucleophilic addition of hydrogen cyanide HCN (using KCN in dilute acid) to a carbonyl compound makes the carbon chain longer. Cyanide ions CN^- are nucleophiles and so will attack the $C^{\delta+}$ carbonyl carbon atom. This reaction produces a nitrile that has a hydroxyl group attached to the carbon bearing the nitrile group. These molecules are called **hydroxynitriles.** For example, the reaction of hydrogen cyanide with ethanal (a two-carbon molecule) produces a molecule with a three-carbon chain, as shown below.

The addition of HCN to the carbonyl group in the aldehyde ethanal.

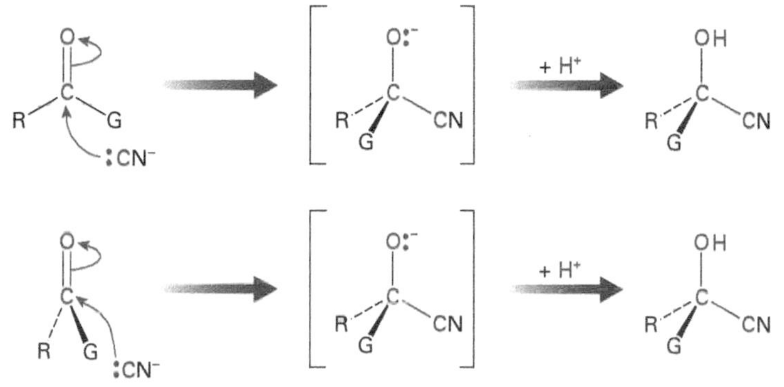

Nucleophilic addition to a carbonyl group leads to an alcohol with a carbon–nucleophile single bond. Note that there is no other product.
The view of the addition of cyanide immediately above gives a more accurate picture of the attack of the ion. Group G is either another R group or a hydrogen.

In general, nucleophilic addition of HCN to an aldehyde or a ketone forms a hydroxynitrile. See spread 28.6 for further details of this reaction. In the next spread we will study another nucleophilic addition reaction, before turning to condensation reactions in spread 26.3.

SUMMARY

- The formula of the carbonyl group is $\supset C{=}O$.
- Aldehydes have the formula RCHO.
- Ketones have the formula RR′CO.
- The carbonyl group is polarized $C^{\delta+}{=}O^{\delta-}$.
- The most characteristic reaction of the carbonyl group is attack by a nucleophile on the carbonyl carbon atom, leading either to a nucleophilic addition reaction or, after elimination, to a condensation reaction.
- HCN adds to aldehydes and ketones to form hydroxynitriles.

C=O and C=C comparison

In the nucleophilic addition to a carbonyl group, an important feature is that *oxygen* carries the negative charge in the species produced when the nucleophile has attacked. A similar reaction at the C=C bond of alkenes would be very unfavourable as a carbon atom is much less stable than oxygen when carrying a negative charge.

Cyanide

Use of cyanides requires carefully controlled conditions because of the cyanide ion's severe toxicity.

Bisulphite addition

Addition of sodium hydrogensulphite (sodium bisulphite, $NaHSO_3$) to an aldehyde or methyl ketone produces a crystalline addition product which can be easily separated; hydrolysis of the addition product in dilute acid regenerates the parent compound. This technique can be used to purify liquid carbonyl compounds.

Aldehydes and ketones in Nature

The aldehyde retinal (see chapter 22 'Alkanes and alkenes') is important to the function of the human eye. The aldehyde CH_2CHCHO (propenal) is a major component of bonfire and barbecue smoke and is partly responsible for making your eyes water. The ketone $CH_3COCOCH_3$ (butanedione) gives butter and stale sweat their characteristic smells. Carvone is the main constituent of oil of spearmint; zingerone is the active ingredient in ginger. The sex hormones progesterone and testosterone are ketones.

PRACTICE

1 Draw structural formulae for each of the following substances:

a Ethanal
b Propanone
c Propanal
d 2-Methylpropanal
e Butanone
f Pentan-2-one
g Pentan-3-one.

2 Give the name and draw the structural formula of the compound that results from each of the following nucleophilic addition reactions:

a Ethanal + hydrogen cyanide
b Propanal + hydrogen cyanide
c Propanone + hydrogen cyanide.

26.2

OBJECTIVES

- $NaBH_4$
- $LiAlH_4$
- Reduction of aldehydes and ketones

Reduction of aldehydes and ketones

The previous chapter discussed the oxidation of alcohols. To summarize the reactions: oxidation of primary alcohols produces aldehydes; oxidation of secondary alcohols produces ketones. Both these oxidation reactions may be carried out using moderately powerful oxidants such as acidified potassium dichromate(VI). This spread shows that the oxidation of alcohols to produce these carbonyl compounds can be reversed by the action of suitable reductants. Sodium tetrahydridoborate $NaBH_4$ is commonly used for reducing carbonyl compounds. An alternative reagent is the more powerful reductant lithium tetrahydridoaluminate $LiAlH_4$.

$NaBH_4$ and $LiAlH_4$

Sodium tetrahydridoborate $NaBH_4$ and lithium tetrahydridoaluminate $LiAlH_4$ are two common reductants that produce the hydride ion H^-. The hydride ion acts as a nucleophile, attacking the $\delta+$ carbonyl carbon atom in aldehydes and ketones. The hydride ion H^- is also present in sodium hydride Na^+H^-, but sodium hydride is too unstable for use with most organic compounds. The two complexes BH_4^- and AlH_4^- supply hydride ions in a steady and controlled manner during reaction; they are sometimes referred to as 'hydride ion carriers'.

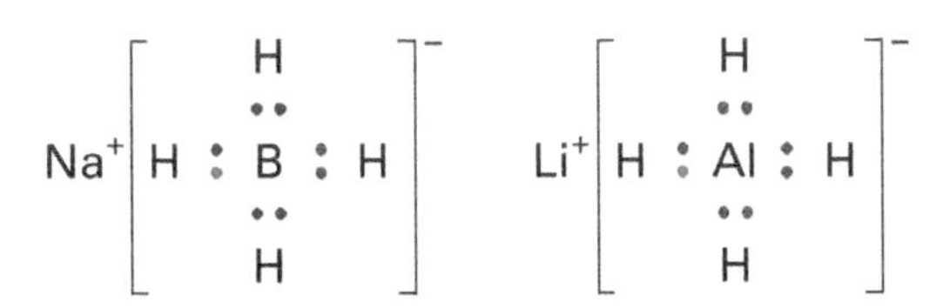

Boron and aluminium are both Group III elements. Their hydrides BH_3 and AlH_3 are Lewis acids (see chapter 12 'Acid–base equilibrium'), and so can accept the electron pair from a hydride ion to form the complexes BH_4^- and AlH_4^-.

Sodium tetrahydridoborate.

Sodium tetrahydridoborate $NaBH_4$

Sodium tetrahydridoborate is a white crystalline solid which reduces aldehydes and ketones. This reductant can be used either in aqueous solution or in alcoholic solution.

- $NaBH_4$ reduces aldehydes to primary alcohols:

  $RCHO \xrightarrow{NaBH_4} RCH_2OH$

For example, ethanal is reduced to ethanol, using [H] to indicate reduction:

$CH_3CHO \xrightarrow{[H]} CH_3CH_2OH$

- $NaBH_4$ reduces ketones to secondary alcohols:

  $RR'CO \xrightarrow{NaBH_4} RR'CHOH$

For example, propanone is reduced to propan-2-ol:

$(CH_3)_2CO \xrightarrow{[H]} (CH_3)_2CHOH$

In both cases, the oxygen atom has not been removed from the molecule, but the carbon–oxygen double bond has been made into a single bond and the molecule has gained two hydrogen atoms. The (simplified) mechanism is shown at the top of the opposite page. To simplify the mechanism, the reagent is considered to be the hydride ion.

Reduction of aldehydes and ketones

The reduction of a carbonyl group by $NaBH_4$ is carried out in two stages: (a) the addition of sodium tetrahydridoborate to the carbonyl compound, followed by (b) protonation to produce the final product.

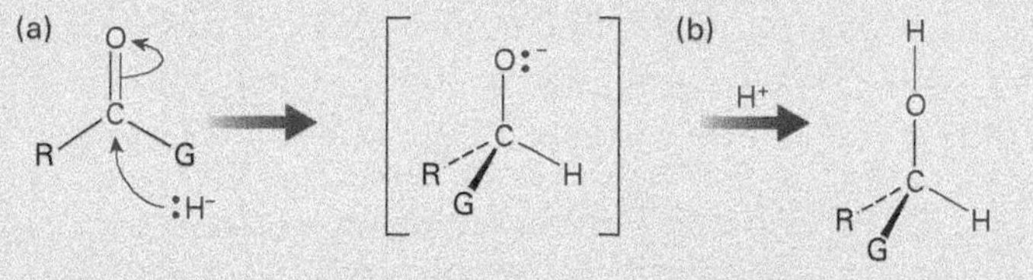

Reduction of a carbonyl group by $NaBH_4$. (a) A hydride ion H^- carries out nucleophilic attack on the $\delta+$ carbonyl carbon atom. (b) Protonation by the solvent forms the product.

Lithium tetrahydridoaluminate $LiAlH_4$

Lithium tetrahydridoaluminate is a more powerful reductant than sodium tetrahydridoborate. $LiAlH_4$ reduces water too vigorously for it to be used in aqueous solution. As a result, anhydrous conditions are needed when reducing aldehydes and ketones using this reagent. Dry ether (ethoxyethane, spread 25.5) is the solvent commonly used; water is added later to complete the reaction. $LiAlH_4$ acts by a similar mechanism to $NaBH_4$, reducing aldehydes to primary alcohols, and reducing ketones to secondary alcohols.

Familiar name

The names 'sodium tetrahydridoborate' and 'lithium tetrahydridoaluminate' are not convenient to use when asking a colleague to hand you a bottle of reagent. The latter is frequently contracted to 'lithal'.

Comparing with other reductants

Compounds with carbon–carbon double bonds are reduced by catalytic hydrogenation using hydrogen passed over the surface of a heated catalyst. For example, ethene is hydrogenated to ethane:

$$CH_2{=}CH_2 + H_2 \xrightarrow[150\,°C]{Ni} CH_3CH_3$$

The most usual catalysts are the transition metals nickel, palladium, and platinum, spread 22.7. The reaction mechanism is a radical process rather than a nucleophilic one because it involves hydrogen *atoms* adsorbed on the surface of the catalyst (see chapter 22 'Alkanes and alkenes').

Other, more specific, reduction processes are also common in organic chemistry. Two examples are sodium metal reacting with alcohols (see the previous chapter), and tin and hydrochloric acid used to reduce nitrobenzene to phenylamine (see chapter 28 'Amines and amino acids').

SUMMARY

- $NaBH_4$ reduces aldehydes and ketones in aqueous or alcoholic solution.
- $LiAlH_4$ reduces aldehydes and ketones in anhydrous solution in ether: water is added to complete the reaction.
- Aldehydes are reduced to primary alcohols.
- Ketones are reduced to secondary alcohols.
- Reduction occurs by hydride ion attack on the carbonyl carbon atom, followed by protonation.

PRACTICE

1 Give the name and structural formula for the product that results from reducing each of the following substances with $NaBH_4$:
 a Propanone
 b Butanal
 c 2-Methylpropanal.

2 Lithium tetrahydridoaluminate $LiAlH_4$ and sodium tetrahydridoborate $NaBH_4$ have similar structures. Explain why $LiAlH_4$ is the more powerful reductant.

3 Suggest a mechanism for the reaction between lithium tetrahydridoaluminate and the carbonyl group.

4 Indicate briefly how you would obtain a sample of ethane from ethanol.

26.3

OBJECTIVES

- Addition–elimination
- DNP

CONDENSATION REACTIONS

The carbonyl carbon atom bears a partial positive charge $C^{\delta+}$ and so is attacked by nucleophiles. As a result, the characteristic reaction of aldehydes and ketones is *nucleophilic addition*. As discussed in an earlier spread, straightforward addition of a nucleophile H—Nu to a carbonyl group ⊃C=O results in the formation of an alcohol with a single carbon–nucleophile bond ⊃C(OH)Nu. However, there is another type of product that can result from the initial nucleophilic attack on a carbonyl group. The addition product can eliminate water, resulting in the formation of a carbon–nucleophile *double* bond.

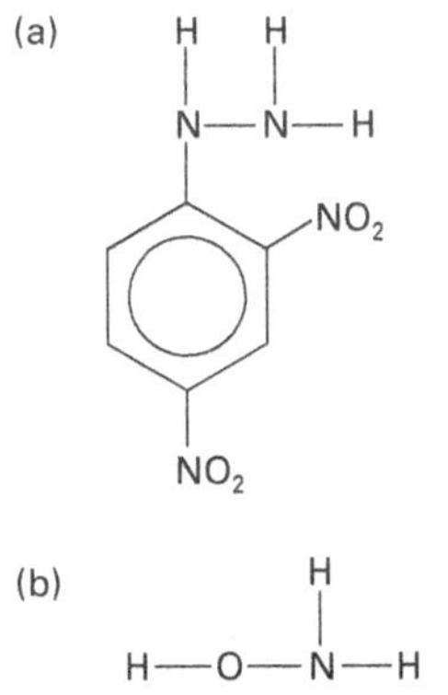

Addition–elimination is condensation

Nucleophilic addition to a carbonyl group may be followed by the elimination of a molecule of water. The oxygen atom in the water molecule originates as the carbonyl oxygen atom and the two hydrogen atoms come from the nucleophile. This class of reaction is called an **addition–elimination** or **condensation** reaction.

This reaction pathway often happens when the electron pair on the attacking nucleophile is the lone pair on a nitrogen atom. The elimination of water results because H—O bonds are stronger than H—N bonds.

There is a series of important nitrogen-containing nucleophiles with the general formula XNH_2. Aldehydes and ketones undergo condensation reactions with these nucleophiles. In general terms:

Aldehyde: $RHC{=}O + XNH_2 \rightarrow RHC{=}NX + H_2O$

Ketone: $RR'C{=}O + XNH_2 \rightarrow RR'C{=}NX + H_2O$

The mechanism for this reaction is outlined below.

Nitrogen-containing nucleophiles with the general formula XNH_2:
(a) 2,4-dinitrophenylhydrazine;
(b) hydroxylamine.

Two possible outcomes of nucleophilic addition to the carbonyl group.

(a) The formation of an alcohol with a carbon–nucleophile *single* bond.

(b) The formation of a carbon–nucleophile double bond by addition and *then* elimination.

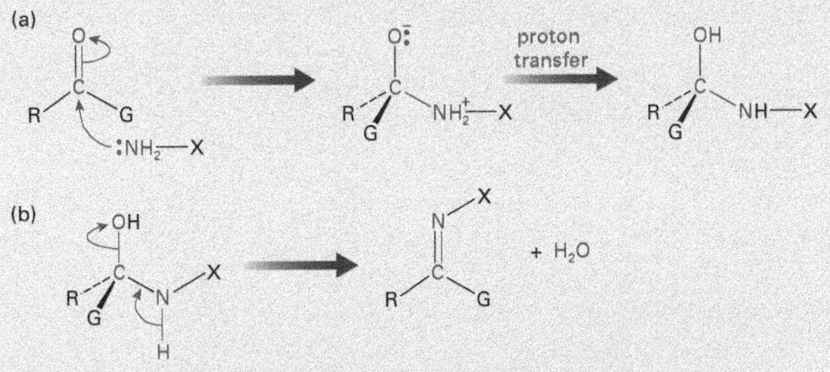

(a) Nucleophilic addition and (b) condensation. G may be R′ or H.

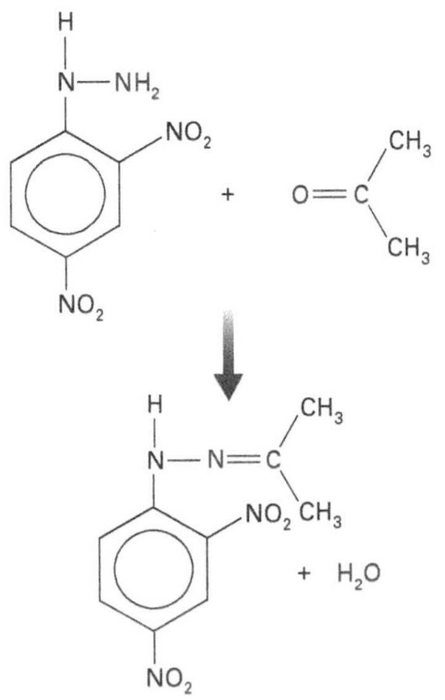

DNP derivatives

DNP is an abbreviation for 2,4-dinitrophenylhydrazine. Aldehydes and ketones undergo rapid condensation reactions in aqueous solution with this substance to form insoluble derivatives called 2,4-dinitrophenylhydrazones. These DNP derivatives are yellow–orange solids. The production of a yellow–orange precipitate on adding an unknown compound to DNP is a positive test for a carbonyl compound (an aldehyde or a ketone). The derivatives also have sharp and characteristic melting points. An aldehyde or ketone may therefore be positively identified by measuring the melting point of its DNP derivative.

The condensation reaction between DNP and propanone. The yellow–orange solid melts sharply at 128 °C. Note that the DNP derivative of propanal (isomeric with propanone) melts at 155 °C.

Ethanol is oxidized by potassium dichromate(VI), which turns green, to ethanal and the ethanal is distilled into DNP solution, which forms a yellow-orange precipitate of ethanal 2,4-dinitrophenylhydrazone.

DNP explosions

Safety organisations have consistently stated the need to ensure that a DNP solution should not be allowed to dry out. In 2016 at least ten schools, concerned about the chance that a sample had dried out, called in bomb disposal squads to destroy the sample with a controlled explosion on the playing fields (occasionally forgetting to warn the neighbours!).

Hydroxylamine derivatives (oximes)

Condensing hydroxylamine $HONH_2$ with an aldehyde or ketone yields the **oxime**. These derivatives usually melt at significantly lower temperatures than the DNP equivalent (e.g. 46 °C for the oxime of ethanal, $CH_3HC{=}NOH$, compared with 168 °C for the DNP derivative). Again, measuring the melting point of an oxime can lead to the identification of the aldehyde or ketone.

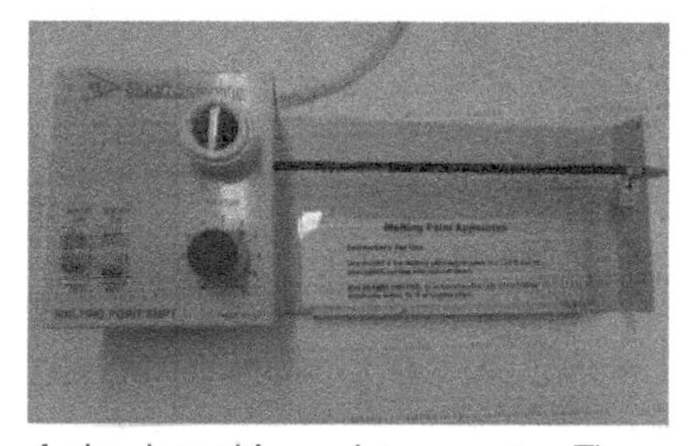

A simple melting point apparatus. The sample is placed in a thin tube very close to the thermometer.

SUMMARY

- Carbonyl compounds undergo condensation reactions by nucleophilic addition of compounds of the form XNH_2 to the carbonyl carbon atom followed by elimination of a water molecule to form the structure ⊃C=NX.
- DNP is 2,4-dinitrophenylhydrazine.
- Formation of a yellow-orange precipitate of a DNP derivative indicates the presence of an aldehyde or ketone.
- The melting points of crystalline DNP derivatives enable specific aldehydes and ketones to be identified.

PRACTICE

1 Using structural formulae, write chemical equations to show the reaction between each of the following pairs of reagents:

a Hydroxylamine and cyclohexanone

b 2,4-Dinitrophenylhydrazine and ethanal.

2 Look at the illustration on the opposite page that shows the mechanism of a general addition–elimination (condensation) reaction. Draw Lewis structures to show how bonds are broken and made during the stages of:

a nucleophilic addition;

b the elimination of water.

3 There are two possible outcomes of nucleophilic addition to the carbonyl group: (i) the formation of an alcohol with a carbon–nucleophile single bond and (ii) the formation of a carbon–nucleophile double bond. Discuss the two sorts of reagent that lead to the two possible outcomes, giving examples of each.

26.4

Oxidation of aldehydes

OBJECTIVES

- Aldehydes: reducing properties
- Tollens' reagent
- Fehling's solution
- Ketones resist oxidation

Reminder: aldehyde and ketone preparation

Aldehydes are prepared by the oxidation of a primary alcohol. For example,

$$CH_3CH_2OH \xrightarrow{Cr_2O_7^{2-}(aq)/H^+(aq)} CH_3CHO$$

Ketones are prepared by the oxidation of a secondary alcohol. For example,

$$(CH_3)_2CHOH \xrightarrow{Cr_2O_7^{2-}(aq)/H^+(aq)} (CH_3)_2CO$$

Practical details are given in the previous chapter.

Tollens' reagent

The test was introduced by Bernhard Tollens (1841-1918), hence the position of the apostrophe. Tollens discovered the molecular formula of glucose (see spread 30.2) in 1888.

Both aldehydes and ketones contain the carbonyl group $\supset C{=}O$. The carbonyl carbon atom in ketones joins with two separate alkyl groups to give the general structure RR′CO. In aldehydes, the carbonyl carbon atom is bonded to a hydrogen atom to give the general structure RCHO. The most important difference between these two structures is the presence of the hydrogen atom in aldehydes. The —CHO grouping gives reducing properties to the aldehyde molecule, and so enables aldehydes and ketones to be distinguished by simple chemical tests. Ketones do not have reducing properties: they resist attempts to oxidize them.

Distinguishing between aldehydes and ketones

Aldehydes have reducing properties: it is possible to oxidize them to carboxylic acids, for example by warming them with acidified potassium manganate(VII):

$$RCHO \xrightarrow{MnO_4^-(aq)/H^+(aq)} RCOOH$$

Ketones do not have reducing properties and so are not oxidized under similar conditions. This distinction forms the basis of two very simple tests that distinguish between the two types of carbonyl compound.

Tollens' reagent: the 'silver mirror' test

Tollens' reagent is sometimes called 'ammoniacal silver nitrate'. It is made as follows. A small quantity of aqueous sodium hydroxide is added to aqueous silver nitrate, forming a precipitate of hydrated silver oxide Ag_2O. Addition of aqueous ammonia causes the precipitate to form a solution containing the silver ammine complex $[Ag(NH_3)_2]^+(aq)$. On warming, aldehydes reduce Tollens' reagent, which is colourless, to a grey precipitate of metallic silver. In a thoroughly cleaned glass container, the precipitate forms as a *silver mirror*. Ketones do not react with Tollens' reagent.

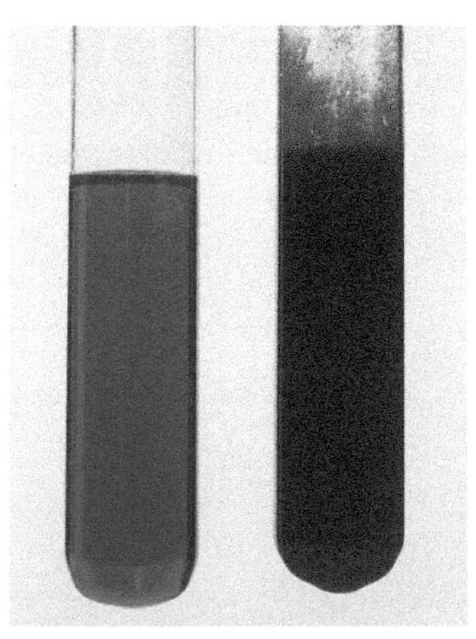

Left: The 'silver mirror' test with Tollens' reagent. Aldehydes reduce the ammine complex of silver $[Ag(NH_3)_2]^+(aq)$ to metallic silver. Ketones do not react with Tollens' reagent. Right: The test with Fehling's solution. Aldehydes reduce a $Cu^{2+}(aq)$ complex to a brick-red precipitate of Cu_2O. Ketones do not react with Fehling's solution.

Fehling's solution

Fehling's solution contains a complex of Cu^{2+}. It is made by mixing 'Fehling's A' (aqueous copper(II) sulphate) with 'Fehling's B' (an alkaline solution of sodium potassium 2,3-dihydroxybutanedioate (sodium potassium tartrate)). On warming, aldehydes reduce Fehling's solution, which is blue, to a brick-red precipitate of copper(I) oxide Cu_2O. (Methanal reduces the solution further to metallic copper.) Ketones do not react with Fehling's solution.

Worked example on identifying aldehydes and ketones

Question: A substance **A** has the molecular formula $C_4H_{10}O$. It oxidizes to the carbonyl compound **B** (C_4H_8O). **B** gives a brick-red precipitate with Fehling's solution. Passing the vapour of **A** over heated aluminium oxide forms **C** (C_4H_8). **C** reacts with hydroiodic acid HI(aq) forming **D** (C_4H_9I). **D** may be hydrolysed and then oxidized to **E** (C_4H_8O). **E** contains a carbonyl group but does not react with Fehling's solution. Identify **A** to **E** and explain the reactions.

Strategy: Find one substance that you can positively identify with no ambiguity and then work from that substance. It is often better to start with the last substance in the sequence, rather than initially concentrating on the first step.

Answer: Compound **E** is a carbonyl compound that contains four carbon atoms. It must be a ketone because it does not react with Fehling's solution. The only four-carbon ketone is butanone. Therefore **E** must be butanone $CH_3CH_2COCH_3$.

E is butanone, so **D** must be the iodoalkane $CH_3CH_2CHICH_3$, which hydrolyses to the corresponding secondary alcohol, $CH_3CH_2CH(OH)CH_3$, which oxidizes to butanone.

C is an alkene to which HI will add to form the iodoalkane **D**. However, the position of the double bond is uncertain at this stage.

B must be an aldehyde because it is a carbonyl compound that reacts with Fehling's solution. The straight-chain four-carbon aldehyde is butanal $CH_3CH_2CH_2CHO$. **B** is formed by oxidation of **A**; therefore, **A** must be the primary alcohol $CH_3CH_2CH_2CH_2OH$ (butan-1-ol).

C is formed by dehydration of **A** and hence the double bond occupies the 1,2- position $CH_3CH_2CH{=}CH_2$ (but-1-ene). When HI adds to **C**, Markovnikov's rule (see chapter 22 'Alkanes and alkenes') states that the product will be $CH_3CH_2CHICH_3$, **D**.

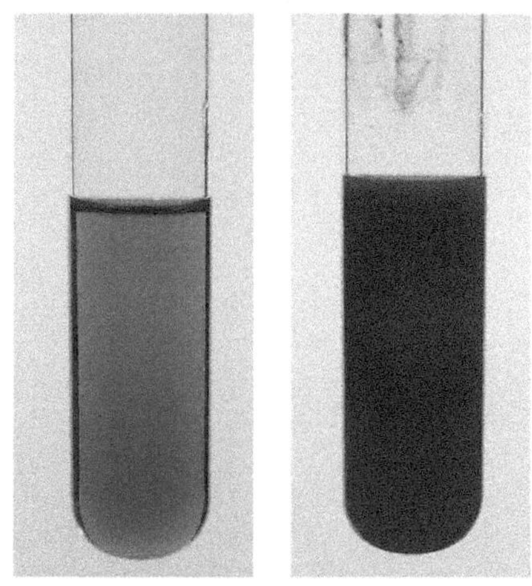

Benedict's solution *(favoured more by biologists than chemists) is very similar to Fehling's solution, but uses citrate ions instead of tartrate ions. On warming, aldehydes reduce the Cu^{2+} complex to brick-red copper(I) oxide. Ketones do not react with Benedict's solution.*

Comment

The ketone **E** (butanone) is an isomer of the aldehyde **B** (butanal). 2-Methylpropanal $(CH_3)_2CHCHO$ is a branched-chain isomer of the aldehyde **B**. This isomer would behave almost identically to butanal. It is therefore a better strategy to identify the carbon chain as unbranched (by identifying **E**) rather than starting at **A** and working forwards, in which case it would be necessary to consider both the straight-chain and branched-chain alternatives.

SUMMARY

- Aldehydes are reductants; ketones are not.
- Aldehydes are readily oxidized to carboxylic acids.
- Aldehydes reduce Tollens' reagent, which is colourless, to a silver mirror.
- Aldehydes reduce Fehling's solution, which is blue, to a brick-red precipitate of copper(I) oxide.
- Ketones do not react with Tollens' reagent or with Fehling's solution.

PRACTICE

1 To what substance is ethanal converted when it reacts with Fehling's solution?

2 Substances **P** and **Q** are carbonyl compounds and both have the molecular formula C_3H_6O. **P** gives a positive result with Fehling's solution; **Q** gives a negative result with Tollens' reagent. Give the names and structural formulae of substances **P** and **Q**.

3 Substance **A** has the molecular formula C_4H_9Br. Hydrolysis with boiling aqueous sodium hydroxide gives **B** ($C_4H_{10}O$). Treating **B** with acidified potassium dichromate(VI) gives a carbonyl compound **C** (C_4H_8O).

   a Give the structural formulae and names of the substances **A**, **B**, and **C**.

   b Give the name and structural formula of an isomer of **C** that would give a positive result with Tollens' reagent.

4 Three compounds are an aldehyde, a ketone, and an alcohol. Identify them from the following information.

   a **A** forms a yellow-orange precipitate on reaction with DNP; **A** does not change the colour of Fehling's solution.

   b **B** does not form a precipitate on reacting with DNP; **B** does not change the colour of Fehling's solution.

   c **C** forms a yellow-orange precipitate on reaction with DNP; **C** deposits a silver mirror on the inside of the test tube when treated with Tollens' reagent.

26.5

OBJECTIVES

- C=C position
- Oxidative cleavage

Some worked examples from analysis

One of the problems often facing an organic chemist is to identify an unknown substance. The task of analysis involves discovering the molecular formula, the structure of the carbon skeleton, the functional groups present, and the positions of those groups on the skeleton. Data come from a wide range of sources, from simple test tube experiments to sophisticated techniques such as mass spectrometry and nuclear magnetic resonance (see the final two chapters in this book). The task is then to arrive at a structural formula that fits the data. While much of the process involves logical analysis of data, chemists also need to trust their intuition and test out hunches.

The position of a double bond

Vigorous oxidation breaks a carbon–carbon double bond, splitting an alkene molecule into two parts. The process is called **oxidative cleavage**. Oxidation of the C=C bond produces two carbonyl groups where the double bond was, i.e.

$$\supset C{=}C\subset \xrightarrow{[O]} \supset C{=}O + O{=}C\subset$$

Oxidation is carried out either by hot acidified potassium manganate(VII) or by ozone O_3. Each of the two carbonyl compounds may then be identified, for example by forming the DNP derivatives and measuring their melting points.

Manganate(VII) summary

Hot acidified potassium manganate(VII) splits an alkene molecule at the C=C double bond. For example,

$CH_2{=}CH_2 \rightarrow 2CH_2O$

Dilute alkaline potassium manganate(VII) changes from purple to green (see photo in spread 13.9) as it oxidizes the C=C bond. For example,

$CH_2{=}CH_2 \rightarrow HOCH_2CH_2OH$

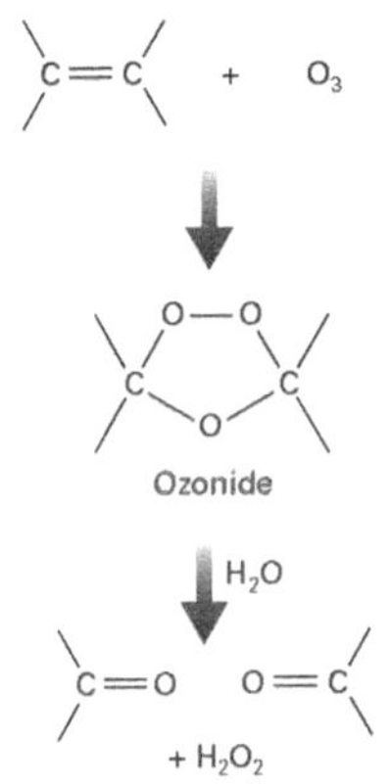

Worked example 1

Question: Identify the alkene that forms propanone and ethanal on heating with hot acidified potassium manganate(VII).

Strategy: The carbon–oxygen double bonds occupy the same positions in the carbonyl compounds as the carbon–carbon double bond occupied in the alkene.

Answer: The diagram below shows the two halves of the molecule being reassembled to form the alkene $(CH_3)_2C{=}CHCH_3$.

Note: Selective cleavage of a double bond can also be achieved by reaction with ozone in a process called **ozonolysis**. The initial product from the addition of ozone is hydrolysed to form the separate carbonyl compounds.

Ozonolysis occurs in two steps: (a) The formation of an ozonide. (b) Hydrolysis of the ozonide by water to form two carbonyl compounds.

Determining the alkene structure by combining the two carbonyl compounds that result from oxidative cleavage.

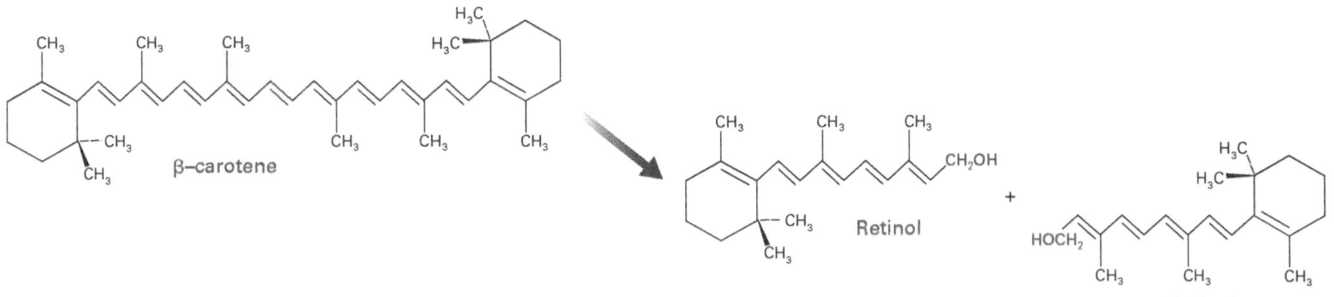

Fission at a double bond occurs in our own human biochemistry. Carrots contain β-carotene, which undergoes enzymatic oxidative cleavage (which produces two alcohols rather than two carbonyl compounds) to produce two molecules of retinol (vitamin A). The oxidized form, retinal, is a constituent of the visual pigment rhodopsin present in the retina, spread 22.6. So – it is actually true that carrots can help you to see better.

Worked example 2

Question: A substance **S** has the molecular formula C_4H_9Br. After boiling with aqueous sodium hydroxide, a product **T**, of formula $C_4H_{10}O$, was formed. On oxidation, **T** formed **U** (C_4H_8O), which reacted to form a DNP derivative but gave no reaction with Fehling's solution. Identify **S**, **T**, and **U**. Name a reagent that can convert **U** into **T**.

Strategy: Find one substance that you can positively identify with no ambiguity and then work from that substance. It is often better to start with the last substance in the sequence, rather than initially concentrating on the first step.

Answer: Because it reacted to form a DNP derivative, **U** is a carbonyl compound. A carbonyl compound that does not react with Fehling's solution must be a ketone. There is only one ketone of formula C_4H_8O: therefore **U** is butanone $CH_3COCH_2CH_3$. A ketone is formed by oxidation of a secondary alcohol. Therefore **T** is butan-2-ol $CH_3CH(OH)CH_2CH_3$. **S** must be the *secondary* halogenoalkane, $CH_3CHBrCH_2CH_3$. A suitable reagent to convert butanone (**U**) to butan-2-ol (**T**) is $NaBH_4$.

S: $CH_3CHBrCH_2CH_3$ —OH^-(aq)→ T: $CH_3CH(OH)CH_2CH_3$ —[O]→ U: $CH_3COCH_2CH_3$ —DNP→ ✓; —Fehling's→ ✗

U —$NaBH_4$→ T

*A flow scheme of the reactions of substances **S**, **T**, and **U**. It is good practice always to summarize your answers to analysis questions in this form.*

SUMMARY

- Oxidation by hot acidified potassium manganate(VII) splits an alkene double bond (by oxidative cleavage) to form two separate carbonyl compounds.
- Ozone adds across a double bond to form an ozonide; hydrolysis results in two separate carbonyl compounds.
- The structure of an alkene is determined by fitting together the structures of the carbonyl compounds formed on oxidative cleavage of the original C=C bond.

PRACTICE

1 An alkene **N** (of formula C_6H_{12}) gave two compounds **M** (of formula C_3H_6O) and **P** (of formula C_3H_6O) on oxidative cleavage. On warming with Fehling's solution, **M** produced a brick-red precipitate but there was no reaction with **P**. Name the compounds **N**, **M**, and **P** (giving your reasoning), and draw the structural formula of each.

2 An alkene **P** (of formula C_6H_{10}) gave a single compound **Q** (of formula $C_6H_{10}O_2$) on oxidative cleavage. **Q** formed an orange precipitate on warming with DNP. It also gave a positive result with Fehling's solution and Tollens' reagent. Suggest possible names for the compounds **P** and **Q** (giving your reasoning), and draw the structural formula of each.

3 Compound **A** is an iodoalkane. Warming **A** with dilute aqueous sodium hydroxide forms **B**, which is oxidized to **C** by acidified potassium dichromate(VI). **C** does not react with Fehling's solution but does form a DNP derivative. Boiling **A** with alcoholic potassium hydroxide solution forms **D** by elimination. Oxidative cleavage of **D** results in **E** (HCHO) and **F** (C_3H_6O), both of which give a silver mirror with Tollens' reagent. Name the compounds **A** to **F** (giving your reasoning), and draw the structural formula of each.

26.6

OBJECTIVES

- Reaction at the α-carbon atom
- The triiodomethane (iodoform) test

THE TRIIODOMETHANE (IODOFORM) REACTION

The reactions of the carbonyl group described so far include nucleophilic addition and condensation (nucleophilic addition–elimination). These reactions are confined to the carbon atom and the oxygen atom that make up the carbonyl group ⊃C=O. However, the influence of the carbonyl group reaches further along the molecule than you might suppose. The carbon atom, and its associated hydrogen atoms, *next to* the carbonyl group have greater reactivity than expected for a hydrocarbon CH_2 group. The hydrogen atoms may be substituted with halogen atoms. This behaviour is the basis of the triiodomethane reaction, a useful test that establishes the presence of two particular molecular structures.

The α-carbon atom

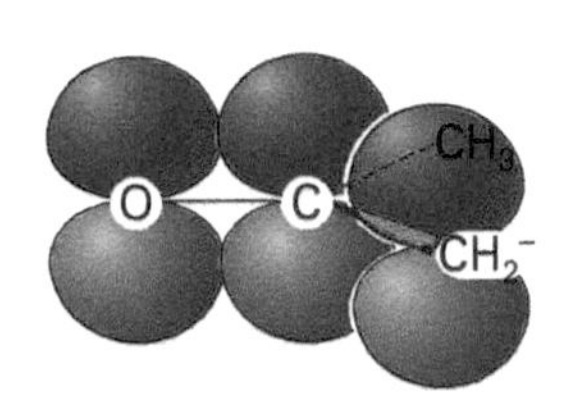

Delocalization of the lone pair on carbon with the π bond of the carbonyl group.

The **α-carbon atom** in a carbonyl compound is the carbon atom *next to* the one bearing the functional group. The carbonyl group ⊃C=O has a potentially useful effect on the C—H bonds of the α-carbon atom. When a hydrogen atom is removed from the α-carbon atom, the C=O double bond helps to delocalize the negative charge left on the α-carbon atom. Delocalization lowers the energy of the negative ion and so makes subsequent reaction much more likely to happen. As a result, the C—H bonds of the α-carbon atom are relatively easy to break. The mechanism is shown below for a typical reaction.

(a) (b) (c)

$CH_2(H)COCH_3 + {:}OH^- \xrightarrow{-H_2O} {}^-{:}CH_2COCH_3 \xrightarrow{I_2} CH_2ICOCH_3 + I^-$

Mechanism of iodination of the α-carbon atom.

The iodination of propanone, spread 15.6, is a good example of a reaction that occurs at the α-carbon atom C—H bond. We shall use this particular example to illustrate the general mechanism:

(a) Base causes the C—H bond to break heterolytically.

(b) The negative ion is stabilized by delocalization of the charge by overlap of the p orbitals on the atomic centres C—C=O.

(c) The ion can then react with a halogen, such as iodine. Note that all three hydrogen atoms on the α-carbon may be substituted by iodine as shown below.

$$CH_3COCH_3 \longrightarrow CH_3COCH_2I \longrightarrow CH_3COCHI_2 \longrightarrow CH_3COCI_3$$

All three of the α-carbon hydrogen atoms may be replaced by halogen.

The triiodomethane (iodoform) test

The substitution by iodine is the basis of a test for carbonyl compounds that contain the structure —$COCH_3$. Iodine in basic solution reacts with these compounds to form triiodomethane CHI_3. This substance forms a yellow precipitate and has a characteristic antiseptic smell. The reaction is known as the **triiodomethane test** or **iodoform test**. The reaction occurs in two stages. First, the iodine substitutes for all three of the hydrogen atoms in the methyl group as outlined above:

$$CH_3CH_2COCH_3 \xrightarrow{I_2/OH^-} CH_3CH_2COCI_3$$

Then excess base hydrolyses the molecule causing the C—C bond to break, releasing triiodomethane:

$$CH_3CH_2COCI_3 \xrightarrow{OH^-} CH_3CH_2CO_2^- + CHI_3$$

The mechanism is shown at the top of the opposite page.

Alcohols

Secondary alcohols that contain the structure $-CH(OH)CH_3$ are oxidized by iodine to the $-COCH_3$ structure and so they also give a positive triiodomethane test. The triiodomethane reaction is also a method of breaking a C—C bond and removing a methyl $-CH_3$ group. It is therefore a method of shortening a chain by a *single* carbon atom.

The mechanism for the formation of triiodomethane from propanone in the triiodomethane test is as follows. (Note that triiodopropanone forms as shown in the mechanism box on the opposite page.)

(a) Base carries out nucleophilic attack on the carbonyl carbon atom.

(b) The C—CI_3 bond breaks to form ethanoic acid and the CI_3^- ion, the three electronegative iodine atoms helping to stabilize the negative charge.

(c) Ethanoic acid transfers a proton to the CI_3^- ion to form triiodomethane and the ethanoate ion.

(a) $CH_3COCI_3 + {:}OH^- \rightarrow CH_3C(O^-)(OH)CI_3$

(b) $CH_3C(O^-)(OH)CI_3 \rightarrow {:}CI_3^- + CH_3COOH$

(c) ${:}CI_3^- + CH_3COOH \rightarrow CH_3CO_2^- + CHI_3$

Mechanism of the triiodomethane test.

SUMMARY

- The α-carbon atom in a carbonyl compound is the carbon atom *next to* the one bearing the carbonyl group.
- The hydrogen atoms on the α-carbon atom in an aldehyde or ketone are substituted in basic solution by halogens.
- Compounds containing the structure —$COCH_3$ or —$CH(OH)CH_3$ form a precipitate of triiodomethane CHI_3 when treated with a basic solution of iodine.
- The triiodomethane reaction is a useful method in chemical synthesis for shortening a carbon chain by a single carbon atom.

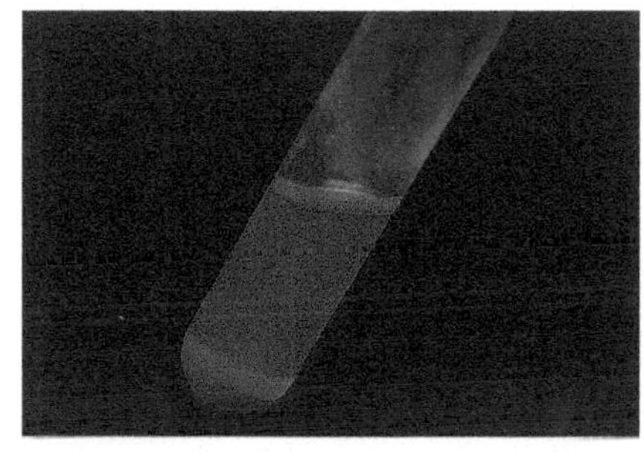

The yellow precipitate of triiodomethane (iodoform).

PRACTICE

1 Draw structural formulae for the two molecular structures, either of which will give a positive triiodomethane reaction.

2 Give examples of three-, four-, and five-carbon compounds that will give a positive reaction to the triiodomethane test.

3 Outline the practical method you would use to produce propanoic acid from butanone.

4 The hydrogen atoms attached to the α-carbon atom of an aldehyde or ketone are sometimes described as being 'acidic'. What evidence can you find in this spread to support this statement?

Chapter 26 Reactions summary

- Oxidation of a primary alcohol to form an aldehyde:
  $CH_3CH_2OH \xrightarrow{Cr_2O_7^{2-}/H^+} CH_3CHO$
- Oxidation of a secondary alcohol to form a ketone:
  $(CH_3)_2CHOH \xrightarrow{Cr_2O_7^{2-}/H^+} (CH_3)_2CO$
- Nucleophilic addition of HCN to ethanal:
  $CH_3CHO + HCN \rightarrow CH_3CH(OH)CN$
- Nucleophilic addition of HCN to propanone:
  $(CH_3)_2CO + HCN \rightarrow (CH_3)_2C(OH)CN$
- Reduction of an aldehyde to a primary alcohol:
  $CH_3CHO \xrightarrow{NaBH_4} CH_3CH_2OH$
- Reduction of a ketone to a secondary alcohol:
  $(CH_3)_2CO \xrightarrow{NaBH_4} (CH_3)_2CHOH$
- Condensation between propanone and DNP:

  2,4-dinitrophenylhydrazine ($C_6H_3(NO_2)_2NH{\cdot}NH_2$) + $O{=}C(CH_3)_2$ → $C_6H_3(NO_2)_2NH{\cdot}N{=}C(CH_3)_2$ + H_2O

  A yellow–orange precipitate on reaction with DNP confirms the presence of a carbonyl compound.
- Oxidation of an aldehyde to a carboxylic acid:
  $CH_3CHO \xrightarrow{MnO_4^-/H^+} CH_3COOH$

  This reaction distinguishes an aldehyde from a ketone.
- Aldehydes reduce Tollens' reagent, which is colourless, to a silver mirror. Ketones do not react.
- Aldehydes reduce Fehling's solution, which is blue, to a brick-red precipitate of copper(I) oxide. Ketones do not react.
- Oxidative cleavage at an alkene double bond:
  $\supset C{=}C \subset \xrightarrow{[O]} \supset C{=}O + O{=}C \subset$

  The oxidant is either hot acidified potassium manganate(VII) or ozone.
- The triiodomethane test is a test for the presence of —$COCH_3$ or —$CH(OH)CH_3$

  A positive result is a yellow precipitate of CHI_3.

PRACTICE EXAM QUESTIONS

1 a Outline the reaction of propanone with the following reagents. Give the equation for the reaction, the conditions, and the name of the organic product.

i Hydrogen cyanide. [3]

ii Sodium tetrahydridoborate (sodium borohydride). [3]

b i Give the mechanism for the reaction in **a i**. [3]

ii What type of mechanism is this? [1]

iii What feature of the carbonyl group makes this type of mechanism possible? Explain how this feature arises. [2]

iv Explain briefly, by reference to its structure, why ethene would not react with HCN in a similar way. [1]

2 a Give a chemical test by which you could distinguish between ethanal and propanone. State the reagent(s) and conditions for the test, describe what you would observe, and give the name or formula of the organic product. [4]

b Consider the following series of reactions involving ethanal, then answer the questions which follow.

$$S \xleftarrow{HCN(l)} CH_3CHO \xrightarrow{\text{2,4-Dinitrophenylhydrazine}} T$$

$$CH_3CHO \xrightarrow{NaBH_4} U$$

i Draw graphical formulae to show the structures of compounds **S**, **T** and **U**. [3]

ii Give the name of compound **T** and describe its appearance. [2]

c Give the name and an outline of the mechanism for the reaction of ethanal with HCN(l) to produce compound **S**. [4]

3 a Give the structural formula to show clearly the organic product formed from each of the following mixtures. If you consider no reaction occurs, you should state 'no reaction'.

i 1-bromobutane with KOH in water;

ii 1-bromobutane with KOH in alcohol;

iii propanal with hydrogen cyanide;

iv 1-bromopropane with potassium cyanide;

v propan-1-ol with ammoniacal silver nitrate;

vi propanone with ammoniacal silver nitrate;

vii propan-1-o1 with acidified potassium dichromate(VI), heated under reflux;

viii propan-2-ol with acidified potassium dichromate(VI), heated under reflux. [8]

b For each of the four reactions **a i–iv**, state what type of reaction is occurring. [4]

c i Give the mechanism for the reaction of hydrogen cyanide with propanal. [3]

ii If the pH of this reaction mixture is too low or too high, the reaction is very slow. Explain why in both cases. [2]

4 Propenal, $CH_2{=}CHCHO$, is one of the materials that gives crispy bacon its sharp odour. In the following question assume that the carbon–carbon double bond and the aldehyde group in propenal behave independently.

a Give the structural formulae of the compounds formed when propenal reacts with:

i hydrogen bromide; [2]

ii hydrogen cyanide; [1]

iii 2,4-dinitrophenylhydrazine. [2]

b i Give the mechanism for the reaction between hydrogen cyanide and the aldehyde group. You may represent the aldehyde group as [3]

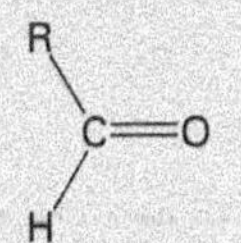

ii The reaction in **i** occurs best in slightly acidic conditions. It is slower if the pH is high or low. Suggest reasons why this is so. [3]

c Explain why lithium tetrahydridoaluminate (lithium aluminium hydride), $LiAlH_4$, reacts only with the $\supset C{=}O$ bond and not with the $\supset C{=}C\subset$, bond, even though these bonds have the same electronic structure. [2]

d Suggest reactions, giving equations and conditions, which would convert propenal into a compound which would react with iodine in the presence of sodium hydroxide solution. [4]

5 a Compound **W** can be converted into three different organic compounds as shown by the reaction sequence below. Give the structures of the new compounds **X**, **Y**, and **Z**.

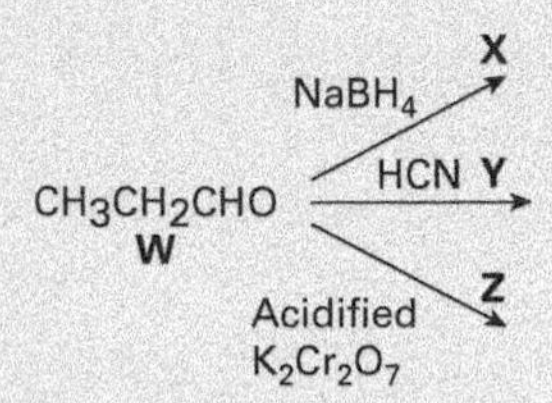

[3]

b Outline a mechanism for the formation of **Y**. [3]

c The infrared spectra shown opposite are those of the four compounds **W**, **X**, **Y** and **Z**. [4]

i Using the table of infrared absorption data given opposite, identify which compound would give rise to each spectrum by labelling each spectrum with the letter **W**, **X**, **Y**, or **Z**.

ii Suggest the wavenumber of the absorption caused by the C≡N bond. (The wavenumber of this absorption is outside the fingerprint region.) [5]

Table of infrared absorption data

Bond	Wavenumber/cm^{-1}
C—H	2850–3300
C—C	750–1100
C=C	1620–1680
C=O	1680–1750
C—O	1000–1300
O—H in alcohols	3230–3550
O—H in acids	2500–3000

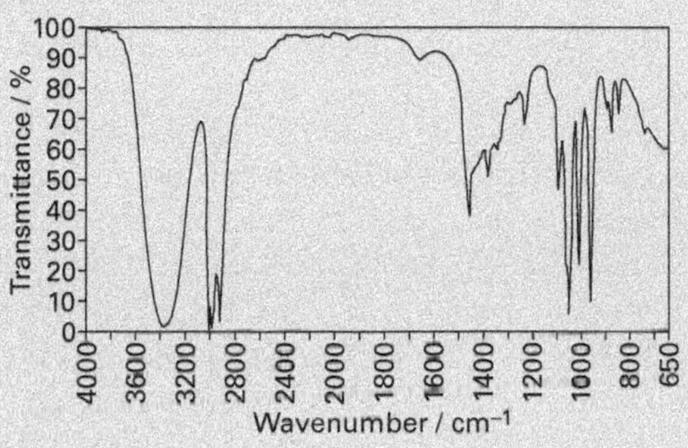

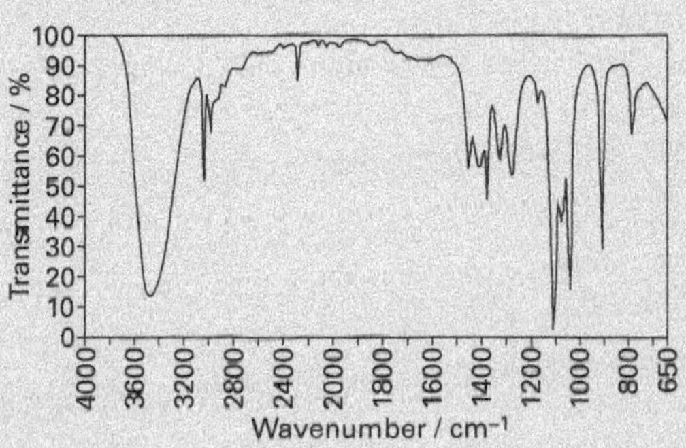

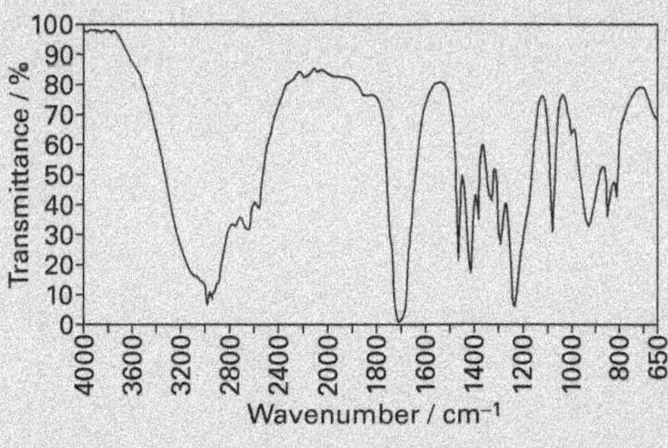

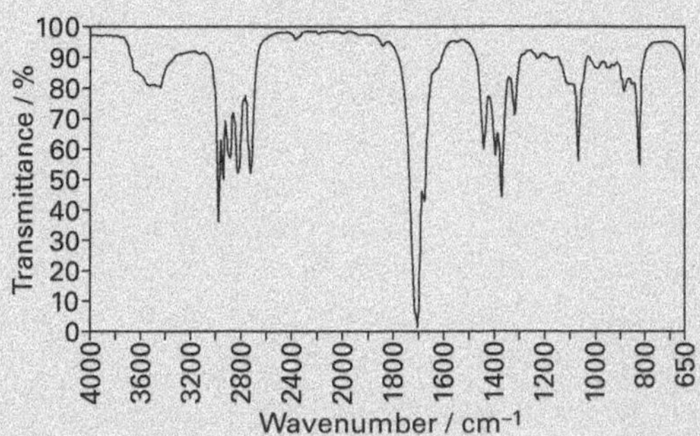

(See also chapter 32.)

6 a i Copy and complete the table below which shows some of the reactions of propanal. [6]

Reactant	Reagent(s)	Organic product	
		Name	Graphic formula
CH_3CH_2CHO	Fehling's solution		
CH_3CH_2CHO	$NaBH_4$		
CH_3CH_2CHO		2-hydroxybutanenitrile	

ii Describe and explain what you would see in the reaction of propanal with Fehling's solution. [3]

iii Give another example of this type of reaction with propanal. [2]

b For each of the following reactions, give the name of the type of mechanism involved and the formula of the attacking inorganic species.

i $CH_3CH_2CH_2Br + NaOH \rightarrow CH_3CH_2CH_2OH + NaBr$ [2]

ii $CH_2CH_2 + HBr \rightarrow CH_3CH_2Br$ [2]

7 The polymer poly(methyl 2-methylpropenoate) (Perspex) can be made by a process which involves the following reactions.

$(H_3C)_2C{=}O \xrightarrow[\text{Step 1}]{\text{Reagent A}} CH_3{-}C(CH_3)(CN){-}OH \xrightarrow{\text{Step 2}} H_2C{=}C(CH_3)COOH$

$H_2C{=}C(CH_3)COOH \xrightarrow[\text{Step 3}]{\text{Reagent B}} H_2C{=}C(CH_3)COOCH_3 \rightarrow \text{Perspex}$

a i Identify Reagent **A**.

ii Name and outline the mechanism for the reaction in Step 1. [6]

b i Name Reagent **B**.

ii Name the type of reaction occurring and give a substance which would act as a catalyst for the reaction in Step 3. [3]

c Draw the repeating unit of Perspex.

d Write an equation for the reaction between the product of Step 2 with sodium hydroxide, showing clearly the structure of the new product. [2]

27 Carboxylic acids and their derivatives

Carboxylic acids RCOOH contain the functional group —COOH. Aldehydes and ketones contain the carbonyl group ⊃C=O, and alcohols contain the hydroxyl group —OH. You might assume that carboxylic acids have a blend of the properties of aldehydes, ketones, and alcohols, but this is not the case. The carboxylic acid group has its own distinctive and unique properties. **Carboxylic acid derivatives** include acyl chlorides RCOCl, esters RCOOR′, and amides $RCONH_2$. In acyl chlorides, a chlorine replaces the —OH group of the parent acid. In esters, the H atom is replaced by an alkyl or aryl group. The first spread of this chapter is devoted to the properties of carboxylic acids; the remaining spreads consider their derivatives.

An introduction to carboxylic acids

27.1

OBJECTIVES

- Preparation of carboxylic acids
- Carboxylic acids as weak acids
- Charge is delocalized in the carboxylate ion
- Comparison with ethanol and phenol

Methanol carbonylation

Ethanoic acid is manufactured using the reaction between methanol and carbon monoxide:

$CH_3OH(g) + CO(g) \rightarrow CH_3COOH(g)$

which has a high atom economy, spread 9.1.

The original process used a catalyst containing cobalt. In the mid 1960s it was found by Monsanto that rhodium catalysts were more efficient: they act at 50 atm and 200 °C. Since 1996 efficiency has been further improved by using an iridium catalyst called Cativa™ by BP Chemicals, with approximate formula $[Ir(CO)_2I_2]^-$.

Preparing benzoic acid

Note that methylbenzene, ethylbenzene, or any other benzene compound with an alkyl side chain (however long) will produce the *same* product, benzoic acid (see chapter 23 'Arenes'). The rest of the side chain is destroyed, forming carbon dioxide and water.

H₃C—C(=O)(—O—H···O=)C—CH₃ (O---H—O ; O—H---O) Hydrogen bonding

Two hydrogen bonds form between molecules of ethanoic acid, holding the molecules together in a dimer.

Carboxylic acids are the most important class of organic acids. They neutralize bases to form salts, and their aqueous solutions have pH values below 7. Acid behaviour is described in chapter 12 'Acid–base equilibrium': Brønsted–Lowry theory defines acids as proton donors. You will see that Brønsted–Lowry theory describes the behaviour of organic acids in exactly the same way as it describes the behaviour of inorganic acids.

The preparation of carboxylic acids

Aliphatic carboxylic acids are prepared by the oxidation of primary alcohols (see chapter 25 'Alcohols'). For example, refluxing ethanol with acidified potassium dichromate(VI) produces the aldehyde ethanal, which is then oxidized to ethanoic acid:

$$CH_3CH_2OH \xrightarrow{[O]} CH_3CHO \xrightarrow{[O]} CH_3COOH$$

Aromatic carboxylic acids consist of a carboxylic acid group attached *directly* to a benzene ring. They are prepared by the oxidation of an alkyl group attached to a benzene ring. For example, refluxing methylbenzene with alkaline potassium manganate(VII) produces benzoic acid (after neutralization):

$$C_6H_5CH_3 \xrightarrow{[O]} C_6H_5COOH$$

Carboxylic acids can also be prepared by acidic hydrolysis of nitriles under reflux, as explained in spread 29.1:

$RCN + H_3O^+ + H_2O \rightarrow RCOOH + NH_4^+$

Physical properties

The lower aliphatic carboxylic acids are colourless liquids, usually with sharp or distinctive smells. Boiling points are higher than would be expected by considering their relative formula masses only; the molecules form dimers by means of hydrogen bonding. Hydrogen bonding also improves the water solubility of the lower carboxylic acids. Pure ethanoic acid CH_3COOH boils at 118 °C and melts at 17 °C. On cold days, it freezes to form needle-shaped crystals, hence the name 'glacial ethanoic acid'.

Benzoic acid (systematically called benzenecarboxylic acid) C_6H_5COOH is the simplest aromatic carboxylic acid. It is a white crystalline solid (T_m = 122 °C; T_b = 249 °C). Compared with ethanoic acid, a larger proportion of this molecule consists of a non-polar hydrocarbon skeleton. It has only limited solubility in water at 25 °C but dissolves fairly readily in hot water.

Acid behaviour

Carboxylic acids are weak acids. For example:

$$CH_3COOH(aq) + H_2O(l) \rightleftharpoons H_3O^+(aq) + CH_3CO_2^-(aq)$$

Aqueous ethanoic acid of concentration 1 mol dm^{-3} has a pH of 2.4 (corresponding to an acid ionization constant K_a of 2×10^{-5} mol dm^{-3}). The negative charge on the resulting **ethanoate ion** ($CH_3CO_2^-$) is stabilized by delocalization, the two oxygen atoms now being identical (see the diagram top right and spread 6.4).

Benzoic acid acts as a proton donor in the same way as ethanoic acid:

$$C_6H_5COOH(aq) + H_2O(l) \rightleftharpoons H_3O^+(aq) + C_6H_5CO_2^-(aq)$$

It has an acid ionization constant of 6×10^{-5} mol dm^{-3}.

(a)

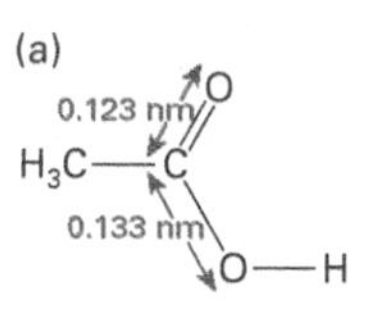

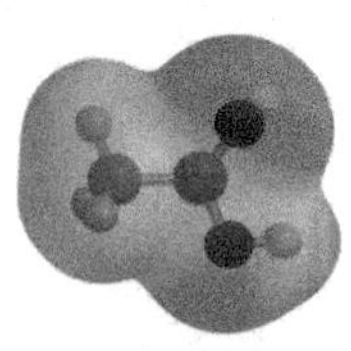

(b)

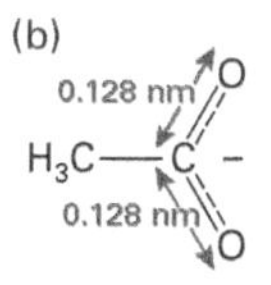

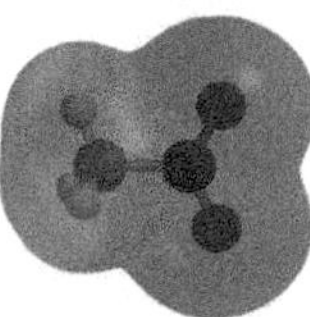

Electrostatic potential maps. (a) The un-ionized ethanoic acid molecule: the C—O bond is longer than the C=O bond. (b) The ethanoate ion: the negative charge in the ethanoate ion is delocalized. Both C—O bonds are equivalent, having the same bond lengths and electron densities (intermediate between C—O and C=O).

Carboxylic acids, alcohols, and phenols

Brønsted acidity is associated with proton transfer from a molecule, usually to water. Wherever there is a C—O—H group in an organic molecule, the potential exists for proton transfer. However, the detailed structure of the organic molecule affects how readily the proton transfers. Remember that water self-ionizes slightly in the liquid state:

$$2H_2O(l) \rightleftharpoons H_3O^+(aq) + OH^-(aq); \quad pK_w = 14.0 \text{ at } 25\,°C$$

Water provides a useful comparison for the acidity of organic compounds. The order of acid ionization is alcohols < water < phenols < carboxylic acids. The reasons for this order can be seen in the stabilities of the ions formed, as shown below.

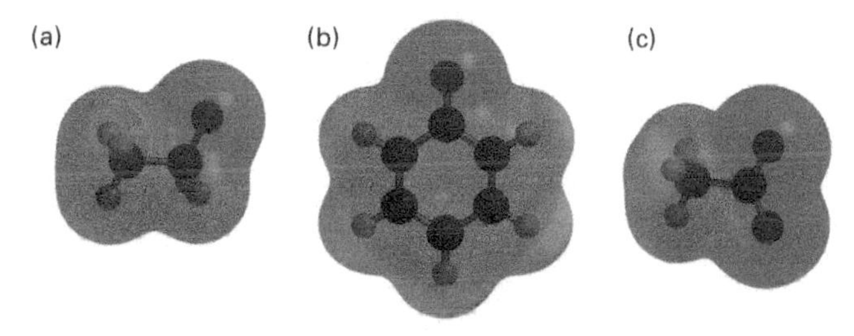

(a) The ethanol molecule has a hydroxyl group attached to an alkyl group. Alkyl groups do not delocalize the negative charge on the alkoxide ion, RO⁻. So, alcohols are not acidic because the alkoxide ion does not form readily (see spread 25.2). (b) The phenol molecule has a hydroxyl group attached to a benzene ring, which involves delocalized electrons. The negative charge on the phenoxide ion $C_6H_5O^-$ becomes partially delocalized with the benzene ring. As a result, phenols are much more acidic than alcohols because the phenoxide ion forms readily. (c) The ethanoic acid molecule has a hydroxyl group attached to a carbonyl group. As explained above, the negative charge on the carboxylate ion RCO_2^- is delocalized with the C=O double bond. Carboxylic acids are therefore more acidic than phenols.

The major consequence of this order of acidities is that ethanoic acid is acidic enough to produce carbon dioxide when reacted with sodium carbonate or sodium hydrogencarbonate: phenol is not acidic enough to react.

Salts

Salts form when carboxylic acids neutralize bases. For example, sodium ethanoate is a white crystalline solid, which may be isolated by evaporation following the reaction:

$$CH_3COOH(aq) + NaOH(aq) \rightarrow Na^+CH_3CO_2^-(aq) + H_2O(l)$$

—COOH and —CO_2^-

The formula for an un-ionized carboxylic acid group is written as —COOH (and not as —CO_2H) to show that the two oxygen atoms are *not* equivalent. The formula for the carboxylate ion is written as —CO_2^- (and not as —COO^-) to show that the two oxygen atoms *are* equivalent.

pK_a values

$$CH_3CH_2OH(aq) + H_2O(l) \rightleftharpoons H_3O^+(aq) + CH_3CH_2O^-(aq)$$

pK_a for ethanol = 16.0

$$C_6H_5OH(aq) + H_2O(l) \rightleftharpoons H_3O^+(aq) + C_6H_5O^-(aq)$$

pK_a for phenol = 9.9

$$CH_3COOH(aq) + H_2O(l) \rightleftharpoons H_3O^+(aq) + CH_3CO_2^-(aq)$$

pK_a for ethanoic acid = 4.8

SUMMARY

- The carboxylic acid functional group has the formula —COOH.
- Carboxylic acids are weak acids: $RCOOH + H_2O \rightleftharpoons H_3O^+ + RCO_2^-$
- The charge on the carboxylate ion is delocalized, the two oxygen atoms being equivalent.

PRACTICE

1 Explain, with the aid of equations, why ethanoic acid is described as a 'weak acid'.

2 Carbon dioxide gas is evolved when sodium hydrogencarbonate reacts with aqueous ethanoic acid. With respect to this reaction:

a Write a balanced chemical equation.

b Show how ethanoic acid acts as a Brønsted acid.

c Name and give the formula of the species acting as a base.

27.2

OBJECTIVES

- Acyl chlorides
- Acid anhydrides
- Acylating agents

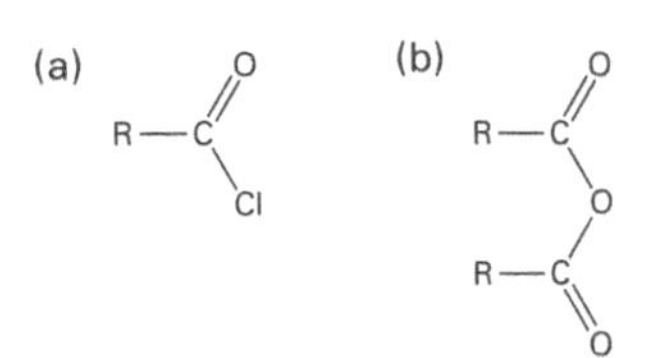

Carboxylic acid derivatives: (a) acyl chlorides; (b) acid anhydrides.

A drop of water on a glass rod causes the vapour of ethanoyl chloride to 'fume' as it is hydrolysed by the water. The fumes are hydrogen chloride gas.

ACYL CHLORIDES AND ACYLATION

Carboxylic acid derivatives include **acyl chlorides** RCOCl. The **acid anhydrides** $(RCO)_2O$ are another group of carboxylic acid derivatives, consisting of two acid molecules condensed together. The uses of acyl chlorides and acid anhydrides as acylating agents are discussed in this spread.

Acyl chlorides: preparation

The chlorinating agents (see spread 25.2 in the chapter on 'Alcohols') PCl_3, PCl_5, and $SOCl_2$ convert carboxylic acids into acyl chlorides:

$$RCOOH + PCl_5 \rightarrow RCOCl + PCl_3O + HCl$$

$$RCOOH + SOCl_2 \rightarrow RCOCl + SO_2 + HCl$$

For example, ethanoic acid CH_3COOH reacts with PCl_5 or $SOCl_2$ to give **ethanoyl chloride** CH_3COCl. Note that the chlorinating agents for carboxylic acids are the same as those for alcohols.

Acid anhydrides: preparation

Acid anhydrides are prepared by refluxing an acyl chloride with the sodium salt of the corresponding carboxylic acid. For example, ethanoyl chloride and sodium ethanoate react to form ethanoic anhydride:

$$CH_3COCl + NaCH_3CO_2 \rightarrow (CH_3CO)_2O + NaCl$$

Acid anhydrides are liquids with high boiling points and limited solubility in water. Their reactions are similar to those of acyl chlorides, but are less violent, and so acid anhydrides are often chosen in industry to avoid excessively vigorous reaction.

Acylating agents

Substitution of RCO— for a hydrogen atom in a molecule is called **acylation**. For example, substitution of CH_3CO— for a hydrogen atom is called **ethanoylation** (traditionally called 'acetylation'). There are two main types of acylating agents: acyl chlorides (e.g. ethanoyl chloride CH_3COCl) and acid anhydrides (e.g. ethanoic anhydride $(CH_3CO)_2O$).

The C—Cl bond in acyl chlorides is more reactive than that in chloroalkanes. Both the chlorine atom *and* the oxygen atom withdraw electron density from the carbon atom, increasing its partial positive charge δ+. Also, the double bond allows the new bond to form before the old one is broken. The chlorine atom is removed from the acyl chloride during acylation. The chloride ion is referred to as a 'good leaving group'. The result is that acyl chlorides engage in *nucleophilic addition–elimination* reactions, which replace the chlorine atom with a nucleophile.

The **nucleophilic addition–elimination** mechanism of acylation involves the following stages:

(a) Nucleophilic attack at the carbonyl carbon atom to form a **tetrahedral intermediate**.

(b) Loss of a chloride ion as the double bond reforms.

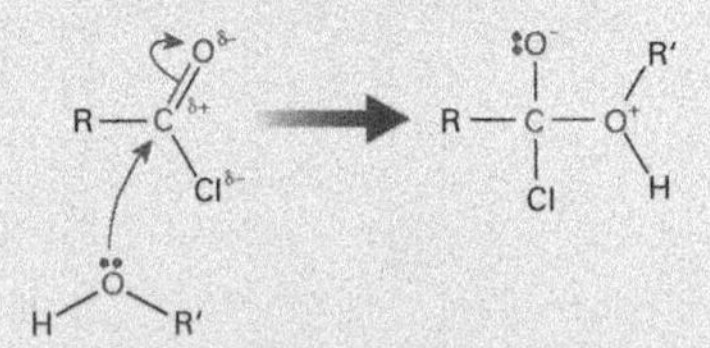

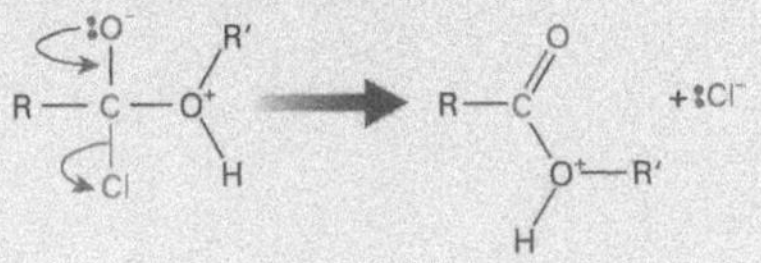

(c) Loss of a proton.

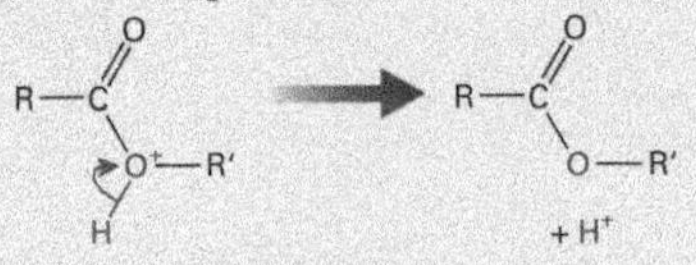

General mechanism of acylation. The reaction occurs by nucleophilic addition of the alcohol, followed by elimination of HCl (in two stages).

Acylation reactions

Acyl chlorides (such as ethanoyl chloride) perform the following reactions:

- Acyl chlorides react with water to form carboxylic acids

  $CH_3COCl(l) + H_2O(l) \rightarrow CH_3COOH(aq) + HCl(g)$

This reaction is described as **hydrolysis** of the acyl chloride, see opposite, because the C—Cl bond has been split by water. Viewed from the point of view of the *water*, a hydrogen atom has been substituted by CH_3CO—.

- Acyl chlorides react with alcohols to form esters

  $CH_3COCl(l) + CH_3CH_2OH(l) \rightarrow CH_3COOCH_2CH_3(l) + HCl(g)$

Viewed from the point of view of the *alcohol*, a hydrogen atom has been substituted by CH_3CO—.

- Acyl chlorides react with phenols to form esters

  $CH_3COCl(l) + C_6H_5OH(aq) \rightarrow CH_3COOC_6H_5(s) + HCl(g)$

- Acyl chlorides react with ammonia to form amides

  $CH_3COCl(l) + 2NH_3(aq) \rightarrow CH_3CONH_2(aq) + NH_4Cl(aq)$

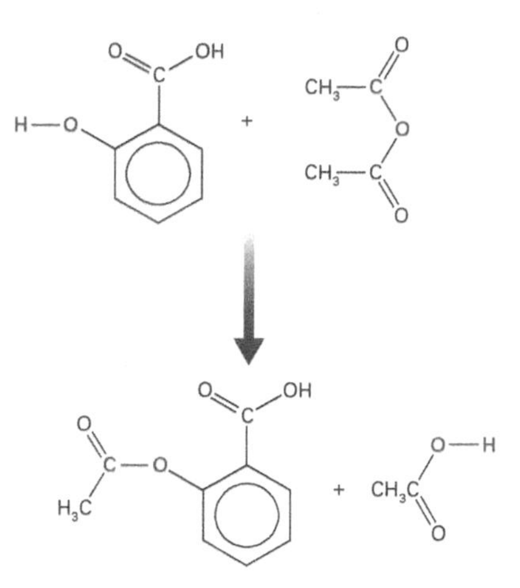

The acylation of 2-hydroxybenzoic acid by ethanoic anhydride (which reacts in a more controlled fashion than ethanoyl chloride). Note that only half of the ethanoic anhydride molecule takes part in the acylation; the other half forms the leaving group.

Aspirin

The manufacture of aspirin involves an example of acylation. The starting material is 2-hydroxybenzoic acid (salicylic acid), which contains a carboxylic acid group and a hydroxyl (phenol) group. Ethanoic anhydride reacts to substitute an ethanoyl group for the hydrogen atom on the hydroxyl group. The resulting compound is 2-ethanoyloxybenzoic acid (acetylsalicylic acid) or 'aspirin'.

The Salix alba *tree (here, in Tealham Moor, Somerset) and the history of aspirin. In 1763, the Reverend Edward Stone noticed that an extract from the bark of the English willow* Salix alba *was able to reduce fever. Fifty years later, the active ingredient was isolated, and after a further fifty years it was successfully synthesized. The active ingredient was called salicylic acid. Unfortunately, treatment often caused severe stomach pains for the patients. (Salicylic acid is now used in creams for removing warts.) Felix Hoffman's father suffered these stomach pains, and his son decided to search for a less acidic derivative. Hoffman found in 1898 that the ester formed by ethanoylating salicylic acid was even more effective at reducing inflammation and yet was better tolerated. He named this derivative 'aspirin': a- for the acetyl (ethanoyl) group, and -spir- for spirsäure, the German word for salicylic acid.*

SUMMARY

- Carboxylic acid derivatives include RCOCl (acyl chlorides), RCOOR′ (esters), $RCONH_2$ (amides), and $(RCO)_2O$ (acid anhydrides).
- Acyl chlorides are more vigorous acylating agents than acid anhydrides.
- Acylation substitutes the acyl group RCO— for a hydrogen atom on alcohols, phenols, and ammonia.
- Acylation of an alcohol gives an ester; acylation of phenol gives an ester; and acylation of ammonia gives an amide.

PRACTICE

1 Draw a structural formula for each of the following acylating agents:
  - **a** Ethanoyl chloride
  - **b** Benzoyl chloride
  - **c** Propanoic anhydride.

2 Each of the acylating agents in question 1 reacts with ethanol to form an ester.
  - **a** Use molecular formulae to write a chemical equation for each of the three reactions.
  - **b** Repeat part (a) using structural formulae.
  - **c** Which of the three acylating agents reacts most slowly?

3 Suggest a mechanism for the hydrolysis reaction between ethanoyl chloride and water.

27.3

OBJECTIVES

- Structure
- Occurrence in Nature
- Laboratory preparations

ESTERS – 1

Esters are carboxylic acid derivatives. They are found widely in Nature in fats and oils (see chapter 30 'Biochemistry'), and are responsible for the aroma of many fruits. Esters are formed in the laboratory by the reaction of alcohols with carboxylic acids, acyl chlorides, or acid anhydrides. This spread covers the production and uses of esters; the following spread looks at hydrolysis reactions, which split esters into their component alcohol and carboxylic acid fragments.

ethyl | ethanoate

The name of the ester ethyl ethanoate derived from ethanol and ethanoic acid.

Ester structure

Esters may be regarded as consisting of two parts, one deriving from an alcohol and the other from a carboxylic acid. For example, the structure of ethyl ethanoate is $CH_3CH_2OCOCH_3$ (or $CH_3COOCH_2CH_3$, see box). This ester results from reacting the alcohol ethanol CH_3CH_2OH with the carboxylic acid ethanoic acid CH_3COOH. The names of esters consist of the name of the alcohol's alkyl group (methyl, ethyl, etc.) followed by the name of the ion corresponding to the carboxylic acid (methanoate, ethanoate, etc.).

Structural formulae of esters

Note that the structure $CH_3CH_2OCOCH_3$ (ethyl ethanoate) presents the alcohol group first, followed by the acid group. This is often reversed, so that the structure of ethyl ethanoate may be written as $CH_3COOCH_2CH_3$. In general, you may find esters written as ROCOR′ or as R′COOR, where R is the alcohol alkyl group and R′ is the alkyl group of the carboxylic acid. In this book we shall generally use the second of these alternatives.

The occurrence and uses of esters

Esters occur widely in Nature. Look down the list of compounds in a mango, and you will see that it is easy to pick out the esters. Simply look for the names that have the form ...yl ...oate. Volatile esters of low relative formula mass usually have distinctive and pleasant fruity aromas, which add to the taste and flavour of fruits. Many of the 'artificial flavours' listed as the ingredients of processed foods, snacks, and confectionery are synthesized esters. For example, the aroma of a chocolate 'rum truffle' is courtesy of a few milligrams of ethyl methanoate $HCOOCH_2CH_3$. Esters are also used as solvents in perfumes and in plasticizers.

Some of the compounds present in an average mango. Spot the esters.

(a) propane-1,2,3-triol

(b) $H_2C(OCO(CH_2)_{14}CH_3)$–$HC(OCO(CH_2)_{14}CH_3)$–$H_2C(OCO(CH_2)_{14}CH_3)$

Fats and oils in both plants and animals are esters of the trihydric alcohol propane-1,2,3-triol (glycerol) (see chapter 25 'Alcohols'). (a) The structure of propane-1,2,3-triol. (b) One of the constituents of palm tree oil.

Esterification: alcohol and carboxylic acid

Esterification is the formation of an ester and water from an alcohol and a carboxylic acid, spread 11.6. For example, ethanol and ethanoic acid react in the presence of a concentrated sulphuric acid catalyst to form the ester ethyl ethanoate:

$$CH_3CH_2OH(l) + CH_3COOH(l) \rightleftharpoons CH_3COOCH_2CH_3(l) + H_2O(l)$$

In this reaction, it is the C(O)—OH bond in the acid that breaks and *not* the C—OH bond in the alcohol. This reaction mechanism is confirmed by using an alcohol that is labelled with the isotope ^{18}O. On separation of the reaction products, it is found that it is the ester that contains the labelled oxygen and not the water:

$$CH_3CH_2{}^{18}OH(l) + CH_3COOH(l) \rightleftharpoons CH_3CO^{18}OCH_2CH_3(l) + H_2O(l)$$

The detailed mechanism is shown below.

The reaction mechanism involves a lone pair on the oxygen atom of the alcohol. The alcohol acts as a nucleophile and attacks the carbonyl carbon atom of the carboxylic acid. The rate at which the reaction reaches equilibrium is increased by using concentrated sulphuric acid as a catalyst. The equilibrium mixture still contains considerable amounts of the reactants. Notice that the bonds *made* are the same as the bonds *broken*, so a *very* small enthalpy change can be expected for the reaction.

The steps shown are as follows:

(a) Protonation of the carboxylic acid.

(b) Nucleophilic attack by the alcohol oxygen lone pair.

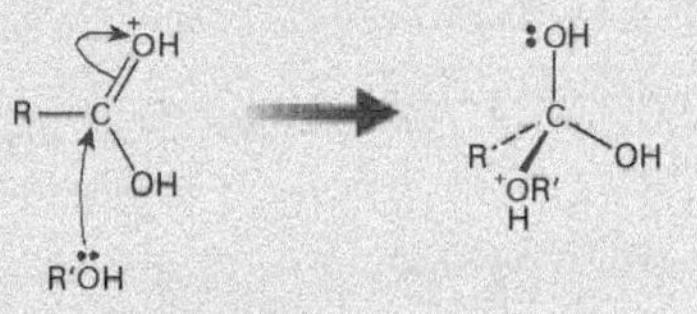

(c) Proton transfer.

(d) Elimination of water.

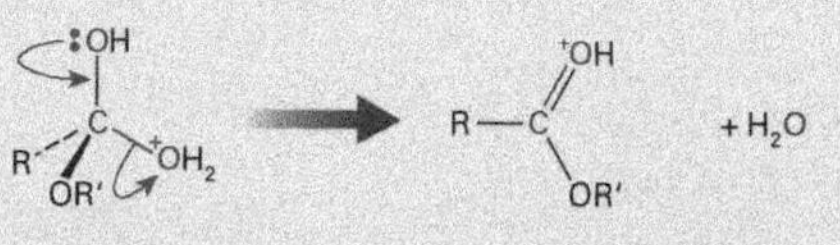

(e) Deprotonation of the ester.

Detailed mechanism of esterification.

Esterification: alcohol and acyl chloride

The synthesis of an ester starting from a carboxylic acid such as ethanoic acid is bound to give a small yield because the equilibrium constant is very close to 1 ($K_c \approx 4$ for ethyl ethanoate at 25 °C). It is much better practice to use an acyl chloride, as the reaction is then much faster and more complete. This behaviour is to be expected because the carbon–halogen bond is being replaced by a stronger carbon–oxygen bond:

$$ROH(l) + R'COCl(l) \rightarrow R'COOR(l) + HCl(g)$$

Ethanoyl chloride ethanoylates ethanol to form ethyl ethanoate:

$$CH_3CH_2OH(l) + CH_3COCl(l) \rightarrow CH_3COOCH_2CH_3(l) + HCl(g)$$

Benzoyl chloride C_6H_5COCl benzoylates ethanol to form ethyl benzoate:

$$CH_3CH_2OH(l) + C_6H_5COCl(l) \rightarrow C_6H_5COOCH_2CH_3(l) + HCl(g)$$

SUMMARY

- An ester molecule consists of an alcohol and a carboxylic acid condensed together.
- Many fats and oils are esters of the trihydric alcohol propane-1,2,3-triol (glycerol).
- Esterification between a carboxylic acid and an alcohol is an equilibrium reaction.
- Esterification of an alcohol with an acyl chloride results in a much higher yield than with a carboxylic acid.
- Esters are used in flavourings, as solvents in perfumes, and in plasticizers.

Acid anhydrides

Esters also form when acid anhydrides $(RCO)_2O$ react with alcohols. The reaction is slower than when acyl chlorides are used.

Esterification: phenol and benzoyl chloride

Phenols are much weaker nucleophiles than alcohols and do not form esters so readily with carboxylic acids. The preferred reactants to produce phenyl benzoate $C_6H_5OCOC_6H_5$ are phenol in aqueous sodium hydroxide with benzoyl chloride C_6H_5COCl. Phenol reacts with the aqueous base to form the phenoxide ion:

$C_6H_5OH(s) + OH^-(aq) \rightarrow C_6H_5O^-(aq) + H_2O(l)$

The phenoxide ion is a stronger nucleophile than phenol, and reacts readily with the benzoyl chloride:

$C_6H_5O^-(aq) + C_6H_5COCl(l) \rightarrow C_6H_5COOC_6H_5(s) + Cl^-(aq)$

27.4

OBJECTIVES

- Ester hydrolysis
- Soap-making
- Mechanism and isotopic labelling

Biodiesel

Biodiesel is formed by a **transesterification** reaction in which a vegetable oil (such as oilseed rape, soya, or palm oil) reacts with methanol, under acid or base catalysis, to form a mixture of the methyl esters of whatever long-chain carboxylic acids were present in the oil. Biodiesel has also been made from *Jatropha curcas* which can grow on marginal arable land irrigated with waste water, in the Philippines and India for example. Production at the world's largest factories now exceeds 100 million gallons per year.

Rudolf Diesel, who originally designed his engine to work on peanut oil, said in 1912 that "The use of vegetable oils for engine fuels may seem insignificant today. But such oils may become in the course of time as important as the petroleum and coal tar products of the present time." Transesterification can also be used to make low-fat spreads from vegetable oils and to make poly(ethenol), spread 22.10.

Liberating the acid

The carboxylic acid itself can be formed from its salt by adding a strong acid such as hydrochloric acid after the hydrolysis is complete:

$Na^+CH_3CO_2^-(aq) + HCl(aq) \rightarrow CH_3COOH(aq) + NaCl(aq)$

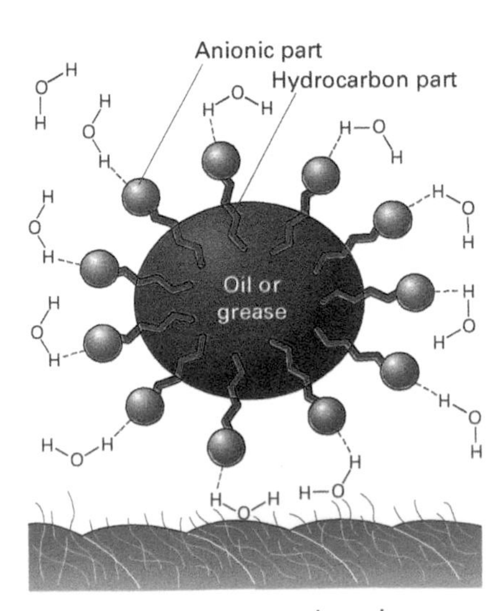

Soap functions because the polar anionic part of the species is solvated by water; the non-polar hydrocarbon part of the species dissolves in oil or grease.

Esters – 2

One of the methods of preparing esters is to react together an alcohol with a carboxylic acid. This process can be reversed by a reaction called **ester hydrolysis**. Esters are hydrolysed to alcohols and carboxylic acids or their salts. The sodium salts of long-chain carboxylic acids are used to make toilet soap and soap flakes. The main purpose of this spread is to describe the mechanism of the hydrolysis reaction and to compare it with the mechanism for esterification.

Ester hydrolysis

Ester, water, alcohol, and acid exist together in the following equilibrium:

$$CH_3COOCH_2CH_3(l) + H_2O(l) \rightleftharpoons CH_3CH_2OH(l) + CH_3COOH(l)$$

The equilibrium mixture (see chapter 11 'Chemical equilibrium') is established by mixing either alcohol and acid, or ester and water. Concentrated sulphuric acid catalyses ester formation (the reverse reaction) and therefore must also catalyse ester hydrolysis (the forward reaction). However, ester hydrolysis gives a much higher yield when a *base* is used as a catalyst. Hydroxide ions are better nucleophiles than water, and will attack the carbonyl carbon more readily. More importantly, with base the equilibrium will shift to the right because the OH^- ion will deprotonate the acid to form the salt:

$$CH_3COOH(l) + NaOH(aq) \rightarrow Na^+CH_3CO_2^-(aq) + H_2O(l)$$

The products of ester hydrolysis using sodium hydroxide are the alcohol and the sodium salt of the acid:

$$CH_3COOCH_2CH_3(l) + NaOH(aq) \rightarrow CH_3CH_2OH(l) + Na^+CH_3CO_2^-(aq)$$

Soap-making

Soaps are the sodium salts of long-chain carboxylic acids. Soft 'toilet' soap is generally made by hydrolysing the esters present in vegetable oils. The basic hydrolysis of esters is sometimes called **saponification** (derived from sapo, the Latin word for 'soap'). During soap-making, a mixture of vegetable oils and aqueous sodium hydroxide or potassium hydroxide is heated with steam. Hydrolysis yields propane-1,2,3-triol (glycerol) and salts of long-chain carboxylic acids.

(a)

$$\underset{\text{Fat}}{\begin{matrix} H_2C-O-CO(CH_2)_{16}CH_3 \\ HC-O-CO(CH_2)_{16}CH_3 \\ H_2C-O-CO(CH_2)_{16}CH_3 \end{matrix}} \xrightarrow{3NaOH} \underset{\text{Propane-1,2,3-triol (glycerol)}}{\begin{matrix} H_2C-OH \\ HC-OH \\ H_2C-OH \end{matrix}} + \underset{\text{Sodium octadecanoate (Sodium stearate, a soap)}}{3Na^+CH_3(CH_2)_{16}CO_2^-}$$

(b)

(a) A soap produced by base-catalysed hydrolysis of a fat. (b) The electrostatic potential map of sodium octadecanoate shows the polar 'head' of the octadecanoate ion.

Base-catalysed ester hydrolysis

Hydrolysis is the opposite reaction to esterification. However, esterification between an alcohol and a carboxylic acid is an *acid-catalysed* process. The mechanism of *base-catalysed* hydrolysis is not simply the reverse of the esterification mechanism (see previous spread). As an introduction to the mechanism of hydrolysis, look at the structure of the ester shown earlier. There are two carbon–oxygen bonds that could be broken, RCOO—R′ *or* RCO—OR′.

The mechanism of base-catalysed ester hydrolysis starts with the attack of the (nucleophilic) hydroxide ion on the carbonyl carbon atom. The carbonyl group is polar due to the presence of the electronegative oxygen atom. Note that this first step is the rate-limiting step. The result is an intermediate that has a tetrahedral shape (a **tetrahedral intermediate**) and a negatively charged oxygen atom. This rate-limiting step is consistent with the order of reaction (see chapter 15 'Chemical kinetics'), which is second order overall, thus showing that *two* species take part in the rate-limiting step. In the next step, the tetrahedral intermediate breaks down by fission of the RCO—OR′ bond to form an alkoxide ion and a carboxylic acid. Finally, a proton transfers from the acid to the alkoxide ion to form the alcohol.

(a) Nucleophilic attack by hydroxide ion on the δ+ carbon atom forming a tetrahedral intermediate.

(b) Breakdown of tetrahedral intermediate.

(c) Proton transfer from carboxylic acid to alkoxide ion.

$RCOOH + R'O^- \longrightarrow RCO_2^- + R'OH$

Mechanism of base-catalysed ester hydrolysis.

More evidence from isotopic labelling

In 1951 the American chemist Myron Bender investigated the hydrolysis of ethyl benzoate in which the carbonyl oxygen was the isotope ^{18}O, $CH_3CH_2OC^{18}OC_6H_5$. The tetrahedral intermediate (see above) had a labelled carbonyl oxygen but the incoming OH^- was not labelled. If the tetrahedral intermediate forms as suggested, the hydrogen of the OH group could swap to the negatively charged, labelled, oxygen very easily. The swap would result in an identical intermediate except that the hydroxyl oxygen would now be labelled rather than the carbonyl oxygen. This swap could not happen if this intermediate did not form.

Bender's flash of genius was to recognize that if this swap happened and the intermediate then broke down to reform the starting materials rather than the products (the reaction is reversible), the reaction mixture would contain some *unlabelled* ester. The hydroxide ion would contain the labelled oxygen instead. Bender therefore looked for *unlabelled* ester at various stages during the reaction, and he found it, thus confirming the suggested mechanism.

Labelling

A molecule containing a radioactive (e.g. ^{14}C) or a stable (e.g. ^{18}O) isotope of an element is called **a labelled compound.** A radioactive isotope in a reaction product may be detected by a Geiger counter; a stable isotope is usually detected by a mass spectrometer. Bender measured the density of the ester.

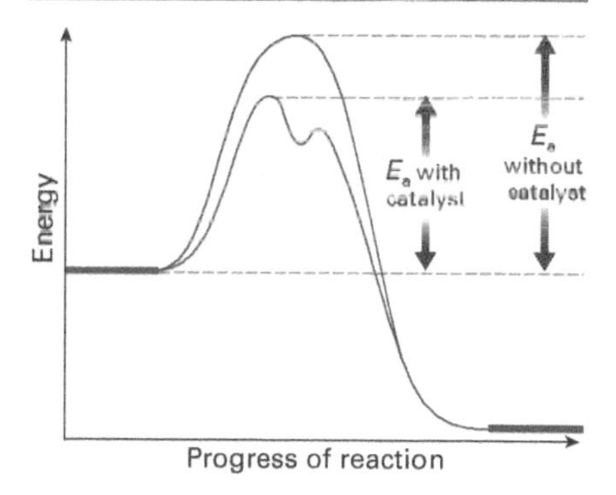

Acid-catalysed ester hydrolysis has a reaction profile in which the catalysed route involves an intermediate, the protonated ester, which then reacts with water. The reaction profile for base-catalysed ester hydrolysis resembles the figure in spread 15.9.

SUMMARY

- Esterification of an alcohol and a carboxylic acid is an acid-catalysed reaction; ester hydrolysis is base-catalysed.
- During base-catalysed hydrolysis, fission occurs at the RCO—OR′ bond.
- A reaction mechanism may be explored by labelling a reactant molecule with an isotope such as ^{18}O. Products are separated and analysed for the presence of the isotope.

PRACTICE

1 Draw a structural formula for each of the following esters:

 a Methyl propanoate
 b Phenyl ethanoate
 c Ethyl benzoate
 d Phenyl benzoate.

2 Give the names and structural formulae of the products that result from the base-catalysed hydrolysis of the four esters listed in question 1.

3 Give the name and structural formula of an acyl chloride and the substance that reacts with it to form each of the esters listed in question 1.

4 Explain with the help of structural formulae and chemical equations the meanings of the three points listed in the 'Summary' section above.

5 Does the work of Myron Bender confirm that hydrolysis involves fission of the RCO—OR′ bond? Explain your answer with the help of structural formulae.

27.5

OBJECTIVES

- Preparation
- Acid–base properties
- Typical reactions

Odours, aromas, stenches, and smells

People differ in their descriptions of smells. However, the penetrating aroma of ammonia reminds parents of babies' nappies long overdue for changing (bacteria metabolize urea in urine to ammonia). Amines smell distinctly fishy, the amine from rotting fish being called 'putrescine'. People who keep pet mice will recognize the smell associated with amides – a sure sign that the cage is due for cleaning out.

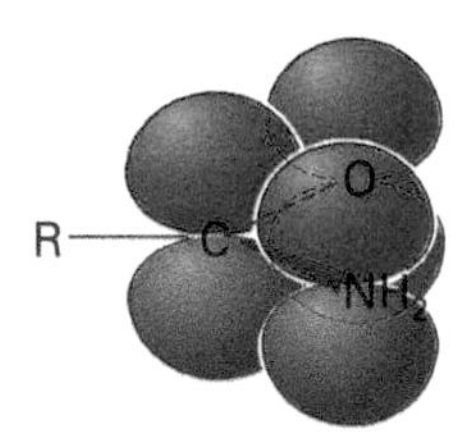

Delocalization in the amide group. Amides are not basic because the nitrogen lone pair is no longer available for protonation. Instead, it is delocalized with the carbonyl group. An amide such as ethanamide CH_3CONH_2 has a delocalization enthalpy (see chapter 23 'Arenes') of about 45 kJ mol^{-1}, i.e. the measured standard enthalpy change of formation is 45 kJ mol^{-1} more exothermic than that calculated from bond enthalpy terms.

The planar shape and bond angles of the amide group.

Amides

Amides contain the functional group —$CONH_2$. Amides may also be regarded as carboxylic acid derivatives: they are named after the corresponding carboxylic acid with the ending '...oic acid' replaced by '... amide'. Hence CH_3CONH_2 is called *ethanamide*. Whilst the amide group would appear to consist of a carbonyl group ⊃C=O attached to an amine group —NH_2, it is more useful to look on —$CONH_2$ as a separate functional group. The amide group has its own unique chemistry, which is not simply a blend of carbonyl and amine group reactions.

Physical properties

Methanamide $HCONH_2$ is a liquid with a high boiling point, whereas all other amides are white solids. The low volatility is due to hydrogen-bonded dimers forming, as in carboxylic acids. The lower members of the amide homologous series are water-soluble.

Preparation of amides

- Ammonia reacts with an acyl chloride (e.g. ethanoyl chloride) to form an amide:

$$CH_3COCl + 2NH_3 \rightarrow CH_3CONH_2 + NH_4Cl$$

- Primary amines react similarly with ethanoyl chloride to form a substituted amide. For example:

$$CH_3COCl + 2CH_3NH_2 \rightarrow CH_3CONHCH_3 + CH_3NH_3Cl$$

The C—N bond in the —CONH— group of a substituted amide is the peptide bond, discussed more fully in the next chapter. Amides are used to protect amine groups during synthesis reactions, as mentioned in the following chapter.

- Amides also result from heating the ammonium salts of carboxylic acids:

$$NH_4^+C_6H_5CO_2^- \rightarrow C_6H_5CONH_2 + H_2O$$

The yield is poor unless the ammonium salt is heated with excess of the parent acid to reduce the degree of ionization:

$$NH_4^+C_6H_5CO_2^- \rightleftharpoons C_6H_5COOH + NH_3$$

Acid–base character

Amides are extremely weak acids, but show greater acidic character than ammonia:

$$RCONH_2(aq) + H_2O(l) \rightleftharpoons H_3O^+(aq) + RCONH^-(aq); \quad pK_a \approx 16$$

$$NH_3(aq) + H_2O(l) \rightleftharpoons H_3O^+(aq) + NH_2^-(aq); \quad pK_a \approx 33$$

Amides are at least 10^{10} times weaker bases than ammonia:

$$RCONH_2(aq) + H_2O(l) \rightleftharpoons RCONH_3^+(aq) + OH^-(aq); \quad pK_b \approx 15$$

$$NH_3(aq) + H_2O(l) \rightleftharpoons NH_4^+(aq) + OH^-(aq); \quad pK_b \approx 4.8$$

Amides are therefore regarded as being essentially neutral in water. For example, aqueous ethanamide has a pH of 7.

Chemical properties

Amides are much weaker bases than amines. Amides are also much weaker nucleophiles than amines. For example, primary amines react with halogenoalkanes to produce secondary and tertiary amines and quaternary ammonium salts; there is no reaction between amides and halogenoalkanes.

- *Hydrolysis*

Amides hydrolyse when refluxed in acidic or in basic solution. Acidic hydrolysis results in the carboxylic acid and the ammonium ion.

For example

$$CH_3CONH_2 + H_3O^+ \rightarrow CH_3COOH + NH_4^+$$

Hydrolysis by base results in the carboxylate ion and ammonia. For example

$$CH_3CONH_2 + OH^- \rightarrow CH_3CO_2^- + NH_3$$

(a) Attack by hydroxide ion on the amide, forming a tetrahedral intermediate.

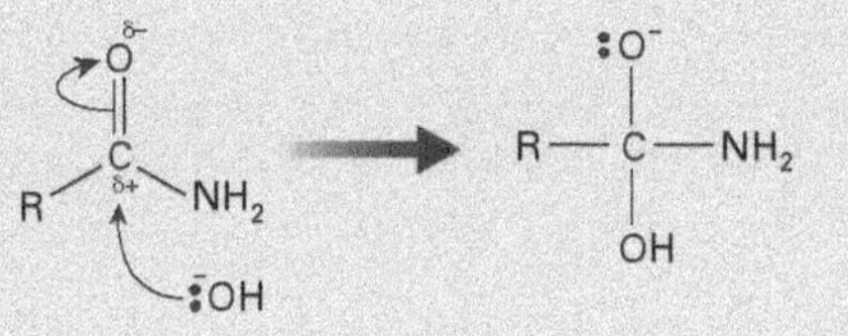

(b) Breakdown of the tetrahedral intermediate, by loss of the amide ion leaving group (NH_2^-).

(c) Proton transfer from carboxylic acid to amide ion.

$$RCOOH + NH_2^- \rightarrow RCO_2^- + NH_3$$

Hydrolysis of an amide by base.

- *Dehydration*

Heating an amide with a dehydrating agent such as phosphorus(V) oxide produces a nitrile. For example

$$C_6H_5CONH_2 \xrightarrow{P_4O_{10}} C_6H_5CN + H_2O$$

- *Reduction*

Powerful reductants such as lithium tetrahydridoaluminate reduce amides to primary amines (see spread 27.8). For example

$$CH_3CONH_2 \xrightarrow{[H]} CH_3CH_2NH_2$$

- *Amine production*

Amides react with a mixture of bromine in aqueous potassium hydroxide to produce amines containing *one less carbon atom*. For example

$$\underset{\text{ethanamide}}{CH_3CONH_2(aq)}+Br_2(aq)+4KOH(aq)\rightarrow \underset{\text{methylamine}}{CH_3NH_2(aq)}+K_2CO_3(aq)+2KBr(aq)+2H_2O$$

This reaction is sometimes called the *Hofmann degradation reaction*.

SUMMARY

- Amides are formed by reacting acyl chlorides with ammonia, or by heating the ammonium salts of carboxylic acids.
- Amides are essentially neutral in aqueous solution: the nitrogen lone pair is delocalized and is not available for protonation.
- Amides are hydrolysed to carboxylic acids by refluxing with either acid or base.
- Amides are dehydrated to nitriles and reduced to primary amines.
- Reaction with bromine in aqueous potassium hydroxide produces an amine with one less carbon atom (Hofmann degradation).

PRACTICE

1 Give the name and structural formula of all of the substances that result when propanamide reacts with each of the following reagents:

a NaOH(aq)

b $H_2SO_4(aq)$

c P_4O_{10}

d $LiAlH_4$

e Br_2/KOH.

27.6

OBJECTIVES

- Polyesters
- Copolymers
- PET, Terylene, and Mylar
- Polycarbonates

Polycotton

Look at the wash labels attached to shirts, trousers, skirts, sheets, and duvet covers. You will see that one of the commonest modern textiles is polycotton, a mixture of cotton and polyester fibres. The strength and crease-resistance of the polyester adds to the 'feel' and durability of cotton.

Biodegradable polymers

Poly(lactic acid), PLA, is a polyester made ultimately from sugar cane or corn starch (in the US). The repeating unit is $-CH(CH_3)COO-$. Plastic bags made from PLA can be broken down in the environment in a matter of months. However, some UK councils have banned customers from putting biodegradable plastics in with their garden rubbish. Further concern is that decomposition in landfill sites is slower and causes emission of methane, a greenhouse gas a couple of dozen times more damaging than carbon dioxide.

Plastarch Material, PSM, is another biodegradable polymer made from corn starch: it can be made into bags, tubing, and food packaging. When disposed of, it can be used as fertilizer. Less than 1 per cent of packaging currently is made from PSM.

Biodegradable polymers made from poly(*cis*-isoprene), PIP, spread 29.5, are used to make absorbent foams for disposable nappies.

The polyester Mylar is used to make the sails on sailboards. It does not 'wet' and is light and 'rip-proof'.

CONDENSATION POLYMERIZATION: POLYESTERS

Polymer molecules consist of long chains of repeating units. The simplest example of a polymer is polythene (also called poly(ethene)). It is made by a process of *addition polymerization* (see chapter 22 'Alkanes and alkenes') in which thousands of ethene molecules $CH_2{=}CH_2$ add together to form a long-chain polymer. Addition polymerization involves *one functional group* only, the carbon–carbon double bond. **Condensation polymerization** forms polymers by condensation reactions (see previous chapter) between *different functional groups*. One of the simplest examples of condensation polymerization is the reaction used to form polyesters.

Polyesters

A condensation reaction involves the combination of two molecules to make a larger molecule, accompanied by the elimination of a small molecule, spread 26.3. Esterification may be classed as a condensation reaction, in which an alcohol and a carboxylic acid react to form an ester, with the elimination of water. A single molecule bearing both an alcohol group and a carboxylic acid group can act as a monomer and undergo a condensation polymerization reaction.

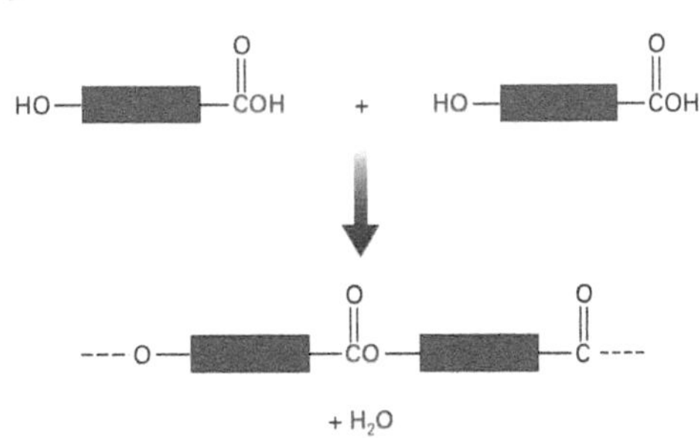

The monomer molecule is a difunctional molecule. A difunctional molecule *bears two functional groups. The alcohol group on one molecule condenses to form an ester linkage with the carboxylic acid group on another molecule. Polymer chains many thousands of monomer units long can result.*

Copolymers

A **copolymer** results from the reaction between two different monomer molecules. A polyester may be formed by condensation between two different difunctional molecules, one of which bears *two* alcohol groups and the other *two* carboxylic acid groups. The majority of commercial polyesters are copolymers.

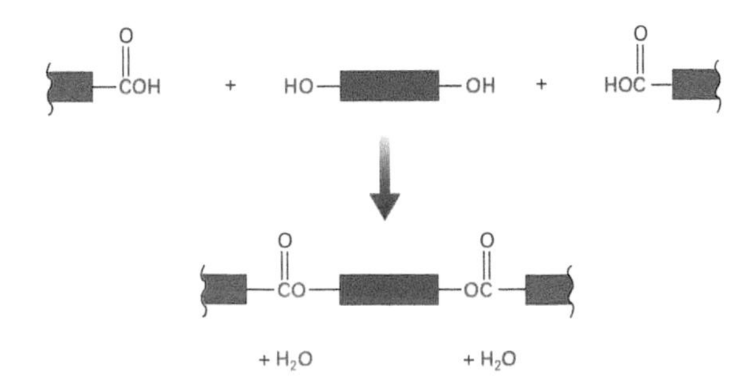

A polyester forms from the reaction between two different monomer molecules.

PET, Terylene, Dacron, and Mylar

Polyesters are made industrially from diacids (or, more usually, the corresponding but more reactive diacyl chlorides) together with dialcohols. PET was first produced in 1941 as a result of the work of the British chemist John Whinfield. PET can be made by heating ethane-1,2-diol ('ethylene glycol', spread 25.6) with benzene-1,4-dicarboxylic acid ('terephthalic acid'). PET (or PETE) is an abbreviation of the traditional name polyethylene terephthalate. It is now the most widely used polyester: for example, the plastic containers for fizzy drinks are made of PET; drawn into filaments or blown into films, PET is also used in the form of yarn and textiles (Terylene in the UK, Dacron in the USA) and as the backing film for audio tapes (Mylar).

$$HO{-}CH_2{-}CH_2{-}OH \quad + \quad CH_3OC(=O){-}C_6H_4{-}C(=O)OCH_3$$

$$\downarrow$$

$$\ldots\ldots O{-}CH_2{-}CH_2{-}OC(=O){-}C_6H_4{-}C(=O)\ldots\ldots$$

$$+\ CH_3OH \qquad + \ CH_3OH$$

PET

The industrial production of PET often uses the dimethyl ester instead of the free acid. The section shown above is the repeating unit of PET.

Polycarbonates

A special class of polyester, the polycarbonates, results from the reaction between dialcohols and derivatives of carbonic acid, such as carbonyl chloride (phosgene) $COCl_2$. Polycarbonates are exceptionally hard materials, which are used to make bullet-proof windows, riot shields, bicycle safety helmets, and car bumpers. Some polycarbonates are sufficiently optically clear to be used for spectacle lenses, in which application they are at least ten times more resistant to breakage than conventional glass.

The polycarbonate Lexan was used to make this helmet visor. The Space Shuttle is reflected in the visor. The 'safety specs' you wear during chemistry practical work are most likely made from a polycarbonate material, as are American football helmets.

SUMMARY

- Condensation polymers result from the polymerization of *difunctional* molecules – molecules bearing two functional groups.
- A copolymer is the result of a polymerization reaction between two different molecules.
- Dialcohols react with diacids to form polyesters.
- PET is a common commercial polyester formed from the reaction between ethane-1,2-diol and benzene-1,4-dicarboxylic acid.
- Polycarbonates are formed from the reaction between dialcohols and derivatives of carbonic acid, such as carbonyl chloride $COCl_2$.

The human-powered Gossamer Albatross *first flew across the English Channel in 1979. Its wings were covered with a thin layer of Mylar polyester film.*

Intermolecular bonding

Polyesters have polar bonds included in their structures. Interaction between these permanent dipoles results in attractive forces between separate polymer chains. Chemists modify the properties of polyesters by choosing carefully the positions and types of polar bond in the polyester chain.

27.7

O B J E C T I V E S

- Polyamides
- Nylon
- Aramid fibres

CONDENSATION POLYMERIZATION: POLYAMIDES

Condensation polymers form between molecules bearing functional groups that undergo condensation reactions. As discussed in the previous spread, polyesters form between diacyl chlorides and dialcohols. Polyamides are another common class of condensation copolymer. They result from the condensation of diacyl chlorides and diamines to form the substituted amide group —CONH—. Nylons are the most common polyamides. Proteins, spread 30.5, are natural polyamides.

The substituted amide group

As discussed in spread 27.5, amides have the general formula $RCONH_2$. A substituted amide has the general formula RCONHR′; an alkyl group R′ is attached to the nitrogen atom in place of one of the hydrogen atoms. A polyamide can be formed from the reaction between a diacyl chloride and a diamine, as shown below.

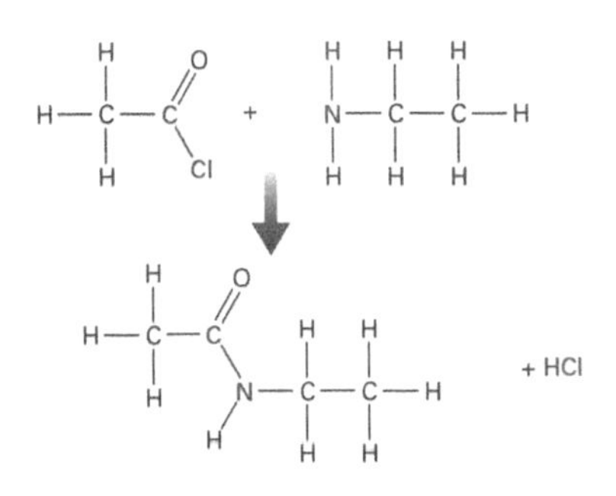

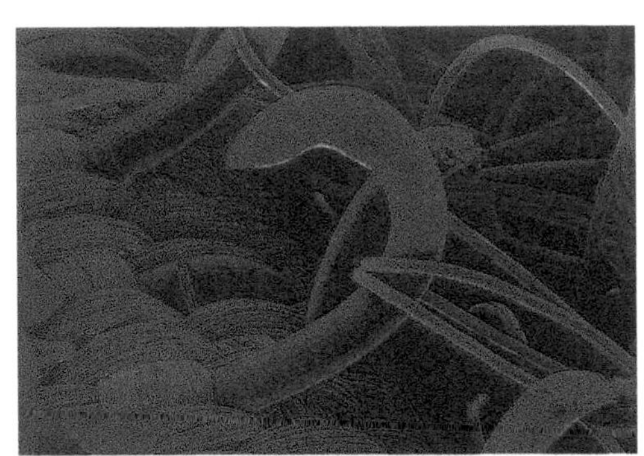

Condensation reaction between ethanoyl chloride and ethylamine to give the substituted amide N-ethylethanamide. A false-colour scanning electron micrograph of a Velcro fastener. One half has a surface with nylon loops; the other has a surface with hooks.

Nylon uses

One of the first uses of nylon was women's sheer 'silk stockings', introduced to critical acclaim at the 1939 New York World's Fair. The all-nylon 'drip-dry' shirt of the 1950s did not require ironing, unlike its conventional cotton counterpart. However, it felt like wearing a plastic bag! Modern all-nylon products include rope and twine, as well as the Velcro fastener. Nylon is frequently blended with other fibres to increase their wear-resistance. Mixed with wool staple, it makes hard-wearing carpets.

Nylon 6,6

A copolymer results when a diacyl chloride reacts with a diamine to form repeating substituted amide groups. The monomers each have two functional groups – they are difunctional molecules. For example, the reaction between hexane-1,6-dioyl chloride and 1,6-diaminohexane forms the polymer known as nylon 6,6 (pronounced 'six-six'). This name signifies that both monomers contain six carbon atoms; hence there are six carbon atoms between each of the repeating substituted amide groups in the polymer. A range of different nylon polymers results from using diamines and diacyl chlorides containing different numbers of carbon atoms.

$$Cl-CO-(CH_2)_4-CO-Cl + H-NH-(CH_2)_6-NH-H \rightarrow \text{----}CO-(CH_2)_4-CO-NH-(CH_2)_6-NH\text{----} + HCl$$

The condensation reaction between hexane-1,6-dioyl chloride and 1,6-diaminohexane forms nylon 6,6. The section shown above is the repeating unit of nylon 6,6.

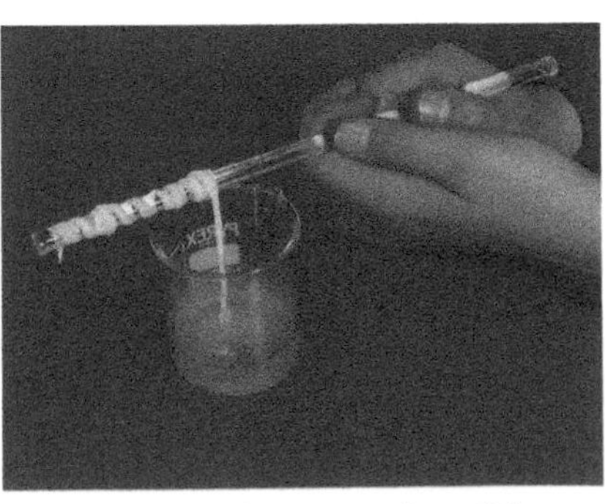

The laboratory demonstration of the formation of nylon 6,6. The reaction occurs at the interface between two immiscible liquids. The lower liquid is tetrachloromethane containing dissolved hexane-1,6-dioyl chloride; the upper liquid is an aqueous solution of 1,6-diaminohexane.

Wallace Carothers and the naming of nylon

Wallace Hume Carothers was born in 1896 in Iowa. After obtaining degrees from the Universities of Illinois and Harvard, in 1928 he took up the post of 'Head of Organic Research' at the DuPont company in Delaware. By 1931 his team had developed the first synthetic condensation polymers.

The raw product of the polymerization reaction was of little commercial use. However, younger members of the research team chanced upon an accidental discovery that led to the material becoming a huge commercial success. While Carothers was away from the laboratory, they engaged in some high-spirited horseplay, which centred on the question: 'Just how far will this stuff stretch?' Pulling on the ends of a sample – technically called 'cold drawing' – they found that it became very silky in appearance and increased in strength. Pulling the fibre aligned the polymer chains; hydrogen bonding between the chains gave a degree of crystallinity similar to that in silk.

The original name of DuPont's first polyamide was '66 polyamide', but this did not sound appealing enough to catch the popular imagination. One reason for the huge commercial success was that stockings made from the material did not 'run' (form 'ladders' when snagged). The name 'norun' was therefore suggested, but this did not satisfy the DuPont management. Spelling the name backwards as 'nuron' eventually led to 'nylon' – the greatest commercial success the company ever had.

Hydrogen bonding between the polyamide chains in nylon 6,6.

Aramid fibres

Aramids are particularly useful variants of polyamides in which the —CONH— group is directly attached to benzene rings. The aramid fibre with the amide groups attached to the 1 and 3 positions of the benzene molecule (Nomex) has exceptional heat- and flame-resistant qualities. Woven Nomex undergarments are worn by fire-fighters, racing drivers, and fighter aircraft pilots. Attaching the amide groups to the 1 and 4 positions gives an exceptionally strong material (Kevlar, made by reaction of benzene-1, 4-dicarboxylic acid and benzene-1,4-diamine) used to make bullet-proof vests, and combined with carbon fibre (as Carbon–Kevlar), the fuel tanks of some F1 racing cars.

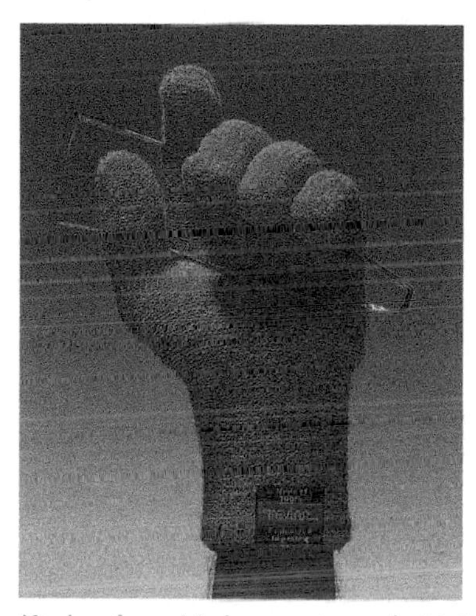

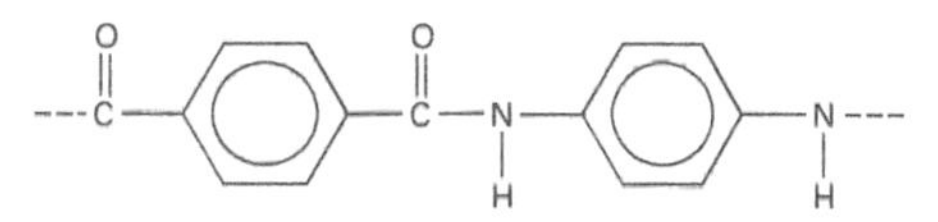

Kevlar. Aramids have extremely high tensile strengths, coupled with toughness, wear- and temperature-resistance, and low density. The 1,4-disubstituted aramids are particularly strong materials. The section shown above is the repeating unit of Kevlar.

SUMMARY

- Polyamides contain the substituted amide (or peptide) linkage —CONH—.
- Polyamides form from condensation reactions between diacyl chlorides and diamines.
- Nylon 6,6 is the polyamide resulting from the reaction between 1,6-diaminohexane and hexane-1,6-dioyl chloride.
- The name 'nylon 6,6' signifies that both monomer molecules contain six carbon atoms.

Biodegradable polymers

Both polyesters and polymides can be hydrolysed; see spreads 27.4 and 27.5. This allows them to be broken down in the environment more readily than any poly(alkene).

PRACTICE

1 Explain the following terms, giving examples:
  a Polyester
  b Polyamide
  c Copolymer
  d Difunctional molecule
  e Condensation polymerization.

2 Give the structural formulae of two monomers suitable for the production of nylon 4,4.

3 Draw the structural formula of a section of the polymer molecule that results from the reaction between propane-1,3-diol and benzene-1,3-dioyl chloride.

27.8

Reduction of acid derivatives

OBJECTIVES

- Reduction of acyl chlorides
- Reduction of esters
- Reduction of amides

The three major classes of carboxylic acid derivatives are the acyl chlorides, the esters, and the amides. All of these can be reduced by strong reductants.

Reduction reactions

Carboxylic acid derivatives have the general structure

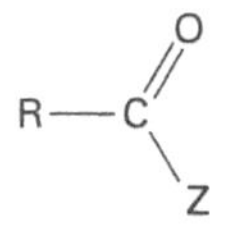

R—C(=O)—Z

where Z is Cl (acyl chlorides), OR′ (esters), or NH_2 (amides). All these derivatives are reduced by lithium tetrahydridoaluminate. A more powerful reductant than $NaBH_4$ (see the previous chapter) is required due to the delocalization of the lone pair on the **Z** group with the carbonyl group.

- An acyl chloride is reduced to an alcohol:

$$\underset{\text{ethanoyl chloride}}{CH_3COCl} \xrightarrow{LiAlH_4} \underset{\text{ethanol}}{CH_3CH_2OH}$$

Note that the parent acid is also reduced:

$$CH_3COOH \xrightarrow{LiAlH_4} CH_3CH_2OH$$

- An ester is reduced to *two* alcohols:

$$\underset{\text{propyl ethanoate}}{CH_3CH_2CH_2OCOCH_3} \xrightarrow{LiAlH_4} \underset{\text{propan-1-ol}}{CH_3CH_2CH_2OH} + \underset{\text{ethanol}}{CH_3CH_2OH}$$

- An amide is reduced to an amine:

$$\underset{\text{ethanamide}}{CH_3CONH_2} \xrightarrow{LiAlH_4} \underset{\text{ethylamine}}{CH_3CH_2NH_2}$$

Lithium tetrahydridoaluminate is a useful reagent, but it is expensive and requires absolutely anhydrous conditions being used in solution in ethoxyethane (ether, spread 25.5). Sodium in ethanol is frequently a suitable alternative:

$$ROCOR' \xrightarrow{Na/CH_3CH_2OH} ROH + R'CH_2OH$$

The reaction products are the same as with lithium tetrahydridoaluminate, but with the added complication that the liberated alcohols will react with the sodium to form their respective alkoxide ions; see spread 25.2. At the end of the reaction, the excess sodium is removed and water is then added to reform the alcohols. Fractional distillation can then be used to separate the three alcohols from the reaction mixture.

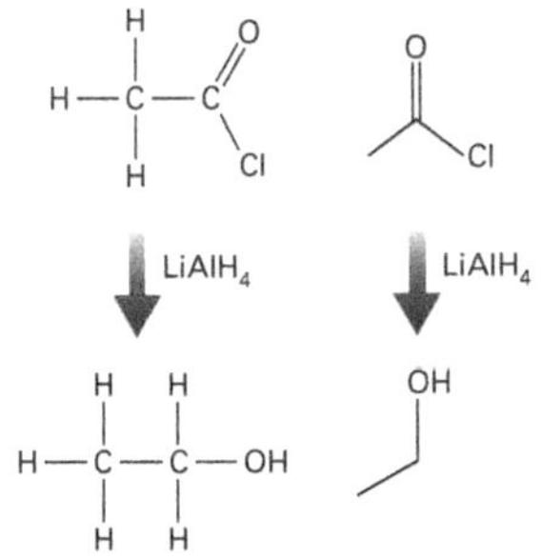

Reduction of ethanoyl chloride to ethanol.

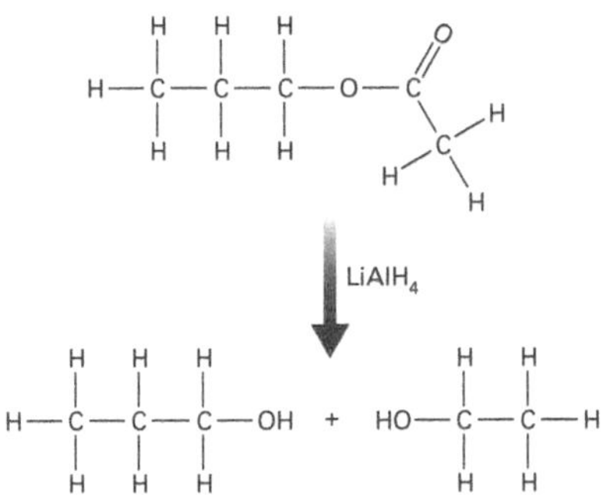

Reduction of propyl ethanoate to propan-1-ol and ethanol.

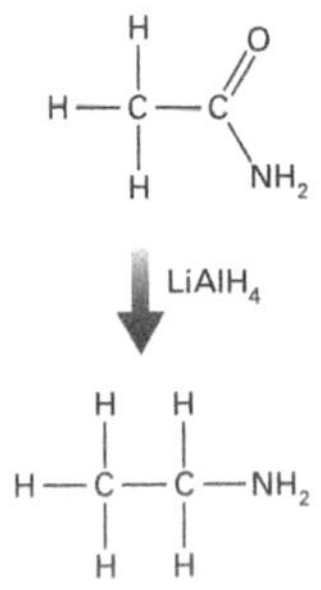

Reduction of ethanamide to ethylamine.

Worked example on reduction

Question: A, $C_4H_8O_2$, is reduced by lithium tetrahydridoaluminate to **B**, $C_4H_{10}O$. **B** is dehydrated to form **C**, C_4H_8. **C** reacts with HBr(aq) to form $(CH_3)_3CBr$. Draw structural formulae for **A**, **B**, and **C**.

Strategy: The loss of one oxygen on reduction suggests the reduction of an acid (an ester would produce *two* products).

Answer: C must be an alkene to add HBr; it must be $(CH_3)_2C{=}CH_2$. The alcohol **B** therefore must be $(CH_3)_2C(OH)CH_3$ or $(CH_3)_2CH$—CH_2OH. **B** has to be formed by reduction of an acid; only the second suggestion fits that piece of data.

Hence **B** is $(CH_3)_2CH$—CH_2OH and
A is $(CH_3)_2CH$—COOH.

SUMMARY

- Esters, acids, and acyl chlorides are reduced by $LiAlH_4$ to alcohols.
- Amides are reduced by $LiAlH_4$ to amines.

PRACTICE

1 Write chemical equations for the reduction of each of the following substances. Name the products formed and give their structural formulae:
 a Ethyl ethanoate
 b Methyl propanoate
 c Ethyl benzoate
 d Phenyl benzoate
 e Ethanoyl chloride
 f Benzoyl chloride.

2 Give the structural formula of ethanamide, together with the name and formula of the product of its reduction.

Chapter 27 Reactions summary

- Aliphatic carboxylic acids – from oxidation of primary alcohols via aldehydes by acidified potassium dichromate(VI):

  $CH_3CH_2OH \xrightarrow{[O]} CH_3CHO \xrightarrow{[O]} CH_3COOH$

- Aromatic carboxylic acids – from oxidation of any alkylbenzene by alkaline potassium manganate(VII):

  $C_6H_5CH_3 \xrightarrow{[O]} C_6H_5COOH$

- Carboxylic acids – protonation of water:

  $CH_3COOH(aq) + H_2O(l) \rightleftharpoons H_3O^+(aq) + CH_3CO_2^-(aq)$

- Carboxylic acids – salt formation:

  $CH_3COOH(aq) + NaOH(aq) \rightarrow Na^+CH_3CO_2^-(aq) + H_2O(l)$

- Preparation of acyl chlorides:

  $RCOOH + SOCl_2 \rightarrow RCOCl + SO_2 + HCl$

- Acyl chloride with water (hydrolysis):

  $CH_3COCl + H_2O \rightarrow CH_3COOH + HCl$

- Acyl chloride with alcohol to form ester:

  $CH_3COCl + CH_3CH_2OH \rightarrow CH_3COOCH_2CH_3 + HCl$

- Acyl chloride with ammonia to form amide:

  $CH_3COCl + 2NH_3 \rightarrow CH_3CONH_2 + NH_4Cl$

- Esterification – alcohol with carboxylic acid to form ester (conc. sulphuric acid catalyst):

  $CH_3CH_2OH(l) + CH_3COOH(l) \rightleftharpoons CH_3COOCH_2CH_3(l) + H_2O(l)$

- Base-catalysed hydrolysis of ester:

  $CH_3COOCH_2CH_3(l) + NaOH(aq) \rightarrow CH_3CH_2OH(l) + Na^+CH_3CO_2^-(aq)$

- Substituted amide from acyl chloride and amine:

  $RCOCl + 2R'NH_2 \rightarrow RCONHR' + R'NH_3Cl$

- Amide from heating the ammonium salt of a carboxylic acid:

  $NH_4^+C_6H_5CO_2^- \rightarrow C_6H_5CONH_2 + H_2O$

- Acidic hydrolysis of an amide:

  $CH_3CONH_2 + H_3O^+ \rightarrow CH_3COOH + NH_4^+$

- Basic hydrolysis of an amide:

  $CH_3CONH_2 + OH^- \rightarrow CH_3CO_2^- + NH_3$

- Dehydration of an amide:

  $C_6H_5CONH_2 \xrightarrow{P_4O_{10}} C_6H_5CN + H_2O$

- Reduction of an amide by $LiAlH_4$:

  $CH_3CONH_2 \xrightarrow{[H]} CH_3CH_2NH_2$

- Amine production by Hofmann degradation reaction:

  $CH_3CONH_2(aq) + Br_2(aq) + 4KOH(aq) \rightarrow CH_3NH_2(aq) + K_2CO_3(aq) + 2KBr(aq) + 2H_2O(l)$

- Reduction of acyl chloride:

  $CH_3COCl \xrightarrow{LiAlH_4} CH_3CH_2OH$

- Reduction of ester:

  $CH_3CH_2CH_2OCOCH_3 \xrightarrow{LiAlH_4} CH_3CH_2CH_2OH + CH_3CH_2OH$

PRACTICE EXAM QUESTIONS

1 The molecular formulae of some compounds that can be prepared from ethanoic acid are given in the scheme below.

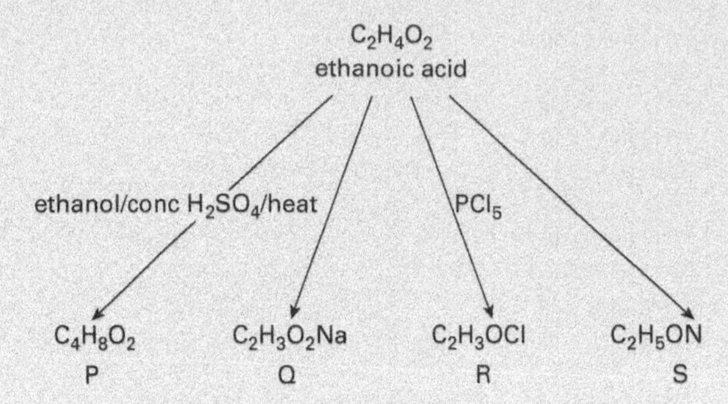

a **i** Give the name and graphical formula of **P**. [2]

ii Give the name of the type of reaction which occurs when **P** is formed from ethanoic acid. [1]

b Ethanoic acid can be obtained from **P**.

i Give the name of the reagent(s) and state the conditions required. [2]

ii Write a balanced equation for the reaction. [1]

c **i** State the reagent and reaction conditions that could be used for converting ethanoic acid into **Q**. [2]

ii Give the name and graphical formula of the organic product of the reaction between anhydrous samples of **Q** and **R**. [2]

iii State how the product formed in **c ii** could be converted into ethanoic acid and write an equation for the reaction. [2]

d **i** Give the name and graphical formula of the amide, **S**. [2]

ii State the reagent(s) and reaction conditions that could be used for converting ethanoic acid into **S**. [2]

iii Write a balanced equation for the reaction between **S** and aqueous hydrochloric acid. [1]

2 **a** Write balanced equations for the following hydrolysis reactions:

i ethanamide and aqueous sodium hydroxide; [2]

ii propanenitrile and aqueous hydrochloric acid; [2]

iii ethanoyl chloride and water; [2]

iv ethanoic anhydride and water. [2]

b What difference is there in the conditions needed for the hydrolyses shown in **a iii** and **a iv**? [1]

c Give the formulae of **two** organic compounds that would react together to give each of the following products:

i CH_3COOCH_3 [2]

ii $CH_3CH_2NHCOCH_3$ [2]

3 Citric acid is used in foodstuffs as an antioxidant and, together with its sodium salt, as an acidity regulator. It occurs naturally in fruit juices.

The formula of citric acid is

```
      CH2COOH
      |
HO — C — COOH
      |
      CH2COOH
```

a **i** Assuming citric acid behaves in aqueous solution as a monoprotic acid:

$$RCOOH + H_2O \rightleftharpoons RCO_2^- + H_3O^+$$

write an expression for K_a for this acid. [1]

ii Calculate the pH of lemon juice which contains citric acid at a concentration of $0.200\ mol\ dm^{-3}$. [K_a for citric acid = $7.4 \times 10^{-4}\ mol\ dm^{-3}$.] [3]

b The use of citric acid together with its salt, sodium citrate, as an acidity regulator depends on the ability of this mixture to act as a buffer.

i What is the function of a buffer solution? [2]

ii Describe how the mixture of citric acid and sodium citrate achieves this buffering action. Give equations for the TWO reactions you describe. [3]

iii Calculate the pH of a buffer solution containing $0.200\ mol\ dm^{-3}$ of citric acid and $0.400\ mol\ dm^{-3}$ of sodium citrate. [2]

c Citric acid forms a liquid ester which has the structural formula

```
      CH2COOC2H5
      |
HO — C — COOC2H5
      |
      CH2COOC2H5
```

i Describe a test you could use to show that the ester contains an —OH group. [2]

ii What reagent would you use to hydrolyse the ester? [1]

iii Treatment of the products of the reaction in **c ii** leads to the production of a pure sample of citric acid. How would you show the presence of the —COOH group in the citric acid other than by the use of an indicator? [2]

4 *Terylene* has the following repeat unit.

—CO—(benzene ring)—$CH_2CH_2CH_2$—O—

a Draw the displayed formulae of the two monomers used to make *Terylene*.

b State the type of polymerization which occurs when *Terylene* is made.

c Explain why holes form when aqueous sodium hydroxide is spilled on a *Terylene* shirt. [4]

5 **a** An organic compound **A** has a molar mass of $46\ g\ mol^{-1}$ and the following elemental composition by mass:

C 52.13%   H 13.15%;   O 34.72%

Determine the molecular formula of **A**. [2]

b The organic compound **A** has an infrared spectrum which shows the following features:

Absorption at 2900 cm^{-1}
C—H stretching frequency

Absorption at 3300 cm^{-1}
O—H stretching frequency

Absorption at 1050 cm^{-1}
C—O stretching frequency

Absorption at 1400 cm^{-1}
C—H bending frequency

The shape and position of the —OH peak indicates substantial hydrogen bonding.

i Explain, briefly, what is meant by the term *hydrogen bonding*. [1]

ii By reference to the infrared spectrum and your answer to **a** deduce the structure of **A**.

Give three reasons in support of your answer. [4]

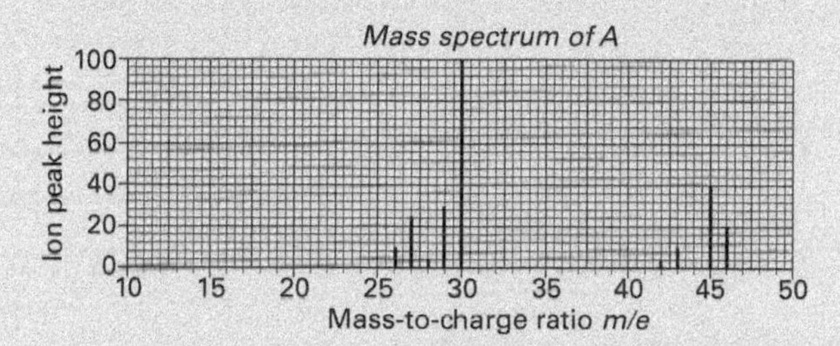

c The mass spectrum of **A** is given above.

Using the structural formula of **A** deduced above, suggest formulae for the positive ions responsible for *m/e* peaks at 45, 31, and 29. [3]

d i When **A** is treated with ethanoyl chloride, a compound **B** is formed containing 4 carbon atoms.

Give the name and structure of **B** and state to which class of organic compounds it belongs. [3]

ii State how you would carry out the addition of ethanoyl chloride to the compound, **A**, paying particular attention to safety. [2]

6 a The figure below shows the infrared absorption spectrum for 'Nylon-6,6'.

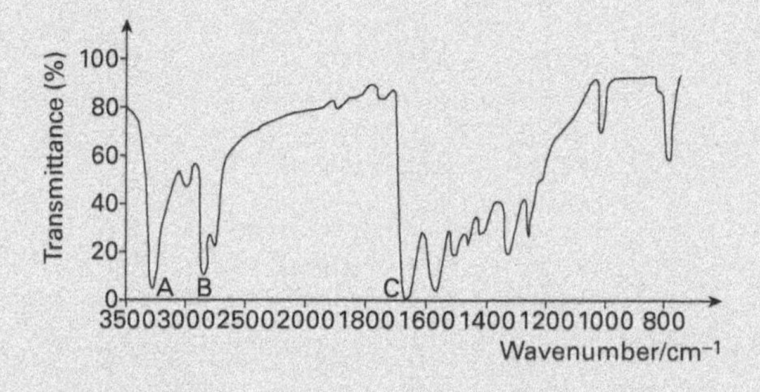

The table below shows some characteristic infrared absorptions.

Molecule or group	Bond absorbing	Wavenumber/cm–1
Alkyl	C–H	2960–2850
	C–H	1460–1370
Aldehyde	C–H	3250–3200
	C=O	1740–1650
Alkene	C–H	3095–3075
	C–H	990–3300
Amide	N–H	3500–3300
Alcohol	C–O	1200–1050
	O–H	3650–3590

i Using the data in the table, identify the cause of the absorption peaks **A**, **B** and **C** in the figure. [3]

ii Draw graphical formulae of two monomers that can be used to produce 'Nylon-6,6'. [4]

b In an experiment involving enzymes, a student used urease, which is found in plants. Urease converts urea to ammonia. The student took some urea solution and added 3 drops of litmus solution, followed by drops of dilute hydrochloric acid until the solution just changed to a red colour. A small quantity of 1% urease solution was then added and the solution quickly changed colour to blue.

i Explain why the colour of the solution becomes blue. [2]

ii Complete and balance the following equation for the reaction in which urea is hydrolysed. [2]

$$(H_2N)_2C=O + H_2O \rightarrow$$

iii The student repeated the experiment, using ethanamide (CH_3CONH_2) in place of urea. There was no change in the colour of the solution when the urease was added. Explain this observation in terms of the activity of the enzyme. [3]

iv Suggest how the enzyme urease could be denatured in this reaction. [1]

(See also chapters 30 and 32.)

7 Nylon 6,6 has the formula

$—(CH_2)_4—CONH—(CH_2)_6—NHCO—(CH_2)_4—CONH—(CH_2)_6—NHCO—$

a Give the formulae of the two monomers which combine to make nylon 6,6. [2]

b What is the name of the other product in this polymerization? [1]

c Suggest, including an equation, what will happen if nylon 6,6 is boiled with dilute acid. [4]

28 Amines and amino acids

Amines and amino acids are organic compounds that contain one or more nitrogen atoms. These different classes of compounds have some underlying properties that derive from the presence of nitrogen. However, the similarities are no more marked than those seen in the alcohols, ketones, and carboxylic acids, which all contain oxygen. This chapter concentrates on amines and amino acids. The study of aromatic amines includes a treatment of diazo compounds; the study of amino acids leads into the topic of optical activity.

Organic nitrogen compounds

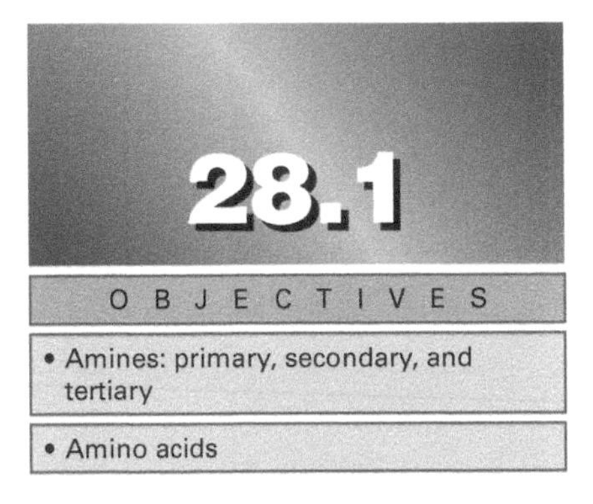

You have already met amides, which are carboxylic acid derivatives that contain the functional group —$CONH_2$. Central to this chapter are two other classes of nitrogen-containing organic compounds – amines and amino acids. **Amines** have an amine group —NH_2 (sometimes one or more of the hydrogen atoms is substituted). **Amino acids** have an amine group —NH_2 *and also* a carboxylic acid group —COOH attached to the molecule. We shall also look at nitriles, which have a carbon–nitrogen triple bond —C≡N, and diazo compounds, which contain a nitrogen–nitrogen double bond —N=N—.

Some molecules that contain nitrogen. From left to right, the molecules are methylamine CH_3NH_2, dimethylamine $(CH_3)_2NH$, trimethylamine $(CH_3)_3N$, ethanamide CH_3CONH_2, and ethanenitrile CH_3CN.

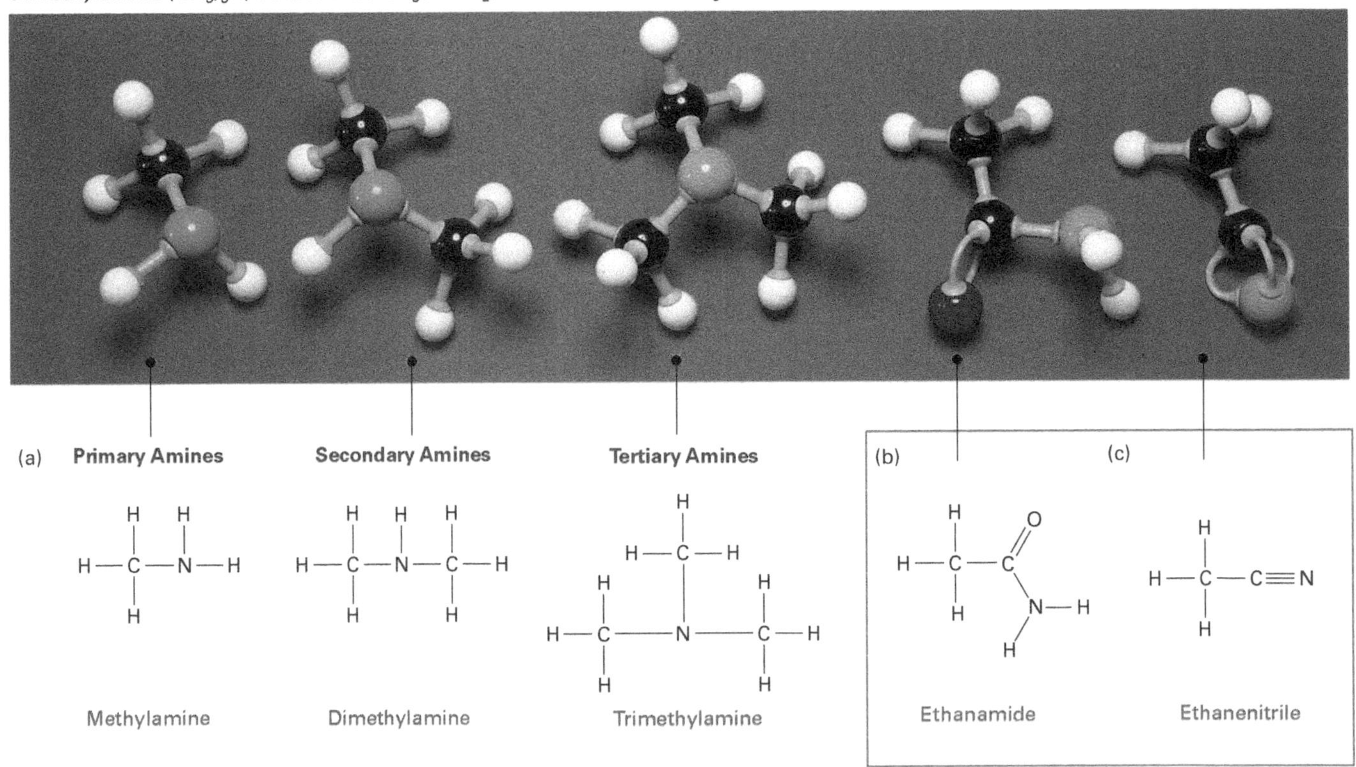

There are other nitrogen-containing molecules apart from amines. On the right are two examples: ethanamide CH_3CONH_2, and ethanenitrile CH_3CN.

Amines

The name of an amine depends on the alkyl or aryl groups present.

Formula	Name
CH_3NH_2	methylamine
$C_6H_5NH_2$	phenylamine
$(CH_3CH_2)_2NH$	diethylamine
$(CH_3CH_2)_3N$	triethylamine

Amines

Amines are organic compounds derived from ammonia in which alkyl (or aryl) groups substitute for some of the hydrogen atoms of the NH_3 molecule. **Primary**, **secondary**, and **tertiary amines** correspond to the substitution of one, two, and three hydrogen atoms respectively. Methylamine, ethylamine, and dimethylamine are gases; other common amines are liquids.

The organic equivalents of the ammonium ion are the **quaternary ammonium salts** $R_4N^+A^-$ in which four groups are attached to a central positively charged nitrogen atom. A^- is an anion such as OH^-, Cl^-, SO_4^{2-}, etc. Quaternary ammonium salts with two long-chain alkyl groups can act as cationic surfactants. Fabric softeners contain cationic surfactants to give a soft feel to cleaned garments; they must be cationic because detergents are usually anionic; see spread 23.5.

Remember that the structures of primary, secondary, and tertiary *alcohols* depend on the number of alkyl groups *attached to the carbon atom* bearing the —OH functional group. Hence, $CH_3CH_2CH_2CH_2OH$ (butan-1-ol) is a primary alcohol, $CH_3CH(OH)CH_2CH_3$ (butan-2-ol) is a secondary alcohol, and $(CH_3)_3COH$ (2-methylpropan-2-ol) is a tertiary alcohol. The structures of primary, secondary, and tertiary amines depend on the number of alkyl groups *attached directly to the nitrogen atom* of the functional group. So, CH_3NH_2 (methylamine) is a primary amine, $(CH_3)_2NH$ (dimethylamine) is a secondary amine, and $(CH_3)_3N$ (trimethylamine) is a tertiary amine.

Quaternary Ammonium Salt

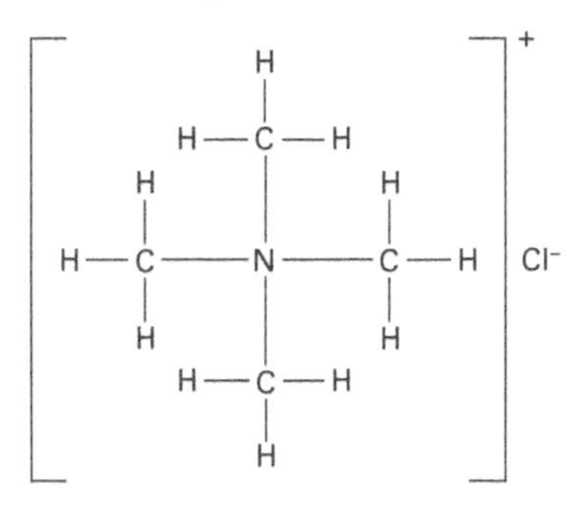

Tetramethylammonium chloride

The biochemically important species acetylcholine is $CH_3COOCH_2CH_2N^+(CH_3)_3$.

Amino acids

The amine group is also present in amino acids. Amino acids contain both the amine group $—NH_2$ and the carboxylic acid group —COOH. Naturally occurring amino acids are the chief constituents of proteins. These α-amino acids have the general structure

```
H   H    O
|   |   //
N — C — C
|   |    \
H   R     O — H
```

The prefix 'α' signifies that the amine group is attached to the carbon atom that itself is attached *directly* to the carboxylic acid group.

The amino acid γ-aminobutanoic acid (also known as GABA) is a neurotransmitter found in nerve synapses in the brain. This is a γ-amino acid and has the structure

```
 O                          H
  \\    α     β     γ       |
   C — CH2 — CH2 — CH2 — N
  /                         |
HO                          H
```

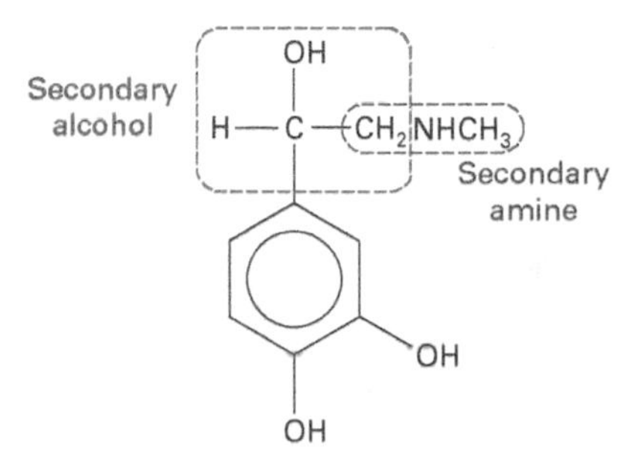

Adrenaline (also known as epinephrine) is a secondary amine. (It is also a secondary alcohol.) Adrenaline is a hormone produced by the adrenal glands, which are situated above the kidneys. Adrenaline increases heart rate as part of the 'fight or flight' response to danger or stress. It is used to treat anaphylactic shock.

SUMMARY

- Nitrogen-containing organic compounds include:
  - primary amines RNH_2
  - secondary amines R_2NH
  - tertiary amines R_3N
  - quaternary ammonium salts $R_4N^+A^-$
  - amides $RCONH_2$
  - nitriles RCN
  - α-amino acids $H_2NCHRCOOH$
  - diazo compounds RNNR
- Some quaternary ammonium salts can be used as cationic surfactants.

PRACTICE

1 Draw structural formulae for each of the following nitrogen-containing compounds:

a Ethylamine
b Methylethylamine
c Dimethylethylamine
d Trimethylethylammonium chloride
e Tetramethylammonium sulphate
f Propanamide
g Ethanenitrile
h 2-Aminopropanoic acid (alanine).

28.2

Aliphatic amines

OBJECTIVES

- Amines as bases
- Amines as nucleophiles

Pharmaceuticals

Many pharmaceuticals, for example the local anaesthetic procaine

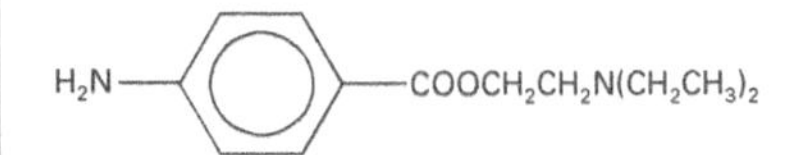

are manufactured as their 'hydrochlorides': this is usually because an amine group within the molecule is protonated.

The complex formed between (left) Cu^{2+} ions and an amine (here, ethylamine) is very similar in appearance and properties to that formed between (right) Cu^{2+} ions and ammonia.

Test for ammonia

The familiar test for ammonia, turning moist litmus paper blue, is not definitive; amines are volatile and their vapours also turn litmus blue.

Aliphatic amines are organic compounds that contain at least one alkyl group bonded directly to a nitrogen atom. The lone pair on the nitrogen atom results in amines and ammonia having similar chemical behaviour. Amines are described as Brønsted bases when they accept a proton. They are described as Lewis bases or nucleophiles when they engage in reactions that involve the donation of this lone pair.

Amines as bases

Ammonia acts as a Brønsted base when it accepts a proton from a Brønsted acid. For example, the neutralization of aqueous ammonia by dilute acid:

$$NH_3(aq) + H_3O^+(aq) \rightarrow NH_4^+(aq) + H_2O(l)$$

Amines can act in a similar manner. For example, a primary amine such as ethylamine $CH_3CH_2NH_2$ acts as a Brønsted base when it accepts a proton from a Brønsted acid. For example, ethylamine reacts with hydrochloric acid to form a substituted ammonium salt. The amine group is protonated:

$$CH_3CH_2NH_2(aq) + H_3O^+(aq) \rightarrow CH_3CH_2NH_3^+(aq) + H_2O(l)$$

Substituted ammonium salts may be isolated in the solid state, e.g. ethylammonium chloride $CH_3CH_2NH_3^+Cl^-(s)$. Strong bases displace weak bases from their salts. For example, the free amine is regenerated by adding aqueous sodium hydroxide to a solution of ethylammonium chloride:

$$CH_3CH_2NH_3^+Cl^-(aq) + NaOH(aq) \rightarrow CH_3CH_2NH_2(aq) + NaCl(aq)$$

Ammonia acts as a Lewis base when it donates a pair of electrons to a Lewis acid. For example, the formation of the nickel(II) complex:

$$6NH_3(aq) + Ni^{2+}(aq) \rightarrow [Ni(NH_3)_6]^{2+}(aq)$$

Primary amines also react with nickel(II) ions to form a solution that contains the nickel(II) ion complexed by the amine:

$$6CH_3CH_2NH_2(aq) + Ni^{2+}(aq) \rightarrow [Ni(NH_2CH_2CH_3)_6]^{2+}(aq)$$

pK_b values

There are small differences in the basicity of primary, secondary, and tertiary amines, but they all have pK_b values within one or two units of that of ammonia:

ammonia	NH_3	$pK_b = 4.8$
methylamine	CH_3NH_2	$pK_b = 3.4$
ethylamine	$CH_3CH_2NH_2$	$pK_b = 3.2$

For aliphatic amines, base strength increases with the size and number of alkyl groups attached to the nitrogen atom. Alkyl groups donate electron density towards the nitrogen atom, increasing the ability of the nitrogen atom to donate its lone pair of electrons.

(a) $H_2N{-}CH_2{-}CH_2{-}NH_2$

(b) $H_2N{-}CH_2{-}CH_2{-}CH_2{-}CH_2{-}NH_2$

The difference a couple of $-CH_2-$ groups can make! (a) 1,2-Diaminoethane, traditionally called 'ethylenediamine' and abbreviated to en, is a common bidentate ligand (see chapter 20 'The transition metals'). It has a very slight odour, but is otherwise inoffensive as a laboratory reagent. (b) Putrescine has the systematic name 1,4-diaminobutane. It has the clinging smell of putrescent rotting fish.

Amines as nucleophiles

The presence of the lone pair on the nitrogen atom results in amines being nucleophiles. For example, primary and secondary amines are acylated by acyl chlorides. The following reaction mechanism makes clear the nucleophilic attack by the lone pair on the acyl chloride δ+ carbon atom; see spread 27.2.

(a) The carbonyl carbon atom in ethanoyl chloride has a δ+ charge due to the electronegative oxygen and chlorine atoms. This carbon atom undergoes nucleophilic attack by the lone pair of electrons on the amine nitrogen atom.

(b) The C—Cl bonding pair of electrons reverts to the chlorine atom, which leaves as the chloride ion Cl^-.

(c) The N—H bonding pair of electrons reverts to the nitrogen atom. Hydrogen leaves as the hydrogen ion H^+.

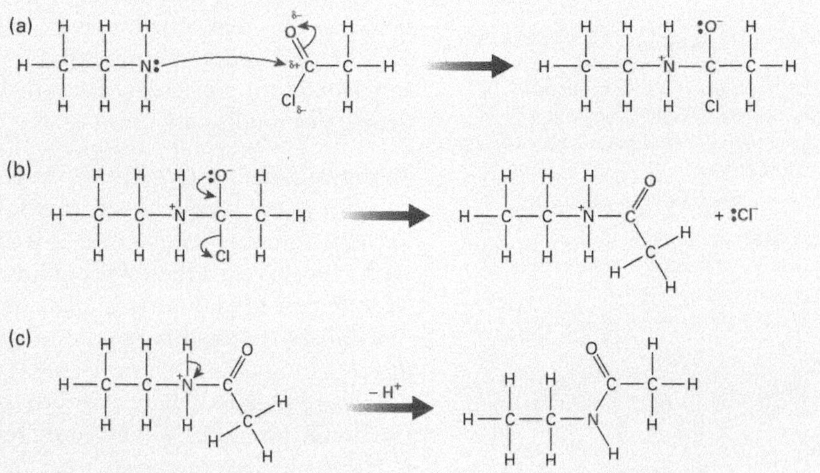

Acylation of ethylamine by ethanoyl chloride.

Amines can be produced by heating excess ammonia in ethanol with a chloroalkane. The primary amine forms as follows:

$$RCl + NH_3 \rightarrow RNH_2 + HCl$$

Unfortunately, the product is itself a nucleophile and can react with more chloroalkane to form successively secondary and tertiary amines and ultimately a quaternary ammonium salt:

$$RCl + RNH_2 \rightarrow R_2NH + HCl$$
$$RCl + R_2NH \rightarrow R_3N + HCl$$
$$RCl + R_3N \rightarrow R_4N^+Cl^-$$

At each step the lone pair on the nitrogen atom carries out nucleophilic attack on the δ+ carbon atom of the chloroalkane. Separation of the mixture is difficult; this reaction is therefore rarely used for the preparation of amines.

A more controlled method of producing *primary* aliphatic amines is to reduce a nitrile with lithium tetrahydridoaluminate; see spread 27.8:

$$CH_3C{\equiv}N \xrightarrow{[H]} CH_3CH_2NH_2$$

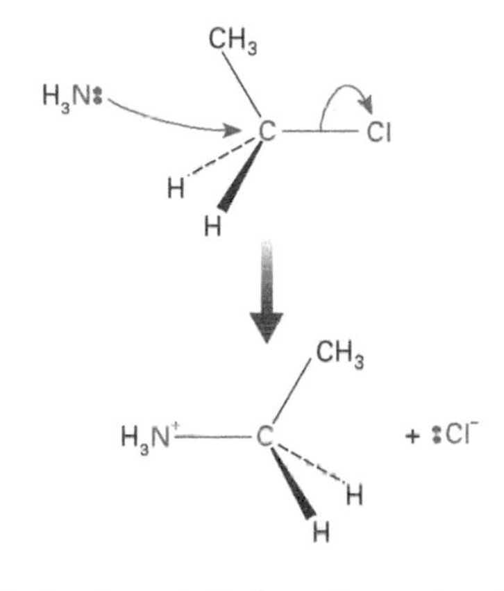

Ethylamine results from the nucleophilic attack of ammonia on chloroethane.

SUMMARY

- Amines react as substituted ammonia molecules.
- Amines act as Brønsted bases when they neutralize strong acids to form substituted ammonium salts.
- The free amine is regenerated when a substituted ammonium salt reacts with a strong base.
- Amines are weak bases; primary amines have slightly lower pK_b values than ammonia.
- Amines act as nucleophiles: the nitrogen lone pair attacks the δ+ carbon atom on halogenoalkanes and acyl chlorides.
- Amines react as Lewis bases when their nitrogen atom donates its lone pair of electrons, e.g. in transition metal complex formation.

PRACTICE

1 Interpret the reaction
$CH_3CH_2NH_2 + H_3O^+ \rightarrow CH_3CH_2NH_3^+ + H_2O$
in terms of a Lewis acid–Lewis base reaction.

2 Explain why the reaction
$6CH_3CH_2NH_2(aq) + Ni^{2+}(aq) \rightarrow [Ni(NH_2CH_2CH_3)_6]^{2+}(aq)$
can be interpreted by Lewis acid–base theory but not by Brønsted-Lowry acid–base theory.

28.3

OBJECTIVES

- Phenylamine preparation
- Aliphatic and aromatic amines compared
- Phenylamine as a nucleophile

A reminder

Aliphatic amines can be produced by the reaction of ammonia with a halogenoalkane (see previous spread), e.g. chloroethane:

$CH_3CH_2Cl + NH_3 \rightarrow CH_3CH_2NH_2 + HCl$

Note that secondary and tertiary amines and ultimately a quaternary ammonium salt also form.

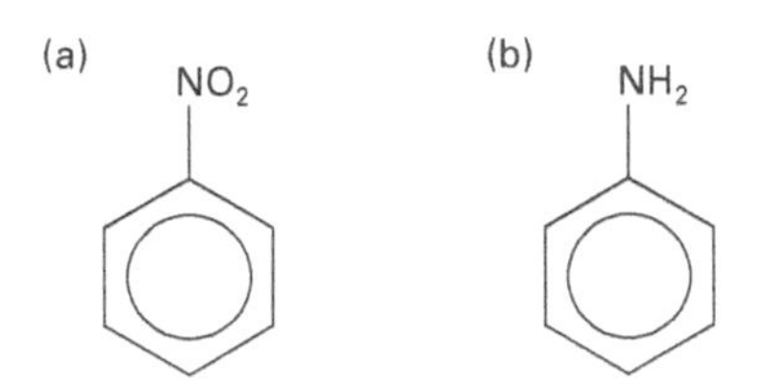

(a) Nitrobenzene has a nitrogen atom attached directly to the benzene ring. (b) The structure of phenylamine suggests that it will result from reduction of nitrobenzene.

Aromatic amines

You should recognize aliphatic amines as organic bases. These substances act as Lewis bases when they donate the lone pair of electrons on their nitrogen atom. The presence of the nitrogen lone pair also makes aliphatic amines act as nucleophiles, which attack centres of positive charge. Aromatic amines consist of an amine —NH_2 group attached to a benzene ring. The ring incorporates a system of delocalized π electrons. This delocalized system is capable of delocalizing the lone pair on the nitrogen atom. For this reason, aromatic amines do not have similar properties to aliphatic amines.

Aromatic amine synthesis

It is not possible to make aromatic amines in a similar manner to aliphatic amines. The reason is that ammonia reacts much more slowly with chloroarenes than with chloroalkanes. Ammonia is a nucleophile: its lone pair of electrons is attracted to centres of positive charge. The delocalized π electron cloud of a chloroarene *repels* the lone pair of an approaching ammonia molecule. The C—Cl bond is also stronger in the chloroarene due to delocalization of electron density from the ring with the bond, spread 24.1. The route to aromatic amines such as phenylamine $C_6H_5NH_2$ is therefore less direct.

Phenylamine: laboratory preparation

Instead of using chlorobenzene, nitrobenzene $C_6H_5NO_2$ (see chapter 23 'Arenes') is chosen as the starting material for the synthesis of phenylamine $C_6H_5NH_2$ (traditionally called aniline). Studying the structures of these molecules shows that nitrobenzene must lose oxygen and gain hydrogen. This change is a reduction and the most effective laboratory reductant is a mixture of granulated tin and concentrated hydrochloric acid. Phenylamine is very weakly basic and the reduction medium is acidic. The product is the salt phenylammonium chloride $C_6H_5NH_3^+Cl^-$:

$$C_6H_5NO_2(l) \xrightarrow{Sn/HCl} C_6H_5NH_3^+Cl^-(aq)$$

(a)

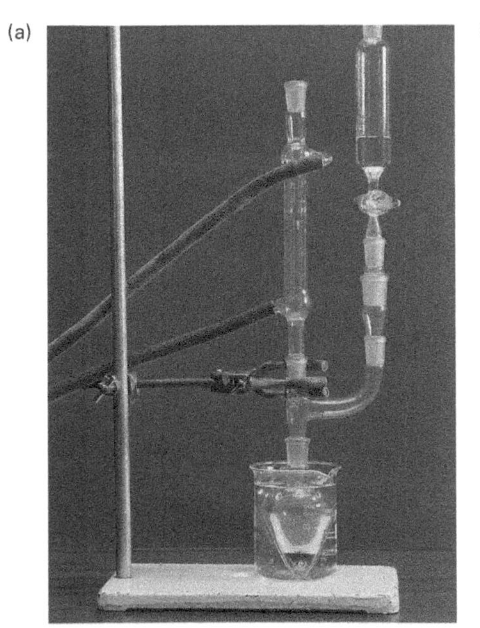

(b)

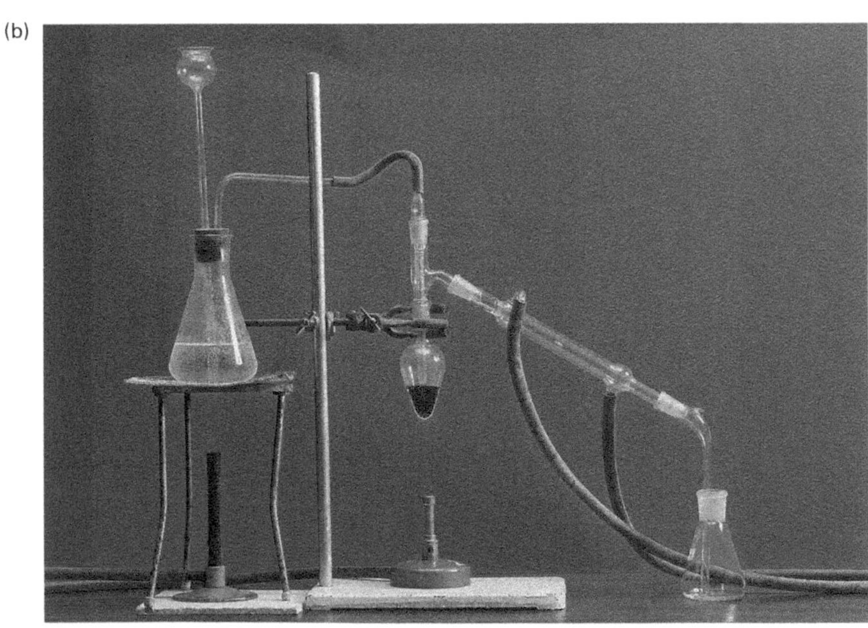

Stages in the preparation of phenylamine. (a) Place granulated tin and nitrobenzene in a flask. Fit a reflux condenser and slowly add concentrated hydrochloric acid, cooling the flask if necessary. Remove the condenser and heat the open flask in a boiling water bath for about one hour. Cool the mixture and add concentrated aqueous sodium hydroxide to produce the free amine by removing a proton from the nitrogen atom:

$C_6H_5NH_3^+Cl^-(aq) + NaOH(aq) \rightarrow C_6H_5NH_2(l) + H_2O(l) + NaCl(aq)$

(b) Steam distil the mixture until the milky emulsion of phenylamine no longer appears in the distillate. The mixture of phenylamine and steam (bubbled through the reaction mixture) boils just below 100 °C, at which temperature phenylamine decomposes much less than it does at its boiling point.

Phenylamine: properties

Aromatic amines are far less soluble in water than are aliphatic amines. Solubility depends partly on the polarity of a molecule. Aromatic amines are less polar because charge is delocalized around the benzene ring.

Phenylamine $C_6H_5NH_2$ (pK_b 9.4) is a much weaker base than ammonia (pK_b 4.8), by a factor of about 10^5. The reason for this difference is that the nitrogen lone pair is delocalized with the benzene ring and so is substantially more difficult to protonate. Phenylamine is, however, sufficiently basic to be protonated by strong acids to form salts containing the **phenylammonium ion** $C_6H_5NH_3^+$.

As for aliphatic amines, phenylamine reacts with acyl chlorides.

- Phenylamine reacts with ethanoyl chloride:

$$C_6H_5NH_2 + CH_3COCl \rightarrow C_6H_5NHCOCH_3 + HCl$$

The reaction is carried out by mixing the reagents. The product is called *N*-phenylethanamide (acetanilide) and hydrogen chloride forms as a by-product.

- Phenylamine reacts with benzoyl chloride:

$$C_6H_5NH_2 + C_6H_5COCl \rightarrow C_6H_5NHCOC_6H_5 + HCl$$

The reaction is carried out by shaking the reagents with aqueous sodium hydroxide. The product is called *N*-phenylbenzamide (benzanilide); water and sodium chloride form as by-products.

The presence of the amine group in phenylamine activates the benzene ring to electrophilic substitution (see chapter 23 'Arenes'). For example, shaking phenylamine with bromine water rapidly produces 2,4,6-tribromophenylamine:

$$C_6H_5NH_2 + 3Br_2(aq) \rightarrow Br_3C_6H_2NH_2 + 3HBr$$

Industry

The production of phenylamine is particularly important in industry. It is the precursor of a range of important dyes, some of which were crucial to the development of antibiotics. The following spread discusses the *diazotization reaction*, which is fundamental to the production of these substances.

SUMMARY

- In aromatic amines, electron density from the nitrogen lone pair is delocalized with the benzene ring.
- Aliphatic amines have lower pK_b values and are more powerful nucleophiles than aromatic amines.
- Aromatic amines react more slowly with acyl chlorides than do aliphatic amines.
- The amine group activates the benzene ring to electrophilic substitution into the 2, 4, and 6 positions.

Phenylamine and medicines

The acylation of a derivative of phenylamine leads to the pain-relieving medicine paracetamol (acetaminophen). Its formula is:

p-$HOC_6H_4NHCOCH_3$

The 'para-' signifies the 4-isomer.

(a)

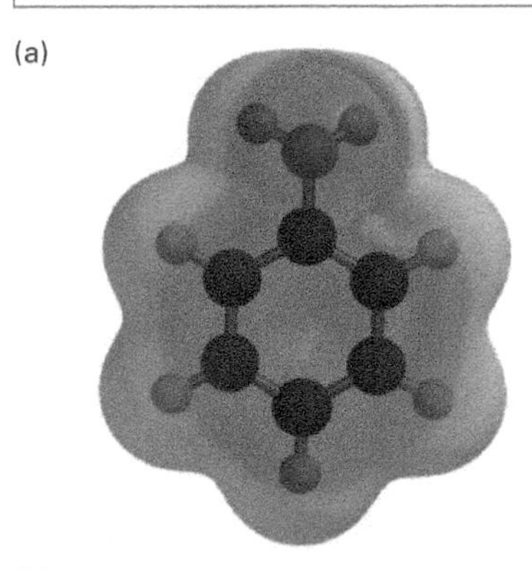

(b)

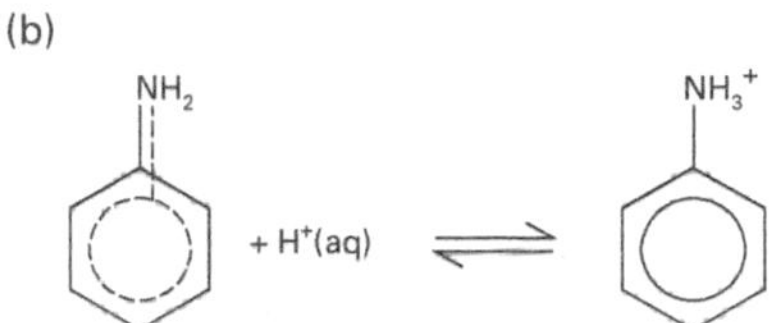

(c)

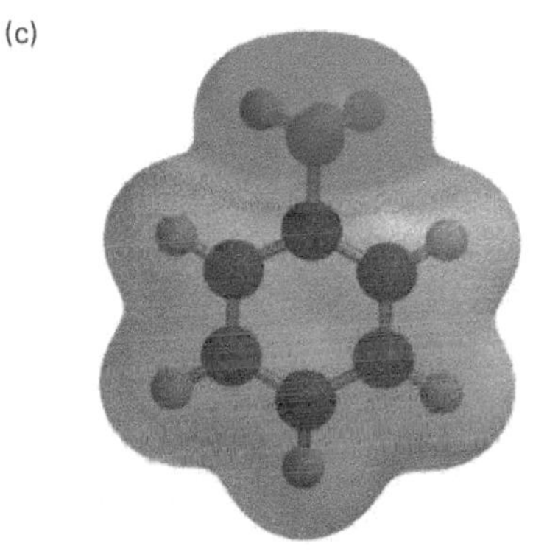

Phenylamine as a base.
(a) Phenylamine (an aromatic amine) is a weaker base than ethylamine (an aliphatic amine) due to delocalization of the nitrogen lone pair with the benzene ring. (b) The phenylammonium ion forms when phenylamine acts as a base (donating its nitrogen lone pair). (c) Delocalization stabilization is absent in the phenylammonium ion because there is no lone pair.

PRACTICE

1 Look at the photographs and captions on the stages in the preparation of phenylamine (opposite page).

a Why is a reflux condenser fitted during the addition of the concentrated hydrochloric acid?

b Why is no condenser required during heating on the water bath?

c Why is concentrated aqueous sodium hydroxide added at the end of the reduction?

2 Suggest a mechanism for the reaction between phenylamine and ethanoyl chloride to produce *N*-phenylethanamide:

$$C_6H_5NH_2 + CH_3COCl \rightarrow C_6H_5NHCOCH_3 + HCl$$

3 Phenylamine reacts with benzoyl chloride to produce *N*-phenylbenzamide:

$$C_6H_5NH_2 + C_6H_5COCl \rightarrow C_6H_5NHCOC_6H_5 + HCl$$

Explain why this reaction is slower than the acylation reaction in question 2, and why aqueous sodium hydroxide is included in the reaction mixture.

Phenylamine and diazotization

OBJECTIVES

- Benzenediazonium salts
- Substitution reactions
- Coupling reactions
- Diazo dyes and drugs

(a)

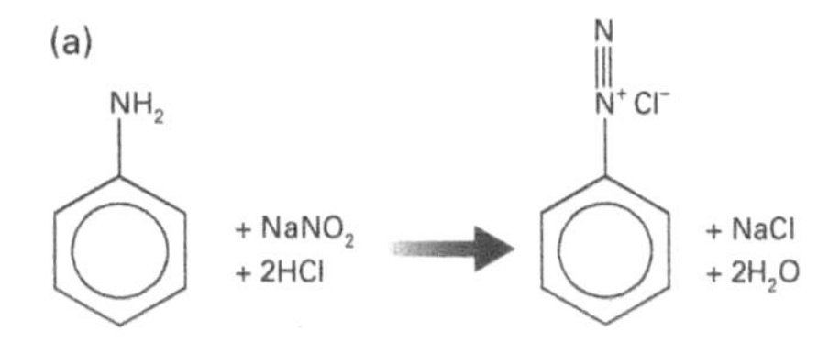

(b)

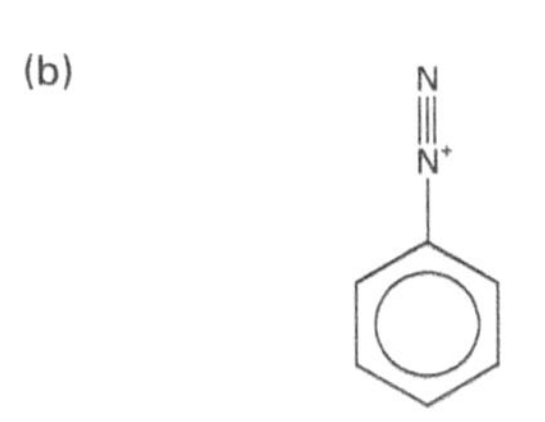

(c)

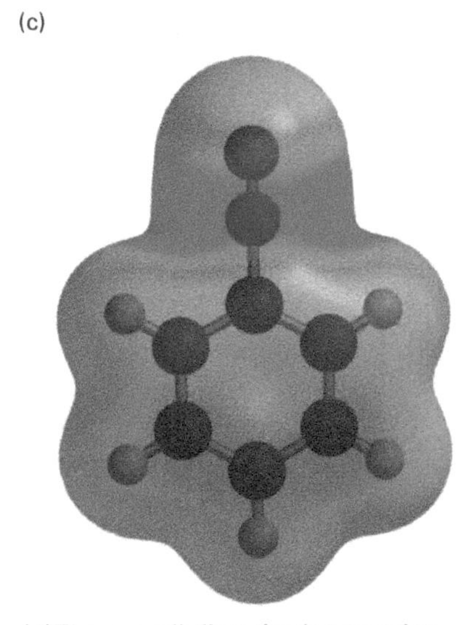

(a) The overall diazotization reaction may be represented by this chemical equation. (b) Classical bonding considerations indicate this structure for the diazonium ion. (c) The electrostatic potential map confirms that the nitrogen atom nearer the ring carries most of the positive charge (blue colour).

Diazonium salts

Aromatic diazonium salts are reasonably stable in solution if they are kept cold. This is because the benzene ring delocalizes the charge on the ion. It is not possible to make aliphatic diazonium salts, because the lack of delocalization makes them extremely unstable. They decompose immediately, releasing nitrogen gas.

Phenylamine (aniline) was originally manufactured from benzene obtained from the distillation of coal tar. This tar has a similar constitution to crude oil and was one of the by-products from heating coal to produce coke and coal gas. The synthetic dyestuff industry began in 1856 when William Perkin discovered a mauve dye while attempting to synthesize quinine, a cure for malaria. Many of the dyes subsequently developed became known as 'aniline' or 'coal tar' dyes. These dyes contain the diazo —N=N— group and are called 'diazo' dyes (or azo dyes). The dyestuffs industry that produces them has a total annual turnover of billions of pounds world-wide, a substantial proportion of which depends on just one chemical reaction – *diazotization*.

Diazotization

The **diazotization reaction** uses nitrous acid HNO_2 (HONO) to convert phenylamine $C_6H_5NH_2$ into the diazonium salt **benzenediazonium chloride** $C_6H_5N_2^+Cl^-$. Nitrous acid decomposes very readily and so cannot be stored. It is prepared *in situ* (within the reaction mixture) from a mixture of sodium nitrite $NaNO_2$ and an acid, i.e.

$$NaNO_2(aq) + HCl(aq) \rightarrow HNO_2(aq) + NaCl(aq)$$

The temperature must be kept below 5 °C in an ice-bath during diazotization otherwise the main product decomposes, losing its nitrogen atoms as nitrogen gas N_2. Temperature control is critical because below 0 °C the rate of diazotization becomes very slow.

Replacing the nitrogen atoms

- *Phenol*

When the temperature rises to around 50 °C, the diazonium ion reacts with water to produce phenol:

$$C_6H_5N_2^+Cl^-(aq) + H_2O(l) \rightarrow C_6H_5OH(aq) + N_2(g) + HCl(aq)$$

Note that, in this reaction, water is acting as a nucleophile.

- *Chlorobenzene*

Other nucleophiles may be used to attack diazonium salts. For example, chloride ions from hydrochloric acid (with copper(I) chloride CuCl as a catalyst) react to form chlorobenzene in the **Sandmeyer reaction** (named after the Swiss Traugott Sandmeyer):

$$C_6H_5N_2^+Cl^- \xrightarrow{HCl/CuCl} C_6H_5Cl + N_2$$

Similar reactions occur with bromide ion and iodide ion (iodide ion does not need a catalyst because it is a more powerful nucleophile). This reaction is the best way of attaching a halogen atom to a benzene ring.

Coupling to the diazo group

Diazonium salts also undergo a type of reaction known as **coupling**. A simple example is the reaction of a diazonium salt (kept below 5 °C) with an aromatic compound such as phenol C_6H_5OH in alkaline solution. The reaction produces an intense orange–yellow precipitate of (4-hydroxyphenylazo)benzene:

$$C_6H_5N_2^+Cl^- + C_6H_5OH \rightarrow HOC_6H_4N{=}NC_6H_5 + HCl$$

In a similar manner, an aromatic amine couples via its carbon atom at position 4 to form a diazo compound, as shown below. Similar diazo compounds are much used in the dyestuffs industry.

Coupling phenylamine with the benzenediazonium ion to produce (4-aminophenylazo) benzene. In all reactions of this type, the attacking molecule couples to the diazo group by the carbon atom in position 4 with respect to its functional group.

Diazo dyes and antibiotics

Before 1900, the vast majority of diseases simply had to be left to run their course because no chemotherapy treatment was available. The first breakthrough came in 1904 when Paul Ehrlich used the diazo dye 'Trypan Red' to treat sleeping sickness and other diseases caused by parasitic trypanosomes. He was awarded the 1908 Nobel Prize in Medicine for this life-saving discovery. In 1909 he introduced Salvarsan, an arsenic-containing molecule designed to treat syphilis.

Ehrlich did not rest on his laurels. He wished to find a compound that would be effective against a wider range of diseases. As a medical student, he had noticed that several synthetic dyes selectively stained biological tissue. He wondered if he could alter such a dye to deliver a 'magic bullet' to kill certain disease-causing pathological microorganisms.

Other scientists joined in the search, and nowhere was activity more intense than at the huge German dye manufacturers I. G. Farben. One particular dye, Prontosil, had achieved impressive results on mice that had been infected with *Streptococcus* bacteria. As the research was at the early stage, it was not felt acceptable to test the compound on humans. However, in 1935 the youngest daughter of Gerhard Domagk, a doctor at I. G. Farben, contracted a severe case of streptococcal infection from a pin prick. With his daughter close to death, Domagk gambled that Prontosil might possibly cure her. It did – she became the first of many to owe their lives to the effects of Prontosil.

Active metabolites

Prontosil is a diazo dye. It was initially suspected that this structure was responsible for its antibacterial activity. Ernest Fourneau of the Pasteur Institute later showed that human metabolism breaks down Prontosil to 4-aminobenzene-sulphonamide (sulphanilamide). This breakdown product (metabolite) is the chemical substance responsible for the destruction of the *Streptococcus* bacterium. (We shall discuss sulpha drugs more fully in the last spread in this chapter.)

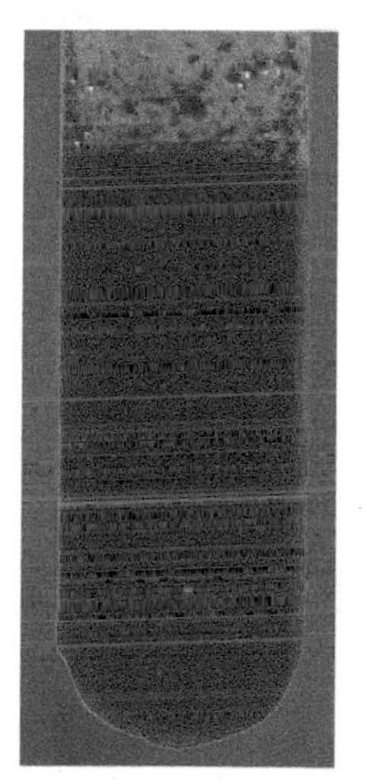

Diazo dyestuffs are used to dye natural and synthetic fibres. Compared to earlier natural dyestuffs, they have greater wash- and light-fastness, and more intense colours. They can also be used in plastics and paints. Their use in food is diminishing.

SUMMARY

- Diazotization forms the benzenediazonium ion $C_6H_5N_2^+$ by the action of nitrous acid HNO_2 on phenylamine $C_6H_5NH_2$.
- The diazo group has the formula —N=N—.
- Nucleophiles such as halide ion and water replace the $—N_2^+$ group on the benzenediazonium ion.
- Phenol and phenylamine couple to the benzenediazonium ion $C_6H_5N_2^+$ to form diazo compounds $C_6H_5N{=}NC_6H_4X$. The X group is in the 4 position relative to the diazo linkage.

PRACTICE

1. Most of the chemical equations given in this spread are written as condensed structural formulae. Rewrite them using *displayed* formulae.
2. A solution of benzenediazonium chloride reacts with the secondary amine *N*-methylphenylamine $C_6H_5NHCH_3$. Draw the structural formula of the product of this coupling reaction.
3. Suggest why iodide ion is a more powerful nucleophile than chloride ion in the reaction with the benzenediazonium ion.

28.5 Amino acids

OBJECTIVES

- The α-amino acids
- Zwitterions
- Condensation reactions
- Polypeptides and proteins

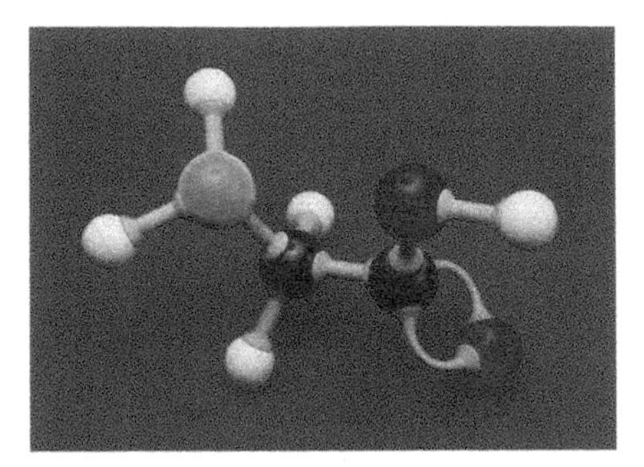
A molecular model of the simplest amino acid, glycine.

Amino acids are substances whose molecules contain both a basic —NH_2 group and an acidic —COOH group. They are very important in biochemistry as the monomers that make up the natural polyamide polymers called proteins (see chapter 30 'Biochemistry'). Almost all naturally occurring amino acids are α-amino acids in which the amine group is attached to the α-carbon atom of a carboxylic acid. As a result, they have the general formula

H H O
N—C—C
H R O—H

Whilst amino acids contain both a basic amine group and a carboxylic acid group, the overall acid–base nature depends on the balance of acidic and basic groups in the R group.

General physical properties

Amino acids have high melting points and low solubility in non-polar solvents. For example, glycine (aminoethanoic acid) NH_2CH_2COOH is a crystalline solid. Propanoic acid CH_3CH_2COOH and butylamine $CH_3CH_2CH_2CH_2NH_2$ have similar relative formula masses to aminoethanoic acid but are liquids. These properties point to amino acid molecules having a significantly polar nature. In both the solid state and in solution, they exist as 'inner salts' or '**zwitterions**'. The hydrogen atom from the carboxylic acid group protonates the basic amine group, i.e.

$$NH_2CH_2COOH \rightarrow {}^{+}NH_3CH_2CO_2^{-}$$

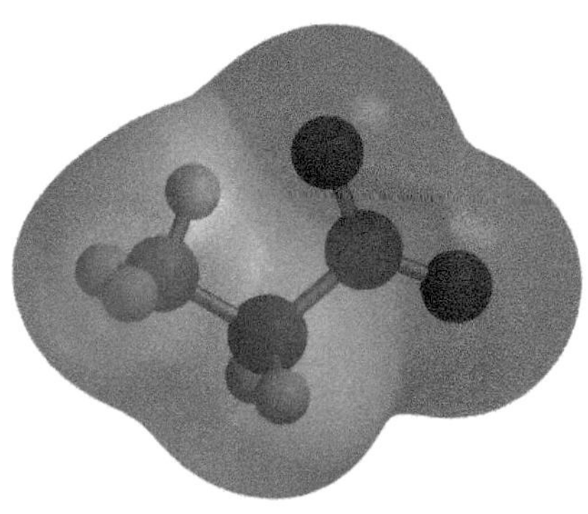
Amino acids typically exist as zwitterions: the protonated amine group is blue and the carboxylate group is red.

Natural amino acids

There are about 20 naturally occurring amino acids. As stated above, almost all are α-amino acids with the general formula $RCH(NH_2)COOH$. The only part of the molecule that varies is the R group. If the R group does not contain any acidic or basic groups, the amino acid is neutral overall. If there are more acidic groups, the amino acid is acidic (aspartic acid, glutamic acid); if there are more basic groups, the amino acid is basic (asparagine, arginine, lysine).

Glycine: $H_2N—CH(H)—COOH$
Alanine: $H_2N—CH(CH_3)—COOH$
Serine: $H_2N—CH(CH_2OH)—COOH$
Threonine: $H_2N—CH(HC(OH)CH_3)—COOH$
Phenylalanine: $H_2N—CH(CH_2C_6H_5)—COOH$
Glutamic acid: $H_2N—CH(CH_2CH_2COOH)—COOH$
Histidine: $H_2N—CH(CH_2\text{-imidazole})—COOH$
Cysteine: $H_2N—CH(CH_2SH)—COOH$

Some of the amino acids occurring in proteins.

Zwitterions

The charges on the zwitterions mean that amino acids move in an electric field. This property is the basis of their separation by electrophoresis (see chapter 31 'Techniques of preparation...').

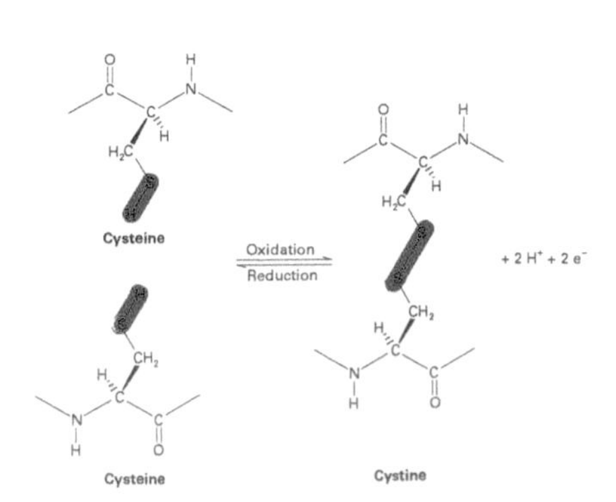

The formation of a disulphide bridge from two cysteine residues is an oxidation reaction.

Amino acids, polypeptides, and proteins

Amino acids undergo condensation reactions (see chapter 26 'Aldehydes and ketones') to form a substituted amide R—CONH—R′. For example, for glycine and alanine

$NH_2CH_2COOH + NH_2CH(CH_3)COOH \rightarrow$
$NH_2CH_2CONHCH(CH_3)COOH + H_2O$

The C—N bond in the —CONH— **peptide link** is called the **peptide bond**. The result of the reaction is a **dipeptide** molecule, which retains an amine group at one end and a carboxylic acid group at the other. Further amino acid molecules may condense with this product to extend the chain length. Chains consisting of up to 20 amino acids are called **oligopeptides**; longer chains are called **polypeptides**. Proteins consist of one or more polypeptide chains coiled into a complex and distinctive structure (see chapter 30 'Biochemistry').

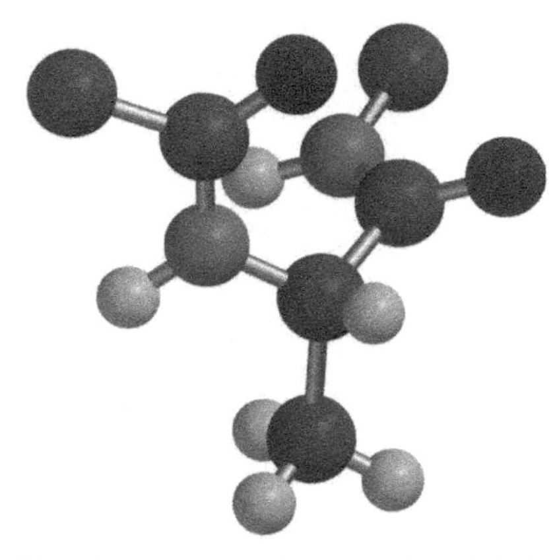

The planar shape at two peptide links in a portion of a protein. Remember that Spartan intentionally does not show the number of bonds; see spread 6.4.

The amino acids glycine and alanine condense to form a dipeptide, with the elimination of water. Note that there are two possible products (Gly–Ala and Ala–Gly). In natural systems, the choice of product is controlled by the structure of DNA.

General chemical properties

Amino acids undergo the typical reactions of primary amines and carboxylic acids:

- Salt formation with acids
  $NH_2CH_2COOH(aq) + HCl(aq) \rightarrow [^+NH_3CH_2COOH]Cl^-(aq) + H_2O(l)$
- Salt formation with base
  $NH_2CH_2COOH(aq) + NaOH(aq) \rightarrow Na^+ \ NH_2CH_2CO_2^-(aq) + H_2O(l)$

One distinctive feature of α-amino acids is that the carbon atom highlighted in the formula R<u>C</u>H(NH_2)COOH is attached to four different groups. As a result, these substances can show optical activity. This topic forms the focus of the following spread.

Crystals of phenylalanine. Amino acids may be isolated as crystalline solids. They have good solubility in water.

Catabolism and anabolism

Digesting the proteins in food involves an overall process called *catabolism*. Protein structure is broken down to liberate the component amino acids. The synthesis, in living systems, of proteins from amino acids is an example of *anabolism*. See chapter 30 'Biochemistry'.

SUMMARY

- Amino acids contain both a primary amine group and a carboxylic acid group.
- Naturally occurring amino acids are usually α-amino acids of general formula $RCH(NH_2)COOH$.
- The carboxylic acid group protonates the amine group to form a zwitterion $^+NH_3CHRCO_2^-$.
- Two amino acid molecules condense with the elimination of water to form a dipeptide containing a peptide link —CONH—, i.e. a dipeptide is a substituted amide.
- Polypeptides contain more than 20 amino acids condensed together. Proteins are natural polypeptides.

PRACTICE

1 Explain the relationship between amino acids, polypeptides, and proteins.

2 Suggest a mechanism for the condensation reaction between two molecules of glycine (aminoethanoic acid) to form a dipeptide containing the peptide link —CONH—.

3 Give the structural formulae of the species you would expect to be present on dissolving solid glycine (aminoethanoic acid) in the following:

a Dilute hydrochloric acid
b Pure water
c Dilute aqueous sodium hydroxide.

4 Suggest approximate pH values for solutions of the following amino acids in pure water:

a Glycine (aminoethanoic acid)
b Aspartic acid (2-aminobutanedioic acid)
c Serine (2-amino-3-hydroxypropanoic acid)
d Lysine (2,6-diaminohexanoic acid).

28.6 Amino acids and optical activity

OBJECTIVES

- Plane-polarized light
- Chiral molecules
- Enantiomers

Stereoisomerism

Stereoisomers have the same structural formulae but differ in their spatial arrangement. There are two types: optical isomers and *E-Z* isomers, spread 22.6.

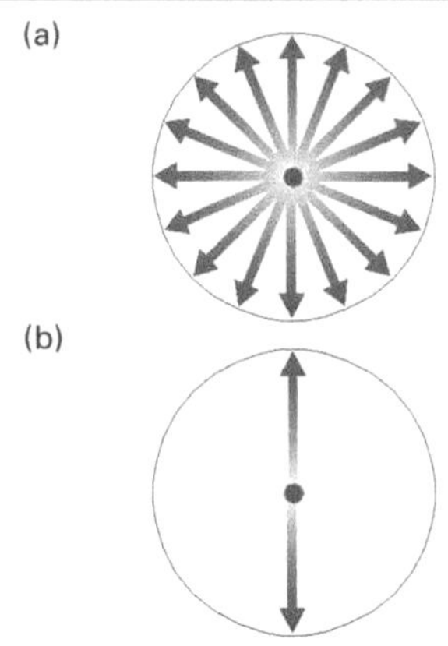

(a) A cross-sectional diagram of a ray of ordinary light travelling directly towards you. The electric field vibrates in all planes. (b) A ray of plane-polarized light. The electric field vibrates in one plane only.

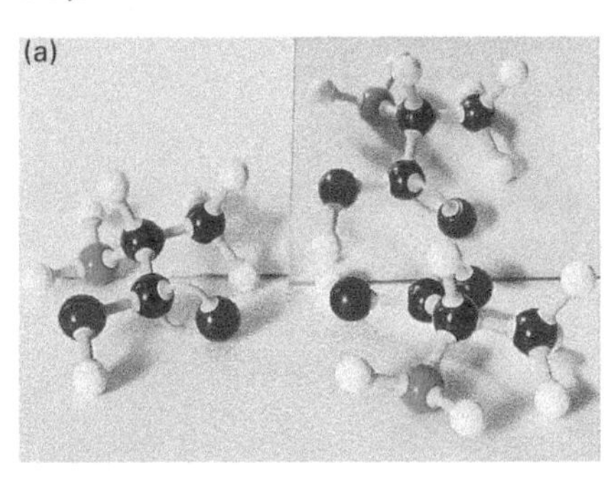

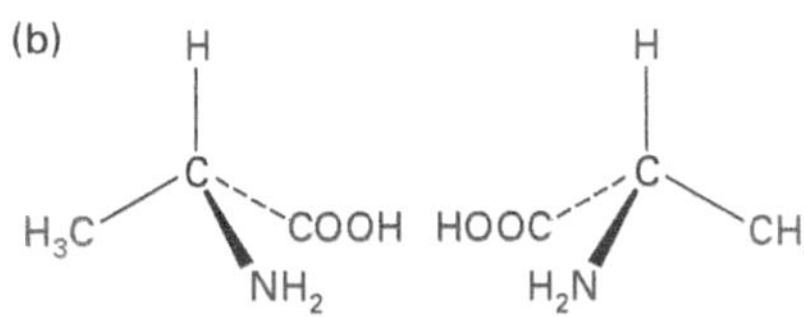

(a) Two non-superimposable mirror images (enantiomers). (b) The two enantiomers of the amino acid alanine (2-aminopropanoic acid) $H_2NCH(CH_3)COOH$. Under identical conditions, samples of the two enantiomers rotate the plane of plane-polarized light through the same angle, but in opposite directions.

Properties of enantiomers

Enantiomers have the same physical properties, with the exception that they rotate the plane of plane-polarized light in opposite directions. Their chemical properties are also identical, unless they are reacting with other chiral molecules.

The general formula of the naturally occurring α-amino acids is $RCH(NH_2)COOH$. With the exception of glycine (where R = H), all these substances contain a carbon atom that is attached to four different groups. These four groups are situated at the corners of a tetrahedron, and are bonded to a carbon atom placed at its centre. As you saw with the inorganic examples in chapter 20 'The transition metals', it is possible for such a compound to exist in two forms that are non-superimposable mirror images of each other. These two optically active isomers will rotate the plane of plane-polarized light in opposite directions to each other.

Polarized light

Light (a form of electromagnetic radiation) is a transverse wave motion, which means that its electric field vibrates at right angles to the direction of motion. In ordinary light, there is an infinite number of planes of vibration for the electric field. Passing this light through a Polaroid filter or certain types of prism results in **plane-polarized light**. The vibration is in one direction (plane) only, depending on how the filter is aligned. Note that laser light is plane-polarized.

The asymmetric carbon atom

A carbon atom is said to be an **asymmetric carbon atom** if it is joined to four different atoms or groups. In the terminology introduced in spread 20.9, such a carbon atom is referred to as a '**chiral centre**' or '**stereogenic centre**'. A compound containing one chiral centre can exist in two forms that are non-superimposable mirror images of each other. These two isomeric forms are called **optical isomers** or **enantiomers**.

Demonstrating optical activity

Optical activity is measured by an instrument called a **polarimeter**. Rays of light of a single frequency (monochromatic light) are polarized as they pass through a polarizing filter. A similar filter is fitted to the eyepiece. This filter is rotated until the light is extinguished. The polarizers are then said to be 'crossed'; their polarizing effects are at right angles to each other. A solution of the substance under investigation is then placed between the two polarizers. If the solution is optically active, then the plane of polarization will be rotated, and light will be seen through the eyepiece. Rotating the eyepiece polarizer can again extinguish the light.

If the eyepiece polarizer must be rotated clockwise – to the right – when viewed looking towards the light source, then the substance is said to be '**dextrorotatory**'; a '**laevorotatory**' substance requires rotation anticlockwise – to the left. The two isomers are known as the + (plus) form and the – (minus) form. The extent of rotation depends on the concentration of the solution, its temperature, the frequency of the light, and the distance the beam travels through the sample. Standard conditions are usually 1 g cm^{-3}, 25 °C, sodium light (589 nm), and 10 cm respectively.

The tube is filled with a saturated solution of cane sugar in water. When plane-polarized light is shone through the tube, bands of light of different colours can be seen along the length of the tube: blue, yellow-green, and red-brown can be seen.

The significance of optical activity

Reaction mechanisms

A reaction such as

$$CH_3\text{—}C(=O)\text{—}H + HCN \longrightarrow CH_3\text{—}C(OH)(CN)\text{—}H$$

results in a product that has an asymmetric carbon atom. However, the reaction always gives a mixture that contains equal amounts of the two enantiomers (a racemic mixture). The cyanide group can attack the planar carbonyl carbon atom from either above or below the molecule. Hydrolysis in acid produces racemic 2-hydroxypropanoic acid (lactic acid); see spread 29.1. An optically active product would indicate attack from one side only. Lactic acid in Nature is almost exclusively one enantiomer.

Racemic mixtures

A solution that contains equal amounts (in moles) of the two enantiomers does not show optical activity. It is called a **racemic mixture**. The rotating effect of one form exactly cancels out the effect of the other.

Frances Kelsey. The drug 'thalidomide' was introduced in Europe in the late 1950s to relieve morning sickness during pregnancy. Evidence slowly emerged that it was teratogenic, causing malformations of the foetus. Children were born with greatly shortened and distorted limbs. The thalidomide molecule is chiral. One enantiomer is teratogenic; the other alleviates morning sickness and is not teratogenic. The drug sold in Europe was a racemic mixture. Frances Kelsey was a pharmacologist working for the US Federal Drug Administration. Every drug used in the USA has to pass the scrutiny of the FDA. Kelsey was concerned about initial reports of nervous disorders in animals, and hesitated to grant a licence for thalidomide, instead requesting further evidence. Before the manufacturers could respond, the human disaster in Europe had unfolded. US mothers who gave birth in the early 1960s have reason to give thanks for Kelsey's diligence. President Kennedy presented her with the Distinguished Federal Civilian Service Award in 1962. Dr Kelsey, a Canadian citizen, died in 2015 at the age of 101, less than 24 hours after being inducted as a member of the Order of Canada.

Chiral synthesis

There are two ways in which a preponderance of one of the two enantiomers can be formed. The first way starts with an already chiral species (**chiral pool synthesis**), for example a natural amino acid. If each subsequent reaction avoids the chiral centre, the final product will itself be chiral. Enzymes are chiral and so often the products of their reactions are also chiral, as for the biosynthesis of lactic acid.

The second way is to use a chiral catalyst (**asymmetric catalysis**). The first example, discovered by William Knowles in 1968, involved a chiral rhodium complex as a homogeneous catalyst, spread 15.11, for hydrogenation. This eventually led to the industrial synthesis of L-DOPA, a drug used clinically for management of Parkinson's disease. Ryoji Noyori introduced a chiral ruthenium complex hydrogenation catalyst, which is used industrially to manufacture the non-steroidal anti-inflammatory drug Naproxen and beta-lactam antibiotics. Barry Sharpless found a chiral titanium catalyst for the formation of epoxides, spread 25.6, so oxidation as well as reduction reactions can be used to introduce chirality. Knowles, Noyori, and Sharpless were awarded the 2001 Nobel Prize.

Living systems

With the exception of glycine, all amino acids obtained from proteins are optically active because the α-carbon atom is a chiral centre. Naturally occurring amino acids all have the amine group, the carboxylic acid group, the hydrogen atom, and the R group arranged in space in the same way (see chapter 30 'Biochemistry'). Organisms cannot metabolize the isomeric substances that are arranged differently at the chiral centre.

When chiral molecules of different substances react together, the situation can be likened to a hand trying to fit into a pair of gloves. The right hand fits snugly into a right glove, but not so comfortably into a left glove. The chiral nature of amino acids (and of many other natural compounds) can give rise to some surprising physiological effects.

- The odours of spearmint and caraway seed are caused by the two enantiomers of carvone. Although infrared and ultraviolet spectrometers (see chapter 31 'Techniques of preparation, ...') cannot distinguish the difference, our noses can!
- One enantiomer of the amino acid asparagine tastes bitter whereas the other tastes sweet.
- Only one of the enantiomers of LSD causes hallucinations.
- One enantiomer of morphine relieves pain strongly and is not greatly addictive, whereas the other is strongly addictive and not nearly so effective at relieving pain.

SUMMARY

- An asymmetric carbon atom is bonded to four different atoms or groups, forming a chiral centre.
- Compounds with an asymmetric carbon atom exist as isomeric pairs called enantiomers; they are non-superimposable mirror images of each other.
- A (+) enantiomer rotates the plane of plane-polarized light clockwise; a (–) enantiomer rotates the light anticlockwise.
- All naturally occurring amino acids (except glycine) are optically active.

28.7

SOME ADVANCED IDEAS

OBJECTIVES

- Cahn–Ingold–Prelog nomenclature
- Sulpha drugs
- Phase transfer catalysis

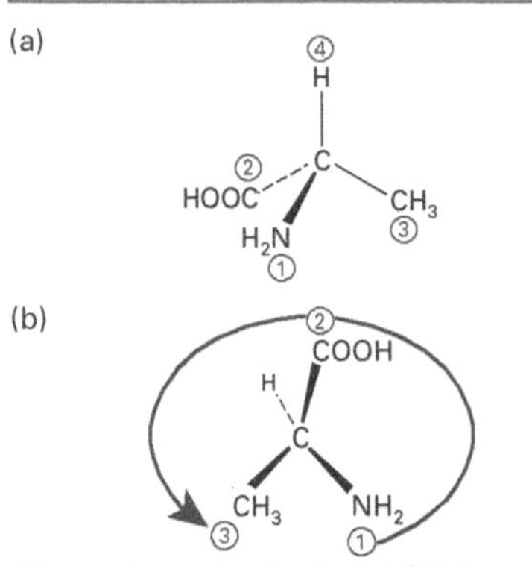

The amino acid alanine. (a) This enantiomer has the four substituents labelled by priority. (b) The hydrogen atom, which has the lowest priority, is drawn pointing into the plane of the paper. The movement from substituent 1 to 2 to 3 is anticlockwise and hence the enantiomer is labelled S*: (*S*)-alanine. This is the enantiomer that can be extracted from living organisms.*

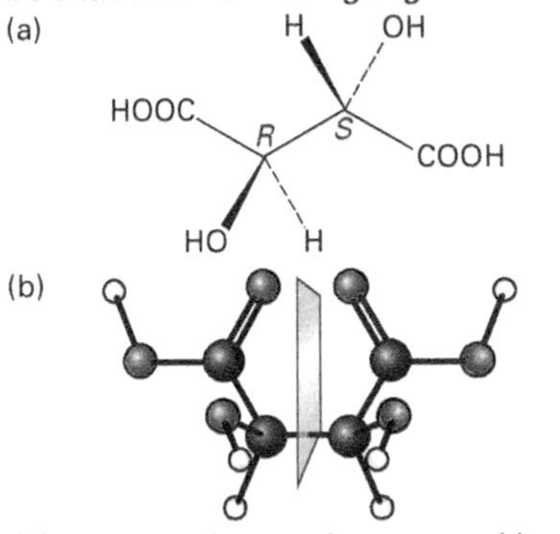

(a) The opposite rotations caused by the two identical chiral centres in tartaric acid cancel each other out: this meso *isomer does not show optical activity. (b)* Meso *tartaric acid has a plane of symmetry and so is not chiral.*

Optical activity, as we saw in the previous spread, is of great importance in living systems. A molecule with a single chiral centre can be labelled as dextrorotatory (+) or laevorotatory (–) on the basis of the direction of rotation of the plane of polarization. It is also helpful to identify the structure of the chiral molecule.

Cahn–Ingold–Prelog nomenclature

To label the two enantiomers of a molecule with a single chiral centre follow a set of rules devised by Robert Cahn, Christopher Ingold, and Vladimir Prelog:

1. Assign a **priority** number to each of the four substituents around the chiral centre; atoms with higher atomic number have higher priority. (This is the same procedure used for labelling *E* and *Z* isomers, spread 22.6.) If two atoms attached to the chiral centre are identical, look further along the two substituents to any attached atoms. For alanine (see previous spread), $H_2NCH(CH_3)COOH$, NH_2 has the highest priority, then COOH, then CH_3, and finally H.
2. Redraw the molecule with the lowest priority substituent pointing into the plane of the paper.
3. Imagine moving from substituent 1 to 2 to 3. If that movement is clockwise, use the label *R* (Latin *rectus*). (Imagine turning a steering wheel clockwise; you would be turning Right.) If that movement is anticlockwise, use the label *S* (Latin *sinister*).

Meso isomers

Some molecules have two (or more) chiral centres. A molecule with two chiral centres would be expected to have four enantiomers and this is the case unless the two chiral centres are identical. In this special case there are only *three* enantiomers as one isomer (called the *meso* isomer) does not show optical activity because the rotation caused by one centre is exactly cancelled by the opposite rotation of the other centre. An important example occurs in tartaric acid (2,3-dihydroxybutanedioic acid), a common ingredient in fizzy drinks.

The synthesis of a sulpha drug

The industrial manufacture of a sulpha drug requires a four-stage synthesis, shown in the box below. The first (and last) steps are required because the amine group is generally too reactive.

The steps in the synthesis of sulphanilamide.

(a) The amine group is **protected** in the first step by reaction with ethanoyl chloride (or ethanoic anhydride) to form a substituted amide, which is much less reactive, e.g. $RNH_2 + CH_3COCl \rightarrow RNHCOCH_3 + HCl$

Phenylamine $\xrightarrow{(CH_3CO)_2O}$ $C_6H_5NHCOCH_3$ + CH_3COOH

(b) The second step is similar to sulphonation of the benzene ring (see chapter 23 'Arenes').

$C_6H_5NHCOCH_3 \xrightarrow[80\ °C]{HOSO_2Cl} ClO_2SC_6H_4NHCOCH_3 + H_2O$

(c) In the third step, nucleophilic substitution replaces a halogen atom by the nitrogen atom of an amine group.

$ClO_2SC_6H_4NHCOCH_3 \xrightarrow{NH_3} H_2NO_2SC_6H_4NHCOCH_3 + HCl$

(d) The protecting CH_3CO— group is finally removed by acidic hydrolysis, to give sulphanilamide, with the structure:

$H_2N{-}C_6H_4{-}SO_2{-}NH_2$

$H_2NO_2SC_6H_4NHCOCH_3 \xrightarrow{(1)\ dil.\ HCl,\ heat\ (2)\ OH^-}$ Sulphanilamide

The synthesis of sulphanilamide.

The molecule sulphanilamide (4-aminobenzenesulphonamide) itself is too toxic for widespread use. The modern group of sulpha drugs are derivatives of this substance, and have a common structure that involves the replacement of one hydrogen in the sulphonamide group $—SO_2NH_2$ to give $—SO_2NHR$, as in sulphapyridine where R is [pyridyl ring structure]

Phase transfer catalysis

An increasingly common strategy for encouraging reactions to occur between organic compounds and ionic compounds involves a process called phase transfer catalysis. Immiscible liquids are mixed, carrying the inorganic nucleophile from the aqueous phase into the organic phase. For example, stirring a mixture of 1-chlorooctane with aqueous sodium cyanide for several days gives a negligible yield of the nitrile. Adding a small quantity of a quaternary ammonium salt (which contains the R_4N^+ ion) causes the mixture to produce a yield of more than 90% of the nitrile within 2 hours, according to the equation:

$$R—Cl + Na^+CN^- \rightarrow R—CN + Na^+Cl^-$$

Without the quaternary ammonium salt, the sodium ion of the sodium cyanide is hydrated and the cyanide ion is forced to remain in the aqueous layer to preserve electrical neutrality. In the presence of quaternary ammonium ions, the rigorous demarcation between the two liquid phases is removed. The quaternary ammonium ions can migrate from the inorganic layer (the aqueous phase) into the organic layer, partly because they are more weakly hydrated than sodium ions, and partly because the organic groups are attracted by the dispersion forces of the organic solvent. To maintain electrical neutrality within each phase, each quaternary ammonium ion takes a cyanide ion with it and so allows the reactants (cyanide and halogenoalkane) to meet in the organic phase.

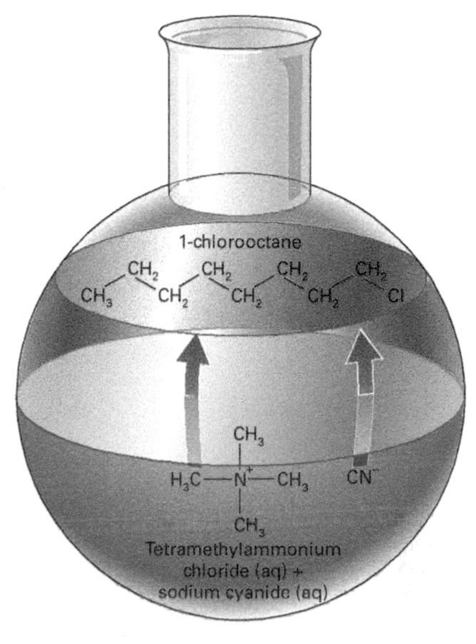

1-Chlorooctane and aqueous tetramethylammonium chloride with sodium cyanide form two immiscible layers. The ion pair tetramethylammonium cyanide can migrate into the organic layer; this is the basis of phase transfer catalysis in this reaction.

SUMMARY

- Amine groups may be protected during multi-stage syntheses by reaction with an acyl halide to form a substituted amide (and hydrolysed back to the amine at the completion of the synthesis).
- Phase transfer catalysis uses a quaternary ammonium salt to carry a nucleophile from the aqueous phase into the organic phase, where reaction takes place.

Chapter 28 Reactions summary

- Amine as a base:
  $CH_3CH_2NH_2 + H_3O^+ \rightarrow CH_3CH_2NH_3^+ + H_2O$
- Regeneration of free amine:
  $CH_3CH_2NH_3^+Cl^-(aq) + NaOH(aq) \rightarrow$
  $CH_3CH_2NH_2(aq) + NaCl(aq)$
- Amine as ligand in complex formation:
  $6CH_3CH_2NH_2(aq) + Ni^{2+}(aq) \rightarrow [Ni(NH_2CH_2CH_3)_6]^{2+}(aq)$
- Successive formation of primary, secondary, and tertiary amine, and quaternary ammonium salt:
  $RCl + NH_3 \rightarrow RNH_2 + HCl$
  $RCl + RNH_2 \rightarrow R_2NH + HCl$
  $RCl + R_2NH \rightarrow R_3N + HCl$
  $RCl + R_3N \rightarrow R_4N^+Cl^-$
- Amine from the reduction of a nitrile by $LiAlH_4$:
  $CH_3C{\equiv}N \xrightarrow{[H]} CH_3CH_2NH_2$
- Aromatic amine from the reduction of a nitro compound:
  $C_6H_5NO_2(l) \xrightarrow{Sn/HCl} C_6H_5NH_3^+Cl^-(aq)$
  $C_6H_5NH_3^+Cl^- + NaOH \rightarrow C_6H_5NH_2 + H_2O + NaCl$
- Phenylamine reacts with ethanoyl chloride:
  $C_6H_5NH_2 + CH_3COCl \rightarrow C_6H_5NHCOCH_3 + HCl$
- Formation of 2,4,6-tribromophenylamine from phenylamine:
  $C_6H_5NH_2 + 3Br_2 \rightarrow Br_3C_6H_2NH_2 + 3HBr$
- Diazotization of phenylamine:
  $C_6H_5NH_2 + NaNO_2 + 2HCl \rightarrow C_6H_5N_2^+Cl^- + NaCl + 2H_2O$
- Substitution reactions with the benzenediazonium ion:
  $C_6H_5N_2^+Cl^-(aq) + H_2O(l) \rightarrow C_6H_5OH(aq) + N_2(g) + HCl(aq)$
  $C_6H_5N_2^+Cl^- \xrightarrow{HCl/CuCl} C_6H_5Cl + N_2$
- Coupling reactions with the benzenediazonium ion:
  $C_6H_5N_2^+Cl^- + C_6H_5OH \rightarrow HOC_6H_4N{=}NC_6H_5 + HCl$
- Amino acid zwitterion formation:
  $NH_2CH_2COOH \rightarrow {}^+NH_3CH_2CO_2^-$
- Formation of peptide link between amino acids:
  $2NH_2CHRCOOH \rightarrow NH_2CHR—CONH—CHRCOOH + H_2O$

PRACTICE EXAM QUESTIONS

1 **a** Explain why ethylamine is a Brønsted–Lowry base. [2]

b Why is phenylamine a weaker base than ethylamine? [2]

c Ethylamine can be prepared from the reaction between bromoethane and ammonia.

i Name the type of reaction taking place and outline a mechanism.

ii Give the structures of **three** other organic substitution products which can be obtained from the reaction between bromoethane and ammonia. [8]

d Write an equation for the conversion of ethanenitrile into ethylamine and give one reason why this method of synthesis is superior to that in part **c**. [2]

2 (Phenylmethyl)amine, $C_6H_5CH_2NH_2$, can be prepared from (bromomethyl)benzene, $C_6H_5CH_2Br$, and also from benzenecarbonitrile, C_6H_5CN.

a **i** Write an equation for the conversion of (bromomethyl)benzene into (phenylmethyl)amine. Name the type of reaction taking place and explain why a low yield of product is obtained.

ii Name the type of reaction involved in the conversion of benzenecarbonitrile into (phenylmethyl)amine. Write an equation for this reaction and suggest a suitable reagent or a combination of reagent and catalyst. Explain why this method of preparation gives a high yield of product. [4]

b State which of the two amines, (phenylmethyl)amine and phenylamine, $C_6H_5NH_2$, is the weaker base, and explain your choice. [3]

3 Consider the compound **A**

$C_6H_5-CH(OH)-CH(CH_3)-NH_2$

which is related to the hormone adrenaline.

a Draw the structures of the organic product(s) which you would expect from the reaction of **A** with

i phosphorus pentachloride; [1]

ii dilute hydrochloric acid; [1]

iii ethanoyl chloride; [2]

iv hot alkaline potassium manganate(VII); [1]

v hot concentrated sulphuric acid. [1]

b Suppose that you have to purify a sample of **A** by recrystallization from trichloromethane. This solvent is toxic by inhalation and skin absorption but is not flammable.

i What safety precautions would you take in using this solvent? [2]

ii Describe in detail how you would recrystallize a sample of about 5 g of **A**. [5]

iii What simple test would you use to determine the purity of your recrystallized material? [2]

4 **a** Explain the term *polymerization*. [1]

b Polymers found in natural materials can be formed by the reaction between amino acids.

i Draw the graphical formula of the product formed when two molecules of alanine, $CH_3CH(NH_2)COOH$, react together. [1]

ii Give the name of the important linkage formed and draw a ring round it on the formula drawn in **b i**. [2]

iii Give the name of the type of naturally occurring polymer containing this linkage. [1]

c Poly(ethene) is an example of a synthetic polymer. It is manufactured in two main forms, low density poly(ethene) and high density poly(ethene). [1]

i Write an equation to represent the polymerization of ethene. [1]

ii What is the main structural difference between the polymer chains in the two main forms of poly(ethene)? Explain how this difference affects the densities of the polymers. [3]

iii Give **one** further physical property that is affected by the structural difference given in **c ii**. [1]

iv Low density poly(ethene) is manufactured via a free radical mechanism. Draw a graphical formula to represent the free radical formed between a free radical, **R•**, and a molecule of ethene in the reaction. [1]

v What type of catalyst is used in the manufacture of high density poly(ethene)? [1]

d Poly(ethene) is non-biodegradable.

i Explain the term *non-biodegradable*. [1]

ii Give **one** environmental benefit of using biodegradable plastics. [1]

iii Developing biodegradable plastics involves compromise. Suggest **one** factor that requires careful consideration and explain your choice. [2]

5 Consider the following reaction scheme:

$$\underset{\mathbf{A}}{C_2H_5COCl} \xrightarrow{NH_3} \underset{\mathbf{B}}{C_3H_7ON} \longrightarrow \underset{\mathbf{E}}{C_2H_7N}$$

$$\underset{\mathbf{B}}{C_3H_7ON} \xrightarrow{P_4O_{10}} \underset{\mathbf{C}}{C_3H_5N} \longrightarrow \underset{\mathbf{D}}{C_3H_9N}$$

Compounds **D** and **E** have the same functional group.

a Identify, using structural formulae, the compounds **B** and **C** and the functional group in both **D** and **E**. [3]

b i Give the reagents and conditions necessary for the conversion of compound **C** into compound **D**. [2]

ii Give the reagents and conditions necessary for the conversion of compound **B** into compound **E**. [2]

c i Give an equation to represent the reaction that takes place when compound **C** is boiled with dilute hydrochloric acid. [2]

ii State the type of reaction taking place in **i**. [1]

d i What structural feature of the functional group present in both **D** and **E** enables them each to react with dilute hydrochloric acid. [1]

ii Using the structural formulae, give an equation to represent the reaction of compound **D** with dilute hydrochloric acid. [2]

e A compound C_6H_7N having the same functional group as **D** and **E** reacts with nitrous acid in the presence of hydrochloric acid. The resulting solution reacts with alkaline 2-naphthol to give a red precipitate.

i Give the equation representing the reaction of C_6H_7N with nitrous acid, using structural formulae. [2]

ii Show the structure of the product which is formed with 2-naphthol. [1]

iii What is the significance of compounds of this type? [1]

6 a Give the structural formulae, showing all covalent bonds, for all the isomers of C_4H_9Br, and name them. [3]

b The compounds in **a** all react on heating with aqueous sodium hydroxide by a nucleophilic substitution reaction. For one of these isomers, the reaction at a given temperature was found to be first order with respect to the organic molecule and first order overall.

i Explain what is meant by the term *nucleophile*.

ii Identify the nucleophile in this reaction.

iii Write the rate expression for this reaction.

iv Give the mechanism for this reaction using the isomer most likely to react in this way. Explain briefly your choice of isomer. [9]

c i Identify the isomer in **a** that is chiral and explain briefly why it is chiral.

ii Assuming that one of the optical isomers of the compound given in **c i** reacts with sodium hydroxide by the same mechanism as in **b iv**, explain how the nature of the intermediate results in a product which is not optically active. [4]

7 *N*-Phenylethanamide can be prepared from benzene in three steps:

benzene —Step 1→ $C_6H_5NO_2$ —Step 2→ $C_6H_5NH_2$ —Step 3→ $C_6H_5NHCOCH_3$

a Give the reagents required to carry out Step 1 and write an equation for the formation of the reactive inorganic species present. Name and outline the mechanism for the reaction between this species and benzene.

b Name the type of reaction taking place in Step 2 and suggest a suitable reagent or combination of reagents.

c Write an equation for the reaction occurring in Step 3. Name and outline the mechanism for this reaction. [7]

8 a Define the following terms, and illustrate them by drawing graphical formulae for the stated examples.

i Structural isomerism, showing the appropriate isomers of C_4H_{10}. [4]

ii Geometrical isomerism, showing the appropriate isomers of C_4H_8. [4]

iii Optical isomerism, showing the appropriate isomers of $CH_3CH(OH)COOH$. [4]

b The figure below shows the chain structure of a dipeptide made from alanine and cysteine.

$H_2N-CH(CH_3)-C(=O)-NH-CH(CH_2SH)-CHO$

i Copy the figure and mark any chiral centres with an asterisk (*). [2]

ii Draw the graphical formulae of the two amino acids alanine and cysteine. [2]

iii Describe how you would show the optical activity of an aqueous solution of one of the optical isomers of alanine. [3]

29 Organic synthesis: changing the carbon skeleton

The aim of organic synthesis is to make a required product from readily available precursors. Most simply described, organic synthesis starts with a chosen substance, and adds or removes carbon atoms and associated functional groups. A complete synthesis may involve many steps, and so the final yield may be very small. For example, a four-step synthesis with a reasonable 30% conversion efficiency per step will produce less than 1 g of product from 100 g of starting material. Such a synthetic route may look feasible on paper, but the economic considerations of the pharmaceutical or dyestuffs industry would make it a complete 'non-starter'. However, the task at this stage of your chemical understanding is simply to suggest synthesis methods based on the most straightforward route possible.

OBJECTIVES

- Friedel–Crafts acylation
- Reduction by $NaBH_4$
- Bromination
- CN^- as nucleophile: chain lengthening
- Nitrile hydrolysis to carboxylic acid

The synthesis of ibuprofen

Ibuprofen is a hugely profitable and useful medicine. It is classified in pharmacology as a non-steroidal anti-inflammatory drug (or NSAID in short). Ibuprofen is mostly used to treat pain resulting from inflammation, e.g. rheumatism, pulled muscles, and back-ache. The side-effects of ibuprofen are far less significant than those of many other NSAID compounds. For this reason, ibuprofen along with aspirin and diclofenac (Voltarol) are the only NSAIDs licensed in the UK for non-prescription 'over-the-counter' sale. The synthesis of ibuprofen is straightforward, and you are already familiar with most of the reactions involved.

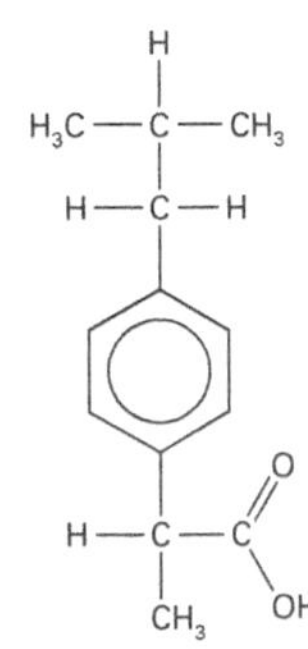

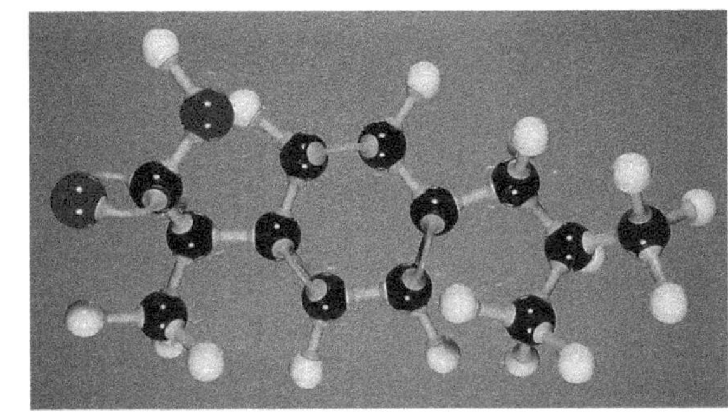

More than 30 types of preparation containing ibuprofen are currently available from pharmacists. They range from gels and sprays for muscle sprains to tablets for the relief of period pains and head-ache. Ibuprofen is a carboxylic acid with a relatively simple molecular structure.

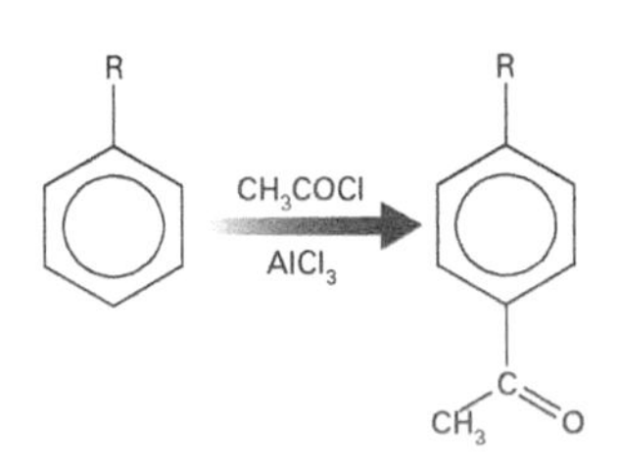

Step 1: Introducing an ethanoyl group in the 4 position to the R group on the benzene ring.

Step 1 Friedel–Crafts acylation

The starting material for the synthesis of ibuprofen is (2-methylpropyl)benzene $(CH_3)_2CHCH_2C_6H_5$. The $(CH_3)_2CHCH_2$— group remains unchanged throughout the synthesis and will therefore be represented by the symbol R– throughout the five steps.

The first step involves Friedel–Crafts acylation (see chapter 23 'Arenes') to introduce an ethanoyl CH_3CO— group in the 4 position on the benzene ring:

$$RC_6H_5 + CH_3COCl \xrightarrow{AlCl_3} RC_6H_4COCH_3 + HCl$$

The acyl group deactivates the benzene ring to further substitution after one group has joined.

R

R

$NaBH_4$

H—C—OH

CH_3

Step 2: Reducing the ketone group to a secondary alcohol.

Step 2 Reduction

Reduction of the carbonyl group ⊃C=O by sodium tetrahydridoborate produces a secondary alcohol group ⊃CHOH (see chapter 26 'Aldehydes and ketones'):

$$RC_6H_4COCH_3 \xrightarrow{NaBH_4} RC_6H_4CH(CH_3)OH$$

Step 3 Bromination

The hydroxyl group can be brominated by reaction with phosphorus tribromide PBr_3 in a similar way that PCl_3 reacts (see chapter 25 'Alcohols'). PBr_3 is produced *in situ* by the addition of red phosphorus and bromine to the reaction mixture:

$$RC_6H_4CH(CH_3)OH \xrightarrow{P(red)/Br_2} RC_6H_4CH(CH_3)Br$$

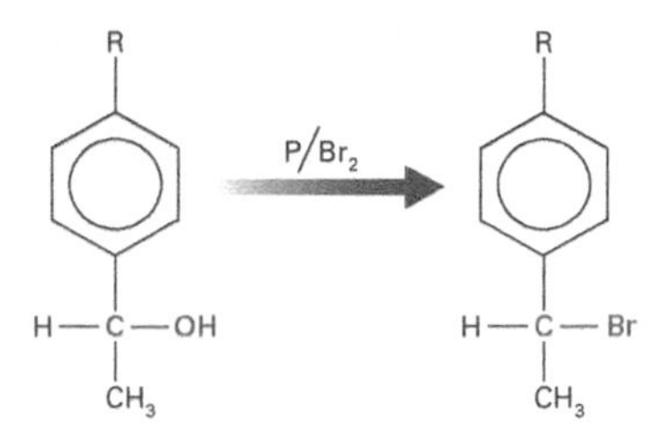

Step 3: Bromination of the alcohol group.

Step 4 Chain lengthening with the cyanide ion

The aim of the *overall* synthesis is to join a carboxylic acid group —COOH to the carbon atom that was attached in the first step to the 4 position on the benzene ring. This fourth step introduces an extra carbon atom at that point in the molecule. Cyanide ion CN^- substitutes for Br^- in a nucleophilic substitution reaction (see chapter 24 'Organic halogeno compounds'). The reaction is carried out using potassium cyanide in ethanol. Ethanol is a polar organic solvent and so will dissolve both the organic halogenoalkane and the ionic cyanide. The mixture is then heated under reflux; the product is a nitrile:

$$RC_6H_4CH(CH_3)Br \xrightarrow{KCN/ethanol} RC_6H_4CH(CH_3)CN$$

Note that this reaction is suitable for use with phase transfer catalysis techniques (see the previous chapter).

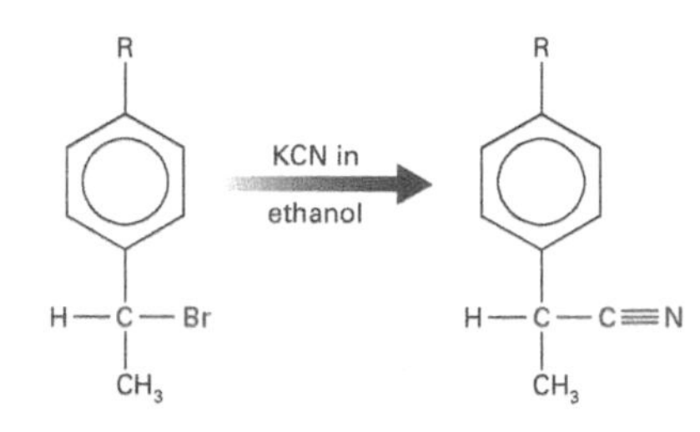

Step 4: Nucleophilic substitution of bromine by cyanide ion.

Step 5 Producing the carboxylic acid group

The most important reaction of nitriles is hydrolysis to form a carboxylic acid. The reaction is carried out by refluxing the nitrile with aqueous acid. Hydrolysis causes the nitrile group to lose nitrogen in the form of ammonia. In acidic solution, the ammonia is protonated to form the ammonium ion. For example:

$$CH_3CH_2CN + H_3O^+ + H_2O \rightarrow CH_3CH_2COOH + NH_4^+$$

(Note that, when hydrolysis by *base* is carried out, the carboxylic acid reacts to form a salt and ammonia *gas* is released.)

This reaction forms the final step of this synthesis, in which the ibuprofen molecule is produced:

$$RC_6H_4CH(CH_3)CN \xrightarrow{H^+(aq)} RC_6H_4CH(CH_3)COOH$$

Each batch of ibuprofen then goes through many stages of purification, recrystallization, and testing before being incorporated into the final packaged product ready for use. Note that the carbon atom next to the benzene ring is a chiral centre; see spread 28.6.

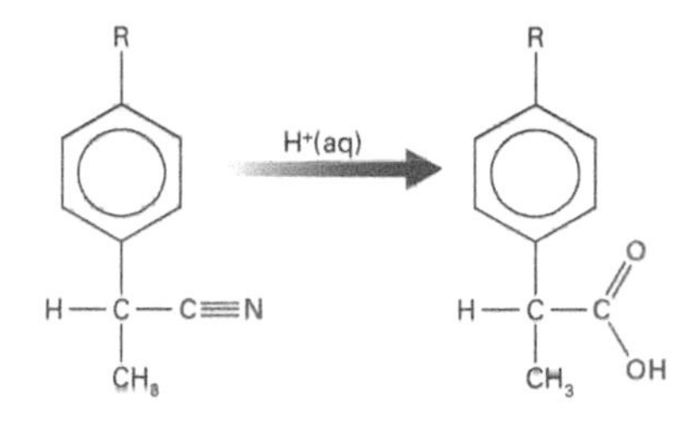

Step 5: Acid hydrolysis of the cyanide group to produce a carboxylic acid.

Some tropical millipedes of the genus Apheloria *have a defence mechanism that relies on the production of hydrogen cyanide HCN from the hydrolysis of the hydroxynitrile $C_6H_5CH(OH)CN$. A millipede can produce enough hydrogen cyanide to kill a small mouse. Note that the laboratory preparation of this hydroxynitrile would involve the addition of HCN to the aldehyde C_6H_5CHO. Millipedes reverse this reaction using a completely different enzyme-mediated biochemical pathway.*

Lactic acid

Addition of hydrogen cyanide HCN to ethanal CH_3CHO followed by hydrolysis in acid produces 2-hydroxypropanoic acid (lactic acid) $CH_3CH(OH)COOH$.

Chiral drugs

The (+) enantiomer of ibuprofen is a much more effective painkiller, so pharmaceutical companies actively search for ways of making this enantiomer preferentially.

Nucleophilic carbon

The cyanide ion is a nucleophile that contains a carbon atom bearing a full negative charge. Cyanide is therefore a useful reagent for adding one carbon atom to a carbon chain. How can more than one carbon atom be added? The answer is to produce a nucleophile consisting of a carbon chain in which one of the carbon atoms bears a negative charge. Grignard reagents provide such nucleophiles; see the following spread.

SUMMARY

- Acidic hydrolysis of a nitrile produces a carboxylic acid:

$$RCN \xrightarrow{H^+(aq)} RCOOH$$

29.2

OBJECTIVES

- Carbon as a nucleophilic centre
- The Grignard reaction
- Organometallics

Grignard reagents and other organometallics

The use of the cyanide ion to form nitriles adds just *one* extra carbon atom to a carbon skeleton. **Grignard reagents** are compounds that have the general formula RMgX, where X is a halogen. They are classed as **organometallic compounds** because they contain a metal (magnesium) bonded to a carbon atom in an alkyl group. During the Grignard reaction, the organometallic reagent adds across the carbonyl group of aldehydes or ketones. The R group of the Grignard reagent attaches to the carbonyl carbon, thus joining together the two carbon skeletons. There are some limitations to the types of carbon skeletons that may be joined using Grignard reagents. However, Grignard reagents are a very powerful tool for the synthesis of more complex organic molecules. They can add more than one carbon atom to the skeleton.

(a)

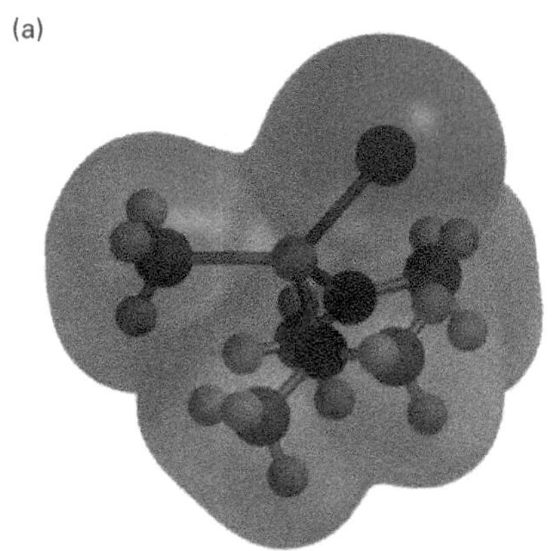

(b)

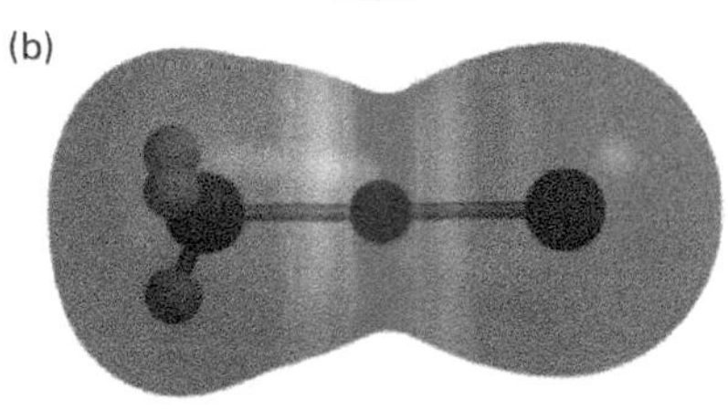

There is evidence to suggest that (a) Grignard reagents are chemically associated by coordinate bonding with the ether solvent in which they are prepared. However, the structure shown in (b) is adequate to explain the reagent's behaviour. Note the high partial positive charge on the magnesium atom shown by the blue colour.

Preparation of Grignard reagents

Grignard reagents are not very stable, and so they must be freshly prepared and used immediately. A Grignard reagent is made by dissolving a halogenoalkane in dry ethoxyethane (ether) and allowing it to stand over turnings of magnesium metal. A vigorous exothermic reaction takes place in which the magnesium disappears as the mixture turns cloudy and boils. The following equation illustrates the reaction for iodoethane:

$$CH_3CH_2I + Mg \xrightarrow{\text{ether}} \underset{\text{ethylmagnesium iodide}}{CH_3CH_2MgI}$$

Although the formula is written in the form RMgX, there is evidence that the actual structure is more complex. The bonds to the magnesium atom are highly polar covalent bonds. Grignard reagents may be thought of as forming a free **carbanion** R^-, an ion in which a carbon atom carries a full negative charge. However, in reality Grignard reagents do not dissociate fully to form $R^-[^+Mg\ I]$, and the situation is more complex than this.

The Grignard reaction

Grignard reagents include a carbon atom bonded to a significantly *less* electronegative atom. The C—Mg bond is very polar, the electron density being pulled towards the carbon, giving the carbon atom a partial *negative* charge δ–. Grignard reagents therefore contain a powerfully *nucleophilic* carbon atom which attacks δ+ carbon atoms on other molecules. The **Grignard reaction** specifically adds a Grignard reagent to the carbonyl group of an aldehyde or ketone. In the carbonyl group ⊃C=O, a carbon atom is joined to a *more* electronegative oxygen atom. This carbon atom has a partial positive charge δ+ and is liable to attack by incoming nucleophiles. The reaction happens in two steps: addition of the Grignard reagent, followed by hydrolysis in acid. Yields of these reactions can be up to 90% or even greater.

- With methanal, the product is a primary alcohol:

$$RMgI + HCHO \xrightarrow{H_2O} RCH_2OH + Mg(OH)I$$

- With other aldehydes, the product is a secondary alcohol:

$$RMgI + R'CHO \xrightarrow{H_2O} RR'CH(OH) + Mg(OH)I$$

- With ketones, the product is a tertiary alcohol:

$$RMgI + R'R''CO \xrightarrow{H_2O} RR'R''COH + Mg(OH)I$$

- With carbon dioxide, the product is a carboxylic acid:

$$RMgI + CO_2 \rightarrow RCOOMgI \xrightarrow{H_2O} RCOOH + Mg(OH)I$$

Notice how the reaction between the Grignard reagent and the organic compound reduces its functional group level (spread 21.6) by one (e.g., a carbon atom doubly bonded to oxygen in an aldehyde or ketone is at FGL 2, whereas a carbon atom singly bonded to oxygen in an alcohol is at FGL 1).

R'
R^-: → C=O
H
(a)
R'
R—C—O^-
H
(b) H^+ transfer from acid
R'
R—C—OH
H

(a) An addition reaction of a Grignard reagent. (b) Hydrolysis of the addition product.

Note that reaction between the Grignard reagent and CO_2 effectively converts a halogenoalkane into a carboxylic acid with one extra carbon atom in the chain. This conversion may also be achieved by reacting the halogenoalkane with potassium cyanide to form the nitrile, followed by acid hydrolysis to form the carboxylic acid (see the previous spread).

(a) $H_2C{=}O$ —(i) CH_3CH_2MgI (ii) H^+ from acid→ CH_3CH_2–CH_2–OH Propan-1-ol

(b) $CH_3(H)C{=}O$ —(i) CH_3MgI (ii) H^+ from acid→ CH_3–$C(CH_3)(H)$–OH Propan-2-ol

(c) $(CH_3)_2C{=}O$ —(i) CH_3MgI (ii) H^+ from acid→ CH_3–$C(CH_3)_2$–OH 2-Methylpropan-2-ol

(d) $O{=}C{=}O$ —(i) CH_3CH_2MgI (ii) H^+ from acid→ CH_3CH_2–C(=O)OH Propanoic acid

Examples of the addition reactions of Grignard reagents.

Other organometallic compounds

Two common organometallic compounds are tetraethyllead $(CH_3CH_2)_4Pb$ and triethylaluminium $(CH_3CH_2)_3Al$. Tetraethyllead was added to some grades of petrol (UK 'four star') to improve its octane rating. Its use has now been stopped, because of fears about the adverse effects of high levels of lead in the environment.

Triethylaluminium forms part of the Ziegler–Natta catalyst used in the industrial polymerization of substituted alkenes (see chapter 22 'Alkanes and alkenes'). It is made by reacting chloroethane with aluminium:

$$3CH_3CH_2Cl + 2Al \rightarrow (CH_3CH_2)_3Al + AlCl_3$$

SUMMARY

- Grignard reagents are a class of organometallic compounds with the general formula RMgX.
- Grignard reagents add across carbonyl groups: an organic product results from subsequent hydrolysis.
- RMgX + methanal → a primary alcohol.
- RMgX + other aldehydes → a secondary alcohol.
- RMgX + ketones → a tertiary alcohol.
- RMgX + CO_2 → a carboxylic acid.

Degree of chain extension

In all these reactions, the product has more carbon atoms than the original carbonyl compound. The number of carbon atoms added depends on the Grignard reagent used.

Victor Grignard

Victor Grignard was born in Cherbourg, the son of a sailmaker. He was 'a simple man with much common sense and a practical mind'. He described his reagent in his doctoral thesis of 1901.

His technique caught on quickly. In 1908, 500 papers using Grignard reagents were published. The Nobel Prize was awarded to him in 1912.

Organolithium compounds

Organolithium compounds react in similar ways to Grignard reagents, spread 16.8. The oral contraceptive ethynyloestradiol is made from the female sex hormone oestrone, a ketone, by reaction with ethynyllithium HC≡CLi and H^+ from acid. The reaction converts C=O into C(OH)C≡CH.

Organometallic complexes

Metals can form complexes that contain metal-carbon bonds, most notably in **metal carbonyls**. A simple rule for rationalizing their formulae is the **eighteen-electron** rule, namely that the number of valence electrons in the metal together with two electrons per carbonyl group should add up to 18. Thus iron carbonyl has the formula $Fe(CO)_5$: $8 + 2 \times 5 = 18$. Nickel carbonyl, $Ni(CO)_4$, with $10 + 2 \times 4 = 18$ electrons, is used in the Mond process for the extraction of nickel. See spread 27.1 for a novel carbonyl complex catalyst.

More exotic complexes exist such as **ferrocene**, which is the compound $Fe(C_5H_5)_2$, where the iron is sandwiched between two cyclopentadienyl ligands.

PRACTICE

1 Give the name and structural formula of the compound that results from reacting ethylmagnesium bromide with water.

2 Give the names and structural formulae of the compounds that result from hydrolysis following the addition of ethylmagnesium bromide to the following substances:

a Propanal
b Cyclohexanone
c Carbon dioxide
d Butanone.

3 Outline the synthesis of butan-2-ol from bromoethane, involving the use of standard inorganic reagents only.

EXTENSION

29.3

OBJECTIVES

- Conjugated dienes
- Dienophile
- 1,4-Cycloaddition

Otto Diels and Kurt Alder

The Diels–Alder reaction was developed by Otto Diels and Kurt Alder in 1928. It proved to be of such value in the synthesis of cyclic compounds that they were awarded the 1950 Nobel Prize in Chemistry.

Conjugated dienes

Conjugated dienes have their π bonds delocalized; see spread 23.1.

THE DIELS–ALDER REACTION

Two of the three most useful reactions for making carbon–carbon bonds have already been described: the Friedel–Crafts reaction (see chapter 23 'Arenes') and the Grignard reaction (see previous spread in this chapter). The third is called the Diels–Alder reaction. It has the added advantage of making *two* carbon–carbon bonds. The **Diels–Alder reaction** involves addition of a compound that has two double bonds, called a **diene**, to a reagent called a **dienophile**. A cyclic product is formed. For example, the reaction joins together the diene butadiene with the dienophile ethene to form the cyclic adduct cyclohexene. The diene must be a *conjugated* diene, which means that the two C=C double bonds are separated by exactly *one* C—C single bond.

1,4-Cycloaddition

The simplest example of a Diels–Alder addition is that between butadiene and ethene. (Butadiene is more correctly called buta-1,3-diene.) However, the yield is low (20%) and a high temperature is required. The reaction is referred to as a '1,4-cycloaddition', because carbon atoms 1 and 4 of the diene add across the double bond of the alkene to produce a cyclic product.

The 1,4-cycloaddition of butadiene to ethene to form cyclohexene.

The yield of this addition reaction is higher when the dienophile has electron-withdrawing groups attached, such as —COOH, —COR, —CN, etc. Notice that the double bond of the dienophile reverts to a single bond.

The mechanism of the Diels–Alder 1,4-cycloaddition reaction is as follows:

(a) The mechanism may be regarded as a migration of the π electrons from the three C=C double bonds. Note that the reaction does not depend on the creation of a charge centre on a carbon atom; it involves a concerted interaction of the orbitals in the reacting molecules.

(b) The electron-withdrawing group X on the dienophile increases the yield.

(c) Ball-and-stick models show the stereochemistry of the reaction.

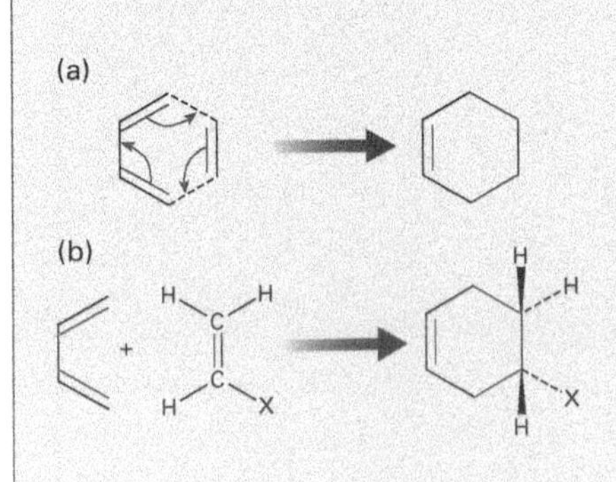

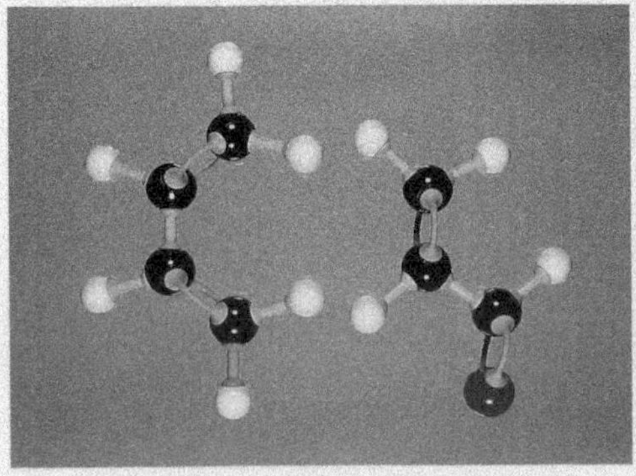

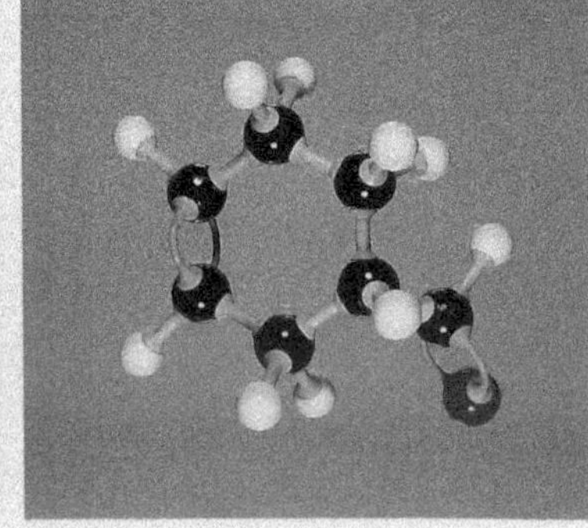

The Diels–Alder 1,4-cycloaddition reaction.

Diels–Alder reactions and natural products

Many naturally occurring or physiologically active substances contain ring structures. The Diels–Alder reaction is extremely useful during the synthesis of such molecules. For example, animals excrete substances called pheromones into the atmosphere in order to affect the social behaviour of other members of the same species that detect them. Female fruit flies produce a pheromone that is a sexual attractant. It is effective at concentrations as low as a few tens of molecules per cubic centimetre of air. Artificially produced pheromones synthesized using the Diels–Alder reaction are used to bait traps for insect pests.

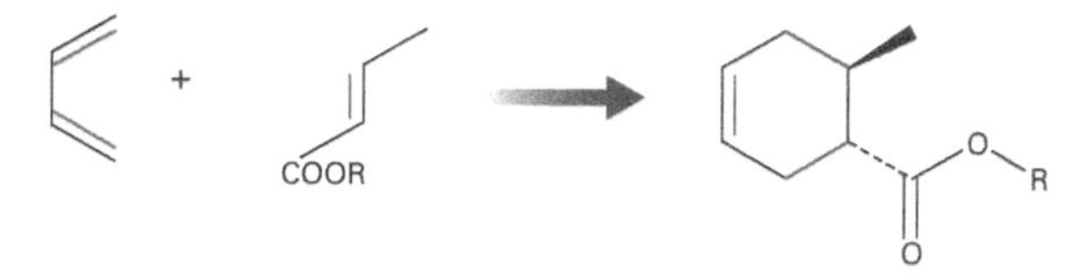

The synthesis of siglure, which mimics the pheromone secreted by the Mediterranean fruit fly. R = $-CH(CH_3)CH_2CH_3$

Robert Burns Woodward

R. B. Woodward is generally regarded as the greatest synthetic chemist of the twentieth century. Together with Roald Hoffmann, he established the mechanism of the Diels–Alder reaction, and was then able to use it for crucial steps of the syntheses of enormous biologically active molecules. In 1951, he produced an intermediate from which he finally prepared cortisone. Woodward synthesized the alkaloid reserpine (then used in the treatment of high blood pressure) in 1956. He achieved a total synthesis of chlorophyll in 1960, and of vitamin B_{12} in 1972.

E.J. Corey and Tamiflu

One of Woodward's colleagues at Harvard, E.J. Corey, responded to the prediction in 2005 by the World Health Organization (WHO) of the possibility of a flu *pandemic* by designing a novel synthesis of Tamiflu, an effective treatment for the disease. He avoided the starting material used in the patented Roche synthesis, starting instead from a Diels-Alder reaction on butadiene. (Corey chose not to patent his synthesis.) Thankfully the WHO's fears did not materialize.

A pheromone-baited trap for a bark beetle in Zurich, Switzerland.

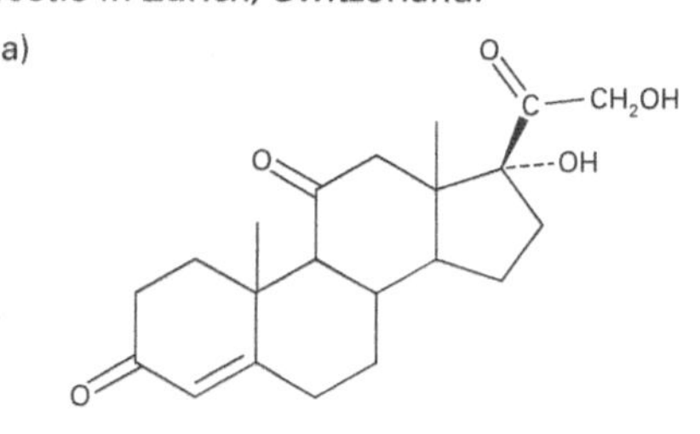

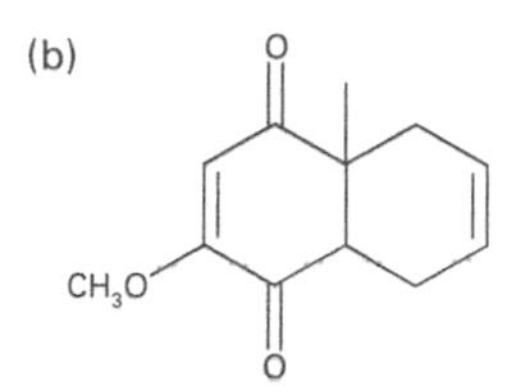

(a) The complex fused ring structure of the steroid hormone cortisone may be synthesized with the help of Diels–Alder cycloaddition reactions. (b) The intermediate synthesized by Woodward that enabled cortisone to be produced synthetically. Cortisone treatment reduces inflammation in severe allergic and rheumatic diseases such as asthma and rheumatoid arthritis.

This is the Harvard Mallinckrodt Chemistry Laboratory where Woodward and Corey worked. Woodward won the 1965 Nobel Prize in Chemistry (and would also have shared the 1981 Prize had he not died two years earlier): Corey won the 1990 Prize.

SUMMARY

- A conjugated diene has two C=C double bonds separated by one C—C single bond.
- Diels–Alder 1,4-cycloaddition adds a conjugated diene to a dienophile to form a ring structure consisting of six carbon atoms.
- An example of the Diels–Alder reaction is the addition of butadiene to ethene to form cyclohexene.
- Yields are increased if the dienophile has electron-withdrawing groups.
- The Diels–Alder reaction is of great use in producing ring structures for synthetic hormones.

PRACTICE

1 Give the structural formulae of the following dienes and dienophiles. In each case give the structural formulae of the Diels–Alder adduct formed by the reaction.

a Buta-1,3-diene; ethene
b Penta-1,3-diene; ethene
c Buta-1,3-diene; propenenitrile $CH_2{=}CH{-}CN$
d Penta-1,3-diene; propenenitrile
e Cyclohexa-1,3-diene; propenenitrile.

2 The structures shown below were formed by Diels–Alder cycloaddition. Give the structural formulae and names of the diene and dienophile from which each was formed.

(a) (b) CH_3 / CH_3 (c) CH_3

29.4

Revisiting polymerization – 1

OBJECTIVES

- Polymerization reactions
- Polymer properties
- Thermosoftening polymers
- Thermosetting polymers

The general principle of addition polymerization: the monomer phenylethene (styrene) $CH_2{=}CHC_6H_5$ polymerizes to give the polymer polystyrene.

An amine group $-NH_2$ on a diamine molecule condenses with a carboxylic acid group $-COOH$ on a diacid. A molecule of water is eliminated and the molecules become connected by a substituted amide group $-CONH-$.

'Tactic'

Giulio Natta's wife coined the term *tactic* to describe the arrangement of side groups along a polymer carbon chain:

- **Atactic** = a random arrangement
- **Isotactic** = side groups all along the same side
- **Syndiotactic** = side groups alternate

When the pattern is regular (iso- and syndiotactic), the polymer is described as being 'stereoregular'.

Polymers consist of long-chain molecules containing bonds formed by polymerization reactions. The two main classes of polymers are natural polymers and synthetic polymers. Natural polymers include the structural proteins and enzymes which are fundamental to the function of living systems. These polymers will be discussed in the following chapter. Synthetic polymers include addition polymers such as polythene and PVC, together with condensation polymers such as polyesters (e.g. Terylene) and polyamides (e.g. nylon). The chemical reactions responsible for the formation of addition polymers were introduced in chapter 22 'Alkanes and alkenes'; condensation polymerization was discussed in chapter 27 'Carboxylic acids and their derivatives'. This spread and the next one develop these ideas within the context of organic synthesis reactions and the design of molecules that show desired properties.

Revision: addition polymers

Typical addition polymers result when the π electrons in substituted alkene monomers form C—C σ bonds between the molecules. Thus a monomer $CH_2{=}CHG$ gives rise to a polymer of general formula $-\!\!\!+ CH_2CHG +\!\!\!-_n$: the structure within the brackets is called the **repeating unit** (or **repeat unit**). Examples include polythene (G = H), polypropylene (G = CH_3), polystyrene (G = C_6H_5), and PVC (G = Cl).

Revision: condensation polymers

Condensation copolymers result from monomer molecules that possess two functional groups. During the polymerization process, condensation reactions take place between groups on adjacent molecules. The principle is illustrated by the formation of a polyamide, in which a diacid molecule condenses with a diamine molecule to form a peptide link —CONH— between the molecules.

Controlling polymer properties

Two very important properties of polymers are tensile strength and softening temperature. Tensile strength determines the structural uses to which a material can be put; softening temperature affects the way in which it can be moulded or extruded. These properties are largely controlled by the forces of attraction between the polymer chains, which in turn are affected by the atoms or groups attached to the carbon skeleton and the way they are arranged along the chain.

For example, there are two distinct forms of polystyrene. Using a Ziegler–Natta catalyst (spread 22.11) for the polymerization of phenylethene (styrene) $CH_2{=}CHC_6H_5$ gives an *isotactic* polymer with all the phenyl groups on *one* side of the chain. Radical polymerization gives *atactic* polystyrene in which the arrangement of the phenyl groups is *random*.

Isotactic PVC. The regular arrangement of the chloro side groups in the isotactic form allows the chains to approach each other more closely.

The regular structure of the isotactic polymer allows dispersion forces to attract the molecules closer to each other. Intermolecular forces are therefore weaker in the atactic form, which can be softened and moulded at a lower temperature than the isotactic form, and which is more soluble in most solvents.

Intermolecular forces

The intermolecular forces between hydrocarbon polymer chains such as polythene and polystyrene are limited to weak dispersion forces. The presence of polar groups allows for stronger dipole–dipole interactions, an example being the commercial fibre Orlon, which is an addition polymer made from propenenitrile $CH_2{=}CHCN$. Much stronger intermolecular forces are produced by hydrogen bonding. This type of force is more common in condensation polymers such as nylon.

Thermoplastics and thermosets

All the polymers described so far are called '**thermosoftening polymers**', or 'thermoplastics'. Increasing the temperature overcomes the intermolecular forces; the chains are able to slide past each other and the material softens. Thermoplastics can be repeatedly heated and cooled without changing their properties.

Another class of synthetic polymeric materials are the '**thermosetting polymers**', or 'thermosets'. Thermosetting polymers are usually made from liquids that are mixed and heated in a mould. The monomer molecules link with each other by forming bonds between the different chains; such bonds are called **cross-links**. A hard product results that has a three-dimensional network of covalent bonds interconnecting the polymer chains. Thermosets can be shaped only once.

Melamine   Methanal

Monomers

The monomers join by cross-links
—NHCH$_2$NH—
formed in condensation reactions between NH_2 groups on neighbouring rings with the methanal

Melamine (2,4,6-triamino-1,3,5-triazine) reacts with methanal to form the thermoset material commonly called 'Melamine'. It can be moulded to produce durable and shatterproof cups, saucers, and plates.

(a) (b)

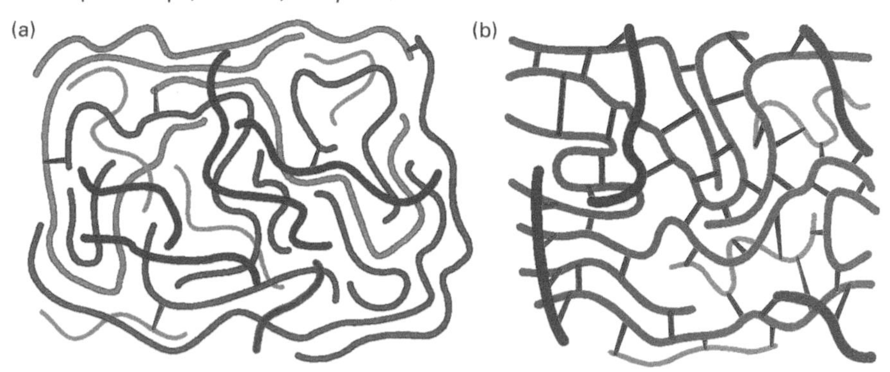

(a) Thermoplastics have polymer chains with weak intermolecular forces between them. There are few covalent cross-links between chains. (b) Thermosets have a covalently bonded three-dimensional network. There is extensive cross-linking.

SUMMARY

- Polymers may be natural or synthetic.
- Synthetic polymers may be formed by addition or condensation polymerization.
- Polymers may be thermosoftening or thermosetting.

(a)

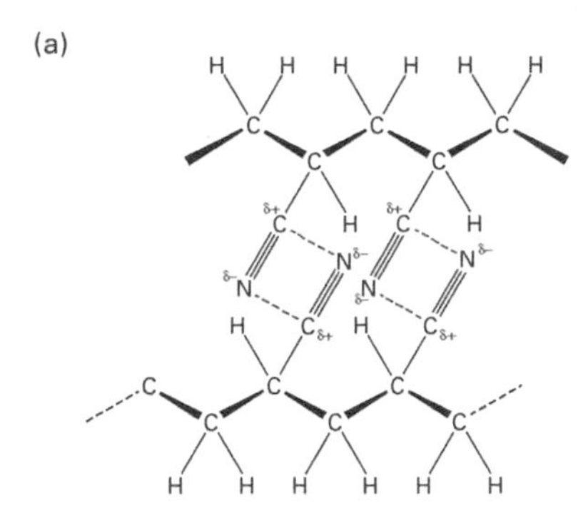

(b)

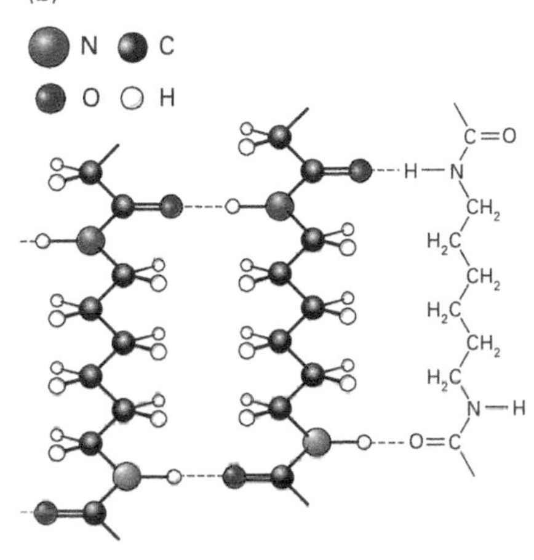

Interactions between polymer chains.
(a) Dipole–dipole forces in Orlon.
(b) Hydrogen bonding in nylon.

Chief distinctions

- **Thermoplastics** soften when heated, are permanently deformed when a force is applied, and harden when cooled.
- **Thermosets** are hard polymeric materials that do not soften when heated, and can be shaped only once.

Recycling plastics

Thermosets are very difficult to recycle. The task of recycling thermoplastics has been simplified by coding individual plastics, thus making their identification easy.

For example:

see spread 22.11.

29.5

OBJECTIVES

- Staudinger's vision
- Chain and step polymerization
- Copolymers
- Rubber

REVISITING POLYMERIZATION – 2

A synthetic polymer can be classified as an addition polymer or as a condensation polymer, according to the type of reaction that formed it. A synthetic polymer may also be classified as thermosoftening or thermosetting, according to the extent of cross-linking and the consequent behaviour on heating. This spread introduces the idea of classifying a polymer according to the mechanism of its formation; it leads to the categories of *step polymers* and *chain polymers*. The vision of Hermann Staudinger links the ideas underlying this chapter to those of the next one.

Hermann Staudinger's foresight

The idea that polymers consist of long-chain molecules was first put forward in the 1920s by Hermann Staudinger, to the disbelief of his contemporaries. Subsequent studies have shown Staudinger to be correct, but he was two years into his retirement before his insight led to him being awarded the Nobel Prize in Chemistry. Turning his attention to the natural polymers in living systems, Staudinger made the prediction in 1936 that: 'Every gene macromolecule possesses a definite structure which determines its function in life.' Modern molecular biology describes in precisely these terms the role of DNA in the synthesis of proteins and the inheritance of characteristics. The next chapter will focus on biochemistry.

Chain polymerization

An example of chain polymerization is the reaction between ethene molecules to produce polythene. During the reaction, the monomer concentration decreases steadily with time. Polymer chains of high molar mass form very rapidly as there is normally only one reactive site (see diagram below). The reaction mixture contains monomer, polymer chains of high molar mass, and a small number of growing polymer chains. The polymer yield increases with reaction time, but the polymer molar mass does not change significantly.

The process of chain polymerization most often occurs when monomers add together. 'Addition polymerization' is often used as an alternative name, but the two terms are not synonymous. Chain polymerization can only produce thermoplastics (and not thermosets).

The chain reaction mechanism

Chapter 22 'Alkanes and alkenes' discussed the chain reaction involved in the chlorination of methane. The reaction involves radicals (species with unpaired electrons) and proceeds by steps: chain initiation; chain propagation; chain termination. Chain polymerization follows the same overall reaction mechanism. The most common types of monomer to polymerize in this manner are derivatives of ethene such as chloroethene and phenylethene.

Chain polymerization is readily started by heating an initiator such as dibenzoyl peroxide $(C_6H_5CO)_2O_2$. The molecule breaks down to yield two radicals of formula $C_6H_5CO_2^{\bullet}$, which loses carbon dioxide to form the phenyl $C_6H_5^{\bullet}$ radical. Polymerization then proceeds as shown below. The resulting chains are atactic and mostly linear, with a certain amount of chain branching. The process of chain termination finally halts the growth of the chain.

(a) $C_6H_5^{\bullet}$ + $CH_2{=}CH(C_6H_5)$ → $C_6H_5{-}CH_2{-}\dot{C}H(C_6H_5)$

(b) $C_6H_5{-}CH_2{-}\dot{C}H(C_6H_5)$ + $CH_2{=}CH(C_6H_5)$ → $C_6H_5{-}CH_2{-}CH(C_6H_5){-}CH_2{-}\dot{C}H(C_6H_5)$

(c) $C_6H_5{-}(CH_2CH(C_6H_5))_n{-}CH_2\dot{C}H(C_6H_5)$ $\xrightarrow{+\,C_6H_5^{\bullet}}$ $C_6H_5{-}(CH_2CH(C_6H_5))_n{-}CH_2CH(C_6H_5){-}C_6H_5$

The radical polymerization of phenylethene. (a) Initiation: the initiator joins with the monomer to form a radical. (b) Propagation: as the chain grows, the unpaired electron continues to be located at the end of the chain. (c) Termination: one possible termination step occurs when a growing chain combines with an initiator radical.

Step polymerization

Step polymerization builds up a polymer in a number of stages. Any two monomers or short chains can react. The monomer concentration decreases rapidly with reaction time; the average molar mass also increases steadily with time. Long reaction times are needed to produce large polymer chains. The reaction mixture contains polymer molecules with a wide range of chain length, from dimers up to those of very high molar mass.

The process of step polymerization most often happens when monomers condense together. 'Condensation polymerization' is often used as an alternative name, but the two terms are not synonymous. For example, polyurethanes (much used for furniture foam) are made by step polymerization but the reaction does not involve condensation as no small molecule is formed as a byproduct. Step polymerization can produce thermoplastics *and* thermosets.

Three monomers – acrylonitrile, (propenenitrile), butadiene, and styrene – polymerize to make the copolymer acrylonitrile–butadiene–styrene, known as ABS. It is a very tough material, used to make suitcases and children's toy bricks.

Copolymers

Copolymers are made from two or more monomers. The polyester Terylene and the polyamide nylon-6,6 described earlier are examples of copolymers, as are many other condensation polymers. If the monomers are A and B, then the polymer chain has the structure ...ABABAB.... Copolymers can be designed to have particular physical properties by using monomers that can polymerize with themselves or with each other. Polymer structure then takes the random form ...AABBBBAABBBA.... An example is Saran, a copolymer of PVC and poly(1,1-dichloroethene) used as a film for wrapping food.

Elastomers

An elastomer is a material that can rapidly recover its original shape after being deformed. The most important example is vulcanized rubber. The structure of natural rubber illustrates the requirements for elastomeric properties: long flexible chains, weak intermolecular forces, and some cross-linking. The all-*cis* configuration limits the approach of adjacent chains and results in low dispersion forces. The all-*trans* isomer occurs naturally as gutta percha, which is hard and non-elastic.

Vulcanization, heating rubber with sulphur, forms disulphide cross-links. Vulcanized rubber is used to manufacture car tyres.

(a)

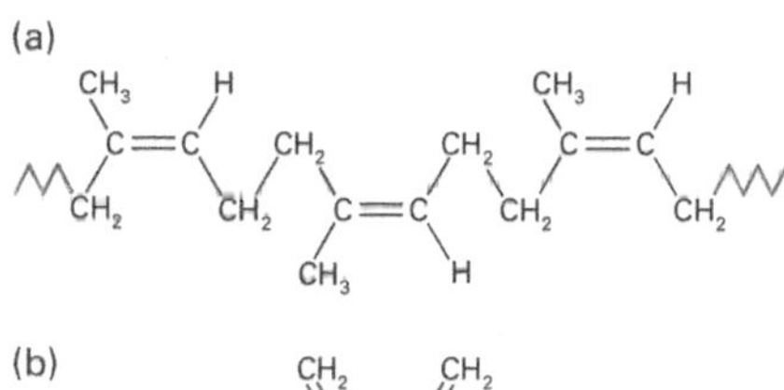

(b)

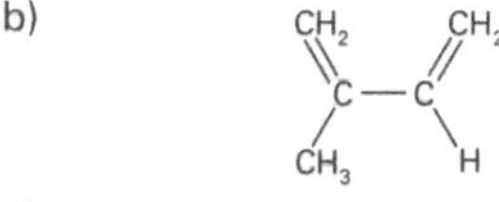

(c)

(a) Natural rubber is poly(cis-2-methylbuta-1,3-diene); the monomer was formerly known as isoprene.
(b) The monomer unit of natural rubber.
(c) Vulcanized rubber is used to make tyres. The extent of cross-linking is controlled carefully to give the product the required hardness, wear resistance, and elasticity.

SUMMARY

- Chain polymerization: proceeds by chain initiation, propagation, and termination; the reaction mixture contains monomer, polymer chains of high molar mass, and a small number of growing chains.
- Step polymerization: chain length increases steadily with time; the reaction mixture contains polymer molecules with a wide range of chain length.
- Copolymers are made from two or more monomers.
- Vulcanization of rubber introduces disulphide cross-links between polymer chains.

Chapter 29 Reactions summary

- Preparation of Grignard reagent:
  $CH_3CH_2I + Mg \xrightarrow{\text{ether}} CH_3CH_2MgI$
- Grignard reagent with carbon dioxide:
  $RMgI + CO_2 \xrightarrow{H_2O} RCOOH + Mg(OH)I$
- Grignard reagent with methanal:
  $RMgI + HCHO \xrightarrow{H_2O} RCH_2OH + Mg(OH)I$
- Grignard reagent with other aldehydes:
  $RMgI + R'CHO \xrightarrow{H_2O} RR'CH(OH) + Mg(OH)I$
- Grignard reagent with ketones:
  $RMgI + R'R''CO \xrightarrow{H_2O} RR'R''COH + Mg(OH)I$
- Diels–Alder 1,4-cycloaddition:
  diene + alkene → cyclohexene

Practice exam questions

1 **a** Normal electric wiring consists of copper wire surrounded by polyvinyl chloride (PVC). In one type of electric wiring used in fire alarm systems, a copper wire is surrounded by solid magnesium oxide to act as insulator, the whole being encased in a copper tube covered with PVC.

i Describe the bonding in copper metal and hence explain how it conducts electricity.

ii What type of bonding is present in PVC? Hence explain why it can be used as an insulator.

iii What type of bonding is present in magnesium oxide? Hence explain how it can act as an insulator.

iv Suggest why magnesium oxide is preferred to PVC alone as an insulator in fire alarm systems. [12]

b Magnesium reacts with bromoethane to form a Grignard reagent. Write the equation for this reaction and give the necessary conditions. [3]

c Name and give the structures of the products when the Grignard reagent from **b** is reacted with:

i propanone;

ii butanal. [4]

d **i** Describe a chemical test which would distinguish between the products of **c i** and **c ii**.

ii Explain briefly why in each case neither of the final liquid products of the reactions in **c i** and **c ii** has any effect on the plane of plane-polarized light which is passed through it. [6]

2 Propranolol is a chiral compound used in some 40 pharmaceutical preparations for the treatment of high blood pressure and cardiac pain. It is a base and is usually used as its hydrochloride salt, which is a white powder soluble to the extent of 50 g dm^{-3} in cold water and much more so in hot.

Propranolol is manufactured from glycidyl butanoate

$CH_3CH_2CH_2{-}C({=}O){-}O{-}CH_2CH{-}CH_2$ (with O bridging CH and CH_2 in an epoxide ring)

which is a chiral ester. This is made from glycidol, which has a boiling point of 56 °C, and butanoyl chloride, the latter being made from butan-1-ol via butanoic acid.

The esterification gives a racemic mixture of the ester but propranolol requires only one of the optical isomers in this mixture.

Butan-1-ol is made commercially from natural gas; it can also be made from an aldehyde and a Grignard reagent, but this is not economic. The alcohol is oxidized to butanoic acid, which is then converted to the acid chloride and then the ester.

a **i** What is meant by the term **chiral**?

ii Draw the two stereoisomers of glycidol,

$CH_2{-}CHCH_2{-}OH$ (with O bridging CH_2 and CH in an epoxide ring)

iii How is chirality detected experimentally? [5]

b Write the equations, stating briefly the necessary conditions, to show how you would bring about the following:

i conversion of butan-1-ol to butanoic acid;

ii conversion of butanoic acid to butanoyl chloride;

iii reaction of butanoyl chloride with glycidol. [6]

c **i** Suggest how butan-1-ol could be prepared using a Grignard reagent and an aldehyde.

ii State how the Grignard reagent you have suggested can be prepared from a halogenoalkane. [5]

d Suggest why only one of the optical isomers of propranolol is effective as a drug. [1]

e Give experimental details of how you could purify propranolol hydrochloride by recrystallization. How would you assess its purity? [7]

f How would you liberate the base propranolol from its hydrochloride salt? [1]

3 Poly(phenylethene) is obtained by a chain polymerization reaction.

a Explain what is meant by chain polymerization. [2]

b The degree of polymerization is governed by physical changes taking place during the polymerization process.

i Describe the physical changes that take place during the polymerization.

ii Explain the principles underlying suspension polymerization in terms of the changes outlined in **i**.

Briefly describe the use of suspension polymerization in the manufacture of poly(phenylethene). [8]

4 **a** Polymers are extensively used in the manufacture of synthetic fibres during which the relative molecular mass must be carefully controlled.

Explain the importance of the relative molecular mass of such polymers in the production and end-use of fibres. [4]

b Several polyamides are in current production in the synthetic fibre industry, including nylon 6, nylon 66, nylon 610, and nylon 11.

i Suggest the structural formula of the monomer that could be used in the manufacture of nylon 11.

ii Predict, with reasons, how you would expect the crystalline melting point of nylon 11 to compare with that of nylon 66. [4]

c After cleaning out some dilute acid bottles, a laboratory technician noticed that small holes had appeared in her nylon stockings.

Suggest a reason for this. [2]

5 **a** Describe briefly the types of bonding present in polymers. [5]

b How do the bonds and molecular interactions within a polymer contribute to;

i the properties of the polymer,

ii the relative ease of processing thermoplastics? [5]

6 This question concerns the addition polymer poly(propene), and the condensation polymer nylon.

a **i** Draw a structural formula for part of the poly(propene) chain showing clearly the repeating unit. [2]

ii Suggest why, when poly(propene) is heated, it softens over a range of temperature rather than melting sharply at a particular temperature. [2]

b Nylon-6,6 is made from a diamine, $H_2N(CH_2)_6NH_2$, and a diacid chloride, $ClOC(CH_2)_4COCl$.

i Draw the structural formula of a representative length of the polymer chain. [2]

ii Using nylon-6,6 and poly(propene) as examples, explain the essential difference between condensation and addition polymerization reactions. [3]

iii Suggest why a diacid chloride is employed to make nylon rather than the corresponding dicarboxylic acid. [1]

iv Nylon fibre is about twice as strong as poly(propene) fibre. Suggest in terms of the intermolecular forces in the polymers why this is so. [2]

c Nylon is an extremely good electrical insulator. Conducting polymers can however be made by polymerising ethyne, $HC{\equiv}CH$. Poly(ethyne) shows stereoisomerism. A section of the cis- form is shown below:

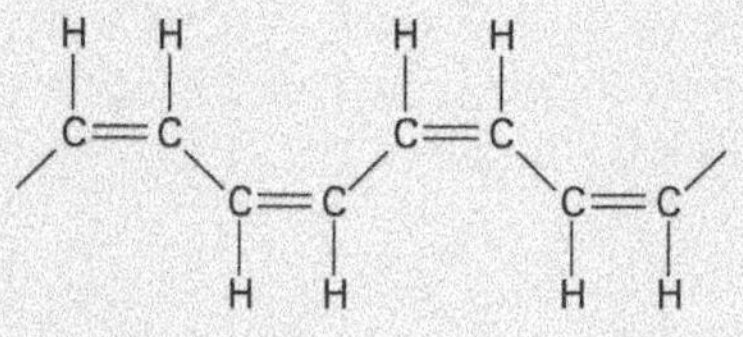

```
  H   H           H   H
  |   |           |   |
 /C===C           C===C
       \         /     \
        C===C           C===C/
        |   |           |   |
        H   H           H   H
```

i Draw a diagram of a section of the trans- form of poly(ethyne). [2]

ii Suggest why poly(ethyne) conducts electricity. [2]
(See also chapter 32.)

7 **a** **i** Describe the type of bonding which holds together the monomer units in a large polymer molecule.

ii Separate molecules or segments of a polymer chain are held together by secondary bonding.

Give **two** examples of secondary bonding found in polymer systems.

iii Give **two** physical properties of polymers which involve the making or breaking of secondary bonding. [5]

b **i** What is meant by the terms *thermosetting* and *thermoplastic* when applied to polymer systems?

ii Give **one** advantage and **one** disadvantage connected with the use of thermoplastics.

iii What is commonly the limiting factor in the use of thermosetting polymers? [5]

30 Biochemistry

Biochemistry studies the chemistry of life. Biochemists seek to describe the chemical structures of organisms and to explain the reactions that underlie the processes of life. Organisms are as much a part of the chemical world as any other class of object. Everything in the world, living or non-living, is composed of an assortment of atoms selected from the known elements. Organisms therefore consist of a collection of chemical substances that work together to produce the familiar set of properties and functions we recognize as 'life'. We can move towards understanding biochemistry by studying the structure and function of the types of molecules found in organisms. These substances fall into four broad categories: **lipids**, **carbohydrates**, **proteins**, and **nucleic acids**. The major part of this chapter treats each of these categories in turn. The final spread shows how these substances work together to support the metabolism of the living cell.

LIPIDS

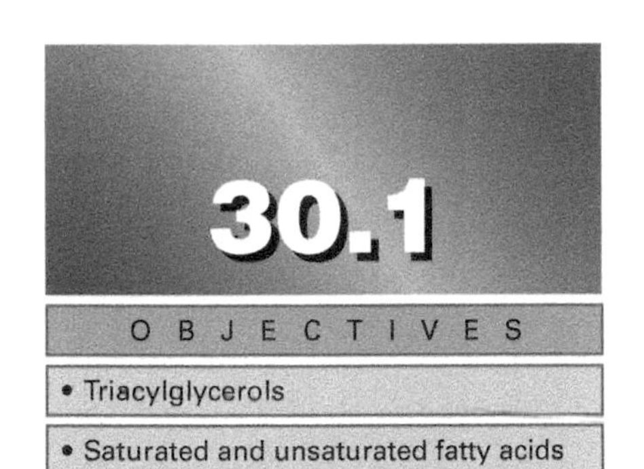

30.1

OBJECTIVES

- Triacylglycerols
- Saturated and unsaturated fatty acids
- Phospholipids
- Fluid mosaic model of the plasma membrane

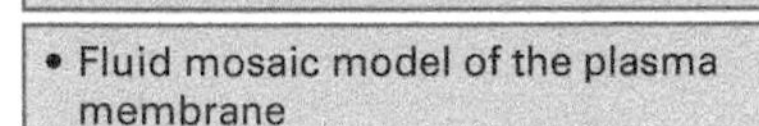

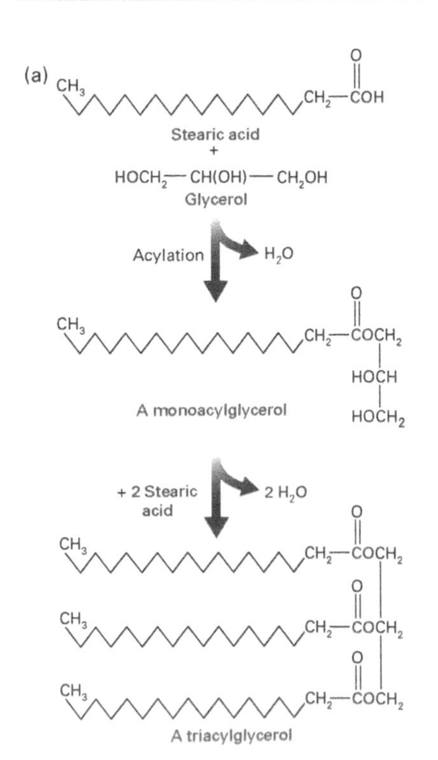

The formation of a triacylglycerol from long-chain fatty acids and the triol glycerol. Note that the three acyl groups could in principle be different; for simplicity, they are shown here to be identical.

Lipid is the general name given to a wide range of substances, found in organisms, that are largely insoluble in water but soluble in non-polar organic solvents. This spread looks at two types of lipid: **acylglycerols** and **phospholipids**. Examples of acylglycerols include the fats and oils used as food stores and the waxes used for waterproofing. Chemically, they are *esters* formed from long-chain carboxylic acids and the alcohol **glycerol**. The long-chain carboxylic acids are often referred to as 'fatty acids'. Note that glycerol has the systematic IUPAC name *propane-1,2,3-triol*, spread 25.6. The shorter familiar name 'glycerol' will be used here. The phospholipids are a major component of the **plasma membranes** of cells and play a pivotal role in their structure and function.

Acylglycerols

Acylglycerols are fatty acid esters of the triol glycerol $HOCH_2CH(OH)CH_2OH$. Esterification can occur at one, two, or all three of the hydroxyl groups of the triol, producing mono-, di-, and triacylglycerols respectively. The most common fatty acid esters found in living systems are **triacylglycerols** (or **triglycerides**), in which long-chain fatty acids attach by acylation reactions (see chapter 27 'Carboxylic acids and their derivatives') to all three hydroxyl groups on the triol.

Saturated and unsaturated fatty acids

Different lipids contain various different fatty acids. The carbon chain of the fatty acid may be *saturated*, as is common in animal lipids, or *unsaturated*, as is more common in plant lipids. The phrase 'high in polyunsaturates, low in saturates' is often used in advertisements to describe butter substitutes (margarine, etc.) made from vegetable oils. A polyunsaturated fatty acid contains more than one carbon–carbon double bond. An excess of saturated fats in the diet can cause a high concentration of the substance cholesterol.

Cholesterol

Cholesterol has a structure based on four alicyclic hydrocarbon rings fused together. It is present in food and is also manufactured in the liver. It is an essential component of plasma membranes and of blood plasma. In the body it is converted to bile acids and many hormones such as the male and female sex hormones testosterone, oestrogen, and progesterone. There is medical evidence to show that increased levels of blood cholesterol are associated with a condition called *atherosclerosis*, in which lipids accumulate on the inner walls of arteries, ultimately restricting or even obstructing blood flow (see spread 22.7).

Phospholipids

Phospholipids are a major constituent of plasma membranes. In a **phospholipid**, two of the hydroxyl groups of glycerol are esterified by fatty acids, but the third is esterified by phosphoric acid. The phosphate group is frequently bound to a nitrogen-containing group. The illustration to the right shows the molecular structure of a lecithin (phosphatidylcholine, the most abundant animal phospholipid).

A phospholipid has two distinct regions. The phosphate group constitutes a polar and hence **hydrophilic** ('water-loving') region, and the fatty acid chains form a **hydrophobic** ('water-hating') region. One important result of this is that, in aqueous environments, phospholipid molecules arrange themselves as a double layer called a **bilayer**, in which all the hydrophobic ends point inwards *away from* the water phase.

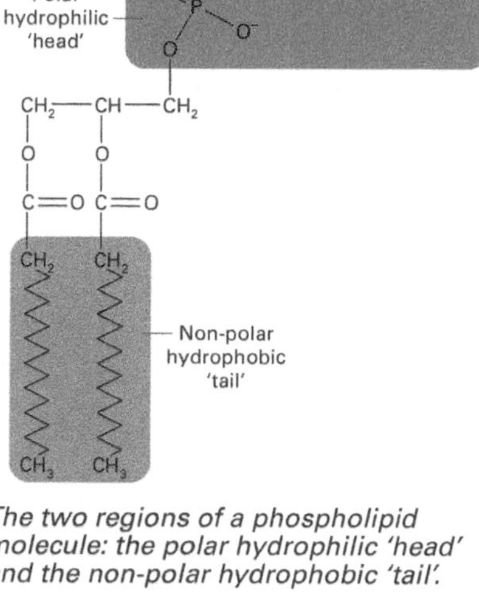

The two regions of a phospholipid molecule: the polar hydrophilic 'head' and the non-polar hydrophobic 'tail'.

Phospholipids and plasma membranes

The fundamental structure of the plasma membranes of cells consists of a phospholipid bilayer in which proteins and other molecules float, thus forming what is described as a **fluid mosaic**. This extremely flexible structure includes all the molecules responsible for the complex functions of the membrane. These functions include controlling the transport of substances across the membrane, enabling the cell to be 'recognized' as part of the organism, and various membrane-bound reactions such as the electron transport chain.

The plasma membrane of a red blood cell.

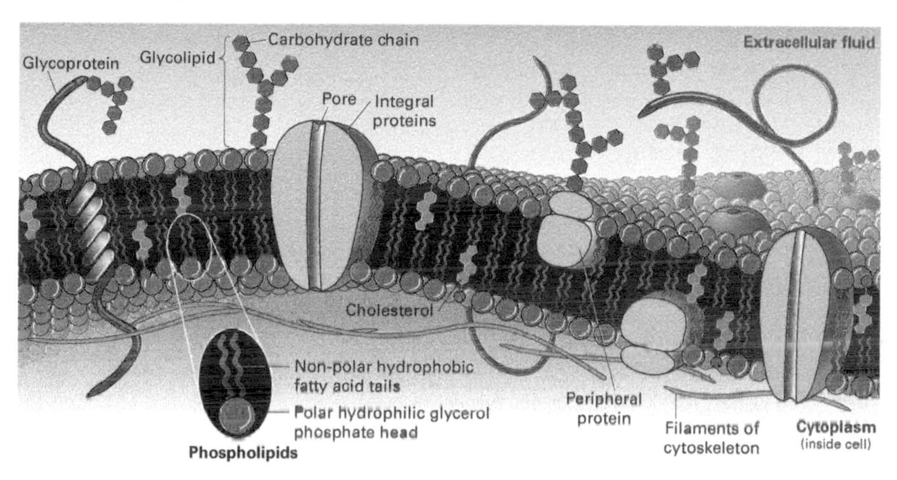

The fluid mosaic model of the plasma membrane. The sheet-like surface of the membrane consists of a phospholipid bilayer (shown alongside in greater detail). Globular proteins are embedded in the bilayer with tree-shaped carbohydrate structures joined to the external surface.

SUMMARY

- Two main classes of lipids are the acylglycerols and the phospholipids.
- Acylglycerols are fats, oils, and waxes consisting of glycerol esterified with fatty acids.
- The fatty acids in acylglycerols may be saturated, unsaturated, or polyunsaturated.
- Phospholipids are diacylglycerols with a phosphate group attached to the third hydroxyl group of the glycerol molecule.
- Plasma membranes have a fluid mosaic structure.

PRACTICE

1 Draw a displayed structural formula for each of the following fatty acids:

a $CH_3(CH_2)_{10}COOH$ (lauric acid)

b $CH_3(CH_2)_4CH{=}CHCH_2CH{=}CH(CH_2)_7COOH$ (*cis,cis*-linoleic acid).

2 Draw the structural formula of the triacylglycerol that results from the esterification of lauric acid at two of the hydroxyl groups of glycerol and *cis,cis*-linoleic acid at the third.

3 A common group found in phospholipids is the ethanolamine group: Sketch the interaction expected between this hydrophilic group and water molecules.

$$-\overset{O}{\overset{\|}{\underset{O^-}{\underset{|}{P}}}}-OCH_2CH_2NH_3^+$$

30.2

OBJECTIVES

- Monosaccharides
- D and L nomenclature
- Ring structures
- α and β anomers

Instant energy

Because carbohydrates are partially oxidized compared with lipids, lipids release more energy per gram than carbohydrates do in complete oxidation. Carbohydrates provide more 'instant access' energy.

(a)	Carbon number	(b)
CHO	①	CH_2OH
H—C—OH	②	C=O
HO—C—H	③	HO—C—H
H—C—OH	④	H—C—OH
H—C—OH	⑤	H—C—OH
CH_2OH	⑥	CH_2OH

(a) Glucose: an aldohexose. Note that the molecule is drawn with the carbonyl group at the top. The carbon atoms are numbered from the top to the bottom of each molecule.
(b) Fructose: a ketohexose.

L-Glucose	Carbon number	D-Glucose
CHO	①	CHO
HO—C—H	②	H—C—OH
H—C—OH	③	HO—C—H
HO—C—H	④	H—C—OH
HO—C—H	⑤	H—C—OH
CH_2OH	⑥	CH_2OH

L- and D-glucose.

L-Ribose	Carbon number	D-Ribose
CHO	①	CHO
HO—C—H	②	H—C—OH
HO—C—H	③	H—C—OH
HO—C—H	④	H—C—OH
CH_2OH	⑤	CH_2OH

L- and D-ribose.

CARBOHYDRATES: MONOSACCHARIDES

The simplest carbohydrates are the monosaccharides such as glucose and fructose, which are important intermediates in the respiration of food to release energy. These two substances are classed as **monosaccharides** because they are not hydrolysed by hydrochloric acid. Monosaccharides however can join together (by a condensation reaction) to form longer chains. Sucrose is a **disaccharide** because dilute acid splits it by hydrolysis into glucose and fructose (both monosaccharides). Disaccharides and polysaccharides are described in the next two spreads. Whatever their complexity, all carbohydrates contain the elements carbon, hydrogen, and oxygen, and have the general formula $C_x(H_2O)_y$.

Monosaccharides

Chemically, the monosaccharides are aldehydes or ketones that also have several hydroxyl groups attached to their carbon skeletons. For example, glucose is known as an **aldose** and has an aldehyde structure at carbon atom number 1 (C1 for short). On the other hand, fructose is known as a **ketose** and has a ketone structure at carbon atom C2.

Glucose and fructose molecules both have the molecular formula $C_6H_{12}O_6$. They both have six carbon atoms and so are called **hexoses.** Monosaccharides that have three carbon atoms are called **trioses**, those with four are **tetroses**, those with five are **pentoses**, and so on. Ribose is the most important pentose. Monosaccharides dissolve easily in water due to the presence of a large number of polar hydroxyl groups, which hydrogen-bond with water molecules.

Chiral centres and optical activity

Look at the structural formula of glucose. You should be able to point out chiral centres (see spread on 'Amino acids and optical activity' in chapter 28 'Amines and amino acids') where carbon atoms are attached to four different groups. These molecules can therefore exist in different enantiomeric forms, some of which may be optically active. The actual shape (stereochemistry) of a monosaccharide molecule may be described by comparing it to a reference standard, glyceraldehyde (2,3-dihydroxy-propanal). This substance has only one chiral centre, which means that it can exist as two enantiomers (see below). The enantiomer with the —OH group on the left is given the label L and the enantiomer with the —OH group on the right is given the label D.

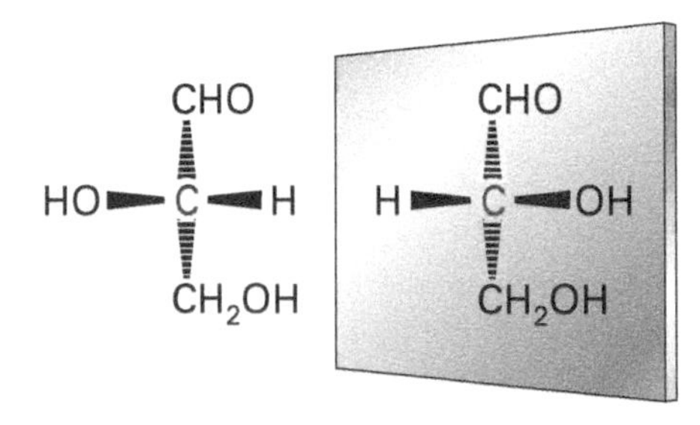

L- and D-glyceraldehyde.

This definition leads to two 'families' of compounds. The L carbohydrates are structurally related to L-glyceraldehyde (at the chiral centre closest to the bottom of the carbohydrate molecule, shown by a coloured carbon atom), and the D carbohydrates are structurally related to D-glyceraldehyde. The two forms of glucose are shown to the left. It is important to note that the labels L and D do *not* relate to the direction of rotation of plane-polarized light (which is described as (+) for dextrorotatory, and (-) for laevorotatory). *Most naturally occurring carbohydrates are D-carbohydrates:* so the label 'D' will be dropped from now on.

Open-chain and ring structures

The glucose and fructose molecules in the illustrations opposite are shown as open-chain structures. However, there is evidence that these molecules exist as *ring* structures when in solution. This happens because the shape and flexibility of the molecule bring the aldehyde or ketone carbonyl group close to one of the hydroxyl groups, and they react to form a cyclic structure. Glucose ($C_6H_{12}O_6$) forms a six-membered **pyranose** ring structure; fructose ($C_6H_{12}O_6$) and ribose ($C_5H_{10}O_5$) both form five-membered **furanose** ring structures.

Sugars

Sugars have a definite, fixed molar mass: they may be monosaccharides, disaccharides, etc. The molar mass of a polysaccharide may vary.

(a)

α–glucose ⇌ (±H⁺) glucose: open-chain aldehyde form ⇌ (±H⁺)

Axial

(b)

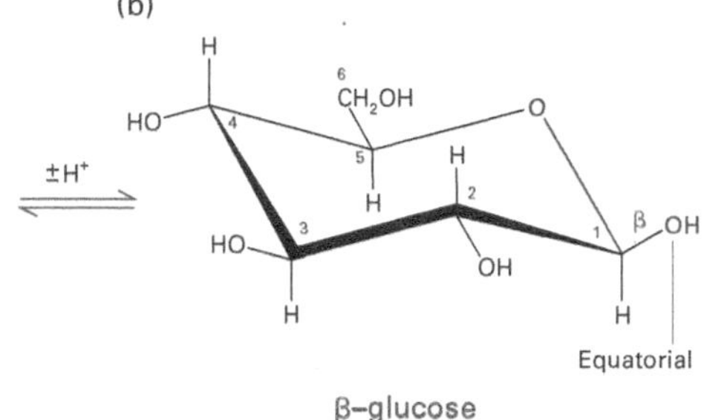

A molecule of glucose undergoes an internal reaction that results in a ring structure. A new chiral centre is formed after reaction at carbon atom number 1 (C1) (called the anomeric centre*). Two possible isomers (called* anomers*) result: (a) α-glucose: attack from above means that the hydroxyl group attached to C1 is in the axial position (pointing out from the plane of the ring). (b) β-glucose: attack from below means that the hydroxyl group attached to C1 is in the equatorial position (pointing into the plane of the ring). The β isomer is more stable for glucose (64% exists in this form). The α isomer has the hydroxyl group on the carbon atom on the 'opposite' side of the ring from the end CH_2OH group, numbered 6. The β isomer has them on the 'same' side.*

Haworth projection for β-glucose.

Haworth projections

Haworth projections show the structures of carbohydrates in a way that makes comparisons easier. The conventions for D and L and α and β isomers are as follows:

- The —CH_2OH group attached to C5 is written *up* for D sugars and *down* for L sugars.
- The —OH group attached to C1 is written *down* for the α isomer and *up* for the β isomer.

Note that Haworth projections give the false impression that the six-membered pyranose ring is flat; in reality it resembles a chair.

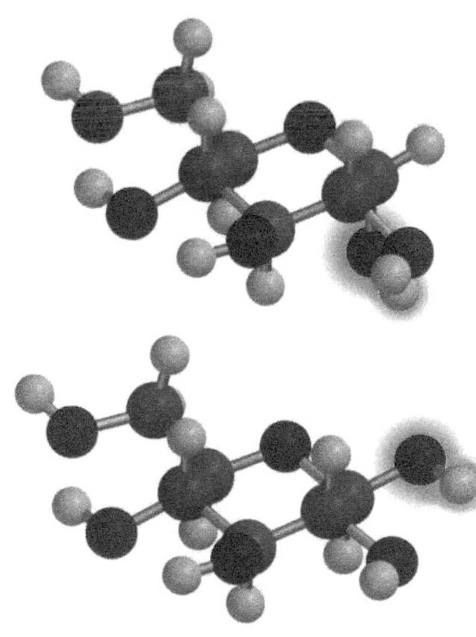

α- and β-glucose. The blue background highlights the position of the hydroxyl group on C1.

SUMMARY

- Carbohydrates have the general formula $C_x(H_2O)_y$.
- The simplest carbohydrates are the monosaccharides.
- Carbohydrates contain chiral centres. The structures of D and L isomers are related to the defined structures of D- and L-glyceraldehyde.
- Most naturally occuring carbohydrates are D-carbohydrates.
- Monosaccharides form ring structures in solution. α and β isomers result that differ in the position of the hydroxyl group at carbon atom number 1.

PRACTICE

1. Draw structural formulae to illustrate the differences between the chain structures of D- and L-glucose.
2. Draw structural formulae and Haworth projections to show the differences between the four ring structures: α-D-glucose, β-D-glucose, α-L-glucose, and β-L-glucose. Give your reasoning.

30.3

Carbohydrates: disaccharides

OBJECTIVES

- The glycosidic link
- Reducing and non-reducing sugars
- Sucrose

The simplest carbohydrates are the monosaccharides such as glucose and fructose. Two monosaccharide molecules can link together by a condensation reaction to form a disaccharide. Examples include sucrose, which is cane sugar, and lactose, which is a constituent of milk. This spread investigates disaccharides and discusses the nature of the linkage that joins the monosaccharide units together.

Disaccharides

Disaccharides consist of two monosaccharide units joined together by a **glycosidic link**. A glycosidic link is made by the reaction between the carbonyl group of one molecule and a hydroxyl group of another molecule. In the case of the disaccharide maltose, two molecules of glucose condense together, forming a glycosidic link between carbon atom 1 on the first glucose molecule and carbon atom 4 on the second.

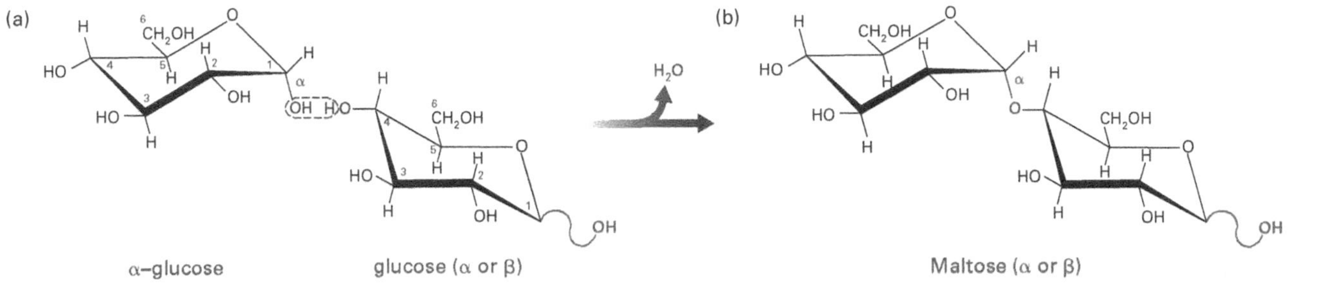

(a) α-glucose glucose (α or β) H_2O (b) Maltose (α or β)

As we saw in the previous spread, the isomers of glucose can interconvert by the ring opening and reclosing, so the wiggly green line on the second glucose molecule shows that either isomer can form maltose. Exactly as for glucose, the right-hand ring in maltose can open and α and β isomers can interconvert. However, the left-hand ring cannot open: it is locked into the α position. Maltose is an α-glycoside.

Enzyme hydrolysis

Enzymes can bring about hydrolysis of disaccharides. Enzyme-catalysed hydrolysis is far more specific than the acid-catalysed reaction. For example, the enzyme maltase catalyses the hydrolysis of α-glycosides much more rapidly than that of β-glycosides.

Acid hydrolysis

Disaccharides hydrolyse when warmed with dilute hydrochloric acid. For example, 1 mole of the disaccharide maltose yields 2 moles of the monosaccharide glucose.

α- and β-glycosides

A glycosidic link can be arranged in different ways, giving rise to a difference in biochemical properties. Look again at the displayed structure of maltose. The oxygen atom of the glycosidic link lies *outside the plane* of the left-hand glucose ring. Maltose is therefore said to have an α-glycosidic link; maltose is an α-glycoside.

Now look at the structure of cellobiose below. Like maltose, it is also a disaccharide built from two glucose molecules. However, the glycosidic link lies *in the plane* of the left-hand glucose ring. Cellobiose is therefore said to have a β-glycosidic link; cellobiose is a β-glycoside.

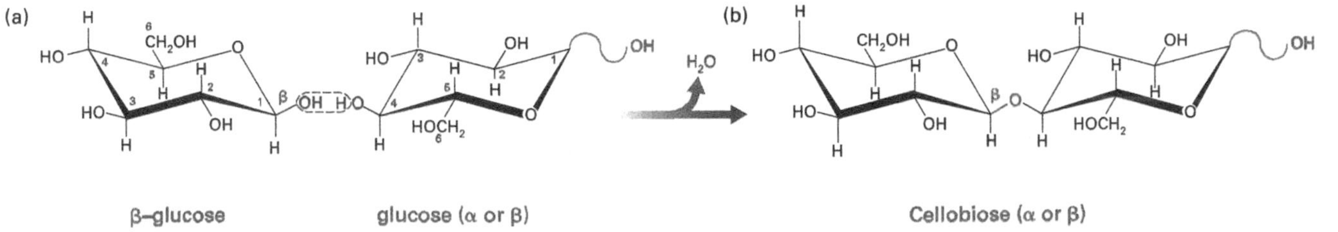

(a) β-glucose glucose (α or β) H_2O (b) Cellobiose (α or β)

The isomers of glucose can interconvert by the ring opening and reclosing, so the wiggly green line on the second glucose molecule shows that either isomer can form cellobiose. Exactly as for glucose, the right-hand ring in cellobiose can open and α and β isomers can interconvert. However, the left-hand ring cannot open: it is locked into the β position. Cellobiose is a β-glycoside.

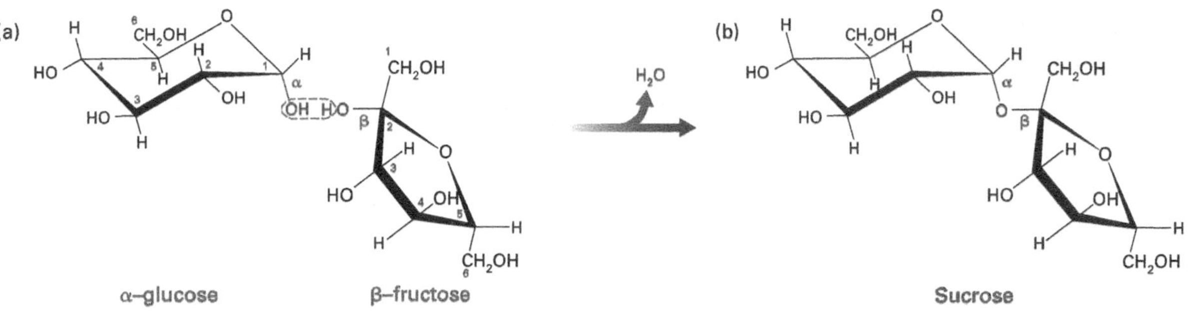

(a) α–glucose β–fructose $\xrightarrow{H_2O}$ (b) Sucrose

Unlike the situation with maltose or cellobiose, the sucrose molecule has the glycosidic link between the anomeric centres of both rings. The molecule results from a condensation reaction between α-glucose and β-fructose. The carbon atom that is the centre for interconversion between the α and β isomers in fructose (carbon atom 2) is no longer able to do so once permanently locked into a β-glycosidic link. The sucrose molecule is unique (there are no α and β isomers).

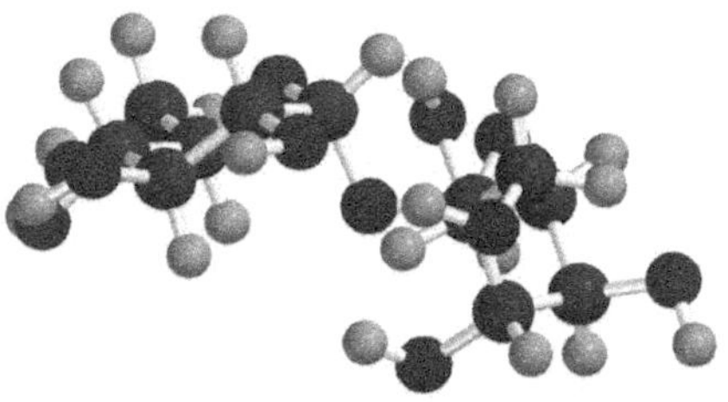

The sucrose molecule as produced by Spartan. Representing as complicated a molecule as this in two dimensions using either artwork or a screen shot is exceptionally difficult.

Reducing and non-reducing sugars

In solution, the ring forms of monosaccharides are in equilibrium with the open-chain forms, which bear aldehyde or ketone groups. Carbonyl groups in aldehydes act as reductants (being oxidized to carboxylic acids), so aldoses would be expected to react positively with Fehling's or Tollens' reagents (see chapter 26 'Aldehydes and ketones'). Ketones do not normally give a positive reaction to these tests because they do not readily oxidize. However, *ketoses* do give a positive result due to the presence of a hydroxyl group attached to the carbon atom next to the carbonyl carbon. Thus, *all* monosaccharides are **reducing sugars**.

Disaccharides may be reducing or non-reducing. To form the open-chain structure from the ring structure, a carbonyl group must be present. If this carbonyl group was involved in forming the glycosidic link of a disaccharide, then that disaccharide will be a **non-reducing sugar**. Look at the structural formulae above and you will see that maltose is a reducing disaccharide but sucrose is non-reducing. You should note that:

- **Reducing sugars** have formed the glycosidic link using a hydroxyl group present in the open-chain structure, and so a carbonyl group remains on one monosaccharide. This can therefore form an anomeric centre, and *two forms* exist: there is an α-maltose and a β-maltose.
- **Non-reducing sugars** have formed the glycosidic link using a hydroxyl group formed on making the ring structure of the molecule, and so no carbonyl group remains. This cannot therefore form anomers, and only *one form* exists: sucrose is a unique molecule.

Benedict's solution

Benedict's solution, spread 26.4, is a more sensitive test for reducing sugars than Fehling's solution. The reagent is a mixture of aqueous copper(II) sulphate with a basic solution of sodium citrate. When boiled with a reducing sugar, the pale blue solution forms a red precipitate. Detection of glucose in urine is a common test for diabetes.

SUMMARY

- A glycosidic link forms by the reaction of a carbonyl group on one monosaccharide with a hydroxyl group on another monosaccharide.
- The α-glycosidic link forms when the bridging oxygen atom is outside the plane of the two monosaccharide rings.
- The β-glycosidic link forms when the bridging oxygen atom is in the plane of the two monosaccharide rings.
- All monosaccharides are reducing sugars.
- Disaccharides are reducing if a carbonyl group remains on one monosaccharide.

PRACTICE

1 Prepare a 10-minute talk (with PowerPoint presentation and other visual aids) to explain:

a the similarities and differences between the α- and β-glycosides;

b the structure that a disaccharide needs to have for it to reduce Benedict's solution.

30.4

Carbohydrates: polysaccharides

OBJECTIVES

- Starch
- Amylose and amylopectin
- Cellulose
- Glycogen
- Amylase

Polysaccharides are natural condensation polymers made up of monosaccharides joined by glycosidic links to form long chains. They are important as carbohydrate energy stores and are fundamental to the structure of plants and fungi. They are also present in the plasma membranes of animal and bacterial cells. This spread concentrates on three glucose-based polysaccharides: starch, cellulose, and glycogen. You will see that, in common with disaccharides, the properties of these polysaccharides depend to a great extent on the type of glycosidic link (α or β) between the monosaccharide units from which they are constructed.

Some polysaccharide sources. (a) Bananas are a source of starch. Digestion in animals breaks down starch into glucose, which circulates in the bloodstream. The human brain requires the energy from two teaspoonfuls of glucose each hour. This glucose comes mainly from sugars (mono- and disaccharides) and starch-based foodstuffs. (b) Chitin (pronounced ki-tin) is a polysaccharide found in the body shells (exoskeletons) of many invertebrate animals, especially arthropods. Chitin is probably the second most abundant organic compound on Earth. The copepods (aquatic crustaceans about 1 mm long) alone produce 10^9 tonnes of chitin per year.

Starch: amylose and amylopectin

Starch from plant sources is a major direct or indirect energy source in the diets of animals. Plants make glucose by photosynthesis, and this may be converted to starch for storage. A typical starch polymer chain consists of around 2500 glucose molecules. Starch is used as a carbohydrate store in roots, tubers, seeds, and fruits. Plant cells store starch in the form of *starch grains*, which usually contain variable proportions of the two polysaccharides *amylose* and *amylopectin*.

The α and β angles

The α-glycosidic link lies outside the plane of one of the monosaccharide rings that it joins; the β-glycosidic link lies in the plane of the rings. As a result, the α link holds the rings at a more acute angle to each other than does the β link.

Amylose and **amylopectin** are both polymers of glucose. Amylose has a structure consisting of glucose units joined by α-(1–4)-glycosidic links. Amylopectin has a branched structure: the chains have the same arrangement as in amylose, branches forming by means of α-(1–6)-glycosidic links. The α-glycosidic link in amylose and amylopectin results in a coiled molecule most suited for storage in starch grains. This contrasts with the β-glycosidic link present in cellulose.

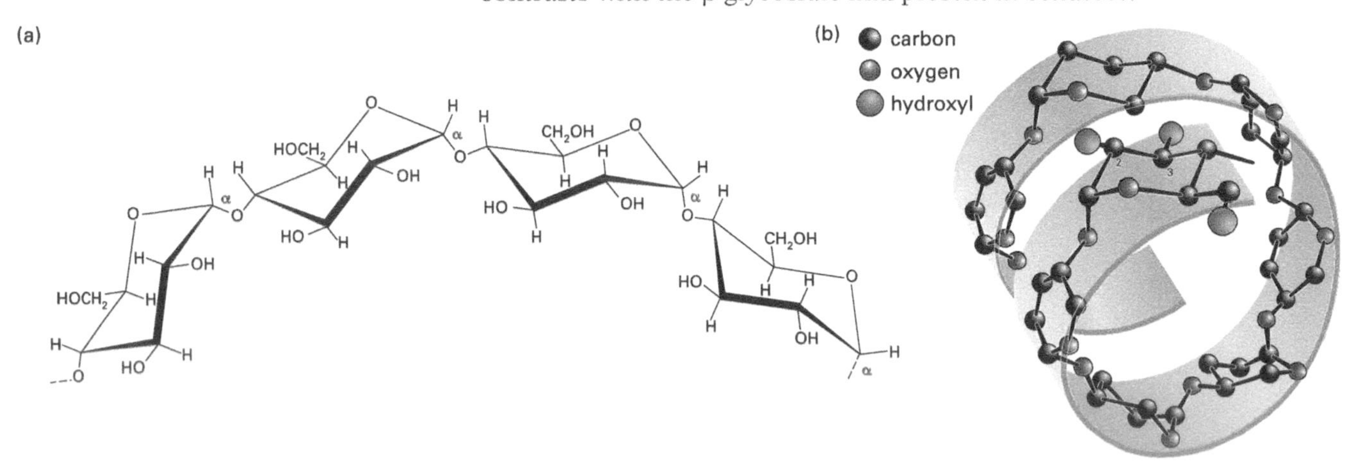

(a) Part of the structure of amylose. (b) The coiled structure of starch results from the angle of the α-glycosidic link. This enables intramolecular hydrogen bonding between the hydroxyl group on C2 with the glycosidic link and the hydroxyl group on C3 with the oxygen atom in the ring.

Cellulose

Cellulose, the main structural component of plant cell walls, is the most abundant organic compound on Earth. It accounts for more than half the carbon in the biosphere. In common with starch, cellulose is a polymer of glucose. However, the (1–4)-glycosidic links between the glucose molecules are at the β angle, resulting in a linear chain structure. Compared with the spiral structure of starch, this shape is more suited to the formation of the rigid walls of plant cells.

Cellulose fibres in the cell wall of the alga Chaetomorpha, *which grows in freshwater pools.*

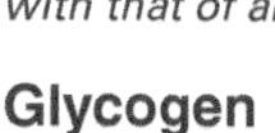

Part of the structure of cellulose. The linear structure of cellulose results from the angle of the β-glycosidic link. Alternate rings are rotated by 180°, which allows hydrogen bonding between the hydroxyl group on C2 with that on C6 and between the hydroxyl group on C3 with the oxygen atom in the ring. Compare this structure with that of amylose.

Glycogen

Starch and cellulose are produced by plants: cellulose is a structural material and starch acts as a food store. Glycogen is the storage carbohydrate of animals. It is very similar to amylopectin, but has a more extensively branched structure. In mammals, starch is digested to glucose. Excess glucose is polymerized and stored in the liver as glycogen. The balance between stored glycogen and free glucose in the blood is controlled largely by the two hormones insulin (promotes conversion of glucose to glycogen) and glucagon (promotes conversion of glycogen to glucose).

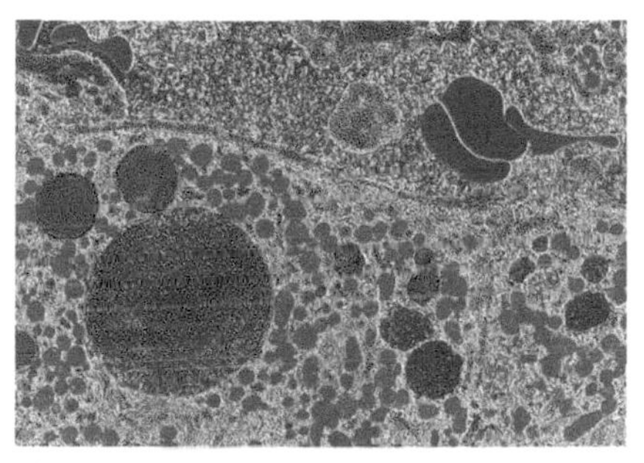

Glycogen granules (pink/red) in the cytoplasm of a liver cell. The glucose units making up these granules probably originated in the starch grains of a plant cell.

Enzyme hydrolysis

Mammals such as humans digest starch by means of a group of catalytic enzymes called *amylases*. These enzymes split the α-glycosidic link to produce a mixture of glucose and the disaccharide maltose. However, human amylases cannot hydrolyse β-glycosidic links. Because of this, we cannot digest cellulose. Herbivores such as cattle and sheep have bacteria in their stomachs that produce an enzyme that digests the cellulose in plant material, breaking it down to the mono- and disaccharides that the body can use.

Almost all enzymes are proteins – this is the subject of the following spread.

SUMMARY

- Polysaccharides are natural polymers consisting of monosaccharides joined by glycosidic links.
- Starch, cellulose, and glycogen are all polysaccharides made up from glucose.
- Starch consists of glucose units linked by α-(1–4)-glycosidic links to form amylose (straight chain) and amylopectin (branched chain).
- The α angle in amylose and amylopectin results in a coiled molecule whose structure is maintained by intramolecular hydrogen bonding.
- Cellulose is a linear polymer consisting of glucose units linked by β-(1–4)-glycosidic links.
- Glycogen is the storage carbohydrate of animals. It has a highly branched structure similar to that of amylopectin.

PRACTICE

1 Compare and contrast the structures and uses of starch, cellulose, and glycogen.

2 Suggest reasons for the presence of 4 mol dm^{-3} hydrochloric acid in the digestive fluids in the stomachs of mammals.

30.5

OBJECTIVES

- Primary structure
- Secondary structure
- Tertiary structure
- Quaternary structure
- Fibrous and globular proteins

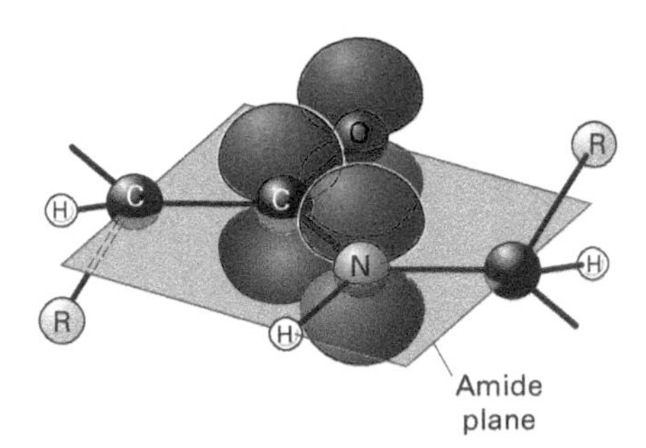

The p orbitals on carbon and oxygen that form the π bond can also overlap with the p orbital on nitrogen, when this is arranged as in the illustration to cause a structure delocalized over all the atoms. This delocalization is the reason why the peptide link must be planar.

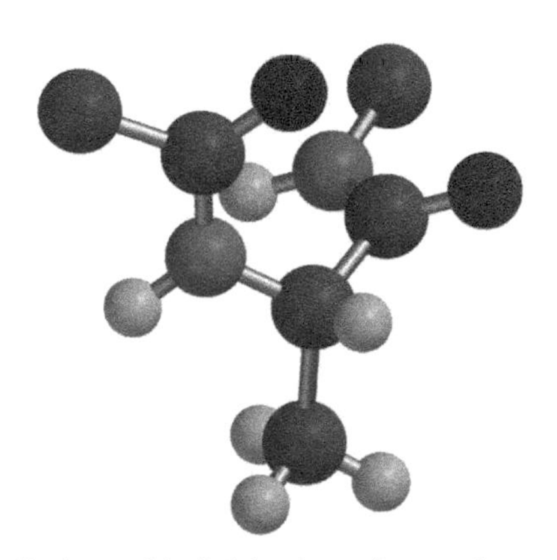

Each peptide link is planar (remember that Spartan uses ball-and-stick models to show atoms that are bonded; the number of bonds is not shown).

Proteins

Proteins are naturally occurring condensation polymers (see chapter 27 'Carboxylic acids and their derivatives'), consisting of chains of amino acid groups. As noted in chapter 28 'Amines and amino acids', there are about 20 naturally occurring amino acids. The human body can synthesize some of these, while others, called the **essential amino acids**, must be present in our food. There are four levels of structure used to describe proteins: primary, secondary, tertiary, and quaternary. The illustrations included in this spread show how these structures relate to the proteins myoglobin and haemoglobin, the oxygen-carrying components of cells.

Primary structure

The **primary structure** of a protein describes the sequence of amino acids present. Each amino acid is joined to the next via a **peptide link**, which forms between the carboxyl group of one amino acid and the amine group of the next. Chemists would describe a protein as a polyamide, see spread 27.7. Writing out the full amino acid structure for each amino acid in a protein would be very tedious, and so primary structure is usually depicted by a three-letter short-hand notation, e.g. **ala** for alanine, **leu** for leucine, and so on, arranged in the appropriate sequence.

The sequence of amino acids in a protein is determined experimentally by reacting the amino acid at one end of the molecule with a substance that creates a coloured or fluorescing derivative. The protein is then hydrolysed by a highly specific enzyme that removes only the terminal amino acid. The marked amino acid is identified by chromatography (see next chapter). Repeating the process many times determines the full primary structure.

H_2N—Val (1)—Leu—Ser—Glu—Gly (5)—Glu—Trp—Gln—Leu—Val (10)—Leu—

—Ala—Ala—Lys (145)—Tyr—Lys—Glu—Leu—Gly (150)—Tyr—Gln—Gly—COOH

This sequence of amino acids gives the primary structure of sperm whale myoglobin.

Secondary structure

The chain of amino acids that makes up the primary structure of a protein can fold itself in two ways, depending on the sequences of amino acids that are next to each other. Hydrogen bonds hold the folded structures in place. This folding of the primary structure is called the **secondary structure**.

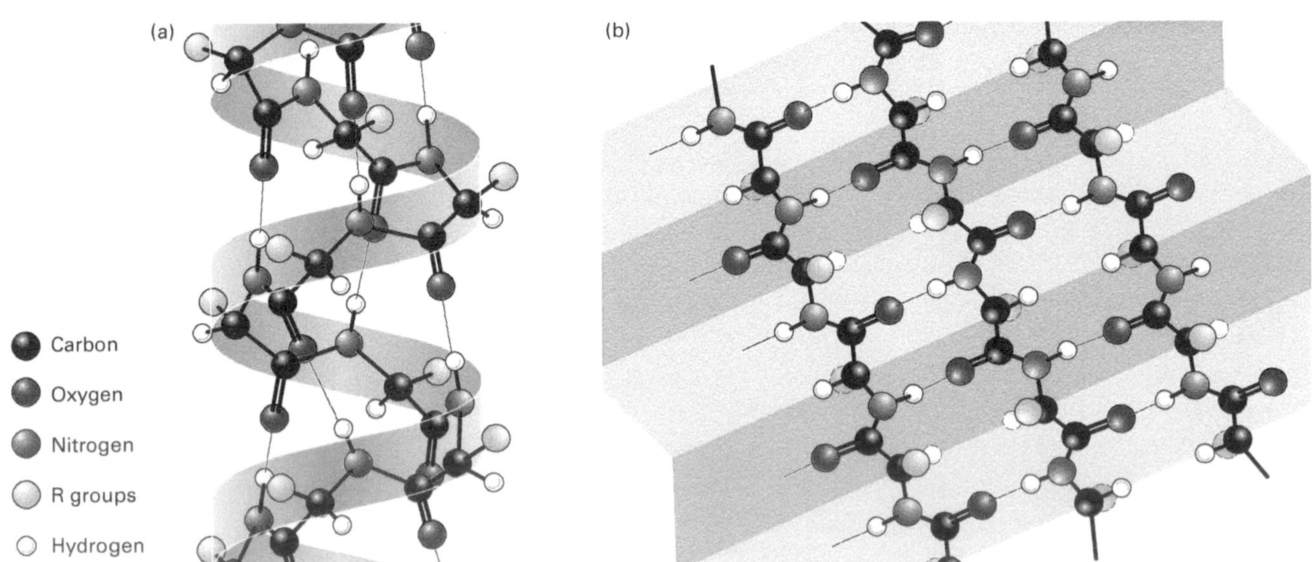

*Hydrogen bonding gives rise to the two forms of secondary structure: (a) the **alpha helix**; and (b) the **beta pleated sheet** (here, three-stranded).*

Tertiary structure

There is overall folding of the chain held by interactions between more distant amino acids. This is called the **tertiary structure**. These interactions include hydrogen bonds and **disulphide bridges** (covalent bonds that form between sulphur atoms on the oxidation of two cysteine amino acids, spread 28.5) as well as ionic interactions and intermolecular forces.

Quaternary structure

Finally, some structure results from interaction between separate protein chains. This is called **quaternary structure**. Not all proteins have a quaternary structure; myoglobin, for example, has only one chain. Haemoglobin has a quaternary structure that includes four protein chains – it has two alpha chains and two beta chains. (Note that the terms 'alpha and beta chains' should not be confused with the terms 'alpha helix' and 'beta pleated sheet' used to describe secondary structure.)

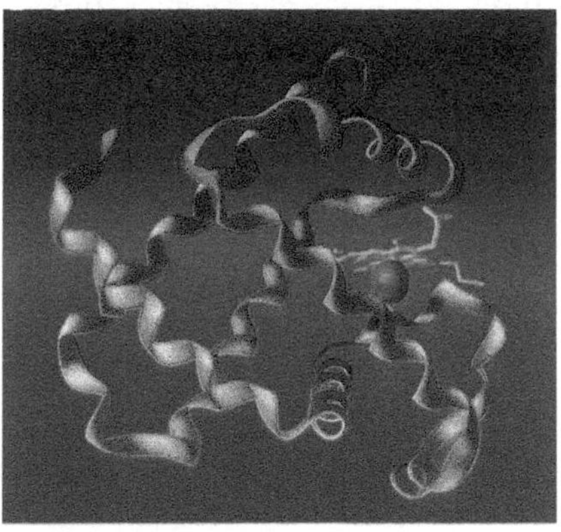

This ribbon diagram model shows the tertiary structure of myoglobin. The position of the oxygen molecule when it binds to the haem group is shown in red (and at exaggerated scale).

Fibrous and globular proteins

The proteins present in organisms may conveniently be divided into two classes. **Fibrous proteins** have long molecules, which are strengthened by many cross-links between the chains. Fibrous proteins form structures such as muscle fibres. **Globular proteins** are smaller and are much more round and compact. One example of a globular protein is ferritin, which acts as the primary store of iron inside cells in animals and many plants. Globular proteins have many roles, particularly as **enzymes** and certain **hormones** (carriers of biochemical messages). The structure of proteins, and hence their function, can be disrupted by a number of factors, such as high temperature. The proteins are then said to be **denatured**; this will be considered in detail in the next spread on enzymes.

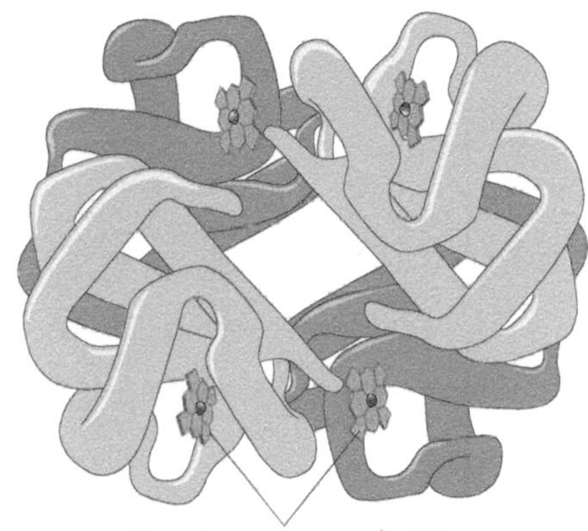

The quaternary structure of haemoglobin consists of two alpha chains and two beta chains arranged at the corners of a tetrahedron.

SUMMARY

- Proteins are condensation polymers of amino acids.
- The structure of proteins is considered at four levels: primary, secondary, tertiary, and quaternary.
- Primary structure: the amino acid sequence in the polypeptide chain.
- Secondary structure: the alpha helix and beta pleated sheet held by hydrogen bonding between adjacent sections of polypeptide chain.
- Tertiary structure: folding of a protein molecule held by interactions between distant amino acids.
- Quaternary structure: fitting together two or more separate protein chains to give the final physiologically active protein.
- Proteins may be fibrous or globular.
- Proteins may be denatured.

Hydrolysis

Acidic hydrolysis (see spread 27.5) of proteins, for example by hot hydrochloric acid, forms amino acids, see spread 28.5.

PRACTICE

1 Refer to the amino acid structures in chapter 28 'Amines and amino acids'.

  a Show how threonine and alanine condense together to make a dipeptide.

  b Explain why lysine, a basic amino acid, and glutamic acid may together be responsible for the tertiary structure of a protein.

2 The atoms in the peptide link are in the same plane. There is limited rotation about the C—N bond. Explain why the carbonyl ($\supset$C=O) oxygen atom and the imino ($\supset$N—H) hydrogen atom do not align *cis* to each other. [Hint: refer to the R groups on each α-amino acid.]

3 Which of the three structural considerations – secondary, tertiary, or quaternary – determines whether a protein is globular or fibrous? Give the reason for your answer.

30.6

OBJECTIVES

- Lowering of activation energy
- Active site
- Lock-and-key and induced-fit models
- Denaturation

The word 'enzyme'

The word 'enzyme' describes any molecule that acts as a biological catalyst by accelerating specific biochemical reactions without itself being permanently altered. Nearly all enzymes are proteins; the remainder are either ribozymes that consist of pure RNA (see spread 30.8) or RNA-protein complexes. The 1989 Nobel Prize was awarded jointly to Sidney Altman and Thomas Cech for their discovery of the catalytic properties of RNA.

Equilibrium constants

In common with other catalysts, enzymes do not alter the position of an equilibrium, i.e. they do not change the value of the equilibrium constant. Enzyme action causes equilibrium to be achieved more rapidly.

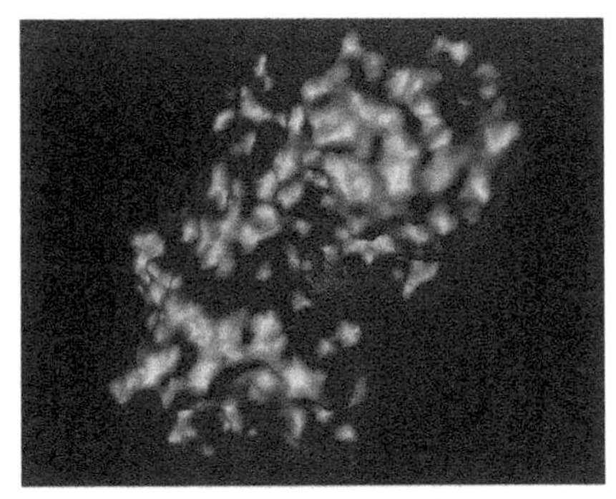

Computer graphics model of hexokinase. When glucose (orange) enters the active site, it causes the protein to change shape, illustrating the induced-fit model.

Stereospecific active site

The active site of the digestive enzyme α-chymotrypsin presents three loci to its substrate: a site containing a serine residue (see spread 28.5) that bonds to the substrate, a hydrophobic area (see spread 30.1) for amino acid side-chain binding, and a hydrogen-bond acceptor site. The active site can therefore bind one enantiomer (see spread 28.6) better than the other.

Enzymes – 1

Enzymes are biological catalysts. In these two spreads we focus exclusively on proteins, because nearly all enzymes are proteins. They increase the rate of biochemical reactions, which would otherwise be so slow that life would not be possible. For example, respiration releases energy from the oxidation of glucose. The overall reaction may be represented as:

$$C_6H_{12}O_6 + 6O_2 \rightarrow 6CO_2 + 6H_2O$$

At the temperature of 37 °C found in the human body, the reaction as written proceeds at a negligible rate. However, glucose is oxidized in organisms by means of a complex metabolic pathway called **glycolysis**, which involves a large number of different enzymes, see final spread in this chapter. Each enzyme is specific to one reaction in the pathway. The overall process of glycolysis produces as much energy per molecule of glucose as does the combustion of glucose in air (following Hess's law; see chapter 10 'Thermochemistry').

The active site

In common with all catalysts, enzymes speed up a reaction by providing an alternative route of *lower activation energy* (see chapter 15 'Chemical kinetics'). They do this by binding the reactant molecule, called the **substrate**, and holding it in a favourable orientation for reaction. The mechanism may be summarized as:

enzyme + substrate → enzyme/substrate complex → enzyme + product

Note that the enzyme, like any other catalyst, is released *unchanged* at the end of the reaction.

The part of the enzyme molecule that binds the substrate is known as the **active site**. There are two models used to describe how the active site works. In one, the *lock-and-key model*, the active site accurately fits its particular substrate molecule (as shown in the illustration below). Alternatively, the *induced-fit model* visualizes the active site changing shape slightly in response to its particular substrate. In both cases the active site is *very specific* to the substrate.

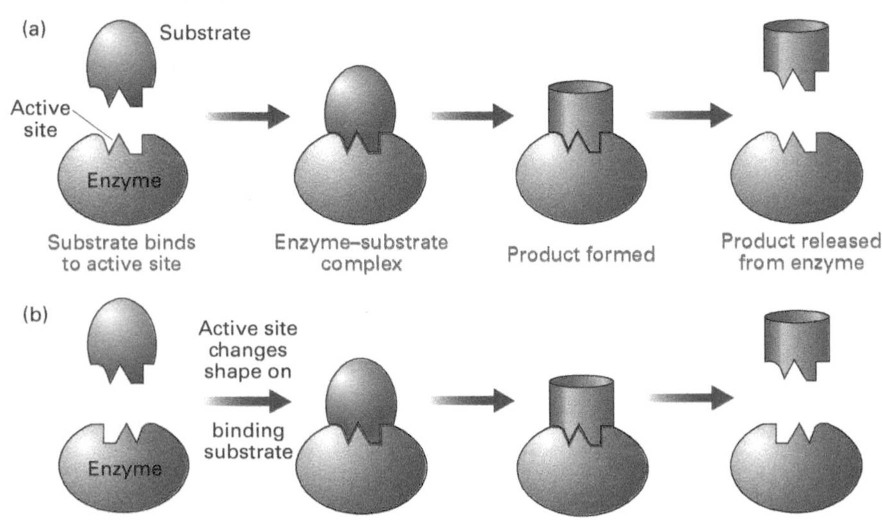

Two models for binding a substrate (reactant) to an enzyme: (a) lock-and-key model; and (b) induced-fit model. Note that the reaction occurs after the substrate has bound to the enzyme; release of the product leaves the enzyme in its initial state and shape.

TIM

The enzyme triose-phosphate isomerase (abbreviated name TIM) interconverts two triose monosaccharides. TIM is spectacularly efficient: one molecule (M_r = 43 000) can catalyse the reaction of 400 000 substrate molecules per second, the limiting factor being the rate of diffusion of substrate to the enzyme. TIM is a component of a variety of biochemical pathways. In muscle, it is involved in the respiration of carbohydrate to produce energy; in green plants, it is involved in the photosynthetic reactions that produce carbohydrates. Its role in the photosynthetic pathway is the reverse of its role in respiration. TIM illustrates many of the general properties of enzymes:

- Enzymes greatly increase the rate of reactions that otherwise would be too slow.
- Enzymes are globular proteins consisting of coiled polypeptide chains.
- Enzymes are highly specific: a given enzyme catalyses a single reaction or a limited class of reactions.
- Substrate and product bind reversibly to the active site.
- The names of most enzymes end in *-ase*, this suffix often being added to the name of the substrate.

TIM catalyses just one reaction out of many hundreds in the biochemical pathway that connects carbohydrate with carbon dioxide, water, and energy. In animals, this pathway is catabolic; in plants, anabolic (see spread 30.11). TIM interconverts glyceraldehyde 3-phosphate (an aldotriose, spread 30.2) with dihydroxypropanone phosphate (a ketotriose).

Denaturation

Most enzymes are destroyed by high temperature because the increased kinetic energy causes the protein structure to break down. Loss of tertiary structure means that the enzyme no longer has the specific shape required for its correct function. The loss of biochemical activity through structural change is called **denaturation.**

The graph to the right plots reaction rate against temperature for an enzyme-catalysed reaction. Note that the optimum temperature for this enzyme is about 45 °C. Above this temperature, the reaction rate falls as the enzyme starts to denature and lose its function. Different enzymes have their own optimum temperature; for human enzymes, the optimum is around 37 °C (body temperature).

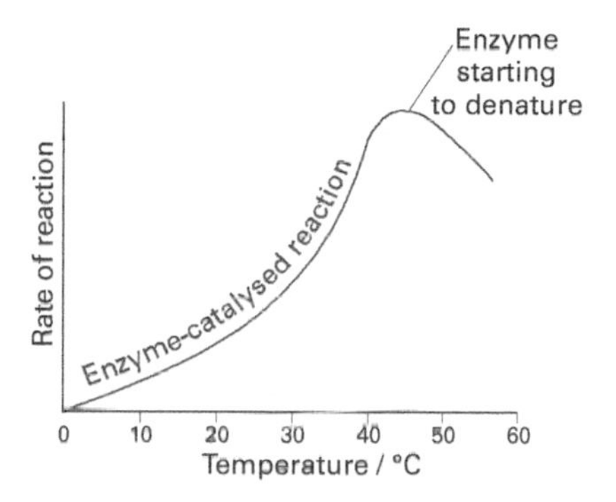

Temperature dependence of enzyme-catalysed reactions. Lactase is an enzyme that breaks the disaccharide lactose ('milk sugar') into the monosaccharides galactose and glucose. The enzyme is denatured at temperatures above about 45 °C, causing the reaction rate to decrease.

Extremes of pH also cause denaturation, and each enzyme also has an optimum pH at which it works best. The illustration below shows a graph of reaction rate against pH for the enzyme fumarase.

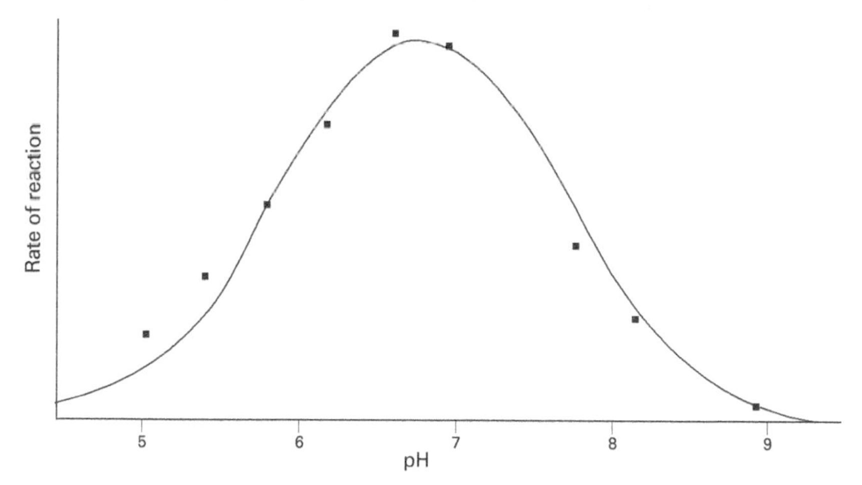

Fumarase has an optimum pH of about 6.5.

The process of cooking (heating) a protein food denatures the protein. For example, boiling an egg causes the protein albumin in the white portion to coagulate and harden. Decreasing the pH also brings about denaturation, as when pickling in vinegar. Note that denaturation destroys an enzyme *irreversibly*.

SUMMARY

- Enzymes are biological catalysts.
- Enzymes have an active site that binds to the substrate.
- Enzyme action is described by the lock-and-key and induced-fit models.
- Enzymes are denatured by high temperature and extremes of pH.

30.7

OBJECTIVES

- Enzyme inhibition
- Immobilized enzymes
- Cofactors

Enzymes – 2

Enzymes are specific to the reactions they catalyse. They are sensitive to temperature and pH, high or low values of these causing denaturation and hence loss of activity. Enzymes can also be *poisoned* in the same manner as non-biological catalysts (see chapter 15 'Chemical kinetics'), and they can be *inhibited*, which decreases activity. Some enzymes are active only in the presence of a substance called a **cofactor**, which acts to 'switch on' the enzyme. Despite the sensitivity of enzymes to their environment, they are used increasingly in industrial processes. Most enzymes act only on a specific substrate, and this specificity as well as excellent conversion efficiencies makes enzymes ideal for carrying out the conversion of a given organic substance in industry.

Enzyme inhibition

Enzyme action can be **inhibited** by the presence of another molecule, with the result that activity decreases. Inhibition may be irreversible, in which case the enzyme is permanently 'poisoned'. This happens when the inhibitor forms a covalent bond with the enzyme which is difficult to break. Conversely, reversible inhibition is an important mechanism for controlling enzyme reactions. A **competitive inhibitor** is a molecule that competes with the substrate for position at the active site. The enzyme molecule 'recognizes' the inhibitor molecule as substrate and so binds to it, but cannot convert it into product. Competitive inhibitors are often chemically very similar to the substrate molecule. A **non-competitive inhibitor** does not bind to the active site, but changes the shape of the active site when it binds to another part of the enzyme. A non-competitive inhibitor is often a heavy metal ion. The enzyme is thus prevented from binding to its substrate. The illustration below shows the distinguishing features of competitive and non-competitive inhibition.

Computer-aided design

Computer-modelling software (see appendix B.1) can be used to investigate the binding of a potential drug to the active site of an enzyme, with the aim of blocking the active site by competitive inhibition.

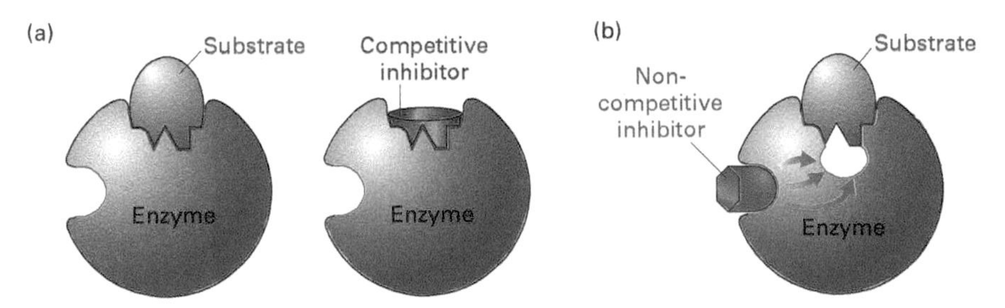

The mechanism of inhibition. (a) Competitive inhibition: both substrate and inhibitor can bind to the active site on the enzyme and hence compete for it. (b) Non-competitive inhibition: the inhibitor binds to a site elsewhere on the enzyme.

Immobilized enzymes

Many enzyme-catalysed reactions have become industrially important. For example, corn syrup is a cheap source of the sugar glucose. Glucose can be converted into the much sweeter (and therefore more valuable) sugar fructose by the enzyme glucose isomerase. A major problem lies in separating the contaminating and expensive enzyme from the product after the reaction.

Immobilization is an industrial process whereby an enzyme is attached to a solid support that does not interfere with its catalytic activity. There are several methods of attaching the enzyme to its support. For example, the enzyme may be trapped within the framework of a polymer, or it may be adsorbed onto the surface of an inert substance such as stainless steel, glass, or cellulose. The support is usually in the form of small beads, which can easily be separated from the product after reaction. The enzyme may be re-used many times. Immobilizing enzymes in this way mirrors the situation in living cells. The majority of enzymes in organisms are anchored within organelles – minute structures inside cells that have a specific specialized function.

Commercial uses

Enzymes are used in a number of commercial enterprises. Prominent among these are the industries based on fermentation (spread 25.1) and the use of enzymes in biological washing powders.

Cofactors

Some enzymes cannot function without the presence of a cofactor. **Vitamins** are substances that must be present in small amounts in food to maintain health, and many vitamins are cofactors. Two important cofactors are **NAD^+** (nicotinamide adenine dinucleotide) and **$NADP^+$** (nicotinamide adenine dinucleotide phosphate), which are synthesized from the B vitamin **niacin**. NAD^+ and $NADP^+$ are examples of a type of cofactor known as a **coenzyme**, because they must be present for the enzyme to work, but they are not bound to it, or are bound reversibly. NAD^+ and $NADP^+$ are essential to the functioning of certain dehydrogenase enzymes that catalyse redox reactions. They are involved in the transport of electrons and hydrogen ions during the oxidation of food to release energy (respiration).

FAD (flavin adenine dinucleotide) is synthesized from the B vitamin **riboflavin**. FAD is a type of cofactor known as a **prosthetic group**, because it binds tightly to the enzyme with which it works. FAD is associated with the enzyme succinate dehydrogenase, for example. **Coenzyme A** is a cofactor synthesized from the B vitamin **pantothenic acid**, which is present in cereal grains, egg yolk, liver, and peas. Coenzyme A is involved in reactions that transfer acyl groups, and it occupies a central position in metabolism (see last spread in this chapter). Some cofactors are metal ions, for example Ca^{2+}, Mg^{2+}, and Fe^{2+}. They form complexes with the lone pair on some amino acids, thus sustaining the tertiary structure of the protein.

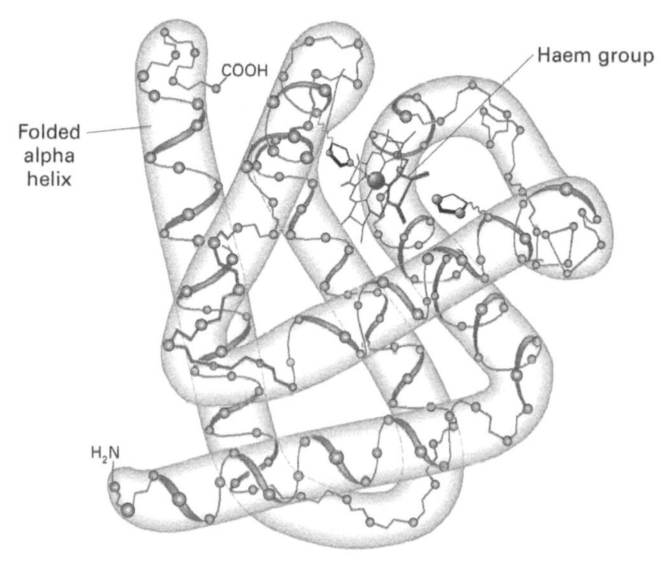

Myoglobin acts as a store for oxygen. An oxygen molecule attaches directly to the Fe^{2+} ion at the centre of a haem group. This group holds the myoglobin polypeptide chain in its tertiary structure.

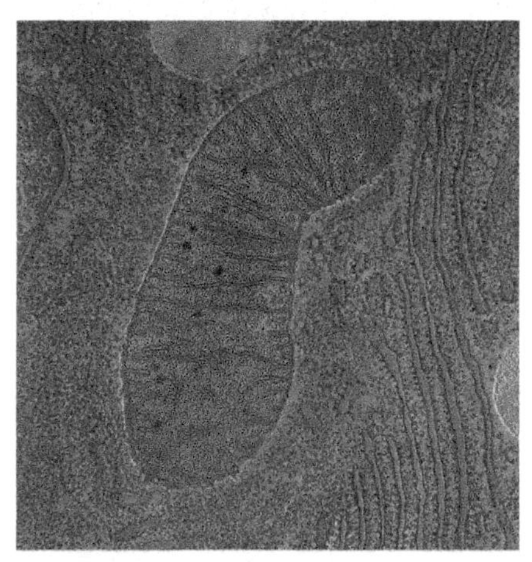

Mitochondrion: the 'powerhouse' of the cell. Mitochondria are constructed mostly from lipids and proteins. They contain the enzymes and coenzymes that control the energy-releasing reactions of respiration. The reactions take place in the numerous micrograms that coat the highly-folded inner membrane.

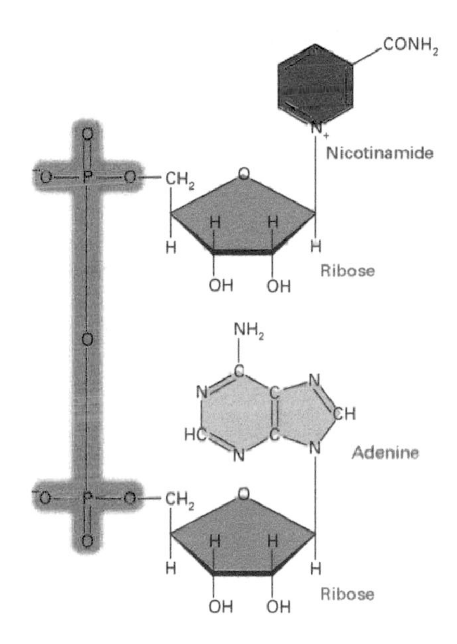

The structure of NAD^+. The molecule consists of the two nitrogen-containing bases nicotinamide and adenine, each bonded to the monosaccharide ribose. Nucleotides are discussed in detail in the next spread. (The lengths of two P—O bonds on the left are exaggerated.)

SUMMARY

- Enzyme inhibitors may be competitive or non-competitive.
- Enzymes used in industrial reactions are often immobilized.
- Some enzymes cannot work without the presence of a cofactor.
- Important cofactors are NAD^+, $NADP^+$, and FAD.

PRACTICE

1 Pectin is a polysaccharide that forms a gel with sucrose. It is widely used as an additive in commercial jam-making. Pectin is present in the germinated grains used for brewing beer. Explain why brewers add the digestive enzyme pectinase to beer during fermentation. Suggest the action of this enzyme.

2 Explain the relationship between the terms 'cofactor' and 'prosthetic group'.

3 An inhibitor lowers the rate at which an enzyme converts substrate to product. Why might a biological system need an inhibitor?

4 Explain the differences between:

a Reversible and non-reversible inhibition

b Competitive and non-competitive inhibition.

30.8

OBJECTIVES

- Nucleosides and nucleotides
- Nucleic acids as polynucleotides
- RNA and DNA
- Ribose and deoxyribose
- Nitrogenous bases

Nucleotides

Nucleotides are widespread throughout biological systems. For example, you have already met NAD^+ (nicotinamide adenine dinucleotide) and FAD (flavin adenine dinucleotide). These two substances each consist of just *two* nucleotides bonded together. Nucleic acids are polymers consisting of chains of *thousands* of nucleotides.

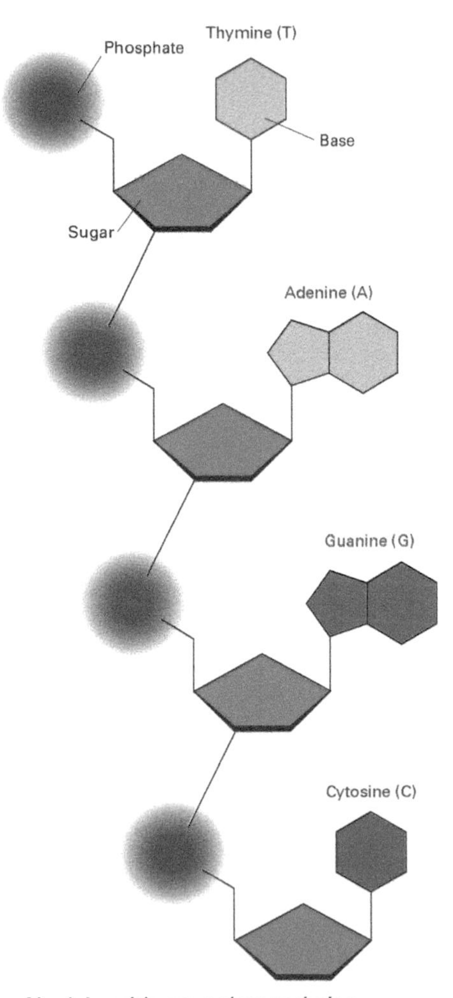

Nucleic acids are polymer chains consisting of phosphate groups alternating with sugar residues. Nitrogenous bases project from the sugar residues of the sugar–phosphate backbone.

NUCLEIC ACIDS – 1

Nucleic acids are the molecules responsible for storing information in a cell and for determining inherited characteristics in an organism. The structure of nucleic acids encodes the information necessary to synthesize all the proteins in an organism. The nucleic acid DNA controls inheritance as its structure is copied and passed on through successive generations. This spread introduces the fundamental structure of nucleic acids; the following two spreads discuss their function.

Outline structure

Nucleic acids are **polynucleotides** – they are polymer chains made up from repeating nucleotide units. A **nucleotide** consists of a phosphate group attached to a nucleoside; a **nucleoside** consists of a nitrogenous (nitrogen-containing) organic base attached to a sugar residue.

The sugar residues in DNA and RNA

There are two types of nucleic acid: **deoxyribonucleic acid** (**DNA**) and **ribonucleic acid** (**RNA**). Both nucleic acids have the same general structure but contain different sugar residues. As the names imply, DNA contains deoxyribose and RNA contains ribose.

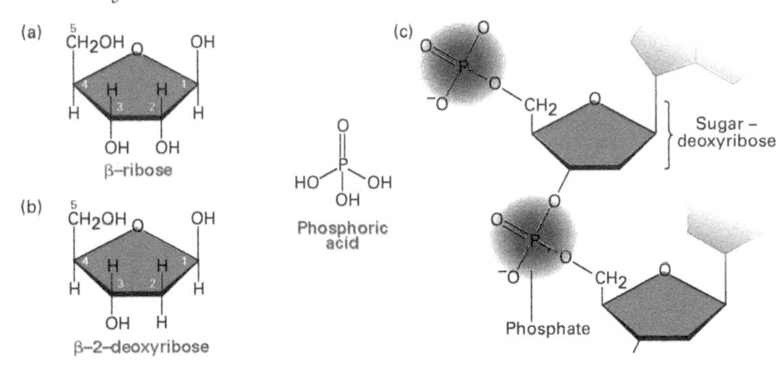

Ribose and deoxyribose. (a) Ribose normally exists as a five-membered ring; the base forms a β-glycosidic link with ribose and so the ring drawn has the systematic name β-ribose. (b) Deoxyribose has the same structure as ribose, with the exception that the hydroxyl group on carbon atom 2 is replaced by a hydrogen atom. Systematic name is β-2-deoxyribose. (c) Deoxyribose bonds to two phosphate groups by means of ester links involving the hydroxyl groups on carbon atoms 3 and 5 to form the sugar–phosphate backbone of DNA.

The nitrogenous bases

Each type of nucleic acid has *four* different nitrogenous bases. DNA contains **adenine** (A), **guanine** (G), **cytosine** (C), and **thymine** (T). RNA has adenine, guanine, cytosine, and **uracil** (U). Cytosine, thymine, and uracil belong to a class of compounds called the **pyrimidines**, whose structures are based on a single nitrogen-containing ring. Adenine and guanine are **purines**, based on two fused nitrogen-containing rings.

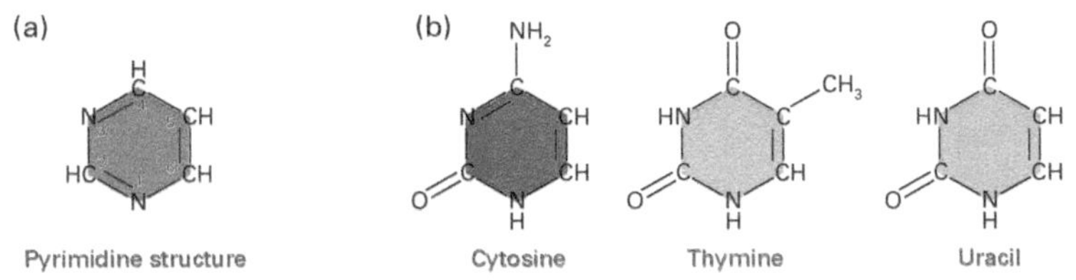

(a) Pyrimidine structure: in a nucleic acid, pyrimidines bond via the nitrogen atom at position 1 to carbon atom 1 on the sugar. (b) The chemical structures of the pyrimidines cytosine, thymine, and uracil.

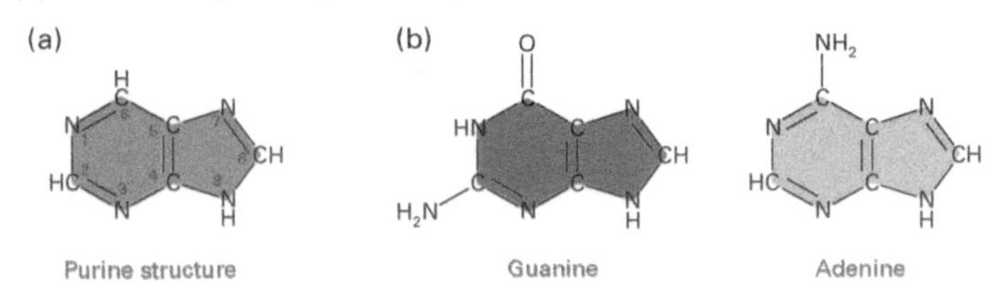

(a) Purine structure: in a nucleic acid, purines bond via the nitrogen atom at position 9 to carbon atom 1 on the sugar. (b) The chemical structures of the purines guanine and adenine.

Detailed chemical structure

As you will see in the following spread, the full structure of DNA involves *two* polynucleotide chains twisted together. The diagrams below show the structures of the single polynucleotide chains of RNA and DNA. The actual bases present in a portion of DNA depend on the function of that portion.

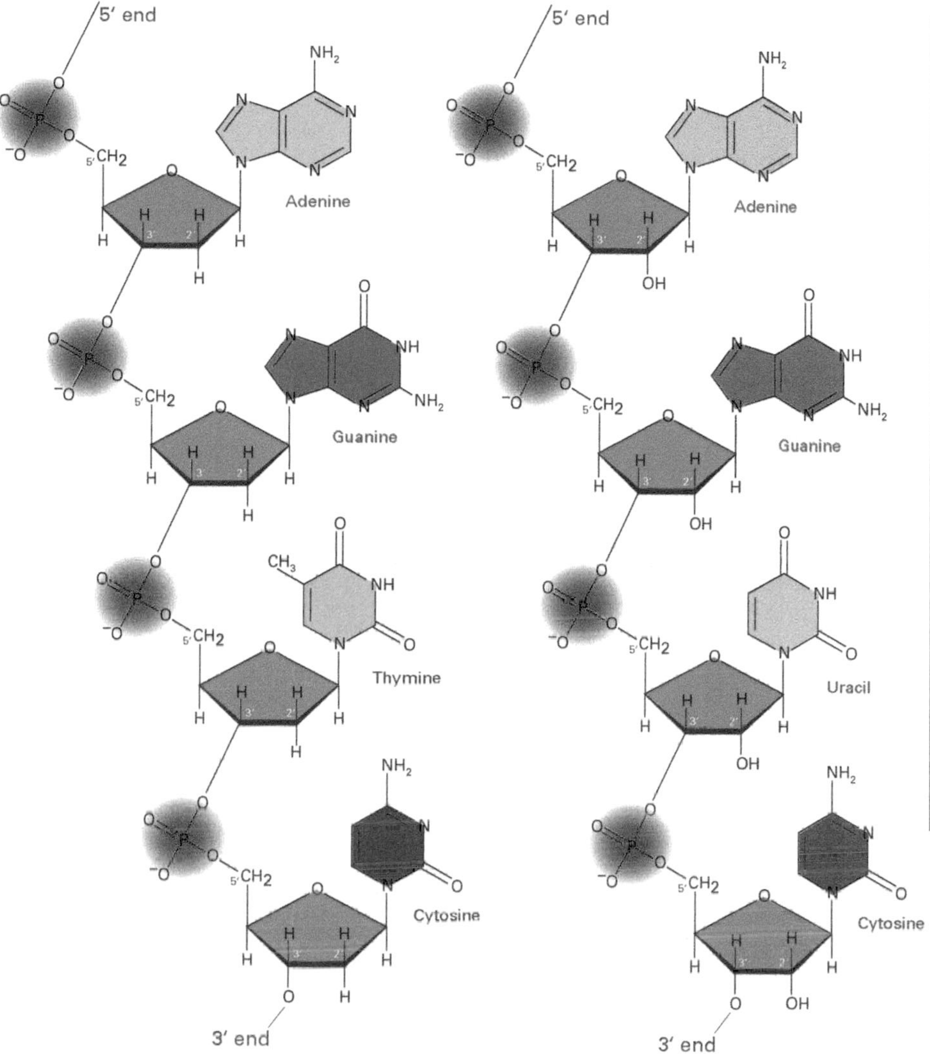

The chemical structures of the polynucleotides making up DNA (left) and RNA (right). The primes refer to positions in the sugar rings (numbers without primes refer to the position in the bases).

The unravelling of DNA

In 1950, the underlying control mechanism for life's processes was not understood. It *was* clear that lipids and carbohydrates were not sufficiently complex to be responsible. Most research workers, including Max Perutz at Cambridge (see spread 20.10), suspected that proteins held the key to life, as they obviously *did* have sufficient complexity.

In 1951, a young post-doctoral student, Jim Watson, asked to work in Perutz's group. Watson was a 'believer' that the molecules that control life are the nucleic acids. The next couple of years saw Watson successfully prove his hunch. He and Francis Crick were crucially aided by the diffraction photograph (shown on the next spread) taken by Rosalind Franklin. Watson and Crick came up with a structure for DNA which accounted in detail for the X-ray pattern. Immediately they had found the structure, they were able to suggest a mechanism for self-replication and hence explain genetic inheritance.

SUMMARY

- Nucleic acids have a backbone of alternating phosphate and sugar groups from which nitrogenous bases project.
- Nucleoside = nitrogenous base + sugar; nucleotide = nucleoside + phosphate.
- DNA (deoxyribonucleic acid) and RNA (ribonucleic acid) differ in the sugar residue that each contains.
- DNA contains the nitrogenous bases adenine, guanine, cytosine, and thymine. RNA contains uracil instead of thymine.

PRACTICE

1 Explain the difference between a nucleoside and a nucleotide.

2 Draw the structures of β-ribose and β-2-deoxyribose.

30.9

Nucleic acids – 2

OBJECTIVES

- Complementary base pairing
- Double helix
- Replication

The nucleic acid DNA contains polynucleotide chains that consist of phosphate groups alternating with deoxyribose sugar residues bearing nitrogenous bases. The structures of these nitrogenous bases are absolutely fundamental to the functioning of the DNA molecule because they form hydrogen bonds with each other in a very specific manner. These hydrogen bonds hold *two* polynucleotide chains together to give DNA molecules their distinctive shape of a *double helix*.

Interactions between the bases

In DNA, the shapes of and partial charges on the molecules dictate that adenine can *only* hydrogen-bond effectively with thymine, and guanine can *only* hydrogen-bond with cytosine. This restriction is called **complementary base pairing**.

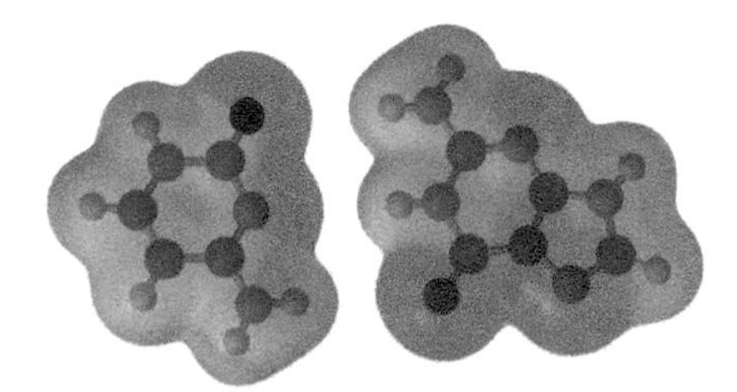

The electrostatic potential map of a cytosine–guanine pair showing how the complementary base pair hydrogen-bonds. The base pair thymine–adenine (T—A) forms only two hydrogen bonds, unlike the three hydrogen bonds formed by the C—G pair.

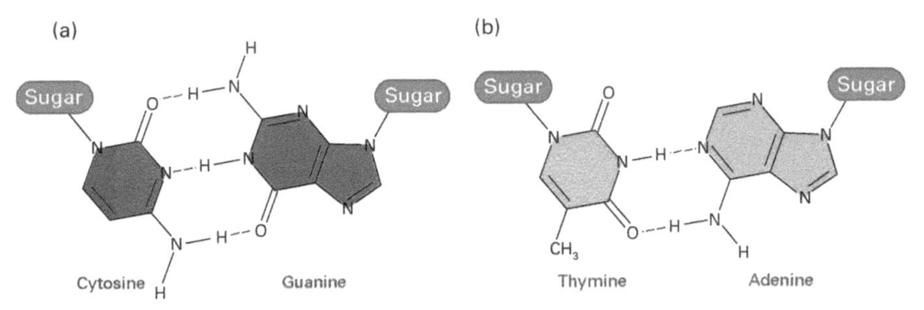

Specific hydrogen bonding between nitrogenous bases: (a) cytosine–guanine (three hydrogen bonds); and (b) thymine–adenine (two hydrogen bonds).

The chemical structure of DNA

DNA consists of two strands of polynucleotide chain held together by hydrogen bonds between complementary bases. The resulting structure of DNA is a double helix in which each strand spirals about the other. DNA can make exact copies of itself. This is how it passes on hereditary information to successive generations. Copies of DNA are made during cell division.

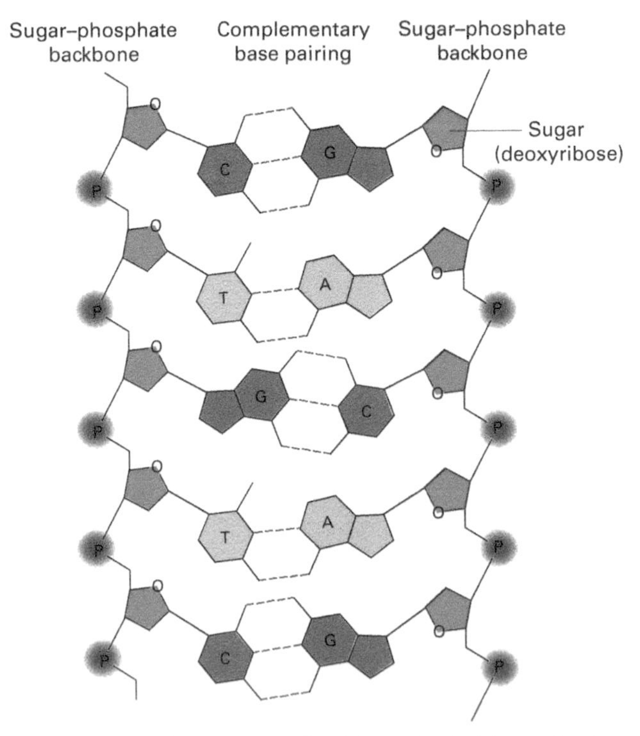

A DNA molecule consists of two polynucleotide chains held together by complementary base pairing. There are 10 base pairs per turn of the sugar–phosphate helix. On the opposite page, we show how the shape of the phosphate group and the carbon atom C5 outside the ring causes the bonds to successive bases (shown in red) to be twisted relative to each other by 36°, forming one of the two helical strands.

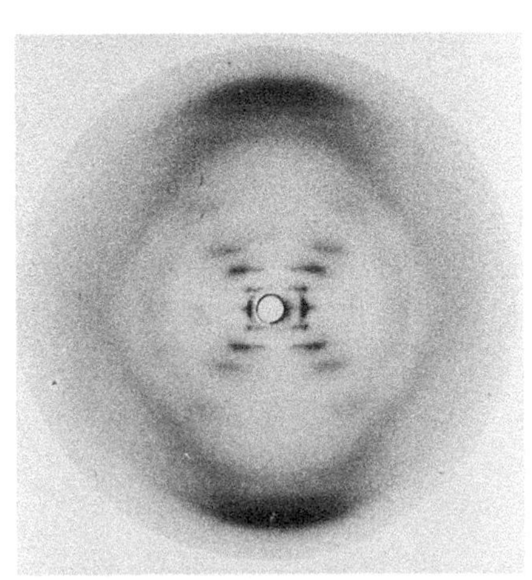

An X-ray diffraction photograph of crystalline DNA prepared by Rosalind Franklin in 1953. This is the photograph that provided Francis Crick and James Watson with the necessary information to deduce their structure for DNA, which was awarded the 1962 Nobel Prize in Medicine. The X-shaped pattern at the centre indicates the presence of a helix. The arcs at the top and bottom of the picture show that the structure repeats every 3.4 nm along the axis of the helix.

Replication

The complete **genetic code** of an organism is called its **genome.** Genetic information is stored in DNA in the sequence of base pairs along the polynucleotide chains. The DNA is packaged in structures called **chromosomes** in the nucleus of each cell. Each cell contains an identical copy of the genome, which carries all the information needed to determine the structure of the entire organism and to control the function of its cells. Sections of the code are switched 'on' or 'off' depending on the type and function of each specific cell. The key property that enables a nucleic acid molecule to perform its function is the ability to make exact copies of itself. This process is called **replication.**

Complementary base pairing allows replication to take place by a mechanism first proposed by Meselson and Stahl. In this mechanism, the double helix of DNA 'unzips', revealing unpaired bases. Free nucleotides present in the cell then pair with these bases and form a chain exactly like the original complementary chain. The mechanism is termed '**semi-conservative**'. This term indicates that the two daughter double-stranded DNA molecules each contain *one* of the parent strands.

In the following spread, you will see how the information stored in the structure of DNA is used to control protein synthesis within cells.

The Meselson and Stahl experiment

Matthew Meselson and Franklin Stahl grew the bacterium *E. coli* in a medium in which the only source of nitrogen was the heavy isotope ^{15}N. After several generations, the bacterial DNA was shown to contain only ^{15}N in its bases. These bacteria were then transferred to a medium containing ordinary ^{14}N. The bacteria were allowed to reproduce, and their DNA was analysed from generation to generation. It was noted that, after a single generation, the DNA was a hybrid of light and heavy forms. In subsequent generations, the light form gradually became dominant. These results were consistent with the semi-conservative mechanism.

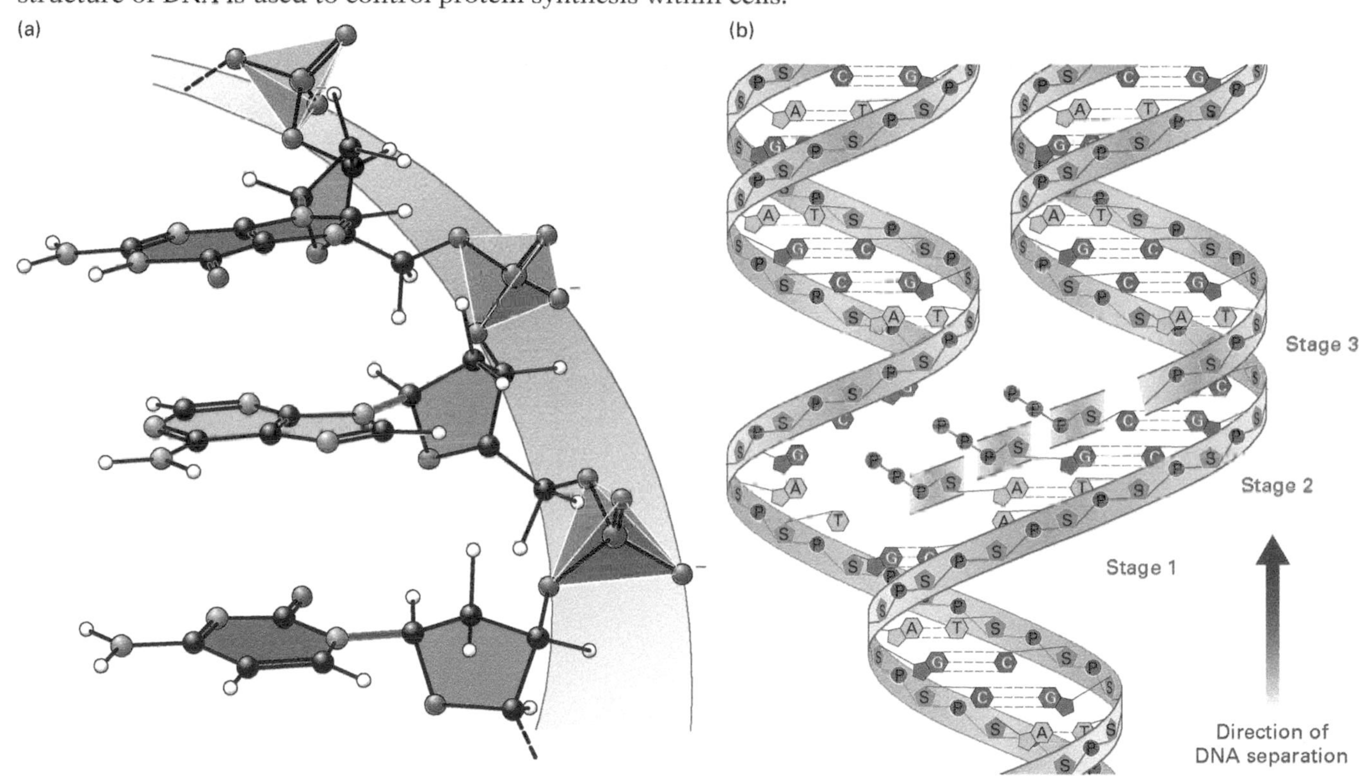

(a) An accurate represention of the three-dimensional structure, showing how the bases twist relative to each other. (b) An outline of DNA replication shown in three stages. Stage 1: The double helix of DNA uncoils and separates. Stage 2: Exposed bases act as a template, which selects complementary bases (as nucleoside triphosphates) from the surroundings. Stage 3: The nucleotides join in sequence to make a new strand of nucleic acid.

SUMMARY

- DNA consists of two polynucleotide chains that are held together by complementary base pairing.
- In complementary base pairing, adenine forms hydrogen bonds with thymine, and guanine with cytosine.
- The two polynucleotide chains in DNA are held in the shape of a double helix.
- DNA replicates by (a) breaking the hydrogen bonding between the complementary bases and then (b) using the two polynucleotide strands as templates: complementary bases (as nucleotides) join in sequence along each strand to form two identical strands of DNA.

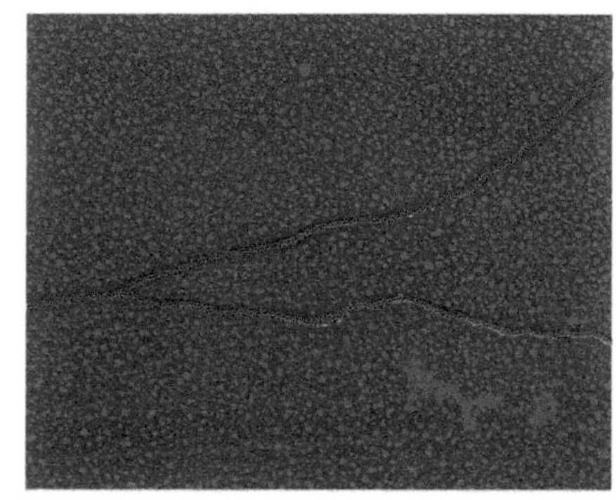

The double helix of DNA uncoils to produce two strands in this false-colour image.

30.10 NUCLEIC ACIDS – 3

OBJECTIVES

- Transcription into RNA
- Translation of RNA into protein
- Triplet code
- Protein synthesis
- Anti-viral drugs

A genetic disorder

Sickle-cell anaemia is a disease in which red blood cells are distorted and have a reduced ability to transport oxygen. In normal haemoglobin, the amino acid at position 6 in the β-polypeptide chain is glutamic acid. In sickle-cell anaemia, this amino acid is valine. This change in one amino acid has come about because of a faulty sequence of nucleotides in the section of DNA that codes for haemoglobin synthesis. This disease is hereditary: it can be passed from parents to their children. An *advantage* of this faulty sequence of nucleotides is that it protects carriers of sickle-cell anaemia from the worst effects of malaria, which is constantly present in the areas of Africa where sickle-cell anaemia is most common.

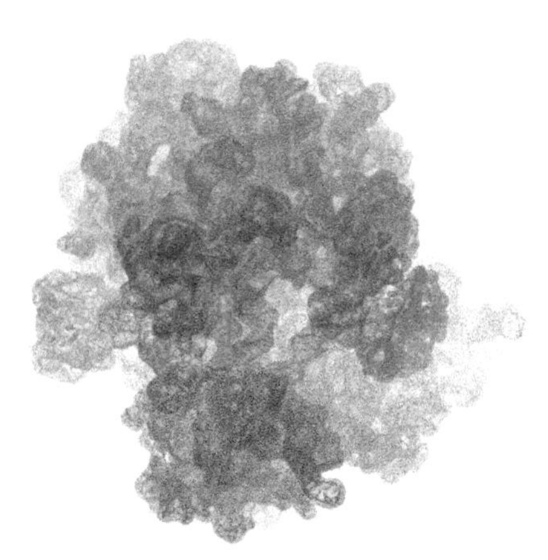

This recent (2009) cryo-electron microscopy image, interpreted using molecular dynamics, shows that the E. coli *ribosome consists of two subunits, the larger coloured cyan and the smaller yellow. Three tRNA molecules (orange, green, and pink) plus a complex incorporating the elongation factor Tu (red) are bound to the ribosome. [This image is owned by the Theoretical and Computational Biophysics Group, NIH Resource for Macromolecular Modeling and Bioinformatics, at the University of Illinois.]*

The previous two spreads discussed the structure of DNA. They also described the ability of DNA to make exact copies of itself and so pass on hereditary information. This spread looks at how the information is stored in a genetic code, and how that code is deciphered into instructions that control the synthesis of proteins. Remember that the architecture of an organism depends mainly on the presence of structural proteins; the metabolism of an organism is controlled by enzymes, which are also proteins. The form and function of all organisms depend on their proteins, which in turn depend on the message encoded in their DNA.

DNA and RNA

DNA is the molecule that carries the genetic code. It is located in the chromosomes contained in the nuclei of cells. RNA is the form into which the code is copied. RNA is transported out of the nucleus into the cytoplasm, where it controls the amino acid sequence during protein synthesis.

Messenger RNA

RNA is a single-stranded molecule that is formed by **transcription** from DNA. The DNA molecule 'unzips' to reveal its bases, as in replication. However, in transcription, it is free *ribo*nucleotides (and not *deoxyribo*nucleotides) that base-pair to it and form an RNA molecule. The RNA molecule thus formed, called **messenger RNA (mRNA)**, then moves out of the nucleus of the cell and attaches to another cell organelle called a ribosome. **Ribosomes** are the sites at which proteins are assembled, during a process called **translation**. RNA is responsible for deciphering the genetic code into protein.

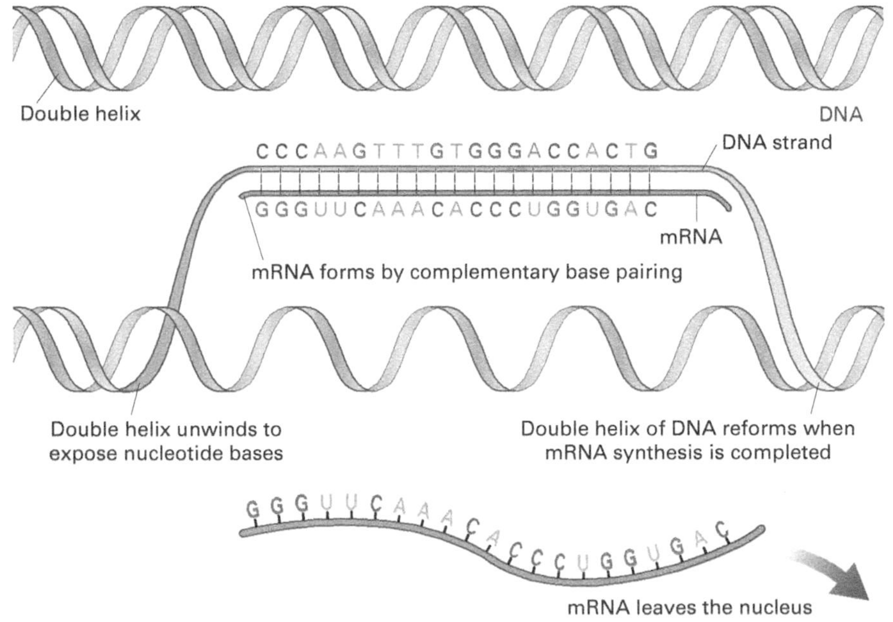

Transcription. All body cells contain identical DNA (with the exception of sperm and ova, the cells that join together to form a new individual in reproduction). However, complex organisms such as humans contain cells that are specialized, arranged in tissues with differing functions. During transcription, the portion of DNA relevant to the function of that cell uncoils and acts as a template for the formation of a complementary chain of messenger RNA.

The triplet code

The primary structure of a protein consists of a chain of amino acids connected by peptide links. There are about 20 naturally occurring amino acids. The structure of DNA includes the four nitrogenous bases adenine, guanine, cytosine, and thymine. The code for each amino acid (called a **codon**) is a *sequence of three bases*.

There are 64 different *triplets* (sequences of three bases) that can be made up by the four bases. As a result, some amino acids are encoded by more than one codon. The codons for some amino acids are given on the right. Of the 64 codons, 61 code for amino acids and 3 act as 'stop' signals that switch off the protein synthesis when the end of the polypeptide chain has been reached.

Amino acid	Codons
Alanine	GCU, GCC, GCA, GCG
Arginine	AGA, AGG, CGU, CGC, CGA, CGG
Glutamic acid	GAA, GAG
Glycine	GGU, GGC, GGA, GGG
Histidine	CAU, CAC
Lysine	AAA, AAG
Phenylalanine	UUU, UUC
Proline	CCU, CCC, CCA, CCG

The codons in mRNA that code for a few amino acids.

Transfer RNA, messenger RNA, and protein synthesis

Protein synthesis takes place at ribosomes located in the cytoplasm. One end of a messenger RNA molecule attaches to a ribosome, which moves along it three bases at a time. Molecules of another type of RNA, called **transfer RNA (tRNA)**, bind to free amino acids in the cytoplasm. Transfer RNA molecules each carry a specific amino acid. They also each have their own base triplet, and this binds to the complementary triplet on the messenger RNA. In this way the messenger RNA determines the order of amino acids. Peptide links form between adjacent amino acids, and the protein chain steadily grows. The illustration below summarizes the process of protein synthesis.

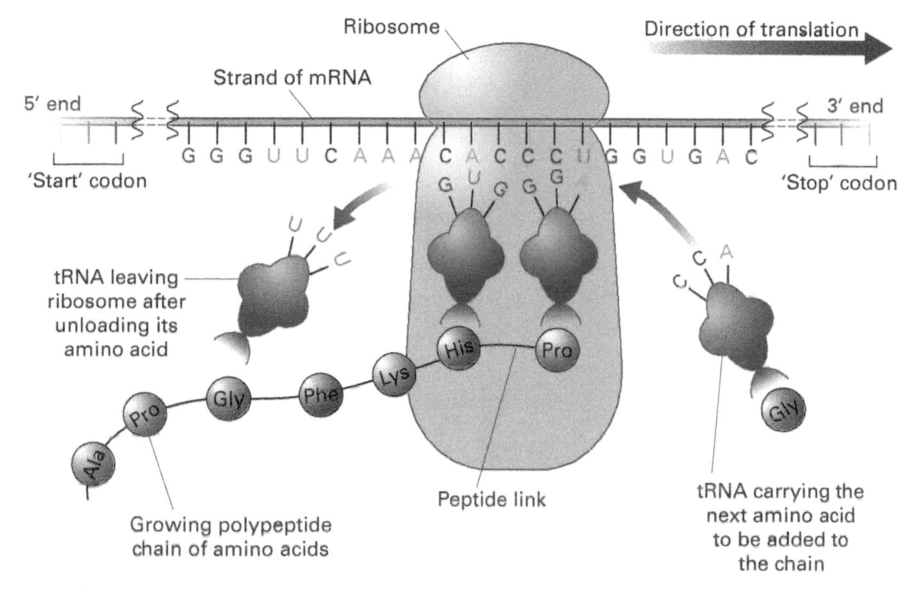

Protein synthesis taking place at a ribosome.

SUMMARY

- The genetic code for each amino acid carried on DNA and messenger RNA is a sequence of three bases – a codon.
- Transcription of DNA forms messenger RNA.
- Translation of the code on messenger RNA results in protein synthesis, which takes place at the ribosomes.

Viruses and anti-viral drugs

Bacteria are microscopic organisms. Some bacteria reproduce in the body and cause disease, for example by producing toxins (e.g. typhoid fever, *Salmonella* 'food poisoning'). Antibiotics are drugs that act against bacteria by disrupting their life processes. Viruses are thousands of times smaller than bacteria, and consist mainly of nucleic acid inside a protein coat. Viruses cause diseases such as the common cold and chickenpox. They inject DNA or RNA into a human (or other host) cell. This viral nucleic acid then overrides the normal function of the cell and instructs it to make more virus particles. The cell ultimately bursts and releases thousands of new viruses, which invade more host cells. Outside host cells, viruses are completely inert; unlike bacteria, viruses do not have life processes or an internal metabolism, and are therefore not affected by antibiotics. A recently developed range of anti-viral drugs are based on the acyclovir molecule. Acyclovir closely resembles the structure of guanine plus ribose, lacking part of the ring and most importantly one of the two hydroxyl groups used to make the polynucleotide chain. A virus incorporating acyclovir cannot continue to build a chain and dies.

PRACTICE

1 Explain the meaning of each of the following terms:
   a Genome
   b Replication
   c Complementary base pairing
   d Nucleic acid.
2 Explain the differences between:
   a A nucleoside and a nucleotide
   b RNA and DNA
   c The functions of transfer RNA and messenger RNA
   d The processes of transcription and replication.
3 Explain why adenine pairs with thymine across the two nucleic acid strands in DNA, whereas adenine and guanine do not pair.
4 Write down the amino acid sequence for the peptide corresponding to the following set of DNA codons: TTC/CGA/CCC/AAG

EXTENSION

30.11

Metabolism

OBJECTIVES

- Anabolism and catabolism
- Photosynthesis
- Glycolysis
- Krebs cycle

Metabolism is the term that refers to the sum total of all the chemical reactions taking place within an organism. There are two broad categories of chemical reactions: **catabolism** refers to any process in which substances are broken down; **anabolism** refers to processes that build up molecules. The catabolic and anabolic processes together make up an organism's metabolism. Reactions usually take place in a series of steps called a **metabolic pathway**. A major concern of any organism is the supply and handling of energy resources. The concluding spread of this chapter considers some of the metabolic pathways associated with this highly important function.

Photosynthesis

It was shown in chapter 14 that, for a reaction to be spontaneous, the standard Gibbs energy change $\Delta G^{\ominus}$ must be negative. Biochemists call such a reaction **exergonic**. Anabolism involves reactions which are endergonic, for which $\Delta G^{\ominus}$ is positive. These reactions can only occur if Gibbs energy is supplied by coupling them so that an endergonic reaction is 'driven' by an exergonic one.

The process of photosynthesis is fundamental to life on Earth because it harnesses the energy of the Sun to drive an endergonic process. This metabolic pathway starts in small green organelles called **chloroplasts** in the cells of plants. Overall:

$$6CO_2 + 6H_2O \rightarrow C_6H_{12}O_6 + 6O_2$$

This simple equation represents two very complex processes: splitting water to produce oxygen and fixing carbon dioxide to produce sugars.

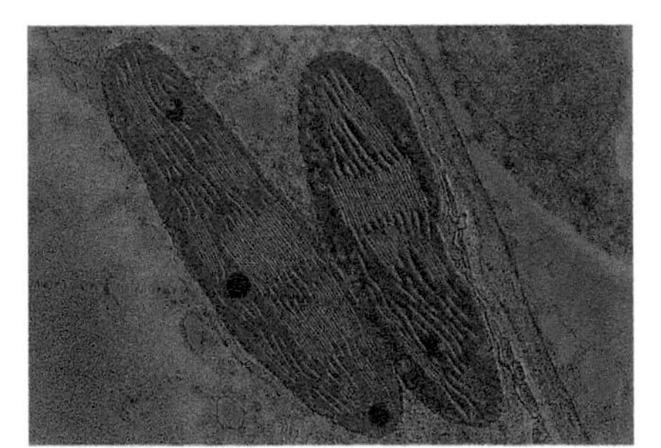

Coloured transmission electron micrograph of two chloroplasts seen in the leaf of a pea plant Pisum sativum.

The splitting of water is called the *light reaction* because it requires the energy of sunlight. The first reaction of photosynthesis is the absorption of light by the green pigment **chlorophyll** which is a complex of magnesium. Activation of chlorophyll by light provides the energy required to synthesize the reductant NADPH.

$$NADP^+ + H_2O \xrightarrow{\text{light/chlorophyll}} NADPH + H^+ + \tfrac{1}{2}O_2$$

The *dark reaction* fixes carbon dioxide to form a series of more-complex organic molecules such as sugars.

Respiration

A major catabolic process is respiration. **Respiration** takes place inside the cells, the net effect being the oxidation of glucose to carbon dioxide and water:

$$C_6H_{12}O_6 + 6O_2 \rightarrow 6CO_2 + 6H_2O$$

This equation is the same as the one written for the burning of glucose in oxygen when the energy released by the reaction appears as heat.

Glycolysis

The first stage of respiration is called **glycolysis** which means 'glucose-splitting'. Glycolysis occurs in the cytoplasm of the cell. The six-carbon molecule of glucose is converted into two three-carbon molecules of **pyruvate** $CH_3COCO_2^-$. The pyruvate is then combined with a very important coenzyme (spread 30.7) called coenzyme A to form **acetyl coenzyme A** (acetyl CoA).

Acetyl coenzyme A holds a key position in metabolism. It is the product not only of the oxidation of carbohydrate by glycolysis, but also of the oxidation of fatty acids and glycerol obtained from the digestion of fats. The metabolic uses of carbohydrate, lipid, and protein from the diet therefore all converge on acetyl CoA as a common intermediate. Acetyl CoA, from whatever source, then enters the Krebs cycle where it is catabolized to carbon dioxide and water in a cyclic series of reactions.

Krebs cycle

The Krebs cycle, discovered by Hans Krebs, is simplified by counting the number of carbon atoms involved at each stage. Two carbon atoms come into the cycle at stage ①; one carbon atom leaves the cycle at stage ③ and another leaves at stage ④.

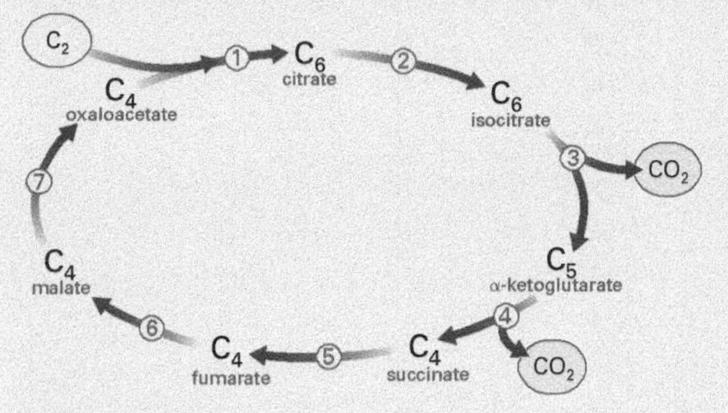

The Krebs cycle

① Incorporation of the two carbon atoms

This stage is the most unfamiliar, involving the addition of acetyl coenzyme A ($CH_3COSCoA$) across the carbonyl bond, followed by hydrolysis

Oxaloacetate $\xrightarrow{CH_3COSCoA}$ intermediate $\xrightarrow[-HSCoA,\ -H^+]{+H_2O}$ Citrate

② Isomerization

This stage occurs by dehydration followed by hydration, with the oxygen attaching to the carbon on the opposite side of the double bond.

Citrate $\xrightarrow{-H_2O}$ intermediate $\xrightarrow{+H_2O}$ Isocitrate

③ Oxidative decarboxylation

This stage occurs by oxidation (loss of hydrogen atoms) followed by loss of carbon dioxide.

Isocitrate $\xrightarrow{NAD^+ \to NADH + H^+}$ intermediate $\xrightarrow[+H^+]{-CO_2}$ α-ketoglutarate

④ Oxidative decarboxylation

This stage is the most complex, despite the simple change to the molecule and involves a complex of enzymes. It is specifically the carboxyl group attached to the carbonyl that is lost. The resulting succinate molecule is symmetrical. Up to this point the acetyl group is intact: in succinate the distinguishing green colour is absent.

α-ketoglutarate $\xrightarrow{NAD^+ + H_2O \to NADH + H^+ + CO_2}$ Succinate

⑤ Oxidation

This stage involves oxidation (loss of hydrogen atoms).

Succinate $\xrightarrow{FAD \to FADH_2}$ Fumarate

⑥ Hydration

This stage involves hydration of a double bond; as the molecule is symmetrical, it does not matter which carbon is attacked.

Fumarate $\xrightarrow{+H_2O}$ Malate

⑦ Oxidation

This stage involves oxidation (loss of hydrogen atoms).

Malate $\xrightarrow{NAD^+ \to NADH + H^+}$ Oxaloacetate

SUMMARY

- The processes that break down molecules are called catabolic.
- The processes that build up molecules are called anabolic.

PRACTICE EXAM QUESTIONS

1 **a** The graphical formulae of alanine and glycine, two important amino acids, are given in the figure below.

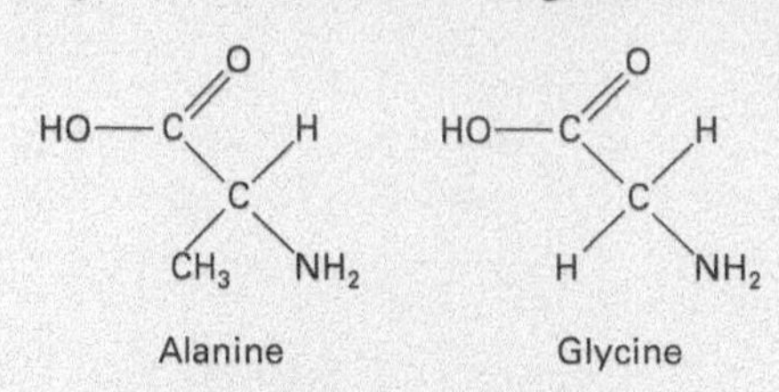

i State, with a reason, which of these two amino acids can exist as optical isomers. [2]

ii Show the formation of a peptide link between these two amino acids. [2]

b Outline the procedure for determining the primary structure of a peptide. [4]

c Suggest two reasons why fluothane, $CF_3CHBrCl$, replaced ether, $C_2H_5OC_2H_5$, as an anaesthetic in the 1950s. [2]

d Chlorofluorocarbons have found widespread use as refrigerants, aerosol propellants, and foam producers. Recently they have been replaced by other compounds, and their manufacture has been restricted.

i Why did chlorofluorocarbons find widespread use as aerosol propellants? [2]

ii Explain the dangers to the environment of chlorofluorocarbons. [3]

2 Diagrams **A**, **B**, **C** and **D** represent carbohydrates.

a Name and draw a displayed formula for the molecule drawn as **A**. [1]

b **B** can be converted into **A**.

i Write a molecular equation for the conversion of **B** into **A** and state what type of reaction it is.

ii Two different methods can be used for this conversion. For each, state the reagents used and the conditions necessary. [5]

c Explain the structural differences between **C** and **D** and the consequent roles that each has in nature. [4]

3 The base pairing in DNA is A ... T and G ... C.

The base pairing in RNA is A ... U and G ... C.

a Draw a block diagram of the double helix of DNA, showing three repeat units. [2]

b Explain how DNA both replicates itself and also leads to the formation of RNA. [4]

c A sequence of bases in one strand of DNA is

—T—G—G—A—C—T—A—A—C—

State the corresponding base sequence in

i the complementary strand of DNA,

ii the RNA derived from this complementary strand of DNA. [2]

d There are about 3×10^9 base pairs in one strand of human DNA. About 10^5 sections of the DNA strand are involved in protein synthesis and, on average, 10^3 base pairs are required for each protein.

What percentage of the DNA strand is involved in protein synthesis? [2]

4 Integral proteins and bilayers are important components in cell membranes.

a What is the function of integral proteins in membranes? Explain why the bilayers cannot perform this function.

A research team in California has synthesized 'molecular channels' which have uniform diameters of only a few nanometres. These molecular channels are called nanotubes and could be used to inject drug molecules or metal ions into malignant cells or into bacteria. Nanotubes are made from rings of cyclic peptides stacked on top of each other. Alternate rings consist of natural amino acids and synthetic amino acids which have the opposite optical activity. The nanotubes form when rings of cyclic peptides stack up from a solution. [1 nanometre = 10^{-9} metres]

b Draw a displayed formula for a cyclic tripeptide, using R to represent the side chain of each amino acid. Label the chiral atoms of the ring with an asterisk(*). [4]

c **i** Suggest how nanotubes of differing diameters can be made.

ii Suggest how the surface characteristics of the nanotubes might be varied for use in different cell walls. [2]

5 Recent techniques have allowed DNA from various sources to be broken down by enzymes (called restriction endonucleases). These enzymes cut DNA molecules at specific places in the sequence of base pairs.

The various fragments that result from the enzyme action are separated by gel electrophoresis carried out in a buffer of pH 7 and constant potential difference.

a Describe, using a block diagram, the structure of the nucleotides which make up nucleic acids. State briefly how these are arranged in DNA. [4]

b Describe how the enzyme functions in this example and state what sort of reaction occurs. [2]

c Explain, in terms of the structure of the nucleotide, why the DNA fragments move towards the positive electrode in the electrophoresis separation. [2]

6 a Explain what enzymes are and how they function in biochemical processes. [5]

b Several deaths occur each year by poisoning due to the accidental swallowing of ethane-1,2-diol (glycol) which is widely used as a solvent or as anti-freeze for car radiators. The glycol is oxidized in the body to ethanedioic acid (oxalic acid) which is poisonous.

If the patient can be treated quickly enough, large doses of brandy (which contains a high percentage of ethanol) are given.

Explain how the ethanol from the brandy acts as a competitive inhibitor and how its action suppresses the oxalic acid poisoning. Include in your answer the formulae of the organic substances involved. [5]

7 a Plant material is composed largely of carbohydrates, an important constituent of the human diet.

Explain why some plants, e.g. potatoes, may be utilized as a source of energy whereas others, e.g. cabbage, provide mainly roughage. [4]

b During jam-making, fruit is boiled with sugar solution.

Some jam was added to water and warmed: the resulting mixture was filtered. A sample of the resulting solution gave a red-brown precipitate on treatment with alkaline Cu^{2+} ions and warming. A sample of the original fruit/sugar mixture gave no reaction before boiling but, after acidification, the resulting solution also produced a red-brown precipitate on warming with alkaline Cu^{2+} ions.

Suggest the possible identities of the two sugars present in the jam. Explain your reasoning. [4]

c Suggest why frozen vegetables may have a higher nutritional value then so-called 'fresh vegetables'. [2]

8 a Describe the β-pleated sheet structure of the protein which constitutes the main material of meat. Your answer should include a suitable diagram, which also shows the bonding between the strands. [5]

b i What is the function of haemoglobin and of myoglobin in the living animal?

ii Explain why unwrapped minced beef for sale at a meat counter is bright red, whereas a slice of raw beef cut from the centre of a joint may be purplish-red.

iii Uncooked minced beef, prepacked in supermarket displays, is wrapped in a plastic film which is permeable to air. Explain why air is allowed to reach the meat. [5]

9 The major components of membranes are phospholipids and proteins. The phospholipids form a bilayer which is interspersed with protein molecules. Transport proteins extend across the membranes and catalyse the transport of specific molecules and ions into and out of the cell. Transport can be passive or active.

a Give the full structural formula of a typical phospholipid. Describe the nature of the different parts of a phospholipid. (If you are unable to draw a structural formula, marks can be scored by a simple labelled diagram of a phospholipid.) [3]

b Explain, using your answer to **a**, how bilayers are constructed. State the nature of the chemical attractions at each end of a phospholipid. [3]

c Explain what is meant by active transport. Describe briefly one common example of active transport. [4]

10 What is meant by each of the terms *primary, secondary, tertiary* and *quaternary* as applied to the structure of proteins? Where possible, illustrate your answer with examples of the chemical bonding involved in each structure. [10]

11 a 'Terylene' is a polymer derived from benzene-1,4-dicarboxylic acid and ethane-1,2-diol.

i Draw graphical formulae to represent benzene-1,4-dicarboxylic acid and ethane-1,2-diol. [2]

ii What type of polymer is 'Terylene' ? [1]

iii Draw the structure of the repeat unit of this polymer. [2]

b i Explain the term *biodegradable*. [1]

ii State **one** environmental benefit of using biodegradable plastics. [1]

c Starch is an important constituent in foodstuffs. It can be hydrolysed by boiling with mineral acid in a test tube.

i Give the name of a source of starch. [1]

ii What is the product of the hydrolysis of starch? [2]

d i Amylase is an enzyme that brings about the same hydrolysis in the body. Suggest why amylase cannot be substituted for the mineral acid in the procedure described in **c**. [2]

ii Suggest the pH environment within which amylase can function best as a catalyst. [1]

31 Techniques of preparation, separation, and identification

It is often easy to overlook the fact that chemistry is a practical subject. Much current information about the properties of elements and their compounds is the result of careful experimentation and observation. The early chemists used primitive apparatus and had few measuring instruments. Modern apparatus is much more convenient to use, and instruments are available today that can probe to the heart of a molecular structure. This chapter looks at some of the practical techniques used to investigate molecules. The first spread discusses some of the techniques you can use to prepare and separate a pure sample of a substance of interest. The rest of this chapter describes three important techniques for identifying substances – electrophoresis, chromatography, and mass spectrometry.

Using simple apparatus

OBJECTIVES

- 'Quickfit' and distillation
- Separatory funnels
- Solvent extraction
- Buchner funnel
- Recrystallization

Distillation and condensers

- Heating 'under reflux' – see spread 25.4 'Oxidation of alcohols'.
- Fractional distillation – see spread 22.1 'The oil industry: fractionation'.
- Steam distillation – see spread 28.3 'Aromatic amines'.

Purifying a liquid product

A liquid product is first dried by leaving it in contact with a small quantity of a substance that absorbs water, such as anhydrous sodium sulphate, for about 30 minutes. The final purification of a liquid product involves redistilling it and collecting the fraction that emerges from the condenser very close to the boiling point of the product (typically ±2 °C).

You should be familiar with carrying out chemical reactions in test tubes where your concern is to observe accurately the reaction taking place and make a record of the results. In preparative chemistry, the object is to make something safely, to obtain a good yield, and to make the sample pure enough for its intended use. 'Test tube' chemistry in the laboratory will rarely achieve these aims, and is certainly not suitable for industrial preparations. The most convenient apparatus for small-scale laboratory and industrial preparations is glass apparatus known as Quickfit™, which is made from borosilicate glass and has standard-size ground glass joints, so that different pieces of equipment can be fitted together easily.

Condensers

The function of a condenser is to provide a cool surface on which vapours produced in a reaction may condense. An air condenser relies on the surrounding air to cool the vapour inside. You will more commonly use a Leibig condenser, in which tap water passes through an outer jacket to make the inner surface colder.

Separatory funnels

Once a reaction has been completed, the product must be separated from the reaction mixture. If an organic product is a liquid, a separatory funnel (traditionally called a 'separating funnel') can be used, for example, to wash out excess acid. This technique relies on the organic product and the aqueous phase used for washing being immiscible.

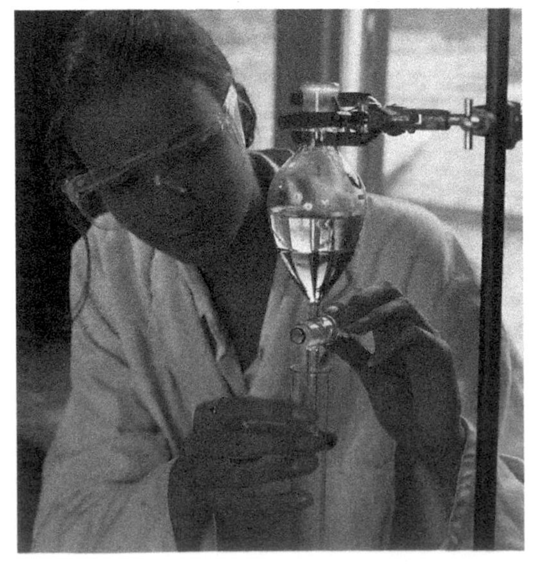

Using a separatory funnel.
(a) With the stopper out to allow the escape of any evolved gases, initially mix the liquids by gentle swirling. This step is particularly important when washing with a carbonate to remove excess acid.
(b) Now give the funnel about 20 vigorous shakes, pausing half way to allow any gases to escape.
(c) Support the funnel while the contents settle and two layers form. One layer is the organic product and the other is the aqueous washing solution. Pour off the lower layer, but always keep both liquids until the organic product has been isolated. Even experienced chemists sometimes discard the wrong layer!

Solvent extraction

If the organic product is more soluble in an organic solvent than it is in water, a technique called solvent extraction can be used to separate the product from its aqueous solution. Under these conditions, the dissolved substance (the solute) becomes unequally distributed between the two solvents. It is **partitioned** between the two solvents.

The two liquids are shaken together in a separatory funnel, and the solute dissolves preferentially in the organic layer. The layers are separated and the solute is recovered by evaporation of the organic solvent.

During the manufacture of penicillin, the product is extracted from its aqueous solution using solvent extraction by trichloromethane. Evaporation of the highly volatile trichloromethane gives pure crystalline penicillin. The large-scale manufacture of penicillin was developed by Norman Heatley, who was awarded the first ever honorary Doctorate in Medicine from Oxford University.

The apparatus for vacuum filtration: a Buchner flask fitted with a Buchner funnel.

Filtration under reduced pressure

Evaporating the solvent from a solution will make the solution more concentrated. A small quantity of solid crystallizes when the solubility of the solid is exceeded. More solid appears as the volume of the solution decreases further. The solvent will also be able to dissolve many of the impurities produced in the reaction along with the desired product. It is therefore not wise to evaporate *all* of the solvent. Evaporation is usually carried out until there is a suspension of the solid product in a small volume of its saturated solution.

Filtration under reduced pressure is the quickest method of separating the solid from its suspension. The apparatus includes a Buchner funnel: the Buchner flask has thick walls to withstand the partial vacuum and a side-arm for attaching a pump to suck the air out. The mixture of crude product is poured gently onto the centre of the filter paper. The partial vacuum in the flask results in rapid filtration.

The ideal solvent

The ideal solvent for recrystallization should:

- not react with the compound;
- have a boiling point lower than the melting point of the compound to be recrystallized;
- be non-toxic and non-flammable.

Above all, the solid product should be very soluble in the hot solvent and *much* less soluble in it when cold.

Recrystallization

A solid product can be purified by recrystallization. The impure solid is dissolved in the minimum volume of a hot solvent and filtered to remove insoluble impurities. The resulting hot, saturated solution of the product, together with any soluble impurities, is then allowed to cool *slowly*, whereupon crystals of pure compound will separate from the solution. These can be filtered out of the solution. The impurities have a much lower concentration and so remain in solution. The crystals of product are tested for purity by melting-point determination (see box).

Lead(II) iodide, recrystallized from hot acidic solution, forms yellow crystals which give an appearance of spangles to the suspension.

Melting-point determination

A pure substance will have a *sharp* melting point: it does not melt slowly over a range of temperature (as does an impure substance).

SUMMARY

- A condenser provides a cold surface on which vapours may condense.
- Condensers are used for reflux and for distillation.
- A separatory funnel is used for separating immiscible liquids, generally during a purification process involving washing with an aqueous solution.
- Solvent extraction separates an organic product from an aqueous solution.
- Solid products are separated from suspensions in liquids by filtration under reduced pressure.
- Solid products are purified by recrystallization.
- The purity of a solid is checked by melting-point determination.

31.2

ELECTROPHORESIS

OBJECTIVES

- Electrophoresis
- PAGE
- Isoelectric point
- DNA 'fingerprints'

Electrophoresis is a method of separation and identification that separates molecules based on how easily they form ions. The ions migrate in a buffered solution under the influence of an applied electric field. The electric field causes positive ions to move towards the cathode and negative ions to move towards the anode. The different ions in a mixture migrate at different rates, and so can be separated. Electrophoresis is particularly useful for separating molecules of high molar mass such as proteins and nucleic acids.

Amino acids and pH

Protein structure is discussed in the previous chapter. Proteins consist of chains of amino acids bonded together by peptide links. As shown to the left, an amino acid contains a carbon atom bonded to an amine group, a carboxylic acid group, and an R group. The R groups are different for different amino acids, and may themselves contain groups that can ionize. The overall charge on a protein depends on the balance of these ionizable groups at a particular pH.

(a) R—C(H)(NH$_3^+$)—COOH

(b) R—C(H)(NH$_2$)—CO_2^-

Amino acid structure at (a) low pH and (b) high pH.

At low pH, amine groups —NH_2 are protonated to form —NH_3^+, but carboxyl groups are not ionized. The result is a net positive charge. Conversely, at high pH, amine groups are not protonated, but carboxyl groups —COOH are ionized to form —CO_2^-. The result is a net negative charge. At pH values between these extremes, the balance of protonation and ionization depends on the particular pK_a values of the R groups.

Polyacrylamide gel electrophoresis

The medium on which electrophoresis is carried out is commonly a **polyacrylamide gel**. This form of the technique is known as **PAGE**, which stands for **p**oly**a**crylamide **g**el **e**lectrophoresis. A gel is a network of polymer molecules with a random structure. The aqueous solution moves through the spaces, or pores, within the network of the gel. Gels of different pore sizes are available for separating molecules of different sizes. The gel acts as a sort of sieve; molecules larger than the average pore size are held back, whereas molecules smaller than the average pore size move through more quickly.

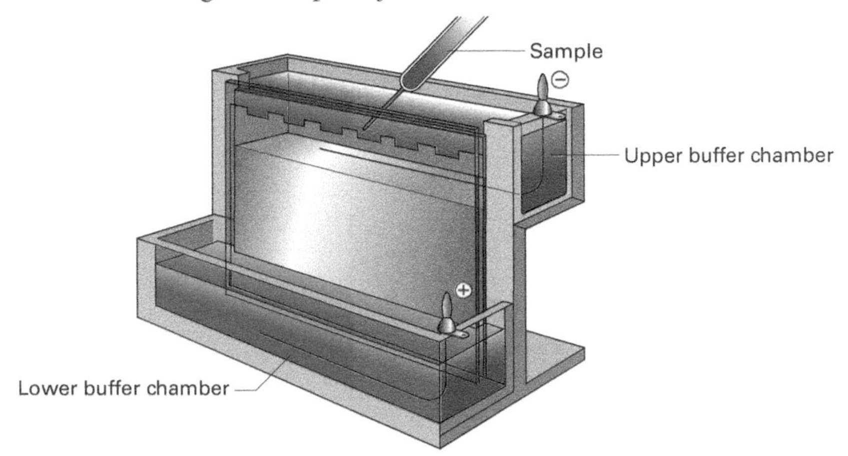

A PAGE cell. Samples are applied in slots in the top of the gel.

The mixture under examination is placed on the centreline of the gel plate. Ions travel towards the anode or the cathode depending on their charge and at speeds depending on their size. Once separation has been achieved, the gel is stained to reveal the individual constituents. A commonly used staining reagent is ninhydrin, which is available in aerosol canisters. On spraying and heating, amino acids and proteins show up as a blue-purple stain. Alternatively, sometimes the components fluoresce under ultraviolet light: in this case a permanent copy is obtained by blotting with radiographic paper. A typical example is shown opposite.

The isoelectric point

At one particular pH, a protein will have no net charge because it contains an equal number of positively and negatively charged groups. This pH is called the **isoelectric point** of the protein, and its characteristic value, found by electrophoresis, may be used to identify the protein. Electrophoresis can also be used to *separate* proteins, because at any given pH the speed with which a protein migrates depends on its particular net charge. Different proteins will travel at different rates through the gel.

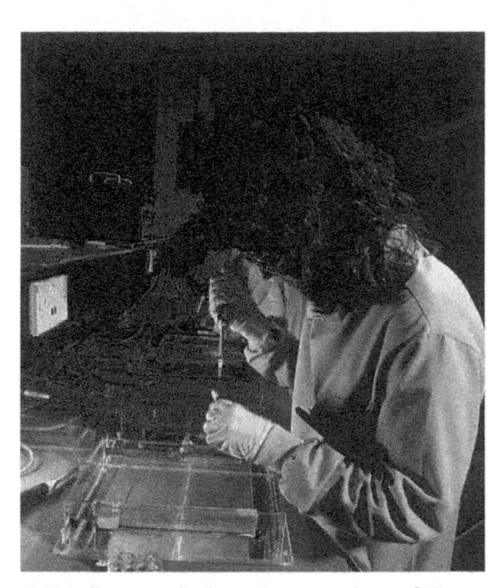

DNA fingerprinting. Preparation of the agarose electrophoresis gel used to separate fragments of DNA into bands.

Genetic fingerprinting

One of the major modern uses of electrophoresis is in separating the components of DNA in the technique of **DNA fingerprinting**.

The DNA of one person is different from the DNA of any other person. The pattern produced by the electrophoresis of DNA taken from a tissue sample can be used to identify the individual from whom the tissue came, just as a fingerprint can be used to identify a particular person. The sample can be extremely small: a smear of blood, the root of a hair, or a flake of skin is enough.

The technique depends on a group of enzymes known as restriction endonucleases. These enzymes act as a sort of 'molecular scissors' that break up a chain of DNA in a predictable way. DNA is extracted from a sample of tissue and is then treated with restriction endonucleases. Electrophoresis of the resulting fragments of DNA gives a pattern unlike that from any other individual except an identical twin. This pattern can therefore be used to identify an individual from tissue inadvertently left at the scene of a crime. Also, common features between the DNA fingerprints of two people can be used to establish hereditary relationships including, for example, the paternity of a child.

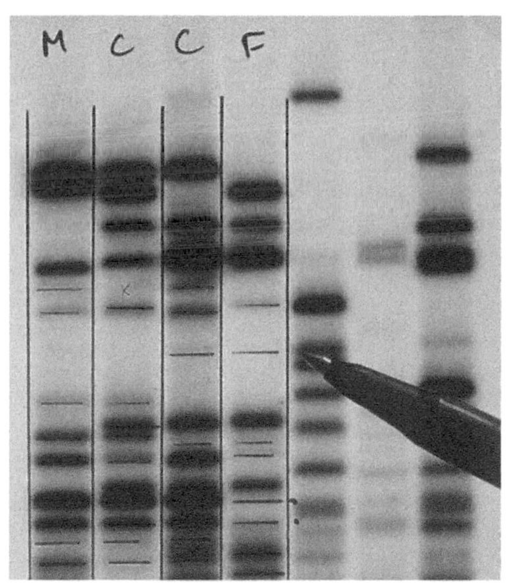

DNA fingerprints of two children (C). Red matches bands common with their mother (M); blue matches bands common with their father (F).

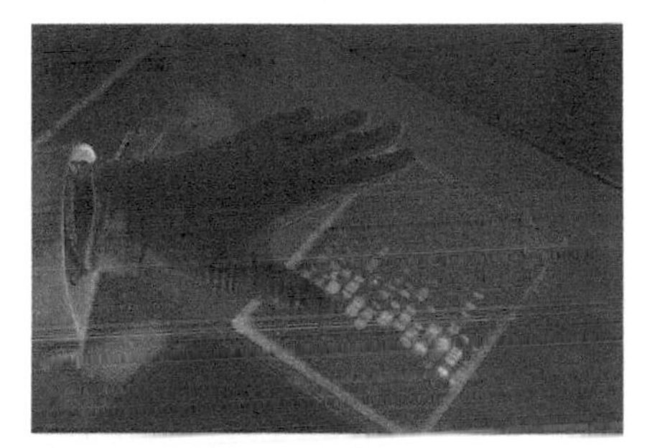

An electrophoresis gel fluorescing to reveal the components of the original mixture. The positions of the constituents are noted. The constituents may be isolated by cutting up the gel and washing each section.

SUMMARY

- Electrophoresis depends on the migration of ions in a buffered solution under the influence of an electric field.
- Electrophoresis is used for separating large molecules such as proteins, nucleic acids, and their breakdown products.
- The isoelectric point is the pH at which a protein has no net charge.
- Electrophoresis may be used for both identification and separation.

PRACTICE

1 Explain why the amino acid glycine NH_2CH_2COOH migrates towards the anode at high pH and towards the cathode at low pH.

2 Would you expect the isoelectric point for the amino acid aspartic acid $HOOCCH_2CH(NH_2)COOH$ to be greater or less than 7? Explain your answer.

31.3

Chromatography – 1

OBJECTIVES

- General principles
- Paper chromatography
- TLC
- Column chromatography

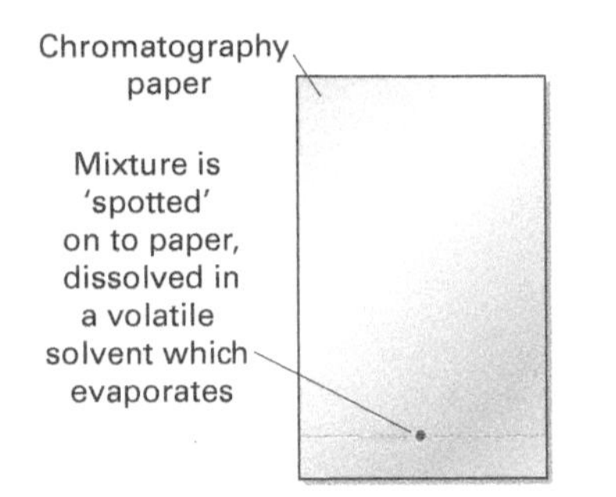

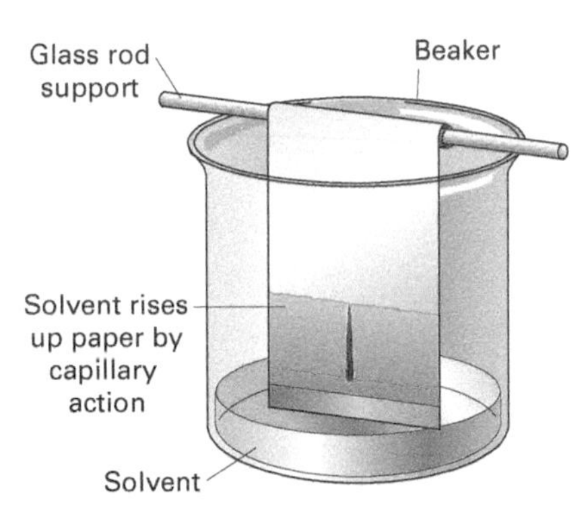

Chromatography paper is a thick absorbent paper rather like filter paper. (a) The mixture to be separated is applied as a spot a short distance from one end of the paper. (b) The end below the spot is dipped into the solvent. Each component of the mixture is carried a different distance by the moving solvent. When the solvent nears the top, the level the solvent has reached is marked on the paper, which is removed from the solvent and dried. The separate components may be directly visible, or are made visible either by staining or by fluorescing under a UV lamp.

Filtration, distillation, and solvent extraction are all useful techniques to separate a significant quantity of product. Chromatography can be used to separate complex mixtures containing very *small* quantities of different substances. There are many different types of chromatography, but in each case there are two phases: a **mobile phase** that moves and a **stationary phase** that stays still. The different components of the mixture become *partitioned* (see first spread in this chapter) or *adsorbed* to different extents between the two phases.

Paper chromatography

A sheet of paper consists largely of cellulose fibres. Cellulose is a polysaccharide (see previous chapter) composed of glucose molecules, which have a large number of hydroxyl groups. Water molecules hydrogen-bond to these groups, so that a sheet of 'dry' paper contains around 10% by mass of water. This water acts as the stationary phase in the technique of paper chromatography. The mobile phase is a solvent consisting of an aqueous solution or an organic solvent such as ethanol or ethanoic acid. The mixture to be separated is dissolved in this mobile phase, which moves along the paper. The movement comes about by capillary action, which results from the forces between the solvent and the solid fibres of the paper.

Thin-layer chromatography

Thin-layer chromatography (**TLC**) is a technique similar to paper chromatography. The stationary phase is a solid; the mobile phase is a liquid. TLC uses a thin layer of a material such as silica (silicon dioxide SiO_2) or alumina (aluminium oxide Al_2O_3) coated onto a glass, plastic, or aluminium plate. The separated substances may be recovered for further analysis or reaction by selectively scraping patches from the plate and dissolving the substance in a suitable solvent. The detection by TLC of the molecule pregnanediol in urine is a positive test for pregnancy.

The R_f value

The R_f value compares the distance moved by each component of the solute with the distance moved by the solvent during the experiment. The **R_f value** (or **retention factor**) of a component of the solute is given by

$$R_f = \frac{\text{distance moved by component of solute}}{\text{distance moved by solvent}}$$

The R_f value may be used to identify a component by comparing it with the R_f values of known substances. The R_f value is dependent on the solute and on the nature of the stationary and mobile phases (such as the solvents used); it is not dependent on the distance travelled by the solvent.

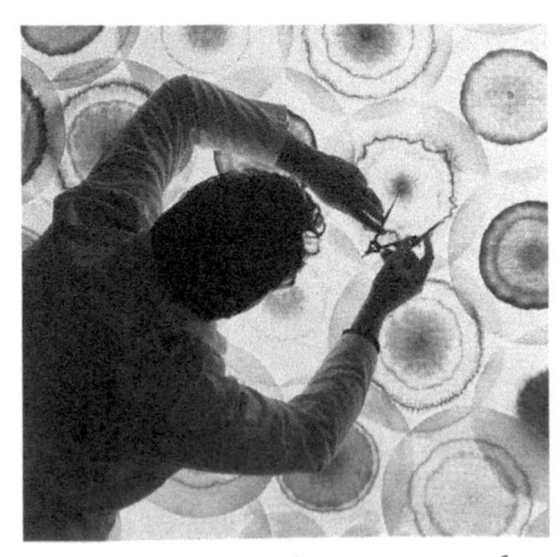

Paper chromatography uses paper of varying shapes and sizes!

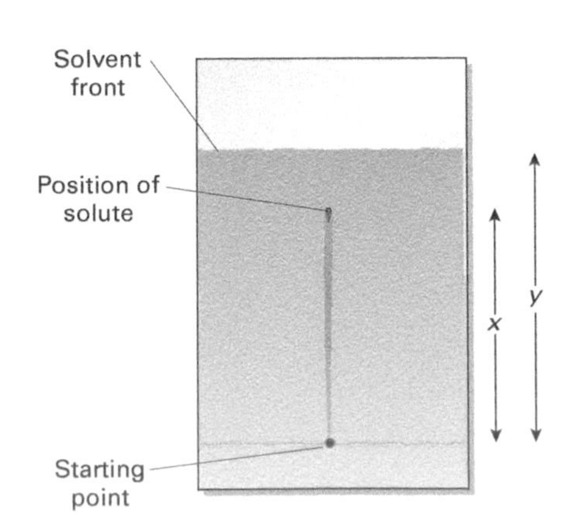

The chromatography paper after removal when the solvent has reached the top.
The R_f value is here given by

$$R_f = \frac{\text{distance moved by solute}}{\text{distance moved by solvent}} = \frac{x}{y}$$

Column chromatography

Column chromatography is a convenient technique for physically separating the components of a mixture for further use, rather than for identification. The stationary phase is a powder packed into a vertical column of quite large diameter (about 1–2 cm) and wetted with solvent. The mixture is applied to the top of the column, followed by solvent, which runs through under gravity. The components of the mixture adsorb onto the surface of the solid to different extents. The different components emerge from the bottom of the column at different times.

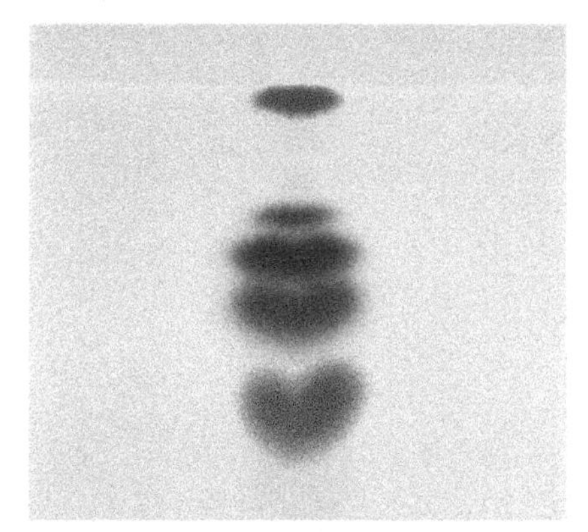

TLC of the pigments in Poa annua *annual meadow grass: the second and third green bands are chlorophylls.*

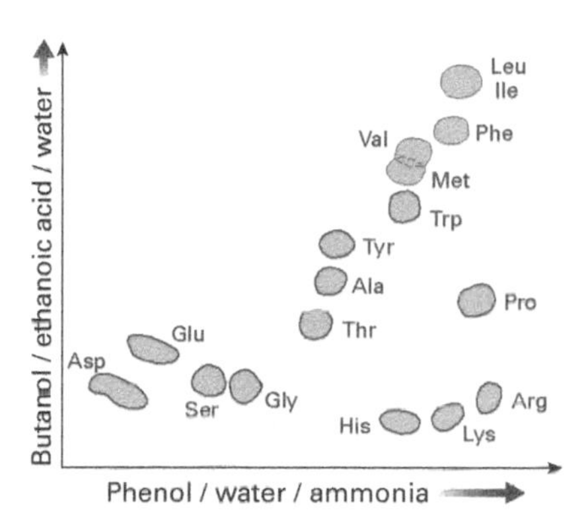

A sophisticated version of thin-layer chromatography is two-dimensional TLC. In this technique, separation is carried out in two stages. The plate is first dipped into one solvent mixture, left for a time, removed, and dried. It is then turned through 90° and dipped into the second solvent mixture, left for a time, removed, and dried. Separation is particularly effective using this method: the amino acids are almost completely separated.

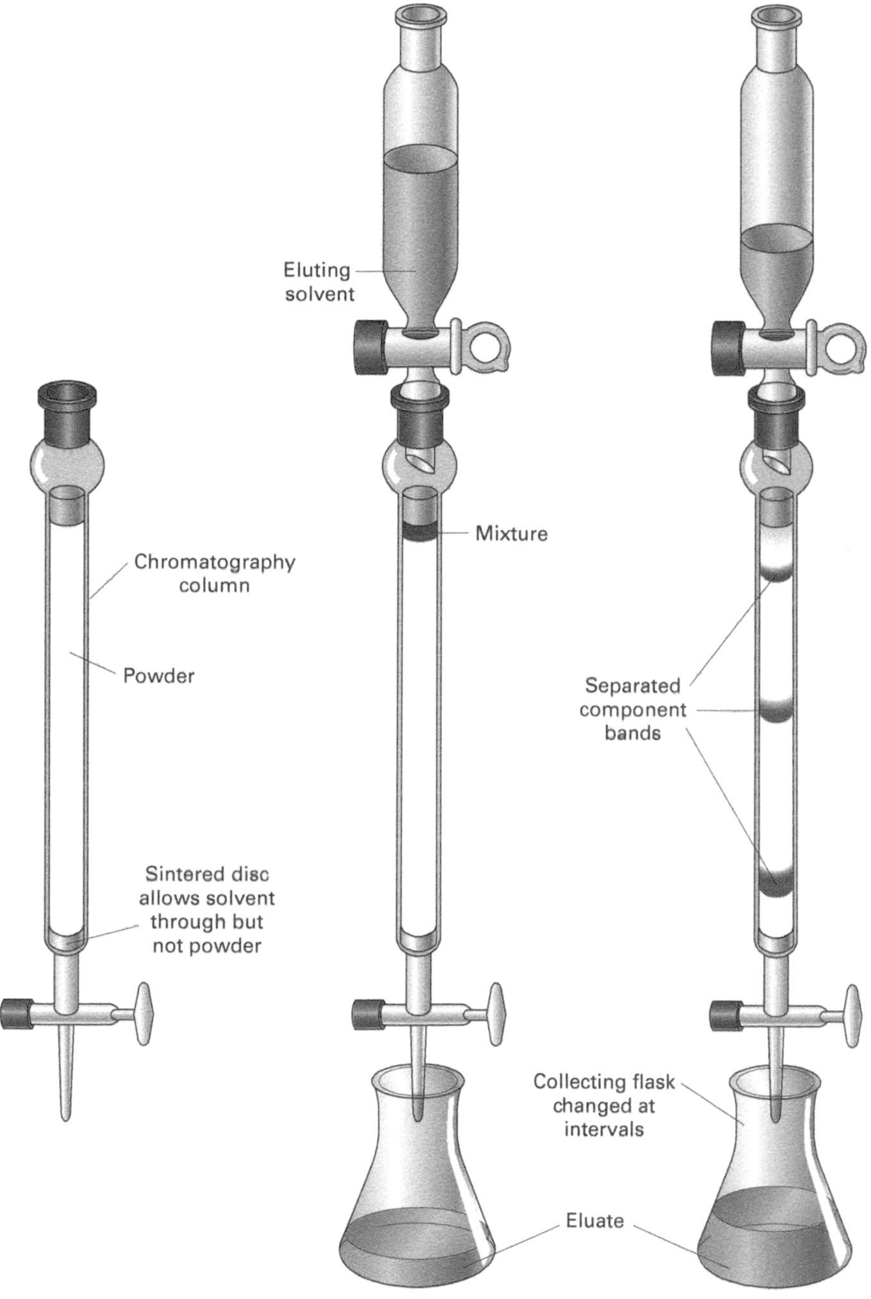

Column chromatography. The mixture in this illustration contains coloured components that appear as coloured bands at different places down the column.

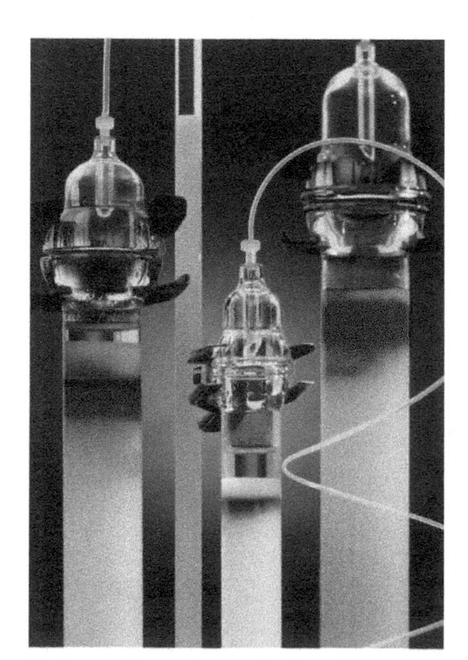

Column chromatography in action.

SUMMARY

- All types of chromatography involve a mobile phase and a stationary phase.
- The different compounds separate because they are partitioned or are adsorbed to different extents.
- The ratio of the distance moved by a component of the solute to the distance moved by the solvent is called the component's R_f value.

OBJECTIVES

- HPLC
- GLC

Chromatography – 2

This spread introduces two important chromatographic techniques: high-performance liquid chromatography (HPLC) and gas–liquid chromatography (GLC). You will see that HPLC is a development of column chromatography described in the previous spread. GLC works by a similar principle, but uses an inert gas to carry the components of a gaseous mixture through the apparatus.

High-performance liquid chromatography (HPLC)

Although it is useful for demonstrating liquid chromatography, column chromatography operating by the force of gravity is no longer used much in laboratories. Instead, chemists use **high-performance liquid chromatography**; in this technique, the stationary phase is held in a column and the mobile phase is forced through under pressure. Separation is much faster. A common stationary phase consists of silica particles with long-chain alkanes adsorbed onto their surfaces. A common mobile phase is methanol.

The separated components of the mixture are usually detected by UV spectroscopy (see next chapter) as they pass through a flow cell between a UV source and detector. The output of the detector is recorded as a **chromatogram**.

HPLC can be used for identification as well as for separation. The components of a mixture are identified by the time they take to pass through the system. The time between injection and the appearance of a peak on the chromatogram is called the **retention time**. Identical substances will have the same retention times under the same conditions.

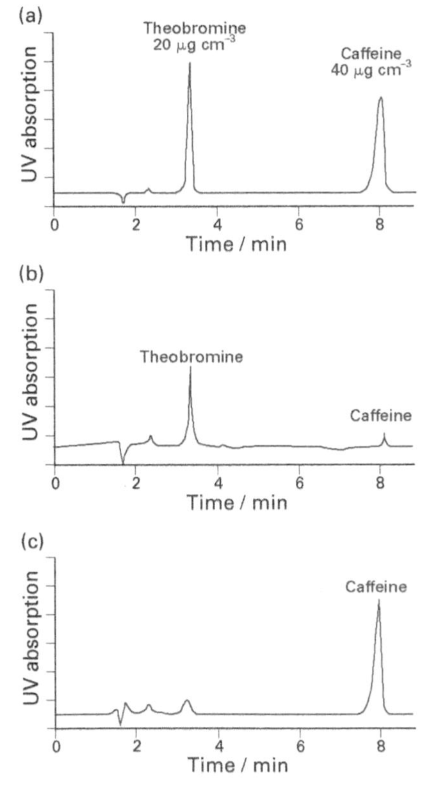

HPLC may be used commercially for the identification of the stimulants theobromine and caffeine. Calibration with standard solutions (i.e. of known concentration) allows the concentrations of theobromine and caffeine in drinking chocolate and tea to be worked out from the chromatograms. (a) Calibrating the machine with standard solutions of theobromine and caffeine. (b) Drinking chocolate extract. (c) Tea extract.

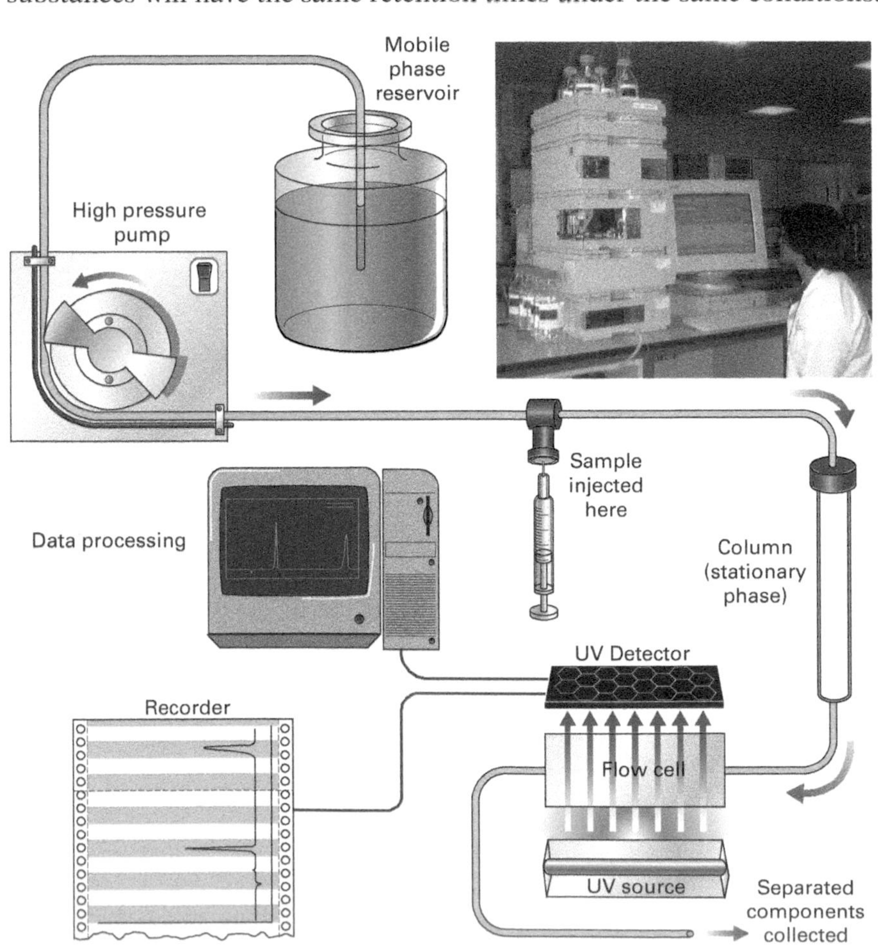

An HPLC system. The photo shows an HPLC system in use. One particular use of this technique is chiral separation, which is able to separate the (+) and (–) forms of optical isomers. Although these isomers have the same functional groups and molecular formula, their differing structures around a chiral centre (their stereochemistry) affects the strength of their adsorption onto a chiral stationary phase.

Gas–liquid chromatography

In common with all the chromatographic techniques described so far, gas–liquid chromatography (**GLC**) uses a stationary phase and a mobile phase. This method is used to separate and identify volatile liquids that do not decompose at temperatures around their boiling points. GLC is generally used for identifying the components of a mixture, and measuring their concentrations. For example, evidence given in Court during prosecutions of drunk drivers usually comes from a gas chromatogram of the defendant's breath.

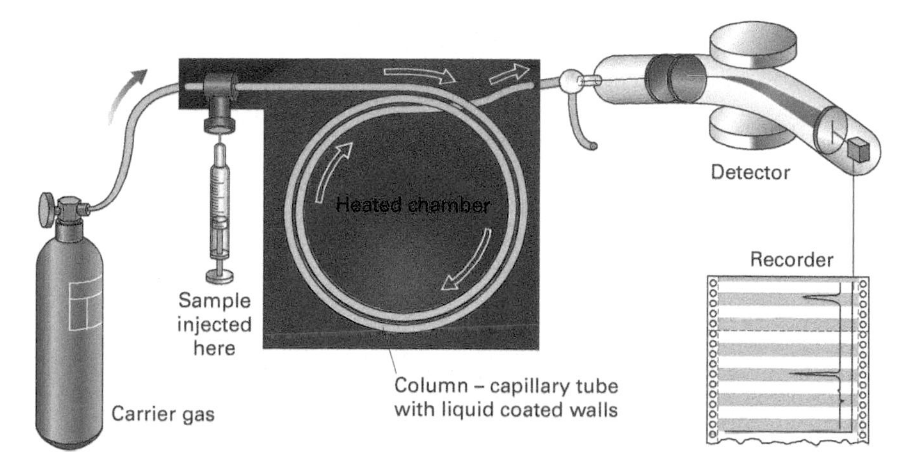

The stationary phase in a GLC apparatus consists of a liquid coated onto the walls of a long, thin capillary tube. The mobile phase is an unreactive gas such as helium or nitrogen. The sample is injected into a heated entrance port, where it immediately vaporizes. The vapour is carried into the column by the mobile phase, which is usually referred to as the *carrier gas*. The carrier gas does not play any part in the separation except to carry the sample along. At the end of the column, the separated components of the mixture pass through a detector, typically a mass spectrometer, see following spread. As in HPLC, the components are identified by their retention times. The relative quantity of each constituent is proportional to the area under its peak in the chromatogram.

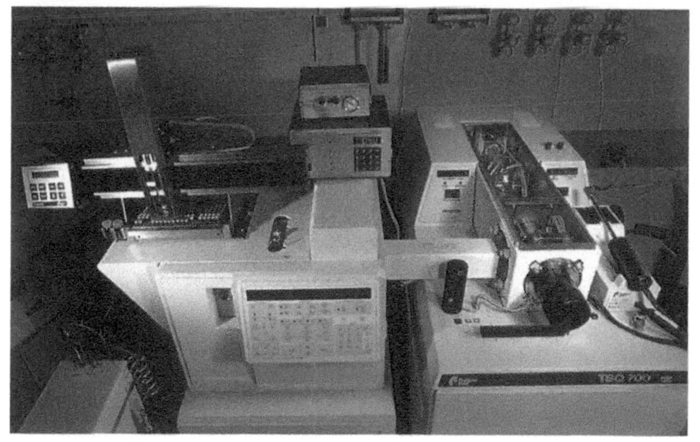

Gas-liquid chromatography/mass spectrometry (GCMS) used in forensic analysis. Mass spectrometry is discussed further in the following spread.

SUMMARY

- In HPLC the solvent is pumped through a column under pressure.
- The time between sample injection and the appearance of a peak in HPLC is called the retention time.
- In GLC the stationary phase is a liquid coated onto the walls of a long, thin capillary tube.
- In GLC the mobile phase is an unreactive carrier gas such as helium or nitrogen.

Nomenclature

Chromatography is the technique.

Chromatograph is the apparatus.

Chromatogram is the result that can be seen – e.g. coloured bands or spots; a graph drawn by a plotter driven by a detector.

A gas–liquid chromatograph. The detector here is a mass spectrometer. GLC columns used for analysis are usually narrow, and can analyse very small samples down to 10^{-7} dm^3 (0.1 μl). GLC can also be used for separation, in which case the columns tend to be wider and are packed with a powdered inert solid.

LC–MS

Current HPLC systems are also frequently interfaced with mass spectrometers: LC–MS provides a powerful and robust addition to the analyst's arsenal, with a number of commercial instruments reducing the price of this once specialized equipment. HPLC can also be interfaced with NMR spectrometers (spread 32.5). Highly integrated instruments are valuable in biomedical applications.

Catching drug cheats

GCMS is used to detect drugs in the urine samples routinely taken from athletes in major competitions. It was the identification by GCMS of the anabolic steroid stanozolol in the urine of Ben Johnson, spread 25.1, which caused his disqualification. The introduction in 1996 of the high resolution mass spectrometer (HRMS) has allowed the detection of lower levels of anabolic steroids.

PRACTICE

1. Describe three different techniques involving chromatography, including one example of its use.
2. Use the chromatogram opposite to comment on the remark that there is more caffeine in tea than in coffee.

31.5

MASS SPECTROMETRY

OBJECTIVES

- Fragmentation
- The molecular ion peak
- Interpreting mass spectra
- Fragmentation patterns

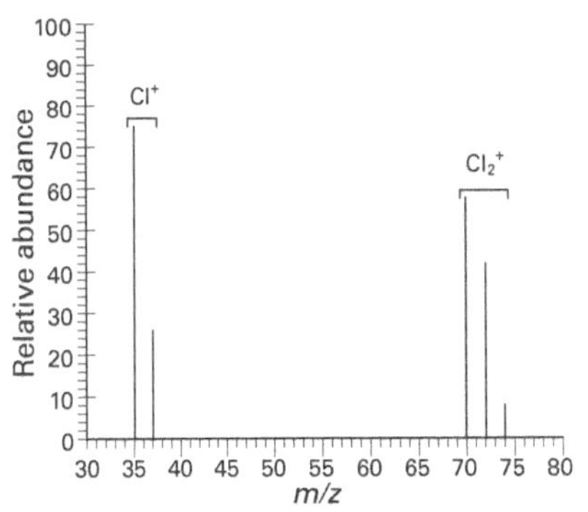

Chlorine is diatomic and so there are peaks for the molecular ion Cl_2^+. However, Cl_2^+ also fragments to produce Cl^+ ions. Chlorine has two isotopes (35 and 37 in a 3:1 ratio), so the peak heights for 70 and 74 are in a 9:1 ratio. The peak height for 72 is 6 because $^{35}Cl^{37}Cl$ and $^{37}Cl^{35}Cl$ both contribute. (It is not possible to compare the relative heights of the Cl_2^+ and Cl^+ peaks.)

High-resolution mass spectrometers

Modern mass spectrometers can measure masses to four or five decimal places, so molecules with almost exactly the same mass can be distinguished. For example, a compound whose relative formula mass M_r is 123 (to the nearest whole number) could be $C_6H_5NO_2$ or $C_8H_{13}N$. Precise values of the relevant relative isotopic masses are:

$^{12}C = 12.00000$ (by definition)

$^{1}H = 1.0078$

$^{14}N = 14.0031$

$^{16}O = 15.9949$

Using these figures,

$M_r(C_6H_5NO_2) = 123.032$

and

$M_r(C_8H_{13}N) = 123.105$

the two substances can therefore be easily distinguished.

Molecular and structural formulae

Note that identifying the molecular ion leads only to the *molecular* formula. Modern instruments, using specialized ionization techniques (such as ionspray), can be tuned to emphasize the molecular ion peak. Obtaining the structural formula requires further analysis; see the following chapter.

In spread 3.2 we looked at mass spectrometry as a means to determine the relative *atomic* masses and relative abundances of isotopes. Mass spectrometry is also used to analyse *molecules*, particularly organic molecules. The mass spectrometer breaks molecules into ionized fragments. These may be detected and displayed as a spectrum, analysis of which gives information about *molecular* mass and structure.

Fragmentation

When a *molecule* is bombarded with high-energy electrons, it loses an electron to form a positive ion and absorbs energy, causing it to break into fragments. These positively charged fragments appear in the mass spectrum. The *tallest* peak in the spectrum is due to the fragment produced in greatest quantity. This peak is called the **base peak**. The relative abundances of all other peaks in the spectrum are measured as a percentage of the abundance of the base peak. The base peak may not be the molecular ion peak (see below).

The molecular ion peak

The **molecular ion peak** is produced by the loss of just one electron from the *complete* molecule. So the mass at which the molecular ion peak is found represents the relative formula mass of the molecule. The molecular ion peak will usually be found within the cluster of peaks of greatest mass in the spectrum, but sometimes it may be very weak or even absent. To determine which one of the cluster of highest-mass peaks represents the molecular ion, the various isotopes that may be present in the molecule must be considered.

The *m/z* ratio for the singly charged molecular ion is labelled M. Peaks of mass 1 or 2 units heavier than the molecular ion are called M+1 and M+2 peaks respectively. They are due to the presence of isotopes that are 1 or 2 mass units heavier than the most abundant isotope.

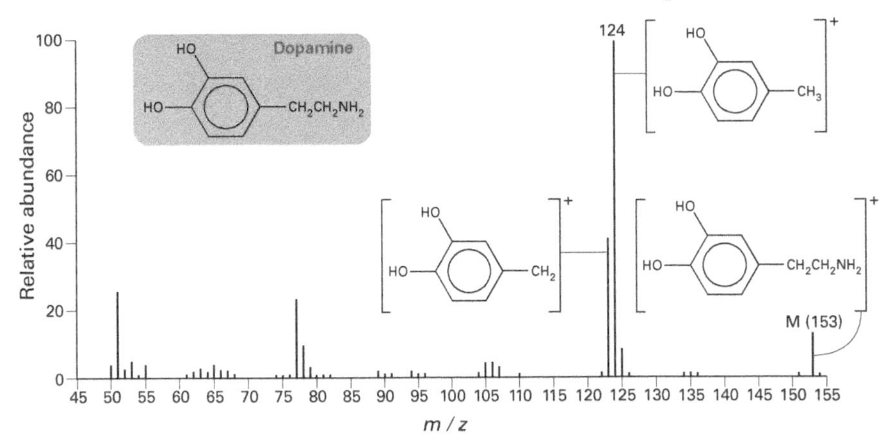

A (simplified) mass spectrum of dopamine, which is a neurotransmitter in the brain. Knowledge of its structure has led to the synthesis of L-dopamine, used to treat some of the symptoms of Parkinson's disease. The molecular ion peak at m/z = 153 corresponds to an ion of formula $C_8H_{11}O_2N^+$. The peak at m/z = 124 (the base peak) corresponds to the ion $C_7H_8O_2^+$. This ion results from fission of the $C_6H_3(OH)_2CH_2{-}CH_2NH_2$ bond (producing $C_6H_3(OH)_2CH_2^+$) followed by capture of a hydrogen atom from other molecular fragments. This structure allows the delocalization of the positive charge with the benzene ring π cloud.

The M+1 peak

An M+1 peak in a molecule containing carbon is probably due to one of the ^{12}C atoms being replaced by a ^{13}C atom. The relative abundance of ^{13}C in Nature is 1% of that of ^{12}C, so a ^{12}C atom will be replaced by a ^{13}C atom 1% of the time. For a molecule with two carbon atoms, a molecule 1 mass unit heavier will therefore occur 2 × 1 or 2% of the time. So, you may expect to see an M+1 peak with an abundance of 2% of the molecular ion peak abundance. The relative abundance of the M+1 peak increases linearly with the number of carbon atoms.

The M+2 peak

The heavier isotopes of chlorine and bromine are each 2 mass units heavier than the lighter isotopes. The natural abundance of ^{81}Br is almost equal to that of ^{79}Br. The mass spectrum of a compound containing a bromine atom therefore has M and M+2 peaks of *almost equal intensity*.

The natural abundance of ^{37}Cl is 33% that of ^{35}Cl. The mass spectrum of a compound containing a chlorine atom therefore has an M+2 peak that is one-third the intensity of the molecular ion peak.

Fragmentation patterns

The molecular ion has an unpaired electron, and so is both a carbocation and a radical; a radical contains an unpaired electron and is unstable. When a molecular ion fragments, the most common result is another carbocation fragment (which will appear in the mass spectrum) together with a neutral fragment with an unpaired electron (which will not be detected), e.g. a methyl radical $^{\bullet}CH_3$.

Fragmentation of the molecular ion occurs most easily at positions that give the most stable carbocation fragments. The stability of carbocations increases with the number of alkyl groups attached to the positive carbon atom, spread 22.8. This can be to used to predict how, for example, an alkane is likely to fragment. The **fragmentation pattern** produced is characteristic and acts as a fingerprint for the parent molecule. With the *m/z* values for various fragments determined in this way, the various peaks in a mass spectrum may be allocated to specific fragments. A substance can be identified by comparing its mass spectrum against a library held on a computer database.

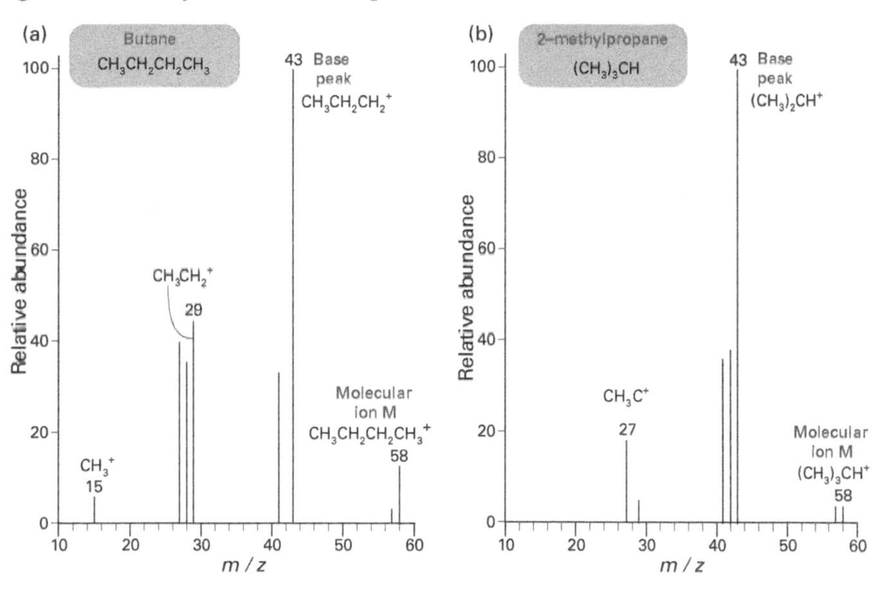

The (simplified) mass spectra of two isomers. (a) Butane. The main fragments are identified. (b) 2-Methylpropane. Note the following 2 major changes. The peak at 29 is very much smaller, as the fragment $CH_3CH_2^+$ cannot be formed. The relative heights of 57 and 58 are the same because breaking the $(CH_3)_3C—H$ bond creates a tertiary carbocation, which is relatively very stable. As for butane, the base peak is formed by the loss of one methyl group.

SUMMARY

- The molecular ion peak is produced by loss of one electron from the complete molecule.
- The M+1 peak shows the number of carbon atoms present in a molecule.
- The M+2 peak identifies compounds containing chlorine and bromine.
- A molecular ion of odd mass suggests the presence of nitrogen.
- The fragmentation pattern produced is characteristic of the molecular structure.
- Fragmentation occurs at positions that give stable carbocations.

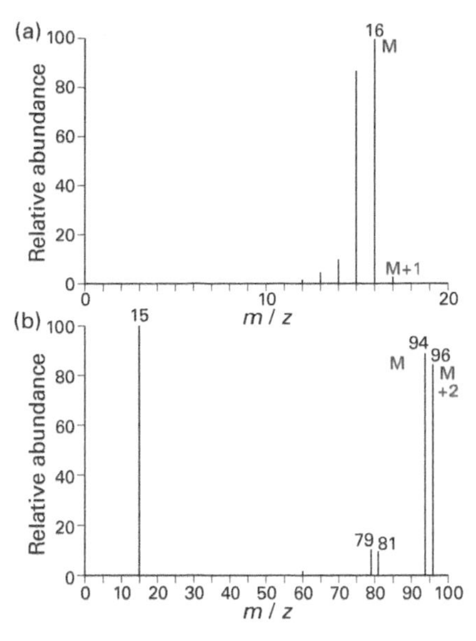

(a) The mass spectrum of methane CH_4 showing the molecular ion peak at m/z = 16 and the (much smaller) M+1 peak at m/z = 17. (b) The (simplified) mass spectrum of bromomethane showing the M+2 peak.

Carbocation stability

A tertiary carbocation R_3C^+ is more stable than the secondary carbocation R_2HC^+, which is more stable than the primary carbocation RH_2C^+, which is more stable than H_3C^+. The fragment RCO^+ is also generally more stable.

Nitrogen

Most organic compounds have masses that are even numbers. A clue to the presence of nitrogen in a molecule arises because, although nitrogen has an even mass (14), it forms an *odd* number of bonds (3). Therefore, the presence of a single nitrogen atom will result in an *odd*-numbered relative formula mass. See box opposite.

Practical techniques

It is possible to predict the fragments likely to form from a given structure. Conversely, deducing the likely identity of the fragments observed gives information about the original molecular structure from which they derive. In practice, a single analytical technique can rarely be used alone to obtain detailed structural information. Mass spectrometry and infrared spectroscopy (see next chapter) are so commonly used together that they are now frequently combined in one single instrument. Sensor outputs are linked to a computer, which displays a three-dimensional pictorial representation of the molecule.

PRACTICE EXAM QUESTIONS

1 **a** The organic compound **M** is introduced into a mass spectrometer.

i What information can be obtained from the precise value of the mass of the molecular ion $M^{+\cdot}$?

ii Suggest why it is usually possible to detect a small peak at one mass unit higher than that of the molecular ion.

iii Write a general equation for the fragmentation of a molecular ion $M^{+\cdot}$ into two new species. Explain briefly why only one of these species can be detected. [4]

b **i** The relative intensities of the peaks at $m/z = 50$ and $m/z = 52$ present in the mass spectrum of chloromethane are in the ratio of approximately 3 to 1, respectively. Suggest why two molecular ion peaks are found.

ii Write an equation for the fragmentation of $CH_3Cl^{+\cdot}$, giving rise to a peak at $m/z = 15$. [4]

2 Mass spectrometry is an important analytical technique used in a variety of applications.

a The table below shows the relative abundances of isotopes of strontium obtained from the mass spectrum of a sample of the element.

Nucleon (mass) number	Relative abundance
84	0.60
86	9.90
87	7.00
88	82.50

i Calculate a value for the relative atomic mass of strontium from the data, showing your working.

ii Suggest why a sample of strontium from a different source may have a slightly different relative atomic mass. [3]

b The high resolution mass spectrum of the products of combustion of a plastic shows peaks at *m/e* values of 27.0109, 29.9980, and 30.0105.

Using the table below identify the components of the mixture responsible for these peaks. Show how you arrive at your conclusions.

Element	Relative atomic mass
Hydrogen, ^{1}H	1.0078
Carbon, ^{12}C	12.0000
Nitrogen, ^{14}N	14.0031
Oxygen, ^{16}O	15.9949

i *m/e* 27.0109

ii *m/e* 29.9980

iii *m/e* 30.0105 [3]

c The mass spectrum shown below was obtained from compound **E**, which contains carbon, hydrogen, and oxygen only.

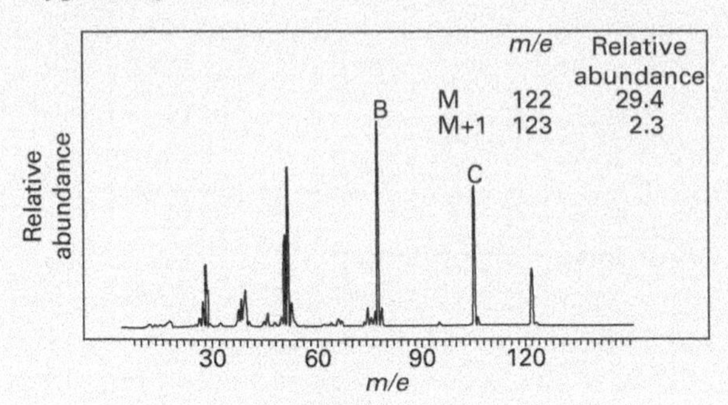

i Using the data provided, show how you can confirm that compound **E** contains seven carbon atoms.

ii Suggest the formulae of the molecular ions responsible for the peaks labelled B and C. [4]

d Suggest a structure for **E**, and identify the functional group present. [2]

3 This question concerns the compounds in the following reaction scheme:

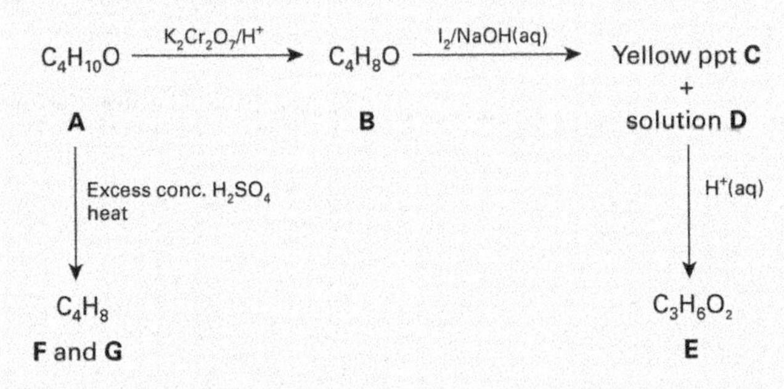

a **F** and **G** are compounds which both decolourize bromine water. **F** has two stereoisomers.

i What functional group is present in both **F** and **G**? [1]

ii Give the structural formulae of both stereoisomers of **F**. [2]

iii Explain how these two isomers arise. [2]

iv Write the structural formula of **G**. [1]

b **B** cannot be oxidized by acidified potassium dichromate(VI) solution.

i Write the structural formula of **B**. [1]

ii Draw the general structural features of molecules which can be detected by the reaction with iodine and alkali. [2]

iii Give the structure of the substance in solution **D**, and of the product **E**. [2]

c The mass spectrum of **A** gives peaks at *m/e* 29 and 45, amongst others. That at 45 is the largest.

This spectrum shows no molecular ion peak, which would be expected at *m/e* 74. **A** is chiral.

i Give the structural formula for **A**. [1]

ii Identify the ions responsible for the peaks at *m/e* 45 and 29, and hence suggest why the molecule shows no molecular ion peak. [3]

d **A** is miscible with benzene and the mixture formed shows a positive deviation from Raoult's law and forms an azeotrope.

i **A** boils at 99 °C, benzene at 80 °C. Sketch a possible boiling point/composition diagram for a mixture of **A** with benzene. [3]

ii Explain why a mixture of **A** with benzene shows a large positive deviation from Raoult's law. [3]

iii The infrared spectrum of **A** is shown below. The very broad peak at 3500 cm^{-1} is due to the presence of an —OH group. This peak becomes much narrower when diluted with benzene and moves to 3600 cm^{-1}.

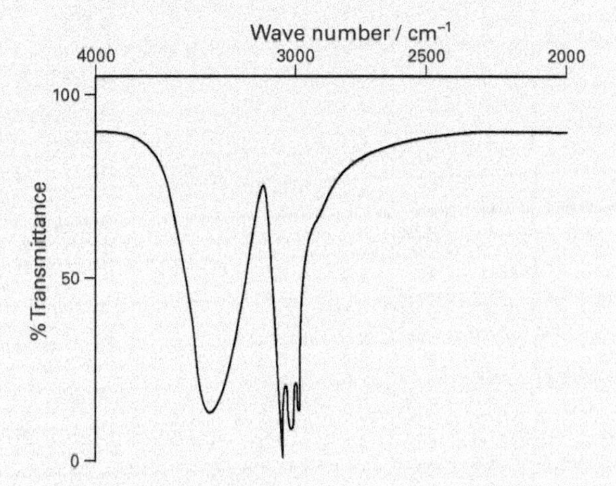

Suggest why the —OH absorption peak changes as **A** is diluted with benzene. [2]

4 a The figure below shows part of the mass spectrum of an organic compound with the molecular formula $C_2H_8N_2$.

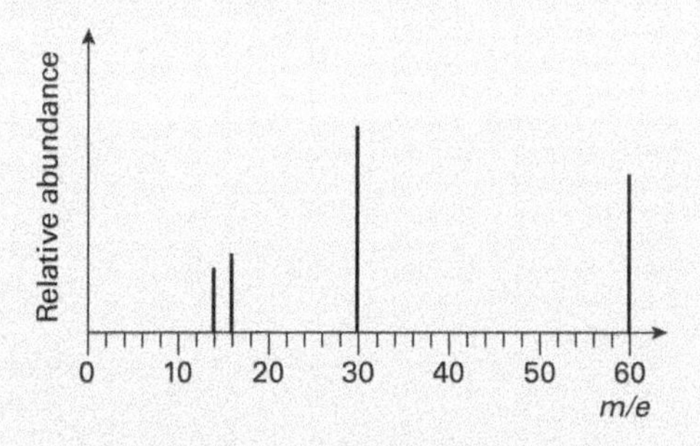

i What is the relative molecular mass of the compound? [1]

ii Suggest the formula of the fragment that corresponds to each of the following masses:

30 mass units; [1]

16 mass units; [1]

14 mass units. [1]

iii Give the name of a functional group in the original molecule. [1]

b Compounds can exhibit structural, geometrical, and optical isomerism. Explain what is meant by *geometrical isomerism*. [2]

c The following formulae represent compounds which exist as geometrical isomers. Draw formulae to show **both** isomers for each compound

i C_4H_8 [2]

ii $[Cr(H_2O)_4Cl_2]^+$ [2]

5 a i Define the terms:

Relative molecular mass

Isotope.

ii Calculate the relative atomic mass of carbon which contains two main isotopes ^{12}C(98.9%) and ^{13}C (1.10%). Give your answer to an appropriate number of significant figures. [6]

b The diagram below shows a simplified mass spectrum for an organic compound **X** which contains carbon, hydrogen, and oxygen only. The infrared spectrum of **X** indicates the presence of a C═O group.

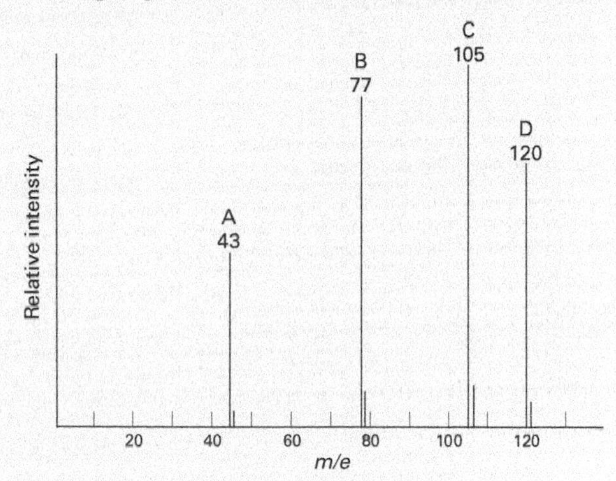

i Deduce from the mass spectrum the relative molecular mass of **X**.

ii Account for the very small peaks that occur next to each of the peaks A, B, C, and D at masses of 44, 78, 106 and 121 respectively. [3]

c i Estimate the number of carbon atoms present in ion D and hence explain why compound **X** is likely to be aromatic and to have the formula $C_6H_5COCH_3$.

ii Deduce the formulae of the ions responsible for peaks B and C.

iii Suggest a reason for the absence of a recorded peak at *m/e* = 15. [7]

d Give the reagents and the result of a simple chemical test to show the presence of the carbonyl group in **X**. [2]

32 Spectroscopy and structure

Once a substance has been separated from a reaction mixture and purified, we then need to identify its structure. The previous chapter introduced mass spectrometry for finding molecular formulae. This chapter describes the use of (i) atomic emission and absorption, (ii) ultraviolet and visible, (iii) infrared, and (iv) nuclear magnetic resonance spectroscopy. Spectroscopy involves the interaction of electromagnetic radiation with matter. These techniques allow the structural formulae of molecules to be determined.

Atomic spectroscopy

32.1

OBJECTIVES

- Spectroscopy: general principles
- Emission and absorption spectra
- Atomic spectra: AES and AAS

Flame photometry

A flame photometer is a simple instrument for detecting emission spectra. It is used in hospital laboratories to detect Na^+ and K^+ in blood serum and urine. The flame burns natural gas, and the sample to be analysed is sprayed into it as an aerosol. Optical filters are placed between the flame and the detector to isolate the strong emission lines, spread 16.9, from Na, K, Li, Ca, or Ba.

Earlier we explained how the emission spectrum of hydrogen provides valuable information about the arrangement of electrons in atoms. Obtaining an *emission* spectrum, however, requires special conditions such as high temperature or low pressure. The instruments for obtaining and analysing emission spectra are often expensive. Information about energy levels can be obtained more cheaply and for a much wider range of substances using absorption spectroscopy. This technique detects the *absorption* of electromagnetic energy that accompanies a transition to a higher energy level. The overall result is an absorption spectrum.

Atomic emission spectroscopy (ICP-AES)

Although the instruments remain very expensive, atomic emission spectroscopy (AES) has been revolutionized by a device called the **inductively coupled plasma (ICP) torch**. A plasma, see below, is a gas containing a high proportion of cations and electrons. In the ICP torch, argon gas is made into a plasma by exciting it with a powerful radio-frequency source. The energy heats the argon by induction to temperatures around 10 000 K. The sample to be analysed is made into an aerosol and is carried into the plasma by the argon flowing through the central tube of the torch.

As the sample enters the torch, it atomizes and is excited into various higher energy levels. The radiation emitted as these excited energy levels return to the lowest energy level is analysed as in any other form of spectrometer. The very significant difference between ICP-AES and atomic absorption spectroscopy (see opposite page) is that no radiation source is required, and so the spectra of many different elements may be analysed together.

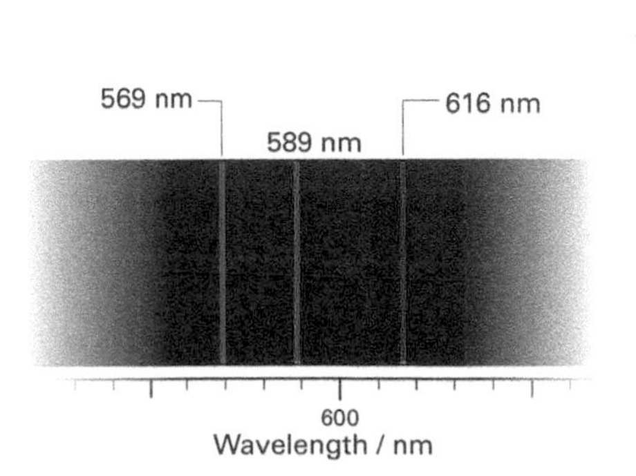

The atomic emission spectrum of sodium showing the intense yellow line at 589 nm. This emission is due to electronic transitions from the 3p energy level down to the 3s energy level. The green line at 569 nm is the result of 4d to 3p transitions; the orange line at 616 nm results from 5s to 3p transitions. See spread 4.3.

(a)

(b)

(a) A hollow-anode nitrogen plasma source used to manufacture light-emitting-diodes (LEDs). (b) Nitrogen plasma around a titanium electrode.

Atomic absorption spectroscopy (AAS)

If you heat a compound to a sufficiently high temperature, it will usually break apart to form atoms in the gas phase. Atomic absorption spectroscopy atomizes a sample at high temperatures (around 3000 K) in an oxygen–ethyne flame. When electromagnetic radiation is shone through the sample, the electrons in the atoms absorb energy and are promoted to higher energy levels. The energy absorbed in each transition corresponds to a particular wavelength of radiation.

Atomic absorption spectroscopy is able to detect just one element at a time. The instrument shines radiation from a hollow cathode discharge tube at the flame containing the vaporized sample. The cathode is made from the element being investigated, so it emits radiation of wavelengths characteristic of that element. Any atoms of that element present in the flame absorb energy at these wavelengths, and the radiation emerging from the flame is measured by the instrument. A typical atomic absorption spectrum shows a series of sharp lines. The extent of the absorption depends on the relative quantity of the element in the sample, allowing the concentration of the element in the sample to be determined.

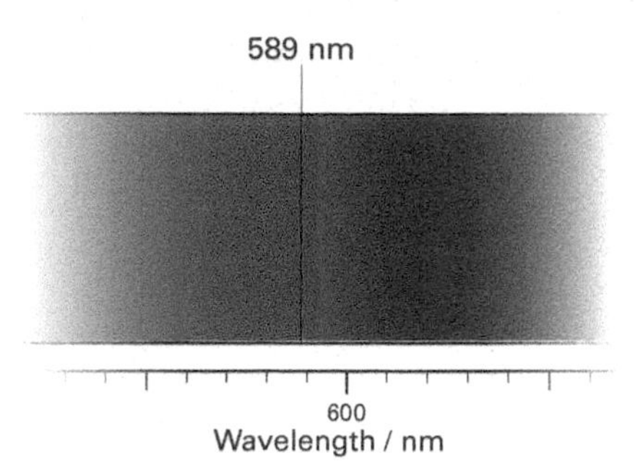

The atomic absorption spectrum of sodium, extending over the same region of wavelengths as the emission spectrum shown opposite. Compare the two spectra and note that both emission and absorption occur at 589 nm.

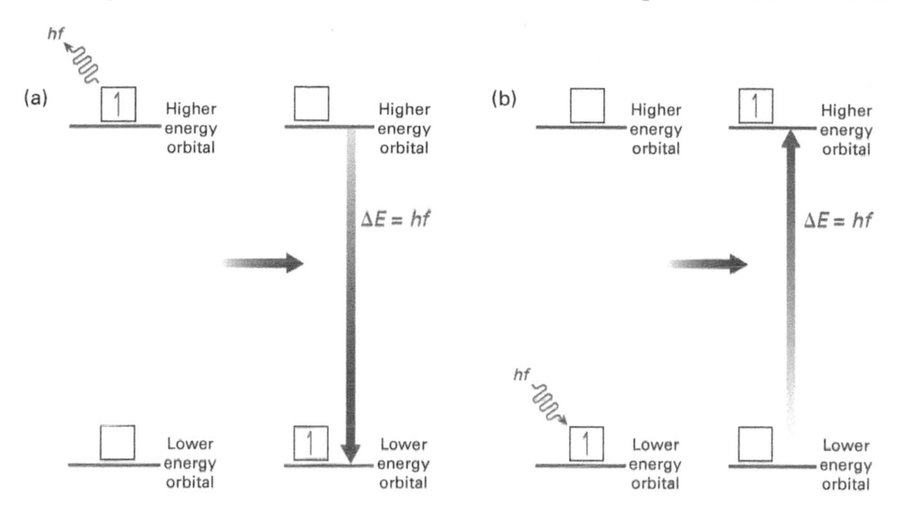

Comparing the mechanisms of emission and of absorption. (a) Emission. An atom emits a quantum of electromagnetic radiation when an electron in a higher energy level makes a transition to a lower energy level. The frequency f of the radiation is given by $\Delta E = hf$, where ΔE is the difference between the two energy levels and h is the Planck constant. (b) Absorption. An atom absorbs a quantum of electromagnetic radiation of the correct energy, causing an electron to make a transition to a higher energy level. The frequency f of the radiation is also given by $\Delta E = hf$.

Photon

A quantum of electromagnetic radiation is called a photon.

SUMMARY

- Atomic emission spectroscopy depends on an atom emitting a quantum of electromagnetic radiation when an electron in a higher energy level makes a transition to a lower energy level.
- Atomic absorption spectroscopy depends on an atom absorbing a quantum of electromagnetic radiation when an electron in a lower energy level makes a transition to a higher energy level.
- The frequency of emitted or absorbed radiation is given by $\Delta E = hf$.
- In atomic absorption spectroscopy only one element is detected at a time.

PRACTICE

1 Explain why the absorption spectrum for atomic sodium contains fewer lines than its emission spectrum.

32.2

OBJECTIVES

- Transitions between molecular energy levels
- Chromophores
- Conjugation

UV/Vis and metal ions

Ultraviolet and visible spectroscopy can also be used to determine the concentration of metal ions in solution after addition of an appropriate ligand to intensify the colour, spread 20.11.

ULTRAVIOLET AND VISIBLE SPECTROSCOPY

Atomic emission and absorption spectra involve transitions of electrons between energy levels (orbitals) in atoms. **Ultraviolet and visible (UV/Vis) spectroscopy** probes the same region of the electromagnetic spectrum, but the spectra result from electronic transitions between *molecular* energy levels. Chapter 5 'Chemical bonding' introduced the idea of molecular orbitals forming from the redistribution of electron density when atomic orbitals overlap. The various possible molecular orbitals represent different quantized energy levels for molecules in the same way as atomic orbitals do for atoms. An electron in a molecule can absorb a quantum of radiation and make a transition between these molecular energy levels, just as an electron in an atom can between atomic energy levels. Analysis of the resulting spectra can lead to identifying the bonds present in a molecule.

Energy level transitions in molecules

A single covalent bond results from the formation of a sigma (σ) molecular orbital, and a double bond also involves the formation of a pi (π) molecular orbital. An unfavourable energy level called an **antibonding orbital** is always formed along with a **bonding orbital**, spread 5.6. Antibonding orbitals are given the symbols σ^* and π^*. Orbitals containing lone pairs of electrons are called **non-bonding orbitals** and are given the symbol n. Stable, unexcited molecules generally have electrons in bonding and non-bonding orbitals. Absorption of energy can promote these electrons to antibonding orbitals.

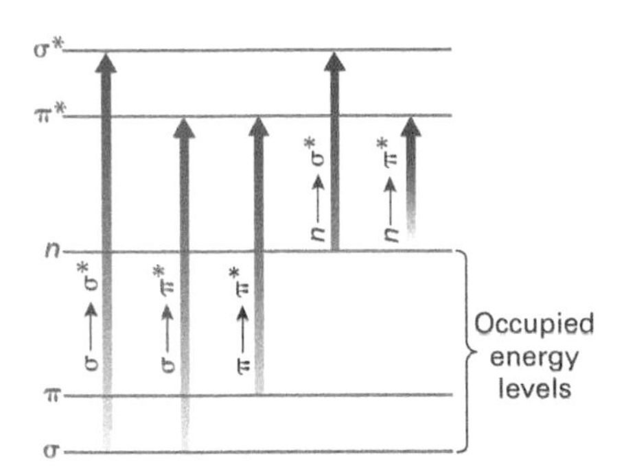

The typical relative energy levels of molecular orbitals and the electronic transitions that can occur between them.

Chromophores

The energy required to make an electronic transition in a molecule depends on how strongly the electrons are attracted by the nuclei of the bonded atoms. So the characteristic energy of a transition is a property of a *group of atoms*. A group of atoms producing such a characteristic absorption is called a **chromophore**.

Different transitions are possible in different chromophores. In molecules such as alkanes that only contain sigma bonds and do not contain any lone pairs, only one electronic transition is possible, i.e. $\sigma \rightarrow \sigma^*$. The energy required for this transition is relatively large. It represents the absorption of ultraviolet radiation of wavelengths too short for detection by most instruments.

Where lone pairs are present as well as sigma bonds, $n \rightarrow \sigma^*$ transitions become possible. The energy of these transitions comes within the range of ultraviolet wavelengths that *can* be measured experimentally: $—NH_2$ and —OH groups absorb in this way in the wavelength range 175–200 nm. (Remember that the frequency f of the radiation is given by $f = \Delta E/h$; and the wavelength λ is given by $\lambda = c/f$, where c is the speed of light.)

If a double bond is present, a $\pi \rightarrow \pi^*$ transition becomes possible. In alkenes this transition occurs at an energy corresponding to a wavelength around 175 nm; and in carbonyl compounds at around 188 nm.

In an unsaturated molecule that also contains a lone pair, an $n \rightarrow \pi^*$ transition is also possible. C=O typically absorbs in this way at wavelengths around 275–290 nm, —N=N— at around 340 nm, and $R—NO_2$ at around 270 nm.

(a)

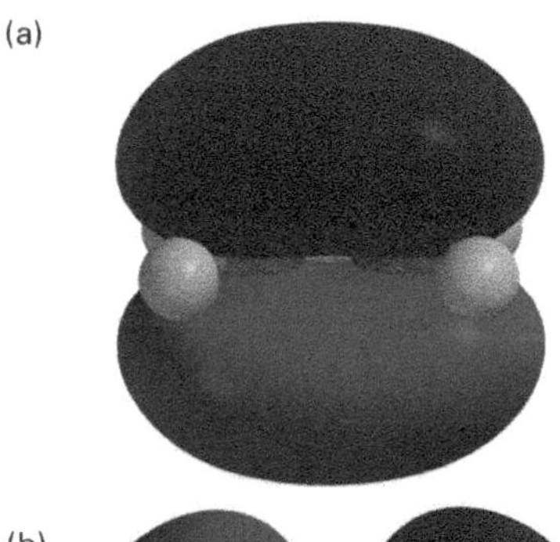

(b)

Electron density diagrams for ethene's molecular orbitals: (a) π, and (b) π^*.

This ultraviolet spectrometer, enclosed in a glove box for safe manipulation, is used in the optics industry. The spectrometer generates a beam of radiation that scans through each wavelength in turn. The output, the ultraviolet spectrum, is typically a plot of absorption against wavelength (from about 180 nm). Comparison of the spectrum with a library of known spectra is routine.

Class	Transition	λ_{max} / nm
ROH	$n \rightarrow \sigma^*$	180
RCHO	$\pi \rightarrow \pi^*$	190
	$n \rightarrow \pi^*$	290
R_2CO	$\pi \rightarrow \pi^*$	180
	$n \rightarrow \pi^*$	280
RCOOH	$n \rightarrow \pi^*$	205
RCOOR′	$n \rightarrow \pi^*$	205
$RCONH_2$	$n \rightarrow \pi^*$	210

Typical absorptions of single isolated chromophores.

Conjugation

Conjugation occurs where there are alternating double and single bonds in a molecule (see chapter 29 'Organic synthesis: ...'). In a conjugated molecule, electron density is delocalized over molecular orbitals stretching across all the atomic centres involved. Conjugation decreases the energy required for a $\pi \rightarrow \pi^*$ transition, and so causes a shift to longer wavelength; it also increases the intensity of absorption. The effect of conjugation is important because it shifts the absorption wavelength to a region of the spectrum much more readily detected using simple instruments. In fact, in extreme cases it shifts it into the *visible* region and the substance is coloured.

Conjugation effects

The C=C double bond in ethene absorbs at 175 nm; whereas the C=C−C=C conjugated bonds in butadiene absorb at 217 nm and in 2-methylbutadiene at 220 nm. Conjugation changes the wavelength by 42 nm; adding the methyl group causes a change of only 3 nm.

(a)

(b)

Organic compounds that are coloured by virtue of conjugated double bonds. (a) β-Carotene. (b) Methyl orange.

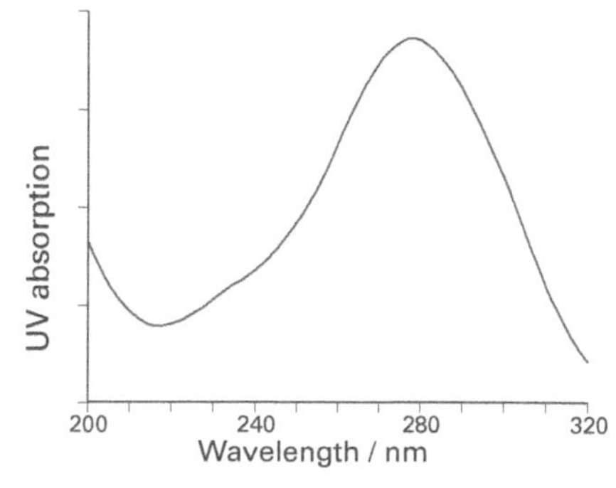

The ultraviolet spectrum of propanone in cyclohexane.

SUMMARY

- UV/Vis spectroscopy is used to detect electronic transitions between molecular energy levels.
- A group of atoms producing a characteristic absorption is called a chromophore.
- Conjugation progressively decreases the energy required for a $\pi \rightarrow \pi^*$ transition.

PRACTICE

1 Explain how absorption spectra could be used to determine atmospheric composition, as for example in the planet revolving around 51 Pegasus.

32.3 INFRARED SPECTROSCOPY – 1

OBJECTIVES

- Vibrational and rotational energy levels
- Stretching and bending vibrations
- Active and inactive bonds
- The infrared spectrometer

Electronic transitions between molecular orbitals giving rise to UV/Vis absorption are not the only types of quantized transition that can occur in molecules. Molecules can interact with electromagnetic radiation in more ways than atoms can. The absorption of energy can cause bonds to stretch, bend, or twist in characteristic ways that distort the electron density. These transitions take place at lower energies than electronic transitions, with the result that vibrational and rotational energy changes are detected by infrared (IR) spectroscopy.

Rotational transitions

- **Electronic transitions** involve energies in the region of 100–400 kJ mol^{-1} and correspond to UV/Vis spectra.
- **Vibrational transitions** involve energies in the region of 20 kJ mol^{-1} and correspond to IR spectra.
- **Rotational transitions** involve energies in the region of 0.01 kJ mol^{-1} and correspond to the microwave region. These spectra are difficult to observe and are not discussed further in this text.

Vibrational transitions

Some molecules absorb radiation of infrared wavelengths, which causes changes in the **modes of vibration** (the various ways in which the molecule vibrates). The wavelengths absorbed correspond to transitions between quantized energy levels. The simplest modes of vibration are **stretching** and **bending**. These modes and some more complicated vibrations are shown in the illustration below. The wavelength of absorption is characteristic of the type of bond present, so infrared absorption spectra give two sorts of structural information:

- Certain groups of atoms give rise to characteristic absorptions, which can be used to identify specific functional groups. It is thus a *'bond-spotting'* technique.
- Molecules with different structures have different vibrational energy levels, and so the infrared spectrum of a compound is like a *'fingerprint'* which is not shared by any other compound. Compounds with identical infrared spectra are identical.

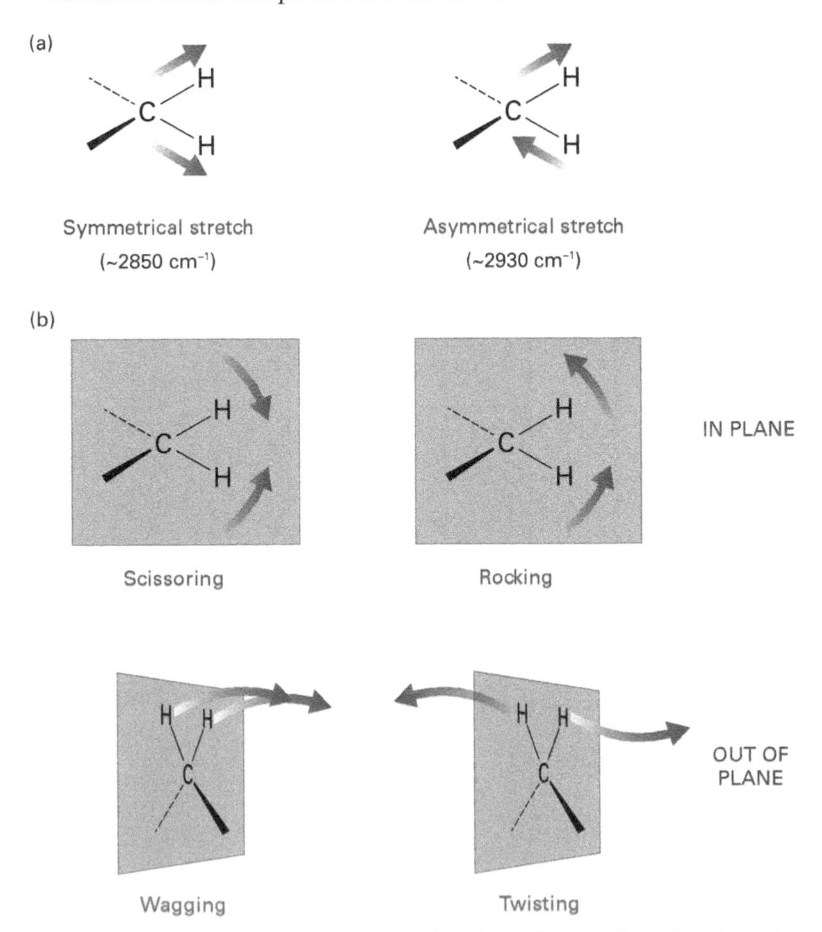

*The modes of vibration. (a) Stretching vibrations. (b) Bending vibrations. Note that the traditional unit used in infrared spectroscopy is cm^{-1}. This measures the **wavenumber** and is the reciprocal of the wavelength in centimetres (1/cm).*

IR absorption and the greenhouse effect

Absorption of infrared radiation by carbon dioxide (and methane and water vapour, for example) is the reason that it causes an enhanced greenhouse effect, spread 22.5.

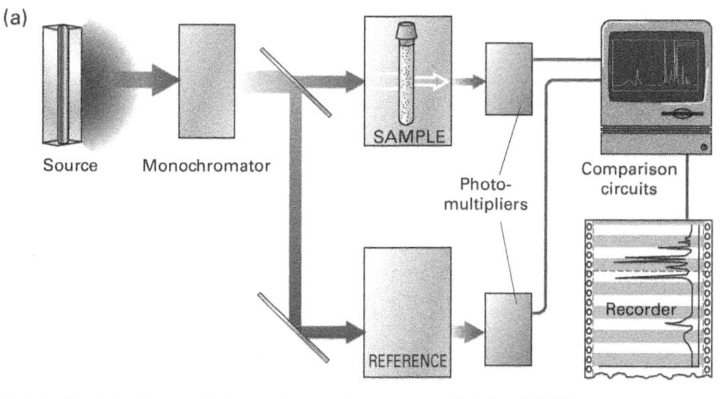

(a) A double-beam infrared spectrometer. (b) An FT-IR spectrometer.

Active and inactive bonds

Not all bonds absorb energy at infrared wavelengths. For energy to be transferred from the infrared radiation, a bond must have an electric dipole, spread 5.9, that changes as it vibrates. Symmetrical bonds, such as those in H_2 or Cl_2, will not absorb, nor will symmetrical bonds that are also symmetrically substituted either side of the bond, e.g. the C=C bond in ethene. These bonds are '**IR-inactive**'. As shown below right, carbon monoxide is a linear diatomic molecule with one **IR-active** mode of vibration. In comparison, carbon dioxide is a linear triatomic molecule: two IR-active modes of vibration are shown with one IR-inactive one.

Sample preparation

For IR spectroscopy (and also for UV/Vis spectroscopy), special materials must be used to contain the sample in order to avoid distortion of the absorption spectrum by the material of the 'container'. **Potassium bromide** is the most commonly used material, because it is transparent at infrared wavelengths. Windows made from a single crystal of potassium bromide are available, which can be used to make 'cells' to contain gases or liquids. Alternatively, a thin film of liquid can be sandwiched between two windows.

FT-IR spectroscopy

Older infrared spectrometers work on the same basic principle as UV/Vis machines, scanning wavelengths from about 2500 nm (4000 cm^{-1}) to 25 000 nm (400 cm^{-1}) and comparing sample and reference beams. Because of the time taken to scan the large range of wavelengths, these instruments have now been largely superseded by **Fourier transform (FT-IR) spectrometers**, which pass several wavelengths through the sample at the same time and analyse the results using the mathematical technique of Fourier analysis. This takes a complex waveform and works out the amplitudes of each single-frequency component in it. The analysis is carried out by a computer (inside the instrument), which then plots the spectrum.

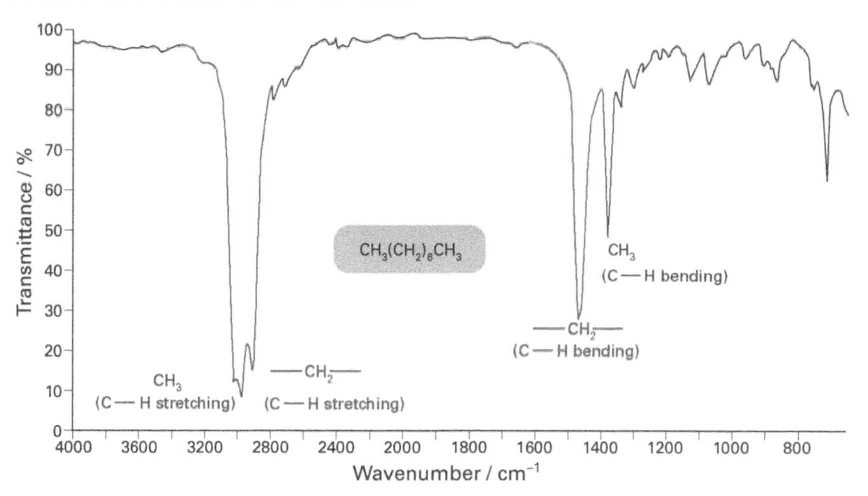

*The infrared absorption spectrum for octane. The standard method of display shows percentage transmission, rather than absorption. This produces downward 'peaks' of the curve at the wavenumbers where a molecule absorbs radiation. The pattern within the **'fingerprint region'** (1500–400 cm^{-1}) is specific to each molecule, and can be used for identification purposes.*

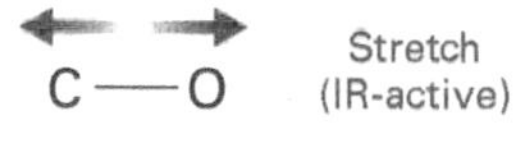

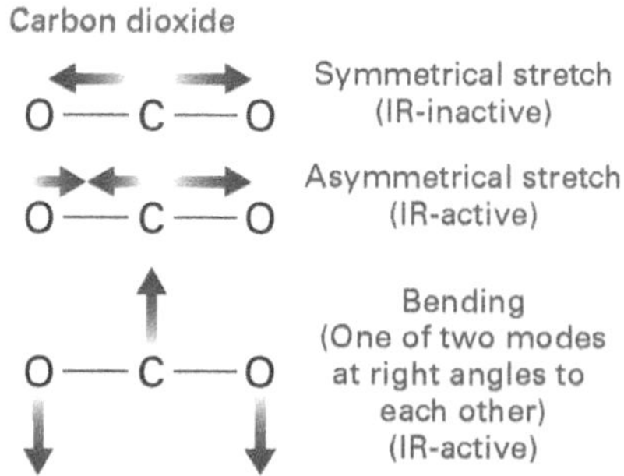

Symmetrical stretching of the CO_2 molecule does not alter the overall dipole moment of the molecule. This vibrational mode is therefore IR-inactive. The nature of the bonding in these molecules is not represented (see spread 10.10).

SUMMARY

- Infrared spectroscopy is used to detect vibrational transitions.
- For a bond to be infrared-active, it must have an electric dipole that changes as it vibrates.
- Infrared spectra can be used to identify specific bond vibrations and to identify complete molecules by using their infrared 'fingerprint'.
- A material transparent to infrared, such as potassium bromide, must be used to contain the sample.

32.4

Infrared spectroscopy – 2

OBJECTIVES

- Correlation charts
- Interpretation of IR spectra

To interpret the infrared spectra of substances and deduce their molecular structures, spectroscopists apply their analytical skills together with an almost intuitive 'feel' for complex spectra gained from experience. Automatic data processing techniques are available that compare measured spectra with libraries of known spectra. Remember that the pattern within the '**fingerprint region**' (1500–400 cm^{-1}) is specific to each molecule, and may be used to identify an unknown substance by comparison of its spectrum with those of previously identified substances. To aid the novice, **correlation charts** list the characteristic absorptions of different isolated functional groups. A simplified correlation chart is shown below. The aim of this spread is to show the interpretation of a number of different IR spectra, and then to help you to make your own interpretations of the spectra of unnamed substances.

A simplified correlation chart.

Bond	Type of compound	Range / cm^{-1}
C—H	Alkanes	2850 – 2960
	Alkenes	3010 – 3095
	Arenes	3030 – 3080
	Aldehydes	2710 – 2730
O—H	Alcohols (H-bonded)	3230 – 3550
	Carboxylic acids (H-bonded)	2500 – 3000
N—H	Amines	3320 – 3560

Bond	Type of compound	Range / cm^{-1}
C=C	Alkenes	1620 – 1680
C=C	Arenes	1500 – 1600
C—O	Alcohols, ethers, carboxylic acids, esters	1000 – 1300
C=O	Aldehydes, ketones, carboxylic acids, esters	1680 – 1750
C—N	Amines	1180 – 1360
C≡N	Nitriles	2210 – 2260

Bond-spotting

Alcohols are identified from the O—H absorption and carbonyl compounds from the C=O absorption.

Bonds to hydrogen

Notice how the bonds that vibrate at the highest wavenumbers are those where atoms are bonded to hydrogen, the element with the lowest mass.

Interpreting IR spectra

Look at each of the following four IR spectra and compare the absorptions with the values given in the correlation chart. Note that the characteristic frequencies of some of the functional groups are not always clearly defined.

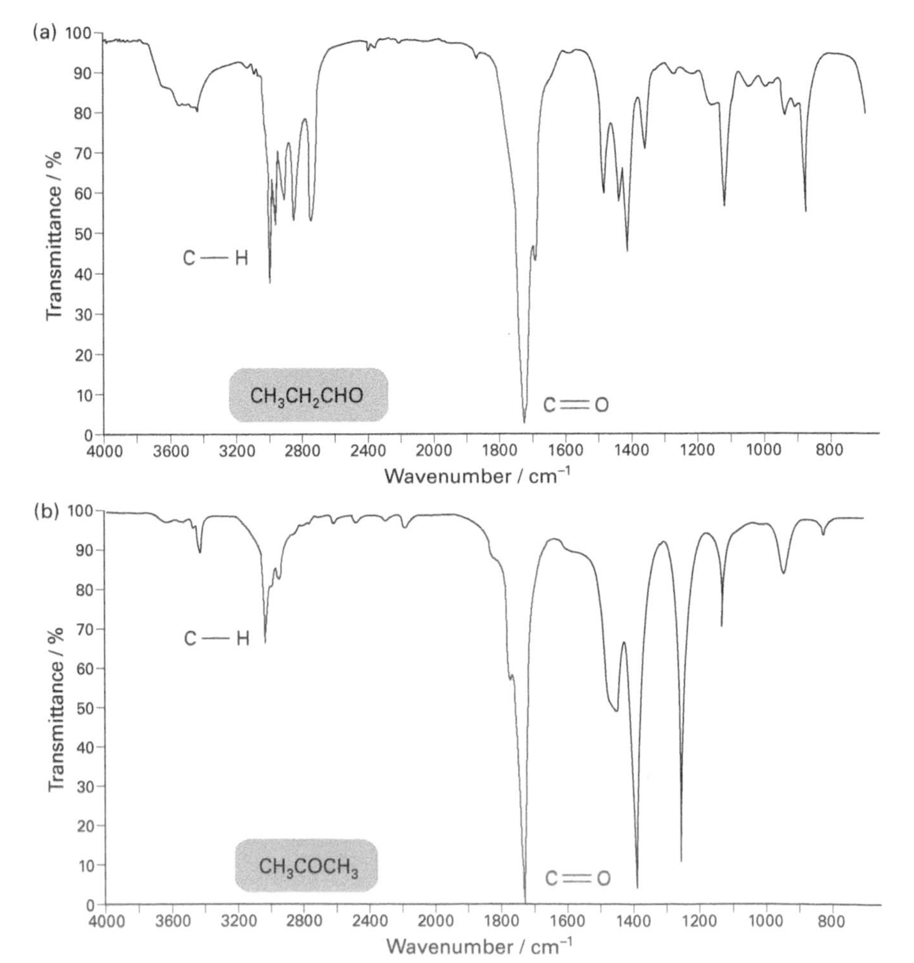

IR spectra of (a) propanal and (b) propanone.

Reflectance IR

The advent of single beam Fourier transform IR spectrometers has enabled the development of a very fast and simple method of handling samples, especially samples which are difficult to prepare for transmission IR (where the beam must pass right through the specimen). Reflectance IR bounces the beam off the surface of the sample instead of passing it right through. At its simplest it is known as diffuse reflectance in which the IR beam is bounced off the rough surface of the sample itself, or one produced by rubbing an abrasive sampling pad on the sample. In attenuated total reflectance (ATR), the IR beam from the spectrometer is directed through a crystal, known as the ATR element, in intimate contact with the sample. Internal reflection of the beam by the crystal causes it to extend into the sample and be selectively absorbed in the usual way. The beam then passes out of the sampling device and into the detector of the instrument where it is analysed. The most famous ATR sampling attachment is known as the 'Golden Gate™' produced by Specac. In this the sample is held in high pressure contact with a diamond ATR element.

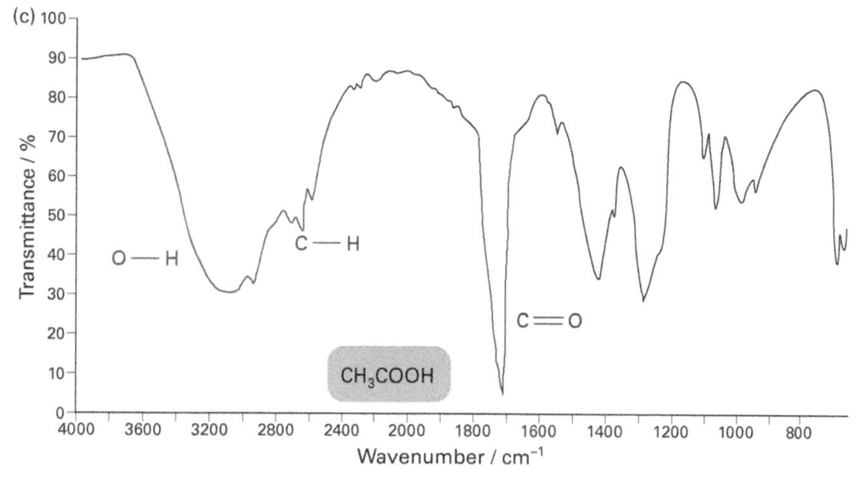

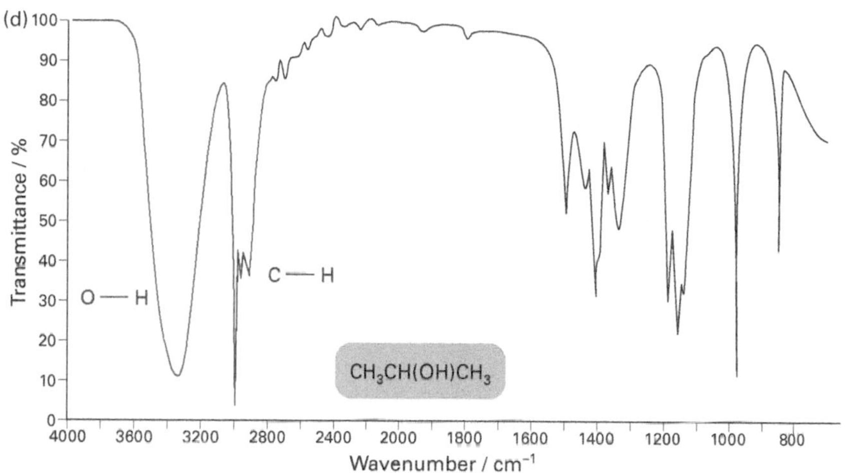

IR spectra of (c) ethanoic acid and (d) propan-2-ol.

Broad O—H

Notice the broad O—H absorption in a carboxylic acid.

Uses of infrared spectrometers

Infrared spectroscopy is a very widely used analytical technique. Infrared spectra have been used to measure levels of ethanol, spread 25.4, and to monitor air pollution. They have also been used to follow the progress of reactions in which the functional group changes, as is the case for the oxidation of alcohols to carbonyl compounds.

SUMMARY

- Inspect the region 3600–1500 cm^{-1} for the characteristic absorptions of O—H, N—H, C=O, C=C, etc.
- The region 1500–400 cm^{-1} is the fingerprint region.
- The region 2500–1800 cm^{-1} frequently contains no peaks.

PRACTICE

1 Describe the modes of vibration in the following:

a Carbon monoxide

b Carbon dioxide

c Water

d Sulphur dioxide.

State which modes are IR-inactive, giving reasons for your choices.

2 Describe as fully as you can the structures of the molecules giving rise to the IR spectra shown below.

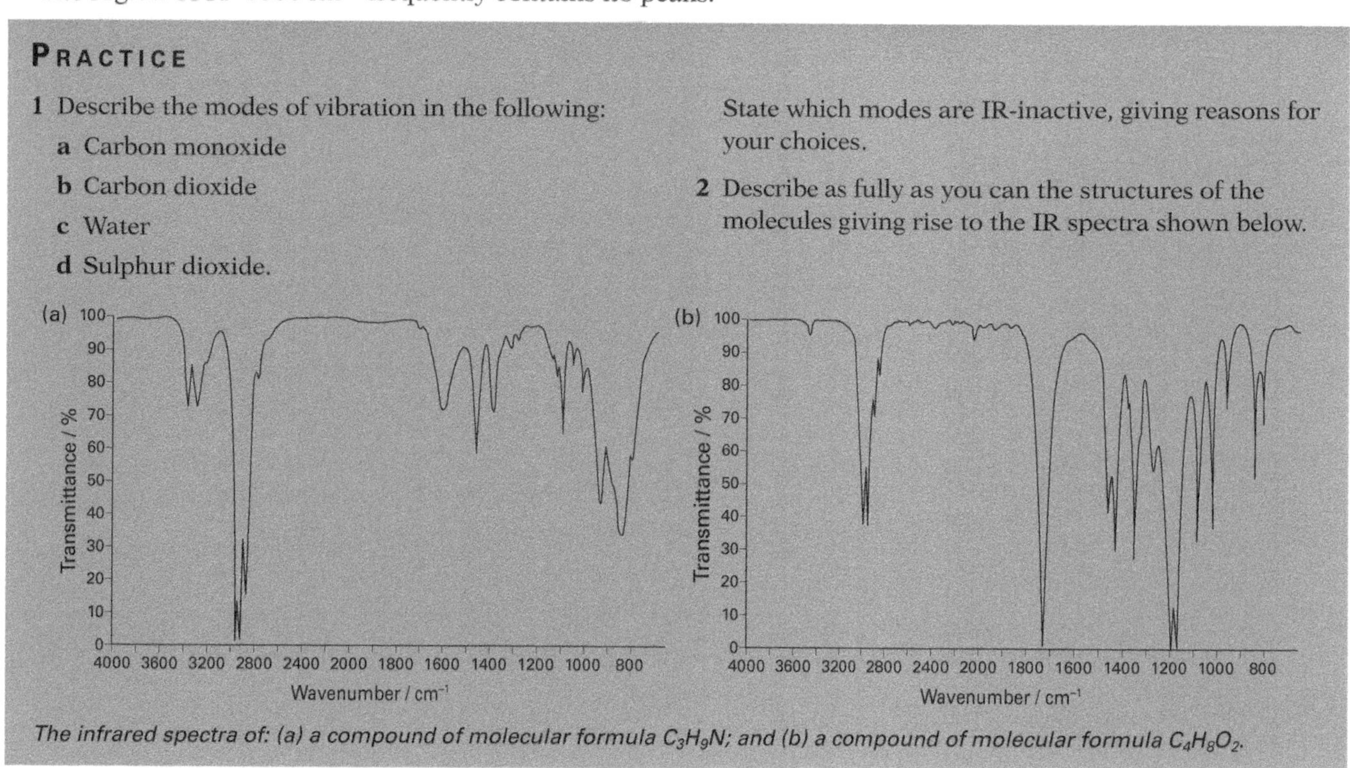

The infrared spectra of: (a) a compound of molecular formula C_3H_9N; and (b) a compound of molecular formula $C_4H_8O_2$.

32.5

OBJECTIVES

- Nuclear spin
- NMR spectra
- Use of TMS

An NMR spectrometer.

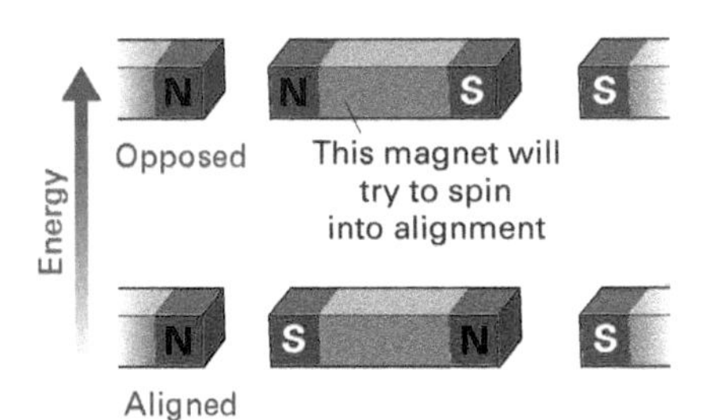

Aligned (lower-energy) and opposed (higher-energy) arrangements of bar magnets.

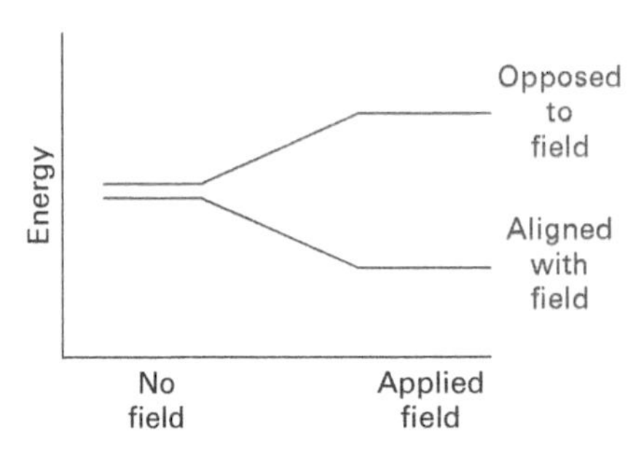

The spin states of a proton both in the absence and the presence of an applied magnetic field.

Proton-free solvents

Proton NMR are obtained using samples dissolved in solvents that do not contain protons, such as CCl_4 or deuterated solvents (such as $CDCl_3$), in which all H atoms are the deuterium isotope D (^2H).

Nuclear magnetic resonance (NMR) – 1

Whereas IR spectroscopy provides information about the types of bonds in a molecule, nuclear magnetic resonance (**NMR**) spectroscopy gives different information about the arrangement of specific atoms in a molecule. The sample is placed in a very strong magnetic field and its absorption of electromagnetic radiation is measured. The nuclei of certain atoms, most importantly hydrogen, absorb strongly under these circumstances. The actual magnetic field experienced by each hydrogen nucleus in a molecule depends on the extent to which electrons shield the nucleus from the applied field. This shielding in turn depends on the arrangement of the electrons around each nucleus. Thus, the magnetic field experienced by an individual hydrogen nucleus depends on its position within the molecule. As with IR spectra, the full interpretation of NMR spectra can be as much an art as an exact science, but NMR spectroscopy is of great value in the fields of research and analytical chemistry.

Nuclear spin

The phenomenon of *electron* spin (either 'up' or 'down') was introduced in spread 4.6. The nuclei of atoms that have an odd mass number also possess spin. The two nuclei that are most often studied in NMR spectroscopy are hydrogen (^1H), as mentioned above, and carbon-13 (^{13}C), because they both possess spin. NMR applied to the hydrogen nucleus is often called 'proton magnetic resonance' (**PMR**). Carbon-13 NMR was useful in identifying buckminsterfullerene, spread 19.4.

Proton NMR spectroscopy

Like the electron, the proton (the hydrogen nucleus ^1H) has two possible spin states available to it. These spin states have the *same energy* until a magnetic field is applied. The spin of the nucleus may then align itself either with the field or opposed to it, as shown on the left. The spin state that is aligned with the field is now at a lower energy than the spin state that is opposed to the field. The precise energy difference between the spin states depends on the strength of the magnetic field experienced by the nucleus.

The nuclei aligned with the applied field can change their spin to the opposite state if they absorb energy that matches the energy required for the transition (in the same way that energy of the correct magnitude is required for an electronic transition from a lower energy level to a higher one). As with other types of spectroscopy, the energy absorbed can be measured. Because the energy difference is very small, the wavelength of the electromagnetic radiation absorbed is *much* longer than for UV/Vis or for IR spectroscopy, and lies in the radio-frequency range, around hundreds of MHz.

Producing NMR spectra

The sample substance, either a pure liquid or in solution, is placed in a cylindrical sample tube, which is spun on its axis between the poles of a powerful electromagnet (typically between 2 and 10 tesla). Around the sample tube is a probe coil, which is connected to a radio-frequency generator and a receiver. To produce a spectrum, the applied magnetic field is varied by means of extra electromagnet coils.

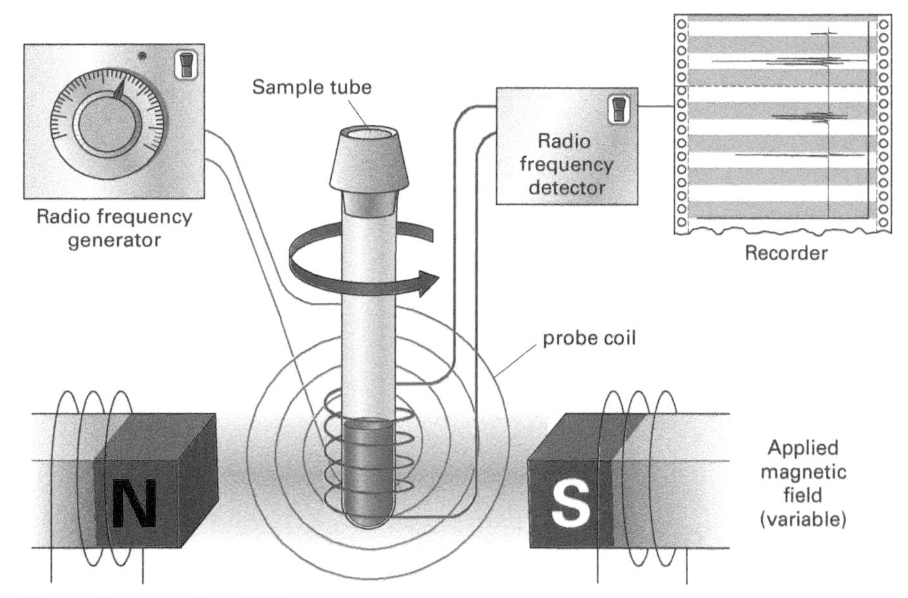

The main parts of an NMR spectrometer.

Chemical shift and TMS

A particular proton within a molecule will absorb electromagnetic radiation at a particular frequency called its **resonance frequency**. The resonance frequencies of protons in different environments are different but very close together. It is very difficult to measure frequencies with enough precision to make out these very small differences. So, instead of trying to measure the *exact* resonance frequency of any proton, a **reference compound** is added to the sample, and the resonance frequency of each proton in the sample is measured *relative to the resonance frequency of the protons in the reference compound*. This technique enables a frequency *difference* to be measured (which is much easier) rather than an exact frequency. Tetramethylsilane (**TMS**), $(CH_3)_4Si$, is universally used as the reference compound for proton NMR because its methyl groups are particularly well shielded (and TMS is non-toxic). The resonances of the protons in the sample are described in terms of how far they are 'shifted' from those of TMS.

The size of this shift varies with the magnitude of the applied magnetic field, which itself depends on the design of the spectrometer and, in particular, on the frequency corresponding to the applied magnetic field. To enable data from different workers and different instruments to be compared, the chemical shift has been *defined* to be independent of applied magnetic field strength (and thus spectrometer radio-frequency used). The **chemical shift** δ is the shift in Hz of the proton in question divided by the operating frequency in MHz of the spectrometer used, i.e.

$$\delta = \frac{\text{shift in Hz}}{\text{spectrometer frequency in MHz}}$$

The units of chemical shift are parts per million (ppm).

SUMMARY

- Nuclei of atoms that have an odd mass number such as hydrogen or carbon-13, possess a spin, which can align with or against an applied magnetic field.
- NMR measures the energy needed to change the spin state, and this energy depends on the environment of the atom.
- Tetramethylsilane (TMS) is used as a reference compound for proton or ^{13}C NMR spectra.
- The chemical shift is the shift in Hz of the proton in question (relative to TMS) divided by the frequency in MHz of the spectrometer used.

Body scanners

NMR using other nuclei, such as ^{13}C and ^{31}P, has found widespread use in medicine. ^{31}P NMR can be used, for example, to determine the extent of damage following a heart attack.

A recent development is the whole-body scanner. The whole-body scanner allows visualization of the entire body by placing the patient inside the magnet of a very large NMR machine. (There is no need to spin the sample!) Molecules in different tissues of the body are in different environments, and so it is possible to distinguish them using NMR. With the aid of a computer, the results can be displayed as a colour map of the organs in the body. In medicine, the technique is known as **magnetic resonance imaging** (MRI) because people are apprehensive of the word 'nuclear'.

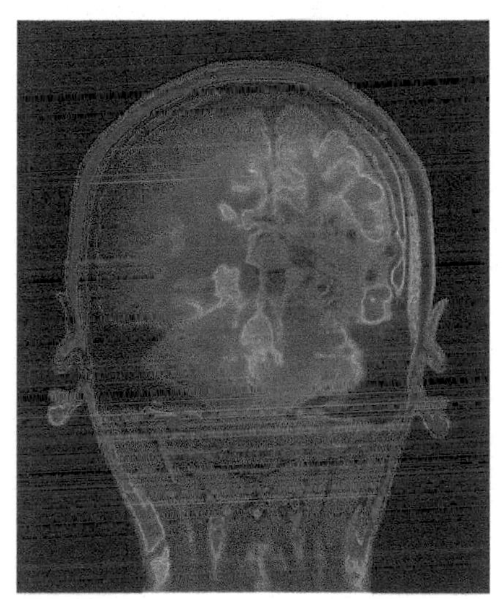

An MRI image of a human head. False-colour imaging (produced by software in the machine) allows the different tissues to be distinguished.

NMR spectra

You are now familiar with the principles underlying the production of NMR spectra. The following two spreads introduce examples of actual spectra and explain how to interpret them.

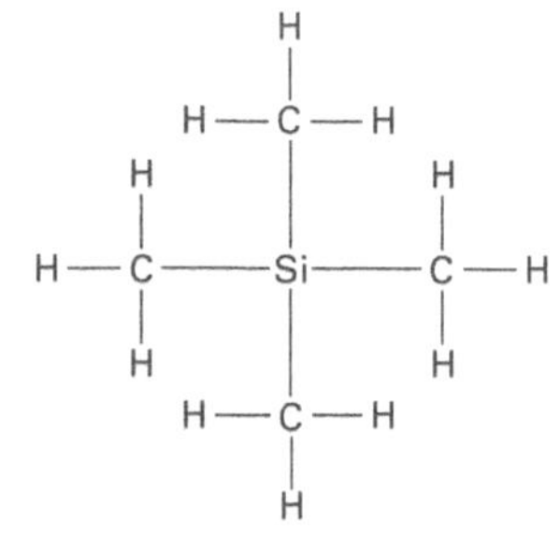

Tetramethylsilane.

32.6

OBJECTIVES

- Chemical equivalence
- Peak area and proton numbers

NUCLEAR MAGNETIC RESONANCE (NMR) – 2

The previous spread covered the basic principles of nuclear magnetic resonance spectroscopy. This spread looks in detail at how the environment of protons or ^{13}C atoms in a molecule can vary, and how to derive structural information from a proton or ^{13}C NMR spectrum.

Chemical equivalence

Protons that are in the same environment in a molecule will have identical chemical shifts. They are said to be **chemically equivalent** and give rise to a single peak in the PMR spectrum of the compound. For example, a benzene molecule contains six hydrogen atoms attached to the six carbon atoms in a regular planar hexagonal ring, spread 23.1.

The nucleus of each of these atoms is in an identical environment, as shown by the single peak in the PMR spectrum given below. The protons of the six hydrogen atoms in benzene are chemically equivalent.

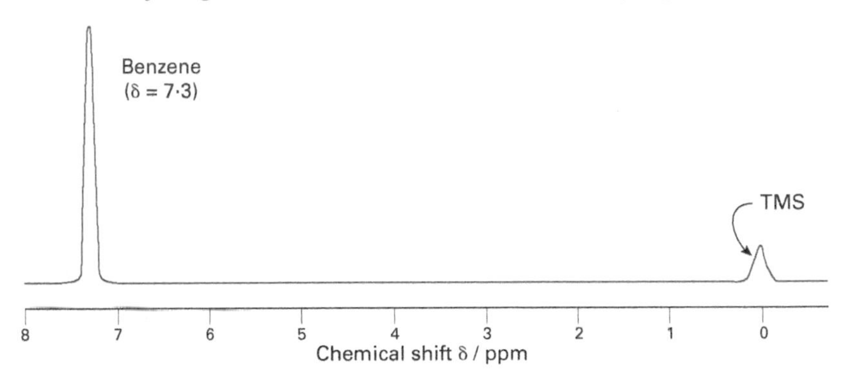

The PMR spectrum of benzene C_6H_6.

Table of approximate chemical shifts and corresponding proton environments. Here R is an alkyl group.

Type of proton	Chemical shift δ/ppm
RCH_3	0.9
R_2CH_2	1.3
R_3CH	1.5
$R_2C{=}CH_2$	~5
C_6H_5–H (benzene ring–H)	7.3
C_6H_5–CH_3 (benzene ring–CH_3)	2.3
R–C(=O)–CH_3	2.3
R–C(=O)–H	9.7

Values for chemical shifts

The value of the chemical shift of a peak in a PMR spectrum tells us about the particular type of proton environment. A table of values for chemical shifts gives us information about the types of molecular structure next to the proton that has that particular chemical shift. Note that chemical shifts alone do not often let us find the precise structure of a molecule.

Interpreting PMR spectra

Protons in different environments will have different chemical shifts, and so each peak in the PMR spectrum represents a different 'type' of proton. Ethyl ethanoate $CH_3COOCH_2CH_3$ has *three* different types of proton corresponding to the two methyl groups —CH_3 and the —CH_2— part of molecule. The two methyl groups are not equivalent, so the PMR spectrum of ethyl ethanoate will have three separate peaks, as shown below.

Exchangeable protons

There is no chemical shift value given above for O–H or N–H protons because they vary greatly. However, these bonds are easy to identify because on shaking the NMR sample with deuterium oxide (heavy water), D_2O, the signals disappear as D is exchanged for H.

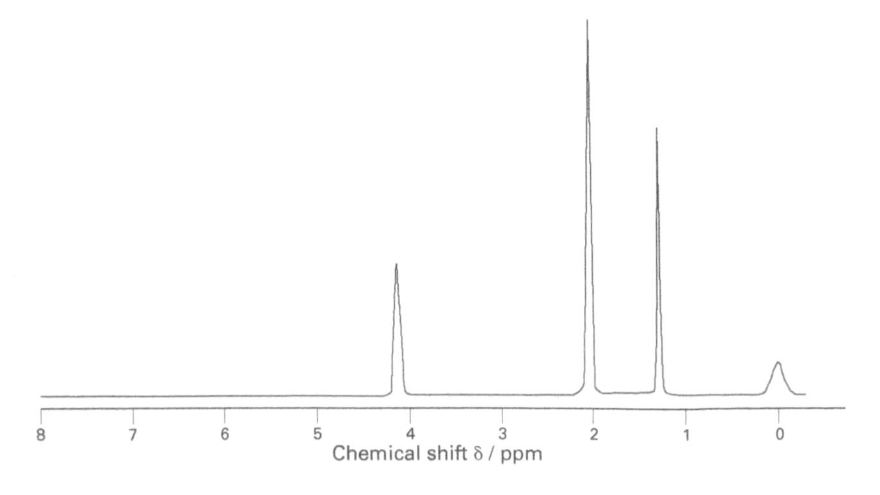

The PMR spectrum of ethyl ethanoate $CH_3COOCH_2CH_3$. There are three different groups of protons, each in its own magnetic environment and each chemically equivalent. See following spread for more details.

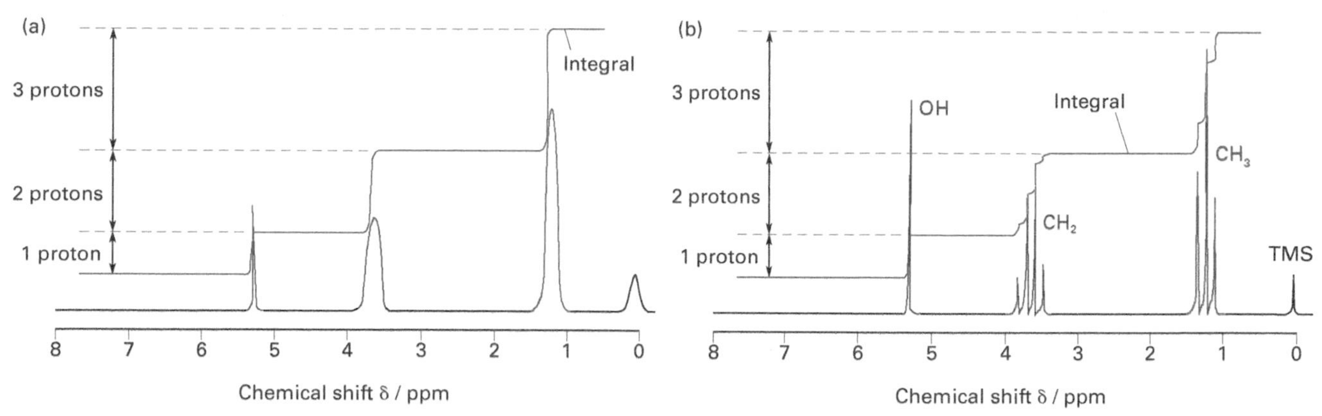

The PMR spectrum of ethanol (a) at low resolution and (b) at high resolution. The additional trace with vertically rising sections 'integrates' the area under each peak. The height of each vertical section is proportional to the area under the peak.

Peak areas

The area under each peak in a PMR spectrum is proportional to the number of protons generating that peak. The area is calculated automatically by the instrument by numerical integration, and can be displayed as a line above each peak, the height of which is proportional to the area. Note that the height of the line does not give the exact number of hydrogen atoms, just the *relative number* within that molecule. It is therefore necessary to look at the areas under all the peaks and work out the simplest ratio between them.

For example, the three peaks in the PMR spectrum for ethanol (above) have relative areas of 1, 2, and 3 respectively. The *true* number of protons must be deduced from other information about the molecule, some of which may be deduced from high-resolution PMR spectra, which will be introduced in the next spread.

The three peaks in ethanol correspond to the proton of the hydroxyl (—OH) group, the two protons of the $—CH_2—$ group, and the three protons of the end methyl ($—CH_3$) group from left to right of the spectrum above. Confirmation of this interpretation requires the high-resolution spectrum discussed in detail in the following spread.

High resolution

Even more details can be extracted from the spectra at high resolution, as explained in the following spread.

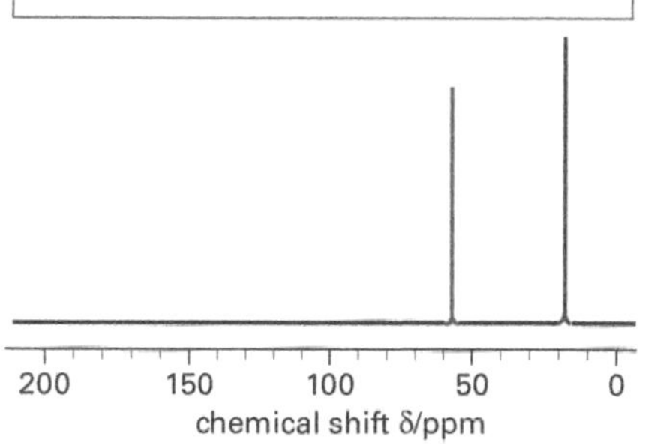

The ^{13}C NMR spectrum of ethanol CH_3CH_2OH.

Carbon-13 NMR spectra

Because of the measurement technique used (broadband proton decoupling), ^{13}C NMR spectra do not show the fine details present in high-resolution PMR spectra and so are easier to interpret. For example, the ^{13}C NMR spectrum for ethanol is shown above right. Note the two peaks expected for two different carbon atoms: the one bonded to oxygen shows a chemical shift above 50 ppm, see table below. The chemical shift value identifies the carbonyl group in paracetamol, spread 28.3, as the origin of the line near 170 ppm in its ^{13}C NMR spectrum (see next spread).

Each chemically different carbon gives a different peak. The four peaks for the four chemically different carbon atoms in butan-1-ol are very clear from the ^{13}C NMR spectrum: the carbon attached to oxygen has a chemical shift above 50 ppm. Note that in ^{13}C NMR spectra the peak area usually does *not* scale with the number of equivalent atoms.

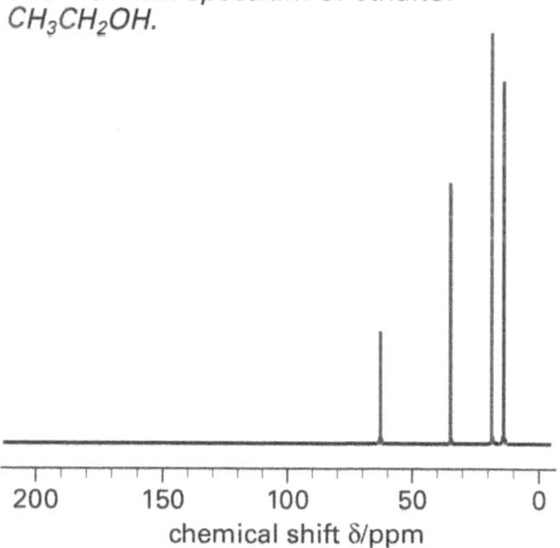

The ^{13}C NMR spectrum of butan-1-ol $CH_3CH_2CH_2CH_2OH$.

SUMMARY

- Chemically equivalent atoms give rise to a single peak in the NMR spectrum.
- Values of chemical shift can be used to indicate the chemical environment of the atoms in a molecule.
- The area under each peak in the PMR spectrum is proportional to the number of protons generating that peak.

δ/ppm	
0–50	Saturated C
50–100	Saturated C next to O
100–150	Unsaturated C
150–200	Unsaturated C next to O

Table of approximate ^{13}C chemical shifts.

PRACTICE

1 Identify the type of proton giving rise to each of the *three* peaks in the PMR spectrum of ethanol. Give reasons for your choices.

2 Explain why propanone gives a *single* PMR peak.

3 Explain why 1,4-dimethylbenzene gives *two* separate PMR peaks.

4 Sketch the PMR spectra for the following:
a 1,1,2-Trichloroethane
b 1,1,1-Trichloroethane.

32.7

OBJECTIVES

- High-resolution spectra
- Spin–spin splitting
- *n* + l rule

Nuclear magnetic resonance (NMR) – 3

All the PMR spectra described so far (except the last) have been of low resolution; they do not show the fine detail of the high-resolution spectra produced by modern instruments. You already know that the area under each peak in a PMR spectrum is proportional to the number of protons generating that peak. Comparing the areas gives the ratios of the different types of proton in a molecule. This section shows how analysis of the fine structure leads to detail about the relative positions of the protons within the molecule.

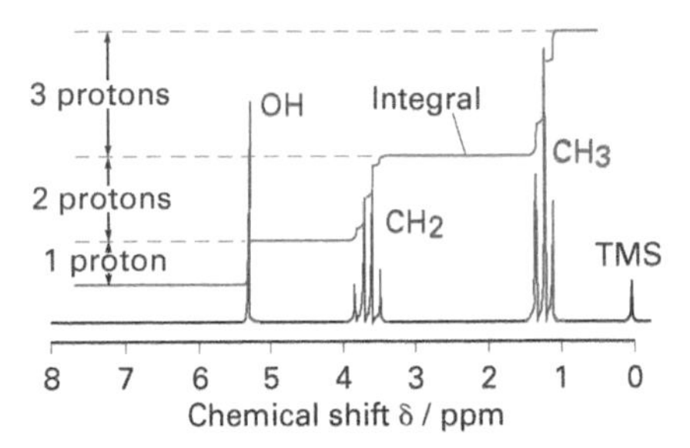

The high-resolution PMR spectrum of ethanol.

Ethanol at high resolution

The $-CH_2-$ peak is split into a quartet by the adjacent $-CH_3$ group.
The $-CH_3$ peak is split into a triplet by the adjacent $-CH_2-$ group.

Spin–spin splitting

The chemical shift of protons in a molecule is slightly altered by other adjacent non-equivalent protons because their respective magnetic fields are close enough to interact. A neighbouring proton with a field aligned with the applied field will have a slightly deshielding effect because its field will add to the applied field. If the field of a neighbouring proton is opposed to the applied field, it will effectively reduce the applied field and so have a shielding effect. The protons are said to be **coupled**.

The effect of coupling on the PMR spectrum is that each type of proton does not necessarily give a single resonance peak. This phenomenon is known as **spin–spin splitting** and becomes evident at high resolution.

- The number of peaks arising from splitting indicates how many protons are *adjacent* to a particular given proton.
- If the number of adjacent non-equivalent protons is n, the peak will split into **(n + 1)** peaks:
  - **one** adjacent proton produces two closely spaced peaks (a **doublet**),
  - **two** adjacent protons produce three closely spaced peaks (a **triplet**),
  - **three** adjacent protons produce four closely spaced peaks (a **quartet**).

Each line within a peak has a relative intensity that can be worked out by a simple mathematical technique called Pascal's triangle.

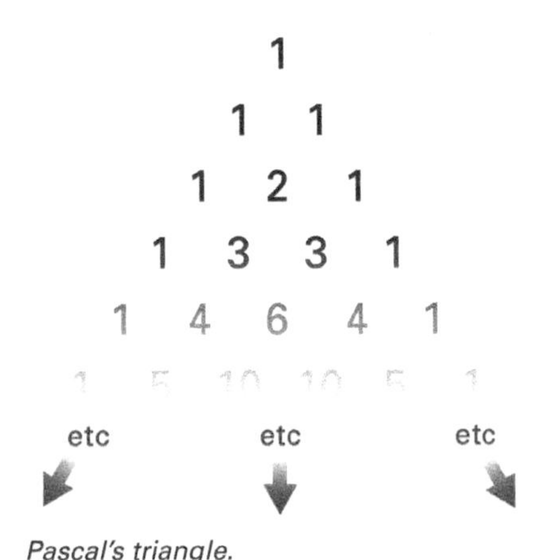

Pascal's triangle.

Each number in the triangle is the sum of the two integers above it to the left and to the right. A doublet therefore has two equal peaks (relative intensities 1:1); a triplet has three peaks (relative intensities 1:2:1) and a quartet has four peaks (relative intensities 1:3:3:1).

Interpreting a high-resolution PMR spectrum

Look at the PMR spectrum of 1,1,2-trichloroethane shown below. The low-resolution spectrum consists of two peaks, indicating that there are two types of proton in the molecule. Integrating each peak to obtain the area under it indicates that the two types of proton are present in the ratio 1:2.

The high-resolution spectrum (below right) shows that the signal arising from the single proton $\mathbf{CHCl_2}CH_2Cl$ is split into three (1:2:1) peaks by the pair of equivalent protons $CHCl_2\mathbf{CH_2}Cl$. The signal arising from the pair of equivalent protons $CHCl_2\mathbf{CH_2}Cl$ is split into two (1:1) peaks by the single proton $\mathbf{CHCl_2}CH_2Cl$.

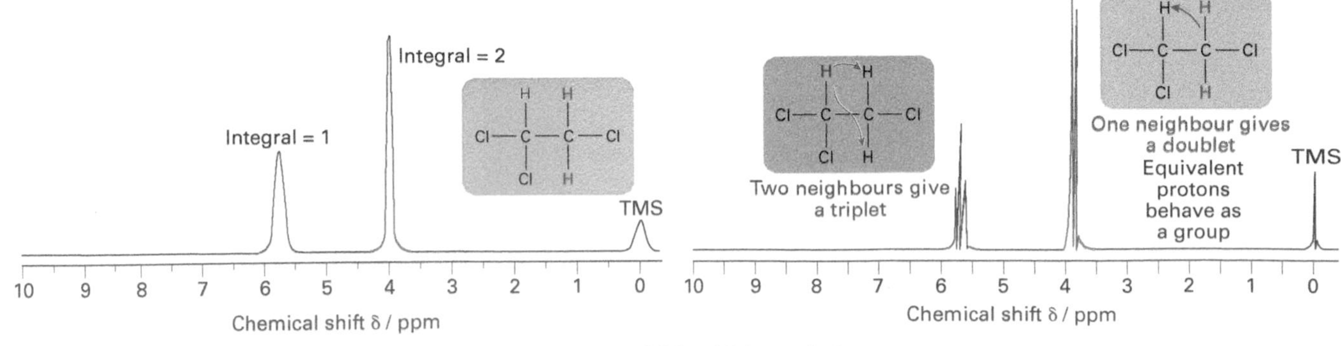

PMR spectrum of 1,1,2-trichloroethane (a) at low resolution and (b) at high resolution.

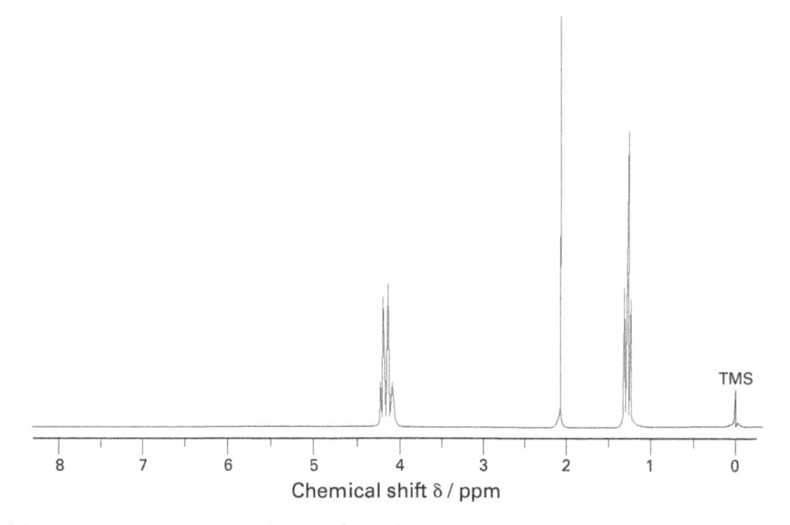

The high-resolution PMR spectrum of ethyl ethanoate.

Carbon-13 NMR spectra show no spin–spin splitting

The figure below shows the ^{13}C NMR spectrum for paracetamol, spread 28.3. Carbon atoms 2 and 6 are identical, as are carbon atoms 3 and 5; hence there are six peaks from the eight atoms.

The relative simplicity of carbon-13 spectra means that more complex molecules can be investigated. The complete biosynthetic pathway for vitamin B_{12} was worked out by Alan Battersby using ^{13}C NMR. Furthermore the technique is non-destructive unlike earlier techniques which required break-up of the molecule into fragments before identification.

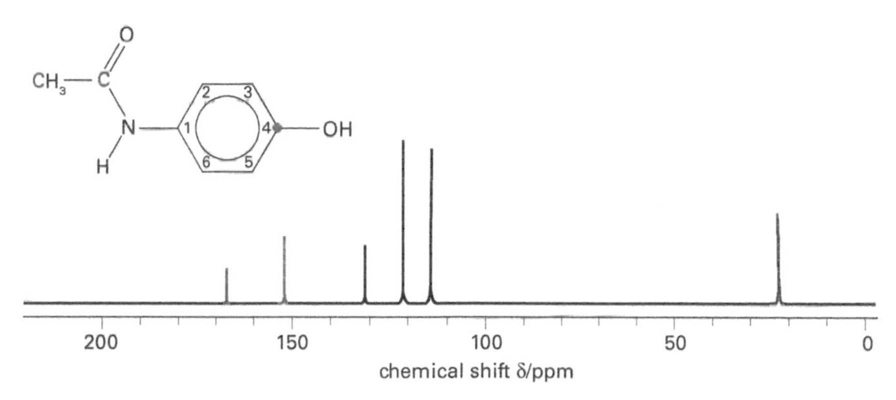

The ^{13}C NMR spectrum of paracetamol. Owing to the symmetry of the positions round the ring, carbon atoms 2 and 6 are identical, as are carbon atoms 3 and 5, hence there are only 3 peaks in the region of 110 to 135 ppm

Ethyl ethanoate

The single peak near 2.1 ppm must be due to the CH_3 connected to the carbonyl group, as it is a singlet (which means there are no adjacent protons).

The peak near 4.1 ppm is a quartet, owing to the $-CH_2-$ group being attached to a $-CH_3$.

The peak near 1.3 ppm is a triplet, confirming that the $-CH_2-$ group is the neighbour of this $-CH_3$ group.

NMR

NMR is one of the most powerful techniques currently used for structure determination in organic chemistry. Modern NMR spectrometers are fully computer controlled and, when coupled with automatic sample changers, provide continuous data acquisition and processing. This high level of automation, together with easy to use software packages, allows wider access to the instruments and frees the chemist to focus on data *interpretation* rather than mere data acquisition and processing.

SUMMARY

- Spin–spin splitting yields information about neighbouring protons.
- If the number of adjacent protons is n, the peak will split into $(n + 1)$ peaks.
- Pascal's triangle can be used to predict the relative intensities of split peaks.

PRACTICE

1 The PMR spectrum shown to the right was obtained from iodoethane. Explain the splitting pattern observed.

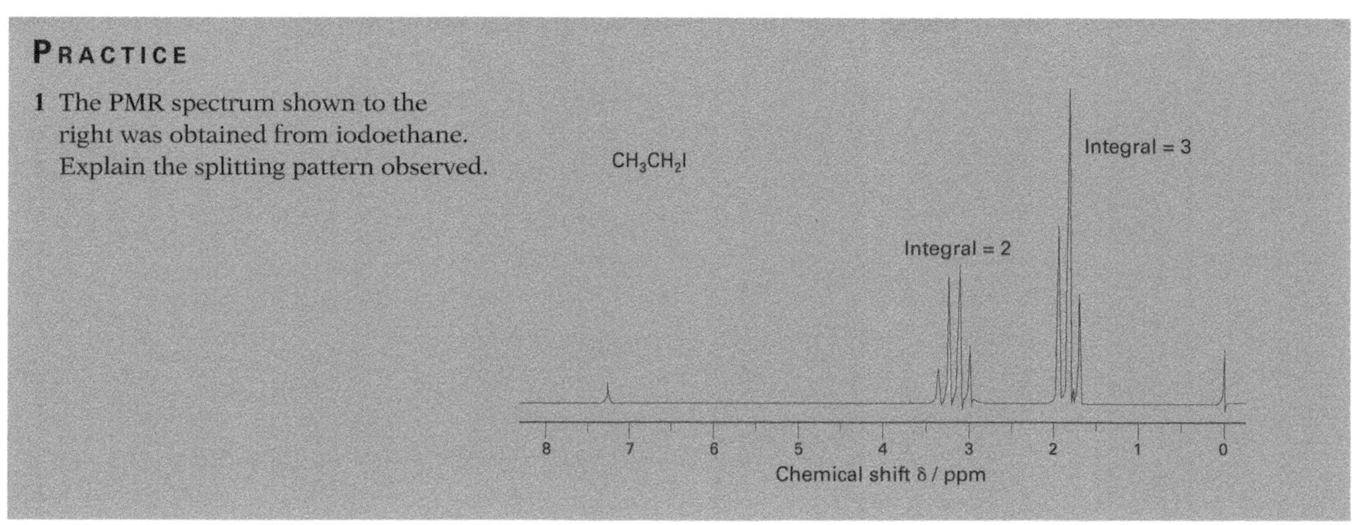

PRACTICE EXAM QUESTIONS

1 a Outline the principles on which atomic absorption spectroscopy is based. [3]

b Briefly describe the main differences in the principles on which atomic absorption spectroscopy and flame emission spectroscopy are based. [2]

c The urine of a patient admitted to hospital with suspected 'heavy-metal' poisoning was analysed by using atomic absorption spectroscopy. The spectrum shown below was obtained.

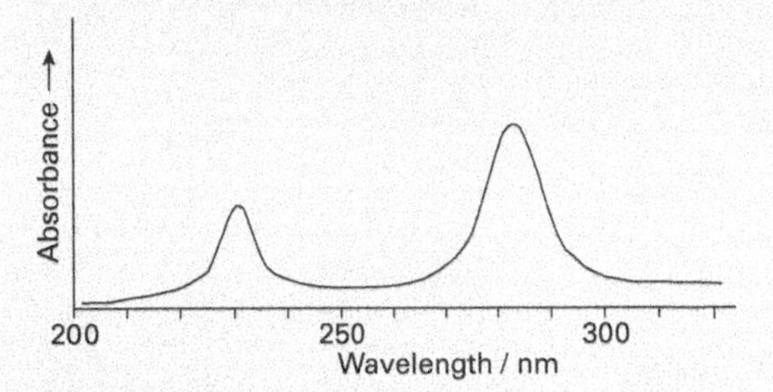

Wavelength of absorption / nm	Element	Possible source
229	Cadmium	Coloured pigments for pottery
254	Mercury	Seafood from contaminated water
283	Lead	Old water pipes, pottery glazes, old paint

Use the data in the table above to identify the likely source of the poisoning, indicating the evidence you have used. [3]

2 a The infrared spectrum shown below is that of compound **X** which has the molecular formula $C_4H_8O_2$.

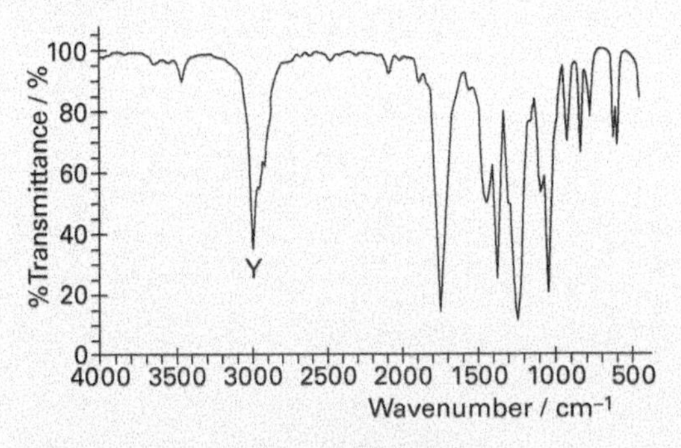

Table of infrared absorption data

Bond	Wavenumber/cm^{-1}
C–H	2850–3300
C–C	750–1100
C=C	1620–1680
C=O	1680–1750
C–O	1000–1300
O–H in alcohols	3230–3550
O–H in acids	2500–3000

i Use the table of infrared data to help you identify the bond responsible for the absorption marked **Y**.

ii Draw the structures of the two carboxylic acids having the molecular formula $C_4H_8O_2$ and explain why **X** cannot be either of these. [5]

b The fingerprint regions of the infrared spectra of **X** and of three other compounds are shown below labelled **I**, **II**, **III**, and **IV**.

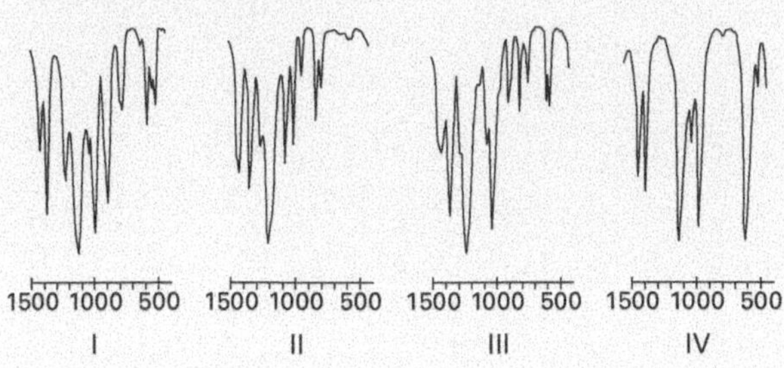

Which one of the fingerprint regions above is that of **X**? [1]

Compound **X** has three peaks with ratio of areas 3:2:3 in its low-resolution proton NMR spectrum. Draw two possible structures for compound **X**. [2]

3 a For each of the following molecules, state how many absorptions you would expect to see in its infrared spectrum:

i H_2O

ii CO_2

iii CO. [3]

b i State two analytical uses of infrared spectroscopy.

ii Give a reason why infrared spectroscopy is suitable for each use. [4]

c The infrared spectrum shown below was obtained from compound **A**, known to contain carbon, hydrogen, and oxygen only. The M_r of compound **A** is 88. Identify the functional groups found in the compound and suggest a structure for **A**, indicating how you arrive at your conclusion. [3]

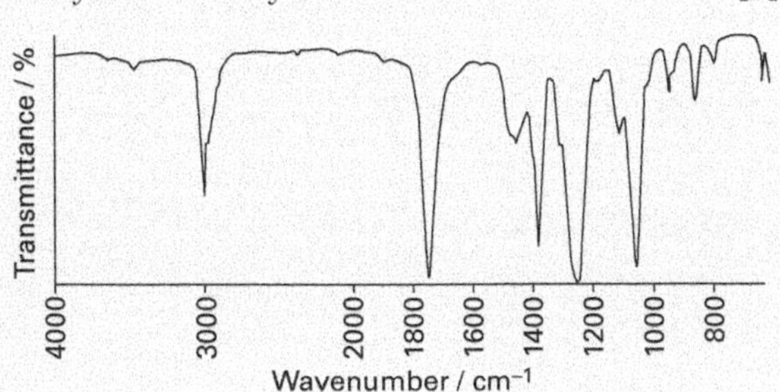

4 a i Outline why a proton can occupy two different energy states when subjected to an external magnetic field.

ii The NMR spectrum of ethanal, CH_3CHO, contains two absorptions, a 1:1 doublet and a 1:3:3:1 quartet. Outline why these splitting patterns occur. [4]

b The two NMR spectra shown below were obtained from compounds F and G, which have the same empirical formula, $C_3H_6O_2$.

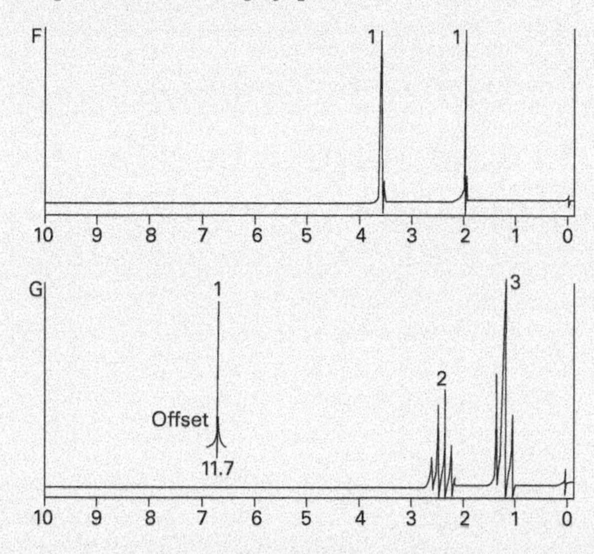

Identify the groups responsible for the absorptions in each spectrum, and hence deduce the structures of the two compounds. [6]

5 The proton NMR spectrum of an alcohol, **A**, $C_5H_{12}O$, is shown below.

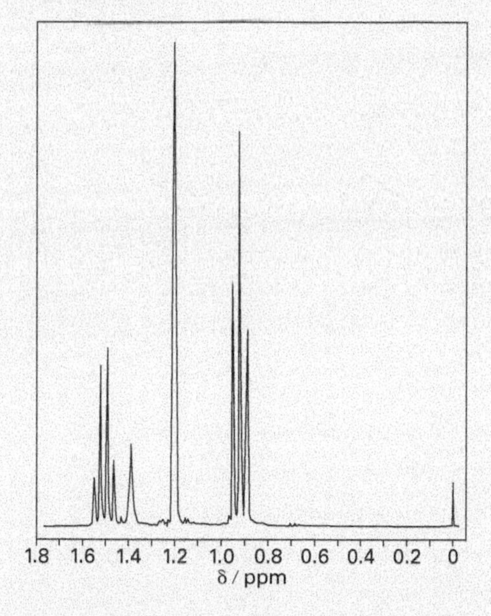

The measured integration trace gives the ratio 0.90 to 0.45 to 2.70 to 1.35 for the peaks at δ 1.52, 1.39, 1.21, and 0.93, respectively.

a What compound is responsible for the signal at δ = 0? [1]

b How many different types of proton are present in compound **A**? [1]

c What is the ratio of the numbers of each type of proton? [1]

d The peaks at δ 1.52 and δ 0.93 arise from the presence of a single alkyl group. Identify this group and explain the splitting pattern. [3]

e What can be deduced from the single peak at δ 1.21 and its integration value? [1]

f Give the structure of compound **A**. [1]

6 a i Explain what is meant by the term *mull* in infrared spectroscopy.

ii Ethanol is a good solvent for many organic compounds. Explain why ethanol is not a suitable solvent for use in producing infrared spectra of such compounds. [3]

b The infrared spectrum shown was obtained from a compound **J** of formula $C_5H_7O_2N$.

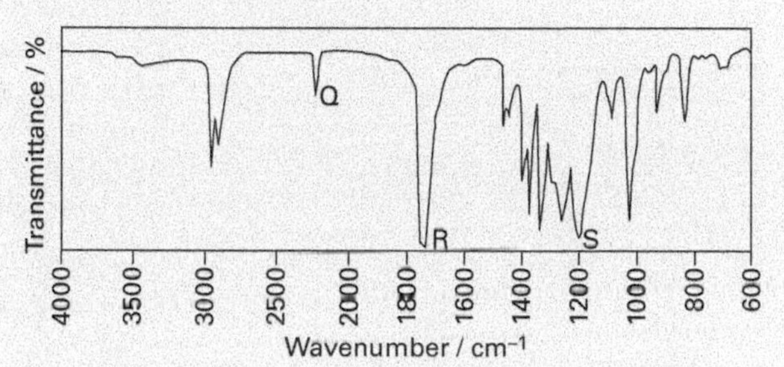

Identify the bonds responsible for the absorptions labelled **Q**, **R**, and **S**, and hence suggest a structure for **J**. [4]

c The mass spectrum shown was obtained from a compound **K**.

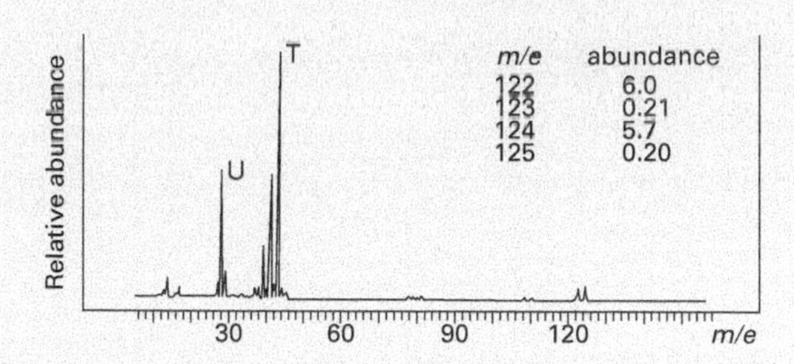

i Deduce what fragment has been lost in forming the peak labelled **T**.

ii Deduce what fragment has been lost from **T** in forming peak **U**.

iii Suggest a formula for **K**. [3]

Appendices

Mathematics toolbox: Standard form, uncertainty, and significant figures

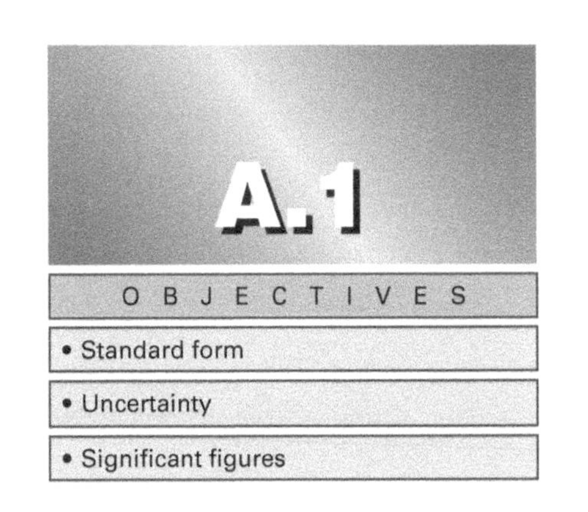

Examples of standard form

101 is 1.01×10^2

0.00000103 is 1.03×10^{-6}

126.5 is 1.265×10^2

Communicating data numerically is an important skill. You need to know the conventions for writing down the information. You also need to understand just what it is you are communicating, the certainty of the information, and the significance of figures.

Standard form

In chemistry, we often meet very large or very small numbers. A method of writing numbers that allows this very wide range to be covered is known as **standard form** or **scientific notation.** In standard form a number is written in two parts that are multiplied together. The first part is a number that is always greater than or equal to 1 and less than 10, and this is multiplied by 10 raised to any power that is an integer.

$$\text{standard form} = \pm a \times 10^p$$

The power of 10 (p) represents the number of moves of the decimal point in a that we need to make if the number we want is less than 1 or greater than 10. If the number is less than 1, the value of p is negative and the decimal point is moved to the left. For a number greater than 10, we move the decimal point to the right in the same way:

$$0.000\,012\,34 = 1.234 \times 10^{-5}$$

$$918 = 9.18 \times 10^2$$

Uncertainty

When scientists say that there is an **uncertainty** in a measurement, they do not mean that they are unsure of its value. Uncertainties, which are sometimes less helpfully called errors, are not mistakes. An uncertainty is a natural variation in a measurement that comes about for a variety of reasons:

- No instrument is exactly accurate.
- Different people may be using different types of instrument.
- No two people read an instrument in exactly the same way.
- The instrument's adjustment may have changed.

No matter how carefully we set up experiments, problems like these always arise. Scientists try to estimate how big an effect such problems may have on an experiment, and they may quote an uncertainty in the measurement.

If you read a scientific paper you will find that the numerical results include a second number and a '±' symbol. For example, a time may be quoted as $4.6 \pm 0.6 \times 10^{-8}$ s. This means that the author believes that the measurement result is 4.6×10^{-8} s but suspects that it may be as large as 5.2×10^{-8} s ($5.2 = 4.6 + 0.6$) or that it could be as small as 4.0×10^{-8} s ($4.0 = 4.6 - 0.6$). This defines the range of uncertainty that is unavoidably part of the experiment.

Worked example on uncertainty

Question: A burette may be read to the nearest 0.05 cm^3 (see spread 9.7). Given that the titre in an experiment is 25 cm^3, estimate the percentage uncertainty in the answer caused by the uncertainty in this titre.

Answer: The percentage uncertainty is

$(0.05\,cm^3)/(25\,cm^3) \times 100 = 0.2$

This particular reading causes a 0.2% uncertainty in the final answer.

Significant figures

The digits reported in a measurement are called the **significant figures** (sig. figs or s.f.). For example, 2.5 cm^3 has 2 sig. figs; 25.2 cm^3 has 3 sig. figs. Calculators quote answers to as many figures as the display will allow. For example, if you divide 10 by 6 on your calculator, it may report the answer 1.6666667. If you are dealing with experimental data, however, it is important to think about how many significant figures it is reasonable to quote.

Generally, it is a good rule to quote calculated values to the same number of significant figures as the data used in the calculation. So, for example, it would be wrong to go from data quoted to two significant figures to a result (calculated from those data) given to three or more significant figures. Equally, though, sometimes whole numbers are given to just one figure because it is obviously exact. Say, for example, you were dealing with four discrete objects. It would be unnecessary to limit the results of calculations using that 4 to just one significant figure. Counting is exact; measurements are always uncertain. You must always consider the certainty of any measurement.

Some confusion can arise over the significance of zeros in a measurement. It is important to distinguish between legitimately measured zeros and ones that are just used to mark the position of the decimal point. The last zeros after the decimal point, as in 25.0 cm^3, are significant because they were measured. If the measurement could be made to ±0.05 cm^3, 25.0 means that the volume is somewhere between 24.95 and 25.05 cm^3. However, some zeros are 'captive'. These may or may not be significant. If you measure 30.7 g, then the zero is significant because it was measured. The zeros in the value 0.035 g are not significant; they are just used to indicate powers of 10. So 0.035 g has two significant figures, but 0.0350 g has three significant figures because the last zero was measured (if it was).

Using significant figures in calculations

You must not quote more significant figures in your answer than the minimum number you had in your data. This means that you often have to round off your result. Round *up* if the last digit is above 5 and round *down* if it is below 5. For numbers ending in 5, always round to the nearest even number to avoid compounding rounding-off errors in your answer.

Do not round off as you work through a calculation, only at the end. In practice, carry all digits in the memory of your calculator until you get the final answer and then round this off in a single step. For example, the result 30.348 rounds to 30.3 to 3 sig. figs but 30.35 to 4 sig. figs.

Significant figures and standard form

A number like 782 000 000 might have been rounded to three significant figures or might be accurate to nine significant figures. It is impossible to tell which just by looking at it.
If, however, the number is given in standard form, then all the figures are significant. So

1.60×10^{-19} is 3 sig. figs

1.602×10^{-19} is 4 sig. figs

1.60218×10^{-19} is 6 sig. figs

Decimal places

The decimal places are the digits that follow the decimal point. Again, we must be consistent in the number of decimal places we quote. In addition or subtraction, the number of decimal places quoted in the result should be the same as the *smallest number of decimal places* in the data.

When multiplying or dividing, revert to the rule that the number of *significant figures* should be the same as the *smallest number of significant figures* in the data.

A.2

OBJECTIVES

- Substituting into equations
- Manipulating equations
- What is a graph?
- Drawing graphs
- Information from graphs
- The equation of a straight line

Mathematics toolbox: equations and graphs

An equation is characterized by the equals sign. For example,

$$5y = x - 3$$

Equations are algebraic, with both numbers and letters. Usually you are in the position of knowing all the terms except one in an equation. You will then substitute the correct value for each letter before you can carry out the calculation. For example, if $x = 18$ in the above equation, then:

$$5y = 18 - 3$$

$$5y = 15$$

$$y = 3$$

Manipulating equations

An equation demands that the relationship denoted by the equals sign is maintained. However, you often need to move quantities about to obtain the answer you need, for example to change the subject of an equation. The important thing to remember is that *you must do the same thing to both sides of the equation* so that the equals sign is not violated. The following steps are commonly used in manipulating equations.

- Add the same quantity to both sides
  For
  $$x - 3 = 5y$$
  adding 3 to both sides gives
  $$x = 5y + 3$$

- Subtract the same quantity from both sides
  For
  $$x + 5 = 7y$$
  subtracting 5 from both sides gives
  $$x = 7y - 5$$

- Multiply the whole equation by the same quantity
  For
  $$\frac{x + 9}{5} = 8y$$
  multiplying both sides by 5 gives
  $$x + 9 = 40y$$

- Divide the whole equation by the same quantity
  For
  $$x + 3 = 2y$$
  dividing both sides by 2 gives
  $$\frac{x + 3}{2} = y$$

- Raise both sides to the same power
  For
  $$x + 3 = 2y$$
  squaring both sides gives
  $$(x + 3)^2 = (2y)^2$$
  $$(x + 3)^2 = 4y^2$$

- Take roots of both sides

  For

  $x^2 = 4y^2 + 9$

  taking the square root of both sides gives

  $x = \pm\sqrt{(4y^2 + 9)}$

If you are unsure whether you have performed an operation correctly, a good check is to substitute simple numbers into the equation. *If one side is not equal to the other, you have done something wrong.*

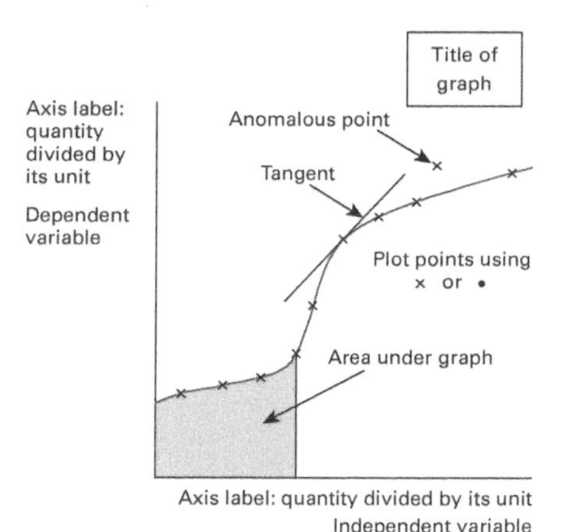

The parts of a graph.

Why use graphs?

A graph is a visual aid. When the results of an experiment are presented as a graph, it is immediately clear if they show a pattern. A graph will immediately reveal the scatter there is of the points about a smooth curve. The scatter is an indication of the size of the random errors involved in the experiment. Any gaps in your evidence will be plain as well. A graph also makes it easy to make estimations between measured points.

Drawing graphs

- Plot the **independent variable** on the horizontal axis or x-axis. You select values for the independent variable in an experiment.
- Plot the **dependent variable** on the vertical axis or y-axis. The dependent variable is the result of your alterations to the independent variable.
- Look at the range of values for both variables and choose a scale for both axes so that the plot fills most of the available space. (The scale need not be the same for both axes.) In some cases, this may mean leaving out the zero or origin from one or both axes.
- Choose a simple scale that will be straightforward to use. Make 10 small divisions represent 1, 2, or 5 (or these numbers multiplied by factors of 10 like 10, 0.2, and 500).
- Do not expect numbers from experimental measurements to give you a perfectly smooth curve or straight line. Draw the line of best fit for the plotted points, ignoring any obviously anomalous point. It is often a good idea to sketch your graph as you take your readings. This means you can take extra readings to fill in any missing detail or repeat the reading of an obviously anomalous point. In case you need to make an alteration, plot the points and draw the best-fit line with a soft lead pencil.

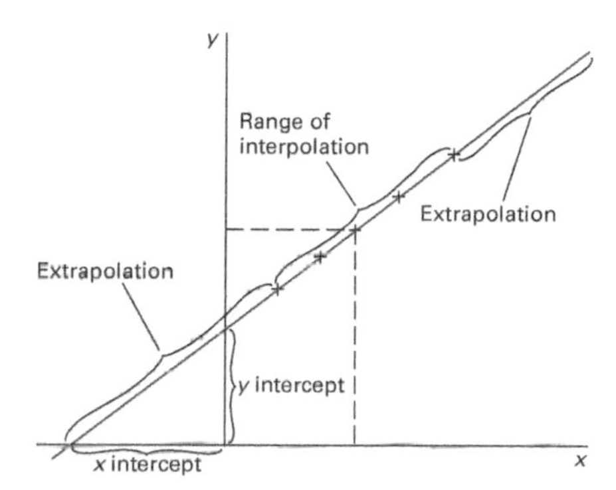

Graph showing interpolated and extrapolated points and intercepts.

> **The equation of a straight line**
>
> If the dimensions of x and y are related by the expression
>
> $y = mx + c$
>
> and you plot y (vertical axis) against x (horizontal axis), you will obtain a straight line. The gradient of the straight line is m, and the intercept on the y axis is c.

Information from graphs

Once the line of best fit is drawn, you can read any value of x in terms of y and vice versa by reading from the value on one axis, crossing to the line of best fit, and reading off the value of the other axis at that point. Points between the experimental readings may be found by **interpolation**. Extending the curve outside the experimental range is called **extrapolation**.

The **gradient** is the slope of a line at a point. To calculate the gradient of a straight-line graph, draw the largest convenient triangle using the line. The gradient is found from the equation

gradient = change in y coordinate/change in x coordinate

If the best-fit line is curved, you can only obtain the gradient for one point at a time. You will need to construct a **tangent** to the curve at that point, and then use the triangle method.

A tangent has the slope of the line at the point where it touches the curve. There are several methods of constructing tangents. To draw a tangent by eye, hold your ruler against the curve at the point where you want to construct the tangent so that it does not obscure the curve. Adjust the slope of the ruler so that the curve falls away equally on both sides of the point of contact.

The gradient of anything plotted against time will give you a rate of change.

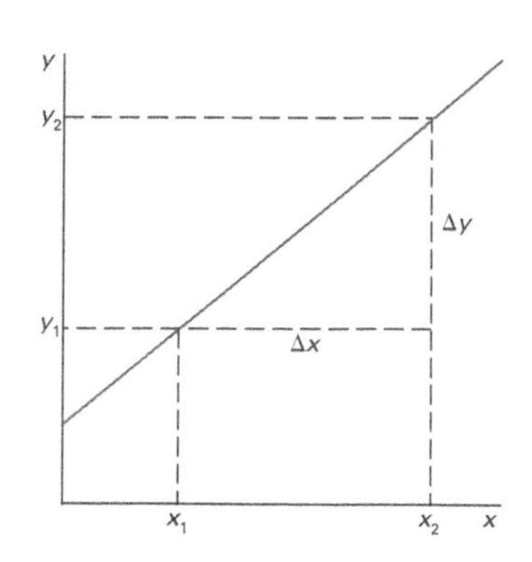

Determining the gradient: gradient = $(y_2 - y_1)/(x_2 - x_1)$, gradient = $\Delta y/\Delta x$

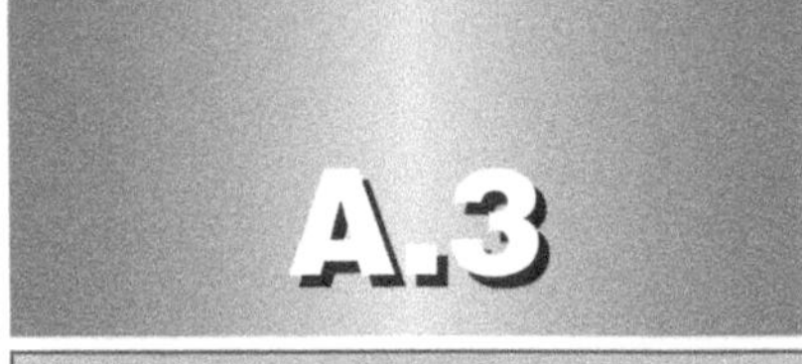

OBJECTIVES

- Common logarithms
- Natural logarithms
- Calculations using logarithms
- The exponential function
- Graphs of exponential functions

Mathematics toolbox: Logarithms and exponentials

Theory of logarithms

If a number y can be written in the form a^x, then the power x is called the **logarithm** of y to the base a. For example, $1000 = 10^3$, so $3 = \log_{10} 1000$. You can check this using the 'log' function on your calculator.

Common logarithms

The logarithm of a number to base 10 is the power that the number 10 has to be raised to in order to equal that number. Logarithms having a base of 10 are called **common logarithms**, and $\log_{10}$ is often abbreviated to 'log' or 'lg'.

Natural logarithms

Logarithms having a base of e, where e is a mathematical constant approximately equal to 2.718, are called **natural logarithms** or sometimes Napierian or hyperbolic logarithms. Log_e is often abbreviated to 'ln'.

Rules of logarithms

- To multiply two numbers, we add their logs:

  $\log(a \times b) = \log a + \log b$
- To divide two numbers, we subtract their logs:

  $\log(a/b) = \log a - \log b$
- To raise a number to a power, we multiply its log by the power:

  $\log a^n = n \log a$

To do the opposite operation, to find the number from its log, we use the antilogarithm or 10^x log key, which is usually found as the second (shifted) function on the log key.

The examples below and alongside show how one particular current calculator works. NB Older calculators may require step 3 to be done first (possibly with the minus sign put in after the number), in which case step 4 is probably not needed. For instance to find the antilog of –3.2

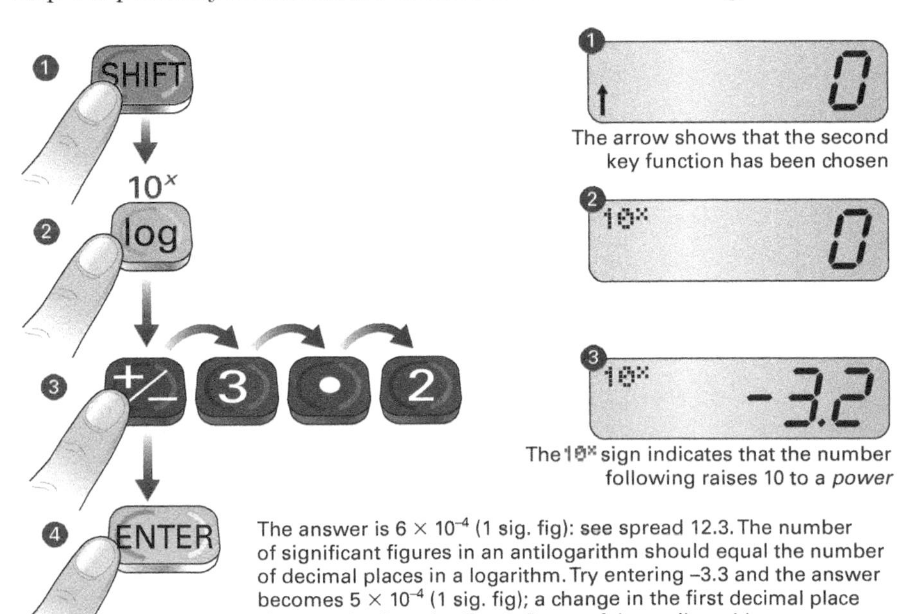

The arrow shows that the second key function has been chosen

The 10^x sign indicates that the number following raises 10 to a *power*

The answer is 6×10^{-4} (1 sig. fig): see spread 12.3. The number of significant figures in an antilogarithm should equal the number of decimal places in a logarithm. Try entering –3.3 and the answer becomes 5×10^{-4} (1 sig. fig); a change in the first decimal place changes the first significant figure of the antilogarithm.

The exponential function

The mathematical constant e, which has a value of approximately 2.718, is called the exponent. A function that contains e^x is called an exponential function. e^x is a function that increases at a rate proportional to its own magnitude.

Know your calculator

There are many different calculators on the market and they often execute functions such as logarithms in slightly different ways. It is well worth checking how your *particular* calculator executes logarithms and exponentials. The examples below and alongside show how one particular current calculator works.

NB Older calculators may require steps 1 and 2 below to be done *in reverse*, in which case step 3 is probably not needed.

Say you want to find the common logarithm of 6.0×10^{23}, as is needed in spread 12.3:

1. Press the key labelled 'log' or 'LOG':

2. Enter your given number, remembering to press the 'enter exponent' key (usually labelled EE or EXP):

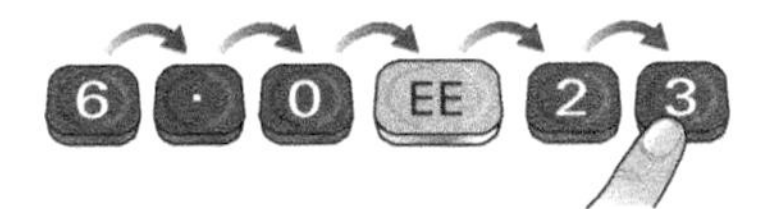

3. Press the key which causes the calculation to be performed (usually labelled ENTER or EXE or =):

The answer you will find is 23.78 (2 d.p.). In spread 12.3 we stated that the number of *decimal places* in a pH value (which uses logs) should equal the number of *significant figures* in your given number: repeat the exercise above using 5.9 in place of 6.0 and the final answer changes to 23.77 (2 d.p.); a change in the second significant figure changes the second decimal place of the logarithm.

The exponent is important because all the natural laws of growth and decay are of the form $y = ae^x$

Logarithms to the base e, natural logarithms, were developed to simplify calculations involving the exponential function.

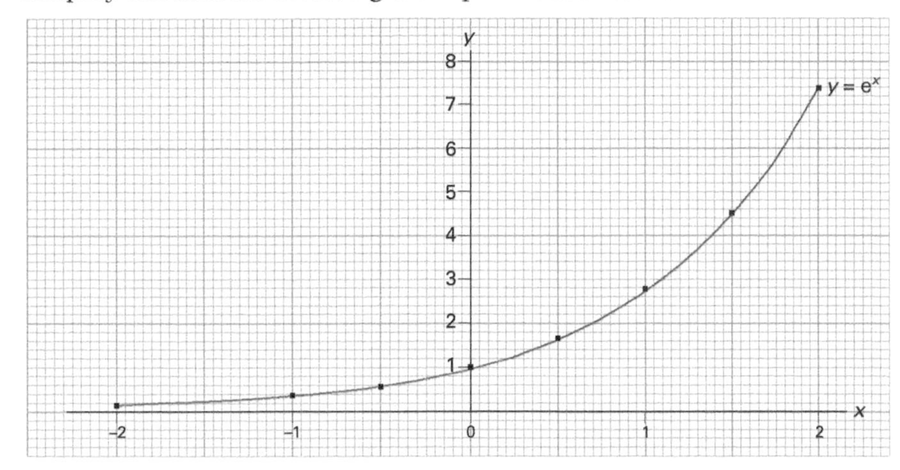

A graph of the curve $y = e^x$. This is called a growth curve.

You can find the value of e^x using the e^x function key on a scientific calculator. The degree of precision given by the calculator is often far greater than is appropriate. The selection of the appropriate number of significant figures is harder than in the case of a common antilogarithm; a rough rule of thumb is to use one more significant figure than there are decimal places in the given number. For example, $e^{-3.2}$ is 0.042 (2 sig. figs), whereas $e^{-3.3}$ is 0.037 (2 sig. figs).

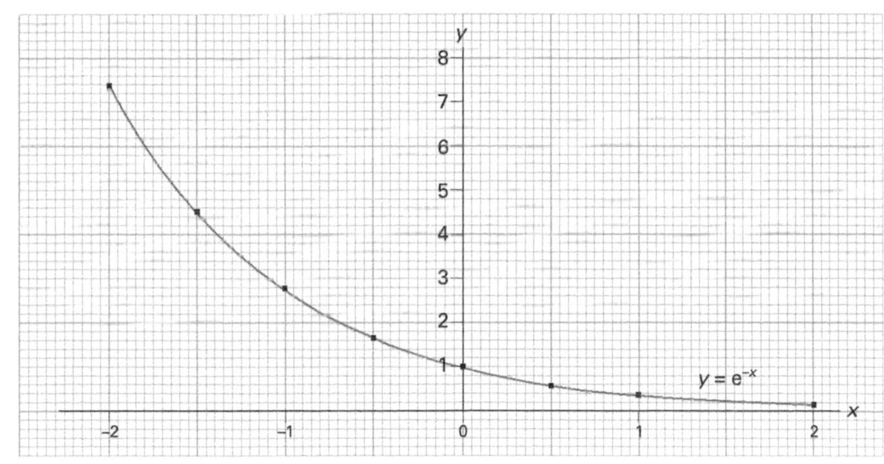

A graph of the curve $y = e^{-x}$. This is called a decay curve.

Graphs of exponential functions

For graphs of the form $y = e^{kx}$, where k is a constant and can be positive or negative, 'k' has the effect of altering the scale of x. For graphs of the form $y = Ae^{kx}$, where A is a constant, 'A' has the effect of altering the scale of y. Hence, every curve of the form $y = Ae^{kx}$ has the same general shape, and A and k are called scale factors of the graph.

Practical uses of growth and decay curves

Experimental data start from a given time, so time has only positive values. Hence, only the terms to the right of the y axis are used.

ln graphs

If $y = Ae^{kx}$ then, by taking natural logarithms, we find that

$$\ln y = \ln(Ae^{kx})$$
$$= \ln A + \ln(e^{kx})$$
$$= \ln A + kx$$

B.1

O B J E C T I V E S

- Spreadsheets
- Databases
- Computer-generated graphs
- Visualization software
- Modelling

THE USES OF COMPUTERS IN CHEMISTRY

The uses of computers in chemistry are particularly widespread. We introduce several of the areas where they are helpful in this spread.

Details on the availability of the various different pieces of software mentioned here may be obtained from one of the authors (MJC), care of the publishers. (MJC gratefully acknowledges the help of former students – most notably Paul Collins and James Whyte – with programming some of the software demonstrated here.)

Spreadsheets

One simple use of a computer involves a spreadsheet to take the tedium out of routine calculations. The example shown below relates to the esterification equilibrium discussed in spread 11.6.

Expt	Titre	OrigVol acid	OrigVol water	OrigVol alc	OrigVol ester
1	18.6	6.0	12.0	6.0	6.0
2	17.1	9.0	3.0	9.0	9.0
3	34.4	9.0	12.0	0.0	9.0
4	15.1	6.0	9.0	9.0	6.0

n(acid) start	*n*(water) start	*n*(alc) start	*n*(ester) start	*n*(acid) eq	*n*(water) eq	*n*(alc) eq	*n*(ester) eq	K_c
0.105	0.664	0.103	0.061	0.102	0.666	0.100	0.064	4.12
0.157	0.166	0.154	0.092	0.094	0.229	0.091	0.155	4.19
0.157	0.664	0.000	0.092	0.197	0.624	0.040	0.052	4.14
0.105	0.498	0.154	0.061	0.082	0.521	0.131	0.085	4.13

Spreadsheet for the esterification equilibrium.

Significant figures

The volumes are given to the nearest 0.1 cm^3; this means that the volumes are correct to either 3 sig. figs (for the titres) or 2 sig. figs (for the original volumes). The third significant figure is retained throughout to avoid rounding errors. The final figure for K_c is shown to 3 sig. figs; the third figure cannot be regarded as correct.

The various different starting compositions are entered into the spreadsheet and the necessary formulae are specified in the various cells. Entering the titres found in a particular experiment can update the sheet instantaneously. The values above represent such an application. Note how the final column, the value for K_c, is almost constant. Providing a shortcut compared with the long-winded and tedious calculation steps (helpfully avoiding the chances of errors cropping up!) frees the student to focus on the important conclusion about the chemistry involved.

Databases

Computers have large memory capacities, allowing storage of huge quantities of data, with fast access available. It is, for example, possible to access a database with essentially all known crystal structures or spectral data. This research tool can be used to find out whether a newly made compound has been fully characterised before. Suitable software is routinely provided with newer spectroscopic equipment.

On a smaller scale, but more useful in educational situations, the periodic table program in ChemSoft is able to provide all the usual data on the elements (melting point, boiling point, ionization energy, and so on). Graphs can be drawn of the variation across periods and down groups. An example is shown below.

(a)

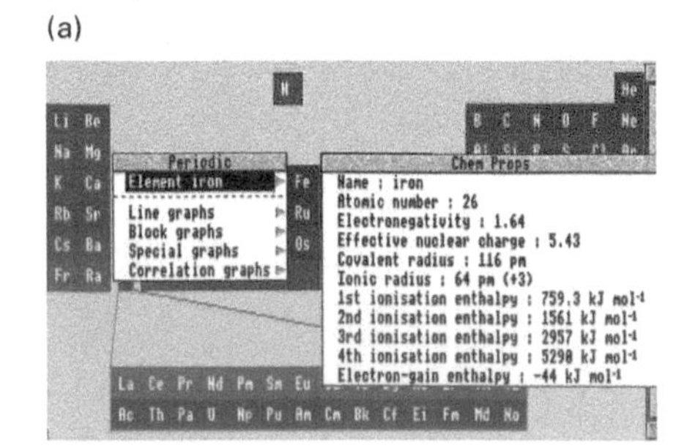

(b)

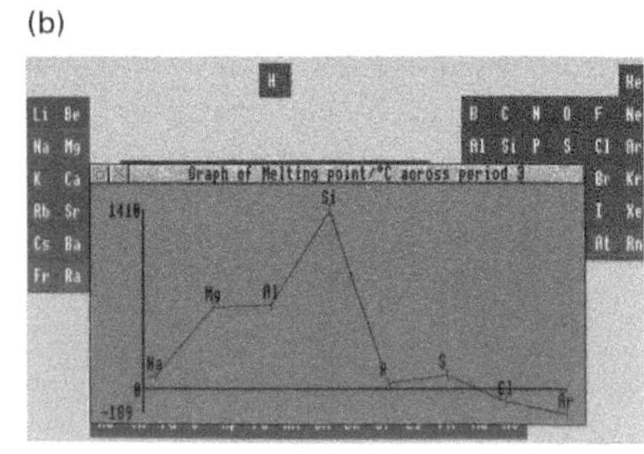

(a) One selection of the data available for an element. (b) Graph of the melting points across Period 3.

Lanthanides and actinides

The correct positioning of the lanthanides and actinides is controversial. In this text, we follow the exam Boards' traditional placing. The program follows current international opinion at research level.

In addition, the correlation between data sets may be investigated. Some of the possible permutations will produce little additional understanding (discovery date and density would be a pair unlikely to warrant study). On the other hand, relative atomic mass and atomic number show close correlation.

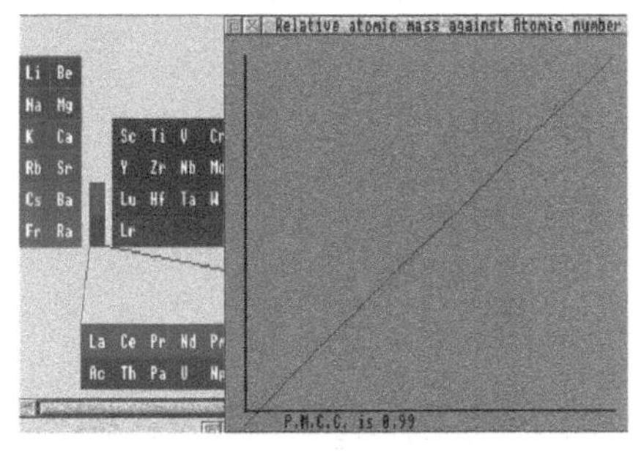

The correlation between relative atomic mass and atomic number.

'Calculate and display' software

Computer-generated graphs are more accurately drawn than are easily possible by hand, assuming the program has been coded correctly!

The graphs of pH against volume shown on spreads 12.8 and 12.9 were originally calculated by a computer program written by one of the authors (MJC). The program allows the user to choose any particular weak acid for example; on specification of the pK_a value, the program will then accurately calculate and display the pH curve in a matter of seconds. It is encouraging that specifications suggest using computers to generate such graphs.

Similar programs for the Maxwell–Boltzmann distribution, Boyle's law curves, shapes of atomic orbitals, and the extent of ionization of a weak acid (among others) are also available.

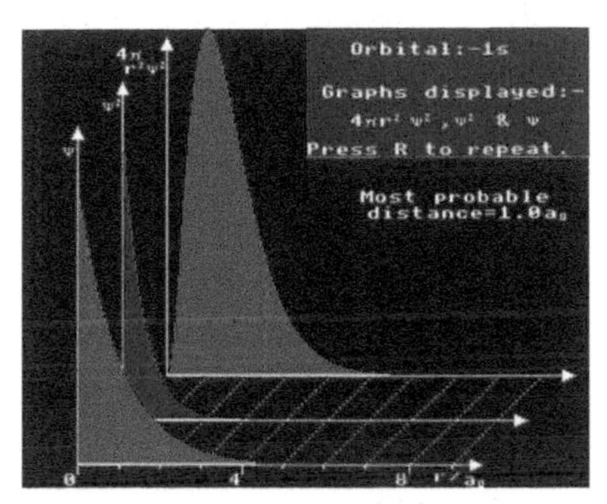

Plots of the wavefunction ψ, its square ψ^2, together with the electron density, which is equal to $4\pi r^2\psi^2$.

Visualization software

One of the most important applications of computers relies on their unrivalled ability to display objects in three dimensions, enabling objects such as crystal structures or molecules to be rotated in real time.

We mentioned in spread 6.7 that the very impressive structure program in ChemSoft enables the most important crystal structures (including f.c.c., h.c.p., rock-salt, fluorite, zinc blende, and several others) to be displayed, resized, and rotated. The octahedral and tetrahedral holes may be highlighted to allow the structure around a specific atom or ion to be investigated. This makes the structures come alive in a visually appealing way.

Modelling

Modelling also aims to take advantage of the display facility of computers, applied mainly to organic molecules. Organic molecules may be rotated to view the detailed electron densities and to inspect the various reactive sites (as explained in spread 21.5).

Such software may be used (most especially by biotechnology companies) to investigate the match of a potential drug to the receptor site on its chosen target.

We have been very fortunate to benefit from the kindness of Wavefunction Inc, and its co-founder and CEO Warren Hehre, who provided many images from their Spartan software package. A number of other companies provide similar packages, mostly for the university sector. We are also grateful to Oxford Molecular, and its founder and CEO Graham Richards, for providing other useful images from software such as CAChe.

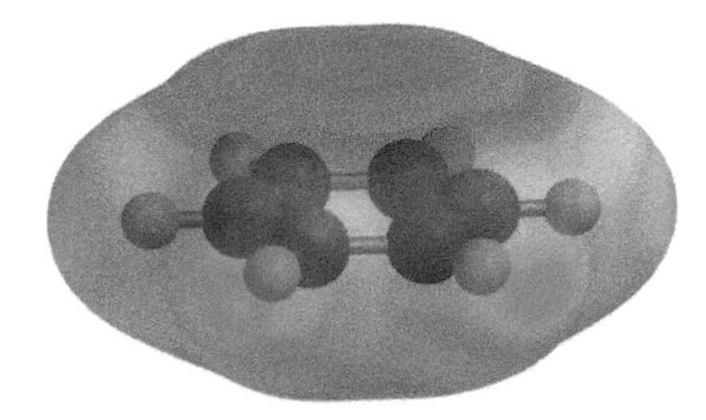

The electrostatic potential map for benzene confirms the high electron density, shown by the red colour, above (and below) the plane of the ring.

A wide range of different molecules may be displayed and manipulated on screen. These may be tailored to show a number of properties. For example, the electron density may be displayed as an electrostatic potential map, spread 5.8. This image may be examined to identify centres of partial charge which will attract nucleophiles or electrophiles for example.

Alternatively, the shape of different molecular orbitals for a selection of the most important molecules can be shown. This can be used to decide which atoms have substantial involvement in for example the so-called frontier orbitals, which are the ones most important in molecular collisions. The frontier orbitals are the highest occupied molecular orbital (HOMO) and lowest unoccupied molecular orbital (LUMO). See for example the π and π^* orbitals for ethene shown in spread 32.2. The shape of the lowest-energy bonding molecular orbital has particular significance for the concept of delocalization, spread 6.4.

The lowest energy bonding molecular orbital for methane is delocalized over all five atoms, see spread 22.12.

A particularly effective use of the technology is to allow visualization of the steps in a mechanism, such as that for nucleophilic substitution shown in spread 24.3.

C.1

OBJECTIVES

- Origin of some names
- Name and symbol
- Atomic number and relative atomic mass

The periodic table

Chemistry is all about the elements and the compounds formed between them. Some elements (such as carbon and gold) have been known since prehistoric times, whereas others (such as fermium and lawrencium) were discovered in the second half of the twentieth century. This spread gives a little historical background to the discovery and naming of the elements.

Element	Atomic no.	Symbol	Isolation	Origin of name and symbol (if different)
aluminium	13	Al	1825, Oersted	Latin *alumen*, alum
argon	18	Ar	1894, Rayleigh/Ramsay	Greek *argos*, inert
barium	56	Ba	1808, Davy	Greek *barys*, heavy
boron	5	B	1808, Davy	Arabic *buraq*, borax, + -on, as in carbon
bromine	35	Br	1826, Balard	Greek *bromos*, stench
caesium	55	Cs	1860, Bunsen/Kirchhoff	Latin *caesius*, sky blue
calcium	20	Ca	1808, Davy	Latin *calx*, lime
carbon	6	C	prehistoric	Latin *carbo*, charcoal
chlorine	17	Cl	1774, Scheele	Greek *chloros*, yellow-green
cobalt	27	Co	1735, Brandt	German *Kobold*, goblin
copper	29	Cu	prehistoric	Latin *cuprum, cyprium*, Cyprus
curium	96	Cm	1944, Seaborg/Ghiorso	Pierre and Marie Curie
europium	63	Eu	1901, Demarçay	Europe
fermium	100	Fm	1952, Ghiorso	Enrico Fermi
fluorine	9	F	1886, Moissan	Latin *fluere*, flow
germanium	32	Ge	1886, Winkler	Latin *Germania*, Germany
gold	79	Au	prehistoric	Anglo-Saxon *gold*; Au: Latin *aurum*
hafnium	72	Hf	1923, von Hevesy	Latin *Hafnia*, Copenhagen
helium	2	He	1895, Ramsay	Greek *helios*, Sun
hydrogen	1	H	1766, Cavendish	Greek *hydr* + *gen*, water-forming
iodine	53	I	1811, Courtois	Greek *iodes*, violet
iron	26	Fe	prehistoric	Anglo-Saxon *iron*; Fe: Latin *ferrum*
krypton	36	Kr	1898, Ramsay/Travers	Greek *kryptos*, hidden
lawrencium	103	Lr	1961, Ghiorso	Ernest Lawrence
lead	82	Pb	prehistoric	Anglo-Saxon *lead*; Pb: Latin *plumbum*
magnesium	12	Mg	1808, Davy	Greek *Magnesia*, district in Greece
mendelevium	101	Md	1955, Ghiorso/Seaborg	Dmitri Mendeleyev
mercury	80	Hg	prehistoric	the planet Mercury; Hg: Latin *hydragyrum*
neon	10	Ne	1898, Ramsay/Travers	Greek *neos*, new
oxygen	8	O	1774, Priestley/Scheele	Greek *oxys* + *gen*, acid-forming
phosphorus	15	P	1669, Brandt	Greek *phosphoros*, light bringer
platinum	78	Pt	prehistoric	Spanish *platina*, silver
plutonium	94	Pu	1940, Seaborg	the planet Pluto
polonium	84	Po	1898, Marie Curie	Poland
potassium	19	K	1807, Davy	English *potash*; K: Latin *kalium*
radium	88	Ra	1898, Curie	Latin *radius*, ray
rubidium	37	Rb	1861, Bunsen/Kirchhoff	Latin *rubidus*, red
selenium	34	Se	1817, Berzelius	Greek *selene*, Moon
silver	47	Ag	prehistoric	Anglo-Saxon *silver*, Ag: Latin *argentum*
sodium	11	Na	1807, Davy	English *soda*; Na: Latin *natrium*
strontium	38	Sr	1808, Davy	Strontian, place in Scotland
sulphur	16	S	prehistoric	Sanskrit *sulvere*
tellurium	52	Te	1783, von Reichenstein	Latin *tellus*, Earth
tungsten	74	W	1783, d'Elhuyar	Swedish *tung sten*, heavy stone; W: *wolfram*
uranium	92	U	1841, Peligot	the planet Uranus
vanadium	23	V	1801, del Rio	*Vanadis*, Scandinavian goddess of beauty
xenon	54	Xe	1898, Ramsay/Travers	Greek *xenos*, stranger
ytterbium	70	Yb	1878, de Marignac	Ytterby, place in Sweden
zinc	30	Zn	prehistoric	German *Zink*

① $_{1}$H Hydrogen 1.0

Period	Group I	II											III	IV	V	VI	VII	VIII
																		$_{2}$He Helium 4.0
②	$_{3}$Li Lithium 6.9	$_{4}$Be Beryllium 9.0											$_{5}$B Boron 10.8	$_{6}$C Carbon 12.0	$_{7}$N Nitrogen 14.0	$_{8}$O Oxygen 16.0	$_{9}$F Fluorine 19.0	$_{10}$Ne Neon 20.2
③	$_{11}$Na Sodium 23.0	$_{12}$Mg Magnesium 24.3											$_{13}$Al Aluminium 27.0	$_{14}$Si Silicon 28.1	$_{15}$P Phosphorus 31.0	$_{16}$S Sulphur 32.1	$_{17}$Cl Chlorine 35.5	$_{18}$Ar Argon 39.9
④	$_{19}$K Potassium 39.1	$_{20}$Ca Calcium 40.1	$_{21}$Sc Scandium 45.0	$_{22}$Ti Titanium 47.9	$_{23}$V Vanadium 50.9	$_{24}$Cr Chromium 52.0	$_{25}$Mn Manganese 54.9	$_{26}$Fe Iron 55.8	$_{27}$Co Cobalt 58.9	$_{28}$Ni Nickel 58.7	$_{29}$Cu Copper 63.5	$_{30}$Zn Zinc 65.4	$_{31}$Ga Gallium 69.7	$_{32}$Ge Germanium 72.6	$_{33}$As Arsenic 74.9	$_{34}$Se Selenium 79.0	$_{35}$Br Bromine 79.9	$_{36}$Kr Krypton 83.8
⑤	$_{37}$Rb Rubidium 85.5	$_{38}$Sr Strontium 87.6	$_{39}$Y Yttrium 88.9	$_{40}$Zr Zirconium 91.2	$_{41}$Nb Niobium 92.9	$_{42}$Mo Molybdenum 96.0	$_{43}$Tc Technetium (97)	$_{44}$Ru Ruthenium 101.1	$_{45}$Rh Rhodium 102.9	$_{46}$Pd Palladium 106.4	$_{47}$Ag Silver 107.9	$_{48}$Cd Cadmium 112.4	$_{49}$In Indium 114.8	$_{50}$Sn Tin 118.7	$_{51}$Sb Antimony 121.8	$_{52}$Te Tellurium 127.6	$_{53}$I Iodine 126.9	$_{54}$Xe Xenon 131.3
⑥	$_{55}$Cs Caesium 132.9	$_{56}$Ba Barium 137.3	$_{57}$La Lanthanum 138.9	$_{72}$Hf Hafnium 178.5	$_{73}$Ta Tantalum 180.9	$_{74}$W Tungsten 183.8	$_{75}$Re Rhenium 186.2	$_{76}$Os Osmium 190.2	$_{77}$Ir Iridium 192.2	$_{78}$Pt Platinum 195.1	$_{79}$Au Gold 197.0	$_{80}$Hg Mercury 200.6	$_{81}$Tl Thallium 204.4	$_{82}$Pb Lead 207.2	$_{83}$Bi Bismuth 209.0	$_{84}$Po Polonium (209)	$_{85}$At Astatine (210)	$_{86}$Rn Radon (222)
⑦	$_{87}$Fr Francium (223)	$_{88}$Ra Radium (226)	$_{89}$Ac Actinium (227)	$_{104}$Rf Rutherfordium (267)	$_{105}$Db Dubnium (270)	$_{106}$Sg Seaborgium (269)	$_{107}$Bh Bohrium (270)	$_{108}$Hs Hassium (270)	$_{109}$Mt Meitnerium (278)	$_{110}$Ds Darmstadtium (281)	$_{111}$Rg Roentgenium (281)	$_{112}$Cn Copernicium (285)	$_{113}$Nh Nihonium (286)	$_{114}$Fl Flerovium (289)	$_{115}$Mc Moscovium (289)	$_{116}$Lv Livermorium (293)	$_{117}$Ts Tennessine (293)	$_{118}$Og Oganesson (294)

$_{58}$Ce Cerium 140.1	$_{59}$Pr Praseodymium 140.9	$_{60}$Nd Neodymium 144.2	$_{61}$Pm Promethium (145)	$_{62}$Sm Samarium 150.4	$_{63}$Eu Europium 152.0	$_{64}$Gd Gadolinium 157.3	$_{65}$Tb Terbium 158.9	$_{66}$Dy Dysprosium 162.5	$_{67}$Ho Holmium 164.9	$_{68}$Er Erbium 167.3	$_{69}$Tm Thulium 168.9	$_{70}$Yb Ytterbium 173.0	$_{71}$Lu Lutetium 175.0
$_{90}$Th Thorium 232.0	$_{91}$Pa Protactinium 231.0	$_{92}$U Uranium 238.0	$_{93}$Np Neptunium (237)	$_{94}$Pu Plutonium (244)	$_{95}$Am Americium (243)	$_{96}$Cm Curium (247)	$_{97}$Bk Berkelium (247)	$_{98}$Cf Californium (251)	$_{99}$Es Einsteinium (252)	$_{100}$Fm Fermium (257)	$_{101}$Md Mendelevium (258)	$_{102}$No Nobelium (259)	$_{103}$Lr Lawrencium (262)

C.2 THERMODYNAMIC PROPERTIES

OBJECTIVES

- Standard enthalpy change of formation
- Standard entropy
- Standard Gibbs energy change of formation

Aluminium Al(s)

Barium Ba(s)

Copper Cu(s)

Iron Fe(s)

Substance	$\Delta H^{\ominus}$(298 K)/kJ mol^{-1}	$S^{\ominus}$(298 K)/J K^{-1} mol^{-1}	$\Delta G^{\ominus}$(298 K)/kJ mol^{-1}
Ag(s)	0	+43	0
AgCl(s)	−127	+96	−110
Al(s)	0	+28	0
$AlCl_3$(s)	−704	+111	−629
Al_2O_3(s)	−1676	+51	−1582
Ar(g)	0	+155	0
BF_3(g)	−1136	+254	−1119
Ba(s)	0	+63	0
$BaCl_2$(s)	−859	+124	−810
BaO(s)	−554	+70	−525
Br_2(l)	0	+152	0
C(s, graphite)	0	+6	0
CH_4(g)	−74	+186	−50
C_2H_6(g)	−84	+230	−32
C_2H_4(g)	+53	+220	+68
C_2H_2(g)	+228	+201	+211
CH_3CHO(l)	−192	+160	−128
CH_3CH_2OH(l)	−278	+161	−175
CH_3COOH(l)	−485	+160	−390
CH_3NO_2(l)	−113	+172	−14
CO(g)	−111	+198	−137
CO_2(g)	−394	+214	−394
Ca(s)	0	+42	0
$CaCO_3$(s)	−1208	+92	−1129
$CaCl_2$(s)	−795	+105	−749
CaH_2(s)	−182	+41	−143
CaO(s)	−635	+38	−603
Cl_2(g)	0	+223	0
CsCl(s)	−443	+101	−415
Cu(s)	0	+33	0
$CuCl_2$(s)	−220	+108	−176
CuO(s)	−157	+43	−130
$CuSO_4$(s)	−771	+109	−662
F_2(g)	0	+203	0
Fe(s)	0	+27	0
$FeCl_2$(s)	−342	+118	−302
$FeCl_3$(s)	−400	+142	−334
Fe_2O_3(s)	−824	+87	−742
Fe_3O_4(s)	−1118	+146	−1015

Substance	$\Delta H^{\ominus}$(298 K)/kJ mol^{-1}	$S^{\ominus}$(298 K)/J K^{-1} mol^{-1}	$\Delta G^{\ominus}$(298 K)/kJ mol^{-1}
H_2(g)	0	+131	0
HBr(g)	−36	+199	−53
HCl(g)	−92	+187	−95
HF(g)	−273	+174	−275
HI(g)	+27	+207	+2
HNO_3(l)	−174	+156	−81
H_2O(l)	−286	+70	−237
H_2O(g)	−242	+189	−229
H_2SO_4(l)	−814	+157	−690
I_2(s)	0	+116	0
K(s)	0	+65	0
KCl(s)	−437	+83	−409
KOH(s)	−425	+79	−379
Mg(s)	0	+33	0
$MgCl_2$(s)	−641	+90	−592
MgO(s)	−602	+27	−569
N_2(g)	0	+192	0
NO_2(g)	+33	+240	+51
N_2O_4(g)	+9	+304	+98
NH_3(g)	−46	+193	−16
Na(s)	0	+51	0
NaCl(s)	−411	+72	−384
Na_2O(s)	−414	+75	−376
NaOH(s)	−426	+64	−380
O_2(g)	0	+205	0
O_3(g)	+143	+239	+163
P_4(s, white)	0	+41	0
PCl_3(l)	−320	+217	−272
Pb(s)	0	+65	0
$PbCl_2$(s)	−359	+136	−314
PbO(s)	−217	+69	−188
S_8(s, rhombic)	0	+32	0
SF_6(g)	−1221	+291	−1117
SO_2(g)	−297	+248	−300
SO_3(s)	−455	+71	−374
$SiCl_4$(l)	−687	+240	−620
SiO_2(s)	−911	+41	−856
TiO_2(s)	−944	+51	−889
V_2O_5(s)	−1551	+131	−1420
Zn(s)	0	+42	0
$ZnCl_2$(s)	−415	+111	−369
ZnO(s)	−351	+44	−321

Potassium K(s)

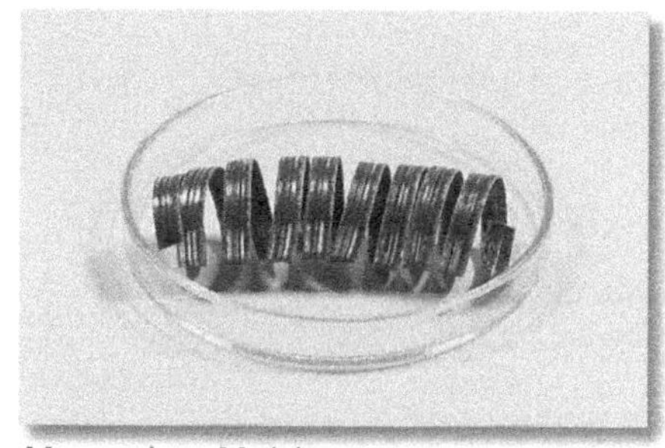
Magnesium Mg(s)

Lead Pb(s)

Sulphur S_8(s)

Suggested answers to selected practice and exam questions

Please note: the following suggested answers and solutions are intended as a guide for the reader. The respective exam boards have not supplied the answers and solutions to past exam questions nor are they in any way responsible for their accuracy or correctness.
In some answers, one extra unjustified significant figure is retained to avoid rounding error at the end.

Chapter 1

Practice questions

Spread 1.4

1 Lithium, magnesium, silicon, bromine, potassium, silver, tin, fluorine.
2 Ge, S, Ca, Au, Kr, Sb, Pb
5 Vanadium, V; aluminium, Al; barium, Ba; krypton, Kr; strontium, Sr; nitrogen, N

Chapter 2

Practice questions

Spread 2.1

2 Simplistically, the ratio of the areas is equal to the square of the ratio of the radii; hence 10^{12} (NB A more sophisticated answer would recognize that close-packed spheres occupy only 74% of space; see spread 6.5)

Spread 2.4

1 a 17 protons, 18 neutrons, 17 electrons
b 12 protons, 12 neutrons, 12 electrons
c 92 protons, 143 neutrons, 92 electrons

2 a Atomic number: 17, mass number: 35
b Atomic number: 12, mass number: 24
c Atomic number: 92, mass number: 235

3 28.1

Chapter 3

Practice questions

Spread 3.1

1 9.0, 10.8, 40.1, 12.0, 55.8, 54.9, 31.0, 23.0, 131.3

2 a 28.0, 44.0, 100.1, 98.1; b 17.0, 64.1, 16.0, 63.0

Spread 3.2

1 72.7

Spread 3.3

1 a 24.0 g; b 12.15 g; c 0.27 g; d 115 g; e 115 g

2 a 55.9 kg; b 1839 kg; c 1.0×10^{-6} kg

3 a 1.00 mol; b 2.00 mol; c 0.249 mol

Spread 3.4

1 a 17.0 g mol^{-1}; b 46.0 g mol^{-1}; c 142 g mol^{-1}; d 48.0 g mol^{-1}

2 a 34.0 g; b 9.2 g; c 257.3 g

3 a 0.010 mol; b 0.33 mol; c 0.050 mol

Exam questions

1 a Neutrons: 10; electrons: 9
b A: ionization; B: acceleration; C: deflection; D: detection
c i For $^{21}Ne^{+}$ the path would curve above the line shown (and hit the top wall of the magnet)
ii The extent of deflection depends on the mass of the ion – lighter ions are deflected more
d $(91.0 \times 20 + 9.0 \times 22)/100 = 20.2$ to 3 sig. figs (NB The question should read 9.0% ^{22}Ne)

2 a See spread 2.4
b i The path of neutral atoms is not affected by an electric or a magnetic field
ii An electrostatic field (charged plates)
c i Boron (atomic number 5); ^{10}B and ^{11}B
ii $(25 \times 10 + 100 \times 11)/125 = 10.8$
d Peaks would be at 5 and 5.5 with the same ratio of heights; the x-axis is mass/charge, and the charge would double

3 a ^{12}C is the standard against which other atoms are compared – it is exactly 12 by definition
b Mass of 1 mol H atoms = 1.0078 g; mass of 1 mol electrons $= 9.1091 \times 10^{-28} \times 6.0225 \times 10^{23} = 5.4860 \times 10^{-4}$ g; mass of 1 mol H^{+} ions $= (1.0078\ g) - (5.4860 \times 10^{-4}\ g) = 1.0073$ g
c i An electron is removed by electron bombardment
ii To avoid multiply charged ions being formed
iii They are accelerated by an electrostatic field (charged plates)
d The mass of the electron lost is very small compared to the mass of the atom

4 a The electron; it has the lowest mass/charge ratio
b See **2 b i**
c $(a \times 10 + b \times 11)/100 = 10.8$; $a + b = 100$; $a = 20$, $b = 80$; $a{:}b = 1{:}4$
d i In 100 g of sample, B = 81.2 g and H = 18.8 g; in moles, B = 81.2/10.8 = 7.52 and H = 18.8/1 = 18.8; simplest ratio = 1:18.8/7.52 = 1:2.5 or 2:5; empirical formula = B_2H_5
ii Mass of empirical formula is 26.6; 54 is approx. 2×26.6, so molecular formula is B_4H_{10}

5 a See spread 2.3
b Electron gun: ionizes atoms; magnet: deflects and separates ions of different masses
c i $^{20}_{10}Ne$ and $^{22}_{10}Ne$
ii $^{22}Ne^{2+}$
iii $M_r = (17.8 \times 20 + 1.7 \times 22)/(17.8 + 1.7) = 20.2$

6 a See spread 3.3
b i $Vc = (0.0236\ dm^3) \times (0.150\ mol\ dm^{-3}) = 0.00354$ mol
ii (0.00354 mol)/2 = 0.001 77 mol
iii $M = (0.245\ g)/(0.00177\ mol) = 138\ g\ mol^{-1}$, so $M_r = 138$
iv $2M + 12 + 3 \times 16 = 138$, so M = 39; M is potassium
c A: magnetic field; B: detector; electron gun: ionizes sample; electric field: accelerates ions

7 a

Particle	Relative mass	Relative charge
proton	1.0	1
neutron	1.0	0
electron	1/1840	–1

b See spreads 2.3 and 2.4
c i So that they will be affected by the electric and magnetic fields
ii Charged plates (electric field)
iii Magnetic field
d See spread 2.4
e i $(78.6 \times 24 + 10.1 \times 25 + 11.3 \times 26)/100 = 24.3$ (NB An answer to 2 d.p. is unjustified, given integer values for the isotopes: compare the answer to spread 3.2, question 1 with the periodic table)
ii Peaks at 24, 25, 26 in the ratio 78.6:10.1:11.3

Chapter 4

Practice questions

Spread 4.7

1 a s; b d; c s; d p; e d; f p

2 a 2s, 2p; b 3s, 3p; c 4s, 3d, 4p

Exam questions

1 a N: atomic number 7, ionization energy slightly higher than that of O; Al: atomic number 13, ionization energy slightly lower than that of Mg
b Na has an electron in a higher shell than the electrons in Ne, which is further from the nucleus and less tightly bound to the atom
c Na: an electron removed from the same shell as for Ne but an extra proton in the nucleus means there is a stronger attractive force on the electrons
d i Ne does not form any bonds with any atoms; hence the value for the electronegativity is not measurable

ii Na; the lowest first ionization energy and the least affinity for electrons

e On dissolving the sodium oxide, ions are separately hydrated; the small oxide ion bonds strongly with the H atoms in the water causing the O—H bonds to break, forming OH^- ions; the excess of these ions increases the pH

f A non-metal oxide, or the oxide of an element in a high positive oxidation state, e.g. SO_3

2 a Spin

b 2p is slightly further from the nucleus

c Spin pair in 2s and two up arrows in two 2p orbitals

d See spread 4.8

e O has additional repulsion from a second electron in one of the 2p orbitals
(Note also that a *half-filled* subshell (as for the 2p in N) is particularly stable; the origin of this stability is complex)

3 a See **2 d**

b Ne has an extra proton, so the attraction for the electrons in the 2p is greater than for F

c See **2 e**

d Na has an electron in the 3s that is further away from the nucleus than the 2p of Ne; hence less energy is required to remove it

e The average difference between O, F, and Ne is 385, so C should be approx. 1400 – 385 = 1015

4 a 3p (Al) is at slightly higher energy than 3s (Mg) and so requires less energy to remove it

b Si has an extra proton in the nucleus, so there is extra force on the outer electrons, so more energy is required for their removal

c The same nuclear charge attracts fewer electrons more strongly

d $Al^{2+}(g) \rightarrow Al^{3+}(g) + e^-(g)$

e The 3rd ionization energy of Mg removes an electron from an inner shell (2p), which requires much more energy; the 3rd electron from Al comes from 3s

5 a Lines occur because the atoms have fixed energy levels and transitions between an upper energy level and a lower one cause emission of light; lines in the ultraviolet end on the energy level with $n = 1$; the energy levels get closer together as the energy increases, so the lines converge

b The convergence limit, where the lines merge, corresponds to an electron at the edge of the atom; the ionization energy is the energy corresponding to that frequency: $E = hf$

6 a The minimum energy required per mole to remove one electron from singly positively charged fluorine ions in the gaseous state

b Generally upward with a steeper upward kink after the 7th electron

7 a i $O(g) \rightarrow O^+(g) + e^-(g)$

ii There is only a pair of electrons in the lowest shell and one more proton in the nucleus compared to H, so He has the greatest attraction for (outer) electrons

iii See **2 e**

iv The outer electron is in progressively higher shells down the group, which are further from the nucleus; hence progressively less energy is required to remove it

b i $O(g) + e^-(g) \rightarrow O^-(g)$; $O^-(g) + e^-(g) \rightarrow O^{2-}(g)$

ii The second electron is repelled by the negatively charged O^- ion, so energy has to be supplied to push it into the ion

c The energy given out when the oppositely charged ions are attracted together to form an ionic crystalline lattice more than compensates for the energy required to form the ions

d The arrangement is 2, 8, 1 (the 2nd ionization energy is much greater than the 1st, so there is one electron in the outer shell; the 2nd to 9th show a steady increase and are in the 2nd shell; the 10th and 11th are much higher and so are in the 1st shell)

8 a i Mg has an extra proton (as before)

ii Al: 1st electron in higher subshell (as before)

iii Electrons from the same subshell but Al has an extra proton

iv The 4th electron is removed from an inner shell, so much more energy is required

b i See **7 b i**

ii See **7 b ii** (NB Values given in this question are different from those in **7**)

iii They are stable because of the ionic bond – strong electrostatic attraction means a lot of energy is given out when the lattice forms (see **7 c**)

9 a See **7 a i**

b i There is increasing charge on the nucleus, which implies increasing attraction for electrons in the same shell

ii B: the electron in 2p has higher energy than the 2s of Be; see **2 e**

c Group IV: the first four ionization energies show a steady increase; then there is a large jump

Chapter 5

Exam questions

1 a Possible answers include SF_6, CH_4, $AlCl_3$, $BeCl_2$

b See spread 5.5

c The lone pairs are more repulsive than bond pairs are

d See spread 7.5

2 a A covalent bond is the attraction between the nuclei of the bonded atoms and a shared pair of electrons

b A coordinate bond is a normal covalent bond where the shared pair of electrons is donated by one of the atoms

c There is no difference: attraction is between N and identical H nuclei; electrons are indistinguishable

d Ammonium ion: 109°; ammonia molecule: 107°; the lone pair occupies more space closer to the nucleus so there is a greater repulsive force, which pushes the bonds closer together

e H bonding. N—H bonds are polar; δ+ on H is attracted to the lone pair on the adjacent δ– N atom

3 a Ionic; $KBr(s) + aq \rightarrow K^+(aq) + Br^-(aq)$; pH 7

b i 57.2% I, 42.8% F; moles: 57.2/126.9 = 0.451, 42.8/19.0 = 2.25; simplest ratio: 0.451:2.25 = 1:5; empirical formula: IF_5

ii $I_2 + 5F_2 \rightarrow 2IF_5$

c i T-shaped (2 lone pairs)

ii BrF_4^-; square planar, 90°

4 a Ionic bond: electrostatic attraction between oppositely charged ions; covalent bond: see **1 a**; dative covalent bond: see **1 b**

b No, no, no, yes, yes, yes

c i Mg has an extra proton

ii Ion has lost the outer shell electron and is left with an inner core of electrons; hence it is smaller (also, the remaining electrons feel a stronger attraction to the nucleus, as there is one fewer electron overall)

iii I^- is large and has high polarizability; Al^{3+} has high charge density, so Al shares the I electrons forming a covalent molecule; F^- is small with low polarizability

5 a Two nuclei attract a shared pair of electrons; one nucleus (more electronegative) has a greater attraction, so the electrons are shared unequally

b When the two atoms have different electronegativities

c When it is close to a cation with a high charge density

d In an unpolarized anion the distribution of electrons is symmetrical, which is not the case in a polarized anion

e i **I** linear; **II** pyramidal; **III** octahedral

ii NCl_3, the only one with a lone pair

6 a A dative or coordinate bond is a normal covalent bond in which both the shared electrons are donated by one of the atoms

b i $H_3N \rightarrow BF_3$; tetrahedral

ii There are four pairs of electrons around the N; mutual repulsion moves them to the tetrahedral positions; there are only three pairs of electrons around the boron atom in BF_3; mutual repulsion pushes them as far apart as possible i.e. to bond angles of 120°, trigonal planar

c i Trigonal bipyramidal

ii BrF_3 is T-shaped; KrF_2 is linear

d i Geometrical or *cis–trans* isomerism; Cl atoms are either adjacent (*cis*) or opposite (*trans*)

ii Only one of the isomers is an active drug; some methods of preparation produce a mixture of the isomers while some selectively produce one isomer or the other

7 a VSEPR theory: the shape is determined by the number of electron pairs in the valence shell of the central atom; repulsion between electron pairs decides the final shape since lone pair/lone pair repulsion > lone pair/bond pair repulsion > bond pair/bond pair repulsion

b Pyramidal

c The greater repulsion of the lone pair pushes the bond pairs closer together

d Nitrogen has a higher electronegativity than the hydrogen atoms, so the electron density in the bonds is uneven, the N having the bigger share *and* the shape is pyramidal

8 a i CH_4: tetrahedral, 109°; NH_3: pyramidal, 107°; H_2O: V-shaped, 104°

ii Methane: dispersion forces; ammonia and water: hydrogen bonds plus dispersion plus dipole–dipole

iii Repulsion from the lone pair(s) pushes the bond pairs together in ammonia and water – more so in water where there are two lone pairs

iv Water: O is the most electronegative, so O—H bonds are the most polar, hence the strongest are hydrogen bonds; also since the Os have two lone pairs, each water molecule can hydrogen bond to four other molecules

b i See spreads 5.2 and 5.3

ii See **6 a** (the ammonium ion is a poor example)

iii All the N—H bonds are identical and indistinguishable

9 a i Vibrations around fixed lattice positions

ii Ions are able to move away from lattice positions, but they remain fairly close together; movement is random

b i Covalent bond – see **6 a**; in LiI the small lithium ion attracts and shares some of the valence electrons in the polarizable iodide ion; there is thus a partial covalent character to the bond between the ions

ii Li^+ is smaller than Na^+ and thus it has a higher charge density

10a A measure of an atom's ability to attract shared electron pairs

b i Ionic; **ii** Covalent (polar); **iii** Covalent

Chapter 6

Practice questions

Spread 6.5

1 a 6; **b** 8; **c** 12; **d** 12

Exam questions

1 a Copper: $1s^22s^22p^63s^23p^63d^{10}4s^1$, d block
Gallium: $1s^22s^22p^63s^23p^63d^{10}4s^24p^1$, p block
Phosphorus: $1s^22s^22p^63s^23p^3$, p block

b Graph shows steady increase with steep rises between 3 and 4, and 11 and 12

c Sodium: body-centred cubic, 8; magnesium: hexagonal close-packed, 12

2 a i On ions next to the + shown and diagonally opposite through the cube

ii 6

b It conducts electricity when molten

c i Electron pair donated by one of the atoms sharing the bond

ii Opposite on the central 'square' of bonds

iii If it is largely covalent, then there are only weak forces *between* molecules

3 a i Ionic

ii Vibration of ions increases as temperature rises; at melting point ions move out of lattice sites and move randomly

iii There are strong electrostatic forces between ions, hence a lot of energy is required to overcome them; there is strong attraction for polar water molecules; the formation of hydrated ions compensates for the lattice energy

b i A shared pair of electrons with high electron density between nuclei

ii A shared pair of electrons with electron density in two lobes above and below the axis between the two nuclei; see figures on spread 5.7

4 a Metallic bonding; ions packed close together, ions held together by mutual attraction for the valence shell electrons; the valence shell electrons are relatively free to move from one atom to another and hence conduct electricity

b i Covalent: $AlCl_3$; ionic: AlF_3

ii Al and F: there is a large difference in electronegativity; Al and Cl: there is a smaller difference in electronegativity, so that a partial covalent bond forms with Al^{3+} and polarizable Cl^-

iii Strong electrostatic attraction between ions requires a large input of energy to overcome forces

5 a i See **2 c i**; $H_3N \rightarrow BF_3$

ii Electrostatic attraction between oppositely charged ions: NaCl

b i Solid turns to gas state on heating with no intervening liquid state

ii $AlCl_3$ attracts the polar water molecule strongly; iodine is non-polar so there is no strong attraction for water molecules

c i Valence electrons are delocalized and relatively free to move between ions

ii Moving electrons carry energy from one ion to the next; also ions are close packed, so, when ions vibrate more and hit adjacent ions, they pass on energy

6 a i I–I is a simple molecule loosely packed in the crystal lattice; see spread 7.4

ii Tetrahedral structure: each atom bonded to four others; giant covalent: see spread 6.3

b Iodine sublimes at low temperature because there are weak dispersion forces between the molecules; diamond has a very high melting point because there are strong covalent bonds between all the atoms

c NaCl: interlocking face-centred cubic, coordination 6,6; CsCl: interlocking simple cubic, coordination 8,8; both ionic; coordination numbers are determined by radius ratios: Na^+ is smaller than Cs^+ so has a smaller ratio (see **8 a iii**)

d See **4 a**

7 a i See **4 a**

ii Covalent bonds between atoms in the PVC molecule; weak intermolecular forces between molecules; all electrons are localized; hence it is a non-conductor

iii Ionic, Mg^{2+} and O^{2-}, electrons are localized; hence it is a non-conductor

b i High melting point; non-flammable

ii MgO is brittle and needs to be protected from fracture by bending or impacts

8 a i The arrangement of ions in the crystal; electrostatic attraction between oppositely charged ions

ii See the figures in spread 6.6; NaCl: 6,6; CsCl: 8,8

iii Radius ratios: $Na^+/Cl^- = 0.52$; $Cs^+/Cl^- = 0.93$; for NaCl it is much smaller, hence NaCl has a lower coordination number

iv X-ray crystallography

b i P $[Ne]3s^23p^3$

ii Trigonal bipyramidal, angles 90° and 120°

iii 5 bond pairs, 0 lone pairs, total 5; the repulsion between pairs is equal

Chapter 7

Practice questions

Spread 7.4

4 70

Exam questions

1 a i $C^{\delta+}$—$Cl^{\delta-}$; $Cl^{\delta+}$—$F^{\delta-}$; $N^{\delta+}$—$H^{\delta-}$
ii See spread 5.8
b i $(CH_3COOH) \times 2 = (24 + 32 + 4) \times 2 = 120$
ii Hydrogen bond
iii Molecules not dimerized and are hydrated by forming H-bonds with water molecules, and there is some dissociation into ethanoate and oxonium ions

2 a i Graph shows steady rise to 80 °C; mixture starts to melt at 80 °C, forming liquid A, but a solid remains until the temperature reaches 120–140 °C, when the solid mixture completes melting
ii Graph shows steady rise to 80 °C then remains horizontal for some time before then rising steadily; the mixture melts at 80 °C, temperature remains constant until all solid has melted
iii A eutectic mixture
b In 100 g of mixture, $n(A) = (28\,g)/(127\,g\,mol^{-1}) = 0.220\,mol$; $n(B) = (72\,g)/(181\,g\,mol^{-1}) = 0.398\,mol$; mole fraction of A = 0.220/(0.220 + 0.398) = 0.36

3 a See **1 a ii**
b i Number of electrons in the molecule increases from S to Te; thus the size of the dispersion forces increases; this results in increasing attraction between the molecules
ii O is the smallest and most electronegative of the Group VI elements and its bonds with H are highly polar, thus the δ+ on the H atom is strongly attracted to the δ– O on adjacent molecules; dotted lines between O of one atom and H of adjacent molecule (chain of three)
c Hydrogen bonds between N—H and O═C; see figures in spread 30.5

4 a Graph
b (What does 'surface' mean for Jupiter and Saturn?) Assuming temperature given is the same at all depths of atmosphere (a false assumption), Mars – low pressure;methane: gas; ammonia: gas; water: solid; Jupiter – pressure increases with depth; methane – all three states possible; gas at high altitudes, then liquid, possibly solid very deep where pressure greatest; ammonia: gas at high altitudes; solid lower; water largely solid; some vapour at top of atmosphere and possibly liquid at very high pressures deep in the atmosphere; Saturn – methane: gas at high altitudes, then solid (no liquid); ammonia: as Jupiter; water: as Jupiter, except liquid less likely

5 a i See spread 4.8
ii $Si^{+}(g) \rightarrow Si^{2+}(g) + e^{-}(g)$
iii Same nuclear charge as more electrons removed means that there is more attraction for the remaining electrons in the shell
iv The 5th electron is taken from a lower energy level than the 4th, so considerably more energy must be supplied
b i Giant covalent; simple molecular
ii In silicon dioxide there are strong covalent bonds linking all the atoms together which require a lot of energy to break them; hence it has a high melting point; silicon tetrachloride has only weak forces between molecules; little energy is required to overcome these forces, hence a low melting point

6 a i See **1 a ii**
ii Increasing the number of electrons in valence shells means that dispersion forces increase down the group; hence the force between molecules increases and so the energy required increases
iii Strong H bonding in HF means that there is a more powerful force between molecules, and a higher temperature is required
b i Shared pair of electrons with high electron density between nuclei
ii Shared pair of electrons with electron density in two lobes above and below the axis between the two nuclei; see figures in spread 5.7
c i Trigonal planar: all angles 120°; delocalized π bond
ii Both have a skeleton of sigma bonds with delocalized π bonding

7 a i See **1 a ii**
ii Dispersion forces
b Methane has more electrons; hence dispersion forces are stronger
c i See **1 a ii**
ii HCl has a bigger difference in electronegativity, hence bond is more polar and molecule has larger dipole, hence there is greater force between molecules; methane has no overall dipole because of tetrahedral arrangement of H atoms

Chapter 8

Practice questions

Spread 8.1

1 a 2 atm **2** 425 cm³

Spread 8.2

1 0.011 m³ **2** 0.015 m³
3 0.18 Pa

Spread 8.3

1 $84.0\,g\,mol^{-1}$ **2** C_6H_{12}

Spread 8.4

1 16; CH_4

Exam questions

1 a i Polar covalent bonds; Cl is more electronegative than H, so Cl δ– and H δ+
ii Polar covalent bonds in the ammonium ion between the N and H atoms (N δ–, H δ+); ionic bonds between ammonium ions and chloride ions
iii Polar covalent bonds between O and H (O δ–, H δ+)
b Hydrogen bonding (see spread 7.5)
c Sodium chloride: there is strong electrostatic attraction between oppositely charged ions; hence a lot of energy is required to separate ions; in iodine, there are only relatively weak dispersion forces between molecules, so much less energy is required to separate the molecules
d $n(I_2) = (4.509\,g)/(253.8\,g\,mol^{-1}) = 0.01777\,mol$; volume = $(0.01777\,mol) \times (2.24 \times 10^4\,cm^3\,mol^{-1}) \times (343\,K)/(273\,K) = 500\,cm^3$ (3 sig. figs)

2 a i Volatile means that it readily evaporates and forms a vapour; ideal gas equation: n = amount of gas in moles, R = the gas constant
ii So that the syringe and contents uniformly reach the temperature of the boiling water bath
iii Particles in constant motion; volume of particles is negligible; there are no interactions between particles other than collisions; collisions are perfectly elastic
b i $n(CH_3CH_2OH) = (0.167\,g)/(46.0\,g\,mol^{-1}) = 3.63 \times 10^{-3}\,mol$; $V = nRT/p = (3.63 \times 10^{-3}\,mol) \times (8.314\,Pa\,m^3\,K^{-1}\,mol^{-1}) \times (373\,K)/(101\,300\,Pa) = 1.11 \times 10^{-4}\,m^3 = 111\,cm^3$
ii Volume of the gas exceeds volume of the syringe

3 a i Gas particles impact the walls of the containing vessel and rebound; change in velocity means there is change in momentum; with many particles hitting a specific area every second, the combined force of the particles exerts a pressure
ii As temperature increases, ions vibrate more violently about the lattice sites; at 801 °C the ions start to move away from their lattice sites
iii A lot of energy is required to overcome the strong electrostatic forces holding ions in their lattice site
b $pV = nRT$; $n = pV/RT = (1.01 \times 10^5\,Pa) \times (63.0 \times 10^{-6}\,m^3)/((8.31\,Pa\,m^3\,K^{-1}\,mol^{-1}) \times (373\,K))$; $n = 0.00205\,mol$; molar mass = $(0.148\,g)/n = 72.2\,g\,mol^{-1}$

4 a i A straight line through the origin
ii Particles do not have a negligible volume: there are forces acting between particles; collisions are not perfectly elastic
b i Slightly concave rising curve (almost parabolic) not starting from the origin
ii A liquid boils at the temperature for which its saturated vapour pressure is equal to the external pressure
iii Some liquids decompose at temperatures below their boiling point at normal atmospheric pressure
c $p_1V_1/T_1 = p_2V_2/T_2$ for the same amount in moles, so $V_1 = (T_1/T_2)V_2 = (273/373) \times 153\,cm^3 = 112.0\,cm^3$; $n = (112.0\,cm^3)/(2.24 \times 10^4\,cm^3\,mol^{-1}) = 5.00 \times 10^{-3}\,mol$; $M = (0.597\,g)/(5.00 \times 10^{-3}\,mol) = 119\,g\,mol^{-1}$ (3 sig. figs)

5 a $pV = nRT$
b $n(CO_2) = (10.0\,g)/(44.0\,g\,mol^{-1}) = 0.2273\,mol$; $p = (nRT/V) = (0.2273\,mol) \times (8.31\,Pa\,m^3\,K^{-1}\,mol^{-1}) \times (273\,K)/(5.00 \times 10^{-3}\,m^3) = 1.03 \times 10^5\,Pa$
c In real gases there are attractive forces between molecules that reduce the pressure

6 a i Separate ions and delocalized sea of electrons
ii Delocalized electrons allow conduction of electricity; since all ions are the same, layers of ions can roll over each other to new stable positions; thus metal is malleable/ductile
b In the vapour, sodium exists as diatomic molecules, Na—Na, with one pair of shared electrons
c i $n(Na) = pV/RT = (25\,Pa) \times (50 \times 10^{-6}\,m^3)/((8.31\,Pa\,m^3\,K^{-1}\,mol^{-1}) \times (293\,K)) = 5.13 \times 10^{-7}\,mol$; $m(Na) = (5.13 \times 10^{-7}\,mol) \times (23.0\,g\,mol^{-1}) = 1.2 \times 10^{-5}\,g$ (2 sig. figs)
ii The assumption that sodium vapour behaves as an ideal gas
d i $1s^22s^22p^63s^1$
ii The energy levels increase from the 1s to the 3s
iii $1s^22s^22p^63p^1$

7 a $pV = nRT$
b i n(acid) in $1\,dm^3$ of vapour $= (101 \times 10^3\,Pa) \times (1 \times 10^{-3}\,m^3)/((8.31\,Pa\,m^3\,K^{-1}\,mol^{-1}) \times (400\,K)) = 3.04 \times 10^{-2}\,mol$;
molar mass $= (2.74\,g)/(3.04 \times 10^{-2}\,mol) = 90\,g\,mol^{-1}$ (2 sig. figs)
ii Some of the molecules are dimers, with a mass of 120; the calculated value is an average for the molecules in the gas
c i A horizontal line
ii Atoms do not have a negligible volume (especially at high pressures) (NB There are also attractions, so the shape in the question is incorrect at low pressures)

Chapter 9

Practice questions

Spread 9.1
1 19.7 g **2** 1.54 kg

Spread 9.3
1 Fe_2O_3 **2** FeS_2

Spread 9.4
1 **a** $1.8\,dm^3$; **b** 2.7 g **2** $3.7\,dm^3$

Spread 9.5
1 $3.5 \times 10^{-3}\,mol$ **2** $0.50\,dm^3 = 500\,cm^3$

Spread 9.6
1 2.93 g **2** $0.04\,mol\,dm^{-3}$

Spread 9.7
1 **a** $0.0816\,mol\,dm^{-3}$; **b** $4.58\,g\,dm^{-3}$
3 $30.0\,cm^3$

Exam questions

1 a i $n(H_2SO_4) = (50\,000\,g)/(98.1\,g\,mol^{-1}) = 510\,mol$
ii $n(NaOH) = 2 \times (510\,mol) = 1020\,mol$
iii $V(NaOH) = (1020\,mol)/(5\,mol\,dm^{-3}) = 204\,dm^3$
b i $n(CaCO_3) = 510\,mol$; M_r $CaCO_3 = 40 + 12 + 16 \times 3 = 100$; mass $= (510\,mol) \times (100\,g\,mol^{-1}) = 51\,kg$ (2 sig. figs)
c NaOH is strongly alkaline: local concentrations could do more damage than the acid; NaOH produces a soluble product that could damage water, plants, and animals; solid calcium carbonate is less dangerous to transport and spread

2 a i M_r(acid) $= 14 + 2 + 32 + 3 \times 16 + 1 = 97$; n(acid) used $= (5.210\,g)/(97\,g\,mol^{-1}) = 0.0537\,mol$; concentration $= (0.0537\,mol)/(0.25\,dm^3) = 0.215\,mol\,dm^{-3}$
ii $n(NaOH) = n$(acid), so $c(NaOH) = (0.215\,mol\,dm^{-3}) \times (22.6/25) = 0.194\,mol\,dm^{-3}$
b $M_r(NaOH) = 23 + 16 + 1 = 40$; $n(NaOH) = (5.0\,g)/(40\,g\,mol^{-1}) = 0.125\,mol$; n(salt) formed $= (0.125\,mol)/2 = 0.0625\,mol$; M_r(salt) $= 2 \times 23 + 32 + 4 \times 16 + 10 \times 18 = 322$; mass of salt $= (0.0625\,mol) \times (322\,g\,mol^{-1}) = 20.1\,g$

3 a i c(acid) $= (0.121\,mol\,dm^{-3}) \times (32.4/25.0) = 0.1568\,mol\,dm^{-3}$
ii $n(HCl) = (0.1568\,mol\,dm^{-3}) \times (4 \times 10^3\,dm^3) = 627.2\,mol$
b n(lime) $= (627.2\,mol)/2 = 313.6\,mol$; M_r(lime) $= 40.1 + 2 \times 16.0 + 2 \times 1.0 = 74.1$; mass required $= (313.6\,mol) \times (74.1\,g\,mol^{-1}) = 23.2 \times 10^3\,g = 23.2\,kg$
c n(lime) $= (1000\,g)/(74.1\,g\,mol^{-1}) = 13.50\,mol$; n(limestone) needed $= 13.50\,mol$; M_r(limestone) $= 40.1 + 12.0 + 3 \times 16.0 = 100.1$; mass of limestone $= (13.50\,mol) \times (100.1\,g\,mol^{-1}) = 1.35 \times 10^3\,g = 1.35\,kg$

4 a Ag: 71.05; C: 7.89; O: 21.06; moles: 71.05/107.9 = 0.658; 7.89/12.0 = 0.658; 21.06/16.0 = 1.32; ratio: 1:1:(1.32/0.658) = 2; empirical formula = $AgCO_2$
b Empirical formula; mass = 108 + 12 + 32 = 152; $M_r = 304 = 2 \times 152$; molecular formula = $Ag_2C_2O_4$
c i $n(X) = (5.00\,g)/(304\,g\,mol^{-1}) = 0.01645\,mol$
ii n(gas) $= pV/RT = (100 \times 10^3\,Pa) \times (8.14 \times 10^{-4}\,m^3)/((8.31\,Pa\,m^3\,K^{-1}\,mol^{-1}) \times (298\,K)) = 0.032\,87\,mol$
iii Ratio of X:CO_2 = 0.016 45:0.032 87 = 1:2.00 = 1:2; $Ag_2C_2O_4 \rightarrow 2Ag + 2CO_2$; residue is silver; proof: mass of $CO_2 = (0.032\,87\,mol) \times (44.0\,g\,mol^{-1}) = 1.45\,g$; mass of residue = 5.00 g − 1.45 g = 3.55 g; mass of Ag in X $= 2 \times (0.01645\,mol) \times (107.9\,g\,mol^{-1}) = 3.55\,g$

5 a i $n(O_2) = (10\,dm^3)/(24\,dm^3\,mol^{-1}) = 0.417\,mol$
$n(H_2O_2) = 2n(O_2) = 0.83\,mol$ (2 sig. figs)
ii $n(H_2O_2) = (0.100\,mol\,dm^{-3}) \times (0.250\,dm^3) = 0.0250\,mol$;
volume $= (0.0250\,mol)/(0.833\,mol\,dm^{-3}) = 0.0300\,dm^3 = 30\,cm^3$ (2 sig. figs)
b $n(KMnO_4) = (0.200\,dm^3) \times (0.020\,mol\,dm^{-3}) = 0.0040\,mol$;
$m(KMnO_4) = (0.0040\,mol) \times (158\,g\,mol^{-1}) = 0.63\,g$ (2 sig. figs)
c i $n(KMnO_4) = (0.0400\,dm^3) \times (0.020\,mol\,dm^{-3}) = 8.0 \times 10^{-4}\,mol$
ii $n(H_2O_2) = (0.0200\,dm^3) \times (0.100\,mol\,dm^{-3}) = 2.0 \times 10^{-3}\,mol$
iii $n(H_2O_2)/n(KMnO_4) = 2.0 \times 10^{-3}/8.0 \times 10^{-4} = 2.5$
iv $2KMnO_4 + 5H_2O_2 + 3H_2SO_4 \rightarrow 2MnSO_4 + K_2SO_4 + 8H_2O + 5O_2$
(NB This question has inconsistent precision)
d i K^+ and Mn^{2+}
ii Manganese(II) hydroxide
e i $M_r(Na_2O_2) = 2 \times 23.0 + 2 \times 16.0 = 78.0$; $n(Na_2O_2) = (0.39\,g)/(78.0\,g\,mol^{-1}) = 0.0050\,mol$; $n(O_2) = \frac{1}{2}n(Na_2O_2) = 0.0025\,mol$; $V(O_2) = (0.0025\,mol) \times (24\,dm^3\,mol^{-1}) = 0.060\,dm^3 = 60\,cm^3$
ii Number $= (0.0025\,mol) \times (6.02 \times 10^{23}\,mol^{-1}) = 1.5 \times 10^{21}$

iii Ox(O) in peroxide = –1; Ox(O) in oxygen = 0; Ox(O) in carbonate = –2; hence oxygen has been oxidized and reduced, i.e. disproportionation

6 a $H_2X + 2NaOH \rightarrow Na_2X + 2H_2O$

b i $n(NaOH) = (0.025\,dm^3) \times (0.100\,mol\,dm^{-3})$ = 0.0025 mol

ii $n(H_2X) = \frac{1}{2}n(NaOH) = 0.00125\,mol$

iii This amount is in 25 cm^3; hence the amount in 500 cm^3 = (500/25) × (0.00125 mol) = 0.025 mol; 2.25 g is equivalent to 0.025 mol; M_r = 2.25/0.025 = 90

iv Mass of H_2X in 1 dm^3 = 2 × 2.25 = 4.5 g; hence, mass of water = 6.3 g – 4.5 g = 1.8 g; $n(H_2O) = (1.8\,g)/(18\,g\,mol^{-1}) = 0.10\,mol$; $n(H_2X) = 2 \times (0.025\,mol) = 0.05\,mol$; $n(H_2O)/n(H_2X) = 0.10/0.05 = 2$

c Shake a measured sealed volume of air with a measured mass of water; titrate the solution formed with standard aqueous sodium hydroxide

7 a See spread 9.3

b From spread 8.3, $M = mRT/pV$ = (0.130 g) × (8.31 Pa $m^3\,K^{-1}\,mol^{-1}$) × (373 K)/ ((101 × 10^3 Pa) × (85.0 × $10^{-6}\,m^3$)) = 46.9 g mol^{-1} (The data in the question is slightly out: 46.0 was meant)

c i Divide by mass of atom: 52.2/12.0 = 4.35; 13.0/1.0 = 13.0; 34.8/16.0 = 2.175; 4.35:13.0:2.175 = 2:6:1; empirical formula = C_2H_6O

ii Ethanol, CH_3CH_2OH, or methoxymethane, CH_3OCH_3

8 a i $^1H = 1.6734 \times 10^{-24} \times 6.0225 \times 10^{23} = 1.0078\,g$
$^{12}C = 1.9925 \times 10^{-23} \times 6.0225 \times 10^{23}$ = 12.000 g (5 sig. figs)

ii ^{12}C given the value 12 is the standard against which other atoms are compared; it is useful as it is a component of a large number of compounds

b i Naturally occurring carbon has a small percentage of other isotopes of carbon such as ^{13}C

ii $C + O_2 \rightarrow CO_2$; use $pV = nRT$: $n(CO_2) = pV/RT$ = (98.0 × 10^3 Pa) × (1.85 × $10^{-3}\,m^3$)/ ((8.31 Pa $m^3\,K^{-1}\,mol^{-1}$) × (293 K)) = 0.07446 mol; $n(C) = n(CO_2)$; $m(C)$ = (0.07446 mol) × (12.0 g mol^{-1}) = 0.894 g

c $CO_2 + 2NaOH \rightarrow Na_2CO_3 + H_2O$; $n(CO_2)$ = (1.54 g)/(44.0 g mol^{-1}) = 0.0350 mol; $c(Na_2CO_3)$ = (0.0350 mol)/(0.0500 dm^3) = 0.70 mol dm^{-3}

CHAPTER 10

Practice questions

Spread 10.4

1 15.7 kJ

Spread 10.6

1 –66 kJ mol^{-1}

Spread 10.9

1 a –84 kJ mol^{-1}; –105 kJ mol^{-1}

2 –760 kJ mol^{-1}

Exam questions

1 a See spread 10.5

b $\Delta_f H^{\ominus}$(nitro) – 1540 = 3 × (–394) + 5/2 × (–242) + 3 × 34 = – 1685, so $\Delta_f H^{\ominus}$(nitro) = –1685 + 1540 = –145 (all in kJ mol^{-1})

c $\Delta_f H^{\ominus}$(reaction) – 145 = 3 × (–394) + (5/2) × (–242) = –1642 (all in kJ mol^{-1})

d A decomposition reaction is more exothermic than combustion

e $\Delta_f H^{\ominus}(l \rightarrow g)$ = –242 – (–286) = +44 kJ mol^{-1}; the process is endothermic, so heat needs to be added

2 a Hydrazine has polar bonds and hydrogen bonding between molecules; hence more energy is needed to separate the molecules than in ethane, which only has weak dispersion forces

b i $N_2H_4 + O_2 \rightarrow N_2 + 2H_2O$

ii $\Delta_f H^{\ominus}$(hydrazine) –624 = 2 × (–286), so $\Delta_f H^{\ominus}$= +52 kJ mol^{-1}

c i $C_2H_6 + 7/2O_2 \rightarrow 2CO_2 + 3H_2O$

ii $2\Delta_f H^{\ominus}(CO_2) + 3\Delta_f H^{\ominus}(H_2O) - \Delta_f H^{\ominus}(C_2H_6)$
2(–394 kJ mol^{-1}) + 3(–286 kJ mol^{-1}) – (–85 kJ mol^{-1}) = –1561 kJ mol^{-1}

d Hydrazine uses less oxygen; the energy given out per mole of oxygen used is greater for hydrazine than for ethane; liquids are easier to handle

3 a See spread 10.5

b See spread 10.5

c i $\Delta_f H^{\ominus}$– 46 = 3 × (–269) – 114, so $\Delta_f H^{\ominus}$= –875 (all in kJ mol^{-1})

ii Bonds broken: 3(N—H): 3 × 388; 3(F—F): 3 × 158; total = 1638; bonds made: 3(H—F): 3 × (–562); 3(N—F): 3 × (–272); total = –2502; $\Delta_f H^{\ominus}$= bonds broken + bonds made = 1638 – 2502 = –864 (all in kJ mol^{-1})

d Bond enthalpies are averages for the bonds of a particular type in a compound and are only approximate, e.g. breaking the three N—H bonds in ammonia in turn requires three different quantities of energy; enthalpies of formation can be obtained either directly by experiment or from enthalpy of combustion measurements and are thus more accurate

4 a i Gaseous ions Na^+ and Cl^- and liquid water

ii NaCl crystal lattice and liquid water

iii Hydrated ions $Na^+(aq)$ and $Cl^-(aq)$ in liquid water

b $\Delta_f H^{\ominus}$(solution) =788 – 784 = +4 (all in kJ mol^{-1})

c Flame tests; clean flame wire by heating and dipping in conc. HCl; dip hot wire in solution and hold in flame; flame colour yellow: sodium, lilac: potassium, red: lithium; silver nitrate test for chloride: white precipitate when nitric acid and aqueous silver nitrate added

5 a i $\Delta_6 H$ (lattice formation enthalpy)

ii $\Delta_1 H + \Delta_2 H + \Delta_3 H + \Delta_4 H + \Delta_5 H + \Delta_6 H = \Delta_7 H$ =–635 – (193 + 590 + 1150 + 248 – 3513) = +697 (all in kJ mol^{-1})

iii $\Delta_3 H$ = 1st E.A. + 2nd E.A. + 697 = 1st E.A. + 844; 1st E.A. = 697 – 844 = –147 kJ mol^{-1}

b i 1st ionization enthalpy

ii Larger for magnesium

iii Magnesium is a smaller atom with the valence shell closer to the nucleus and hence bound more tightly

c i Flame test: calcium gives a brick-red flame

ii Dissolve the solid in dilute nitric acid (the carbonate reacts and effervesces), then add a few drops of aqueous silver nitrate; a white precipitate suggests chloride; add excess aqueous ammonia: the precipitate dissolves, confirming the presence of chloride ions

6 a

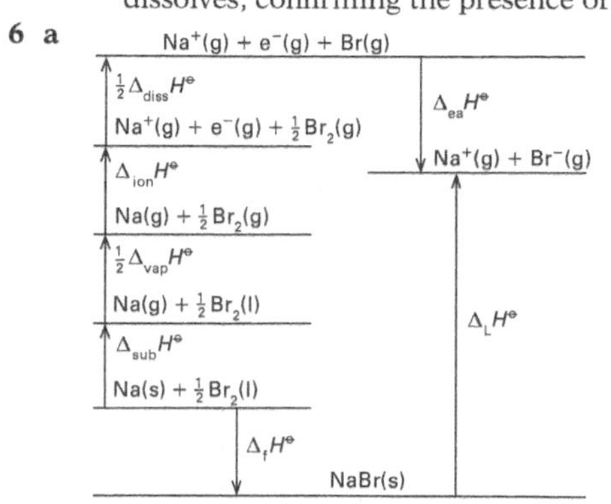

b $\frac{1}{2}\Delta_{vap}H^{\ominus}$ = –361 – (107 + 498 + 194/2 – 325 – 753) = +15 so $\Delta_{vap}H^{\ominus}$ = +30 (all in kJ mol^{-1})

7 a $CH_3CH_2OH + 3O_2 \rightarrow 2CO_2 + 3H_2O$

b i See spread 10.5

ii Ethanol: M_r = 46.0; hence the energy released by 1 g = 1370/46 =29.8 kJ; glucose: M_r = 180; hence the energy released by 1 g = 3000/180 = 16.7 kJ

iii Total energy from ethanol = 20 × 29.8 = 596 kJ;
total energy from glucose = 20 × 16.7 = 334 kJ;
total energy from 1 dm^3 of beer = 930 kJ (2 sig. figs)

CHAPTER 11

Practice questions

Spread 11.7

1 3.8 mol dm^{-3}

2 0.13 mol H_2 and 0.13 mol I_2

Exam questions

1 a See spread 11.1

b The position of equilibrium is independent of the rate at which equilibrium is reached

c B; $K_c = [NO_2]^2/[N_2O_4]$

d i x: **B**; y: **A** or **D**; z: **C**

ii Reaction will not be fast enough

iii It may become too expensive to make the containment vessels

iv z is the ammonia synthesis, which is exothermic, so the yield is lower at higher temperature

2 a Using $pV = nRT$, $n = pV/RT$; n(total) = ((1.59 × 10^6 Pa) × (1.04 × 10^{-3} m^3))/((8.31 Pa m^3 K^{-1} mol^{-1}) × (380 K)) = 0.524 mol; $n(CH_3OH)$ = 0.524 mol – 0.122 mol – 0.298 mol = 0.104 mol

b $[CH_3OH]$ = (0.104 mol)/(1.04 dm^3) = 0.100 mol dm^{-3};
$[H_2]$ = (0.298 mol)/(1.04 dm^3) = 0.2865 mol dm^{-3};
[CO] = (0.122 mol)/(1.04 dm^3) = 0.1173 mol dm^{-3};
$K_c = [CH_3OH]/[H_2]^2[CO]$ = (0.100 mol dm^{-3})/((0.2865 mol dm^{-3})^2 × (0.1173 mol dm^{-3}))
= 10.4 dm^6 mol^{-2} (3 sig. figs)

c i $K_p = p(CH_3OH)/p(CO)p(H_2)^2$

ii x(CO) = 0.122/0.524 = 0.2328;
$x(H_2)$ = 0.298/0.524 = 0.5687;
$x(CH_3OH)$ = 0.105/0.524 = 0.1985

iii $p(H_2)$ = 0.5687 × (1.59 MPa) = 0.904 MPa

iv p(CO) = 0.2328 × (1.59 MPa) = 0.370 MPa;
$p(CH_3OH)$ = 0.1985 × (1.59 MPa) = 0.316 MPa;
K_p = (0.316 MPa)/((0.370 MPa) × (0.904 MPa)^2)
= 1.0 MPa^{-2} (2 sig. figs)

3 a Increasing temperature → lower yield of sulphur trioxide; the reaction is exothermic, so an increase in temperature favours the backward reaction and reduces the equilibrium constant

b Increasing p → higher yield of sulphur trioxide; the number of gas moles decreases in the reaction, so the reaction shifts in the direction that minimizes the increase in pressure

c $p(O_2)$ = 120 kPa – 33 kPa – 39 kPa = 48 kPa;
$x(O_2)$ = 48/120 = 0.4

d i $K_p = p^2_{SO_3}/p^2_{SO_2}\, p_{O_2}$

ii K_p = (39 kPa)^2/((33 kPa)^2 × (48 kPa)) = 0.029 kPa^{-1}

4 a $n(Cl_2)$ = (11.1g)/(71.0 g mol^{-1}) = 0.156 mol;
$n(PCl_3)$ = 0.156 mol; PCl_5 used up = 0.156 mol;
$n(PCl_5)$ at start = (83.4 g)/(208.5 g mol^{-1}) = 0.400 mol;
$n(PCl_5)$ at equilibrium = 0.400 mol – 0.156 mol
= 0.244 mol

b i $K_c = [PCl_3] \times [Cl_2]/[PCl_5]$

ii $[PCl_3] = [Cl_2]$ = (0.156 mol)/(9.23 dm^3)
= 0.016 90 mol dm^{-3}; $[PCl_5]$ = (0.244 mol)/(9.23 dm^3)
= 0.026 44 mol dm^{-3}; K_c = (0.016 90 mol dm^{-3})^2/(0.026 44 mol dm^{-3}) = 0.011 mol dm^{-3} (2 sig. figs)

c i $K_p = p(PCl_3) \times p(Cl_2)/p(PCl_5)$

ii $x(Cl_2)$ = 0.156/(2 × 0.156 + 0.244) = 0.281

iii $p(PCl_5)$ = (0.244/(2 × 0.156 + 0.244)) × (250 kPa)
= 110 kPa

iv K_p = (0.281 × 250 kPa)^2/110 kPa = 45 kPa

5 a Forward and backward reactions still occurring, but the concentrations of the reactants and products do not change

b The number of gas moles decreases in the forward reaction so there is a decrease in pressure; equilibrium favours the direction that minimizes change in external pressure, so high pressure favours the forward reaction

c i $n(NH_3)$ = (24.0 g)/(17.0 g mol^{-1}) = 1.41 mol;
$n(H_2)$ = (13.5 g)/(2.0 g mol^{-1}) = 6.75 mol;
$n(N_2)$ = (60.3 g)/(28.0 g mol^{-1}) = 2.15 mol;
n(total) = 10.31 mol;
$x(NH_3)$ = (1.41 mol)/(10.31 mol) = 0.137;
$x(H_2)$ = (6.75 mol)/(10.31 mol) = 0.655;
$x(N_2)$ = (2.15 mol)/(10.31 mol) = 0.209;

ii Pressures are 1.37 atm NH_3, 6.55 atm H_2, 2.09 atm N_2;
$K_p = p^2_{NH_3}/p_{N_2}\, p^3_{H_2}$ = (1.37 atm)^2/((2.09 atm) × (6.55 atm)^3)
= 3.2 × 10^{-3} atm^{-2} (2 sig. figs)

iii Reaction is exothermic, so increasing the temperature favours the backward reaction, decreasing the yield and reducing K_p

d Thermodynamic stability refers to the enthalpy change (strictly Gibbs energy change) for the reaction, which in this case is negative, i.e. the products are at a lower energy level (more stable) than the reactants; kinetic stability refers to the rate at which the change occurs; if the reaction is slow (i.e. there is a high activation energy barrier for the reactants to surmount) then the reactants are said to be kinetically stable

6 a i See **1 c**

ii 0.5 mol N_2O_4 forms 1 mol NO_2; volume is 10.0 dm^3, so $[N_2O_4]$ = 0.05 mol dm^{-3}; $[NO_2]$ = 0.1 mol dm^{-3}; so K_c = (0.1 mol dm^{-3})^2/(0.05 mol dm^{-3}) = 0.2 mol dm^{-3}

iii $\Delta_r H^{\ominus}$ = 2 × (+33.9 kJ mol^{-1}) – (+9.70 kJ mol^{-1})
= +58.1 kJ mol^{-1}

iv The dissociation will increase because the reaction is endothermic

b Decrease the pressure

c i Faster

ii None

iii Faster

d The bonds in the reactants are stronger than those in the products

7 a i Bonds broken: H—H and I—I = (+436 kJ mol^{-1}) + (+151 kJ mol^{-1}) = +587 kJ mol^{-1}; bonds made: 2 × H—I = 2 × (–299 kJ mol^{-1}) = –598 kJ mol^{-1}; so ΔH = (+587 kJ mol^{-1}) + (–598 kJ mol^{-1})
= –11 kJ mol^{-1}

ii Energy goes up, reaches a transition state, and then falls to just below the original energy level

b i Energy down from reactants to products;

ii Energy up from reactants to transition state

c Higher activation energy because the Cl—Cl bond is stronger than the I—I bond

d i $K_c = [HI]^2/[H_2][I_2]$

ii 1.5 mol H_2 and 1.5 mol I_2 react to form 3.0 mol HI, so the amount in moles left of H_2 and I_2 = 1.9 mol – 1.5 mol = 0.4 mol; $[H_2] = [I_2]$ = (0.4 mol)/(0.25 dm^3)
= 1.6 mol dm^{-3}; [HI] = 3.0 mol/(0.25 dm^3)
= 12 mol dm^{-3}; $K_c = [HI]^2/[H_2][I_2]$
= (12 mol dm^{-3})^2/(1.6 mol dm^{-3})^2 = 56

CHAPTER 12

Practice questions

Spread 12.7

1 11.3 **2** 11.1

Spread 12.9

3 0.579 mol dm^{-3}

Exam questions

1 a i $C_2H_5COOH(aq) \rightleftharpoons H^+(aq) + C_2H_5COO^-(aq)$

ii pK_a = 4.91

b i $C_2H_5COOH(aq) + NaOH(aq) \rightarrow C_2H_5COONa(aq) + H_2O(l)$

ii Phenolphthalein

c i A solution that resists change in pH despite small additions of acid or base

ii Addition of acid or base shifts the equilibrium, so counteracting the addition; e.g. if acid is added, the excess hydrogen ions cause the equilibrium to shift to the left, the propanoate ions combine to form more propanoic acid molecules, and the hydrogen ion concentration remains almost constant
iii Volume of NaOH added neutralizes the same volume of acid = 15 cm^3; volume of acid remaining = 15 cm^3, since amounts of acid and salt are equal; pK_a= pH
iv Carbon dioxide/hydrogencarbonate ions; maintains a constant pH – structure of proteins (e.g. enzymes) sensitive to changes in pH

2 a See spread 12.2
b HCl, Cl^-, H_2SO_4, HSO_4^-; NH_4^+, NH_3
c i and **ii** See spread 12.4
iii CH_3COOH **A1**, $CH_3CO_2^-$ **B1**, H_3O^+ **A2**, H_2O **B2**
d i $CH_3COOH + NH_3 \rightarrow NH_4^+ + CH_3CO_2^-$
ii Ammonia is a base, encouraging the deprotonation to occur

3 a X: strong base (e.g. sodium hydroxide)
Y: strong acid (e.g. hydrochloric)
b Curve B – X: weak acid, pH higher at start than for strong acid in A, equivalence point above pH 7; Y: strong base, final pH similar to A; curve C – X: weak base, pH lower at start than it would be for a strong base of same concentration, Y: strong acid, final pH same as start in A, equivalence point below pH 7
c $[H^+] = 5.0 \times 10^{-3}$ mol dm^{-3};
$K_a = (5.0 \times 10^{-3}\text{ mol dm}^{-3})^2/(0.1\text{ mol dm}^{-3})$
$= 2.5 \times 10^{-4}$ mol dm^{-3}

Chapter 13

Exam questions

1 a Reduction is the gain of electrons by a species
b The standard hydrogen electrode; difficult to maintain standard conditions (e.g. pressure of hydrogen gas) and hence reliable cell emf; risk of ignition of hydrogen gas
c i $Ag^+(aq) + e^- \rightarrow Ag(s)$; $Cu(s) \rightarrow Cu^{2+}(aq) + 2e^-$; the silver ions are reduced
ii $S_2O_8^{2-}(aq) + 2e^- \rightarrow 2SO_4^{2-}(aq)$;
$Mn^{2+}(aq) + 4H_2O(l) \rightarrow$
$MnO_4^-(aq) + 8H^+(aq) + 5e^-$; the $S_2O_8^{2-}$ is reduced

2 a Hydrogen electrode; 0 V
b No current drawn from cell; solutions of 1 mol dm^{-3}; temperature: 298 K; pressure: 1 bar
c To provide electrical contact betwen the two half-cells; potassium chloride
d Table of redox half-equations (usually written as reduction reactions) in order of their standard electrode potentials (most negative at the top)
e Iron(II) ions reduced to Fe: Zn has the greater potential to lose electrons and donates electrons to the iron(II) ions

3 a $Fe(s)|Fe^{2+}(aq)||Fe^{3+}(aq)|Fe(s)$
emf = (–0.04 V) – (–0.44 V) = + 0.4 V
$Fe(s) + 2Fe^{3+}(aq) \rightarrow 3Fe^{2+}(aq)$
b Manganate(VII); the manganate(VII) standard half-cell has a more positive potential than the chlorine half-cell and will oxidize chloride ions to chlorine, thus interfering with the estimation of the iron
c i The manganate(VII) will oxidize the chromium(III)
ii $6MnO_4^-(aq) + 10Cr^{3+}(aq) + 11H_2O(l) \rightarrow$
$6Mn^{2+}(aq) + 5Cr_2O_7^{2-}(aq) + 22H^+(aq)$

4 a Method 1: reduction by carbon (blast furnace); the metal ore; the reducing agent; high temperature, e.g. iron; method 2: electrolysis of molten metal compound; metal compound; high temperature, e.g. aluminium; method 3: reduction by stronger reducing agent, e.g. Thermit process
b i Carbon oxidized by oxygen
ii Raw materials cheaper, less energy required (lower temperatures)
iii Titanium able to withstand higher temperatures

5 a i Digital voltmeter; high resistance
ii 1 mol dm^{-3}
b i Provides electrical contact between the solutions in the two half-cells
ii Aqueous potassium chloride
iii It is the ions in the solution which are the charge carriers, not free electrons as in the metal wire
c i emf = $E^\ominus$(r.h. electrode) – $E^\ominus$(l.h. electrode) = + 0.38 V
= +0.34 V – $E^\ominus(Fe^{3+}/Fe)$, so $E^\ominus(Fe^{3+}/Fe)$
= +0.34 V – 0.38 V = –0.04 V
ii $Fe(s)|Fe^{3+}(aq)||Cu^{2+}(aq)|Cu(s)$
iii $2Fe(s) + 3Cu^{2+}(aq) \rightarrow 2Fe^{3+}(aq) + 3Cu(s)$
d Zinc has a more negative standard electrode potential than iron, i.e. it will reduce any iron(II) formed and will be oxidized itself, so even if the surface is scratched to expose the iron, the zinc will prevent rusting occurring; tin has a more positive standard electrode potential than iron; if tinplate is scratched, the iron will oxidize and prevent the tin from being oxidized, so rusting is accelerated

6 a $\mathbf{A}|\mathbf{A}^{n+}||\mathbf{B}^{n+}|\mathbf{B}$
b **C** least reactive, then **B**, **D**, **A**
c Used a digital voltmeter
d $E^\ominus$ = (+1.10 V) – (+0.47 V) = +0.63 V

7 a i Pb
ii PbO_2
iii H_2SO_4
b $Pb(s) + PbO_2(s) + 2H_2SO_4(aq) \rightarrow 2PbSO_4 + 2H_2O(l)$
c $E^\ominus$= (+1.69 V) – (–0.36 V) = +2.05 V
d Opposite of the equation above
e They are heavy
f i Oxygen also needed
ii Rusting more severe where there is already an indentation
iii Corrosion more likely where this is stress
iv Magnesium corrodes preferentially as Mg is more reactive than Fe

8 a II
b $4Al(s) + 3O_2(g) + 6H_2O(l) \rightarrow 4Al(OH)_3(s)$
c The electrodes get coated in the non-conducting aluminium

9 a i $Cl_2 + 2KI \rightarrow I_2 + 2KCl$
ii $Cl_2 + H_2 \rightarrow 2HCl$
iii See spread 13.4
b i The reactions at the electrodes involve the gain and loss of electrons
ii See spread 19.2

Chapter 14

Exam questions

1 a Conservation
b $\Delta H^\ominus - T\Delta S^\ominus = -RT\ln K_c$; so $\ln K_c = \Delta S^\ominus - \Delta H^\ominus/T$
c i As T increases, if $\Delta H^\ominus$ is negative, the second term gets less positive and so the value of K_c falls
ii If $\Delta H^\ominus$ is positive, the second term above gets less negative and so the value of K_c rises

2 a i A: digital voltmeter; B: salt bridge; C: copper wire
ii Zn^{2+}; 1 mol dm^{-3}
iii Zero current
b i $Zn(s) + Cu^{2+}(aq) \rightarrow Zn^{2+}(aq) + Cu(s)$
ii $E^\ominus$ = (+0.34 V) – (–0.76 V) = +1.10 V
iii $\Delta G^\ominus = -zFE^\ominus = -2 \times (96500\text{ C mol}^{-1}) \times (+1.10\text{ V})$
= –212 kJ mol^{-1}
iv From Zn to Cu

3 a i $\Delta_r H^\ominus$ is positive, $\Delta_r S^\ominus$ is positive
ii Below 0, $|T\Delta_r S^\ominus| < |\Delta_r H^\ominus|$ so $\Delta_r G^\ominus$ is positive and the reaction is not spontaneous
b $\Delta_r H^\ominus$ for reaction is positive, and overall $\Delta_r S^\ominus$ is also positive as gas is released; $|T\Delta_r S^\ominus| > |\Delta_r H^\ominus|$, so $\Delta_r G^\ominus$ is negative, and the reaction goes

c $\Delta_f H^\ominus$ is negative, so more energy is released by new bonds made than is required to break bonds in reactants; $\Delta_f S^\ominus$ is positive, so there is an increase in the number of molecules during the reaction, and hence $\Delta_f G^\ominus$ is negative

4 a i –350 kJ mol^{-1} (CO); –420 kJ mol^{-1} (CO_2); –410 kJ mol^{-1} (Fe)

ii FeO + C: $\Delta_f G^\ominus$ = (+410 – 350)/2 = +30 kJ mol^{-1}; FeO + CO: $\Delta_f G^\ominus$ = (+410 – 420)/2 = –5 kJ mol^{-1}

iii Only the reaction with CO is feasible at 800 K because $\Delta_f G^\ominus$ is negative for this reaction but positive for the reaction with C

iv Approx. 1000 K (where the C line crosses and falls below the Fe line)

v Both reactants are solid: low S; one of the products is a gas: high S, so change is positive

5 a $Zn(s) \rightarrow Zn^{2+}(aq) + 2e^-$; $Pb^{2+}(aq) + 2e^- \rightarrow Pb(s)$

b It takes account of heat lost from the calorimeter; you do not have to assume the specific heat capacity of the solution is the same as for pure water

c Energy given out = (25.0 J s^{-1}) × (305 s) = 7625 J; n(lead) used = (0.5 mol dm^{-3}) × (0.1 dm^3) =0.05 mol; energy released per mole lead = (7625 J)/0.05 = 152.5 kJ; $\Delta_f H^\ominus$ = –152.5 kJ mol^{-1}

d i $Zn(s)|Zn^{2+}(aq)||Pb^{2+}(aq)|Pb(s)$

ii Lead; it is the one being reduced (taking up electrons)

iii 1 M aqueous lead nitrate and zinc nitrate

iv See question 8, chapter 13

e i $\Delta G^\ominus = -zFE^\ominus = -2 \times (9.65 \times 10^4\ \text{C mol}^{-1}) \times (0.63\ \text{V})$ = –121.6 kJ mol^{-1} = $\Delta H^\ominus - T\Delta S^\ominus$, so $T\Delta S^\ominus = \Delta H^\ominus - \Delta G^\ominus$ = (152.5 kJ mol^{-1}) – (–121.6 kJ mol^{-1}) = –30.9 kJ mol^{-1}; thus $\Delta_f S^\ominus$ (system) = (–30.9 × 10^3 J mol^{-1})/(298 K) = –103.7 J K^{-1} mol^{-1} (answer correct as given)

ii $\Delta_f S^\ominus$(surr) = $-\Delta_f H^\ominus / T$ = (+152.5 × 10^3 J mol^{-1})/(298 K) = +511.7 J K^{-1} mol^{-1}

iii $\Delta_f S^\ominus$ (sys) + $\Delta S^\ominus$(surr) = –103.7 J K^{-1} mol^{-1} + 511.7 J K^{-1} mol^{-1} = 408 J K^{-1} mol^{-1}; this is equal to $-\Delta_f G^\ominus / T$, so if $\Delta_f S^\ominus$ (total) is positive, the reaction is spontaneous, which it is

6 a Any reaction where a solid becomes a liquid or gas, or liquid becomes gas, or where the number of particles increases, e.g.
$2Na(s) + 2H_2O(l) \rightarrow 2Na^+(aq) + 2OH^-(aq) + H_2(g)$

b 214 – 205 –$S^\ominus$ (graphite) = 3, so $S^\ominus$ (graphite) = 6 J K^{-1} mol^{-1}; graphite is solid, and has a highly ordered lattice with strong bonds between atoms; hence low $S^\ominus$

c i Given in **1**

ii When $\Delta_f G^\ominus$= 0, $T\Delta_f S^\ominus = \Delta_f H^\ominus$, or $\Delta_f S^\ominus = \Delta_f H^\ominus / T$

d $\Delta_f S^\ominus$=6000/273 =22 J K^{-1} mol^{-1}; M_r(water) = 18, hence (54 g)/(18 g mol^{-1}) = 3 mol; total entropy change = (3 mol) × (22 J K^{-1} mol^{-1}) = 66 J K^{-1}

Chapter 15

Exam questions

1 a Reducing the size of the lumps increases the surface area of the marble in contact with the acid, hence providing a greater probability that the acid will collide with and react with the marble and increasing the rate of reaction

b i Expt 2: compared to expt 1, concentration of A and B doubles, hence the rate is multiplied by 2 × 2^2 = 8: 8 × 1.2 × 10^{-4} = 9.6 × 10^{-4}; expt 3: concentration of B is the same as in 2 but the rate is 1/4 of the rate in expt 2, so [A] = 1/4 of 0.04 = 0.01; expt 4: compared to expt 1, we have [A] × 3, [B] × 1.5, so rate = 1.2 × 10^{-4} × 3 × 1.5^2 = 8.1 × 10^{-4}; expt 5: compared to 1, we have [A] doubled, rate × 6, so 6 = 2 × b^2, where b = 1.732, so [B] = 1.732 × 0.02 = 0.035

ii In expt 1, 1.2 × 10^{-4} = kx, 0.02 × 0.022k = 15 mol^{-2} dm^6 s^{-1}

2 a i Increased temperature

ii The total number of molecules

iii There is no change to the number of molecules

b i The minimum energy particles must have for collisions to bring about a reaction

ii Catalyst: causes activation energy to decrease (rate to increase); if volume is reduced, pressure increases, but there is no effect on the activiation energy (but rate increased)

c Few particles move from their lattice positions and so there are few collisions between the reactants and few have sufficient kinetic energy to produce a reaction

3 a Homogeneous: same phase as reactants; heterogeneous: different phase

b See **2 b i**

c i k: rate constant, the constant of proportionality in rate expression (gradient of line of rate against [A]); [A]:concentration of reactant A; n: the order of the reaction with respect to reactant A

ii The order of reaction with respect to B is zero

d i Heterogeneous; reactants adsorbed onto surface of the catalyst; weak bonds formed between reactants and catalyst which weaken bonds in reactants, allowing molecules close together on catalyst surface to form new bonds

ii Homogeneous; reactant molecules fit into and bond with active sites on enzyme molecules; this weakens bonds in the reactant molecules allowing reactions to take place

e Increase the surface area of the catalyst, e.g. by making it into porous pellets (vanadium(V) oxide) or coating it onto an inert substrate (e.g. platinium on ceramic in catalytic converters); recover and re-use the catalyst after use

5 a i k: the rate constant; A: the pre-exponential factor

ii Gradient = –12 500 K

iii E = (–12 500 K) × (–8.31 J K^{-1} mol^{-1}) = 104 kJ mol^{-1}

b i Order: the powers of the concentration in the rate expression, e.g. rate = $k[A]^x$, where x is the order with respect to A

ii Increased concentration means that there are more collisions between reactant molecules and hence a greater probability that the reaction will occur

iii Proportional increase in concentration = 0.54/0.18 = 3; rate increases by factor of 9 = 3^2; reaction is second order; rate is proportional to concentration squared

Chapter 16

Exam questions

1 a i Barium has stronger metallic bonding; because 2 electrons in outer shell are involved in bonding, not 1; barium atoms are smaller than caesium atoms

ii Electrons dropping from higher to lower energy levels

iii Discrete lines

b i NaO; **ii** Na_2O_2

iii Weighed mass of Y reacted with known volume of $H_2SO_4(aq)$; titrate aliquots with $KMnO_4(aq)$ of known concentration

2 a i Molten NaCl; **ii** Aqueous NaCl

b i Cl_2; $2Cl^- \rightarrow Cl_2 + 2e^-$

ii O_2; $4OH^- \rightarrow 2H_2O + O_2 + 4e^-$

c Although OH^- is easier to oxidize ($E^\ominus$ is less positive under standard conditions), concentration of Cl^- is much higher

d White precipitate ($[Al(OH)_3(H_2O)_3]$) formed from colourless solution ($[Al(H_2O)_6]^{3+}$), which redissolves in excess to give colourless solution ($[Al(OH)_4]^-$)

3 a i $1s^2 2s^2 2p^6 3s^2 3p^6 4s^2$

ii Ca atoms lose outer electrons to form cations in a sea of delocalized electrons; held together by strong electrostatic forces of attraction

b i $Ca + 2H_2O \rightarrow Ca(OH)_2 + H_2$
0 +1–2 +2–2+1 0

ii Ca is oxidized; $Ca \rightarrow Ca^{2+} + 2e^-$

iii Calcium hydroxide (limewater); 11

c **i** White precipitate after a few drops
ii No obvious change
d $Mg^{2+}(aq) + 2OH^-(aq) \rightarrow Mg(OH)_2(s)$
$Mg(OH)_2$ is much less soluble than $Ba(OH)_2$, which is a strong base; fewer drops of NaOH(aq) are required

4 **i** Method 1: heat magnesium in oxygen
$2Mg(s) + O_2(g) \rightarrow 2MgO(s)$
Method 2: heat carbonate, nitrate, or hydroxide
$MgCO_3(s) \rightarrow MgO(s) + CO_2(g)$
ii $[Mg]^{2+}$ $[\overset{\times\times}{\underset{\bullet\bullet}{:O:}}]^{2-}$
iii It contains O^{2-} ions, which can act as proton acceptors
iv **I** Magnesium oxidizes when heated in air
II 175 kg

5 **a** **i** White precipitate; $Mg^{2+}(aq) + 2OH^-(aq) \rightarrow Mg(OH)_2(s)$
ii Reagents: add $AgNO_3(aq)$ acidified with $HNO_3(aq)$; observation: white precipitate which does not dissolve in $HNO_3(aq)$; equation: $Ag^+(aq) + Cl^-(aq) \rightarrow AgCl(s)$
b **i** Lattice enthalpies for both compounds are similar; hydration enthalpy of Mg^{2+} is much greater than that of Ba^{2+}
ii High melting point/hard/cleaves/conducts only when liquid or aqueous
c Required for chlorophyll production

6 **a** **i** $Ba + 2H_2O \rightarrow Ba(OH)_2 + H_2$
ii More vigorously; the lower ionization energy of Ba means that electrons are lost more easily
iii $BaCO_3(s) \rightarrow BaO(s) + CO_2(g)$
iv Calcium carbonate
v Flame test; calcium: brick-red; barium: yellow-green
b **i** BaO_2; **ii** O_2^{2-}; **iii** -1; **iv** Na_2O_2
c **i** $Na_2O_2 + H_2SO_4 \rightarrow Na_2SO_4 + H_2O_2$
ii $BaSO_4$ is insoluble and could be removed by filtering

7 **a** **i** $-348\ kJ\ mol^{-1}$; the Born–Haber cycle
ii The theoretical lattice enthalpy is based on a purely ionic model, so the experimental lattice enthalpy would be different because magnesium chloride has some covalent character
b **i** As the cation radius increases (from Ca to Ba), the ions have less attraction to water
ii The sulphates become less soluble (from $CaSO_4$ to $BaSO_4$), because the cationic radius increases, so the hydration enthalpy increases; the decrease in lattice enthalpy is only slight in comparison
iii Barium sulphate is very insoluble, so there are few barium ions free to poison the patient
iv $Ba(s) + 2H_2O(l) \rightarrow Ba(OH)_2(aq) + H_2(g)$
v Add dilute sulphuric acid, filter the barium sulphate precipitate formed, and dry

8 **a** **i** The aqueous magnesium nitrate would immediately form a white precipitate with sodium hydroxide; the aqueous barium nitrate would not form a precipitate with a few drops of NaOH, but when excess is added, a white precipitate would form
ii Magnesium hydroxide is insoluble and hence precipitates immediately; $Mg^{2+}(aq) + 2OH^-(aq) \rightarrow Mg(OH)_2(s)$; barium hydroxide is much more soluble and so would only form a precipitate if NaOH was added in large amounts; $Ba^{2+}(aq) + 2OH^-(aq) \rightarrow Ba(OH)_2(s)$
b **i** As the atomic number increases, the Group II carbonates become more stable, because the cation radius increases, so the cations polarize the carbonate anion less, so the carbonates require higher temperatures to decompose; more importantly, the lattice energy of the oxide formed would be less
ii Sodium carbonate is much more stable to heat because the sodium ion only has a 1+ charge, so the sodium ion does not polarize the carbonate so much
c **i** Electrolysis of molten sodium chloride
ii Magnesium chloride, from dried-up salt lakes
iii High temperature to melt the salt, and electricity; $Mg^{2+} + 2e^- \rightarrow Mg$

iv Chlorine; examples of uses: manufacture of bleach, PVC, insecticides
d **i** The electronic energy levels around a lithium ion are fixed; the flame heat promotes electrons to a high energy level; when the electrons fall down energy levels, light is emitted; as the energy levels around the ions are fixed, the energy released is fixed, so a particular colour is produced
ii Rubidium is much more reactive, so burning it in chlorine would be too dangerous

Chapter 17

Exam questions

1 **a** The first three elements have a relatively high melting point owing to their metallic bonding; silicon in the middle has a very high melting point owing to a giant covalent lattice; the last elements have low melting points owing to the weak intermolecular forces between the simple molecules
b As atomic number increases in Group I, the atomic radius increases, because there is an increased number of electron shells; the ionization energies decrease as the electron shells increase, producing an increased distance from the nucleus to the outer electrons
c **i** $2Na(s) + 2H_2O(l) \rightarrow 2NaOH(aq) + H_2(g)$
ii $Na_2O(s) + H_2O(l) \rightarrow 2NaOH(aq)$
iii $NaH(s) + H_2O(l) \rightarrow NaOH(aq) + H_2(g)$
d 14

2 **a** **i** Cl_2
ii Si, P (red)
b NaCl; MgS
c $AlCl_3$: trigonal planar, 120°; PCl_3: pyramidal, less than 109°
d **i** Sodium: $Na_2O + H_2O \rightarrow 2NaOH$
ii Sulphur: $SO_2 + H_2O \rightarrow H_2SO_3$

3 **a** The way that the properties of the elements (and their compounds) show a repeating pattern when ordered by atomic number
b The number of protons in the nucleus increases and electrons in the same shell shield poorly, so the nucleus increasingly attracts electrons
c **i** Mg has stronger metallic bonding because 1 more electron per atom is involved in the delocalized sea
ii Si has a giant covalent structure; therefore very strong covalent bonds must be broken
iii S_8 has stronger dispersion forces between molecules than P_4
iv In Ar there are only very weak dispersion forces between individual atoms

4 **a** NaCl; $MgCl_2$; $AlCl_3$; Al_2Cl_6; PCl_3/PCl_5; S_2Cl_2
b 16.5%
c **i** :Cl: / :Cl × Si × Cl: / :Cl:
ii
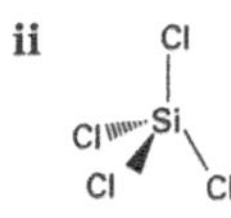

4 bond pairs (and 0 lone pairs) repel each other equally to the corners of a tetrahedron
iii $AlCl_3$ has 3 bond pairs (and 0 lone pairs) which repel each other equally to the corners of a triangle
d **i** $SiCl_4 + 2H_2O \rightarrow SiO_2 + 4HCl$
ii CCl_4 does not react because the outer shell of the C atom does not have a vacant orbital to accept a lone pair from a water molecule (see artwork in spread 19.6)

5 **a** A measure of the attraction of an atom for a shared pair of electrons (in a covalent bond)
b Electronegativity increases, because the number of protons increases and the added electron shields poorly, so the effective nuclear charge increases, so the strength of attraction for electrons increases
c **i** Ionic → covalent
ii As atoms become more electronegative, the difference in electronegativity decreases, so electrons are shared rather than transferred
d $P_4O_{10} + 6H_2O \rightarrow 4H_3PO_4$

6 a Metallic bonding in Mg: strong force of attraction between cations and delocalized electrons; ionic bonding in $MgCl_2$: strong force of attraction between oppositely charged ions
b Covalent/ionic borderline, because Al is more electronegative than Mg; the difference in electronegativity is less, *or* Al^{3+} has a higher charge density than Mg^{2+}, so it polarizes Cl^- more
c $MgCl_2(s) + aq \rightarrow Mg^{2+}(aq) + 2Cl^-(aq)$; the force of attraction between positive Mg^{2+} and polar water molecules (or coordinate covalent bonds $\rightarrow [Mg(H_2O)_6]^{2+}$)

7 a i $Na_2O + H_2O \rightarrow 2NaOH$; pH = 14
ii $SO_2 + H_2O \rightarrow H_2SO_3$; pH = 1
b Ionic oxides give basic solutions, covalent give acidic
c i $MgCl_2(s) + aq \rightarrow Mg^{2+}(aq) + 2Cl^-(aq)$; pH = 7
ii $SiCl_4(l) + 2H_2O(l) \rightarrow SiO_2(s) + 4HCl(g)$; pH = 0

8 a Electronegativity is a measure of the attraction of an atom to a shared electron pair in a covalent bond
b i Ionic; **ii** Covalent (polar); **iii** Covalent
c i The electron pairs around a central atom move as far apart as possible, to decrease the repulsion between them; lone pair – lone pair repulsion is greater than lone pair – bonding pair repulsion, which is greater than bonding pair – bonding pair repulsion
ii Ammonia has a pyramidal shape; draw the lone pair straight up, with one bond in the plane of the paper at 107° from the lone pair, and then the other two bonds off the other way, one into the paper and one out
iii The lone pair repels more effectively than the N—H bonding pairs, so the lone pair pushes the bonding pairs together
iv The N—H bond is polar owing to the difference in electronegativity of the N and H atoms; the whole molecule is polar because the polar bonds do not cancel each other since they are off to one side of the molecule
d i $2Mg + O_2 \rightarrow 2MgO$; $MgO + 2HCl \rightarrow MgCl_2 + H_2O$
ii $Al_2O_3 + 6H^+ \rightarrow 2Al^{3+} + 3H_2O$; $Al_2O_3 + 3H_2O + 2OH^- \rightarrow 2[Al(OH)_4]^-$;
iii CO_2 is more acidic; SiO_2 requires molten NaOH at high temperatures

CHAPTER 18

Exam questions

1 a It decreases; as the number of shells increases, the distance from the nucleus increases
b It increases; more electrons increase the strength of the dispersion forces
c Br_2 is non-polar, while ICl is polar, because Cl is more electronegative than I; therefore dipole–dipole forces exist as well as dispersion forces

2 a i $2P + 3I_2 \rightarrow 2PI_3$
ii $PI_3 + 3H_2O \rightarrow H_3PO_3 + 3HI$
b i Yellow precipitate then no change
ii Black solid/purple vapour (rotten egg smell)

3 a i $NaCl(s) + H_2SO_4(l) \rightarrow NaHSO_4(s) + HCl(g)$; at room temperature
ii I^-/HI is a stronger reductant than Cl^-/HCl, therefore $H_2SO_4 + 8HI \rightarrow 4I_2 + H_2S + 4H_2O$
b Ionization: $HF + H_2O \rightarrow H_3O^+ + F^-$; it is less with HF, because the H—F bond is strongest
c Add $AgNO_3(aq)$; white precipitate with Cl^-, yellow precipitate with I^-
d i $2Br^-(aq) + Cl_2(g) \rightarrow Br_2(l) + 2Cl^-(aq)$
$E^\ominus$ for the reaction is positive ((1.36 – 1.09) V) and therefore it is energetically feasible
ii It allows the production of Br_2 from seawater

4 a i Covalent bonding; the diagram should show the three covalent bonds in the PCl_3 as three pairs of dots and crosses; the P also has a lone pair of electrons
ii The diagram should show a pyramidal shape like that of ammonia; the electron pairs repel until they are as far apart as possible
iii Drip PCl_3 through dry chlorine gas using ice-cold apparatus
iv PCl_4^+ and PCl_6^-
v PCl_5 has an unexpected ionic structure, which causes an increased melting point
vi PI_5 is unknown because it is not possible to fit five large iodine atoms around the relatively small phosphorus atom
b i Propanoyl chloride, CH_3CH_2COCl; the graphical formula would show all the bonds
ii There is a vigorous production of HCl (seen as white fumes) which can be violent; the HCl fumes would form concentrated hydrochloric acid, which could harm the experimenter
iii Graphical formulae would show all the bonds; the diagrams would show propan-1-ol, $CH_3CH_2CH_2OH$ and propan-2-ol, $CH_3CH(OH)CH_3$; R is propan-1-ol
iv Add R to potassium manganate(VII) (or potassium dichromate) with dilute sulphuric acid and heat under reflux; the product would be purified by fractional distillation

5 a Chlorine is produced by electrolysis of brine using a steel cathode and titanium anode
b i $3Cl_2(g) + 6NaOH(aq) \rightarrow NaClO_3(aq) + 5NaCl(aq) + 3H_2O(l)$; the green chlorine gas produces a colourless solution, by disproportionation
ii $Cl_2(g) + 2Br^-(aq) \rightarrow Br_2(l) + 2Cl^-(aq)$; the chlorine oxidizes the bromide ions to bromine; the green chlorine produces a brown solution; a redox reaction
iii $Cl_2(g) + C_2H_4(g) \rightarrow ClCH_2CH_2Cl(l)$; the green chlorine gas reacts with ethene to make a colourless oily liquid; the chlorine adds to the ethene by electrophilic addition

6 a Main oxidation states: Na: +1; Mg: +2; Al: +3; Si: +4; P: +3 and +5; ionic chlorides dissolve in water;
$NaCl(s) \rightarrow Na^+(aq) + Cl^-(aq)$;
$MgCl_2(s) \rightarrow Mg^{2+}(aq) + 2Cl^-(aq)$;
covalent chlorides hydrolyse and produce white hydrogen chloride fumes;
$Al_2Cl_6(s) + 3H_2O(l) \rightarrow 2Al(OH)_3(s) + 6HCl(g)$;
$SiCl_4(l) + 2H_2O(l) \rightarrow SiO_2(s) + 4HCl(g)$;
$PCl_3(l) + 3H_2O(l) \rightarrow H_3PO_3(aq) + 3HCl(g)$;
$PCl_5(l) + 4H_2O(l) \rightarrow H_3PO_4(aq) + 5HCl(g)$;
b i $2S_2Cl_2(l) + 3H_2O(l) \rightarrow 3S(s) + H_2SO_3(aq) + 4HCl(g)$
ii 22.2 cm^3

7 a i The Born–Haber cycle should include 2 × $\Delta H^\ominus_{at}$(F), and 2 × E.A. of fluorine
ii –2640 kJ mol^{-1}
b i $CaF_2 + H_2SO_4(l) \rightarrow CaSO_4(s) + 2HF(g)$
ii +58 kJ mol^{-1}
iii The reaction should be heated; the equilibrium will shift right with increasing temperature, so the forward reaction will be favoured, because the reaction is endothermic

8 a i Aqueous NaCl (brine)
ii $2Cl^- \rightarrow Cl_2 + 2e^-$
iii $2H^+ + 2e^- \rightarrow H_2$, or $2H_2O + 2e^- \rightarrow H_2 + 2OH^-$
iv $2NaCl + 2H_2O \rightarrow 2NaOH + Cl_2 + H_2$
b i Remove the diaphragm so that Cl_2 and NaOH can react
ii $2NaOH + Cl_2 \rightarrow NaCl + NaOCl + H_2O$
iii Bleach
c i Simultaneous oxidation and reduction of the same species in a reaction
ii $3ClO^- \rightarrow 2Cl^- + ClO_3^-$
+1 –1 +5
iii $ClO^- + 4OH^- \rightarrow ClO_3^- + 2H_2O + 4e^-$
$ClO^- + H_2O + 2e^- \rightarrow Cl^- + 2OH^-$

9 a As the atomic number increases, the halogen atomic radius increases, so the hydrogen halide bonds become weaker, so the hydrogen halides become less stable to heat
b i Hydrogen sulphide, H_2S, and some sulphur dioxide, SO_2
ii The iodide ions have more electron shells, so there

is a greater distance from the nucleus to the outer electrons, so the outer electrons are more weakly held, so the iodide ions are easily oxidized by sulphuric acid (Using $E^{\ominus}$ values would be better)

c Bromide ion
d 0.015mol I_2

10a Iodide ions
b Standard means that the solution is made accurately to a particular concentration; this is only possible because the thiosulphate salt can be made pure
c Starch, blue-black to colourless.
d $2S_2O_3^{2-}(aq) + I_2(aq) \rightarrow S_4O_6^{2-}(aq) + 2I^-(aq)$
e 0.84 mol dm^{-3}

Chapter 19

Exam questions

1 a i Al_2O_3 has a very high melting point, so energy costs would be high
ii $Al^{3+} + 3e^- \rightarrow Al(l)$
iii $2O^{2-} \rightarrow O_2 + 4e^-$; hot oxygen reacts with graphite anodes to form gaseous CO_2 and CO
b Electrode reactions, heating the melt
c i Conservation of resources, less environmental damage
ii Labour

2 a i Reaction is vigorous, exothermic, with steamy fumes
$Al_2Cl_6 + 6H_2O \rightarrow 2Al(OH)_3 + 6HCl$ (simplified)
ii $[Al(H_2O)_6]^{3+}(aq) + H_2O \rightleftharpoons [Al(H_2O)_5(OH)]^{2+}(aq) + H_3O^+(aq)$
iii Effervescence, gas turns limewater cloudy, white precipitate of $Al(OH)_3$
$2H^+(aq) + CO_3^{2-}(aq) \rightarrow H_2O(l) + CO_2(g)$
b i White precipitate forms, which does not react with excess; $Al(OH)_3(H_2O)_3$, *or* $Al(OH)_3$
ii White precipitate would react with excess; forms $[Al(OH)_4]^-$

3 a i CO; **ii** SiO_2; **iii** $SnCl_2$; **iv** CCl_4
b i $x = 4$; **ii** Tetrahedral
c i Diamond: 3D lattice, tetrahedral arrangement of atoms; graphite: 2D lattice/layers of atoms, trigonal planar arrangement of atoms
ii Mass spectrometry

4 a CCl_4, CO; $SiCl_4$, SiO_2; $PbCl_4$, PbO
b $SiCl_4$ hydrolyses to produce HCl fumes;
$SiCl_4 + 2H_2O \rightarrow SiO_2 + 4HCl$
there is no hydrolysis with CCl_4 because there are no vacant orbitals in the outer shell of the C atom (see artwork in spread 19.6)
c $PbO_2 + 4H^+ + 2e^- \rightarrow Pb^{2+} + 2H_2O$; $2Cl^- \rightarrow Cl_2 + 2e^-$
d CO_2 has a simple molecular structure – only dispersion forces have to be broken; SiO_2 has a giant covalent structure, and therefore covalent bonds must be broken

5 a i See spread 5.5
ii See spread 5.5; 3 bond pairs and 1 lone pair repel each other to the corners of the tetrahedron
iii A bond formed by collinear overlap of atomic orbitals
b i In both cases there are dispersion and dipole–dipole forces between molecules; NH_3 forms H bonds as well
ii H bonds can form between H_2O and NH_3 molecules
c 47.6%

6 a CCl_4: liquid/poor/immiscible, no reaction/covalent;
$SiCl_4$: liquid/poor/hydrolyses, produces white fumes and white precipitate/covalent;
$PbCl_2$: solid/good/insoluble solid/ionic
b i $CaO(s) + SiO_2(s) \rightarrow CaSiO_3(s)$ in blast furnace (several answers possible)
ii $PbO(s) + 2HNO_3(aq) \rightarrow Pb(NO_3)_2(aq) + H_2O(l)$ (other reactions of tin and lead oxides are possible)
c i Tin
ii $2MnO_4^-(aq) + 5Sn^{2+}(aq) + 16H^+(aq) \rightarrow 2Mn^{2+}(aq) + 5Sn^{4+}(aq) + 8H_2O(l)$

7 a i It can be moved in pipes; it has to be kept hot
ii $S(l) + O_2(g) \rightarrow SO_2(g)$; $2SO_2(g) + O_2(g) \rightleftharpoons 2SO_3(g)$
b i Exothermic
ii Vanadium(v) oxide
iii To cool the gases (increases yield) and to raise steam ('recycle' energy)
iv The cooler the bed, the greater the yield of SO_3 (Le Chatelier's principle)
v Conc. H_2SO_4; $H_2SO_4 + SO_3 \rightarrow H_2S_2O_7$
c Possibility of loss of harmful gases, e.g. SO_2/ugly/leak of acid could occur

8 a Copper is a good conductor because it has metallic bonding where the electrons are free to move; SiO_2 is a poor conductor because the electrons are held in strong covalent bonds
b SiO_2 has a giant covalent structure, so all the strong Si—O bonds need to break to make the oxide liquid; copper is easier to turn into a liquid because the metallic bonds are not so strong
c SiO_2 is an electrical insulator, has a high melting point, and is hard (scratch resistant)
d Advantage: copper is easy to melt and pour into axe-head moulds; disadvantage: copper is not hard so the axe edge will blunt easily

Chapter 20

Exam questions

1 a $Fe_2O_3 + 3C \rightarrow 2Fe + 3CO$
$Fe_2O_3 + 2Al \rightarrow 2Fe + Al_2O_3$
$Fe_2O_3 + 3H_2 \rightarrow 2Fe + 3H_2O$
b i Carbon; it is cheapest/most plentiful
ii Hydrogen; byproduct is a gas and there is no contamination by a solid reductant
c SO_2, acid rain; CO_2, global warming
d O_2 blown through molten iron; impurities oxidized to gases or acidic solids that can be removed by calcium oxide

2 a High melting point/dense/hard
b i $[Ar]3d^64s^2$; $[Ar]3d^6$
ii Fe^{3+} has a stable half-filled subshell ($3d^5$)
c i Pale green; $[Fe(H_2O)_6]^{2+}$; octahedral
ii It turns brown (depending on pH); Fe^{2+} oxidized to Fe^{3+} by aerial O_2
d i Orange-brown solution is produced; I^- is oxidized to I_2 by Fe^{3+}
ii $2Fe^{3+} + 2I^- \rightarrow 2Fe^{2+} + I_2$
e Associated ligands split the 3d subshell into 2 energy levels; promotion of electrons between levels absorbs visible light of a particular wavelength

3 a i A molecule or ion with a lone pair that can make a coordinate covalent bond with a metal ion
ii 1 molecule or ion that has 2 atoms which can donate an electron pair to the same metal ion
b i Electron pair acceptor
ii $[Co(H_2O)_6]^{2+}$, octahedral; $[CoCl_4]^{2-}$, tetrahedral
iii Cl^- ions repel each other
iv Different ligands/different coordination number
c $[Co(en)_3]^{2+}$

4 a V: $[Ar]3d^34s^2$; V^{2+}: $[Ar]3d^3$
b i Because of the partially filled d subshell (see **2 e**)
ii Coordinate covalent
c i $NH_4^+ + OH^- \rightarrow NH_3 + H_2O$
ii Copper(II) hydroxide
iii $[Cu(NH_3)_4(H_2O)_2]^{2+}$
iv +4; +3; +2
d $2VO_2 + \frac{1}{2}O_2 \rightarrow V_2O_5$
e i Dioxovanadium(v)
ii Tetraamminedichlorochromium(III)
f Octahedral

5 a Heterogeneous: in different phase to reactants; catalyst: speeds up a reaction, recoverable at end

b Variable oxidation states
c See answer to chapter 19, exam question 7
d MnO_4^- is reduced to pale pink Mn^{2+}; Mn^{2+} catalyses the reaction so it gets progressively faster as $[Mn^{2+}]$ increases (= autocatalysis)

6 a Change in oxidation state; ligand exchange; coordination number change
b i Ligand exchange
ii Ligand exchange and coordination number change
c i A catalyst in the same phase as the reactants
ii Variable oxidation state
iii See **4 d**

7 a Coloured ions; variable oxidation state; complex formation; catalytic properties
b Example: V_2O_5, Contact process; variable oxidation state
c i $[Ar]3d^54s^1$; ii $[Ar]3d^3$
d i $+6 \rightarrow +3$; ii 2.00×10^{-3}; iii 16.7 cm^3
e Green precipitate forms which reacts with excess to give a green solution

8 a $CuCl_4^{2-}$
b i $Fe(s) + Cu^{2+}(aq) \rightarrow Fe^{2+}(aq) + Cu(s)$
ii The solid is bronze coloured; the solution becomes green rather than blue
iii $[Fe(H_2O)_6]^{2+}$
c **P**: anionic, tetrahedral; **Q**: cationic, octahedral
d i Green; ii $Fe(OH)_3$;
iii Oxygen in the air; iv Oxidation

Chapter 21

Practice questions

Spread 21.3

3 3,3-dimethylpentane

Chapter 22

Exam questions

1 a Kerosine
b Different boiling points
c Naphtha: cracking; diesel: industrial furnaces/central heating
d i $2C_4H_{10} + 13O_2 \rightarrow 8CO_2 + 10H_2O$
ii Not volatile, viscous liquid
e Decomposes/combusts

2 a i Partial vapour pressure = mole fraction × vapour pressure of pure component
ii Ideal behaviour; draw liquid and vapour lines and use tie lines (horizontal lines joining the liquid and vapour lines) to explain purification
iii Oil has many more than two components; fractions are themselves mixtures; fractions are taken from many levels, not just the top and bottom
b i It is easier/safer to transport liquids/more can be contained in the same space
ii Carcinogenicity
iii The mixture explodes too soon/before the spark/it causes engine damage/fewer mpg
iv Lead compounds cause damage to the central nervous system

3 a Homolytic/(free) radical substitution
b It breaks the Cl—Cl bond to form radicals
c $CH_4 + Cl_2 \rightarrow CH_3Cl + HCl$
d CH_2Cl_2
$CH_3Cl + Cl\cdot \rightarrow \cdot CH_2Cl + HCl$
$\cdot CH_2Cl + Cl_2 \rightarrow CH_2Cl_2 + Cl\cdot$
e $\cdot CH_3$ radicals collide in a termination step
$2\cdot CH_3 \rightarrow CH_3CH_3$
f CCl_4

4 a A: 2-bromobutane $CH_3CH_2CHBrCH_3$
B: butane $CH_3CH_2CH_2CH_3$
C: butan-2-ol $CH_3CH_2CH(OH)CH_3$
D: butane-2,3-diol $CH_3CH(OH)CH(OH)CH_3$
b i Electrophile
ii See spread 22.8
c i $CH_2{=}CH{-}CH{=}CH_2$
ii $C_4H_8 \rightarrow C_4H_6 + H_2$

5 a i $CH_2{=}CH_2$
ii $-CH_2-CH_2-CH_2-CH_2-CH_2-CH_2-$
(C—H bonds not shown)
b High pressure shifts the position of equilibrium to the right, increasing yield/increases rate; low temperature shifts the position of equilibrium to the right, increasing yield, as the reaction is exothermic; compromise so that the rate is not too low
c i $R-CH_2CH_2\cdot + CH_2{=}CH_2$
$\rightarrow R-CH_2-CH_2-CH_2-CH_2\cdot$
ii Any two radicals forming a bond, e.g.
$2R-CH_2-CH_2\cdot \rightarrow R-CH_2-CH_2-CH_2-CH_2-R$
d i The chains are unbranched
ii Higher melting point/stronger
e $-C(CH_3)_2-CH_2-C(CH_3)_2-CH_2-C(CH_3)_2-CH_2-$
f i $H_3C-C(CH_3)(Br)-CH_2Br$ (Not all bonds are shown)
ii $H_3C-C(CH_3)(OH)-CH_2OH$

6 a i The reaction is exothermic
ii Sodium hydroxide/sodium carbonate/limewater
iii The same rate can be achieved at a lower temperature; lower temperatures increase the yield by shifting the position of equilibrium to the right
b i $CH_2ClCH_2Cl \rightarrow CH_2ClCH_2\cdot + Cl\cdot$
ii $CH_2ClCH_2Cl \rightarrow CH_2{=}CHCl + HCl$
iii Fractional distillation
iv Recycled
c i CH_2CHCl
ii Poly(chloroethene) = PVC, raincoats, pipes, etc.
iii HCl is a strongly acidic gas; hydrocarbon–oxygen mixtures are explosive

Chapter 23

Exam questions

1 a Test with ammonia gas; when no more white smoke is produced, the reaction is over
b i Can accept a pair of electrons
ii Electrophilic substitution (see spread 23.5)
c i 1,2-dimethylbenzene
(structures of 1,2-, 1,3- and 1,4-dimethylbenzene: $C_6H_4(CH_3)_2$)
ii Fractional distillation
d i C_3H_4
ii C_9H_{12}; C_6H_5R with $R = C_3H_7$
e $C_6H_5COCH_3$

2 a i $C_6H_6 + CH_3COCl \rightarrow C_6H_5COCH_3 + HCl$ (SF needed)
ii Electrophilic substitution (see spread 23.5)
b i A large surface area to give a fast reaction
ii H_2O hydrolyses $AlCl_3$ and CH_3COCl
iii The reaction is exothermic/could get out of control
iv Harmful: HCl; carcinogenic: benzene fumes
v It removes acids
c i $(2.2\,g) \times (120/78.5) = 3.4\,g$
ii 59%
iii Side reactions/incomplete reactions/losses during purification

3 a i It produces a higher concentration of NO_2^+ than in

just conc. HNO_3
- **ii** The reaction is exothermic and the temperature must be kept low
- **iii** To prevent further nitration
- **iv** For a good rate
- **v** Because acids and benzene are immiscible
- **vi** It dilutes/removes excess acid
- **vii** It removes any acid dissolved in nitrobenzene
- **viii** It separates nitrobenzene from other volatile components

b Sn/conc. HCl/heat under reflux

c i

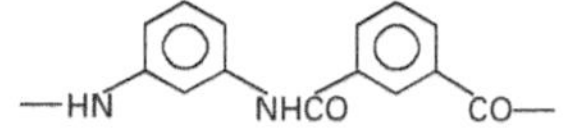

- **ii** Condensation or polyamide
- **iii** The acid chlorides are more reactive and so react faster and give a better yield
- **iv** PCl_5 or SCl_2O at room temperature

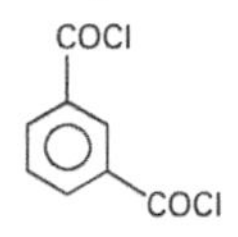

4 a i Cl_2/UV; $C_6H_5CHCl_2$ and $C_6H_5CCl_3$
- **ii** $Cl_2/AlCl_3$/dry +

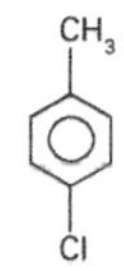

b See spread 22.4 for radical substitution

c Electrophilic substitution

d i

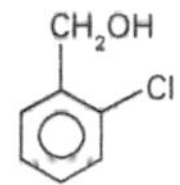

- **ii** Cl in ring is attached and cannot be substituted owing to overlap of the Cl p orbital with the delocalized π cloud

CHAPTER 24

Exam questions

1 a Replacement of an atom or group by another in a reaction initiated by an electron-rich species donating a pair of electrons

b The electronegative Br atom makes the C atom electron deficient

c $CH_3CH_2CH_2Br + NH_3 \rightarrow CH_3CH_2CH_2NH_2 + HBr$
nucleophilic substitution (S_N2)

d CH_3CH_2Br/KCN dissolved in ethanol
heat under reflux

2 a $CH_3CHBrCH(CH_3)_2$

b $CH_3CHBrCH(CH_3)_2 + NaOH$
$\rightarrow CH_3CH(OH)CH(CH_3)_2 + NaBr$
substitution
S_N1 or S_N2 (see spread 24.3)

c Elimination; base;
$CH_2{=}CHCH(CH_3)_2$ and $CH_3CH{=}C(CH_3)_2$

3 a Nucleophilic substitution, $(CH_3)_2CHCH(OH)CH_3$; the mechanism is based on nucleophilic substitution; S_N1 or S_N2 would be acceptable

b i Elimination; the KOH is acting as a base (proton acceptor)
- **ii** See spread 24.4

4 a i

```
                     CH3
                     |
CH3CH2CH2CH2Br ; CH3CHCH2Br
```

- **ii** 2-bromomethylpropane

b i Nucleophilic substitution
- **ii** $CH_3CH_2CH(NH_2)CH_3$
- **iii** Isomer 2; it is a tertiary bromide

c i NaOH dissolved in ethanol/heat
- **ii** $CH_3CH_2CH{=}CH_2$ and $CH_3CH{=}CHCH_3$

d i Electrophilic addition
- **ii** See spread 22.8

5 a See spread 24.3 for nucleophilic substitution with the ammonia as the nucleophile (S_N2). Repeated nucleophilic substitution occurs, resulting in a succession of products: $CH_3CH_2CH_2NH_2$, $(CH_3CH_2CH_2)_2NH$, $(CH_3CH_2CH_2)_3N$, and $(CH_3CH_2CH_2)_4N^+$

b i IR would only differ in the fingerprint region
- **ii** $CH_3CH_2CH_2Br$ would show 3 peaks; $(CH_3)_2CHBr$ would show 2 peaks

6 a i Electrophilic addition (see spread 22.8)
- **ii** CH_2BrCH_2OH or CH_2BrCH_2Cl
- **iii** The first step involves an electrophile; Cl^- is a nucleophile and can react only after electrophilic addition has occurred

b i Heat under reflux with NaOH(aq)
- **ii** Faster because the C—I bond is weaker

7 a i Decolorized
- **ii** Mole ratio of A:Br_2 = 1:3; this is consistent with 3 C=C bonds

b i NaOH dissolved in ethanol/heat under reflux
- **ii** H atom on any adjacent carbon could be eliminated, producing different products
- **iii** Optical – there are two chiral C atoms in B

8 a i Substitution; nucleophile
- **ii** $(CH_3)_2CHCH(OH)CH_2CH_3$
(see spread 24.3 for mechanism)

b i $(CH_3)_2CHCOCH_2CH_3$
- **ii** 57: $CH_3CH_2CO^+$
71: $(CH_3)_2CHCO^+$

c i Elimination; base
- **ii** $(CH_3)_2C{=}CHCH_2CH_3$
see spread 24.3 for mechanism
- **iii** $(CH_3)_2CHCH{=}CHCH_3$
geometric

9 a i $CH_3CH_2CHBrCH_2CH_3$
- **ii** Elimination (see spread 24.4)
- **iii** Geometric

b i $CH_3CH_2{}^+CHCH_2CH_3$; $CH_3{}^+CHCH_2CH_2CH_3$
- **ii** 3-bromopentane; 2-bromopentane
- **iii** They are both formed from similarly stable secondary carbocations

CHAPTER 25

Exam questions

1 a i $CH_3CH_2CH_2CH_2OH + [O] \rightarrow CH_3CH_2CH_2CHO$
orange → green
- **ii** Butan-1-ol produces butanal; if Fehling's solution, blue solution → brick-red precipitate
butan-2-ol produces butanone – no change with Fehling's solution

b $CH_3CH_2CH_2COOH$

c Isomer 1 = methylpropan-1-ol

```
CH3
|
CH3CHCH2OH
```

Isomer 2 = methylpropan-2-ol

```
    CH3
    |
CH3 C CH3
    |
    OH
```

2 a C—C

b i 700–1500
- **ii** Complex pattern is unique to that compound

c i Too broad/not specific for ethanol
- **ii** 1700
- **iii** Several possibilities: PCl_5, $KMnO_4/H_2SO_4$(aq), $K_2Cr_2O_7/H_2SO_4$(aq); DNP, (see text for details)

3 a Higher temperatures would increase loss of $CH_2{=}CH_2$ as CO_2 and H_2O or reaction is exothermic; therefore lower temperatures favour better yields

b i See 1

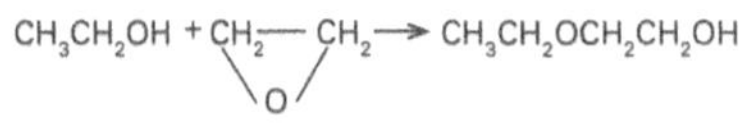

$CH_3CH_2OH + CH_2(-O-)CH_2 \rightarrow CH_3CH_2OCH_2CH_2OH$

ii $CH_3CH_2(OCH_2CH_2)_nOH$

c $H_2C(-O-)CH_2 + NH_3 \rightarrow H_2NCH_2CH_2OH$

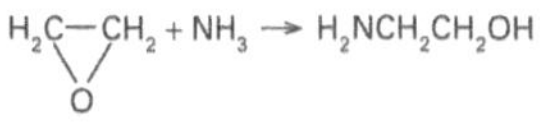

4 a Reaction I: KBr + conc. H_2SO_4 + heat *or* PBr_3 (P + Br_2) room temperature; reaction II: named oxidant e.g. $Cr_2O_7^{2-}$(aq) + H_2SO_4(aq) + heat; reaction III: conc. H_2SO_4, 180 °C *or* pass vapour over hot Al_2O_3

b Reaction I: $(CH_3)_3CBr$; reaction II: no reaction; reaction III: $(CH_3)_2C{=}CH_2$

5 a $(CH_3)_2CHCH_2OH$

b i Elimination

ii $CH_3CH_2CH{=}CH_2$; $CH_3CH{=}CHCH_3$ but-1-ene; (*cis-* and *trans-*) but-2-ene

c i Butanal

ii Butanoic acid

iii $CH_3CH_2COCH_3$

iv Oxidation

d i $CH_3COOCH(CH_3)_3$

ii Ethanoic acid

iii Conc. H_2SO_4

e i The rate of formation of a white precipitate increases in the order: butan-1-ol → butan-2-ol → 2-methylpropan-2-ol

ii 2-chlorobutane/$CH_3CHClCH_2CH_3$

6 a i 4 peaks: 3:2:2:3

ii $CH_3CH_2{-}O{-}CH_2CH_3$

b i Nucleophilic substitution (S_N2); see spread 25.5

ii CH_3O^-

c Tertiary $(CH_3)_3COH$

d $C_4H_{10}O + 6O_2 \rightarrow 4CO_2 + 5H_2O$

e Broadened peak between 3230 and 3550 only in alcohols

Chapter 26

Practice questions

Spread 26.4

2 **P**: propanal, CH_3CH_2CHO; **Q**: propanone, CH_3COCH_3

3 a **A**: $CH_3CH_2CHBrCH_3$; **B**: $CH_3CH_2CH(OH)CH_3$ **C**: $CH_3CH_2COCH_3$

b $CH_3CH_2CH_2CHO$

4 a ketone; **b** alcohol; **c** aldehyde

Spread 26.5

1 **N**: $CH_3CH_2CHC(CH_3)_2$; **M**: CH_3CH_2CHO; **P**: CH_3COCH_3

3 **A**: 2-iodobutane, $CH_3CH_2CHICH_3$;
B: butan-2-ol, $CH_3CH_2CH(OH)CH_3$;
C: butanone, $CH_3CH_2COCH_3$;
D: but-1-ene, $CH_3CH_2CHCH_2$; (assuming no but-2-ene forms);
E: methanal, HCHO;
F: propanal, CH_3CH_2CHO

Exam questions

1 a i $CH_3COCH_3 + HCN \rightarrow CH_3COH(CN)CH_3$ aqueous/pH 5 2-hydroxy(-2-)methylpropanenitrile

ii $CH_3COCH_3 + 2[H] \rightarrow CH_3CH(OH)CH_3$ aqueous propan-2-ol

b i See spread 26.1

ii Nucleophilic addition

iii Electron-deficient C atom C is bonded to an electronegative O atom

iv Ethene is susceptible to electrophilic not nucleophilic attack due to 'electron-rich' ⊃C=C⊂

2 a Any named *oxidant* + conditions + observations, e.g. Fehling's or Tollens'; only ethanol reacts

b i S = $CH_3CH(OH)CN$

T = $CH_3(H)C{=}NNH{-}C_6H_3(NO_2)_2$ (2,4-dinitrophenyl)

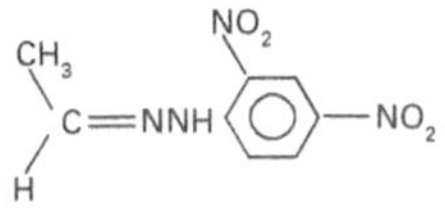

U = CH_3CH_2OH

ii Ethanol-2,4-dinitrophenylhydrazone yellow–orange precipitate

c Nucleophilic addition; (see spread 26.1)

3 a i $CH_3CH_2CH_2CH_2OH$

ii $CH_3CH_2CH{=}CH_2$ **iii** $CH_3CH_2CH(OH)CN$

iv $CH_3CH_2CH_2CN$

v No reaction

vi No reaction

vii CH_3CH_2COOH

viii CH_3COCH_3

b i Nucleophilic substitution

ii Elimination

iii Nucleophilic addition

iv Nucleophilic substitution

c i Nucleophilic addition (see spread 26.1)

ii See **4 b ii**

4 a i $CH_3CHBrCHO$; **ii** $CH_2CHCH(OH)CN$

iii $CH_2CH(H)C{=}NNH{-}C_6H_3(NO_2)_2$ (2,4-dinitrophenyl)

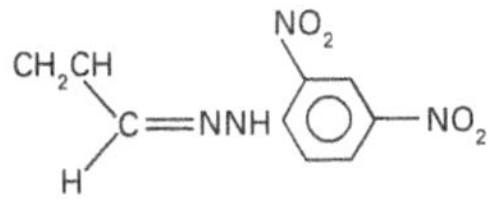

b i Nucleophilic addition (see spread 26.1)

ii High pH [H^+] is low, therefore 2nd step becomes very slow; low pH CN^- is protonated, therefore [CN^-] is low and 1st step becomes very slow

c $LiAlH_4$ (contains H^-) is a nucleophilic reagent; only >C=O is susceptible to nucleophilic attack as carbon–carbon bond has no polarity

d I_2/NaOH = iodoform reagent, therefore convert it to $CH_3CH(OH)CHO$ by, e.g., using H_2O/H^+(aq)

5 a X = $CH_3CH_2CH_2OH$
Y = $CH_3CH_2CH(OH)CN$
Z = CH_3CH_2COOH

b See spread 26.1

c i Y, X, Z, W; **ii** 2900–3000

6 a i Propanoic acid CH_3CH_2COOH propan-l-ol $CH_3CH_2CH_2OH$ HCN $CH_3CH_2CH(OH)CN$

ii Blue solution → brick-red precipitate propanal reduces Cu^{2+} to Cu_2O

iii Any named oxidant e.g. Tollens'

b i Nucleophilic substitution (S_N2); OH^-

ii Electrophilic addition; H^+

7 a i HCN

ii Nucleophilic addition (see spread 26.1)

b i Methanol

ii Esterification; conc. H_2SO_4

c $[-CH_2-C(CH_3)(COOCH_3)-]$ (repeat unit: H, H on first C; CH_3, $COOCH_3$ on second C)

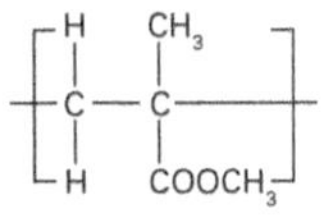

d $CH_2{=}C(CH_3)COOH + NaOH \rightarrow CH_2{=}C(CH_3)CO_2^-Na^+ + H_2O$

Chapter 27

Exam questions

1 a i Ethyl ethanoate $CH_3COOCH_2CH_3$ (linear abbreviation)

ii Esterification

b i H_2SO_4(aq) / heat

ii $CH_3COOCH_2CH_3 + H_2O \rightleftharpoons CH_3COOH + CH_3CH_2OH$

c i NaOH(aq)/room temperature

ii Ethanoic anhydride $(CH_3CO)_2O$ (linear abbreviation)

iii Add water and warm $(CH_3CO)_2O + H_2O \rightarrow 2CH_3COOH$

d i Ethanamide CH_3CONH_2 (linear abbreviation)

ii Add NH_3(aq), heat to evaporate water and heat solid (dehydrates) iii $CH_3CONH_2 + H_2O + HCl \rightarrow CH_3COOH + NH_4^+ + Cl^-$

2 a i $CH_3CONH_2 + NaOH \rightarrow CH_3CO_2^-Na^+ + NH_3$
ii $CH_3CH_2CN + 2H_2O + HCl \rightarrow CH_3CH_2COOH + NH_4^+ + Cl^-$
iii $CH_3COCl + H_2O \rightarrow CH_3COOH + HCl$
iv $(CH_3CO)_2O + H_2O \rightarrow 2CH_3COOH$
b iii Vigorous at room temperature, whereas **iv** slow at room temperature
c i (CH_3COOH or) $CH_3COCl + CH_3OH$
ii $CH_3COCl + CH_3CH_2NH_2$

3 a i $K_a = [RCO_2^-][H_3O^+]/[RCOOH]$
ii pH = 1.9
b i A buffer solution resists a change in pH on the addition of small amounts of acid or alkali
ii The citric acid reacts with alkali;
$RCOOH(aq) + OH^-(aq) \rightarrow RCOO^-(aq) + H_2O(l)$;
the citrate ion reacts with acid;
$RCOO^-(aq) + H^+(aq) \rightarrow RCOOH(aq)$
iii pH = 3.4
c i Add PCl_5 and there will be vigorous production of smoky fumes (of HCl)
ii Aqueous sodium hydroxide
iii Sodium hydrogencarbonate would effervesce

4 a $HOOCC_6H_4COOH$ and $HOCH_2CH_2OH$ (linear abb.)
b Condensation
c NaOH hydrolyses ester

5 a C_2H_6O
b i See spread 7.5; **ii** CH_3CH_2OH
c 45: $CH_3CH_2O^+$; 31: CH_2OH^+; 29: $CH_3CH_2^+$
d i Ethyl ethanoate; $CH_3COOCH_2CH_3$; ester
ii Add slowly to dry ethanol at room temperature

6 a i A: N—H; B: C—H (alkyl); C: C=O
ii $H_2N(CH_2)_6NH_2$ and $ClOC(CH_2)_4COCl$ (linear abbreviation)
b i NH_3 is produced, which is basic
ii $(NH_2)_2CO + H_2O \rightarrow 2NH_3 + CO_2$
iii Enzymes are specific and cannot hydrolyse ethanamide at an appreciable rate
iv H bonds are disrupted by ethanamide

7 a $H_2N(CH_2)_6NH_2$
$ClOC(CH_2)_4COCl$
b Hydrogen chloride
c Hydrolyses $—OC(CH_2)_4CONH(CH_2)_6NH— + 2H_2O \rightarrow H_2N(CH_2)_6NH_2 + HOOC(CH_2)_4COOH$

Chapter 28

Exam questions

1 a Can act as a proton acceptor owing to lone pair on N atom
b Lone pair is delocalized by overlap with benzene π system
c i Nucleophilic substitution (see spreads 24.3 and 28.2)
ii $(CH_3CH_2)_2NH$; $(CH_3CH_2)_3N$; $(CH_3CH_2)_4N^+$
d $CH_3CN + 4[H] \rightarrow CH_3CH_2NH_2$
only the primary amine is formed

2 a i $C_6H_5CH_2Br + NH_3 \rightarrow C_6H_5CH_2NH_2 + HBr$
nucleophilic substitution
(possibly $NH_3 + HBr \rightarrow NH_4Br$)
secondary amines, tertiary amines, and quaternary salt are also produced
ii Reduction
$C_6H_5CN + 4[H] \rightarrow C_6H_5CH_2NH_2$
$LiAlH_4$ (or H_2/Ni)
other amines cannot be formed
b Phenylamine
lone pair on N atom is delocalized with the benzene π system

3 a i $C_6H_5CHClCH(CH_3)NH_2$
ii $C_6H_5CH(OH)CH(CH_3)NH_3^+Cl^-$
iii $C_6H_5(OCOCH_3)CH(CH_3)NHCOCH_3$
iv C_6H_5COOH
v $C_6H_5CH=C(CH_3)NH_2$

b i Perform in fume cupboard/wear gloves
ii See spread 31.1
iii Determine melting point

4 a Joining many small molecules (monomers) to form a long chain polymer
b i $H_2NCH(CH_3)CONHCH(CH_3)COOH$ (linear abbreviation)
ii Peptide (amide)
iii (Polypeptides) proteins
c i $n\ CH_2=CH_2 \rightarrow [CH_2—CH_2]_n$
ii Low density has more branched chains which cannot pack so closely
iii Melting point
iv $R—CH_2CH_2•$
v Ziegler–Natta
d i Not broken down in the environment (by organisms) / does not rot
ii Will not 'clutter' the environment
iii More prone to hydrolysis

5 a B = $CH_3CH_2CONH_2$
C = $CH_3CH_2C \equiv N$
functional group = $–NH_2$
b i Reduce with e.g. $LiAlH_4$
ii NaOH(aq), Br_2(l)
c i $CH_3CH_2CN + 2H_2O + HCl \rightarrow CH_3CH_2COOH + NH_4^+ + Cl^-$
ii Hydrolysis
d i Lone pair on N atom
ii $CH_3CH_2CH_2NH_2 + HCl \rightarrow CH_3CH_2CH_2NH_3^+ + Cl^-$
e i $C_6H_5NH_3^+ + HNO_2 \rightarrow C_6H_5N_2^+ + 2H_2O$
ii See spread 28.4
iii Used as dyes

6 a $CH_3CH_2CH_2CH_2Br$, $CH_3CH_2CHBrCH_3$, $(CH_3)_3CBr$, $(CH_3)_2CHCH_2Br$
b i A nucleophile is an electron pair donor
ii Hydroxide ion
iii Rate = $k[C_4H_9Br]$
iv The mechanism is S_N1, so the halogenoalkane loses the Br^- in the first step, the rate-determining step, and then the hydroxide ion reacts with the carbocation produced in the second, fast step; tertiary halogenoalkanes tend to undergo S_N1, so $(CH_3)_3CBr$ should appear in the mechanism
c i $CH_3CH_2C^*HBrCH_3$ is chiral, as indicated by the '*'; this C has four different groups attached and so is optically active
ii The intermediate forms a carbocation that is trigonal planar; the hydroxide ion may attack from either side of the carbocation and so produces two optically active isomers in equal amounts; the isomers rotate the plane of polarized light equally in opposite directions, so the mixture is not optically active

7 a Conc. H_2SO_4 + conc. HNO_3
$2H_2SO_4 + HNO_3 \rightarrow NO_2^+ + H_3O^+ + 2HSO_4^-$
electrophilic substitution (see spread 23.3)
b Reduction; Sn/conc. HCl
c $C_6H_5NH_2 + CH_3COCl \rightarrow C_6H_5NHCOCH_3 + HCl$
condensation (nucleophilic addition then elimination); see spread 28.2

8 a i Arrangement of C atoms in backbone differs
$CH_3CH_2CH_2CH_3$ and $(CH_3)_3CH$ (linear abbreviation)
ii When a pair of groups is on the same or opposite sides of a double bond

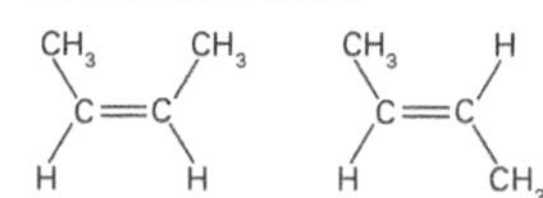

iii When a pair of molecules are non-superimposable mirror images

COOH COOH
C H H C
CH3 OH HO CH3

b i N—C*—C—N—C*—C
ii See spread 28.5
iii Use a polarimeter; (see spread 28.6)

CHAPTER 29

Exam questions

1 a i Metallic, delocalized electron sea conducts
ii Covalent, no delocalized electrons
iii Ionic, therefore cannot conduct as a solid
iv MgO has a much higher melting point than PVC (ionic lattice structure versus molecular)
b $CH_3CH_2Br + Mg \rightarrow CH_3CH_2MgBr$
(dry ether solvent, I_2 catalyst, heat under reflux)
c i $CH_3CH_2C(CH_3)_2OH$ 2-methylbutan-2-ol
ii $CH_3CH_2CH(OH)CH_2CH_2CH_3$ hexan-3-ol
d i test with any common oxidant e.g. $KMnO_4$ (see spread 25.4 for details)
ii For **c i** the product is not chiral
for **c ii**, although products are chiral, a racemic mixture is produced (carbonyls are planar and can be attacked by the Grignard from either side)

2 a i Mirror image is not superimposable
iii By using a polarimeter to detect/measure optical activity
b i $CH_3CH_2CH_2CH_2OH + 2[O] \rightarrow CH_3CH_2CH_2COOH + H_2O$; heat under reflux with $KMnO_4(aq)/H_2SO_4(aq)$ or similar
ii $CH_3CH_2CH_2COOH + PCl_5 \rightarrow CH_3CH_2CH_2COCl + HCl + POCl_3$; room temperature
iii $H_2C—CH_2$ (epoxide, O bridging) $+ NH_3 \rightarrow H_2NCH_2CH_2OH$
room temperature
c i React $CH_3CH_2CH_2MgBr$ with HCHO then HCl(aq)
ii See **1 b**
d Enzymes (the target) are stereospecific.
e See spread 31.1 for details of recrystallization assess purity by determining melting point
f Add NaOH(aq)

3 a See spread 29.5
b i The liquid mixture of monomers thickens as the monomer molecules react in the presence of an initiator to produce long-chain molecules; the reaction is exothermic, so the mixture becomes hotter
ii The monomer, poly(phenylethene), is immiscible with water, so when the two liquids are mixed, a suspension is formed; the heat generated is absorbed by the water, so the temperature does not get too high

4 a If the molecule chains are too short, the softening point of the product would be too low, and the product would be too soft and weak; if the chains are too long, then the softening point may be too high, making it difficult to melt the polymer and mould it; it also may become too brittle
b i A ring structure of 11 carbon atoms and one nitrogen atom, including an amide link
ii Nylon 11 would have many more amide links than nylon 66, so there would be more hydrogen bonding between the nylon 11 chains than the nylon 66 chains, so the crystalline melting point of nylon 11 would be much higher than that for nylon 66
c The nylon amide links are hydrolysed by acid

5 a Covalent bonds; dispersion forces between chains; polyamides (nylon) hydrogen bond
b i The covalent bonds make the polymers insulators and quite resistant to chemical change; the hydrogen bonds make the polymers have high melting points
ii The intermolecular forces between the polymer molecules influence the processing of thermoplastics because if the intermolecular forces are weak, then the plastic would soften easily when heated; if the polymer molecules hydrogen bond, then the melting point may be so high that the polymer would not soften easily

6 a i See spread 22.10
ii Chain length varies/behaves as a mixture
b i See spread 27.7
ii Condensation: small molecule like H_2O or HCl is formed as a byproduct (= addition followed by elimination) Addition: no byproduct
iii Acid chloride is more reactive
iv H bonds between —CONH— links on neighbouring chains
c i Copy the diagram but with each bond having the *trans* geometry (groups across the bond) rather than *cis*
ii The conjugated double bonds allow delocalization and hence conduction

7 a i Covalent bonding, in which atoms share electrons
ii Polyamide chains are held together by hydrogen bonding; polyesters and polyhalogenoalkenes (i.e. PVC) are held by dipole–dipole attraction
iii Melting and mechanical strain, which pulls the polymer chains apart
b i Thermosetting plastics are polymers that solidify as they are formed and will not melt when heated; thermoplastics are polymers that will soften after being made
ii Advantage: thermoplastics may be extruded and moulded; disadvantage: they have low softening points and so cannot be used at high temperatures and are often weak
iii Whether the thermosetting plastic can be made in the mould

CHAPTER 30

Exam questions

1 a i Alanine is chiral
ii $CH_3CH(NH_2)CONHCH_2COOH$ or other way around
b Edman degradation (1) modify N terminus; (2) mild acid hydrolysis releases modified N terminus; (3) released amino acid identified by chromatography; (4) repeat process
c Less toxic, not flammable
d i Volatile, low toxicity, inertness
ii Broken by UV in stratosphere to chlorine radicals which lead to destruction of ozone etc.

2 a A is glucose (see spread 30.2 for graphical formula)
b i $B + H_2O \rightarrow 2A$; this is hydrolysis
ii Either an enzyme (a yeast, at 35 °C) or aqueous hydrochloric acid could be used and heated under reflux
c C is starch, used as a food store; D is cellulose; see spread 30.4 for the consequences of the α and β glycosidic links

3 a and **b** See spreads 30.8–10
c i —A—C—C—T—G—A—T—T—G—
ii —U—G—G—A—C—U—A—A—C—
d 3.3%

4 a Active transport of substances across the cell membrane; the bilayers repel most substances
b The side chains of the peptides could be made non-polar or polar, depending on the membrane
c i Increase the number of C atoms between amino and carboxyl groups
ii Change the side chains, i.e. vary the proportion with hydrophilic and hydrophobic groups

5 a See spreads 30.8–10
b The enzyme fits certain combinations of base pairs; when it contacts the correct sequence it holds on to the chain and breaks it
c Phosphate groups are negatively charged

6 a See spreads 30.6 and 30.7
b Ethanol competes for active sites of oxidative enzymes which oxidize glycol, giving the body time to eliminate the glycol before it is oxidized

7 a Potatoes contain starch for which humans have the digestive enzyme; cabbage has cellulose for which humans do not have the enzyme

b The first jam had a sugar with an aldehyde group which it is likely to be glucose; the second jam contained a disaccharide which contained no aldehyde group, but when boiled in acid, the disaccharide split to make glucose

c Vitamins break down naturally in fresh vegetables; the low temperature of the frozen vegetables slows the breakdown of the vitamins

8 a See spread 30.5

b i Haemoglobin: transfer of O_2 from lungs to tissues; myoglobin: storage of O_2 in tissues

ii Residual respiration in tissues uses O_2 bound to myoglobin; myoglobin near surface is recharged with O_2 from air

iii To keep it red/attractive

9 a and b See spread 30.1

c Active transport is the movement of a solute across a biological membrane from a region of low to high solute concentration, requiring the input of energy and the involvement of enzymes and specific transport proteins; sodium pump involves Na^+ ion

10 The primary protein structure is the sequence of covalently bonded amino acids; the secondary structure is the chain of amino acids forming a helix held by hydrogen bonds; the tertiary structure is the helix bent to form globular or planar structures held by hydrogen bonds or —S—S— bridges; the quaternary structure is the folds in the tertiary structure, held by hydrogen bonds

11 a i The graphical formula should show all bonds; $HOCO—C_6H_4—COOH$ and $HOCH_2CH_2OH$

ii polyester

iii $[—CO—C_6H_4—CO—OCH_2CH_2O—]_n$

b i Biodegradable means it will break down naturally in a short time

ii The plastics would break down so would not fill landfill sites, or they could even be used as a source of energy

c i Potatoes, wheat, rice, maize

ii Glucose

d i Mineral acid in a food stuff would harm the person.

ii Around pH 7

Chapter 31

Exam questions

1 a i M_r; **ii** due to ^{13}C; **iii** $M^{+\bullet} \rightarrow P\bullet + Q^+$ only one carries a positive charge and is accelerated towards detector: the other is a radical

b i There are two isotopes ^{35}Cl and ^{37}Cl whose abundance ratio is 3:1

ii $CH_3Cl^{+\bullet} \rightarrow CH_3^+ + Cl\bullet$

2 a i 87.7; **ii** Different relative abundances of isotopes

b i HCN; **ii** NO; **iii** HCHO

c i M+1 peak is approx 7% of M peak

ii B = $C_6H_5^+$ C = $C_6H_5CO^+$

d C_6H_5COOH; —COOH

3 a i C=C; **ii** *Cis-* and *trans-*

iii Restricted rotation around C=C;

iv $CH_3CH_2CH=CH_2$

b i $CH_3CH_2COCH_3$; **ii** $—COCH_3$ or $—CH(OH)CH_3$

iii D: $CH_3CH_2CO_2^-Na^+$ E: CH_3CH_2COOH

c i $CH_3CH_2CH(OH)CH_3$; **ii** 29: $CH_3CH_2^+$ 45: $CH_3CH(OH)^+$ most parent molecules fragment

d i See spread 22.2, which assumes Raoult's law

ii Positive deviation means that there is more vapour than expected from Raoult's law, because the forces between the two molecules are poorly matched; they have stronger forces with their own kind (such as hydrogen bonding for the alcohol)

iii Broadening caused by hydrogen bonding, which gets less prevalent on dilution

4 a i 60; **ii** 30: CH_2NH_2; 16: NH_2; 14: CH_2

iii Amine

b See spread 22.6

c i See spread 22.6

ii This is very similar to $[Co(NH_3)_4Cl_2]^+$, as in spread 20.9

5 a i Relative molecular mass; the weighted mean of the mass numbers of a sample of an element, using 1/12 of a carbon-12 atom as one unit; isotopes have the same atomic number (proton number) but different mass number (number of neutrons)

ii 12.0

b i 120

ii The small peaks are due to the molecular ions containing one ^{13}C atom

c i A molecule with mass 120 (which must contain 1 oxygen atom) must contain at least 8 carbon atoms; with one less C-atom it would be difficult to think of a possible molecule; with one more carbon atom there would not be spare mass for enough hydrogen atoms

ii B: 77; $C_6H_5^+$; C: 105; $C_6H_5CO^+$

iii The CH_3^+ ions would be removed/would not be detected as the analysis concentrated on the larger fragments

d Add 2,4-dinitrophenylhydrazine, and a brightly coloured solid would be made if the compound contained a C=O group

Chapter 32

Practice questions

Spread 32.4

2 a $CH_3CH_2CH_2NH_2$ **b** $CH_3CH_2COOCH_3$

Exam questions

1 a See spread 32.1 **b** See spread 32.1

c Cadmium and lead from pottery glazes

2 a i C—H

ii $CH_3CH_2CH_2COOH$
$(CH_3)_2CHCOOH$
no broadened O—H peak evident

b III

c $CH_3COOCH_2CH_3$ and $CH_3CH_2COOCH_3$

3 a H_2O: 3
CO_2: 2
CO: 1

b See spreads 32.3 and 32.4

c Any ester $C_4H_8O_2$, such as ethyl ethanoate (the actual molecule requires comparison with known IR fingerprints). Peaks are due to C=O, C—O—, C—H, etc.

4 a i See spread 32.5

ii See spread 32.7

b F: CH_3COOCH_3
G: CH_3CH_2COOH

5 a $Si(CH_3)_4$ **b** 4 **c** 2:1:6:3

d $CH_3CH_2—$
splitting of δ 1.52 due to neighbouring $—CH_3$; splitting of δ 0.93 due to neighbouring $—CH_2$

e Single peak, therefore no spin–spin coupling, no H atom on adjacent group; 6 of these H atoms

f $CH_3CH_2C(CH_3)_2—OH$

6 a i Paste made from solid sample

ii Interactions (e.g. H bonding) between ethanol and test substance complicate interpretation

b Q: C≡C or C≡N
R: C=O
S: C—O, C—N
J is $NCCH_2COOC_2H_5$

c i T has a mass of 43, which suggests $C_3H_7^+$

ii U has a mass of 28, which suggests $C_2H_4^+$

iii The M+2 peak of similar height to the M peak suggests that bromine is present: this confirms that T is mass 43, because 43 + 79 = 122; K may be $(CH_3)_2CHBr$

Index

Acknowledgements

Photographs and computer-generated images

(FPNY = Fundamental Photographs New York; GSF = GeoScience Features; LGPL = Leslie Garland Picture Library; SPL = Science Photo Library; (t) = top; (b) = bottom; (l) = left; (r) = right; (c) = centre.)

10–11 IBM Corporation, Research Division, Almaden Research Centre; 12 Crown Copyright/Health and Safety Laboratory/SPL; 13(l) Derby Museum and Art Gallery, Derbyshire UK/Bridgeman Art Library; 13(r) The Metropolitan Museum of Art, Purchase, Mr and Mrs Charles Wrightsman Gift, in honour of Everett Fahy, 1977 (1977.10), photo: © 1989 Metropolitan Museum of Art (detail); 14(bl) Andrew Lambert/LGPL; 14(tl, tc, tr, br) GSF; 18(l) Statens Historiska Museum Stockholm/Werner Forman Archive; 18(r) Andrew Lambert/LGPL; 19(t) Lawrence Migdale/SPL; 19(b) Lawrence Berkeley National Laboratory/Science Photo Library; 20 Science Museum/Science and Society Picture Library; 21(bl, br) IBM Corporation, Research Division, Almaden Research Center; 21(t) scan courtesy of Purdue University, image courtesy of Digital Instruments, Veeco Metrolology Group, Santa Barbara, CA, USA; 22 Jean-Loup Charmet/SPL; 23 Cavendish Laboratory, University of Cambridge; 25 The Oxford Story exhibition; 26 Cavendish Laboratory, University of Cambridge; 28 Thermit Welding (GB) Ltd; 30 Peter Gould; 31 Tek Image/SPL; 32 GSF; 34 GSF; 35 Charles D. Winters/Timeframe Photography Inc.; 39 Dept of Physics, Imperial College/SPL; 43 Jerry Mason/SPL; 44(bl) Mr Fred Wondre, Dept of Physics, University of Oxford; 44(br) Dr Tongguang Zhai, Department of Materials Science, University of Oxford; 44(t) Peter Gould; 51 GSF; 58 GSF; 59 Charles D. Winters/SPL; 60 Bancroft Library, University of California Archives no. 13.596; 62 Bochsler Photographics+Imaging, Burlington, Canada; 64 Martin Sookias; 68 IBM Corporation, Research Division, Almaden Research Center; 71 © Richard Megna/FPNY; 76(l) © Andy Washnik ñ CORPRICOM; 76(r) © Richard Megna/FPNY; 77 Dr Jeremy Burgess/SPL; 79 Andrew Lambert/LGPL; 83(l) Nubar Alexanian/Corbis; 83(tr) Rover, Oxford; 83(br) Andrew Syred/Microscopix; 84(t) J C Revy/SPL; 84(r) Michael W Davidson/SPL; 84(b) GSF; 85(b) Charles D. Winters/Timeframe Photography; 85(t) Andrew Lambert/LGPL; 86 Fred Ward, Black Star / Colorific!; 87(t) Tony Craddock/SPL; 87(b) Robert Harding Picture Library; 89 Will Biddle/Michael Clugston/John Clugston; 90 Charles D. Winters/Timeframe Photography; 94 ChemSoft, with thanks to James Whyte; 95 Mike Clugston; 98(tc) Yoav Levy/Phototake Inc/Robert Harding Picture Library; 98(tr) © Richard Megna/FPNY; 98(r) Charles D. Winters/SPL; 98(b) Adam Hart-Davis/SPL; 99 Dan Guravitch/SPL; 100(t) from Mark Ladd: Introduction to Physical Chemistry 3rd edn, Cambridge University Press 1998; 100(b) Geoff Tompkinson/SPL; 101 Mike Clugston; 106 Mehan Kulyk/SPL; 108 Runk/Schoenberger from Grant Heilman; 109(t) Tim Clayton/Sidney Morning Herald; 109(b) D. Wells/The Image Works; 110 Rob Friedmann/iStock.com; 114 Andrew Lambert/LGPL; 115(t) Charles D. Winters/Timeframe Photography; 117 L. S. Stepanowicz/The Picture Cube; 118 Martin Sookias; 119 Ken McVeigh/Tony Stone Worldwide; 120(l) Bochsler Photographics+Imaging, Burlington, Canada; 120(r) Andrew Lambert/LGPL; 126 Andrew Lambert/LGPL; 128(t) MIRA; 129 David Nunuk/SPL; 128(b) Chris Fairclough/Image Select; 130 Lester V. Bergman/Corbis; 131 Alfred Pasieka/SPL; 133 Auto Express/Quadrant Picture Library; 135(t) Martin Sookias; 135(b) © Richard Megna/FPNY; 137(tl, b) Andrew Lambert/LGPL; 137(tr) Ken Karp; 138 British Steel plc; 142 NASA/SPL; 143 Chris Barry/Action-Plus Photography; 145(l) Charles D. Winters/Timeframe Photography; 145(r) Lawrence Migdale/SPL; 151 Robert Smith/Rex Features; 153(t) NASA Goddard Institute for Space Studies/SPL; 153(b) Crown Copyright/Health and Safety Laboratory/SPL; 154 Neil Tingle/Action-Plus Photography; 156 GSF; 158 © Richard Megna/FPNY; 164 GSF; 165 Simon Fraser/SPL; 167(t) Hulton Getty; 167(b) BASF; 168 James Scherer/Houghton Mifflin Company; 169 © Richard Megna/FPNY; 170(t) Bochsler Photographics+Imaging, Burlington, Canada; 170(b) Andrew Lambert/LGPL; 174 Andrew Lambert/LGPL; 178 St Bartholomew's Hospital/SPL; 180 Bryan Pickering/Corbis; 181 David Taylor/SPL; 182 © Richard Megna/FPNY; 183(l) Dr E. R. Degginger; 183(r) Andrew Lambert/LGPL; 186 Leslie Garland/LGPL; 187 Alfred Pasieka/SPL; 190(t) Adam Hart-Davis/SPL; 190(bl) Corel Professional Photos; 190(bc) Oxford University Press; 190(bc) Martin Sookias; 191 Andrew Lambert/LGPL; 192 Andrew Lambert/LGPL; 193 Andrew Lambert/LGPL; 195(b) Dr E. R. Degginger; 195(t) Andrew Lambert/LGPL; 196 Andrew Lambert/LGPL; 200 Metrohm; 205 Jason Burke/Eye Ubiquitous; 206 Mike Clugston; 207 Andrew Lambert/LGPL; 208 Ken Karp; 209 Andrew Lambert/LGPL; 212 Bochsler Photographics+Imaging, Burlington, Canada; 213(b) Andrew Lambert/LGPL; 213(t) Robert Harding Picture Library; 214 Andrew Lambert/LGPL; 215 Andrew Lambert/LGPL; 216 Andrew Lambert/LGPL; 217 GSF; 218(t) Andrew Lambert/LGPL; 218(b) Paul Silverman/FPNY; 220(tl) Charles D. Winters/Timeframe Photography; 220(tr) Charles D. Winters/SPL; 220(b) Ken Karp; 221 Jerry Mason/SPL; 222 Andrew Lambert/LGPL; 224 © Andy Washnik – CORPRICOM; 225 Andrew Lambert/LGPL; 226 Andrew Lambert/LGPL; 227 Andrew Lambert/LGPL; 228(t) Peter Gould; 228(b) Andrew Lambert/LGPL; 229 Dr Dennis Kunkel/Phototake NYC/Robert Harding Picture Library; 231 NASA/Glenn Research Center; 232 Andrew Lambert/LGPL; 234(t) Kevin Wisniewski/Rex Features; 234(b) Andrew Lambert/LGPL; 235(t) Yu Xiang; 235(b); Dr

Jeremy Burgess/SPL; 235(m) Michael Clugston; 236(tr) Kunsthistorisches Museum Vienna/photo: Erich Lessing/AKG London; 236(tl) GSF; 236 (bl) Michael Clugston; 237 Sascha Burkhard/iStock.com; 237(br) Philip Craven/Robert Harding Picture Library; 240 Manley Prim Photography; 241 Andrew Lambert/LGPL; 244 Dean Conger/Corbis; 246 SPL; 250 Bochsler Photographics+Imaging, Burlington, Canada; 251 Paolo Koch/ Robert Harding Picture Library; 254 Andrew Lambert/LGP; 255(tl, tr) Charles D. Winters/Timeframe Photography; 255(b) © Richard Megna/FPNY; 257 Physics Dept, Imperial College/SPL; 262 © Richard Megna/FPNY; 267(t) Geoff Tompkinson/SPL; 267(b) US Dept of Energy/SPL; 268 © Richard Megna/FPNY; 269 Dr Ivan Polunin/ NHPA; 271(t) Thomas Eisner, Daniel Aneshansley, Cornell University; 271(bc) Johnson Matthey; 272 Oxford University Press; 275 Clive Freeman/The Royal Institution/SPL; 278–279 SPL; 281 (lithium, sodium, potassium, calcium, magnesium , strontium) GSF; 281 (beryllium, rubidium) Russ Lappa/SPL; 281 (cesium, barium) Lester V. Bergman/ Corbis; 283 GSF; 284(t) (all except cesium) Andrew Lambert/LGPL; 284(t) (cesium) still from BBC Open University video The elements organised in the periodic table; 284(b) Bochsler Photographics+Imaging, Burlington, Canada; 284(bl) Mike Clugston; 285 © Andy Washnik – CORPRICOM; 286 GSF; 288(t) Phototake/Robert Harding Picture Library; 288(b) © Richard Megna/FPNY; 290 James Davis Worldwide Photographic Travel Library; 292 The Egyptian Museum, Cairo/Werner Forman Archive; 293 CNRI/ SPL; 296(a) David Taylor/SPL; 296(b) Jerry Mason/SPL; 296(c) E. R. Degginger/SPL; 296(d) Lester V. Bergman/Corbis; 297 A. R. Lampshire/Robert Harding Picture Library; 301 (all except chlorine) GSF; 301 (chlorine) Charles D. Winters/SPL; 303(l) T. Waltham/Robert Harding Picture Library; 303(r) © Richard Megna/FPNY; 306 Andrew Lambert/LGPL; 307 Hartmann/Sachs/Phototake NYC/Robert Harding Picture Library; 308 Andrew Lambert/LGPL; 309 GSF; 311 David Taylor/SPL; 312 Andrew Lambert/ LGPL; 313 Andrew Lambert/LGPL; 314 Charles D. Winters/Timeframe Photography; 318 Ken Karp; 319 (fluorine, iodine) GSF; 319 (bromine) Andrew Lambert/LGPL; 319 chlorine Charles D. Winters/SPL; 321 Srulik Haramaty/Phototake NYC/Robert Harding Picture Library; 322 GSF; 323 Bochsler Photographics+Imaging, Burlington, Canada; 324(t) Andrew Lambert/LGPL; 324(b) Frank Herholdt/Tony Stone Images; 326(t) James Scherer/Houghton Mifflin Company; 326(b) Andrew Lambert/LGPL; 328 Andrew Lambert/LGPL; 333 GSF; 334 Oberlin College Archives, Ohio; 335(t) Andrew Lambert/ LGPL; 335(b) Digital imagery copyright 1999 PhotoDisc, Inc.; 337(l) © Richard Megna/ FPNY; 337(r) Andrew Lambert/LGPL; 338(b) AT&T Archives; 338(t) Lucara Diamond Corp; 339(tl) John Mead/SPL; 339(tr) Geoff Tompkinson/SPL; 339(b) Donald Huffman, University of Arizona; 340 © Richard Megna/FPNY; 342 Bochsler Photographics+Imaging, Burlington, Canada; 343 Niall MacLeod/Corbis; 344 GSF; 346 Andrew Lambert/LGPL; 347 Andrew Lambert/LGPL; 349 James L. Amos/Corbis; 350 NASA; 351 GSF; 353 Bochsler Photographics+Imaging, Burlington, Canada; 353(b) Will Biddle/Michael Clugston/John Clugston; 354 Runk/Schoenberger from Grant Heilman; 355(t) Keenan & Wood: General College Chemistry, 5th edn, Harper & Row, New York 1976; 355(br) Argonne National Laboratory; 359(t) (all except cobalt, nickel and chromium) GSF; 359(t) (nickel) Lester V. Bergman/Corbis; 359(t) (chromium) Klaus Guldbrandsen/SPL; 359 (cobalt) Russ Lappa/SPL; 359(b) Andrew Lambert/LGPL; 360(t) GSF; 360(b) J Allan Cash; 361(t) British Steel plc; 361(r) David Cumming/Eye Ubiquitous; 361(b) Leslie Garland/LGPL; 363(tl) Courtesy of Micromed Technology, Inc.; 363(tr) Atlantide Phototravel/Corbis; 363(cr) GSF; 363(b) Leslie Garland/LGPL; 364 Houghton Mifflin Company; 364 Richard Megna/FPNY; 365 James Scherer/ Houghton Mifflin Company; 366 GSF; 367 Mike Clugston; 368 Andrew Lambert/LGPL; 369 Eye of Science/SPL; 371 GSF; 372 Jerry Mason/SPL; 374 NIH/Phototake; 376 Andrew Lambert/LGPL; 378 © Richard Megna/FPNY; 379(t) Martin Sookias; 379(b) Andrew Lambert/LGPL; 384 Geoff Tompkinson/SPL; 386 Andrew Lambert/LGPL; 387 Andrew Lambert/LGPL; 388 Volker Steger/SPL; 389 Andrew Lambert/LGPL; 390 Andrew Lambert/LGPL; 396(l) Sarah Leigh Lewis/Panos Pictures; 396(r) Mike Taylor/ LGPL; 397 Andrew Lambert/LGPL; 400 Ian McKinnell/Ace; 403 James Holmes/SPL; 404 Warren Gretz/NREL/US Dept of Energy/SPL; 407 Prof. P. Motta/Dept of Anatomy/ University La Sapienza Rome/SPL; 409(t, c) Andrew Lambert/LGPL; 409(b) Custom Medical Stock Photo/SPL; 413 Vauxhall Motors Ltd; 414(l) Martin Sookias; 414(r) Courtesy of Ice-Nook/MNS Holdings, Ontario, Canada, web site: http://www.pathcom.com/~mns; 415(tl) Courtesy of European Council of Vinyl Manufacturers; 415(tr, bl) BASF; 415(br) DuPont; 417 John Clugston; 424 Andrew Lambert/LGPL; 427(t,c) Andrew Lambert/LGPL; 427(b) Craig Hammell/The Stock Market; 429 Andrew Lambert/LGPL; 430 Colin Smedley/Skyscan; 431 CNRI/SPL; 433 BSIP/SPL; 438 Andrew Lambert/LGPL; 439 Peter Gould; 442 BSIP Leca/SPL; 443 Camera Press; 446 Jonathan Blair/Corbis; 447 Gail Mooney/Corbis; 449 Andrew Lambert/LGPL; 450 Andrew Lambert/LGPL; 451 Anthony Blake Photo Library; 453(l) Peter Gould; 453(r) Jerry Mason/SPL; 455 Mike Clugston; 457 Craig Aurness/Corbis; 462 Andrew Lambert/LGPL; 465(t) Bochsler Photographics+Imaging, Burlington, Canada; 466(r) Mike Clugston; 466(l) James Scherer/ Houghton Mifflin Company; 466(r) Andrew Lambert/LGPL; 467 Andrew Lambert/LGPL; 471 Peter Gould; 476 Andrew Lambert/LGPL; 477 David Woodfall/NHPA; 478 Corel Professional Photos; 484 Corel Professional Photos; 485(t) Jim Sugar/Corbis; 485(b) Digital imagery copyright 1999 PhotoDisc, Inc.; 486(t) Dr Jeremy Burgess/SPL; 486(b) Andrew Lambert/LGPL; 487 DuPont; 492 Andrew

Lambert/LGPL; 493 David Parker/SPL; 494 Andrew Lambert/LGPL; 496 (l,r) Andrew Lambert/LGPL; 499(l) Andrew Lambert/LGPL; 499(r) Jeremy Hofner/Panos Pictures; 500 Andrew Lambert/LGPL; 501 Alfred Pasieka/SPL; 502(t) GSF; 502(t) Mike Clugston, Andrew Worrall, David Braybrook, Jill Jordan; 503 Topham Picturepoint; 508 GSF; 509 K. G. Preston-Mafham/Premaphotos Wildlife; 512 GSF; 513 Eric Soder/NHPA; 513(b) Michael Clugston; 514 Andrew Lambert/LGPL; 517(t) Mike Clugston; 517(b) Barnabas Bosshart/Corbis; 521 NIBSC/SPL; 526(l) Ecoscene; 526(r) J. C. Revy/SPL; 527(t) Biophoto Associates; 527(b) CNRI/SPL; 529 Ken Eward/SPL; 530 Ken Eward/SPL; 533 Bill Longcore/SPL; 536 Kings College London; 537 Dr Gopal Murti/SPL; 540 Dr Kari Lounatmaa/SPL; 544 Andrew Lambert/LGPL; 545(l) Mike Clugston; 545(r) Andrew Lambert/LGPL; 546 James Holmes/Cellmark Diagnostics/SPL; 547(t) Phillippe Plailley/SPL; 547(b) David Parker/SPL; 548 Geoff Tompkinson/SPL; 549(t) Sinclair Stammers/SPL; 549(b) Digital imagery – copyright 1999 PhotoDisc, Inc.; 550 Hichrom; 551(t) Unicam Chromatography; 551(b) SmithKline Beecham; 556(t) Lawrence Berkeley National Laboratory/SPL; 556(b) Jean Collomber/SPL; 559 Chris Rogers - Rainbow/Getty; 561 Spectronic Unicam; 565 Simon Fraser/Royal Victoria Infirmary, Newcastle on Tyne/SPL.

In a few cases we have been unable to trace the copyright holder prior to publication. If notified, the publishers will be pleased to amend the acknowledgements in any future edition.

Picture research by Anne Lyons

We gratefully acknowledge the generosity of Dr Warren Hehre of Wavefunction, Inc. in creating and providing images throughout the book generated with Spartan software.

Exam questions

We are grateful to the following examination boards for their kind permission to reproduce past examination paper questions:

AQA (Assessment and Qualifications Alliance):
- AEB (Associated Examination Board): p37q6,7; p96q1,2; p97q6–8; p112–13q1,3,5,6; p124q2,3; p162q3; p210q2; p252q2; p277q5; p298q3; p316–17q1; p330q4; p357q7; p380–1q2,8; p435q4; p458–9q5; p472–3q2,6; p490–1q1,2,6; p506–7q4,8; p542q1; p543q11; p555q4
- NEAB (Northern Examinations and Assessment Board): p36–7q1–5; p56q1–4; p80–1q1–3,5,9; p113q7; p140–1q1–4,8; p162–3q1,2,6; p184–5q1–4; p210–11q3; p238q1–4; p252–3q1,3,4,6; p276q1–3; p316q3,5–7; p330q1; p356q1,2; p380–1q1,3,5,6; p420q1,3; p444q1–5,8,9; p458–9q1–3,6; p472–3q5,7; p506–7q1,2,7; p519q6; p554q1; p570–1q,2,5

NICCEA (Northern Ireland Council for the Curriculum, Examinations, and Assessment):
- p56–7q5; p112q2; p162–3q4; p185q5; p210q1; p331q10; p434q1; p491q7

EDEXCEL Foundation, including London Examinations:
- p57q6–9; p80q4; p81q6–8,10; p96–7q3–5; p140–1q5,6; p163q5; p185q6,7; p238–9q5,7,8; p253q5; p277q6; p298–9q1,2,6–8; p316–17q4,8; p330–1q3,8; p356–7q4,5; p380q4; p420–1q2,5; p434–5q2,3; p444–5q6,7; p458q4; p472q1,3,4; p490q3; p506–7q3,5,6; p518q1,2; p554–5q3,5

OCR (Oxford, Cambridge and RSA Examinations):
- UCLES (University of Cambridge Local Examinations Syndicate): p112–13q4; p141q7; p316q2; p163q7; p239q6,9; p330q2; p330–1q5,6,7; p331q9; p357q6,8; p420q4; p490–1q4,5; p518–19q3,4,5,7; p542–3q2–10; p554q2; p570–1q1,3,4,6
- UODLE (University of Oxford Delegacy of Local Examinations): p125q5,6,7

WJEC (Welsh Joint Education Committee)
- p124–5q1,4; p211q4–6; p276–7q4; 298–9q4,5; p356q3; p381q7; p421q6.